Proceedings of
The Dynamic Flow Conference 1978

on

Dynamic Measurements in Unsteady Flows

I.M.S.T., Marseille, France, September 11-14, 1978
Johns Hopkins University, Baltimore, USA, September 18-21, 1978

Scientific Editorial Committee
Prof. Leslie S.G. Kovasznay
Department of Mechanical Engineering
University of Houston, Texas 77004, USA

Prof. A. Favre
Membre de l'Academie des Sciences
Institut de Mécanique Statistique de la Turbulence
Université d'Aix-Marseille II, France

Dr. P. Buchhave
Manager, SRE Research & Development Dept.
DISA Elektronik A/S
DK-2740 Skovlunde, Denmark

Dr. Louis Fulachier
Institut de Mécanique Statistique de la Turbulence
Université d'Aix-Marseille II, France

Technical Editor
Bengt Wulff Hansen, M.Sc.
Manager, Information Dept.
DISA Elektronik A/S
DK-2740 Skovlunde, Denmark

Sijthoff & Noordhoff 1979
Alphen aan den Rijn, The Netherlands

ISBN-13: 978-94-009-9567-3   e-ISBN-13: 978-94-009-9565-9
DOI: 10.1007/ 978-94-009-9565-9
Sijthoff & Noordhoff
International Publishers bv

Contents   Page

Contents   Page

Measurements in Two-Phase Flow   287

Contents

Contents   **Page**

Contents   Page

Preface

This volume contains the Proceedings of the 1978 Twin Conference on Dynamic Measurements in Unsteady Flows (the "Dynamic Flow Conference 1978"). The first of the two conferences was held at the Institut de Mécanique Statistique de la Turbulence (the "I.M.S.T.") of the Université d'Aix-Marseille II, September 11-14, 1978. The second took place directly afterwards at the Johns Hopkins University, Baltimore, USA, September 18-21, 1978.

The subjects discussed at the Twin Conference were not, primarily, results of measurements in dynamic flows but, rather, the measuring methods and signal and data processing techniques employed. Though some of the methods used in fluid flow research have been known for more than 50 years, the entire field of flow instrumentation is in a state of extreme flux. New methods are being developed, and rapid advances in signal and data processing are having an impact on all types of instrumentation and measuring techniques.

Such conferences, combining the interests of fluid mechanics researchers, industrial development and laboratory engineers, and representatives of the specialized instrumentation industry, have often triggered off further progress in the field. As a participant in this ongoing trend DISA Elektronik has previously initiated two such conferences in co-operation with leading universities. The first of these was the 1972 Leicester Conference at Leicester University, England. The second was the 1975 LDA-Symposium in Copenhagen, Denmark. The present Twin Conference was co-sponsored by the Institut de Mécanique de la Turbulence of the Université d'Aix-Marseille II, the Johns Hopkins University Baltimore, and DISA Elektronik.

This year's conference differed slightly in scope from the previous ones as more emphasis was placed on invited state-of-the-art reviews and tutorial lectures than on contributed research papers. Considerable time and effort were invested by the host institutions and organizers and by the authors of review lectures, invited papers, and contributed papers. The high quality of the contributions and the success of the conference are ample proof of the existence of a real need for such interdisciplinary meetings to discuss measuring techniques and equipment, if such meetings are held at the proper times. The composition of the attendance — approx. 50 % from universities and 50 % from the industries — shows that the desire for this kind of communication is equally shared by both groups.

It is our belief that the 1978 Twin Conference on Dynamic Measurements in Unsteady Flows will contribute to the further development of fluid flow research instrumentation and that the Proceedings Volume will be accepted as a lasting contribution to the literature on the subject.

Thanks are due to the Organizing Secretariat, in particular Mr. N.J. Madsen and Mrs. Bente G. Christensen, for their efficient organization of the Conference and the staffs of the subsidiary companies of DISA Elektronik in France, the United Kingdom, Germany, Italy, Canada, and the United States for their assistance in providing the framework for a successful conference.

The Scientific Editorial Committee

Probes for Multivariant Flow Characteristics

Probes for Multivariant Flow Characteristics

Anthony Demetriades
Supervisor, Fluid Mechanics Section, Aeronutronic Division
Ford Aerospace & Communications Corporation
Newport Beach, California

Abstract

Diagnostic methods in unsteady, multivariant flows can aim either at measuring one flow property regardless of variations in others, or at measuring each of two or more properties at the same time. This review lecture primarily addresses Eulerian detectors in a continuum, chemically inert environment undergoing a stationary random process, and assumes that the detector is simultaneously affected by two or more flow variables. An outline of the basic theory is given which examines the quasi-steady detector response and its statistical moments in terms of the statistics of the flow variable fluctuations. This theory reveals the need for reinterpretation of classic detecting techniques and sensors in regimes of current technological interest involving non-linear and high frequency phenomena. From this point of view the lecture addresses the current status of the most popular detectors and evaluates their potential in isolating the individual history of the flow velocity, density, pressure, temperature and concentration.

1. Introduction

This review lecture concerns itself with classical aerodynamics in which the moving fluid is a chemically inert gas and the temperature does not distort grossly the molecular or atomic structure. In the field of such classical flows, the demand for experimental diagnostics is presently concentrated in two main areas: (a) modeling of turbulent or unsteady flows, almost wholly for constant-density cases, and mostly limited to a need for higher moments of the velocity components and (b) a combination of theoretical modeling and empirical descriptions for turbulent flows of chemically inert but multi-component fluids and of compressible flows. The questions affecting the velocity fluctuations of incompressible fluids are now well-posed and the instrumentation refined to a high degree. A large variety of the higher moments of the fluctuations Δu, Δv and Δw of the velocity components can be obtained by a skillful experimenter equipped with laser or hot wire anemometers, modern data acquisition, storage and computation gear and the freedom to design an experiment limited to low speeds and room temperatures with controllable boundary conditions. This area will not be considered here, since it will be ably dealt with in other papers of this volume.★

We will concern ourselves with multivariant unsteady flows in which two or more of the aerodynamic properties (u, v, w), ρ, p, T and concentration c vary with time and space. We are concerned mostly with stationary phenomena, primarily in the Eulerian frame, although the techniques discussed are equally applicable to transient flows; in spatial resolution which allows field mapping, i.e., measurement in a single point in the body of the flow; and in unsteadiness of gas flows as opposed to liquid flows.

Multivariant diagnostics play an important role in an increasing number of modern technologies. In pollution studies, for example, binary mixing involves the simultaneous development of cross-correlation of velocity and concentration fluctuations. In noise abatement, one is interested in measuring the velocity and pressure fluctuations; in optical communications the density and temperature fluctuations driven by atmospheric disturbance. In the supersonic aerodynamics of boundary layer turbulence there are intermingled velocity, density, temperature and pressure fluctuations (such a multivariant flow is pictured in Figure 1); if to those one adds binary mixing such as occuring, for instance, in transpiration-cooled missile components, one begins to appreciate the difficulties in the measuring techniques needed.

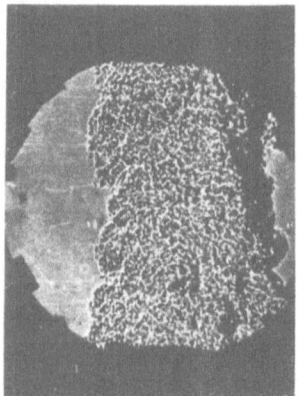 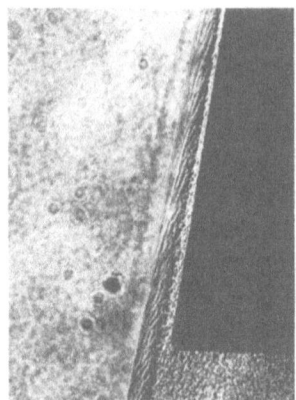

Fig. 1. Density fluctuations as viewed for a wake, at left (driven mainly by temperature spots) and for a Mach 20 boundary layer (right), (driven by temperature and pressure fluctuations).

★For the purpose of this paper, a "multivariant" flow will be defined to exclude the measurement of $\Delta u \equiv u'$, $\Delta v \equiv v'$ and $\Delta w \equiv w'$ alone and will, instead, focus on physical properties additional to flow speed.

Regardless of the form and complexity of the sensors employed in each case, one outstanding feature of multivariant unsteady flows is the difficulty of estimating, in advance, the correct hierarchy of the fluctuating terms. For example, in low speed turbulent flows, it has been customary to (essentially) invoke isotropy for truncating the equations of motion, e.g., to retain second-degree terms $\overline{(\Delta u_i)(\Delta u_j)}$ in favor of third-degree terms $\overline{(\Delta u_i)(\Delta u_j)^2}$ or $\overline{(\Delta u_i)^2 (\Delta u_j)}$. In a compressible hypersonic flow, on the other hand, terms such as $\overline{(\Delta u_i)(\Delta T)^2}$ can be shown to be important and in not-so extreme cases, of the same order as terms of the type $\overline{(\Delta u_i)^2}$ This has created certain disorder in the priorities with which diagnostics of hypersonic unsteady flow have been developed. The same can be said of the hierarchy of terms arising in binary mixing where the fractional concentration fluctuations $(\Delta c/\bar{c})$ can be close to unity or larger; in this case, correlations of the type $\overline{(\Delta u)(\Delta c)^2}$ are of the second rather than the third order; i.e., they approach the magnitude of $\overline{(\Delta u)(\Delta c)}$.

Such considerations prompt us to re-state, for purposes of reference, the rudiments of what might be called the multivariant measurement method in the following section.

2. The Multivariant Measurement Method

The measurement of a time-dependent property $x(t)$ (say, the velocity, density or pressure) at a point in a multivariant flow is done by detecting the fluctuations of a "sensed variable" Q, which usually is affected by properties other than x as well. For example, to detect temperature with a hot wire we detect, as sensed variable Q, the wire heat transfer rate which also depends on velocity; for detecting Mach number variations we can use the Q of a pitot tube which is the pitot pressure and thus depends also on the static pressure; and in the electron-beam fluorescence technique Q, the radiated intensity, depends equally on gas temperature and density.[★] Such sensing methods follow in common the mathematical rules outlined below, which can serve as departure points for the full understanding of sensor response. In the general case the quantities affecting Q, written as $x_i(t)$, are independent variables although physical links among them usually exist (e.g., by the equation of state). In the quasi-steady viewpoint, the dependence of x_i on the time t is neglected, at the expense of increasing the number of unknowns when time integrals (averages) are later formed. The key to success of the quasi-steady approach is the existence of a function Q stated as a physical law or discovered by calibration, and the target of the experimentalist is to measure the individual histories $x_i(t)$ by measuring $Q(t)$.

★The detector actually converts Q to a voltage which is usually only a matter of calibration.

2.1 Direct Methods for Measuring Fluctuations

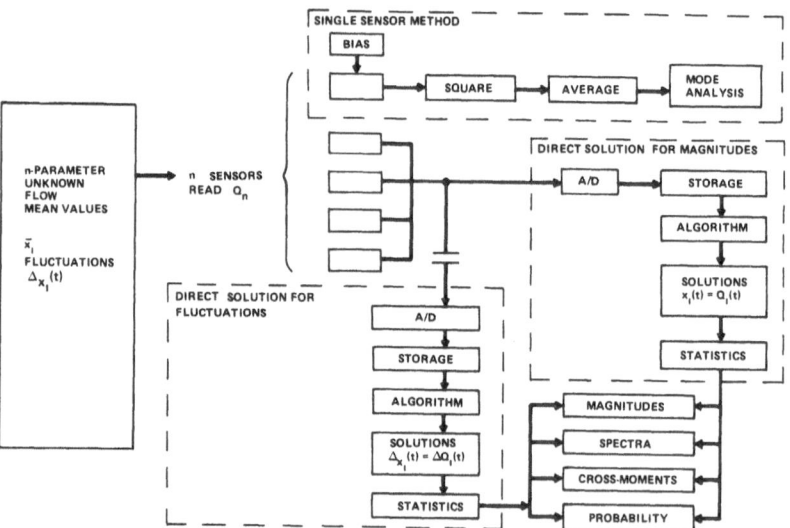

Fig. 2. Various direct and indirect methods exist for multivariant unsteady flows, depending on whether the number of sensors at a point can match the number of unsteady properties.

The recent rapid advances in data acquisition equipment and computer systems make it possible to make direct measurements of time-varying flows (see Fig. 2) In univariant unsteady or fluctuating flows of the type $Q(x)$, high frequency sampling of $Q(t)$ will give the time history of $x(t) = x(Q)$. For multivariant flows, however, of the type $Q(x_1, x_2,....x_n)$, n *simultaneous* measurements are necessary in order to obtain the solutions $x_i = x_i(Q)$.

This measurement is not easy to make, either conceptually or in practical laboratory conditions. To illustrate, consider a "sensed variable" $Q(x_i)$ linearly dependent on n independent quantities:

$$Q(x_i) \equiv C_1 x_1(t) + C_2 x_2(t) + \ldots \ldots C_n x_n(t) \tag{1}$$

where the C_i are numerically known in advance or determined by calibration. The multivariant method requires that, to measure $x_i(t)$, we obtain the n equations at the same instant t_o :

$$Q_1(t_o) = C_{11} x_1(t_o) + C_{12} x_2(t_o) + \ldots \ldots + C_{1n}(t_o)$$
$$C_2(t_o) = C_{21} x_1(t_o) + \ldots \ldots \qquad\qquad + C_{2n}(t_o) \tag{2}$$
$$C_n(t_o) = C_{n1} x_1(t_o) + \ldots \ldots \qquad\qquad + C_{nn} x_n(t_o)$$

The coefficients C_{ij} are obtained by "biasing" n times the original coefficients C_i. In this case "bias" means a change of C_i by some controllable change in the operating characteristics of the system. It could take the form of the placement at the same point in the flow (or as close together as possible!) of two or more sensors which are of the *same type* but *non-identical;* for heated film sensors, for example, this would mean the placement in proximity of such sensors each of different Nusselt-Reynolds number characteristic, or each held at different heating current. For fluorescent excitation techniques, it would mean separate recording of the radiation from two or more spectral lines. Obviously, for the system [2] to have a solution, biasing must affect each of C_1, C_2, ... C_n by unequal amounts; for example, in a good system, a bias change may double or triple C_1 but leave C_2 unaffected.

Multivariant detection schemes may, therefore, be defined as those for which biasing is possible and ranked according to the effect which bias has on the co-effecients of the functional $Q(x_i)$. Similar concepts apply when the latter depends on x_i in a way more complex than shown in eq. [1] (e.g., non-linear or implicit dependence); in this case computer algorithms must be devised to solve the system of simultaneous equations, which may be prohibitive for some schemes of detection. If successful, however, this method will give directly the history of $x_i(t)$ from which all manner of statistical information of the variables can be obtained.

2.2 Expansion of the Multivariant Sensed Variable $Q(x_i)$

Historically, diagnostic instrumentation of stationary fluctuating flows has dealt with the "a.c. component" of the flow and it is convenient to think in terms of the Taylor expansion of $Q(x_i)$.

For n independent variables x_1, x_2, ... x_n, each of which has a mean and fluctuating component in time:

$$x_i(t) = \bar{x}_i + \Delta x_i(t) \qquad [3]$$

this approach is also best suited to teach us practical things about our experiment (non-linearity thresholds, sensitivities, etc.), as well as to enrich our intuition of the measured events especially when the functional $Q(x_i)$ is non-linear in x_i. Thus,

$$Q(x_1, x_2, \dots x_n) \equiv Q(x_i) = Q(\bar{x}_1, \bar{x}_2, \dots \bar{x}_n) + \Delta x_1 \frac{\partial Q}{\partial x_1}\Big|_{\bar{x}_i} + \Delta x_2 \frac{\partial Q}{\partial x_2}\Big|_{\bar{x}_i}$$

$$+ \dots + \frac{1}{2}\left[(\Delta x_1)^2 \frac{\partial^2 Q}{\partial x_i^2}\Big|_{\bar{x}_i} \dots\right] + \dots + \frac{1}{k}\left[\sum_{i=1}^{n}(\wedge x_i)\frac{\partial}{\partial x_i}\right]^k Q(x_i)\Big|_{\bar{x}_i} + \dots [4]$$

or,

$$Q(x_i) = \sum_{k=0}^{\infty} \frac{1}{k!} \left[\sum_{i=1}^{n} (\Delta x_i) \frac{\partial}{\partial x_i} \right]^k Q(x_i) \, |\bar{x}_i \qquad [5]$$

where the notation $|\bar{x}_i$ implies derivatives evaluated at $\bar{x}_i$.

2.3 Linearization
Linearization (truncation) of [5] is permissible for a Eulerian observer in a moving fluid if the fluctuations are small, as often in turbulence:

$$Q(x_i) \cong Q(\bar{x}_i) + \sum_{i=1}^{n} (\Delta x_i) \frac{\partial Q}{\partial x_i} \, |\bar{x}_i \qquad [6]$$

It is a fundamental difficulty of multivariant unsteady flow measurements that the statement $(\Delta x_i)^{m+1} \ll (\Delta x_i)^m$ (or rather $(\Delta x/x)_i^{m+1} \ll (\Delta x/\bar{x})_i^m$) is *not* sufficient for truncation (it is insufficient in univariant flows also).

The true necessary and sufficient condition is of the form

$$\left| \left(\frac{\Delta x_i}{\bar{x}_i}\right)^{m+1} \frac{\partial^{m+1} Q}{\partial x_i^{m+1}} \, |\bar{x}_i \right| \ll \left| \left(\frac{\Delta x_i}{\bar{x}_i}\right)^m \frac{\partial^m Q}{\partial x_i^m} \, |\bar{x}_i \right| \qquad [7]$$

In other words, it is insufficient to say that "the fluctuations are small"; one needs to consider the magnitude of the *sensitivity coefficients*

$$\frac{\partial^k Q}{\partial x_i^k} \, |\bar{x}_i, \; \frac{\partial^k Q}{\partial x_i^{n-k} \partial x_j^k} \, |\bar{x}_i \; \text{etc.,} \qquad [8]$$

as well. For some highly non-linear variables Q, the second degree terms containing $(\Delta x_i)^2$ can be larger than those containing (Δx_i) even for $(\Delta x_i)/\bar{x}_i$ of order 0.1 or less. The real difficulty, of course, is that the truncation requires an advance estimate of the fluctuation magnitude, as well as knowledge of the functional $Q(x_i)$.

2.4 Averaging
If time averages (means) are denoted by an overbar, then

$$Q(x_i) = Q(\bar{x}_i) + \sum_{0}^{\infty} \frac{1}{k!} \left[\sum_{i=1}^{n} (\Delta x_i) \frac{\partial}{\partial x_i} \right]^k Q(x_i) \, |\bar{x}_i \qquad [9]$$

with the operator in the bracket supplying the familiar "moments" $\overline{(\Delta x_1)}$, $\overline{(\Delta x_1)^2}$, $\overline{(\Delta x_1)(\Delta x_2)}$, $\overline{(\Delta x_1)(\Delta x_2)(\Delta x_3)}$ and so on. It is evident from [9] that the mean value of Q and its value at the mean values of x_i are not the same; only in valid linearized cases is $\overline{Q(x_i)} = Q(\bar{x}_i)$. In the general case where the fluctuation magnitudes are arbitrary and if Q is highly non-linear, the disparity between $\bar{Q}$ and $Q(\bar{x}_i)$ can be serious; this author has found disparities by orders of magnitude in a flow example not distant from the classical flows considered here (1973). Smith (1975) has had the same experience in hypersonic turbulence. More on this subject will be said below.

2.5 Fluctuation Measurement by the Taylor Expansion

In analogy with the direct method of measuring the time history of $x_i(t)$ by recording the history of $Q(t)$ (see Section 2.1) we can find the fluctuations Δx_i by recording the excursions ΔQ of Q about its true mean $\overline{Q(x_i)}$. In the linearized multivariant case, the fluctuation of Q about its mean is

$$Q(x_i) - \overline{Q(x_i)} \equiv \Delta Q(x_i) = \sum_{i=1}^{n} (\Delta x_i) \frac{\partial Q}{\partial x_i} \Big|\overline{x}_i \qquad [10]$$

The fundamental problem is to recover the quantities Δx_i from the fluctuation $\Delta Q(x_i) = \Delta Q(t)$ in the Eulerian system, as we recover, for example, the velocity and temperature fluctuations from the heat flux fluctuation of a hot wire or hot film. We can do this by producing again, a linear system of n equations:

$$\Delta Q_A = (\Delta x_1) \frac{\partial Q_A}{\partial x_1} \Big|\overline{x}_i + (\Delta x_2) \frac{\partial Q_A}{\partial x_2} \Big|\overline{x}_i + \dots (\Delta x_n) \frac{\partial Q_A}{\partial x_n} \Big|\overline{x}_i$$

$$\Delta Q_B = (\Delta x_1) \frac{\partial Q_B}{\partial x_1} \Big|\overline{x}_i + (\Delta x_2) \frac{\partial Q_B}{\partial x_2} \Big|\overline{x}_i \dots \qquad [11]$$

$$\dots\dots$$

$$\Delta Q_n = (\Delta x_1) \frac{\partial Q_n}{\partial x_1} \Big|\overline{x}_i + \dots\dots$$

This system can then be solved for each instant t by recording $\Delta Q(t)$, once the sensitivity coefficients are known (which is possible since $Q(\overline{x}_i)$ in the linear case will yield $\overline{x}_i$) and their determinant is again non-zero, consistent with the comments in Section 2.1. This technique has practical advantages over that of Section 2.1 when $Q(x_i)$ is very complicated algebraically.

In exchange, however, the situation with the system [11] becomes difficult when the independent variables x_i make large excursions (fluctuations) in time, which give us reasons for suspecting the linear representation of eq. [10] as unsuitable. The system [11] is then replaced by one of non-linear equations, each containing a mixture of powers of all the unknowns, for which formal solutions do not exist in the general case. For univariant functions non-linearity is usually no problem since the sensed variable $Q(x)$ becomes a polynomial in (Δx) for which algebraic or numerical solutions can be found. It is a characteristic of multivariant sensed variables, then, that the time history of the fluctuation in the contributing quantities cannot be generally constructed from $\Delta Q(t)$ when non-linearity is present. *Specific* exceptions occur, however, depending on the algebraic form of $Q(x_1, x_2 . . .)$, which can be used to advantage by the experimentalist; or algorithms can be set up to provide numerical solutions to some degree of accuracy. In other words, the problems then are identical to solving simultaneous non-linear algebraic equations in more than one variable.

2.6 Measurement by Detection of the Higher Moments

In stationary flows the demand placed on the experimentalist usually concerns the lower order moments such as $\overline{(\Delta x_i)^2}$ or $\overline{(\Delta x_i)(\Delta x_j)}$. For this reason, and also because steady voltages have been historically easier to measure, the great majority of unsteady but stationary flow phenomena reported in the literature were obtained by electronic averaging of the mean-square:

$$(\Delta Q)'^2 \equiv \overline{(\Delta Q)^2} = \overline{(Q(x_i) - \overline{Q(x_i)})^2} = \tag{12}$$

$$\overline{\left[\sum_0^\infty \frac{1}{k!} \left[\sum_{i=1}^n (\Delta x_i) \frac{\partial}{\partial x_i} \right]^k Q(x_i) \, |\bar{x}_i \; - \; \overline{\sum_0^\infty \frac{1}{k!} \left[\sum_{i=1}^n (\Delta x_i) \frac{\partial}{\partial x_i} \right]^k Q(x_i) \, |\bar{x}_i} \right]^2}$$

where ( )$'$ refers to the root-mean-square. For example, if we have valid grounds for linearizing the response of a thermal sensor to velocity (u), temperature (T) and pressure (p) fluctuations, then [12] reduces to the following well-known form of the sensor voltage fluctuation $\Delta e \equiv \Delta Q$:

$$\overline{(\Delta e)^2} = \overline{(\Delta u)^2} \, S_u^2 + \overline{(\Delta T)^2} \, S_T^2 + \overline{(\Delta p)^2} \, S_p^2 + 2 S_u S_T \overline{\Delta u \Delta T}$$

$$+ \, 2 S_u S_p \, \overline{(\Delta u)(\Delta p)} + 2 S_T S_p \, \overline{(\Delta T)(\Delta p)} \tag{13}$$

The sensitivity coefficients S_u, S_T, etc. have the form given by [8] and are biased by changing the heating current.

If the full non-linear form of the squared $(\Delta Q)'$ is processed as written in eq. [12], it would provide all the higher order moments $\overline{(\Delta x_i)^k}$ and cross-moments $\overline{(\Delta x_i)^\alpha (\Delta_j)^\beta}$ which seem to be needed, say, in preparing an empirical mode of turbulent flow. This hope originates in the possibility of a "mode diagram" which is an experimental curve in the $\overline{(\Delta_e)^2}$ vs. b plane, where b is the biasing parameter controlling the magnitudes of the sensitivity coefficients; the unknown moments $\overline{(\Delta x_i)^k}$ etc., could then be determined from the shape parameters of this curve. Unfortunately, the biasing parameter does not always produce independent sensitivity coefficients. In the example of eq. [13], there are six unknowns so that the $\overline{(\Delta_e)^2}$ vs b curve must be of the fifth degree — which it is not for the hot wire and hot film sensors.

2.7 Application to a Practical Example

To illustrate the rules outlined above, we will discuss an example of how they can be applied to a typical turbulence-measuring technique. Since the thermal sensor technique (hot wire, hot film) is amply documented in the past and probably quite familiar to the reader, we will choose as an example the electron beam fluorescence technique. In doing so, we will deal with the elementary principles of this method, leaving its constrains and details aside for the moment and basing out discussion on the higher density cases such as studied experimentally, for instance, by Smith (1975). Our exposition below can then be condensed, and refinements added later without upsetting the main results.

In this technique, the sensed variable is the intensity I of some line in the emission spectrum of a gas bombarded by electrons, and the time-dependent quantities are the gas density ρ and temperature T:

$$I = \frac{A\rho}{1+C\rho T^D} \qquad [14]$$

where the constants A, C and D are characteristic of the particular spectral line observed with a particular experimental set-up. The bias enters through the optical filter selecting the spectral line (Smith gives A = 5 $\times$ 10^{-18} and 15.5 $\times$ 10^{-18}, C = 1.83 $\times$ 10^{-18} and 8.6 $\times$ 10^{-18} and D = 0.33 and 0.35 for the helium 5016- and 3965-angstrom lines, respectively) but it can also conceivably enter through the electron beam current and other causes. Here we shall first assume that $\Delta\rho/\bar{\rho}$ and $\Delta T/\bar{T}$ are small, and that digital data are acquired at the proper rate to recover the form of I(t) in detail.

We can find $\rho(t)$ and $T(t)$ at the point of observation first directly by observing $I_1(t)$ and $I_2(t)$ at two spectral lines 1 and 2 and applying the method of Section 2.1. Then the two equations

$$I_1(t) = \frac{A_1\rho}{1+C_1\rho T^{D_1}}$$

$$I_2(t) = \frac{A_2\rho}{1+C_2\rho T^{D_2}} \qquad [15]$$

can be solved simultaneously for $\rho(t)$ and $T(t)$ from which $\bar{\rho}$ and $\bar{T}$ can be determined (these, if inserted in eq. [15] will give an $I_{1,2}$ $(\bar{\rho}, \bar{T})$ agreeing with the measured $\bar{I}_1, \bar{I}_2$ if the fluctuations are small). The fluctuations are also possible to extract at this point by the method of Section 2.5 using the linear system [11]. The sensitivity coefficients will be:

$$\frac{\partial I}{\partial \rho})_{1,2} = \frac{\overline{I}_{1,2}}{\overline{\rho}} [1-\psi_{1,2}]$$

$$\frac{\partial I}{\partial T})_{1,2} = (-D_{1,2}\psi_{1,2})\frac{\overline{I}_{1,2}}{\overline{T}}$$

[16]

where $\psi \equiv \dfrac{C\overline{\rho}\overline{T}^D}{1+C\overline{\rho}\overline{T}^D}$ $(0 < \psi < 1)$ [17]

The system of equations to be solved at each t consists of

$$\frac{\Delta I_1}{\overline{I}_1} = \frac{\Delta\rho}{\overline{\rho}}(1-\psi_1) - \frac{\Delta T}{\overline{T}} D_1\psi_1$$ [18]

$$\frac{\Delta I_2}{\overline{I}_2} = \frac{\Delta\rho}{\overline{\rho}}(1-\psi_2) - \frac{\Delta T}{\overline{T}} D_2\psi_2$$

which is easy to do since $\overline{I}_1$, $\overline{I}_2$ and ΔI_1(t), ΔI_2 are measured directly, since A_1, A_2, C_1, C_2, etc. are calibration constants and since $\overline{\rho}$, $\overline{T}$ (and hence ψ_1, ψ_2) were already determined. Thus, we find the time history of the fractional fluctuations $\Delta\rho(t)/\overline{\rho}$, $\Delta T(t)/\overline{T}$ from which Fourier, probability and cross-correlation analysis can proceed in a straightforward manner.

In the preceding, it has been assumed that equipment to perform fast sampling and A/D conversion exists for each channel. When no such equipment is available, the historically older "mode analysis" of Section 2.6 can be used. The equivalent of eq. [12] will require only mean-square measurement of the fluctuations:

$$\overline{(\frac{\Delta I}{\overline{I}})^2} = \overline{(\frac{\Delta\rho}{\overline{\rho}})^2}(1-\psi)^2 + \overline{(\frac{\Delta T}{\overline{T}})^2}D^2\psi^2 - \overline{(\frac{\Delta\rho}{\overline{\rho}})(\frac{\Delta T}{\overline{T}})}D\psi(1-\psi)$$ [19]

Since there are three unknowns involved here, additional bias is needed (the biasing parameter b here being the spectral wavelength) in the form of at least a third set $\overline{I}$, A . . D, which can be obtained by intensity observations at a third spectral line or some other means. It would be interesting, for example, to obtain sets of $\overline{I}$, A, C, D and hence ψ(which, however, must produce a non-vanishing determinant) for different values of the electron beam current, since in this method the observations do not have to be done simultaneously as before.

Assuming the experimental feasibility of thus acquiring the rms $(\Delta I)'$, the average I and the calibration factors for three or more cases, we have at our disposal sets (ΔI), $\overline{I}$, D and ψ; and in the $(\Delta I/\overline{I})'$, ψ plane we will get a second-degree curve, formed by data points as follows:

$$\overline{\left(\frac{\Delta I}{I}\right)}^{2} = \psi^{2}\left[\overline{\left(\frac{\Delta\rho}{\rho}\right)}^{2} + 2\overline{\left(\frac{\Delta\rho}{\rho}\right)\left(\frac{\Delta\theta}{\theta}\right)} + \overline{\left(\frac{\Delta\theta}{\theta}\right)}^{2}\right]$$ [20]

$$- 2\psi\left[\overline{\left(\frac{\Delta\rho}{\rho}\right)}^{2} + \overline{\left(\frac{\Delta\rho}{\rho}\right)\left(\frac{\Delta\theta}{\theta}\right)}\right] + \left[\overline{\left(\frac{\Delta\rho}{\rho}\right)}^{2}\right]$$

where the rms $(\Delta\theta/\theta)' = D(\Delta T/T)'$. Least-squares curve-fitting will then produce three numerical values for each of the brackets above, from which the
three unknowns $(\Delta\rho/\rho)^{2}$, $(\Delta T/T)^{2}$ and their correlation can be determined. The
following comments are in order regarding eq. [20]:

— For a temperature-independent line $D = 0$ and $\Delta I/\bar{I}$ is affected only by
    $\Delta\rho/\bar{\rho}$ as it should.
— The (extrapolated) intercept of the curve with the $\Delta I/\bar{I}$ axis (i.e., $\psi = 0$)
    gives directly the mean-square density fluctuations.
— The (extrapolated) value of the curve at $\psi = 1$ gives directly the temper-
    ature fluctuations within the known factor D; i.e., $(\Delta I/\bar{I})'$ $(\psi = 1) =$
    $D(\Delta T/\bar{T})'$.
— For zero pressure fluctuations $\Delta\rho/\bar{\rho} = -\Delta T/\bar{T}$ and

$$\overline{\left(\frac{\Delta I}{\bar{I}}\right)}^{2} = \overline{\left(\frac{\Delta\rho}{\rho}\right)}^{2}[\psi(1 - D) -1]^{2}$$ [21]

We see that the limits $\psi = 0$ and $\psi = 1$ (the latter attainable at the shorter wave-
lengths and larger $\bar{\rho}$ and $\bar{T}$) give us independently the $(\Delta\rho/\bar{\rho})'$ and $(\Delta T/\bar{T})'$,
respectively. Thus the modal analysis for the linear case, although giving us
only three of the many possible moments, gives us a priori valuable hints about
the experiment design and the magnitude of the fluctuations at play. The mode
diagram for this technique is shown in Figure 3.

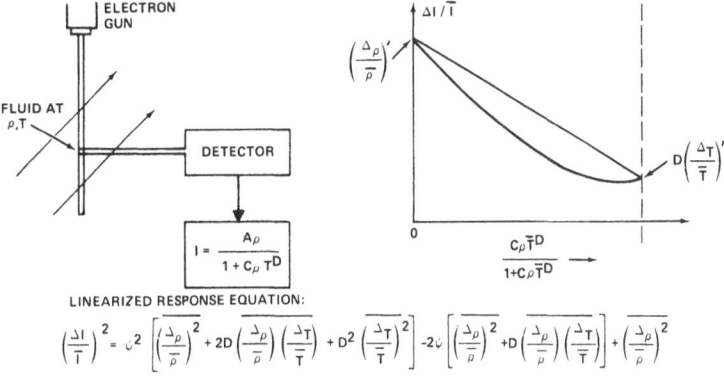

Fig. 3. Mode diagrams are not the exclusive province of thermal sensors. On the
right is the mode diagram for the electron-beam fluorescence technique; the
two curves are for two values of D.

We can exploit the present example further by a brief mention of what happens for large fluctuations $\Delta\rho/\rho$ and $\Delta T/T$. Because the algebra consumes too much space we will mention only the distorting effect on the mean value and then only when we retain nothing beyond second-degree terms. Then, according to [9]:

$$\overline{I(\rho,T)} = I(\bar{\rho},\bar{T}) + \frac{1}{2}\overline{(\Delta\rho)^2}\frac{\partial^2 I}{\partial\rho^2}\Big|_{\bar{\rho},\,\bar{T}} +$$

$$\frac{1}{2}\overline{(\Delta T)^2}\frac{\partial^2 I}{\partial T^2}\Big|_{\bar{\rho},\,\bar{T}} + \overline{(\Delta\rho)(\Delta T)}\frac{\partial^2 I}{\partial\rho\partial T}\Big|_{\bar{\rho},\bar{T}} \qquad [22]$$

which can be shown to be

$$\frac{\overline{I(\rho,t)}}{I(\bar{\rho},\bar{T})} = 1 - \overline{\Big(\frac{\Delta\rho}{\bar{\rho}}\Big)^2}\,\psi^2 - \overline{\Big(\frac{\Delta T}{\bar{T}}\Big)^2}\,\frac{D}{2}\,[\psi(D-1) + D(1-2\psi)]$$

$$- \overline{\Big(\frac{\Delta T}{\bar{T}}\Big)\Big(\frac{\Delta\rho}{\rho}\Big)}\,2D\psi\,(1-\psi) \qquad [23]$$

Thus, for a particular spectral line emitted from the bombarded gas of mean properties $\bar{\rho}$ and $\bar{T}$ the average measured intensity $\bar{I}$ will differ from that at the mean values of ρ and T by the amount of the last three terms on the right of eq. [23]. It appears, considering that $0 < \psi < 1$, that these "distortion" co-efficients $D(\psi(D-1) + D(1-2\psi))/2$ and $2D\psi(1-\psi)$ of the temperature fluc-tuations and temperature-density cross-correlation are of order $D/2$, and for the examples quoted by Smith (1976), this is a small quantity since $D \sim 0.3$. For a choice of spectral line and ρ, T, however, where ψ can approach unity, the distortion can approach $((\Delta\rho)/\bar{\rho})^2$. Since large values of the latter are expected in certain interesting flows, the distortion there can be serious indeed.

3. Some Current Trends in Multivariant Sensing Methods
Multivariant sensors can be thought of as capable of measuring one property independently of others varying along with it; or of measuring cross-products of two or more variables by bias changes for purposes of correlation detection, or for increasing the efficiency and economy of the experiment in general. The following list, also shown in outline in Figure 4, presents the most publicized examples of both types of probes, to which the analysis of Section 3 can be applied.

PROPERTY	METHOD (GENERAL)	SPECIFIC
VELOCITY	TIME-OF-FLIGHT	COHERENT SCATTERING (LASER DA)
	THERMAL EFFECT	HOT-WIRE, HOT FILM
TEMPERATURE	THERMAL EFFECT	HOT WIRE, HOT FILM
	ATOM / MOLECULE EXCITATION	ELECTRON BEAM
DENSITY	THERMAL EFFECT	HOT-WIRE, HOT FILM
	ATOM / MOLECULE EXCITATION	ELECTRON BEAM, LASER
	LIGHT REFRACTION	SCHLIEREN, SHADOW (INTERFEROMETRY)
PRESSURE	THERMAL EFFECT	HOT WIRE, HOT FILM
	MICROPHONE	PRESSURE TRANSDUCERS
CONCENTRATION	THERMAL EFFECT	HOT WIRE, HOT FILM
	ATOM / MOLECULE EXCITATION	LASER FLUORESCENCE

Fig. 4. Current diagnostic trends in multivariant flows

3.1 Unsteady Velocity Measurements

No single instrument capable of measuring unsteady velocity in multivariant flows has yet been invented which can improve on the time-of-flight method of the laser anemometer, much about which appears in subsequent papers. The main advantage of the LDA, within the present context, is its direct insensitivity to other variables. For many practical situations, on the other hand, the LDA is difficult or impossible to use. These include confined or obstructed fields of view (e.g., three dimensional cavities) or large and awkward wind tunnels where vibrations are expensive to suppress or provide for. In such cases, it is possible to return to the thermal sensors (hot wire or hot film), about which further is said in Section 3 and in the paper by Gaviglio (1978). The thermal sensors are especially useful in flows with variable density and temperature, in which the correlations $\overline{(\Delta u)_i (\Delta \rho)}$, etc., can be obtained in principle. In fact, the thermal anemometer variety of probes remains the only type of instrument widely used for correlations of this type. However, it should be strongly stressed that the measurement of multiple-point correlations with thermal sensors in multivariant media has its own rigorous rules (Demetriades 1973b) of which the classic mode analysis is only a special case. This "correlation mode" discipline is illustrated in Figure 5.

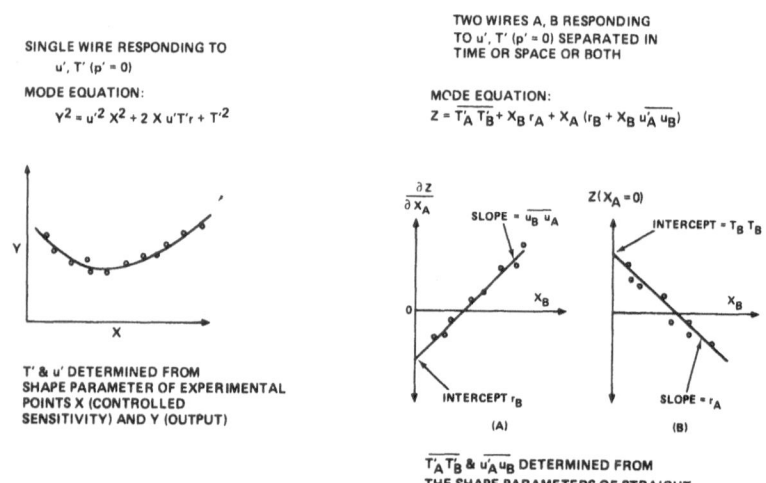

Fig. 5. Space-time correlation measurements with thermal sensors (at right) literally provide another dimension to the classic mode diagram (at left).

Note should be made of the comparison of $\Delta u/\bar{u}$ as measured with the LDV and a hot wire anemometer in Figure 6. Such long overdue comparisons have been made by now at several laboratories and have increased confidence in the use of thermal-type sensors in multivariant flows.

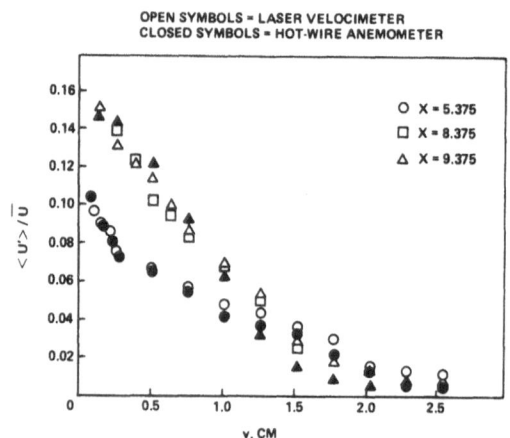

Fig. 6. The advent of the LDA provided a long-needed cross-check of the hot wire response in multivariant flows, in this case a supersonic boundary layer. (After Johnson and Rose).

In certain situations, direct velocity measurements can be done with an alternative time-of-flight technique developed by Bauer (1965). This consists of two parallel wires held at some distance apart, with the first wire emitting heat pulses. Velocity is measured by the time interval needed for the heated pulse to convect to the second wire; a simple mechanism allows the two elements to be placed on the same streamline. Modern signal processing methods would allow this probe to measure all the velocity moments.

3.2 Unsteady Temperature Measurements

Indirect measurement of fluctuating temperatures (i.e., measurement of a Q which depends on other parameters as well), is usually done with a biased thermal sensor. This measurement would appear to be simplest in low speed heated gases, by taking the limiting slope of the sensor ac output vs its heating current parameter at small current i (Corrsin 1947). However, this technique cannot be applied when fluctuations or changes in the velocity or Reynolds number result in recovery-factor fluctuations. The user of such a method will also discover that the benefit of the limiting approximation ($i \rightarrow 0$) is difficult to achieve for slightly heated, high turbulence media. At higher speeds, the $i \rightarrow 0$ limit gives only the fluctuations in total temperature T_o, but the extraction of ΔT from such data is relatively straightforward by means of the mode diagram. There is also the possibility in compressible unsteady or turbulent flows with small Δp, of setting the heating current at a value linearly proportional to $(\Delta T)'$ (Kovasznay 1950); This is an important operating scheme because it allows a direct recording of the $\Delta T(t)$ waveform with many desirable dividents (e.g., simple measurement of probability properties such as $\overline{\Delta T^k}$). Of course, the velocity-temperature $\overline{(\Delta u)(\Delta T)}$ or velocity-density $\overline{(\Delta u)(\Delta \rho)}$ cross-correlation is available from the mode method, but again, only if restrictive assumptions (such as $\Delta p = 0$) can be made.

It is quite ironic that in combustion research, where high temperatures and multiple species are involved, there is considerably more optimism about measuring temperature histories (Goulard 1976). In "classical" flows, we should keep in mind that we are usually discussing $(\Delta T)'$ of order 10 % around mean room temperature levels or lower.

In a large number of flows of interest (e.g. free shear flows), the absolute shear is small and so is the sound field intensity. In these cases the temperature fluctuations are numerically equal (and anti-correlated with) the density fluctuations and can thus be inferred from the latter, which are discussed below.

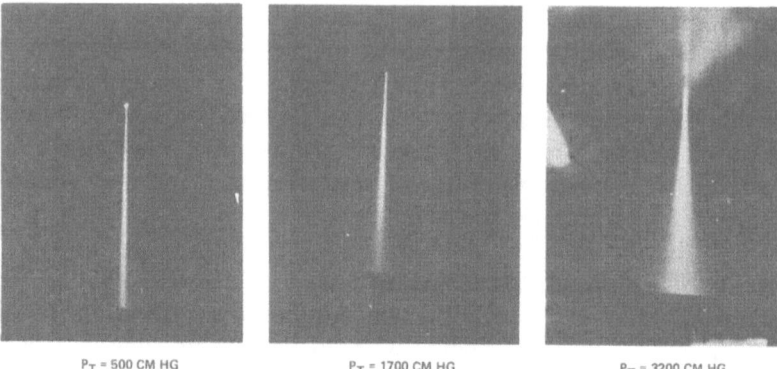

P_T = 500 CM HG P_T = 1700 CM HG P_T = 3200 CM HG

Fig. 7. Capable of simultaneous measurement of gas density and temperature fluctuations, the electron-beam fluorescence technique is severely limited by beam spreading at the higher pressures (after McRonald).

At low gas pressures, the local gas temperature history can be measured from the fluorescence induced by energetic electron beams such as discussed in the form of an example in Section 2.7. Workers at several laboratories have measured the moments of such time histories (mean and fluctuating values, spectra, correlations, etc.); see Muntz (1968), Smith (1975), McRonald (1975). To the extent that ionic excitation processes (Ashkenas 1967) are understood, and subject to the hardships listed typically by Smith (1976), this technique has great promise in hypersonic or low-density flows or combinations of both. A typical upper threshold in pressure for the use of this technique is about 15 torr-inch (McRonald 1975) where the length refers to the portion of the electron beam immersed in the flow; Figure 7 shows what happens at higher pressures. Finally, although the emission from the electron-excited gas is a multi-variant which follows the general analysis of Section 2, it is apparently possible to pick emission lines for which the signal is nearly univariant, i.e., which depends only on the rotational gas temperature, for example.

3.3 Unsteady Density Measurements

In contrast to temperature measurements, density fluctuation measurements are endowed with many candidate schemes, some of them quite old, in aerodynamic flows of widely varying conditions:

(a) Schlieren and Shadowgraph Methods (Optical Refraction)

These are among the earliest "direct" methods; the fundamentals were explained a long time ago by Kovasznay (1954). The measuring equipment is commonplace, and an advantage is offered in picturing the entire flowfield, which eliminates preliminary diagnosis and accelerates concentration on interesting regions of the flow. The disadvantages include the necessity for some key assumptions (e.g., for isotropic turbulence, its three-dimensional spectrum is simply related to the microdensitometer correlations). Also, stationarity is marginal because of the finite number of eddies in the processed region, so that for most flows (e.g., jets, shear layers) many spark pictures (or shadow movies) may be needed. Typical application of such techniques can be found in Slattery (1962). In this connection, the laser Schlieren (crossed-beam technique of Fisher and Krause (1967)) and its subsequent refinements, deserve attention as a potential method for density fluctuation measurements.

(b) Fluorescence Methods (Atomic/Molecular Excitation)

The bombardment of a gas by particles (electron beam approach) or waves (e.g., laser fluorescence approach) offers the possibility of instantaneous density measurements with focused optics, since the emitted radiation depends on the number density of atoms or molecules in the sampled volume; the electron beam technique is discussed in 2.7, and the laser fluorescence method is mentioned in 3.4. The credentials of these techniques are strongest in low density, high-temperature flows, or where an additive can be conveniently chosen to resonate with the exciting radiation. There are very few attempts to use such techniques in "classical" flows of moderate (e.g. room) temperature and densities, where the type of working fluid is not at the disposal of the experimenter, or when additive fluctuations do not safely reflect fluctuations in the density of the carrier gas. Also, note that in the fluorescence technique emission lines can be picked so that $(\Delta\rho)$ can be reportedly measured directly (i.e., where the analysis of Section 2 favors the density fluctuations). Figure 8 shows a comparison between electron beam and thermal sensor measurements of density in a "classical" flow; the correspondence is good and the apparent lower frequency response of the thermal sensor in this case can of course, be vastly improved.

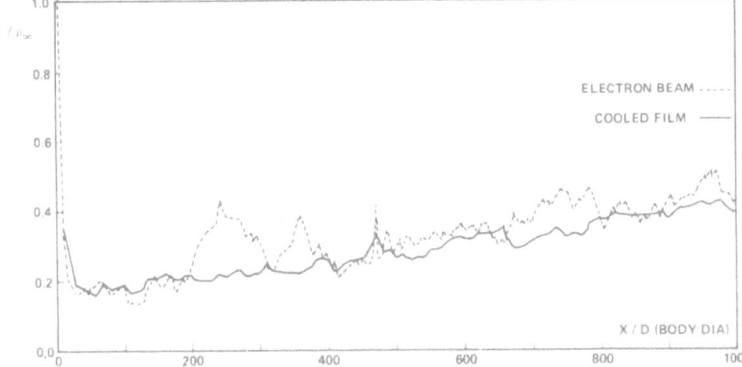

Fig. 8. Simultaneous demonstration of unsteady density measurement with a thermal sensor and the electron beam (after Ellington).

(c) Electroresistive Methods

The class of diagnostic probes often known as "glow anemometers" has probably been permanently laid to rest because of prohibitive practical problems such as inadequacies of understanding electric discharges in moving, multi-variant media, capricious electrode behavior, etc. However, untapped advantages may exist in this method which is a direct one, for example, by the repetitive measurement of breakdown voltage (instead of the steady-state current fluctuations) or the measurement of the normalized fluctuation $\Delta\rho/\bar{\rho}$.

(d) Thermal Methods

Indirect measurements of density fluctuations with hot wire and hot film anemometers are routine for compressible and other similar flows, if the assumption can be made that the term carrying the pressure fluctuations is much smaller than that carrying $(\Delta T)/\bar{T}$. In this case, the equation of state

$$\frac{\Delta p}{\bar{p}} = \frac{\Delta\rho}{\bar{\rho}} + \frac{\Delta T}{\bar{T}} = 0 \tag{24}$$

is invoked, and the response equation ends up as a bivariant, containing only $\Delta u/\bar{u}$ and $\Delta\rho/\bar{\rho}$ (the same argument can be made for detecting $\Delta T/\bar{T}$). This technique remains the most common, at this writing, for the measurement of density fluctuations.

3.4 Measurement of Concentration Fluctuations

The still-outstanding problem of binary mixing in turbulent flow has given rise to a multitude of experiments to measure turbulent diffusion of additives. This particular problem gives research workers much latitude and opens up many diagnostic possibilities, since additives can be arbitrarily picked for laboratory study which also make the diagnostic easier. Thus, there is a large number of suggested techniques which have so far fallen into certain broad categories:

(a) **Thermal Methods.** These are based on differences of thermal conductivity between the working fluid and the additive, and are especially suited to low-speed gas flows. Corrsin (1947) was first to indicate that thermal fluctuations due to concentration fluctuations $c(t)$ produce fluctuations in a hot wire voltage:

$$e(t) = S_u u(t) + S_c c(t) \tag{25}$$

where S_c, the sensitivity to concentration fluctuations is, according to [6], found from $\partial e(u,c)/\partial c|_{\bar{u},\bar{c}}$. Devillers and Diep (1973) showed subsequently that the Corrsin technique is feasible even in supersonic flow because of the good sensitivity of the hot wire to changes in the gas composition. Two interesting modern variations of this technique deserve mention. In one, Brown and Rebollo (1972) immersed the hot wire in a microscopic wind tunnel whose "stagnation chamber" ahead of the supersonic nozzle throat is the flow region to be sampled. The choking characteristics of their probe inlet suppressed the sensed velocity fluctuations so that concentration fluctuations (to 1 % of helium in nitrogen) could be detected with a frequency response of at least 5 kHz.

In another interesting application, Libby and co-workers (1970) installed an array of hot wires and hot films in close proximity in various mixtures; the essential element of this array was a "detector" placed in the wake of a "heater", for which these workers observed higher sensitivities to Δc, as would be expected, than obtained for a single hot wire or single hot film. Experimenting with low speed helium-air mixtures, these authors report numerical results on no less than 31 different moments, such as $(\Delta u)'/\bar{u}$, $\overline{(\Delta v)(\Delta c)}/\bar{c}\bar{u}$, $\overline{(\Delta c)(\Delta \rho)(\Delta v)}/\bar{\rho}\bar{c}\bar{v}$, etc.

(b) Optical Methods

The widespread use of laboratory lasers has given rise to a class of concentration fluctuation techniques which are based on additives fluorescing under laser excitation. The set-up features the classic lines of a narrow illumination beam, observed along one of its points by an apertured optical detector. Proper matching of the laser and additive will produce the desired light emission, while proper matching between the additive and the collecting optics will register the proper filtered voltage at the photomultiplier output. In the applications discussed in the literature, the fluorescent intensity is found to be linearly proportional to the additive concentration; the sensed variable is thus univariant ($Q(x_i) = Q(x)$ from the standpoint of Section 2) and also unaffected by nonlinearity since all terms in eq. [5] vanish except those containing $\partial Q/\partial x$. Many of the laser-fluorescence works report on liquids: Dewey (1976) reports using rhodamine 6G dye as an additive, and collecting light at 5600-5800 angstroms; Owen (1976) used uranin dye and managed to measure velocity in the process by switching to the LDA mode.

Another method for measuring concentration fluctuations depends on light absorption and is exemplified by the work of Batt and co-workers (1970). Unlike other optical methods, this technique requires an immersible probe featuring a gap across which light, not necessarily coherent, travels from an emitter to a receiver. As the concentration of the additive streaming through the gap fluctuates, so does the received light intensity by Lambert's law, and so does the generated voltage at the PM tube. In Batt's experiment, the additive used was nitrogen dioxide and the filter aperture centered at 4000 angstroms.

32 Anthony Demetriades

(c) Electroresistive Methods

If the additive has different electrical conductivity (resistivity) properties than the surrounding fluid, then its fluctuations will produce a variable emf at an electrode immersed in the flow of interest. Gibson (1968), using such a conductivity probe, measured concentration fluctuations of a weak electrolyte in water.

In summary, there seems to be a wealth of methods for measuring fluctuations of concentration; some of these are "non-interfering"★, quite easy to set up, simple in principle, and oblivious to simultaneous fluctuations of other properties. On the other hand, nearly all experiments reported so far with these techniques took place either in liquids or in low speed flows of gases and seem more pre-occupied with the excellent flow visualization obtained than with quantitative measurement. Their quantitative utility in certain broad regimes of classical aerodynamics (e.g., higher temperature or high-speed gas flows) remains to be demonstrated.

★The "non-interfering" feature of optical techniques sometimes provides an exaggerated influence on the choice of measuring technique. The claim that all immersible probes provide local flow distortion which invalidates the data is simply not true. Immersible probes do distort the flow greatly in certain situations (e.g. separated regions) and are awkward to work with in certain others, but in many cases the probe-flow interaction is either nil or forms part of the measuring system (e.g. pitot tube shock wave in unsteady supersonic flow, see Section 3.5). For most cases, the interference is either measurable or calculatable in advance. Strictly speaking, optical or other remote methods also have the potential for interference, such as streamline distortion by material addition in the flow (light-scattering particle or fluorescent dye addition), local distortion due to heat absorption by radiation, etc. The sensor-flow interaction issue should, therefore be addressed "from the ground up" for each particular measurement.

3.5 Unsteady Pressure Measurements

(a) Direct Static Pressure Measurements

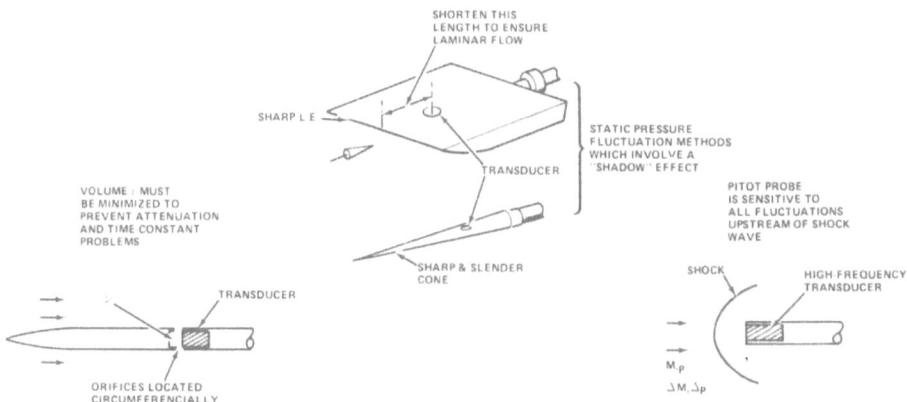

Fig. 9. Various schemes of unsteady pressure measurements.

Direct measurement of fluctuating pressure at a point of a flow is a field in which no innovations have occurred for decades. Such measurements are still done by transducers (capacitance, piezoelectric) supplied by specialist manufacturers rather than constructed by the researchers. Good progress toward reliability, frequency response and miniaturization have been made (it is now possible to obtain transducers with active surfaces in the 0.02 - 0.1 cm diameter range) but usually at the expense of sensitivity and tolerance to high temperatures. Furthermore, since static pressure fluctuations, unlike other fluctuations, do not convect along streamlines (Kovasznay 1954), there are intrinsic difficulties in measuring the former; as stressed in Figure 9, there is a "shadow" effect created by the solid structure supporting the transducer, by which a portion of the pressure waves radiated to the transducer are obscured.

(b) Indirect Static Pressure Measurements
These have already been mentioned incidentally as by-products of other techniques in cases where Δp exists simultaneously with other variable properties (e.g., the "hot wire microphone" is an old idea, see Tucker 1921). If the electron beam technique (Section 2.7) is perfected at higher pressures than heretofore possible, it offers a unique method of Δp measurement, since the ΔT and $\Delta \rho$ can be measured independently by it.

(c) Unsteady Pitot Measurements
Because of the physical simplicity of the Rankine-Hugoniot relations, the pitot probe can be used to measure certain unsteady properties, if it is combined

with a miniature pressure transducer such as those available at present. In supersonic flow, the fluctuating pitot pressure Δp_T is connected to fluctuations
Δp in the static pressure and Mach number ΔM, for example if $\gamma = 1.4$, by

$$\frac{\Delta p_T}{\bar{p}_T} = \frac{\Delta p}{\bar{p}} + \left(7 - \frac{35\bar{M}^2}{7\bar{M}^2 - 1}\right)\frac{\Delta M}{\bar{M}} \qquad [26]$$

in the linearized case. As can be seen, there is no possibility of bias in this
relation by which a system such as [11] can be obtained. However, this type of
"unsteady pitot probe" could give additional algebraic relations among the unknown fluctuations if used in conjunction with other probes. The Mach number function in paranthesis has a numerical limit (=2) as soon as M exceeds a
value of 2 or so; since

$$\frac{\Delta M}{\bar{M}} = \frac{\Delta u}{\bar{u}} - \frac{1}{2}\frac{\Delta T}{\bar{T}} \qquad [27]$$

and since the equation of state $\Delta p/p = \Delta \rho/\rho + \Delta T/T$ can be invoked, eq. [26]
then gives

$$\frac{\Delta p_T}{\bar{p}_T} = \frac{\Delta p}{\bar{p}} + 2\frac{\Delta M}{\bar{M}} = \frac{\Delta(\rho u)}{\overline{\rho u}} + \frac{\Delta u}{\bar{u}} \qquad [28]$$

This further simplifies the use of the pitot probe as a complement to other sensors; that is, it provides an additional equation to an undetermined system if
the pitot tube can be placed at the same point as another, "underterminate"
sensor.

4. Current Hot Wire Anemometer Methods

Perhaps a full one-half of the multivariant unsteady measurements in fluids reported in the recent years have been done with some of the hot wire anemometer probes or its variants involving one or more hot wires or hot films simultaneously; it, therefore, seems fitting to elaborate on this technique, especially as
it applies to multivariance generated by high speeds. In his paper, Dr. Gaviglio
will elaborate further on this type of sensing technique.

4.1 The Hot Wire in Supersonic and Hypersonic Flow

With the exception of Kistler's study of supersonic boundary layer turbulence,
it took close to fifteen years before the hot wire technique came into general
use in compressible flows. Among the serious obstacles causing this delay were
the problems associated with high speeds; large dynamic stress, high temperatures, high-frequency requirements and the difficulty of controlling dirt particle content in supersonic wind tunnels. Solutions to these hardships seemed
always contradictory; for example, decreasing the wire diameter to reach the

higher frequencies weakens the element structurally. The most serious obstacle, however, had to do with the signal processing needed to "split" the signal into the mode contributions. The high cost of operating supersonic and hypersonic facilities is incompatible with the lengthy operations needed for hot wire measurements such as calibration, compensation and bias current changes. For example, modal analysis of a spectrum (e.g., 100 Fourier components) utilizing ten overheat currents would require 1000 readings — an already very risky proposition in a particle-laden 3 or 4-psia flow; and we well know that once the wire breaks (or simply shifts resistance) a change of probes (and a repetition of the measurement) is necessary since the heat transfer characteristic differs from one wire to another.

Hugh Dryden and his contemporary pioneers would probably be quite impressed to observe a modern hot wire experiment in progress. Imagine a 50-inch (127 cm) hypersonic wind tunnel continuously operating at Mach 8 and T_o = 850° F, and an airlock on top of it, the size of a small room. Inside the airlock sits a large probe holder/actuator with a shielded hot wire anemometer mounting a 0.00002"-diameter (0.000051 cm) wire of length 0.004" (0.01 cm) attached to nickel prongs by gold paint (soldering or spot-welding is impossible for such dimensions). As soon as desired, controllers set in motion machinery which open a trap door linking the air lock to the test section, lower the probe holder to the model and "fly" it to the desired position on it. Further commands remove the probe shield and, while closed-circuit television monitors display a microscope view of the probe tip, position the hot wire relative to the model surface with a precision of 0.0001" (0.00025 cm). The Nusselt number and recovery factor characteristics are then measured quickly with the aid of an analog computer. For the turbulence measurement itself, the heating current is rapidly stepped fifteen times from a push-button panel. The wire ac signal, together with its mean voltage, current, local flow conditions, etc. is continuously stored in analog and/or digital form, the human operator being limited to monitoring recorders and alarming the system to anomalies or failures★. Figure 10 gives some idea of the differences between "older" and "modern" hot wire anemometer detection systems.

The experimenter still has the task of critical analysis of the data, but the great bulk of the reduction process labor next belongs to the computer. Routine digital operations on the stored signals provide "instant" mode diagrams for the broad band; fast Fourier transform programs and/or passage through digital/averaging wave analyzers provide mode diagrams at a great number of

★This seemingly utopian system is in operation at the U.S. Air Force's Arnold Engineering Development Center (AEDC), Tullahoma, Tennessee, U.S.A.

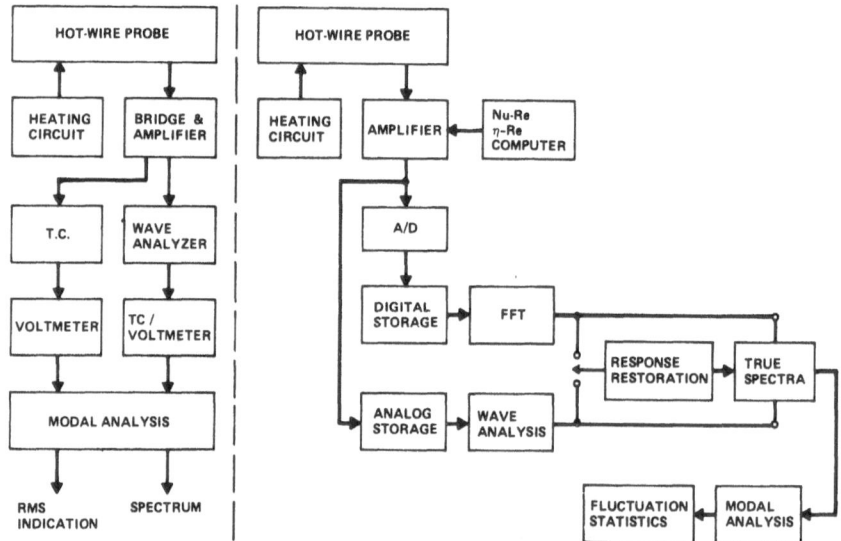

Fig. 10. Modern uses of thermal sensors in multivariant flows generate complex measuring circuits of hardware and software.

Fourier components just as readily. The hot wire spectrum at each current is first "response-restored" by computing the amplitude transfer function of each circuit component — the wire response, for example, is corrected by computing its time constant from the known mean flow conditions and comparing it with the time-constant setting of the compensator. The result is a mode diagram (in the broadband or at any Fourier cell) which comes close to one taken with "infinite bandwidth" electronics. By least-squares fitting of the mode diagram the computer provides the velocity and temperature fluctuations, their cross-correlation and their spectra.

Hot wire anemometry in multivariant flows has, therefore, come of age again thanks to the giant steps in electronic and computational technology taken in the last quarter century. But this advance does much more than free the time of the experimenter; it makes the application of the hot wire technique to multivariant flows possible. For example, experience shows that problems in the probe behavior often do not become apparent until the final stages of data reduction; or, the reduction process itself requires approximations which can be made only after mode-diagram inspection. In cases like these, repetition of the measurement or the data reduction process is necessary — a prospect made palatable only by the modern techniques outlined above.

On the other hand, certain key aspects of hot wire anemometry for high speed, multivariant applications have changed little over the years, and these are listed below in the form of problem areas and possible solutions.

4.2 Heat Transfer Calibration: The Case for the "Non-Ideal" Wire

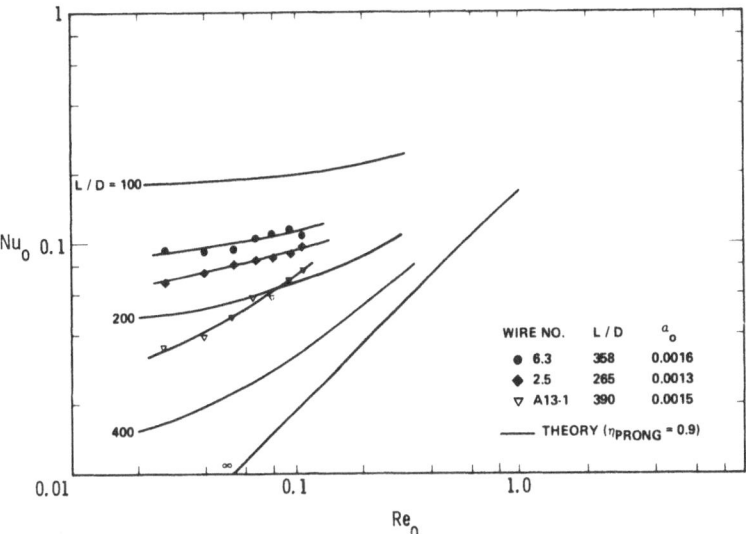

Fig. 11. Heat transfer characteristics of 0.00001 inch (0.000025 cm) diameter hot wires in hypersonic flow compared with theoretical predictions for finite aspect ratio L/D.

King's Law (1914) is invalid at the Reynolds numbers encountered in high-speed hot wire experimentation, where the reduced wire diameters required for high frequency measurements place the sensor in the slip or free-molecular regime (see Dewey 1965). Dewey and later Behrens (1973) have derived formulas for the wire Nusselt-Reynolds and recovery-factor behavior to account for this effect and also for the finite-length effect. In principle, these theories, illustrated in Figure 11, can be used to find the wire sensitivity coefficients to fluctuations (Morkovin 1956) and thus to forego the need for calibrating individual wires. As Figure 11 shows, however, individual hot wires specifically built for high-speed applications depart greatly from the theoretical predictions. The reason for this is that the reduced wire length (necessary to produce a strong moderate-aspect-ratio structure) also produces a "non-ideal" wire geometrically, instead of the straight cylinder normal to the flow vector. Such hot wires require individual calibration in the flow prior to the fluctuation measurement.★ The practice of individual calibration which is in anyway indicated by good laboratory practice in all flows, is thus mandatory for high-speed measurements.

4.3 Frequency Response Requirements

What frequency range should one design for use at supersonic or hypersonic speeds? This question is crucial since, in affecting the wire diameter, it affects the validity of most arguments presented here. If one is only interested in fluctuation magnitudes (integrals of the spectrum), then one is content with a system low-passing signals in the so-called "energy-containing" region. This lies roughly below a frequency, for example for a boundary layer, corresponding to the convected integral scale Λ, i.e., below $f\Lambda/u = 1$. Since $\Lambda \cong \delta/5$ and if we conservatively take $u = u_e$, then an acceptable upper limit would be $f = 5u_e/\delta$; for typical values of $u_e = 1000$ m/sec and $\delta = 1$ cm, this implies $f = 500$ kHz. For spectrum studies, a factor 2 to 4 beyond that is desirable (to 2 MHz).

4.4 Constant-Current Compensation in High-Speed Turbulent Flows

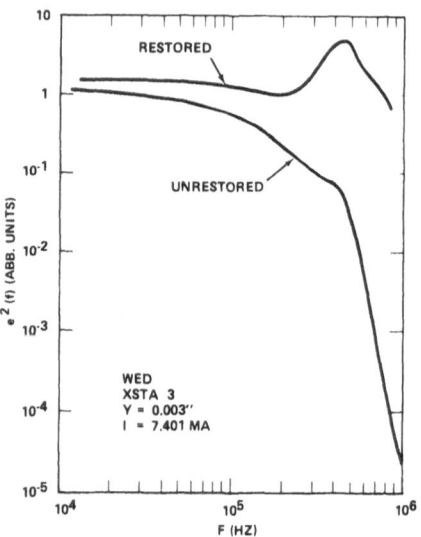

Fig. 12. Computer-aided response restoration reveals spectral features (in this case at 450 kHz) of this hot wire application to supersonic flow, which are not discernible in the unrestored output.

★In Laderman and Demetriades (1974) wires of diameter 0.00001" (.000025 cm) and lengths as small as 0.002" (0.005 cm) were used at Mach 10, $T_o = 1000°$F.

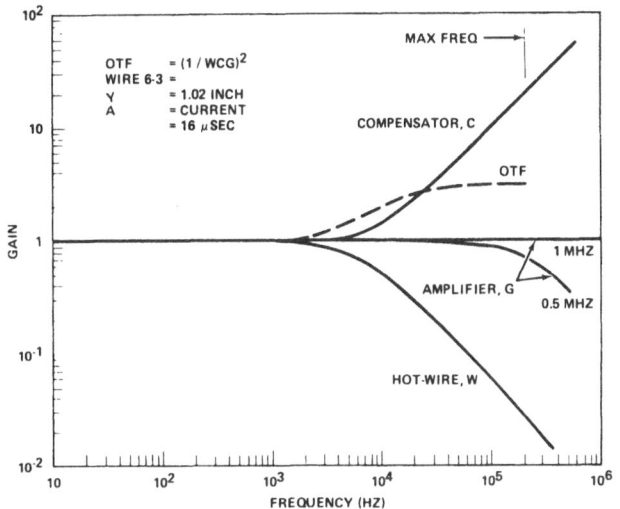

Fig. 13. Automatic computation of the overall transfer function (OTF) corrects for mismatched frequency response of the hot wire anemometer components (solid lines).

The widespread use of constant-current hot wire amplifiers has always dictated the use of electronic response compensation by non-uniform amplification. Figures 12 through 14 give an intimation of the penalty to be paid by compensation which is too low or too high. (Even for 0.00001'' dia wires which have a time constant τ of order 20 microseconds the maximum uncompensated frequency of use is only of order 10 kHz which compares unfavorably with the 2 MHz bandwidth quoted above). The difficulty in high speed shear flows is that the wire constant varies greatly across the flow (and with the bias current). Regardless of the compensation method (Kovasznay 1954, Sandborn 1974) the point-by-point adjustment of the compensation network would make the process prohibitively time consuming. In the AEDC system, the compensation network remains at one setting; the wire time-constant variation is calculated from the flow properties and compensation is introduced digitally in the data reduction process.

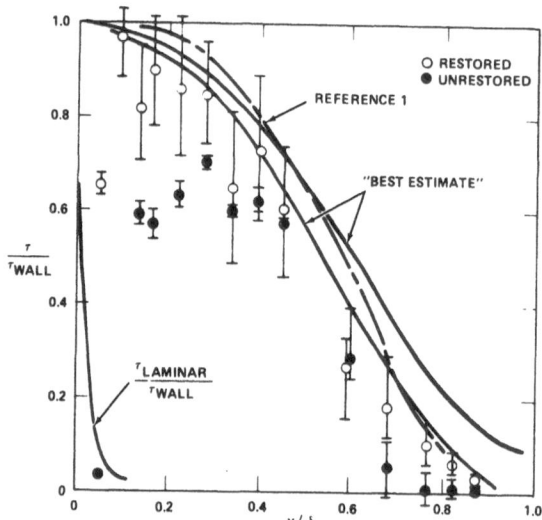

Fig. 14. Response restoration of hot wires causes substantial changes in mea-sured Reynolds stresses at supersonic speeds (after Laderman).

4.5 On Certain Approximate Methods For Interpreting The Data

Often, statements are found in the literature to the effect that the hot wire, operated in a multivariant flow where both velocity and temperature fluctuations are active, will respond only to the latter at small currents and only to the former at high currents. This practice is unjustified. At low currents, the wire responds to total temperature fluctuations in which the velocity fluctuations play a role. In fact, there are circumstances where the wire responds only to density/temperature fluctuations (when $\Delta p = 0$) as indicated earlier by the mode-diagram method (Kovasznay 1950) but this occurs (if at all) primarily at high overheats! However, the most serious faults of such approximations is that they are wrongly based on a comparison of the magnitude of the sensitivity co-efficients, while they should be based on the magnitude of the *product* of the fluctuation and the sensitivity coefficients (see Section 2.3).

4.6 Non-Linearity

A likely difficulty to be encountered by hot wire and hot film probes has to do with the fluctuation magnitude at high speeds. Whereas at low speeds the velocity fluctuations in ordinary turbulence are truly second-oder (1 % - 10 %), in flows with shear-driven thermal effects (e.g. supersonic or hypersonic bound-ary layers) the temperature fluctuations are of order $M^2 \Delta u$. Despite the ob-served decrease of Δu in the hypersonic regime, the potential of very large ΔT or $\Delta \rho$ (of order 100 %) is real. The usual linearized approach to the response equation is, therefore, invalid as is the mode diagram method of the hot wire as we know it today.

4.7 On the Unresolvability of the Hot Wire Output

Although it is usually mentioned that the $\Delta p \neq 0$ condition makes it impossible to separate the hot wire output into individual components, the problem is basically algebraic with consequences which are more general. The response equation connecting the voltage output $\Delta e(t)$ with any set of three independent variables (e.g., entropy σ, vorticity τ and sound π) via the sensitivity coefficients e_σ, e_τ and e_π is, in the mean square (analogous to eq. [13] in Section 2.6),

$$\overline{e'^2} = e_\sigma^2 \, \overline{\sigma^2} + e_\tau^2 \, \overline{\tau^2} + e_\pi^2 \, \overline{\pi^2} + 2e_\sigma e_\tau \, \overline{\sigma\tau} + 2e_\sigma e_\pi \, \overline{\sigma\,\pi} + 2e_\tau e_\pi \, \overline{\tau\,\pi} \qquad [29]$$

The classic hot wire technique realizes that the current-biased controllable variable $e_\tau/e_\sigma = \chi$ can be used to put this into the form:

$$\overline{e'^2}/e_\sigma^2 = \overline{\sigma^2} + \overline{\tau^2}\chi^2 + f_1(\chi) \, \overline{\pi^2} + 2\chi \, \overline{\sigma\tau} + f_2(\chi) \, \overline{\sigma\pi} + f_3(\chi) \, \overline{\tau\pi} \qquad [30]$$

The fundamental difficulty here is that none of the coefficients $f_1(\chi)$, $f_2(\chi)$, etc., involve powers of χ higher than 2, so that at best, the above formula can be re-written as

$$Y \equiv \frac{\overline{e'^2}}{e_\sigma^2} = C_1 + C_2\chi + C_3\chi^2 \qquad [31]$$

The three coefficients C_1, C_2, C_3 can be found from the shape parameters of the curve (χ, Y), but since they depend on the six unknown fluctuation parameters, they provide only half of the equations needed for determining the fluctuations.

The single most frequent assumption made to resolve this difficulty is that $\Delta p = 0$ (more correctly, that the term including Δp or π in the response equation [29] is negligible compared to the other terms). This assumption seems justified at lower Mach numbers (say below 5) but not at high Mach numbers for which it is known that acoustic radiation from turbulence is high. The indeterminacy is sometimes circumvented if many points on the mode diagram are available since the degree of "sound content" of the flow is then recognizable on this diagram. Furthermore, using alternative assumptions rather than $\Delta p = 0$, Laderman and this author (1974) have found that the measured velocity and temperature fluctuations are not changed much. In the opinion of this author, however, this difficulty remains severe and perhaps prohibitive for high hypersonic Mach numbers.★

4.8 On the Use of Multiple Hot Wires in Multivariant Flows

In previous statements, we noted the inherent physical (mathematical) limitations of hot wires in multivariant flows, namely the indeterminacy arising in the processing of the mean-square ac signal. The historical reason for using the mean-square signal is that the signal acquisition and processing technology was, comparatively speaking, at its infancy during the formative years of the hot wire technique. In those days, a stationary flow produced a single signal at each heating current which could be easily recorded by an ac/dc converter.

Prospects are quite bright, however, for multiple hot wire or hot film probes positioned at the same point and operated at different bias settings to give instantaneous outputs of the type represented by the system [11]. Ellington (1970) and more recently Libby (1974) and his co-workers have already taken steps in that direction. The skills and systems required for such a "multiprobe" are as follows:

1. The mounting and survival of individual elements within a volume in space which insures spatial resolution. In view of practical element lengths which can now be made of order 0.005 cm, this becomes a not-insurmountable question of manual skill. Hot films can replace hot wires since the frequency response of the former can be predicted and since computer response-restoration will be in anyway necessary.

2. Data A/D conversion to handle up to 1 MHz and storage to handle 100,000 words per channel or better.

3. Computer algorithms for solving the simultaneous system of equations obtained.

With these improvements, it is not unlikely that an immersible probe capable of recovering the time history of all the vector and scalar properties of the unsteady flow will be feasible.

★This argument applies to flows with high shear such as boundary layers. For multivariant low shear flows such as wakes, the sound radiation decreases greatly.

References

The following list contains references cited in the text, as well as other sources useful to a reader looking for the historical evolution, physical explanation or typical application of each technique mentioned.

1. **Hot Wire Anemometry for Multivariant Flows**
   Behrens, W. (1973): Private communication
   Corrsin, S. (1947): *Rev. Sci. Instr.,* Vol. **18**, No. 7, p. 469
   Demetriades, A. (1968): *Physics of Fluids,* Vol. **11**, No. 9, p. 1841
   Demetriades, A. (1973b): *Trans. of ASME, Series E, J. App. Mech.* Vol. **40**, No. 3, p. 822
   Ellington, D., Picard, A.G. and Park, K.R. (1970): *DREV TN-1901/70,* Quebec, Canada
   Gaviglio, J. (1978): Paper M1.1, *Proceedings of The Dynamic Flow Conference 1978*
   Johnson, D.A. and Rose, W.C. (1975): *AIAA J.* Vol. **13**, No. 4, p. 512
   King, L.V. (1914): *Phil. Trans. Roy. Soc. (London)* Vol. **214**, No. 14 Sect. A, pp. 373-432
   Kovasznay, L.S.G. (1950): *J. A. Sci.* Vol. **17**, p. 565
   Kovasznay, L.S.G. (1954): "Turbulence Measurements, F.2: Hot Wire Method" in Physical Measurements in Gas Dynamics and Combustion, Vol. **IX**, *High-Speed Aerodynamics and Jet Propulsion Series* (R.W. Ladenburg, ed.) Princeton Univ. Press
   Laderman, A.J. and Demetriades, A. (1974): *JFM* Vol. **63**, part 1, p. 121
   Laderman, A.J. (1976): *AIAA J.* Vol. **14**, No. 9, pp 1286-1291
   Morkovin, M.V. (1956): *AGARDograph* No. **24**, NATO
   Mikula, V. and Horstman, C.C. (1975): *AIAA J.* Vol. **13**, No. 2, p. 1607

2. **Velocity Methods (Other Than LDV or Typical Hot Wire)**
   Bauer, A.B. (1965): *AIAA J.* Vol. **3**, No. 6, p. 1189
   Bradbury, L.J.S. and Castro, I.P. (1971): *JFM* Vol. **29**, Part 4, p. 657
   Donaldson, J. (1976): *AEDC TR 76-88,* Tullahoma, Tenn.

3. **Unsteady Temperature and Density Measurements (Excluding Thermal Sensors)**
   Ashkenas, H. (1967): *Phy. of Fluids,* Vol. **10**, No. 12, p. 2509
   Fisher, M.J. and Krause, F.R. (1967): *JFM* Vol. **28**, Part 4, pp 705-717
   Funk, B.H. Jr., and Cikanek, H.A. Jr. (1971): *NASA TN D-6077,* Washington, D.C.
   Kovasznay, L.S.G. (see under Section 1 of References)
   Krause, F.R., Davies, W.O. and Cann, M.W.P. (1969: *AIAA J.* Vol. **7**, No. 4, p. 587
   Ladenburg, R., Winkler, J. and Van Voorhis, C.C. (1948): *Phys. Review,* Vol. **73**, p. 1359
   Maguire, B.L., Muntz, E.P. and Thomas, K.M. (1972): *AIAA Paper 72-118,* New York, N.Y.
   McRonald, A.D. (1975): Ph.D. Thesis, Univ. of Southern California, Los Angeles, CA
   Muntz, E.P. (1962): *Phy. of Fl.* Vol. **5**, No. 1, p. 80
   Muntz, E.P. (1968): *AGARDograph,* No. **132**, Paris

Smith, J.A. and Driscoll, J.F. (1975): *JFM* Vol. **72**, Part 2
Smith, J.A. and Driscoll, J.F. (1976): *AGARD CP-193,* Paper No. 16
Slattery, R.E. and Clay, W.G. (1962): *Phy. of Fl.* Vol. **5**, No. 7, p. 849

4.   **Unsteady Concentration Measurement Techniques**
Batt, R.G., Kubota, T. and Laufer, J. (1970): *AIAA Paper 70-721,* New York
Blackshear, P.L. and Lingerson, L. (1962): *ARS J.* Vol. **32**, No. 1, p. 1709
Brown, G.L. and Rebollo, M.R. (1972): *AIAA J.* Vol. **10**, No. 5, p. 649
Demetriades, A. and Doughman, E.L. (1973a): *J. of Plasma Phy.* Vol. **9**, Part 3, p. 367
Devillers, J.F. and Diep, G.B. (1973): *DISA Information* No. **14**, p. 29
Dewey, C.F. jr., (1976): Paper No. 17, *AGARD CP-193*
Gibson, C.H., Chen C.C. and Lin, S.C. (1968): *AIAA J.* Vol. **6**, No. 4, p. 642
Lapp, M., Penney, C.M. and Asher, J.S. (1973): General Electric Company, Report No. SRD-72-085
Lapp, M. and Penney, C.M. (1973): Laser Raman Gas Diagnostics, Plenum Press
Owen, F.K. (1976): Paper No. 27, *AGARD CP-193*
Stanford, R.A. and Libby P.A. (1974): *AIAA J.* Vol. **17**, No. 7, p. 1353
Way, J. and Libby, P.A. (1970): *AIAA J.* Vol. **8**, No. 5, p. 976

5.   **Pressure Methods (Other Than Hot Wire Mode Analysis)**
Rasmussen, C.G. (1965): *DISA Information*, No. **2**, p. 5
Tucker, W.S. and Paris, E.T. (1920): *Phil. Trans. Roy. Soc.* **A221**

6.   **General Reading**
Andersen, O.K. (1966): *DISA Information,* No. **3**, p. 21
Davies, P.O.A.L. and Fisher, M.J. (1964): *JFM* Vol. **18**, Part 1, pp 97-116
Goulard, R. (1976): In "Applications of Non-Intrusive Instrumentation In Fluid Flow Research", *AGARD CP 193*
Sandborn, V. (1974): *Resistance Temperature Transducers,* Metrology Press

Quelques aspects de l'anémomètrie par fil chaud dans les écoulements turbulents de gaz présentant de forts gradients de température★

J. Gaviglio
Institut de Mécanique Statistique de la Turbulence
12, Avenue General Leclerc - 13003 Marseille

1. Introduction

1.1 Rappels

Les propriétés d'écoulements turbulents à basse vitesse de gaz non réactifs dans lesquels existent des sources de chaleur, et d'écoulements à vitesse élevée avec ou sans sources de chaleur peuvent être souvent étudiées au moyen de l'anémomètre à fil chaud. Un travail de synthèse a été récemment consacré à ce sujet (GAVIGLIO, 1978). On rappelle ici quelques progrès récents obtenus dans le domaine des mesures, et on expose quelques problèmes non résolus et le sens des efforts qui leur sont consacrés.

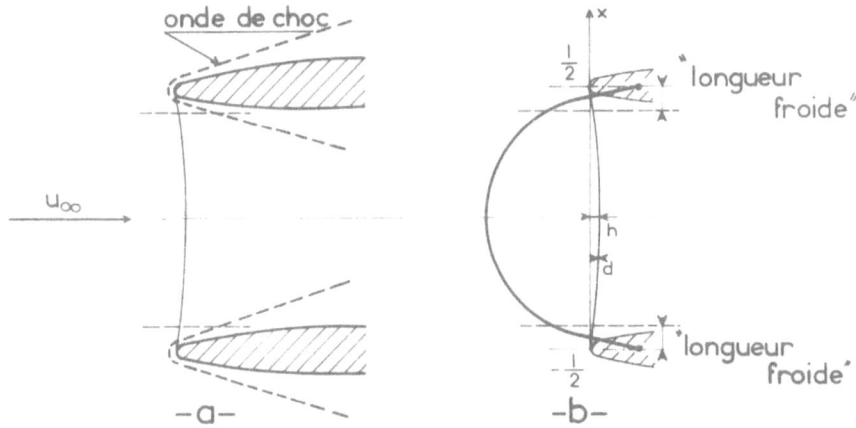

Fig. 1 Géométrie du fil chaud

L'anémomètre étudié est du type "à courant constant". Le fil (Fig. 1) est un petit cylindre de métal, de longueur ℓ, de diamètre d. Il est placé perpendiculairement à l'écoulement ou, lorsque le texte le spécifiera, incliné d'un angle $\varphi = \pi/2$ sur la direction moyenne locale. Le courant de chauffage est i, la

★ Some Aspects of Hot-Wire Anemometry in Turbulent Flows of Gases, with Strong Temperature Gradients

résistance du fil est R_w, et lorsque i tend vers o, R_w tend vers une limite R_e. Les propriétés de l'anémomètre dépendent des lois des pertes de chaleur. Comme l'a montré MORKOVIN (1956), leur expression en fonction des nombres de NUSSELT et de REYNOLDS correspondant à la température d'arrêt locale θ_t constitue la forme la plus aisément exploitable par l'expérimentateur. θ_t est relié à la température θ d'une particule, à la vitesse $\vec{q}$ et au nombre ou Mach M par la relation:

$$\theta_t = \theta + q^2/2c_p = \theta \ (1 + \frac{\gamma-1}{2}M^2) \tag{1}$$

$\vec{q}$ a pour composantes u, v, w; c_p est la chaleur spécifique du gaz (ici l'air) à pression constante. On définit encore la pression génératrice locale p_t par sa relation à la pression p:

$$p_t = p(1 + \frac{\gamma-1}{2}M^2)^{\frac{\gamma}{\gamma-1}} \tag{2}$$

On sait que deux méthodes sont utilisées, principalement, pour séparer les grandeurs moyennes, ($\sim$) ou ($^-$), et les fluctuations, (') ou (''). La distinction ne sera introduite qu'au §4. Jusque-là les moyennes seront notées ($^-$) et les fluctuations (').

1.2 Sur le choix des caractéristiques des fils chauds

1.2.1. Le fil peut subir des contraintes mécaniques élevées dues principalement: aux régimes transitoires correspondant à l'amorçage et au désamorçage des tuyères dans lesquelles sont faites les mesures; à la traînée due à l'écoulement moyen, et aux vibrations qu'engendrent ou qu'entretiennent les perturbations présentes dans l'écoulement, notamment la turbulence. Le choix du tungstène s'impose alors. Une fine couche de platine accroît sa résistance à l'oxydation. Au contraire, lorsque les efforts sont modérés mais les températures élevées, le platine est préféré.

   a) Les *vibrations* provoquées par l'*amorçage* et le *désamorçage* peuvent être réduites en abaissant la masse spécifique du fluide, donc la pression génératrice; en diminuant le plus possible les temps d'amorçage et de désamorçage, et, lorsque cela est technologiquement possible, en abritant le fil dans une couche limite, ou mieux derrière un obstacle. Dans le cas des mesures faites à l'I.M.S.T. en écoulements supersoniques, p_t est réduit à 0,13 bar pendant ces régimes critiques dont les temps, réglés par la manoeuvre d'une vanne, sont seulement de 2,5 sec.

   b) La traînée du fil dans l'écoulement moyen a pour expression:

$$R_x = \frac{1}{2}\,\gamma\overline{p}M^2\,\ell dc_x \cong \overline{p}_t\frac{0,7M^2}{(1+0,2M^2)^{7/2}}\,\ell dc_x$$

où c_x est le coefficient de traînée. Si le fil prend la forme d'une chaînette la tension mécanique unitaire du matériau,

$$T_u = \frac{R_x}{\ell}\,\frac{\ell^2}{2h}\,\frac{4}{\pi d^2} \cong \frac{2}{\pi}\,\overline{p}_t\,\frac{0,7\,M^2}{(1+0,2M^2)^{7/2}}\,\frac{\ell}{d}\,\frac{\ell}{h}\,c_x$$

dépend de ℓ, d et h par les rapports ℓ/h et ℓ/d qui caractérisent la géométrie du fil. Lorsque $\overline{p}_t$ est maintenu constant, et compte tenu de la variation de c_x en fonction de Rd_t qui dépend lui-même du nombre de MACH, T_u passe par un maximum, obtenu pour $M_\infty > 1,4$. Pour une tension unitaire T_u à ne pas dépasser, plus le nombre de MACH est élevé, et plus la pression génératrice peut être fixée à une valeur élevée.

c) Les vibrations susceptibles d'être engendrées ou entretenues par le caractère instationnaire de l'écoulement (turbulence, ondes de choc instables . . .) augmentent avec la pression génératrice et les tensions de REYNOLDS. L'auteur a observé, à l'occasion de l'étude d'un proche sillage (GAVIGLIO et al. 1977), des ruptures par vibrations intervenant systématiquement lorsque le fil était placé dans le voisinage d'une surface sur laquelle le frottement moyen par la turbulence était maximum. Ces ruptures furent supprimées après réduction de la pression génératrice. En ce qui concerne l'influence du nombre de MACH, dans le cas d'une couche limite en quasi-équilibre les tensions de REYNOLDS restent à peu près proportionnelles au frottement pariétal.

$$\tau_w = 1/2\,\gamma\overline{p}M_\infty^2\,c_f$$

où γ est le rapport des chaleurs spécifiques. En raison de la diminution du produit pM_∞^2 (lorsque p_t est maintenu constant) et de c_f lorsque M_∞ augmente, l'effet de l'accroissement du nombre de MACH nominal est encore favorable. Les effets sur le fil ou sur les broches, ou sur le support de sonde, des instationnarités, sont à craindre plus que les contraintes dues au champ moyen, à cause des résonances mécaniques possibles.

1.2.2. Pour étudier les écoulements hypersoniques, DEMETRIADES, LADERMAN, DOUGHMAN (Cf. réf. citées) utilisent des fils de platine fortement miniaturisés: d = 0,25 microns, ℓ = 50 microns par exemple, ce qui procure un haute résolution spatiale et fréquentielle. Mais à cette échelle, il est malaisé de construire des sondes identiques, chacune d'elles doit donc être tarée, ce qui implique l'utilisation d'un système puissant d'acquisition de données. La solution adoptée à l'I.M.S.T. est plus classique. (Fig. 1). Le fil est un tungstène recouvert de platine, d = 2,5 microns, ℓ = 0,7 à 0,9 mm selon le modèle. Les sondes construites sur un même modèle ont des sensibilités adimensionnelles données par des tarages d'une sonde-type. En ce qui concerne les effets mécaniques et électriques des ondes de choc qui enveloppent les broches, ces dernières sont très aigues de façon à réduire l'intensité des chocs qui rencontrent le fil à une distance des broches comparable à la "longueur froide" $\ell_c/2 = \ell(S_t/4)$, où S_t est le "paramètre d'effets de bouts" défini au §3.3.1. Pour les exemples donnés au §3.3.3., $\ell_c/2 \leqslant 0,1$ mm. La "longueur active.. $1-\ell_c$ du fil est faible si la conductivité thermique k_w du fil est élevée. De ce point de vue, l'emploi de tungstène dont la conductivité thermique est supérieure à celle du platine, est favorable, puisque la résolution spatiale est meilleure, mais ceci accroît l'importance d'une correction liée à cet effet (§3.3.). Enfin, le fait que l'effet d'inertie thermique soit plus important qu'avec des fils miniaturisés oblige à corriger les mesures (§3.2.).

De telles sondes, équipées de fil de platine, sont également utilisables dans les écoulements à basse vitesse chauffés.

2. Progrès récents dans la détermination des grandeurs moyennes

La détermination des grandeurs moyennes en tout point d'un écoulement turbulent nécessite la connaissance locale du nombre de MACH M et de la température d'arrêt $\overline{\theta}_t$. M est déduit de mesures de pression. $\overline{\theta}_t$ est déterminé à l'aide d'un fil dont on fait tendre la surchauffe vers zéro, car alors sa température prend une valeur moyenne $\overline{\theta}_e = \eta \overline{\theta}_t$ dite "d'équilibre", voisine de $\overline{\theta}_t$. Le coefficient de récupération η tient compte de l'influence des broches, dont la température $\overline{\theta}_b = \eta_b \overline{\theta}_t$ peut être différente de celle du fil. Il dépend des nombres de MACH, de REYNOLDS, des pertes de chaleur à l'écoulement qu'exprime le nombre de NUSSELT Nu_t, et de l'évacuation de la chaleur vers les broches par conduction thermique qui altère d'ailleurs Nu_t:

$$\eta = \eta[Rd_t, M, Nu_t, (d/\ell)\sqrt{k_w/k_t}]$$

k_w et k_t sont les coefficients de conductibilité calorifique du métal du fil et de l'air, supposés constants. Pour un fil chauffé, Nu_t a pour expression:

$$Nu_t = (R_w i^2)/M\ell k_t(\overline{\theta}_w - \overline{\theta}_e) = Nu_t[Rd_t, M, (d/\ell)\sqrt{k_w/k_t}, \overline{a}_w]$$

où $\overline{a}_w \cong \overline{R}_w - \overline{R}_e/\overline{R}_e$ est le coefficient de surchauffe. Dans l'expression de η, intervient la valeur numérique limite de Nu_t lorsque $\overline{a}_w \to 0$.

LAUFER et McCLELLAN (1956) ont développé une méthode de détermination de η pour le cas des écoulements supersoniques où η est indépendant de M. Par un tarage fait dans l'écoulement à potentiel d'une tuyère, ces auteurs établissent une liaison entre le nombre de REYNOLDS correspondant à la température du fil non chauffé, le nombre de NUSSELT et le coefficient de récupération: la mesure du NUSSELT en tout point d'un écoulement fournit ainsi la valeur de η.

Avec BISSONETTE, BURNAGE (1971, 1973) a proposé une méthode qui accroît la précision, surtout aux faibles nombres de REYNOLDS où η varie rapidement. Cette méthode remplace la connaissance de Nu_t par celle de $\overline{p}_t$, mieux déterminée. Une loi de variation $\eta = \eta(Rd_e)$ est établie à partir de l'identité:

$$\rho u = \sqrt{\frac{\gamma}{R} \frac{p_t}{\theta_t}} \; M \left(1 + \frac{\gamma-1}{2} M^2\right)^{-\frac{\gamma+1}{2(\gamma-1)}}$$

où R est la constante des gaz parfaits. On obtient:

$$\eta = \frac{R\mu_e^2}{\gamma \overline{p}_t^2 \left[M\left(1 + \frac{\gamma-1}{2} M^2\right)^{-\frac{\gamma+1}{2(\gamma-1)}} \right] d^2} \; \theta_e \, Rd_e^2 \qquad [3]$$

η et le nombre de REYNOLDS local sont déterminés par l'intersection de cette fonction avec la courbe de tarage $\eta(Rd_e)$, Fig. 2.

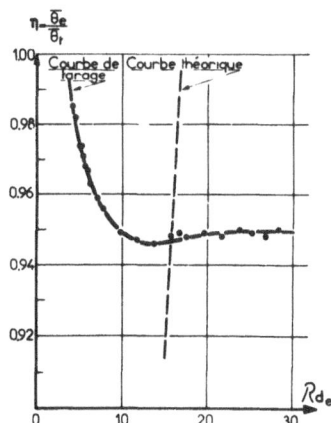

Fig. 2. Détermination de $\eta = \dfrac{\overline{\theta}_e}{\theta_t}$

Lorsque η correspond au minimum de cette dernière, la méthode s'applique de manière optimale. La relation [3] et la méthode de LAUFER et McCLELLAN forment un ensemble cohérent, permettant de vérifier que mesures et calculs sont aussi cohérents. Par ailleurs, la relation [3]peut être transformée en une relation directe entre η et le nombre de REYNOLDS Rd_t correspondant à la température d'arrêt. Ceci est possible, la représentation $\eta(Rd_t)$ conservant l'indépendance de η vis-à-vis de M_∞ (Fig. 3, Cf. GAVIGLIO, 1978). De la relation

$$(Rd_e/Rd_t) = \mu_t/\mu_e \cong (\theta_t/\theta_e)^m \text{ où } m \cong 0,765, \text{ on déduit:}$$

$$\eta^{2m+1} = \frac{R\mu_e^2}{\gamma\overline{p}_t^2 \left[M(1+\gamma_2-1M^2) ^{-\frac{\gamma+1}{2(\gamma-1)}} \right] d^2} \theta_e Rd_t^2 \qquad [4]$$

En fait, Rd_e et Rd_t peuvent être souvent confondus.

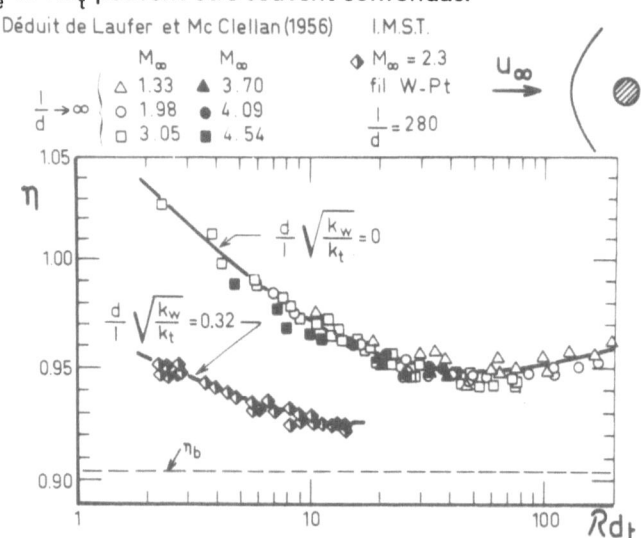

Fig. 3. Variation de η à vitesse supersonique

3. Progrès récents dans la caractérisation des grandeurs fluctuantes moments d'ordre 2.

3.1 Rappel de la méthode. Difficultés d'application.
3.1.1. D'une étude faite par KOVASZNAY à propos des écoulements super-soniques, on déduit que les perturbations turbulentes présentes dans un écoule-ment de gaz homogène peuvent être, quel que soit le nombre de MACH, rat-tachées aux trois ''modes'' suivants: fluctuations de vitesse q'_ω à caractère rotationnel; fluctuations non isentropiques de température θ'_s, et fluctuations de pression p' souvent exprimées, lorsqu'un schéma d'ondes s'applique, au

moyen des fluctuations associées de vitesse $u'_{\to p}$, irrotationnelles, et de température, θ'_p, isentropiques. Les fluctuations *totales* de vitesse et de température
s'écrivent:

$$q'_{\to} = q'_{\to \omega} + q'_{\to p} \qquad \theta' = \theta'_s + \theta'_p \qquad\qquad [5]$$

avec $p' = \rho a |q'_r|$ où est célérité du son;

$$\theta'_p = \overline{\theta} \frac{\gamma - 1}{\gamma} \frac{p'}{\overline{p}} \quad \text{et} \quad q'_{\to} = u'_{\to} + v'_{\to} + w'_{\to}$$

u'_{ω} étant la composante de $q'_{\to \omega}$ dans la direction de la vitesse moyenne, $\overline{q}_{\to}$, la
tension relative $e'/\overline{e}$ aux bornes d'un fil supposé sans inertie thermique s'écrit
sous forme linéarisée:

$$(e'/\overline{e}) = F_{\omega} \frac{u'_{\omega}}{\overline{q}} + F_{\theta} \frac{\theta'_s}{\overline{\theta}} + F_p \frac{p'}{\overline{p}} \qquad\qquad [6]$$

où F_{ω}, F_{θ}, et F_p sont les "sensibilités" à u'_{ω}, θ'_s et p'. Sauf aux basses vitesses, on ne sait pas déterminer ces dernières directement par l'expérience. Il est
nécessaire d'écrire $e'/\overline{e}$ en fonction des fluctuations *totales* de u, de ρ et de
θ_t:

$$(e'/\overline{e}) = -F_q \frac{u'}{\overline{q}} - F_{\rho} \frac{\rho'}{\rho} + G \frac{\theta'_t}{\theta_t} \qquad\qquad [7]$$

F_q, F_{ρ} et G sont sensibilités à u', à ρ' et à θ'_t. On peut les déduire de l'expérience directe. F_{ω}, F_{θ} et F_p en sont des combinaisons linéaires. Il y a 6 inconnues:

$\sqrt{\overline{u'^2_{\omega}}}$, $\sqrt{\overline{\theta'^2_s}}$, $\sqrt{\overline{p'^2}}$, et les corrélations $r_{u\theta}$, $r_{\theta r}$, et r_{pu}

3.1.2. Aux vitesses supersoniques, l'équation [7] se simplifie car F_{ρ} et F_u sont
pratiquement égaux. Les méthodes classiques des "diagrammes de fluctuations"
(KOVASZNAY, 1953, MORKOVIN, 1956), équivalentes à un système redon-
dant d'équations déduites de [7] permettent de déterminer trois des carac-
téristiques du champ turbulent, par exemple $(\sqrt{\overline{u'^2_{\omega}}}/q)$, $(\sqrt{\overline{\theta'^2_s}}/\overline{\theta})$ et le coefficients de corrélation $r_{u\theta}$ entre u'_{ω} et θ'_s. Ces méthodes supposent que les
fluctuations de pression ont un niveau assez faible pour que leur effet sur le
tracé des diagrammes soit négligeable: $(p'/\overline{p}) \ll (u'/\overline{q})$, $(\theta/\overline{\theta})$, il n'y a donc plus
alors que 3 inconnues, $\sqrt{\overline{u'^2_{\omega}}}$, $\sqrt{\overline{\theta'^2_s}}$ et $r_{u\theta}$.

C'est la une hypothèse "d'isobarie" (LAUFER, 1969). Hormis le cas de certains écoulements, tels des écoulements séparés au sein desquels les fluctuations de pression mesurées à la paroi peuvent être d'un ordre de grandeur supérieures à ce qu'elles sont dans une couche en quasi équilibre (KISTLER, 1964), cette hypothèse d'applique si le nombre de MACH nominal est modéré, M<4. Aux vitesses hypersoniques par contre, $(\sqrt{\overline{p'^2}}/\overline{p})$ peut avoir des valeurs non négligeables, à la paroi (KISTLER et CHEN, 1963) et à l'intérieur des écoulements (LAUFER, 1964). Pour interpréter les résultats des mesures en réduisant à 3 le nombre des inconnues, il faudrait introduire dans les calculs des hypothèses physiques supplémentaires, prenant en compte d'une part les mécanismes liés à la génération des fluctuations de pression, et d'autre part à l'action sur l'anémomètre des "sources équivalentes" de pression qui sont comme transportées par l'écoulement et environnent le fil chaud. Cela rend difficile une schématisation. Néanmoins, des hypothèses "de fermeture" existent. LADERMAN et DEMETRIADES (1973, 1974) en ont donné un exemple dans le cas d'une couche limite au nombre de MACH 9,4. La solution suppose notamment qu'au sein des écoulements les fluctuations u'_ω et θ'_s sont en corrélation −1, et que chacune est en corrélation nulle avec les fluctuations p'. Ceci ramène à 3 le nombre des inconnues. Les résultats sont raisonablement en accord avec ceux que permettent de prévoir les lois établies par LAUFER et par KISTLER et CHEN, aux vitesses supersoniques.

3.1.3. D'autres problèmes posés par l'anémométrie des écoulements compressibles ont été bien analysés et peuvent souvent être résolus par des corrections. Ce sont notamment l'insuffisance de résolution spatiale des sondes dont la longueur n'est pas petite vis-à-vis des "échelles caractéristiques" de la turbulence à mesurer, la non linéarité de l'anémomètre pour les signaux de forte amplitude, les "effets de jauge" pour les fils vibrants. On peut consulter là-dessus LOWELL, 1950; MORKOVIN, 1956; CORRSIN, 1963; BRADSHAW, 1971; COMTE-BELLOT, 1976. On mettra ici l'accent sur la correction de deux erreurs liées au choix des caractéristiques des sondes (§1.2).

3.2 Corrections aux mesures anémométriques dans les écoulements "minces"
On-sait (KISTLER, 1959) que l'étude des couches limites cisaillées aux vitesses supersoniques et aux grands nombres de REYNOLDS nécessite que l'ensemble fil chaud-amplificateur-filtre compensateur de l'inertie thermique, ait une bande passante assez élevée pour ne pas altérer trop le spectre de fréquences. Par exemple, dans le cas d'une couche limite, lorsque le nombre de REYNOLDS Re_θ formé avec l'épaisseur de quantité de mouvement est supérieure à 3.10^4, la bande passante F_c à 3 dB doit être supérieure à $5(u_\infty/\delta)$ où u_∞ est la

vitesse de l'écoulement libre et δ l'épaisseur de la couche limite. Cette condition est d'autant moins sévère que le nombre de Reynolds est moins élevé. La fréquence de coupure F_c est déterminée: 1° par l'amplificateur non compensé dont on limite souvent la bande passante, f_c, pour réduire, à un niveau très inférieur à celui des signaux turbulents, des perturbations dues au bruit de fond ou à des vibrations par effet de jauge, ou des parasites radio-électriques, 2° par le filtre compensateur de l'inertie thermique, bande f'_c. Citée par KOVASZNAY (1954) par CORRSIN (1963), cette source d'erreurs a été récemment analysée et corrigée (GAVIGLIO et DUSSAUGE, 1977; GAVIGLIO, 1978). On résume ici cette analyse.

Mis à part les effets de conduction thermique (§3.3.) la réponse du fil chaud à une excitation harmonique $s' = s.e^{j\omega t}$, où s représente l'un quelconque des trois "modes", est telle que

$$e'(t) \, a \frac{s.e^{j\omega t}}{1+j\mathcal{M}\omega} \qquad [8]$$

a indique la proportionnalité, $\mathcal{M}$ désigne la constante de temps du fil. Le filtre électrique incorporé à l'amplificateur ne permet de corriger que partiellement le signal. Le calcul de la fonction de transfert T_r d'un tel filtre (GAVIGLIO et DUSSAUGE, 1977; GAVIGLIO, 1978) fournit en effet la forme adimensionnelle, applicable aux circuits de compensation couramment utilisés:

$$T_r = (1+\mathcal{M}\omega)(1+j \frac{g_0}{g_\infty}\mathcal{M}\omega)^{-1} \qquad [9]$$

où g_0 et g_∞ sont les gains réels du quadripole aux fréquences tendant vers zéro et vers l'infini. Ainsi, la réponse du fil "compensé" à une oscillation sinusïdale,

$$[e'(t)]_c \, a \, s. \, e^{j\omega t} \, (1+j \frac{g_0}{g_\infty} \mathcal{M}\omega)^{-1}$$

approche la réponse idéale avec une atténuation $\sqrt{1+(g_0/g_\infty)^2\mathcal{M}^2\omega^2}$ en amplitude et un retard de phase $- \arctan(g_0/g_\infty)\mathcal{M}\omega$. La fréquence f'_c d'atténuation à 3 dB est donc $f'_c = (g_0/g_\infty)(1/2\pi\mathcal{M})$. Souvent négligeable lorsqu'on utilise des fils chauds fortement miniatures (LADERMAN et DEMETRIADES, 1974), l'erreur oblige à des corrections, pour des fils de 2,5 microns par exemple.

Si la bande passante de l'amplificateur non compensé f_c est assez large pour que l'erreur provienne seulement du filtre compensateur, l'énergie $\overline{e'^2_m}$ du signal mesuré et celle du signal corrigé, $\overline{e'^2_c}$, sont dans le rapport

$$\frac{\overline{e'^2_c}}{\overline{e'^2_m}} = \int_0^\infty F_m(n) \left[1 + \left(\frac{g_0}{g_\infty} \right)^2 \mathcal{M}^2 \omega^2 \right] dn \qquad [10]$$

où $F_m(n)$ est la fonction spectrale normée du signal mesuré telle que $\int_0^\infty F_m(n)dn$ = 1.   Lorsqu'on ne peut mesurer $F_m(n)$, en raison par exemple du volume considérable d'information qu'implique les tracé d'un diagramme de fluctuations, la correction peut être faite si on a déterminé au moins une "fréquence équivalente" N_m, telle que $N_m^2 = \int n^2 F(n) dn$. Cette fréquence, introduite en 1964 par A. MARTINOT LAGARDE, est celle de l'oscillation sinusoïdale dont la courbe d'autocorrélation est osculatrice au sommet à la courbe d'autocorrélation du signal aléatoire aux bornes du fil. Cette définition fournit la méthode de mesure. La relation [10] s'écrit encore:

$$\frac{\overline{e'^2_c}}{\overline{e'^2_m}} = 1 + \left(\frac{g_0}{g_\infty} \right)^2 (2\pi N_m \mathcal{M})^2 \qquad [11]$$

La correction est d'autant meilleure que la bande passante F_c est plus élevée. Pratiquement, elle est toujours faite par défaut. On doit donc vérifier, en comparant certains des résultats auxquels elle conduit, à d'autres obtenus dans des conditions plus favorables, que l'erreur résiduelle est faible. Si tel n'est pas le cas, la valeur de f_c doit être accrue, et l'on doit s'efforcer de minimiser les erreurs en réduisant $\mathcal{M}$ le plus possible.
Des applications de la méthode sont détaillées dans les publications citées en référence. Leur importance est indiquée sur des exemples, Fig. 5.

3.3 Corrections liées à la conduction thermique

3.3.1. Conditions d'application des corrections. Dans un article récent (1978), MILLON, PARANTHOEN et TRINITE ont analysé les erreurs qu'introduisent, dans les mesures de *fluctuations θ' de température* d'un écoulement à basse vitesse, les échanges de chaleur entre un fil peu chauffé et ses supports. Cette analyse, qui suppose linéaires les pertes de chaleur par convection avec l'air en fonction de la différence de température, reprend dans un cas simple une étude générale, non linéaire et relative à un fil chauffé, d'une grande complexité, due à BETCHOV (1948, 1949). Le calcul de MILLON et al. a été adapté, (travail effectué en collaboration avec J.F. DEBIEVE) au cas des écoulements supersoniques, en ce qui concerne les seules fluctuations θ'_t de température d'arrêt qui jouent un rôle analogue à celui de θ' pour les faibles vitesses, mais dont la signification physique est évidemment différente (relation [2]).
Si on note $\overline{\theta}_{e\infty} = \eta_\infty \overline{\theta}_t$, la température *moyenne dans le temps* d'un fil non

chauffé de longueur infinie, l'équation qui régit la température instantanée θ (x, t) en tout point d'abcisse x d'un fil de longueur finie et chauffée, s'écrit:

$$\frac{\pi d^2}{4} \rho_w c \frac{\partial \theta_w}{\partial t} + \pi k_t (\theta_w - \theta_{e\infty}) Nu_t - \frac{\pi d^2}{4} k_w \frac{\partial^2 \theta_w}{\partial x^2} = R(x, t)i^2 \qquad [12]$$

où ρ_w et c sont la masse volumique et la chaleur spécifique du matériau du fil, et R_{w1} la résistance par unité de longueur. Cette équation, dans laquelle on fait tendre i vers o, permet de calculer la répartition au long du fil de la température notée $\overline{\theta}_w$ (x), prise en moyenne dans le temps, qui est évidemment différente de la température prise en moyenne au long du fil notée

$$\overline{\theta}_w = \frac{1}{\ell} \int \theta_w (x) \, dx \Big]_{-\ell/2}^{+\ell/2}$$

On montre que, bien que la température des broches $\overline{\theta}_b = \eta_b \, \theta_t$ soit différente de celle du fil (Fig. 3), avec les approximations usuelles:

a) $\overline{\theta}_w$ (x) est représentée par une loi en cosinus hyperbolique comme aux basses vitesses (CORRSIN, 1963).

b) le coefficient de récupération moyen du fil, η, s'exprime en fonction du paramètre

$$S_t = \frac{d}{\ell} \sqrt{\frac{k_w}{k_t}} \frac{1}{\sqrt{Nu_t}}$$

qui est la limite, pour i → o, du paramètre introduit par KOVASZNAY et TORMARCK (1950) pour caractériser les effets de conduction. Ceci est conforme à un résultat de DEWEY (1961, MEMO).

3.3.2. Influence de la conduction thermique sur les mesures de θ'_t aux vitesses supersoniques.

a) "Sensibilités" d'un fil non chauffe: Lorsque $\overline{a}_w \to 0$, le fil chaud est surtout sensible à θ'_t. En effet, les coefficients F_ρ et F_u de la relation [5] pratiquement égaux si M⩾1,2 s'annulent si $\partial \eta / Rd_t = 0$ (GAVIGLIO), 1978). Sinon, θ_r est contaminé par $(\rho u)'$, de même qu'aux faibles vitesses θ' est contaminé par u', (VEROLLET, 1972; FULACHIER, 1972).

Selon la Fig. 3, récemment publiée (GAVIGLIO, 1978), la condition $\partial \eta / \partial R \cong 0$ est remplie dans l'intervalle des nombres de REYNOLDS $40 < R < 80$, si $\ell/d \to \infty$. Pour un fil d'allongement fini, la limite inférieure est plus basse, de l'ordre de 10 par exemple.

b) Erreur sur la mesure de $\overline{\theta'^2_t}$. Le calcul tenant compte de la conduction thermique montre que bien que les températures moyennes $\overline{\theta}_b$ et $\overline{\theta}_w$ des broches et du fil soient différentes, la fonction de transfert globale du fil a même forme qu'à basse vitesse:

$$Tr_w = (1+j\mathcal{M}\omega)^{-1} H_c (\mathcal{M}\omega)$$

$(1+j\mathcal{M}\omega)^{-1}$ serait la fonction de transfert du fil en l'absence de conduction; $H_c (\mathcal{M}\omega)$ traduit l'effet de cette dernière. On a négligé ici l'influence[*] de la variation de Nu_t en fonction de θ_t, ce que justifie un examen d'ordres de grandeur, et on a supposé infinie la capacité thermique des broches, dont l'effet est effectivement négligeable. On a exprimé $H_c (\mathcal{M}\omega)$ en fonction de S_t et de $\mathcal{M}\omega$.

$$H_c (\mathcal{M}\omega) = 1 - \frac{\left(\text{th} \ (S_t\sqrt{2})^{-1}\left\{ \left[(1+\mathcal{M}^2 \omega^2)^{\frac{1}{2}} +1 \right] + j\left[(1+\mathcal{M}^2 \omega^2)^{\frac{1}{2}} -1 \right] \right\} \right)}{(S_t\sqrt{2})^{-1}\left\{ \left[(1+\mathcal{M}^2 \omega^2)^{\frac{1}{2}} +1 \right] + j\left[(1+\mathcal{M}^2 \omega^2)^{\frac{1}{2}} -1 \right] \right\}} \quad [15]$$

le carré H_e^2 du module de H_c est représenté sur la Fig. 4. Les courbes A, B, D, F correspondent à celles de la Fig. 2 de la publication de MILLON et al. L'erreur diminuant de façon monotone lorsque $\mathcal{M}\omega$ augmente, on peut s'attendre à ce qu'aux vitesses supersoniques, en raison de l'étendue des spectres d'énergie l'erreur globale sur $\overline{\theta'^2_t}$ soit relativement moins importante qu'aux basses vitesses. L'expérience confirme ce point de vue. (Fig. 5, a et b). Mais, la correction requiert la connaissance pratiquement complète du spectre d'énergie.

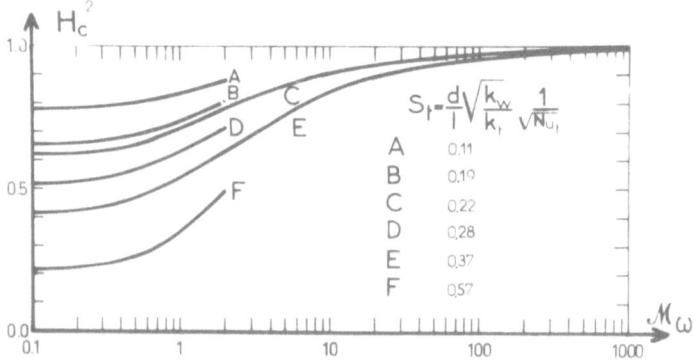

Fig. 4. *Influence de la conduction thermique sur les répartitions spectrales d'énergie θ'^2.*

[*] qui porte sur le terme $Nu_t (\theta_w - \theta_{e\infty})$ de la relation [12]

c) Erreurs à craindre sur la mesure de la constante de temps des fils. Aux vitesses supersoniques, une erreur sur $\mathcal{M}$ entraîne une erreur plus grande que dans le cas des écoulements à basse vitesse, car elle intervient sur un intervalle beaucoup plus étendu des fréquences non-dimensionnelles $\mathcal{M}\omega$. Elle pourrait résulter de l'influence sélective de la conduction thermique, qu'indique la Fig. 4. Lorsqu'on mesure $\mathcal{M}$ par une méthode dite "de signaux carrés" (BETCHOV, 1968; GAVIGLIO, 1958) on superpose au courant continu de chauffage du fil des impulsions périodiques dont l'action peut être prévue lorsque la fonction de transfert Tr_w est connue. A l'effet d'inertie thermique qui peut être compensé (§3.2.) par le réglage convenable d'un filtre, s'ajoute celui de la conduction — terme $H_c (\mathcal{M}\omega)$ — qui, privilégiant l'amplitude des harmoniques de l'oscillation fondamentale, pourrait conduire à sous-estimer $\mathcal{M}$ si les relations de phases étaient convenables. Des essais effectués avec des fils communément utilisés (tungstène recouvert de platine d = 2,5 microns; ℓ/d = 300; S_r = 0,35; $\bar{a}_w$ = 0,05; $\mathcal{M}$ = 0,27 10^{-3} sec.) montrent qu'un tel effet existe, mais qu'il est inférieur à l'écart type des erreurs aléatoires. Il est donc ici tenu pour négligeable.

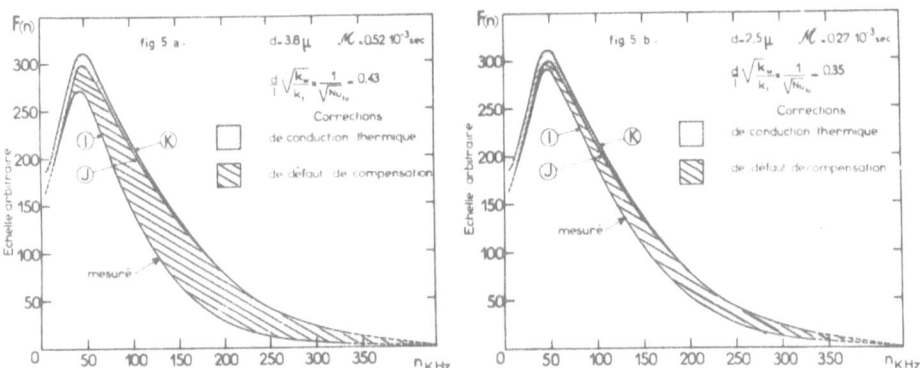

Fig. 5. Importance des corrections

3.3.3. Résultats concernant l'importance des différentes corrections. La Fig. 5, a et b, montre, sur deux exemples, une application à l'étude d'une couche limite "mince" des corrections analysées aux §§ 3.2 et 3.3.1. Les conditions sont les suivantes: δ = 4 mm; $\mathcal{M}$ = 2,3; distance à la paroi (y/δ) = 0,25. Les courbes I correspondent à des mesures. Les courbes J sont déduites de I par correction de compensation d'inertie thermique. Les courbes K sont déduites de J par correction d'effet de conduction thermique.

Les fils, en tungstène recouvert de platine ont les caractéristiques suivantes:

Cas (a)
d = 3,8 microns, 1 = 0,85
$\mathcal{M}$ = 0,52 10^{-3} sec., S_r = 0,43

La correction de l'erreur due aux défauts du filtre compensateur est de 30 % de l'énergie totale. Celle due à la conduction thermique est de 4 % dans l'intervalle des fréquences 5 à 400 kHz.

Cas (b)
d = 2,5 microns, ℓ = 0,75 mm
$\mathcal{M}$ = 0,27 10^{-3} sec., S_r = 0,35

La correction de l'erreur de compensation est seulement ici de 13 %, celle due à la conduction est de 3,6 %.

Il y a un bénéfice évident à utiliser des fils de faible diamètre. En ce qui concerne l'effet de conduction thermique, qui intervient quelle que soit la surchauffe du fil et quel que soit le nombre de MACH, son étude serait avantageusement étendue au cas du fil chauffé.

4. Remarques concernant l'intérêt, pour les mesures, de la pondération par la masse des variables de base

On connaît (FAVRE, GAVIGLIO, cf. FAVRE et al., 1976, parties II et V) l'intérêt qui s'attache à pondérer par la masse, dans le cas des écoulements compressibles, les variables de base: réduction du nombre des termes des équations et clarification du sens physique de ces termes. Il en résulte aussi une simplification des mesures (GAVIGLIO, 1978). On rappelle comment intervient cette simplification, et on formule quelques observations sur son application au cas du fil incliné.

4.1 Simplification des mesures

Les équations régissant les propriétés des écoulements turbulents compressibles de gaz sont obtenues en décomposant chaque grandeur instantanée u, θ, p, ρ .., en une valeur moyenne et une fluctuation. Deux décompositions sont principalement utilisées. L'une est celle de REYNOLDS.

(a) $u_i = \overline{u}_i + u_i'' \Leftrightarrow \overline{u_i''} \equiv 0 \qquad \theta = \overline{\theta} + \theta'' \Leftrightarrow \overline{\theta''} \equiv 0$ [16]

L'autre est celle des "grandeurs pondérées par la masse" (cf. FAVRE, p. ex., 1958, 1976):

(b) $u_i = \tilde{u}_i + u'_i \Leftrightarrow \overline{\rho u'_i} \equiv 0$ $\theta = \tilde{\theta} + \theta' \Leftrightarrow \overline{\rho \theta'} \equiv 0$ [17]

Les relations $p = \overline{p} + p' \Leftrightarrow \overline{p'} \equiv 0$ et $\rho = \overline{\rho} + \rho' \Leftrightarrow \overline{\rho'} \equiv 0$ sont communes aux deux décompositions. u'' et θ'', de même que ρ' et p' sont des "variables centrées". Les équations de type (b) comportant moins de termes que les équations de type (a) (sauf lorsque les effets moléculaires sont importants), leur emploi implique un nombre de mesures moins élevé. Les équations qui expriment les bilans des variations de quantités transportables (masse, quantité de mouvement, énergie interne, entropie, énergie cinétique d'agitation) comportent des termes ayant même forme avec les deux décompositions, et ceci aussi bien en ce qui concerne les grandeurs moyennes que les grandeurs turbulentes, mais les équations (a) contiennent, par rapport aux équations (b), des termes supplémentaires qui peuvent être importants. En outre, ces termes sont souvent difficiles à mesurer: il s'agit, par exemple, du terme $\overline{v\rho'\theta''} = \overline{\rho}\overline{\theta}\tilde{v}\sqrt{\overline{\theta''^2}}/\overline{\theta}$ qui figure dans l'équation de l'énergie et dont la connaissance implique celle de $\tilde{v}$, composante de la vitesse moyenne perpendiculaire à la vitesse $\overline{u}$ dans la direction de l'écoulement. On montre en effet que $\tilde{v}$ se calcule *comme en incompressible:*

$$\tilde{v} = \frac{1}{\overline{\rho}} \int_o^y \frac{\partial}{\partial x} (\overline{\rho}\tilde{u}) dy$$

dans le cas d'une couche limite. Les lignes ou surfaces de courant sont inclinées d'un angle $\tilde{v}/\tilde{u}$ aisément déterminé. L'expression de $\overline{v}$ est par contre plus compliquée:

$$\overline{v} = \tilde{v} - \tilde{u}\, r_{\rho v} (\sqrt{\overline{\rho'^2}}/\overline{\rho})(\sqrt{\overline{v'^2}}/\overline{q})$$

Donc, il est beaucoup plus difficile de connaître l'inclinaison v/u des lignes de courant définies en méthode (a). On pourrait objecter que si la méthode (b) paraît s'imposer parce que le fil est sensible à ρq et non à q, les signaux fournis par un amplificateur d'anémomètre de type "à courant constant" sont centrés, comme ceux que permet de traiter la méthode (a). Néanmoins, la méthode (b) s'applique avec plus d'intérêt pour les raisons déjà indiquées, et parce qu'il est plus facile d'introduire dans les équations (b) les termes mesurés auxquels on applique les corrections rendues explicites par FAVRE (dans FAVRE et al., 1976, partie II) que d'utiliser les équations (a) dont certains termes supplémentaires ont une valeur beaucoup plus élevée que les termes correctifs.

4.2 Remarques sur l'orientation des sondes inclinées, utilisées pour mesurer les flux de quantité de mouvement et de chaleur

4.2.1. Des sondes à deux fils croisés, ou à un seul fil incliné sur la direction moyenne de l'écoulement d'un angle φ_+ puis $\varphi_- = \pi - \varphi_+$, sont utilisées pour mesurer les flux de quantité de mouvement, quel que soit le nombre de MACH, et aussi des flux de chaleur aux vitesses supersoniques. (MORKOVIN-PHINNEY, 1958; ROSE, 1972; DEMETRIADES - LADERMAN, 1973). Lorsque les intensités de turbulence sont faibles, l'angle φ est repéré par rapport à une direction que l'on détermine comme dans le cas bien connu des écoulements incompressibles à basse vitesse: elle est telle que le refroidissement du fil est le même en moyenne pour les incidences φ_+ et φ_- .

Par contre, la détermination de cette direction repère peut soulever des problèmes lorsqu'il s'agit d'étudier des écoulements présentant des gradients importants de masse spécifique, et lorsque cette détermination ne peut être obtenue par un tarage angulaire fait "in situ". En effet, le refroidissement du fil dépend, quel que soit le nombre de Mach, des variables, ρ, q, θ_t et φ, et l'on peut penser que, comme dans le cas des écoulements supersoniques, le groupement ρq puisse jouer en rôle important dans les lois des pertes de chaleur. Or la direction du vecteur $\overrightarrow{\rho q}$, est différente de celle du vecteur vitesse moyenne $\overrightarrow{q}$ des particules qui sert de référence dans le cas des écoulements incompressibles.

Une solution simple apparait lorsque le nombre de Mach formé avec la composante normale au fil de la vitesse est supérieur à l,2. Fig. 6. Alors le refroidissement du fil ne peut être le même en moyenne pour les deux positions symétriques φ_+ et φ_- que si les composantes normale et tangentielle du vecteur $\overrightarrow{\rho q}$ ont même valeur dans les deux cas. La direction repère est donc celle de ce vecteur, qui est tangent aux lignes ou surfaces de courant macroscopiques qui sont les lignes d'iso-débit de masse étanches en moyenne à tout échange de masse par la turbulence. La direction $\tilde{u}/\tilde{v}$ de ces lignes se détermine aisément (§4.1.) par l'emploi des vitesses pondérées par la masse.

En l'absence de sources de chaleur, $\Delta\alpha$ n'excède pas ici 1 degré, donc reste de l'ordre des erreurs faites sur φ lorsque l'orientation est déterminée par l'expérience. Mais on peut avoir à étudier des situations telles que $\Delta\alpha$, dont l'expression est indépendante de M, ait des valeurs plus élevées, notamment dans le cas d'écoulements à des vitesses non supersoniques corportant des sources de chaleur.

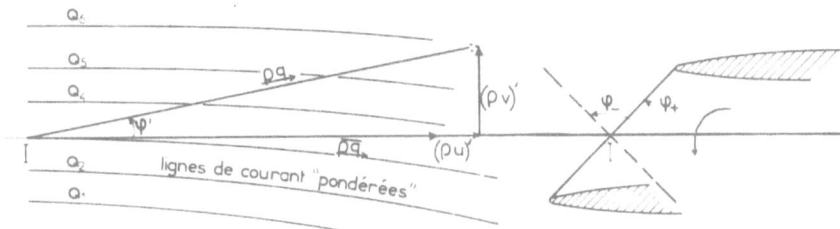

Fig. 6. Fil incliné

5. Conclusions

5.1. Un certain nombre de problèmes qui se posent à l'expérimentateur étudiant, au moyen de l'anémomètre à fil chaud, les propriétés des écoulements compressibles, à haute ou à basse vitesse, avec ou sans sources de chaleur, ont été examinés. Leurs solutions sont basées sur un choix, que l'on justifie, de la géométrie des sondes et du matériau du fil, et sur la connaissance des lois des pertes de chaleur.

5.2. On a rappelé des progrès récents effectués dans la détermination des grandeurs moyennes, par des mesures conjuguées de pressions et du coefficient de récupération d'un fil non chauffé.

5.3. En ce qui concerne la mesure des fluctuations turbulentes de vitesse et de température, l'attention a été attirée sur les différences qu'introduit la prise en compte des fluctuations de pression, et sur l'intérêt que présenterait une meilleure connaissance des mécanismes liés à ces fluctuations au sein des écoulements, et de leur action sur l'anémomètre.

5.4. Deux sources d'erreurs ont été analysées de façon détaillée. L'une provient de l'insuffisance de la définition fréquentielle des anémomètres résultant de l'imperfection des filtres compensant les effets d'inertie thermique. La formulation adimensionnelle de la fonction de transfert de ces filtres a permis d'apporter une correction qui peut être très importante dans le cas des écoulements supersoniques ou hypersoniques. La seconde source d'erreurs est due à la conduction de la chaleur au long du fil, vers les broches. Elle a été étudiée dans le cas de fils peu chauffés, sensibles aux fluctuations de température d'arrêt, placés dans un écoulement supersonique. La fonction de transfert équivalente à l'effet de conduction a été exprimée en fonction d'un paramètre non-dimensionnel caractérisant les pertes de chaleur par les broches, et appliquée au cas où la température moyenne du fil peu chauffé est différente de celle des supports. La correction est plus faible qu'aux basses vitesses, mais elle est plus difficile à appliquer. L'importance de cette correction suggère d'en étendre l'étude au cas d'une surchauffe quelconque. La conduction pourrait encore provoquer une erreur sur la détermination de l'inertie thermique du fil. Cette erreur paraît négligeable dans les conditions usuelles d'étude des écoulements supersoniques.

5.5. Enfin, on a montré à l'aide d'exemples, comment la formulation, en termes de grandeurs pondérées par la masse, des équations statistiques des gaz turbulents compressibles simplifie la tâche de l'expérimentateur et accroît la précision, ce qui indirectement aide à l'application de cette formulation aux problèmes de prédétermination d'écoulements. Sur ces bases, on a posé le problème de l'orientation, dans les écoulements compressibles non supersoniques, des sondes inclinées servant à mesurer des flux de chaleur et de quantité de mouvement.

5.6. Après plus de six décennies, l'anémomètre à fil chaud continue de faire l'objet de perfectionnements et d'applications nouvelles, notamment à l'étude des écoulements turbulents comportant des sources de chaleur. Il ne saurait être supplanté par l'anémomètre laser puisque ce dernier ne fournit pas de renseignement sur les champs de température et de masse spécifique. Mais les deux types d'appareils sont mutuellement complémentaires, car le laser permet de connaître, sans troubler l'écoulement, le champs des vitesses particulaires dans des zones "interdites" au fil chaud. D'autre part, il apporte par des moyens très différents, des informations qui peuvent être avantageusement comparées à celles de l'anémomètre dans les domaines communs d'investigations.

Remerciements

L'auteur remercie vivement M. DEBIEVE, avec qui certains calculs ont été faits et interprétés, et MM. DUSSAUGE et ELENA pour la discussion et la lecture qu'ils ont bien voulu faire du texte.

Références

BEGUIER, C.L.: 1971, Th. Doct. ès Sc., Univ. de Provence, Marseille.
BETCHOV, R.: 1948, *Proc. of Royal Dutch Acad. of Sc.,* Vol. **VI**, no. 2.
BETCHOV, R.: 1949, *Proc. of Royal Dutch Acad. of Sc.,* Vol. **LII**, no. 3.
BRADSHAW, P.: 1971, *An Intr. to Turb. and Its Measurements,* Pergamon Press, Oxford
BURNAGE, H.: 1973, Th. Doct. ès Sc., Univ. de Provence, Marseille
COMTE-BELLOT, G.: 1976, *Ann. Rev. of Fl. Mech.,* Vol. **8**, p. 209.

CORRSIN, S.: 1963, *Handbuch der Physiks,* Truesdale ed., Baltimore, p. 524.

DEMETRIADES, A.; LADERMAN, A.J.: 1973, *A.I.A.A.J.,* Vol. **11**, no. 11, p. 1594.

DEWEY, C.F., Jr.: 1961, *J. Amer. Rocket Soc.,* Vol. **28**, p. 1709; et Memo Hypers. Res. Proj., G.A.L.C.I.T., App. A.

DEWEY, C.F., Jr.: 1965, *J. Heat and Mass Transf.,* Vol. **8**, p. 245.

DOUGHMAN, E.L.: 1972, *Rev. of Sci. Instr.,* Vol. **43**, p. 1200.

FAVRE, A.J.: 1958, C.R. Acad. Sci., t. 246, no. 2576, 2723, 2839, 3216, Paris.

FAVRE, A.J.: 1965, *Journ. Méca.,* Vol. **4**, no. 3, p. 361 et Vol. **4**, no. 4, p. 391.

FAVRE, A.J.; KOVASZNAY, L.S.G.; DUMAS, R.; GAVIGLIO, J.; COANTIC, M.: 1976, *Turb. en Méca. des Fl.,* Gauthier Villars, Paris.

FULACHIER, L.: 1972, Th. Doct. ès Sc., Univ. de Provence, Marseille.

GAVIGLIO, J.: 1958, Th. Doct. ès Sc., Fac. Sc. Aix-Marseille; 1962, P.S.T. Minist. Air 385, Paris.

GAVIGLIO, J.: 1978, *J. de Méca. Appl.* Vol. **2**, no. 4, 1978.

GAVIGLIO, J.; DUSSAUGE, J.P.: 1977, *Proc. Symp. N.B.S.,* Gaithersburg, publ. 484, p. 649, T.P. O.N.E.R.A. 1977-168.

GAVIGLIO, J.; DUSSAUGE, J.P.; DEBIEVE, J.F.; FAVRE, A.J.: 1977, *Phys. of Fl. Suppl.* Vol. **20**, no. 10, p. S179, T.P. O.N.E.R.A., 1977-171.

HORSTMAN, C.C.; ROSE, W.C.: 1977, *A.I.A.A.J.,* Vol. **15**, no. 3, p. 395.

KING, L.V.: 1914, *Phil. Trans. Roy. Sc.,* A, Vol. **214**, p. 373.

KISTLER, A.L.: 1964, *J. Acoust. Sc. of Amer.,* Vol. **36**, p. 543.

KISTLER, A.L.; CHEN, W.S.: *J. Fl. Mech.,* Vol. **16**, no. 1, p. 41.

KOVASZNAY, L.S.G.: 1950, *J. Aero. Sc.,* Vol. **17**, no. 9, p. 565.

KOVASZNAY, L.S.G.: 1953, *J. Aero. Sc.* Vol. **20**, no. 10, p. 657.

KOVASZNAY, L.S.G.: 1956, N.A.C.A. Rept. 1209.

KOVASZNAY, L.S.G., TORMACK, S.I.A.: 1950, J.H.U., Rpt. 127, Bumblebee Series

LADERMAN, A.J.; DEMETRIADES, A.: 1973, *Phys. of Fl.,* Vol. **16**, no. 2, p. 179.

LADERMAN, A.J.; DEMETRIADES, A.: 1974, *J. Fl. Mech.,* Vol. **63**, part 1, p. 121.

LAUFER, J.: 1964, *Phys. of Fl.,* Vol. **7**, p. 1191.

LAUFER, J.: 1969, The RAND Corp. and Univ. of CALIF., MEMO 5946-Pr.

LAUFER, J.: McCLELLAN, R.: 1956, *J. Fl. Mech.,* Vol. **1**, p. 276.

LOWELL, H.: 1950, N.A.C.A., T.N. 2117.

MARTINOT-LAGARDE: 1946, N.T. G.R.A. no. 55, PARIS, Non publiée

MILLON, F.; PARANTHOEN, P.; TRINITE, M.: 1978, *Int. J. Heat and Mass Transf.,* Vol. **21**, p. 1, Pergamon Press.

MORKOVIN, M.V.: 1956, Agardograph 24.

MORKOVIN, M.V.; PHINNEY, R.E.: 1958, J.H.U. A.F.O.S.R. T.N. 58469.

ROSE, W.C.: 1972, Th. Doct., Univ. of Washington.

ROSE, W.C.; CHILDS, M.E.: 1974, *J. Fl. Mech.,* Vol. **65**, part 1, p. 177.

VEROLLET, E.: 1972, Th. Doct. ès Sc., Univ. de Prov. - Rpt. CEA-R-4872, 1977.

Local and Instantaneous Measurements in Liquid Metal MHD

by

Réne Moreau
Institut National Polytechnique de Grenoble

1. Introduction

In liquid metals used for laboratory experiments in magneto-hydrodynamics, such as mercury, the techniques of local and instantaneous measurements are known to be much more delicate than in classical fluids like water and air. Whatever diagnostic method is used, two categories of difficulties must be overcome: those due to the presence of a magnetic field and those due to the nature of the fluid. The former generally appear as fundamental difficulties, since the magnetic field acts like a new parameter which modifies the physics of the measurement apparatus; we will see further on that this can be put to use for imagining new methods which have no equivalent in classical hydrodynamics. The latter appear rather like secondary disturbances which would not exist in perfect experiments and which can (at least theoretically) be reduced by the care taken by the experimenter.

These diagnostic methods consists almost always in the introduction into the fluid of a probe on which the outflow generates a measurable phenomenon. This phenomenon can be either a stagnation pressure when the probe is a simple obstacle, an electric potential difference when the probe is formed by electrodes between which the moving fluid in the presence of the magnetic field induces an electric field, or a flux of heat when the probe is an object cooled by the fluid. We shall examine these three types of sensors successively, we shall analyse their specific difficulties, and we shall try to bring out the means of getting around them.

2. Measurement of the Stagnation Pressure

Pitot tube can, of course, be used in liquid metals, on the one condition that it be made out of materials which cannot be attacked by the liquid metal used. In the case of mercury, stainless steel, glass and plastics open wide possibilities. It is therefore the presence of a magnetic field that brings the main causes of difficulties, and these are of two types.

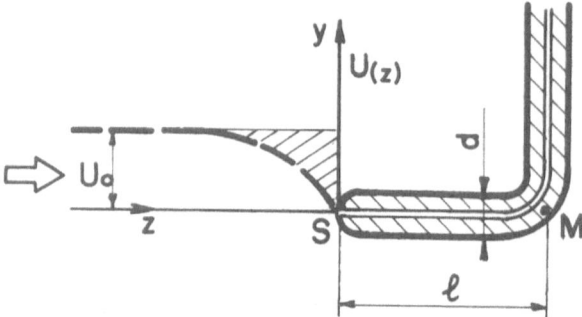

Fig. 1. Schema of a Pitot tube in the presence of a transverse magnetic field

First it appears that the stagnation pressure is not exactly equal to the fluid's loss of kinetic energy, since some work is generally furnished to the fluid particles by the electromagnetic forces which tend to contrary their slowing down. Euler's equation projected on the stream line arriving at the stagnation point (direction z) is written, in the presence of a transverse magnetic field B assumed to be uniform:

$$\frac{d}{dz}(p + \frac{1}{2} \rho U^2) = - \sigma B(E+UB) \tag{1}$$

In the majority of cases, if the Pitot tube is small enough the electric field can be written $E = -BU_o$ (to roughly a certain constant) where U_o is the velocity to be measured. Then the over-pressure at the stagnation point S (Fig. 1) is written:

$$P_s - P_o = \frac{1}{2}\rho U_o^2 + \sigma B^2 \int_{-\infty}^{o} (U_o - U) \, dz \tag{2}$$

It can also be put in the form:

$$P_s - P_o = \frac{1}{2}\rho U_o^2 [1 + 2\alpha N_d] \tag{3}$$

in introducing the interaction parameter built with the diameter of Pitot tube

$$N_d = \frac{\sigma B^2 d}{\rho U_o}$$

where σ is the electrical conductivity of the fluid and ρ its density, and a coefficient α which is a characteristic of the tube used and which is generally smaller thań one. With a magnetic field of 0.5 Tesla and mercury, large velocities ranging about 1 m/s must be used in order that N_d be less than 1 %; on the contrary, in regions of small velocity ranging about 1 cm/s it can reach value near one. It is therefore clear that a correction is often necessary and requires a calibration of each probe in order to determine its coefficient α (GNATYUK and PARAMONOVA, 1969).

The second risk of error comes from another variation of pressure in the inside of the Pitot tube and its support, situated like an electromagnetic pump in the crossed electric and magnetic fields. This effect is of course present only if the Pitot tube is an electric conductor and has good electric contact with the liquid metal. It can be estimated by integrating the equation of the hydrostatic equilibrium of the fluid between points S and M on Fig. 1, which leads to:

$$p_M - p_S \cong \sigma B^2 U_o L \cong \frac{1}{2} \rho U_o^2 \cdot N_\ell \qquad [4]$$

where

$$N_\ell = \frac{\sigma B^2 \ell}{\rho U_o}$$

is an interaction parameter built with the length ℓ necessarily much longer than d. This second correction, if it must be considered, is the most important by far. But it must be noted that it is extremely easy to do away with this risk of error by utilizing a probe that is insulating or insulated. In most laboratory experiments with slightly polluted mercury, a thin layer of impurities (oxydes or grease particles) is always present on the Pitot tube and is enough to insulate it electrically in order to practically annul this cause of error, even when the tube is of stainless steel.

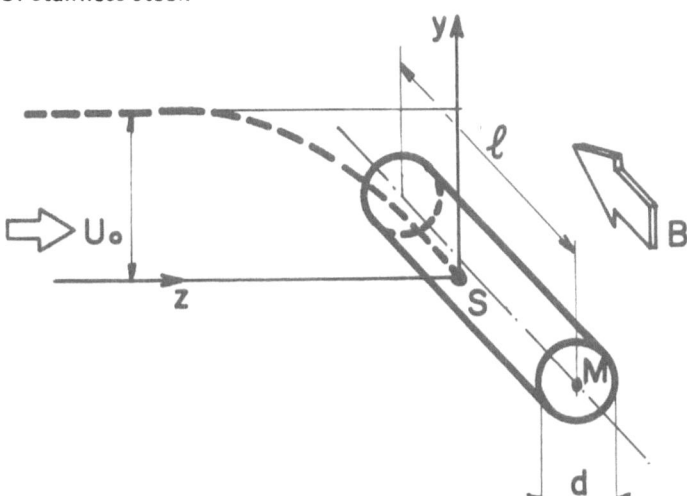

Fig. 2. Schema of a two-dimensional sensor for stagnation pressure in the presence of a transverse magnetic field

In Grenoble, we have worked out and used, in particular to calibrate our Pitot tubes and to measure with precision the errors mentioned above, a probe which is capable of measuring the true stagnation pressure. The sensor is a cylinder whose axis is parallel to the magnetic field and perpendicular to the velocity (Fig. 2). If this cylinder was infinitely long, the outflow in the neighbourhood of the aperture would be two-dimensional and the measurement would then be perfect since no electric current would be induced in the fluid, (SCHERCLIFF, 1965, p. 86). With these probes, built such that $\ell/d \cong 10$ to 20, the modification of the stagnation pressure is not exactly zero but is easy to estimate and appears d/ℓ times smaller than in the case of the classical Pitot tube (ROSANT, 1976). The pumping effect along side the tube is also suppressed. This probe thus appears completely rid of the risks of error to which the Pitot tube is submitted. Its only inconveniency is that it presents an obstruction often too important; for this reason we continue to use Pitot tubes after having them calibrated with this probe under suitable conditions.

3. Measurement of the Local Electric Potential and Field

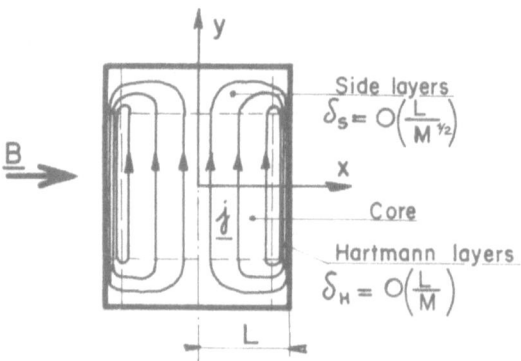

Fig. 3.   *Organisation of fully established flow in insulating rectangular duct with Hartmann number much larger than one*

This method is particularly interesting when the magnetic field is transverse, since in that case an electric field is induced perpendicular at the same time to B and U. To illustrate this, let us consider the example of a fully established duct flow with a Hartmann number $M = (\frac{\sigma}{\rho \nu})^{\frac{1}{2}} BL$ much larger than unity. This outflow is organized in three regions with quite distinct characteristics (Fig. 3):

— a core where the velocity is constant and only slightly larger than the average velocity $\frac{V_c - V_m}{V_m} = 0(M^{-1})$ and where the electric field $\underline{E}$ and the density of the current $\underline{j}$ are also uniform,

— the Hartmann layers, extremely thin since their thickness is of the same order as L/M, where all the electric current passing through the core confines itself (if the walls of the duct are insulating),

— the side layers, where the electric current lines have a component parallel to the magnetic field which permits them to bend in from the core towards the Hartmann layers, and whose thickness is about $L/M^{1/2}$.

It results from this organisation so well characterized and essentially imposed by the conservation of the electric current, that to the large Hartmann numbers, the three terms of the Ohm's law:

$$\frac{\underline{j}}{\sigma} = \underline{E} + \underline{V} \wedge \underline{B} \qquad \qquad [5]$$

are of the same order only in the Hartmann layers. Everywhere else the left hand side is negligible (at least in the direction y) before each of the other terms which balance almost exactly. Consequently, a measure of the component E_y of the electric field gives the velocity profile $V(y)$, the only one which interests us since $V(x)$ is uniform. The relative error made by neglecting j_y/σ is of the order of M^{-1} in the core and $M^{-1/2}$ in the side layers. The realization of sensors of this component E_y presents no difficulty.

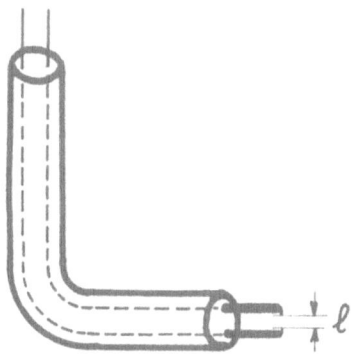

Fig. 4. Schema of a sensor for local electric field

In Grenoble we use bended cylindrical hot film probes which allow us to have a distance between the two electrodes varying from 1 to 3 mm (Fig. 4). Owing to a gold deposit obtained by vacuum evaporation, and which form an amalgan with the mercury, the electric contact is entirely excellent and stable. The signal obtained with a magnetic field of 0.5 Tesla and a velocity of 0.2 m/s ranges about 10^{-1} mv, i.e. very easy to measure after amplification, as long as one is interested only in the average value of the signal. Indeed in certain experiments where the electromagnets are fed by turning machines (ROSANT, 1976) the background noise is very important (it can attain 20 %). It can be considerably reduced by using storage batteries as the source of direct current (BRANOVER and GERSHON, 1976).

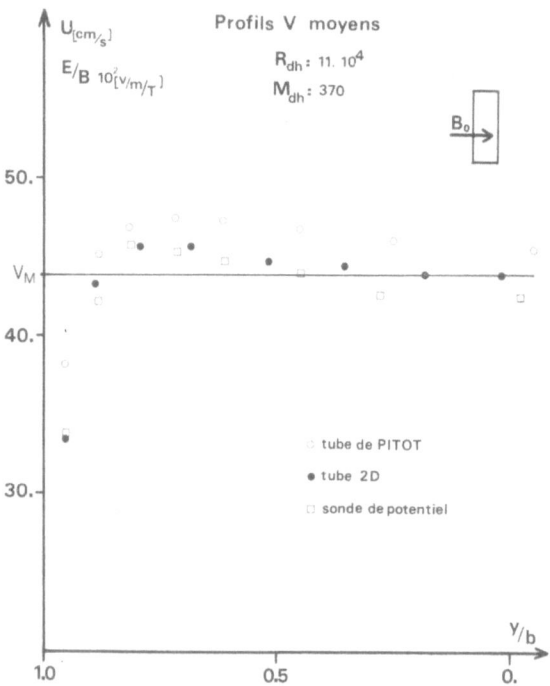

Fig. 5. Comparison of the profiles of the mean velocity in rectangular duct with insulating walls and transverse magnetic field, obtained with a Pitot tube, a two-dimensional pressure probe and a sensor for the electric field (after RO-SANT, 1976)

Fig. 5 allows the comparison of the velocity profiles V(y) obtained with the three types of sensors previously described. The error by excess of the measurements made with Pitot tube is perfectly clear although the corrective terms $2\alpha N_d$ of [3] remains about 4 %. The error by defect of measurements of the electric field is equally clear, but a bit slighter (about 2 %).

An important question cannot fail to be brought up at this point, although it has not yet received a satisfactory answer: that is to know if the fluctuation of the difference of instantaneous potential measured between the two-point electrode of a sensor such as that of Fig. 4 when the background noise is slight, can form the subject-matter of a precise interpretation and can serve as a diagnostic of certain characteristics of turbulence. The conservation of electric current (divj = 0) imposes the relation $\nabla^2\varphi = \underline{B}$ curl $\underline{v}$ between the electric potential φ ($\underline{E} = -\underline{\nabla}\varphi$) and the velocity $\underline{V}$.

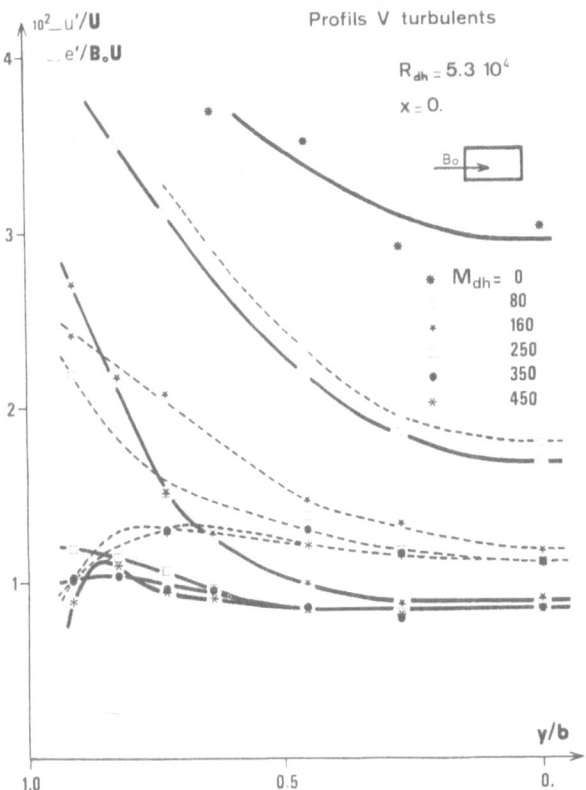

Fig. 6. Comparison of turbulent intensities in direction of the mean flow, measured with hot film and with sensor for electric field (after ROSANT, 1976)

If, to simplify, we limit ourselves to homogeneous turbulent fields, the solution to the Poisson equation can be written (MOREAU, 1969):

$$\varphi(\underline{x}) = -\frac{1}{4\pi}\int \underline{B}\cdot\underline{\omega}(\underline{x}') \frac{dr}{r} \qquad [7]$$

where $\underline{r} = \underline{x}' - \underline{x}$ and $\underline{\omega} = $ curl $\underline{V}$, and where the integral is extended to the entire space. It seems possible to use this expression [7] several ways, for example by deducing from it an expression of the value of local electric energy (MOREAU, 1969):

$$\overline{e^2} = \overline{\frac{\partial \varphi}{\partial x_p} \cdot \frac{\partial \varphi}{\partial x_p}} = \frac{B_i B_j}{4\pi} \int \overline{\omega_i(\underline{x}) \cdot \omega_j(\underline{x}')} \, \frac{dr}{r} \qquad [8]$$

These relations [7] and [8] which link the electric quantities to the vorticity seem extremely interesting. However, it must be noted that through the integral a global dependence appears and that these electric quantities cannot serve directly as measurements of local values of the vorticity or of certain correlation of the vorticity. Nevertheless, it seems that these relations have not yet been completely exploited and that capable experimenters could perhaps use them to obtain pertinent information on essential aspects such as intermittency and structure of small eddies. It must, however, be yet noted that the quantity measured by a probe like that of Fig. 4 is a difference of potential between two small electrodes and not a local electric field. This implies a second global effect which leads us to think that the exploitation of these ideas would necessitate a considerable refinement of these probes.

Here should also be mentioned another method of deducting the distribution of average velocity in a duct from measurements of electric potential, used by LECOCQ (1964) but never used since then. LECOCQ utilises the relation [6] to deduce the distribution of velocity $V(x,y)$ from a distribution of electric potential $\varphi(x,y)$ which itself is obtained by displacing a point electrode in all the cross section of the duct and by measuring the difference of potential between an electrode attached to the wall and this moving electrode. The local value of $\nabla^2 \varphi$ is obtained numerically by a double derivation, and the velocity profiles then follow from an integration of the relation [6]. It seems that LOCOCQ arrived at relatively precise results by using techniques sufficiently sophisticated to reduce the risk of error on $\nabla^2 \varphi$.

A variant of this method, suggested by GROSSMAN, LI and EINSTEIN, (1957) and later by SCHERCLIFF (1965), often discussed at colloquia (HUNT and MOREAU, 1976) and tested only at Riga to my knowledge (KIT, 1970) consists in using a probe with five point electrodes separated by a step h rather small compared to the characteristic length scales of the flow. If one assigns the suffix 0 to the central electrode and 1 to 4 to the other four electrodes situated at the summit of a square, the quantity $\nabla^2 \varphi$ could be obtained instantaneously:

$$\nabla^2 \varphi = \frac{\varphi_1 + \varphi_2 + \varphi_3 + \varphi_4 - 4\varphi_0}{h^2} \qquad [9]$$

with a greater precision than that obtained by LECOCQ.

4. Measurement of the Heat Flux Extracted from a Hot Film

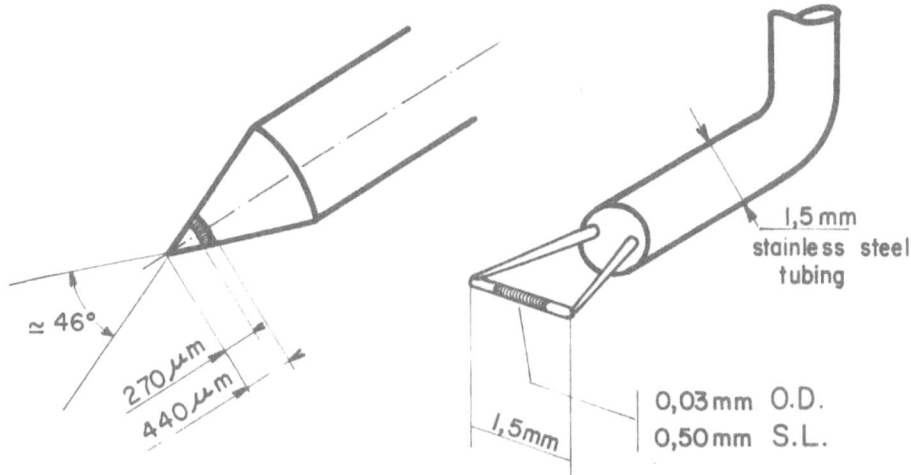

Fig. 7. Examples of hot film probes realized with a platinum film deposited on a quartz support (conical or cylindrical) and insulated from the liquid by a thin quartz coating

Hot film anemometry, a measurement technique now well mastered in air, both for local average velocities and for its fluctuations, is evidently usable in liquid metals, in spite of the two categories of difficulties mentioned in the introduction. The probes used are made with a platinum film deposited on a quartz support and electrically insulated from the surrounding fluid with the help of a thin layer of quartz. Fig. 7 shows two examples of realizations of such probes to carry out measurements in mercury; the constructors (Thermo Systems Inc. and DISA Electronics) now propose a large enough variety of probes so that one can always be found which is adaptable to the experiment foreseen. The anemometers used are generally of the constant temperature type.

All those who utilize this method deduce from the fact that the interaction parameter built with the diameter (or another characteristic length in the case of non-cylindrical films) of the sensitive element is very small

$$N_d = \frac{\sigma B^2 d}{\rho U} \leqslant 10^{-2}$$

that the flow around the film can be considered as not being disturbed by the magnetic field. Several among them, however, have some misgivings about this subject (MALCOLM, 1969). It must be said here that the influence of the magnetic field on the outflow around an obstacle is still poorly known and that on the theoretical level the misgivings seem justified (suppression of vorticity, etc.) However, it is evident from the entire measurements made with hot film (LYKOUDIS, 1976, ALEMANY et al., 1979) that this influence certainly remains confined in the acceptable incertitudes of the turbulence measurements.

Among all the difficulties which arise from the nature of the fluid, the most oftenly occurring is certainly the presence of impurities which can settle on the sensor and constitute a thermal contact resistance, which varies with time. Only ROBINSON and LARSSON (1973) using a small enough quantity of mercury so that the experiment might be made in a vacuum, seem to have managed to annul this difficulty. All the other experimenters seem to be reduced to accepting this deposit of impurities and imagining a method for eliminating its influence on the measurements. In fact, two methods have been proposed, which both consist in admitting that this thermal contact resistance depends very little on fluid velocity and in trying to eliminate it by effecting a difference between the measurement at zero velocity and the measurement at the actual velocity. SAJBEN (1965) and MALCOLM (1969) admit that the total thermal resistance is made of two resistances in series, the one which is due to the conduction in the fluid, in the supports and through the impurities, and which is supposed to be independent of the velocity (noted K below), and the convection resistance, inversely proportional to the Nusselt number Nu, which is supposed to depend only on the Peclet number. The relation:

$$\frac{\pi k L \Delta T}{Q} = K + \frac{1}{Nu} \qquad [10]$$

where k designates the thermal conductivity of the fluid, L the length of the film, ΔT the difference in temperature between the film and the fluid, and Q the flux of energy furnished to maintain ΔT constant, permits the elimination of K by the difference:

$$\pi k L \Delta T \left[\frac{1}{Q(o)} - \frac{1}{Q(P\acute{e})} \right] = F(P\acute{e}) \qquad [11]$$

The measurements of MALCOLM (1969) in the range $0.01 \leqslant P\acute{e} \leqslant 1$ seem to justify quite well the hypothesis of resistances in series, and also confirm that the quantiy F depends almost exclusively on Pé.

In Grenoble, with experiments generally bigger and with higher velocities, we have Péclet numbers comprised between 1 and 20, and we use the schema opposed to that of SAJBEN and MALCOLM, supposing that the two thermal resistances are in parallel. (ALEMANY, 1978; ALEMANY et al., 1979). This leads us to effect another difference in order to obtain a quantity supposed to be independent of the thermal contact resistance:

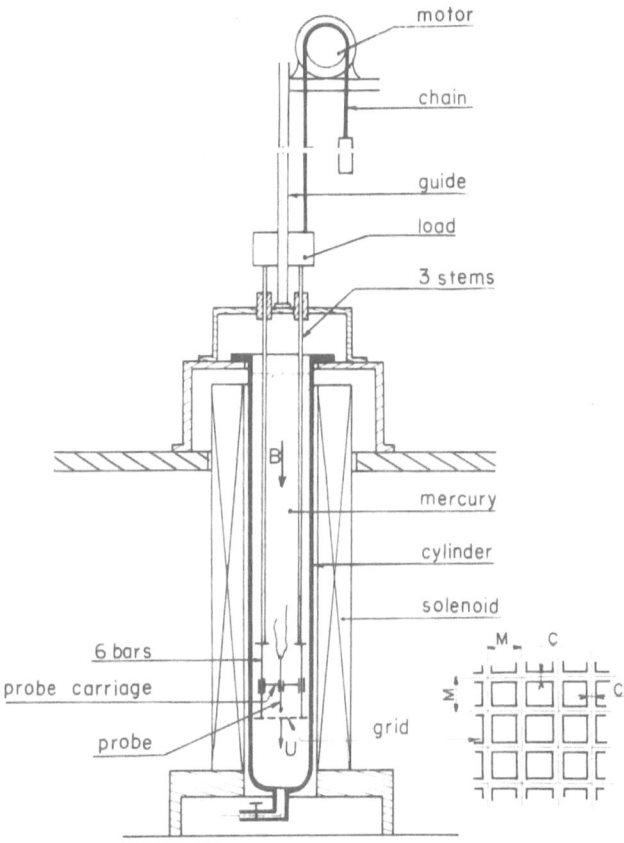

Fig. 8. Experiment carried out in Grenoble to investigate MHD homogenous turbulence (after ALEMANY, 1978)

$$\pi k L \Delta T \cdot Nu(P\acute{e}) = Q(P\acute{e}) - Q(o) \qquad [12]$$

My own opinion on this question consists in considering that the reality is much more complex than these schematic dispositions in series or in parallel. To me, it even seems questionable that the contact resistance be independent of the fluid velocity which must certainly modify the disposition of impurities on the probe. But one has to search for a way of getting around this difficulty, and I believe that the idea of reducing the influence of the impurities by means of a difference is to be retained. I also believe that each experimenter can chose empirically between the two differences [11] and [12] or can imagine a better combination of $Q(o)$ and $Q(P\acute{e})$. Indeed, if the influence of the impurities is weak and only slightly dependent on the velocity, the differences [11] and [12] lead to some improvement of the signal. It is evident, of course, from

these remarks that it is essential to clean the mercury and the probes as carefully as possible and to reduce to a minimum the risks of contaminating during
the experiments, for example by never crosssing a free surface with a probe.
HOFF (1969) tried to reduce this thermal contact resistance by depositing a
very thin gold coating on the insulating film of quartz, thus forming a superficial amalgam; we have also tried to use gold coated probes in Grenoble
(ROSANT, 1976) but the disturbance due to the mechanical erosion of the
gold coating revealed themselves to be as harmful as those due to impurities, so
that we did not continue to use these probes.

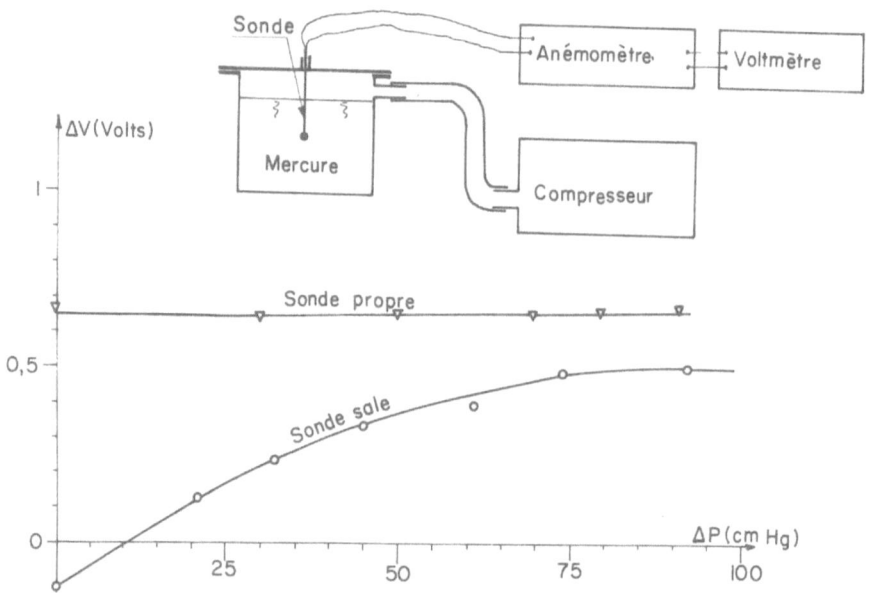

*Fig. 9. Experiment carried out to demonstrate the influence of the pressure
on the thermal contact resistance of the deposit of impurities on hot films
(after ALEMANY, 1978)*

An experiment illustrated by Fig. 9 was carried out in Grenoble to bring to
light the sensibility of thermal contact resistance to pressure. It very clearly
shows that this influence of pressure on the signal delivered by the probe after
having functioned for some ten hours, is due to the deposit of impurities.

The other difficulties in thermoanemometry in mercury have been the object
of a very complete synthesis by MALCOLM (1976). They fall into three categories:

A.   The risk of drift of the liquid metal temperature during the experiment.
MALCOLM (1969) shows that an error of 5 % on ΔT leads to an error of
24 % of F(Pé) in [11], and consequently to an important error in the
measurement of Pé and of the velocity. He indicates that at very low
velocities (a few cm/s) one must obtain temperatures which are stable to
approximately 10^{-1} °C in order to obtain sufficiently precise measurements. The main rule which becomes apparent from this remark, is that
experiments must be conceived in such a way as to have a Péclet number
built with the diameter of the hot film clearly superior to unity.

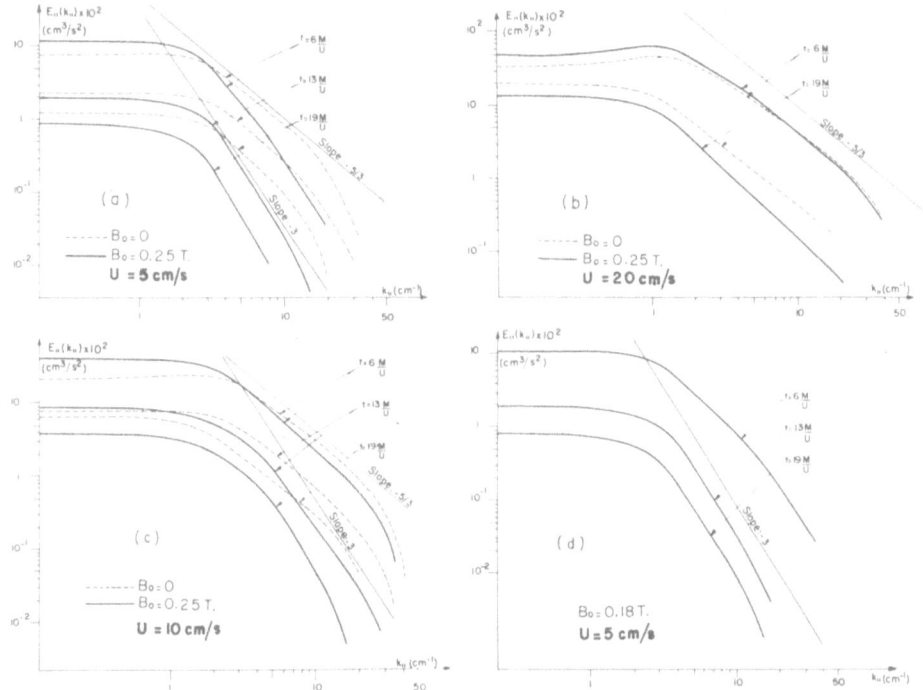

*Fig. 10. Examples of energy spectra obtained with hot films in mercury in the
presence of a uniform magnetic field (after ALEMANY et al., 1979)*

B.   The directional sensitivity of cylindrical probes. HILL and SLEICHER
(1971) have demonstrated that in liquid metals at low velocity this sensitivity is not a simple function of the angle between the directions of
the film and of the velocity. Two other variables, the fluid velocity and
the elongation of the cylinder have an important influence on the signal.

This difficulty makes particularly arduous the utilization of X probes for separating two components of the velocity especially when the latter is small. Such measurements, however, have been carried out recently at Purdue University by REED and LYKOUDIS (1978) and by PATRICK and LYKOUDIS (PATRICK, 1976; LYKOUDIS, 1976) with sufficient reproductibility and precision, but at rather high velocities (20 to 50 cm/s).

C. The frequency response at low Péclet numbers. MALCOLM (1976) shows that when Pé $<$ 1 a reduction of amplitude of 10 % can be expected at frequencies of the order of 0.02 V^2/α Hz, and of 90 % at frequencies of 4 V^2/α Hz. With mercury in laboratory experiments the frequencies reduced to 10 % range about 4000 Hz, and the main part of the kinetic energy is situated in frequencies inferior to 1000 Hz (see Fig. 10). It seems, therefore, that as long as the measurements pertain the turbulent large scales and concern the turbulent intensities, $\overline{u'^2}$ and $\overline{v'^2}$ and correlations such as $\overline{u'v'}$, this difficulty is not a real handicap.

In conclusion, two rules seem to impose themselves on all experimenters using hot film anemometers in mercury: clean mercury and high velocities ($\geqslant$ 10 cm/s) should be employed.

References

ALEMANY, A., MHD à l'échelle du laboratoire, quelques résultats et quelques applications, Thèse de Doctorat d'Etat, Univ. de Grenoble, 1978

ALEMANY, A., MOREAU, R., SULEM, P.L. and FRISCH, U., Influence of an external magnetic field on homogenous MHD turbulence, *J. de Mécanique*, Vol. **18**, to appear in 1979

BRANOVER, H. and GERSHON, P., *Proc. of the Bat-Sheva International Seminar on MHD Flows and Turbulence*, John Wiley, p. 81, 1976

GNATYUK, V.V. and PARAMONOVA, T.A., Magni. Gidrod., no. 4, p. 143, 1969

GROSSMANN, LI and EINSTEIN, *Proc. Amer. Soc. Civ. Engrs*, Vol. **83**, p. 1394, 1957

HILL, J.C. and SLEICHER, C.A., *Rev. Sci. Inst.*, Vol. **42**, p. 1461, 1971

HOFF, M. *Instruments and Control Systems*, Vol. **42**, p. 83, 1969

HUNT, J.C.R. and MOREAU, R., *J.F.M.*, Vol. **78**, p. 261, 1976

KIT, L.G., Magnit. Gidrod., no. 4, p. 41, 1970

KIT, L.G., PETERSON, D.E., PLATNIEKS, I.A. and TSINOBER, A.B., Magnit. Gidrod., no. 4, p. 47, 1970

LECOCQ, P., *Bull. du Centre de Rech. et d'Essais de Chatou*, Suppl. au no. 8, 1964

LYKOUDIS, P.S., *Proc. of the Bat-Sheva International Seminar on MHD Flows and Turbulence*, John Wiley, p. 103, 1976

MALCOLM, D.G., *J. Fluid Mech.*, Vol. **37**, p. 701, 1969

MALCOLM, D.G., *Proc. of the Bat-Sheva International Seminar on MHD Flows and Turbulence*, John Wiley, p. 119, 1976

MOREAU, R., *Proc. of the Symp. on Turbulence of Fluids and Plasmas*, Polytechn. Inst. Brooklyn, MRI Series, Vol. **18**, p. 359, 1969

PATRICK, R.P., Magneto-fluid Mechanic Turbulence, Ph.D. Thesis, Purdue Univ., 1976

REED, Cl., An Investigation on Shear Turbulence in the Presence of a Magnetic Field, Ph.D. Thesis, Purdue Univ., 1976

REED, Cl. and LYKOUDIS, P., *J. Fluid Mech.*, to appear 1979

ROBINSON, T. and LARSSON, K., *J. Fluid Mech.*, Vol. **60**, p. 641, 1973

ROSANT, J.M., Écoulements hydromagnetiques turbulents en conduites rectangulaires, Thèse de Docteur-Ingénieur, Univ. de Grenoble, 1976

SAJBEN, M., *Rev. Sci. Inst.*, Vol. **36**, p. 945, 1965

SCHERCLIFF, J.A., *A Text Book of Magnetohydrodynamics*, Pergamon, 1965

On Data Acquisition in Heated Turbulent Flows

by

H. Fiedler
Hermann-Föttinger-Institut für Thermo- und Fluiddynamik
Technische Universität Berlin

1. Introduction and Specification of Problem

The problem of taking measurements in non-isothermal flows is encountered in many practical cases as well as laboratory investigations. Typical examples are:

— combustion flows (chemical reactions)
— compressible flows ($M > (\gg)\ 1$)
— flows with heat transfer

The first two cases depict genuinely coupled thermal- and velocity fields. The temperature is in general not separately adjustable and temperature differences are relatively high — as in the case of a flame. Thus in these cases the thermal field is a substantial and "active" component of the problem.

On the other hand, temperature-contamination is often profitably used for diagnostics of as such isothermal flows. There the thermal field in its structure is fully related and determined by the velocity field, its magnitude, however, is adjustable where the upper and lower limits may be determined by:

1. The demand for passivity of the contaminant for which the Richardson number is a criterion, and

2. The resolution of the temperature sensor, i.e. its signal-to-noise ratio.

For the experimental analysis of any non-isothermal flow it is of essence to determine certain quantities purely, i.e. to separate them from other parameters. These may be single physical quantities, like velocities, temperatures, etc. or combinations like correlations etc. allowing for an unimpaired interpretation of the separate fields and their possible interactions.

Of the great number of different principles to obtain temperature and velocity data in non-isothermal turbulent flow the most common ones today are:

1.   Heat transfer methods (extended hot-wire principle) and
2.   Laser optical methods.

While in the following this paper will only concentrate on a few main problems of the heat transfer method as applied to non-isothermal turbulent flow fields, leaving those interested in the more advanced Laser-Raman-diagnostics behind (however, referring them e.g. to the paper by S. Lederman (1)) we shall at least try to list a few points of comparison of the two methods:

	Heat transfer methods	Laser methods
Temperature field	**CC-Resistance probe**	**Laser-Raman scattering**
range of temperature	limited by probe material only	particularly for high temperatures (flames)
calibration	easy	difficult
economy	low price simplicity	high costs complicated technique
specificity	approximate	yes
space resolution	good	good
time resolution	good	problematic
non intrusiveness	no	yes
Flow field	**Hot wire anemometry**	**Laser Doppler anemometry**
calibration	simple	not necessary
economy	medium price relatively simple	high costs complicated
specificity	no	yes
space resolution	good	good
time resolution	good	problematic
non intrusiveness	no	yes

2. Heat Transfer Sensors

2.1 General Principles

The influence of various flow variables on the heat transfer from a heated wire in a flow and thus on its resistance has been discussed in great detail by many authors e.g. Corrsin (2, 3), Kovasznay (4, 5). I shall, therefore, restrict myself to a mere consideration of the global influence of the species temperature and velocity on a hot-wire heated at constant current I_o. Making use of King's equation we write:

$$\frac{R_w \cdot I_o^2}{R_w - R_M} = A + Bc^n, \text{ and}$$

$$\frac{R_w \cdot I_o^2}{R_{w_o} - R_M} = A = \frac{R_{w_{oo}} \cdot I_o^2}{R_{w_{oo}} - R_o}$$

where

R_w = Resistance of wire

R_{w_o} = Resistance of wire for $c = o$

$R_{w_{oo}}$ = Resistance of wire for $c = \Delta T = o$

R_M = Resistance of wire at medium-temperature T_M

ΔT = $T_M - T_o$

T_O = Reference (calibration) temperature

n ≈ 0.5; A, B $\approx$ const

These equations may be combined to yield a relation for c and ΔT, which for the case

$$E_B = I_o (R_w - R_{w_{oo}}) = 0 \text{ (bridge voltage)}$$

allows for evaluation of the influence of the actual parameter for a given I_o. The result is shown in Fig. 1 for a specific probe configuration (d = 0.65 μm). It is obvious that for a heating current of, say, 0.1 mA the maximum velocity influence corresponds to a temperature difference of approx. 0.3°C. For a velocity of 10 m/s

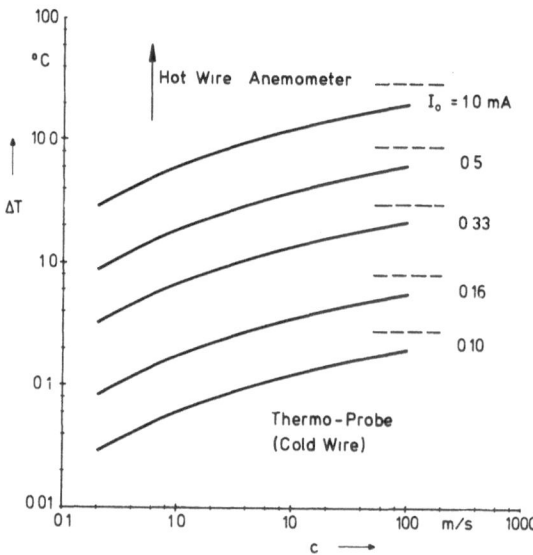

Fig. 1. Velocity- and temperature-influence for hot-wire systems at different wire heatings (R_o = 600 Ω, d = 0.65 μm)

$$\Delta T_{error} \approx 0.12°C$$

Thus a wire heated at such low current may well serve as a temperature sensor with negligible velocity influence on the output. It is of interest to note, that for I_o = 1 mA the maximal possible velocity influence on the wire reading already corresponds to 30°C of flow temperature difference! For equal overheat ratios the heating current is related to the wire diameter by $I_o \sim d^{3/2}$.

To investigate a non-isothermal flow field by hot-wire anemometry one thus may separate velocity c from temperature ΔT by obtaining two outputs at different heating currents I_o, which then may be combined to yield f_1 (ΔT) and f_2 (c). The two readings, when obtained simultaneously, may also be evaluated to yield T' and c', the fluctuating quantities in a turbulent flow. If, however, a great number of data is needed of the temperature field alone, it might be of advantage to use a simple resistance probe (i.e. a sensor with I_o = 0 (0.1) mA) where the velocity influence is negligible and which for small temperature differences may easily be combined with an ordinary hot-wire system and thus allow for temperature-compensated measurements as well.

In the following those temperature sensors and compensation circuits will be discussed.

2.2 The Thermo-probe

2.2.1 Electronic Circuitry
The sensor is part of a Wheatstone Bridge which — for reasons of simplicity — is a DC-bridge with 1:10 resistance ratio. The electronic circuit used in our investigation, which was developed by E. Froebel (6) is shown in Fig. 2.

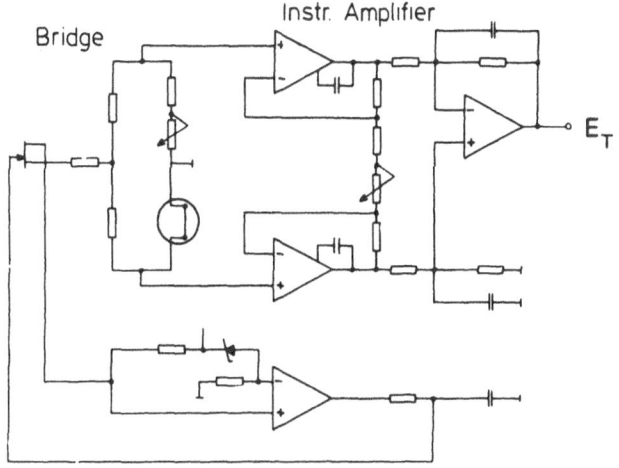

Fig. 2. Thermo-probe bridge circuit

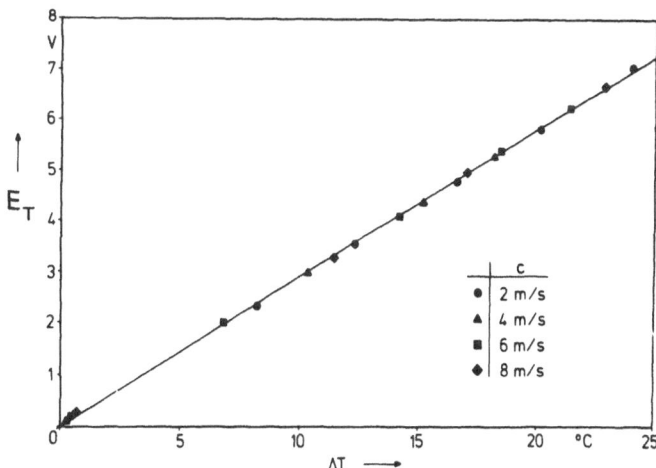

Fig. 3. Calibration curve of thermo-probe for various flow velocities. Arbitrary ordinate scale.

Changes in ambient reference temperature are easily compensated by a second probe, the bridge output of which is subtracted from the measuring bridge. A heating current of 0.25 mA for the full bridge was found to be a good compromise to yield sufficient signal voltage at still negligible influence of velocity in the signal. This for the problem at hand — a heated turbulent shearlayer — where the maximum velocity was $c = 10$ m/s and the maximum temperature difference $\Delta T = 15°C$. Fig. 3 shows the independence on velocity and on the other hand a sufficient linearity between E_B and ΔT which is best around the balanced bridge case.

2.2.2 Sensor Geometry

The most serious shortcoming for this kind of sensor is its restricted frequency response. The time constant of the wire due to its thermal inertia is a function of its diameter. There then the demand goes for small wire diameters d. The smallest diameter available on the market is $d = 0.5$ μm. In our investigation we used $d = 0.65$ μm. In addition to a high frequency resolution a high signal-to-noise ratio is asked for to enable the measurement of very small temperature increments. Since $E_B/\Delta T = \alpha\, I_o \cdot R_w$ a high resistance of the sensor is needed to obtain a signal voltage E_B well beyond the electronic noise (the thermal noise of the probe wire $E_R = (4k\, T_o\, f \cdot R)^{1/2}$ is negligible).

For a freely suspended wire (unlike film probes as used by Schmidt et al. (7) which, however, cannot be used for combination probes), the upper limit for the wire resistance is given by its tensile strength.

$$\text{Let } \sigma = \frac{F}{A} = \frac{C_D \cdot \ell \cdot d \cdot \frac{\rho}{2} c^2}{\frac{\pi d^2}{4}}$$

$$\text{where: } C_D = C_{D_{oseen}} = \frac{8\pi}{Re(2 - \ell n\, Re)} \approx \frac{1.6}{Re}$$

$$\text{for } Re = \frac{c \cdot d}{\nu} \approx 10^{-6}$$

$$\text{Since } R \sim \frac{\ell}{d^2}$$

$$\curvearrowright \sigma \sim c \cdot \eta \cdot R$$

Thus for a certain R, chosen for optimum signal-to-noise ratio the wire diameter is of no consequence to the mechanical strength and durability and may thus be chosen from considerations of time constant and spacial resolution (wire length ℓ) only.

Experience has led us to using sensors with d = 0.65 μm and ℓ = 1 mm and a wire resistance of R $\approx$ 600 Ω. Sensors of these dimensions have also been used by other workers in this field, e.g. Sreenivasan, Antonia and Danh (8).

With these probes the RMS-value of the noise (high frequency and 50 Hz hum) in the system described corresponds to approx. 0.10°C and is thus of the same (negligible) order as the velocity contamination of the signal.

2.2.3 Time constants and frequency resolution
The actual time constant of the probe system may be impaired by the bridge system in addition to the thermal inertia of the wire itself (see Bremhorst and Krebs (9). Considering only the latter it is approximately described by the relationship

$$\tau_w \approx k \cdot d^n \cdot f(c),$$

where theoretically   n = 1.5 for c $\rightarrow \infty$ and n = 2.0 for c $\rightarrow$ o.

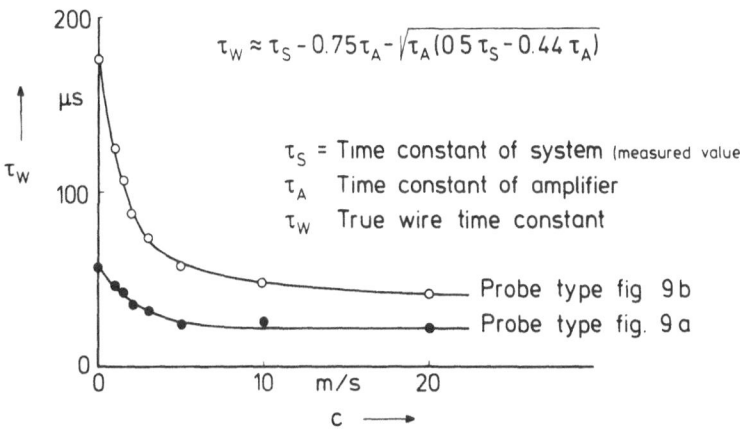

$$\tau_w \approx \tau_S - 0.75\tau_A - \sqrt{\tau_A(0.5\,\tau_S - 0.44\,\tau_A)}$$

τ_S = Time constant of system (measured value)
τ_A Time constant of amplifier
τ_w True wire time constant

Probe type fig. 9 b
Probe type fig. 9 a

Fig. 4. Wire time constants for two probe configurations. d = 0.63 μm

Fig. 4 shows time constants τ_w versus flow velocity c for two different probe configurations (measurements by P. Mensing). The measurements were done by external heating of a wire portion using a chopped laser beam (see Fig. 6) and corrected for zero amplifier time constants τ_A. The approximate correction formula in Fig. 4 was derived for a system with two time constants in series. The drastic differences for the two probe types under identical circumstances illustrate the need for individual calibration since f(c) in the above relation is not universal. Further investigations will be conducted on this question. Apparently probe type Fig. 9b is inferior to type a of Fig. 9 for small scale high frequency fluctuations.

The time constant can be effectively reduced, that is, the limiting frequency increased if necessary by electronic compensation circuitry, as was discussed e.g. by Kovasznay (1) and by Freymuth (10). With a 0.63 μm wire-probe this may not be necessary, unless measurements in the dissipative frequency range of high velocity turbulent flow are to be performed. In this case it may also be useful to check more closely into the velocity contamination of the signal (see Mimaud-Lacoste (11)).

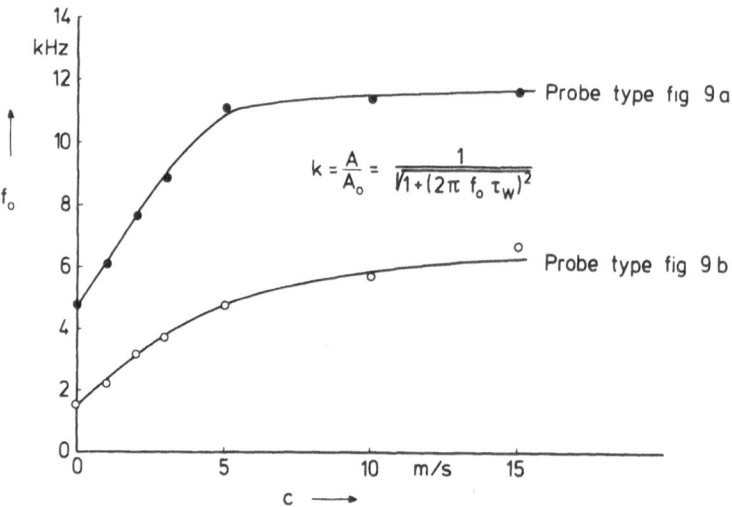

Fig. 5. Limiting frequencies (A/A$_o$ = 0.5, φ_{Phase} = –60°) versus velocity for two different probe configurations

Fig. 5, showing limiting frequencies for the d = 0.63 μm wire probes as evaluated from Fig. 4, may serve for better interpretation.

To measure τ_w a number of different methods are possible, e.g.:

1. Electric "pulsing" of the wire exposed to cold flow. The electric pulse may be a pure step-pulse. A somewhat more elegant way is to use high frequency bursts, which will only heat the wire, however, not be felt "directly" by the slow bridge-amplifier-system (see also Bremhorst and Krebs (9)).

2.   External heating of the probe, e.g. by a pulsed laser beam.

3.   Exposing the wire to an air-stream with a sudden change in temperature (moving the sensor from a cold into a hot air-stream and vice versa).

Methods 1 and 2 are similar in that the physical process of heating is different from the real case and thus only the time constant of the cooling phase may be considered.

Method 3 poses technical difficulties due to finite thickness of the separating layer between the two air-streams.

Another, quite important point is, that by method 1 only the wire will be heated, while the prongs remain cold. By applying method 2 the heating may also be applied to the prongs, however, there is no control as to whether the prongs will assume the same temperature as the wire. In the real case the prongs will assume the average temperature of the medium following to some extent the low frequency content of the temperature fluctuations according to their time constants which typically is of the order of 0.2 sec.

With the pulsed-laser method applied in different ways − as shown in Fig. 6 − it was found that as long as the prongs are also subjected to temperature fluctuations their time constant influences the probe response to a step pulse. It became, however, obvious that it is not the heat conduction into and from the prongs (see Maye (12) that is responsible for this disturbing effect, but the radiation between prongs and wire. (Cooperation with P. Mensing).

The time constant of the prongs was measured and is shown in Fig. 7. The existence of this second time constant may pose a serious problem, since it implies different, i.e. reduced sensitivity of the probe for "fast" fluctuating temperatures as compared to its response at constant or slowly varying temperatures, or alternating temperature levels of long duration e.g. in the intermittency regime of a heated turbulent jet. Prong influence on the wire response has also been observed and reported by Hojstrup et al. (21), who also proposed an interesting method for time-constant calibration where temperature fluctuations are produced by sound (21, 22).

Since the above method does not supply a quantitatively correct value of the prong influence A_2/A_1

Test arrangement:

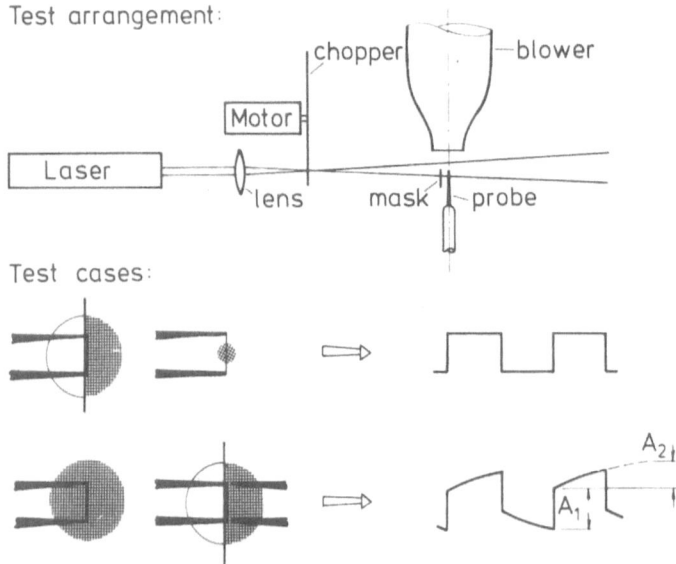

Test cases:

Fig. 6. Time constant calibration with laser heating

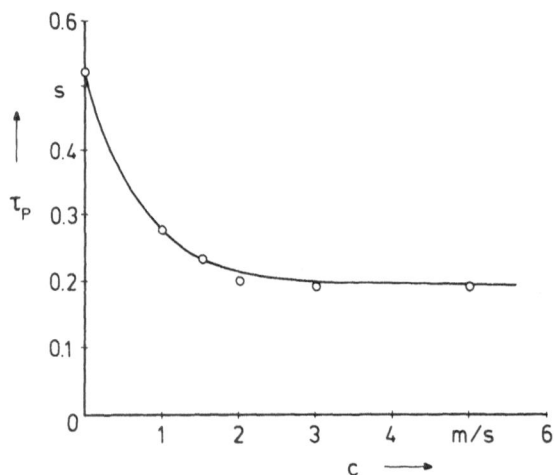

Fig. 7. Time constant of wire prongs versus velocity

the third method was revived. The apparatus is shown in Fig. 8. Designed as a fluidic amplifier its two streams of different temperature can be "switched" to either side of the diffusor exit, leaving the probe tip in the cold or the warm stream. This method is too slow to measure the wire time constant. It serves, however, well for measuring time constant and amplitude ratio A_2/A_1 of the prong influence.

This is shown in Fig. 9, which shows two simultaneously recorded step signatures from two probes of different design: While the common probe design (a) clearly suffers from prong influence, for the modified type (b) this prong influence is essentially suppressed. There the radiative influence is eliminated by increasing the distance between element and supports. This method is superior to electronic compensation of τ_p since it is independent of the flow velocity, as well as of its structural scale.

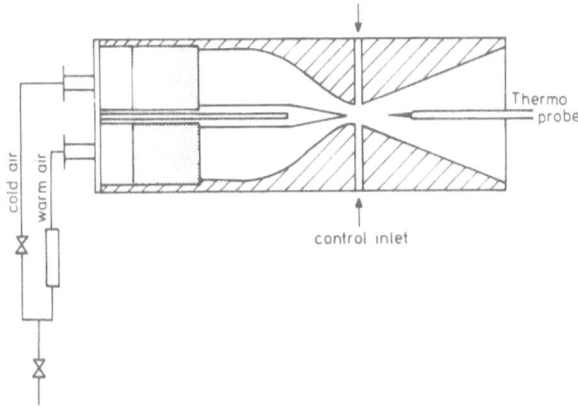

Fig. 8. Time constant calibration unit with cold and hot air-streams. Fluidic principle

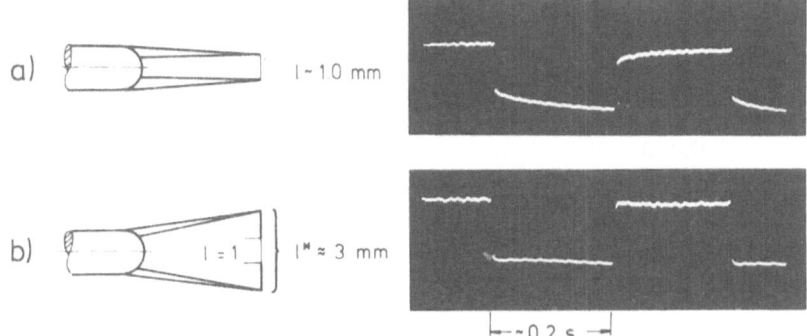

Fig. 9. Temperature step response of two different probe configurations

2.2.4 Probes
The probes were manufactured of Wollaston wire with 0.65/40 μm diameter. The sensing material is Pt/10 % Rh with a silver coating. Wollaston wire is soldered (80 % gold, 20 % tin) on Nimonic 90-prongs of 40 μm diameter at their tips. Of the 3 mm wire length a central portion of 1 mm length is then etched down to the core. Thermal resistance coefficient was measured as: $\alpha = 0.0013$.

The great simplicity of the temperature sensing circuitry makes the use of multi-probes as a tool for studies in turbulence structure very easy and attractive. Fig. 10 shows — as an example — a multi-probe signature of temperature traces in a heated free shear layer. The figure shows clearly defined large scale structures and serves as a good example of a powerful diagnostic method in turbulence research.

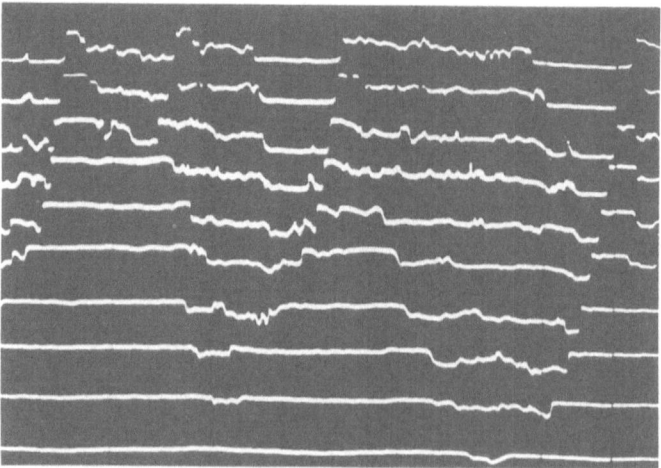

Fig. 10. Simultaneous temperature traces from a 10-probe rake in a heated shear layer

3. Velocity-temperature Separation Methods
Many workable methods to separate temperature- and velocity signals in non-isothermal fluids have been proposed (Sakao (13), Chevray and Tutu (14), Fabris (15), Artt and Brown (16) etc.), and an interesting analysis of these methods was published by Drubka et al. (17). In the following two very simple methods are described.

3.1 Combination method
This method was developed during our work on the heated shear layer (e.g. Fiedler, Korschelt, Mensing (18)). The details are published elsewhere (Fiedler

and Mensing (19)). I shall, however, summarize the underlying principle: The following Fig. 11 shows the bridge voltage of a CT-hot-wire anemometer as a function of velocity c and temperature T. The temperature-compensated bridge voltage E_{B_o} may be described as

$$E_{B_o} = E_B + m \cdot \Delta T$$

where the index o stands for ambient temperature. To a good degree of accuracy m is a constant for velocities $c > 8$ m/s in the range $20°C \leqslant \Delta T \leqslant 50°C$. For $c < 8$ m/s m = const leads to overcompensation. It was then found that full compensation is reached by choosing

$$m = K_1 (E_B + K_2), \text{ where } K_{1,2} = \text{constants}$$

thus

$$E_{B_o} = E_B + K_1 (E_B + K_2) \cdot \Delta T$$

The electronic circuit for this compensation is shown in Fig. 12. The compensation voltage is added to the bridge voltage ahead of the linearizer, thus a linearization need be performed only once at ambient temperature.

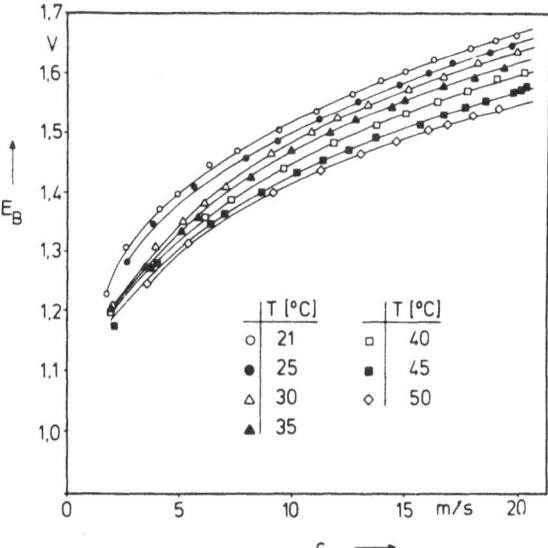

Fig. 11. Bridge voltage of constant temperature. Hot-wire anemometer for different air temperatures versus velocity.

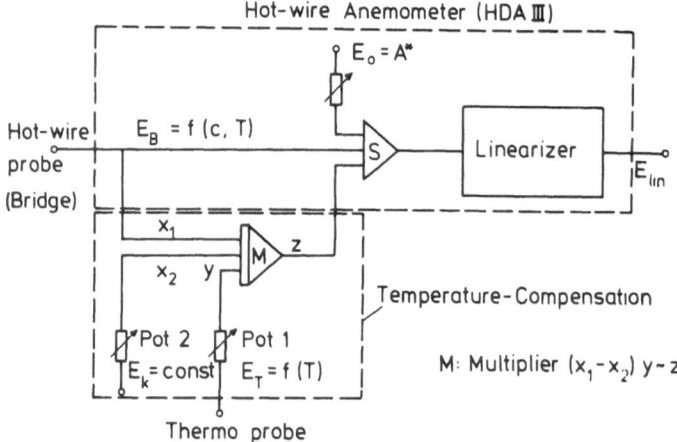

Fig. 12. Block-circuit for temperature compensation of CT-hot-wire signal

Fig. 13 shows a linearized temperature-compensated calibration curve of the hot-wire anemometer which was obtained after 3 iteration steps of compensation.

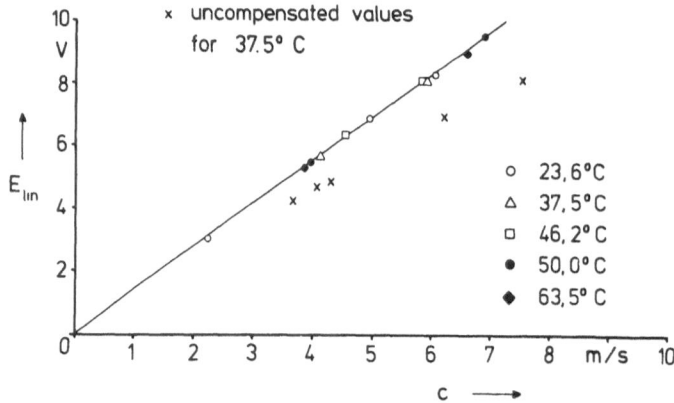

Fig. 13. Temperature compensated calibration curve of linear hot-wire output

More complicated and powerful compensation schemes similar to what is proposed for instance by Chevray and Tutu (14) or others have also been developed and tested but were then abandoned in favour of the simplicity of the above method. Fig. 14 shows typical combination probes consisting of an X-wire and a thermal wire which were used with this compensation. For the design according to Fig. 14a care should be taken that the thermal probe is at no time reached by the wake of a hot wire pushed sideways by a cross fluctuation.

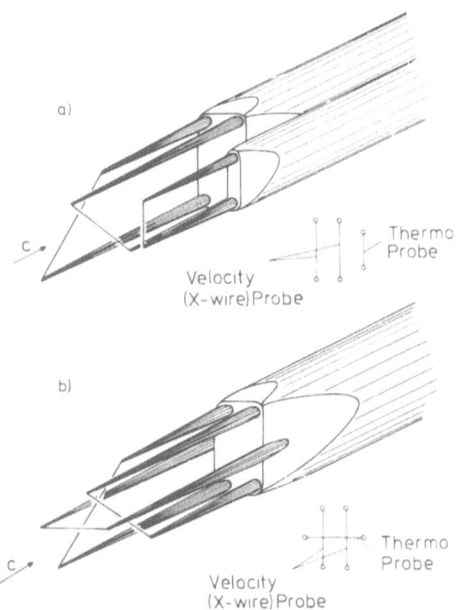

Fig. 14. Velocity temperature combination probes

3.2 Single wire method

U. Michel in cooperation with E. Froebel (not published) devised a dynamic method of quasi-simultaneous measurement of temperature and velocity by evaluating the output of a CT-hot-wire anemometer whose overheat ratio periodically changed every 0.5 ms (see also Artt and Brown (16). The block diagram in Fig. 15 shows the principle of the "switched" hot-wire anemometer. Fig. 16 shows a typical bridge voltage trace versus time, where the maximum frequency resolution is $f_{max} = \frac{1}{4 \cdot \Delta t}$ (Δt = switching period).

$$1/\Delta t = f_s \sim P_w = Ew^2/Rw$$

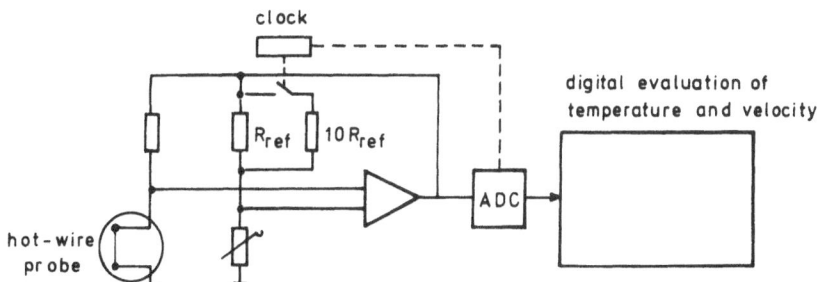

Fig. 15. CT-hot-wire anemometer with alternating (switched) bridge resistance

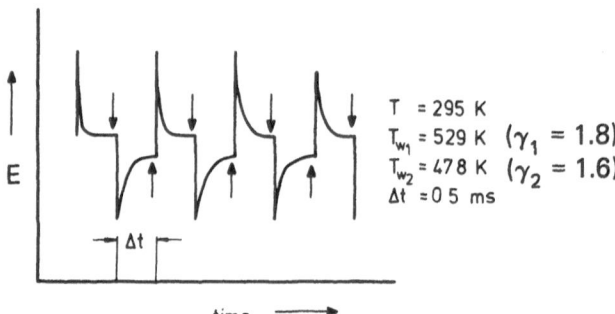

Fig. 16. Voltage of switched hot-wire bridge for constant flow velocity and temperature. Arrows denote positions of sampling

Evaluation of the hot-wire signals is done digitally based on the equation:

$$P_w = \frac{E_w^2}{R_w} = F(c) \, (T_w - T_M) \, (1 + \alpha(T_w - T_M))$$

for the electrical power P_w of the wire as function of the velocity c, medium temperature T_M and wire temperature T_w. Solutions for F(c) and $T_M = T$, with $| \alpha(T_w - T_M) | \ll 1$ are obtained as:

$$F = \frac{P_{w1} - P_{w2}}{T_{w1} - T_{w2}} - \alpha(P_{w1} - P_{w2})$$

and

$$T = T_{w1} - \frac{P_{w1}/F}{1 + P_{w1}/F}$$

The inverse function of F(c) is approximated by

$$c = a_1 G + a_2 G^2 + a_3 G^3,$$

where $G = (F - F_o)^2$

and $a_2 = a_3 = o$ For King's law.

The calibrations F_o, a_1, a_2, a_3, α, T_{w1}, T_{w2}, R_{w1}

are determined by a calibration program which requires a minimum of 9 measurements of E_{w1}, E_{w2} at different velocities c and temperatures T_M. The relative errors of the approximate inverse functions of F(c) are:

$$\frac{\Delta F}{F} < 0.01 \text{ and } \frac{\Delta(T_M - T_w)}{T_M - T_w} < 0.0002$$

if $| \alpha(T_{w_1} - T_M) | < 0.04$. To avoid oscillating mechanical stresses in the hot-wire as a consequence of their alternating temperature they have to be attached to the prongs loosely.

This very promising method has so far been tested only with ordinary 5 μm sensors yielding to frequency resolution of $f_{max} = 500$ Hz. Further investigations, however, are under way.

3.3 On the need for compensation
It has often been argued (e.g. Arya and Plate (20)) that for certain measurements in non-isothermal flow temperature compensation of hot-wire signals may be neglected, particularly where — with double wire probes — hot-wire signals are subtracted. There a word of caution seems appropriate: Assumptions of this kind are based on a relation like:

$$E'_{\ell in} = K_1 \cdot c' + K_2 \cdot T'$$

assuming that $K_1 \neq f(T)$

and $K_2 \neq f(c)$,

which at best may hold for very small temperature differences. For the very real case of a turbulence field with $u_o = c_{max} \approx 10$ m/s and $\Delta T_{max} \approx 15°C$ errors of up to 25 % were observed for the "subtraction" term

$$\frac{\overline{v'^2}}{u_o^2}$$

when measured without compensation. Similar errors were found for most velocity- and velocity-temperature-combinations.

The above relation follows from

$$dE_{\ell in} = \frac{\partial E_{\ell in}}{\partial T} \cdot dT + \frac{\partial E_{\ell in}}{\partial c} \, dc$$

which may be written as ($E = E_{\ell in}$)

$$E' \approx \frac{\partial E}{\partial T} \, T' + \frac{\partial E}{\partial c} \, c' \qquad \text{for small fluctuations.}$$

Putting $E = E_{\ell in} \approx [\sqrt{(A + B\sqrt{c})}\,(\Delta T_{w_o} - \Delta T) - \sqrt{A\,\Delta T_{w_o}}\,]^4$

we obtain

$$\frac{\partial E}{\partial c} \approx \frac{[\sqrt{(A+B\sqrt{c})}\,(\Delta T_{w_o} - \Delta T) - \sqrt{A\Delta T_{w_o}}\,]^3 \, B(\Delta T_{w_o} - \Delta T)^{\frac{1}{2}}}{[(A+B\sqrt{c})\,c]^{\frac{1}{2}}}$$

and

$$\frac{\partial E}{\partial T} \approx \frac{-2\,[\sqrt{(A+B\sqrt{c})}\,(\Delta T_{w_o} - \Delta T) - \sqrt{A\,\Delta T_{w_o}}\,]^3 \,(A+B\sqrt{c})^{\frac{1}{2}}}{(\Delta T_{w_o} - \Delta T)^{\frac{1}{2}}}$$

where

$$\Delta T = \Delta \overline{T} + T'$$

$$c = \overline{c} + c'$$

$$\Delta T = T_w - T_o = \text{const.}$$

Indeed for small fluctuations (linearized) and small temperature differences we obtain

$$E' \approx f_1\,(\Delta T_{w_o}, \overline{c})\, c' - f_2\,(\Delta T_{w_o}, \overline{c})\, T'$$

corresponding to the above assumption.

For "realistic" fluctuations of, say $T'_{max} = 0(0.1\ T_{w_o})$ and $c'_{max} = 0(c_o)$ these simplifications are no longer justified. There not only the non-linear terms might (and will) assume values which no longer can be neglected. An additional error will also evolve by the fact that for the wires I and II in an X-wire

$$\frac{\partial E}{\partial T}\bigg|_{I} \neq \frac{\partial E}{\partial T}\bigg|_{II}$$

and thus the temperature term will not cancel upon subtraction.

P. Mensing, HFI/Technische Universität Berlin; E. Fröbel and U. Michel, DFVLR, Inst. für Turbulenzforschung Berlin, have contributed to this paper, which is based on work sponsored by Deutsche Forschungsgemeinschaft.

References

1. Lederman, S.: The use of Laser Raman diagnostics in flow fields and combustion. *Prog. Energy Combust. Sci.,* Vol. **3**, 1977. Pergamon Press
2. Corrsin, S.: Extended applications of the hot-wire anemometer. Techn. Note No. 1864 April 1949, Nat. Adv. Committee for Aeronautics
3. Corrsin, S.: Turbulence: Experimental methods. In *Handbuch der Physik,* Bd. **VIII/2**. Herausgegeben v. S. Flügge, C. Truesdell. Springer Verlag 1963
4. Kovasznay, L.S.G.: Hot wire method. In *Phys. Measurements in Gas Dynamics and Combustion,* Vol. **IX** High Speed Aerodynamics and Jet Propulsion. Princeton Univ. Press 1954
5. Kovasznay, L.S.G.: The hot-wire anemometer in supersonic flow. *Journ. of the Aeronautical Sciences,* Sept. 1950
6. Froebel, E.: not published
7. Schmidt, D.W.; Schmidt, W.; Wagner, W.J.: Eine Metallfilmsonde zur Messung von kleinen und schnellen Strömungstemperaturschwankungen. *Wärme- u. Stoffübertragung 6* (1973). Springer Verlag
8. Sreenivasan, K.R.; Antonia, R.A.; Danh, H.Q.: Temperature dissipation fluctuations in a turbulent boundary layer. *The Physics of Fluids,* **20**, 8, Aug. 1977
9. Bremhorst, K.; Krebs, L.: Reconsideration of constant current hot wire anemometers for the measurements of fluid temperature fluctuations. *Journ. of Physics E: Scientific Instruments 1976,* Vol. **9**
10. Freymuth, P.: Compensation for the thermal lag of a thin wire resistance thermometer by means of a constant temperature hot-wire anemometer. *Journ. of Physics E: Scientific Instruments 1969,* Vol. **2**
11. Mimaud-Lacoste, F.A.: An experimental study of the measurement of fluctuating temperature. M.S. thesis, Sept. 1972, Dept. of Aerospace Engineering, The Penn. State Univ.
12. Maye, J.P.: Fehler bei Temperaturmessungen mit einem als Widerstandsthermometer benutzen Drahtanemometer durch die thermische Ableitung zwischen dem Fühlerdraht und seinen Haltern. *DISA Information Nr. 9,* Feb. 1970
13. Sakao, F.: Constant temperature hot wires for determining velocity fluctuations in an air flow accompanied by temperature fluctuations. *Journ. of Physics E: Scientific Instruments 1973,* Vol. **6**
14. Chevray, R.; Tutu, N.K.: Simultaneous measurements of temperature and velocity in heated flows. *Review of Scientific Instruments,* Vol. **43**, No. 10, Oct. 1972
15. Fabris, G.: Probe and method for simultaneous measurements of "true instantaneous temperature and three velocity components in turbulent flow. *Review of Scientific Instrum.,* Vol. **48**, No. 11, Nov. 1977
16. Artt, D.W.; Brown, A.: The simultaneous measurement of velocity and temperature. *Journ. of Physics E: Scientific Instrum.* Vol. **4**, (1971)
17. Drubka, R.E.; Tan-atichat, J.; Nagib, H.M.: Analysis of temperature compensating circuits for hot-wires and hot-films. *DISA Information No. 22,* Dec. 1977

18. Fiedler, H.; Korschelt, D.; Mensing, P.: On transport mechanism and structure of scalar field in a heated plane shear layer. *Lecture Notes in Physics,* Vol. **76** Structure and Mechanisms of turbulence, Proceedings 1977

19. Fiedler, H.; Mensing, P.: Eine Methode zur Kompensation des Temperatureinflusses bei Hitzdrahtmessungen. *Z. Flugwiss. Weltraumforsch.* 1 (1977), Heft 2

20. Aray, S.P.S.; Plate, E.J.: Hot-wire measurements in non-isothermal flow. *Instr. and Control Systems,* March 1969

21. Höjstrup, J.; Rasmussen, K.; Larsen, S.E.: Dynamic calibration of temperature wires in still air. *DISA Information No. 20,* Sept. 1976

22. Höjstrup, J.; Rasmussen, K.; Larsen, S.E.: Dynamic calibration of temperature wires in moving air. *DISA Information No. 21,* Apr. 1977

The Temperature Sensitivity of Hot Wires

by

F.H. Champagne
University of Arizona
Tucson, Arizona 85721

Abstract

Significant errors in measurements of some velocity and joint velocity-temperature statistics arise from the non-negligible contamination of the measured hot-wire velocity signal by concomitant temperature fluctuations. Measurements of the temperature sensitivity of hot wires operated at high overheat ratios are considerably more difficult to obtain than measurements of the velocity sensitivity. Therefore to provide a check on the measured values, an analytical expression for the temperature sensitivity is derived which can be evaluated from parameters readily obtained from the velocity calibration data. The analysis, based on the convective heat transfer relation of Collis and Williams (1959), gives values of the temperature sensitivity which compare well with directly measured values. The present analysis is extended to show that the temperature and velocity sensitivities can be expressed entirely in terms of the linearized voltage output obtained by direct calibration for each hot-wire or hot-film sensor. Estimates of the errors in some measured statistics caused by the concomitant temperature fluctuation contamination of the hot-wire signal are determined.

1. Introduction

An important result from our many atmospheric boundary layer experiments and laboratory studies of a heated jet using hot-wire anemometers and platinum resistance thermometers is the non-negligible contamination of some measured velocity and joint velocity-temperature statistics by the concomitant temperature fluctuations. This contamination is present for hot-wire or hot-film sensors even when operated at high overheat ratios in air. Correction of the measured hot-wire velocity signal requires knowledge of the temperature fluctuations and the temperature sensitivity of the hot-wire, β. The former can be obtained from a resistance thermometer or cold wire located in close proximity to the hot wire and the latter can be obtained from careful calibration of the hot wire over the range of ambient temperatures expected. Measurements of the temperature sensitivity of hot wires operated at high overheat ratios are considerably more difficult to obtain than measurements of the velocity sensitivity. Thus to provide a check on the measured values, an analytical expression for β is derived which involves parameters readily obtained from the velocity calibration data. The desired analytical expression is determined from an analysis of the instantaneous response of a constant-temperature hot-wire anemometer based upon the convective heat transfer relation of Collis and Williams (1959). The analysis also yields an analytical expression for the velocity sensitivity and gives the effects of variations in mean ambient temperature on the estimate of mean velocity. Estimates of the errors in some measured statistics cased by the concomitant temperature fluctuation contamination of the hot-wire signal are determined. The results are shown to agree well with those of Rose (1962), who performed a somewhat similar derivation. Rose's results, although useful, are expressed in terms of a parameter K_1 which varies with the type of wire material, the aspect ratio of the wire and the overheat ratio. The present analysis is extended to show that the velocity and temperature sensitivities can be expressed entirely in terms of the linearized voltage output *obtained by direct calibration* for each hot-wire or hot-film sensor, and thus is more general and usable.

Other studies concerned with the use of hot wires in non-isothermal flows including those by Aray and Plate (1969), Freymuth (1970), Chevray and Tutu (1971), Bearman (1971), Hollasch and Gebhart (1972), Koch and Gartshore (1972), and Drubka, Nagib, and Tan-atichat (1977) emphasize different aspects of the problem from those presented here, so will not be discussed. The present study is confined to flows of air at low subsonic speeds.

2. Background Considerations
For a given fluid, the instantaneous linearizer output voltage E_L is a function of the wire or film average temperature T_w, the instantaneous ambient fluid temperature T_a, the calibration temperature of the fluid T_c, and, to first approximation, the instantaneous normal component of the fluid velocity Q_N. It will be assumed that the sensor is aligned with its axis perpendicular to the mean flow direction and that the turbulence intensities are sufficiently low that second and higher order fluctuations in the binomial series expansion of Q_N can be neglected so that $Q_N = U$, the instantaneous streamwise velocity component. Absolute temperatures will be used for all temperatures. The linearizer output voltage can therefore be expressed in the form

$$E_L = g(U, T_w, T_c, T_a) = f(U, T_c - T_a) \qquad [1]$$

as T_w and T_c are constant. The resulting hot-wire response equation for velocity, temperature and linearized voltage fluctuations can be written

$$e_L = \frac{\partial E_L}{\partial U}\bigg|_{T_c-T_a^u} - \frac{\partial E_L}{\partial(T_c-T_a)}\bigg|_U \theta = \alpha u - \beta\theta. \tag{2}$$

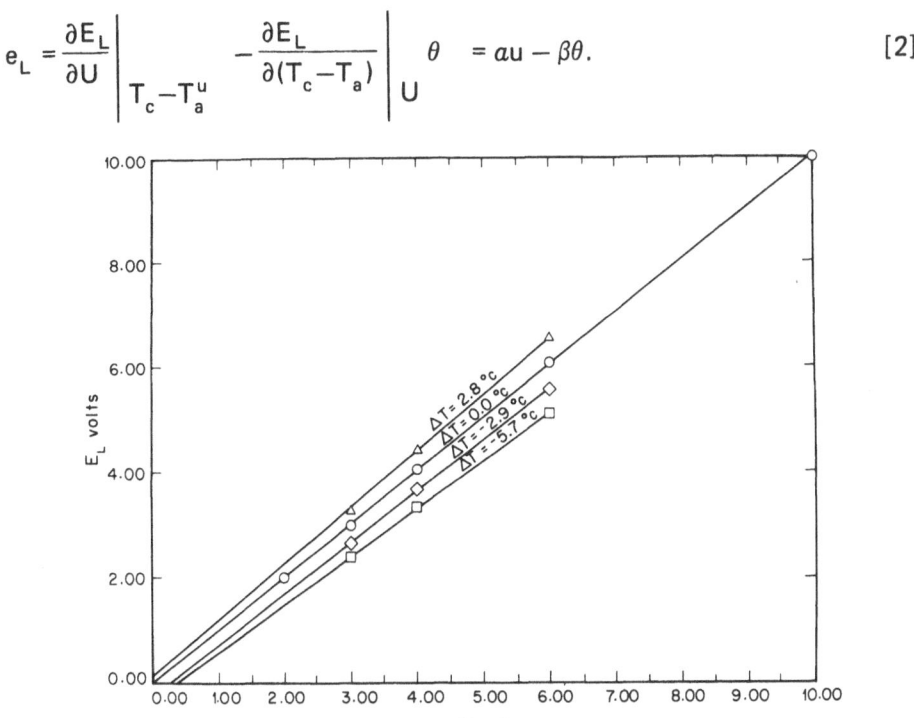

Fig. 1. *The velocity sensitivity of a hot-wire anemometer at various values of* $T_c - T_a$ *for an overheat ratio of 0.80.*

Typical calibration results for a wire aligned normal to the calibration airstream are shown in Figs. 1 and 2. Direct evaluation of β from the linearized output is not considered in some of the earlier studies possibly because of the difficulties involved in controlled variation of the calibration airstream temperature.

To determine the hot-wire or hot-film probe response characteristics to velocity and temperature analytically, let us consider the dominant heat loss mechanism, the convective heat loss. The rate of convective heat loss from a heated wire or cylindrical film sensor with its axis aligned perpendicular to the flow can be expressed in the form (Collis and Williams 1959 and Corrsin 1963)

$$Nu = \left[A_1 + B_1\left(\frac{Ud_w}{\nu_m}\right)^n\right]\left[1 + 1/2\left(\frac{T_w-T_a}{T_a}\right)\right]^{n_1} \tag{3}$$

where $Nu = hd_w/k_m$ is the Nusselt Number, h is the convective heat transfer

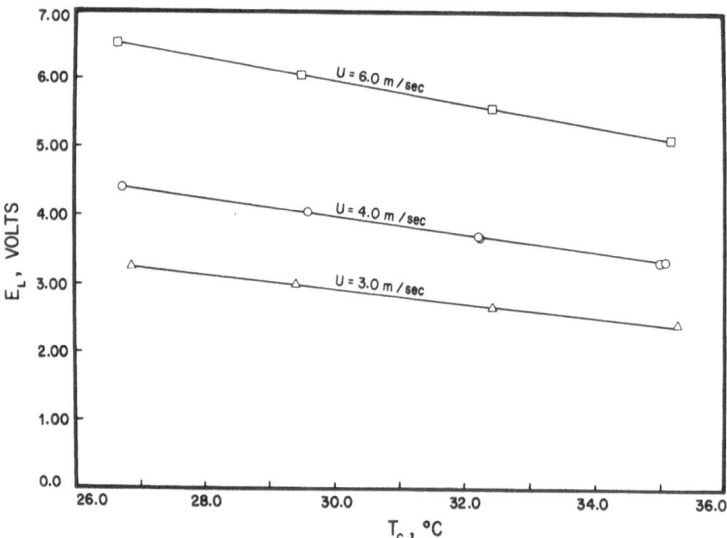

Fig. 2. The temperature sensitivity of a hot-wire anemometer operated at an overheat ratio of 0.80.

coefficient, d_w is the sensor diameter, and ν_m and k_m are the viscosity and thermal conductivity of the fluid evaluated at the mean temperature $(T_w + T_a)/2$. A_1, B_1, n_1, and n are empirical constants. Collis and Williams found n to be 0.45 rather than 0.50 as reported in the early work by King (1914). Collis (1956) points out that three-dimensional effects caused by the influence of wire support geometry become significant for aspect ratios less than 10^3. The main effects reported are on the values of A_1 and B_1, the value of n remaining constant. Koch and Gartshore (1972) report that their heat transfer data for wires with aspect ratios of about 200 agreed well with the Collis and Williams relation n = 0.45, although with different values of the empirical constants A_1, B_1, and n_1 which is attributed to aspect ratio effects. The Collis and Williams relation will be used in the subsequent analysis and to assess the importance of temperature loading effects on the value of the heat transfer coefficient. Values of the constants n and n_1 of 0.45 and 0.17 will be assigned wherever their evaluation is necessary.

The instantaneous ambient fluid temperature, T_a, can be written as

$$T_a = T_c - \Delta T + \theta \qquad [4]$$

where $\Delta T = T_c - \overline{T}_a$, the deviation of the mean ambient temperature from the calibration temperature, T_c and θ is the temperature fluctuation. A reasonable linear fit to the kinematic viscosity and thermal conductivity of air for 0°C to 200°C are given by (Batchelor 1967 and Collis and Williams 1959)

$$\nu = 0.150 + 1.00 \times 10^{-3}(T-293) \text{ cm}^2/\text{sec} \tag{5}$$

$$k = 5.76 \times 10^{-5}[1 + 3.17 \times 10^{-3}(T-273)] \text{ cal/cm-sec-°K}.$$

These variables evaluated at the mean fluid temperature can be expressed as

$$\nu_m = \nu_c - 0.50 \times 10^{-3}(T_c - T_a) \tag{6}$$

$$k_m = k_c - 9.130 \times 10^{-8}(T_c - T_a)$$

where the subscript c refers to the respective values at calibration conditions. From equations [3] and [6], and using a binomial series expansion for the temperature loading term in [3], the heat transfer coefficient can be shown to be

$$h = A_c \left[1 + (T_c - T_a)\left(\frac{n_1 T_w}{2 T_c^2 \left(1 + \frac{n_1 a_w}{2 a_c T_c}\right)} - \frac{9.13 \times 10^{-8}}{k_c} \right) \right] \tag{7}$$

$$+ B_c U^n \left[1 + (T_c - T_a)\left(\frac{n_1 T_w}{2 T_c^2 \left(1 + \frac{n_1 a_w}{2 a_c T_c}\right)} + \frac{0.5 \times 10^{-3} n}{\nu_c} - \frac{9.13 \times 10^{-8}}{k_c} \right) \right]$$

where

$$A_c = \frac{k_c A_1}{d_w}\left(1 + \frac{n_1 a_w}{2 T_c}\right), \quad B_c = \frac{k_c B_1}{\nu_c^n d_w^{1-n}}\left(1 + \frac{n_1 a_w}{2 T_c}\right).$$

The overheat ratio a_w is defined by $a_c(T_w - T_c)$ where a_c is the temperature coefficient of resistance of the wire. To estimate the order of magnitude of the various terms, let $a_w \doteq 0.8$, $T_w = 443°K$, and $T_c = 293°K$ so then

$$h = A_c[1 - 7.99 \times 10^{-4}(T_c - T_a)] + B_c U^{0.45}[1 + 1.20 \times 10^{-4}(T_c - T_a)]. \tag{8}$$

For measurement situations to be considered here, that is typical laboratory or atmospheric surface layer experiments in air, a value of $T_c - T_a$ of 10°K will be considered large. The resulting variation in either term is less than about 1 % so the temperature dependence of the heat transfer coefficient can be neglected and equation [7] can be approximated by

$$h = A_c + B_c U^n \tag{9}$$

The equation of thermal equilibrium for a heated wire maintained at constant temperature by a system such as that shown in Fig. 3 is

$$NI_w^2 R_w = \pi d_w \ell_w h(T_w - T_a) + Q_E \qquad [10]$$

where I_w = the wire current, R_w = the wire resistance, N is the conversion constant from watts to calories/sec, Q_E is the end conduction loss, and the radiation loss has been neglected. Q_E for a typical ℓ_w/d_w = 200 wire operated at an overheat ratio of 0.8 is about 8 percent of the convective loss (Champagne, Sleicher and Wehrmann 1967). When T_a varies, the convective loss term changes linearly with $T_w - T_a$, providing $\Delta T < 10°K$.

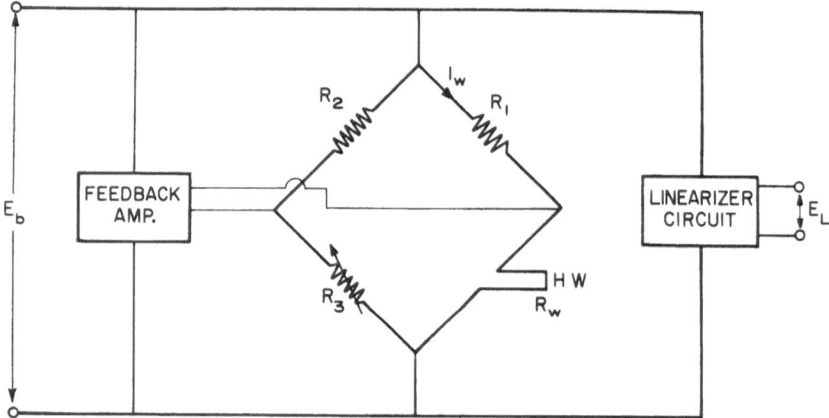

Fig. 3. Schematic diagram of a linearized constant temperature hot wire anemometer system.

The end conduction loss also varies linearly with $T_w - T_a$ if one assumes that the temperature gradient at the wire ends scales accordingly. Thus, the variation in the value of the end conduction loss with $T_c - T_a$ or ΔT is an order of magnitude smaller than the corresponding changes in the convective loss. Q_E is normally tacitly incorporated into the convective loss term.

3. Instantaneous Response Equation
The constant temperature anemometer bridge voltage output can be expressed in the form

$$E_b^2 = \frac{(R_1 + R_w)^2}{N R_w} [\pi d_w \ell_w (A_c + B_c U^n)(T_w - T_a)] = (A_a + B_a U^n)(T_w - T_a) \qquad [11]$$

where $A_a = (R_1 + R_w)^2 \pi d_w \ell_w A_c/N R_w$ and $B_a = (R_1 + R_w)^2 \pi d_w \ell_w B_c/N R_w$.

The instantaneous linearizer voltage output can thus be expressed as

$$E_L = K_o U \left[1 + \frac{T_c - T_a}{T_w - T_c} \left(1 + \frac{A_a}{B_a U^n} \right) \right]^{1/n} \qquad [12]$$

where $K_o = K B_a^{1/n} (T_w - T_c)^{1/n}$ is a calibration constant and K is the constant linearizer gain. The parameters A_a and B_a can be directly determined for each probe from velocity calibration data holding T_c constant. If it is assumed that the measured linearizer signals are caused by velocity alone, i.e. ignoring the fact that T_c and T_a may differ, then

$$E_L = K_o U_m \qquad [13]$$

where the subscript m refers to assumed measured conditions (i.e. $E_L (U_m, T_c) = E_L (U, T_a)$). Thus, from [12] and [13]

$$U_m = U \left[1 + \frac{\alpha_c (T_c - T_a)}{a_w} \left(1 + \frac{A_a}{B_a U^n} \right) \right]^{1/n} \qquad [14]$$

Substituting [4] and

$$U_m = \bar{U}_m + u_m, \quad U = \bar{U} + u$$

into [14], using a binominal series expansion to first order in the argument

$$x = \frac{\alpha_c \Delta T}{a_w} \left(1 + \frac{A_a}{B_a U^n} \right)$$

for the bracketed term, and using a binominal expansion to first order in $u/\bar{U}$ for the $(\bar{U} + u)^{-n}$ term, one obtains

$$\frac{\bar{U}_m}{\bar{U}} = 1 + b_1 \Delta T \qquad [15]$$

and

$$u_m = u \left[1 + \left(\frac{\alpha_c}{a_w} - (1-n) b_1 \right) \Delta T \right] - b_1 \bar{U} \theta \qquad [16]$$

where $b_1 = \frac{\alpha_c}{n a_w} \left(1 + \frac{A_a}{B_a \bar{U}^n} \right)$.

Equation [16] can be written in the form of [2]

$$e_L = \alpha u - \beta \theta \qquad [17]$$

where $\alpha = K_o + \left[\frac{\alpha_c}{a_w} - (1-n) b_1 \right] K_o \Delta T$ and $\beta = b_1 \bar{U} K_o$.

For the experimental conditions considered here, the argument x is on the order of 0.15 or less so the expansion in equation [14] is valid. If n = 0.50, then b_1 is identical to Rose's parameter K_1. It is readily shown that equations [15] and [16] agree with Rose's results to first order in fluctuating quantities with the exception of the bracketed term in equation [16]. Rose gives a value of unity for this term as he evaluates all his sensitivity coefficients at calibration conditions rather than at prevailing ambient conditions. This term accounts for the slope with ΔT indicated in Fig. 1.

The parameters A_a and B_a can be readily obtained from velocity calibration data where the fluid temperature is maintained constant. Knowledge of A_a/B_a, n and the mean flow conditions allows evaluation of both the velocity and temperature sensitivity coefficients, a and β, respectively. The parameters a_c and a_w are known from the physical properties of the wire and the sensor operating conditions. It should be noted that variations in the value of a_c from the value cited in the International Critical Tables can occur and are probably caused by effects of impurities, the wire or film manufacturing process, etc. Variations in the wire diameter from the nominal manufacturers value may also occur. Values of A_a/B_a and n have been determined by the present author for many 5μ diameter wires with ℓ_w/d_w of 200 and an overheat ratio of 0.80. The results are $1.75 \pm 0.11 (m/sec)^{0.45}$ and 0.45, respectively, which agree well with the values obtained by Koch and Gartshore (1972) of 1.79 and 0.45, respectively, for similar wires and overheat ratios. A_a and B_a vary with ℓ_w/d_w and A_a/B_a varies with the wire diameter and fluid viscosity as $(\nu/d_w)^{0.45}$. In view of possible variations and uncertainties in d_w, a_c, and ℓ_w/d_w, it is desirable to calibrate each wire *directly* to determine the velocity and temperature sensitivities from the linearized output. As the linearized output voltage can be expressed in the form presented by equation [1], the sensitivities a and β can be determined from equations [2] and [12] to be

$$a = \frac{\partial E_L}{\partial U}\bigg|_{T_c - T_a} = K_o \left(1 + \frac{T_c - T_a}{T_w - T_c}\right)\left[1 + \frac{T_c - \overline{T}_a}{T_w - T_c}\left(1 + \frac{A_a}{B_a \overline{U}^n}\right)\right]^{1/n-1} \qquad [18]$$

and

$$\beta = -\frac{\partial E_L}{\partial T_a}\bigg|_{\overline{U}} = \frac{K_o \overline{U}}{n(T_w - T_c)}\left(1 + \frac{A_a}{B_a \overline{U}^n}\right)\left[1 + \frac{T_c - \overline{T}_a}{T_w - T_c}\left(1 + \frac{A_a}{B_a \overline{U}^n}\right)\right]^{1/n-1}. \qquad [19]$$

Binominal series expansion of the bracketed terms, which can be written in the form $[1 + y]^{1/n-1}$, and retaining terms to first order in the argument y, and then evaluating the results at the steady calibration conditions equivalent to the mean ambient conditions of the experiment, gives the following:

$$a = K_o \left[1 + \frac{a_c \, /^{\cdot}}{na_w} \cdot \left(1 + (1-n) \frac{A_a}{B_a \bar{U}^n} \right) \right] \qquad [20]$$

$$\beta = \frac{K_o \bar{U} a_c}{na_w} \left(1 + \frac{A_a}{B_a \bar{U}^n} \right) \qquad [21]$$

which agrees with the results presented in [17].

Therefore, the sensitivities a and β can be determined *directly* from the linearized output as originally indicated by equation [2]. Further, if A_a/B_a, a_c, a_w, n and the mean flow conditions are accurately known, equations [20] and [21] allow evaluation of a and β. As measurements of β are much more difficult to obtain than those of a, equation [21] provides a means of checking the measured values of β or to estimate its value if circumstances prevent its direct measurement. Also equations [20] and [21] allow determination of the effect of variations in ΔT on a and the effect of variations in $\bar{U}$ on β.

Note that here the partial derivatives defining the sensitivity coefficients are evaluated at the true mean velocity and at the actual mean ambient temperature, $\bar{T}_a$. It is assumed that the mean conditions such as $\bar{T}_a$ are known, so that $\bar{U}$ may be correctly obtained from $\bar{E}_L$, and finally a and β may be evaluated at the true mean conditions of velocity and temperature. One minor problem with the present technique is that initially only an implicit relationship exists between $\bar{U}_m$ and $\bar{U}$ as equation [15] can be written

$$\bar{U}_m = \bar{U} - \frac{1}{K_o} \left. \frac{\partial E_L}{\partial T_a} \right|_{\bar{U}} \Delta T. \qquad [22]$$

However, this can be solved iteratively and the correction to $\bar{U}_m$ converges rapidly.

5. Comparison of Measured and Computed Sensitivities

Details of the experimental facility used to obtain the calibration data are presented in Champagne (1978). Evaluation of the velocity and temperature sensitivities from the data shown in Figs. 1 and 2 gives the results shown in Table 1. The computed values were determined by evaluating A_a, B_a, a_c/a_w, and n from the bridge voltage calibration data, obtained simultaneously with the linearized voltage data, and substituting into equations [20] and [21]. The simultaneously obtained bridge voltage data is shown in Figs. 4 and 5. Fig. 4 presents the variation of the squared bridge voltage with velocity to the 0.45 power for the various values of ΔT. Straight lines are drawn through the data and appear to fit the data quite well. Similarly, a linear fit of the bridge voltage squared versus calibration temperature data for fixed velocity seems reasonable

TABLE 1

U, m/sec	β, volts/°C		ΔT, °C	a, volts/m/sec	
	Measured	Computed		Measured	Computed
3.0	0.10	0.11	2.8	1.07	1.07
4.0	0.13	0.14	0	1.00	1.00
6.0	0.17	0.19	−2.9	0.95	0.93
			−5.7	0.89	0.86

as shown in Fig. 5. Thus equation [11] would appear to be a good approximation over the range of temperature and velocity tested with n = 0.45. A value
of $\overline{U}$ = 4 m/sec was used in evaluating a at the various values of ΔT, but the
value of a is not very sensitive to variations in $\overline{U}$, especially for the limited
range used here.

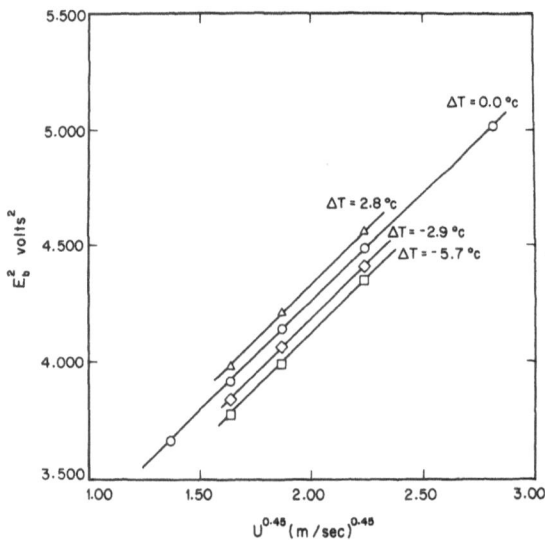

Fig. 4. *The variation of squared bridge voltage with velocity for various values
of ΔT. Overheat ratio is 0.80.*

The measured response equation for a 2.3 μ diameter tungsten wire operated at
a_w = 0.80 with K_o = 1.00 volt/m/sec is

$$e_L = 1.00\,u - 0.17\,\theta \qquad\qquad [23]$$

for U = 6 m/sec and $\Delta T = 0°K$. For these conditions $b_1 = 2.8 \times 10^{-2}°K^{-1}$ and Rose's data gives $K_1 = 2.5 \times 10^{-2}°K^{-1}$ for a 3.8 μ tungsten wire, which is good agreement.

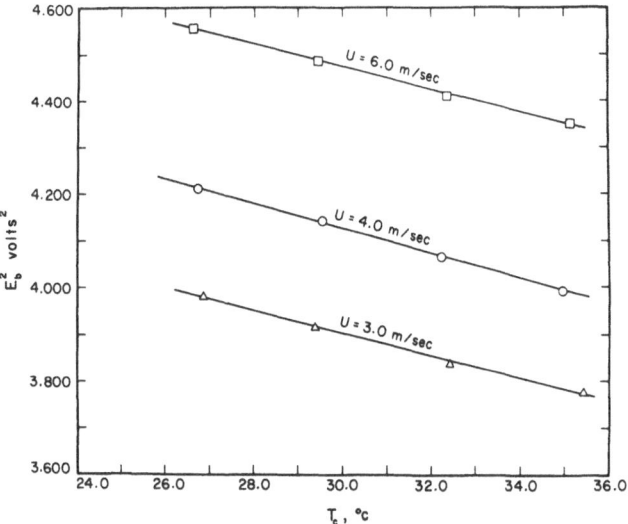

Fig. 5. The variation of squared bridge voltage with calibration stream temperature for an overheat ratio of 0.80.

6. Error Estimates for some Measured Statistics

The importance of errors caused by the concomitant temperature fluctuation contamination of the hot-wire signal depends on the statistic measured and the experimental conditions. A joint velocity-temperature statistic that can be strongly contaminated by the temperature sensitivity of the hot wire is

$$\frac{\partial u}{\partial x} \frac{\partial \theta}{\partial x} ,$$

which is important in studying the fine scale structure of turbulent velocity and temperature fields. This can be demonstrated as follows. To estimate

$$\frac{\partial u}{\partial x} \frac{\partial \theta}{\partial x}$$

one would measure $\overline{u \dot{\theta}}$, where the dot refers to time differentiation, and use Taylor's hypothesis. Measurement of $\overline{u \dot{\theta}}$ requires simultaneously using a hot wire and cold wire which are separated by less than a Kolmogorov length scale. In such a measurement the velocity sensitivity of the cold wire must be acacounted for. Wyngaard (1971) demonstrated the importance of the velocity sensitivity of cold wires on measurements of the skewness of the temperature derivative signal. The response equations of the wires can be written

$$u_m = \frac{a}{K_o} u - \frac{\beta}{K_o} \theta \qquad [24]$$

$$\theta_m = \theta - cu \qquad [25]$$

where c is the velocity sensitivity of the cold wire. It can be easily shown from [24] and]25] that

$$\frac{\overline{\dot{u}_m \dot{\theta}_m}}{[\overline{\dot{u}^2 \dot{\theta}^2}]^{1/2}} = \left(\frac{a}{K_o} + \frac{\beta c}{K_o}\right) \frac{\overline{\dot{u}\,\dot{\theta}}}{[\overline{\dot{u}^2\,\dot{\theta}^2}]^{1/2}} - \frac{\beta}{K_o} \left[\frac{\overline{\dot{\theta}^2}}{\overline{\dot{u}^2}}\right]^{1/2} - \frac{ac}{K_o} \left[\frac{\overline{\dot{u}^2}}{\overline{\dot{\theta}^2}}\right]^{1/2} \qquad [26]$$

Typically $\overline{\dot{u}^2}_m$ and $\overline{\dot{\theta}^2}_m$ are negligibly contaminated. For $\Delta T = 0$, then assigning typical values from atmospheric surface layer results (see Champagne, Friehe, LaRue, and Wyngaard 1977), $\overline{\dot{u}^2} = 3 \times 10^3$ m²/sec⁴, $\overline{\dot{\theta}^2} = 1.6 \times 10^3$ °K/sec², $a = 1.00$ v/m/sec, $\beta = 0.17$v/°K, $K_o = 1.00$ v/m/secm $c = 8 \times 10^{-2}$ U^{-1}°K/m/sec, $\overline{U} = 6$ m/sec, equation [26] gives

$$\frac{\overline{\dot{u}_m \dot{\theta}_m}}{[\overline{\dot{u}^2 \dot{\theta}^2}]^{1/2}} = 1.00 \frac{\overline{\dot{u}\,\dot{\theta}}}{[\overline{\dot{u}^2 \dot{\theta}^2}]^{1/2}} - 0.124 - 0.018 \qquad [27]$$

If the fine scale velocity and temperature fields are isotropic, then the $\overline{\dot{u}\dot{\theta}}$ correlation should be zero, thus showing the possible importance of the temperature sensitivity of the hot-wire in such a measurement.

Extending the analysis to x-wire probes, one would obtain equations of the form

$$e_1 = au + kw - \beta\theta \qquad [28]$$

$$e_2 = au - kw - \beta\theta \qquad [29]$$

where the subscripts 1 and 2 refer to the two wires which are assumed identical and inclined at the same angle to the mean flow direction. Standard manipulation of these equations yields

$$\overline{(e_1 + e_2)^2} = 4a^2 \overline{u^2} - 8a\beta\overline{u\theta} + 4\beta^2 \overline{\theta^2} = 4a^2 \overline{u_m^2} \qquad [30]$$

$$\overline{(e_1 - e_2)^2} = 4k^2 \overline{w^2} = 4k^2 \overline{w_m^2} \qquad [31]$$

$$\overline{(e_1 + e_2)(e_1 - e_2)} = 4ak \overline{uw} - 4k\beta \overline{w\theta} = 4ak \overline{u_m w_m} \qquad [32]$$

As $\overline{u\theta}$ is negative for the unstable atmospheric boundary layer, the ratio of $\overline{w^2}_m/\overline{u^2}_m$ would be smaller than that of $\overline{w^2}/\overline{u^2}$ because of the θ-contamination. It is readily shown from equations [30] and [31] that

$$\frac{\overline{w^2_m}}{\overline{u^2_m}} = \frac{\overline{w^2}/\overline{u^2}}{1 - 2\dfrac{\beta}{a} R_{u\theta}\left[\dfrac{\overline{\theta^2}}{\overline{u^2}}\right]^{\frac{1}{2}} + \dfrac{\beta^2\overline{\theta^2}}{a^2\overline{u^2}}} \qquad [33]$$

where the correlation coefficient $R_{u\theta}$, defined as

$$R_{u\theta} = \frac{\overline{u\theta}}{\left[\overline{u^2}\ \overline{\theta^2}\right]^{\frac{1}{2}}}$$

is introduced. To estimate the magnitude of the correction for typical atmospheric surface layer results   as before, let $\overline{u^2} = 0.18\overline{U^2}$, $\overline{\theta^2} = 0.10°K^2$, $R_{u\theta} = -0.6$, $a = K_o \cos 45° = 0.707$ v/m/sec, $\beta = 0.17$ v/°K, and $\overline{U} = 6$ m/sec. Substituting these values into [33] gives

$$\frac{\overline{w^2_m}}{\overline{u^2_m}} = 0.92 \frac{\overline{w^2}}{\overline{u^2}} \qquad [34]$$

The spectral analog of these equations indicate the effect of this contamination on spectral tests of local isotropy and the 8 % bias cannot be neglected when searching for 4/3 spectral ratio.

The θ-contamination of hot-wire signals may be significant in many experiments where temperature is used to tag the fluid as usually temperature fluctuations much larger than those naturally occurring in the atmospheric boundary layer are used. Finally, it should be emphasized that the θ-contamination increases as the overheat ratio of the wire or film decreases and as the mean velocity decreases. Thus a careful analysis of the probe response equations should be undertaken for each statistic to be measured, especially for low overheat ratios.

Acknowledgement
This work was supported by AFOSR Grant 77-3172.

References

Arya, S.P.S. and Plate, E.J., 1969, *Instr. and Control Sys., **42**, 87*

Batchelor, G.K., 1967, *An Introduction to Fluid Dynamics.* Cambridge University Press

Bearman, P.W., 1971, *DISA Information No. 11*, 25

Champagne, F.H., Sleicher, C.A., and Wehrmann, O.H., 1967, *J. Fluid Mech., **28**, 153*

Champagne, F.H., Friehe, C.A., LaRue, J.C., and Wyngaard, J.C., 1977, *J. Atmos. Sci., **34**, 515*

Champagne, F.H., 1978, *J. Fluid Mech., **86**, 67*

Chevray, R. and Tutu, N.K., 1971, *Proceedings Second Symposium on Turbulence Measurements in Liquids,* Rolla, Missouri

Collis, D.C., 1956, *J. Aero Sci., **23**, 697*

Collis, D.C. and Williams, J.J., *J. Fluid Mech., **6**, 357*

Corrsin, A., 1963, Turbulence: Experimental Methods. *Handbuch der Physik,* VIII/2, 524

Drubka, R.E., Nagib, H.M., and Tan-atichat, 1978, *J. ITT Fluids and Heat Transfer,* Report R77-1

Freymuth, P., 1970, *Instr. and Control Sys.* 43, 82

Hollasch, K. and Gebhart, B., 1972, *J. Heat Transfer ASME,* **94**, 17

Koch, F.A. and Gartshore, I.A., 1972, *J. Physics E: Scientific Instr.,* **5**, 58

Rose, W.G., 1962, *J. Applied Mech.,* **29**, 554

Hot-wire Anemometry for Turbulence Measurements in Helium-Air Mixtures★

by

Paul A. Libby and John C. LaRue
Department of Applied Mechanics and Engineering Sciences
University of California, San Diego
La Jolla, California 92093

Abstract

The use of extended hot-wire anemometry involving an interfering probe is shown to permit measurements of variable density turbulence such as arises in the mixing of helium and air. The methods of calibration and data reduction leading to time series in one or more velocity components, in the mass fraction of helium, and in the mixture density are described. Typical results in various flows to which the technique has been applied are discussed.

1. Introduction

For a number of years we have been making measurements in turbulent flows involving the mixing of helium and air by application of hot-wire anemometry and digital techniques. Our motivation for this effort resides in the need for high quality data on variable density turbulence. Many turbulent flows of practical interest involve variations in density due to temperature and/or composition inhomogeneities; in the theoretical treatment of these flows the effect of variations of density is usually assumed to be confined to the mean density and the models effecting closure of the describing equations are usually transcribed without alteration from those which are more or less successful in constant density turbulence. Until there are available detailed experimental data on suitable variable density flows permitting comparison of prediction and experiment, this casual treatment of density variations can go unchallenged and potentially important new effects associated with density inhomogeneities can be disregarded. We report briefly on our efforts to supply such data.

The notion of using extended hot-wire anemometry to measure composition as well as velocity is not new. To facilitate discussion we consider the measurement of the principal velocity component u and the mass fraction of a foreign species denoted c in air with the understanding that a second velocity component can be measured by suitable extension of the ideas we discuss. Corrsin

★ This contribution is an abbreviated but up-dated version of Libby (1977).

(1949) suggested that two wires with suitable dissimilar responses could be used to obtain composition and velocity. Straight-forward application of this idea seems to be possible for some gases but not for others, in particular not for helium. In fact, there have been several unsuccessful attempts at applying hot-wire anemometry to helium-air mixtures.

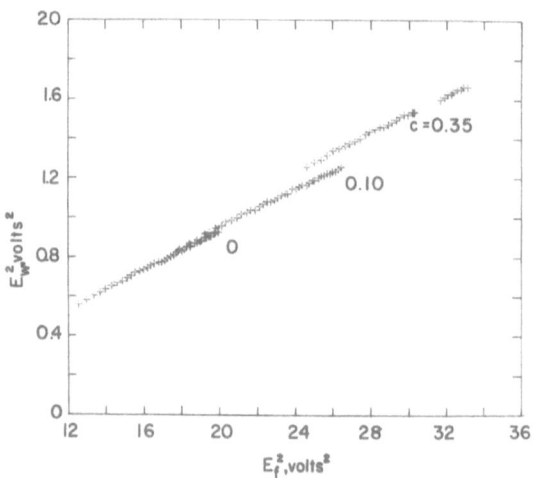

Fig. 1. The $E_w^2 - E_f^2$ map for a non-interfering wire and film (from Way and Libby (1970)).

Our own early experiments (cf. Way and Libby (1970)) indicate that there is a severe ambiguity problem which must be overcome; it is difficult to distinguish between a low velocity in a high concentration of helium from a high velocity in pure air. The problem is shown graphically in Fig. 1; the squared voltages from a wire and a film, i.e.; E_w^2 and E_f^2, are plotted so that various velocities in mixtures of fixed helium concentration form the lines shown. The ambiguity cited earlier is clearly seen to prevail.

2. An Interfering Probe and Its Utilization

The solution is to place the wire within the thermal field of the film with the result that the same presentation of squared voltages leads to Fig. 2. The basic idea of this interfering probe is that a portion of the energy required to maintain the wire at a constant average temperature is supplied by the thermal field around the film. If the velocity is so high and the helium concentration is so low that none of the energy needed by the wire is due to the film, we have the non-interfering, ambiguous case indicated in Fig. 1. If, on the contrary, all of the heating to the wire is due to the film, there will be no signal from the wire and again the probe fails; this latter situation tends to arise in low velocities with high helium concentrations. Thus, we note that the probe must be designed and operated so as to avoid the upper left and lower right corners of the $E_w^2 - E_f^2$ map. This can be done for a variety of flow situations of interest by adjusting the operating temperatures of the wire and film and the spacing between the two sensors.

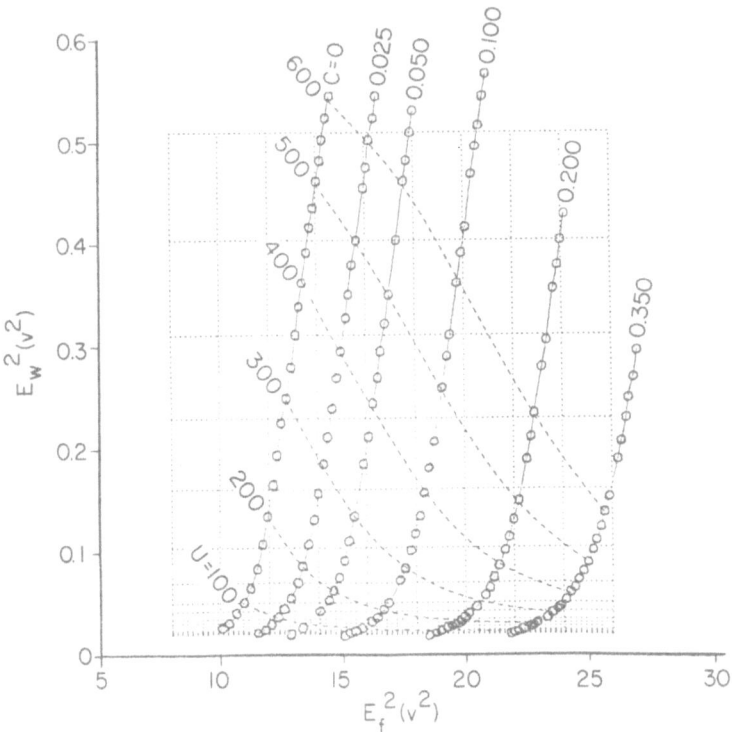

Fig. 2. The $E_w^2 - E_f^2$ map for a typical interfering probe (from Way and Libby (1971)).

Fig. 2 shows lines of both constant helium concentration and of constant velocity obtained from a typical interfering probe. As the velocity and helium concentration increase, the voltage required by the film increases as expected. However, for a fixed velocity an increase in helium concentration leads to a decrease in voltage required by the wire, presumably because the influence of the thermal field of the film increases as the thermal conductivity of the mixture increases and the Reynolds number decreases. On the contrary, for a fixed helium concentration increases in velocity result in increases in wire voltage as the influence of the film on the thermal balance of the wire decreases.

Fig. 2 displays calibration data and is generated by carrying out the probe calibration much as in standard hot-wire anemometry, but in this case in five to eight helium-air mixtures of known concentration. The first step in data reduction involves inversion of a voltage pair, E_w^2 and E_f^2, to the corresponding u-c pair. Over the years we have used a variety of techniques making changes in each case in order to reduce computing time and costs. One of our early techniques involves the use of the overlaid grid shown in Fig. 2. Values of u and c are interpolated from the calibration lines onto the mesh points; a further interpolation involving a polynomial approximation to the surfaces $u(E_w^2, E_f^2)$ and $c(E_w^2, E_f^2)$ is used.

At present we use a more direct, less cumbersome strategy as follows: The film behaves in the various helium-air mixtures as a conventional anemometer with calibration constants dependent on the mass fraction of helium. Thus we can write for constant temperature operation in isothermal mixtures

$$E_f^2 = A(c) + B(c)\,(\rho u)^{1/2} \qquad\qquad [2.1]$$

where $\rho = \rho(c)$ is the mass density of the mixture. The coefficients A(c) and B(c) are obtained initially at the discrete concentrations used in calibration, but are subsequently represented by polynomials in c. The result is an explicit equation for $u = u(E_f^2, c)$.

The second relation involves polynomial approximation to the surface $c = c(E_w^2, E_f^2)$ in the form

$$c = \sum_n \sum_m a_{nm}\,(E_w^2)^n\,(E_f^2)^m \qquad\qquad [2.2]$$

with a least square approximation used to select the a_{nm} coefficients. Thus the two voltages are used in [2.2] to compute the mass fraction of helium and the result used in the solution of [2.1] to determine the velocity. Even with an old, relatively slow computer, a CDC 3600, this technique requires one second of computing time to invert one second of data.

In discussing the exploitation of this probe to make measurements in turbulent flows involving helium-air mixing it is useful to distinguish between calibration data and turbulence data. The former corresponds to voltage pairs for known velocity-concentration pairs, i.e., to the data used to generate Fig. 2 and the constants A(c), B(c) and a_{nm} in [2.1] and [2.2]. Turbulence data correspond to voltage pairs which are to be inverted by use successively of [2.2] and the equation for u derived from [2.1] to obtain the related u-c pair.

Both calibration and turbulence data are recorded digitally. Our recorder limits us to a sampling rate of two to four thousand samples per second per channel. Accordingly, we adjust the flow velocities in our experiments appropriately; typical velocities are five-ten meters per second. Data are collected for sufficient time to accumulate a quarter of a million samples, a number which permits satisfactory statistics to be established even with heavy conditioning.

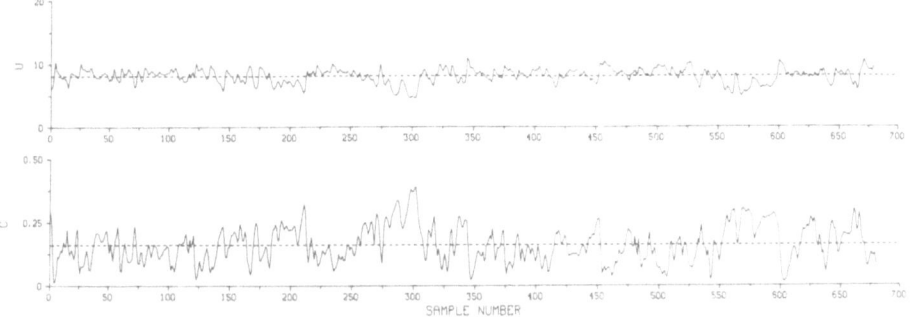

Fig. 3. A computer plot of a segment of the time series in u and c (from Libby (1977)).

With time series in voltages inverted into corresponding time series in velocity component and helium concentration, a third series in the density can be computed from the relation

$$\rho/\rho_o = (1 + wc)^{-1} \qquad [2.3]$$

where ρ_o is the mass density of air at the pressure and temperature conditions of the experiment and w is a molecular weight parameter equal to 6.22★. These three time series can be subjected to the usual methods of time series analysis and conditioned sampling. We shall review some of the results from this technique later. For the present it is illuminating to display as Fig. 3 a typical computer plot of a short section of typical time series in u and c. One appreciates from the previous discussion that considerable effort precedes the display of data in this form; it is not the same as an on-line oscilloscope display!.

★It is interesting that [2.3] is identical with a frequently employed expression for the density in premixed combustion; w corresponds to a heat release parameter τ and c to the concentration of product normalized to unity when the reaction is complete (cf. Bray and Libby (1976)). A value of τ = 6.22 is of practical interest in combustion.

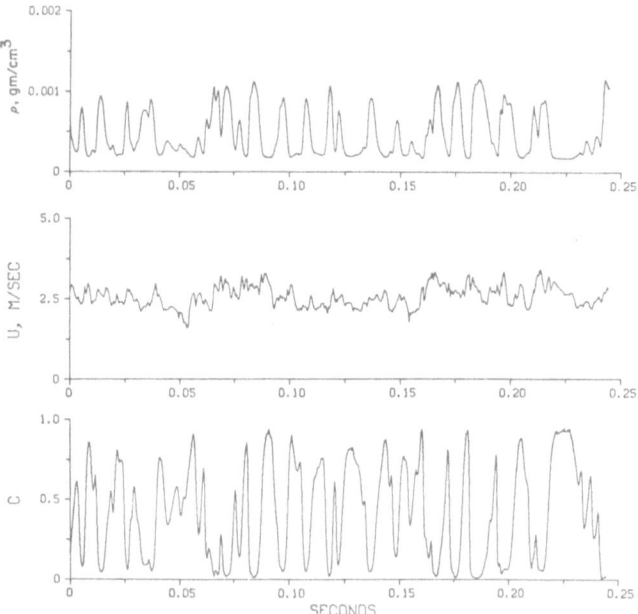

Fig. 4. A computer plot of a segment of time series involving intense turbulence (from LaRue and Libby (1977)).

Fig. 4 is a similar presentation of u-c data but is selected to indicate the intensity of density fluctuations which can arise in extreme circumstances. Here we show the time series in the density as well. These data are obtained in the turbulent boundary layer downstream of a slot which introduces helium into an air boundary layer. At this station the sheet separating the helium and air and originating at the edge of the splitter plate has not been destroyed by accelerated molecularity and is oscillating across the probe. The result is large excursions in concentration and density and significantly smaller velocity fluctuations. We shall discuss later the consequence of this high intensity turbulence relative to a three sensor probe for the measurement of two velocity components and helium mass fraction. Full details are in LaRue and Libby (1977).

3. Assessment of Accuracy
The most obvious and the first check on accuracy involves treating the calibration data as turbulence data and comparing the predicted and known velocity-concentration pairs. Another check involves the collection of turbulence data in a helium-air mixture of fixed and known helium concentration. For example in Way and Libby (1971) we applied our probe to measurements of a helium jet discharging into quiescent air; by mounting a mailing tube over the jet orifice and passing a mixture of 10 % helium through the tube we established a

turbulent pipe flow with fixed helium concentration. Finally, we must face the reality of non-negative helium concentrations. We do not legislate values of $c < 0$ from our data reduction scheme so that the number of negative entries in a probability density function for c at a given spatial location also provides a crucial measure of accuracy. The same applies for values of $c > 1$.

Our experience in making these measurements is that we have an accuracy of $\pm 10^{-3}$ in concentration at low values of c and of 1 % in velocity and in helium concentration at high values of c. Another indication is that for $\bar{c} = 0.014$, and $\overline{(c'^2)}^{1/2} = 0.0133$ we obtain negative values for c in one sample per 400. We therefore believe that insofar as u-c data are concerned we achieve our goal of high quality data in variable density turbulence.

4. Supplementary Measurements

In Stanford and Libby (1974) and LaRue and Libby (1977) an additional sensor, a swept film off to the side of the wire and film pair giving u-c data, is used to measure a second velocity component. Under some circumstances the displacement of the swept film is such as to lead to problems of spatial resolution. Typically the wire and film have a spatial resolution of 0.5 by 0.25 mm, adequate for most laboratory investigations. However, in order to avoid unwanted interference between the two films, the swept film is placed to the side with the result that the three sensor probe has a spatial resolution corresponding to a cube 0.63 mm on a side. In one region of the flow involving the slot injection of helium into an air boundary layer described in LaRue and Libby (1977) and discussed briefly in connection with Fig. 4, this resolution is inadequate insofar as the v-velocity component is concerned. At other positions in this slot flow and in fact in all of the other flows we have studied the number of default values of the v-velocity component are acceptably small. However, our experience with the slot flow indicates to us that a probe with better spatial resolution than the present three sensor configuration is indicated. We have in mind a split film with a wire mounted upstream and in the thermal field of the two films. In this same vein we note that we shall attempt to develop a skin-friction meter for helium-air mixtures in order to obtain surface data in our slot experiment.

We have completed an experiment involving heated helium discharging from a two-dimensional jet into a moving airstream and are reducing the data at present. A cold-wire has been mounted upstream of a standard u-c probe consisting of a wire and a film in order to permit measurements of u, c and θ. Our principal motivation for this extension of our previous experiments on a two-dimensional jet which we discuss later relates to the different Prandtl and

Schmidt numbers for dilute helium air mixtures, namely 0.75 and 0.25 respectively. The extent to which this difference will alter the superlayer and the detailed statistics within the turbulent fluid is of interest.

5. Some Representative Results

We have applied the technique outlined here to several flow configurations. In Way and Libby (1971), the first application of the technique, a circular jet of helium discharging into quiescent air is studied. This flow has the disadvantage of having low velocities in the outer edges of the mixing region so that after these initial results this flow has been abandoned. In Stanford and Libby (1974) a second film is added to the original u-c probe and measurements are made of u-v-c in a porous tube with helium injected through the cylindrical surface. This flow has the advantage that the technique developed for helium and air can be applied to other binary mixtures with higher molecular weight ratios, e.g., to helium and carbofluoride mixtures so that the molecular weight parameter in [2.3] can take on values of roughly thirty.

One of the interesting results from the tube experiments is the observation of counterfluxes of helium, i.e., the mean flux of helium $\overline{v'c'}$ has the same sign as the gradient $\partial\bar{c}/\partial y$. Although this result has not been fully clarified, the explanation appears to reside in small radial pressure gradients, gradients which have a negligible effect in constant density flows but which are significant when the density is variable. The effect seems to be analogous to the influence of small density fluctuations on the behaviour of turbulence in stratified flows. In this connection we note that Bilger (1976) suggests the importance of small radial pressure gradients on the balance of turbulent kinetic energy in variable density turbulence. The same considerations appear to apply to the flux of concentration. We also note that the same counterfluxes are found in the slot flow (LaRue and Libby (1977)).

One instructive means for displaying the results of our measurements is in terms of probability density functions. Multivariable pdf's are difficult to display, although we have done so in commenting on the inappropriateness of a theoretical model for a joint probability density function in the literature; thus single variable pdf's are generally considered. As an example of data in this form we show in Figs. 5 and 6 the pdf's of the axial velocity and the helium concentration at several radial positions at the exit plane of our porous tube. Particularly interesting are the increasing skewness of the velocity as the wall is approached and the reality of the $c \geqslant 0$ barrier for the helium concentration.

In recent years we have focussed on two flows which would appear to have more general interest; we refer to the slot flow involving the injection of helium into an air boundary layer (LaRue and Libby (1977)) and the two-dimensional helium jet in a slower moving airstream (Anderson et. al. (1977)). We discuss some representative results from these flows.

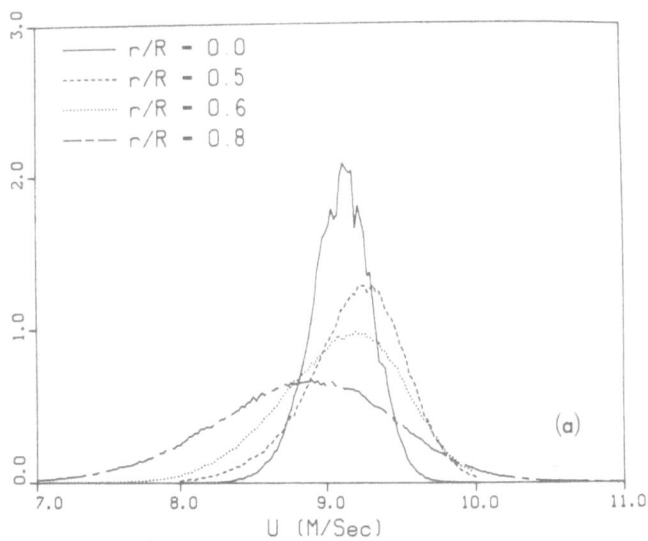

Fig. 5. The probability density distribution of axial velocity (from Stanford and Libby (1974)).

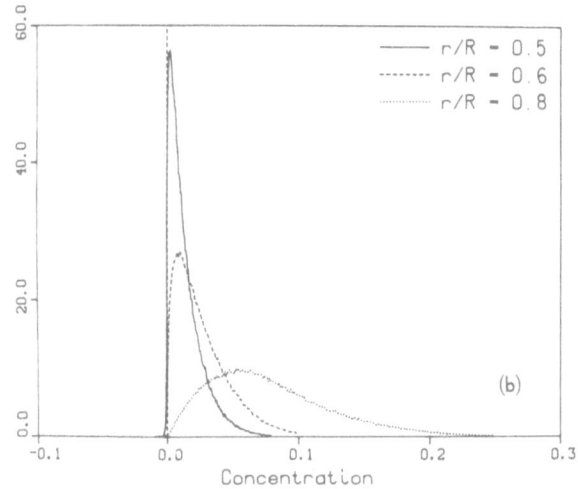

Fig. 6. The probability density distribution of helium concentration (from Stanford and Libby (1974)).

In the slot flow we encounter the full range of helium concentrations from pure helium to pure air, zero helium. The original presentation of the data involved conventional, unconditioned averaging, i.e., mean values, $\bar{\rho}$, $\bar{u}$, $\bar{c}$; intensities $\overline{\rho'^2}$, $\overline{u'^2}$, $\overline{v'^2}$, $\overline{c'^2}$; and fluxes $\overline{v'u'}$, $\overline{v'c'}$. We are currently reanalyzing these same time series in terms of conditioned, Favre-averaged statistics. To our knowledge this is the first time the two notions, conditioned sampling and Favre-averaging, have been combined. We follow this line of development with the conviction that Favre-averaging will become increasingly accepted for the description of turbulent flows involving variable density; this acceptance is already wide-spread in combustion circles.

It is perhaps useful at least from the point of view of notation to define conditioned Favre-averaging by means of a few examples. We assume that on the basis of some discrimination, in our case on the basis of a level of helium concentration, an intermittency function $I(\underline{x},t)$ is established. Then the conditioned Favre-averaged u-velocity component and related quantities are

$$(\tilde{u})_1 = \overline{\rho u I}/\bar{\rho}\,\bar{I}$$

$$(\overline{\rho u''^2}/\bar{\rho})_1 = \overline{\rho(u - \tilde{u})^2\,I}/\bar{\rho}\,\bar{I} \qquad\qquad [5.1]$$

$$(\overline{\rho v''u''}/\rho)_1 = \overline{\rho(v - \tilde{v})(u - \tilde{u})I}/\bar{\rho}\,\bar{I}$$

where the overbar indicates the usual time averaging. Note that we define the fluctuations, u'' and v'' for example, in terms of the deviation from the relevant unconditioned Favre-averaged quantity. We find little utility in introducing the fluctuations relative to conditioned means. For our particular purposes it is convenient to display the results in terms of distributions of the unconditioned quantities and the appropriate product $\bar{I}(\ )_1$; to illustrate consider the u-velocity component, namely

$$\tilde{u} = \bar{I}(\tilde{u})_1 + (1-I)(\tilde{u})_0 \qquad\qquad [5.2]$$

where $(\tilde{u})_0 = \overline{\rho u(1 - I)}/\bar{\rho}(1 - \bar{I})$ is the Favre-averaged velocity when $I = 0$.

The conditioned data we show involves discrimination based on a value of zero for the helium concentration, a "gate" value determined by examination of the probability density function for c at several spatial locations; there is no hold time used in the discrimination except that implicit in the digitization. Figs. 7 - 9 show respectively the mean concentration, the mean u-velocity component, and the Reynolds stress at several downstream stations. From Fig. 7 we see that at the most upstream station considered, namely at x/s = 1.94 where s is the height of the slot, there is a well defined layer adjacent to the surface of nearly pure helium and that at the two stations further downstream this layer dissipates. However, near the surface at the most downstream station considered, there is sufficient helium concentration so that the mean density is only 70 % of air, implying the possibility of significant effects.

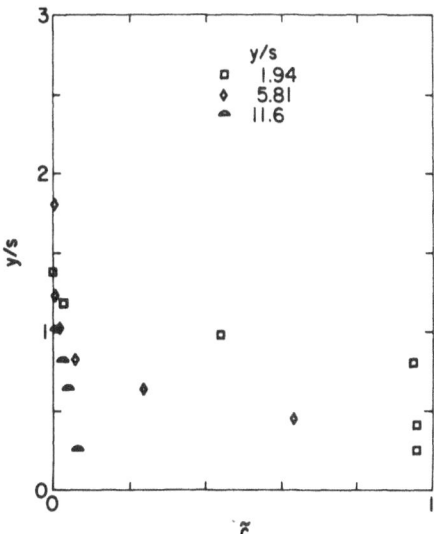

Fig. 7. Distributions of Favre-averaged helium concentration in a turbulent boundary layer involving slot-injection of helium.

In Figs. 8 and 9 both unconditioned and conditioned distributions are shown; the unconditioned data are the open symbols and the product of the conditioned quantities as defined by [5.1] and the intermittency $\bar{I}$ are represented by solid symbols. Points involving coincidence of the two quantities, unconditioned and conditioned, are denoted by a flag. Comparison of these two figures with Fig. 7 indicates that despite the small concentrations of helium for y/s > 1 the helium does contribute significantly to the mean velocity and the Reynolds stress, presumably because high helium concentrations correspond to large velocity defects.★

★ The data indicate that $\overline{\rho' u'}$ is positive as expected.

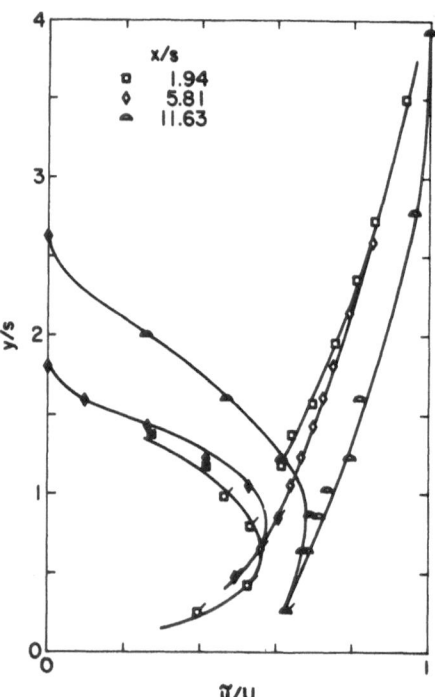

Fig. 8. Distributions of unconditioned and conditioned streamwise velocity component according to Favre-averaging in a turbulent boundary layer involving the slot-injection of helium.

Fig. 10 shows the distribution of the mean normal flux of helium, again Favre-averaged. Comparison with Fig. 7 indicates that when the helium concentration is sufficiently low so that the flow is nearly of constant density, i.e. for $y/s \geqslant 1$, the fluxes are positive and in accord with gradient transport. Near the surface, for all streamwise stations, the fluxes are negative, contrary to gradient transport. This result clearly warrants further study since it has significant implications regarding the widespread use of gradient transport in the phenomenology of turbulent flows with variable density, in particular for turbulent reacting flows.

The final data we discuss as representative of the results we have obtained with our hot-wire anemometry are given by Figs. 11 and 12, data from Anderson et al. (1978), i.e. from a two-dimensional helium jet in a slower-moving airstream.

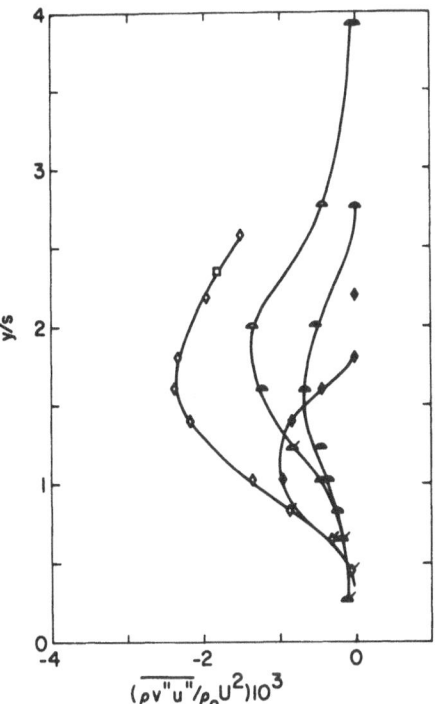

Fig. 9. Distributions of unconditioned and conditioned Reynolds stress according to Favre-averaging in a turbulent boundary layer involving the slot-injection of helium. See Fig. 8 for symbols.

Our measurements in this case are all made so far downstream that the helium acts as a passive scalar. The data in Anderson et. al. (1978) involve the usual streamwise and transverse distributions of the unconditioned u-velocity component and the concentration and zone averages based on discrimination relative to a level of helium concentration. Comparison is made with previous measurements of the velocity distributions; there appear to be no previous data regarding passive scalars for two-dimensional jets in moving streams.

The most interesting results from our experiment relate to what we term range conditioned point statistics. Briefly, at a given spatial location all turbulent passages within a specified tolerance of a specified mean duration are selected; ensemble statistics relative to the upstream and downstream crossing for this subset are established. Thus this technique provides a statistical picture of turbulent burst of a given scale.

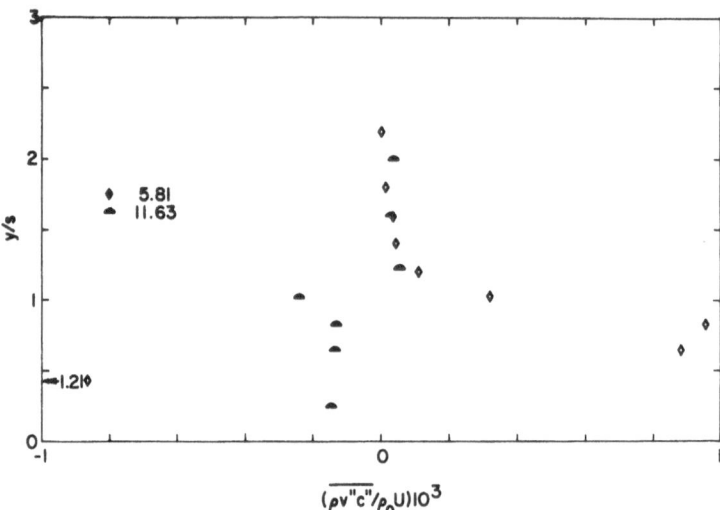

Fig. 10. Distributions of the unconditioned flux of helium according to Favre-averaging in a turbulent boundary layer involving the slot-injection of helium.

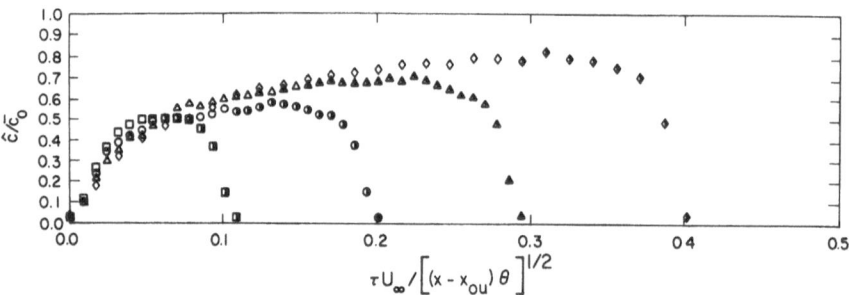

Fig. 11. Distributions of helium concentration given by range conditioned point statistics (from Anderson et al. (1977)).

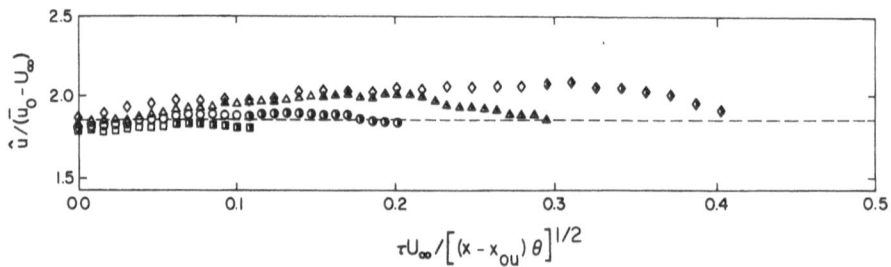

Fig. 12. Distributions of streamwise velocity given by range conditioned point statistics (from Anderson et al. (1977)).

Note, however, that we are unable to discriminate between passages through the center of a small turbulent structure from grazes off the centerline of a large structure.

In Figs. 11 and 12 τ is the mean duration, U_∞ the velocity in the airstream, θ the momentum increment of the jet, and $\bar{c}_0$ is the mean centerline concentration at the streamwise station involved. The caret implies ensemble average. Four different values of τ are shown in these figures with all passages lined up so that the upstream crossing is at the origin. We see from Fig. 11 that there are three distinct regions in these structures: upstream and downstream regions involving large gradients in helium concentration, and a central region involving a gradual increase in helium concentration in the downstream direction. The two edge regions have a thickness on the order of 55 times the Batchelor scale; this is somewhat thicker than we would expect on the basis of the results of similar measurements by LaRue and Libby (1974) in the heated wake of a cylinder. Nevertheless, we attribute these edge regions to the superlayer. The gradual increase in the mean helium concentration as the downstream edge is approached is consistent with preferential entrainment at the upstream edges of the turbulent structures. This is also indicated in Fig. 12 by the higher velocity near the downstream edges. It must be recalled that in this jet the turbulence moves faster than the surrounding airstream so that the upstream edges are the leeward edges; preferential entrainment there leads to lower helium concentrations and slower moving fluid. This result is consistent in terms of preferential entrainment at leeward edges with previous results in which the turbulent fluid moves slower than the surrounding fluid.

6. Concluding Remarks

We have reviewed our work on extended hot-wire anemometry and digital techniques for making measurements in turbulent flows involving the mixing of helium and air. The success of the technique depends on a special hot-wire probe which removes an essential ambiguity. Typical results from a variety of flows are described with particular emphasis on the turbulent boundary layer which results from the slot injection of helium into an air boundary layer and on a two-dimensional helium jet in a moving airstream.

Future work will involve further measurements relative to slot injection and will utilize a skin-friction meter for helium-air mixtures and to the two-dimensional jet permitting measurements sufficiently close to the discharge so that significant dynamic effetcs arise.

Our research on the turbulent mixing of helium-air is supported by the Office of Naval Research as part of Project SQUID under Contract NOO14-75-C-1143 NR-098-038 and by the National Aeronautics and Space Administration under Grant NSG 1386. The support of these organizations is gratefully acknowledged.

References

Anderson, P., LaRue, J.C. and Libby, P.A., 1977 Project SQUID TR UCSD-9-PU

Bilger, R.W., 1976 *Prog. Energy Combust. Sci.* **1**, 87

Bray, K.N.C. and Libby, P.A., 1976 *Phys. Fluids,* **19**, 1687

Corrsin, S., 1949 NACA TN 1864

LaRue, J.C. and Libby, P.A., 1974 *Phys. Fluids,* **17**, 1956

LaRue, J.C. and Libby, P.A., 1977 *Phys. Fluids,* **20**, 192

Libby, P.A., 1977, in *Studies in Convection,* Vol. **2**, ed. Launder, B.E., Academic Press, London, 1-43

Stanford, R.A. and Libby, P.A., 1974, *Phys. Fluids* **17**, 1353

Way, J. and Libby, P.A., 1970 *AIAA J.* **8**, 976

Way, J. and Libby, P.A., 1971, *AIAA J.* **9**, 1567.

Measurements in Intermittent and Periodic Flow

Measurement in Intermittent and Periodic Flow

by

Leslie S.G. Kovasznay★
The Johns Hopkins University
Baltimore, Maryland, USA

Abstract

For the purpose of the present discussion, it may be taken for granted that the signals from the various individual probes (hot-wire, LDA, microphone) follow faithfully the physical variables at a given point, consequently the concern here is shifted to the signal processing techniques that enable the extraction of relevant information about the nature of the flow field.

The methods to be discussed here all deal with the full, essentially unfiltered, signals. On the other hand, sampling and averaging are performed subject to some criterion developed in real time. The earliest and also the most developed among the methods is conditional sampling in combination with ensemble averaging. Originally introduced by the author in the 1960's with this method in an intermittent turbulent flow, one may obtain separate average values for the turbulent state and for the non-turbulent state. Further refinements include the introduction of the point average. In that case, samples are taken only at certain specific instants, subject to some well-defined criteria, such as the instant when the turbulent interface passes over the detector probe.

In the simplest case when the sampling criterion is obtained from a periodic signal, the above method is called periodic sampling. The applications are quite numerous. When studying the flow field in rotating machinery, the instantaneous position of the rotor blades may be used for a sampling criterion; then by averaging over a large number of periods the strictly periodic or "deterministic" portion of the phenomenon can be fully recovered.

The most advanced signal processing offered today is termed as "pattern recognition". In its simplest form, this involves only rescaling typical events, both in amplitude and in time. The first events detected this way were the so-called "bursts" and "sweeps" observed near the wall in turbulent flows. As the

★Present address, University of Houston, Texas 77004

use of large scale computers is advancing rapidly, this new method will be used more extensively to "recognize" the passage of typical large-scale coherent turbulent structures.

But, at this point, one must also give a word of caution. The conditional sampling and pattern recognition methods always give some spurious response when applied to a random signal. The only proper protection against such "false alarm" is to check-out the circuit, or the computer program, first on an artificial random signal, designed to possess the same overall statistical properties (spectra and probability density function) as the flow field and see how much "false alarm" will be detected by using the original criteria on this synthetic signal.

1. Introduction
In unsteady flow measurements, especially in a turbulent flow, there are two basic problems. The first is the appropriate response of a probe and for the sake of present discussion, it will be assumed that this problem has been solved. The second problem, the conceptually more difficult one, is how to understand complex unsteady flows through a limited number of probe signals, especially when the flow properties are only statistically definable. This is the reason why signal processing plays such a central role in unsteady flow measurements. The particular subject of discussion here will be the detection and measurement of intermittent or periodic features occurring in an otherwise random, irregular flow. Naturally, the first step is to recognize that the flow is indeed intermittent or periodic. Signal processing in such a situation generally proceeds in the real time domain as the use of Fourier transforms per-se does not offer many advantages. One must remember here that in turbulence research the power spectra of the fluctuations were measured mostly for the convenience of using only a single probe and thus avoid the necessity of measuring at two or more points in space. A convenient rationalization of this practice was offered by invoking the so-called Taylor's hypothesis, for the exchange of space and time coordinates in the interpretation of the measurements. On the other hand, today there is an impressive arsenal of turbulence measuring techniques: correlation is being measured routinely at two or more points, higher order spectra are obtained, but more of that subject will be given in one of the later sessions. The main subject here today is a different one: It involves the selection of some identifiable condition in time or in some cases a selection in amplitude. The selection in time is based always on some "master signal", typically a train of pulses that provides the appropriate condition. This master signal may be obtained from external sources in case of "driven" periodic flows or it may be obtained from the flow field itself by defining some condition for conditional sampling.

The aim here is to understand periodic or intermittent flows by extracting from the continuous probe signal some recurring features by proper selection and proper averaging. In all these techniques, whatever are the selection criteria, after a selection is made ensemble average is performed on the selected samples. First these operations were performed by analog circuits. Later most signal processing became digital so instead of analog circuits, computer programs were developed. As far as selection criteria are concerned, they will be described here in an evolutionary sequence.

2. Periodic Phenomena

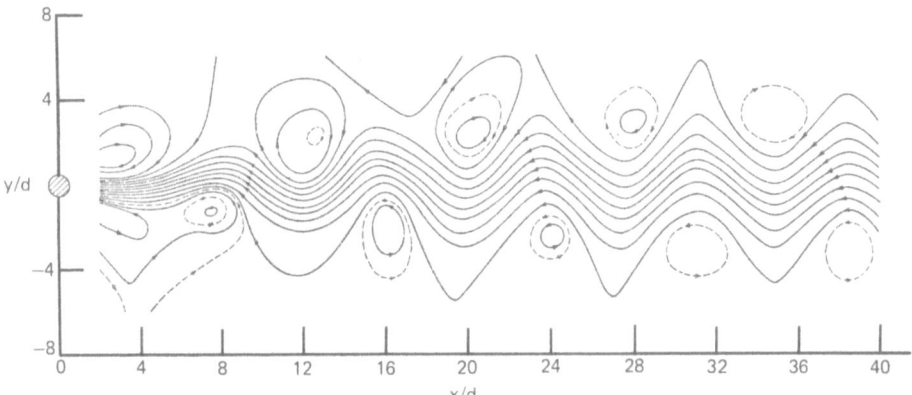

Fig. 1. Periodic wake (Karman vortex street) at Re = 56. Reconstruction from hot-wire measurement. From Kovasznay (1949).

Periodic phenomena are quite common. A function f is periodic if, and only if

$$f(t + T) = f(t)$$

where T is the period.

If the signal is strictly periodic, or in other words, there is no additive noise, recording and analyzing one single period is sufficient. If the periodic velocity fluctuations are recorded, the entire periodic flow field may be reconstructed and the only practical problem is the accurate determination of phase. The Karman vortex street is strictly periodic in the Reynolds number range $40 < Re < 100$ so Kovasznay (1949) reconstructed the entire instantaneous flow field from the signals of two hot-wires. One probe was used at a fixed location outside the vortex street and it provided the phase reference, while the other was moved to various x, y locations (Fig. 1).

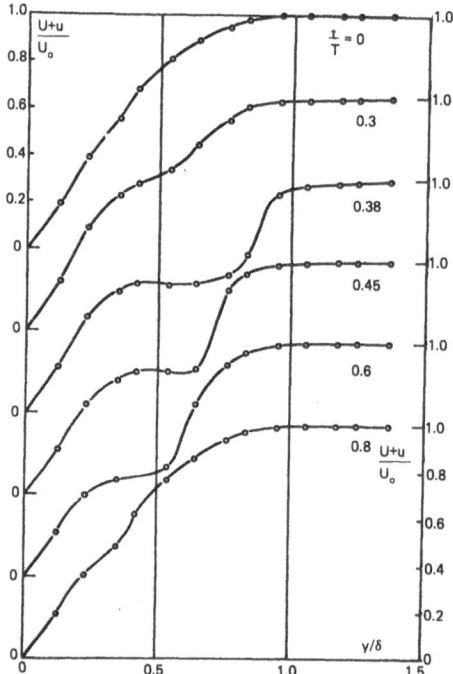

Fig. 2. Instantaneous velocity profiles during transition in boundary layer. From Kovasznay, Komoda & Vasudeva (1962).

Similarly, the non-linear development of the periodic instability waves in the transition of laminar boundary layers was studied by Kovasznay, Komoda and Vasudeva (1962) using the periodic excitation signal as a phase reference. Fig. 2 shows the instantaneous velocity profiles at different times during one full period.

2.1 Ensemble Average, Noise Suppression

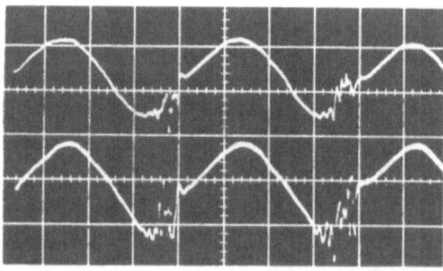

Fig. 3. Boundary layer transition in periodic flow. Traces taken at two span-wise stations. From Obremsky & Fejér (1967).

The underlying assumption here is that there is a strictly periodic component in the phenomenon, even if it is buried in the noise. In most practical cases the periodicity is imposed by the experimenter, the flow is "driven" period-ically. In such a case, the "master signal", a sine wave or a set of periodically occurring timing pulses can be derived from the driving mechanism. Early ex-amples are periodically driven flows such as those by Obremski and Fejer (1967). They studied the transition in a periodically driven boundary layer. Fig. 3 shows two oscilloscope traces taken at two spanwise locations showing transition only during a small fraction of the period. Periodically "driven" turbulent boundary layer was studied by Karlsson (1959), as well as Nakagawa, Hayakawa and Kobashi (1977). In all these cases, specially constructed wind tunnels provided a periodically varying mean flow and the timing signals were obtained from the rotating shaft of the periodical drive. The main interest here is to obtain the periodic or repetitive component in the presence of turbulence.

Let us define two kinds of averages for an arbitrary function g(t), the first one is the conventional time average defined as

$$\overline{g} \equiv \underset{T \to \infty}{Lt} \frac{1}{T} \int_0^T g(t)dt$$

and the second, the periodic average is defined as

$$\widetilde{g}(t) \equiv \underset{K \to \infty}{Lt} \frac{1}{K} \sum_{n=1}^{K} g(t + nT)$$

where T is a period and K is a large integer.

Now let us assume that a function $f(t)$ consists of three parts, first the time average F (or the D.C. component), second a zero mean (A.C.) strictly periodic component $f_p(t)$, and third a random noise $f_N(t)$:

$$f(t) = F + f_p(t) + F_N(t)$$

The periodic component obeys the condition

$$f_p(t) = f_p(t + nT)$$

for any integer n.

By definition, the time averages of the three components are

$$\overline{f_N} = \overline{f_p} = 0 \quad , \quad \overline{f} = F$$

The periodic average of the random component is then

$$\widetilde{f}_N = 0$$

and

$$\widetilde{f}(t) = F + f_p(t)$$

Clearly by taking the periodic average we suppress the random signal component, and only the D.C. average and the deterministic periodic components are
retained. In actual experiments, however, K the number of the samples averaged cannot increase indefinitely, so the contribution from the random component will decrease proportional to $\sqrt{K}$ for large K, and for sufficiently large K
this component will become negligible compared to F and f_p. In the experiments typically $100 < K < 2000$ is used.

Important examples where periodic averaging is useful are flows driven periodically by imposing perturbations on the flow by thin ribbons placed near a
solid surface in a permanent magnetic field and driven periodically by an electric current. There are many examples of this technique. Instability of the
laminar boundary layer was originally studied by this method in the 1940's at
the National Bureau of Standards and many more researchers followed it later.
Kovasznay, Komoda and Vasudeva (1962) perfected the technique in studying
laminar instability and the transition to turbulence in a laminar boundary
layer.

The same technique was used by Hussain and Reynolds (1970) in a turbulent
channel flow. They have driven a turbulent flow periodically by placing thin
ribbons near the wall. They studied the fate of the imposed periodic disturbances within the fully turbulent flow regime. Other groups have excited flows
periodically by sound waves. In these cases, vortices are shed from the solid
edges. The flow so excited were usually of very low Mach number so that the
direct cause of the excitation was not compressibility, but rather the induced
velocity fluctuation near the solid wall.

Excitation by loudspeaker offers also the possibility of using two or more different frequencies and to study the nonlinearity of the phenomenon, such as
by Sato (1970).

Another important application of this technique is for the periodic sampling of
velocity and pressure signals and averaging them in flows behind passing blades
in rotating machinery. For a stationary observer, the passage of an upstream

blade represents a strictly periodic disturbance and the accurate timing signals must be obtained from the rotating shaft. In case of rotating machinery, the number of blades may be large (typically 30-40 blades) so one must resolve the phase angle so well that the timing error would be only a small fraction of a blade passage time. As a result, the time resolution expressed as a shaft angle must be better than one half degree, consequently the use of electronic pulse shaping circuits are a necessity. As a good practical test, the accuracy of such periodic master pulse train can be easily estimated by driving a stroboscopic light source from the pulse train then observe the blade row appear as stationary. Then, by changing the speed of rotation the apparent position of the blade row must remain unchanged. On the other hand, if with changing r.p.m., the blade row appears to change position in the "strobe" light, this is inadequate for many applications. Author was able to achieve great stability by developing appropriate pulse shaping circuit. More details about measurements in rotating machinery can be found in Okiishi (1978).

Periodic sampling is always advantageous where there is a genuine periodic mechanism to drive the experiment. On the other hand, it is less useful in approximately periodic phenomena, where conditional sampling is more advantageous, e.g. in the Karman vortex street at a somewhat higher Reynolds numbers (Re > 100).

2.2 Recurrent Phenomena
The difference between periodic sampling and recurrent sampling is subtle. Let us suppose that the phenomenon is driven not from a strictly periodic source, but from a "deterministic source", meaning that the driving mechanism is always the same although the actual time of occurrence is arbitrary. Often this is the case when there is a short transient phenomenon which is externally triggered, and it will occur again and again in the same manner, although the time interval between occurrences is not constant but arbitrary, even random. Sometimes one may deliberately drive a flow in such a manner, not periodically but according to some other program for some other advantages, as it will be explained later. Three examples will be given here. In the first case, there is a transient phenomenon (e.g. passage of a shock wave; sweeping by of a wake, etc.) that may be studied under highly turbulent conditions by performing the transient experiment not once, not ten times, but hundreds of times in order to obtain a reliable ensemble average. In all such cases, one wishes to repeat the experiment always under identical conditions, and a simple solution is to drive the phenomenon repetitively by choosing a slow repetition rate, so that during the time interval, the previous event will decay and steady state will be reached before the start of the next event. The actual technology is very similar to that of periodic sampling, except here the master signal consists of a train of pulses

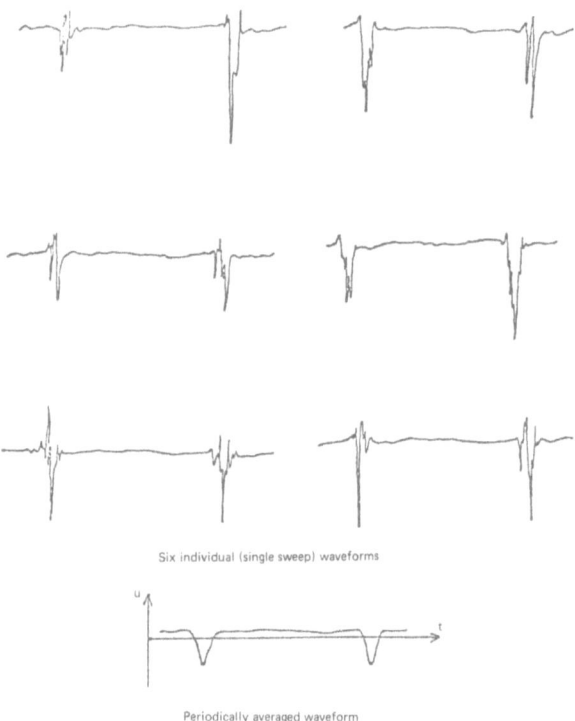

Fig. 4. Wakes behind passing rods (d = 1 cm; U 38 m/sec; V = 13 m/sec). From
Kovasznay & Fujita (1971).

with unequal intervals that signal the initiation or arrival of the phenomenon.
Fig. 4 shows six samples of individual records of a wake passage. A circular rod
of 1 cm diameter is cutting across the open jet working section of the wind tunnel as discussed by Fujita & Kovasznay (1974). The wakes are all different in
detail but they form a statistically reproducible ensemble. The bottom trace
shows the ensemble average taken over 600 samples. In such an experiment,
the periodicity of the phenomenon is not important as long as the interval between events is long enough for the event to decay. It is important, however,
that the triggering of the sampling circuit would be synchronized with the
passage of the wakes. The above technique was used to study "wake cutting"
by mounting an airfoil, or even a small cascade of airfoils into the working section of the open jet type wind tunnel. By rotating a pin-wheel made of rods in
front of the airfoil, an individual wake passage causes an unsteady aerodynamic and acoustic response that can be studied by taking ensemble averages of
all flow variables. The time interval between wake passages was chosen to be

sufficiently long so that each "wake cutting" could be regarded as an independent experiment. Details are given by Fujita and Kovasznay (1974); Ho and Kovasznay (1976) and Kovasznay (1977).

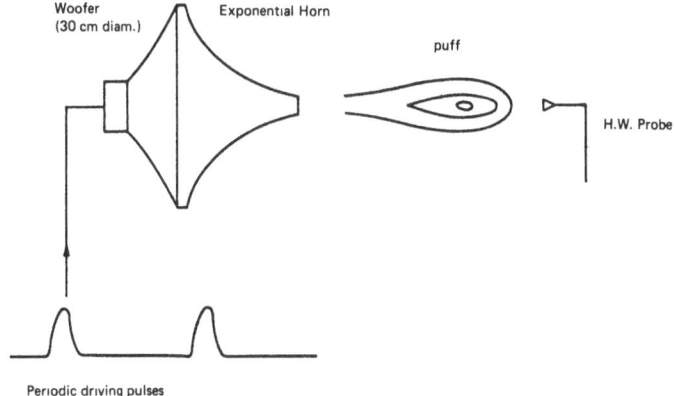

Woofer
(30 cm diam.)   Exponential Horn

puff

H.W. Probe

Periodic driving pulses

Fig. 5. Puff experiment. From Kovasznay; Fujita & Lee (1974).

Another example of averaging over recurrent single events, is the emission of turbulent puffs described by Kovasznay, Fujita and Lee (1974). Such puffs can be generated by large, powerful loudspeakers, when provided with a nozzle attachment, as shown in Fig. 5. Both laminar vortex rings and turbulent puffs can be emitted (by placing a grid at the nozzle exit) at pre-determined times by imposing an electric pulse on the driving coil of the speaker. Turbulent puffs were studied using this technique by Kovasznay, Fujita and Lee (1974). Both ensemble averaged mean velocity and ensemble averaged instantaneous turbulence level were measured. Fig. 6 shows the value of $\tilde{U}(t)$ and $\tilde{u}'(t)$ on the axis of $x/d = 2$. The definitions of those quantities are:

$$\tilde{U}(t) = N^{-1} \sum_{K=1}^{N} u(t_K + t)$$
$$[\tilde{u}'(t)]^2 = N^{-1} \sum_{K=1}^{N} [u(t_K + t) - \tilde{U}(t)]^2$$

where the t_K —s are the instants of the timing pulses, obtained from the pulse generator (possibly delayed by a fixed amount of time) and N is the total number of puffs ensemble averaged.

The same method was extended to produce turbulent puffs made of a combustible mixture by Oshima, Kovasznay and Oshima (1978), where the ensemble averaged signature of the combustion generated sound pulses were also obtained.

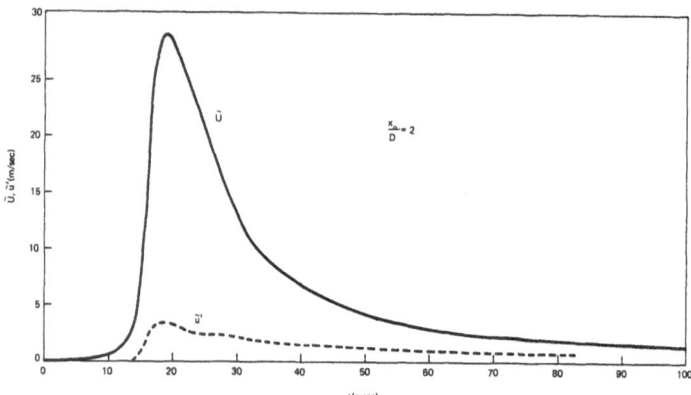

Fig. 6. Ensemble averaged mean velocity and rms velocity fluctuation in turbulent puff (x/d = 2). From Kovasznay, Fujita & Lee (1974).

Another example where recurrent sampling and averaging has some unique advantages is in the area of wave propagation. In acoustical propagation experiments, in the absence of an expensive anechoic chamber, simple and multiple reflections plague many experiments. If the reflecting surfaces cannot be eliminated, one may choose to study short solitary phenomena such as pulse trains by "keying" the detection to the master signal corresponding to the emission of the pulse modulated wave trains and to "average out" the response to earlier pulses so effectively "blur out" the reflections by initiating the pulses not periodically but at random intervals. This can be accomplished by having an approximate period and then distributing the intervals at random around the average interval (by a large random jitter). This way, suppression of reflections from previous events is possible by randomizing their cummulative effect. Naturally, one must first make a statistical estimate about the suppression of reverberation to be achieved in this manner.

What are the actual technologies necessary in all these cases? Periodic or recurrent sampling can be obtained by so-called "signal averaging" or "eductor" equipment. These in general contain a large array of storage devices. In the older analog form, these were high quality condensers that were charged up and accumulated the averaged values of the enhanced signal. In the more modern versions, storage devices are digital registers that accumulate the averages. In addition, there is always an appropriate switching mechanism that initiates the sampling mode and delays the signal so that each appropriate storage device receives the signal with the proper delay. After certain time in the sampling mode, covering hundreds or thousands of events, the storage registers can be read and the enhanced average signal reconstructed.

2.3 Flow Field Reconstruction

An important application of periodic and recurrent sampling should be mentioned here: any unsteady flow that is periodic, two-dimensional and incompressible can be fully reconstructed from the samples of the two velocity components over a coordinate grid. If the two velocity components $u(t)$ and $v(t)$ are obtained at every point on a two-dimensional x,y plane, one may reconstruct the instantaneous stream function by simple integration.

Let us define a stream function $\psi(x,y,t)$ so that

$$u(x,y,t) = \frac{\partial \psi(x,y,t)}{\partial y}$$

$$v(x,y,t) = - \frac{\partial \psi(x,y,t)}{\partial x}$$

Since only the derivatives of ψ occur in the physical quantities, one can assign, without losing any generality, $\psi = 0$ at a point $x = x_0$ and $y = y_0$ for all t. Let us consider the flow field at a particular fixed time, then the stream function can be obtained by integrating the measured velocities as follows.

By assigning, the value

$$\psi(x_0, y_0) = 0$$

first, $v(x,y)$ is integrated with respect to x so $\psi(x,y_0)$ is obtained as

$$\psi(x,y_0) = - \int_{x_0}^{x} v(x', y_0) \, dx'$$

Next, by integrating $u(x,y)$ with respect tu y, $\psi(x,y)$ is obtained as

$$\psi(x,y) = \int_{y_0}^{y} u(x, y') \, dy' + \psi(x, y_0)$$

or

$$\psi(x,y) = - \int_{x_0}^{x} v(x', y_0) \, dx' + \int_{y_0}^{y} u(x, y') \, dy'$$

The instantaneous streamlines can be drawn by connecting points having the same values $\psi = $ constant.

Fig. 7 (Fujita and Kovasznay 1974) shows two different streamline patterns representing the same flow field at the same instant of time, but presented as observed from two different coordinate systems, one from the "Laboratory"

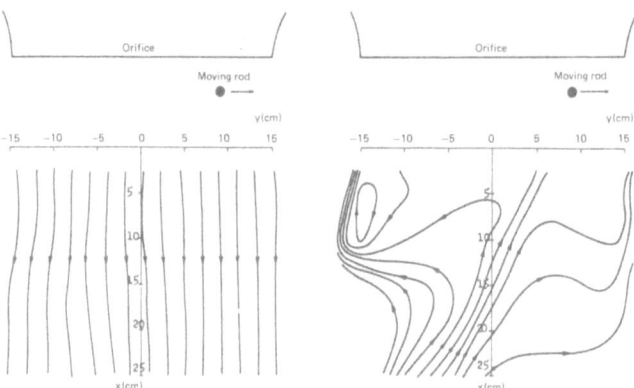

Fig. 7. Instantaneous flow pattern reconstructed from hot-wire data. Left: in laboratory coordinates; right: coordinate system convected with mean flow. From Fujita & Kovasznay (1974).

coordinates fixed to the apparatus, and the other from a coordinate system moving in the downstream direction with the nominal mean velocity U_o. The position of the wake producing rod is indicated by the solid black circle. The flow pattern as seen from the laboratory coordinate system is, in a way, disappointing because it reveals very little of the flow (left) as the streamlines appear nearly parallel and uniform. On the other hand, the pattern seen from the moving coordinate system (as shown on the right), accentuates the flow features quite clearly. The velocity defect in the wake appears as a jet directed towards the wake producing rod, and a large vortex, is found near the left edge of the jet at the upper left corner. The undisturbed region moving approximately with the nominal mean velocity U_o here appears nearly at rest. The rod entering and leaving the shear layer always leaves large vortices.

In such a flow field, the reconstruction of the periodic or ensemble averaged flow is a large advantage. On the other hand, if one would wish to reconstruct the instantaneous flow field, in a non-periodic flow, one would need simultaneously a very large number of probes in order to obtain the full velocity field. Since in case of periodic samplings, any one periodically sampled and ensemble averaged record is equivalent to any other one, the measurements can be done sequentially, one at a time, at each point, by maintaining identical conditions in the flow field. Naturally, by such a technique only the periodic or recurrent, or as quite often referred to as the "deterministic", portion of the signal is recovered.

3. Intermittent Phenomena

It was first discovered in the 40's that the turbulent-non-turbulent interface is sharp, not diffuse. The outer edge of a turbulent flow appears diffuse only in the distribution of statistical averages, but when observed instantaneously, it is quite clear and distinct. When placing a probe into the intermittent zone, the flow will be found turbulent for a finite fraction of the time and non-turbulent over the rest of the time. Townsend (1949) defined the intermittency factor γ, giving the fraction of time when the flow is turbulent at a point in space. He measured γ by measuring the kurtosis (normalized fourth moment) of the hot-wire signal and by assuming that the turbulent portion is gaussian. Corrsin and Kistler (1955) obtained the intermittency function $I(x,y,z,t)$ from the structure of the streamwise vorticity signal. Fiedler and Head (1966) used optical detection. They introduced thin smoke into a turbulent boundary layer and illuminated it with a thin sheet of light. The presence of the smoke, or more specifically, the scattered light from the smoke was taken as an indication of the turbulent state, so the properly clipped phototube output signal was regarded as the intermittency signal and γ was determined by simple averaging in time.

The intermittency at the turbulent-non-turbulent interface is rather clear. Less clear is the problem of the so-called "internal intermittency". Following Kolmogorov's suggestion, it was suspected that even within homogeneous turbulence there are high vorticity and low vorticity regions. Kuo and Corrsin (1971) made an attempt to measure this internal intermittency and even to make a rough estimate of the shape of the high vorticity regions within a homogeneous shear flow. From an experimental point of view, intermittency can be defined if there are two experimentally distinct states in the flow.

3.1 Detector Functions

The detection of the turbulent and non-turbulent states has an extensive literature: Corrsin and Kistler (1955) based it on a vorticity meter giving ω_x the instantaneous vorticity in the mean flow direction. Kovasznay, Kibens & Blackwelder (1970) used the velocity derivative $\partial u/\partial y$ that may be regarded as the approximate span-wise vorticity ω_z as a basis of the detection in boundary layers. Using digital techniques, Kaplan and Laufer (1969) used the magnitude of the short term (typically 8-point) average variance of the streamwise velocity fluctuation as a basis of detecting intermittency. There were efforts to base detection on the instantaneous uv product, whose average value is the Reynolds stress, as well as many other combinations, e.g. Sunyach and Mathieu (1969), Hedley and Keffer (1974), LaRue (1974). The influence of the numerical parameters was examined by Kibens, Kovasznay and Oswald (1974) as well as by Antonia and Atkinson (1974).

The problem of conditional sampling, of course, is always centered on the criteria used to define the detector function and one may appreciate the difficulty of defining the criteria if one moves a little away from the one's technical field at hand. Let us suppose that we monitor a cocktail party with a microphone and associated signal processing equipment. If our aim is to find the characteristics of the human coughing pattern, we must be able to identify, align and ensemble average all the coughing that occurs at the cocktail party, knowing quite well that the signals are all deeply embedded in the overall noise of the chatter. In order to define proper criteria for the detection of turbulent state in general, there are two minimum requirements. One is to define a detection threshold to be applied. This requires that some variable or combination of variables must exceed a certain predetermined threshold level. The second is the time window, namely, the threshold criteria must be applied for a given finite time interval in order to make a valid decision. This is absolutely necessary in order to avoid spurious responses that would be called a "false alarm", In case of coughing, there would be a minimum amplitude, but also a minimum time to qualify as a cough. As the flow properties vary from case to case, the numerical values of the detection criteria must be adjusted to the flow variables. The threshold value of velocity or vorticity fluctuations must be adjusted to represent a given fraction of the corresponding average values in the fully turbulent portion of the flow. The time window or, as it is often called, the "holding time" must be chosen proportional to the characteristic time of the turbulence structure, most typically to the quantity $\lambda/\bar{U}$ where λ is the Taylor microscale and $\bar{U}$ is the local mean velocity. After the detector parameters are set, it is worthwhile to verify that the criteria chosen really do correspond to our intuitive notion of what the turbulent and the non-turbulent flow should be. In a way, this can be used as an a posteriori justification. Such a check can be best made by comparing the simultaneous record of the fluctuations and the derived intermittency function $I(t)$ either on a print-out or, for faster phenomena, on a storage scope. Nevertheless, the setting of detection criteria only by "eyeball" is no longer considered acceptable. In connection with this problem, there is another difficulty, namely the Reynolds number of the turbulent flow. If the Reynolds number of the experiment decreases toward the point when the flow barely remains turbulent, the ratio between the scales of the largest and the smallest eddies approaches unity. Correspondingly, the detection becomes increasingly difficult and increasingly arbitrary.

In case the flow is heated (sometimes it may be heated slightly just to facilitate detection), temperature may be used as a turbulence indicator, e.g. if the wall is heated, the turbulent portion of the boundary layer will be heated while the entrained non-turbulent fluid will remain cold. Extensive research in air concluded that in a heated turbulent flow, the interface is the same for turbulent velocity fluctuations and for turbulent temperature fluctuations.

In heated flows the temperature signal properly clipped and normalized can serve as the intermittency function $I(t)$ because the turbulent transport of heat is very much larger than the molecular transport, and one may assume that all parts of the turbulent zone will become heated, while the non-turbulent fluid will remain at ambient temperature. In such a case, the turbulence detector is merely a "cold wire" probe. This is a hot-wire probe, used with very low over-heat, followed by a comparator. The only problem is that a low-drift D.C. amplifier is needed to amplify the "cold-wire" temperature signal in order to obtain a reliable threshold. With only A.C. coupling, the threshold level would be floating therefore unreliable.

All detector circuits must be calibrated by "synthetic turbulence" signals. Those are intermittently modulated noise type of signals, where the inter-mittency statistics, as well as the noise spectrum, are predetermined and known quantities. Antonia & Atkinson (1974), as well as Kibens, Kovasznay & Oswald (1974) give detailed examples.

3.2 Zone Average, Point Average
Conditional sampling, at least in principle, is a generalization of the periodic or recurrent sampling, to the case when the master signal is random and what is more important, it is obtained from some property of the flow field itself. Historically, the first conditional sampling experiments were done in or around the turbulent, non-turbulent interface, Kovasznay, Kibens & Blackwelder (1970) defined both the conditional zone average and the conditional point average.

Since there appears to be clearly two distinct states, one may immediately define an intermittency function $I(x,y,z,t)$ so that

$$I(x,y,z,t) = \begin{cases} 1 \text{ for turbulent flow,} \\ 0 \text{ for non-turbulent flow.} \end{cases}$$

The time average value of $I(t)$ is the intermittency factor, γ, which gives the fraction of the time that the probe spends in turbulent flow:

$$\gamma = \overline{I} \equiv \underset{t_o \to \infty}{Lt} \frac{1}{t_o} \int_{t_i}^{t_i+t_o} I(t)\, dt$$

The intermittency function $I(t)$ at a point is a random square wave and the in-stants when it changes from 0 to 1 and from 1 to 0 represent the condition that the turbulent non-turbulent interface is located at the turbulence detector probe. Since it is assumed here that the interface is a continuous curved surface that separates the two types of flow, for a fixed observer there are two types of transitions. One is when the probe enters the turbulent regime (Front). The other is when the probe leaves the turbulent regime (Back).

It is clear to the experimenter that, in attempting to determine the intermittency function $I(t)$, there will be always some arbitrariness in the definition, so details will depend on the magnification, if you like, on the coarseness of the method of detection. Nevertheless, if one assumes that $I(x,y,z,t)$ can be suitably defined, it becomes possible to define other new kinds of averages.

If $Q(t)$ is a fluctuating quantity (e.g. a velocity component at a given point), then the 'conventional time average', denoted here by a bar, is given by

$$\bar{Q} \equiv \underset{t_o \to \infty}{Lt} \frac{1}{t_o} \int_{t_1}^{t_1+t_o} Q(t)\, dt$$

Utilizing the intermittency function, $I(t)$, conditional averages may be formed.

If $Q(t)$ is averaged only during the time intervals when $I = 1$, this is termed a 'turbulent zone average':

$$\tilde{Q} = \frac{\overline{IQ}}{I} \equiv \underset{t_o \to \infty}{Lt} \frac{1}{\gamma t_o} \int_{t_1}^{t_1+t_o} I(t)Q(t)\, dt$$

Conversely, if $Q(t)$ is averaged only during the non-turbulent time intervals when $I = 0$, the non-turbulent zone average is obtained.

$$Q = \frac{\overline{(1-I)Q}}{(1 - I)} \equiv \underset{t_o \to \infty}{Lt} \frac{1}{(1-\gamma)t_o} \int_{t_1}^{t_1+t_o} [1-I(t)]\, Q(t)\, dt$$

The notation used for zone averages is as shown above ($\tilde{\ }$) for the turbulent and ($\bar{\ }$) for the non-turbulent zone average. It is clear from the definition that the conventional average is the weighted sum of the two zone averages.

$$\bar{Q} = \gamma\tilde{Q} + (1-\gamma)\tilde{Q}$$

$Q(t)$ and $I(t)$ of course, do not need to be taken from the same point in space, and, in that case, the zone average will depend on the relative position of the detector probe located at x_o, y_o, z_o giving $I(x_o, y_o, z_o, t)$ and the signal probe located at x, y, z, t giving $Q(x, y, z, t)$.

Another entirely different type of sampling and averaging can be constructed based on the instants when the interface passes over the detector probe. As mentioned above, one can distinguish here two kinds of transitions (Fronts and Backs). If the intermittency function $I(t)$ is differentiated with respect to time, a pulse train, consisting of positive and negative delta functions, results:

$$\dot{I} \equiv \frac{\partial I}{\partial t} = (-1)^n \, \delta(t-t_n)$$

where each t_n is the instant of change in $I(t)$. For an even integer, $n = 2i$, the monitor probe is entering the turbulent region (Front); for an odd integer, $n = 2i-1$ the monitor probe is leaving the turbulent region (Back). The two pulse trains can be defined separately as $\hat{P}(t)$ and $\check{P}(t)$. For sampling at the 'Fronts',

$$2P(t) \equiv \dot{I}(t) + |\dot{I}(t)| = \delta(t-t_{2i})$$

For sampling at the 'Backs',

$$2P(t) \equiv -\dot{I}(t) + |\dot{I}(t)| = \delta(t-t_{2i-1})$$

The conditional sampling of the continuous variable $Q(t)$ can now be performed and each pulse train of conditional samples can be represented as one of the two generalized functions of time:

(Fronts) $\hat{p}(t) = Q(t) \, \hat{P}(t)$ and   (Backs) $\check{p}(t) = Q(t) \, \check{P}(t)$

and the 'conditional point averages' can be defined as:

$$\hat{Q} = \underset{t_o \to \infty}{LT} \; \frac{\int_{t_1}^{t_1+t_o} Q(t) \, \hat{P}(t) \, dt}{\int_{t_1}^{t_1+t_o} \hat{P}(t) \, dt}$$

$$\check{Q} = \underset{t_o \to \infty}{LT} \; \frac{\int_{t_1}^{t_1+t_o} Q(t) \, \check{P}(t) \, dt}{\int_{t_1}^{t_1+t_o} \check{P}(t) \, dt}$$

All integrations are performed in the sense of generalized functions such as the Dirac delta function. In practice, of course, the pulse width used in the electronic equipment will be finite but still small compared to the time intervals during which $Q(t)$ changes appreciably. The point average is in general a function of both sets of co-ordinates: x_o, y_o, z_o and x, y, z.

As far as the function Q is concerned, it may be any flow variable, such as a velocity component, a velocity derivative or any square of the instantaneous velocity fluctuation; another is the instantaneous velocity product, such as the instantaneous value of the Reynolds stress.

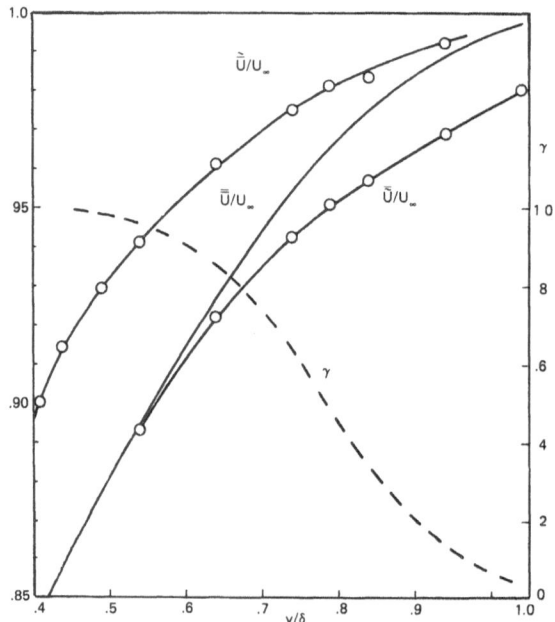

Fig. 8. Zone averages of the streamwise velocity component in a turbulent boundary layer. (Intermittency factor also given). From Kovasznay; Kibens & Blackwelder (1970).

Figs. 8 and 9 are taken from Kovasznay, Kibens & Blackwelder (1970), giving zone average of both the streamwise velocity component and of the stream-wise velocity fluctuation level (r.m.s.) in the turbulent boundary layer. The distribution of the intermittency factor is also given as reference.

Conditional point averages are especially useful to determine the flow distribution relative to the turbulent non-turbulent interface. Blackwelder and Kovasznay (1972) were able to reconstruct a qualitative picture of the interface bulges (Fig. 10). The composite picture is drawn for an observer moving with the average convection velocity of the turbulent "bulge" and the figure clearly indicates that the turbulent region has a whirl character and the outside potential flow rushes over the "bulges".

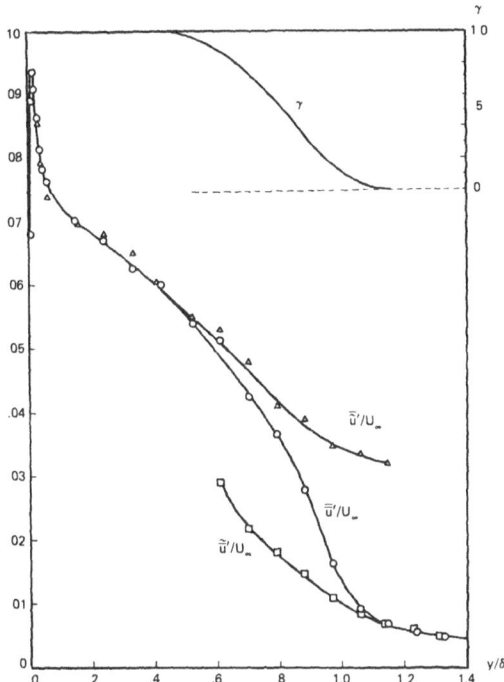

Fig. 9. Zone averaged turbulence level in a turbulent boundary layer. Intermittency factor given above. From Kovasznay; Kibens & Blackwelder (1970).

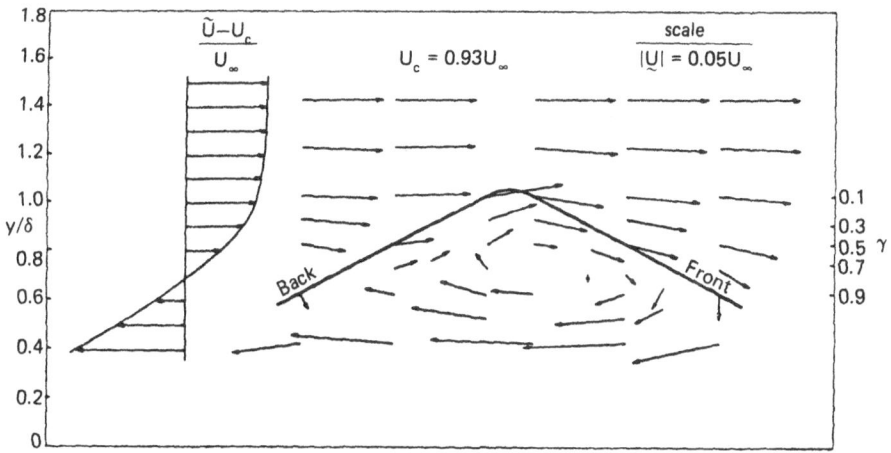

Fig. 10. Flow field around turbulent "bulges" reconstructed from conditional
point averages. Shown in coordinate system moving with average convection
velocity. From Blackwelder & Kovasznay (1972).

3.3 Multiple Conditional Sampling

Naturally, one may use not only one, but two or even more conditions to select the subset of instants when sampling should be performed. The only penalty to do so is that the more restrictive are the criteria, the fewer are the samples,
so that the total averaging time needed to obtain stable averages will keep increasing. Each condition is represented by a detector function $P_1(t)$, $P_2(t)$,
$P_3(t)$, etc. The multiple condition is equivalent to a single detector function

$$P(t) = P_1(t) \, P_2(t) \, P_3(t)$$

Naturally if each detector function is constructed from very narrow pulses, the
resulting coincidences will be few, and $P(t)$, will have only rare pulses.

A few interesting examples will be mentioned here. One is the double conditional sampling in a quasi-periodic phenomenon, e.g., in a Karman vortex
street behind an obstacle. The flow is sampled by a rotating arm mounted hot-
wire or "flying hot-wire", Coles (1978). The first sampling is essentially a periodic sampling that provides the accurate definition of the position of the
rotating arm, it defines the coordinates of the point in space where the flow
field is sampled. But, the second sampling is different. It fixes the appropriate
phase of the vortex street. The master signal for the second sampling is derived
from the nearly periodic reference signals taken from a stationary hot-wire
probe outside the vortex street. Coles (1978) was able to reconstruct the detailed flow in the wake of a cylinder or airfoil at high Reynolds numbers.

Another example of double conditioning is the use of two detector probes to
detect those moments when the turbulent non-turbulent interface crosses
both probes. Imaki (1968), was able to make conditional sampling where not
only the position but also the slope of the interface was thus specified.

It is appropriate to mention here that a large number of interface detectors
was used by Paizis & Schwarz (1974) to determine the convolutions of the turbulent non-turbulent interface and the statistics of the multiple valuedness of
the interface height over the solid wall.

4. Special Problems

There are extensions and modifications of the basic periodic or conditional
sampling technique that are worth mentioning separately. An especially astute
conditional sampling and averaging was performed by Browand (1975). He was
able to capture the different states of development in the vortex merger of a
two-dimensional mixing layer. Here the conditioning criteria were derived from

the signals of the two hot-wire probes placed outside of the shear layer. The vortex merging itself was identified by a shift in the relative phase difference observed on the two sides of the layer.

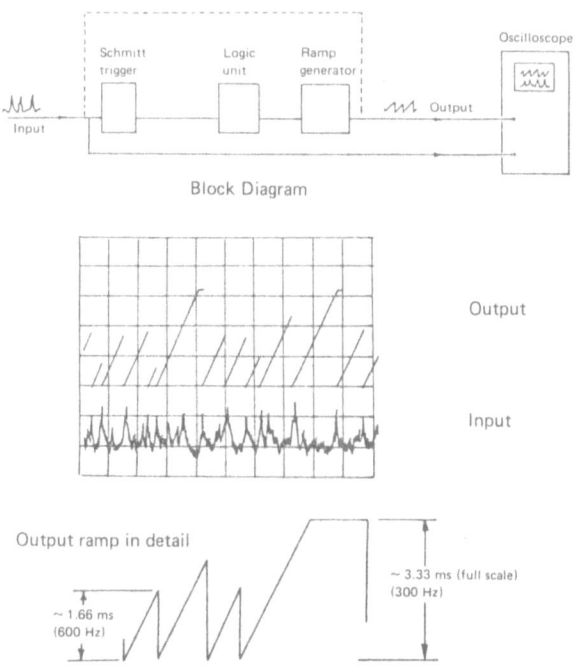

Fig. 11. *Sampling of recurrent spikes obtained in turbulent jet. From Lau &*
Fisher (1975).

Another conditional sampling type of detection was used by Lau & Fisher (1975). In the study of the structure of a turbulent jet, they observed passage of discrete vortices identified as "spikes". The spikes occurred in a somewhat random manner, and the hot-wire signal was converted to a "ramp" signal (random, sawtooth wave, Fig. 11). The recurrent sampling relative to the spikes and ensemble averaging gave results that were displayed as a function of time.

4.1 Iteration

The ultimate goal of using the conditional sampling ensemble averaging techniques is the recognition of some semi-permanent flow structures within the turbulent flow. Just three well known examples are cited. First: the large scale structures near the turbulent non-turbulent interface. Second: the "burst" and "sweep" sequence near the viscous sublayer in wall flow. Third: transition "spots" in laminar boundary layers.

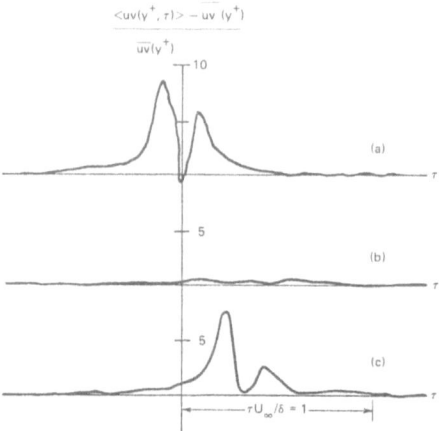

Fig. 12. Conditionally averaged Reynolds stress near wall (y_+ = 15). (a) x = 0 (b) Δx = δ/4 simple average (c) Δx = δ/4 corrected for random convection velocities. From Blackwelder & Kaplan (1976).

In all three cases, the identified structures can be detected and followed for a relatively long distance during their downstream journey. On the other hand, in addition to their uniform parallel translation, they also execute a random walk, so they are not found exactly at the location one would except them. Blackwelder and Kaplan (1976) show it for the local Reynolds stress near the viscous sublayer (y_+ = 15). It is surprising how well one can recover the travelling disturbance if the detection is corrected for random convection velocity (Fig. 12). The same effect is discussed in Blackwelder (1977) and a sample is shown in Fig. 13 .

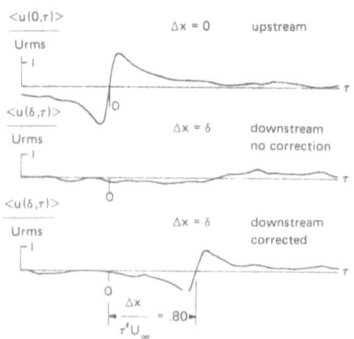

Fig. 13. Ensemble averaged streamwise velocity fluctuation near wall (y_+ = 15). From Blackwelder (1977).

Wygnanski (1978) describes the iteration procedure in the detection of transition spots within the fully turbulent layer. The procedure always consists of the following steps: first calculate the conditional average over the whole ensemble. Then each member is correlated with the ensemble average and the optimal time shift for each one is determined. Finally, using the realigned samples, the ensemble average is performed again. For a few iterations the sharpness of the results improves significantly.

4.4 Pattern Recognition

When one mentions pattern recognition, it immediately evokes the thought of flow visualization. In fact this is quite far from reality. Flow visualization gives information that is of a different nature than obtainable from probe signals (hot-wire, LDA, etc.). The latter are all providing time records in an Eulerian framework, while flow visualization shows the history of marked particles in a Lagrangian framework. When interpreting pattern recognition of probe signals, one must remember the difference between these two points of view, otherwise it is very easy to confuse the two types of information. Pattern recognition in this context involves the use of a computer to select typical events or flow structures recognized by some more complex criteria than simple conditional sampling.

The difference may be more quantitative than qualitative, e.g. Antonia (1972) first used rescaling of the time and the amplitude of hot-wire signals in the turbulent portion of an intermittent flow and obtained the average velocities and the mean square fluctuations during the passage of a normalized turbulent "bulge".

The best known work termed as pattern recognition is the result of a group cooperation, centered in Goettingen. Wallace, Brodkey & Eckelmann (1977), followed by Eckelmann and Nychas (1977) also Eckelmann (1978). They studied the burst in and near the viscous sublayer, and were able to recognize these events. They identified the burst, sweep sequence. The computer program consists of searching for the following scenario: in a full cycle first the streamwise u velocity gradually decreases and then rapidly increases again. A typical set of criteria is that the rapid increase after a decrease be at a rate at least twice as rapid as the original gradual decrease was. Using some pre-filtering and putting rather broad thresholds on the rate of increase and decrease, they were able to select and renormalize (change the amplitude and the time scale) such events. Finally they were able to account for about 65 % of the total time by using these recognized events.

In all recognition and detection methods, the important point is that one should not develop criteria that would result in the collection of a very small number of events because in that case there is a danger that one would find such "events" even in a completely random synthetic signal obtained by using a random noise generator. A very good verification of the soundness of the detection method is to produce a synthetic signal of the same r.m.s. value, power spectrum and probability density distribution as the hot-wire signal to be processed, then apply the pattern recognition detector circuit or computer program to it in order to see how many "events" it would "catch" in such a completely random signal. Such a detection level would serve as a reference or a control to see how efficient is the discrimination. In case of very rare events, one must have serious doubts about the value of the method because very rarely even some odd combinations can occur and that may not be statistically significant at all.

Pattern recognition certainly will be used in the future to recognize the passage of various large-scale structures but its success will depend largely whether or not a statistically significant portion of the total phenomenon can be accounted for by such recognized structures.

4.3 Conclusions

As one may observe, new signal processing techniques immediately became applied to the detection of phenomena in unsteady flows, especially in unsteady turbulent flows, although the actual details may vary from case to case. The field of signal processing has progressed very far from the original state when the measurement of two-point correlation and of a power spectrum was considered the most advanced technique.

The handling of periodic and intermittent flows is already well established as evidenced by the important results in the literature. The availability of advanced signal processing techniques made the study of turbulence in unsteady mean flows quite attractive. It may be that the greatest advance in the near future will be in the technologically important unsteady flows, e.g. reciprocating engines, starting flows or periodically modulated flows.

On the other hand, one must caution here against the over-elaboration of such detection systems because the more complex schemes necessarily introduce arbitrariness, and the results become dependent on too many parameters, especially of hidden parameters not sufficiently understood and not very well under rational control.

References and Bibliography

ANTONIA, R.A., 1972. Conditionally sampled measurements near the outer edge of a turbulent boundary layer. *J. Fluid Mech.* **56**, 1

ANTONIA, R.A. & ATKINSON, J.D. 1974. Use of a pseudo-turbulent signal to calibrate an intermittency measuring circuit. *J. Fluid Mech.* **64**, 679-699

ANTONIA, R.A.; PRABHU, A. & STEPHENSON, S.E., 1977. Conditionally sampled measurements in a heated turbulent jet. *J. Fluid Mech.* **72**, 455

BILGER, R.W.; ANTONIA, R.A. & SREENIVASAN, K.R., 1976. Determination of intermittency from the probability density function of a passive scalar. *Phys. Fluids* **19**, 1471-1474

BLACKWELDER, R.F. & KOVASZNAY, L.S.G., 1972, Time scales and correlation in a turbulent boundary layer. *Phys. Fluids,* **15**, 1545

BLACKWELDER, R.F. & KAPLAN, R.E., 1976. On the wall structure of the turbulent boundary layer. *J. Fluid Mech.* **76**, 89-112

BLACKWELDER, R.F., 1977. On the role of phase information in conditional sampling. *Phys. Fluids* **20**, S 232-242

BROWAND, F.K., 1975. "Ensemble-averaged large scale structure in the turbulent mixing layer", *Turbulent Mixing in Reactive and Nonreactive Flows* (Ed. S.N.B. Murthy), 316, Plenum Press, New York and London

COLES, D., 1978. On-line control and data processing for the flying hot-wire. *Proceedings of the Dynamic Flow Conference 1978,* Marseille and Baltimore

CORRSIN, S. & KISTLER, A.L., 1955. Free-stream boundaries of turbulent flows. NACA Rep. 1244, Washington, D.C.

ECKELMANN, H. & NYCHAS, S.G. 1977. Vorticity and turbulence production in pattern recognized turbulent flow structures. *Phys. Fluids* **20**, S 225

ECKELMANN, H., 1978. The structure of turbulence in the near wall region. *Workshop of Coherent Structures of Turbulent Bounday Layers,* Lehigh University & AFOSR, May 1-3, 1978

FIEDLER, H. & HEAD, M.R., 1966. Intermittency measurements in the turbulent boundary layer. *J. Fluid Mech.* **25**, 719-735

FUJITA, H. & KOVASZNAY, L.S.G., 1974. Unsteady lift and radiated sound from a wake cutting airfoil. *AIAA Jour.* **12**, 1216

HANDA, N. 1976. Ph.D. Thesis, Department of Mechanical Engineering, Nihon University, Tokyo, Japan (in Japanese only)

HEDLEY, T.B. & KEFFER, J.F. 1974. Turbulent/non-turbulent decisions in an intermittent flow. *J. Fluid Mech.* **64**, 625-644

HEDLEY, T.B. & KEFFER, J.F., 1974. Some Turbulent/non-turbulent properties of the outer intermittent region of a boundary layer. *J. Fluid Mech.* **64**, 645-678

HIRSCH, C. & KOOL, P. 1973. Application of a periodic sampling and averaging technique to flow measurements in turbo machines. Rep. V.U.B.-STR-4, Department of Fluid Mechanics, Free University, Brussels

HO, C.M. & KOVASZNAY, L.S.G., 1976. Sound generation by a simple cambered blade in wake cutting. *AIAA Jour.* **14**, 763

HUSSAIN, A.K.M.F. & REYNOLDS, W.C., 1970. The mechanics of an organized wave in turbulent shear flow. *J. Fluid Mech.* **41**, 241-258

IMAKI, K., 1968. Structure of the superlayer in the turbulent boundary layer I & II. *Institute of Space and Aeronautical Science, University of Tokyo,* Vol. 4, No. 3, 348-367; No. 4, 536-563. In Japanese with short English abstract

KAPLAN, R.E. & LAUFER, J., 1969. The intermittently turbulent region of the boundary layer. *Proc. 12th. International Congress of Appl. Mech.* Stanford, 236, Springer-Verlag

KARLSSON, S.K.F., 1959. An unsteady turbulent boundary layer. *J. Fluid Mech.* **5**, 622

KIBENS, V.; KOVASZNAY, L.S.G. & OSWALD, L.J., 1974. Turbulent-nonturbulent interface detector. *Rev. Sc. Instr.* **45**, 1138-1144

KOVASZNAY, L.S.G., 1949. Hot-wire investigation of the wake behind cylinders at low Reynolds numbers. *Proc. Roy. Soc.* A **198**, 74

KOVASZNAY, L.S.G.; KOMODA, H. & VASUDEVA, B.R., 1962. Detailed flow field in transition. *Proc. of the 1962 Heat Transfer and Fluid Mechanics Institute,* Stanford University Press

KOVASZNAY, L.S.G., 1968. "Should we still use hot-wires?" *Advances in Hot Wire Anemometry,* (Eds. W.L. Melnich and J.R. Weske) AFOSR Rep. 68-1492, 211, University of Maryland

KOVASZNAY, L.S.G.; KIBENS, V. & BLACKWELDER, R.F., 1970. Large scale motion in the intermittency region of a turbulent boundary layer. *J. Fluid Mech.* **41**, 283

KOVASZNAY, L.S.G., & FUJITA, H., 1971. Unsteady boundary layer and wake near the trailing edge of a flat plate. *IUTAM Symposium on Unsteady Boundary Layers.* 806-833, Laval University Press, Quebec

KOVASZNAY, L.S.G.; FUJITA, H. & LEE, R.L., 1974. "Unsteady turbulent puffs", *Advances in Geophysics* (Ed. F.N. Frenkiel) Academic Press, New York, **18B**, 253

KOVASZNAY, L.S.G., 1977. "Wake cutting experiments". *Turbulence in Internal Flows* (Ed. S.N.B. Murthy) Hemisphere Publishing Company. Washington, D.C. 463

KUO, A.Y.S. & CORRSIN, S. 1971. Experiments on internal intermittency and fine-structure distribution function in fully turbulent fluid. *J. Fluid Mech.* **50**, 285-319

LaRUE, J.S. 1974. Detection of the turbulent-nonturbulent interface in slightly heated turbulent shear flows. *Phys. Fluids,* **17**, 1513-1517

LAU, J.C. & FISHER, M.J. 1975. The vortex-street structure of "turbulent" jets. Part I. *J. Fluid Mech.* **67**, 299-337

LU, S.S. & WILMARTH, W.W., 1973. Measurements of the structure of the Reynolds stress in a turbulent boundary layer. *J. Fluid Mech;* **60**, 481-511

NAKAGAWA, K.; HAYAKAWA, M. & KOBASHI, Y., 1977. Boundary layer transition in unsteady flow. *Trans. JSME* **43**, 1005-1014

OBREMSKI, H.J. & FEJER, A.A., 1967. Transition in oscillating boundary layer flows. *J. Fluid Mech.* **29**, 93-111

OKIISHI, T.H., 1978. Measurement of the periodic variation of turbomachine flow fields. *Proceedings of the Dynamic Flow Conference 1978,* Marseille and Baltimore

OSHIMA, K.,; KOVASZNAY, L.S.G. & OSHIMA, Y., 1978. "Sound emission from burning puffs". *Structures and Mechanisms of Turbulence,* (Ed. H. Fiedler) Lecture Notes in Physics, **76**, 219-230. Springer Verlag, Berlin, Heidelberg and New York

PAIZIS, S.T. & SCHWARZ, W.H. 1974. An Investigation of the topography and motion of the turbulent interface. *J. Fluid Mech.* **63**, 315-343

RAO, K.N.; NARASIMHA, R. & BADRI NARAYANAN, M.A., 1971. The "bursting" phenomena in a turbulent boundary layer. *J. Fluid Mech.* **48**, 339-352

SATO, H. 1970. An experimental study of non-linear interaction of velocity fluctuations in the transition region of a two-dimensional wake. *J. Fluid Mech.* **44**, 741-765

SUNYACH, M. & MATHIEU, J. 1969. Zone de melange d'un jet plan. *Int. Heat & Mass Transfer,* **12**, 1679

TOWNSEND, A.A., 1949. The fully developed turbulent wake of a circular cylinder. *Austr. J. of Sci. Res.,* A **2**, 451-468

WALLACE, J.M. BRODKEY, R.S. & ECKELMANN, H., 1977. Pattern-recognized structures in bounded turbulent shear flows. *J. Fluid Mech.* **83**, 673-693

WYGNANSKI, I.; SOKOLOV, M. & FRIEDMAN, D. 1975. On transition in a pipe. Part 2. The equilibrium puff. *J. Fluid Mech.* **69**, 283-304

WYGNANSKI, I., 1978. On the possible relationship between the transition process and the large coherent structures in turbulent boundary layers. *Workshop in Coherent Structures of Turbulent Boundary Layers,* Lehigh University & AFOSR, May 1-3, 1978.

Pattern Recognition,
a Means for Detection of Coherent Structures
in Bounded Turbulent Shear Flows
by

Helmut Eckelmann, James M. Wallace[*]
and Robert S. Brodkey[**]
Max-Planck-Institut für Strömungsforschung
D 3400 Göttingen, Fed. Rep. of Germany

Abstract

In the last decade, many publications have shown that there is now an abundance of evidence for the existence of coherent structures in turbulent shear flows. The authors have worked out a technique to recognize these structures in the turbulence signals. A digital computer was programmed to extract and ensemble average the patterns which repeatedly occur in the streamwise fluctuating velocity signal, u, of a bounded turbulent shear flow. It is believed that these patterns are the signatures of the coherent structures. Only simple criteria were used to recognize these patterns, which can be characterized by a relatively weak deceleration of the flow followed by a strong acceleration. The acceleration was, on the average, over twice as strong as the deceleration in the region near the wall. In the region where turbulence production is most intense, $10 \leqslant y^+ \leqslant 30$, the patterns occurred in more than 65 % of the total sample of the u-signal. The lengths of such patterns varied over quite an extensive range (1:25). Thus to obtain a meaningful ensemble average, a normalizing scheme had to be used. This scheme normalized each pattern to an arbitrarily chosen length in time. During the same signal interval that the u pattern was recognized, the simultaneously measured signals, v, w, uv, uw, $\partial U/\partial y$, $\partial U/\partial z$, etc. were also normalized and ensemble averaged with no predetermined criteria applied to these signals.

The new pattern recognition technique can detect and obtain ensemble averages of simple wave forms that are characteristic of a sequence of events producing Reynolds stress in the wall region. The pattern recognition technique also provides insight into the vorticity dynamics of organized structures. By using two probes simultaneously at different locations, this technique can give additional information about the spatial extension of the coherent structures.

[*] University of Maryland, USA, [**] Ohio State University, USA.

I. Introduction

Given that coherent structures exist in turbulent shear flows, one is interested in the most optimal means for their detection. It has become clear that overall average values cannot provide the details necessary for understanding the mechanism of turbulent shear flows. By overall measurements we mean such values as the mean velocity, the fluctuation rms value, and the spectra or probability density distribution of these. There is simply not enough information available in such measurements to allow one to extract the details necessary to picture what is happening with regard to individual events in the flow. Considerably more information is available in the instantaneous signals; however, this is unfortunately too much information to process. Therefore, what is needed is something in between to reduce the data to a more manageable degree, and which corresponds to the events which occur in the flow.

The signal processing technique that we have developed to obtain detailed information about individual events that exist in the flow field involves the use of pattern recognition. This technique was first introduced for use in detecting individual events in turbulent shear flows by Wallace, Brodkey, and Eckelmann (1977). Further applications of the technique were presented by Eckelmann, Nychas, Brodkey, and Wallace (1977) and Kastrinakis, Wallace, Willmarth, Ghorashi, and Brodkey (1978). In these articles the pattern recognition technique was developed and applied to turbulence signals that were obtained in bounded turbulent shear flows. The main emphasis of these efforts was to obtain information to elucidate the nature of the structures that were believed to exist in these flows. Furthermore, an effort was made to integrate the information obtained with earlier visual results. In the sequel, we will give some example results from the application of this technique; but, in the present presentation we wish to emphasize the application of the pattern recognition technique to the signals obtained from complex probe configurations.

II. The Pattern Recognition Technique

Fig. 1a is an example segment of the u signal. This is a part of the raw digitized data with every second point shown. One can see the striking characteristic of a weak deceleration and the strong acceleration. In Wallace at al. (1977), the nature of the signal was verified by measurements of the skewness factors of both u and its time derivative. In this reference are also given the details of the pattern recognition technique. For the present purposes, only the briefest of details are presented here. Several key ideas were incorporated into the technique and it is those ideas that we wish to emphasize.

The first of these novel ideas was the concept of using a short-time temporal average, which we will refer to as TPAV. This is an average taken over the

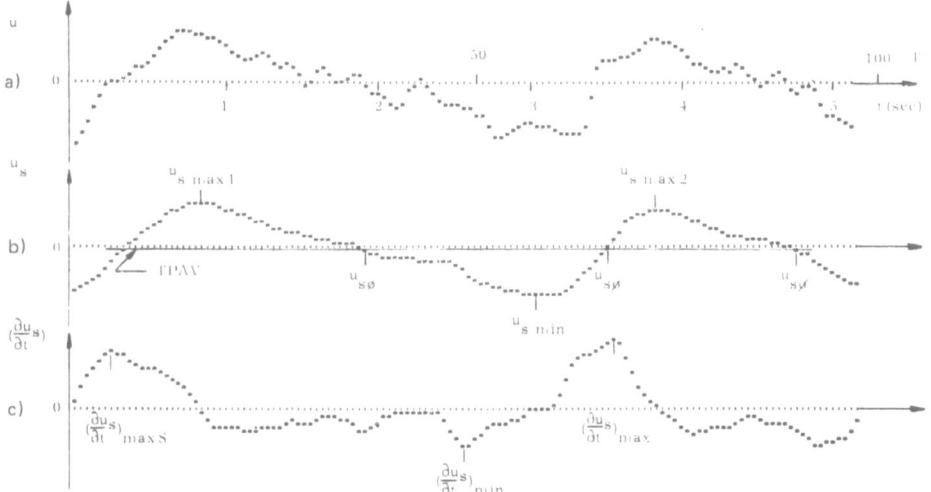

Fig. 1. a)u signal (raw data) digitized with every second point shown.
b) smoothed u signal.
c) time derivative of the smoothed u signal.

length of the pattern. Since it is necessary to know the pattern in order to determine the TPAV, clearly an iterative process is necessary. *Fig. 2* is a sketch
that illustrates the use of TPAV in the recognition program. As can be seen
from the sketch, TPAV is taken to be the period from one maximum in the u
signal time derivative to the next. Note that if the long-time mean velocity as
noted in Fig. 2 had been used, these patterns would both have been missed.

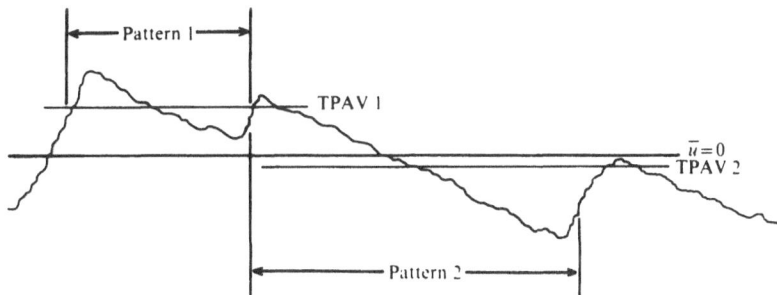

Fig. 2. Sketch illustrating the use of TPAV in the recognition program.

A second key idea in the pattern recognition scheme involved a decision on
when a recognized event should or should not be accepted. We wanted a technique that did not involve an arbitrarily chosen discriminator level. Since it can
be seen in Fig. 1a that the deceleration slope is much less in absolute value than

the acceleration, we simply require that the positive slope be greater in absolute value than the negative slope. These then became the accepted events to be analyzed and all others were also analyzed but called unaccepted events. In Wallace et al. further tests are cited where it was required that the positive slope be either two or three times greater than the negative slope.

Another key idea was that the recognition was done on a smoothed u signal. The smoothed signal was used only for the recognition process; the raw data itself was used when ensemble averages were formed. Smoothing of the recognized signal is necessary so that the pattern recognition technique will not recognize jitter in the turbulence signal. In Wallace et al. the effect of different degrees of smoothing was tested and reported. *Fig. 1b* shows the smoothed signal corresponding to the raw data in *Fig. 1a,* and *Fig. 1c* shows the corresponding time derivative signal. We must emphasize that it is the unsmoothed signals that are ensemble averaged and not the smoothed signals on which the pattern recognition operates. Furthermore, this was done for both accepted and unaccepted patterns, since we wished to study both.

Since the patterns that are recognized can vary greatly in their length, one should normalize them to an arbitrary length. If this is not done, then short events would be arbitrarily averaged into only some small part of the large event. Consequently, all events small or large, were normalized to the same unit length.

A final point to emphasize is that the recognition criteria and normalization technique actually set the shape of the averaged recognized u pattern. However, any other signal measured simultaneously with the u signal has no criteria imposed upon it which predetermines its shape. This is an important feature of our pattern recognition procedure.

The computer flow diagram for the pattern recognition technique is shown in *Fig. 3.* The flow diagram is self-explanatory when used with the notation provided in Figs. 1b and 1c. Complete details can be found in Wallace et al. One particularly difficult point is illustrated in the sketch of *Fig. 4* which shows the iteration procedure used in the program for finding sub-patterns.

III. Pattern Recognized u, v, w, and uv Signals

As an example of the potential of the pattern recognition technique, *Fig. 5* shows the normalized and ensemble averaged u, v, w, and uv patterns in a wall bounded turbulent flow field at a non-dimensional wall distance of $y^+ = 15$. There are several characteristics that are noteworthy. For example, although

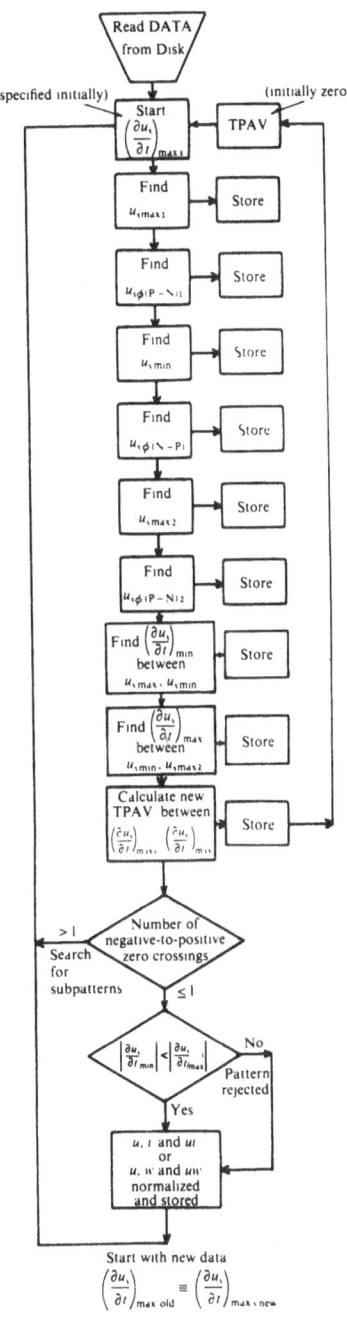

Fig. 3. Flow diagram showing the operation of the recognition program.

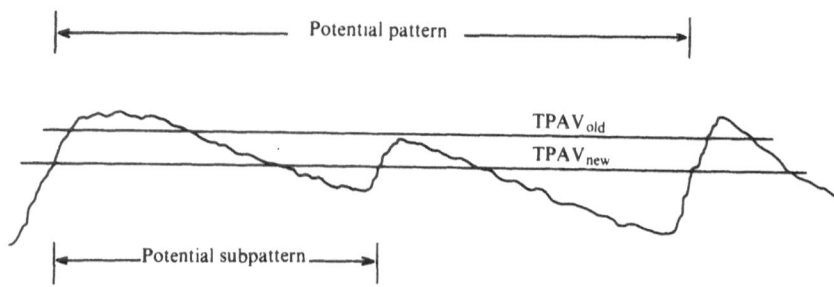

Fig. 4. Sketch illustrating the iteration procedure in the recognition program for finding subpatterns.

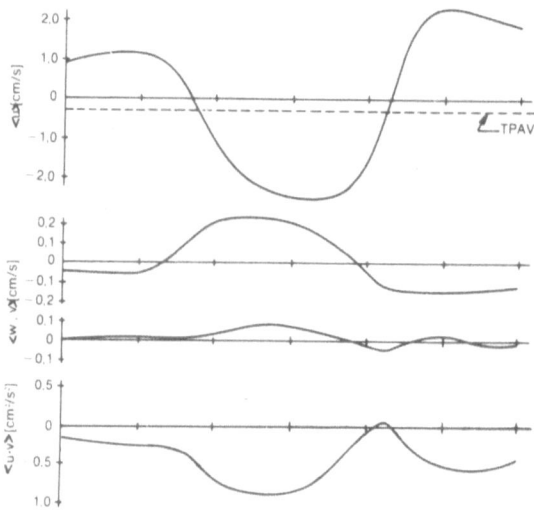

Fig. 5. The normalized and ensemble-averaged u, v, w, and uv patterns.

the u pattern is predetermined, we see that the acceleration is considerably larger than the deceleration. Furthermore, the v signal is largely out of phase with the u pattern, which results in large negative pulses in the uv pattern. Further details are again given in Wallace et al. Remember that these patterns are normalized in length and that the abscissa scale is in arbitrary time units. In reality there is a wide distribution of lengths as shown in *Fig. 6.*

In order to verify further the pattern recognition technique, Wallace et al. tested the method on computer-generated random signals that were designed to simulate a turbulence signal. The test was done on pairs of both uncorrelated and correlated signals. For uncorrelated signals, the ensemble averaged v and uv patterns were zero across the entire normalized time interval.

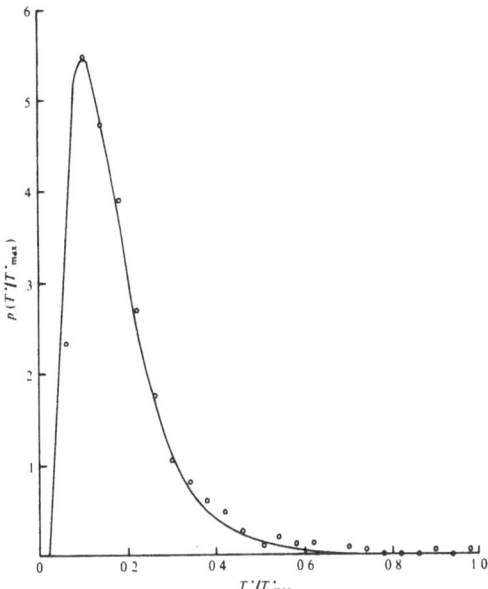

Fig. 6. Probability density distribution of the normalized pattern length.

IV. Signals from Complex Probes

The signals from an X- or V-probe provides only a limited amount of information. If one wants information about other stresses or about the vorticity field, much more complex probe configurations are necessary. However, when such complex probes are utilized, a great deal of precaution must be exercised to be sure that the probes are providing the information desired. One would not expect simple overall average criteria such as rms fluctuations, spectra, etc., to be adequate tests for a probe that is being used to provide ensemble averages by pattern recognition. Clearly, a more detailed verification of the adequacy of the probe is necessary. One method to accomplish this is to use the pattern recognition technique on separate probes that reproduce a part of the complex probe that we wish to test. To illustrate this in some detail let us take two examples that involved efforts at measuring the vorticity field in a bounded shear flow.

5-Sensor Probe

The 5-sensor probe configuration described by Eckelmann et al. (1977) is shown in *Fig. 7*. Sensors 4 and 5 form an X-probe configuration used to obtain the u and v components, sensors 2 and 3 form a V-configuration to obtain the u and w components, and sensor 1 was oriented into the flow to measure the u component. Thus, the u component was measured at three individual locations, which provided spatial gradients in the u signal. Basically, within the small volume of the probe we were able to measure the three instantaneous components in the velocity and gain enough derivative information (also requiring Taylor's hypothesis) to provide both the spanwise and normal components of the vorticity. In addition the instantaneous turbulence production, which is the product of the Reynolds stress and the spanwise velocity gradient, $uv(\partial U/\partial y)$ could also be obtained.

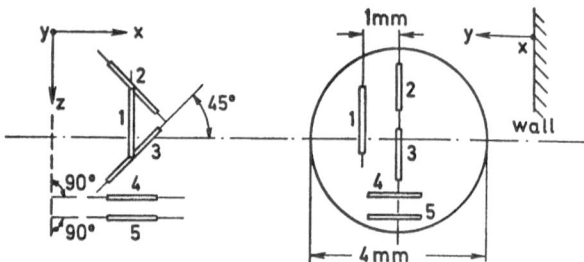

Fig. 7. Sketch of five-sensor probe. V probe, 2 and 3; X probe, 4 and 5; U probe, 1.

All of the normal tests that could be performed to check the adequacy of the probe were done. For example, tests showed that there was no thermal interference between the sensors. A static calibration of the sensors was adequate because experiments with similar hot-film sensors when dynamically calibrated in oil showed that the static and dynamic frequency responses were identical in the range for which the data was obtained (Hofbauer (1975)). However, all of these tests, including the values for the mean velocity and rms fluctuations of the various fluctuating components, do not prove that the probe is really adequate for use in pattern recognition work.

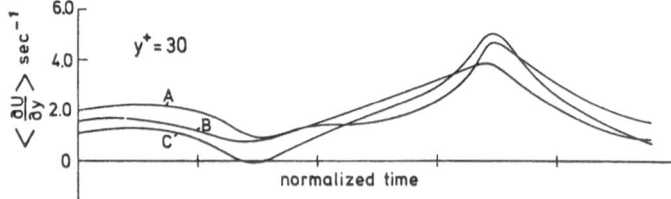

Fig. 8. Pattern recognized $(\partial U/\partial y)$ at $y^+ = 30$ for A: 30,000 data pairs from gradient probe, B: same as A, but 60,000 data pairs, C: 128,000 data pairs from five-sensor probe.

The means used to check the adequacy of the 5-sensor probe was to compare results obtained with the 5-sensor probe with similar results obtained with probes that were simpler in design and easier to use that would at least in part give some of the same information. For example, data can be taken using a simple gradient probe which consists of two single sensors separated by the same distance as in the 5-film probe (e.g., between sensors 1 and 2, 3). These can be directly compared with the same gradient obtained from the 5-sensor probe. However, since the measurements cannot be done simultaneously with the two probes in the same location, the instantaneous velocities cannot be compared. But since the results are going to be used for pattern recognition, the pattern recognized gradients from the two sensors could be compared. Such a comparison is given in *Fig. 8* for two sets of data from gradient probes (A and B) and one full set of data from the 5-sensor probe (C). As indicated by Eckelmann et al., measurements of velocity gradients are extremely difficult when the separation of the sensors is of the order of the Kolmogorov microscale, as they were for the experiments reported there; but to obtain true velocity gradients, the separation distance must be about of this order.

The problem in obtaining an accurate measurement of the gradient is the result of trying to take small differences between two large values of the u velocities at the two points, especially when in two of the cases the u velocity is calculated from either an X- or V-probe configuration. Thus it should not be surprising that the absolute values of the velocity gradients measured over such small scales are difficult to obtain accurately. The differences in the absolute values of the mean velocity gradients, $\partial U/\partial y$, measured were large indeed. However, when these data were processed using the pattern recognition technique, it was found that the *shapes* of the patterns for the streamwise velocity gradient were very similar, although not the same in magnitude. As can be seen in Fig. 8, at any point along the normalized axis of the pattern recognized gradients, the difference between the ensemble averaged values can be quite large; however, the shapes of the patterns are very similar. Remember that in pattern recognition analysis, one is concerned primarily with the shapes of the signals being recognized because these shapes give us insight into the turbulent mechanism. Errors in the absolute amplitudes are of less concern since they do not significantly effect the shape of the signals. Remember also that the absolute amplitudes have less meaning for pattern recognized signal patterns, since signals of all different sizes have been normalized in time and added together.

A similar analysis can be done by using the 5-film probe to obtain pattern recognized X-probe information and comparing that to the pattern recognized results that were obtained with a simple X-probe and previously presented as Fig. 5. This comparison is presented in *Fig. 9* where the pattern recognized ensemble averaged u, v, and uv from the 5-sensor probe are shown as the solid lines and the previous X-probe results as dots. The data from the 5-sensor probe compares well with that obtained with the earlier single X-probe.

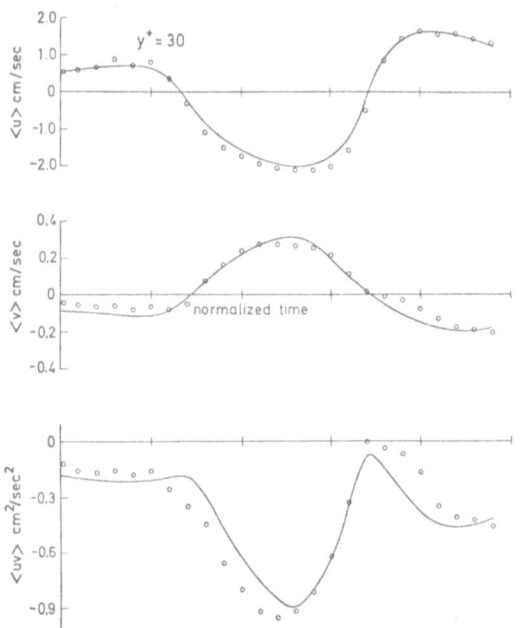

Fig. 9. Pattern recognized u, v, and uv from five-sensor probe (solid curves) compared to previous X probe results (o).

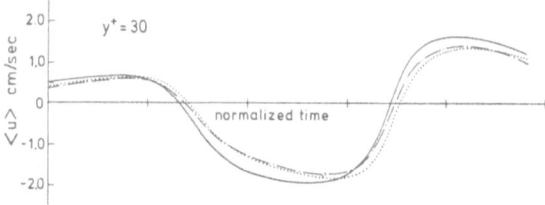

Fig. 10. Pattern recognized u for three simultaneous u signals available from the five-sensor probe: ——— pattern recognized signal from X probe; ······ from V probe; —·—·— from single U probe.

In *Fig. 10* are shown the pattern recognized results for the three simultaneous u signals that can be obtained from the 5-sensor probe. The signal from the X-probe configuration was pattern recognized and is shown as a solid line. The two additional u signals that are simultaneously ensemble averaged (but not pattern recognized) are also shown. The patterns are very similar, although there is a slight shift in phase indicating that the probe is small enough so that all the sensors do respond to essentially the same structures in the flow.

Axial Vorticity Probes

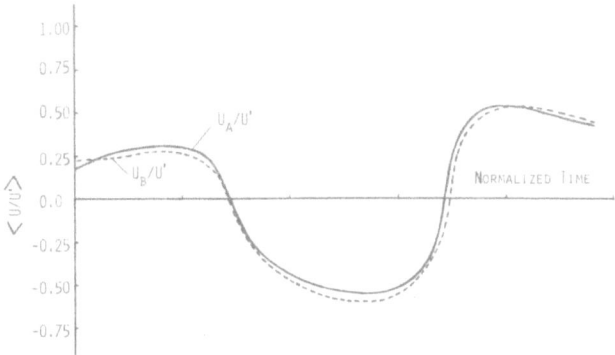

Fig. 11. Pattern recognized axial velocity normalized with rms values: solid line recognized on Probe A; dashed line recognized on probe B.

Kastrinakis et al. (1978) reported on results that were obtained with a pair of probes whose design was originally proposed by Kovasznay (1954) and which are sensitive to both the u component of the velocity and the streamwise vorticity, ω_x. Details about the nature of this probe can be found in Kastrinakis, Eckelmann, Wallace and Nychas (1978), which is a paper being presented at this conference and which will be published elsewhere. When using two probes, either u signal pattern can be recognized and the simultaneously occurring other signals from the probe can be ensemble averaged. *Fig. 11* shows the normalized and ensemble averaged recognized pattern obtained from each of the probes in this work. Note that they are very similar to each other and similar to those that have been reported previously in Figs. 5 and 10. Clearly, the comparison shows that this pattern is a distinguishing characteristic of the flow, since the two flow systems were considerably different. *Fig. 12* gives the ω_x-pattern that was obtained from the same probe for which the u-pattern recognition was performed.

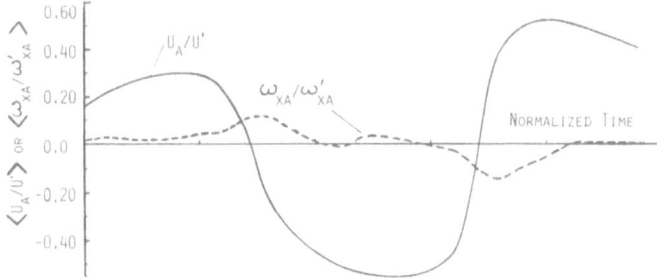

Fig. 12. Simultaneously averaged axial vorticity (normalized) from same probe as pattern recognized axial velocity.

V. Discussion

Pattern recognition is a technique that has been well documented in the references cited herein. In the present article, the emphasis has been on the use of the pattern recognition technique to verify the adequacy of signals obtained from complex probe configurations. It has been shown for two examples, the 5-sensor probe and the axial vorticity probe, that these probes can provide adequate pattern recognized signals when one realizes that it is only the shape of those signals that is of concern. Once this has been demonstrated, the probes and the pattern recognition technique can be used for further investigation of the mechanism that is believed to be occurring in bounded turbulent shear flows.

Indeed, the articles already cited in this paper were directed towards a better understanding of the mechanism of turbulent shear flows. The analysis by Wallace et al. (1977) was carried further than that shown in Fig. 5 by incorporating the quadrant splitting technique along with pattern recognition. Eckelmann et al. (1977) investigated both the spanwise and normal vorticites as well as the instantenous production term. They also used quadrant splitting in conjunction with the pattern recognition technique. Furthermore, they deduced a model for the flow by using extensively the quadrant split and pattern recognized information for the u velocity gradients. Finally, Kastrinakis et al. (1978) showed that useful information with regard to the axial vorticity can be obtained from a pair of probes. They also pattern recognized the u signal on one probe and simultaneously ensemble averaged the u signal on a probe at different positions in the z direction and verified the z^+ separation 100.

References

Eckelmann, H., Nychas, S.G., Brodkey, R.S. and Wallace, J.M., 1977 *Phys. Fluids,* **20**, S225

Kovasznay, L.S.G., 1954, Physical measurements in gas-dynamics and combustion. Princeton University Press, Princeton, New Jersey

Kastrinakis, E.G., Wallace, J.M., Willmarth, W.W., Ghorashi, E. and Brodkey, R.S., 1978 *Lecture Notes in Physics* **75**, 175, Springer Verlag Berlin, Heidelberg, New York

Hofbauer, M., 1975, Diplomarbeit, Georg-August-Universität, Göttingen (also Max-Planck-Institut für Strömungsforschung Report No. 118/1975)

Wallace, J.M., Brodkey, R.S. and Eckelmann, H., 1977 *J. Fluid Mech.* **83**, 673.

Pattern Recognition of Coherent Eddies

by

Ron F. Blackwelder
Department of Aerospace Engineering
University of Southern California
Los Angeles, California 90007

Abstract
Within the past couple of decades, considerable effort has been devoted to-wards describing turbulent flow fields as a random distribution of coherent eddies (i.e., patterns). The motivation for this approach is to explain the experimental observation of definite flow patterns and to obtain a more tractable analytical formulation of the flow field. Simultaneously the field of mathematical pattern recognition has been developed from earlier work in communication and detection theory. Although it most often finds application in analyzing photographs, written alphabetic characters and other two-dimensional figures, the mathematical tools are much more universal and can be utilized to study transducer outputs and other electronic signals. A brief introduction to pattern recognition is given and its application to the detection of coherent eddies in turbulent flow fields is outlined.

Pattern Recognition
The capability of recognizing and classifying different patterns is one of the most fundamental skills acquired in our learning process. Considerable effort in our formative years is devoted to recognizing different objects and events. Although the details of how this is mentally accomplished are unclear, we can represent this learning process as the ability to perceive and categorize a set of parameters which accurately describe the object or pattern. Some of the basic parameters that one uses are the size, shape, color, etc. However, armed with these crude parameters, it is impossible to discern the difference between most physical objects. Thus, other quantitative parameters such as relative shape, sound, types of motion, etc., must be considered before classification can be obtained. In an example closely related to fluid mechanics, if we are asked to define a cloud, we would use features which are most efficient in recognizing it; e.g., color, size, location, etc. But if we are asked to describe the different types of clouds, we must be more exacting and use additional parameters such as length scales, aspect ratio, height, etc.

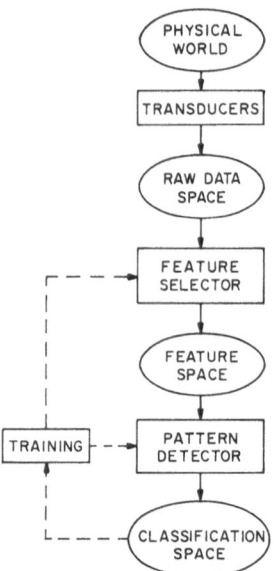

Fig. 1. Flow chart representation of a pattern recognition scheme.

To begin describing pattern recognition, means for acquiring and handling information must be established. This proceeds as shown in Fig. 1. The mathematical theory of pattern recognition assumes that of an infinite number of variables that can be measured, a finite number of transducers sample M of these variables. The choice of the measured quantatives is usually a compromise between the desired variables and easily measured ones. These variables are usually obtained by sampling M different quantatives at each point in time. Depending upon how the series and/or parameters are chosen, the M variables may not be completely independent because of a correlation between the variables. Indeed, often this correlated relationship between parameters defines the desired pattern. Mathematically each set of M points can be considered as a vector in an M-dimensional raw data space. This space forms a basis for choosing the features that best describe the patterns of interest.

The feature selector in Fig. 1 is used to choose those characteristics that are the most desirable and efficient in describing the patterns. A finite number, N, of these characteristics, called features, are used to define the N-dimensional feature space. In practice, the feature selector is a digital algorithm, or its analog equivalent, constructed to compute the desired features. The design of the feature selector is the most important aspect of the pattern recognition and presents the greatest challenge. Essentially one desires to retain the minimum number of features without sacrificing the loss of essential information so that correct classification of the patterns is still possible. To accomplish this requires a good deal of physical insight and experience to judge which features are optimum.

Often the M dimensional raw data space contains information that is redundant or is irrelevant to the problem of interest. If the raw data consists of visual observations, e.g., photographs, films, etc., the amount of information is very large although the information density is often low. For recognizing patterns, much of this data is repetitious and/or unnecessary; for example, the outline of an object is often sufficient to identify it without knowing its internal details. In this case, the feature selector should be designed to discard the excess information and retain only the relevant features. Then the number of features, N, would be significantly smaller than the dimension of the raw data space, M, which greatly reduces the difficulty and computation time. Alternatively, the information in the data space may consist of a single time series of t_m points. Depending upon the sought pattern, this series could be treated as one-dimensional with M realizations, or as groups of M points forming a multi-dimensional space but having fewer time points. In this case the feature selector should seek to emphasize those aspects of the time series that are most relevant in describing the pattern, e.g., high localized energy, strong coherence, etc.

The pattern detector is the final algorithm which categorizes the features into the K different classes of patterns or events. Mathematically, it used an N−1 dimensional surface in the feature space so that each different pattern is ascribed to a distinct region in that space. If there are K possible patterns, then the classification space has K dimensions and the pattern detector maps the N-dimensional vector in feature space into the appropriate pattern. The simplest classification space is one in which we are seeking a given event or the lack of it. For example, in the interfacial region of free turbulent shear flows, we may be interested in knowing only if the flow is turbulent or nonturbulent as studied by Townsend (1949), Corrsin and Kistler (1955) and others. On the other hand, four different types of flow patterns were described by the pattern detector of Wallace, Eckelmann and Brodkey (1972) and Willmarth and Lu (1972).

Training is a process by which the feature selector and/or pattern detector are "taught" to recognize a set of patterns. This is accomplished by using a data set with known classifications in a feedback mode; i.e. the embedded patterns have been determined a priori and this information is used to "teach" or "train" the machine to make the correct decisions. Several different mathematical techniques exist to diagnose which of the features are most informative in the detection process. In its most sophisticated version, the computer can study the data and train itself. This involves dynamic programming and is evident in the advanced versions of computer games, such as chess. Usually a human operator is injected into the loop to study the results and modify the algorithms as required for more efficient recognition of the patterns. The detailed description of these topics can be found in Hancock and Wintz (1966), Nilsson (1970), Andrews (1972) and other standard texts on the subject.

Identifying Coherent Eddy Structures

Different flow fields are known to often contain various coherent motions. Perhaps the best known example of such eddies is the Karman vortex street behind a circular cylinder. Many different types of eddies are known to exist and, hence a general definition is difficult. Nevertheless, for purposes here, a coherent eddy structure will be defined as a volume of fluid occupying a confined region in space and time which has distinct phase relationships between the variables associated with its constituent parts. These structures are usually vortical; however, the pressure fields associated with the eddies can extend their influence over considerable distances and into irrotational regions. The coherent structures must satisfy the equations of motion to some suitable approximation. Since they must also conform to the boundary conditions of the flow field, different types of eddies exist in different flow fields. There is no guarantee that a single eddy structure is sufficient to fully describe a flow field. Two or more coherent events may be required depending upon the initial conditions, Reynolds number, etc.

To design and implement a pattern recognition scheme, the variables most descriptive of the eddy must be ascertained. Often the desired information cannot be obtained and, hence, other sensors and signals must be used to recognize the eddies. For example, the vorticity may be the most descriptive variable, but it is difficult to measure with sufficient spatial resolution. Hence, a compromise must be reached in which different variables, such as one or more components of the velocity are used. Tracer gases and other passive contaminents such as temperature or dye are often used. Since a turbulent eddy involves definite phase relationships in space and time, these should be considered when choosing the number of transducers to be used and their location in space. If they are separated by distances significantly smaller than the length scale of the eddy in that direction, they will probably measure redundant information. On the other hand, if they are separated by too great a distance, the phase relationship between the measured variables may be so random, that they are independent and no eddy structure can be detected. To reduce the computation, it is desirable to measure quantities that can be used directly as features. This is often impossible. However, the measured quantities should be chosen so that the feature selector is as simple as possible. Otherwise the pattern recognition scheme may become unduly complicated and obscure the detected structure. Examples of features that require minimal computation include simple filters, derivatives, short time average, etc. Finally, since training is often used, the acquisition and utilization of the additional information required for training must be considered in designing a pattern recognition scheme.

Detection of Turbulent/Non-Turbulent Fluid
The most widely developed use of detection techniques in experimental fluid mechanics has been to discriminate between turbulent and irrotational fluid as found in unbounded turbulent flow fields. Corrsin (1943) first noticed that the turbulent zones in a jet were clearly separated from the non-turbulent regions by a thin interface. This phenomenon is now known to characterize the interfacial region of all free shear flows. Townsend (1949) developed an analogue circuit capable of discriminating between these two types of flow in a wake and was able to measure the intermittency, i.e. the per cent of time that the flow was turbulent. Subsequent studies by Corrsin and Kistler (1955), Bradbury (1964), Demetriades (1968) and others have extended this type of measurement to boundary layers and other flows at various Reynolds and Mach numbers. The concept of conditional sampling was introduced by Kaplan and Laufer (1969) using digital techniques and Kovasznay, Kibens and Blackwelder (1970) with analogue methods. Using a simplified version of pattern recognition, they detected the turbulent fluid and obtained averages of the flow field variables separately in the turbulent and irrotational zones to study the large scale eddies. Numerous other techniques have since been devised and used to study the same phenomenon and to extend this type of exploration to other free shear flows.

The determination of the intermittency at a point in space has typically utilized the velocity signal obtained from one or two transducers. These velocity signals are used to define the raw data space as discussed earlier and from it various features could be computed. Kaplan and Laufer (1969) used the streamwise velocity and calculated the short time average of the temporal derivative to derive a single feature. Kovasznay et al. (1970) computed a spatial and temporal derivative with filtering to obtain two features. In both cases, the pattern detector was designed so that whenever the features exceeded a threshold, the flow field was considered to be turbulent. In these cases, a training technique was used that visually compared the output of the detection scheme with traces of the raw signals.

Since this early work, many experimenters have utilized different transducers to measure other characteristic signals. Hedley and Keffer (1974) and others have used two simultaneous velocity components to form a detection signal and Kibens et al. (1974) have used a vorticity component. The addition of a slight amount of heat has proven useful in flow fields where it can be used as a passive contaminant to mark either the turbulent or non-turbulent fluid; e.g., Sunyach (1971) in a mixing layer, LaRue (1974) in a wake, Antonia, Prabbu and Stevenson (1975) in a jet, Chen (1975) in a boundary layer and others. When properly introduced, the temperature signals are often so free of noise that they can be used directly as a feature rather than requiring additional data processing.

To discuss the turbulent/non-turbulent recognition problem, consider the probability density function of a variable q_1 shown in Fig. 2. This bimodal distribution could possibly have been attained in an interfacial region which has a higher value of q_1 in the turbulent zone than in the irrotational region. The mathematical theory of pattern recognition does not require that there be bimodal distributions and they are usually not found, However, if one or more features can be obtained that have such a distribution, a clearer demarcation between the different patterns results with lower errors of discrimination. The total probability density is represented by

$$p = \gamma p^t + (1-\gamma) \, p^n$$

where p^t and p^n are the densities in the turbulent and non-turbulent regions and γ is the actual intermittency at this spatial location. All of the probability densities have been normalized so their integrals are unity. The finite width of p^n is assumed due to background noise in the system whereas the width of p^t is due to the turbulent fluctuation of q_1. In general, features should be chosen so that the distance between the peaks (or mean values), called the interclass distance, is as large as possible. In addition, the width of each distribution, as measured for example, by its rms value, is called the intraclass distance and should be as small as possible. In the irrotational flow field, velocity fluctuations are possible which give a relatively large intraclass distance to p^n. For this reason a component of vorticity or a passive contaminant provides a better feature than the velocity signal.

Given the underlying probability distributions in Fig. 2, a simple pattern detector would be a threshold value set at the location q_D, as shown. Any threshold will result in some statistical error in the decision process. If the distributions are known, the statistical error associated with false classifications can be

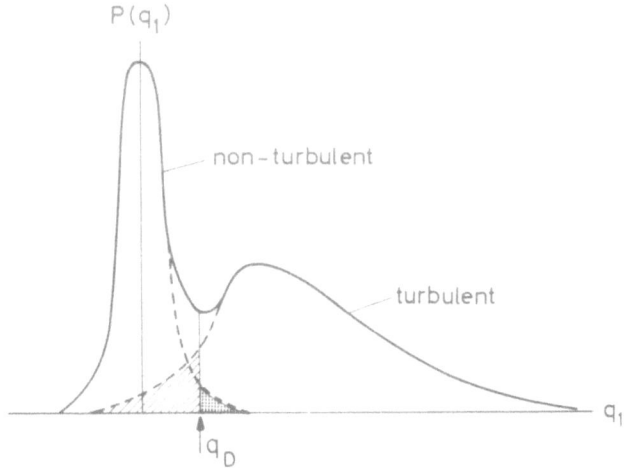

Fig. 2. Bi-modal probability distribution of a possible feature q_1 in the tur-bulent/non-turbulent region of a shear flow.

computed. For example, the probability that a turbulent region is indicated by the algorithm when actually non-turbulent fluid is present is given by the con-ditional probability to be

$$\epsilon = \text{Prob}\{q > q_D \mid \text{nonturb}\} = \int_{q_D}^{\infty} p^n(q)\, dq$$

This is represented by the stippled region in Fig. 2. A similar expression gives the error of detecting non-turbulent fluid in the turbulent zone and is repre-sented by the crosshatched area. These errors are analytic functions of the threshold q_D, and hence the relative merits or varying the threshold can be ascertained. For example, Chen and Blackwelder (1978) have shown that under the above conditions, it is impossible to obtain a region where the indicated intermittency is independent of the threshold. The threshold may be varied to minimize one type of error at the expense of the other. The relative merits of this risk or loss analysis are independent of the distribution shape and are dis-cussed by Andrews (1972). If the detection scheme is being used with con-ditional sampling, the above statistical approach can be used to analytically compare and/or predict the values of different threshold setting. For example, the experimental turbulent zone average of q_1 is

$$<q_1>^t = \int_{q_D}^{\infty} q_1\, p(q_1)\, dq_1$$

which is differentiable as discussed by Chen and Blackwelder.

To improve upon a pattern detection scheme, either additional data, i.e., more information in the data space, and/or better features must be supplied to the detection algorithm. For example, instead of measuring only a single variable such as the temperature in the interfacial region, a component of the vorticity could possibly be measured giving the joint probability distribution sketched in Fig. 3. In this case, the higher temperature and the vorticity are assumed to be associated with only the turbulent fluid. Since these two signals are good indicators of turbulence, this data space can be used directly as the feature space. The pattern detector could be a simple threshold as shown. This linear threshold can be replaced by a nonlinear segmented or curved line if more sosphistication is required, although this can increase the required computations immensely.

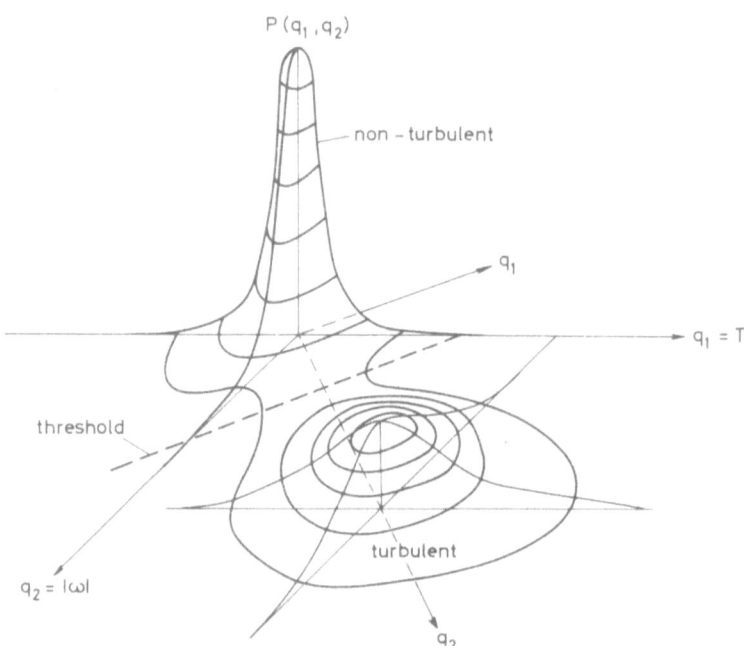

Fig. 3. Bi-modal joint probability distribution of the feature q_1 and q_2 in the intermittent region of a turbulent shear flow.

Several techniques are available to determine which features or combinations thereof contain the most information and are the most useful in detecting a pattern. When guided by the physics of the problem, these methods are also helpful in choosing the features. For example, assume a single feature were desired from the distributions in Fig. 3. A rotational transformation from the

(q_1, q_2) axes to the (q'_1, q'_2) coordinates defines two new features q'_1 and q'_2 as linear combinations of the old ones. An examination of the probability distributions indicates that the feature q'_1 would be almost useless in discriminating between the turbulent and non-turbulent fluid. The single feature q'_2 is better than either q_1 or q_2 because the transformation has increased the interclass distance as can be seen. The Karhunen-Loeve transformation is optimum for reducing the number of features while maintaining a minimum error. Other linear transformations that have been used include the Fourier and Walsh transforms, which are popular because they can both be implemented by fast algorithms. The divergence and Bhattacharyya transforms are useful when the underlying statistics are Gaussian. These and other non-linear transforms are discussed by Andrews (1972).

Detection of Eddies Embedded in Turbulent Fluid

The detection of coherent structures completely embedded within turbulent shear flows has proven to be much more of a challenge to experimentalists than the interfacial problem. There seems to be no universally identifiable characteristic associated with these eddies and consequently no simple demarcation between them and the background turbulence exists. This has thwarted efforts to define distinguishing features and often, but not always, a more subjective criteria must be incorporated into the detection scheme. Initially, most of the information acquired about these eddies was obtained by visualization techniques in which the observers found a coherent motion involving well defined parcels of fluid in space. The dynamics of these parcels were sufficiently repeatable that they could be identified as recurrent similar events although they appeared randomly in space and time. Further studies have shown that the most unique features of these eddies is the phase relationship maintained between the flow variables associated with their structure. The amplitudes of these variables are typically one rms unit depending upon which part of the eddy is recorded. For example, the phase relationship between the velocity fluctuations along the axis of the mean flow and those in the direction of the mean gradient within the coherent structure usually produce a large tangential Reynolds stress in spite of the fact that the amplitudes of the two velocity components are not necessarily large.

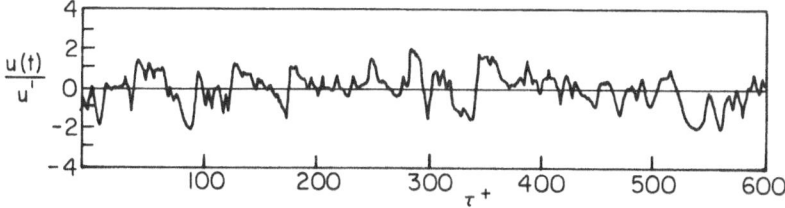

Fig. 4. *The streamwise velocity component at $y^+ = 15$ in a turbulent boundary layer at $R_\theta = 2550$.*

The difficulty of finding these eddies using Eulerian techniques is readily seen by observing a transducer signal as given in Fig. 4. The fluctuating streamwise velocity component from a turbulent boundary layer at $y^+ = 15$ is plotted. Based upon previous studies, a sufficient amount of time has elapsed in the figure so that several eddies should have passed the probe. However, it is difficult to determine a priori from the velocity trace what characteristic signature they possess and when they occur.

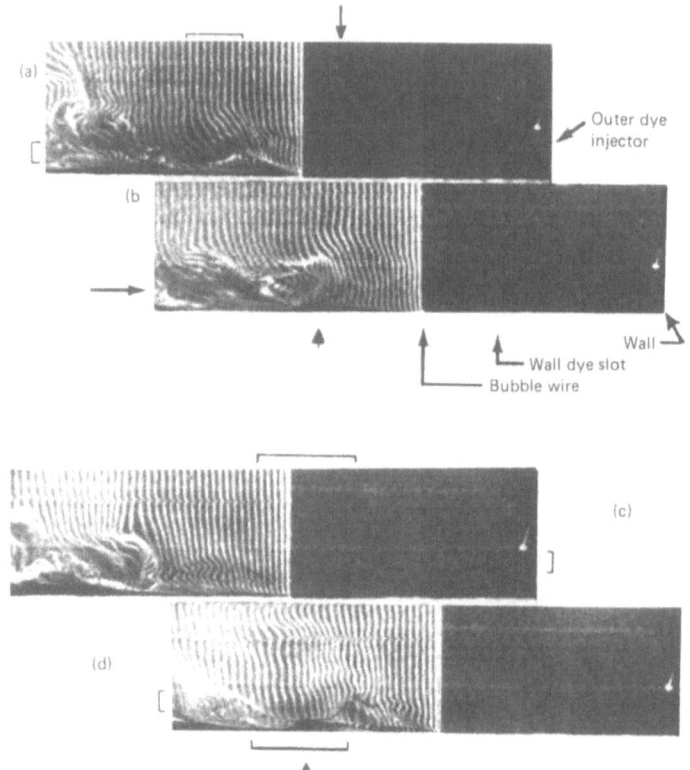

Fig. 5. Photographic sequence of a ragged sweep (a and b) and a smooth sweep (c and d). The flow is to the left and the photos are displaced by the distance a particle with velocity U_∞ travels during the time interval $\Delta t = 1.5$ s between views. Reproduced from Offen and Kline (1974).

A photo of this region of the boundary layer taken from Offen and Kline (1974) is shown in Fig. 5. Obtained from a movie film, these sequences depict a characteristic series of events collectively called the bursting phenomenon that disturb the wall region. Since the human eye is able to recognize spatial structures quite well, it can easily detect these visual events, especially when the temporal information is available as in the movie. However, it is not a trivial task to use this Lagrangian information in order to detect these structures in the Eulerian frame as demonstrated by Offen and Kline (1973). The reasons are several. In the motion pictures of the Lagrangian studies, the eye is allowed to follow the particles in space and time. Whenever a relative coherent group of particles associated with an eddy structure follows a similar trajectory, the eye can easily discriminate between this and the random background motion. After detection, this motion can be followed until it either loses its coherency or it is out of the field of view. None of these advantages exist in the Eulerian frame. Not only can the transducers not follow the eddy structure, but the mere detection of the eddy is more difficult. In addition, the motion pictures contain a tremendous amount of data, but have low information densities. On the other hand, an Eulerian signal has a high information density and is more quantitative.

This fundamental difference between low density photographic information and high density electronic signals leads to a major difference in the data processing. When analyzing a two-dimensional object, as in alphabetical character recognition, the dimensionality of the raw data space in Fig. 1 is quite large, e.g. 256×256. One of the major tasks of the feature selector in this case is to greatly reduce this dimensionality during the mapping into the feature space. On the other hand, when analyzing electronic signals it is often advantageous to construct several features from each time sequence so that different aspects of the signal can be emphasized. Thus, the dimension of the feature space may be greater than that of the raw data space.

Since the characteristic eddies of turbulent shear flows are often described by their spatial coherence, it has proven advantageous to record Eulerian data simultaneously at several locations in space. One example of this is shown in Fig. 6 of the streamwise velocity signals at ten locations in the wall region of a turbulent boundary layer. Regions of strong coherency in space are evident over large areas of the flow field. For example the structure at $300 < \tau^+ < 350$ extends out to $y^+ > 100$. Possibly because the eddies pass by the probes at different stages of their development and/or because of the three-dimensional nature of these structures, some of the synchronous signals maintain their coherency over a more limited spatial extent. Irrespective, the distinguishing signature of the coherent eddies seems to be their spatial correlation which is most readily observed in the direction of the mean shear. Similar conclusions can be obtained in other shear flows as seen in the data of Sunyach (1971) and Fiedler (1975).

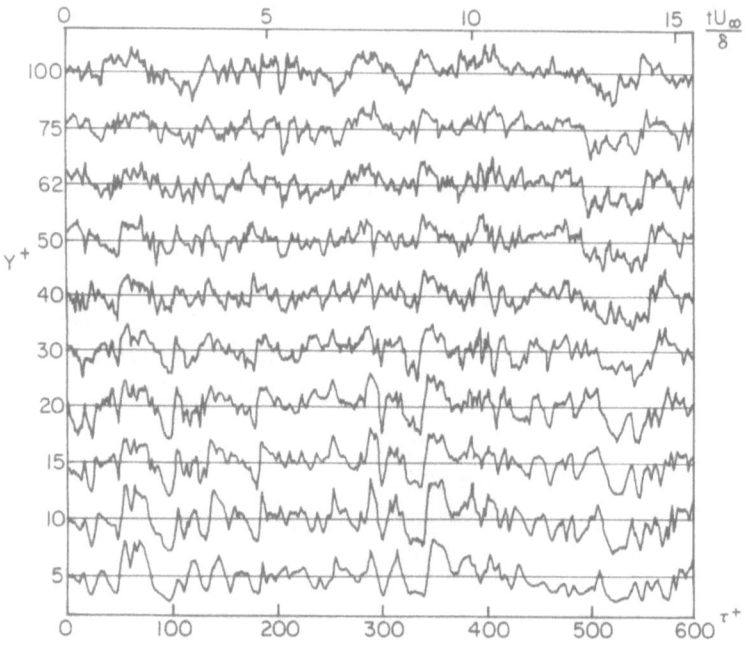

Fig. 6. Simultaneous streamwise velocity signals at ten locations in the wall region of a turbulent boundary layer at R_θ = 2550.

The definition of the features is the most difficult aspect of detecting embedded eddies. The choice of the features and their relative weighting in the decision process must be governed by the physics of the problem. Although the mathematical tools from pattern recognition cannot provide much help in the initial choice of features, they can be useful in determining which features contain information most efficient in classifying different patterns. The features are usually obtained by differentiating, filtering, taking short time averages, etc. of the signals in the raw data space. The N features are often a combination of these functions and have been suitably normalized, for example, by their rms values to form the feature space. Short time cross-correlations and/or coherence functions could be taken from the data in Fig. 6 to locate those regions that have a strong spatial similarity. If a known pattern is being sought in the data, it can be used to define a vector that is compared directly with the raw data. Depending upon the randomness of the problem being studied, it may be desirable to rescale the amplitude and/or time scale of the known pattern to determine the best fit. The minimum variance between the pattern and the data could be used to define the point of detection. This technique has been used to track coherent structures downstream by Zilberman,

Wygnanski and Kaplan (1977) in a transitioning boundary layer and by Black-welder (1977) in a turbulent boundary layer.

A simple detection scheme is usually desired in order to minimize probe interference, to facilitate repeatability between different laboratories, etc. Often signals obtained from a single spatial location are used and several features are derived from them. If an eddy with greater spatial coherence than the surrounding background turbulence passes a transducer, it will produce a signal with greater temporal coherence. Thus, a possible feature, f, given by the integral scale

$$f(t, T) = \int_0^T | R(t, \tau, T) | \, d\tau$$

where $R(t, \tau, T)$ is a short-time auto-correlation function taken over the interval T, should be larger in the presence of the eddy than in the background noise alone. Other features utilizing either powers and/or moments of the auto-correlation function could be defined that emphasize other temporal characteristics of the signal.

Short-time averages of the fluctuating signal have been used by Blackwelder and Kaplan (1976) to detect the structure in Fig. 6. Letting the average over an internal T be

$$\hat{g}(t) = \frac{1}{T} \int_{t-T/2}^{t+T/2} g(s) \, ds$$

several of the short-time averages that can be derived from a single function are shown in Fig. 7. All of these features used an averaging interval of $T^+ = 20$. Although the averaging interval can be varied, they found that the detected eddies were relatively independent of T^+ as it was changed by a factor of two. Blackwelder and Kaplan used only a single feature, $\hat{u}^2 - \hat{u}^2$, which is a measure of the energy fluctuations in time T^+. After further study Chen and Black-welder (1978) added a feature that distinguished between accelerations and decelerations. Willmarth and Lu (1972) used the streamwise velocity signal at $y^+ = 15$ and its derivative with and without filtering to form two features in another study of the turbulent boundary layer. Zaric (1974) used the stream-wise velocity and its time derivative in a different detection scheme and computed the probability distributions due to different events. Thomas (1977) has also used short-time averages of a single streamwise velocity signal and has detected embedded eddy structures whenever a large correlation exists between the $\hat{u}$ and $|u - \hat{u}|$ features.

Several other detection schemes have been utilized in bounded turbulent shear flows. Wallace et al. (1972), Willmarth and Lu (1972) and Eckelmann et al. (1977) used the streamwise and normal velocity signals u and v directly as features. They separated the uv plane into four quadrants and studied the relative contribution to the Reynolds stress from each quadrant at $y^+ = 15$. Blackwelder and Eckelmann (1979) used two velocity gradient signals at two wall locations and formed a four-dimensional feature space which allowed them to detect the streamwise vortices in the wall region.

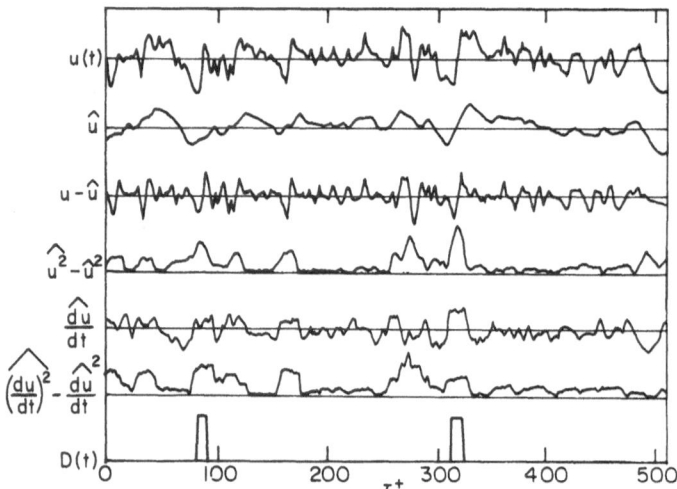

Fig. 7. The velocity signal at $y^+ = 15$ of Figs. 4 and 6 and several features derived by using short-time averages. The detector signal is given at the bottom.

Independently of how the features are defined and computed, ultimately a decision must be made determining whether a coherent structure exists. Also if several different eddies are possible, the correct classification of the eddy needs to be ascertained. This is the function of the pattern detector and consists of defining an N−1 dimensional surface in the feature space for each type of eddy and computing into which subspace the features lie. For example, the Euclidean square metric between an unclassified vector $\underset{\sim}{x}$ and the mean values of the features in the kth class is

$$d^2(\underset{\sim}{x}, \underset{\sim}{\overline{f}}) = \sum_{n=1}^{N} w_n (x_n - \overline{f}_n^{\,k})^2$$

where $\overline{f}_n^{\,k}$ are the most probable or mean values for the kth class and w_n is the weighting function for the nth feature. The unknown vector, $\underset{\sim}{x}$, can be assigned to the class that gives a minimum value to d^2. This is equivalent to deciding on which side of the threshold $\underline{x}$ lies in Figs. 2 or 3. If the fluctuating velocity components are used directly as features, the Euclidean distance from the origin is a measure of the energy. Those eddies with large energy associated with the u and v velocity fluctuations would lie outside the threshold sketched in Fig. 8a. Alternatively a non-linear surface can be defined in the feature space. One such surface, uv = constant, used by Willmarth and Lu (1972) and others is shown in Fig. 8b. If the surface becomes highly nonlinear, it often involves many FORTRAN IF statements or their equivalent which can require considerable computation.

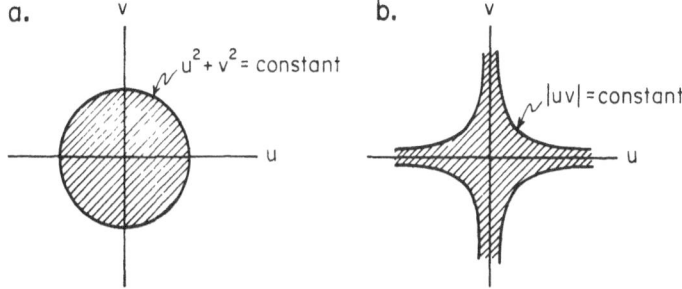

Fig. 8. An Euclidean mean square threshold proportional to the energy is shown in a. A nonlinear decision surface used by Willmarth and Lu (1972) and others is illustrated in b.

When detecting internal embedded eddies, most methods tend to lock onto a particular phase of the coherent structure rather than determining the location and/or duration of the entire eddy. For example, the detector signal in Fig. 7 is typically non-zero only 1 % of the total time, yet the eddy structure as found by conditional sampling is present 25 % of the time. As the detection methodology improves, this may change so that the entire eddy will be located as is usually done in the intermittent regions of free turbulent shear flows.

Calibration
Since the problems that require pattern recognition are necessarily dealing with stochastic data, there is always a finite probability of an error in the detection scheme, as was seen in Figs. 2 and 3. Thus, even though the detection scheme is carefully designed to eliminate false alarms, one has only to wait a finite amount of time before an incorrect decision is made. Random data with no embedded structure will also have a finite probability of finding vectors that lie near the desired pattern. The density of points near the pattern in feature space should be considerably greater when embedded structures lie within the data.

However, this density is obviously a function of the choice of features; the better the feature, the greater the density. The background noise level, i.e., the density when no structure exists, is usually an unknown factor and should always be determined a priori by using a suitably filtered random signal to simulate the uncoherent turbulence.

Artificial signals have been used by many authors to calibrate detection schemes. Kibens, Kovasznay and Oswald (1974) and Antonia and Atkinson (1974) describe the use of suitably filtered analog noise with a known intermittency which they utilized during design of their circuits, Blackwelder and Kaplan (1976) and others have used a random digital number generator and filtering to test their pattern recognition scheme.

The spectra and the probability density of the real and artificial signals should be made nearly identical. This can be accomplished with an analog or digital Gaussian noise generator and appropriate filtering so that the mean and mean square statistics between the two signals are similar. The coherent eddy structures are often characterized by deviations from a Gaussian distribution, hence this distinguishing feature should be incorporated into the detection scheme. In many shear flows, this non-Gaussian character is associated with the existence of sharp gradients of the flow variables along some surface associated with the eddy structure. Consequently, features which utilize some aspects of these sharp gradients have been among the most successful in the search for coherent eddies. For example, the gradient of fluctuating vorticity demarcates the turbulent/non-turbulent interface in free shear flows as used by Kibens et al. (1974). By using a slight amount of heat as a passive contaminate, sharp temperature gradients extending across the turbulent boundary layer were found by Chen and Blackwelder (1978). Similar gradients have been reported in the atmospheric boundary layer by Taylor (1958) and others.

Blackwelder and Kaplan (1972) used a short time average of the streamwise velocity as their feature and found that the bursting phenomenon in bounded shear flows is associated with sharp velocity acceleration in the wall region. This was verified by Wallace, Brodkey and Eckelmann (1977) using a highly nonlinear pattern scheme. In a similar study, Thomas (1977) found that by testing his detection scheme with non-skewed noise, he detected no structure. However, by adding some skewness, his scheme triggered indicating that the non-Gaussian character was being detected.

Although some of the elementary ideas from pattern recognition have been successfully used in fluid mechanics, there is still a wealth of mathematical tools available that can guide and assist those who are searching for coherent eddies in turbulent shear flows.

The support of the Army Research Office - Durham under Grant DA-ARO-DAAG29-76-G-0297 during the period covered by this work is gratefully acknowledged.

References

Andrews, H.C., 1972 *Intro. to Mathematical Techniques in Pattern Recognition,* Wiley-Interscience

Antonia, R.A. & Atkinson, J.D., 1974, *J. Fluid Mech.,* Vol. **64**, p. 769

Antonia, R.A., Prabbu, A. & Stephenson, S.E., 1975, *J. Fluid Mech.,* Vol. **72**, p. 455

Blackwelder, R.F., 1977 *Phys. Fluids,* Vol. **20**, p. 232

Blackwelder, R.F. & Eckelmann, G., 1979 to appear in J. Fluid Mech.

Blackwelder, R.F. & Kaplan R.E., 1972 Intermittent structures in turbulent boundary layers *NATO-AGARD Con. Proc. No. 93,* London: Technical Editing and Reproduction, LTD.

Blackwelder, R.F. & Kaplan, R.E., 1976 *J. Fluid Mech.,* Vol. **76**, p. 89

Bradbury, L.J.S., 1964 *Aeronautical Quart.,* Vol. **15**, p. 281

Chen, C.H.P., 1975 The large scale motion in a turbulent boundary layer: a study using temperature contamination, Ph.D. thesis, U. of Southern California

Chen, C.H.P. & Blackwelder, R.F., 1978 *J. Fluid Mech.,* Vol. **89**, p. 1

Corrsin, S., 1943 NACA Report No. W-94

Corrsin, S. & Kistler, A.L., 1955 NACA Report No. 1244

Demetriades, A., 1968 *J. Fluid Mech.,* Vol. **34**, p. 3

Eckelmann, H., Nychas, S.G., Brodkey, R.S. & Wallace, J., 1977 *Phys. Fluids,* Vol. **10**,

Fiedler, H., 1975 *Turbulent Mixing in Reactive and Non-reactive Flow,* Proc. SQUID Conference, Plenum Press

Hancock, J.C. & Wintz, P.A., 1966 *Signal Detection Theory,* McGraw-Hill, New York

Hedley, T.B. & Keffer, J.F., 1974 *J. Fluid Mech.,* Vol. **64**, p. 625

Kaplan, R.E. & Laufer, J., 1969, *Proc. 12th IUTAM Congress,* Stanford, p. 236, Springer

Kibens, V., Kovasznay, L.S.G. & Oswald, L.J., 1974 *Rev. Sci. Instruments,* Vol. **45**, p. 1138

Kovasznay, L.S.G., Kibens, V. & Blackwelder, R.F., 1970 *J. Fluid Mech.,* Vol. **41**, p. 283

LaRue, J.C., 1974 *Phys. Fluids,* Vol. **17**, p. 1513

Nilsson, Nils, J., (1970) *Learning Machines,* McGraw-Hill, New York

Offen, G.R. & Kline, S.J., 1973 Report MD-31, Stanford University, Department Mech. Eng.

Offen, G.R. & Kline, S.J., 1974 *J. Fluid Mech.,* Vol. **62**, p. 223

Sunyach, M. 1971 "Contribution a l'etude des frontieres d'ecoulement turbulents libres" Ph.D. thesis, Ecole Centrale de Lyon

Taylor, R.J., 1958 *Aust. J. Phys.,* Vol. **11**, p. 168

Thomas, A.S.W., 1977 Organized structures in the turbulent boundary layer, Ph.D. thesis, U. Adelaide

Townsend, A.A., 1949 *Aust. Jour. Sci. Research,* A, Vol. **2**, p. 451

Wallace, J.M., Brodkey, R.S. & Eckelmann, H., 1977 *J. Fluid Mech.,* Vol. **83**, p. 763
Wallace, J.M., Eckelmann, H. & Brodkey, R.S., 1972, *J. Fluid Mech.,* Vol. **54**, p. 39
Willmarth, W.W. & Lu, S.S., 1972, *J. Fluid Mech.,* Vol. **55**, p. 65
Zaric, A., 1974 *Advances in Geophysics,* Vol. **18A**, p. 249, Academic Press, New York
Zilberman, M., Wygnanski, I. & Kaplan, R.E., 1977 *Phys. Fluids,* Vol. **20**, p. S258

The Recognition of an Evoked Large Scale Structure in Turbulent Shear Flows

by

I. Wygnanski

Tel-Aviv University, Ramat-Aviv, Israel

Introduction

The experimental research in turbulent shear flows has been recently influenced by four seemingly unrelated developments.

1. The recognition that large coherent structures exist in most or perhaps all turbulent shear flows.

2. The appearance of the mini-computer and the concommitant digital signal processing as an aid to the experimenter in the laboratory.

3. The development of inexpensive electronic components permitting measurements with arrays of sensors in a low budget operation.

4. The possibility of evoking coherent structures by disturbing the flow in a manner which triggers inherent instabilities.

For many years turbulent flows were described statistically by dividing the velocity and pressure fields into mean and fluctuating parts. The discovery of intermittency (Corrsin and Kistler, 1955) caused reconsideration of the averaging process in flows having a free interface. Conditional sampling and zone averaging was introduced by Kovasznay et al., (1970), which provided a separate statistical description for the quantities in the turbulent and potential zones. It was realized that the turbulent fluctuations are confined to large bulges of turbulent fluid which protrude from the turbulent zone. Some statistical data on the scales, shape, and characteristic celerity of the bulges was compiled leading to the concept of a large eddy at the outer part of the boundary layer. At approximately the same time Kline et al., (1967) observed visually that the boundary layer has a very streaky structure near the surface. They also observed a clearly identifiable process in which a dye streak was slowly lifted from the surface; the "lifting-up" was followed by oscillations and finally by a violent burst. The process is loosely defined in the literature as a "burst".

There is evidence that a bulge observed in the outer part of the layer and the burst originating near the wall are related in some way although no causality was as yet established. The burst and the bulge are clearly recognizable events which occur randomly in space and time. To describe an "average" burst implies averaging events which are at various stages of development. Furthermore, since most measuring devices (like a hot wire probe) provide information at a "point" the averaging is also done over events which originated at different spatial locations in the flow. Thus, if the structure occupies a *volume* of fluid, one might average over different sectors of the volume. There is no clear way of telling how the spatial averaging in a direction transverse to the flow differs from the temporal averaging over different periods in the life span of the structure.

Pattern recognition techniques (Wallace et al. 1977) in which recognized events are stretched and normalized by an arbitrarily chosen time interval, imply that only the scale of the events changes with time. It would be more desirable to have an array of sensors in a plane perpendicular to the flow which would provide simultaneous data across the entire event and its development with time. The task may be a difficult one but not impossible. Inexpensive multi-channel, constant-temperature anemometers are available in a few laboratories and simultaneous measurement with arrays of hot wires containing 10-16 sensors were made (e.g. Blackwelder and Kaplan 1972).

Approximately 16 years ago Kovasznay, Komoda and Vasudeva (1962) used ten-channel linearized constant temperature equipment while studying transition. They had to convert a television set into a 10 channel display oscilloscope, photograph the traces, and evaluate the velocities by projecting the film on a graph paper. The appearance of the mini-computer in the laboratory caused profound changes in the way in which data is acquired. Output from an array of sensors is digitized, linearized digitally, and then written on a magnetic tape or a disc for further processing. The fact that an entire experiment can be stored and processed at leisure and reprocessed whenever the need arises, is significant in itself. It is of particular importance whenever complicated processing, like pattern recognition, is required for identification of special events in the flow.

A major simplification of the problem of mapping and tracking the development of a large coherent structure is achieved whenever the structure can be evoked artificially. The question is whether a momentary perturbation can trigger a large coherent eddy which is independent of the nature of the perturbation but is determined by the flow parameters. This being the case, provides the experimenter with a time reference and an average trajectory of the evoked structure, and since the process can be made repeatable the recognition techniques and sampling criteria are simplified. Attempts to perturb the fully turbulent pipe or channel flow failed to evoke the required large coherent structure. A well documented attempt was made by Hussein and Reynolds (1970), who introduced two-dimensional waves into a fully developed channel flow. The waves simply decayed some distance downstream. Experiments, carried out in a boundary layer in the absence of pressure gradient and in fully developed pipe flows indicated that a typical coherent structure, which is independent of the nature of the perturbation, could be evoked provided the perturbation occurred in the laminar region just prior to transition.

The importance of digital data acquisition, multi-channel sensors and artificial initiation of transitional structures in detecting and mapping the large coherent motions in fully turbulent flow will be demonstrated by way of examples.

The Puffs as the Basic Coherent Structure in Turbulent Pipe Flow

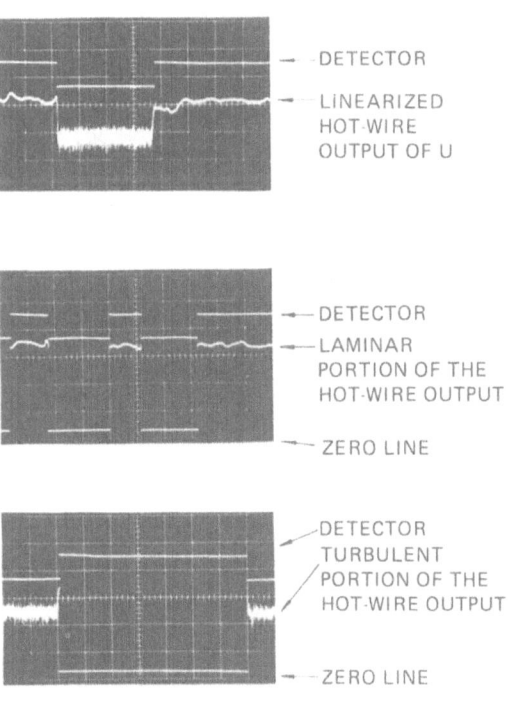

T = 1.0 Sec/cm

Fig. 1a. Oscillograms showing the operation of the turbulence detector and the switching circuits.

The earliest conditionally sampled measurements in pipe flow at Reynolds numbers corresponding to the onset of turbulence consisted of zone averaging and point averaging methods (Wygnanski 1971). The signal used to detect intermittency was $[u'^2 + (\partial u'/\partial t)^2]^2$ filtered between 50 - 50,000 Hz. It was obtained by manipulating the output of a single hot wire sensor using analog circuits. The choice of the signal was rather arbitrary. The addition of u'^2 to $(\partial u'/\partial t)^2$ resulted in a signal which has less zero crossings than either u'^2 or $(\partial u'/\partial t)^2$. Consequently, excessive capactive smoothing was avoided when a telegraph signal representing intermittency was finally obtained.

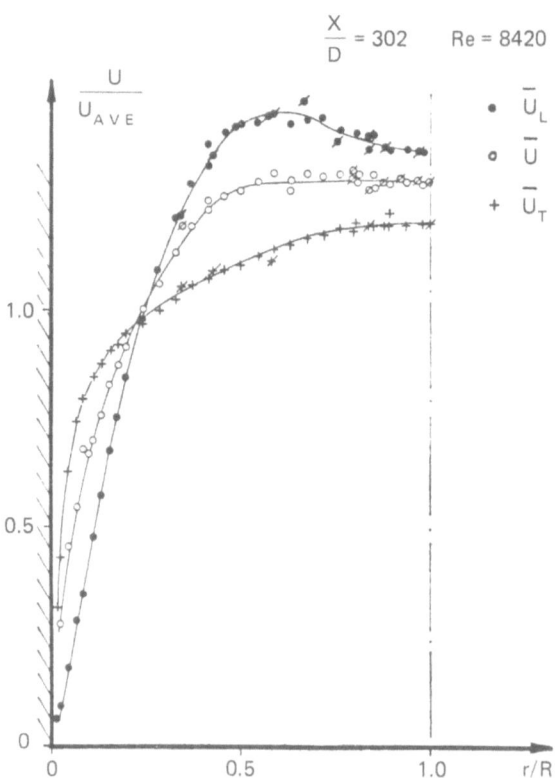

Fig. 1b. Conventional and zone-averaged velocity profiles in a pipe (slash indicates data collected on the other side of the center-line).

The intermittency signal actuated a switch which directed the velocity signal to two different integrators (Fig. 1a) giving turbulent and laminar zone-averaged velocities (Fig. 1b). The results themselves are outdated by todays standards but they indicated that the intensity of the velocity fluctuations in a tur-

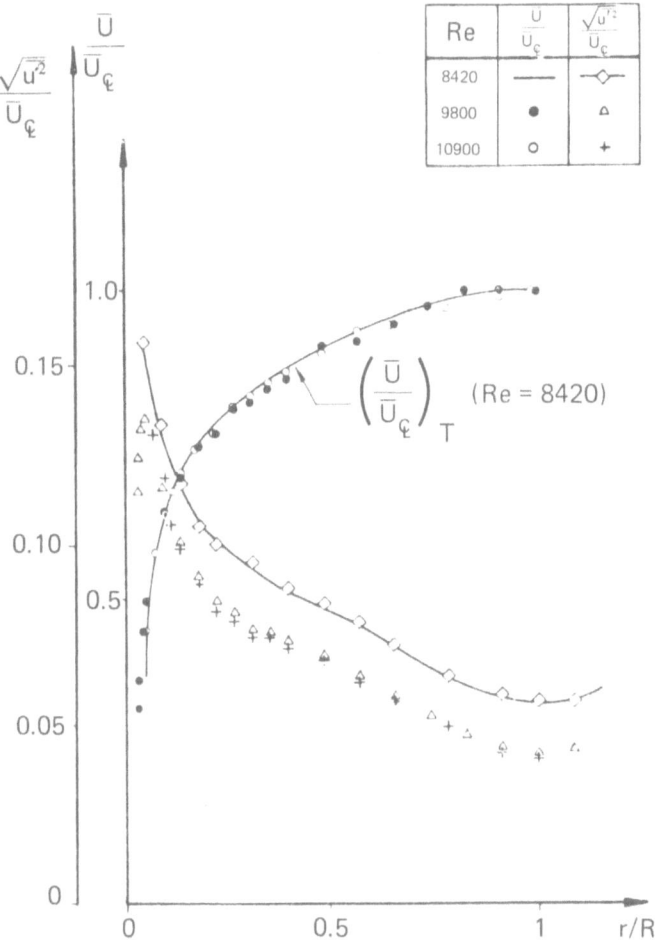

Fig. 2. *The effect of intermittency on the longitudinal component of turbulent zone-averaged intensity.*

bulent zone was much higher than in fully developed pipe flows, although the normalized radial distribution in both cases was almost identical (Fig. 2). The intensity level in the turbulent zone depended on the intermittency γ. In the center of the pipe the intensity of $(u')_T$ dropped from 5.7 % when $\gamma = 0.5$ to 4.2 % when the Reynolds number was increased so that $\gamma \rightarrow 1$. In fully turbulent pipe flow the intensity is approximately 3.3 %. Examining the reasons for this result alerted us to the fact that near the interface of a turbulent slug there is an enhanced turbulent activity which does not disappear at the instant the slugs merge into a continuous turbulent region.

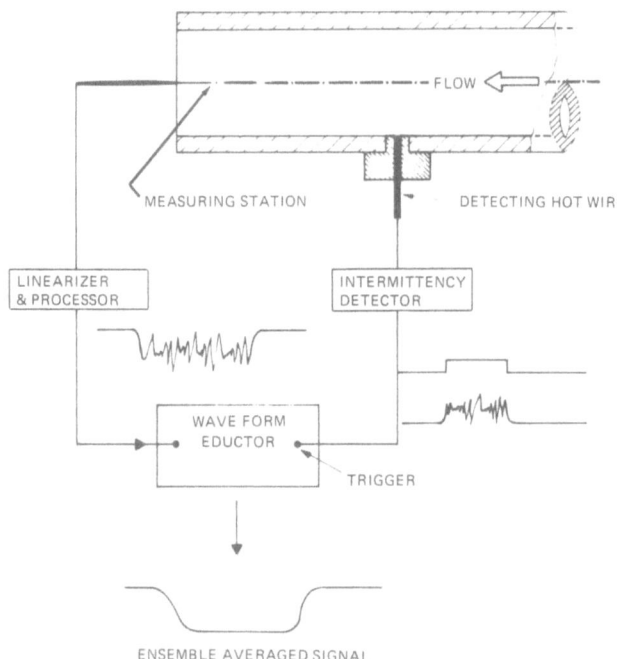

Fig. 3. A Schematic diagram showing the use of a wave-form eductor in pipe flow.

The wave-form eductor permits the ensembling of data at various times from a known occurrence. The use of the instrument is illustrated for the case of transitional flow in a pipe, (Fig. 3). A hot wire embedded flush with the inner surface of the pipe some distance upstream of the measuring station is used as a turbulence detector, thereby providing a trigger signal in advance of the oncoming turbulent-non-turbulent interface. The trigger activates a clock and the instrument waits for a predetermined time before sampling the data. The time delay permits the interface to arrive at the measuring station. The data is sampled during a period of interest and the process is repeated until the ensemble-averaged signal no longer changes. From the data collected at the measuring station one may infer what happens to various flow parameters near the interface. The sampling procedure depends on the assumption that successive interfaces move downstream at the same velocity, at a given Reynolds number, and do not distort while passing from the detector to the measuring station. Since these assumptions are only valid in a statistical sense the ensemble-averaged results are smeared somewhat, limiting the spatial resolution of the ensemble-averaged data to approximately two diameters. To improve the resolution near the interface, a time delay network was designed permitting the use of a single signal for the detection and analysis of turbulence. The signal was sent to two separate channels. One (non-delayed) went to the turbulence detector, while the other (delayed) provided the data to be conditioned on the detection of an interface. By delaying a second channel, one could recapitulate the signal (representing the flow velocity or its fluctuations) which existed just before the interface arrived at the measuring station. The results obtained this way were free from averaging over the speed and shape of the interface, however, the knowledge of these quantities was lost.

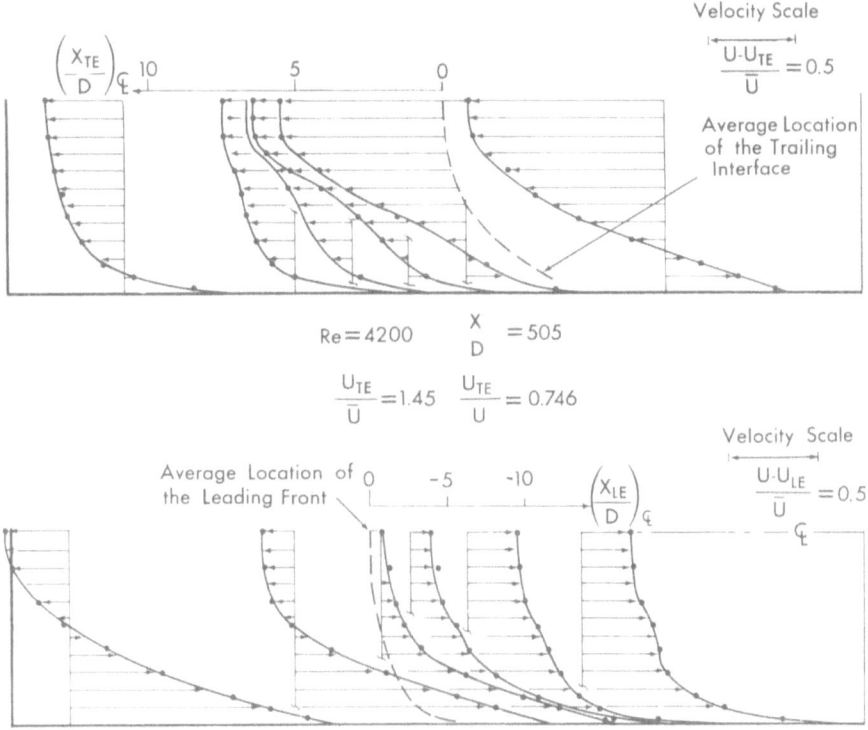

Fig. 4. Velocity profiles in the vicinity of an interface of a slug Re = 4200; x/d = 505.

The results near an interface of a "slug" indicated that the velocity profiles contain points of inflection as the flow became turbulent. (Fig. 4). The turbulent intensities imparted to the freshly entrained fluid exceed the level of fluctuations existing in the interior of the slug, and for that matter in the fully turbulent pipe flow. The turbulent levels encountered in the vicinity of the interface of the slug are, generally speaking, 4 to 7 times higher than in the fully developed turbulent pipe flow (Fig. 5), (see also Wygnanski and Champagne 1973). Furthermore, the turbulent production terms near the interface, averaged over the entire pipe cross section are 4 times higher than the total average production in a fully developed turbulent pipe flow. Preliminary observations of velocity signals indicated that the enhanced turbulent activity persists for quite a long time after adjacent slugs joined together alluding to the possibility that the "burst" may be associated with the activity of a "faded" interface.

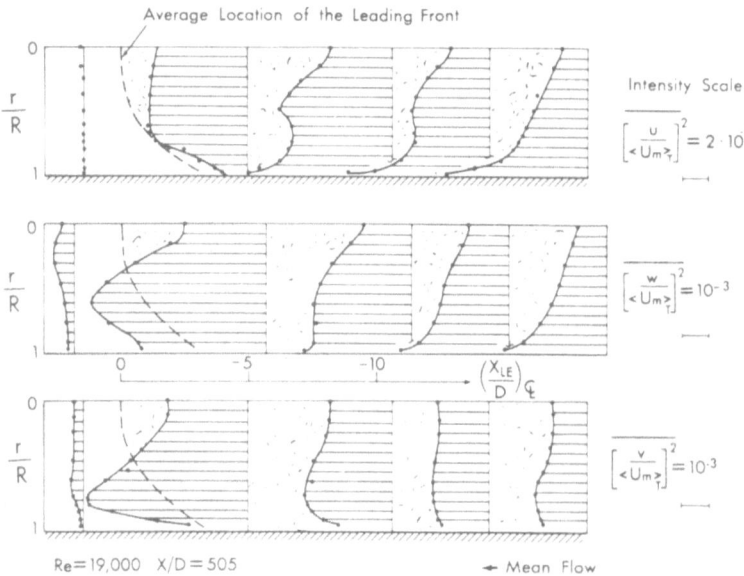

Fig. 5. The three components of turbulent intensity near the leading interface of a slug Re = 1.9 × 10⁴ ; x/d = 505.

The spatial resolution obtained using analog devices is inadequate for measurements at $2000 < Re < 2700$. Turbulent puffs occurring at these Reynolds numbers are approximately 25 pipe-diameters long, and their leading interface is not well defined. For this reason a jitter of 2 diameters in ensembling the data does not represent properly the individual realizations. Furthermore, puffs triggered by a fixed disturbance like an orifice plate tend to propagate in groups except at the lowest Reynolds number; but the frequency of occurence at these Reynolds numbers is so low that it takes a long time to ensemble the data. In the second version of the experiment (Wygnanski, Sokolov and Friedman,

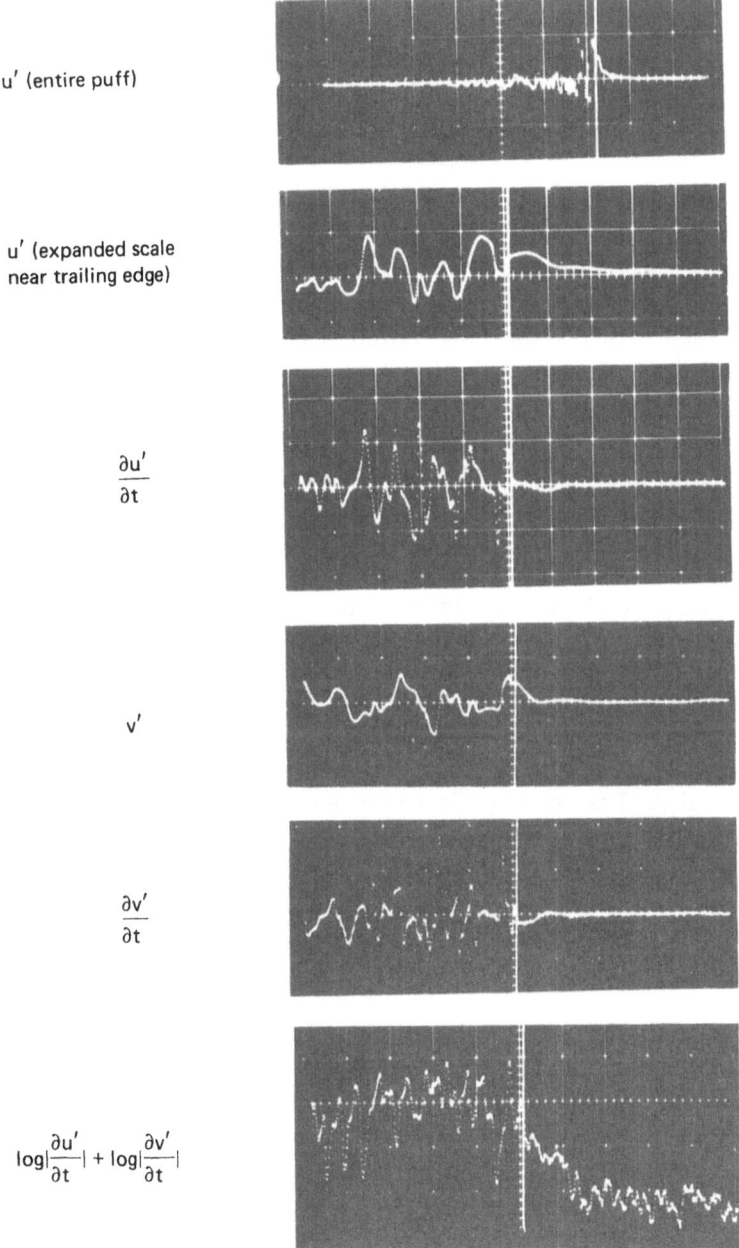

Fig. 6. The detection of an interface in a puff.

1975), a loudspeaker mounted on a funnel produced momentarily a small jet (1 mm in diameter) normal to the direction of the flow and evoked a single puff at Re $\leqslant$ 2300. The structure evoked by a momentary disturbance was compared with randomly occurring puffs resulting from a fixed protrusion in the flow and found to be identical.

The data was digitized and recorded. Ensemble averages were obtained by detecting the interface and shifting the signals representing different events until their respective interface location was brought into mutual alignment. After inspecting many events, a simple function $(\partial u'/\partial t)^2 * (\partial v'/\partial t)^2$ was suggested as one to which a simple threshold criterion could be applied. However, it was noted that the logarithm of this function would have the advantage of showing the full dynamic range of activity. As illustrated in Fig. 6, this function has two plateaus, one indicating the level of the background and the other the level of the turbulent activity. The trailing edge was defined as the point at which the signal dropped *for the last time,* to one decade below its mean turbulent value. Only by using a digital system one could conveniently scan the time axis to determine where the last turbulent fluctuation occurs. Furthermore, histograms of the cumulative distribution of the interface locations for each probe station were recorded to allow the trailing edge of the puff to be mapped by knowing the average time delay relative to the station on the centerline; the latter was used for synchronizing the data-recording operation. In this way, the time delay from the initiating pulse (phase averaging in Prof. Kovasznay's terminology) gave the shape of the interface while conditional sampling yielded the flow field around it. The velocity field with and without realignment is shown in Fig. 7, and the importance of realignment is obvious. An expanded computer output of 101 averaged realization at r/R = 0.363 is shown in Fig. 8, and the temporal derivative of the ensemble-averaged streamwise velocity can be estimated from it. The slope of U near the interface represents a combination of factors:

(i)   The actual deceleration of the fluid as it becomes turbulent which may be related to the thickness of the interface.

(ii) An error in locating the interface.

(iii) Temporal uncertainty resulting from finite digitizing rate (5000 samples/ sec).

If we assume that a discontinuity in velocity is allowed across the interface (i.e. $\partial u/\partial t \rightarrow \infty$) we find that the maximum possible error in locating the interface is 3 msec corresponding to an average spatial resolution of 3 mm.

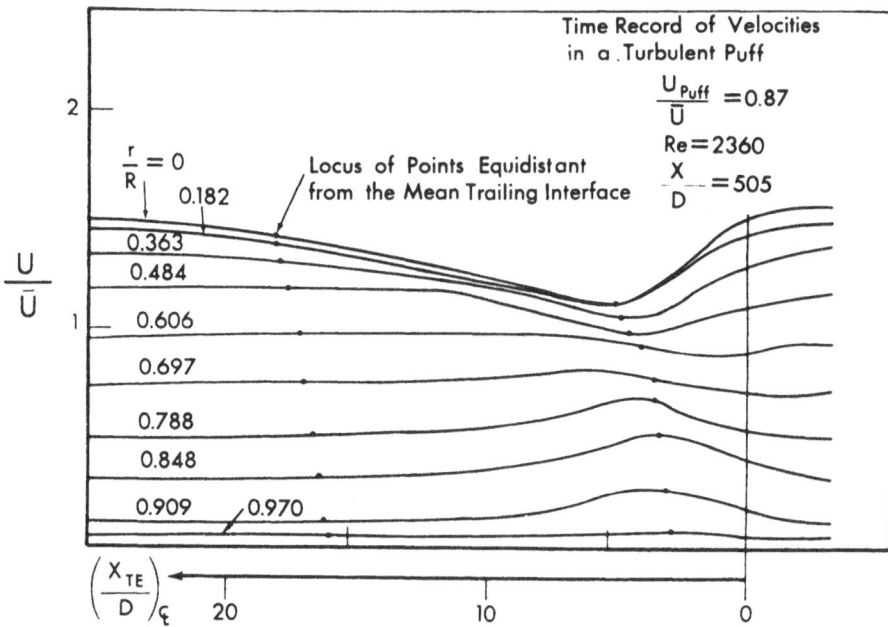

Fig. 7a. Ensemble-averaged time-records of velocities in a puff without re-alignment.

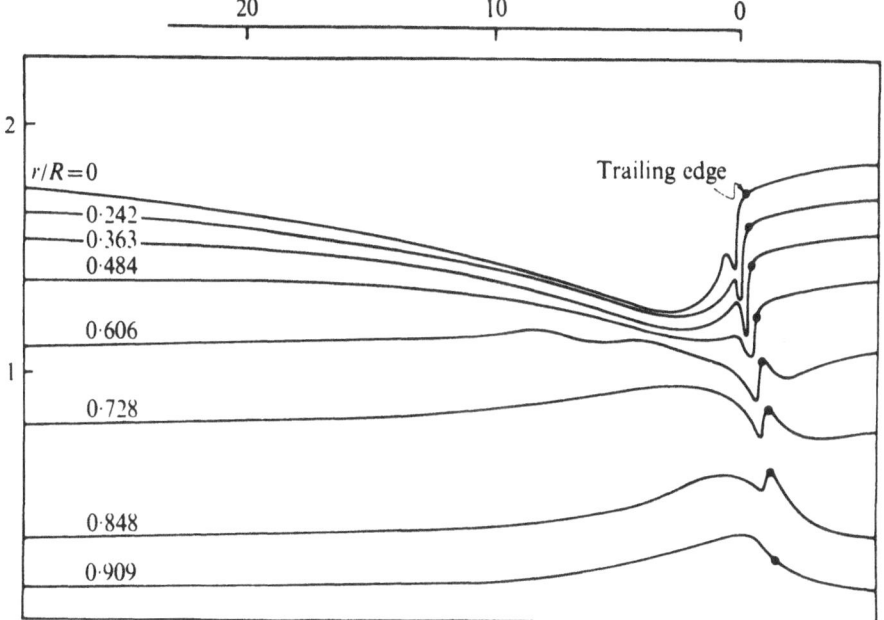

Fig. 7b. The effect of realignment on the ensemble averaged velocities.

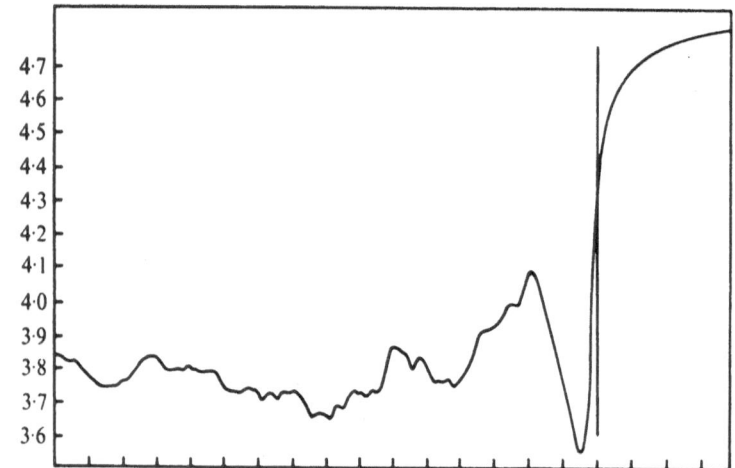

Fig. 8. Expanded record of velocity near the trailing interface of a puff.

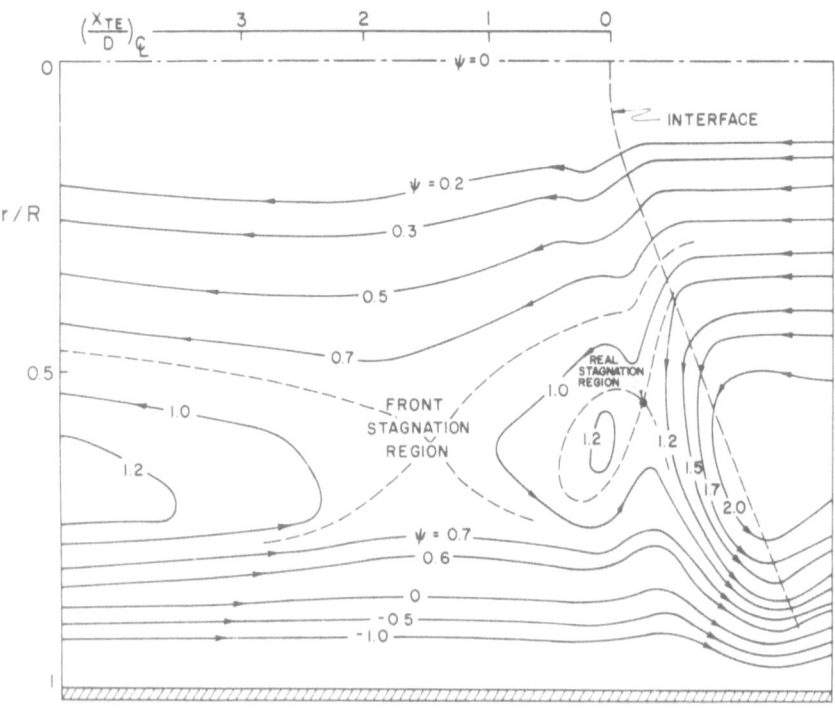

Fig. 9. Streamline pattern in an ensemble-averaged puff.

The kinematics of the flow in the puff is fairly well resolved (Fig. 9). The puff consists of two flattened vortices which rotate in the same direction and a smaller ring vortex in the vicinity of the trailing interface. All the turbulent activity is associated with this ring vortex. The other two flat vortices may result from the fact that the puff is embedded in laminar flow having a parabolic velocity distribution. At Re $\cong$ 2200 the puff is in equilibrium; is does not grow nor shrink as it moves downstream, however, at 2300 < Re < 2600 it generates new puffs most of which retain their typical structure; for Re > 2600 these puffs merge as quickly as they are generated giving rise to slugs. We propose, albeit without definitive evidence, that the slug is an array of these ring vortices (puffs) and the large coherent structure in fully turbulent pipe flows is related to puffs. A "burst" for example, in which fluid is ejected away from the wall may occur between adjacent puffs. Two interacting puffs created artificially at Re = 2100, and a slug generated by a single pulse at Re = 2600 show the similarity between the two structures (Figs. 10a, 10b). Sometimes the velocity traces generated at the same Reynolds numbers with identical disturbances show different degree of interaction between adjacent structures, suggesting that the interaction process cannot be easily controlled even in artificially evoked structures. The large structures which could be identified and ensembled using pattern recognition techniques, are clearly visible in Fig. 10. This is particularly true near the wall where the high frequencies are missing or a bit further away from the surface where most of the turbulent production occurs.

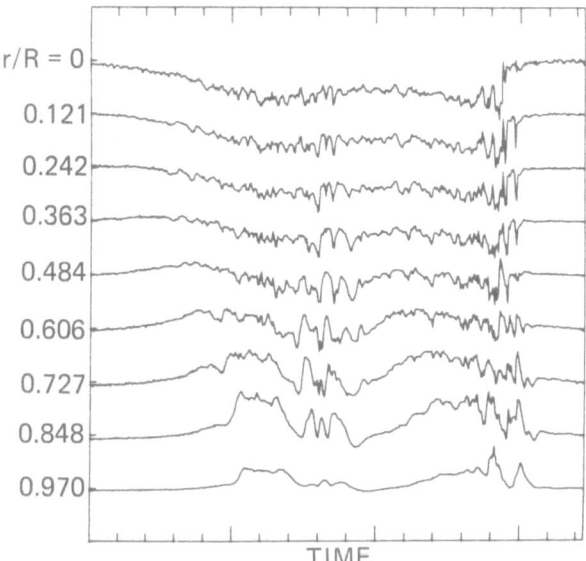

Fig. 10a. Velocity records from two evoked puffs interacting at Re = 2100.

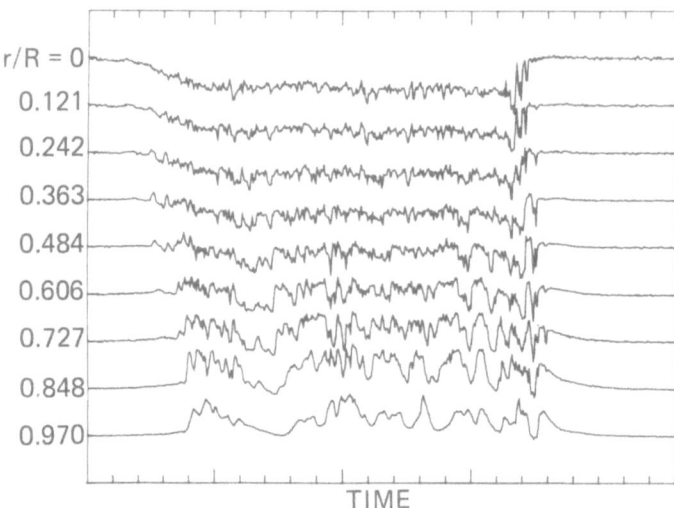

Fig. 10b. Velocity records in a slug at Re = 2600.

The Transitional Spot as a Large Coherent Structure in a Boundary Layer.
The turbulent velocity profiles based on wall parameters in a fully developed
pipe and in the boundary layer are essentially identical. Large coherent structures (i.e. "bursts", sweeps, ejections, etc.,) were seen in both flows so by in-
ference the experimental techniques should be similar. For the time being, the
transitional spot seems to be the primary module in the boundary layer and it
is easily evoked by an electric discharge, or a small jet of fluid ejected from the
surface. Amini (1978) showed that an insipient spot, generated by a jet normal
to the surface, prior to its total breakdown to turbulence may be regarded as an
array of horseshoe vortices which are very orderly. In the initial stages of breakdown the flow field observed by Amini resembles the flow field in a puff. The
spots, after merging in quite random fashion, produce a turbulent boundary
layer, but even an array of spots originating regularly from the same source give
rise to the universal velocity profile (Wygnanski, 1978).

An experiment was carried out in which an evoked transitional spot merged
with a fully developed turbulent boundary layer which was generated in the
usual way by tripping the flow with roughness elements. The purpose of the
experiment was to establish how long can a spot persevere in the turbulent environment before losing its identity. In the range of Reynolds numbers considered, a portion of the spot remained clearly identifiable over extremely long
distances. Conditional pattern recognition techniques were used in the en-
semble-averaging procedure but the educed signatures were not streched or
normalized in any way. The ensembling of the data was done in a few stages.
(Haritonidis et al., 1977; Zilberman et al., 1976).

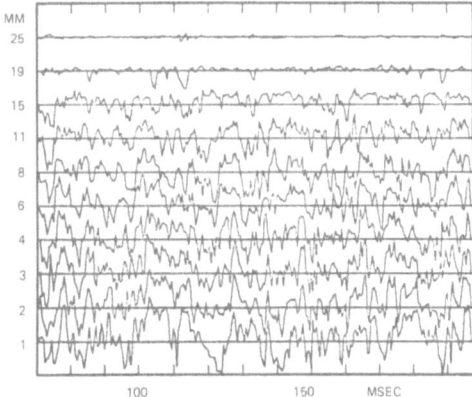

Fig. 11. Velocity signals in a tripped turbulent boundary layer 900 mm downstream of a disturbance. Each trace is displaced 20 % U_∞.

(i)   A spark signal triggered a clock which was used as a basic time reference. The acquisition program was activated by the trigger and a time delay was programmed to allow the evoked spot to reach the measuring station. The signal was recorded on a disc file. Typical velocity signals showing an individual realization are plotted in Fig. 11.

(ii) After recording a large number of events an ensemble average is generated using the original time reference (Fig. 12a). The large eddies arriving at the measuring station at random are averaged out and only the evoked structure represents a perturbation with respect to the local mean velocity. In the case shown 1000 events were considered. The velocity scale in Fig. 12a is four times larger than in Fig. 11.

(iii) The ensemble average generated in the previous step serves as a pattern for aligning other events, or realigning the same set of recorded events. Each event is low-pass-filtered and correlated with the pattern. The time at which the maximum correlation occurs is noted and the individual realizations are shifted in time to bring their large structure into mutual alignment. A new ensemble-averaged pattern is recomputed. The reason for this procedure lies in the fact that the events under consideration do not arrive at the measuring station at precisely the same time, and indeed the new average is in general more representative of the individual large scale events. Only those events which are detected within 10 % of their nominal time arrival are accepted as being the evoked events. By disregarding the events which are associated with larger time shifts the probability of erroneous pattern recognition becomes very small indeed.

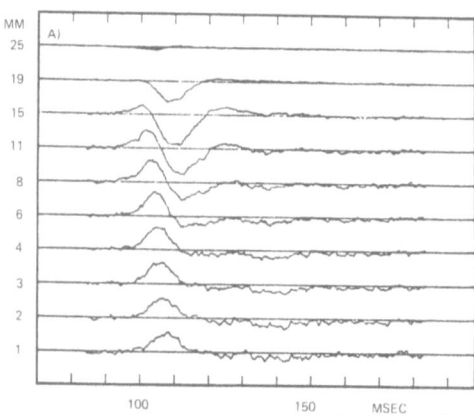

Fig. 12a. Non-aligned ensemble averaged velocity histories in a tripped boundary layer. Each trace is displaced 5 % U$_\infty$.

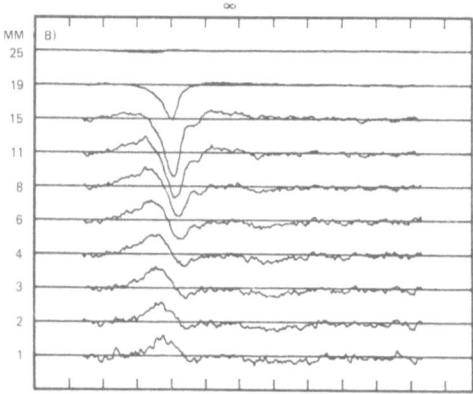

Fig. 12b. Educed and aligned with respect to wire 3 (y = 15 mm from surface).

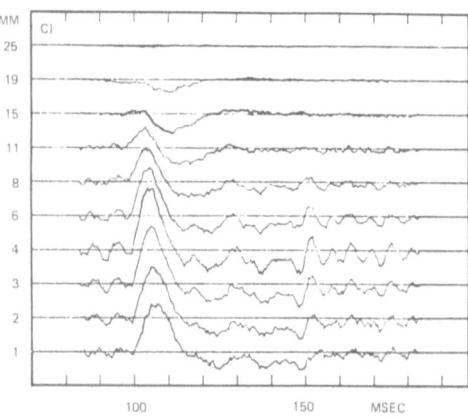

Fig. 12c. Educed and aligned with respect to wire 7 (y = 4 mm).

(iv) The new average can serve again as a pattern for repeating the entire process. It was found empirically that the process converged and the structure stopped changing in most cases after the second iteration (Fig. 13).

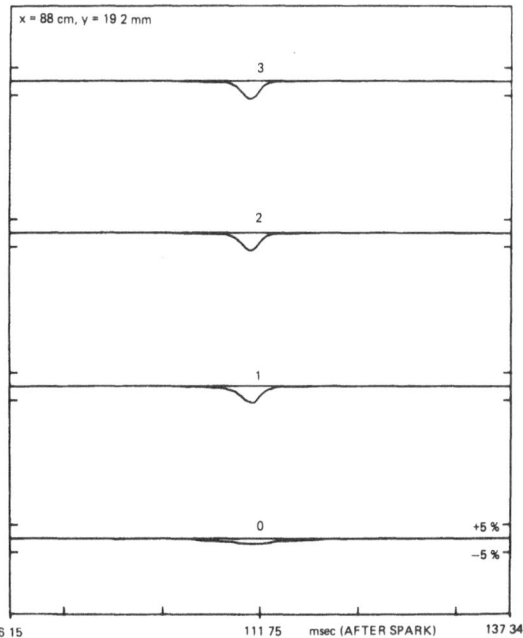

Fig. 13. Development of an educed velocity signature at each stage of the iterative procedure; ±5 % U_∞ departure from local mean is shown.

(v) Whenever an array of sensors is used, the above mentioned procedure may be applied to any or all sensors in the array. By recording the histograms of the time shifts as they are applied to various wires, one may draw some conclusions about the motion of the entire structure. If the application of the procedure to any one wire produces identical results the jitter in the time of arrival of the structure to the measuring station is caused by variations in the convection velocity, and the structure moves in translation like a solid body. Solid body rotation is also easily seen in the histograms, however, in most instances, there is also some internal distortion in the large coherent eddies. For example, applying the pattern recognition technique to the wire located 15 mm from the surface (third trace from the top in Fig. 12b) improved the coherence of the structure in the vicinity of this wire. In particular, the velocity defect observed 15 mm from the surface almost doubled when compared to the unaligned pattern. The effect of alignment at Y = 15 mm was not felt near the surface leaving the velocity history at Y = 1 mm almost unaltered. When a sensor located at Y = 4 mm was used for alignment (Fig. 12c) only the velocity perturbation near the surface was enhanced. When each wire was treated separately, the resulting velocity histories can be almost presented as a combination of Figs. 12b and 12c.

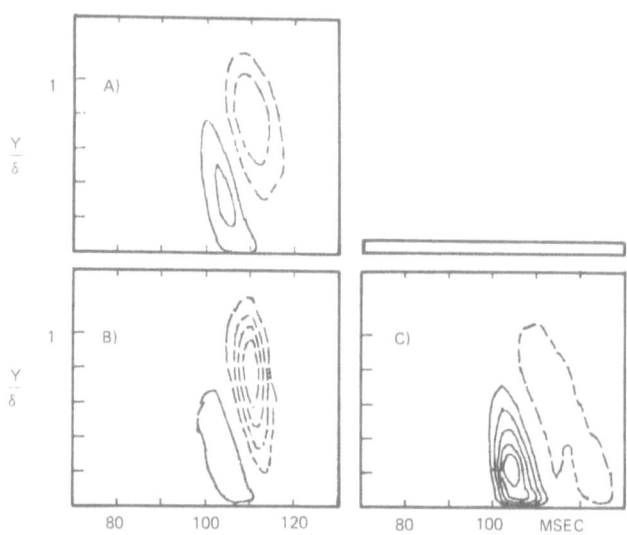

Fig. 14. Countours of velocity perturbation (every 1.5 % U∞). Generated by an evoked structure in a turbulent boundary layer. (a) No alignment (b) Wire 3 aligned (15 mm) (c) Wire 7 aligned (4 mm).

Contours of constant perturbation velocity are generated by cross plotting and interpolating the results in Fig. 12. Fig. 14 a-c corressponds to Fig. 12 a-c, showing contours at intervals of ± 1.5 %. Negative perturbation contours are drawn in broken lines. The narrow rectangle above Fig. 14 is indicative of the distortion of the vertical scale calculated at a convection velocity characteristic of the structure. The contours shown in Fig. 15 a-b are obtained after each wire was realigned individually and are drawn at intervals of ± 3 %. The contours shown in Fig. 15a were measured 900 mm downstream of the disturbance while the contours in Fig. 15b were measured at the 1200 mm station. The perturbation did not weaken in the interval between x_s=900 mm and 1200 mm although the defect region shrank somewhat while the excess region grew in the streamwise (time) direction. In the direction normal to the surface the pattern scales approximately with the boundary layer thickness. The overall scale and character of the pattern does not change by the recognition and re-alignment process.

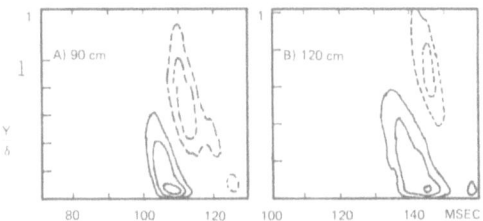

Fig. 15. Velocity perturbation contours (every 5 % U_∞), 900 mm and 1200 mm from the disturbance. Each wire was aligned individually.

There is no doubt that an evoked spot persists in a turbulent boundary layer and it seems to be related to the "Bulge" at the outer part of the boundary layer. Since the evoked spot is a universal structure one may infer that the natural transition spot is also related to the bulge in the same way. Furthermore, there seems to be a relationship between the bursting phenomenon and the outer structure which could be studied in this fashion.

The Two-Dimensional "Vortex" as a Basic Coherent Structure in the Plane Mixing Layer

The evoked structure in a plane mixing layer is obtained by vibrating a ribbon at the trailing edge of a splitter plate or oscillating a flap which is attached to the splitter plate. Both arrangements cause a regular shedding of vortices which triggers an inherent instability in the flow. The two dimensional perturbation was imposed on the flow after it was observed that the naturally generated large eddies are indeed quasi two-dimensional (Oster, Wygnanski and Fiedler, 1976). The structures evoked persisted throughout the length of the test section (which was 1500 mm; $Re_x \cong 10^6$). The data was acquired in the same manner as described before, with a simple phase-averaged pattern serving as a first stage in a pattern recognition program. Velocity histories acquired by a rake of hot wires and the smoke picture shown in the insert exhibit the regularity at which the vortex pattern repeats itself (Fig. 16).

It should be noted that conditional sampling and phase averaging may also be applied to flow visualization. An intermittency signal can trigger an electronic flash or a stroboscope in the same way as an induced perturbation can (e.g. a sine wave generator). Fig. 17 shows an ensemble averaged picture of the large eddies in a two-stream mixing layer. The exposure time was 1 sec. while a stroboscopic light was activated at a frequency of 40 Hz. The light was phase-locked to the perturbation in the flow and hence the pattern appeared to be stationary (Oster et al., 1977).

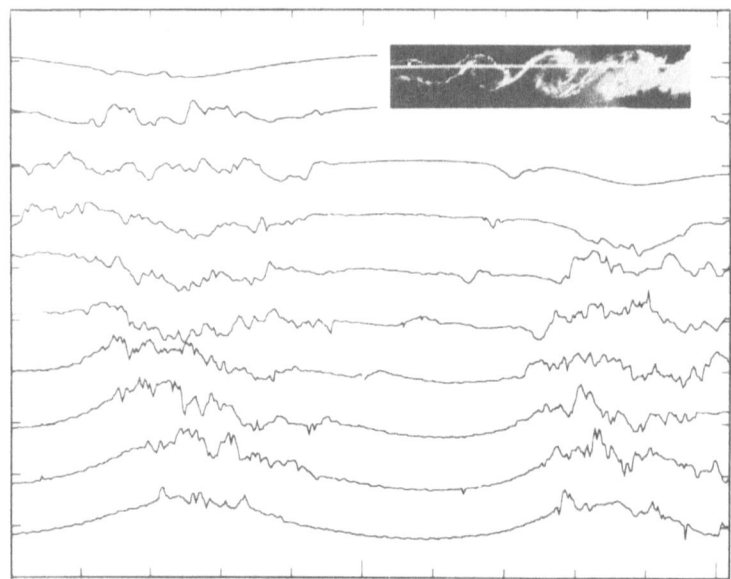

Fig. 16. Velocity histories in a mixing layer and the visualization of the large
coherent structures using smoke.

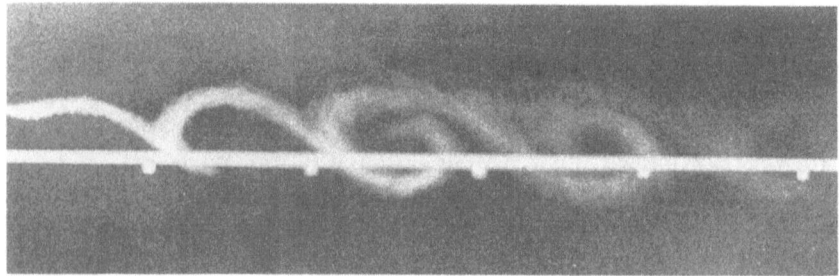

Fig. 17. Phase-averaged photograph of the large coherent structures in a mixing
layer.

In concluding, the use of sensor arrays rather than single probes, digital data
acquisition techniques and evoking coherent structures in the turbulent flow,
seems to be a promising way to help us understand the very difficult problems
at hand.

References

Amini, J. (1978). Transistion controlee en couche limite. Ph.D. Thesis. Universite Scientifique et Medical de Grenoble.

Blackwelder, R.F. and Kaplan R.E. (1972). *Proceedings of 12th IUTAM Congress of Applied Mechanics,* Moscow.

Corrsin, S. and Kistler, A.L. (1955). NACA Rep. 1244.

Emmons, H.W. (1951). *J. Aero Sc.,* **18**, 490.

Haritonidis, J.H., Kaplan, R.E., Wygnanski, I. (1977), in *Lecture Notes in Physics,* Vol. **75**, 234. Springer-Verlag, Berlin (Structure and Mechanisms of Turbulence 1, Proceedings, Berlin 1977).

Hussain, A.K.M.F. and Reynolds, W.C. (1970). *J. Fluid Mech.* **41**, 241.

Kline, S.J., Reynolds, W.C., Schraub, F.A., Runstadler, P.W. (1967). *J. Fluid Mech.* **30**, 741.

Kovasznay, L.S.G., Komoda, H. and Vasudeva, B.R. (1962). *Proceedings of Heat and Transfer and Fluid Mechanics Inst.,* Stanford University Press.

Kovasznay, L.S.G., Kibens, V. and Blackwelder, R.F. (1970). *J. Fluid Mech.,* **41**, 283.

Oster, D., Wygnanski, I. and Fiedler, H. (1976). In *Turbulence in Internal Flows,* SNB Murthy Editor, Hemisphere Press, Washington.

Oster, D., Wygnanski, I., Dziomba, B. and Fiedler, H. (1977). In *Lecture Notes in Physics,* Vol. **15**, 48, Springer Verlag, Berlin (Structure and Mechanisms of Turbulence 1, Proceedings, Berlin, 1977).

Wygnanski, I. (1971). *Israel J. of Technology,* **9**, 105.

Wygnanski, I. and Champagne, F.H. (1973). *J. Fluid Mech.* **59**, 281.

Wygnanski, I., Sokolov, M. and Friedman, D. (1975). *J. Fluid Mech.* **69**, 283.

Zilberman, M., Wygnanski, I. and Kaplan, R.E. (1977). *Phys. Fluids,* **20**, S 258.

Detection of Intermittent Events Maintaining Reynolds Stress

by

Geneviève Comte-Bellot, Jean Sabot and Issal Saleh
Laboratoire de Mécanique des Fluides
Associé au C.N.R.S.
École Centrale de Lyon, 69130 Ecully

1. Introduction

Over the last decade great attention has been paid to the intermittent maintenance of the Reynolds shear stress $- \rho\overline{uv}$ and the possible existence of coherent structures in wall turbulent shear flows. Experimental investigations have been made using different techniques such as:

— flow visualization pictures with fixed or moving camera of the wall region (KLINE, REYNOLDS, SCHRAUB and RUNSTADLER, 1967; CORINO and BRODKEY, 1969; GRASS, 1971) and of the outer part of the flow (NYCHAS, HERSHEY and BRODKEY, 1973; OFFEN and KLINE, 1974; FALCO, 1974 and 1977).

— u, v plane quadrant analysis with classification of events related to the four combinations of signs of u and v (WALLACE, ECKELMANN and BRODKEY, 1972):

quadrant I   : outward interactions ($u > 0$, $v > 0$)
quadrant II   : ejections ($u < 0$, $v > 0$)
quadrant III   : inward interactions ($u < 0$, $v < 0$)
quadrant IV   : sweeps ($u > 0$, $v < 0$).

u is the streamwise component of the velocity fluctuation and v its component along the outward normal to the wall. A detection level $H = |uv|/u'v'$ where $u' = \sqrt{\overline{u^2}}$ and $v' = \sqrt{\overline{v^2}}$ has been later introduced to take into account the "strength" of these events (LU and WILLMARTH, 1973).

— narrow band filtering of u and use of a discriminator level acting on the amplitude of the filtered signal (RAO, NARASIMHA and NARAYANAN, 1971; NARAYANAN, RAJAGOPOLAN and NARASIMHA, 1977).

— The selection of a u velocity pattern (WALLACE, BRODKEY and ECKELMANN, 1977) consistent with observed large skewness factors for the time derivative.

Our previous research has been so far mostly concerned with the second technique. For example, in the core region of a fully developed turbulent pipe flow we have been able to isolate the ejections and sweeps associated with opposite radial directions (SABOT and COMTE-BELLOT, 1976). The case of a very rough wall has also been investigated and clearly illustrates that the intermittent maintenance of $\overline{uv}$ occurs throughout the flow so that external scales rather than wall scales are the relevant parameters (SABOT, SALEH and COMTE-BELLOT, 1977).

In this lecture we shall concentrate on the most recent aspects of our work and the following points will be considered:

— effect of the detection level H on the usual characteristics of ejections, sweeps and interactions (intermittency frequency, intermittency factor, fractional contribution to $\overline{uv}$).

— suggestion of a characteristic level of detection (noted $\tilde{H}_J$ with J = I, II, III, IV according to the u, v quadrant analysis) useful for comparisons in various turbulent shear flows.

— determination of new statistical characteristics related to the intermittent maintenance of $\overline{uv}$ such as:

the amplitude distribution of u and v during ejections and sweeps

the signature of the u, v and uv fluctuations during ejections and sweeps

the mean streamwise size of ejections and sweeps with use of a modified Taylor approximation.

Digital data acquisition along with conditional sampling and ensemble averaging techniques were the main tools of our recent investigations. The number of imposed conditions was often large. For example, 4 conditions were needed when analyzing the amplitude distribution of u and v during ejections or sweeps.

II. Experimental Procedure

II.1 Flows Investigated

Most of the experiments were made in a fully developed pipe flow with a smooth wall at a Reynolds number $U_{max} \cdot D/\nu$ = 135 000 (center line velocity: U_{max} = 20 ms^{-1}, pipe diameter D = 2R = 10 cm). Detailed properties of the turbulence such as turbulence intensities, integral and Taylor length scales, spectra, integral time scales in convected coordinates are available elsewhere (SABOT, RENAULT and COMTE-BELLOT, 1973; SABOT and COMTE-BELLOT, 1976; SABOT, 1976).

In addition, two other flows were investigated when checking that the threshold suggested in Section IV was indeed appropriate:

— a pipe flow with a very rough wall. The geometry of the ring shape roughness, which is similar to that used by HANJALIC and LAUNDER (1972) in an asymmetric channel, is such that the friction velocity is larger by a factor of 2.5 than in a smooth pipe at the same Reynolds number (SABOT, SALEH and COMTE-BELLOT, 1977)

— a conventional boundary layer (smooth plate wall with zero pressure gradient) described by CHARNAY (1974) with an external velocity $U\infty$ of 5 ms^{-1}. At the measurement station the boundary layer thickness δ was 6.1 cm and the Reynolds number $Re_{\delta} = U\infty \; \delta/\nu$ = 20 000.

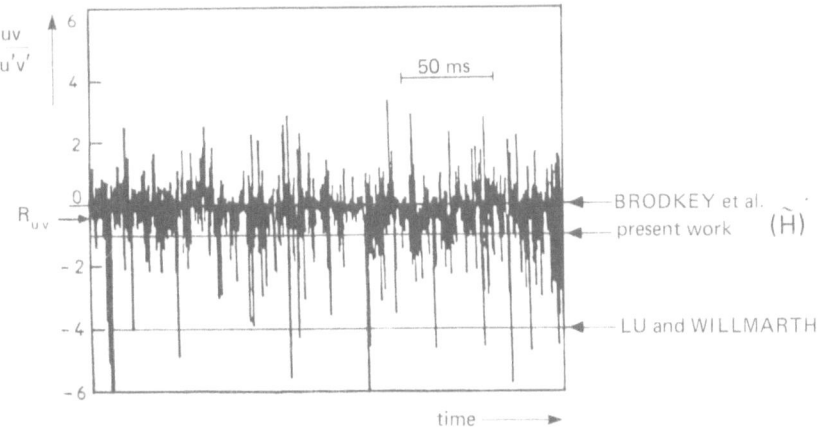

Fig. 1. Example of record of uv fluctuation (smooth pipe; Re = 135 000; distance from the wall y/R = 0.40) and location of the different thresholds used in literature to detect ejections.

II.2 Hot-Wire Anemometry

Standard DISA equipment was used: X-meter probe type 55A38, constant temperature anemometers type 55D01 and linearizers type 55D10. The separation of u and v was accomplished by means of BURR BROWN operational amplifiers type 3114 and 3003.

II.3 Data Acquisition and Data Processing

Digital acquisition of the u and v signals was made at a sampling frequency of 20 kHz and with 12 bits including sign (PRESTON-SERCEL equipment type GMAD 3). For all the flows investigated this sampling frequency was larger than twice the value of the highest frequencies contributing to the spectra of u and v.

The data were then stored on a magnetic disc (5000 k mots) and processed on a HEWLETT PACKARD 2108 computer by means of Fortran programs (SALEH 1978). Each simultaneously stored record of u and v corresponded to a real time of 10 s. which is noticeably larger than the life-time of the most coherent structures (SABOT and COMTE-BELLOT, 1972).

Fig. 1 shows a typical uv signal with which most of this work deals. It corresponds to the inner part of the smooth pipe flow (y/R = 0.40 at which $\overline{uv}$ = 0.40 u'v'). It can be observed that the maintenance of $\overline{uv}$ is related to the existence of many large negative peaks of uv.

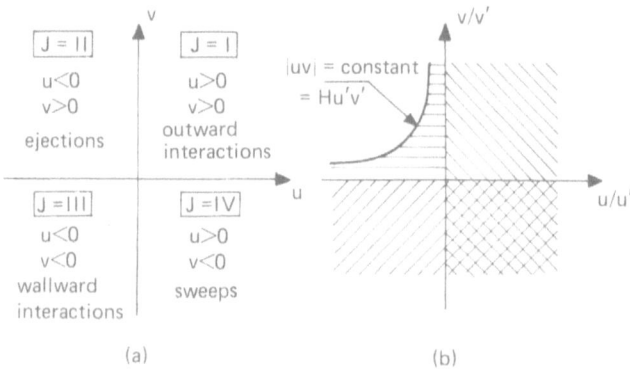

Fig. 2. (a) Classification of the four classes of events in the u, v plane; (b) signal conditioning of the uv fluctuation in a given quadrant of the u, v plane (example: J = II. Three conditions: u < 0, v > 0 and | uv |/u'v' ≥ H).

The sequence of conditions imposed on the signals was as follows (Fig. 2):

- sign of u
- sign of v } hence choice of a u,v plane quadrant (J = I, II, III, IV)

- amplitude of uv larger than an arbitrarily fixed threshold H i.e. | uv | $\geqslant$ H u'v' (hence only the highest peaks of the uv signal are kept when H is large)
- when investigating the amplitude distribution of u (or v) *during* ejections or sweeps an additional condition is introduced on u (or v) i.e. | u | $\geqslant h_u$ u' or | v | $\geqslant h_v$ v' where h_u and h_v are arbitrarily "fixed thresholds (Fig. 3)".

Statistical characteristics of the different events were then analyzed using the corresponding conditional data. For the u, v and uv signatures (i.e. ensemble averages) given in section VI, these three first conditions determined a reference time associated with each sample of the simultaneous u, v and uv traces. Ensemble averages were then made with all the reference times brought into coincidence. The number of samples taken into account to construct ensemble averages is of the order of 10^3.

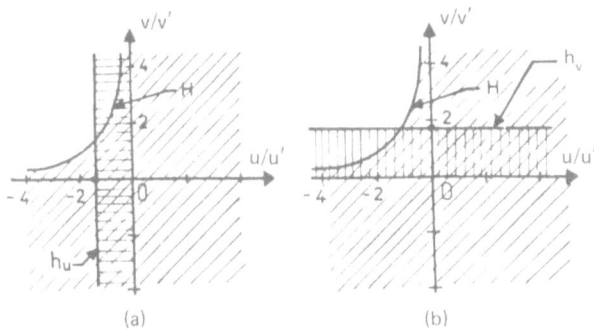

Fig. 3. (a) Signal conditioning of u during ejections detected at the level H. Four conditions: u < 0, v > 0, | uv |/u'v' $\geqslant$ H and u/u' $\leqslant h_u$ (b) signal conditioning of v during ejections detected at the level H. Four conditions: u < 0, v > 0, | uv |/u'v' $\geqslant$ H and v/v' $\geqslant h_v$.

III. Effect of the Threshold H on the Usual Characteristics of Ejections, Sweeps and Interactions

For the ejections, sweeps, inward and outward interactions (i.e. quadrants II, IV, III and I), the statistical characteristics usually considered are:

- the fractional contributions to the Reynolds stress
- the mean period T_J between two successive events of class J always beginning at the leading front

— the intermittency factor γ_j defined as the percentage of time during which events of class J are present.

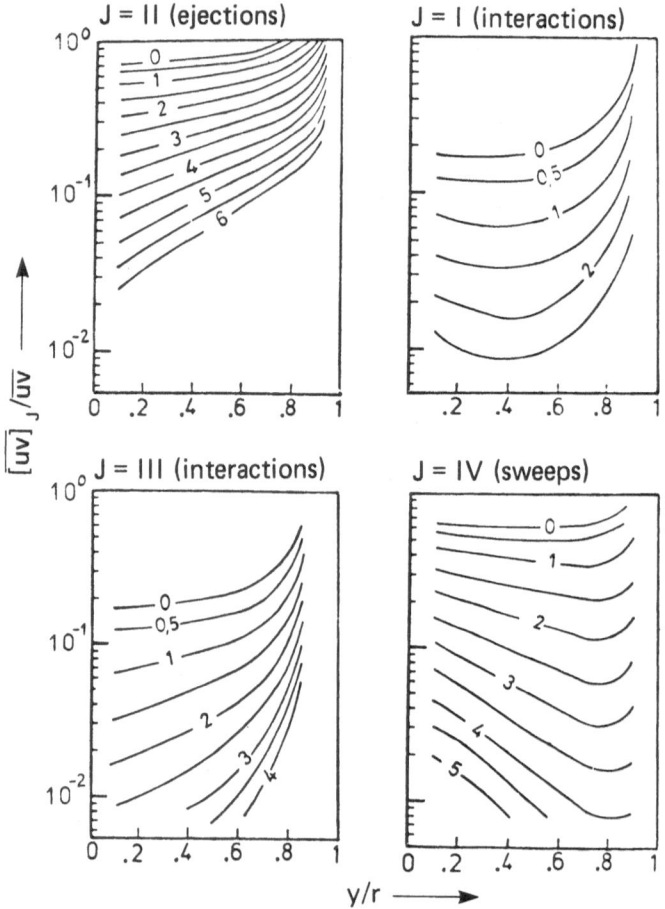

Fig. 4. Contribution to $\overline{uv}$ from the events of class J as function of the distance from the wall for several values of the detection level H (smooth pipe, Re = 135 000).

In the case of the smooth pipe flow, Figs. 4, 5 and 6 show how these characteristics depend on the level of detection H = | uv |/u'v' for various distances from the wall.

In Fig. 4 it can be noted that the relative importance of these different events remains the same for small and moderate values of H in most of the inner part of the flow (0.1 ≤ y/R ≤ 0.7). The ejections are indeed the dominant events.

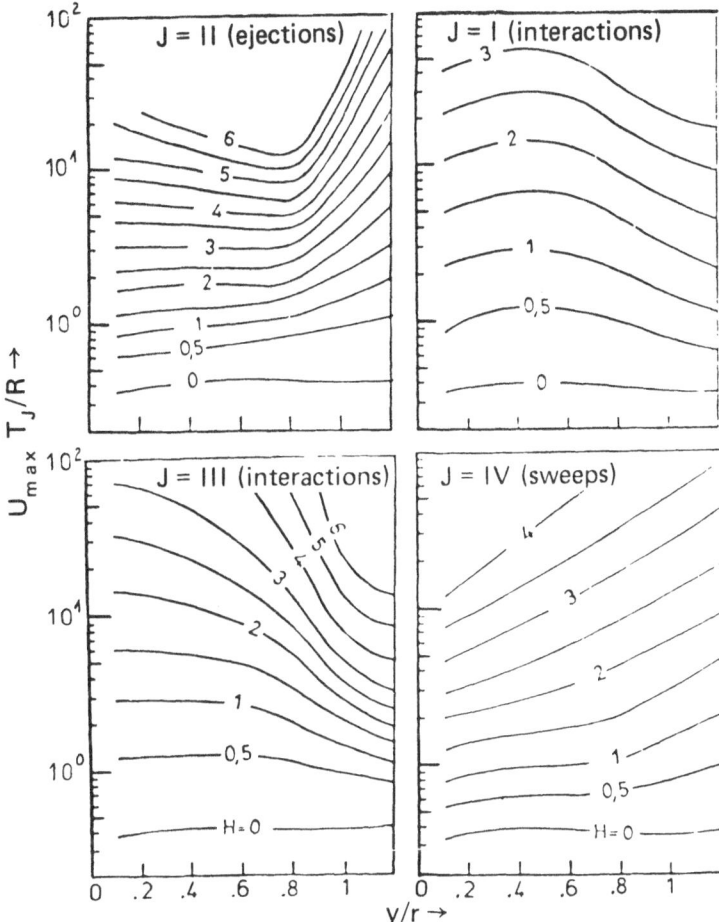

Fig. 5. Effect of the detection level H on the events of class J as function of the distance from the wall (smooth pipe, Re = 135 000).

The sweeps come next, and the interactions are nearly negligible. In the core region $(y/R \geqslant 0.7)$ the inward (and outward) interactions seem to increase in importance, but that comes only from a contamination due to the ejections (and sweeps) which are associated with the radially opposite wall and which extend beyond the pipe axis (SABOT and COMTE-BELLOT, 1976).

On the other hand, the effect of H is quite large on T_J (or $U_{max} T_J/R$) and γ_J as expected (Figs. 5 and 6). However, it can be noted that T_J and γ_J are nearly constant across the inner region for small and moderate values of H. In the core region, the apparent decrease of T_J and γ_J for the interactions is again due to the ejections and sweeps due to the opposite wall.

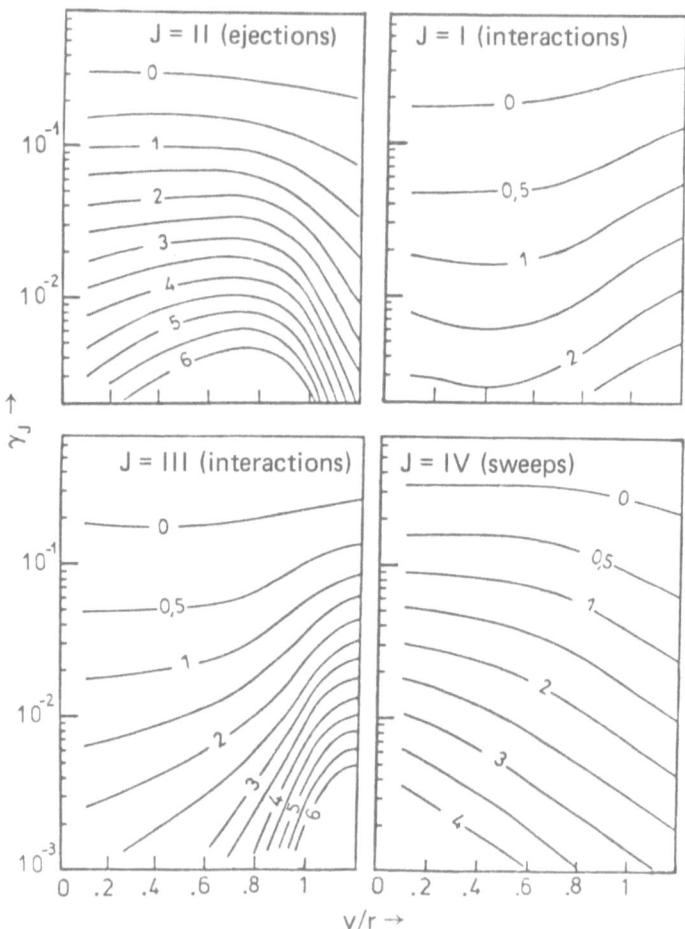

Fig. 6. Effect of the detection level H on the time intermittency factor γ_J of the events of class J as function of the distance from the wall (smooth pipe, Re = 135 000).

IV. Suggestion for a Characteristic Threshold

From the above results it is clear that a relevant level of detection has to be defined for every class J of events. So far only two procedures have been used:

- $H_J = 0$ for any J (BRODKEY, WALLACE and ECKELMANN, 1974)
- $H_{II} = 4$ and $H_{IV} = 2.5$ as derived from a criterion based on the fractional contribution from every quadrant J to $\overline{uv}$ (LU and WILLMARTH, 1973).

These two procedures are not fully satisfactory since in the former the same importance is given to the very "weak" and to the very "violent" events, whereas for the latter only the "violent" events are retained (see Fig. 1). Furthermore, in the latter case, the events J = III and J = I cannot be separated because of their similar fractional contribution to $\overline{uv}$.

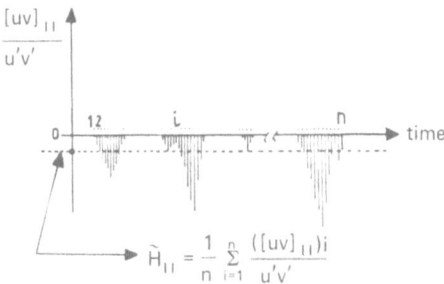

Fig. 7. Sketch showing the definition of the characteristic threshold $\tilde{H}_J$ in the case of ejections (J = II).

In order to introduce an objective procedure we propose to define a threshold $\tilde{H}_J$ based on the conditional average value of uv over the events of class J only, *but over all these events* (Fig. 7). The values of $\tilde{H}_J$ obtained in different flow configurations are given in Fig. 8. At fixed J they do not significantly depend on the flow considered and are of the same order across the inner wall region, i.e.:

$$\tilde{H}_{II} \cong 1$$
$$\tilde{H}_{IV} \cong 0.70$$
$$\tilde{H}_I \cong \tilde{H}_{III} \cong 0.40$$

Use of this new level of detection $\tilde{H}_J$ is illustrated in Fig. 9 which gives the values of $\tilde{T}_J$ (or $U_{max} \tilde{T}_J/R$ or $U_{max} \tilde{T}_J/\delta$) obtained in different flows. The collapse of all the values around:

$$U_{max} \tilde{T}_{II}/R \cong 0.85 - 1.25$$
$$U_{max} \tilde{T}_{IV}/R \cong 0.65 - 0.85$$

indicates that the relevant scales are the outer flow parameters U_{max} and R (or δ) and that the streamwise separation between these events is of the order of the largest integral length scale of the flow, i.e. $L_{11}^{(1)}$ related to the u fluctuation and a streamwise separation. Furthermore the fact that $\tilde{T}_{II} > \tilde{T}_{IV}$ indicates that some ejections could be followed by more than one sweep.

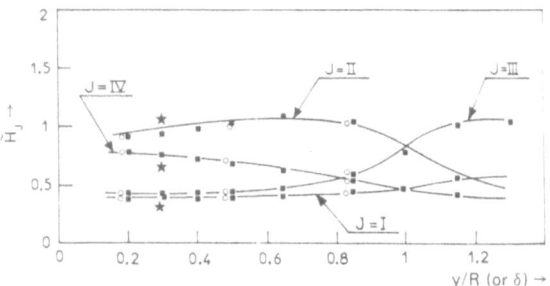

Fig. 8.   Profiles of the characteristic threshold $\tilde{H}_J$
   - (smooth pipe, Re = 135 000);
   ○ (smooth pipe, Re = 68 000);
   ■ (rough pipe, Re = 135 000);
   ★ (boundary layer, Re_δ = 20 000).

Finally, we have observed that $\tilde{T}_{II}/T_{u,0} \cong 1.6 - 1.9$ where $T_{u,0}$ is the mean period between two successive zero crossings of u. From a very different experimental way NARAYANAN, RAJAGOPOLAN and NARASIMHA (1977) have found a mean "burst" period T_b such that $T_b/T_{u,0} \cong 1.66$. We deduce therefore that $\tilde{T}_{II} \cong T_b$. However, extension to other flow configurations including free shear flows such as jets or wakes, and consideration of other Reynolds numbers would be of great interest to confirm the trend of this first set of data.

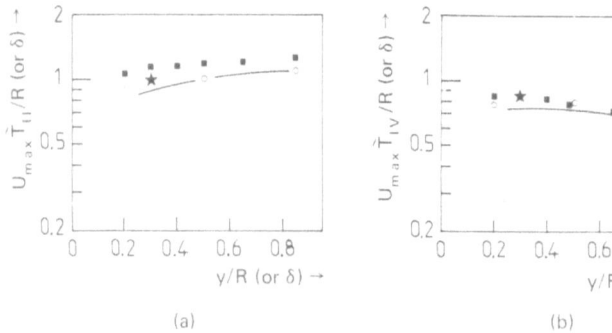

(a)   (b)

Fig. 9. Mean period $\tilde{T}_J$ between two successive ejections or sweeps detected at level $\tilde{H}_J$ as function of the distance from the wall:
   — (smooth pipe, Re = 135 000)
   ○ (smooth pipe, Re = 68 000)
   ■ (rough pipe, Re = 135 000)
   ★ (boundary layer, Re_δ = 20 000).

In what follows, the effect of H will be further systematically considered so that all the characteristics corresponding to the suggested threshold $\tilde{H}_J$ can be easily deduced from the above given values of $\tilde{H}_J$.

V. Amplitude Analysis of u and v During Ejections (or Sweeps)

For the main events contributing to the Reynolds stress $\overline{uv}$ hence providing large negative peaks in the uv signal, it is interesting to investigate the concomitant amplitudes reached by the u and v signal separately. The additional conditions $|u| \geqslant h_u u'$ or $|v| \geqslant h_v v'$ which are needed have been indicated in section II and an example of the experimental findings is given in Fig. 10. It corresponds to the ejections observed in a smooth pipe at the wall distance $y/R = 0.40$. Other data and additional interpretations are available in the work of SALEH (1978).

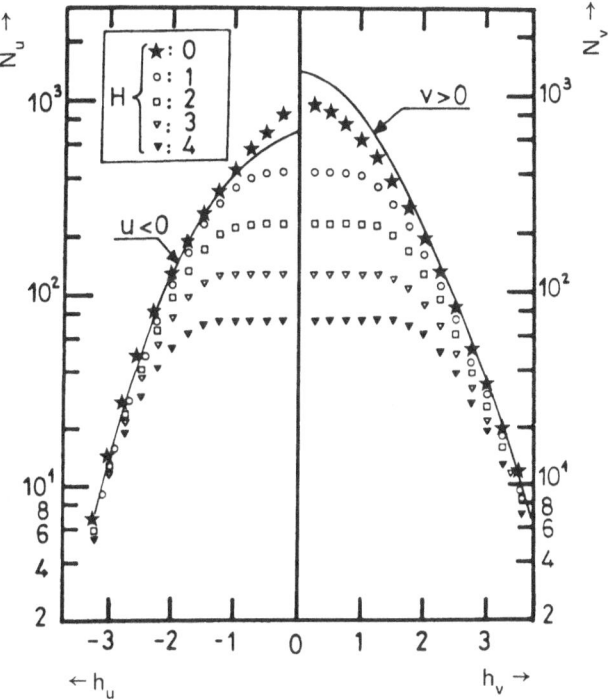

Fig. 10. Frequency distribution N_u of u and N_v of v during ejections detected at the level H as function of the thresholds h_u and h_v. Solid lines correspond to the only truncated fluctuations (smooth pipe, Re = 135 000, y/R = 0.40).

From Fig. 10, where N_u (or N_v) denotes the mean crossing frequency of u (or v) at the threshold h_u (or h_v) two main results can be obtained:

— there is a plateau for N_u (or N_v), at fixed H over a relatively large h_u (or h_v) interval from $h_u = 0$ (or $h_v = 0$). As a consequence, u and v have to be simultaneously relatively large during ejections (i.e. $| u | \cong 1.5\,u'$ and v $\cong 1.5\,v'$ as soon as $H > 3$).

— for large h_u (or h_v), N_u (or N_v) is independent of H and has the same value as that obtained for a merely truncated signal $u < 0$ (or $v > 0$), i.e. not conditioned by ejections. The largest amplitudes of u and v are therefore entirely due to the ejections.

VI. Signature of u, v and uv During Ejections (or Sweeps)

An example of the ensemble average of u, v and uv traces is given for the smooth pipe flow in Fig. 11 for ejections and Fig. 12 for sweeps. Around the instant of detection, a time interval of -1.5 ms, $+1.5$ ms has been chosen in order to retain the whole duration of the selected events.

From Figs. 11 and 12, a specific form, hence a signature, is obtained for u, v and uv whatever the value of H. In particular, an asymmetry exists on the traces and their time derivatives on either side of the instant of detection. In order to sort out results which are pertinent to turbulence (i.e. not observed with randomly generated signals), it might prove useful to compare such a behavior (a) with the asymmetrical pattern assumed by WALLACE, BRODKEY and ECKELMANN (1977) for $\partial u / \partial t$ and (b) with the role attributed to the odd moments of the joint probability distributions of u and v by NAKAGAWA and NEZU (1977).

VII. Mean Streamwise Size of Ejections (or Sweeps)

Because of the large velocity defect which exists during ejections as shown by Fig. 11 (and similarly the large velocity increase durings sweeps, Fig. 12) the mean streamwise size ℓ_J of ejections and sweeps cannot be deduced in a fully satisfactory way from the mean time duration $\Delta T_J = \gamma_J T_J$ of the events through the usual Taylor approximation based on the local mean velocity U. We have therefore used a conditional average of the instantaneous streamwise velocity during the selected events (noted $\tilde{U}_J$, Fig. 13) so that:

$$\ell_{II} = \tilde{U}_{II}\,\Delta T_{II} \text{ for ejections and}$$

$$\ell_{IV} = \tilde{U}_{IV}\,\Delta T_{IV} \text{ for sweeps}$$

Results concerning these sizes are given in Figs. 14 and 15 for different values

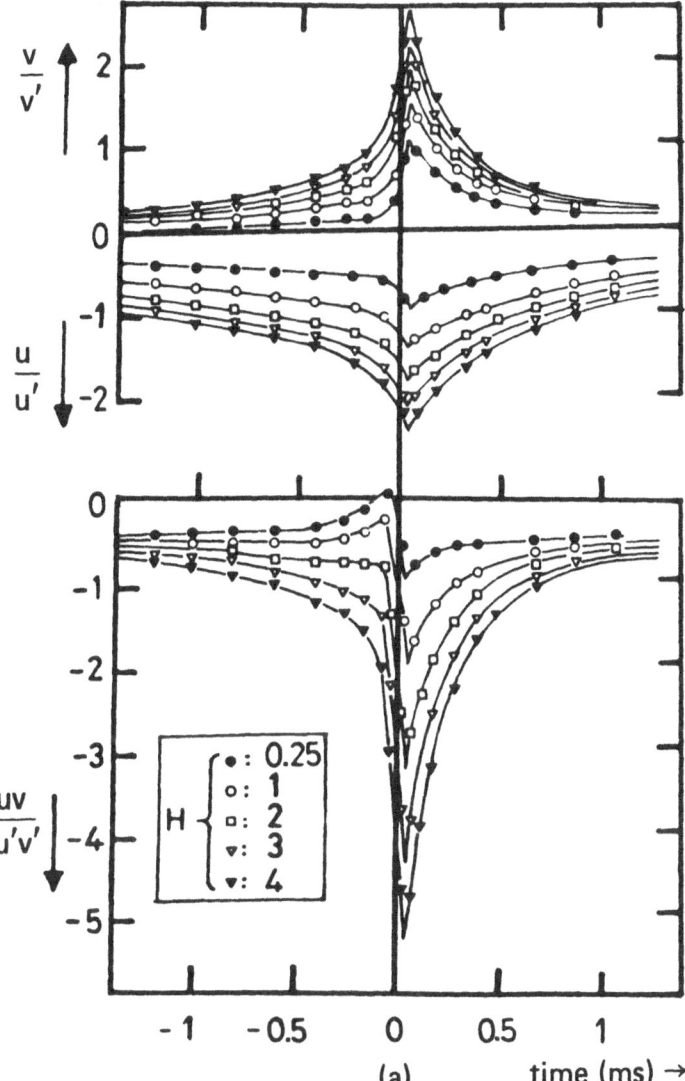

Fig. 11. Signature of ejections: ensemble averages of the u, v and uv signals around the time of detection $t_0 = 0$ of the ejections as function of their level of detection H (smooth pipe, Re = 135 000, y/R = 0.40).

of the level of detection H for the smooth pipe and at the level $\tilde{H}_J$ suggested in section IV for several flow configurations. It is important to note that:

$$\frac{\tilde{U}_{II}\,\widetilde{\Delta T}_{II}}{R \text{ (or } \delta)} \cong \frac{\tilde{U}_{IV}\,\widetilde{\Delta T}_{IV}}{R \text{ (or } \delta)} \cong 0.06 - 0.10$$

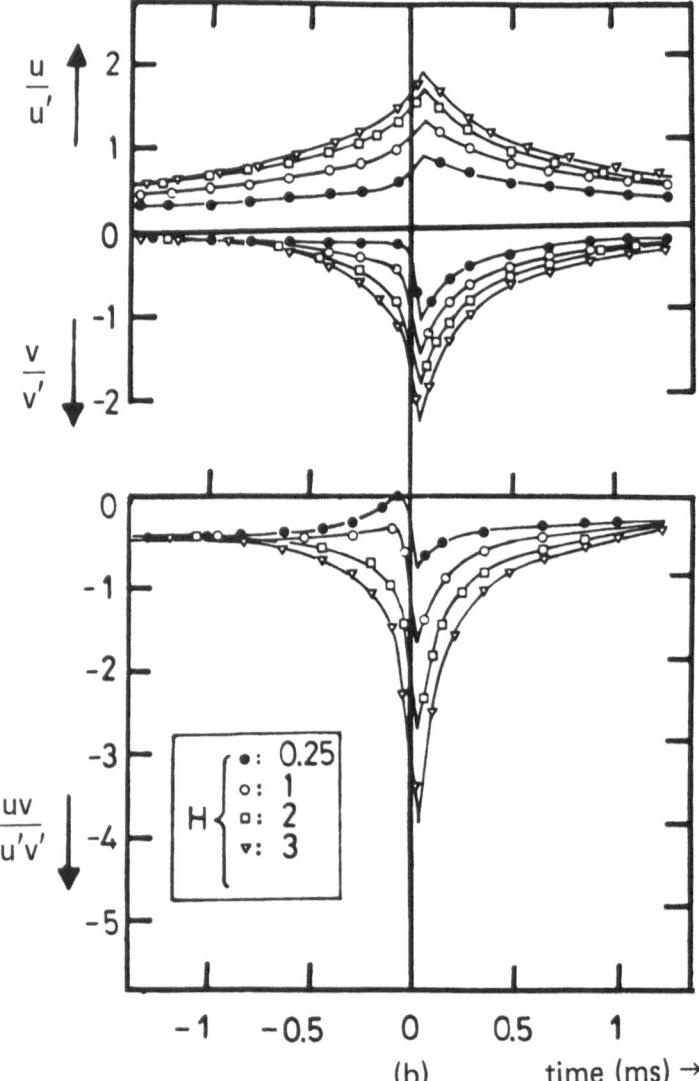

Fig. 12. Signature of sweeps: ensemble averages of the u, v and uv signals around the time of detection $t_0 = 0$ of the sweeps as function of their level of detection H (smooth pipe, Re = 135 000, y/R = 0.40).

For the smooth pipe the mean streamwise size of ejections and sweeps can then be compared with the integral length scales previously obtained (SABOT, 1976). It therefore appears that these sizes are smaller by a factor of 10 than $L_{11}^{(1)}$ considered in section IV. On the other hand these sizes are of the same order as $L_{22}^{(1)}$ or $L_{22}^{(2)}$ which are the integral length scales of the v fluctuation in either a streamwise or a radial separation. This information helps to understand the spatial structure of turbulence in a pipe flow.

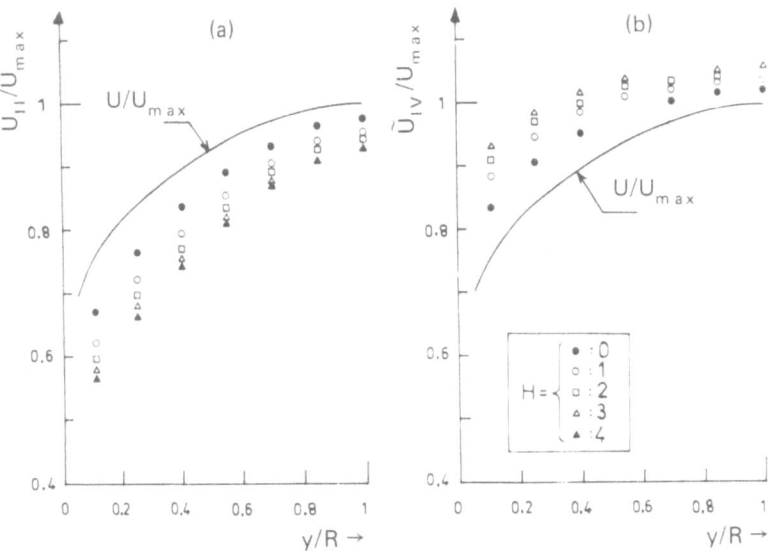

Fig. 13. Conditional average $\tilde{U}$ of the instantaneous streamwise velocity during ejections (a) and sweeps (b) detected at the level H as function of the distance from the wall. (Smooth pipe, Re = 135 000).

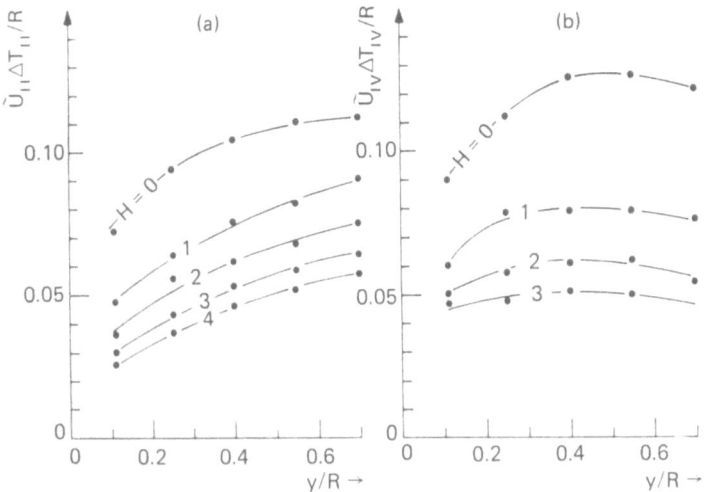

Fig. 14. Mean streamwise size of: (a) ejections and (b) sweeps as function of the distance from the wall for several values of H. (Smooth pipe, Re = 135 000) Data obtained by a modified Taylor approximation.

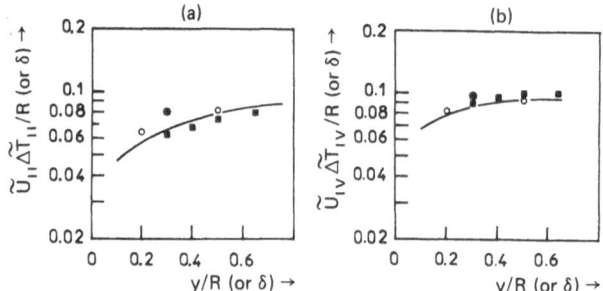

Fig. 15.   Mean streamwise size of: (a) ejections and (b) sweeps detected at level $\tilde{H}_J$ as function of the distance from the wall:
   — *(smooth pipe, Re = 135 000)*
   ○ *(smooth pipe, Re = 68 000)*
   ■ *(rough pipe, Re = 135 000)*
   ★ *(boundary layer, Re_δ = 20 000)*
Data obtained by a modified Taylor approximation.

Conclusion

In this paper we described several signal processing techniques. These techniques allowed us to obtain information relevant to the intermittent events maintaining the Reynolds stress in various bounded turbulent shear flows. Other intermittent phenomena for which two correlated random variables are simultaneously recorded could be investigated in a similar way, for example the transverse heat flux $\overline{v\theta}$ in heated turbulent shear flows.

In addition to the detailed and quantitative description which could be obtained, it is hoped that this amount of information could be useful to elaborate better numerical models of turbulence or to allow comparisons with the intermittent phenomena which may appear when simulating directly the Navier-Stokes equations.

Finally, research on the physical mechanisms which trigger the intermittent phenomena observed is also an important objective for future work with conditional techniques. In order to obtain a dynamical description of the flow organization before, during and after the detected event, one must emphasize that the present signal processing techniques should probably be applied to several points of observation in space and time. In that context the association of flow visualizations with signal processing techniques could prove to be a powerful method for further investigations.

References

BRODKEY, R. WALLACE, J.M., and ECKELMANN, H. 1974 *J. Fluid Mech.*, **63**, 209

CHARNAY, G., 1974 "Caractéristiques d'une couche limite turbulente évoluant en présence d'un écoulement extérieur turbulent". Thèse d'Etat No. 264, Université de Lyon

CORINO, E.R. and BRODKEY, R.S., 1969 *J. Fluid Mech.* **37**, 1

FALCO, R.E., 1974 *AIAA*, 12th Aerospace Sc. Meeting, Washington

FALCO, R.E., 1977 *Phys. Fluids* **20**, S124

GRASS, A.J., 1971 *J. Fluid Mech.* **50**, 233

HANJALIC, K. and LAUNDER, B.E., 1972 *J. Fluid Mech.* **51**, 301

KLINE, S.J., REYNOLDS, W.C., SCHRAUB, F.A. and RUNSTADLER, P.W., 1967 *J. Fluid Mech.* **30**, 741

LU, S.S. and WILLMARTH, W.W., 1973 *J. Fluid Mech.* **60**, 481

NAKAGAWA, H. and NEZU, I., 1977 *J. Fluid Mech.* **80**, 99

NARAYANAN, M.A.B., RAJAGOPOLAN, S. and NARASIMHA, R., 1977 *J. Fluid Mech.* **80**, 237

NYCHAS, S.G., HERSHEY, H.C. and BRODKEY, R., 1973 *J. Fluid Mech.* **61**, 513

OFFEN, G.R. and KLINE, S.J., 1974 *J. Fluid Mech.* **62**, 223

RAO, K.M., NARASIMHA, R. and NARAYANAN, M.A.B., 1971 *J. Fluid Mech.* **48**, 339

SABOT, J., 1976 "Etude de la cohérence spatiale et temporelle de la turbulence établie en conduite circulaire" Thèse d'Etat No. 76-36, Université de Lyon

SABOT, J. and COMTE-BELLOT, G., 1976 *J. Fluid Mech.* **74**, 767

SABOT, J., RENAULT, J. and COMTE-BELLOT, G., 1973 *The Phys. of Fluids,* **16**, 1403

SABOT, J., SALEH, I. and COMTE-BELLOT, G., 1977 *The Phys. of Fluids* **20**, S150

SALEH, I., 1978 "Contribution à l'étude de l'entretien intermittent de la tension de Reynolds dans les écoulements turbulents anisotropes" Thèse de Docteur-Ingénieur No. 295, Université de Lyon

WALLACE, J.M., ECKELMANN, H. and BRODKEY, R., 1972 *J. Fluid Mech.* **54**, 39

WALLACE, J.M., BRODKEY, R.S. and ECKELMANN, H., 1977 *J. Fluid Mech.* **83**, 673

Measurement of Spanwise Distribution of Turbulent Structures

by

James F. Keffer

Mechanical Engineering, University of Toronto, Canada

Abstract

Techniques which can be used to determine the three-dimensional nature of the structure within turbulent flows are described. The methods include spanwise, space-time correlations based on conventional and intermittency signals and evaluation of co-spectra and coherence.

Introduction

It is now generally accepted that turbulent flows are not as random as once thought. Large eddy structures in the outer intermittent regions of free shear flows, (Grant, 1958; Keffer, 1965) have been observed for some time and the effect of these organized eddies upon the processes of spread and entrainment has been incorporated into models of the turbulence motion, (Townsend, 1970). More recently, so-called coherent structures have been detected within the inner regions of turbulent flows, e.g. the vortex pairing phenomenon in the mixing layer reported by Brown and Roshko (1974) and Winant and Browand (1974).

The identification of these structures has been aided in great part by the rapid development over the last decade of a number of signal processing techniques. The conditioning of the raw signal to the presence or absence of turbulent fluid, (Kovasznay, et al. 1970, Hedley and Keffer, 1974a), in intermittent flows has allowed us to distinguish the separate contributions of the potential and turbulent fields. Point statistics have provided us with a view of the turbulent/non-turbulent interface as well as a mapping of the interior of these bursts (Antonia, 1972; Hedley and Keffer, 1974b; LaRue & Libby, 1974). Conditioning upon the sign and the magnitude of the signals has permitted a breakdown of the contribution to the important transport terms. More recently, pattern recognition techniques have been employed to search out the underlying structures of the motion, within fully turbulent flows (Blackwelder, 1977, Wallace, et al., 1977).

All of these approaches have given us considerable insight into the streamwise development of the motion and its associated spread laterally into the ambient environment. Correlation mappings e.g. Grant (1958) and Kovaznay et al. (1970) have revealed spanwise features for the motions as well, but by comparison relatively little is known about the extent of these. The focus of the present paper is upon this latter aspect. Techniques are presented which permit a general description of the 3-dimensional nature of the large eddies of the flow. We shall restrict the examination to nominally plane flows. In some cases, it will be convenient to heat the flows slightly, using temperature as a passive scalar contaminant to trace the structural features. Basic signal conditioning upon either the velocity or temperature will be employed to define the turbulent activity within intermittent regions. Space-time correlations, (Favre et al., 1957, 1958) based on both the turbulent fluctuating quantity and the intermittency function will be used to construct a first order model of the composite 3-dimensional eddy. To complete the picture, further details of the transport and structure will be obtained by examining the co-spectra and coherence of the fields.

Experimental Approach

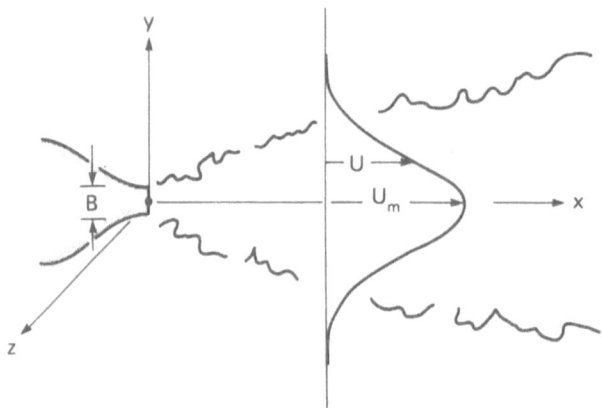

Fig. 1. Definition sketch of the plane jet.

A variety of basic turbulent shear flows has been used to generate the data for illustrating the techniques. Measurements have been taken in both the intermittent and fully turbulent regions of a plane jet discharging into an ambient environment (Moum, 1978) defined in Fig. 1; a heated two-dimensional wake (Barsoum et al., 1978) shown in Fig. 2; a mixing layer with a step jump in temperature (Béguier et al., 1979) in Fig. 3 and heated asymmetric plane jet (Keffer et al., 1979) in Fig. 4.

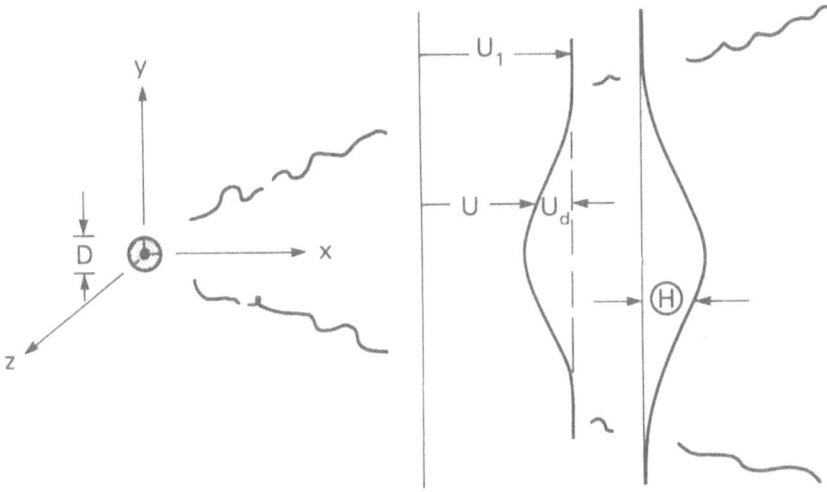

Fig. 2. Definition sketch of the heated two-dimensional wake.

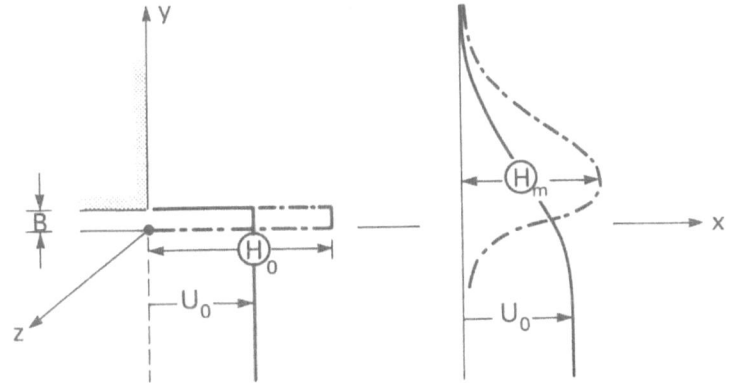

Fig. 3. Definition sketch of the mixing layer with a jump in temperature.

The raw signals representing instantaneous values of the velocity field, u, v and sometimes w, and of temperature θ, were recorded using conventional DISA hot-wire anemometer equipment. In some cases, for the temperature fields, special probes (Béguier et al., 1978) were employed for simultaneous measurement of u,v. and θ and separation of temperature from the velocity, achieved either by appropriate electronic circuitry or by digital techniques (Keffer et al., 1978). The data were generally recorded on analogue tape using an AMPEX PR 2200. Digitization of the data was accomplished at the University of Toronto Computer Centre using a PDP 11 at rates up to 10 kHz per channel and then stored for subsequent processing on an IBM 370-165.

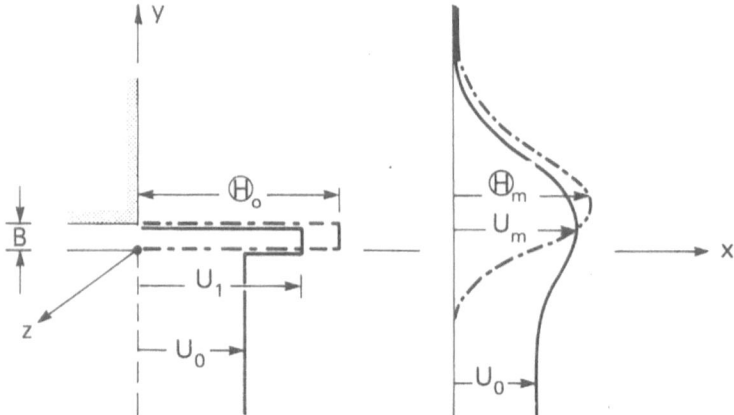

Fig. 4. Definition sketch of the asymmetric heated plane jet.

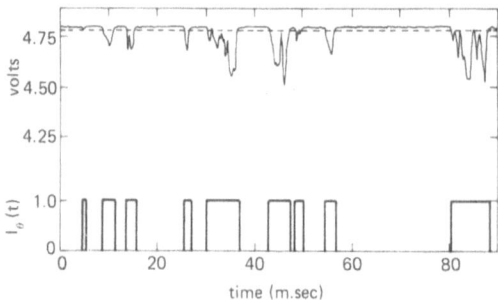

Fig. 5. Typical temperature-based intermittency signal, heated two-dimensional wake.

Analytical Background

For analysing the intermittent flow rather standard signal conditioning techniques were used to generate a turbulence indicator function. These were based on either velocity or temperature. The indicator function is defined as

$$I(x,y,z,t) = \begin{cases} 1 \text{ if turbulent fluid occurs at } (x,y,z) \text{ at time } t \\ 0 \text{ otherwise,} \end{cases}$$

and the mean value of this random square wave, $\overline{I}(x,y,z)$, is γ, the intermittency factor.

The generation of $I(t)$ can be accomplished in a number of ways. Typically, discrimination of the raw signal is based upon the behaviour of a non-negative,

turbulence detector function, D(t). The problem is not straightforward however, since any real detector function will occasionally take on zeros inside a
valid turbulent zone as a simple consequence of the characteristic of its probability density function. This is termed the signal dropout problem. Moreover,
D(t) may be expected to have non-zero values in a valid potential region of the
flow because of background low level turbulence and legitimate velocity fluctuations in the ambient "irrotational" fluid. This is termed the noise problem.
Both effects are discussed more fully in Hedley and Keffer (1974a).

There are basically three methods of reducing the effects of signal dropout and
background noise. The one most commonly employed is to incorporate a hold
time, T_H. Thus an intermediate function H(t) is formed such that

$$H(t) = \begin{cases} 1 \text{ when } D(t) \geqslant 0 \\ 0 \text{ otherwise,} \end{cases}$$

where C is a prescribed threshold level for D(t). Any zero values of H(t) of
duration less than T_H are converted to unity values and the resulting function
is taken to be I(t).

A second method involves a short term integration or smoothing of D(t) over
a time period T_S. This generates a criterion function S(t) and the threshold
level is applied to this, i.e.,

$$I(t) = \begin{cases} 1 \text{ when } S(t) \geqslant 0 \\ 0 \text{ otherwise,} \end{cases}$$

A third method involves the use of back-up detector functions based upon
higher order derivatives of the basic detector function, e.g.,

$$I(t) = \begin{cases} 1 \text{ when } D(t) \geqslant C \text{ or } D_1(t) \geqslant C_1 \text{ etc.} \\ 0 \text{ otherwise.} \end{cases}$$

All of the above techniques have been employed at one time or another in our
analyses. A typical indicator function based upon temperature is shown in Fig.
5 for the two-dimensional heated wake.

Once having the indicator function we can proceed to calculate the usual conditional and point averages such as intermittency factor, burst rate, zone aver-

ages, etc. (Hedley and Keffer, 1974b; Kawall and Keffer, 1977). This aspect will not concern us here however. Rather, the present intent is to focus our attention upon the extent of the large structures of the motion. This can be conveniently accomplished in terms of the space-time correlation of the signal $I(t)$ as will be seen in the next section.

The correlation of a signal, $\alpha(t)$, with zero mean, obtained at a given point in a turbulent flow is defined as

$$\rho_{\alpha\alpha}(\tau) = \overline{\alpha(t)\alpha(t+\tau)} \, / \, \overline{\alpha^2} = \lim_{T\to\infty} \frac{1}{T} \int_0^T \alpha(t)\alpha(t+\tau)dt,$$

where τ is a time lag and $\overline{\alpha^2}$ is the variance of the signal.

The one-dimensional spectrum of $\alpha(t)$, the Fourier transform of $\overline{\alpha^2} \, \rho_{\alpha\alpha}(\tau)$, is given by,

$$E_{\alpha\alpha}(n) = \overline{2\alpha^2} \int_0^\infty \rho_{\alpha\alpha}(\tau) \exp(-i2\pi n\tau) \, d\tau,$$

where n is a frequency such that $n \geqslant 0$. Since

$$\rho_{\alpha\alpha}(\tau) = \frac{1}{\overline{\alpha^2}} \int_0^\infty E_{\alpha\alpha}(n) \exp(+i2\pi n\tau)dn,$$

and $\rho_{\alpha\alpha}(0) = 1.0$, it follows that

$$\overline{\alpha^2} = \int_0^\infty E_{\alpha\alpha}(n)dn.$$

The space-time correlation of $\alpha(t)$ signals measured at two points, A and B, separated by a distance S, is defined as

$$\rho_{\alpha_A \alpha_B}(\tau;S) = \frac{\overline{\alpha_A(t)\,\alpha_B(t+\tau)}}{(\overline{\alpha_A^2}\,\overline{\alpha_B^2})^{\frac{1}{2}}}$$

We next extend the analysis to two signals $\alpha(t)$ and $\beta(t)$. As before the correlation function is

$$\rho_{\alpha\beta}(\tau) = \overline{\alpha(t)\,\beta(t+\tau)} \, / \, \alpha'\beta',$$

where primed quantities are the root-mean-square values of the signals. The cross-spectrum, the Fourier transform of $\rho_{\alpha\beta}(\tau)$, is given by

$$E_{\alpha\beta}(n) = \int_{-\infty}^\infty \rho_{\alpha\beta}(\tau) \exp(-i2\pi n\tau)d\tau.$$

$E_{\alpha\beta}(n)$ is a complex quantity, the real part of which is referred to as the co-spectrum and the imaginary part, the quadrature spectrum. The cospectrum, $C_{\alpha\beta}(n)$, represents the correlation between the in-phase components of $\alpha(t)$ and $\beta(t)$ at frequency n and the quadrature spectrum, $Q_{\alpha\beta}(n)$ represents the correlation between the $90°$ out-of-phase components at n. Since

$$\rho_{\alpha\beta}(0) = \overline{\alpha\beta} \, / \, \alpha'\beta',$$

it follows that the turbulent transport term is given by

$$\overline{\alpha\beta} = \alpha'\beta' \int_0^\infty C_{\alpha\beta}(n)dn.$$

The coherence or degree of correlation between $\alpha(t)$ and $\beta(t)$ at frequency n, is described by the coherence spectrum which is defined as

$$\Gamma_{\alpha\beta}(n) = \frac{C_{\alpha\beta}^2 (n) + Q_{\alpha\beta}^2 (n)}{E_{\alpha\alpha}(n) \, E_{\beta\beta}(n)} .$$

Finally, if measurements of $\alpha(t)$ and $\beta(t)$ are separated a distance in space, S, then the space-time correlation will be

$$\rho_{\alpha_A \beta_B}(\tau;S) = \frac{\overline{\alpha_A (t) \, \beta_B (t+\tau)}}{(\overline{\alpha_A^2} \, \overline{\beta_B^2})^{\frac{1}{2}}}$$

and the coherence function will be dependent upon S, i.e.

$$\Gamma_{\alpha_A \beta_B}(n;S) = \frac{C_{\alpha_A \beta_B}^2 (n;S) + Q_{\alpha_A \beta_B}^2 (n;S)}{E_{\alpha\alpha}(n) \, E_{\beta\beta}(n)} .$$

Results and Interpretation

We consider first the application of the above methods to an isothermal plane jet, (Moum, 1978). The intermittency signal I(t) was based on the streamwise fluctuating component of the turbulent velocity, u and space-time measurements of I(t), viz., $\rho_{I_A I_B} (\tau:S)$, were made in the outer regions of the flow at roughly the half intermittency point.

Typical results at x/B = 10, for a traverse in the spanwise direction, S_z, are shown in Fig. 6. It is seen that $\rho_{I_A I_B}$ is symmetric about $\tau = 0$ and that values of maximum correlation drop monotonically as S_z increases. This is contrasted with a similar correlation taken in the lateral direction, S_y, shown in Fig. 7. Some important differences can be seen. Symmetry about $\tau = 0$ no longer exists. Furthermore, with the exception of $S_y = 0$, the peak in $\rho_{I_A I_B}$

occurs at negative values of τ. The interpretation of these results is straight-forward. The shift in the point of maximum correlation indicates a tilt of the turbulent bursts in the x-z plane. By defining an average convection velocity we estimated the magnitude of the tilt to be approximately 25°. Furthermore, the shift of $\rho_{I_A I_B}$ (max) with the negative τ region tells us that the bursts are tilted backwards as they are ejected from the central portion of the fully turbulent flow, which would be expected from the physics of the motion.

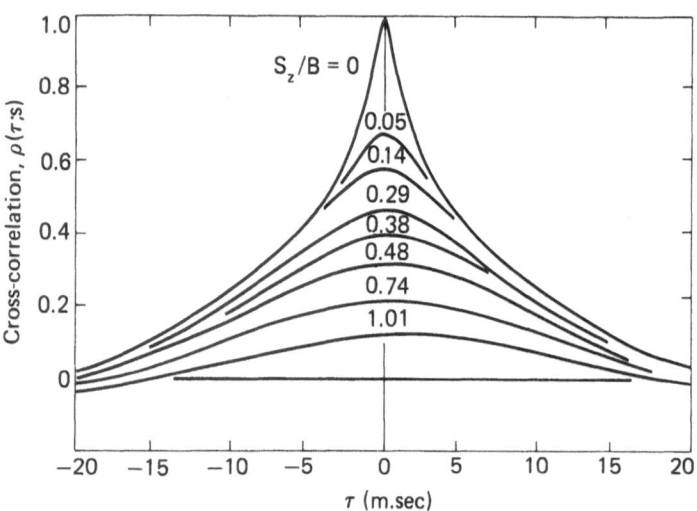

Fig. 6. Spanwise space-time correlation for the plane jet, x/B = 10; $\gamma \approx 0.5$.

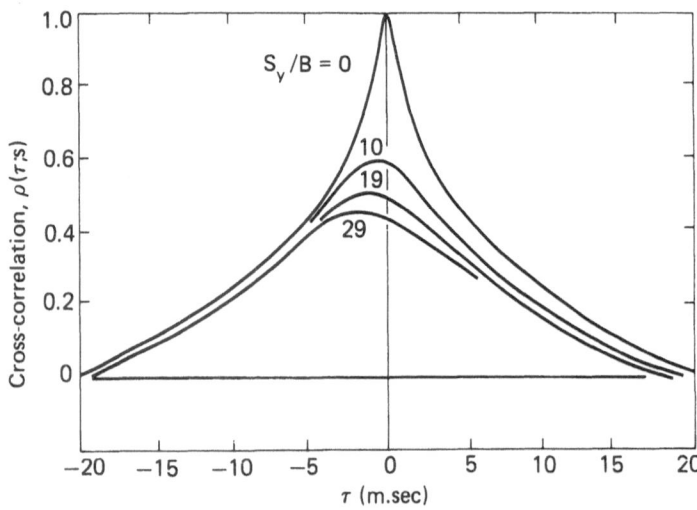

Fig. 7. Lateral space-time correlation for the plane jet, x/B = 10, $\gamma \approx 0.5$.

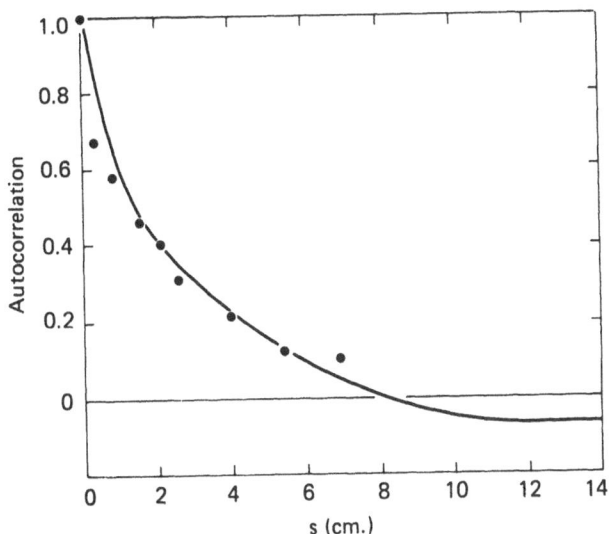

Fig. 8. Autocorrelations for the plane jet; x/B = 10; γ ≈ 0.5; •, spanwise; —, streamwise.

Some measure of the extent of these bursts can be obtained by a comparison of integral scales of the correlation functions. Fig. 8, shows a plot of the peaks of $\rho_{I_A I_B}(\tau; S_z)$ along the span of the flow, compared with the streamwise autocorrelation transformed by Taylor's hypothesis. It is seen that the integral scales and by implication, the lateral and streamwise extent of the bursts, are of the same order. It may be assumed that the lateral extent is of the order of two standard derivations of the probability density function of the interface position (Kawall and Keffer, 1977). With this information, a composite first order picture of the outer intermittent bursts can be constructed. We can picture them as being of roughly equivalent size in the span and streamwise directions, slightly elongated in the lateral plane and tilted backwards at about 25°. In terms of the nozzle width B and integral scales, L, the dimensions are $L_x/B \approx L_z/B \approx 0.75$ while $L_y/B \approx 1.13$.

A similar experiment has been performed for the heated two-dimensional wake (Barsoum et al., 1978). In this case the intermittency function was constructed from the temperature signal and $\rho_{I_A I_B}(\tau; S)$, measured at approximately the half intermittency point at the position x/D = 96. Results for the spanwise space-time correlation are shown in Fig. 9. As before, the functions are symmetric about $\tau = 0$ indicating no tilt of the bulges in the spanwise direction. A comparison of the maximum values of $\rho_{I_A I_B}$ with the transformed streamwise autocorrelation in Fig. 10, is significant. The bursts are evidently narrow in the spanwise direction, L_x/L_z being the order 2 or in terms of the cylinder diameter, $L_x/D \approx 12$, $L_x/D \approx 6$. Further details are given in Barsoum et al., 1978.

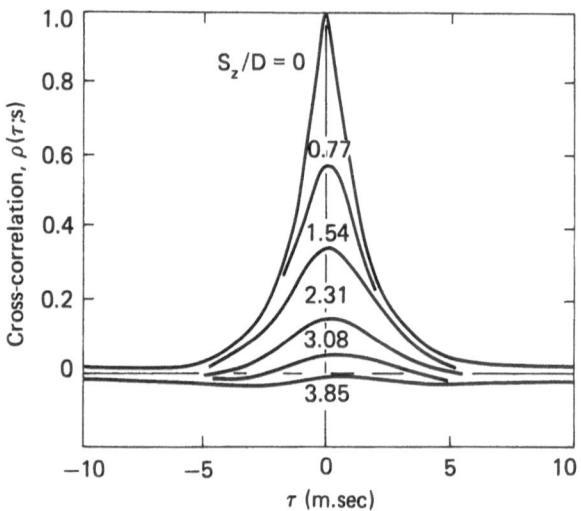

Fig. 9. Spanwise space-time correlations for the heated two-dimensional wake; $x/D = 96$; $\gamma \approx 0.5$.

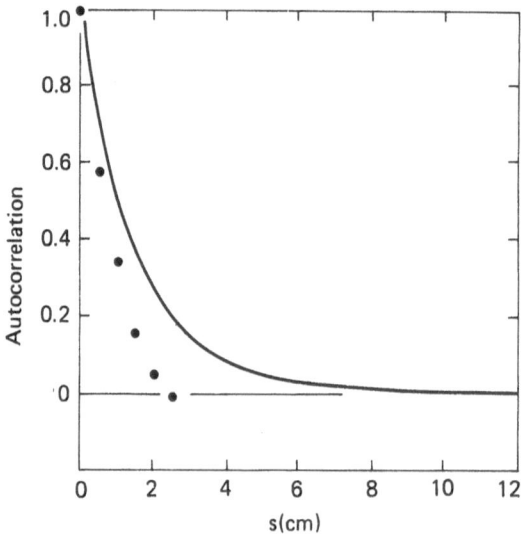

Fig. 10. Autocorrelations for the heated two-dimensional wake; $x/D = 96$; $\gamma \approx 0.5$; ●, spanwise; —, streamwise.

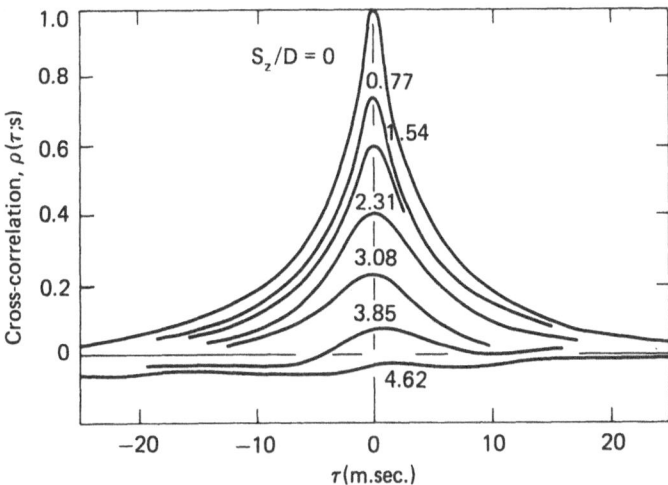

Fig. 11. Spanwise space-time correlations for the heated two-dimensional wake;
x/D = 96; y/D = 0.

Using the temperature signal directly, rather than I(t), an indication of the
structure within the fully turbulent region can be obtained, although the
visualization of "eddy" is less precise than a "burst". Figs. 11 and 12 show the
corresponding space-time and autocorrelations of $\theta(t)$. It can be seen that the
general behaviour is the same and one may conclude that the largest turbulent
eddies at the core of the flow are experiencing a similar elongation in the
streamwise direction as the outer bursts.

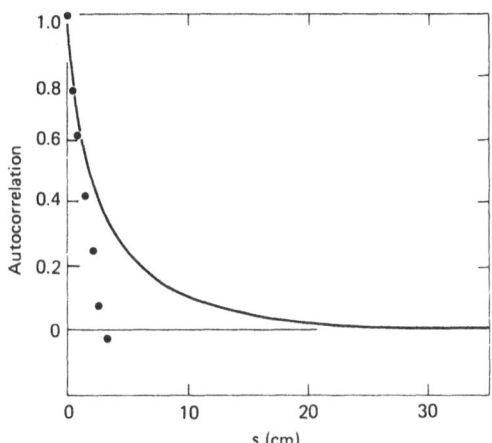

Fig. 12. Autocorrelations for the heated two-dimensional wake; x/D = 96;
y/D = 0; •, spanwise; −, streamwise.

A more definitive picture of the structure may be obtained by an examination of the coherence function, $\Gamma_{\alpha\beta}$. We note that whereas the autospectrum will identify dominant periodic structures within the flow, it can tell us little about those Fourier components which contribute to the correlation. The co-spectrum will reveal those eddy sizes which are instrumental to the transport which comes about as a result of the in-phase components, while the coherence spectrum will describe the effects of the out-of-phase as well as the in-phase components and as such will be a more sensitive indicator of coherent structures.

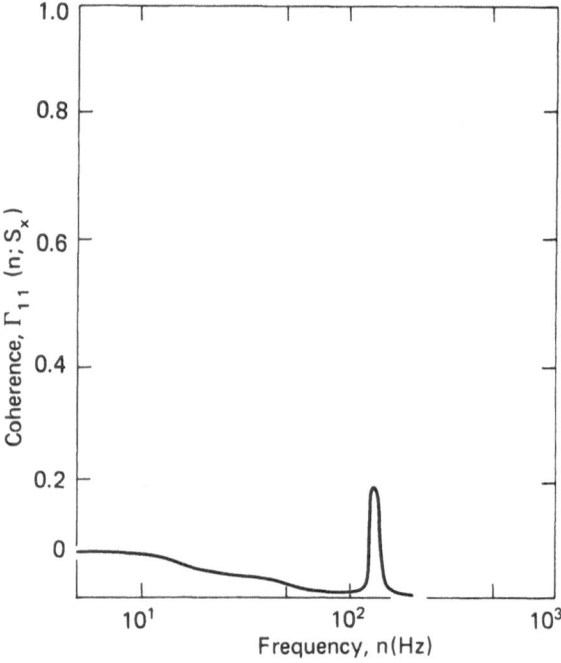

Fig. 13.   Coherence in the vortex street of a heated wake; reference probe at x/D = 6; moving probe at x/D = 40.

A simple illustration of this is seen with respect to experiments carried out on the Karman vortex street shed by a circular cylinder (Budny et al., 1979). In the present situation, the cylinder was heated and the vortices traced by the temperature signal. The coherence function was evaluated with pairs of probes, the fixed probe being located at either x/D of 6 or 8, and the moving probe, at separations up to x/D of 100. The autospectrum taken at x/D = 40 showed no peak at the Strouhal frequency of 135 Hz. In contrast, we see in Fig. 13, a distinct peak in the coherence function obtained with S_x/D = 34, at this frequency. The coherence clearly indicates a measurable residual vortex activity

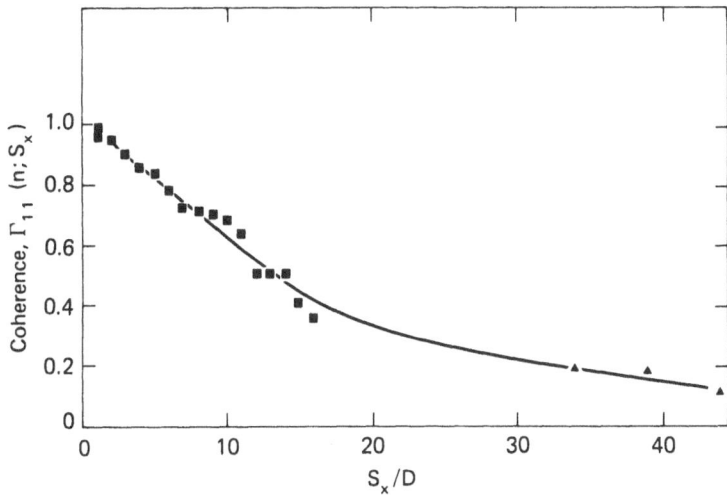

Fig. 14. Streamwise variation of coherence in the vortex street of a heated wake; ■ *- reference probe at x/D = 6;* ▲ *- reference probe at x/D = 8.*

although this is undetectable with conventional single point statistics. Fig. 14 illustrates how the magnitudes of Γ_{11} decreases as the separation increases.

The technique was applied to the interior of the plane jet, (Moum et al., 1979). Streamwise velocity signals were used and measurements made in the spanwise direction. A composite plot of the coherence functions is shown in Fig. 15. At small separations, $S_z/B < 0.02$, the complete spectrum of turbulent eddies contribute to the correlation. As S_z increases, the magnitude of the maximum coherence drops and the high frequency cut-off region shifts to lower frequencies. This is not unexpected since the smallest eddies become statistically independent. A peak in the function eventually occurs, at approximately $S_z/B = 0.30$ thus defining a coherent structure, centered at about 15 Hz for this flow.

More information can be obtained by applying these techniques to complicated flows, e.g. those free turbulent shear flows exhibiting asymmetries in the mean profiles and concomitantly, having regions in which the net production of either turbulent kinetic energy or the thermal equivalent of this, is locally negative (Béguier et al., 1977). In such cases, the lateral transport of the turbulent quantity is significant and co-spectral measurements can be used to discern the relative importance of the large eddy bulk motions, which may be counter-gradient, and the small eddy gradient diffusion processes. This can be seen from experiments carried out on the thermal field of a mixing layer having a jump in temperature superimposed (Béguier et al., 1978). A plot of the co-

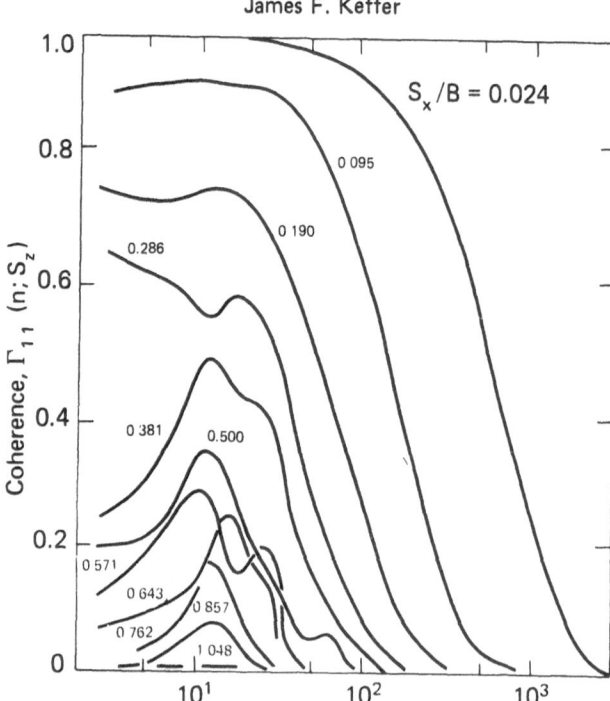

Fig. 15. Spanwise coherence of plane jet; x/B = 20; y/B = 0.

spectra of $v\theta$ is shown in Fig. 16 and it can be seen that the transport activity can be broadly divided into regions of large and small wave number.

This is more clearly seen for the case of an asymmetrically heated plane jet. Co-spectra of $v\theta$ for this situation are plotted in Fig. 17 and show the same bimodal features as for the mixing layer. Furthermore, the corresponding coherence functions are given in Fig. 18 and these show explicitly that these centres of turbulent transport activity are associated with large and small eddy motions.

Acknowledgements

This work was sponsored by the National Research Council of Canada under Grant No. A2746. Experimental work was carried out at the Institut de Méca-nique Statistique de la Turbulence, Marseille and the Department of Mechanical Engineering, University of Toronto. I would like to thank J.G. Kawall for his constructive comments on the manuscript.

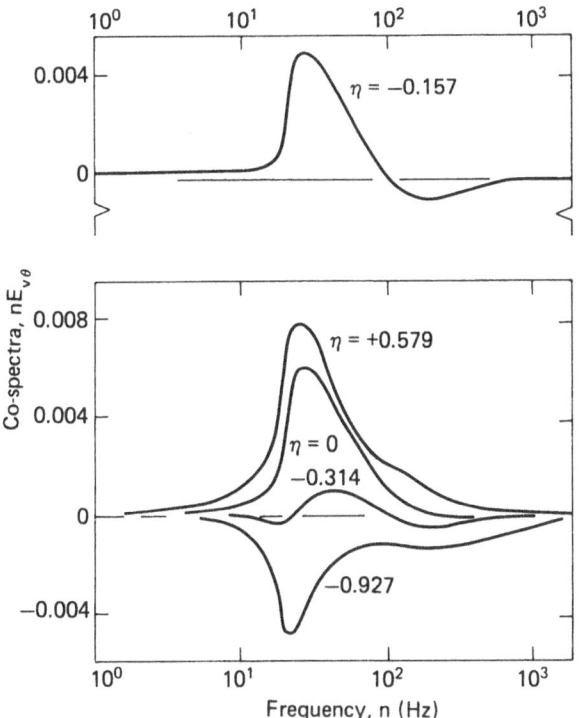

Fig. 16. Co-spectra of vθ for mixing layer with a jump in temperature; x/B = 30.

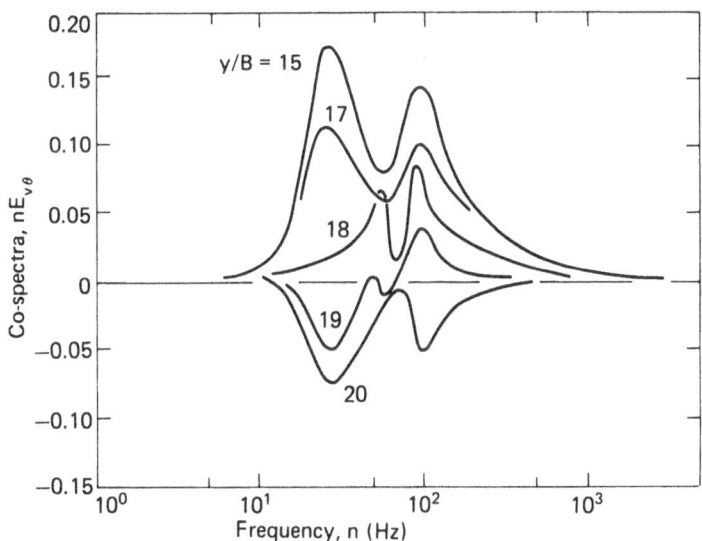

Fig. 17. Co-spectra of vθ for asymmetric heated jet; x/B = 30.

James F. Keffer

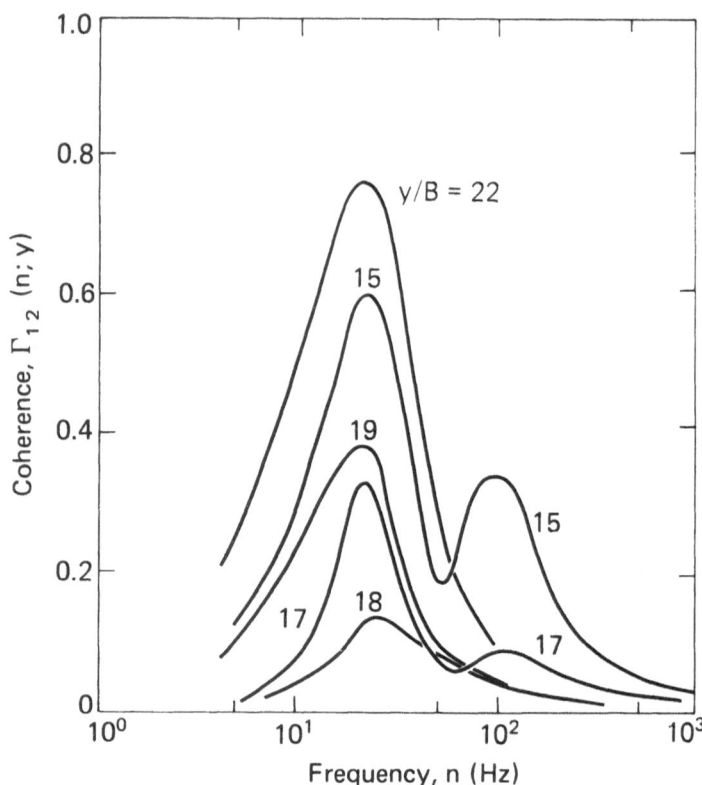

Fig. 18. Coherence of $v\theta$ for asymmetric heated jet; $x/B = 30$.

References
1. Antonia, R.A. (1972), Conditionally sampled measurements near the outer edge of a turbulent boundary layer. *J. Fluid Mech.,* **56**, 1
2. Barsoum, M.L., Kawall, J.G. and Keffer, J.F. (1977), Spanwise structure of the plane turbulent wake. *Physics of Fluids,* **21**, 157
3. Béguier, C., Fulachier, L. and Keffer, (1978). The turbulent mixing layer with an asymmetrical distribution of temperature. *J. Fluid Mech.,* **89**, 561
4. Béguier, C., Giralt, F., Fulachier, L. and Keffer, J.F., (1977), Negative production in turbulent shear flows. *Structure and Mechanisms of Turbulence II,* **23**, Berlin
5. Blackwelder, R.F. (1977), On the role of phase information in conditional sampling. *Physics of Fluids,* **20**, S232
6. Brown, G.L. and Roshko, A. (1974), On density effects and large structure in turbulent mixing layers. *J. Fluid Mech.,* **64**, 775
7. Budny, R.S., Kawall, J.G. and Keffer, J.F. (1979) Vortex street evolution in the wake of a circular cylinder, to be published, 2nd Symp. Turb. Shear Flows
8. Favre, A., Gaviglio, J., and Dumas, R. (1957). Space-time double correlations and spectra in a turbulent boundary layer. *J. Fluid Mech.,* **2**, 313
9. Favre, A., Gaviglio, J., and Dumas, R. (1958). Further space-time correlations of velocity in a turbulent boundary layer. *J. Fluid Mech.,* **3**, 344
10. Grant, H.L., (1958). The large eddies of turbulent motion, *J. Fluid Mech.* **4**, 149
11. Hedley, T.B. and Keffer, J.F. (1974a), Turbulent/non-turbulent decisions in an intermittent flow. *J. Fluid Mech.* **64**, 625
12. Hedley, T.B. and Keffer, J.F. (1974b), Some turbulent/non-turbulent properties of the outer intermittent region of a boundary layer. *J. Fluid Mech.* **64**, 645
13. Kawall, J.G. and Keffer, J.F. (1977), Uniform distortion of a heated turbulent wake. *Structure and Mechanisms of Turbulence II,* 85, Berlin
14. Keffer, J.F., (1965), The uniform distortion of a turbulent wake, *J. Fluid Mech.,* **22**, 135
15. Keffer, J.F., Budny, R.S. and Kawall, J.G. (1978), Digital technique for the simultaneous measurement of velocity and temperature. *Rev. Scientific Instr.,* **49**, 1343
16. Keffer, J.F., Kawall, J.G., Giralt, F. and Béguier, C. (1979), Analysis of turbulent structures in complex shear flows, to be published, 2nd Symp. Turb. Shear Flows
17. Kovasznay, L.S.G., Kibens, V. and Blackwelder, R.F. (1970), Large scale motion in the intermittent region of a turbulent boundary layer. *J. Fluid Mech.* **41**, 283
18. LaRue, J.C. and Libby, P.A. (1974), Temperature fluctuations in the plane turbulent wake *Physics of Fluids* **17**, 1956
19. Moum, J.N. (1978), Structure of the plane turbulent jet, B.A.Sc. Thesis, Mech. Eng. U. of T.
20. Moum, J.N., Kawall, J.G. and Keffer, J.F. (1979), Two-point statistics in the plane turbulent jet, to be published
21. Townsend, A.A. (1970). Entrainment and the structure of turbulent flow, *J. Fluid Mech.,* **32**, 145
22. Wallace, J.M., Brodkey, R.S. and Eckelmann, H., (1977), Pattern-recognized structures in bounded turbulent shear flows. *J. Fluid Mech.,* **83**, 673
23. Winant, C.D. and Browand, F.K. (1974). Vortex pairing: the mechanism of turbulent mixing-layer growth at moderate Reynolds numbers. *J. Fluid Mech.* **63**, 237

Measurement of the Periodic Variation of Turbomachine Flow Fields

by

Theodore H. Okiishi
Department of Mechanical Engineering
Iowa State University, Ames

and

Douglas P. Schmidt
The Trane Company, LaCrosse, Wisconsin 54601

I. Introduction

In turbulent flow measurement situations involving an appreciable periodically varying component of flow, it is possible to separate the periodic flow information from the chaotically varying measurement signal normally involved. This strategy is particularly useful in turbomachine flow studies where the periodic variation of flow, due to relative motion between the stationary and the moving blades, is of interest because of its direct relationship (a relationship not yet well understood) to discrete frequency noise generation, forced blade vibration and energy transfer. The unsteady anemometer output voltage associated with a hot-wire sensor immersed in a turbomachine flow field may be considered to consist of the sum of two components, one related to periodic variations in flow and the other due to turbulence. If the anemometer signal is sampled continuously for an appropriate length of time and an average value is obtained, a time-average or steady-state measurement is obtained. If this signal is sampled periodically (synchronized with a particular rotor sampling position) an appropriate number of times and these periodic samples are averaged, a periodic-average measurement is made.

Several researchers (e.g. Whitfield et al (1972), Evans (1975), Raj and Lakshminarayana (1976), and Hirsch and Kool (1977)) have used this kind of measurement technique to obtain periodic-average flow data for turbomachine rotor flows which, except for fluctuations due to turbulence, were assumed to be steady to a coordinate system moving with the rotor blades. Such an assumption is probably justified for an isolated low-speed flow rotor. In most turbomachines, however, the flows through all blade rows are unsteady in all reference frames because of interaction. Blade rows are normally spaced closely enough axially that a downstream blade row potential field will affect the upstream blade row moving relative to it and cause a periodic variation in the upstream blade row exit flow. The wakes shed by an upstream blade row that are subsequently chopped and transported by a downstream blade row moving relative to them cause a periodic variation in the downstream blade row flow. Thus, detailed information about how turbomachine flows vary with increments of rotor motion seems pertinent.

Use of a periodic-average measurement procedure for obtaining this kind of data is not common, with only a few documents appearing in the open literature (Fujita and Kovasznay (1974, 1975), Ho and Kovasznay (1976, 1976), Gallus (1976), Schmidt and Okiishi (1977), and Wagner et al. (1978)). Details of the periodic-average flow measurement system and procedure used at Iowa State University will be described in this paper.

2. Measurement System

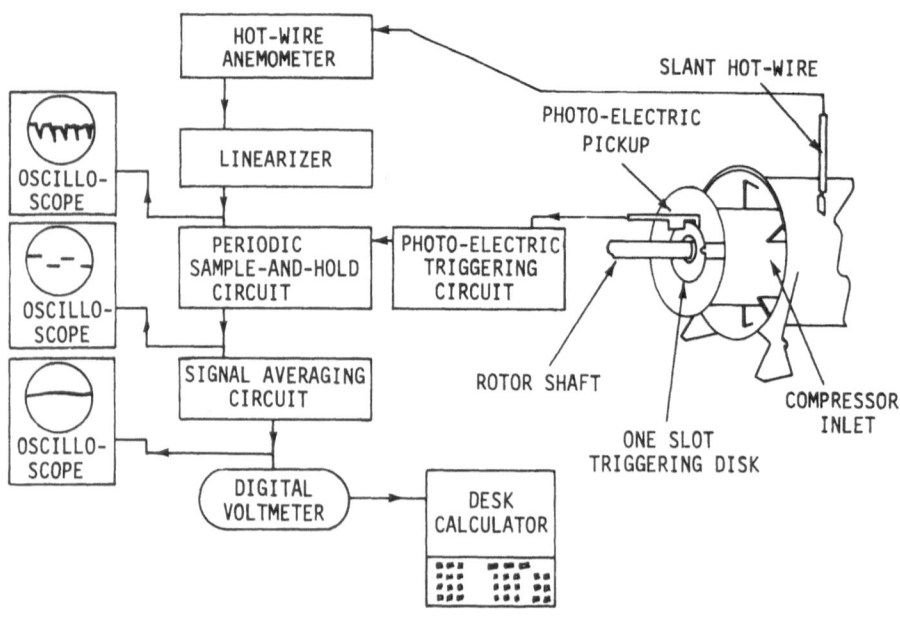

Fig. 1. Schematic set-up diagram of periodic-average flow measurement system.

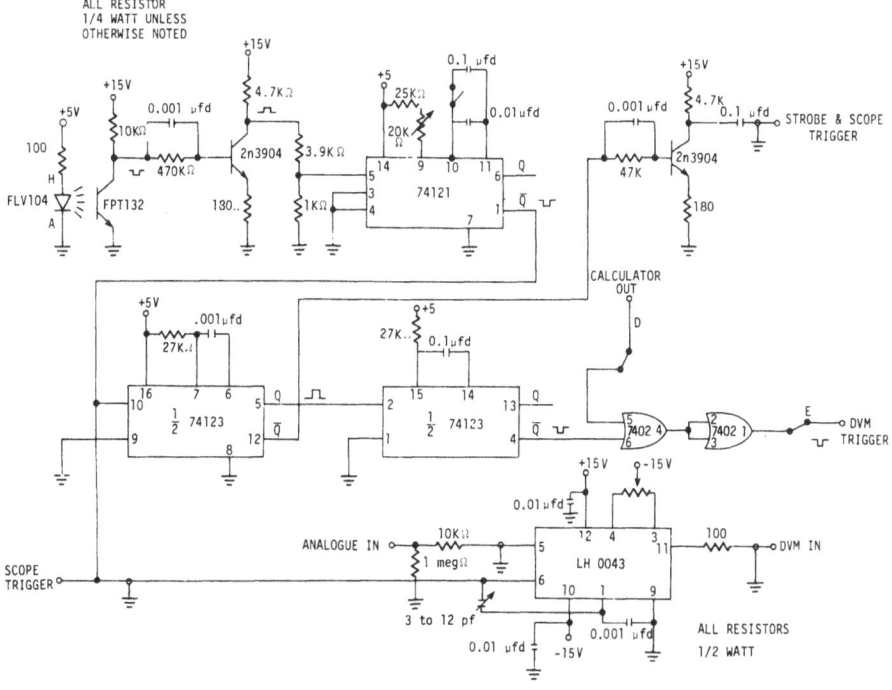

Fig. 2. Circuit diagram of triggering and sample-and-hold circuits.

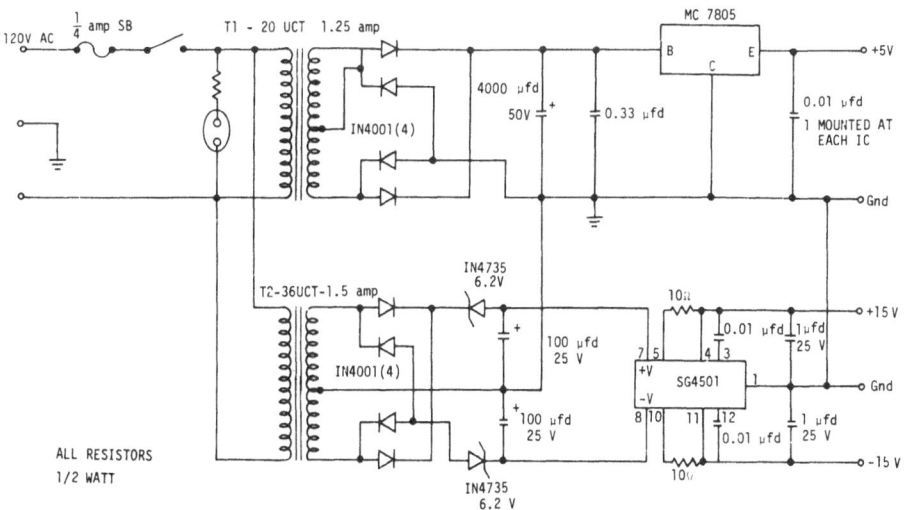

Fig. 3. Power supply for triggering and sample-and-hold circuits.

A schematic sketch of the measurement system used at Iowa State is shown in Fig. 1. Only the important features of the system will be discussed. The hot-wire probe involved a single 5 μm diameter, platinum-plated tungsten wire having a sensing length of 1.25 mm and copper and gold plated ends. The sensor was inclined at an angle with respect to the probe axis.

The periodic sample-and-hold circuit was designed and built on campus. Details of the circuit and interfacing appear in Figs. 2 and 3. A 5 μs sample of the linearizer output voltage was obtained each time the photoelectric pickup was activated (one per revolution of the shaft). The time delay between photoelectric pickup activation and linearizer output voltage sampling could be varied to permit control of the rotor blade sampling position. The photoelectric pickup system could also be mechanically moved to vary the rotor blade sampling position. With the measurements made to date, a variable time constant (1 to 100 s) low pass filter was used to electronically smooth the periodically sampled data. A time constant of 1.0 s was normally used. A sample-and-hold circuit was used to make the voltmeter perform an A/D conversion for the calculator after receiving a "READY" signal from the calculator *and* a trigger signal from the photoelectric cell system, in that order. Because of the relatively low operating speed of the calculator, a value of linearizer voltage could be entered in at about 0.17 s intervals only.

3. Measurement Procedure

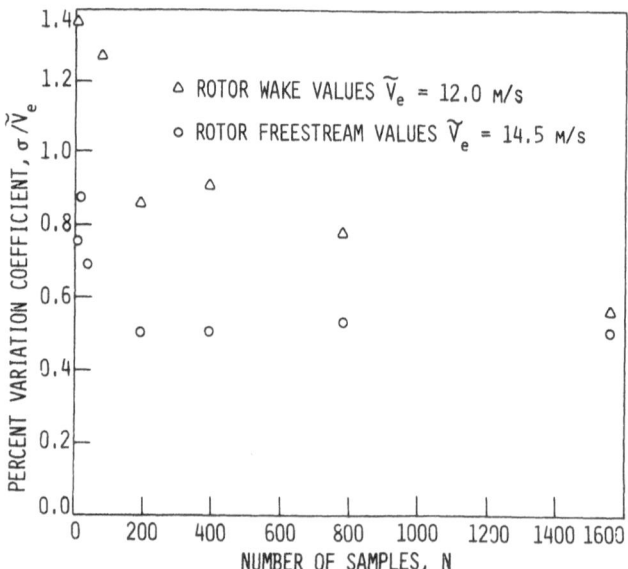

Fig. 4. *Percent variation coefficient of the periodic-sample averages for different values of N.*

The precision of the periodic-sampling and averaging technique is largely dependent on N, the number of periodic samples involved. Hirsch and Kool (1977) suggested that the variance associated with the periodic-average value, σ, is related to the variance associated with turbulence, σ_T, by the relationship

$$\sigma^2 = \sigma_T^2/N \qquad [1]$$

where N is the number of periodic samples involved. During the present study, data were obtained to determine the effect of the number of periodic samples on σ. The results are summarized in Fig. 4 for measurements in and out of a rotor wake. Although 1600 periodic-average samples would have been desirable according to Fig. 4, that choice would have resulted in prohibitively long test times. Therefore, only about 700 samples were actually obtained in the present tests. These 700 samples were electronically smoothed with the low pass filter. From the electronically smoothed data, approximately 180 samples were digitized, entered into the calculator, and arithmetically averaged.

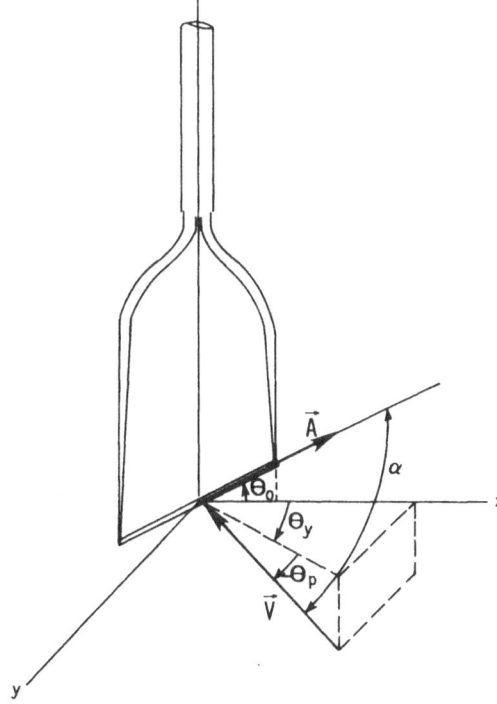

Fig. 5. Hot-wire configuration relating velocity vector, $\vec{V}$, to hot-wire sensor and probe coordinates, x, y, z.

A single hot-wire sensor was used to measure periodic-average, three-dimensional velocity vectors. The sketch in Fig. 5 relates the hot-wire sensor to the probe coordinates x, y, and z and to the velocity vector, $\vec{V}$. The z coordinate was placed along the hot-wire probe axis while the x coordinate was located so that the sensor lay in the x-z plane. The wire sensor is represented in Fig. 5 by the unit vector $\vec{A}$ slanted at the angle θ_0 to the x axis. The direction of the velocity vector with respect to the probe is defined by θ_y, the probe yaw angle, and θ_p, the probe pitch angle. Since the probe coordinates x, y, and z were fixed to the probe, the probe angle θ_y changed by the amount of turning as the probe was rotated about its axis, whereas the pitch angle, θ_p, remained constant. The sensor yaw angle, α, was defined as the angle between the velocity vector $\vec{V}$ and the axis to the hot-wire sensor. It will be useful to note that the sensor yaw angle can be expressed in terms of the angles θ_0, θ_p, and θ_y. The unit vector $\vec{A}$ and the velocity vector $\vec{V}$ expressed in terms of vector components are

$$\vec{A} = \cos\theta_0 \,\vec{i} + \sin\theta_0 \,\vec{k} \qquad [2]$$

and

$$\vec{V} = -V \cos\theta_p \cos\theta_y \,\vec{i} - V \cos\theta_p \sin\theta_y \,\vec{j} - V \sin\theta_p \,\vec{k} \qquad [3]$$

The dot product of $\vec{A}$ and $\vec{V}$ is

$$\vec{A}\cdot\vec{V} = |\vec{A}| \,|\vec{V}| \cos(180-\alpha) = -|\vec{V}| \cos\theta_0 \cos\theta_p \cos\theta_y$$
$$- |\vec{V}| \sin\theta_0 \sin\theta_p \qquad [4]$$

Hence, the sensor yaw angle relationship is

$$\cos\alpha = \cos\theta_0 \cos\theta_p \cos\theta_y + \sin\theta_0 \sin\theta_p \qquad [5]$$

If a hot-wire velocity calibration is made in the typical manner with the sensor yaw angle, α, equal to 90° (wire sensor normal to the flow) and if the hot-wire sensor is then used for velocity measurement at a sensor yaw angle other than 90°, then the velocity indicated by the sensor will not equal the actual velocity and is therefore defined as the "effective cooling velocity", V_e. V_e was related to the linearized anemometer bridge voltage, E_ϱ, by the empirical second order equation

$$V_e = K_1 + K_2 E_\varrho + K_3 E_\varrho^2 \qquad [6]$$

where the three coefficients K_1, K_2, and K_3 were determined from a velocity calibration with the wire sensor normal to the flow. The hot-wire measurement technique used in the present study was based on knowing the precise relationship for the effective cooling velocity/actual velocity ratio, V_e/V. The sine law is a useful relationship for sensor yaw angles near $90°$:

$$\frac{V_e}{V} = \sin\alpha \qquad\qquad [7]$$

Another commonly used relationship is

$$\left(\frac{V_e}{V}\right)^2 = \sin^2\alpha + k^2\cos^2\alpha \qquad\qquad [8]$$

where k is claimed to be dependent on the sensor type and length-to-diameter ratio. The latter relationship, attributed to Champagne et al. (1967), takes into consideration the residual velocity sensitivity when the velocity vector is parallel to the sensor due to the finite length and nonuniform temperature of the sensor. For this experimental investigation, both the sine law and the Champagne et al. (1967) relationship were judged as being inadequate for the inclined hot-wire probe and the measurement conditions involved. The sine law was not appropriate since sensor yaw angles as small as $40°$ were encountered. Experimentally determined values of k for the Champagne et al. (1967) relationship for the slant wire probe were found to vary considerably depending on sensor yaw angle and pitch angle.

Experiments showed that V_e/V was strongly dependent on sensor yaw angle, weakly dependent on pitch angle, and only very slightly dependent on velocity level. The dependence of V_e/V on sensor yaw angle was determined for several combinations of velocities and pitch angles with a probe calibration nozzle. Typical results for a $35.35°$ slant hot-wire sensor are shown in Fig. 6.

A second order empirical correlation was used to express the effective cooling velocity ratio as a function of sensor yaw angle, pitch angle, and velocity, as follows:

$$\frac{V_e}{V} = b_0 + b_1\alpha + b_2\theta_p + b_3 V + b_4\alpha^2 + b_5\theta_p^2 + b_6 V^2 + b_7\alpha\theta_p + b_8\alpha V + b_9\theta_p V \qquad [9]$$

The coefficients b_0 through b_9 were determined with a least squares fit of effective cooling velocity calibration data. Since the sensor yaw angle, α, and the probe yaw angle, θ_y, are geometrically related (Equation [5]), either angle could have been selected as one of the independent variables in Equation [9].

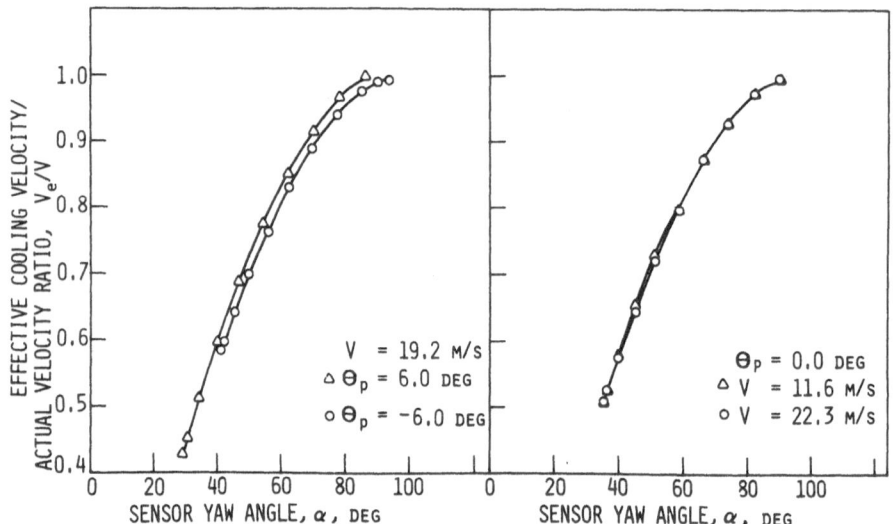

Fig. 6. Typical effective cooling velocity calibration results for a 35-degree inclined hot wire.

However, the sensor yaw angle was chosen since the dependence of V_e/V with respect to α facilitated correlation with a second order empirical type fit. Since hot-wire sensor behavior was not symmetrical about the axis of the probe, two sets of coefficients were used, one for each of the probe yaw angle θ_y ranges of 0° to 90° and 0° to −90°.

Hot-wire measurements were made at compressor flow-field measurement points by positioning the hot-wire sensor and recording data at each of three probe angle orientations (a, b, and c) corresponding to probe yaw angles of $\theta_{y,a}$, $\theta_{y,b}$, and $\theta_{y,c}$ which equal:

$$\theta_{y,a} = \theta_y \qquad\qquad\qquad\qquad\qquad\qquad\qquad [10]$$

$$\theta_{y,b} = \theta_y - m_b \qquad\qquad\qquad\qquad\qquad\qquad [11]$$

$$\theta_{y,c} = \theta_y - m_c \qquad\qquad\qquad\qquad\qquad\qquad [12]$$

where m_b and m_c are constant probe turning angle increments. For each wire orientation, a geometric relationship similar to Equation [5] could be expressed and an effective cooling velocity relationship similar to Equation [9] applied. The resulting six equations are:

For position a

$$\frac{V_{e,a}}{V} = b_{0a} + b_{1a}\alpha_a + b_{2a}\theta_p + b_{3a}V + b_{4a}\alpha_a^2 + b_{5a}\theta_p^2 + b_{6a}V^2$$

$$+ b_{7a}\alpha_a\theta_p + b_{8a}\alpha_aV + b_{9a}\theta_pV \qquad\qquad [13]$$

$$\cos\alpha_a = \cos\theta_0\,\cos\theta_p\,\cos\theta_y + \sin\theta_0\,\sin\theta_p \qquad\qquad [14]$$

For position b

$$\frac{V_{e,b}}{V} = b_{0b} + b_{1b}\alpha_b + b_{2b}\theta_p + b_{3b}V + b_{4b}\alpha_b^2 + b_{5b}\theta_p^2 + b_{6b}V^2$$

$$+ b_{7b}\alpha_b\theta_p + b_{8b}\alpha_bV + b_{9b}\theta_pV \qquad\qquad [15]$$

$$\cos\alpha_b = \cos\theta_0\,\cos\theta_p\,\cos(\theta_y - m_b) + \sin\theta_0\,\sin\theta_p \qquad\qquad [16]$$

For position c

$$\frac{V_{e,c}}{V} = b_{0c} + b_{1c}\alpha_c + b_{2c}\theta_p + b_{3c}V + b_{4c}\alpha_c^2 + b_{5c}\theta_p^2 + b_{6c}V^2$$

$$+ b_{7c}\alpha_c\theta_p + b_{8c}\alpha_cV + b_{9c}\theta_pV \qquad\qquad [17]$$

$$\cos\alpha_c = \cos\theta_0\,\cos\theta_p\,\cos(\theta_y - m_c) + \sin\theta_0\,\sin\theta_p \qquad\qquad [18]$$

Since the effective velocities $V_{e,a}$, $V_{e,b}$, and $V_{e,c}$ were measured values, six unknown variables (α_a, α_b, α_c, θ_p, θ_y, V) remained, and the six nonlinear equations (Equations [13 - 18]) could be solved simultaneously by using the Newton-Raphson numerical method. With V, θ_p, and θ_y determined, the velocity vector is completely specified with respect to the probe coordinate system. The velocity vector with respect to compressor coordinates could be determined when the relationship between the probe and compressor coordinate system is known.

The precision of this measurement technique was largely dependent on the selection of the three probe measurement angles $\theta_{y,a}$, $\theta_{y,b}$, and $\theta_{y,c}$. It was important to select the measurement angles in regions of high sensor yaw angle sensitivity for best resolution. As can be seen in Fig. 6, the sensor yaw angle region near 90° was unfavorable since large variations in α (5°) could result from small changes of effective cooling velocity ratio. Therefore, in this region, substantial measurement errors were likely to occur. In fact, if the sensor yaw angle was near 90° at any one of the three probe angle measurement positions, the solution would not converge. In addition, sensor yaw angles near 90° were also avoided since this position was most susceptible to velocity gradients along the sensor. Since the effective cooling velocity ratio was calibrated over the two distinct probe angle regions 0° to 90° and 0° to −90°, probe angle measurement positions were selected to insure that the varying sensor yaw angle caused by the random variation of the velocity direction at the measurement point always remained in one of these two regions. To avoid probe prong interference, the shorter prong was always positioned into the flow upstream of the longer prong.

The angle of wire inclination, θ_0, is also an important factor in the measurement technique since it affects the sensitivity and uniqueness of solution. An angle of 35.3° was used by Whitfield et al. (1972) so that the sensor would be in three orthogonal orientations when the probe was rotated about its axis in 120° increments. This special angle is, however, not necessarily optimal. The measurement sensitivity to tangential and radial angles increases as θ_0 is decreased. The lower limit of θ_0 is governed by the value of the radial angle and the selection of the probe angle measurement positions. Hirsch and Kool (1977) state that in order to obtain a unique solution, the pitch angle, θ_p, must be smaller than the value computed from

$$\tan\theta_p = \frac{\tan\theta_0}{\cos\theta_y} \qquad\qquad [19]$$

which is dependent on the wire angle and probe measurement yaw angle. In addition, if θ_0 is too small, unsteady probe prong interference effects could cause large errors for probe measurement angles near 0.0°. Of the two hot-wire probes used (θ_0 of 45° and 35°), the results obtained from the 35° probe were more consistent and judged to be more accurate. For this reason, the majority of the measurements were made with the 35° probe. Possibly, the accuracy can be slightly increased by further optimization of the wire angle for the measurement conditions involved.

4. Example Results

Some examples of results obtained with the periodic-sampling data acquisition system are presented in this section. Measurements were made in the first stage of the low-speed research compressor shown in Fig. 7 at measurement stations 3 and 4 (see Figs. 8 and 9). Periodic-average velocity component data for flow behind the first rotor (measurement station 3) and the first stator (measurement station 4) at 50 % span are shown in Figs. 10 and 11 for different rotor sampling positions. The nomenclature is explained in Figs. 9 and 12. A com-

Probe Measurement Stations

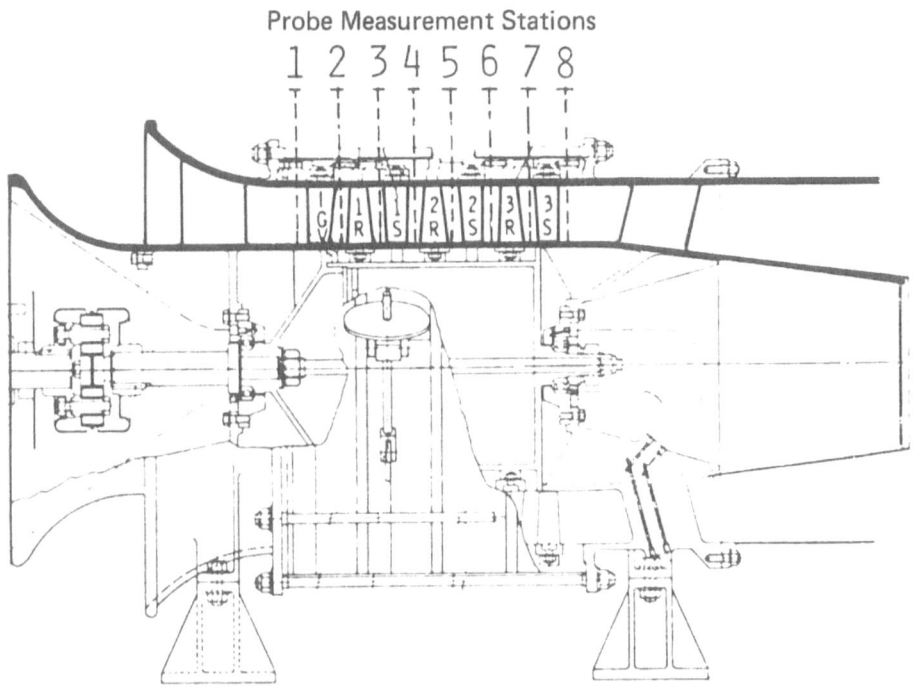

Fig. 7. Research compressor with probe measurement stations.

parison of the axial velocity values obtained with the hot-wire (periodic-aver-age) and with a total-pressure probe (time-average) is shown in Fig. 13. Estima-ted uncertainties for the hot-wire measurements are indicated in Table 1.

From data like those shown in Figs. 10 and 11, blade-to-blade plane flow sketches illustrating wake transport and interaction (such as shown in Fig. 14) could be constructed as explained by Wagner et al. (1978). Three-dimensional representation of velocity vector data, as shown in Fig. 15, efficiently displays a large amount of data. More details about the research compressor measure-ments may be found in papers by Schmidt and Okiishi (1977) and Wagner et al. (1978).

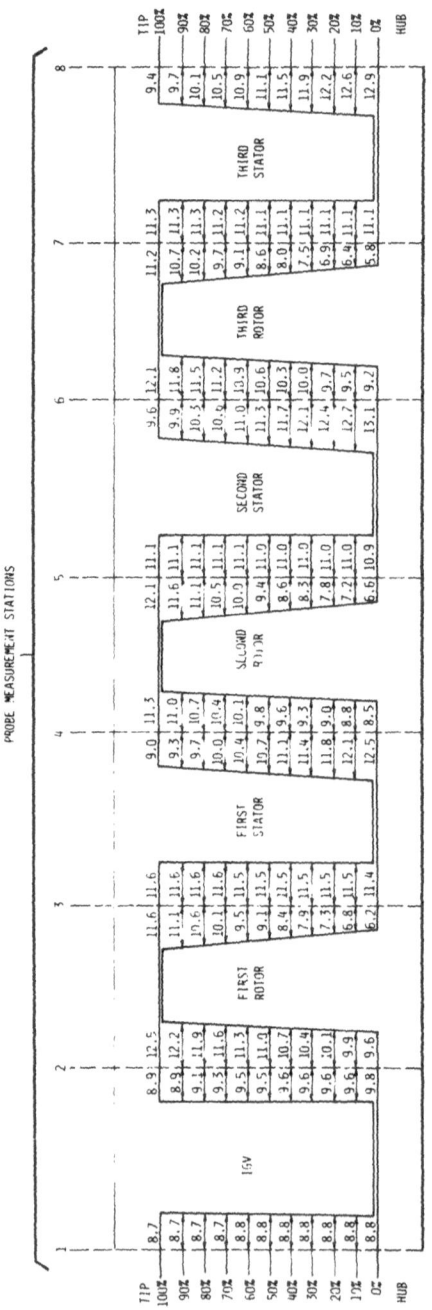

Fig. 8. Schematic diagram showing axial location of probe measurement stations (dimensions in mm).

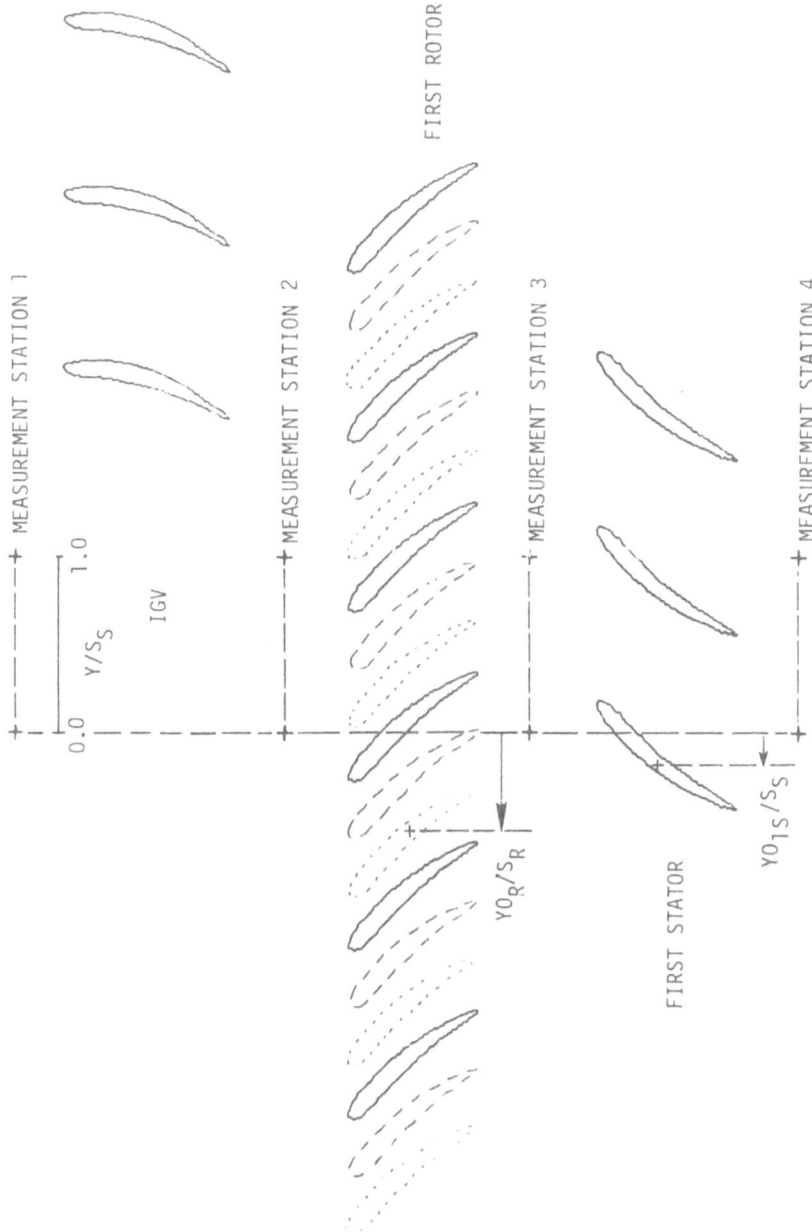

Fig. 9. Blade cascade showing relative positions of blades for several rotor sampling positions. (S_R is rotor blade spacing; S_S is stator blade spacing; Y is measurement circumferential location; YO_R is reference rotor blade circumferential location; YO_{1S} is reference first stator blade circumferential location).

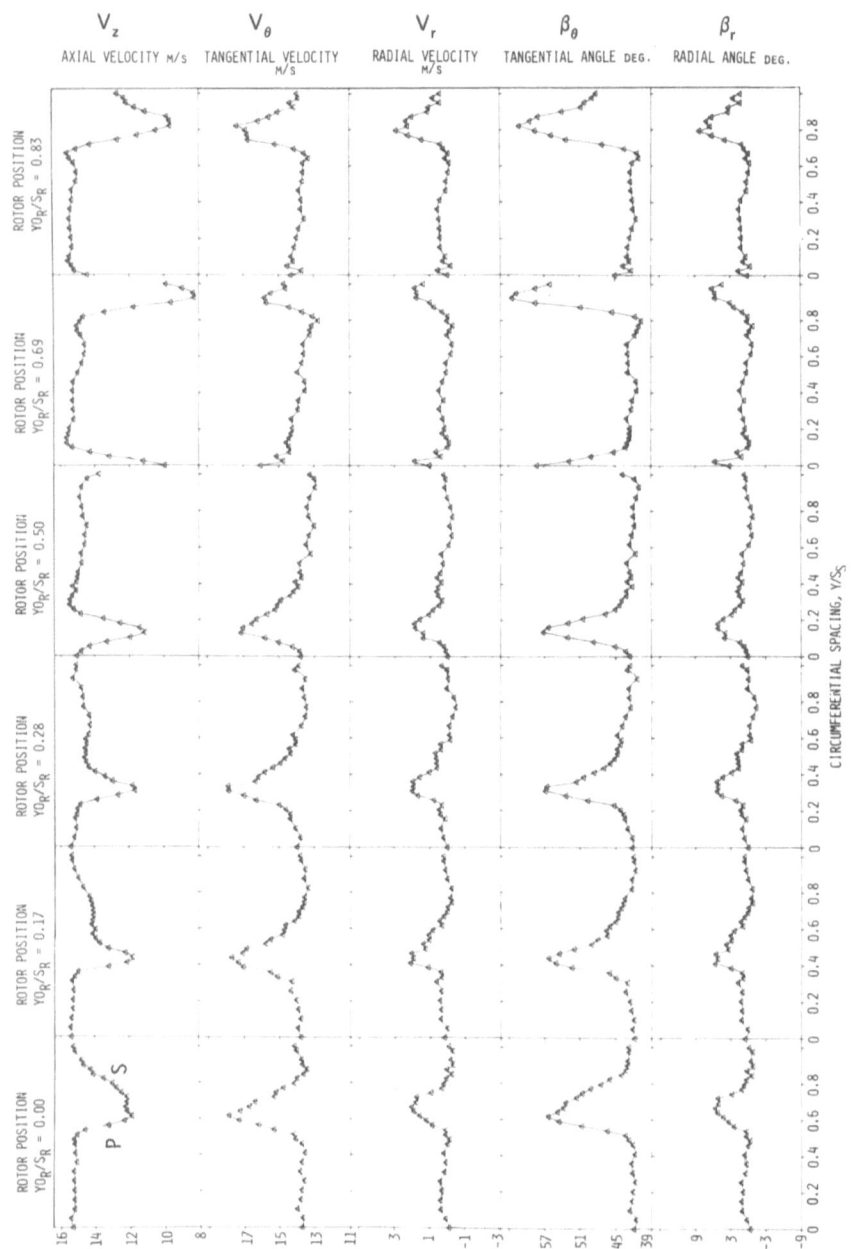

Fig. 10. Periodic-average flow patterns for first rotor exit at 50 % span and different rotor positions.

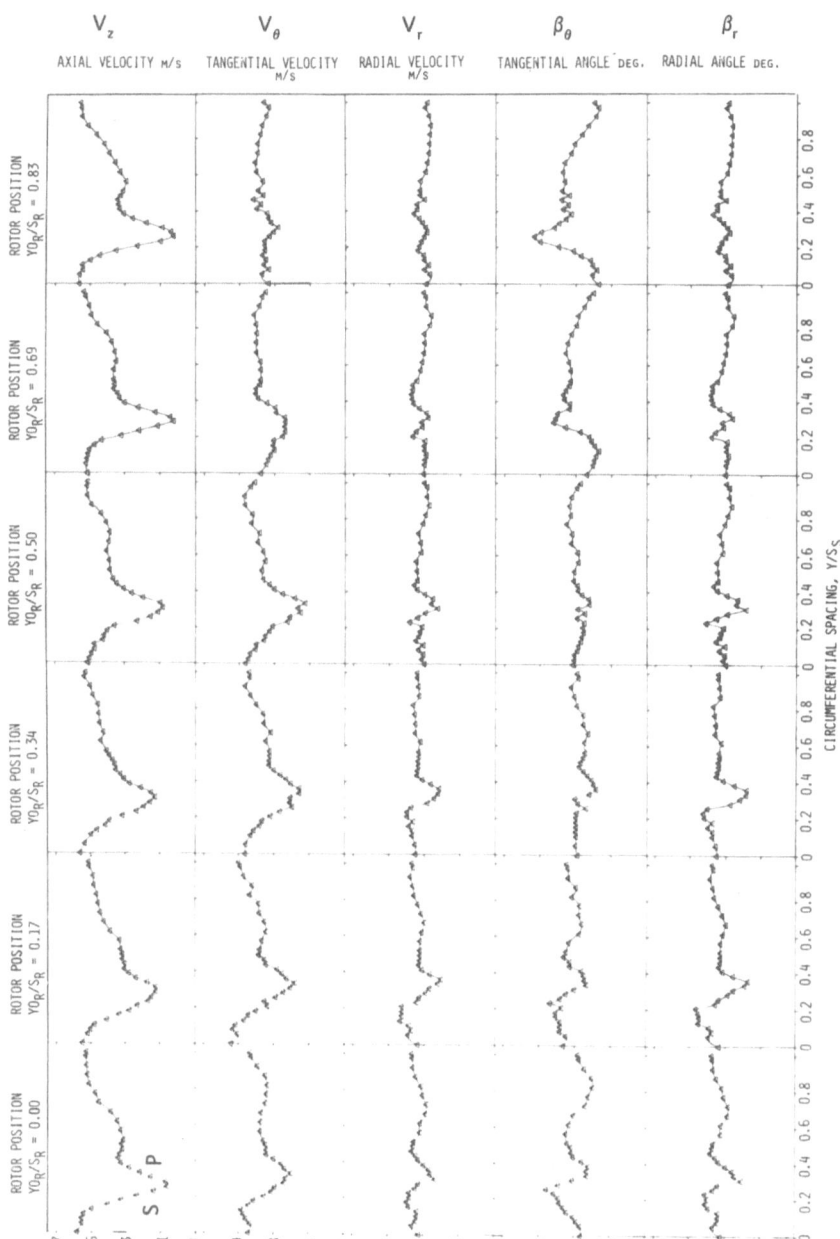

Fig. 11. Periodic-average flow patterns for first stator exit at 50 % span and different rotor positions.

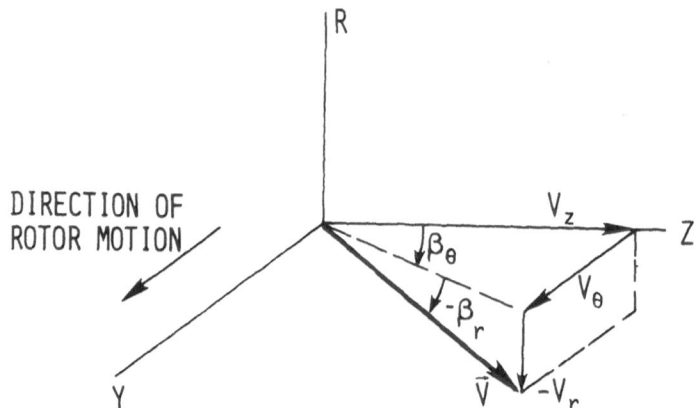

Fig. 12. Compressor coordinate system showing nomenclature and sign convention for three-dimensional periodic-average velocity and angle parameters. Radial coordinate positive direction is outward from the compressor axis.

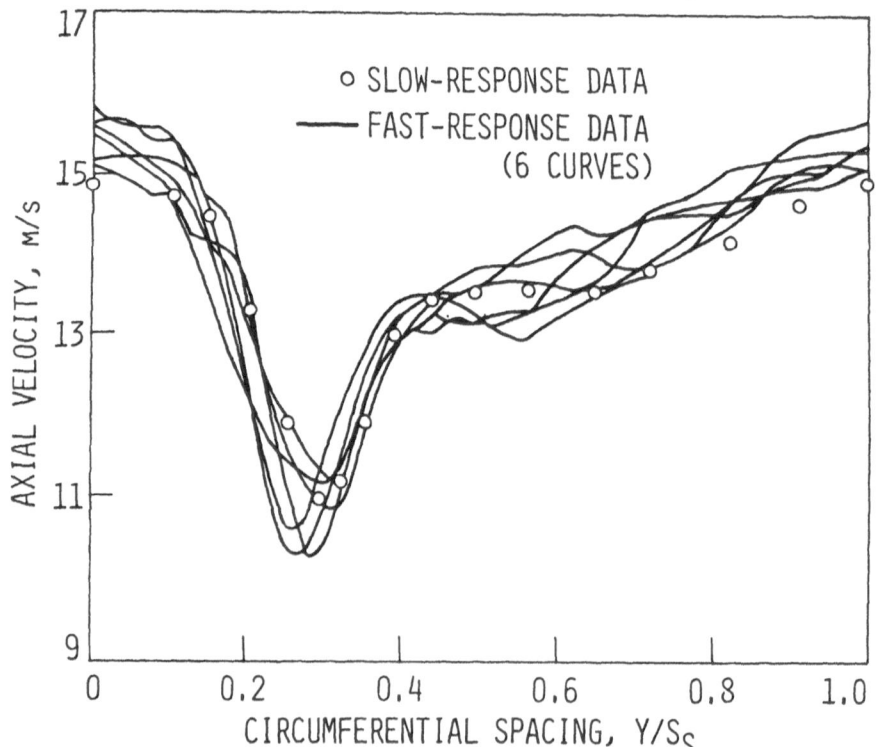

Fig. 13. Comparison of slow-response and fast-response data behind the first stator.

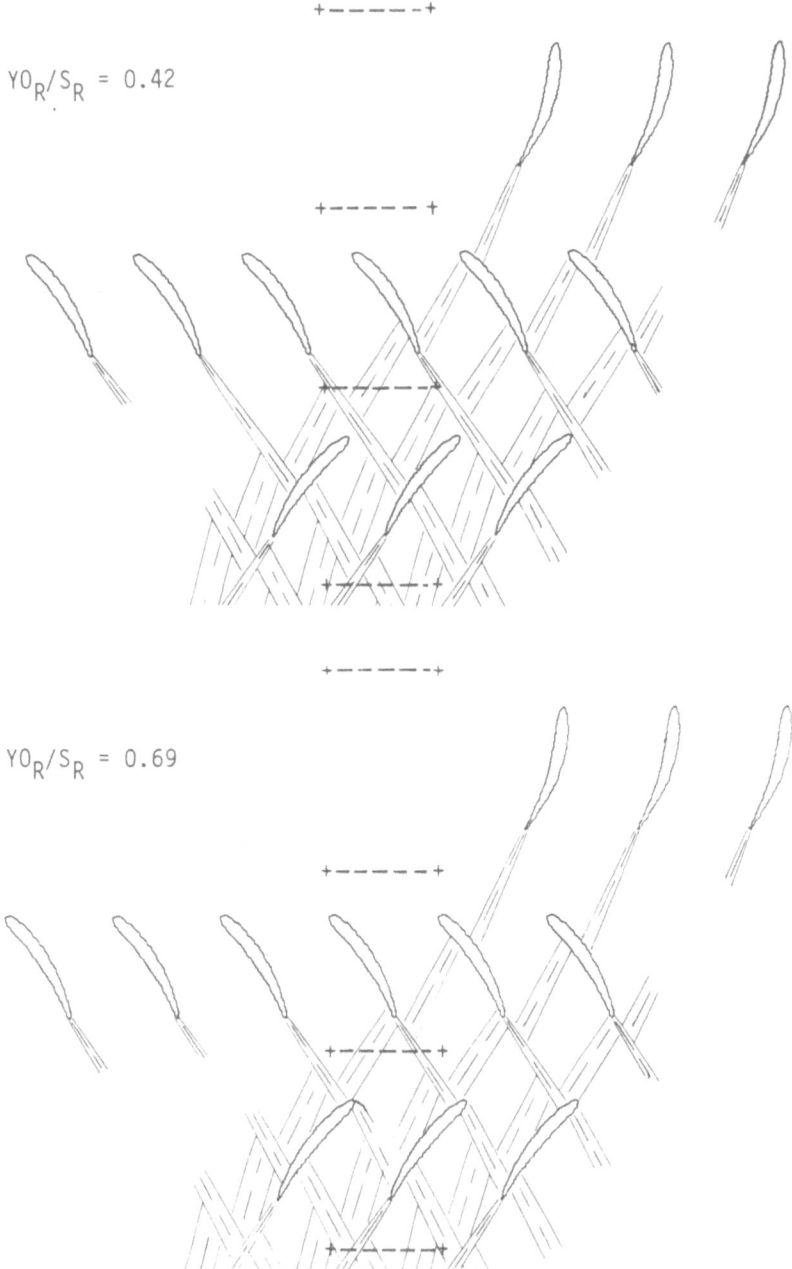

$YO_R/S_R = 0.42$

$YO_R/S_R = 0.69$

Fig. 14. Periodic-average cascade plots for the first stage of the research com-pressor at 30 % passage height.

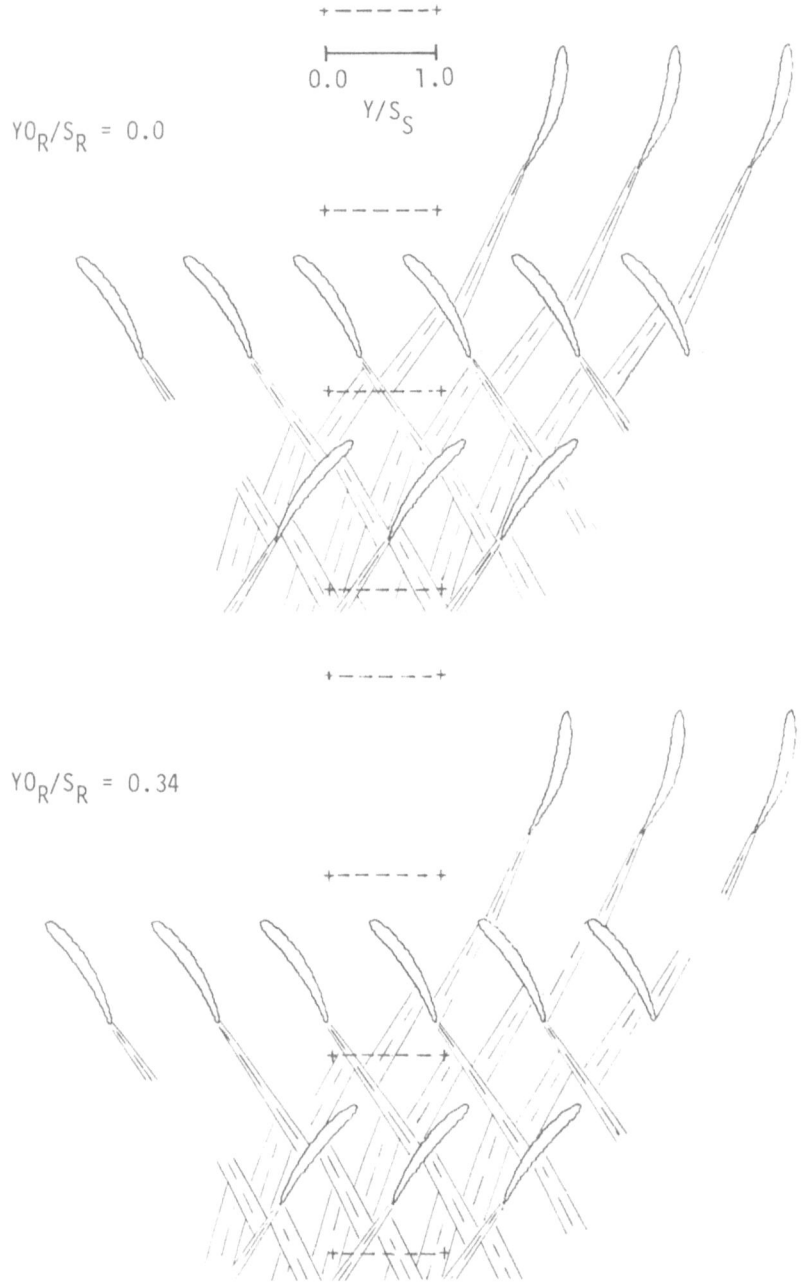

Fig. 14. Concluded.

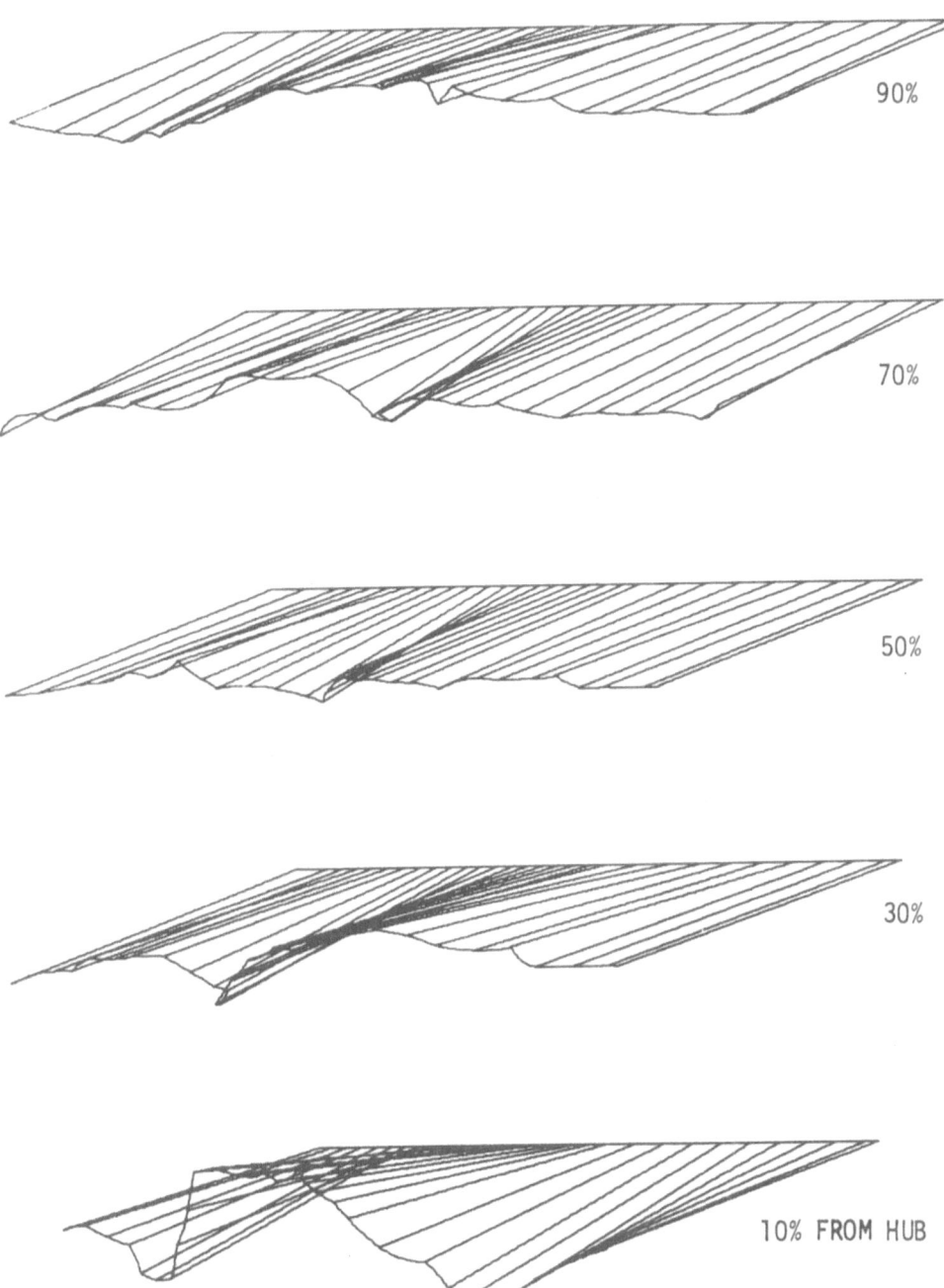

90%

70%

50%

30%

10% FROM HUB

Fig. 15. Hub-to-tip variation of first stator exit flow between two adjacent stator blades. Rotor sampling position, $YO_R/S_R = 0.34$.

Table 1.
Uncertainty and scatter of periodic-average flow field parameters

Flow Parameters	Symbol	Typical Values	Estimated Uncertainty (20 to 1 odds)
Absolute velocity	V	18.0 m/s	0.8 m/s
Absolute tangential angle	β_y	40.0°	1.5°
Radial angle	β_r	0.0°	2.0°
Axial velocity	V_z	13.0 m/s	0.7 m/s
Absolute tangential velocity	V_θ	12.0 m/s	0.6 m/s
Radial velocity	V_r	0.0 m/s	0.5 m/s

Conclusions

A reasonable procedure for obtaining periodic-average velocity vector data in a turbomachine has been developed. By rotating a slanted single-sensor hot-wire probe around its axis, three-dimensional velocities were measured. Synchronization of the data collection with rotor sampling position led to measurements indicating the periodic unsteadiness of flow in turbomachine.

Acknowledgments

The authors are appreciative of the support received under Air Force Office of Scientific Research Grant No. AFOSR-76-2916, with cost sharing participation by the Engineering Research Institute and the Mechanical Engineering Department of Iowa State University, Ames, Iowa. The equipment grants received from the National Science Foundation and the ISU Research Foundation are gratefully acknowledged. The authors are also indebted to their colleagues for their valuable suggestions.

References

Champagne, F.H., Sleicher, C.A. & Wehrman, O.H. 1967 Turbulence measurements with inclined hot-wires, part 1. *J. Fluid Mech.* **28**, 153-175

Evans, R.L. 1975 Turbulence and unsteadiness measurements downstream of a moving blade row. *Trans. ASME, J. Engng. Power* **A97**, 131-139

Fujita, F. and Kovasznay, L.S.G., 1974 Unsteady lift and radiated sound from a wake cutting airfoil. *AIAA J.* **12**, 1216-1221

Fujita, F. and Kovasznay, L.S.G. 1975 Sound generated by wake cutting. *AIAA Progress in Astronautics and Aeronautics* **38**, 35-41

Gallus, H.E. 1976 Results of measurements of the unsteady flow in axial subsonic and supersonic compressor stages. AGARD-CP-177

Hirsch, C. and Kool, P. 1977 Measurement of the three-dimensional flow field behind an axial compressor stage. *Trans ASME, J. Engng. Power,* **A99**, 168-180

Ho, C.-M. and Kovasznay, L.S.G., 1976 Sound generation by single cambered blade in wake cutting. *AIAA J.* **14**, 763-766

Ho, C.-M. and Kovasznay, L.S.G., 1976 Wake cutting by a cascade of cambered blades. *AIAA 2nd Aero Acoustic Conference Proc.,* 43-53

Raj, R. and Lakshminarayana, B. 1976 Three dimensional characteristics of turbulent wakes behind rotors of axial flow turbomachinery. *Trans. ASME, J. Engng. Power,* **A98**, 218-228

Schmidt, D.P. and Okiishi, T.H. 1977 Multistage axial-flow turbomachine wake production, transport, and interaction. *AIAA J.* **15**, 1138-1145

Wagner, J.H., Okiishi, T.H. & Holbrook, G.J. 1978 Periodically unsteady flow in an imbedded stage of a multistage, axial-flow turbomachine. *ASME paper 78-GT-6*

Whitfield, C.E., Kelly, J.C. & Barry, B. 1972 A three-dimensional analysis of rotor wakes. *Aero. Quart.* **23**, 285-300

Measurements of Internal Intermittency and Dissipation Correlations in Fully Developed Turbulence

by

Yves Gagne, E.J. Hopfinger and J. Marechal
Institut de Mécanique (associé au C.N.R.S.)
Université de Grenoble, Grenoble

Relatively little is known about the small scale structure in fully developed turbulence and there is a great lack of experimental data. The reasons for this lack are various inherent difficulties in getting reliable data on the statistics of the small scale structure, and in fact, on this aspect of turbulence, theories seem to be more advanced than experiments.

The proposed contribution is concerned with internal intermittency and flatness factors, where results can be confronted with those of Kuo and Corrsin (1971), and furthermore with spatial correlations of the dissipation rate ϵ. These correlations have a theoretical support (Kolmogorov, 1962). In obtaining them, local isotropy and the Taylor hypothesis are assumed to be valid so that time derivatives can be taken and longitudinal correlations can be replaced by auto-correlations. Intermittency was studied by using a bandpass filter with central frequency f_m and adjustable bandwidth Δf.

Measurements were made in three flows giving a range of Reynolds numbers R_λ between 50 and 500: in grid turbulence, flow I, with and without plane strain ($R_\lambda \leqslant 100$), in a plane channel, flow II, with $R_\lambda \approx 200$ to 500 and in an axisymmetric jet, flow III, with $R_\lambda \approx 500$. Flow I was at a mean velocity of 15 ms^{-1} and the mesh size M was 2.13 cm. Measurements were made at $x_1/M = 39$ where the turbulence was isotropic and $R_\lambda \approx 50$ and further downstream in the plane strain section at $x_1/M = 152$ corresponding to a strain ratio of 11, (Marechal 1972) and a $R_\lambda \approx 100$. Flow II was that studied by Comte-Bellot (1965). The velocities were 17 ms^{-1} and 34 ms^{-1} and measurements were made at $x_1/D = 60$ and $x_2/D = 0.75$ ($R_\lambda \approx 200$) and $x_2/D = 0.4$ ($R_\lambda \approx 300$ and 500), 2D being the channel width. Roughness elements were placed at the channel inlet in order to enhance flow establishment. The jet of nozzle diameter d = 12 cm, was run at 40 ms^{-1} and measurements were taken at $x_1/d = 35$.

A hot wire 1 micron in diameter and of ℓ_w = 0.3 to 0.4 mm in length was used in all measurements by giving in the worst case, flow (III), $\ell_w/\eta \approx 2\star$ (where η is the Kolmogorov scale). A constant current anemometer was specially built for this study which had a noise level at input of 0.15 μVolts in the frequency range of 2 Hz to 30 kHz, and the maximum acceptable signal was 7 mVolts, giving a dynamic range of 5×10^4. The low noise level was necessary in order to obtain a sufficient resolution of the dissipation scales. After amplification the signal was either differentiated and low-pass filtered or band-pass filtered. Two types of filters were used: a Krohn-Hite 3323 (40 dB/oct.) and later on, a convolution type filter with 45 dB/0.2 oct.$\star\star$ and 2 % phase linearity. The steep cut-off is an interesting feature especially in the high frequency range but we have as yet not made a comparative study with the Krohn-Hite. The convolution filter was used in the ϵ correlation measurements whereas all the other results were obtained with the Krohn-Hite. The filter output was transmitted on-line to an analog-digital converter Preston (G.M.A.D. 2 - 15 bits + sign) which is connected with an IBM 7 computer. The sampling frequency was 38.4 kHz and the storage capacity 1.25×10^6 data points, giving a signal length of 32 sec.

When defining the wave number associated with η by $k_K = 1/\eta$ the corresponding frequency is $f_K = $ <U>$/2\pi\eta$ where <U> is the mean velocity. This frequency is used as reference frequency. When plotting the flatness factor F = <u^4>/[<u^2>]^2 as a function of the central frequency f_m/f_k using a nominal bandwidth Δf = 0.4 f_m $\star\star\star$ it is found that F has a maximum at $f_m/f_K \approx 2$. The magnitude of F as well as the location of the maximum are in fairly good agreement with the results of Kuo and Corrsin (Fig. 1) who used an effective bandwidth Δf = 0.52 f_m and analog devices only. The earlier drop-off of Kuo and Corrsin's F is believed to be due to a limitation in frequency response of the analog system. The kurtosis K = <$(\partial u/\partial t)^4$>/[<$(\partial u/\partial t)^2$>]^2 plotted as a function of high cut-off frequency f_c/f_K also shows a maximum at $f_c/f_K \approx 2$. It is therefore desirable that the frequency response is better than $2f_K$ and the sampling frequency $\geqslant 4 \, f_K$.

$\star$Ideally, one would like to have $\ell_w/\eta \leqslant 1$. With ℓ_w/η = 2 the attenuation of the spectrum is according to Wyngaard (J. of Scientific Instruments (J. of Phy. E), 1969, series 2, Vol. 2), less than 10 % at $f_K = 2\pi$<U>$/\eta$.

$\star\star$Developed in our laboratory by Dr. Auchere.

$\star\star\star$This is a nominal filter setting and it corresponds to an effective Δf = 0.57 f_m at 3 dB, a value very close to Kuo and Corrsin's effective Δf = 0.52 f_m.

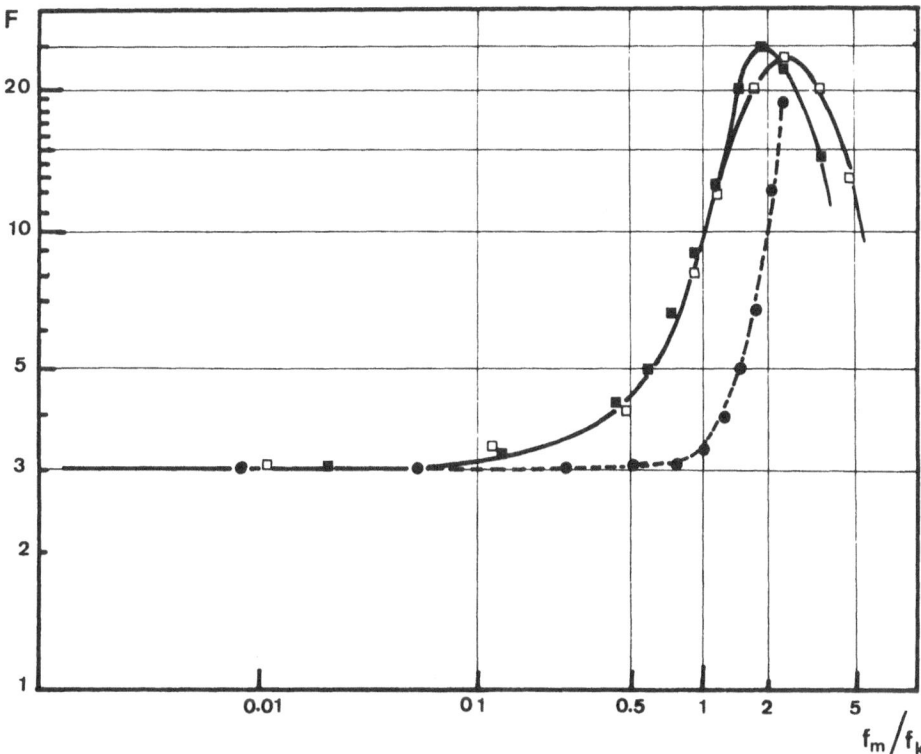

Fig. 1. Flatness factor of band pass signal as function of midband frequency for fixed relative bandwidth $\Delta f/f_m = 0.4$. ■ *Kuo and Corrsin (1971)* □ *present study (Flow I). Flatness factor obtained from intermittency function:* ● $F = 3/\gamma$, $f_K = \langle U \rangle /2\pi\eta$.

For Flow I, an intermittency function was generated from the band-pass filtered signal and this for $\Delta f/f_m = 0.3$, 0.4 and 0.5. The intermittency factor γ obtained in this way is considerably higher than $3/F$ (see Fig. 1). Even though there is some ambiguity about the choice of the threshold and the retardation time, this cannot explain the difference. In fact it is not surprising that $\gamma \neq 3/F$ but it is not clear which assumption to it is more violated, the normal distribution assumption of the signal in its high intensity state or the zero value in its low-intensity state. From the intermittency function the mean width (mean duration) as well as the most probable widths were calculated. The latter coincides fairly well with η (Fig. 2).

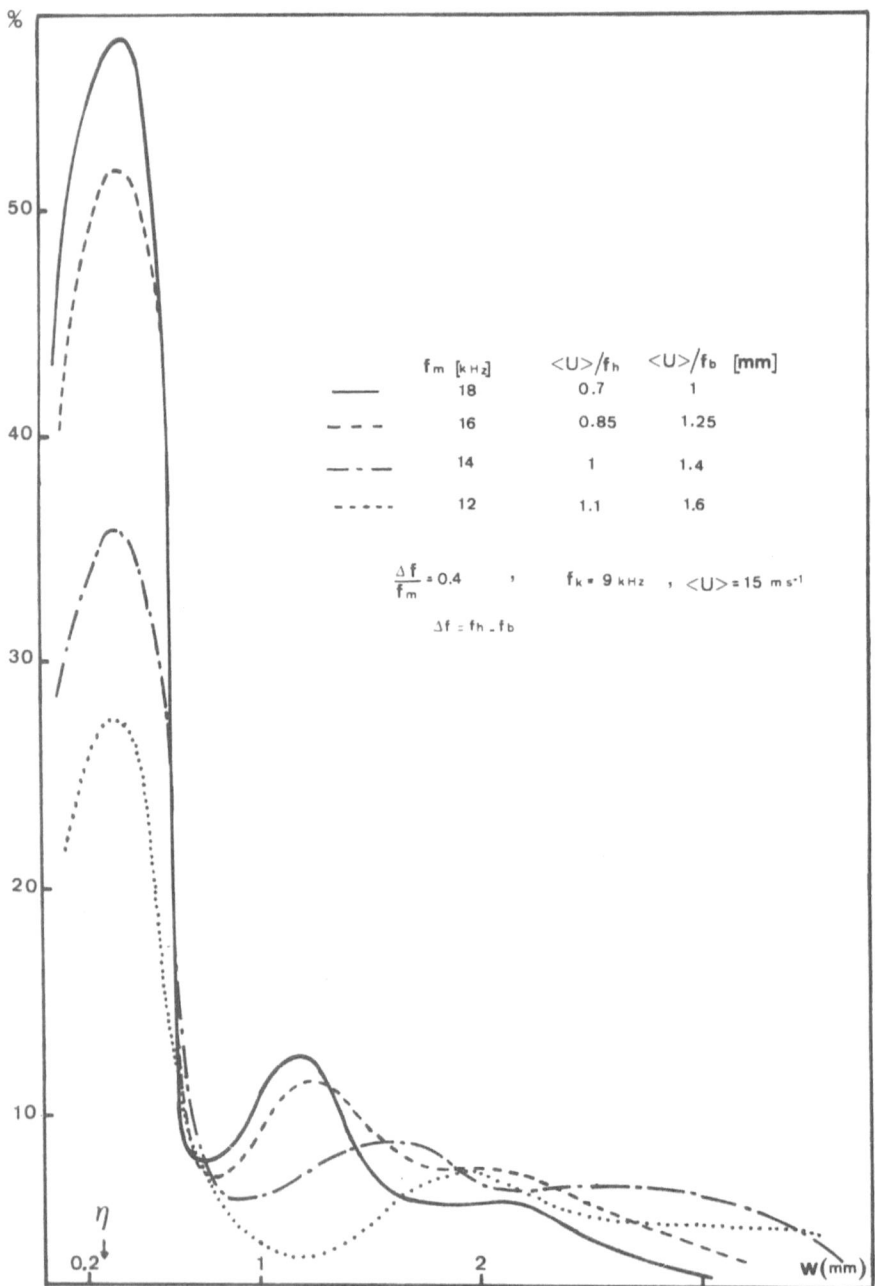

Fig. 2. Distribution of widths (durations) of intermittency function.

The kurtosis K was calculated for values of R_λ between 50 (flow I) and 500 (flow III). At low R_λ the results agree well with published results (see Kuo and Corrsin) whereas at higher values of R_λ our values are generally 20 to 30 % higher (Fig. 3). This can be explained by the difference in f_c/f_K which is less important at low R_λ but has a considerable influence at high R_λ.

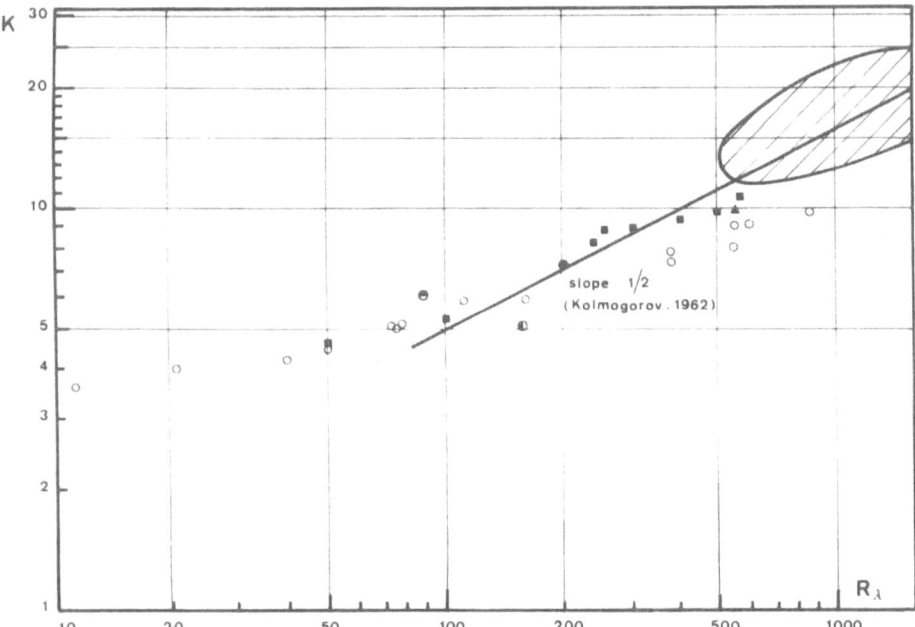

Fig. 3. Kurtosis as a function of Reynolds number R_λ.

■ *Present study $f_c = 2f_K$*

● *Wyngaard (1970) $f_c = 1.4\, f_K$*

○ $f_c = f_K$
◒ $f_c = 1.2\, f_K$ } Kuo and Corrsin (1971)
◓ $f_c = 0.8\, f_K$

▲ *Friehe, Van Atta, Gibson (1971) $f_c = 0.75\, f_K$; Atmospheric data* ▨

The intermittency in dissipation implies that ϵ is a random variable with a distribution which differs considerably from a normal one and Kolmogorov's (1962) theory gives:

$$\langle \epsilon(x_1) \cdot \epsilon(x_1 + r)\rangle \propto (r/\ell)^{-\mu}$$

where ℓ is the integral scale and μ a universal constant whose value must be determined experimentally.★ Spectra of $(\partial u/\partial t)^2$ taken in the atmosphere have shown that $0.30 \leqslant \mu \leqslant 0.65$ (Monin and Yaglom, 1975). In the laboratory similar

★Note that any positive quantity characteristic of the small scale structure should follow the same power law.

$$\left[<\mathcal{E}_{(x_1+r)}\cdot \mathcal{E}_{(x_1)}> - \left(<\mathcal{E}_{(x_1)}>\right)^2 \right] \Big/ <\mathcal{E}^2_{(x_1)}>$$

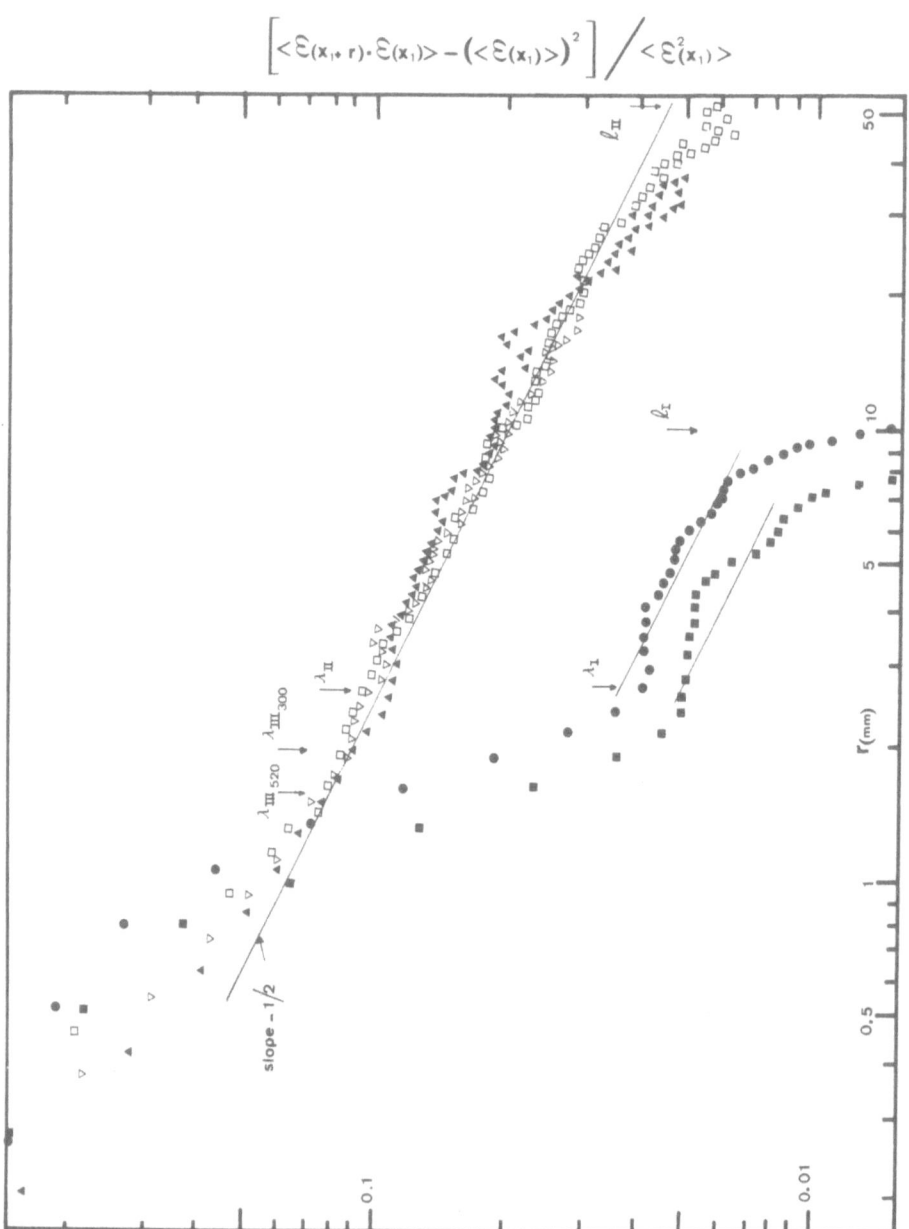

Fig. 4. Dissipation correlations for three different flows: grid turbulence ■, ●, R_λ 50; plane channel flow □, $R_\lambda = 230$; axisymmetric jet, ▲ $R_\lambda = 520$; ▽ $R_\lambda = 300$.

spectra have been taken by Friehe, Van Atta and Gibson (1971) from which a value of 0.5 seems to emerge. We have made correlation measurements in all three flows but only in flow II and III an inertial range exists over about one decade (Fig. 4). The slope can be defined with good precision particularly in flow II and $\mu = 0.5 \pm 0.05$. In flow III the experimental points oscillate. This is most likely due to the fact that in the jet the turbulence level is high and the Taylor hypothesis begins to be violated. The non-linearity in the constant current hot wire signal is probably less problematic since we only work with derivatives of the velocity signal.

The ϵ correlations correspond to moments of 4th order which requires a dynamic range $D = 12Q$ (where Q is the r.m.s. signal to noise ratio, Tennekes and Wyngaard (1972)). We are large in this respect ($D \approx 60Q$) and this will permit us to obtain higher order correlations of ϵ which is of interest to test existing theories.

Acknowledgements
The authors are very grateful to Mr. J-P. Barbier for building the low noise level anemometer and to Mr. M. Lagarde for having written most of the computer programs.

References

Comte-Bellot, G., (1965), Paris, P.S.T., Min. Air, No. 419
Friehe, C.A., Van Atta, C.W., Gibson, C.H., (1971), *Proc. AGARD, Spec. Meeting on Turbulent Shear Flows,* London
Kolmogorov, A.N. (1962), *J. Fluid Mech.* **13**, 81
Kuo, A.Y.S., Corrsin, S., (1971), *J. Fluid Mech.* **50**, 285
Marechal, J., (1972), *J. de Mécanique,* Vol. **11**, No. 2
Monin, A.S., Yaglom, A.M., (1975), *Statistical Fluid Mechanics,* Vol. **2**, M.I.T. Press
Tennekes, H., Wyngaard, J.C., (1972), *J. Fluid Mech.* **55**, 93

Time-Variant Aerodynamic Measurements in a Research Compressor

by

Sanford Fleeter
Mechanical Engineering, Purdue University
West Lafayette, Indiana 47907

and

William A. Bennett and Robert L. Jay
Detroit Diesel Allison, Division of General Motors
Indianapolis, Indiana 46206

Introduction

The fundamental mechanism for energy transfer in a turbomachine is the unsteady through-flow. However, current design systems consider either a time-averaged flow field or assume a steady flow in the relative frame of reference. Hence, extensive empirical correlations are required to compensate for this inadequate understanding of the fundamental unsteady flow phenomena.

To achieve further significant improvements in turbomachinery design and performance, it is necessary to understand and to control these unsteady flows. Fortunately, the required quantitative experimental investigations of these high reduced frequency time-variant flows are now becoming possible with the recent developments and availability of fast response instrumentation and data handling and processing equipment.

Herein is briefly described the experimental techniques, instrumentation, data acquisition and analysis procedures, and subsequent data-theory correlation developed and used in a fundamental investigation to determine the high reduced frequency fluctuating pressure distribution on a compressor stator, with the primary source of excitation being the wakes generated by an upstream rotor.

Instrumentation

The time-variant quantities of fundamental interest include the fluctuating aerodynamic forcing function — the rotor wakes, and the resulting chordwise distribution of the complex time-variant pressure on a downstream stator vane.

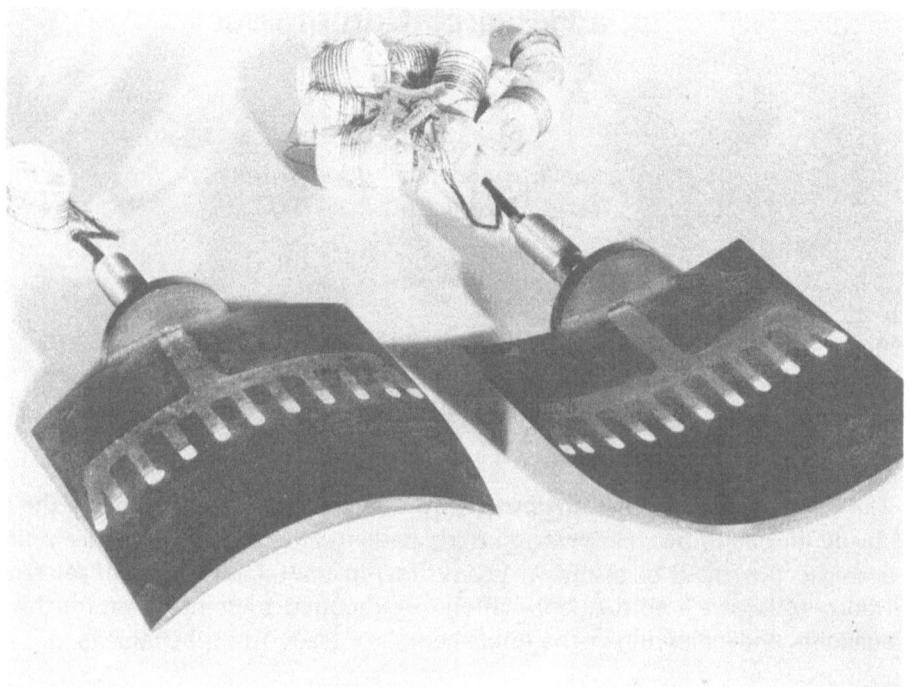

Fig. 1. Research compressor dynamically instrumented stator vanes.

The blade surface dynamic pressure measurements are accomplished with Kulite thin-line design dynamic pressure transducers flush mounted on a pair of the NACA Series 65 stator vanes. These instrumented vanes, seen in Fig. 1, are positioned in the stator row such that one flow passage is instrumented.

The time-variant wake measurements are obtained with a cross-wire probe, calibrated and linearized to 200 feet per second and ±25° angular variation. This probe is located in the rotor-stator axial gap of the flow passage adjacent to the dynamically instrumented airfoils, as seen in Fig. 2. The mean absolute rotor exit flow angle is determined by rotating the cross-wire probe until a zero voltage difference is obtained between the two linearized hot-wire signals. This mean angle is then used as a reference for calculating the instantaneous absolute and relative flow angles. The output for each channel is corrected for tangential cooling effects and the individual fluctuating velocity components parallel and normal to the mean flow angle calculated from the corrected quantities.

Fig. 2. View of crossed hot-wire probe and dynamically instrumented vane suction surface with rotor installed.

Data Acquisition and Analysis

The time-variant data acquisition and analysis technique used is based on a data averaging or signal enhancement concept. The key to such a technique is the ability to sample data at a preset time. For this investigation, the signal of interest is generated at the blade passing frequency. Hence, the logical choice for a time or data initiation reference is the rotor shaft. An optical encoder which delivers a square wave voltage signal having a duration of 1.5 microseconds is mounted on the rotor shaft for this purpose. The computer analog-to-digital-converter is triggered from the positive voltage at the leading edge of the pulse, thereby initiating the acquisition of the time unsteady data at a rate of up to 100,000 points per second. The data is sampled for N blade passages over M rotor revolutions. These rotor revolutions are not consecutive because a finite time is required to operate on the N blade passage data before the computer returns to the pulse acceptance mode which initiated the gathering of the data. For this experimental program, 80 to 100 digitized data points are obtained for each of three blade passages averaged over 400 rotor revolutions (N = 3, M = 400).

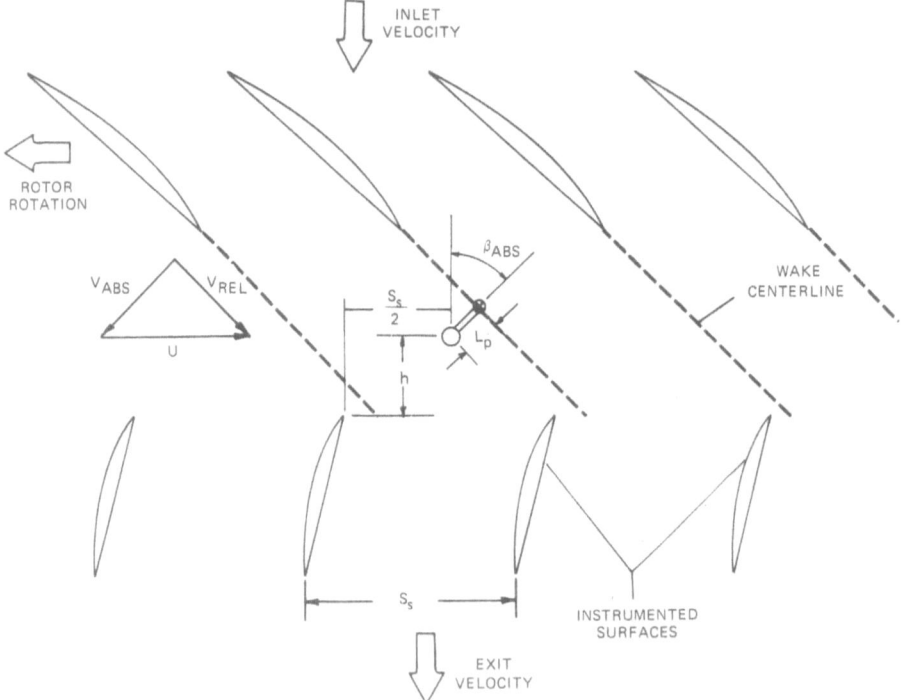

Fig. 3. Schematic of flow field used in dynamic data analysis.

Each of the time-variant signals is digitized and Fourier decomposed into its harmonics. From these, both the magnitude and the phase lag as referenced to the data initiation pulse are obtained. Fig. 3 presents a schematic of the flow field and dynamic instrumentation. The rotor wakes are located relative to the dynamic instrumentation planes through the assumptions that: (1) the wakes are identical at the hot-wire and stator leading edge planes; (2) the wakes are fixed in the relative frame. From these, the times at which the wake is present at various locations are determined. The increment times between occurrences are then related to phase differences between the perturbation velocities and the vane surfaces.

Following this procedure the pressure differences across an equivalent single vane at each transducer location are calculated. The final form of the data consists of a dynamic pressure coefficient, normalized with respect to steady-state properties of the flow and the magnitude of the transverse gust ($C_p = P/\rho V^2$ v/V), and an aerodynamic phase lag, which is referenced to a transverse gust at the leading edge of the instrumented vane.

Results
Two studies were undertaken in the course of this experimental program: (1) a qualitative study of the rotor wake velocity profile as a function of both compressor loading and downstream axial distance; (2) a quantitative investigation of the resulting time-variant surface pressures induced on the stator vanes by the upstream generated rotor wakes as they are convected downstream.

Wake Investigation
Fig. 4 presents the variation of the wake profile with axial distance as measured from the rotor at two levels of loading along the 100 % speed line. As indicated, increased axial distance from the rotor decreases the difference between the wake centerline velocity and the freestream velocity. At the high level of loading, this trend is extremely pronounced, becoming less so at the lower level of loading. An increase in the pressure ratio (a decrease in the mass flow rate) at a constant axial position increases the wake profile width and the deficit between the freestream and the wake centerline velocity.

Unsteady Pressure Differential Data-Theory Correlation
Fig. 5 presents the dynamic pressure coefficient and aerodynamic phase lag for the first harmonic of the unsteady pressure difference across the vane as a function of percent vane chord for a zero incidence angle at a reduced frequency value of 8.30. Also included are the incompressible predictions obtained from the state-of-the-art flat plate cascade transverse gust analysis of Reference I.

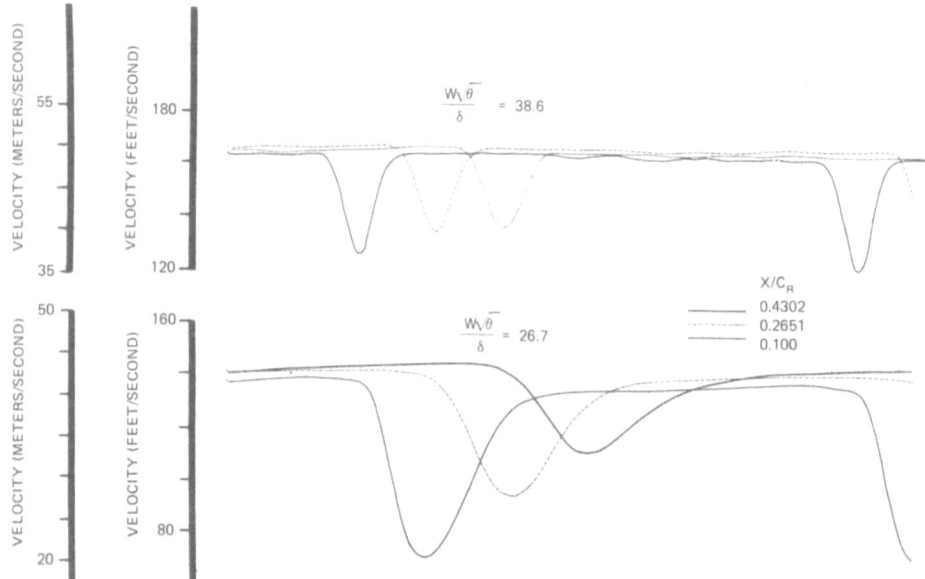

Fig. 4. Variation of the rotor wake profile with axial distance for two levels of loading.

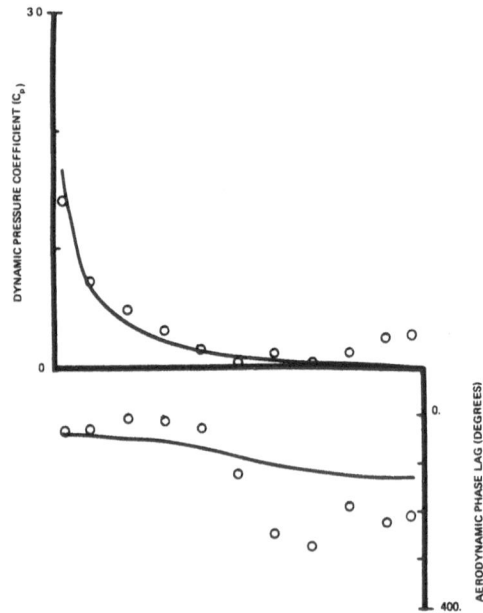

Fig. 5. Chordwise data for first harmonic unsteady pressure difference across the vane and prediction at a low incidence angle and a reduced frequency of 8.30.

As can be seen, the dynamic pressure coefficient data-theory correlation is quite good over the entire vane chord, with both decreasing in value in the chordwise direction. That the magnitude of this unsteady pressure differential data approaches a finite near zero value in the vicinity of the trailing edge is significant in that it reflects upon the validity of the Kutta condition at these high reduced frequency values.

The aerodynamic phase lag data-theory correlation is excellent over the front portion of the vane but becomes less favorable over the rear. This correlation trend is attributable to the differences between the actual vane and the analytical model: the vane has approximately 49° of camber whereas the analysis considers flat plate airfoils. Hence, there is good correlation over the front of the vane where the flow is aligned with the vane, and poorer correlation over the rear, where the camber effects have become important.

Acknowledgment
This research was sponsored, in part, by the Air Force Office of Scientific Research (AFSC), United States Air Force.

Reference
Fleeter, S., "Fluctuating Lift and Moment Coefficients for Cascaded Airfoils in a Nonuniform Compressible Flow" *AIAA Journal of Aircraft*, Vol. **10**, No. 2, February 1978.

Measurements in Two-Phase Flow

Measurements in Two-Phase Flow

Victor W. Goldschmidt

Purdue University, West Lafayette, Indiana

I. Introduction

An experimental fluid mechanicist can easily become awed by the area of measurements in two-phase flow, finding it both challenging and discouraging. Classical measurements of pressure, velocity, and flow rates become more complex and many times inaccurate. Problems dealing with phenomena involving sprays, mists, bubbly flows, dusts, smoke (made up of bubbles, drops, or solid particulates) are brought up by sister disciplines such as nuclear, civil, mechanical, and chemical engineering. They generally call for basic data on the size distribution, spatial distribution (i.e., concentration or density), and kinetic distribution (i.e., velocity or trajectories) of the added phase, adding to the complexity of the measurements.

An additional and growing concern is the corresponding effect of the additional phase (bubbles, drops, or solid particles) on the carrier itself. As an example, a complete quantification of the possible changes in the turbulent structure of the carrier stream due to the carried contaminant is, to this reviewer's judgement, still missing (for an update see, for instance, Danon et al., 1977)[*]. Similarly, for that matter, the measurement and prediction of the corresponding contaminant turbulent transport is also far from resolved. Although neither of those concerns are to be treated now, it is obvious that one of the missing bits of information to address them can be attributed to the lack of accurate experimental data subject to generalization.

The field to be considered in this review will be purposely limited.

1. Measurements of bulk properties (averaged over a surface or a volume considerably larger than the turbulent microscale) and measurements on the confining walls will generally not be considered. This will exclude measure of static pressure drops, measurements with venturi meters, etc. Of concern will be measurements leading to a quantification of the size, velocity, or concentration of the suspended bubble, drop, or solid particulate.

[*]Throughout the discussion only the first author will be noted even when only two are listed.

2. The review will limit itself to turbulent flows, except for particular references describing measuring systems applied to laminar flows but applicable to turbulent flows as well.

3. In addition, the "far deep" and "far high" will be neglected excluding areas such as sediment transport and atmospheric pollution. The references will be limited to those dealing with the type of flows which can generally be modeled through laboratory experiments.

4. Only fully dispersed flows of bubbles, drops, or solid particles in carrier liquids or gases will be considered. The case of two gas phases or solutions in liquids (see, for instance, McQuaid, et al. (1973), Quarmby, et al. (1974), Edwards, et al. (1976) and Torrest, et al. (1969)) and of Hewitt et al. (1970) will not be treated, although many of the experimental techniques apply there as well.

5. A number of excellent reviews on measurements in two-phase flows have already been published. In order to avoid some redundance, the literature referred to now will only be that published after 1964. It is worthy to note that those few references now quoted and not reviewed by this author will so be indicated in the text correspondingly acknowledging their source.

The diversity of disciplines concerned with two-phase flows, and the different means of measuring properties in them, do not provide a clear path for a literature survey. Unlike reporting on measurements of turbulence in pipe flows, which would have Professor Laufer's publication as a common denominator, a review on turbulent diffusion, undoubtedly branching off from Sir G.I. Taylor's contribution, or a review on intermittency and conditional sampling, stemming from the researchers at the Johns Hopkins University, led by Professors Corrsin and Kovasznay, there is no clear-cut path for the starting point of a literature review on measurements in two-phase flows. Simultaneously, the "publication explosion" is evident in two-phase flows as well as in most other areas of engineering. There are, as a result, many omissions of importance. For example, the German literature is not properly represented, and papers in proceedings of a number of conferences (such as, for example, the May 15, 1978 Symposium on "Advances in Particle Sampling and Measurement" and the just past conference on Two Phase Flows, chaired by F. Durst) are not included. The intent is instead to provide the reader with a representative set of references in order to give an entryway to the newcomer and an overview to those already in it.

In the sections that follow, different measurement techniques will first be listed. Following that, their expected applicability to measurements in different types of two-phase flows will be noted. The types of two-phase flows are selected according to contaminant size and state (drops, bubbles, or solid particles).

II. Classification of Measurement Techniques

Measurement techniques will be classified into five major categories. **Removal Methods** (called "mechanical" by Soo (1967)) will include trapping of a given sample, isokinetic sampling, inertial collection (usually on a treated microscopic glass slide), and collection after freezing (an original drop), encapsulating, etc. These methods do not permit a measure of velocities but can lead to measure of some space and time average concentration and contaminant size distribution.

The second category will be that of **Probe Methods** (generally lending themselves to local concentration and sizing measurements and, in some instances, velocity data as well). The methods included are impact probes, optical probes, microthermocouples, hot wire/film anemometry, and electrical probes (such as resistance or conduction probes, capacitance, and probes dependent on charge or inductance). In some instances, an array of two or more probes may be used to give velocity data.

Optical Methods will comprise the third category. In addition to the classical photographic techniques, holograms, light scattering methods, laser dopplers, and interferometers are included. The fourth category is that of **Tracking Methods**, where individual particles are tagged with a radioactive or similar tracer or simply tracked by some optical scheme.

Bulk Methods, leading to measure of void fraction in bubbly flows, including radiation attenuation methods, impedance gages, and sonic and ultrasonic devices, will be placed in the last category. In tabular form, then, the methods will be classified as follows:

TABLE 1

Classification of Techniques

1. Removal Methods	1.1 Trapping
	1.2 Isokinetic Sampling
	1.3 Inertial Collection
	1.4 Freezing, Encapsulating, etc.

2. Probe Methods   2.1 Impact Probe
                                       2.2 Optical Probe
                                       2.3 Microthermocouples
                                       2.4 Hot Wire/Film Anemometry
                                       2.5 Electrical Probes

                                           2.5.1 Resistance or Conductance
                                           2.5.2 Capacitance
                                           2.5.3 Charge
                                           2.5.4 Inductance

3. Optical Methods   3.1 Photographic
                                       3.2 Laser Holograms
                                       3.3 Laser Scattering
                                       3.4 Laser Doppler
                                       3.5 Laser Interferometer

4. Tracking Methods   4.1 Radioactive or other tagging
                                       4.2 Optical (dual beams, etc.)

5. Bulk Methods   5.1 Radiation Attenuation and Light
                                           Transmission
                                       5.2 Sonic and Ultrasonic Devices

The measurement methods can, in turn, be applied to measurements of size, velocity, or concentration in flows with drops, bubbles, or solid particles. The measurement objectives can then be identified through the following:

TABLE 2

Measurement Objectives

1.1 Drop Sizes	2.1 Bubble Sizes	3.1 Solid Particle Sizes
1.2 Drop Velocities	2.2 Bubble Velocities	3.2 Solid Particle Velocities
1.3 Drop Concentration	2.3 Bubble Concentration	3.3 Solid Particle Concentration
	2.4 Void Fraction	

III. Description and References Representative of Measurement Techniques

In the short sections that follow the different measurement schemes, as coded through Table 1, will be briefly described and sample references given. When defined, the applicability of the method will be stated as per the numbering scheme of Table 2.

1.1 Trapping Methods

Concentration measurements may be made simply by quickly closing a section of the flow (Jepson, et al., 1970) or the ends of a cylinder or similar device, which has been inserted in the flow, and trapping a known sample volume. When the trapping device is provided with a glass slide, photomicrographic sizing of the particulates is possible. There must be some flexibility as to the size of the trapped volume to assure that a large enough sample ultimately settles on the glass slide after the sample is captured. In an extreme case, too many particles might be caught resulting in drenching or stacking on the slide. The system lends itself for 1.1, 1.3, 3.1, and 3.3 measurements of drops and solid particles in a gas flow. Treatment of the glass slide with coatings such as immersion oil (Namie, et al., (1972) and Hiroyasu, et al., (1976)), lamp black, or magnesium oxide (see, for instance, Orr (1966)) may be desirable. The particulates in the latter case will leave a crater proportional to their size. Alternatively, when sizing liquid droplets on glass slides, their spread factor has to be accounted for. Both Nash, et al. (1967) and Fujimoto, et al. (1967) describe a slide and shutter scheme for sizing droplets. Goldschmidt (1965) describes a trapping cylinder used for both size and (through count of particles) concentration measurement of liquid droplets. A similar device was used by Onuma (1974) to obtain both a number and size of droplets in a spray burner. Concentration measurements can be obtained more effectively using a tracer dye (and some optical device after trapping the sample (Verhoff, et al. (1977)). An application of trapping in a supersonic flow of zinc particles is described by McBride, et al. (1972), while a trapping device for drops in liquid-liquid systems is used by Mlynek, at al. (1972).

1.2 Isokinetic Sampling

The isokinetic sampler is simply a sampling tube facing the oncoming flows. A sample is withdrawn and analyzed for size and concentration distribution. A rather extensive survey is given by Schraub (1969). Sampling errors (primarily through size biasing when sampling is non-isokinetic) are discussed in Inoya (1976) and clearly described by Boothroyd (1971). Hewitt (1972) and Bankoff

(1965), among many others, refer to the work at CISE Milan for examples of methods employed to assure isokinetic sampling. vanBreugel, et al. (1970) describe a probe and report measurements in dense gas-solid suspensions. Parker, et al. (1970) report the use of isokinetic sampling with submicron size solid particles in air while Brown, et al. (1968) and Jepson, et al. (1970) describe the use of an isokinetic sampler for gas-liquid systems. Suction tubes were also used by Laats (1966) for dust in air, Briller, et al. (1969) and Namie, et al. (1972) for droplets in air, Kubie, et al. (1977) for drops in liquids, and Suneja, et al. (1972) for KCl particles in air. A commercial counter is used after sampling by Bragg, et al. (1975) for dust in air, while a method based on the Coulter counter is described by DeBlois, et al. (1970) and by Sprow (1967). Rao, et al. (1971) and Pchelkin, et al. (1975) extend and compare the use of the isokinetic probe to measure momentum flux or thrust due either to the change in direction or the stagnation of the two-phase flow. The analysis is similar to that employed to interpret the use of pitot-static probes in two-phase flows. (See, for example, Doig, et al. (1967), Mobbs, et al. (1970), Jepson, et al. (1970), Keller (1973), Crane, et al. (1975), Soo (1965), and Michiyoshi (1978)). Isokinetic samplers are best suited to 1.3, 2.3, 3.1, and 3.3 type applications.

1.3 Inertial Collection
These include conventional schemes primarily applicable to droplets or solid particles in air (1.1, 1.3, 3.1, 3.3) where collecting slides, ribbons, or other objects are placed in the stream for collection of a sample. Their application is straightforward but complicated when actual particle trajectories prior to impaction are desired in order to determine concentration via a collection efficiency (Nash, et al. (1967), Householder, et al. (1969)).

1.4 Freezing, encapsulation, etc.
The use of molten wax while determining properties of sprays (Kim, et al. (1971)) and the encapsulation of droplets (Mlynek, et al. (1972) and Karabelas (1978)) are examples of cases where the size distribution of droplets (1.1, 1.3) can easily be determined photographically or microscopically.

2.1 Impact Probe
The resulting forces due to the change in momentum flux of a two-phase flow subject to stagnation already referred to can be used for metering. Boothroyd (1971) describes some of the schemes based on impact on cantilevered bodies, strain gages, etc. They all have the same shortcoming as the inertial collection method when actual particulate trajectories are needed for proper analysis. Neal, et al. (1965), Burick, et al. (1974) and Sato, et al. (1975) give examples when the measured stagnation pressure in an impact probe similar to a pitot total pressure probe is used to measure time average properties such as velocity or mass flux. Sato, et al. apply the concept to measurement of bubbles

in liquids whereas Burick, et al. consider the flow of drops in air (Neal, et al. treat nitrogen-mercury flow). In either case, a calibration dependent on the probe characteristics is desirable.

2.2 Optical Probes

Optical probes can be designed based on their sensitivity to the change in refractive index or light transmission of the surrounding medium. Jones, et al. (1976) classify optical probes as glass rod, fiber bundle, or U-shaped fiber systems. They all operate on the same principle, light is transmitted through a light guide to the probe tip. From there it emerges and according to the sursounding material's refractive index is partly reflected back to the probe and through another light guide to a phototransistor. The output of the phototransistor will be dependent on the presence or absence of drops, bubbles, or solid particulates in the immediate vicinity of the probe. Miller, et al. (1969) describe a glass rod system whereas the U-shaped fiber system as developed by Delhaye is referred to in Jones, et al. (1976). Optical probes of this type were developed, used, and critically evaluated by Hinata (1972), Hinata, et al. (1977), Abuaf, et al. (1978 a,b), and Abuaf, et al. (1978). Dunn, et al. (1976) use two probes, one downstream of the other, to measure velocities. Measurements were also reported by McSweeney, et al. (1972) and Oki, et al. (1975), (1977) who also present an analysis of the probe response. An alternative scheme is to miniaturize a transmitter and receiver of a light beam, mount them on a single probe, and with them sense the crossing of bubbles, drops, or solid particles. This method was used by Fenton, et al. (1976) for particles in air, by Weinstein, et al. (1973) and Schindler, et al. (1968) for drops in liquids, and in a simplified manner by Peskin, as referred to in Boothroyd (1971) and Soo, et al. (1964), (1966). Obvious problems such as particulate trajectories, coincident pulses in dense flows, spatial averaging, and the coating of the optical surfaces are generally not major and can be resolved. (Fenton, et al. (1976) and Boothroyd (1971), for example, describe a purging system to blow out deposits on the optical surfaces). With proper data analysis, the method can be applied to almost all the measurement objectives (a double probe could be used for velocity measurement).

2.3 Microthermocouples

The microthermocouple is primarily suited for measurements of the type 2.4 for boiling two-phase flows. Delhaye, et al. (1973) exhibit how a small thermocouple's signal, when treated in order to obtain the amplitude histogram of the temperature, can be used to determine the void fraction. Afgan and co-workers (1973, 1973, 1978) report similar applications of a microthermocouple in a boiling two-phase flow.

2.4 Hot Wire/Hot Film Anemometry

Hot wire/film anemometry lends itself for measurement of concentration flux and sizing of liquid droplets in air (application 1.1 and 1.3) as noted by Goldschmidt, et al. (1966) and Goldschmidt (1965). The higher frequency of the cooling signals attributed to impaction of the cooler droplet and ensuing vaporization permit discrimination from the lower frequency turbulent signals. In order to size the droplets, Goldschmidt, et al. (1968, 1969) determined that there should be a linear relationship between the peak values of the signal due to impact and size. However, with further data, Bragg, et al. (1974) contradicted these results. (Further insight into the controversy might be gained through a related study by Seki, et al. (1978)). Experience on measurements of liquid droplet transport in pipe flows is documented by Ginsberg (1971) whereas Hetsroni, et al. (1969, 1971) and Goldschmidt, et al. (1972) present extensive data for droplet transport in turbulent air jets.

The use of hot film anemometry in air/liquid flows (primarily for measurement of objectives 2.1, 2.2, 2.3, 2.4) is thoroughly discussed by Delhaye (1969), Chuang, et al. (1971), and Bremhorst, et al. (1976). Delhaye, et al. (1977) note that the overheat of the sensor has to be limited to avoid degassing and apply a scheme to measure local void fraction and velocities (using a conical sensor). (A method for digital analysis of the signal is given by Postaire, et al. (1977). Measurements with bubbles in a rectangular liquid channel flow are presented by Jones, et al. (1975, 1977) whereas data in a hydraulic jump are recorded by Resch, et al. (1974, 1975, 1976). A thorough review on the use of hot film/ wire anemometer is given by Jones, et al. (1976) partly based on Delhaye (1969).

Related reports on the use of heated elements for bubble detection are presented by Hirata, et al. (1977), who use a heated platinum wire and a bridge circuit to discriminate between air and helium bubbles in water and by Kubie (1975) who with a similar set-up determines the frequency of bubble generation.

On the other hand, van Paasen (1974) analyzes and exhibits the measuring of liquid droplet sizes with a thermocouple on which the droplets evaporate (similar in parts to Seki, et al. (1978)), while Marsheck, et al. (1965) present a method for detecting silica-alumina particles in air (application 3.3). They used two thermistors, one as a "heater" and the other as an "indicating" thermistor. The presence of a particle between them would be accompanied by a decrease in the receiver's temperature.

2.5 Electrical Probes

Soo (1967) describes electrical methods as utilizing the measurement of charge, mobility, capacitance, and resistance. Bergles (1969) in his rather extensive review paper classifies electric probes as conductance probes, capacitance probes, and spray analyzers, which measure the charge withdrawn as a drop touches a charged wire.

Electrical probes may lend themselves to measurement of concentration and size whereas the use of two probes could provide velocity data.

2.5.1 Resistance (or conductance) probes are dependent on the varying electrical conductivity of two-phase flows. The voltage between two electrodes (one of which is the tip of the probe and the other a second probe, or the ground of the test section or probe support) will depend on the presence of bubbles, drops, or solid particulates in their proximity. They are best suited for 2.1, 2.2, 2.3, and 2.4 type applications. Kobayasi, et al. (1970) and Akagawa (1964) describe the use of an electrode probe to measure the void fraction in air-water flow. In Akagawa, et al. (1966) and Akagawa, et al. (1971) the concept is expanded (using more than one probe) to measure velocities of the bubbles and slugs. (Uga (1971) also uses more than one probe and applies it to the measure of bubble size).
Serizawa, et al. (1975a) develop a rather elaborate system based on a double-sensor probe with which local void fraction, bubble impaction rate, bubble size and velocity in a liquid stream can be determined. The method is applied to air-water bubbly flow in Serizawa, et al. (1975b,c) and Michiyoshi (1978) as well as to bubbly flow in Michiyoshi, et al. (1977). A similar method was developed and reported earlier by Park, et al. (1969) and Rigby, et al. (1970). Earlier work by Kitayama (1972), also utilizing a single and double probe, does, however, suggest that a single probe may be more applicable to measure bubble sizes in the case of high bubble concentrations. Burgess, et al. (1975), in three consecutive papers, develop and apply a five element array to properly measure bubble size and velocity (eliminating the uncertainty of incorrect alignment to the bubble trajectory in two probe elements).

Sekoguchi, et al. (1975a, 1975b) analyze and measure the expected signal due to a bubble passing past an electric resistivity probe consisting of two electrodes. They utilize both a trapping and a photographic measurement procedure to verify the applicability of the probe to measure void fraction and bubble sizes. Bubble velocity, in addition to size and impaction rate, was measured by Konstantinov, et al. (1976) with an array of fourteen electrodes, whereas, in a completely unrelated effort, Sultan, et al. (1978) use a single conductance probe to detect bubbles in boiling water. Gardner, et al. (1970) use an "electrolytic probe" to quantify void fraction in water and Keller (1973) describes an array of probes for field measurements of aeration in a spillway.

In applications to other than bubble flow in liquids, Hoffer, et al. (1975) report measurements in liquid-liquid dispersions, Steiner, et al. (1974) predict and confirm with measurements the response of a point electrode to solid glass spheres in an electrolyte and Beck, et al. (1973) determine the particle scale of turbulence for preselected sieved sand particles in air.

2.5.2 Capacitance probes, where two electrodes are placed so as to form a capacitor, with the two-phase medium acting as the dielectric (Bergles (1969), Cimorelli, et al. (1967)) are best suited for 2.4 and 3.3 type measurement objectives. The method is outlined by Boothroyd (1971) where he points out the basic limitations due to the usual need for relatively large size plates, although he does refer to earlier measurements in fluidized beds and in a transfer line. The particulates alter the capacity of the plates, which in turn can be used to unbalance a bridge circuit.

The use of capacitance probes was notably modernized by Werther, et al. (1973 a,b) and Werther (1974) through the application of miniaturized capacitance probes. Using either a single or double probe, the mean bubble pulse duration, impact rate, and bubble velocity are determined. (A recent patent disclosure by Dunn (1978) describes a related probe capable of measuring local void fraction via a measure of impedance variation).

2.5.3 Charge probes are dependent on the change of charge of a wire or other sensor placed in the stream as a drop or solid particule comes in contact with it. Gardiner (1964) considers the case of liquid droplets in air. He finds that the charge drawn from a charged wire is a function of droplet size and can hence be used as a size sampling scheme. Pinczewski, et al. (1977) carry the concept further improving the probe design and effectively calibrating for the response of the sampler.

The method can be used for concentration measurements as well as sizing, (see, for instance, Soo, et al. (1964), (1966)), but is generally used for 1.1 type applications.

2.5.4 An inductance probe, supposedly the first of its kind, is described by Cranfield (1972). It was designed in order to detect and measure bubbles in a fluidized bed. The probe consists of an inductive transducer of proprietary nature, supposedly capable of measuring bubble size, impact rate, and velocity.

3.1 Photographic Methods

Photographic methods for measuring both size and velocity of a particle (drop, bubble or solid) can generally be used for particles between 10 and 500 microns. (Soo (1967)). In order to obtain velocities high speed motion pictures or successive exposures are necessary. Although photographic techniques have the virtue of not disturbing the flow they are usually laborious and preclude the acquisition of sufficient data (Carlson et al. (1975)). The major problems appear to be illumination (overcome in some cases with high intensity sources such as lasers) and determining which particles are in focus and which are not. (Orr (1966), Hewitt (1972)). Hsu, et al. (1969) present a survey paper including photographic techniques. They classify these as direct illumination, shadowgraph and schlieren photography. At this time "photographic" methods will be classified as direct photographic, multiple or streak exposures and high speed photographs.

Direct photographs for example, are best applicable to sizing (1.1, 2.1 & 3.1) (see for example Jeffreys, et al. (1970), Card, et al. (1971) and Tsuji, et al. (1978)). The particles may be illuminated with a collimated light source (see for instance Hayashi et al. (1976)) or alternatively viewed with a limited focal plane (in some instances through a microscope (see for instance Graham et al. (1973), Pogson et al. (1970)) or even a periscope (Mlynek et al. (1972)). Collins et al. (1970) summarize earlier comments by Kinter et al. (Can. J. Chem. Eng. Vol. 39 pp 235-241, 1961) and Ward (Ph.D. thesis, Oregon State University, 1964) on the limitations of using photographic techniques. As particle size decreases and/or concentrations increase light transmission decreases and the optical resolution suffers. They also refer to a method through which the final photographs may be analyzed with a narrow beam and photocell sweeping the negatives. Illumination remains the major problem although in backlighting (Temkin et al. (1972), Chen et al. (1978)) the needed illumination is not as much as in forward lighting. The needed intensities are usually met with stroboscopic light sources as done by Chien et al. (1967), Cumo et al. (1974) and Ochi et al. (1978).

The use of stroboscopic light flashes for illumination lends itself to multiple exposures (of at least two sparks) through which particle trajectories can be traced and velocities determined (applications 1.1, 1.2, 2.1, 2.2, 3.1, 3.2). Use of this method is noted in Temkin, et al. (1972), Maneri, et al. (1974) Chigier (1974) Sugimoto et al. (1978) and reviewed in Somerscales (1969). Alternative schemes of chopping the light source or rotating windows on the lens are described by Abuaf, et al. (1974) and Armand (1972) while a double flash and one stereocamera was used by Reddy, et al. (1969).

Streak photography is an option to the multiple flash exposure but is usually limited to larger particles where illumination intensity is not as crucial. The motion of solid particles in a stirred tank was determined in this manner by Schwartzberg et al. (1968) and together with a stereomicroscope by Pogson et al. (1970). The analysis of the data can be tedious although the results can be quite consistent. (See Kramer et al. (1972) where they also refer to a cross-correlation method).

The use of high speed photography can give quantitative as well as qualitative results. The problems of illumination are even more severe and generally limit the use of high speed photography to larger particles. Examples of high speed photography in bubbly and boiling flows are given by Frost et al. (1967), Geldart et al. (1972), Kling et al. (1972), Holmes et al. (1975), Wairegi et al. (1976) and Unal (1977, 1978).

3.2 Laser Holograms
Holographic techniques (usually employing a ruby laser for recording and a He-Ne laser for reconstructing) are capable of giving spatial distribution and velocities of droplets and solid particles in gases or bubbles in liquids like water. Fourney et al. (1969) in a rather descriptive publication demonstrate the use of holographic techniques. Lee et al. (1974) describe and use the method to measure the size and velocity of drops in an air-water mixture.

The literature abounds with recent publications specifically directed to the use of lasers in a scattering, interferometric, or a Doppler frequency analysis mode for measuring particle size, velocity and concentration. The trend appears to be towards simultaneous measurement of size and velocity — usually with an LDA (laser Doppler anemometer system) for measuring particle velocity and with scattered light for measuring size (Durst, et al. (1975)). The dependence on size of the scattered light is based on Mie's theory for particles in the micron and submicron size and extensions thereof for larger size particles.

3.3 Laser Scattering
Laser scattering techniques (applicable primarily to 1.1 and 3.1 objectives) are reported by Ferrara, et al. (1970), Landa (1972) and Holve, et al. (1978). Ho, et al. (1978) show how measuring the scattered light at two angles allows more than one particle in the focal volume at one time whereas Bachalo (1978) describes the development of a dual beam scattering instrument for particle sizing.

The principles of scattering apply to other light sources as well. Rudinger (1970) and Gooderum et al. (1967) are two examples of such. Systems capable of measuring both particle size (via light scattering) and velocity (via a laser Doppler anemometer) are described by Ogden et al. (1978) and by Durst et al. (1976). A similar scheme, where part of the optical arrangement leads to particle sizing and another to velocity measurement is described by Wittig et al. (1978) in what they call a "laser-two-focus" technique.

3.4 Laser Doppler

The development of the laser Doppler anemometer has been exciting to watch by most of us and intriguing to be in on the development by others. The prevalent "claim" is that if the seeding is small enough then the anemometer measures the fluid velocity, whereas if the particles are of sufficient size then indeed we have a two-phase flow device to measure particle velocity, applicable to 1.2, 2.2, and 3.2 objectives. The state of development of the instrumentation and the intensity of involvement on the part of many researchers has led to quite a bit of constructive controversy on the best optics as well as the best method for signal analysis. Reported instances where laser Doppler anemometers were used to measure the velocity of particles are reported (amongst others) by James, et al. as early as 1968 for mist and solid particles in air; Popper, et al. (1974), Crane, et al. (1975) and Voss-Spilker (1977) for droplets in air, by Carlson, et al. (1975), Birchenough, et al. (1976), Jurewicz, et al. (1977) and Stock, et al. (1975) for solid particles in air, by Einav, et al. (1973) for solid particles in water and Sullivan, et al. (1978) for bubbles in a liquid.

Recent literature also considers ways through which a laser Doppler anemometer can, by itself, be extended to measure particle size (as well as velocity). Ben-Yasef et al. (1975) extract size information from the velocity distribution (assuming a Stokes type drag). In a related method — but dependent on the phase lag between the motion of the carrier and the particles — Mazumder, et al (1977, 1978) and Sato et al. (1978 a, b) use a laser Doppler anemometer as a size sampler. The laser Doppler signal itself can be used for sizing. Lee et al. (1978) use the duration of the individual laser-Doppler burst, Chigier et al. (1978) extract size information from the peak values of the filtered anemometer signals while Driscoll et al. (1978) used for their sizing of submicron particles the broadening of the optical spectrum of the laser scattered light.

The availability of laser Doppler anemometers has also extended their use from velocity measurement to concentration measurements. In a rather unusual application, Schmidt et al. (1974) report the measurement of frazil ice concentration with a laser Doppler anemometer.

3.5 Laser Interferometer
In a series of publications Farmer and co-workers describe a method for determining particle size, concentration and velocity based on the nonuniform illumination of a particle by a set of interference fringes generated by crossed laser beams. In Farmer (1972) the feasibility of measuring particle size, concentration and velocity is analyzed, while results on measurements with particles 10 - 120 microns in size are shown in Farmer (1974). The latter two publications, both Farmer et al. (1978) report on calibration and measurements in turbine precombustors and in rocket exhausts.

Within these proceedings F. Durst has a discussion on Studies of Particle Motion by Laser-Doppler Techniques while Farmer has further descriptions and applications of the technique.

4. Tracking Methods
A review on tracking methods is given by Somerscales (1974). Within his broad classification two methods can be uniquely identified now. One of these tracks an individual particle, or sets of particles with a tracer. The other follows an untagged particle through dual beams of light or some similar scheme. (A number of the methods identified in earlier sections had tracking aspects within them, such as for instance, the double laser already referred to). The method is best applied to objectives 1.2 and 3.2.

4.1 Radioactive or other Tagging
Roberts et al. (1971) tagged particles with radioactivity and tracked them with a scintillation counter while observing the motion of neutrally buoyant particles in a liquid. In a similar method the tracking of particles in beds is reported by Van Velzen et al. (1974) and Gatt (1977). Rao et al. (1973) used a pressure sensitive radiopill as a tracer and tracked it with a ferrite antenna system while determining the flow of particles in a hopper. Jones et al. (1968) first used light emitting particles and photomultiplier detectors while later on Howard, et al. (1975) improved the system with radioactive tagging and appropriate detectors. The array of detectors used (mounted on a movable carriage) permitted a continuous tracking of the turbulent motion of the particles.

4.2 Time of Flight
The time of flight method implies the use of two detectors (usually optical) to measure the time for a particle to go between two observation points. It lends itself to measurements of the type 1.2, 2.2 and 3.2. Examples are given by Laderman, et al. (1969), Matthes, et al. (1970), and Lading (1977, 1978).

5.1 Radiation Attenuation & Light Transmission Methods

Radiation attenuation techniques are described by Schrock (1969) and Jones et al. (1976). A collimated beam of radiation when passed through the two-phase flow will be attenuated depending on the phases it passes through. With proper calibration a measure of the void fraction (application type 2.4) can be obtained. Obviously, the interpretation of the attenuated intensity is easier for flow in rectangular channels than in those with a circular cross section. (See for instance Thorpe et al. (1970) and, for a different approach Rowe et al. (1971), (1972) where x-rays were used). Measurements in rectangular channels are reported by St. Pierre, et al. (1967) and Jones, et al. (1975), whereas data for annular channels is given by Evangelisti, et al. (1969) and for circular tubes by Pike, et al. (1965) and El Halwagi, et al. (1967). A rather unique scheme employing three beams is given by Heidrick, et al. (1977), extending the earlier work by Lassahn (1976) using correlation techniques (a dual beam x-ray photometer was used by Jones in his Ph.D. thesis as referred by Jones, et al. (1978).

The light transmission methods are quite similar in concept. Hodkinson (1966) gives a rather extensive discussion of extinction of light by aerosols, whereas Kondic (1970) makes reference to light attenuation results. The measurement of void fractions in air-water flows is exhibited by Tsuji et al. (1978) and in part by El-Kaissy et al. (1976).

5.2 Sonic and Ultrasonic Devices

The velocity of propagation of acoustic waves will be different in fluids of different properties. Consequently, the presence of drops or bubbles will cause a phase shift of the detected waves forced to travel through the medium. Smith (1974) describes the use of that phase shift to determine the volume fraction of drops in a liquid (and also suggests that the variance of that phase shift will give a measure of their size).

Using two parallel ultrasonic beams and cross-correlating their modulations transmitted across the pipe of a gas liquid flow Olszowski et al. (1976) determine the mean velocity and (although not shown) suggest that the quality of the flow could be measured as well.

Also applying the principles of changes in propagation of a wave in two-phase flows, Jones et al. (1978) describe a compact radio wave frequency probe (the emitter and receiver are two wires mounted on the probe tip) capable of measuring both local velocity and void ratio. The scheme is relatively new and may be worthy of further follow-up.

IV. Comments and Conclusions
After skimming over some 300 references (with some 200 finally included in this paper) I have reached the conclusion that the whole field of measurements in two-phase flows is still just as uncertain and shaky as it was while I was deeply involved in it some ten years ago. The advent of the laser Doppler has opened some hope, but unfortunately the cost and individual attention required to make it useful limit its applicability (without mentioning the many 'optimum' schemes for signal retrieval and processing). One of the shortcomings appears to have been the perspective in development of instrumentation. Researchers appear to have been asking themselves, "I have this great scheme for instrumentation. Now what shall I apply it to?" instead of, "I have this specific two-phase flow for which I need data. What instrumentation can I use or adapt that my technician can then use on my behalf?"

In most cases calibration remains a problem. (Unfortunately in some it is also an ignored problem). The use of standard "particle sources" for calibration (a suggestion stemming from the 50's) might be worthy of consideration. Liu (1976) for example, presents a convincing argument on the need for standardization and calibration.

The benefits of centering our sights on the flow field to be measured, rather than the sensor and electronics to apply, is obvious when we recognize the advances in measurements in unique areas such as bubbly flows and liquid air interfaces. I am looking forward to surveys by Resch (Marseille Conference) and Leutheusser (Baltimore Conference) amongst others in this conference where that approach is apparent and fruitful.

In Table 1 the classification for different techniques was presented; in Table 2 possible measurement objectives were listed. The review is not complete without combining those two tables. Such as attempted in Table 3.

Acknowledgements
The author acknowledges the input received from a number of colleagues in the form of reprints, etc. Part of the time dedicated to this effort was included under research activities sponsored in part by the ONR and the NSF.

TABLE 3

Suggested Measurement Techniques	1. DROPS SIZE .1	.2	.3	VEL .1	.2	.3	CONC .1	.2	.3	2. BUBBLES SIZE .1	.2	.3	VEL .1	.2	.3	CONC .1	.2	.3	Void Fract.	3. SOLID PART. SIZE .1	.2	.3	VEL .1	.2	.3	CONC .1	.2	.3
1.1 Trapping	*	*					*	*	*							x	x	x	x	*	*					*	*	
1.2 Isokinetic	x	x					*	*	*											x	*					x	*	
1.3 Inert Collection	x	x					x	x	x											x	*					x	x	
1.4 Freezing, etc.							x	x																				
2.1 Impact Probe							x	x	x							x	x	x	x							x	x	x
2.2 Optical Probe	x						x	*										x	*							x		
2.3 Micro Thermocouple																			x	x								
2.4 Hot Wire/Film	x							*		*						*	*		x									
2.5.1 Resistance											x						*		*									
2.5.2 Capacitance	x								x							x			*							x		
2.5.3 Charge																				x								
2.5.4 Induction	x								x	x						x			x							x		
3.1 Photographic	x	*		x	x		x	x		*					x	x				x	x				x	x		
3.2 Holograms	x	x		x	x		x	x			x									x	x				x	x		
3.3 L Scattering	x	x								x	x									x	x							
3.4 L Doppler				x	*								x	*									x	*				
3.5 L Interferometer	x			x							x			x							x			x				
4.1 Tagging				x	*										*								x	*	*			
4.2 Dual Beams				x	*																			x				
5.1 Attenuation																			*									
5.2 Sonic									x						x				x									

CODE: .1 = under 0.5 μ; .2 = 0.5 to 10 μ; .3 = larger; * = preferable; x = practical. (Assuming no unusual flow configuration or other limiting conditions).

References

1.   Abuaf, N. and Gutfinger, C., "Trajectories of Charged Solid Particles in an Air Jet under the Influence of an Electrostatic Field", *Int. J. Multiphase Flow,* Vol. **1**, pp. 513-523, 1974.

2.   Abuaf, N., Jones, O.C., and Zimmer, G.A., "Response Characteristics of Optical Probes", Scheduled for presentation at the ASME 1978 Winter Annual Meeting, ASME Paper 78-WA/HT-3.

3.   Abuaf, N., Jones, O.C., and Zimmer, G.A., "Optical Probe for Local Void Fraction and Interface Velocity Measurements", *Review Scientific Instruments,* Vol. **44** (To be published), August 1978.

4.   Abuaf, N., Jones, O.C., Zimmer, G.A., Leonhardt, W.J., and Saha, P., "BNL Flashing Experiments: Test Facility and Measurement Techniques", (to be presented at the 1978 OECD Specialists Meeting on Transient Two-Phase Flows), Paris, 1978.

5.   Afgan, N., Jovanovic, L.J., Stefanovic, and Pislar, V., "An Approach to the Analysis of Temperature Fluctuation in Two-Phase Flow", *Int. J. Heat Mass Transfer,* Vol. **16**, pp. 187-194, 1973.

6.   Afgan, N.H. and Jovic, L.A., "Intermittent Phenomena in the Boiling Two-Phase Boundary Layer", *Int. J. Heat Mass Transfer,* Vol. **21**, pp. 427-434, 1978.

7.   Afgan, N., Stefanovic, M., Jovanovic, L.J., and Pislar, V., "Determination of the Statistical Characteristics of Temperature Fluctuation in Pool Boiling", *Int. J. Heat Mass Transfer,* Vol. **16**, pp. 244-256, 1973.

8.   Akagawa, K., "Fluctuation of Void Ratio in Two-Phase Flow , (1st Report, The Properties in a Vertical Upward Flow)" , *Bull. JSME,* Vol. **7**, pp. 122-128, 1964.

9.   Akagawa, K., Hamaguchi, H., Sakaguchi, T., and Ikari, T., "Studies on the Fluctuation of Pressure Drop in the Two-Phase Slug Flow (1st Report, Experimental Study", *Bull. JSME,* Vol. **14**, pp. 447-454, 1971.

10.   Akagawa, K. and Sakaguchi, T., "Fluctuations of Void Ratio in Two-Phase Flow (3rd Report, Absolute Velocities of Slugs and Small Bubbles and Distribution of Small Bubbles in Liquid Slugs)", *Bull. JSME,* Vol. **9**, pp. 111-120, 1966.

11.   Armand, R., "Etude Experimentale de l'Ecoulement Turbulent d'une Suspension Trajectoires et Vitesses des Particules Transferts Thermiques entre les deux Phases", *Int. J. Heat Mass Transfer,* Vol. **15**, pp. 2217-2229, 1972.

12.   Bachalo, W.D., "On Line Particle Diagnostic Systems for Application in Hostile Environments", *Third International Workshop on Laser Velocimetry,* Purdue University July 11-13, 1978.

13.   Bankoff, S.G., "Some Aspects of Gas-Liquid Flows", pp. 53-68, in *Single and Multi-Component Flow Processes,* Ed. by R.L. Peskin and C.F. Chen, Rutgers Univ., 1965.

14.   Bantin, R.A. and Streat, M., "Dense-Phase Flow of Solids Water Mixtures in Pipelines", *Hydrotransport 1,* Paper G1, BHRA, 1970.

15.   Beck, M.S., Lee, K.T., and Stanley-Wood, N.G., "A New Method for Evaluating the Size of Solid Particles Flowing in a Turbulent Fluid", *Powder Technology,* Vol. **8**, pp. 85-90, 1973.

16.   Ben Yasef, N., Ginio, Ol, Mahlab, D., and Wertz, A., "Bubble Size Distribution Measurement by Doppler Velocimeter", *J. Appl. Phys.,* Vol. **46**, pp. 738-741, 1975.

17.   Bergles, A.E., "Electrical Probes for Study of Two-Phase Flows", in *Two-Phase Flow Instrumentation,* ASME, pp 70-81, 1969.

18.   Birchenough, A. & Mason, J.S. "Local Particle Velocity Measurements with a Laser Anemometer in an Upward Flowing Gas-Solid Suspension". *Powder Technology,* Vol. **14**, pp. 139-152, 1976.

19.   Boothroyd, R.G., *Flowing Gas-Solids Suspension,* Chapman and Hall, Ltd., 1971.

20.   Bragg, G.M. and Bednarik, H.V., "Particulate Diffusion Across a Plane Turbulent Jet", *Int. J. Heat Mass Transfer,* Vol. **18**, pp. 443-451, 1975.

21.   Bragg, G.M. and Tevaarwerk, J., "The Effect of a Liquid Droplet on a Hot Wire Anemometer Probe", in *Flow, Its Measurement and Control in Science and Industry,* Vol. **1**, pp. 599-603, ASME Fluids Engineering Conference, ISA, 1974.

22.   Bremhorst, K. and Gilmore, D.B., "Response of Hot Wire Anemometer Probes to a Stream of Air Bubbles in a Water Flow", *J. of Physics E. Scientific Instruments,* Vol. **9**, pp. 347-357, 1976.

23.   Briller, R. & Robinson, M. "A Method for Measuring Particle Diffusivity in Two-Phase Flow in the Core of a Duct", *AIChe J.,* Vol. **15**, pp. 733-735, 1969.

24.   Brown, F.C. & Kranich, W.L. "A Model for the Prediction of Velocity and Void Fraction Profiles in Two-phase Flow" *AIChe J.,* Vol. **14**, pp. 750-758, 1968.

25.   Burgess, J.M. & Calderbank, P.H. "The Measurement of Bubble Parameters in Two-Phase Dispersions — I, II, III" (Three papers). *Chem. Eng. Science,* Vol. **30**, pp. 743-750, 1107-1121 and 1511-1518, 1975.

26.   Burick, R.J., Scheuerman, C.H., and Falk, A.Y., "Determination of Local Values of Gas and Liquid Mass Flux in Highly Loaded Two-Phase Flow", *Flow, Its Measurement and Control in Science and Industry,* Vol. **1**, pp. 153-160, ASME Fluids Engineering Conference, ISA, 1974.

27.   Card, D.C., Sims, G.E., and Chant, R.E., "Ultrasonic Velocity of Sound and Void Fraction in a Bubbly Mixture", *J. of Basic Engineering,* ASME, Vol. **93**, pp. 619-623, 1971.

28.   Carlson, C.R. and Peskin, R.L., "One-Dimensional Particle Velocity Probability Densities Measured in Turbulent Gas-Particle Duct Flow", *Int. J. Multiphase Flow,* Vol. **2**, pp. 67-68, 1975.

29.   Carrard, G. and Ledwidge, T.J., "Measurement of Slip Distribution and Average Void Fraction in an Air-Water Mixture", *Progress in Heat and Mass Transfer,* Vol. **6**, pp. 405-418, 1972.

30.   Chakko, M.K., "Measurement of Aerosol Concentration in Turbulent Flows", Syracuse University Research Institute, Final Report Contract DA-18-108-AMC-49 (A), 1963.

31.   Chandok, S.S. and Pei, D.C.T., "Particle Dynamics in Solids-Gas Flow in a Vertical Pipe", *Progress in Heat & Mass Transfer,* Vol. **6**, pp. 465-474, 1972.

32.   Chen, J.J., Lienhard, J.H., and Eichhorn, R., "A Method for Measuring Transparent Droplet Diameters", *Int. J. Multiphase Flow,* Vol. **4**, pp. 233-235, 1978.

33.   Chien, S.F. and Ibele, W.E., "Photographic Study of the Interfacial Disturbance of Liquid Films in Falling Film Flow, and in Vertical, Downward, Annular Two-Phase Flow", *Int. J. Heat Mass Transfer,* Vol. **10**, pp. 1016-1018, 1967.

34.   Chigier, N.A., "Velocity Measurements of Particles in Sprays", in *Flow, Its Measurement and Control in Science and Industry,* Vol. **1**, pp. 823-831, ISA, 1974.

35. Chigier, N.A., Ungut, A., and Yule, A.J., "Particle Sizing in Flames with Laser Velocimeters", *Third International Workshop on Laser Velocimetry*, Purdue University, July 11-13, 1978.

36. Chuang, S.C. and Goldschmidt, V.W., "The Response of a Hot Wire Anemometer to a Bubble of Air in Water", in *Turbulence Measurements in Liquids*, Ed. by G.K. Patterson and J.L. Zakin, pp. 88-95, Univ. of Missouri, 1971.

37. Cimorelli, L. and Evangelisti, R., "The Application of the Capacitance Method for Void Fraction Measurement in Bulk Boiling Conditions", *Int. J. Heat Mass Transfer*, Vol. 10, pp. 277-288, 1967.

38. Collins, S.B. and Knudsen, J.G., "Drop-Size Distribution Produced by Turbulent Pipe Flow of Immiscible Liquids", *AIChE Journal*, Vol. 16, pp. 1077-1080, 1970.

39. Crane, R.I. and Melling, A., "Velocity Measurements in Wet Stream Flows by Laser Anemometry and Pitot Tube", *ASME J. of Fluids Engineering*, Vol. 97, pp. 113-116, 1975.

40. Cranfield, R.R., "A Probe for Bubble Detection and Measurement in Large Particle Fluidized Beds", *Chem. Eng. Science*, Vol. 27, pp. 239-245, 1972.

41. Comu, M., Farello, G.E., Ferrasi, G., and Palazzi, G., "On Two-Phase Highly Dispersed Flows", *ASME Journal of Heat Transfer*, Vol. 96, pp. 496-503, 1974.

42. Danon, H., Wolfstein, M., and Hetsroni, G., "Numerical Calculations of Two-Phase Turbulent Round Jet", *Int. J. Multiphase Flow*, Vol. 3, pp. 223-234, 1977.

43. DeBlois, R.W. and Bean, C.P., "Counting and Sizing of Submicron Particles by The Resistive Pulse Technique", *The Rev. of Scientific Instruments*, Vol. 41, pp. 909-916, 1970.

44. Delhaye, J.M., "Hot-Film Anemometry in Two-Phase Flow", in *Two-Phase Flow Instrumentation*, ASME, pp. 58-69, 1969.

45. Delhaye, J.M. and Galaup, J.P., "Hot-Film Anemometry in Air-Water Flow" in *Turbulence in Liquids*, Ed. by J.L. Zakin and G.K. Patterson, pp. 83-90, Science Press, 1977.

46. Delhaye, J.M. and Jones, O.C., Jr., "A Summary of Experimental Methods for Statistical and Transient Analysis of Two-Phase Gas-Liquid Flow", ANL, 76-75, 1976.

47. Delhaye, J.M., Someria, R., and Flamand, J.C., "Void Fraction and Vapor and Liquid Temperatures: Local Measurements in Two-Phase Flow Using a Microthermocouple", *ASME, J. of Heat Transfer*, Vol. 95, pp. 365-370, 1973.

48. DISA Elektronik A/S, "Type 55S Two-Phase Flow Equipment", Skovlunde, Denmark, Reg. No. 9150A7311.

49. Doig, I.D. and Roper, G.H., "Air Velocity Profiles in the Presence of Concurrently Transported Particles", *I&EC Fundamentals*, Vol. 6, No. 2, pp. 247-256, 1967.

50. Driscoll, J.F. and Mann, D., "Submicron Particle Size Measurements in an Acetylene/Oxygen Flame", *Third International Workshop on Laser Velocimetry*, Purdue University, July 11-13, 1978.

51. Dunn, P.F. "Patent Securement Proposal: Current-or-Voltage Clamp Impedance-Variation Probe Circuit" Argonne National Laboratories, August 17, 1978.

52. Dunn, W.E., Chao, B.T., and Clausing, A.M., "Simple Optical Detector for Measuring Production Rate and Local Velocity of Drops", *Rev. Sci. Instruments,* Vol. **47**, pp. 321-323, 1976.

53. Durst, F. and Umhauer, H., "Local Measurements of Particle Velocity, Size Distribution and Concentration with a Combined Laser Doppler Particle Sizing System", (presented at the LDA-Symposium 1975, Copenhagen), SFB80/EM/81 Karlsruhe University, February, 1976.

54. Durst, F., and Zari, M., "Laser Doppler Measurements in Two-Phase Flows", SFB80/ TM63, Karlsruhe University, July 1975.

55. Edwards, W.M., Zuniga-Chaves, J.E., Worky, F.L., Jr., and Luss, Dan, "Measurements of Concentration Fluctuations in Gaseous Mixtures", *Ind. Eng. Chem. Fundamentals,* Vol. **15**, pp. 341-343, 1976.

56. Einav, S. and Lee, S.L., "Particles Migration in Laminar Boundary Layer Flow", *Int. J. Multiphase Flow,* Vol. **1**, pp. 73-88, 1973.

57. El Halwagi, M.M. and Gomezplata, A. "An Investigation of Solids Distribution, Mixing and Contacting Characteristics of Gas-Solid Fluidized Beds" *AIChE J.,* Vol. **13**, pp. 503-512, 1967.

58. El-Kaissy, M.M. and Homry, G.M., "Instability Waves and the Origin of Bubbles in Fluidized Beds, Part 1: Experiments", *Int. J. Multiphase Flow,* Vol. **2**, pp. 379-395, 1976.

59. Evangelisti, R. and Lupoli, R., "The Void Fraction in an Annular Channel at Atmospheric Pressure", *Int. J. Heat Mass Transfer,* Vol. **12**, pp. 699-711, 1969.

60. Farmer, W.M., "Measurement of Particle Size, Number Density, and Velocity Using a Laser Interferometer", *Applied Optics,* Vol. **11**, pp. 2603-2612, 1972.

61. Farmer, W.M., "Observation of Large Particles with a Laser Interferometer", *Applied Optics,* Vol. **13**, pp. 610-622, 1974.

62. Farmer, W.M., Harwell, K.E., Hornkohl, J.O., and Schwartz, F.A., "Particle Size Interferometer Measurements in Rocket Exhausts", *Third International Workshop on Laser Velocimetry,* Purdue University, July 11-13, 1978.

63. Farmer, W.M., Hornkohl, J.O., Brand, G.J., and Meier, J., "Design and Calibration of a Laser Interferometer System for Particle Size and Velocity Measurements in Large Turbine Combustors", *Third International Workshop on Laser Velocimetry",* Purdue University, July 11-13, 1978.

64. Fenton, D.L. and Stukel, J.J., "Measurement of the Local Particle Concentration in Fully Turbulent Duct Flow", *Int. J. Multiphase Flow,* Vol. **3**, pp. 141-145, 1976.

65. Ferrara, R., Fiocco, G. and Tonna, G., "Evolution of Fog Droplet Size Distribution Observed by Laser Scattering", *Applied Optics,* Vol. **9**, pp. 2317-2321, 1970.

66. Fourney, M.E., Matkin, J.H., and Waggoner, A.P., "Aerosol Size and Velocity Determination via Holography", *Rev. Scientific Instruments,* Vol. **40**, pp. 205-213, 1969.

67. Frost, W. and Kippenhan, C.J., "Bubble Growth and Heat Transfer Mechanisms in the Forced Convection Boiling of Water Containing a Surface Active Agent", *Int. J. Heat Mass Transfer,* Vol. **10**, pp. 931-949, 1967.

68. Fujimoto, H., Konishi, Y., Hirata, K., Hosoi, Y., and Sato, G.T., "Liquid Atomization in Electrostatic Field", *Bull. JSME,* Vol. **10**, pp. 155-163, 1967.

69. Gardiner, J.A., "Measurement of the Drop Size Distribution in Water Sprays by an Electrical Method", *Instrument Practice,* April, pp. 353-356, 1964.

70. Gardner, G.C. and Neller, P.H., "Phase Distribution in a Flow of an Air-Water Mixture Round Bends and Past Obstructions at the Wall of a 76 mm Bore Tube" *Proc. Instr. Mech. Engrs.,* Vol. **184**, Part 3C, pp. 93-101, 1970.

71. Gatt, F.C., "Flow of Individual Pebbles in Cylindrical Vessels" *Nuclear Eng. and Design,* Vol. **42**, pp. 265-275, 1977.

72. Gauvin, W.H., Katta, S., and Knelman, F.H., "Drop Trajectory Predictions and Their Importance in the Design of Spray Dryers", *Int. J. Multiphase Flow,* Vol. **1**, pp. 793-816, 1975.

73. Geldart, D. and Kelsey, J.R., "The Use of Capacitance Probes in Gas Fluidized Beds" *Powder Technology,* Vol. **6**, pp. 45-50, 1972.

74. Ginsberg, T., "Droplet Transport in Turbulent Pipe Flow", ANL.7694, 1971.

75. Goldschmidt, V.W., "Measurements of Aerosol Concentrations with a Hot-Wire Anemometer", *J. Colloid Science,* Vol. **20**, pp. 617-634, 1965.

76. Goldschmidt, V.W. and Eskinazi, S., "Two-Phase Turbulent Flow in a Plane Jet", *ASME, J. Appl. Mech.,* Vol. **33**, pp. 735-747, 1966.

77. Goldschmidt, V.W. and Householder, M.K. "Measurements of Aerosols by Hot-Wire Anemometry", in *Advances in Hot Wire Anemometry,* Melnik, W.L. and Weske, J.R., Eds., Univ. of Maryland, pp. 134-152, 1968.

78. Goldschmidt, V.W. and Householder, M.K., "The Hot Wire Anemometer as a Droplet Size Sampler", *Atmospheric Environment,* Vol. **3**, pp. 643-651, 1969.

79. Goldschmidt, V.W., Householder, M.K., Ahmadi, G. and Chuang, S.C., "Turbulent Diffusion of Small Particles Suspended in Turbulent Jets", *Progress in Heat and Mass Transfer,* Vol. **6**, pp. 487-507, 1972.

80. Gooderum, P., Bushnell, D., and Huffman, J., "Mean Droplet Size for Cross-Stream Water Injection into a Mach 8 Air Flow", *J. Spacecraft,* Vol. **4**, pp. 534-536, 1967.

81. Graham, C. and Griffith, P., "Drop Size Distribution and Heat Transfer in Dropwise Condensation", *Int. J. Heat Mass Transfer,* Vol. **16**, pp. 337-346, 1973.

82. Gregory, G.A., Nicholson, M.K., and Aziz, K., "Correlation of the Liquid Volume Fraction in the Slug for Horizontal Gas-Liquid Slug Flow", *Int. J. Multiphase Flow,* Vol. **4**, pp. 33-39, 1978.

83. Hayashi, Y., Takimoto, A., and Kanbe, M., "Transport-Reaction Mechanism of Mist Formation Based on the Critical Supersaturation Model", *ASME, Journal of Heat Transfer,* Vol. **98**, pp. 114-119, 1976.

84. Heertjes, P.M., Verloop, J., and Willems, R., "The Measurement of Local Mass Flow Rates and Particle Velocities in Fluid-Solids Flow", *Powder Technology,* Vol. **4**, pp. 38-40, 1970.

85. Heidrick, T.R., Saltvold, J.R., and Banerjee, S., "Application of a 3-Beam Gamma Densitometer to Two-Phase Flow Regions and Density Movements", *AIChE Symposium Series No. 164,* Vol. **73**, pp. 248-255, 1977.

86. Hetsroni, G., Cuttler, J.M., and Sokolov, M., "Measurements of Velocity and Droplets Concentration in Two-Phase Flows", *ASME J. Appl. Mech.,* Vol. **36**, pp. 334-335, 1969.

87.   Hetsroni, G. and Einav, S., "Some Basic Properties of Low Quality Two-Phase Tur-
      bulent Flow" in *Turbulence Measurements in Liquids,* Ed. by G.K. Patterson and
      J.L. Zakin, pp. 81-87, Univ. of Missouri, 1971.

88.   Hetsroni, G. and Sokolov, M., "Distribution of Mass, Velocity and Intensity of Tur-
      bulence in a Two-Phase Turbulent Jet", *ASME J. Appl. Mech.,* Vol. **38**, pp. 315-
      327, 1971.

89.   Hewitt, G.F., "The Role of Experiments in Two-Phase Systems with Particular Refer-
      ence to Measurement Techniques", *Progress in Heat and Mass Transfer,* Vol. **6**,
      pp. 295-343, 1972.

90.   Hewitt, G.F. and Boure, J.A., "Some Recent Results and Development in Gas-Liquid
      Flow: A Review", *Int. J. Multiphase Flow,* Vol. **1**, pp. 139-171, 1973.

91.   Hewitt, G.F. and Hall-Tayler, N.S., *Annular Two-Phase Flow,* Chapter 12, Experi-
      mental Techniques for Annular Flow, Pergamon Press, 1970.

92.   Hinata, S., "A Study on the Measurement of the Local Void Fraction by the Optical
      Fibre Glass Probe, (1st Report on the Local Void Fraction in a Liquid Metal Two-
      Phase Flow)", *Bull. JSME,* Vol. **15**, pp. 1228-1235, 1972.

93.   Hinata, S., Kuga, O., and Kobayasi, K., "Diffusion of Bubbles in Two-Phase Flow
      (The 1st Report, on the Method for the Measurement of the Diffusivity of the
      Bubbles", *Bull. JSME,* Vol. **20**, pp. 1299-1305, 1977.

94.   Hiroyasu, H. and Kadota, T., "Fuel Droplet Size Distribution in Diesel Combustion
      Chamber", *Bull. JSME,* Vol. **19**, pp. 1064-1072, 1976.

95.   Ho, C.W., Tveten, A.B., Chan, P.W., and She, C.Y., "Particle Size Measuring Device
      in Real Time for Dense Particulate Systems", *Applied Optics,* Vol. **17**, pp. 631-634,
      1978.

96.   Hodkinson, J.R., "The Optical Measurement of Aerosols" in *Aerosol Science,* Ed. by
      C.N. Davies, Academic Press, 1966.

97.   Hoffer, M.S. and Resnick, W., "A Modified Electroresistivity Probe Technique for
      Steady- and Unsteady-State Measurements in Fine Dispersions — I, II, III" (Three
      papers), *Chem. Eng. Science,* Vol. **30**, pp. 473-502, 1975.

98.   Holmes, T.L. and Russell, T.W.F., "Horizontal Bubble Flow", *Int. J. Multiphase
      Flow,* Vol. **2**, pp. 51-66, 1975.

99.   Holve, D. and Self, S.A., "An Optical Particle-Sizing Counter for In-Situ Measure-
      ments", *Third International Workshop on Laser Velocimetry,* Purdue University,
      July 11-13, 1978.

100. Householder, M.K. and Goldschmidt, V.W., "The Impaction of Spherical Particles on
      Cylindrical Collectors", *J. Colloid and Interface Sci.,* Vol. **31**, pp. 464.768, 1969.

101. Howard, N.M., Meek, C.C., and Jones, B.G., "Experimental Measurement of Particle
      Dispersion in Turbulent Flow", in *Turbulence in Liquids,* Ed. by G.K. Patterson and
      J.L. Zakin, pp. 259-273, Univ. of Missouri, 1975.

102. Hsu, Y.Y., Simoneau, R.J., Simon, F.F., and Graham, R.W., "Photographic and
      Other Optical Techniques for Studying Two-Phase Flow", in *Two-Phase Flow In-
      strumentation,* ASME, pp. 1-23, 1969.

103. Inoya, K., "Research and Development in Japan on Fine Particle Measurement and
      New Control Devices", in *Fine Particles,* Academic Press, pp. 24-37, 1976.

312 Victor W. Goldschmidt

104. Ivey, H.J., "Relationships Between Bubble Frequency, Departure Diameter and Rise Velocity in Nucleate Boiling", *Int. J. Heat Mass Transfer,* Vol. **10**, pp. 1023-1040, 1967.
105. Jakubowsky, S. and Sideman, S., "A Simulation Model for Two-and Three-Phase, Agitated Systems", *Int. J. Multiphase Flow,* Vol. **3**, pp. 171-180, 1976.
106. James, R.N., Babcock, W.R., and Seifert, H.S., "A Laser Doppler Technique for the Measurement of Particle Velocity", *AIAA Journal,* Vol. **6**, pp. 160-162, 1968.
107. Jepson, J.C. and Ralph, J.L., "Hydrodynamic Studies of Two-Phase Upflow in Vertical Pipelines", *Proc. Inst. Mech. Engrs.,* Vol. **184**, Part 3E, pp. 154-165, 1970.
108. Jeffreys, G.V., Davies, G.A., and Pitt, K., "Rate of Coalescence of the Dispersal Phase in a Laboratory Mix in Settler Unit: Part I", *AIChE J.,* Vol. **16**, pp. 823-827, 1970.
109. Jones, B.G., et al., "An Experimental Study of Small Particles in a Turbulent Fluid Using Digital Techniques for Statistical Data Processing", *Dev. in Mech.,* Vol. **4**, p. 1249 ff, 1968.
110. Jones, O.C., Jr. and Delhaye, J.M., "Transient and Statistical Measurement Techniques for Two-Phase Flows: A Critical Review", *Int. J. Multiphase Flow,* Vol. **3**, pp. 89-116, 1976.
111. Jones, O.C., Jr. and Zuber, N., "The Interrelations Between Void Fraction Fluctuations and Flow Patterns in Two-Phase Flow", *Int. J. Multiphase Flow,* Vol. **2**, pp. 273-306, 1975a.
112. Jones, O.C. Jr. and Zuber, N., "Use of a Cylindrical Hot-Film Anemometer for Measurement of Two-Phase Void and Volume Flux Profiles in a Narrow Rectangular Channel",*AIChE Paper No. 45,* 1975b.
113. Jones, O.C. Jr. and Zuber, N., "Interfacial Passage Frequency for Two-Phase, Gas-Liquid Flows in Narrow Rectangular Ducts", *I. Mech. E.,* C192/77, pp. 5-10, 1977.
114. Jones, O.C. Jr. and Zuber, N., "Slug-Annular Transition with Particular Reference to Narrow Rectangular Ducts", Manuscript, c. 1978.
115. Jurewicz, J.T., Stock, D.E., and Crowe, C.T., "Particle Velocity Measurements in an Electrostatic Precipitator with a Laser Velocimeter", *AIChE Symposium Series 165,* Vol. **73**, pp. 138-141, 1977.
116. Karabelas, A.J., "Droplet Size Spectra Generated in Turbulent Pipe Flow of Dilute Liquid/Liquid Dispersants", *AIChE J.,* Vol. **24**, pp. 170-180, 1978.
117. Keller, R.J., "Instrumentation in Full Scale Self-Aerated Flows", *Journal of Hydraulic Research,* IAHR, Vol. **11**, pp. 325-341, 1973.
118. Kim, K.Y. and Marshall, W.R. Jr., "Drop Size Distribution from Pneumatic Atomizers", *AIChE J.,* Vol. **17**, pp. 575-584, 1971.
119. Kitayama, Y., "Digital Void Velocimeter", *J. Nucl. Science and Technology,* Vol. **9**, pp. 613-617, 1972.
120. Kling, C.L. and Hammitt, F.G., "A Photographic Study of Spark-Induced Cavitation Bubble Collapse", *ASME J. Basic Engineering,* Vol. **94**, pp. 825-832, 1972.
121. Kobayasi, K., Iida, Y., and Kanagae, N., "Distribution of Local Void Fraction of Air-Water Two-Phase Flow in Vertical Channel", *Bull. JSME,* Vol. **13**, pp. 1005-1012, 1970.

122. Kondic, N.N., "Lateral Motion of Individual Particles in Channel Flow-Effect of Diffusion and Interaction Forces", *ASME J. of Heat Transfer,* Vol. **92**, pp. 418-428, 1970.

123. Konstantinov, S.M., Kisurkin, A.K., and Neduzhko, Ye A., "Determination of Bubble Velocity and Other Properties in A Gas-Liquid Bed", *Fluid Mechanics-Soviet Research,* Vol. **5**, pp. 97-101, 1976.

124. Kramer, T.J. and Depew, C.A., "Experimentally Determined Mean Flow Characteristics of Gas-Solid Suspensions", *ASME J. Basic Engineering,* Vol. **94**, pp. 492-499, 1972.

125. Kubie, J., "Bubble Induced Heat Transfer in Two Phase Gas-Liquid Flow", *Int. J. Heat Mass Transfer,* Vol. **18**, pp. 537-551, 1975.

126. Kubie, J. and Gardner, G.C., "Drop Sizes and Drop Dispersion in Straight Horizontal Tubes and Helical Coils", *Chem. Engr. Science,* Vol. **32**, pp. 195-202, 1977.

127. Laats, M.K., "Experimental Study of the Dynamics of an Air-Dust Jet", *Inzhenerno-Fizicheskii Zhurnal,* Vol. **10**, pp. 11-15, 1966.

128. Laderman, A.J., Lewis, C.H., and Byron, S.R., "Time-of-Flight Measurement of Particle Velocity", *AIAA Journal,* Vol. **7**, pp. 556-557, 1969.

129. Lading, L., "The Time-of-Flight Laser Anemometer", *AGARD Conference Proceedings No. 193,* pages 23-1 to 23-20, c. 1977.

130. Lading, L., "The Time-of-Flight Laser Anemometer versus the Laser Doppler Anemometer", *Third International Workshop on Laser Velocimetry,* Purdue University, July 11-13, 1978.

131. Landa, I. and Tebay, E.S., "The Measurement and Instantaneous Display of Bubble Size Distribution Using Scattered Light", *IEEE Trans. on Instr. & Meas.,* pp. 56-59, Feb., 1972.

132. Lassahn, G.D., "Two-Phase Flow Velocity Measurements Using Radiation Intensity Correlation", *ISA Transactions,* Vol. **15**, pp. 297-300, 1976.

133. Lee, S.L. and Srinivasan, J., "Measurement of Local Size and Velocity Probability Density Distributions in Two-Phase Suspension Flows by Laser-Doppler Techniques" *Int. J. Multiphase Flow,* Vol. **4**, pp. 141-155, 1978.

134. Lee, Y.C., Fourney, M.E., and Moulton, Ralph W., "Determination of Slip Ratios in Air-Water Two-Phase Critical Flow at High Quality Levels Utilizing Holographic Techniques", *AIChE J.,* Vol. **20**, pp. 209-219, 1974.

135. Leutheusser, A.J. and Ward, C.A., "Thermodynamic Aspects of Hydraulic Aeration", *Proceedings of the XVIth Congress of the IAHR,* Vol. **3**, pp. 500-508, 1975.

136. Liu, B.Y., *Fine Particles, Aerosol Generation, Measurements, Sampling and Analysis,* Academic Press, 1976.

137. Liu, B.Y.H., "Standardization and Calibration of Aerosol Instruments", in *Fine Particles,* Academic Press, pp. 39-53, 1976.

138. Maneri, C.C. and Zuber, N., "An Experimental Study of Plane Bubbles Rising and Inclination", *Int. J. Multiphase Flow,* Vol. **1**, pp. 623-645, 1974.

139. Marsheck, R.M. and Gomezplata, A., "Particle Flow Patterns in a Fluidized Bed", *AIChE J.,* Vol. **11**, pp. 167-173, 1965.

140. Matthes, W., Riebold, W., and DeLooman, E., "Measurement of the Velocity of Gas Bubbles in Water by a Correlation Method", *Rev. Scientific Instruments,* Vol. **41**, pp. 843-845, 1970.

141. Mazumder, M.K., and Kirsch, K.J., "Single Particle Aerodynamic Relaxation Time Analyzer", *Rev. Sci. Instr.,* Vol. **48**, pp. 622-624, 1977.

142. Mazumder, M.K., Ware, R.E., Wilson, J.D., Sherwood, L.T., and McLeod, P.C., "Application of SPART Analyzer for Monitoring Real-Time Aerodynamic Size Distribution of Stack Emission", *Third International Workshop on Laser Velocimetry,* Purdue University, July 11-13, 1978.

143. McBride, D.D. and Sherman, P.M., "Condensed Zinc Particle Size Determined by a Time Discrete Sampling Apparatus", *AIAA Journal,* Vol. **10**, pp. 1050-1063, 1972.

144. McQuaid, J. and Wright, W., "The Response of a Hot-Wire Anemometer in Flows of Gas Mixtures", *Int. J. Heat Mass Transfer,* Vol. **16**, pp. 819-828, 1973.

145. McSweeney, A. and Rivers, W., "Optical Fiber Array for Measuring Radial Distributions of Light Intensity for Particle Analysis", *Appplied Optics,* Vol. **11**, pp. 2101-2102, 1972.

146. Merilo, M., Dechine, R.L., and Cichowlas, W.M., "Void Fraction Measurement with a Rotating Electric Field Conductance Gauge", *ASME J. of Heat Transfer,* Vol. **99**, pp. 330-332, 1977.

147. Michiyoshi, I., "Two-Phase Two-Component Heat Transfer", Keynote Paper, *6th Int. Heat Transfer Conference,* Toronto, Canada, 1978a.

148. Michiyoshi, I., "Heat Transfer in Air-Water Two-Phase Flow in a Concentric Annulus", *6th International Heat Transfer Conference,* Toronto, Canada, pp. 499-504, Paper FB-39, 1978b.

149. Michiyoshi, I., Funakawa, H., Kuramoto, C., Akita, Y., and Takahashi, O., "Local Properties of Vertical Mercury-Argon Two-Phase Flow in a Circular Tube Under Transverse Magnetic Field", *Int. J. Multiphase Flow,* Vol. **3**, pp. 445-457, 1977.

150. Miller, N. and Mitchie, R.E., "The Development of a Universal Probe for Measurement of Local Voidage in Liquid/Gas Two-Phase Flow Systems", in *Two Phase Flow Instrumentation,* ASME, pp. 82-88, 1969.

151. Mlynek, Y. and Resnick, W., "Drop Sizes in an Agitated Liquid-Liquid System", *AIChE J.,* Vol. **18**, pp. 122-127, 1972.

152. Mobbs, F.R., Bowers, H.M., Riches, D.M., and Cole, B.N., "Influence of Particle Size Distribution on the High-Speed Flow of Gas-Solid Suspensions in a Pipe", *Proc. Inst. Mech. Engrs.,* Vol. **184**, Part 3C, pp. 67-76, 1970.

153. Moeck, E.O., "Measurements of Liquid Film Flow and Wall Shear Stress in Two-Phase Flow", in *Two-Phase Flow Instrumentation,* ASME, pp. 36-46, 1969.

154. Namie, S. and Ueda, T., "Droplet Transfer in Two-Phase Annular Mist Flow (Part 1, Experiment of Droplet Transfer Rate and Distributions of Droplet Concentration and Velocity)", *Bull. JSME,* Vol. **15**, pp. 1568-1580, 1972.

155. Namie, S. and Ueda, T., "Droplet Transfer in Two-Phase Annular Mist Flow (Part 2, Prediction of Droplet Transfer Rate)", *Bull. JSME,* Vol. **16**, pp. 752-764, 1973.

156. Nash, J.H., Lester, G.G., and Grimm, F., "Sampling Device for Liquid Droplets", *Rev. Sci. Instruments,* Vol. **38**, pp. 73-77, 1967.

157. Neal, L.G. and Bankoff, S.G., "Local Parameters in Cocurrent Mercury-Nitrogen Flow: Parts I and II", *AIChE J.,* Vol. **11**, pp. 624-635, 1965.

158. Ochi, M. and Ikemori, K., "Minimum Transport Velocity of Granular Materials at High Concentration in a Horizontal Pipe", *Bull. JSME,* Vol. **21**, Paper 156-10, pp. 1008-1014, 1978.

159. Ogden, D.M. and Stock, D.E., "Simultaneous Measurement of Particle Size and Velocity via the Scattered Light Intensity of a Real Fringe Laser Anemometer", *Third International Workshop on Laser Velocimetry,* Purdue University, July 11-13, 1978.

160. Oki, K., Akehata, T., and Shirai, T., "A New Method for Evaluating the Size of Moving Particles with a Fiber Optic Probe", *Powder Technology,* Vol. **11**, pp. 51-57, 1975.

161. Oki, K., Walawender, W.P. and Fan, L.T., "The Measurement of Local Velocity of Solid Particles", *Powder Technology,* Vol. **18**, pp. 171-178, 1977.

162. Olszowski, S.T., Coulthard, J. and Sayles, R.S., "Measurement of Dispersed Two-Phase Gas-Liquid Flow by Cross Correlation of Modulated Ultrasonic Signals", *Int. J. Multiphase Flow,* Vol. **2**, pp. 537-548, 1976.

163 Onuma, Y., Tsuji, T. , and Ogasawaru, M., "Studies on the Flame Structure of a Spray Burner (1st Report, Measurement of Quantities Related to Droplets and Gas within the Flame, and Their Correlation)", *Bull. JSME,* Vol. **17**, pp. 1296 ff, 1974.

164. Orr, C. Jr., *Particulate Technology,* Macmillan, 1966.

165. Park, W.H., Kang, W.K., Capes, C.E. and Osberg, G.L., "The Properties of Bubbles in Fluidized Beds of Conducting Particles as Measured by an Electroresistivity Probe", *Chem. Eng. Science,* Vol. **24**, pp. 851-865, 1969.

166. Parker, G.J. and Ryley, D.J., "Equipment and Techniques for Studying the Deposition of Sub-Micron Particles on Turbine Blades", *Proc. Inst. Mech. Engrs.,* Vol. **184**, Part 3C, pp. 43-51, 1970.

167. Pchelkin, I.M., Kalakutskaya, N.A., and Parfent'yeva, I.F., "Local Characteristics of Two-Phase Flows at Nozzle Discharges", *Fluid Mechanics-Soviet Research,* Vol. **4**, 1975.

168. Pike, R.W., Wilkins, B., and Ward, H.C., "Measurement of the Void Fraction in Two-Phase Flow by X-Ray Attenuation", *AIChE J.,* Vol. **11**, pp. 794-800, 1965.

169. Pinczewski, W.V, and Fell, C.J.D., "Droplet Sizes on Sieve Plates Operating in the Spray Regime", *Trans. Inst. Chem. Engrs.,* Vol. **55**, pp. 46-52, 1977.

170. Pogson, J.T., Roberts, J.H., and Waibler, P.J., "An Investigation of the Liquid Distribution in Annular-Mist Flow", *ASME J. of Heat Transfer,* pp. 651-658, 1970.

171. Popper, J., Abuaf, N., and Hetsroni, G., "Velocity Measurements in a Two-Phase Turbulent Jet", *Int. J. Multiphase Flow,* Vol. **1**, pp. 715-726, 1974.

172. Postaire, J.G. and Fitreman, J.M., "A Digital Special-Purpose Signal Processor for Two-Phase Flow Real-Time Analysis", *IEEE Transactions on Instrumentation and Measurement",* Vol. **IM-26**, pp. 254-257, 1977.

173. Quarmby, A. and Quirk, R., "Axisymmetric and Non-Axisymmetric Turbulence Diffusion in a Plain Circular Tube at High Schmidt Number", *Int. J. Heat Mass Transfer,* Vol. **17**, pp. 143-147, 1974.

174. Rao, C.S. and Dukler, A.E., "The Isokinetic-Momentum Probe. A New Technique for Measurement of Local Voids and Velocities in Flow of Dispersions", *Ind. Eng. Chem. Fundam.,* Vol. **10**, No. 3, pp. 520-526, 1971.

175. Rao, V.L. and Venkateswarla, D., "Determination of Velocities and Flow Patterns in Mass Flow Hoppers", *Powder Technology,* Vol. **7**, pp. 263-265, 1973.

176. Reddy, K.V.S. and Pei, D.C.T., "Particle Dynamics in Solids-Gas Flow in a Vertical Pipe", *Ind. Eng. Chem. Fundam.*, Vol. **8**, No. 3, pp. 490-497, 1969.

177. Resch, F.J. and Leutheusser, H.J., "Two-Phase Flow", in *Turbulence in Liquids*, Ed. by G.K. Patterson and J.L. Zakin, pp. 243-249, Univ. of Missouri, 1975.

178. Resch, F.J., Leutheusser, H.J. and Alemy, S., "Bubbly Two-Phase Flow in Hydraulic Jump", *ASCE, J. Hydraulic Div.*, HYI, pp. 137-149, 1974.

179. Resch, F.J., Leutheusser, H.J., and Coantic, M., "Etude de la Structure Cinematique et Dynamique du Ressant Hydraulique", *Journal of Hydraulic Research*, Vol. **14**, No. 4, pp. 293-314, 1976.

180. Rigby, G.R., Van Blockland, G.P., Park, W.H. and Capes, C.E.,"Properties of Bubbles in Three Phase Fluidized Beds as Measured by an Electroresistivity Probe", *Chem. Eng. Science*, Vol. **25**, pp. 1729-1741, 1970.

181. Roberts, C.P.R. and Kennedy, J.F., "Particle and Fluid Velocities of Turbulent Flows of Suspensions of Neutrally Buoyant Particles", in *Advances in Solid-Liquid Flow in Pipes and Its Application*, Pergamon Press, pp. 59-72, 1971.

182. Rouhani, Z., "Effect of Wall Friction and Vortex Generation on the Radial Distribution of Different Phases", *Int. J. Multiphase Flow*, Vol. **3**, pp. 35-50, 1976.

183. Rowe, P.N. and Everett, D.J., "Fluidized Bed Bubbles Viewed by X-Rays: Part 1 - Experimental Details and the Interaction of Bubbles with Solid Surfaces", *Trans. Inst. Chem. Engrs.*, Vol. **50**, pp. 52-58, 1972.

184. Rowe, P.N. and Matsaro, R., "Single Bubbles Injected into a Gas Fluidized Bed and Observed by X-Rays", *Chem. Eng. Science*, Vol. **26**, pp. 923-935, 1971.

185. Rudinger, G., "Effective Drag Coefficient for Gas-Particle Flow in Shock Tubes", *ASME J. Basic Engineering*, Vol. **92**, pp. 165-172, 1970.

186. Sato, T., Kishimoto, T., and Sasaki, K., "Laser Doppler Particle Measuring System Using Forced Vibration Synchronized Detection and Power Spectral Analysis", *Applied Optics*, Vol. **17**, pp. 230-234, 1978a.

187. Sato, T., Kishimoto, T., and Sasaki, K., "Laser Doppler Particle Measuring System Using Nonsinusoidal Forced Vibration and Bispectral Analysis", *Applied Optics*, Vol. **17**, pp. 667-670, 1978b.

188. Sato, Y. and Sekoguchi, K., "Liquid Velocity Distribution in Two-Phase Bubble Flow", *Int. J. Multiphase Flow*, Vol. **2**, pp. 79-95, 1975.

189. Schindler, H.D. and Treybal, R.E., "Continuous-Phase Mass-Transfer Coefficients for Liquid Extraction in Agitated Vessels", *AIChe J.*, Vol. **14**, pp. 790-798, 1968.

190. Schmidt, C.C. and Glover, J.R., "A Frazil Ice Concentration Measuring System Using a Laser Doppler Velocimeter", *Journal of Hydraulic Research*, Vol. **13**, pp. 299-314, 1975.

191. Schraub, F.A., "Isokinetic Probe and Other Two-Phase Sampling Devices: A Survey" in *Two-Phase Flow Instrumentation*, ASME, pp. 47-57, 1969.

192. Schrock, V.E., "Radiation Attenuation Techniques in Two-Phase Flow Measurements", in *Two-Phase Flow Instrumentation*, ASME, pp. 24-35, 1969.

193. Schwartzberg, H.G. and Treybal, R.E., "Fluid and Particle Motion in Turbulent Stirred Tanks - Fluid Motion", *Ind. Eng. Chem. Fundamentals*, Vol. **7**, No. 1, pp. 1-6 1968a.

194. Schwartzberg, H.G. and Treybal, R.E., "Fluid and Particle Motion in Turbulent Stirred Tanks - Particle Motion", *Ind. Eng. Chem. Fundamentals,* Vol. **7**, No. 1, pp. 6-12, 1968b.

195. Seki, M., Kawamura, H., and Sanokawa, K., "Transient Temperature Profile of a Hot Wall Due to an Impinging Liquid Droplet", *ASME, Journal of Heat Transfer,* Vol. **100**, pp. 167-169, 1978.

196. Sekoguchi, K., Fukui, H., Matsuoka, T., and Nishikawa, K., "Investigation into the Statistical Characteristics of Bubbles in Two-Phase Flow (1st Report, Fundamentals of the Instrumentation Using the Electric Resistivity Probe Technique)", *Bull. JSME,* Vol. **18**, pp. 391-396, 1975.

197. Sekoguchi, K., Fukui, H., Tsutsui, M., and Nishikawa, K., "Investigation into the Statistical Characteristics of Bubbles in Two-Phase Flow (2nd Report, Application and Establishment of Electric Resistivity Probe Method)", *Bull. JSME,* Vol. **18**, pp. 397-404, 1975.

198. Serizawa, A., Katoka, I., and Michiyoshi, I., "Turbulence Structures of Air-Water Bubbly Flow — I. Measuring Techniques", *Int. J. Multiphase Flow,* Vol. **2**, pp. 221-233, 1975a.

199. Serizawa, A., Katoka, I., and Michiyoshi, I., "Turbulence Structure of Air-Water Bubbly Flow - II. Local Properties", *Int. J. Multiphase Flow,* Vol. **2**, pp. 235-246, 1975b.

200. Serizawa, A., Katoka, I., and Michiyoshi, I., "Turbulence Structure of Air-Water Bubbly Flow - III. Transport Properties", *Int. J. Multiphase Flow,* Vol. **2**, pp. 247-259, 1975c.

201. Smith, T.N., "Measurements of Drop Size in Liquid-Liquid Dispersion", *Chem. Eng. Sciences,* Vol. **29**, pp. 583-587, 1974.

202. Somerscales, E.F.C., "Fluid Velocity Measurement by Particle Tracking", in *Flow, Its Measurement and Control in Science and Industry,* Vol. **1**, pp. 795-821, ISA, 1974.

203. Soo, S.L., *Fluid Dynamics of Multiphase Systems,* Blaisdell, 1967.

204. Soo, S.L., "Gas-Solid Flow", pp. 3-52, in *Single and Multi-Component Flow Processes,* Ed. by R.L. Peskin and C.F. Chen, Rutgers Univ., 1965.

205. Soo, S.L. and Trezek, G.J., "Turbulent Pipe Flow of Magnesia Particles in Air", *I&EC Fundamentals,* Vol. **5**, No. 3, pp. 388-392, 1966.

206. Soo, S.L., Trezek, G.J., Dimick, R.C. and Hohnstreiter, G.F., "Concentration and Mass Flow Distributions in a Gas-Solid Suspension", *I&EC Fundamentals,* Vol. **3**, No. 2, pp. 98-106, 1964.

207. Sprow, F.B., "Distribution of Drop Sizes Produced in Turbulent Liquid-Liquid Dispersion", *Chem. Engr. Sciences,* Vol. **22**, pp. 435-442, 1967.

208. Steiner, L., Shoukry, E., and Hartland, S., "Conductometric Measurement of Particle Size in a Dispersion", *Ind. Eng. Chem. Fundamentals,* Vol. **13**, No. 3, pp. 267-272, 1974.

209. Stock, D.E., Jurewicz, J.T., Crowe, C.T., and Eschback, J.E., "Measurement of Both Gas and Particle Velocity in Turbulent Two-Phase Flow", in *Turbulence in Liquids,* Ed. by J.L. Zakin and G.K. Patterson, pp. 91-102, Science Press, 1975.

210. St. Pierre, C.C. and Bankoff, S.G., "Vapor Volume Profiles in Developing Two-Phase Flow", *Int. J. Heat Mass Transfer,* Vol. **16**, pp. 237-249, 1967.

211. Sugimoto, A. and Kobayashi, K., "Experimental Investigation of Burning Spray (Relation Among Dimensionless Numbers)", *Bull. JSME,* Vol. **21**, Paper 153-18, pp. 494-501, 1978.

212. Sullivan, J.P. and Theofanous, T.G., "The Use of LDV in Two-Phase Bubbly Pipe Flow", *Third International Workshop on Laser Velocimetry,* Purdue University, July 11-13, 1978.

213. Sultan, M. and Judd, R.L., "Spatial Distribution of Active Sites and Bubble Flux Density", *ASME, Journal of Heat Transfer,* Vol. **100**, pp. 56-62, 1978.

214. Suneja, S.K. and Wasan, D.T., "Dispersion of Charged Particles in a Turbulent Air Stream under Transverse Flow Conditions", *Ind. Eng. Chem. Fundamentals,* Vol. **11**, No. 1, pp. 57-66, 1972.

215. Temkin, S, and Reichman, J.M., "A New Technique to Photograph Small Particles in Motion", *Rev. Sci. Instruments,* Vol. **43**, pp. 1456-1459, 1972.

216. Thermo-Systems, Inc., "Particle Research Instruments — A condensed Catalog, T.S.I., Minnesota, 1977.

217. Thorpe, J.F., Funk, J.E., and Borg, T.Y., "Void Fraction and Pressure Drop in a Water Electrolysis Cell", *ASME, J. of Basic Engineering,* Vol. **92**, pp. 183-191, 1970.

218. Torrest, R.S. and Ranz, W.E., "Improved Conductivity System for Measurement of Turbulent Concentration Fluctuations", *I&EC Fundamentals,* Vol. **8**, No. 4, pp. 810-816, 1969.

219. Tsuji, S. and Katakura, H., "A Fundamental Study of Aeration in Oil, 2nd Report: The Effects of the Diffusion of Air on the Diameter Change of a small Bubble Rising in a Hydraulic Oil", *Bull. JSME,* Paper 156-11, Vol. **21**, pp. 1015-1021, 1978.

220. Tsuji, S. and Matsui, K., "On the Measurement of Void Fraction in Hydraulic Fluid with Entrained Bubbles", *Bull. JSME,* Paper 152-9, Vol. **21**, pp. 239-245, 1978.

221. Ueda, T. and Tanaka, H., "Measurements of Velocity, Temperature and Velocity Fluctuation Distribution in Falling Liquid Films", *Int. J. Multiphase Flow,* Vol. **2**, pp. 261-272, 1975.

222. Uga, T., "Determination of Bubble-Size Distribution in a BWR", *Nucl. Eng. and Design,* Vol. **22**, pp. 252-261, 1972.

223. Unal, H.C., "Void Fraction and Incipient Point of Boiling During the Subcooled Nucleate Flow Boiling of Water", *Int. J. Heat Mass Transfer,* Vol. **20**, pp. 409-419, 1977.

224. Unal, H.C., "Determination of Void Fraction, Incipient Point of Boiling, and Initial Point of Net Vapor Generation in Sodium-Heated Helically Coiled Steam Generator Tubes", *ASME, J. of Heat TRansfer,* Vol. **100**, pp. 268-274, 1978.

225. van Breugel, J.W., Stein, J.J.M. and deVries, R.J., "Isokinetic Sampling in a Dense Gas-Solids Stream", *Proc. Instr. Mech. Engrs.,* Vol. **184**, Pt. 3C, pp. 18-23, 1970.

226. vanPaasen, C.A.A., "Thermal Droplet Size Measurements Using a Thermocouple", *Int. J. Heat Mass Transfer,* Vol. **17**, pp. 1527-1548, 1974.

227. Van Velzen, D., Flamm, H.J., Langenkamp, H. and Casile, A., "Motion of Solids in Spanted Beds", *Can. J. of Chem. Engr.,* Vol. **52**, pp. 156-161, 1974.

228. Verhoff, F.H., Ross, S.L., and Curl, R.L., "Breakage and Coalescence Processes in an Agitated Dispersion. Experimental System and Data Reduction", *Ind. Eng. Chem. Fundam.,* Vol. **16**, No. 3, pp. 371-377, 1977.

229. Voss-Spilker, P., "Flow Measurements in Two-Phase Diffuse Flow", *DISA Information,* No. **22**, December, 1977.

230. Wairegi, T. and Grace, J.R., "The Behaviour of Large Drops in Immiscible Liquids", *Int. J. Multiphase Flow,* Vol. **3**, pp. 67-77, 1976.

231. Walmet, G.E. and Staub, F.W., "Pressure, Temperature, and Void Fraction Measurements in Nonequilibrium Two-Phase Flow", in *Two-Phase Flow Instrumentation,* ASME, pp. 89-101, 1969.

232. Weinstein, B. and Traybal, R.E., "Liquid-Liquid Contacting in Unbaffled, Agitated Vessels" *AIChE J.,* Vol. **19**, pp. 304-312, 1973.

233. Werther, J., "Bubbles in Gas Fluidized Beds - Part I", *Trans. Instn. Chem. Engrs.,* Vol. **52**, pp. 149-159, 1974.

234. Werther, J. and Molerus, O., "The Local Structure of Gas Fluidized Beds — I. A Statistically Based Measuring System", *Int. J. Multiphase Flow,* Vol. **1**, pp. 103-122, 1973a.

235. Werther, J. and Molerus, O., "The Local Structure of Gas Fluidized Beds — II. The Spatial Distribution of Bubbles", *Int. J. Multiphase Flow,* Vol. **1**, pp. 123-138, 1973b.

236. Wilcox, R.L. and Tate, R.W., "Liquid Atomization in a High Intensity Sound Field", *AIChE J.,* Vol. **11**, pp. 69-72, 1965.

237. Wittig, S.L.K. and Sakbani, K., "Simultaneous Particle Size and Velocity Measurements in Postflame Gases", *Third International Workshop on Laser Velocimetry,* Purdue University, July 11-13, 1978.

238. Wu, J., "Fast-Moving Suspended Particles: Measurements of Their Size and Velocity" *Applied Optics,* Vol. **16**, pp. 595-600, 1977.

239. Yang, W.J. and Clark, D.W., "Spray Cooling of Air-Cooled Compact Heat Exchangers", *Int. J. Heat Mass Transfer,* Vol. **18**, pp. 311-317, 1975.

Optical Methods in Two-Phase Flow

J.M. Delhaye
Centre d'Etudes Nucléaires de Grenoble
Service des Transferts Thermiques

Abstract
Optical methods constitute an essential diagnostic tool for the elaboration of two-phase flow models.

In the first part of the paper the interface visualization techniques are examined. First the different systems used to illuminate the flow field confined in a test section is characterized. Then the design of this test section is discussed. The final section deals with special shooting techniques employed according to the test section or the flow characteristics.

The second part of the paper is devoted to the visualization of the liquid velocity and temperature fields in two-phase flows.

The review discards the techniques applied to small bubbles or droplets less than one millimeter in diameter and the measurements of space fractions by means of photon attenuation methods.

Introduction
As recognized several years ago by G.F. HEWITT, two-phase flows, boiling and condensation are highly photogenic phenomena. However, it seems doubtful that this feature only has led to the enormous amount of literature displaying two-phase pictures. In fact, two-phase flows are so complex that the easiest way to obtain modeling guidelines is to look at the flow patterns. Moreover, optical techniques can provide quantitative data over a large flow field at a given time, which is impossible with local probes.

A usual statement at the beginning of a review paper is to claim that the survey will not be exhaustive. We will not depart from this attitude by saying that our review will discard the techniques applied to small bubbles or droplets less than one millimeter in diameter. These methods are reviewed in the same conference by F. DURST and F. RESCH. Although we will present the possibilities of x-radiography, we will not handle the quantitative measurements of space fractions by means of photon attenuation techniques which were surveyed in detail by JONES and DELHAYE (1976) and by DELHAYE (1979).

The mores of writing a review paper imply a paragraph on the great advantages and slight disadvantages of the methods which are presented. Here it is! Usually an optical technique is said not to disturb the flow field. Actually an optical method may require a special viewing system such as transparent windows which may modify the boundary conditions. If this drawback can be overcome optical techniques present four definite *advantages* (1) an excellent time resolution which can reach the picosecond range with some electronic high-speed cameras, (2) a good space resolution which can be obtained by means of microscopic systems, (3) a high sensitivity, e.g. in the temperature field determination by means of holographic interferometry, (4) an easy data storage although automatic data processing techniques still require further developments. Beside these appealing features, some *disadvantages* can be put forward. Optical methods require transparent media and a flow confinement of a good optical quality. No general recipes are available for the illumination of the flow field. This problem has to be solved by trial and error and each case leads to a particular solution which requires skill and ingenuity. However, general principles must be recalled and will be given in the first section of this paper. Very often photographs give rise to equivocal and subjective interpretations due to reflection, refraction and absorption effects. In fact, a photograph does not record an interface per se but the image of the light source through this interface. As a result the apparent size of a droplet is not equal to its actual size. Polygonal contours have been photographed and taken for bubbles. In reality these contours are the images of the diaphragm of the camera when operated out-of-focus. Finally, quantitative optical measurements based on the variations of the refractive index need two-dimensional plane or axisymmetric flow fields.

The basic information on the visualization of single-phase flow fields can be found in MERZKIRCH's book (1974) and in the article written by HAUF and GRIGULL (1970). The former covers the tagging techniques based on the addition of foreign materials into gaseous and liquid fluid flows along with the optical methods for compressible flows whereas the latter emphasizes the visualization of temperature fields by means of shadow and schlieren techniques and interference methods. General information concerning applied optics and optical engineering can be found in the series of books edited by KINGSLAKE, the first volume of which appeared in 1965.

Although many papers display photographs of two-phase phenomena, very few articles discuss the techniques which should be used to obtain high quality pictures or reliable quantitative optical information. The major references on these subjects were written about a decade ago, primarily by the people working at Harwell, England. A general review was compiled by COLLIER and HEWITT (1966) who gave the main techniques to illuminate the flow field, to obtain high-speed pictures and to improve visualization. X-ray photography was discussed and applications to adiabatic two-phase flow, pool boiling and convective boiling were considered. Satellite papers were published in a journal specialized in scientific photography (COOPER et al. 1964; ARNOLD and HEWITT, 1967). A second review article was published in 1969 by HSU et al. in an ASME booklet devoted to Two-Phase Flow Instrumentation. Ninety-nine references are quoted concerning experimental techniques and the information which can be extracted from optical data.

The purpose of the present review paper is essentially to complement and update the above surveys.

In a first section we will examine the interface visualization techniques while in the second section we will discuss the visualization methods of the liquid flow field.

1. Interface Visualization
In the first part we will give the characteristics of the different systems used to illuminate the flow field confined in a test section the design of which is discussed in a second part. The last two parts deal with special shooting techniques employed according to the test section or the flow characteristics.

1.1 Lighting of the Flow Field
Very often people think of a bubble or a droplet as an object. In fact this opinion is entirely wrong since the *interfaces* act as *refracting* or *reflecting surfaces* through which the image of the light source is perceived. As a result a linear magnification is introduced and can lead to erroneous conclusions if it is not properly taken into account. A simple example of such a phenomenon is given when measuring the diameter of transparent liquid droplets. In this case the droplets are illuminated from the side with a black background (Fig. 1). CHEN et al. (1978) give the ratio of the actual diameter D to the apparent diameter D_{app} and the ratio of the droplet center displacement d to the apparent diameter D_{app} as a function of the illumination angle θ (Fig. 2).

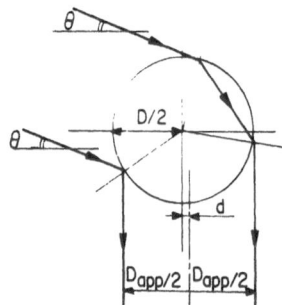

Fig. 1. Light rays through a transparent droplet illuminated at an angle θ (CHEN et al., 1978).

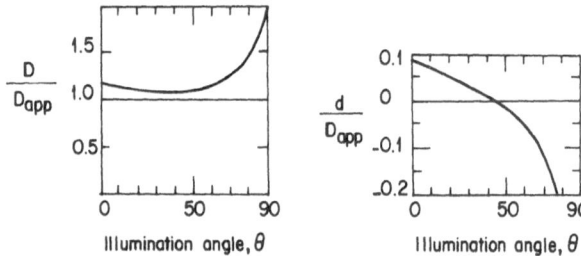

Fig. 2. Apparent diameter and displacement of the center of a droplet (CHEN et al., 1978).

Fig. 2 shows that the lens effect is not negligible. If small particles are to be visualized, a *dark background* is more efficient than a *light background*. This is easily understood by looking at Fig. 3. If a light background is used the diffraction pattern produced by a small particle produces a low contrast whereas this contrast is higher when the background is dark.

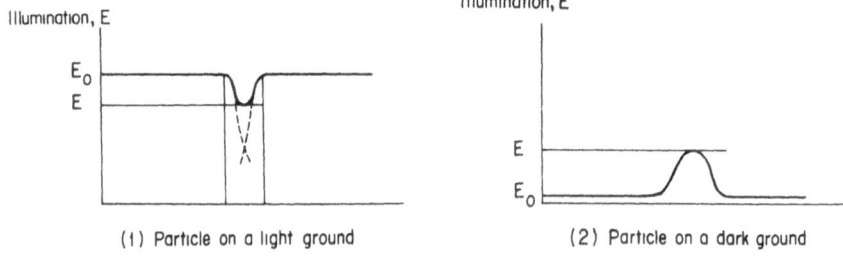

Fig. 3. Light background versus dark background illumination (contrast ≜ |$E_o - E$|/E_o)

A dark background requires a directed illumination system in order to account for a limited zone of the flow field only. Generally *diffused light* is preferred when interface patterns must be recognized. Should *directed light* be used, part of the light-rays reflected and refracted by the interfaces would be stopped by the entrance aperture of the camera. As a result only some portions of the interfaces would be recorded. However, in film flow, directed light is used because of the focusing effect of the interface which enables the ripples to be detected (Fig. 4). Recommended combinations of lights, reflectors, screens, films, and f-stop values for photography of bubbles and drops can be found in the paper by KINTNER et al. (1961).

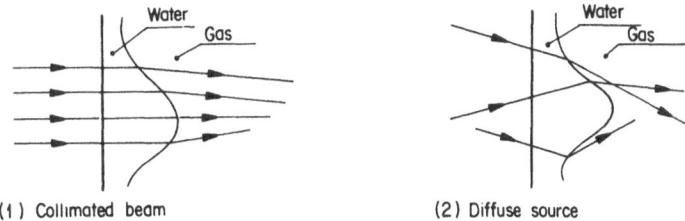

Fig. 4. Directed light versus diffused light for film flow illumination (COOPER et al., 1964).

The contrast can be increased by using a *polarizer* on the incident light beams and an *analyzer* on the beams entering the camera. Proper adjustment of the polarization angles eliminates unwanted reflections.

When small particles, bubbles or droplets, must be detected the light can be brought into the flow field by a *laser scanner* (Fig. 5). A twelve-face rotating mirror reflects a He-Ne laser beam which sweeps the observation plane perpendicular to the camera axis. The luminous energy received by a particle is far higher than in a conventional directed illumination system. So is the contrast. Rotating rates up to 10^5 rpm can be used and enable transient phenomena to be studied.

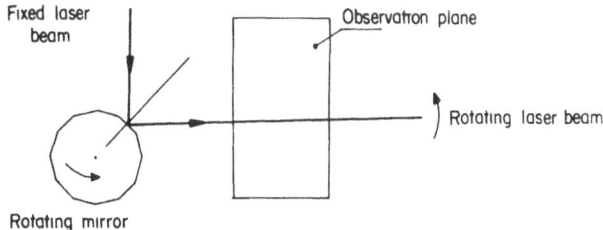

Fig. 5. Laser scanner (AID System)

Very sharp pictures of two-phase flow patterns were obtained by JONES (1973) by means of a *retro-reflex* or *cataphotic screen*. With such a screen the light is returned retrodirectively rather than angularly. The experimental setup used by JONES (1973) is represented in Fig. 6. Some results are given in Fig. 7.

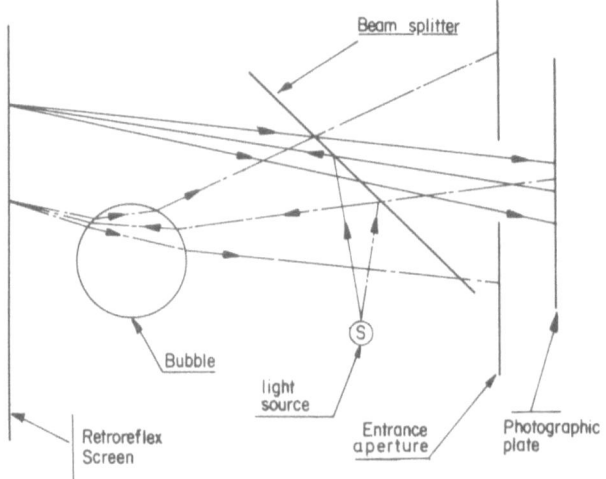

Fig. 6. Retroreflex technique used by JONES (1973).

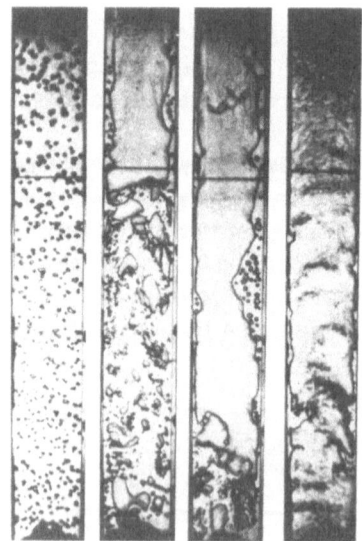

Fig. 7. Photographs of flow patterns obtained by JONES (1973) with the retro-reflex technique. Air-water flow in a vertical channel of rectangular cross section (4.98 × 63.50 mm).

1.2 Test Section Designs

Unlike a test section of rectangular cross section (JONES, 1973) a test section of circular cross section introduces *lens effects* due to the *curvature of the pipe wall*. These effects are easily suppressed by using a circular test section drilled in an acrylic resin block of rectangular cross section (GALAUP, 1975) or by using a parallelepipedic view-box surrounding the circular test section (BROWN and GOVIER, 1961; COOPER et al., 1964; JACOWITZ and BRODKEY, 1964; ROUMY, 1969). Fig. 8 summarizes the different solutions adopted. Note that this method must be employed only if the predominant phase within the test section is the liquid phase. If the predominant phase is the gas phase, the use of such view-boxes would result in a negative lens effect.

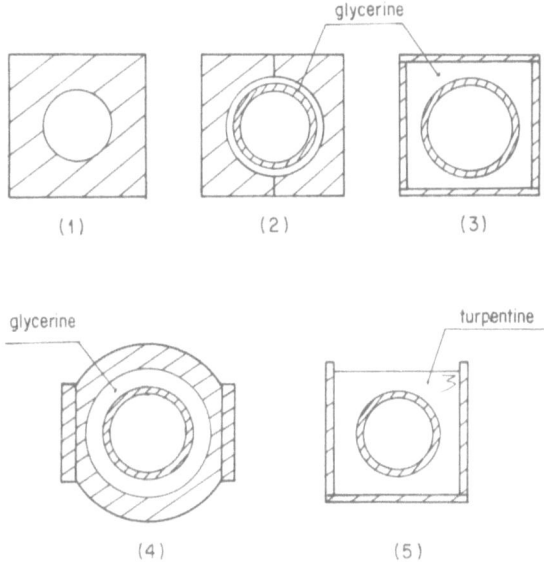

Fig. 8. Different kinds of view-boxes made up of acrylic resin (1) HEWITT and ROBERTS, 1969; GALAUP, 1975; (2) COOPER et al., 1964; (3) ROUMY, 1969, (4) BROWN and GOVIER, 1961; (5) JACOWITZ and BRODKEY, 1964.

Special test section designs are used for the *visualization of convective boiling or pool boiling phenomena*. The rectangular test section designed by HOSLER (1967) permits the simultaneous front and side viewings of a steam-water flow, the steam being generated along the heated wall (Fig. 9).

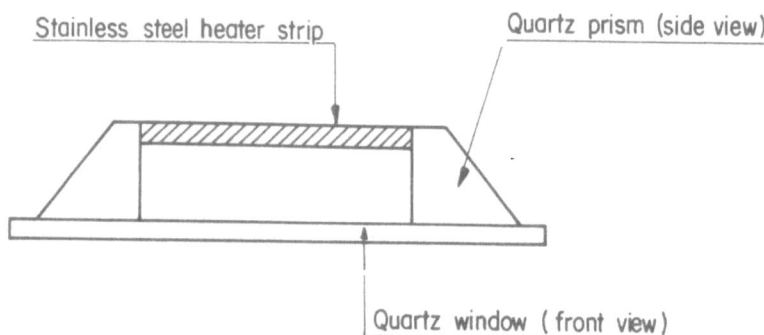

Fig. 9. Schematic of the test section used by HOSLER (1967) for the visualization of steam-water flow.

Convective boiling has been studied also in annular test sections provided with a central heated cylinder (Fig. 10). Transparent heated walls enable boiling phenomena to be observed in circular test sections. Steam-water flow patterns in forced convection boiling were photographed by HSU and GRAHAM in 1963 by using a transparent metallic coating on the external side of a circular glass tube. The maximum heat flux was rather low, about 0.11 W cm^{-2}.

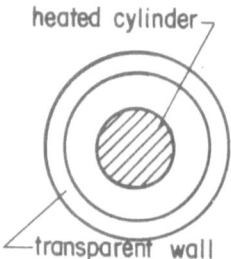

Fig. 10. Visualization of forced convection boiling in an annular test section.

Recently chugging instabilities have been experimentally studied by ACHARD and BESSET (1978) in a Pyrex tube internally coated with indium oxide (Fig. 11). The tube can be heated with heat fluxes up to 25 W cm^{-2}. The 0.2 μm thick coating is obtained by means of a chemical vapor deposition technique. Tubes, as long as 1 m, are coated and present a uniformity of the local resistance better than ±10 % except near the electrical connection zones. Small dimension heater strips were used by KIRBY and WESTWATER (1963) for studying pool boiling characteristics of carbon tetrachloride and methanol. Electrically conducting glass was able to sustain heat fluxes up to 45 W cm^{-2}. However, this device could not be used with water which attacked the coating.

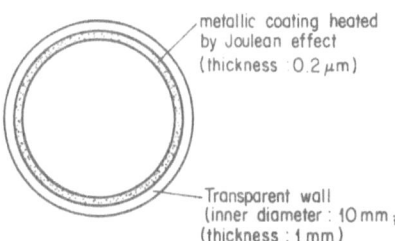

metallic coating heated
by Joulean effect
(thickness : 0.2 μm)

Transparent wall
(inner diameter : 10 mm ;
(thickness : 1 mm)

Fig. 11. Coated heating tube used for the visualization of convective boiling.

High pressures (140 bar), *high temperatures* (330°C) are encountered in several industrial processes involving two-phase flows. As a result, many experiments have been carried out on boiling phenomena at high pressure. Deionized water at high pressure and high temperature has been proved to be corrosive to normal glass and even to quartz (SiO_2) and borosilicate glasses (Pyrex type). In his study of pool boiling under 140 bar, SEMERIA (1963) used two observation windows: the first one in sapphire ($Al_2 O_3$) to withstand the high-temperature, deionized water corrosivity, the second one in quartz to resist the pressure difference.

1.3 Special Shooting Techniques Due To The Test Section Design

Forced convection boiling at high pressure requires opaque, metallic test sections to withstand the mechanical stresses. Although windows could be used, at the expense of serious technological difficulties and size limitations, several investigators preferred to look for a visualization based on photon attenuation techniques.

Fluoroscopy was proposed by JOHANNS (1964) to study steam-water flow patterns. The more severe conditions tested by the author correspond to a stainless steel tube, 54.8 mm of inner diameter, 2.8 mm thick. A collimated x-ray beam is produced by an x-ray tube operating at 130 kV and 6 mA. After being absorbed by the test section wall and the flowing two-phase mixture, the beam (Fig. 12) is collected by an image intensifier tube of gain 3000. The output image of this tube is recorded by a 800 line vidicon TV camera. The low efficiency of this system does not permit the recording of the final image on a photographic film but video tapes can be used to store the information. Despite this drawback, the same technique was employed by BAKER (1965) to study the flow patterns of boiling Refrigerant-11 ($CCl_3 F$) at a pressure of 25 bar In a stainless steel rectangular test section, 26.7 × 9.6 mm.

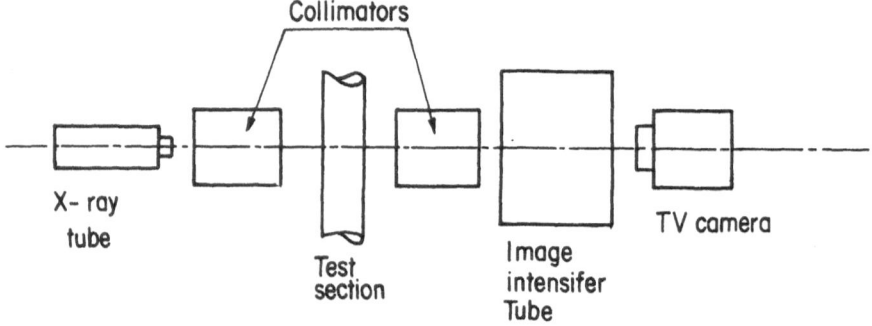

Fig. 12. Fluoroscopic system (JOHANNS, 1964).

X-ray radiography was tentatively investigated by DERBYSHIRE et al. (1964). Their first experiments were carried out in an air-water loop whose test section was an acrylic resin tube, 15.9 mm in diameter. The x-ray pulse, 0.1μs in duration, was produced by a medical x-ray unit operating at 60 kV and 800 mA. The results were encouraging and the same group of investigators decided to apply this technique to the determination of steam-water flow regimes at high pressure (BENNETT et al., 1965-1966). Several requirements had to be met for the channel walls which must exhibit a low absorption coefficient for x-rays, a mechanical strength sufficient to withstand the high pressure, a chemical in-activity with respect to the high temperature water and an electrical resistivity compatible with the power supply of the test section. Considering all these con-straints a ribbed titanium tube was designed as shown in Fig. 13. The results were very interesting and led to their well known flow regime map. However, the definition of the pictures was not so good for high liquid flowrates. These conditions were investigated by HEWITT and ROBERTS (1969) for an air-water mixture flowing in a circular tube, 31.2 mm in diameter at a pressure ranging from 0.41 to 4.48 bar. These authors showed the existence of the so-called wispy-annular flow where clouds of droplets are moving in the central core of an annular flow.

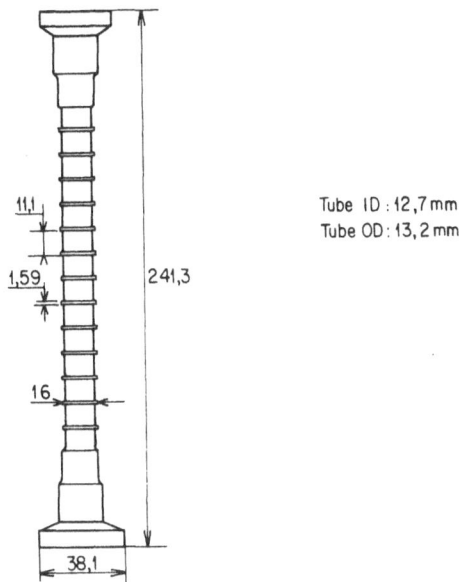

Tube ID : 12,7 mm
Tube OD : 13,2 mm

Fig. 13. Ribbed titanium tube used by BENNETT et al. (1965-1966) for x-ray radiography of steam-water flows at high pressure (up to 69 bar). Dimensions in mm.

1.4 Special Shooting Techniques Due To The Flow Characteristics

In dispersed annular flow the liquid phase moves under two different configurations: a film along the wall and droplets in the central gas core. The interface of the liquid film exhibits different kinds of waves such as ripples or roll waves. The droplets are torn off from the film interface, migrate in the central gas core and may redeposit on the liquid film. The investigation of the effect of phase flowrates on the wave structure and on droplet entrainment was carried out by HEWITT and ROBERTS (1969b) by analyzing high speed cine films frame by frame using an *axial view technique.* An air-water dispersed annular flow was moving upward in a vertical tube 31.75 mm in diameter and a cross section plane of the flow was observed by means of the device depicted in Fig. 14. This device has two purposes, (1) to ensure the return of the two-phase mixture to the separation tank, (2) to provide a clear viewing window by avoiding any droplet impingement. This second purpose is achieved by a countercurrent air flow in the viewing tube. A circumferential illumination slit, 6.35 mm wide, surrounds the plane of focus of the camera unit whose optics is adapted to the 12.7 mm aperture of the viewing tube and to the position of the plane of focus located 114.3 mm from the entrance of the viewing tube.

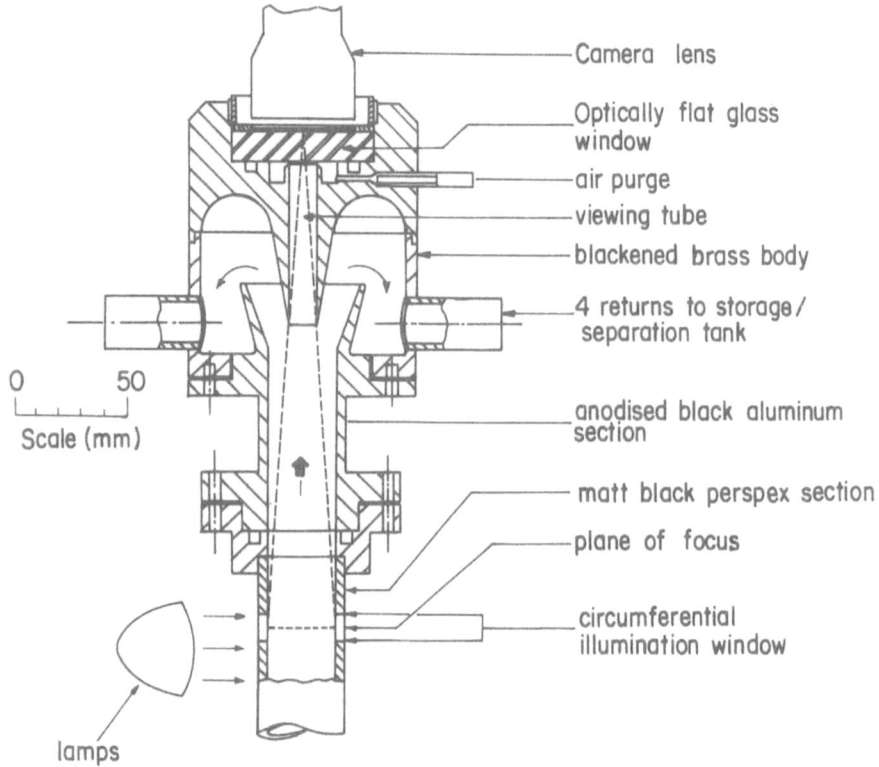

Fig. 14. Axial viewing device used by HEWITT and ROBERTS (1969a) to visualize dispersed annular flow.

The wave structure can be observed by means of *light guide techniques*. Fig. 15 shows the twin-color piped light illumination system designed by COOPER et al. (1964). The light is brought into the liquid film by two acrylic resin conical light guides mounted flush with the external wall of the circular test section. The optical junction between the test section and the light guides is ensured by a thin film of a liquid having the same refractive index as acrylic resin. The pictures are sharply contrasted since the direction of the wave slopes is indicated by the color, red or green, reflected from it. The light guides eliminate the light losses which would have resulted from reflections if the test section has been directly illuminated. Note that the twin-color illumination was also used by BAKER (1965) for flow regime determination. However, no light guides were used.

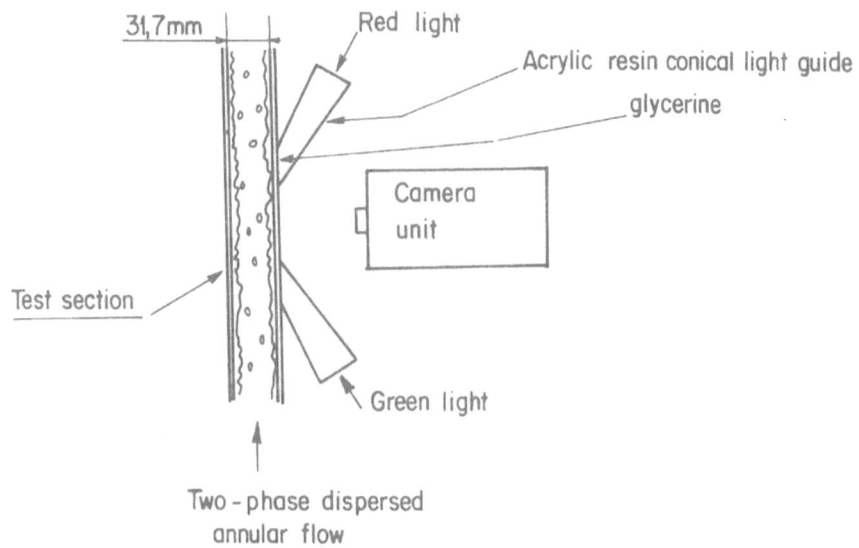

Fig. 15. Twin-color piped light illumination system (COOPER et al., 1964).

Generally heterogeneity is a specific feature of two-phase mixture and it might be worthwhile to follow a given disturbance of the flow such as a bubble, a group of droplets or a roll wave. *Tracking techniques* are rather limited but give extremely valuable information on the flow structure. Rising velocities of Taylor bubbles in a 25.9 mm bore tube, 2.40 m in length, were measured by NICKLIN et al. (1962) by using a cine camera moving on rails and raised by a rope winding on to a drum entrained by an electric motor through a variable speed hydraulic gear box. The maximum rising velocity was about 20 cms^{-1}. The recent interest in transient two-phase flows led to a more sophisticated device (PHAM, 1978) where a TV camera can be displaced at velocities up to 2 ms^{-1}, with acceleration up to 40 ms^{-2}. This device (AID system) enables the position of a bubble, hence its velocity and acceleration, to be detected by processing the video signal of the TV camera. A simpler technique using a rotating mirror was developed by HEWITT and LOVEGROVE (1970) to measure the velocities of disturbance waves and ripple waves in annular flow. The principle is the same as the one used by ROUMY (1969a) which is represented in Fig. 16. The rotating speed of the mirror is set so that a stationary image is obtained with respect to the object of interest i.e. a roll wave, a bubble,...

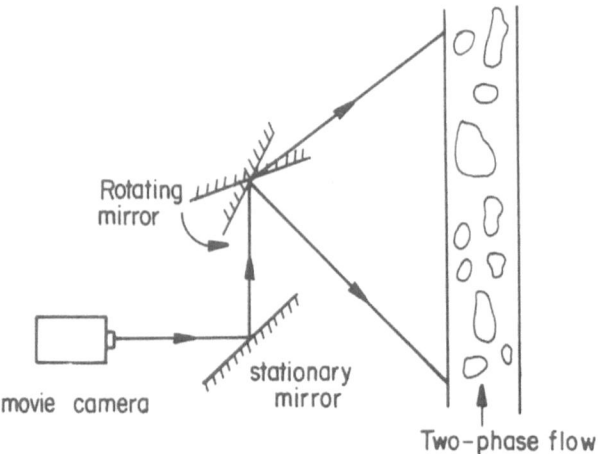

Fig. 16. Bubble tracker used by ROUMY (1969a).

Boiling and cavitation offer very fast events to be visualized. *High-speed cinematography* has been used many times since the first movies on boiling obtained by Fritz in 1935 with a 600 fps camera. A good example of the investigating power of high-speed movies (8000 fps) is given by SEMERIA (1963) in his study of pool boiling at high pressure (140 bar). Recent equipments are thoroughly discussed by MARILLEAU (1972) and HUSTON (1978). Picosecond cameras are now available to study very fast events such as cavitation bubble collapse or vapor explosion.

High-speed cinematography can be avoided when investigating almost periodical events by means of *stroboscopic techniques* (BEHAR, 1963). A stroboscope flash is triggered by the occultation of a reference light beam after a time interval which can be adjusted to any value up to 0.1 s. Bubbling of air bubbles into a liquid and coalescence of bubbles emerging from two hypodermic needles close together were visualized by BEHAR (1963).

2. Liquid Flow Field Visualization
Boiling mechanisms have been investigated for many years in order to understand the heat transfer enhancement due to the generation of vapor bubbles on a heated wall. Such a comprehension requires the knowledge of the kinematic and thermal fields of the liquid phase surrounding the vapor bubble. Although various local techniques are able to provide quantitative information, too many data would be necessary to obtain an overall picture of the flow field.

Conversely, optical techniques are particularly suited to apprehend the complete pattern of such a complex phenomenon. In the following we will first examine the liquid velocity field visualization and finally the liquid temperature field visualization.

2.1 Velocity Field Visualization

The general principle of this visualization consists in tagging the liquid phase either with small electrolytic hydrogen bubbles or with small solid particles. These elements must be small enough not to lag behind the fluid motion. A good discussion of these techniques is to be found in the second chapter of MERZKIRCH's book (1974).

BARAKAT and SIMS (1977) combined the *hydrogen-bubble technique* with high-speed cinematography to study the liquid flow patterns about barbotage bubbles. These bubbles were obtained by injecting air into water through an orifice, 1.16 or 2.52 mm in diameter, drilled into a cast polyester plate. Nine platinum wires, 76 μm in diameter, were moulded into the plate along a diameter and polished flush with the surface. These wires acted as cathodes in a dc circuit whose second electrode was immersed in the water. Tiny hydrogen bubbles were generated on the cathodes and were illuminated against a dark background. The liquid velocity was determined taking into account buoyancy effects on the hydrogen bubbles.

The flow pattern about barbotage bubbles was compared by BARAKAT and SIMS (1977) with the liquid motion around boiling bubbles which had been reported by KUTATELADZE and MAMONTOVA (1973). These authors used *solid markers* consisting in lycopodium powder first wetted in ethyl alcohol. The diameter of the particles (22 μm) and their density (0.963 g cm^{-3}) were such that they could follow the liquid motion without any significant lag. High-speed cinematography was used with a light background.

The liquid motion about a barbotage or a boiling bubble is theoretically an axisymmetric problem. Three-dimensional flow can be visualized quantitatively by means of multi-color illumination as reported by VAN MEEL and VERMIJ (1960). These authors developed an optical technique to determine the direction and magnitude of velocity vectors in three-dimensional flow field. A black and white photograph of a particle following a three-dimensional trajectory permits only the determination of the projection of the path of the particle on a plane normal to the optical axis of the camera, thus giving two coordinates. However, color photography enables the third coordinate to be determined. This is achieved by illuminating the flow field with a system of ten plane parallel slices of light of different colors, directed normal to the optical axis of the camera. An opaque polystyrene sphere traversing these colored slices will be photographed with a color depending on its position whereas if it moves in a plane normal to the optical axis no color change will occur. The colors are arranged in contrasting order (Fig. 17) to make the determination of the third component easier. The color slices were obtained with optical filters glued onto a thick glass-plate support. These filters were illuminated with parallel light coming from a point-shaped light source associated with a parabolic mirror.

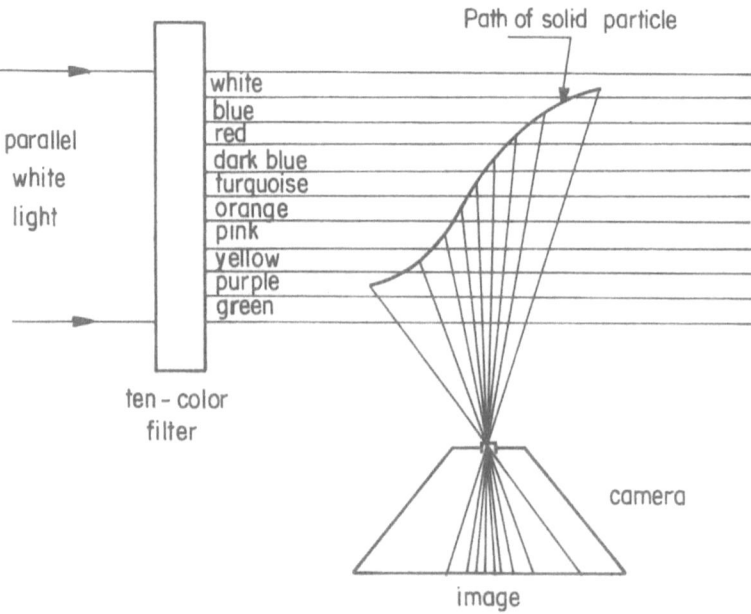

Fig. 17. Multicolor illumination of a flow field for the visualization of the three-dimensional path of a solid particle (VAN MEEL and VERMIJ, 1960).

2.2 Temperature Field Visualization

Although a good account of temperature field visualization techniques can be found in GOLDSTEIN's article (1976), more detailed information is available in the article by HAUF and GRIGULL (1970) and in MERZKIRCH's book (1974). These three general references contain all the material needed for a thorough understanding of the optical methods which can be used for temperature measurements based on a change of the refractive index (interferometry), its first derivative (schlieren) or its second derivative (shadowgraph), in a direction normal to the light beam.

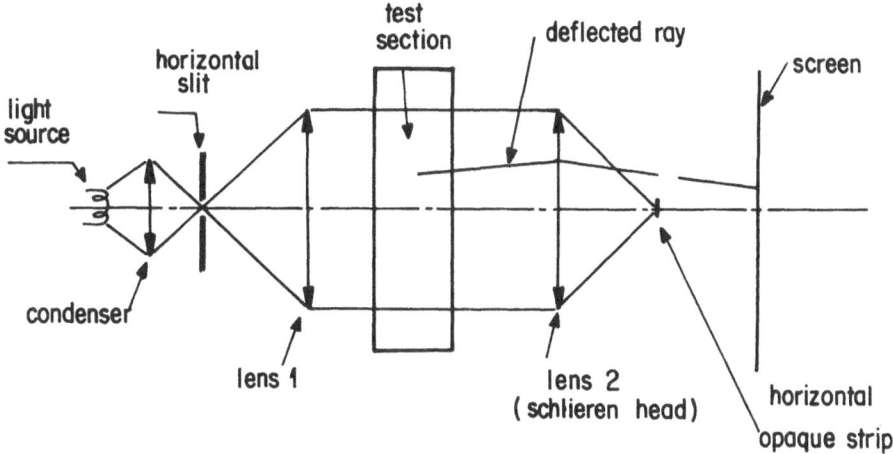

Fig. 18. Schlieren system used by BEHAR and SEMERIA (1963) for pool boiling studies.

Degassing and nucleate pool boiling mechanisms were investigated by BEHAR and SEMERIA (1963, 1963a) by means of black-and-white and color *schlieren-techniques*. They used the Toepler system consisting of two lenses and a horizontal linear source. The lenses were photographic objectives whose focal lengths were 15 cm and f-number 4.5, thus ensuring a large field of view and a narrow depth of field. This last point is an advantage since only the phenomenon of interest is examined notwithstanding the heterogeneities of the bulk of the liquid. In the black-and-white schlieren system an opaque strip was placed at the focus of the schlieren head (Fig. 18). The dimensions of the strip were exactly those of the image of the light source slit through the system. In the color schlieren system the opaque strip was surrounded by several parallel, colored, transparent strips. This experimental technique led to important results concerning the heat transfer mechanism in nucleate subcooled boiling which were confirmed later by determining directly the temperature field by means of microthermocouples. A similar investigation was carried out by JACOBS and SHADE (1969). Subcooled pool boiling of carbon tetrachloride was visualized with a schlieren system consisting of two lenses and an iris diaphragm as analyzer. A knife edge was used by PRICE and BRAMALL (1971) in their observation of vapor column formations in film boiling of carbon dioxide. The diaphragm is less sensitive than the narrow strip or the knife edge but has nearly the same sensitivity in all directions. A schlieren system using two lenses, a knife edge and a He-Ne laser light source was used by BARANENKO et al. (1974) in their investigation of degassing and pool boiling of water, benzene, butyl and isobutyl alcohols at atmospheric pressure. The use of a laser light gives closely spaced fringes instead of a continuous illumination distribution obtained in an ordinary schlieren record. Physically this stems from the fact that, at the focus of the schlieren head, the knife edge interacts with the diffraction pattern of the laser beam rather than with a geometrical point of light (OPPENHEIM et al., 1966). According to SMIRNOV and BARA-NENKO (1971) the interference fringes in a laser schlieren system correspond to isotherms.

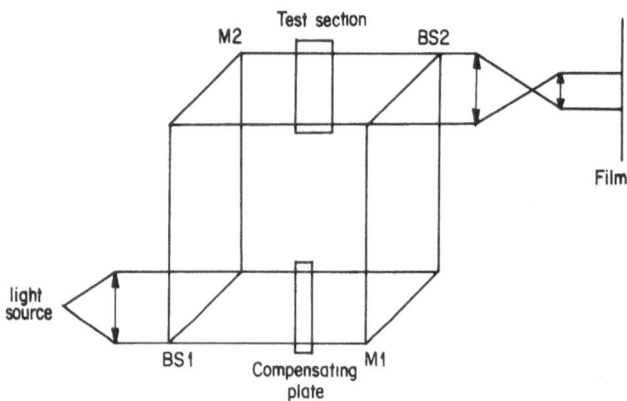

Fig. 19. Principle of a conventional MZI (BS: beam splitter; M: mirror).

Whereas the schlieren techniques make the refractive index field visible, *inter-ferometry* permits the measurement of the refractive index and consequently of the temperature. BEER (1969) and BEER et al. (1977) determined the temperature field about a growing vapor bubble in water and Freon 113 under atmospheric pressure with a Mach-Zehnder interferometer equipped with a Helium-Neon laser light source and associated with high speed cinematography. Fig. 19 shows the principle of a conventional MZI. A monochromatic light source delivers a parallel beam of light which is split into two beams. The first one passes through the test section whereas the second one passes through a uniform medium. The two beams are subsequently recombined producing interference fringes. The test and reference beams must be set parallel to ensure a broad fringe spacing and a vibration free, constant temperature area is used to maintain the alignment of the beams. In some cases, a compensation chamber can be placed on the reference beam to adjust the optical path lengths of the test and reference beams. A clear and practical guide to interferometry equations can be found in GOLDSTEIN's article (1976).

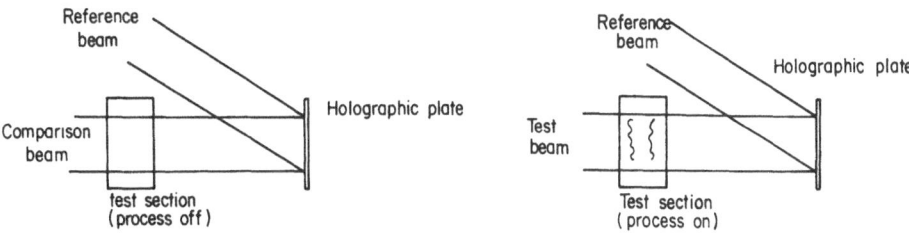

Fig. 20. Holographic interferometry.

Holographic interferometry is less costly than conventional interferometry due to the possibility of using ordinary windows and mirrors. However, it is as accurate as MZI. The first application of holographic interferometry to a two-phase flow problem can be found in a paper by MAYINGER et al. (1974). So far the basic reference remains the invited lecture given by MAYINGER and PANKNIN at the Tokyo International Heat Transfer Conference (1974). The simplest arrangement is the double exposure technique represented in Fig. 20. A comparison beam passes through the test section where the heat transfer process has been turned off, and combines with a reference beam on a holographic plate. A first hologram is recorded which takes account of all the optical imperfections of the optical components. After the heat transfer process has been turned on in the test section, a second hologram is recorded leading to macroscopic interference fringes between the recorded wave patterns of the comparison and test beams. The holographic plate is then developed and fixed. Finally, the interference fringes can be observed or recorded by illuminating the hologram with the same laser light. For a real time observation one uses the single exposure technique.

First a comparison beam passes through the test section where the heat transfer process has been turned off and combines with a reference beam on the holographic plate. The hologram is recorded, then developed and fixed either in place or elsewhere and accurately repositioned. Afterwards the heat transfer is turned on and moving macroscopic interference fringes appear due to the combination of the time dependent test beam wave pattern with the recorded comparison beam wave pattern. The resulting interferogram is then continuously observed or recorded on still or movie films. Holographic interferometry has been extensively applied to investigations on bubble growth and collapse in subcooled liquids (NORDMANN and MAYINGER, 1976; NORDMANN, 1977) and on the effect of liquid viscosity on bubble formation and heat transfer in boiling (MAYINGER and HOLLBORN, 1977).

Optical techniques used to measure temperatures are based on the change of the refractive index along a light beam. As a result they give an average value over a segment, which leads to the design of two-dimensional test sections. However, it is possible to study an axisymmetric flow field by solving an Abel integral equation (MATEKUNAS and WINTER, 1971).

3. Conclusions
Optical methods constitute an essential diagnostic tool for the elaboration of two-phase flow models. An accurate portrayal of two-phase systems can be obtained providing that the measuring technique itself has been modeled. As examples, we would like to recall the need for evaluating the deviation of the light rays through a transparent droplet or for calculating the temperature field in an axisymmetric flow from line averaged information.

Due to page limitations, this review has been restricted to the basic optical methods used in two-phase flow experimental analysis. It should be complemented with the descriptions of infrared photography, photon attenuation technique for void fraction measurements and even with a review of the optical techniques available for liquid film thickness measurements.

Acknowledgements
The author is particularly grateful to the following people for their contribution to the improvement of the review paper: P. BELL (Centrad, Grenoble), M. BOURE (Heat Transfer Laboratory, Grenoble), F. DANEL (AID, Grenoble) LECOFFRE (Neyrtec, Grenoble), R. SEMERIA (Head, Heat Transfer Laboratory, Grenoble), B. VALIBOUSE (Neyrtec, Grenoble).

Nomenclature

D	diameter
d	displacement
E	illumination
θ	illumination angle

Subscripts

app	apparent
o	ground

References

ACHARD, J.L. and BESSET, G., 1978, CENG/STT, Private communication

ARNOLD, C.R. and HEWITT, G.F., 1967, Further developments in the photography of two-phase gas-liquid flow, *The J. of Photographic Science,* Vol. **15**, 97-114

BAKER, J.L.L., 1965, Flow regime transitions at elevated pressures in vertical two-phase flow, ANL-7093

BARAKAT, S.A. and SIMS, G.E., 1977, Liquid flow patterns about barbotage bubbles, *Int. J. Multiphase Flow,* Vol. **3**, 383-397

BARANENKO, V.I., CHICHKAN, L.A., NIKOLAEV, and SMIRNOV, G.F., 1974, Optical investigation of heat transfer mechanism with boiling, *Heat Transfer 1974,* Vol. **4**, Paper B2.4, 50-54

BEER, H., 1969, Beitrag zur Wärmeübertragung beim Sieden, *Progress in Heat and Mass Transfer,* Vol. **2**, IRVINE, T.F., IBELE, W.E., HARTNETT, J.P. and GOLDSTEIN, R.J., Eds., 311-370

BEER, H., BUROW, P. and BEST, R., 1977, Bubble growth, bubble dynamics, and heat transfer in nucleate boiling, viewed with a laser interferometer, *Heat Transfer in Boiling,* HAHNE, E. and GRIGULL, V., Eds., Ch. 2, Academic Press-Hemisphere, 21-52

BEHAR, M., 1963, Méthode stroboscopique à éclairs retardés pour l'étude des bulles, *La Houille Blanche,* No. 6, 692-696

BEHAR, M. and SEMERIA, R., 1963, La strioscopie appliquée à l'étude de l'ébullition et du dégazage, *La Houille Blanche,* No. 6, 687-691

BEHAR, M. and SEMERIA, R., 1963a, Sur la mise en évidence par strioscopie de certains mécanismes d'échanges thermiques dans le dégazage et l'ébullition, *CRAS,* t. 257, 2801-2803

BENNET, A.W., HEWITT, G.F., KEARSEY, H.A., KEEYS, R.F.K., and LACEY, P.M.C., 1965-1966, Flow visualization studies of boiling at high pressure, *Instr. Mech. Engrs., Proc. 1965-1966,* Vol. **180**, Part 3 C, 260-270

BROWN, R.A.S. and GOVIER, G.W., 1961, High-speed photography in the study of two-phase flow, *Canadian J. Chem. Engng.,* Vol. **39**, No. 4, 159-164

CHEN, J.I., LIENHARD, J.H. and EICHORN, R., 1978, A method for measuring transparent droplet diameters, *Int. J. Multiphase Flow,* Vol. **4**, 233-235

COLLIER, J.G. and HEWITT, G.F., 1966, Experimental techniques in two-phase flow, *Brit. Chem. Engng.,* Vol. **11**, No. 12, 1526-1531

COOPER, K.D., HEWITT, G.F. and PINCHIN, B., 1964, Photography of two-phase gas/liquid flow, *The J. of Photographic Science,* Vol. **12**, 269-278

DELHAYE, J.M., 1979, Two-phase flow instrumentation, *Thermohydraulics of two-phase systems applied to industrial design and nuclear engineering,* DELHAYE, J.M., GIOT, M., and RIETHMULLER, M.L., Eds., (to be published by Hemisphere Publ. Corp.)

DERBYSHIRE, R.T.P., HEWITT, G.F. and NICHOLLS, B., 1964, X-radiography of two-phase gas-liquid flow, AERE-M-1321

GALAUP, J.P., 1975, Contribution à l'étude des méthodes de mesure en écoulement diphasique, Thèse de docteur-ingénieur, Université Scientifique et Médicale, Institut National Polytechnique de Grenoble

GOLDSTEIN, R.J. Optical techniques for temperature measurements, *Measurements in Heat Transfer,* Sec. Ed., ECKERT, E.R.G. and GOLDSTEIN, R.J. Eds., Hemisphere-McGraw-Hill, 241-293

HAUF, W. and GRIGULL, V., 1970, Optical methods in heat transfer, *Advances in Heat Transfer,* Vol. 6, HARTNETT, J.P. and IRVINE, T.F., Eds., Academic Press, 133-366

HEWITT, G.F. and LOVEGROVE, P.C., 1970, A scanning device for velocity measurement and its application to wave-studies in annular two-phase flow, *J. Physics E: Scientific Instruments,* Vol. 3, 6-8

HEWITT, G.F. and ROBERTS, D.N., 1969, Studies of two-phase flow patterns by simultaneous x-ray and flash photography, AERE-M-2159

HEWITT, G.F. and ROBERTS, D.N., 1969a, Investigation of interfacial phenomena in annular two-phase flow by means of the axial view technique, AERE-R-6070

HOSLER, E.R., 1967, Flow patterns in high pressure two-phase (steam-water) flow with heat addition, WAPD-TM-658

HSU, Y.Y. and GRAHAM, R.W., 1963, A visual study of two-phase flow in a vertical tube with heat addition, NASA-TN-1564

HSU, Y.Y., SIMONEAU, R.J., SIMON, F.F. and GRAHAM, R.W., 1969, Photographic and other optical techniques for studying two-phase flow, *Two-Phase Flow Instrumentation,* LETOURNEAU, B.W. and BERGLES, A.E., Eds., 1-23

HUSTON, A.E., 1978, High-speed photography and photonic recording, *J. Physics E: Sci. Instrum.,* Vol. 11, 601-609

JACOBS, J.D. and SHADE, A.H., 1969, Measurement of temperatures associated with bubbles in subcooled pool boiling, *J. Heat Transfer,* 123-128

JACOWITZ, L.A. and BRODKEY, R.S., 1964, An analysis of geometry and pressure drop for the annular flow of gas-liquid systems, *Chem. Engng. Sci.,* Vol. 19, 261-274

JOHANNS, J., 1964, Development of a fluoroscope for studying two-phase flow patterns, ANL-6958

JONES, O.C. 1973, Statistical considerations in heteregeneous, two-phase flowing systems, Ph.D. thesis, Rensselaer Polytechnic Institute, Troy, N.Y.

JONES, O.C. and DELHAYE, J.M., 1976, Transient and statistical measurement techniques for two-phase flows: a critical review, *Int. J. Multiphase Flow,* Vol. 3, 89-116

KINGSLAKE, R. Ed., 1965, *Applied Optics and Optical Engineering,* Vol. 1, Academic Press

KINTNER, R.C., HORTON, T.J., GRAUMAN, R.E. and AMBERKAR, S., 1961, Photography in bubble and drop research, *The Canadian J. of Chem. Engng.,* Vol. 39, No. 6, 235-241

KIRBY, D.B. and WESTWATER, J.W., 1963, Photography from below: nucleate boiling on electrically heated horizontal glass plates, *Chem. Engng. Sci.,* Vol. 18, No. 7, 469

KUTATELADZE, S.S. amd MAMONTOVA, N.N., 1973, The nature of motion of a liquid about a vapor bubble, *Heat Transfer - Soviet Research,* Vol. 5, No. 6, 149-153

MARILLEAU, J., 1972, Cinématographie électronique ultra-rapide, *BIST CEA,* No. 169, 3-36

MATEKUNAS, F.A. and WINTER, E.R.F., 1971, An interferometric study of nucleate boiling, Int. Symp. Two-Phase Systems, Haifa, Israel

MAYINGER, F., NORDMANN, D. and PANKNIN, W., Holographische Untersuchungen zum unterkühlten Sieden, *Chemie-Ingenieur-Technik,* Vol. 46, No. 5, 209

MAYINGER, F. and PANKNIN, W., 1974, Holography in heat and mass transfer, *Heat Transfer 1974,* Vol. 6, IL 3, 28-43

MAYINGER, F. and HOLLBORN, E., 1977, The effect of liquid viscosity on bubble formation and heat transfer in boiling, *Heat Transfer in Boiling,* HAHNE, E. and GRIGULL, V., Eds., Ch. 17, Academic Press-Hemisphere, 391-424

MERZKIRCH, W., 1974, *Flow Visualization,* Academic Press

NICKLIN, D.J., WILKES, J.O. and DAVIDSON, J.F., 1962, Two-phase flow in vertical tubes, *Trans. Inst. Chem. Engrs,* Vol. 40, 61

NORDMANN, D., 1977, Interferometric investigations of bubble growth and collapse in subcooled liquids, *Two-Phase Flows and Heat Transfer,* Vol. 3, KAKAC, S. and VEZIROGLU, T.N., Eds., 1151-1167

NORDMANN, D. and MAYINGER, F., 1976, Experimental investigations of bubble growth and collapse during subcooled boiling, European Two-Phase Flow Group Meeting, Erlangen

OPPENHEIM, A.K., URTIEW, P.A., and WEINBERG, F.J., 1966, On the use of laser light sources in schlieren-interferometer systems, *Proc. Roy. Soc., London, A 291,* 279-290

PHAM, P.D., 1978, CENG/STT, Private communication

PRICE, I. and BRAMALL, J.W., 1971, A diametrical effect on vapour column formations in film boiling in carbon dioxide near the critical state, *Int. J. Heat Mass Transfer,* Vol. 14, 1750-1751

ROUMY, R., 1969, Structure des écoulements diphasiques eau-air. Etude de la fraction de vide moyenne et des configurations d'écoulement, CEA–R-3892

ROUMY, R. 1969a, 16 mm silent movie on two-phase flow patterns, CENG/STT

SEMERIA, R., 1963, La cinématographie ultra-rapide et l'ébullition à haute-pression, *La Houille Blanche,* No. 6, 679-686

SMIRNOV, G. and BARANENKO, V., 1971, Experimental study of temperature profiles in a thermal boundary layer during the boiling of a liquid in a free volume, *J. of Engng Physics,* Vol. 21, No. 2, 1003-1006

VAN MEEL, D.A. and VERMIJ, H., 1960, A method for flow visualization and measurement of velocity vectors in three dimensional flow patterns in water models by using color photography, *Appl. Sci. Res.,* Vol. A, No. 10, 109-120

Studies of Particle Motion by Laser Doppler Techniques

by

F. Durst

Sonderforschungsbereich 80
"Ausbreitungs- und Transportvorgänge in Strömungen"
Universität Karlsruhe, Germany

Abstract

The present paper provides an introduction into the application of laser-Doppler anemometers to investigations of particulate flow systems. The basic principles of LDA-measurements in particulate two-phase flows are outlined and it is shown that correct LDA-measurements require the optical system to consist of two parts that permit the information from the fluid velocity field to be separated from the information relating to the properties of the moving particles. The various information on particulate flow systems which is contained in the optical signals is explained and electronic processing systems are introduced that permit this information to be extracted.

The paper describes the application of laser-Doppler anemometers to particulate flows and presents results that were obtained for a laminar bubble driven liquid flow and for a turbulent gas flow of suspended particles. Implications of these results to improve the present knowledge on particle motion in turbulent flows are provided.

1. Introduction

Basic studies of the transport of solid and liquid particles by gaseous flows and the motion of solid and gas particles in liquid flows have been the subject of numerous scientific and engineering investigations, e.g. see refs. (14), (18), (23) and (24). This fact indicates the far-reaching interest in understanding simple flows of this kind as a basis for a deeper insight into the nature of more complex two-phase flows, such as sand and snow storms and the conveyance of plant pollen by winds. In addition, basic information on solid and liquid particle behaviours in turbulent flows will also be valuable to treat more accurately "man made" particulate two-phase flows which are associated with the dispersion of emission products in the atmosphere, in lakes, rivers and the sea. Hence, results obtained in laboratory particulate two-phase systems are likely to find far-reaching applications in treatment of more complex flows.

Carefully conducted experiments with particulate two-phase flows clearly reveal that meaningful results on particle dynamics can be obtained only when related to specific particle size distributions, the shape of particles, the local particle concentration, and to the intrinsic density of the suspended particles relative to the density of the flowing fluid. These properties can only be studied, however, when suitable measuring techniques are available to locally measure the aforementioned parameters that describe particulate flow systems. Hence, attention has been focused for many years on measuring techniques that accurately measure local particle size distributions, particle concentrations and the velocity distributions of the fluid and the suspended particles, e.g. see refs. (1), (2), (4), (6), (12) and (13). Laser-Doppler techniques have been suggested for such measurements and it has turned out that these techniques work satisfactorily in particulate flow systems in which spherical particles are suspended. In systems of this kind information can be obtained on local properties such as:

(a) Instantaneous velocity of the flowing fluid
(b) Instantaneous velocity of suspended particles
(c) Size distribution of suspended particles
(d) Concentration of suspended particles.

First attempts to carry out laser-Doppler measurements in two-phase bubble flows were undertaking by Lading (17) and Davies (5). Davies (5) showed that LDA-signals of high quality can be obtained from particles much larger than the fringe spacing of a conventional laser-Doppler system. This finding was confirmed by Ohba (19) who combined reference beam LDA-measurements with records of light attenuation to obtain velocity and concentration information in turbulent bubble flows. Some LDA-measurements have also been attempted in particulate gas-solid two-phase flows, e.g. see Farmer (11), Riethmueller (22), Carlson and Peskin (3), Popper et al. (21), Kolansky et al. (16), Stümke and Umhauer (26). Similar attempts have been reported in flows with large liquid particles, e.g. in wet steam, which confirm the findings obtained with solid particles.

The aforementioned publications reflect the present state of development of LDA-systems for two-phase flow studies: No extensive measurements of two-phase flow velocity field are presently available; suitable instruments to carry out such measurements are still under development. Guidelines for such developments have been put forward by Durst and Zaré (10) together with explanations on the physics of LDA-signals obtained from particulate two-phase flow systems. The present paper reports on the continuation of this work; improved LDA-optical units for two-phase flow studies are described and suitable electronic processing systems are introduced. Applications of these systems to simple two-phase flows are described that were carried out to verify the performance of all the components prior to a detailed study on a turbulent gas flow with solid particles. Results of this study are provided in this paper together with suggestions of further system improvements.

2. Laser-Doppler Instrumentation for Two-Phase Flow Measurements in Particulate Flow Systems

2.1 Basic Ideas of Two-Phase Flow LDA-Measurements

The basic ideas for LDA-velocity measurements in two-phase flows have been forwarded by Durst and Zaré (10) who showed that the light wave fronts, produced by two laser beams reflected on the smooth surface of a large body, interfere and produce fringes in space. The location and shape of the interference pattern is dependent on the arrangement of the incident laser beams, the shape of the body and its location. Furthermore, the rate at which the fringes cross a detector in space is linearly related to the transverse velocity component of the reflecting body, perpendicular to the symmetry line between the two incident beams.

The reflected light fronts reaching any point in space originate from two near points A and B on the surface of the large body, Fig. 1a, and the movements of these *two points* contribute to the detected Doppler signal:

$$f_D = \frac{1}{\lambda} \left\{ [(\{K_I\}_i - \{K_R\}_i) \, \{U\}_i]_A + [(\{K_I\}_i - \{K_R\}_i) \{U\}_i]_B \right\} \qquad [1]$$

The size of fringe separation distance Δx is determined by the wave length of the laser light and the angle between the two reflected beams reaching the detector position.

$$\Delta x = \frac{\lambda}{2} \left[E_{ijk} \cdot \{K_R\}_{iA} \{K_R\}_{jB} \right] = \frac{\lambda}{2\sin\beta} \qquad [2]$$

β is a function of incident and reflected beam directions, together with the position of the body and that of the detector, and can be calculated for each specific case.

A similar analysis can be carried out for light beams which are refracted at the surface of a transparent particle. In this case, the total Doppler shift of the light frequency in each of the beams is affected by the velocity at a point on the front surface and a point on the back of the body, Fig. 1b. Consequently, the frequency of the intensity variations at the observer, produced by the interference of two refracted light beams, can be formulated as:

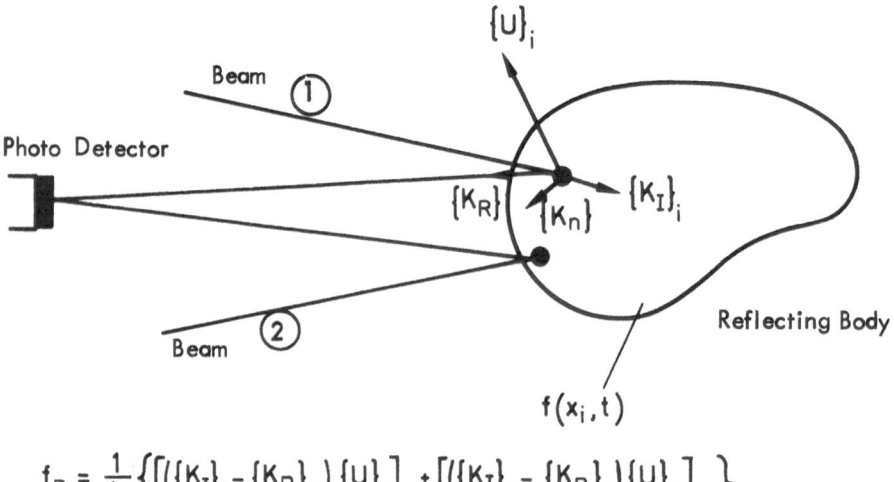

$$f_D = \frac{1}{\lambda} \left\{ \left[\left(\{K_I\}_i - \{K_R\}_i \right) \{U\}_i \right]_A + \left[\left(\{K_I\}_i - \{K_R\}_i \right) \{U\}_i \right]_B \right\}$$

a) Reflecting Body

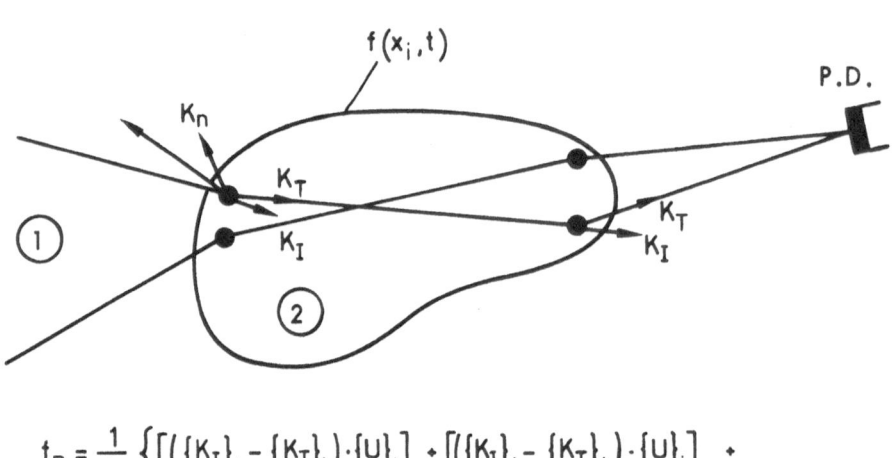

$$f_D = \frac{1}{\lambda} \left\{ \left[\left(\{K_I\}_i - \{K_T\}_i \right) \cdot \{U\}_i \right]_A + \left[\left(\{K_I\}_i - \{K_T\}_i \right) \cdot \{U\}_i \right]_B + \right.$$
$$\left. \left[\left(\{K_I\}_i - \{K_T\}_i \right) \cdot \{U\}_i \right]_C + \left[\left(\{K_I\}_i - \{K_T\}_i \right) \cdot \{U\}_i \right]_D \right\}$$

b) Transparent Body

Fig. 1. General configurations to explain laser-Doppler measurements utilizing refracted and defracted light beams.

$$f = \frac{1}{\lambda} \left\{ [(\{K_I\}_i - \{K_T\}_i) \cdot \{U\}_i]_A + [(\{K_I\}_i - \{K_T\}_i) \cdot \{U\}_i]_B + \right.$$

$$\left. [(\{K_I\}_i - \{K_T\}_i) \cdot \{U\}_i]_C + [(\{K_I\}_i - \{K_T\}_i) \cdot \{U\}_i]_D \right\}$$

[3]

The relations concerning the unit vector of the incident and transmitted light waves include the refractive indices of the two media besides other factors.

The above considerations are applicable to general body shapes, but are only of practical relevance when applied to a specific particle shape. In this section, the relations for reflecting and refracting spherical particles are presented.

a) Reflecting Spheres

The two laser beams, reflecting from a spherical particle, interfere and produce a non-linear fringe system. The shape and spacing of the fringes are functions of the angle between the incident laser beams and their wave length, as well as the sphere size and the direction of observation. The fringes are, however, of simple shape — namely, parallel planes — in the backward direction. The fringe separation distance in this region, for large ratios L/R and small angles φ, can be formulated as:

$$\Delta x = [1 + \frac{2(L-R)}{R \cdot \cos\varphi}] \cdot \frac{\lambda}{2\tan\varphi}$$

[4]

This relationship indicates that the radius of the sphere R can be obtained by measuring the fringe distance Δx in the backward direction. The author has used a double element photodiode, with elements spaced 2 mm apart, to obtain information on the sphere diameter through phase measurements between the two detected signals. Examples of such measurements are given in reference (10).

As shown in Fig. 2a, when the sphere moves across the two laser beams, the fringes appear to originate from a point along the direction of one of the beams, move around the entire space and then disappear into a point along the direction of the second beam.

The rate at which fringes cross any point, when produced by a moving non-deformable sphere, is the same at all points in the surrounding space, and is equal to the rate of fringe appearance at the source line or the rate of their disappearance at the sink line. The Doppler frequency detected at the source, resulting from the movement of the sphere, can be formulated as:

$$f_D = \frac{2(U_\perp \cos\beta \pm U_\parallel \sin\beta) \sin\varphi}{\lambda}$$

[5]

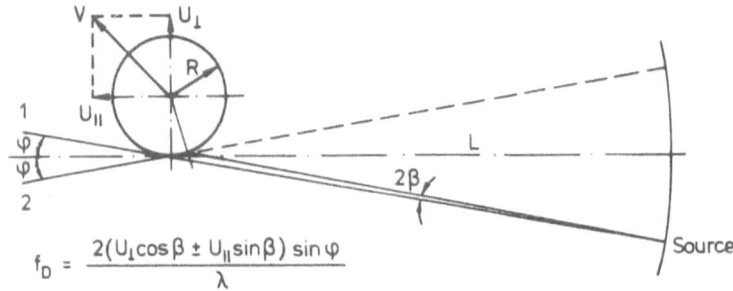

$$f_D = \frac{2(U_\perp \cos\beta \pm U_\parallel \sin\beta)\sin\varphi}{\lambda}$$

a) Reflecting Sphere

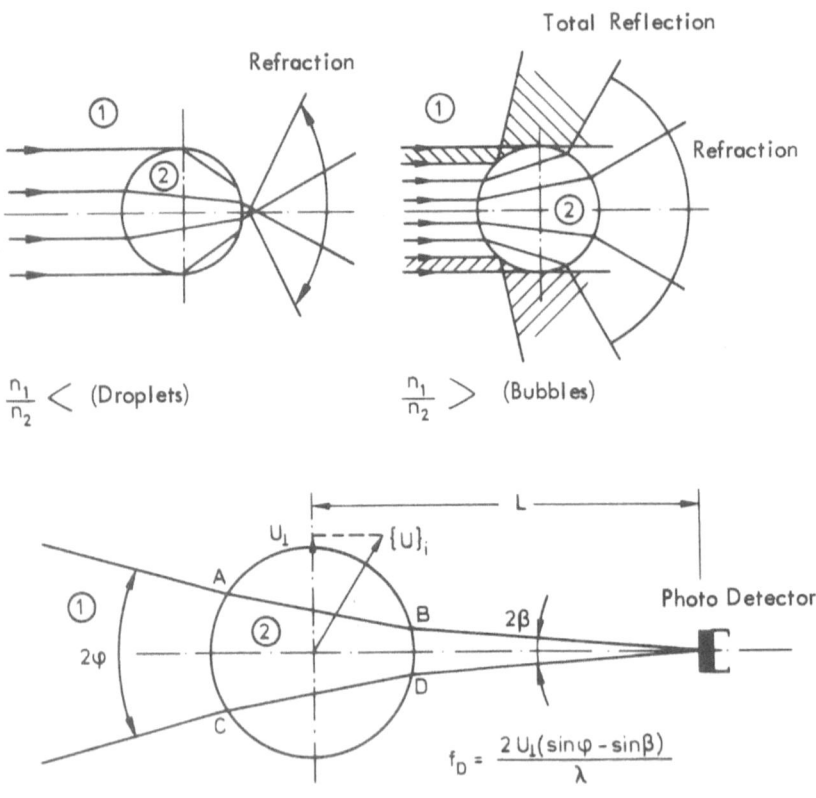

$$f_D = \frac{2\,U_\perp(\sin\varphi - \sin\beta)}{\lambda}$$

b) Transparent Sphere

Fig. 2. Configuration for spherical particle

The angle β is a function of L/R and φ. For large values of ratio L/R and small angles φ, equation [5] tends towards the universal equation of laser-Doppler anemometry for single phase measurements. In the case of a deforming particle the rate of change of fringes would not remain the same all over space, and this fact allows us to measure the particle deformation through the use of proper multi-detection systems.

b) Transparent Spheres

As the light beam passes through a transparent particle, the relative refractive index of the particle plays a role in the characteristics of the propagated light. Two cases are distinguishable, Fig. 2b: (i) $n_1/n_2 < 1$; the light propagates from an optically lighter medium to a denser medium. All of the light incident on the sphere passes through it and the sphere acts as a positive lens. An example of this case occurs in the measurements of water drops falling in air. (ii) $n_1/n_2 > 1$; the light propagates from an optically denser medium to a lighter medium. The sphere acts as a negative lens, and at a certain region total reflection occurs on the sphere surface. Such a case happens in the measurement of gas bubbles-liquid flows and insoluble liquid-liquid flows. The partly or totally reflected light from the interface of the spherical particle can be treated in a manner similar to part (a) of this section.

When two beams are transmitted through a sphere, the shape of the fringes produced by the beams is linear in the vicinity of the optical axis in the forward direction, and for large ratios L/R and small angles φ one may show the fringe distance to be:

$$\Delta x = [1 + (\frac{2n_1}{n_2} - 2) \frac{L}{R}] \frac{\lambda}{2n_1 \sin\varphi} \qquad [6]$$

For constant geometrical and physical conditions the fringe distance are functions of sphere radius R. Consequently, spherical particle size measurements can be performed using a double photo-detector.

When transmitted through a moving particle, each of the beams is refracted at both interfaces (entering and leaving) and therefore Doppler shifted twice. The general motion of the sphere and its surface movements all contribute to the rate of passage of the interference fringes. For a simple case of a solid sphere with no rotational movements (so that the four points A,B,C and D in Fig. 2b have equal velocities), the Doppler frequency detected in the forward direction can be formulated as

$$f_D = \frac{2U_\perp (\sin\varphi - \sin\beta)}{\lambda} \qquad [7]$$

For large ratios L/R and small beam crossing angle φ, equation [7] tends towards the universal equation of laser-Doppler anemometry for single phase
measurements.

2.2 Optical System and Signal Processing Electronics

Continuing the work by Durst and Zaré (10), the author and his collaborators
have persued the development of suitable optical and electronic systems to
carry out LDA-measurements in particulate two-phase flows. This development work resulted in two systems that are presently in use at the Sonderforschungsbereich 80 at the University of Karlsruhe. They represent extensions
of LDA-systems that have been built for single-flow measurements as can be
seen from Figs. 3 and 5. Figs. 4 and 6 show the appropriate signal processing
systems in form of diagrams. These diagrams show automatic filter banks
which are used in conjunction with amplitude discriminators in order to separate those signals that are obtained from the fluid and the particles.

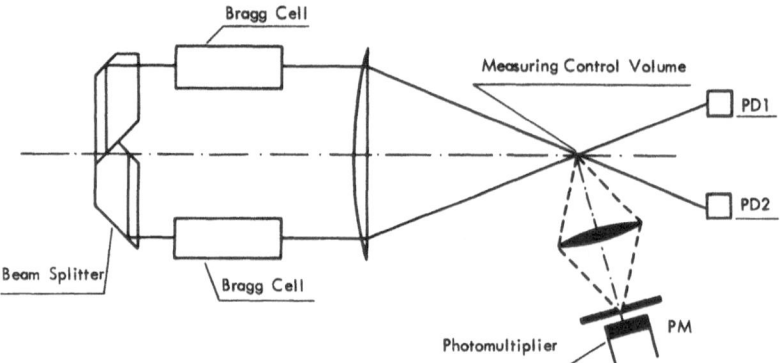

Fig. 3. Optical arrangement I for two-phase flow LDA-measurements.

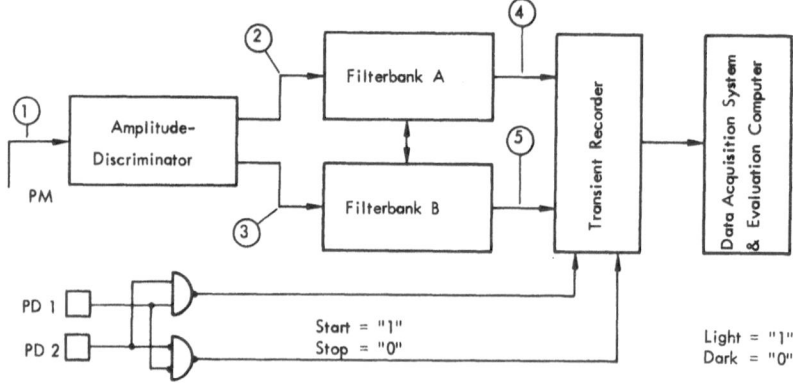

*Fig. 4. Blockdiagram of signal processing system I for two-phase flow LDA-
measurements.*

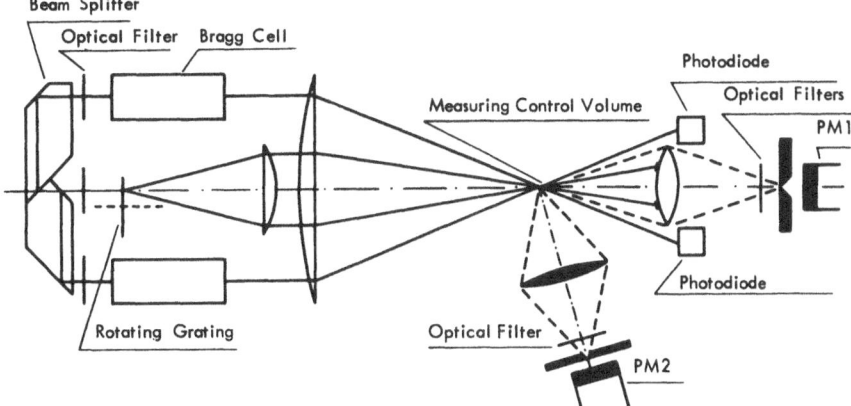

Fig. 5. Optical arrangement II for two-phase flow LDA-measurements.

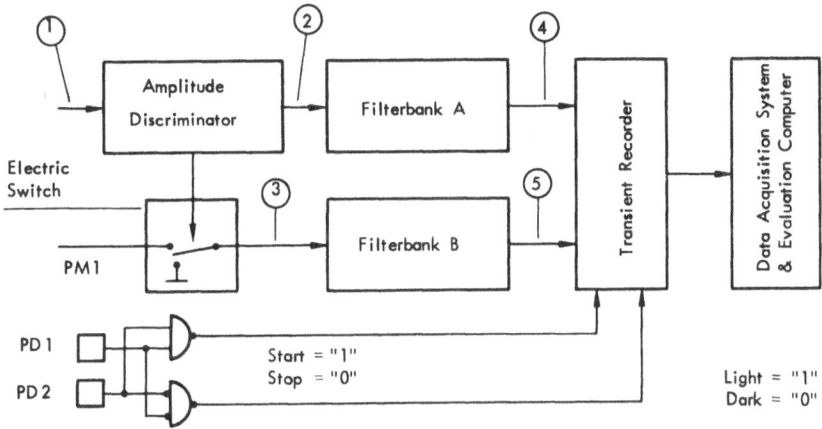

Fig. 6. Blockdiagram of signal processing system II for two-phase flow LDA-measurements.

The aforementioned filter banks have been designed for LDA-measurements in single-phase flows and have been extended to two-phase flow applications as described by Durst and Heidbreder in (7). These filter banks consist of 15 band pass filters that permit automatic selection of the optimum filter for LDA signals within a range between 500 Hz and 50 MHz. An internal logic uses the output of the filters in order to select the optimum filter. This yields filtered signals as indicated in Fig. 8. This figure shows that signals of very low signal-to-noise ratio are recognized by the filter logic and that correct decisions of the optimum band pass filter are made.

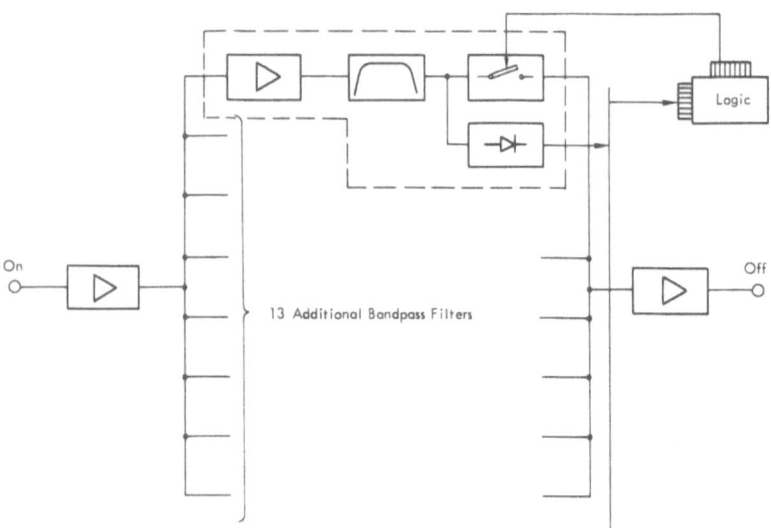

Fig. 7. Blockdiagram of automatic filterbank.

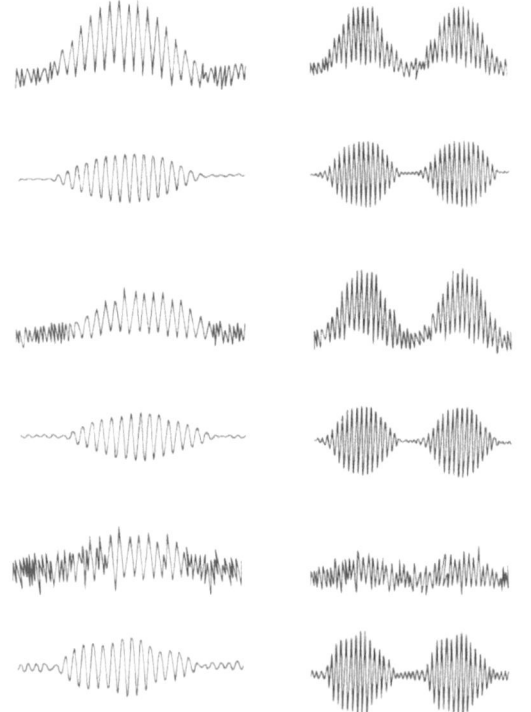

Fig. 8. Noisy signals at input and output of automatic filterbank.

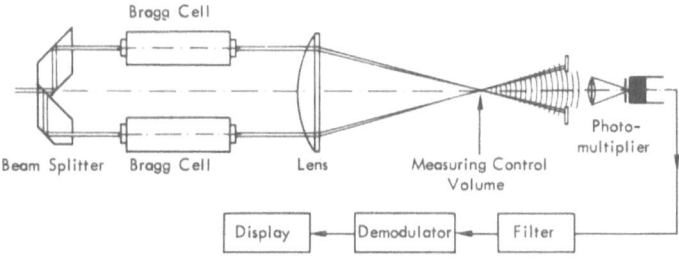

Fig. 9. Laser-Doppler optic and blockdiagram of signal processing electronics for single-phase flow measurements.

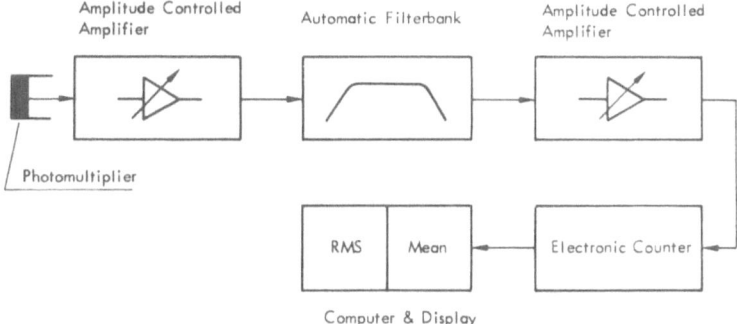

Fig. 10. LDA-signal processing electronics employed for measurements in a single-phase turbulent pipe flow.

The application of the aforementioned filter banks in single-phase flows using the optical system shown in Fig. 9 and the electronics given in Fig. 10 permits velocity profile measurements to be carried out without the usual necessity to switch the band pass filters in the signal processing electronics. Such measurements were carried out in the flow system indicted in Fig. 11 and yielded the velocity distribution given in Fig. 12.

As pointed out in ref. (7), LDA-optical and electronic systems for single-phase flow measurements can be extended for two-phase flow applications. In particulate two-phase flow signals are obtained that differ in amplitude and modulation depth dependent on whether they are obtained from small particles that follow the fluid motion or whether they are recorded from large particulates. The signals indicated in Fig. 13 and Fig. 14 demonstrates the separation of the signals by amplitude discrimination and filtering. This sequence of signals indicates the functioning of the amplitude control in the electronic system

given in Fig. 4 and also shows the functioning of the automatic filter banks A
and B to produce the resulting signals d and e. These signals can be processed
to obtain local information on the velocity fields of the two phases. In addition to the amplitude discrimination the signals from the two photo-diodes
are used to separate the signals of large and small particles. The combined
usage of the amplitude discrimination and the photo-diodes permits the cross
talk between signals from the small and large particles to be reduced.

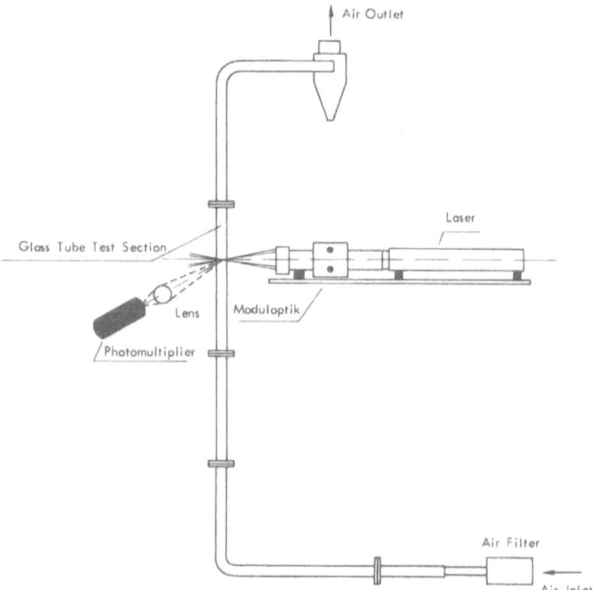

Fig. 11. Test section to verify automatic filter bank for single-phase flow measurements.

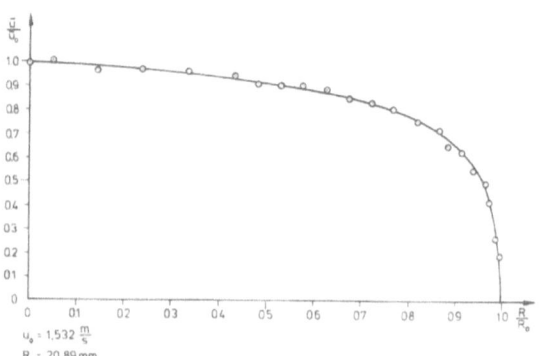

Fig. 12. Mean velocity profile across vertical pipe flow test section.

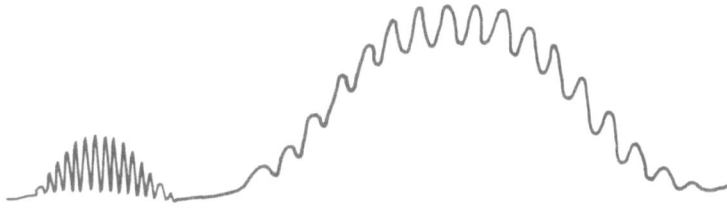

Fig. 13. Typical signals from particulate two-phase flow system.

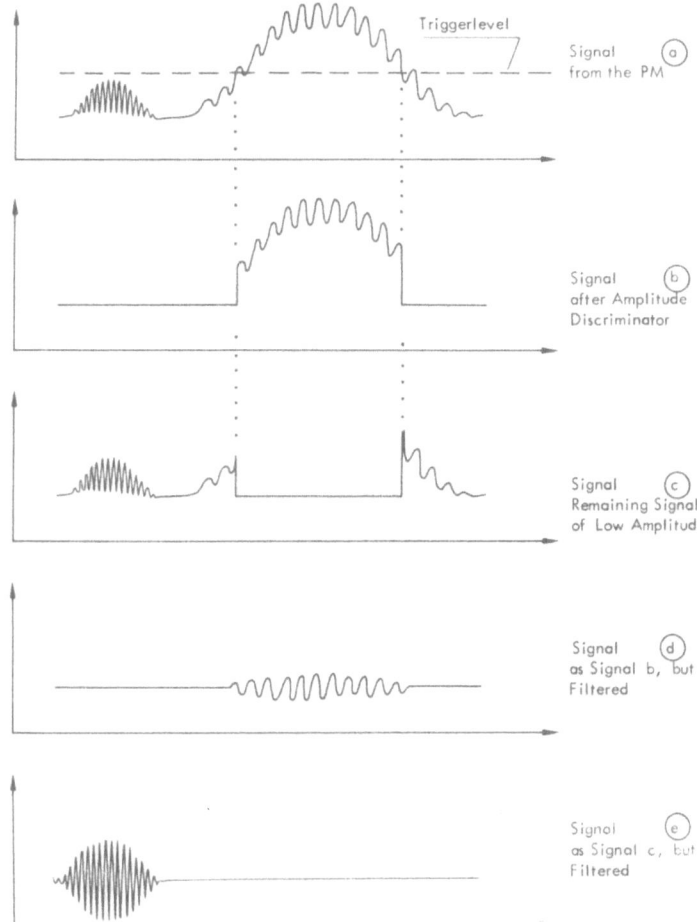

Fig. 14. Separation of signals by amplitude discriminator and automatic filterbanks.

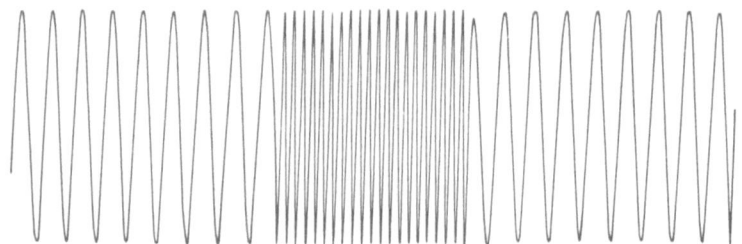

Fig. 15a. Simulated two-phase flow signals; bursts of different frequency.

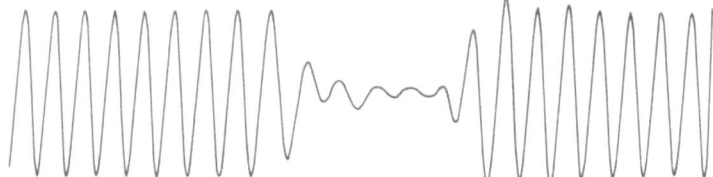

Fig. 15b. Output of automatic filterbank A.

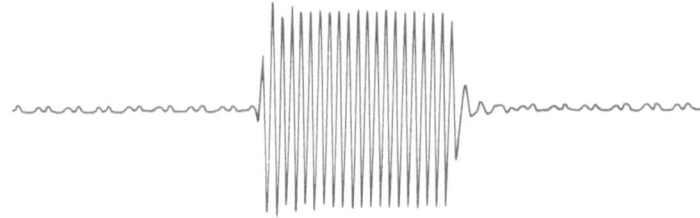

Fig. 15c. Output of automatic filterbank B.

In order to test the signal processing electronics, simulated signals were applied to the input of the signal processing systems. Such simulated signals are shown in Fig. 15a and consisted of bursts of different frequencies but of equal amplitude. In such a case, the amplitude discriminator would not be able to distinguish between the contributions from the two phases but the employment of the automatic filter banks in a so-called "slave-operation" permits the two contributions to be separated. It is only the employment of different controls like amplitude discriminator, "slave-control" of the filter bank, photodiode-selection control, etc. that permits the cross talk between signals from small and large particles to be reduced.

2.3 Signal Processing and Data Acquisition

The processing of laser-Doppler signals by available digital computer became possible with the development of high speed analog-to-digital converters with storage capability and with digitizing rates of 100 MHz and faster. Devices of this kind permit Doppler bursts to be digitized and stored with sufficient accuracy to allow meaningful Doppler frequencies to be calculated by digital computers following the A/D-converter. The evaluation of the Doppler frequency from the recorder signals is controlled by soft-ware programs and, hence, differences in signal qualities can be easily accounted for by introducing suitable data validation schemes.

The A/D converter which was used in the present study was the 8 bit "Biomation Transient Recorder" model 8100. This model is particularly suited to record signal bursts as obtained in laser-Doppler anemometry because of its ability to operate in a "delayed-trigger" mode which allows the recording of the signals that occur prior to the triggering event. Hence, the recorder can be triggered on the high amplitude peaks in the center of the Doppler bursts and still record the entire bursts in the 2048 word memory of the transient recorder. The Biomation 8100 also offers input attenuation and amplification, and proper adjustment of these input parameters of the transient recorder permits optimal usage of the 8 bits of amplitude resolution.

In the present investigations, the contents of the 2048 word memory was directly transferred to the core memory of an available Hewlett Packard computer (HP 2116C). Once in the core, the data can be either directly analysed (on-line operation) or stored on disk for further processing to obtain information on the Doppler frequency. Reading the data in and storing them on disk can be carried out at approximately 10000 words/sec (5 complete memories), and the on-line processing rate depends entirely on the evaluation method used, as discussed in the next paragraph. If the entire 2048 word memory is not required for analysis then a smaller portion of it may be read thus allowing more Doppler bursts to be scanned per second. However, this modification means fewer points to describe the Doppler bursts and its application is therefore restricted to the processing of good quality signals.

The present chapter describes the different computer programs that were developed by the author and one of his collaborators (C. Tropea) to process LDA-signals on an available digital computer after the signal had been transferred from the Biomation transient recorder to the computer core. Employing transient recorders and computers to process LDA-signals yields signals of finite length which can be adjusted to contain information on a single Doppler burst only. Hence, the computer programs were written to accept discrete Doppler bursts from the output of the filter banks and to calculate the Doppler frequency from signal samples of finite length.

The first program to be discussed is the high speed program which has been developed for Doppler signals of high signal-to-noise ratio. This program assumes that there are no multiple zero-crossings in the signal and that every change of the sign of the input signal indicates another halfcycle of the high-pass filtered Doppler signal. The basic approach chosen in the high-speed computer program was therefore to calculate the position of the first and last zero-crossings exactly and to count the numbers of zero-crossings between these end-positions. Knowing the number of signal cycles and also the time for these zero-crossings to occur, permits the Doppler frequency to be calculated.

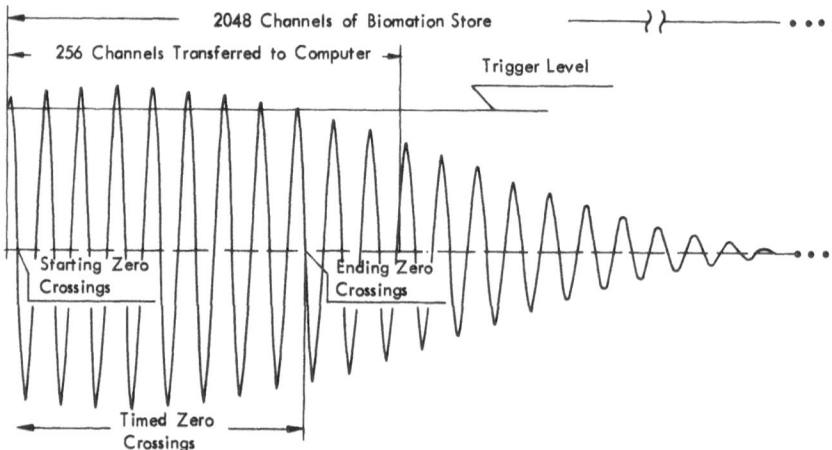

Fig. 16. Data transfer and signal validation using the computer program HISPE.

The program operates directly on-line, that is, each burst is processed before the next burst is read from the Biomation recorder. In order to speed up the data transfer, only the first 256 words of the transient recorder's 2048 word memory are read in by the program for the analysis described above. This requires the Biomation to be triggered in such a manner that the high amplitude regions of the bursts occur in the initial region of the store, see Fig. 16. In addition, an amplitude check in the program ensures that only the high amplitude part of the burst enters the evaluation procedure. The accurate calculations of the zero-crossings for the first and last cycle are carried out using a least square fit of the signal amplitude.

The second program to be described in this chapter is the low speed program designed to handle signals with moderate levels of noise. This program offers a more flexible data input than the high speed program and can read data directly from the transient recorder (on-line operation) or from the disk. The number of points read per burst can also be specified as 256, 512, 1024 and 2048. The number of points read in has to be increased with decreasing signal-to-noise ratio of the Doppler bursts.

The low speed program works sequentially through the different signal cycles in the burst that exceed a certain amplitude level and finds every maximum and minimum by defining these as a change of the sign of the gradient over 4 consecutive points. After finding the first signal amplitude maximum which is over the specified discrimination level, the program calculates the exact position of the next zero-crossing by using a least-squares fit of a straight line through all signal points lying around zero within a fraction of the preceding maximum. In this way the event of multiple zero-crossings is compensated for. Using all data around zero within a fraction of the preceding amplitude maximum, permits the program user to adjust the program in accordance to validation requirements imposed by the signal-to-noise ratio.

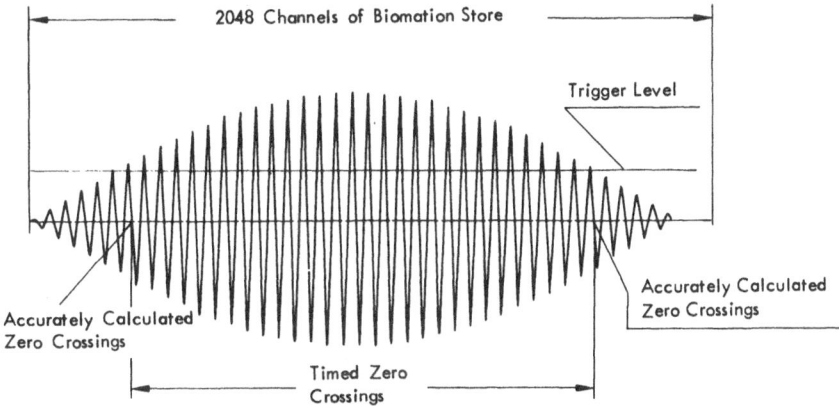

Fig. 17. Data transfer and signal validation using the computer program LOSPE

After the first zero-crossing is determined accurately, the program continues calculating and counting signal peaks until the end of the stored information is reached or the peaks fall below the pre-set discrimination level, see Fig. 17. At this point a second zero-crossing is determined accurately and the frequency calculated via the time between the accurately determined zero-crossings and the number of peaks between these. However, if there are fewer than 4 cycles between the accurately determined zero crossings the burst is disregarded and a new burst is read in for processing.

In order to carry out LDA-measurements with the aforementioned programs it is necessary for the operator to put in the following information: sample time interval of the transient recorder, the input mode of the program (i.e. on-line or from disk operation), the number of the bursts to be analyzed, the length of the bursts to be used in the program and the two discriminator levels used in the data validation scheme. The program returns the number of accepted bursts, the average frequency and the rms-value of the frequency deviations. The individual frequency values can be stored on a disk for further analysis such as obtaining probability density distributions and/or carrying out other statistical calculations.

3. Experimental Verifications

During several stages of their experimental program, the author and his collaborators have verified the correct functionung of the optical and electronic systems of their two-phase flow LDA-instruments. Such verifications are best carried out in simple two-phase flows and/or simulated flows that can be set up to emphasize particular features of the more complex flow situations in which the final instruments have to be applied. The present section describes examples of such verification experiments that clearly demonstrate the correct functioning of the two-phase flow LDA-systems described in this paper.

3.1 Bubble-Driven Liquid Flow

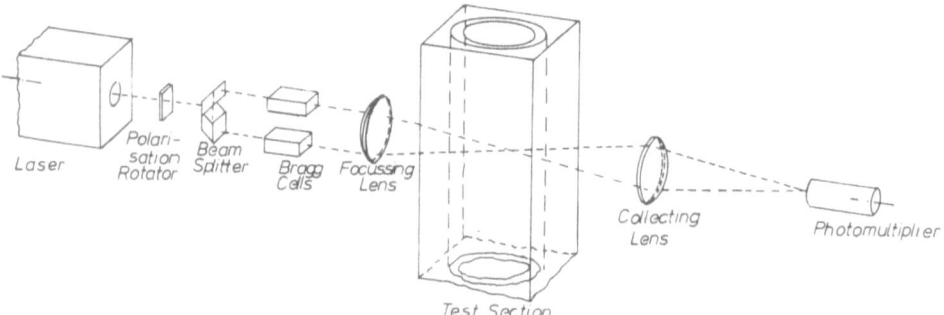

Fig. 18. Schematic diagram of test section and LDA-optical system.

In order to demonstrate that correct liquid velocity can be measured in flow situations in which bubble-driven flow components can exist locally and vary with time, Durst, Taylor & Whitelaw (8) set up a vertical cylindrical test section as shown in Fig. 18. This figure also shows an LDA-system used for liquid flow measurements and also indicates that the cylindrical test section was mounted inside a square outer housing that permitted an outside liquid to be used for refractive index matching. In this way displacement errors due to the cylindrical walls can be avoided. Fig. 19 shows the traversing table that was used to move the test section relative to the optical system in order to obtain complete traverses of the velocity profiles of the bubble-driven liquid flow.

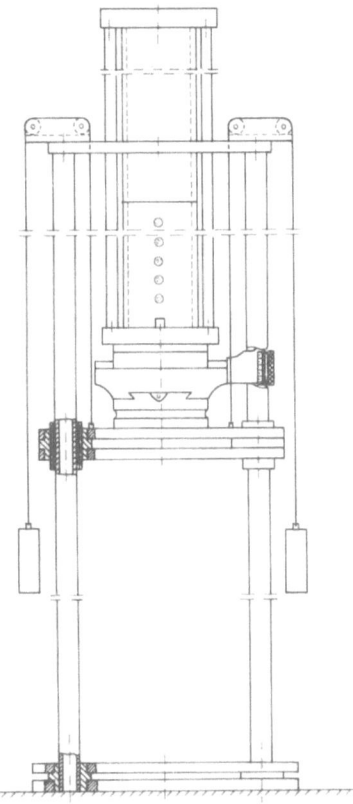

Fig. 19. Test section and traversing table for bubble driven liquid flow.

Fig. 20. Flow pattern of bubble driven liquid flow.

Fig. 20 shows a photograph of the recirculating flow that established itself in the test section. It can be seen that the flow was driven by a stream of air bubbles rising through the liquid inside the cylinder which had an inside diameter of 100 mm. The bubbles were formed by delivering a carefully regulated flow of air to a .5 mm diameter nozzle located at the base of the cylinder. As the bubbles broke away from the nozzle and rose through the liquid, the regulatory flow established itself as shown in Fig. 20. The peak Reynolds number (based on the peak bubble velocity, the measured kinematic viscocity of the oil at 20°C and the bubble diameter) was .1 and guaranteed the laminar flow of equal bubble spacing if the time between the bubbles was chosen to be .55 sec. This time was kept constant for the different experiments carried out for different liquid hights in the circular cylinder.

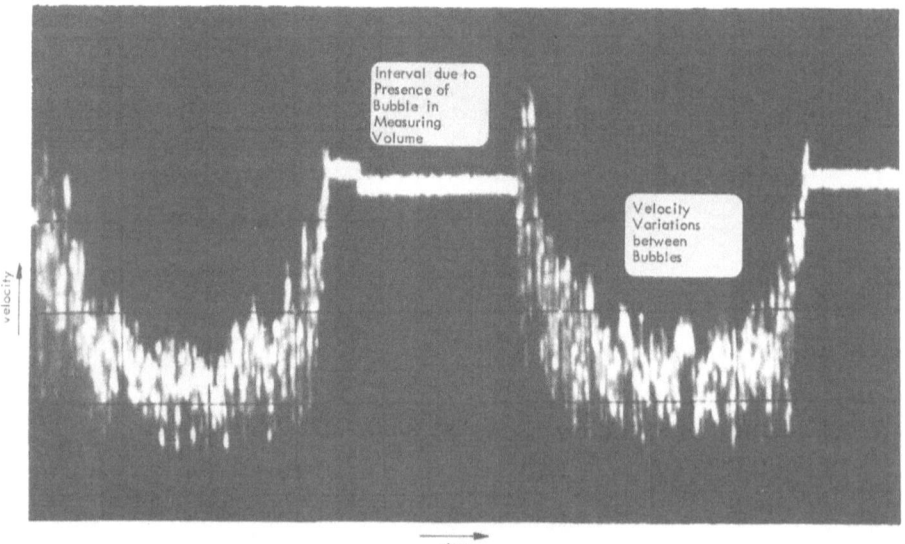

Fig. 21. Variations of liquid velocities between bubbles.

Durst, Taylor and Whitelaw demonstrated in the aforementioned flow situation that conventional frequency trackers do not only correctly measure the instantaneous flow outside the bubble column but are also able to record locally the variations of the liquid velocity between the bubbles. This is indicated in Fig. 21 which shows the tracker output for the liquid velocity variations between subsequent bubbles. This clearly demonstrates that LDA-systems can be used to record local, instantaneous velocity variations caused by bubble motions. Information of this kind is of interest in studying two-phase flows.

3.2 Model-Two-Phase Flows

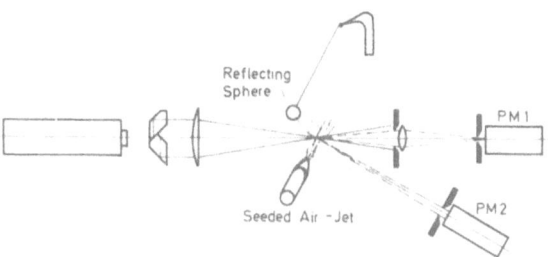

Fig. 22. Experimental arrangement to stimulate laser flow.

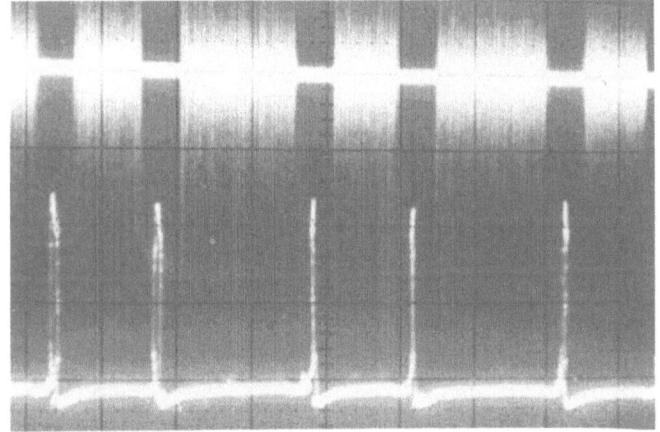

Fig. 23. Laser-Doppler signals from the two photomultipliers of the optical arrangement shown in Fig. 22.

The present section describes a model two-phase flow situation that was set up in the way proposed by Durst and Zaré (10). This model two-phase flow is schematically shown in Fig. 22 and consists of a low velocity air jet. The second phase, i.e. the large particles in the flow, is simulated by a highly reflecting steel sphere which is made to swing through the measuring control volume. The optical and electronic system tested in this particular flow situation are also shown in Fig. 22 and are self-explanatory.

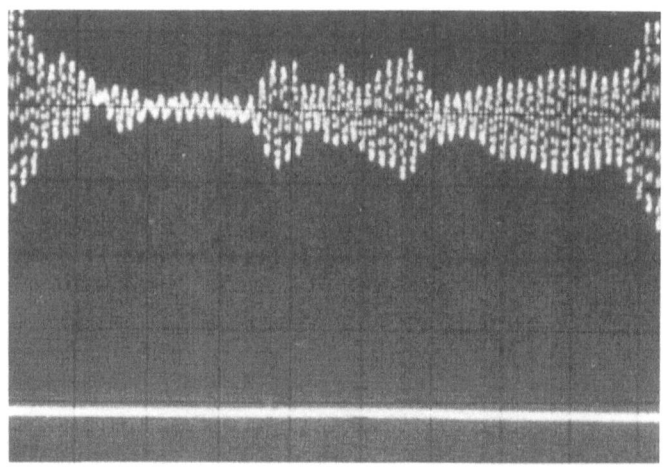

Fig. 24a. Signals from photomultipliers: photomultiplier 1 shows the signal from small silicon oil particles, whereas no signals on photomultiplier 2.

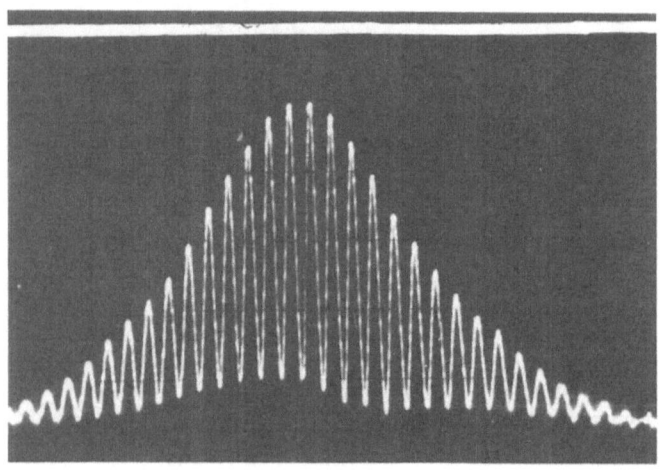

Fig. 24b. Signals from photomultipliers: no signals are shown on photomultiplier 1 when the sphere penetrates the measuring control volume. Photomultiplier 2 shows the LDA-signal obtained from the sphere.

Fig. 24a shows the signals that were recorded by the two channels of the transient recorder and replayed onto a scope for photographing purposes. The two signals originating from the air flow and the reflecting spheres are given on the scope display and it is shown that the signal from the air flow is interrupted when the signal from the reflecting sphere is displayed. There is a time gap between the signal from the small particles in the fluid and the large particles which is due to the spatial displacement between the two photodetectors used to record the signals; when the reflecting spheres enter the measuring control volume, it interrupts the signal on the first photodetector but some time is required until the moving fringe pattern has reached the second detector.

Fig. 24b shows the same signals as provided in Fig. 24a but with a higher time resolution on the transient recorder. It can be clearly seen that there is no signal from small particles in the air flow when the signal from the reflecting sphere is recorded and vice versa. This clearly demonstrates that the electronic system is able to separate the information from the two phases simulated in this way. There can be cross talks between the channels, however, if the particles does not move through the center of the measuring control volume but only touches the two laser beams. The combined usage of the automatic filter bank and the photodiode control circuit helps to avoid this kind of cross-talks. Nevertheless, still some more development work is required to improve the signal separation in complex flow cases.

4. Turbulent Pipe Flow with Solid Particles

The present section describes, as a further demonstration that LDA-measurements can be carried out in particulate two-phase flow systems, measurements in a turbulent air flow with solid particles of different diameters. These measurements were performed in the test section shown in Fig. 25 which is an extension of the system shown in Fig. 11. It consists also of a vertical pipe flow test section through which filtered air was driven which was supplied from a regulated pressure system. The supplied air contained enough particles to measure the air flow in the vertical glass tube so that no seeding problems existed. Glass particles of different diameters were added to this air flow via a venturi orifice that supplied the sub-pressure to suck the particles into the flow. A wire mesh and a flow straightener ensured an equal distribution of the particle and also helped to regulate the disturbed flow from the venturi orifice. A zyclon at the end of the test section permitted the glass particles to be separated from the air flow and re-used as indicated in Fig. 25. Particles of diameters d_1 = 100 μm, d_2 = 200 μm, d_3 = 400 μm and d_4 = 800 μm were employed in the present experiments.

368 F. Durst

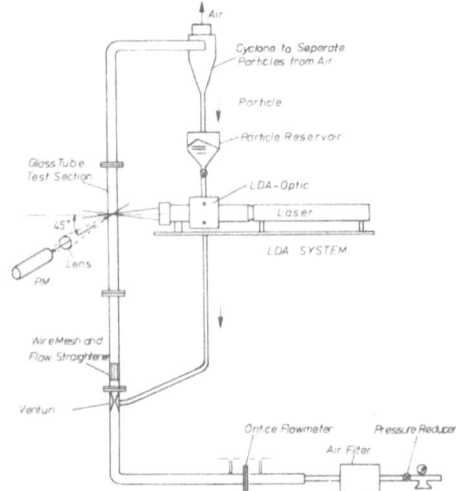

Fig. 25. Schematic diagram of the test section and the LDA-system for two-phase flow measurements.

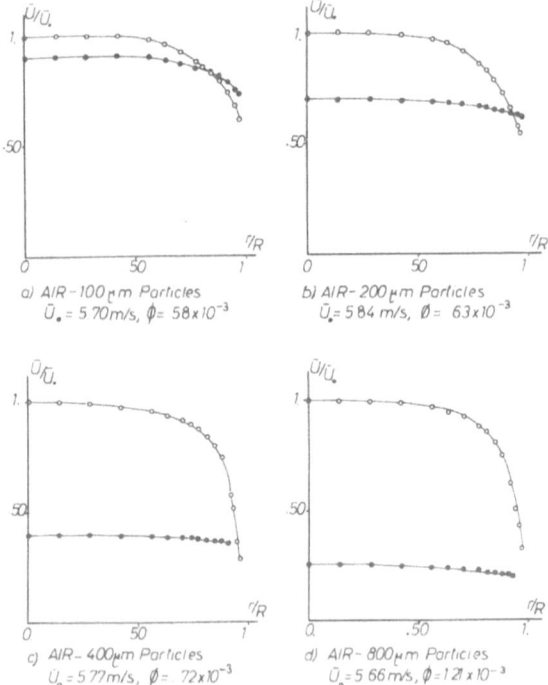

a) AIR - 100 μm Particles
$\bar{U}_o = 5.70 m/s$, $\emptyset = 5.8 \times 10^{-3}$

b) AIR - 200 μm Particles
$\bar{U}_o = 5.84 m/s$, $\emptyset = 6.3 \times 10^{-3}$

c) AIR - 400 μm Particles
$\bar{U}_o = 5.77 m/s$, $\emptyset = 7.2 \times 10^{-3}$

d) AIR - 800 μm Particles
$\bar{U}_o = 5.66 m/s$, $\emptyset = 12.1 \times 10^{-3}$

Fig. 26. LDA-measurements in two-phase flows. Mean velocity profiles of air flow and glass spheres in upward direction pipe flow.

Fig. 26 shows examples of the results obtained in the aforementioned test section. Mean velocity distributions are shown across the entire test section for the air flow and the stream of suspended particles. The increasing particle velocity with decreasing particle diameter is clearly given. With increasing particle diameter, the particle velocity profile becomes constant across the pipe diameter. At smaller particle diameters, the results show lower mean velocities of the particle in the center of the pipe but higher velocities at the pipe walls. This seems to indicate that the action of turbulent flows on solid particles shows basic differences that are dependent on the particle diameter. This is presently under study by the author and one of his colleagues (Professor R.S.L. Lee, State University of New York at Stony Brook).

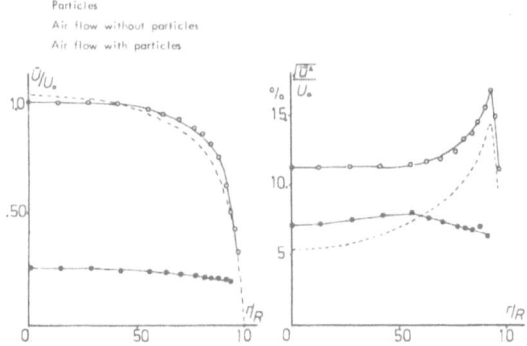

Fig. 27. Variations of the air flow due to the presence of particles.

Fig. 27 shows the effect which solid particles have on the properties of the air flow. The velocity profiles and turbulence intensity profiles for the axial velocity fluctuations are shown in this figure with and without particles. With particles, the mean velocity profile flattens in the center of the pipe but becomes steeper in the region near to the wall. The figure also shows that the turbulence intensities in the air flow are increased due to the particle motion. These are typical results for large particles. For small particles reversed actions can be observed.

5. Conclusions and Final Remarks
The present paper is a result of an attempt to summarize the work that has been carried out to study particulate two-phase flows by laser-Doppler techniques. It is shown that measurements can be carried out by laser-Doppler anemometers in particulate two-phase flows if the shape and the size of the particle can be defined by a single parameter as it is the case for particles of spherical shape. In this case, the basic relationships to carry out size and velocity

measurements are provided and explanations are given that indicate that studies of local particle size, particle concentration and particle velocity can be performed. The paper emphasizes velocity measurements and optical and electronic systems are described that are under development at the Sonderforschungsbereich 80 of the Universität Karlsruhe. Verification experiments are provided that clearly demonstrate the correct functioning of parts of the system. These increase the confidence in the measurements that are provided for the flow consisting of a turbulent air stream containing solid glass particles. Examples of results are presented for a vertical, turbulent pipe flow to which glass particles were added. The particle diameter was varied in these experiments and results are provided for different diameters.

In spite of the descriptions of successful applications, the present paper stresses the fact that LDA-systems for two-phase flow measurements are not completely developed and further developments will be required in order to ensure correct measurements in complex two-phase flow situations. Experiences with existing LDA-systems have shown that different controls have to be employed in two-phase flow signal-processing electronics in order to ensure cross talks between the channels that supply the information on the two phases. Simple amplitude discriminations are insufficient. The present paper suggests amplitude discriminators to be combined with automatic filter banks and photodiode controls to minimize contributions from large particles to the information from small particles and vice versa.

Irrespective of the need for further developments, the present paper shows that laser-Doppler systems have great potential in studying particulate two-phase flows. They are limited, however, to small particle concentrations. If the particle concentration causes frequent beam interruptions by particles outside the measuring volume, the measuring time is increased until the concentration has reached a limited value beyond which no measurements can be made. This value is dependent on the test section size; it can be large for small test sections but has to be small when the test section dimensions increase. Void fractions of a few percent have been studied by the author for test sections in the order of 100 mm in diameter.

Acknowledgements

The present report summarizes work on LDA-measurements that has been carried out in the Sonderforschungsbereich 80 at the University of Karlsruhe. The author is very thankful to the different contributions which his collaborators have made. In particular the work by J. Heidbreder, A. Taylor, C. Tropea and M. Zaré need to be mentioned, Miss G. Bartman and Mr. M. Gerber were of great help in completing the manuscript.

The author likes to acknowledge the financial support of this work by the Deutsche Forschungsgemeinschaft.

Nomenclature

Δx	Fringe separation distance
f_D	Doppler shift frequency
$\{K_i\}$	Unit vector
L	Distance between the probe volume and the detector
n_1, n_2	Refractive indices of the two phases
R	Radius of curvature of the interface in the plane of the two beams
$\{U\}_i$	Velocity vector
U	Velocity component
2β	Angle between two interfering light rays
λ	Wave length of the laser light
φ	Half angle between two crossing laser beams

Subscripts

D	Detector, Doppler
I	Incident
n	Normal
R	Reflected
T	Transmitted
‖	Parallel
⊥	Perpendicular

References

1. ASME - Symposium - Proceedings "Two-Phase Flow Instrumentation", Minneapolis, Minnesota, USA, August 1969
2. BRAND, F.L. "Akustische Verfahren zur Messung von Strömungsgeschwindigkeiten", VOITH Forschung und Konstruktion, Heft 21, Aufsatz 6, 1973
3. CARLSON, C.R. & PESKIN, R.L. "One-Dimensional Particle Velocity Probability Densities Measured in Turbulent Gas-Particle Duct Flow", *International Journal of Multiphase Flows*, Vol. 2, pp. 67-78, 1975
4. CLASS, G., LOEFFEL, R. & REIMANN, J. "Mass Flow Measuring Techniques in Transient Two-Phase Flow", Specialist Meeting on Transient Two-Phase Flow, OECD/ Nuclear Energy Agency, Toronto, Canada, August 1976
5. DAVIES, W.E.R. "Velocity Measurements in Bubbly Two-Phase Flows Using Laser-Doppler Anemometry", Institute for Aerospace Studies, University of Toronto, Part I & II, UTIAS-Technical-Notes No. 184 & 185, 1973
6. DELHAYE, J.M. "Mesures en écoulement diphasiques", B.I.S.T., Commisariat à L'Energie Atomique, No. 187, 1974
7. DURST, F. & HEIDBREDER, J. "Eine automatische Filterbank für LDA-Signale", Sonderforschungsbereich 80, SFB 80/M/140, Universität Karlsruhe, January 1979
8. DURST, F., TAYLOR, A. & WHITELAW, J.H., "Experimental investigations of a bubble driven axisymmetric, low Reynolds number flow", in preparation

9. DURST, F. & ZARE, M. "Bibliography of Laser-Doppler Anemometry", Universität Karlsruhe, Sonderforschungsbereich 80, SFB80/M/44, 1974

10. DURST, F. & ZARE, M. "Laser-Doppler Measurements in Two-Phase Flows", *Proceedings of the LDA-1975 Symposium, Copenhagen,* Denmark, August 1975

11. FARMER, W.M. "Observation of Large Particles with a Laser-Interferometer", *Journal of Applied Optics,* Vol. 13, No. 3, pp. 610-622, 1974

12. GALAUP, J.P. "Contribution à l'étude des Methods de Mesure en écoulement diphasique", Dr.-Ing. Thèse, Université de Grenoble, 1975

13. HEWITT, G.F. & LOVEGROVE, P.C., "Experimental Methods in Two-Phase Flow Studies", Heat Transfer and Fluid Flow Service, Atomic Energy Research Establishment, Harwell, England, 1975

14. INONE, A., AOKI, S. & KOGA, T. "On the Void and the Velocity Profiles of Two-Phase Bubble Flow in a Vertical Pipe", Research Laboratory for Nuclear Reactors, Tokyo Institute of Technology, Japan 1976

15. KOBUS, H. "Bemessungsgrundlagen und Anwendungen für Luftschleier im Wasserbau", Wasser und Abwasser in Forschung und Praxis, Erich Schmidt Verlag, 1973

16. KOLANSKY, M.S., WEINBAUM, S. & PFEFFER, R., "Drag Reduction in Dilute Gas Solid Suspension Flow: Gas and Particle Velocity Profiles", Third International Conference on the Pneumatic Transport of Solids in Pipes, Paper C1, April 1976

17. LADING, L. "Two-Phase Measurements Utilizing Laser Anemometer", Danish Atomic Research Energy Commission, Research Establishment Risö-M-1368, 1971

18. NISHIKKAWA, K., SEKOGUCHI, K., NAKASATOMI, M. & KANENZ, A., "Two-Phase Annular Flow in a Smooth Tube and Groved Tubes", Gas-Liquid-Flow, pp. 47-80, Plenum Press, 1969

19. OHBA, K., KISHIMOTO, I. & OGASAWARA, M., "Simultaneous Measurements of Local Liquid Velocity and Void Fraction in Bubbly Flows Using a Gas Laser: Part I: Principles and Measuring Procedure, Technology Reports of the Osaka University, No. 1328, pp. 547-556, 1976. Part II: Local Properties of Turbulent Bubbly Flow", Technology Reports of the Osaka University, Vol. 27, No. 1358, pp. 229-238, 1977

20. PESKIN, R.L. & KAN, C.J., "Numerical Simulation of Particulate Diffusion", Rutgers University, New Brunswick, N.J., USA, 1972

21. POPPER, J., ABUAF, N. & HETSRONI, G., "Velocity Measurements in a Two-Phase Turbulent Jet", *International Journal of Multiphase Flow,* Vol. 1, pp. 715-726, 1975

22. RIETHMULLER, M.L., "Optical Measurements of Velocity in Particulate Flows", Von Karman Institute for Fluid Dynamics, 1973

23. SATO, Y., & SEKOGUCHI, K., "Liquid Velocity Distribution in Two-Phase Bubble Flows", *International Journal of Multiphase Flows",* Vol. 2, pp. 79-95, 1975

24. SOO, L.S., "Fluid Dynamics of Multiphase Systems", Blaisdell Publishing Company, A Division of Ginn and Company, 1967

25. STEVENSON, W.H., DOS SANTOS, R. & METTLER, S.C., "A Laser Velocimeter Utilizing Laser-Induced Fluorescence", *Applied Optics Letters,* 27, pp. 395, 1975

26. STÜMKE, A. & UMHAUER, H., "Local Particle Velocity Distributions in Two-Phase Flows Measured by Laser-Doppler Velocimetry", *Proceedings of the Dynamic Flow Conference 1978,* Marseille & Baltimore, September 1978.

Measurement of Particle Size and Concentrations Using LDV Techniques

by

W. Michael Farmer
The University of Tennessee Space Institute
Tullahoma, Tennessee

Introduction

Laser velocimeters are devices used to measure particle velocity in fluid flows. A wide variety of applications and increased theoretical understanding have emphasized the operational significance of the physical characteristics of particles measured with these devices. First, either the particle must follow the fluid flow or a well defined correlation between particle velocity and fluid velocity must be known if fluid velocity is to be estimated. Second, the light scattering characteristics of the particle strongly affect measurement accuracy through signal-to-noise ratio effects.

The dynamics of spherical particles in fluids are well understood and a correlation between fluid velocity and particle velocity can be made when particle diameter and density are known. For particles of nonspherical shapes or unknown density, response to fluid velocity can be expressed in terms of an "equivalent" spherical particle diameter of unit density.

Particle scattering characteristics are predictable when particle shape, size, index-of-refraction and wavelength of the illuminating light are known. It can be shown that those particles which scatter to produce the highest signal-to-noise ratio (S/N) in a given instrument are those which are often preferentially measured. In some cases conflicting requirements arise between those particles which can be measured optimally and those which follow the fluid flow velocity with a known degree of correlation. To circumvent this impasse there are obvious alternatives:

1. to artificially seed the fluid with particles of known density and size which have relatively large scattering cross-sections
2. measure the naturally occurring particle sizes and mass densities in addition to velocity or
3. assume that all measurable particles are adequately following the fluid flow.

The first alternative is the most straightforward solution to determining the particle characteristics of the fluid but is often impractical to implement, particularly where fluid volumes are large. The third alternative is most often followed and in many cases can be used as a reasonable approximation. The second alternative can be expected to ultimately yield the most reliable velocity measurements when artificial seeding is impractical or undesirable.

Aside from the fact that a knowledge of particle size and density are necessary for accurate velocity measurements, a large number of cases arise where the measurement of particle size, number density, and population distribution are the required measurements. The techniques to measure particle size are numerous and imposing, ranging from the electron microscope or complicated and elegant light scattering analysis, to the simple mechanical methods of cyclone or seive analysis. When constraints are imposed requiring that dynamic particles be measured in situ, the applicable techniques reduce to a very few direct optical and radiometric light scattering techniques. One approach to the measurement of dynamic particle size which has received considerable attention during the past five years has been the determination of particle size from the shape of a real fringe type laser velocimeter signal. Initial research suggested that for certain optical geometries, the signal visibility (defined as the AC signal amplitude divided by the mean signal amplitude) could be related directly to particle diameter independent of the index-of-refraction of the particle. It was suggested that with this technique 1) a measurement of size could be determined independent of signal magnitude (providing there existed sufficient S/N), 2) typical radiometric calibrations involving the light source and photodetector were unnecessary and 3) since the particle size was measured relative to a parameter which was a direct function of the velocimeter optical geometry — the fringe period — the same instrument could be adjusted to cover a particle size range from submicron to millimeters. Because the measurement was from an LDV signal, velocity and particle size would be measured simultaneously, leaving little doubt concerning the correlation between fluid and particle velocity. Such a technique thus appeared to offer an extremely powerful and useful tool for determining particle size for otherwise impossible applications. Work which has followed from the initial research and reported in the open literature has been primarily analytical due to the lack of signal processors which could measure the required signal shape parameters, although some experimental work which has been reported reflects signal shape measurements obtained for the most part with storage or computing oscilloscopes.

Spectron Development Laboratories Inc. (SDL) has developed and applied a signal processor capable of measuring the laser velocimeter signal visibility and signal time period. A visibility processor of similar basic design has also been developed at the University of Tennessee Space Institute (UTSI). These applications have included water mists in fog simulations, atmospheric dust measurements, measurements of particles in supersonic dust erosion test facilities, solid rocket motor exhausts, and smokes. This instrumentation has been interfaced with microprocessor computer memories making instruments capable of taking large numbers of measurements in histogram formats and at high data rates. Data obtained with these instruments has often agreed (within experimental error) with that simultaneously obtained with other methods. In some cases this data appears at variance with computations and predictions made in some other studies and that the initial visibility computations apply over a much broader range of optical system and particle parameters than these studies predict.

An examination of research pertinent to this particle sizing technique published during the past three years shows studies in which the Lorenz-Mie scatter functions have been computed for a sphere illuminated by two coherent beams. These scatter functions are then integrated over a solid collection angle which is comparable to that subtended by typical laser velocimeter optical systems (a calculation not performed in the initial research). Such calculations must be performed by large digital computers which are capable of evaluating the lengthy series of Bessel functions and Legendre polynominals involved in the Lorenz-Mie solution of light scattered by a sphere. Some general conclusions drawn by these calculations are:

1.   The visibility depends on the size of the solid collection angle subtended by the receiver.
2.   For paraxial receiver geometry and very small solid collection angles the visibility is nearly 1 and independent of particle diameter.
3.   The visibility is a sensitive function of index-of-refraction, being significantly different for highly absorptive and highly refractive particles.
4.   No á priori correlation should be expected for visibility measured in the forward scatter direction with that measured in the backward direction.

Though meager, experimental results reported in the literature have been highly variable. In some cases it has agreed well with that predicted in the original research. In other cases, however, large differences between theory and experiment have been found which cannot be easily explained.

In this lecture, we will review past work which has been performed in this area, present additional unpublished data, and summarize how existing analytical predictions compare with experimental measurements.

Review of Previous Research

References 1 through 30 list those reports and publications which are available for general scrutiny or have been referenced in the open literature. References 31-37 list technical reports which were available for the preparation of this lecture but are yet to be published openly.

The original research which suggested that particle size measurements could be extracted from a measurement of the signal shape from a dual scatter LDV interferometer was developed independently and at about the same time by several different authors [2, 11, 12, 18]. Of this initial work, that of Farmer and Fristrom et al., was perhaps the most detailed and presented the only calculations for the relationship of signal shape to particle size for spheres[12, 18]. Both Farmer and Fristrom suggested that the signal shape parameters be called the signal "visibility" after Michelson's definition of interferometer fringe contrast. While their conclusions were the same, Farmer and Fristrom developed their calculations from two different approaches. Fristrom assumed that the signal was the result of an infinite ensemble of point particles in the shape of a sphere or a cylinder. Fristrom suggested in this work that the particle size be determined by adjusting the fringe period in the interferometer until a zero was reached in the visibility function and presented data from wire measurements that showed that the zero values of the visibility function did indeed fall where the theoretical calculations indicated that they should. Later, Holly apparently unaware of the work by Fristrom was to propose a similar device for the continuous measurement of fiber diameters and precision slits[19]. He showed experimentally that the zeros in the visibility function predicted for fiber diameters or slits widths were within 0.1 % of that measured by other means. Farmer's approach to the calculation of the visibility function was to assume that the light scattered by a particle was observed paraxially along the bisector between the beams and that the observation aperture was of a sufficient size that the functional dependence of the scattered radiation patterns was the same[12, 16]. Under these assumptions, the calculations showed that the scattered light intensity appeared to be similar to that from a single particle but with the incident illumination non-uniformly distributed over the cross-sectional area of the particle. Farmer suggested that the observed intensity scattered by the particle would reflect a spatial average of the incident illumination across the particle's cross-section. The results Farmer obtained using this approach were that the visibility function V for a sphere was

$$V = 2J_1 (\pi D/\delta)/(\pi D/\delta) \qquad\qquad [1]$$

while for a cylinder of length ℓ oriented perpendicular to the fringe pattern it was

$$V = \sin(\pi\ell/\delta)/\pi\ell/\delta \qquad\qquad [2]$$

J_1 is a first order Bessel function of the first kind, D is a spherical particle diameter, ℓ is a cylindrical particle length and δ is the fringe period. Both of these functions were identical to those derived by Fristrom and equation [2] is identical to results obtained by Altman and Durst for similar assumptions[2, 11]. In contrast to Fristrom, Farmer suggested that the signal visibility be measured directly for a fixed fringe period and related to particle size rather than adjusting the fringe period to a zero in the visibility function. In later experimental work for particles ranging in diameter from 20 - 200 μm, Farmer was able to obtain reasonable agreement with theoretical predictions for measurements made paraxially in the forward direction[13, 14]. He did observe that apparently because of the reflective properties of the particles, the visibility function was very sensitive to the paraxial alignment of the scattered light collecting lens (also called the "receiver"). When the axis of the receiver was displaced slightly from the axis of the transmitter it was found that the visibility could be much larger than originally estimated[13, 14]. A calculation performed to account for the off-axis reflection characteristics of spherical particles showed that in contradiction to equation [1], the visibility could be near 1 for particles several δ in diameter if the observation were made off-axis. Durst and Zare have reported similar observations and performed measurements with very large particles[10]. Their analysis suggested that the curvature of the particle could be measured by observing the phase differential in the signal as detected at two different observation positions.

In his experimental work Farmer also found little or no dependence of the visibility on the size of the solid collection angle or the shape of the collecting aperture. These experiments were performed under tightly controlled conditions in terms of known particle size and trajectory through the probe volume. This was achieved by mounting the particles to be measured on AR coated optical flats which were pushed on known trajectories through the probe volume. The particle diameters were initially measured with a microscope. Visibility measurements in these experiments were made from signals recorded using a storage oscilloscope.

Robinson and Chu repeated Farmer's experiments but covered a particle size range down to about 1 micron[29]. They also presented a different approach to the calculation of the visibility function. Using a Fourier optics approach,

Robinson and Chu showed that the visibility computed from the light scattered in the forward half space could be expressed as

$$V = \tilde{T}(S,0)/\tilde{T}(0,0) \qquad\qquad\qquad\qquad\qquad\qquad [3]$$

where $\tilde{T}(S,0)$ is the Fourier transform of the object cross-section evaluated at a spatial frequency $S(S=1/\delta)$ and $\tilde{T}(0,0)$ is the Fourier transform evaluated at $\underline{0}$ spatial frequency. For a receiver of infinite diameter, equation [3] was found to yield results identical to that previously derived[2,11,12,18]. However, it was found that when the size of the collecting aperture was reduced and the scattered light distribution computed from well-known scalar diffraction theory, the visibility tended toward a value of 1, independent of particle size. If the collecting aperture was made large enough to observe both the forward scatter lobes from each of the illuminating beams, then it was found that the visibility agreed with that computed through equation [1]. This led Robinson and Chu to conclude "that the dependence of visibility arises mainly from the interference between the two forward scatter lobes"[29]. Although Robinson and Chu obtained experimental results which seem to verify this conclusion and although for sufficiently large collecting apertures ($F \cong 2$, for a 2-μm particle) their visibility measurements agreed with Farmer's; their conclusion concerning the requirement that the light collection system observe the main forward scatter radiation lobes was inconsistent with Farmer's experiments and appeared to be so for those of Fristrom[12,18]. The scattered light collection geometry of Farmer was such that most of the measurements were obtained by blocking the primary diffraction lobes[12]. This also appears to be the case for the measurements made by Fristrom[18].

Because the aperture effects on the visibility have been predicted in a number of research papers, examples of these results are shown in Fig. 1[29] (see also Ref. 1,6,8,20,21, and 27). Fig. 1 shows the predictions of Robinson and Chu for values of the visibility for different particle sizes as a function of a dimensionless aperture parameter $R = D/2\lambda F$ where F is the F number of the receiver. The dashed line at $R = 1.35$ represents the approximate limiting value for which equation [1] is still accurate. This figure suggest that for very small values of R (i.e. small observation apertures) no light is scattered in the paraxial direction when a particle is centered over a dark fringe and this is independent of the size of the particle. Chu and Robinson also performed similar calculations using the Mie solution for scatter by spheres and again reached the same conclusions[6] (see Fig. 1). This work indicated that the visibility strongly depended on the particles' index-of-refraction. Jones has performed Mie scatter computations for scatter in the plane of the illuminating beams and has reached

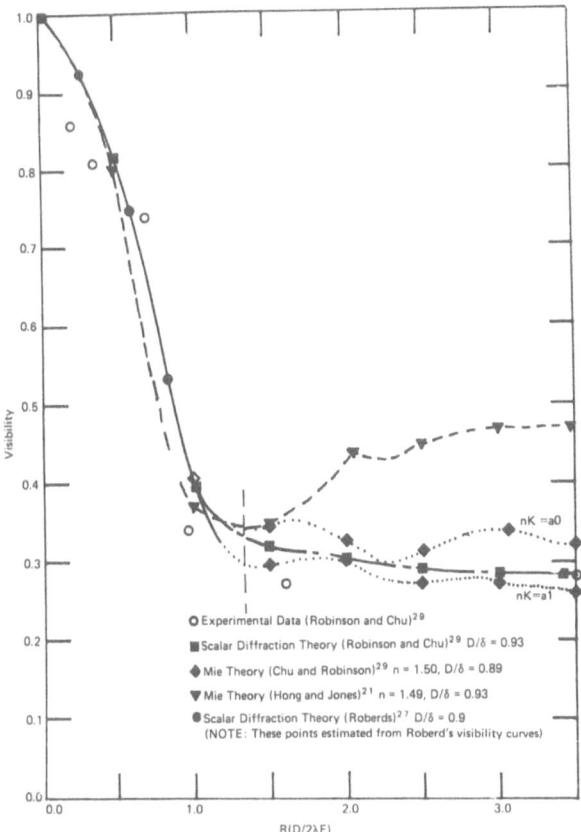

Fig. 1. Visibility as a function of collection aperture size.

similar conclusions[22]. Jones' computations showed the interesting result that when the particle is centered over a bright fringe the scattered radiation is symmetric about the bisector between the beams; when it is half way between a bright and dark fringe it is asymmetric, and when it is centered over a dark fringe it is again symmetric but no light propagates in the direction of the beam bisector. It is interesting to note that Hong and Jones present calculations that suggest that the index-of-refraction ignored by the scalar diffraction theory calculations of Robinson and Chu significantly affects the predicted value of the visibility[21] (see Fig. 1). Experimental data reported by Robinson and Chu do not appear to support such a conclusion[29]. In fact Robinson and Chu noted how remarkably well diffraction theory calculations seemed to support experimental measurements of particles as small as 2 microns.

Roberds independently computed the visibility function using diffraction theory in the same fashion as Robinson and Chu[27,28]. However, his computations included those for a scattered light receiver with a central circular beam stop geometry. This geometry was identical to that used by Farmer[13]. His computations suggest that for a given particle size and a centrally stopped observation geometry, the visibility can be much lower than that predicted by equation [1]. These predictions are in conflict with Farmer's early experimental measurements. Roberd's experimental work was performed using a Bragg Cell beam splitter and thus reflects measurements involving moving interference fringes as opposed to those of, for example, Farmer or Robinson and Chu or Hong and Jones who have reported stationary fringe measurements[13, 16,20,21,29] (in theory there should be no difference). The results of Roberds reflect measurements obtained with an electronic visibility signal processor developed at ARO, Inc. This device was designed to function with a Bragg Cell type beam splitter. Roberds has also reported backscatter measurements which appear to reflect qualitative agreement with Chu and Robinson, i.e. large fluctuations in the visibility for presumably small variations in particle size[6,26, 27]

Hong and Jones have performed calculations on the effects of limiting apertures and particle indices of refraction[20,21]. They have reported experimental results for small aperture-off-axis forward scatter measurements of glass spheres in the size range of 1-10 microns to support their conclusions[20]. Their experimentally determined (through visibility measurements) size distributions were compared with those measured with optical and electron microscopes. The comparisons are favorable but not definitive.

Boiarski et al. have reported visibility measurements of kerosene droplets in a fuel spray[3]. The size range was approximately 20-130 microns. They compared their size distribution measurements (using equation [1]) obtained in this manner with those obtained from MgO impact measurements and mean size determination from forward scatter intensity ratio measurements. Their visibility measurements were made with a computing oscilloscope. Agreement was found to be quite good between the sizing techniques even though the measurements were made 10° off the bisector between the beams in the backscatter direction.

By comparison, backscatter visibility measurements made by Orloff et al. have indicated poor agreement with that predicted by equation [1][26]. In these experiments backscatter visibility measurements were made of polystyrene 6.8-9.8 μm diameter spheres in water. The fringe period was adjusted for various values of D/δ in the neighborhood where equation [1] predicts the first zero in the visibility function. Orloff et al. found in these measurements that the visibility did not pass through zero and was consistently higher than that predicted by equation [1]. In fact, the measured visibilities appeared to be displaced from those predicted by equation [1] by a constant amount with the minimum value apparently falling exactly where the first zero was predicted. When this data is compared to the computation of Chu and Robinson, the large oscillations predicted for the backscatter visibility are not present[6].

Adrian and Orloff published computations and experimental data which seemed to support work which each had done separately and reported in 1975[1,26]. In some cases Adrian's computations seemed to be verified by Orloff's data while in others agreement is not as good as might be surmised by Adrian's detailed calculations. It is interesting to note that in the data presented in this analysis the visibility function is not shown to go through zero — even for forward scatter observations. This appears to be a variance in the results reported by Farmer, Fristrom and Holly for large particles and by Hong and Jones and Robinson and Chu or Roberds for the range 1-50 microns[16,18,19, 21,27,29]. It should also be noted that qualitative comparison of this work with the calculations of Chu and Robinson do not appear to be consistent[6].

In the following section data will be presented which have been obtained by the author with the SDL and UTSI visibility processors which may help in resolving some of the apparent conflicts in the literature which have been discussed.

Summary of Unpublished Experimental Results

The results to be discussed in this section have been reported or referred to in technical   reports or have been obtained during measurements obtained for these reports[31-37]. The previously indicated visibility processors have been used to acquire the data. A variety of optical systems were involved in the measurements — these will be described along with the pertinent experimental results.

Measurements with a Monodisperse Droplet Generator

The geometry for these experiments is shown in Fig. 2. The object of the measurements was to obtain a calibration of the optical system sample space as a function of particle size, and to test the visibility theory as reflected by equation [1]. A monodisperse droplet generator was mounted on a traverse such that it could be positioned with a 25 micron positional accuracy along the X axis and 100 microns along the Z axis (the Z coordinate is parallel to the bisector between the beams, X is perpendicular to the plane of the beams, Y is perpendicular to XZ). The droplet stream could be measured by the interferom-

W. Michael Farmer

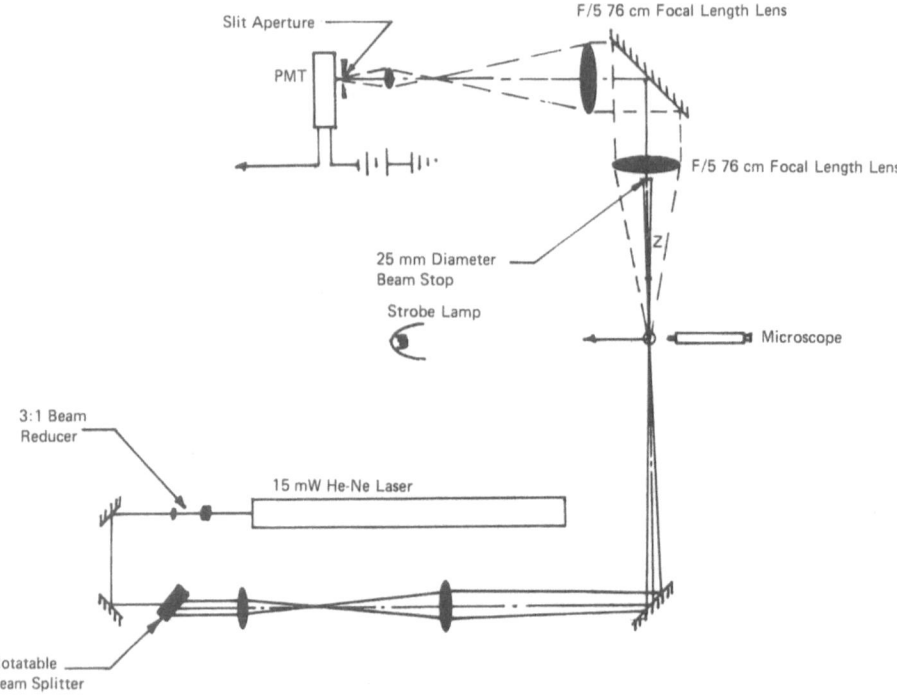

Fig. 2. Interferometer system used with monodisperse particle generator.

eter system and photographed simultaneously with a strobe lamp and microscope system. Note that Fig. 2 shows that a centrally located 25-mm diameter
beam stop was used with an F/5 scattered light collection system. When the
image of the droplet stream was focused at the split aperture, the visibility
processor gave results which were consistently within 2 - 3 % of that determined photographically. As Fig. 3 shows, these results were often better than this
value. The significance of this data is to be found in the fact that it was obtained with the central beam stop in place. The fringe period for these measurements was 180 microns. For a 150 micron particle the diameter of the primary
radiation lobe at the beam stop was approximately 7.82-mm and was centered
about the transmitted beams. The beam separation at the beam stop was approximately 2.6-mm. Hence, the 25-mm beam stop completely eliminated the
primary diffraction lobes. In fact, it eliminated more than the first three forward scatter lobes. This fact seems to contradict the conclusion by Robinson
and Chu and by Roberds that it is the interference of the forward scatter lobes
which produces the visibility function[27,29]. Roberds' diffraction theory calculations for the visibility measured with the previously described geometry, predict a mean particle size of about 100 microns.

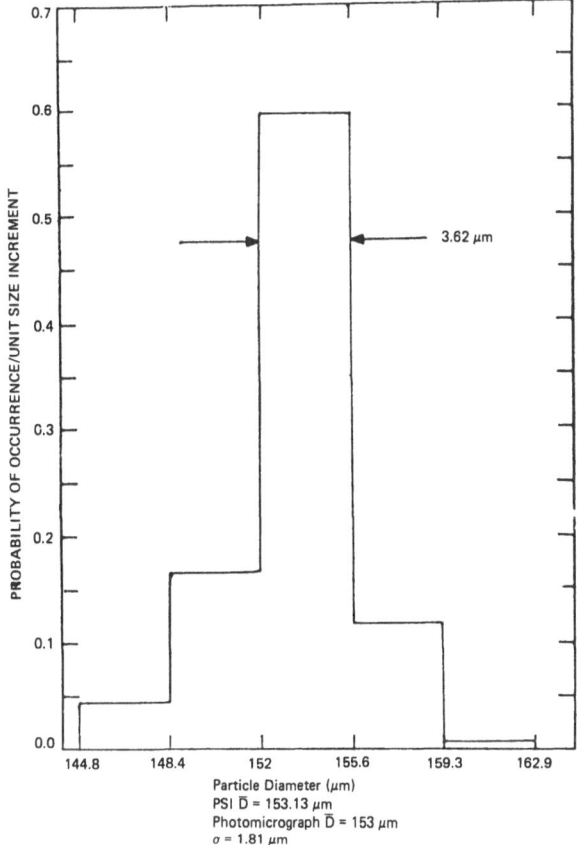

Fig. 3. Particle size distribution of monodisperse particle generator as determined from 100,000 measurements using the particle sizing interferometer.

These experiments also revealed a previously unreported characteristic of the visibility function in relation to the optical system used to obtain the visibility measurement. In these experiments, the light scattered by the droplets was imaged at an adjustable slit (oriented parallel to the image of the particle trajectory) set immediately in front of the PMT. It was initially observed that when the slit width was set for only transmittance of all the image light, the observed visibility was that predicted by equation [1]. As the slit width was reduced removing an increasingly larger fraction of the image light (effectively reducing the solid collection angle), the visibility was observed to decrease. The effect was also the same when the slit width was left open for initially good image transmission and the particle trajectory changed such that the droplet image was misfocused on the slit aperture. A systematic set of measurements was made of this effect. The droplet stream was directed through a set of

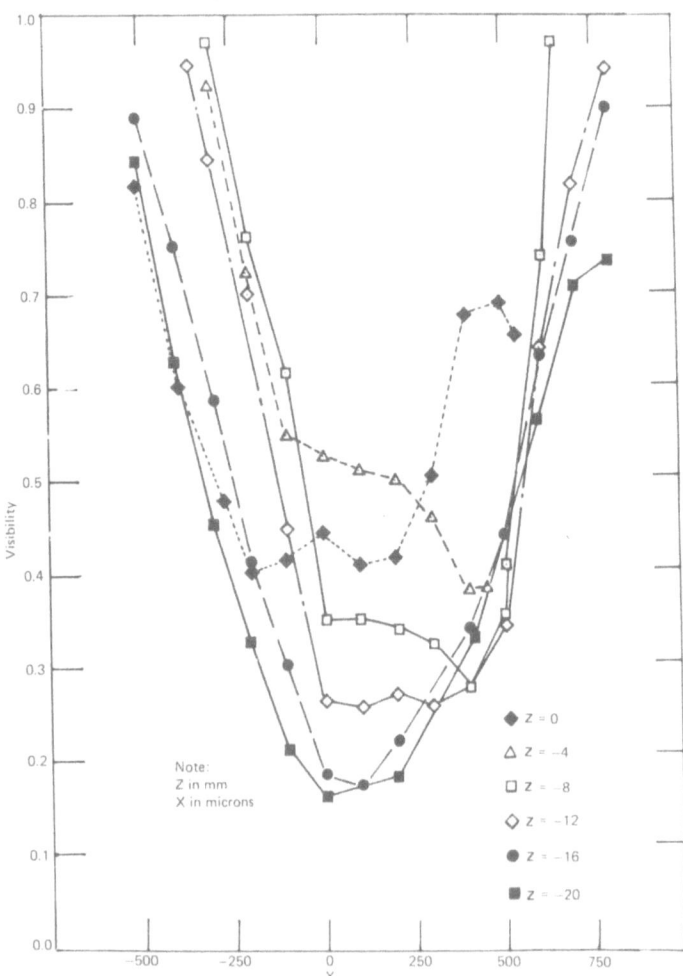

Fig. 4. Visibility as a function of probe volume co-ordinates for 153-μm water droplet.

trajectories normal to the YZ plane as defined in Fig. 2. These trajectories were spaced approximately 200 microns apart for a given XY plane and 4-mm apart along the Z axis. The signal processor required that the input signal has a preset number of cycles above a preset amplitude threshold. Thus, a sample space for a given particle and its possible trajectory set was defined by the signal processor and the signal magnitude[15]. Fig. 4, which summarizes these results, shows that the visibility varies strongly with particle trajectory. When the trajectory is near the edge of the sample space, the visibility is near 1. When the trajectory is near the center of the sample space and its image is well focused, it is near the value predicted by equation [1]. When the trajectory is centered along the middle of the sample space but away from "best focus" the visibility *decreases*. The reduction in visibility in this case appears to follow a *reduction* in the effective solid collection angle. This follows from the fact that the solid collection angle which the detector responds to is identical to that defined by the depth of field and limiting aperture of the optical system. This data was obtained for a particle diameter of approximately 120 microns which should give a visibility as predicted by equation [1] of 0.55. An average of the visibility obtained for all trajectories surveyed within the sample space defined by the signal processor yielded a value which was within 6-10 % of that predicted by equation [1]. It should be emphasized that fringe contrast variation in the probe volume was less than 1 % for these measurements. Thus, the results depend solely on the geometry of the receiver system. These results are at considerable variance with the calculations which have assumed that the solid collection angle subtended by the scattered light collection lens defines the solid collection angle viewed by the light detector. This data also suggests that the imaging properties if the scattered light observation optics must be considered. These particular experiments suggested that the visibility *decreases* rather than *increases* as predicted by previous calculations for decreasing solid collection angle. The results also suggested a possible explanation of the wide variability of experimental results when relatively few particles of a given size are sampled and the experiment is not tightly controlled (particularly where particle trajectory is concerned), i.e. insufficient numbers of particles were measured to provide sufficient smoothing of data with respect to particle trajectory.

Backscatter Visibility Measurements

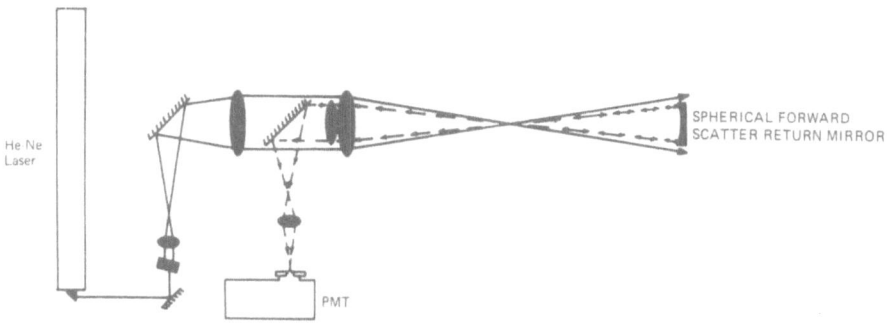

Fig. 5. Optical system geometry for 32-μm fringe period.

Two different sets of measurements but with similar optical systems are described. The optical system is shown in Fig. 5. The instrument was so designed that fringe periods between 2.3 and 500 microns could be obtained by choosing the correct lens combinations. Either forward or backscatter measurements could be made by using a collecting mirror for forward scatter or by removing it for backscatter. The laser produced approximately 15-mW of power at a 632.8 nm wavelength and the photodetector had about a 2 % quantum efficiency at the laser wavelength.

The first set of measurements were made in the cloud simulation tunnel at UCLA. This tunnel has a slow speed, well controlled, airflow. The flow is seeded with a background water mist ranging in size from 1 - 30 microns. This mist is used as the background for studying the growth of suspended ice crystals through collisions with supercooled mist particles. Particle size distributions obtained from backscatter visibility measurements for a fringe period of 30 microns were compared to those obtained by measuring the size distribution with a gelatin impact technique (see Ref. 3). In this method the mist particles were allowed to collide with a gelatin covered rod. The impact craters were measured microscopically and corrected in size to account for aerodynamic flow around the rod and collision efficiency as a function of size. Fig. 6 is a comparison of the data obtained by the two methods. The mean sizes and geometric standard deviations are within 10 % of each other. The data points show biases present in the sampling technique which have been corrected according to a technique described by Cadle[38]. In view of previous analytical predictions and some experimental measurements concerning backscatter visibility, the agreement in particle size between the two techniques is remarkable[1,6,25,27]. The size distribution does not seem to reflect any mode sizes which one might expect due to multiple values of particle size for the same visibility[40].

In the second set of measurements the optical arrangement shown in Fig. 5 was used with and without the forward scatter mirror. The particle source was a water mist generated at controlled pressure conditions from a nebulizer made by the Inspiron Corp.. First, the electronics system was used to acquire a 10^4 count histogram distribution of visibility measurements from the mist. A forward scatter observation geometry and 2.3 micron fringe period was used. No attempt was made to extract the particle size distribution or to weight the visibility increments to obtain a normalized sample space. The forward scatter collection aperture was F/4. The instrument was then changed to a backscatter observation mode with the same observation aperture and fringe period. The phototube and amplifier gains were adjusted until backscatter signals of apparently similar magnitude (as determined by oscilloscope observation) as the forward scatter signals were obtained. A 10^4 count visibility histogram was then

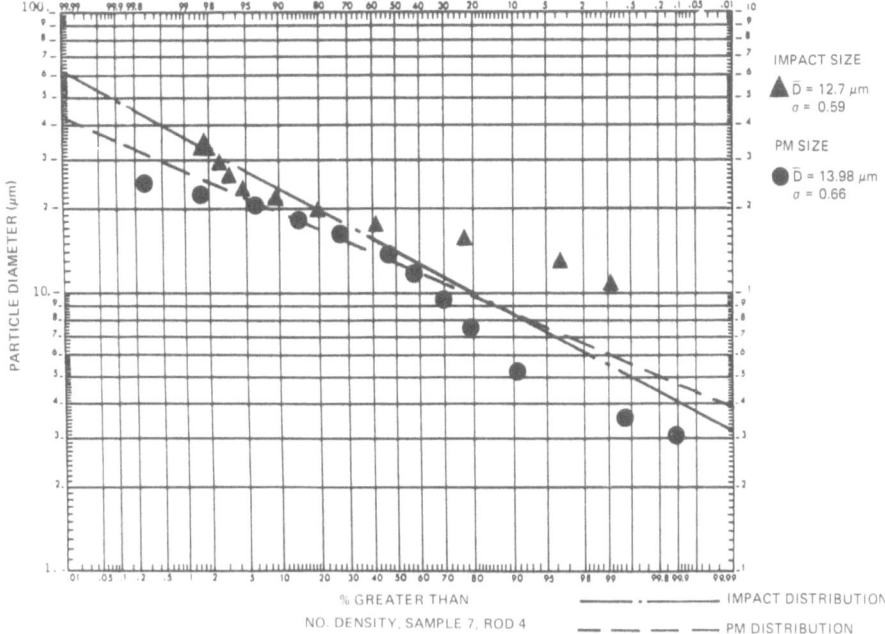

Fig. 6. Comparison of backscatter visibility size measurement with gelatin impact measurements.

taken and compared with that obtained for the forward scatter measurements. The probability distributions as a function of visibility for the two sets of measurements are plotted in Fig. 7. The figure shows virtually no difference in the visibility distributions for the forward scatter or backscatter measurements. This observation does not seem to support the conclusions from Mie theory computations which suggest strong differences in the expected visibility for forward scatter versus backscatter measurements. It is interesting to note that significant variations did appear between the two sets of measurements when the same electronic gains were used for both forward and backscatter observations suggesting possible sample bias effects due to signal-to-noise ratio.

Data such as presented in Fig. 7 cannot be taken as definitive since no experimental control was used to define the size distribution. It might be argued, for example, that a combination of aperture limiting effects and observation of different portions of the mist size distributions due to different electronic gains for the two sets of measurements could yield the same visibility distributions. While such a possibility appears unlikely it should be tested both analytically and experimentally.

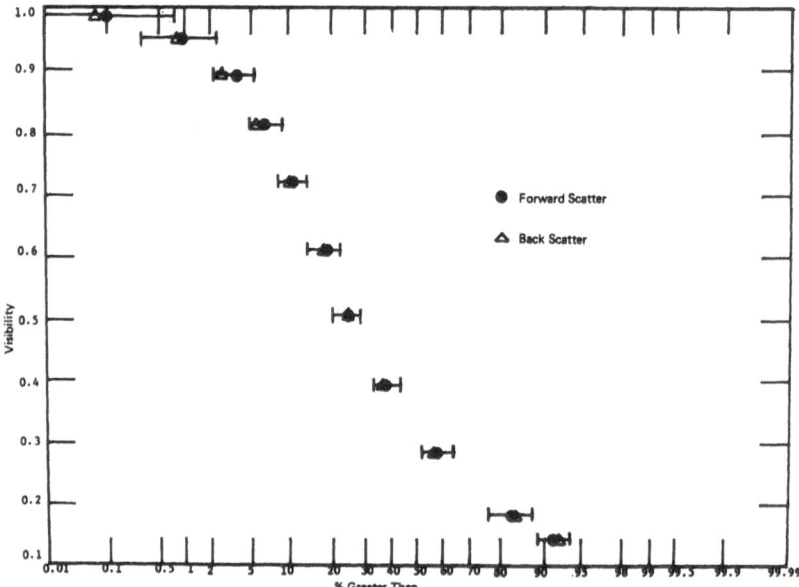

Fig. 7. Comparison of visibility distribution from a water mist for F/4 collection aperture forward and backscatter.

Aperture Limited Visibility Measurements

In these experiments, the same optical system shown in Fig. 5 was used. An aperture stop was employed to change the collection aperture from F/4 to F/60. This aperture stop could be positioned relative to the center of the large F/4 limiting aperture.

Fig. 8. compares the visibility probability distribution functions obtained for F/60 limiting aperture and for the fully opened F/4 aperture in a forward scatter observation mode. The data trends show a definite shift towards higher visibilities for the reduced aperture as other authors have suggested. It is important to note, however, that the magnitude of the shift in these observations it not nearly as large as has been predicted previously. The for maximum particle sizes expected from the nebulizers (approximately 6-7 microns) the smallest value of visibility which has been predicted is about 0.8[6,20,27,29].

It is interesting to note in Fig. 9 that there is a variance in the visibility about the plane of symmetry defined by illuminating beams when the small aperture is placed at different positions in the plane of the F/4 aperture. This effect would be expected qualitatively on the basis of previous calculations[14] with

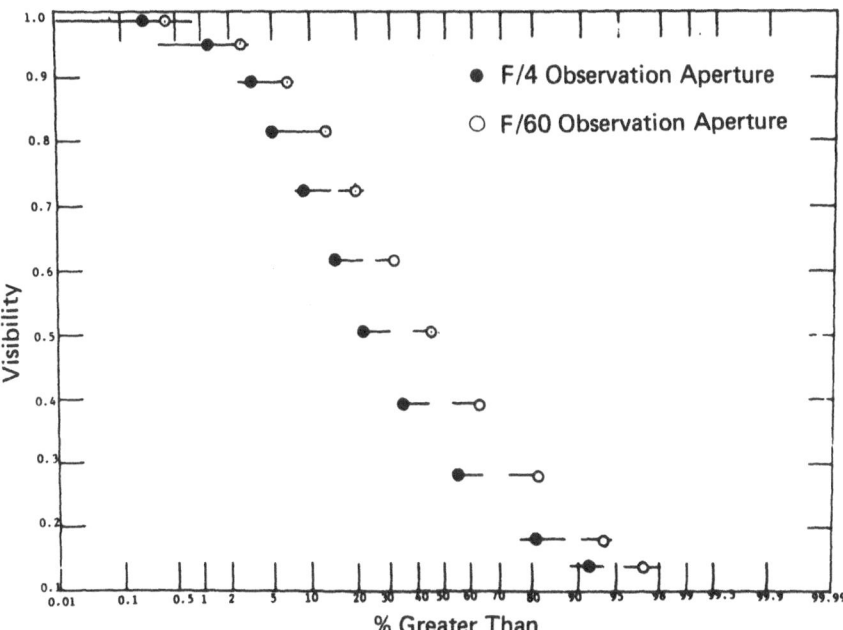

Fig. 8a. Effect of limited aperture on visibility distributions for forward scatter measurements.

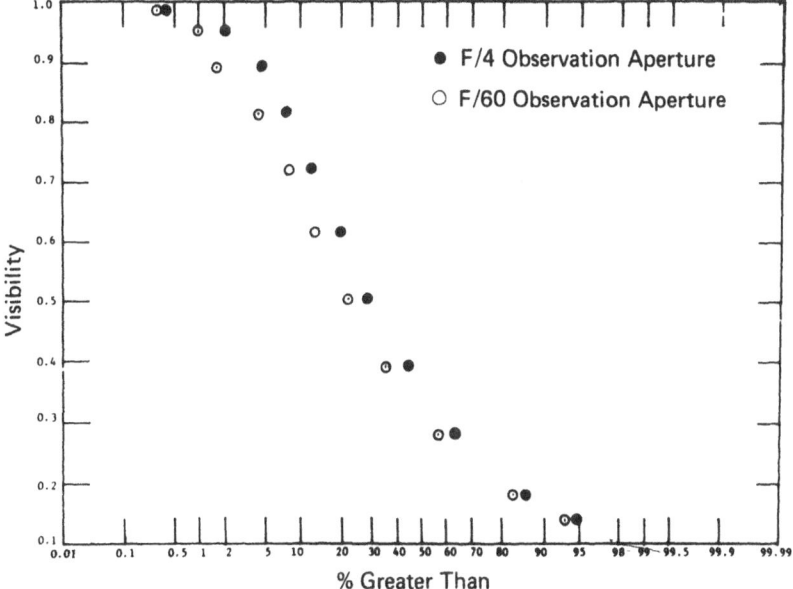

Fig. 8b. Effect of limited aperture on visibility distribution for backscatter measurements.

respect to the polarization characteristics of the signal. It is somewhat surprising that the variance is asymmetric however, and suggests that further research should examine this effect in detail.

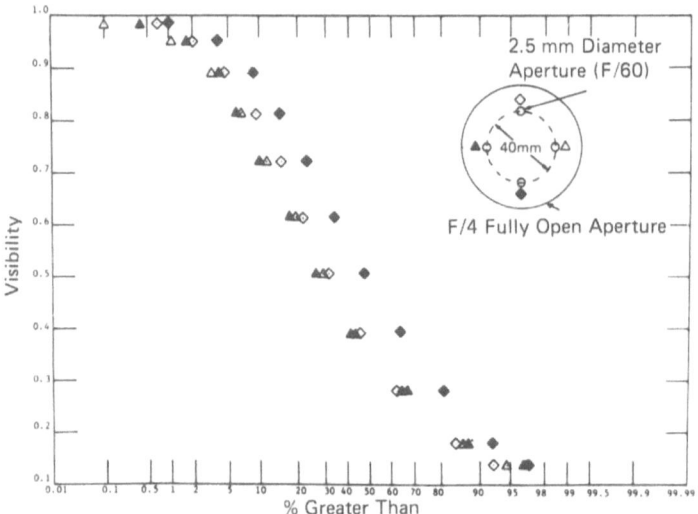

Fig. 9a. Visibility distribution for various F/60 aperture locations in F/4 forward scatter observation aperture.

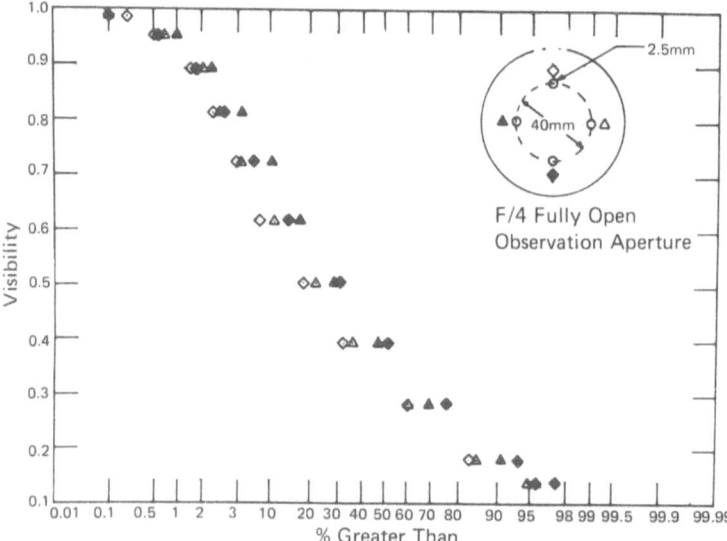

Fig. 9b. Visibility distribution for various F/60 aperture locations in F/4 backscatter observation aperture.

Particle Number Density Estimates

Measurement of particle number density or size distribution with an LDV whether through a set of visibility measurements, signal magnitudes, or particle drag measurements is difficult. The sample space (i.e. the volume in which a particle measurement can be obtained with the LDV) is a function of the scattering cross-section of the particle, measuring system sensitivity to scatter magnitude, and the logic constraints imposed by the signal processor for acceptable signals[15]. Hence, calibration involving all the optical parameters associated with both the particle (wavelength, index-of-refraction, and scatter angle) and LDV system is necessary.

A simple model will illustrate how these parameters tie together to affect the number density estimate. For particles and an optical system where the normalized scattering cross-section is nearly independent of particle diameter (F number of the scattered light receiver is less than about $\pi D/\lambda$), it can be shown that the cross-sectional area A_i of the sample space normal to the flow direction for some particle diameter D_i is given by

$$A_i = A_o D_i / D_o \qquad [4]$$

where A_o is the cross-sectional area for some calibration particle diameter D_o and set of instrument calibration settings. The number density N_i for D_i measured with the LDV is then written

$$N_i = \dot{N} f_i / V_i A_i \qquad [5]$$

where V_i is the set of velocities associated with D_i, $\dot{N}$ is the rate at which D_i is measured and f_i is the measured fraction of particles with diameter D_i. The factor $V_i A_i$ yields a set of weighting factors for a measured particle size histogram that references the set of measurements to a normalized volume. If only the probability density of the particle size distribution is of interest then relative weighting factors may be computed for the LDV system directly using equation [4] and accounting for signal processor constraints. When absolute number densities are required, then A_o for a given D_o must be measured experimentally. Fig. 10 shows a comparison of number density measurements made using a forward scatter PSI system and a commercially available optical counter. The particles are a urea-formaldehyde resin with a mean diameter of about 3-4 micrometers. The commercial optical counter is obtaining its estimate by withdrawing a sample from a large 20 m³ box while the PSI is focused directly inside the box. Differences in the measured number densities are thought to be due to the method in which the particles were extracted and

W. Michael Farmer

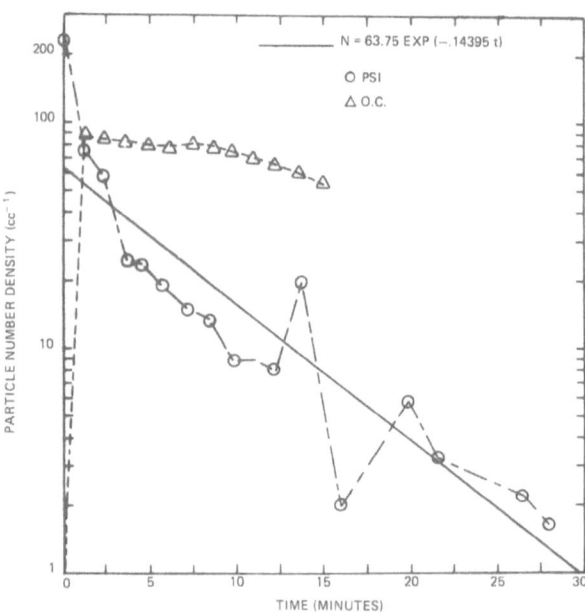

Fig. 10. Comparison of number density measurements using a PSI and a commercial optical counter.

kept in suspension in the box. The PSI measurements show a number density decreasing exponentially with time as predicted by Fuchs[39]. The measured value of A_o for this particular sample space was found to be about $1.7 \cdot 10^{-2}$ mm^2.

Summary and Conclusions
The data reported in the previous sections represents the results from a number of different experiments. These experiments have used an electronic signal processor to obtain the visibility measurements. An integral part of the signal pocessor has been a small computer system which could generate histograms of large numbers of individual particle measurements. Both forward scatter and backscatter measurements have been obtained and compared against independent particle size measuring techniques. Experiments have also been conducted for forward scatter and backscatter measurements for variable aperture observations of a well-defined but not independently calibrated particle size distribution. The results from these experiments were:

1. Excellent agreement between the particle size determined from the originally derived visibility function and microscopic measurements. These

measurements were obtained with an optical system which blocked all of the primary lobes in the forward scatter diffraction pattern.

2. A systematic trajectory scan of the sample space with a large particle stream from a monodisperse generator showed that:
    a) The visibility decreased with decreasing observation solid angle.
    b) The visibility increased for trajectories at the edges of the sample space.
    c) The average visibility for all trajectories was close to that predicted by equation [1].

3. Particle size distributions of a water mist determined from backscatter visibility measurements for the size range 2-30 microns showed reasonably good agreement with those determined from gelatin impact measurements.

4. For a particle size distribution not independently calibrated in the approximate size range 0.1 - 7 microns, experimental results showed:
    a) Identical visibility distributions for an F/4 observation aperture in both forward scatter and backscatter observations.
    b) A slight visibility dependence on aperture size.
    c) A relationship between visibility, PMT gain, and the limiting aperture.

5. Number density estimates using an LDV are possible with careful instrument calibration.

These results raise serious questions pertinent to the applicability, range, and interpretation of previous computational results involving either scalar diffraction theory or Mie scattering theory. For example, if the optical system geometry is such that diffraction contributes only a small part of the total scattered light (small F/No. receiver with a central stop blocking, say the first lobes in the radiation pattern) then results based solely on diffraction theory cannot be expected to apply. Conclusions deduced beyond the range of the model will be erroneous. Similarly, calculations using Lorenz-Mie theory for small particles have assumed perfect spheres which are materially homogeneous. For the most part, with the possible exception of liquid droplets (even then there is a question of homogeneity due to possible air bubbles) or artificially generated particles, naturally occurring aerosols are neither perfect spheres or homogeneous. Hence, while calculations involving diffraction or Lorenz-Mie theory may be perfectly valid, they may not apply to the experiment even for the particle size ranges for which they were computed. In fact the results from simpler theories might often be applied with equal accuracy. The data also indicates a need to be able to process large numbers of measurements and to be able to determine the measurements independently with a signal processor. Without such a device, visibility measurements will continue to reflect the experimentalists' biases in selecting signals for oscilloscope measurements. Furthermore, accuracy in such measurements will be limited to values of perhaps no better than ±5 % and will continue to be very sensitive to moderate noise levels.

A review of the experimental results published in the open literature also emphasizes the need for tighter experimental control when attempting to compare theory with experiment. Accurate in situ particle size measurements which can be compared with the visibility measurements are essential for model verification. A careful analysis of experimental precision and accuracy should also be made for an estimate of the quality of the measurements used to verify the theory. These two factors should make the greatest contribution toward a full understanding of this method of particle size analysis.

References

1. Adrian, R.S. and Early, W.L., "Evaluation of LDV Performance Using Mie Scattering Theory", *Proceedings of The Minnesota Symposium on Laser Anemometry,* Editor E.R.G. Eckert, University of Minnesota (1976) p. 426
2. Altman, W.P., "Incoherent Disdrometer Employing CO_2 Laser Radiation", Masters Thesis (School of Electrical Engineering, University of Pennsylvania, Philadelphia, Pennsylvania 1972)
3. Boiarski, A.A., Barnes, R.H. and Luce, R.G., "Applicability of Laser Interferometry Technique for Droplet Size Determination", Presentation at the 46th Semi-Annual Meeting of the Supersonic Tunnel Association held at Battelle Columbus Laboratories, Columbus, Ohio, September 30 and October 1, 1976
4. Brayton, D.B., "Small Particle Signal Characteristics of a Dual-Scatter Laser Velocimeter", *Applied Optics,* Vol. **13**, p. 2346 (1974)
5. Chigier, N.A., Ungut, A. and Atakan, S., "Particle Size Measurement by Laser Anemometry", Meeting of the AIAA, Washignton, D.C. (1976)
6. Chu, W.P. and Robinson, D.M., "Scattering From a Moving Spherical Particle by Two Crossed Coherent Plane Waves", *Applied Optics,* Vol. **16**, p. 619 (1977)
7. Drain, L.E., "Coherent and Non-Coherent Methods in Doppler Optical Heat Velocity Measurement", *J. Phys. D. Appl. Physics,* Vol. **5**, p. 481 (1972)
8. Durst, F. and Eliasson, B., "Properties of Laser Doppler Signals and Their Exploitation for Particle Size Measurements", The Accuracy of Flow Measurements by Laser Doppler Methods, *Proceedings of the LDA Symposium,* Copenhagen 1975, p. 457
9. Durst, F. and Umhauer, H., "Local Measurements of Particle Velocity Size Distribution and Concentration with a Combined Laser Doppler Particle Sizing System, Ref. 8, p. 430.

10. Durst, F. and Zare, M., "Laser Doppler Measurements in Two-Phase Flows", Ref. 8, p. 403

11. Durst, F., "Development and Application of Optical Anemometers", Ph.D. Thesis (Heat Transfer Section, Dept. of Mechanical Engr., Imperial College, London, S.W. 1, 1972) pp. 170-173

12. Farmer, W.M., "Measurement of Particle Size, Number Density and Velocity Using a Laser Interferometer", Applied Optics, Vol. 11, p. 2603 (1972)

13. Farmer, W.M., "Observations of Large Particles with a Laser Interferometer", Applied Optics, Vol. 13, p. 610 (1974)

14. Farmer, W.M., "Scattering From Small Particles Illuminated by Coherent Beams: Polarization Properties", Journal of Optical Society Am., 63, p. 1563 (1973)

15. Farmer, W.M., "Sample Space for Particle Size and Velocity Measuring Interferometers", Applied Optics, Vol. 15, p. 1984 (1976)

16. Farmer, W.M., "The Interferometric Observation of Dynamic Particle Size, Velocity, and Number Density", Ph.D. Thesis (Dept. of Physics, The University of Tennessee, Knoxville, TN. (1973)

17. Farmer, W.M. and Hornkohl, J.O., "Environmental Aerosol Measurements Using an Airborne Particle Morphokinetometer", EPA Technical Report No. EPA-600/3-76-087 August, 1976

18. Fristrom, R.M., Jones, A.R., Schwar, M.S.R. and Weinburg, F.S., "Particle Sizing by Interference Fringes and Signal Coherence in Doppler Velocimetry", Faraday Symposia of the Chemical Society, Vol. I, p. 183 (1973)

19. Holly, Sandor, "Lateral Interferometry — Its Characteristics, Technology and Applications", Optics Engr., 15, p. 146 (1976)

20. Hong, N.S. and Jones, A.R., "A Light Scattering Technique for Particle Sizing Based on Fringe Anemometry", Journal of Physics D. Appl. Phys. 9, p. 1839 (1976)

21. Hong, N.S. and Jones, A.R., "Light Scattering by Particles in Laser Doppler Velocimeter Using Mie Theory", Applied Optics, Vol. 15, p. 2951 (1976)

22. Jones, A.R., "Light Scattering by a Sphere Situated in an Interference Pattern with Relevance to Fringe Anemometry and Particle Sizing", Journal of Physics, D. Applied Physics, 7, p. 1369 (1974)

23. Li, Ming Chiang, "Scattering of Two Coherent Light Beams", Journal of Optic Society A., 65, p. 586 (1975)

24. Meyers, James F., "Computer Simulation of a Fringe Type Laser Velocimeter", Proceedings of the Second International Workshop on Laser Velocimetry", H.D. Thompson and W.H. Stevenson, Editors, Purdue University, March 27-19, 1974

25. Myer, F. and Mikasa, M., "Interferometric Measurement of Visibilities of Known Particle Size", Engineering Report, Harvey Mudd College, Claremont, CA, Sept., 1975

26. Orloff, K.L., Myer, F.C., Mikasa, M.F. and Phillips, J.R., "Limitations on the Use of Laser Velocimeter Signals for Particle Sizing", Ref. 1, p. 359. See also Adrian and Orloff, "Laser Anemometer Signals: Their Visibility Characteristics and Application to Particle Sizing", Applied Optics, 16, p. 677 (1977)

27. Roberds, D.W., "Particle Sizing Using Laser Interferometry", Submitted to Applied Optics, December 1976

28. Roberds, D.W. "Electronic Instrumentation for Interferometric Particle Sizing", Ph.D. Thesis (Dept. of Electrical Engineering, The University of Tennessee, Knoxville, TN, 1975)

29. Robinson, D.M. and Chu, W.P., "Diffraction Analysis of Doppler Signal Characteristics for a Cross-Beam Laser Doppler Velocimeter", *Applied Optics,* Vol. **14**, p. 2177 (1975)

30. Yule, A.J., Chigier, N.A., Atakan, S. and Ungut, A., "Particle Size and Velocity Measurement by Laser Anemometry", *AIAA Paper No. 77-214,* Presented at the AIAA 15th Aerospace Sciences Meeting, Los Angeles, CA, pp. 24-26, January, 1977

31. Farmer, W.M., Hornkohl, J.O., Tidwell, E.D. and Enis, C.P., "Particle Sizing Interferometer Measurements of Simulated Clouds I: Interferometer Calibration and Evaluation", SDL Technical Report No. SDL-75-6804 (1975)

32. Farmer, W.M., Hornkohl, J.O., Tidwell, E.D., Enis, C.P., and Blanks, J.R., "Particle Sizing Interferometer Measurements of Simulated Clouds II: Application Results", SDL Technical Report No. SDL-75-6805 (1975)

33. Farmer, W.M. and Trolinger, J.D., "Particle Sizing Techniques: Application at the Arnold Engineering Development Center Vol. II Design Study for an Icing Test Facility Particle Morphokinetometer", SDL Technical Report No. SDL 76-6061 (1976)

34. Farmer, W.M. and Hornkohl, J.O., "Particle Morphokinetometer Measurements of Ambient and Injected Water Droplets in the AEDC Dust Erosion Tunnel", SDL Technical Report No. SDL-76-6822 (1976)

35. Farmer, W.M., Schwartz, F.A. and Stallings, E.S., "Water Spray Measurements by a Particle Morphokinetometer at The General Electric Icing Test Facility in Peebles Ohio, Vols. I and II", SDL Technical Report No. SDL-76-6812 (1976)

36. Farmer, W.M. and Stallings, E.S., "Particle Size Calibrations for Nebulizers and Filters Produced by The Inspiron Company, Vols. I and II", SDL Technical Report No. SDL 76-6817 (1976)

37. Farmer, W.M., Dixon, R.L., Trolinger, J.D., and Busch, C.W., "RV Instrument Feasibility Study: Particle Sizing Interferometers", SAI Technical Report No. SAI-73-502-TT.

38. Cadle, R.D., *Particle Size,* Reinhold New York (1965)

39. Fuchs, N.A., *The Mechanics of Aerosols,* Pergamon Press, New York, 1964, pp. 250-257

40. Farmer, W.M., Harwell, K.E., Hornkohl, J.O. and Schwartz, F.A., "Particle Sizing Interferometer Measurements in Rocket Exhaust", presented at the 3rd International Workshop on Laser Velocimetry, Purdue University, July 11-13, 1978.

Experimental Studies of Two-Phase Air-Water Flows

by

Hans J. Leutheusser
Department of Mechanical Engineering
University of Toronto, Toronto, Ontario, Canada

Abstract
The paper presents a brief survey of experimental techniques; and their application to laboratory investigations of, respectively, bubbly two-phase flow in open channels, bubble streams in closed conduits, and pressure effects due to bubble deformation.

Introduction
Multiphase flow is the rule rather than the exception in engineering fluid mechanics, with examples ranging from sediment transport in rivers, to the combustion of air-fuel mixtures in prime movers, and the generation of steam in power plants. In any analysis of multiphase flow the well-known difficulties which are encountered in studies of single-phase problems are multiplied, thereby rendering the whole subject matter highly empirical and almost entirely dependent upon the acquisition of reliable experimental data.

The present paper deals with two-phase air-water flows. Problems of this type are particularly prevalent in hydraulic engineering under such varied guises as the aeration of free-surface water flows, the deployment of bubble screens for ice control and containment of oil spills, and the occurrence of air bubbles in water lines. While much work has been done in the past to clarify the dynamics of air-water flows, many technically important problems are still unresolved. The veracity of this statement is also borne out by the few case studies which are dealt with herein. In their treatment it is the viewpoint of the fluid dynamicists rather than that of the instrumentation specialist which prevails. This particular ordering of priorities should help to underline the magnitude of multiphase flow problems in engineering fluid mechanics and yet, at the same time, demonstrate their susceptibility for experimental analysis with the aid of what are essentially routine laboratory techniques.

Measurement Techniques for Two-Phase Air-Water Flows

From a practical point of view, mixture concentration (or void ratio) is probably the generally most useful information on air-water flows. However, this parameter describes only the fluid density and, hence, the mass transport of the mixture; it does not reveal the internal make-up of the two-phase system. Indeed, the same void ratio may be due to either a large number of small bubbles (to the limit of bubble emulsion), or a small number of large bubbles (to the limit of slug flow). Since the two cases differ both in bubble size and in the extent of interfacial contact area, it becomes essential, particularly in studies related to interphase mass transfer, to complement data on void ratio by measurements of the average equivalent bubble diameter.

The practical realization of the measurement needs for studies of two-phase air-water flows are varied, often intertwined, and development of refined techniques is in a state of constant flux. Nevertheless, certain operating principles have become established over the years as standards. For detailed annotated bibliographies on the subject, reference should be made, for instance, to the works of Rao and Kobus (1975) and Barczewski (1976).

Measurement of Concentration

Mechanical Methods: — A sample of the air-water mixture is withdrawn from the flow and separated into its component parts.

Electrical Methods: — These techniques exploit the differences which exist between air and water of, respectively, conductivity, dielectric constant, resistance, and heat transfer potential.

Absorption Methods: — Concentration is deduced from the attenuation of the intensity of, for instance, gamma-ray or light beams.

Measurement of Bubble Size

Direct Methods: — Individual air bubbles are extracted from the flow and passed through a volume meter.

Indirect Methods: — As exemplified by the hot-film anemometry technique, signal duration in the gaseous phase and velocity information on the liquid phase are combined to compute the diameter of a supposedly exactly bisected spherical bubble.

Photographic Methods: — Bubble size is deduced from still photographs.

Other Measurements

Velocities and turbulence characteristics of air-water mixtures are most conveniently determined by hot-film anemometry. The rapidly evolving method of laser-Doppler anemometry holds comparable potential, but tends to become inapplicable at bubble concentrations in excess of about 3 %. For the measurement of mean and fluctuating pressures, transducers of various types are suitable. An overview of the highly unsteady flow processes characterizing multiphase flow is obtainable with the aid of high-speed cinematography. In conjunction with time-correlated transducer outputs, such a pictorial record can represent a particularly powerful diagnostic tool.

Selected Case Studies

Bubbly Two-Phase Flow in Open Channels

The Problem: — Hydraulic jump is the abrupt transition from supercritical to subcritical open-channel flow. The transformation is characterized by a discontinuity in depth of flow and corresponding formation of a surface roller. Other symptoms are visible air entrainment in the form of bubbles of all sizes, intense internal mixing, and dissipation of kinetic flow energy. Because of its features the hydraulic jump appears well-suited for the purpose of recharging the oxygen content of polluted rivers and streams. However, in order to properly assess its aeration potential, information is needed on the pertinent flow characteristics of turbulence, air bubble concentration, and bubble size distribution.

Method of Solution: — The investigation, spanning several years and exhaustively documented in the technical literature (cf., e.g., Resch, Leutheusser and Coantic, 1976), was carried out in a horizontal laboratory flume. Flow properties were determined with the aid of hot-film anemometry using conical probes. Since a hot-film sensor in two-phase flow operates in each fluid phase on a different voltage level, indications of standard, indiscriminately averaging analog meters become meaningless. In order to separate the signal component corresponding to the liquid phase from the signal component arising from a bubble passage, recourse was taken to a digital method of data treatment. In this, successive signal extremes were computed and compared to a peak-to-peak fluctuation threshold. The applicable threshold level was initially selected on the basis of visual inspection of signal records. Later on in the experiments, a more sensitive discriminator tied to the notion of void ratio was employed. With the phase separation accomplished, the various turbulence parameters of the liquid phase, and the characteristics of the bubbly two-phase flow could then be evaluated numerically.

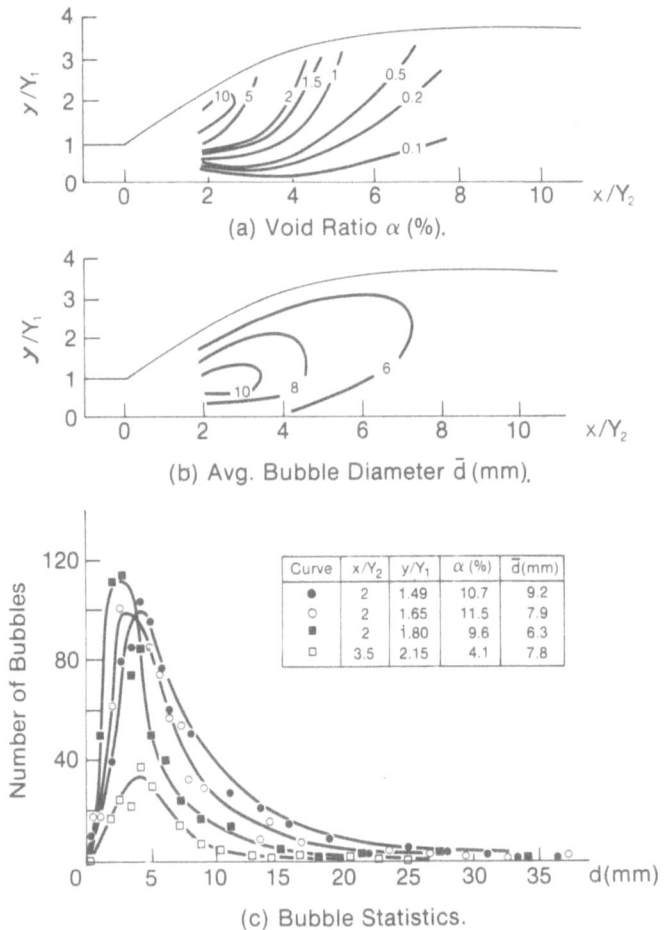

Fig. 1. Characteristics of bubbly two-phase flow in hydraulic jump (Froude number F = 2.85, undeveloped inflow).

Typical Results: — Representative data on void ratio, bubble size, and bubble statistics are compiled in Fig. 1. The results demonstrate the utility of the experimental technique employed for gaining a detailed insight into the structure of bubbly two-phase flow. It may be mentioned that it is only through experimental results of the kind presented in Fig. 1 that there exists now a proper appreciation of the interaction between the fluid dynamic and thermodynamic processes which are involved in the reaeration of water flows by hydraulic jump (Leutheusser and Ward, 1975).

Bubble Stream in Closed Conduits

The Problem: — Flow separation entails the formation of a surface of velocity discontinuity. With increasing distance downstream from the point of separation, this free shear layer has the tendency to become, at first, wavy; then, to roll up into distinct vortexlike eddies; and, eventually, to degenerate into random fluctuations. The occurrence of distinct eddies in the flow causes pressure fluctuations which, if they are of sufficient periodicity, may lead to flow-induced vibrations of the enclosing flow boundary. The danger of this to happen is apparently increased if the fluid flowing happens to be a liquid-gas mixture. This, at least, is the implication of experiments according to which pressure fluctuations in separated flow tend to become both more pronounced and more periodic under cavitation conditions. It has been speculated (Naudascher and Jezdinsky, 1972) that this phenomenon may be caused by some "lubricating" action of minute vapour bubbles in the viscous core of the separation vortices. Experiments are needed to explore whether or not similar effects do also occur in non-cavitating separated flows of gas-liquid mixtures.

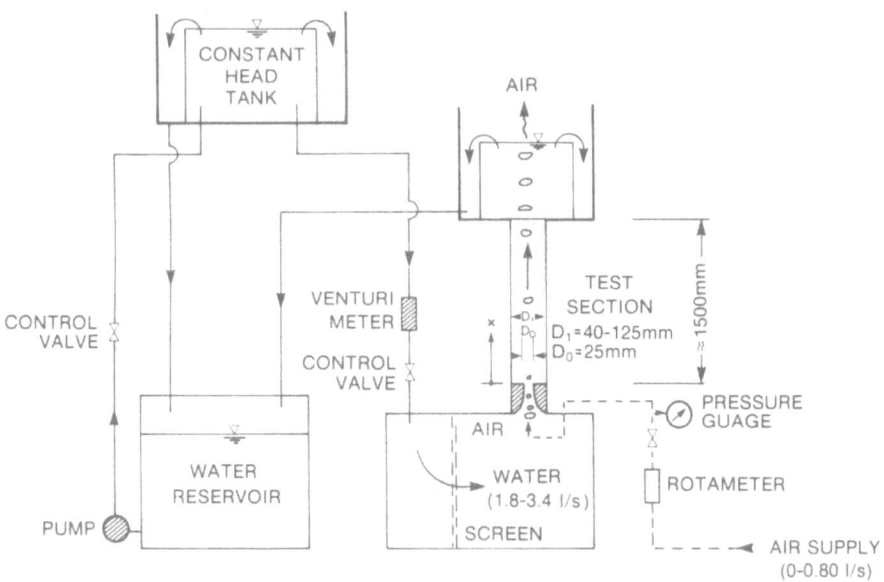

Fig. 2. Experimental apparatus for the investigation of bubble streams in closed conduits.

402 Hans J. Leutheusser

Method of Solution: — The investigation, also spanning several years but not yet documented in the readily available technical literature, was carried out in the separated flow resulting from a sudden concentric expansion of a vertical plastic pipe conveying water in the upward direction. Continuous admixture of air in the form of streams of bubbles of various sizes took place upstream of the sudden expansion, and was achieved in several ways, viz. through the walls of porous feeder units (producing a bubble mist), or the variable-sized discharge nozzle of an immersed air line (emitting a string of discrete bubbles). A schematic diagram of the experimental apparatus is depicted in Fig. 2. Besides bubble size as determined by the mode of air injection, experimental variables included water flow rate, air flow rate, axial position of the air injection point, and diameter ratio of the sudden expansion. The principle measurements taken concerned mean and fluctuating pressures, as well as pressure correlations, prevailing on the wall of the pipe downstream of the sudden expansion. Pressures were determined with the aid of inductive-type pressure transducers. High-speed cinematography was used for the purpose of flow visualization, and still photography formed the basis for evaluating the bubble size distribution.

Typical Results: — In Fig. 3 are combined some typical findings concerning, respectively, the longitudinal distribution of rms-pressure, and the variation of maximum rms-pressure with air-water concentration. The data reveal a pronounced influence of void ratio on the magnitude of pressure fluctuations. However, the presence of undissolved air in water is only significant when it occurs in the form of finite-sized bubbles; very finely dispersed air, such as emanated in the experiments from the porous feeder units, has practically no effect. It may be mentioned that the distribution of bubble diameter downstream of the separation cavity was found to be independent of the mode of air injection, i.e., shear apparently forced the breakup of large bubbles, and the coalescence of small ones.

Pressure Effects Due to Bubble Deformation

The Problem: — The mechanics of the augmentation of pressure fluctuations due to the presence of gas bubbles in liquid systems is not yet understood. Indeed, the investigation described in the foregoing has seemingly demonstrated that, in contrast to experiences with cavitating separated flows, the effects of externally introduced minute air bubbles on the magnitude of eddy-induced fluctuating pressures are negligible. Clearly, the observed large increase in fluctuating pressure due to the presence of finite-sized air bubbles must have a different cause than vortex "lubrication". As it happens, bubbles themselves may be the source of pressure radiation due to volume and shape oscillations. However, most of the limited information presently available on the dynamic behaviour of gas bubbles in liquid environments pertains to small bubbles of spherical shape. Experimental information is required to explore the dynamic behaviour of large gas bubbles whose shape tends toward the form of a spherical cap.

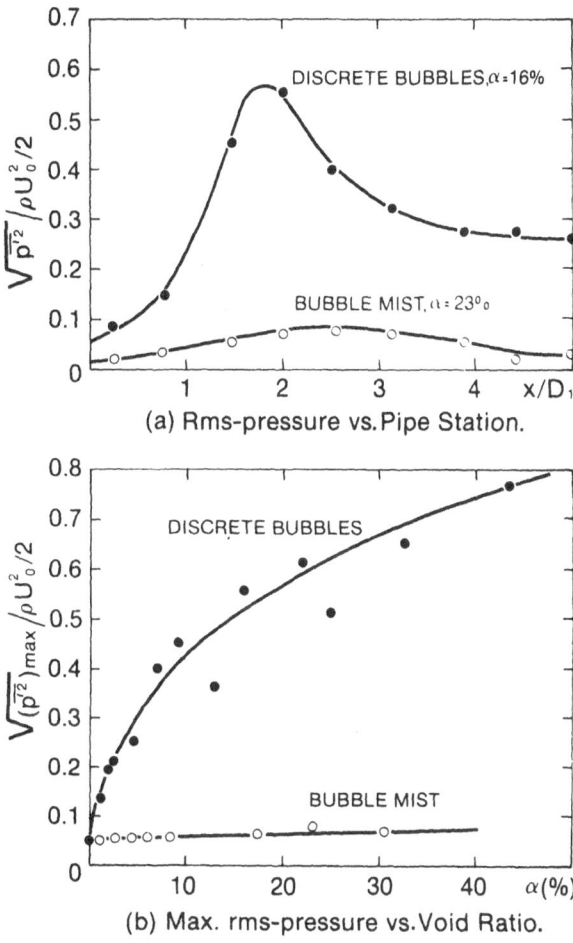

(a) Rms-pressure vs. Pipe Station.

(b) Max. rms-pressure vs. Void Ratio.

Fig. 3. Pressure characteristics downstream of sudden expansion with D_1/D_0 = 2.0.

Method of Solution: — The investigation, currently in progress, is carried out on isolated large air bubbles rising in a vertical plastic pipe filled with quiescent water. The main features of the experimental apparatus are depicted in the schematic diagram of Fig. 4. For the investigation of the behaviour of artificially distorted bubbles, orifice plates of various diameters are placed at different

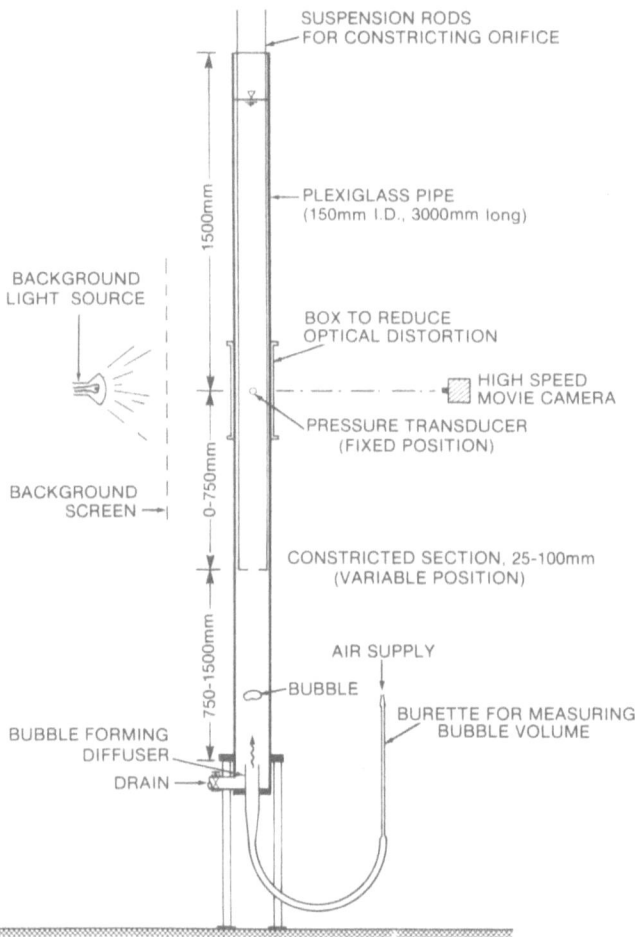

Fig. 4. Experimental apparatus for the investigation of pressure effects due to bubble deformation.

positions inside the pipe. A similar system is employed to force the breakup of large air bubbles. The bubbles are introduced into the water through a diffuser at the bottom of the pipe. The bubble motion is observed and recorded by high-speed cinematography, enabling the determination of mode, amplitude, and frequency of the bubble oscillations. A pressure transducer, installed in the wall of the pipe is used to measure and record the pressure signals.

Typical Results: — Experimental results are presently being assembled, and representative examples were presented at the Conference.

Summary and Conclusions

Problem formulations and solution methods have been sketched in the foregoing for a number of two-phase air-water flow situations of engineering importance. The examples show that significant and revealing results can be obtained with presently available experimental techniques.

Acknowledgements

The writer gratefully acknowledges contributions to this paper by B. Barczewski (Sonderforschungsbereich 80, Universität Karlsruhe, Germany), F. Hara (Department of Mechanical Engineering, Science University of Tokyo, Japan), E. Naudascher (Institut für Hydromechanik, Universität Karlsruhe, Germany), C. Plesko (Department of Mechanical Engineering, University of Toronto, Canada), and F. J. Resch (Institut de Mécanique Statistique de la Turbulence, Université d'Aix-Marseille, France). Portions of the research work described in the paper were supported through Grant A-1541 of the National Research Council of Canada.

References

Barczewski, B., 1976 *Sonderforschungsbereich 80, Universität Karlsruhe, Germany,* Report SFB 80/ME/72

Leutheusser, H.J. and Ward, C.A., 1975 *Proc. XVIth Congr. Intl. Assoc. Hydr. Res., Sao Paulo, Brazil,* 3/500

Naudascher, E. and Jezdinsky, V., 1972 *Institut für Hydromechanik, Universität Karlsruhe, Germany,* Report 510

Rao, N.S.L. and Kobus, H.E. 1975, *Characteristics of Self-aerated Free-surface Flows,* Erich Schmidt Verlag, Berlin, Germany

Resch, F.J., Leutheusser, H.J. and Coantic, M., 1976 *J. Hydr. Res.* 14, 4, 293.

Measurement Techniques in the Two-Phase Flow Region of the Air-Sea Interface Layer

by

François Resch
Institut de Mécanique Statistique de la Turbulence
Marseille, France

1. Introduction

The measurement of size and velocity (if possible) of suspended particles is often required in many engineering fields. This is especially the case for meteorologists and oceanographers who investigate the production of liquid marine aerosols at the sea-surface.

Many diversified methods are available to measure the two-phase flow parameters of these moving particles, but not all of them are suitable for describing the process of spray production through bubble bursting on which we will focus our attention here.

For a better specification of the measurement techniques that have to be selected, let us first define the flow configurations we have to deal with.

II. Two-Phase Flow Aspects of the Particulate Exchanges within the Air-Sea Interface Layer

It is estimated that the air-sea particulate exchange processes are responsible for the transfer of about 10^{10} tons of chemical material — mainly sodium chloride — from the oceans to the atmosphere. It turns out that most (if not all) of this quantity enters the atmosphere as drops ejected from air bubbles bursting at the water surface. This production of marine aerosols can be schematically represented by the following "cascade" process, see table I.

As we are only interested in the production processes our primary aim is to determine the two-phase flow parameters concerning bubbles in water and droplets in air.

TABLE I

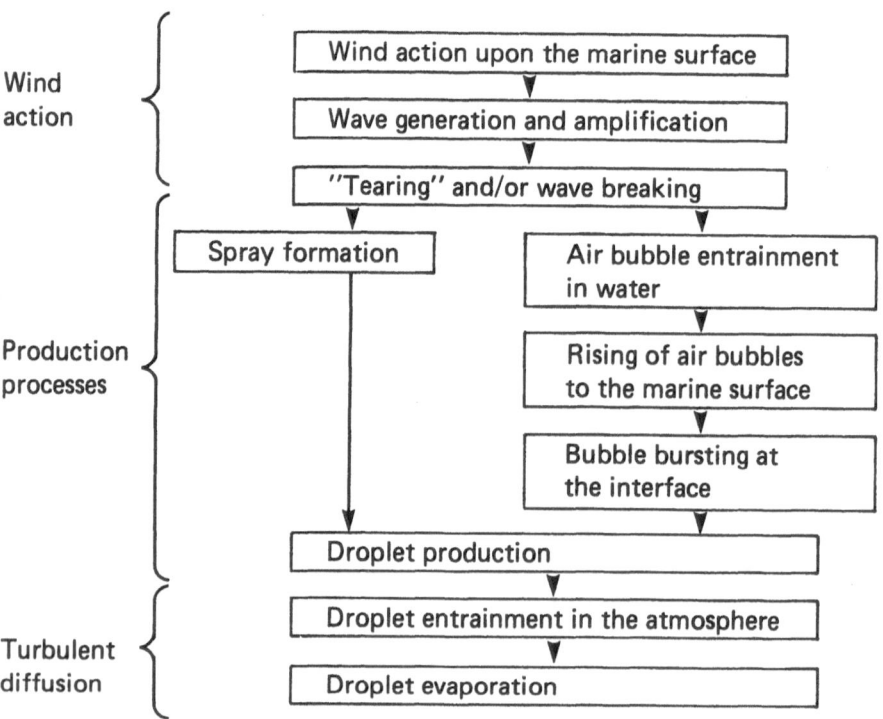

1. **Bubbles in water**

   Bubbles are drawn back into the water by the action of breaking waves, down to a certain depth under the combined action of inertia and gravity forces. The size spectrum is expected to cover diameters from 50 microns up to 5 mm.

2. **Droplets in air**

   Bubble bursting at the interface gives rise to two families of droplets:
   − The film drops, which are due to the rupture of the bubble film that protrudes through the interface. Resulting drop sizes are approximately a few microns.

   − The jet drops, which are produced by a jet of water ejected upwards as the bubble cavity collapses after bubble bursting. Their size ranges from a few microns to 500 microns and they are ejected at variable heights.

Our aim is then to be able to determine:
— Sizes, concentrations and injection depths of air bubbles in water;
— Sizes, concentrations and ejection heights of liquid droplets in air.
Both concentrations are expected to be less than one hundred particles per cubic centimeter. This last consideration, together with the approximate size spectrum (some tenths of microns to several millimeters for bubbles, and a few microns to some hundreds of microns for drops) creates a specific class of two-phase flow configurations for which a special metrology has to be worked out. Whenever feasible, it will be desirable to use the same measurement technique both in air and water.

As usual in two-phase flow measurements, global and local techniques will be considered separately.

III. Global Technique
Let us first recall that a global measurement is defined as the measure of the space averaged value of a quantity over either a segment, a surface area, or a volume.

We are presenting here three kinds of global techniques:

1. **Photographic technique**
    The first global technique to be used was photographic and was the basis for most of the original observations. It can be used not only in both liquid and gaseous phases but also at the interface itself for the recognition of whitecaps, MONAHAN (1951). Although this technique is still extensively used and continuously improved, it presents particular short-comings as already mentioned by HEWITT (1972). The obtaining of high enough light intensities, and the detection of particles which are not in the plane of focus represent the basic difficulties that experimentalists have to face. Use of laser beams and holography can bring partial solutions to these problems; nevertheless, the photographic technique does not seem easy to handle because of the long and tedious data processing it necessitates.

2. **Impaction technique**
    Impactors or impingers have often been used to study the size distribution of atmospheric aerosols, since MAY (1945). When cascade impactors operate, a certain amount of air containing the dispersed phase is directed towards a series of plates so that suspended particles are deposited on the plates through inertia as the flow stream is deviated. Various versions of cascade impactors are available, see RENOUX et al. (1974). For example,

TOBA and TANAKA (1967) have developed a hand-operated impactor with a chloride reagent film fixed on the surface of impact. HOUNAM and SHERWOOD (1965) have developed a cascade centripeter for determining the concentration and size distribution of aerosols.

Impingers having a suitable collecting surface (e.g. magnesium oxide-coated glass) may be used in the presence of an electric field to achieve an electrostatic-collection of suspended particles. If the electrical field is sufficiently high, the drops move rapidly under the influence of an electrical force to impact the plate, BLANCHARD and SYZDEK (1975).

These techniques are widely used for the detection of marine and atmospheric aerosols. They are usually quite precise but processing of data is also long and cumbersome. Moreover, the use of such methods is obviously restricted to the gas phase and not applicable to bubble measurements. Another inconvenience arises from the fact that drops will dry out on the impaction surface, so that only their "signatures" can be studied.

3. Optical techniques

Two kinds of global techniques using light-scattering and light extinction are described.

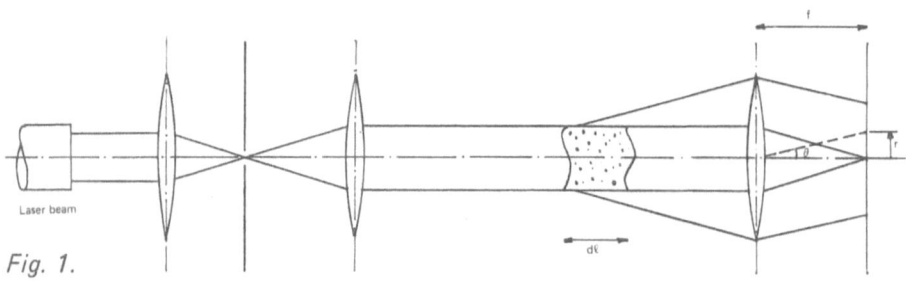

Fig. 1.

The light-scattering, or "Mie"-scattering for particles whose dimensions are not negligible compared to the wavelength, is used to determine the particle size distribution of the dispersed phase. The schematic optical principle is represented in Fig. 1. Starting with the scattering of an individual spherical particle, one can estimate the energy received on a disc of radius r. Integrating a finite volume of suspended particles, the size spectrum is determined from the measurement of the radial distribution of the light energy in the image plane. This is achieved by means of micro-densitometers moved about on this plane, or by taking advantage of the recent development of several ($\cong$ 30) photocells placed along a semi-circular ring path.★

★ Malvern Instruments Ltd. ST1800

This type of measurement is very delicate to perform and the numerical data processing creates difficulties; moreoever, an a priori size distribution (usually a log-normal one) has to be assumed. Nevertheless, this method has the advantage of detecting particles as small as one micron and of being useable in both gaseous and liquid phases.

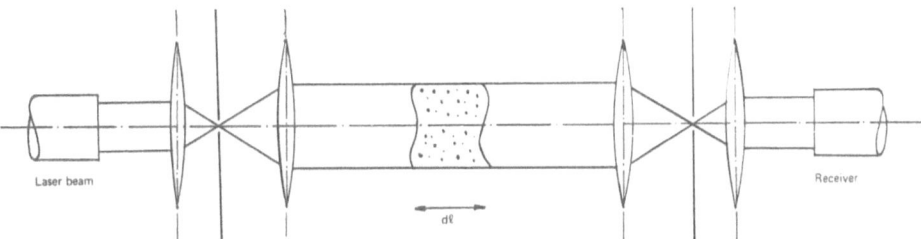

Laser beam   Receiver

dℓ

Fig. 2.

The light-extinction technique allows the determination of the effective superficial area of the dispersed phase. The incident intensity I_o of a collimated laser beam, see Fig. 2, is attenuated by the scattering due to particles, so that the intensity I of the received beam is expressed as:

$$I/I_o = \exp(-\gamma\ell)$$

where ℓ is the optical path and γ the extinction coefficient proportional to the surface area for particles larger than 30 microns. It should be noted that this simple technique gives successful results only if the "pinhole" is in line with the optical axis.

This method has been recently adapted by J. WU (1977) for a low concentration dispersion as is the case for sea spray; he has also developed a dual laser beam attenuation method for measuring particle velocities.

This technique seems promising, since it can be used alternately in the two phases. Actual detected diameters range between 30 microns and 1 millimeter, a fact which is convenient for the air bubble case.

Other global techniques have to be mentioned here as they may be of interest for suspended particles, although not well adapted for our present topic. The chemical reaction method is quite promising for some flow configurations, see CHARPENTIER (1977). The X-ray attenuation method is well developed but not so easy to use. The acoustic method also seems promising but still has to be improved to become efficient. Finally, electrostatic methods involving the measurement of the sensibility distribution of charged particles could eventually be used.

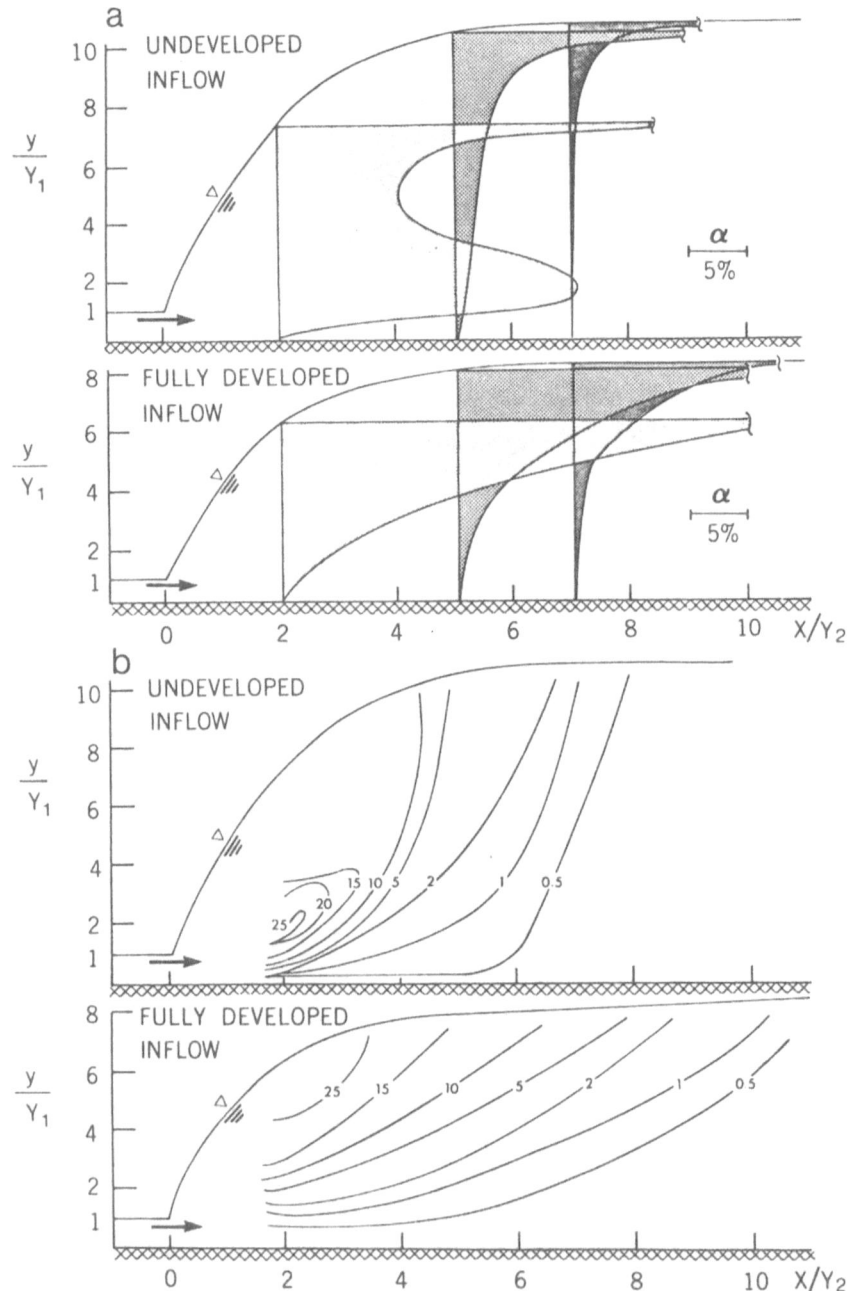

Fig. 3.

IV. Local Techniques

Usual local two-phase flow metrology can be used for bubble and droplet size determination, provided the smallest dimension can be detected by the sensing element of the probes. Although there is now a marked tendency to miniaturize the latter, it seems, for the time being, difficult to measure diameters less than some hundred microns. Present techniques, however, may be of interest for the wave breaking generated bubbles: electrical, optical or thermal probes can then be chosen.

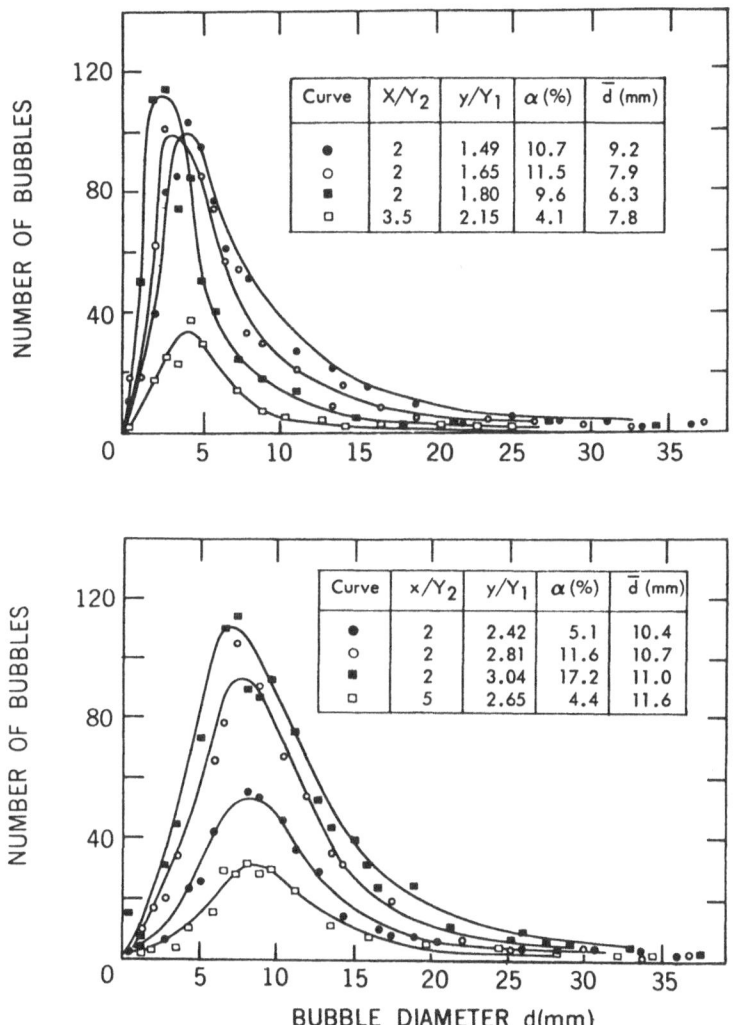

Fig. 4.

A thermal method has been used by RESCH et al. (1974) with the aid of a hot-film probe for determining the two-phase flow parameters in the air-en-training region of the hydraulic jump, a phenomenon similar to the breaking of a wave. Applying a special digital data processing technique (amplitude filtering) to the output signal of a conical hot-film probe, it is possible to evaluate void ratios, bubble diameters, water flow velocities, etc.

Results are displayed here in Fig. 3 for the void ratio distribution and in Fig. 4 for the statistical bubble diameter distribution, see RESCH et al. (1974). Although several difficulties seem to arise the same technique has already been used for droplet granulometry determinations by GOLDSCHMIDT and HOUSE-HOLDER (1968, 1969).

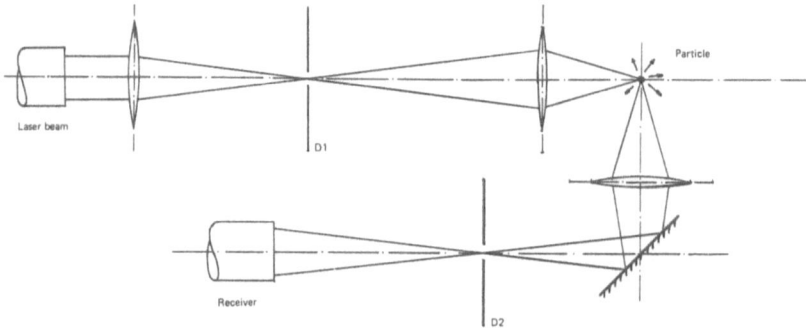

Fig. 5.

For smaller diameters, less than a few hundred microns, another recently developed technique, see e.g. EDERHOF (1976) is presented with an optical probe using light scattering at an angle of ninety degrees from the incident light beam, see Fig. 5. The light flux scattered by a particle present in the small detection volume, is a direct function of its diameter. A calibration of the probe is then sufficient to determine the particle sizes. If the control volume is precisely defined (in optical terms), it is possible to determine the particle velocity by measuring the residence time. Automatic digital treatment of the output signal of the photocell is now being developed.

V. Conclusions

All the above described techniques are shown in Table II, see also F. AVEL-LAN (1978).

Some of them seem especially convenient for the determination of the two-phase flow characteristics of the air-sea interface layer; they are:

— for global measurements: the light attenuation technique
— for local measurements: the light scattering (at 90 degrees) techniques for the smallest bubbles or drops (from 1 to $\cong$ 200 microns) and the hot-film anemometry method for larger bubbles and drops (a few hundred microns to a few millimeters).

VI. Acknowledgements

The work described herein is supported by the "Centre National pour l'Exploitation des Océans" (CNEXO convention no. 78.1869) and the "Centre National de la Recherche Scientifique" (CNRS).

These studies continue to be carried out in cooperation with F. AVELLAN, a doctoral student.

TABLE II	Techniques	Phase	What can be measured ?	Data Treatment
Global Techniques	Photographic	Bubbles, Droplets	Granulometry (1 μ minimum) Velocity	non automatic, long
	Impaction	Droplets	Granulometry (micronic and submicronic)	non automatic, long
	Light scattering	Bubbles, Droplets	Granulometry (1 - 500 μ)	difficult numerical resolution
	Light attenuation	Bubbles (Droplets)	Specific area Granulometry (30 μ - 1 mm)	automatic, fast
Local Techniques	Hot film anemometry	Bubbles, Droplets	Granulometry (200 μ - 10 mm) Void fraction velocity	automatic, complex
	Light scattering at an angle of 90° from the incident light	Bubbles, Droplets	Granulometry (1 μ - 500 μ) Velocity	automatic, fast

VII. References

AVELLAN, F. (1978) "Methodes de mesure des caractéristiques diphasiques au voisinage de l'interface air-mer" *Rapport d'activités D.E.A.,* I.M.S.T., Université d'Aix-Marseille II

BLANCHARD, D.C. and SYZDEK, L.D. (1975) "Electrostatic collection of jet and film drops". *Lymnology and oceanography,* Vol. **20**, No. 5, September 1975, pp. 762-774

CHARPENTIER, J.C. (1977) "Considerations générales sur les contacteurs gaz-liquide" Communication au groupe "Métrologie diphasique" de la section "Écoulements polyphasiques" de la Sociéte Hydrotechnique de France, Grenoble, décembre 1977

EDERHOF, A. (1976) "A light-scattering probe for droplet size and wetness fraction measurement in two-phase flows". *Two-phase Steam Flow in Turbines and Separators,* ed. Moore, M.J. and Sieverding, C.H., Hemisphere Publishing Comp., London, pp. 249-260

GOLDSCHMIDT, V.W. and HOUSEHOLDER, M.K. (1968) "Measurement of aerosols by hot-wire anemometry" *Advances in Hot-Wire Anemometry,* ed. Melnik, W.L. and Weske, J.R., pp. 134-152, Department of Aerospace Eng., University of Maryland

GOLDSCHMIDT, V.W. and HOUSEHOLDER, M.K. (1969) "The hot-wire anemometer as an aerosol droplet size sampler". *Atmospheric Environment,* Vol. **3**, pp. 643-651

HEWITT, G.F. (1972) "The role of experiments in two-phase systems with particular reference to measurement techniques" *Progress in Heat and Mass Transfer,* ed. Hetsroni, G., Sideman, S. and Hartnett, J.P., Vol. **6**, pp. 295-343, Pergamon, Oxford

HOUNAM, R.F. and SHERWOOD, R.J. (1965) "The cascade centripeter: a device for determining the concentration and size distribution of aerosols" *American Industrial Hygiene Association Journal,* Vol. **26**, pp. 122-131

MAY, K.R. (1945) "The cascade impactors. An Instrument for sampling coarse aerosols" *Journal of Scientific Instruments,* Vol. **22**, pp. 187-195

MONAHAN, E.C. (1971) "Oceanic whitecaps". *Journal of Physical Oceanography,* Vol. **1**, pp. 139-144

RENOUX, A., TYMEN, G., BUTOR, J.F., and MADELAINE, G. (1974) "Utilisation de cascades impactors pour l'étude de la répartition granulométrique des aérosols atmosphériques. Application à l'aérosol marin". *Journal Rech. Atmos.,* Vol. **8**, pp. 709-721

RESCH, F.J., LEUTHEUSSER, H.J. and ALENU, S. (1974) "Bubbly two-phase flow in hydraulic jump", *Journal of the Hydraulic Division, ASCE,* Vol. **100**, No. HY1, Proc. Paper 10297, pp. 137-149

TOBA, Y. and TANAKA, M. (1967) "Simple technique for the measurement of giant sea-salt particles by use of hand-operated impactor and a chloride reagent film". Special contributions, Geophysical Institute, Kyoto University, No. 7, pp. 111-118

WU, J. (1977) "Fast-moving suspended particles: measurements of their size and velocity". *Applied Optics,* Vol. **16**, No. 3, pp. 596-600.

Local Particle Velocity Distributions in Two-Phase Flows Measured by Laser-Doppler Velocimetry

by

A. Stümke and H. Umhauer

Institut für Mechanische Verfahrenstechnik

Universität Karlsruhe

Introduction

In many gas-solid flow studies it has to be considered that the particle phase may not move with a uniform one-dimensional velocity, but may show at any place of measuring a particular velocity distribution the mean of which may be distinctly different from the fluid's local mean velocity. Generally this happens with particle sizes exceeding the order of a few microns and because of the turbulence of the fluid flow. The character of those distributions may rarely be assessed theoretically. The complete frequency distributions of the local particle velocities therefore have to be determined by registering the velocities of individual particles passing through an exactly limited probe volume.

The authors set up an appropriate Laser-Doppler velocimeter (LDV) meeting these requirements. The important problem of a "representative" measurement with a wide particle size distribution given in the flow was also discussed and extensively investigated (1). In case any correlation exists between particle size and particle velocity, a correct local particle velocity distribution can only be obtained if one can be sure that the velocities for *all* particle size classes are registered accurately according to their frequencies in the respective intervals. This demand also applies for the measurement of an exact mean value and the scattering range of the velocities.

"Representativity" of the measurement

As is known from literature, the signal quality m and the signal strength A_{ac} of LD-bursts may strongly vary. With increasing particle size x ($x \geqslant d_f$, d_f = fringe distance), decreasing and oscillating modulation degrees can be observed. The definition of signal quality $m = (A_{max} - A_{min})/(A_{max} + A_{min})$ and a graph m(x) (e.g. for a square particle and a fringe distance $d_f = 10 \mu m$) are given in *Fig. 1*; see (2), (3). Given wide particle distributions, e.g. $Q_o(x)$ ranging from x_{min} to x_{max}, signal drop out may occur in certain intervals — x_1 to x_2 — if the signal quality m(x) sinks below a certain value m^*, which is supposed to indicate the sensitivity level of the analysing electronics. Quantitative calculations have been carried out to demonstrate that this may lead to serious errors in the measurement of the velocity distributions, providing any dependences $w(x)$; one example is given in *Fig. 2.*

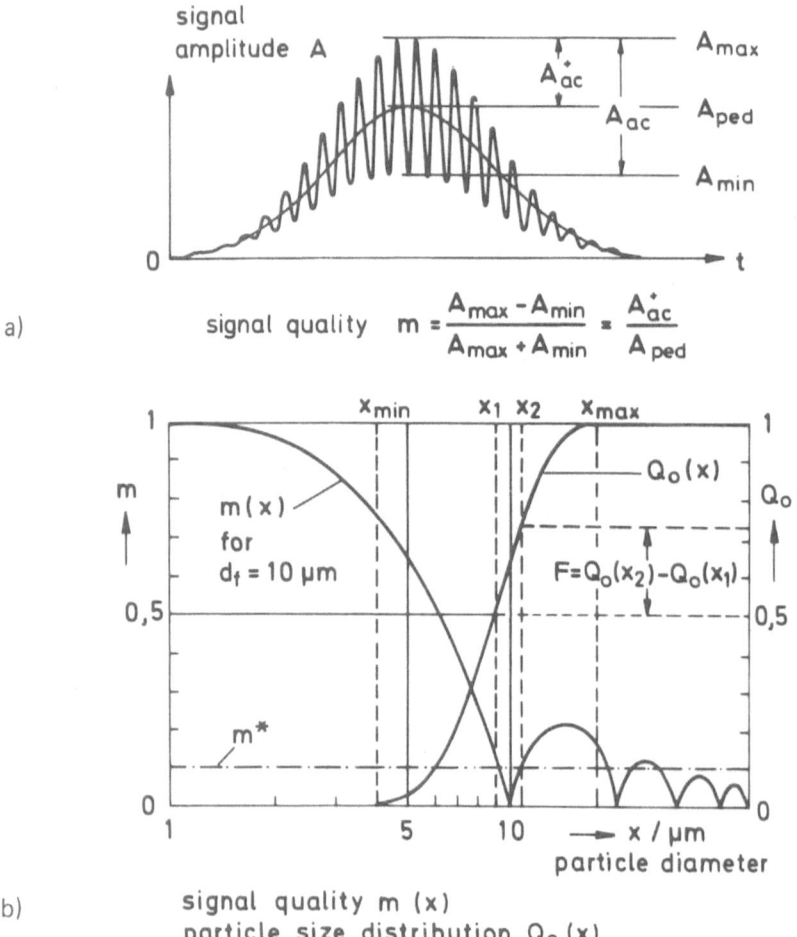

a)

signal quality $\quad m = \dfrac{A_{max} - A_{min}}{A_{max} + A_{min}} \approx \dfrac{A_{ac}^{+}}{A_{ped}}$

b)

signal quality $m(x)$
particle size distribution $Q_o(x)$

Fig. 1. Signal modulation.
a) Definition of signal quality m by signal amplitudes; cp. visibility (2), (3).
b) Example of possible signal loss F between x_1 and x_2 with given signal quality $m(x)$ — valid for a square particle (2) — and a supposed particle size distribution $Q_o(x)$.

In order to avoid signal drop out and to ensure a representative measurement of the velocity distribution, the measuring system has to be optimized according to the given particulate properties.

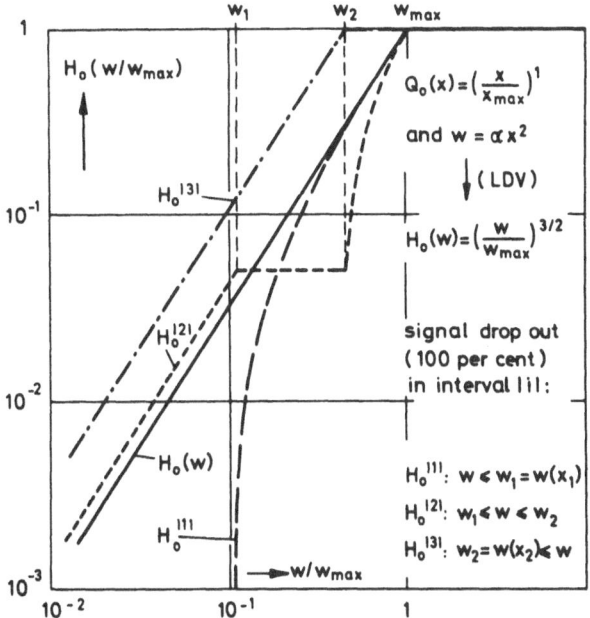

Fig. 2. Calculated examples for incorrect measurement of a given particle velocity distribution $H_o(w)$. Flow situation considered is a sedimentation process in the Stokes' region with settling velocity $w \sim x^2$.

LD-Measuring Device

Following is the most important information on the dual-beam measuring instrument used in this study; *Fig. 3*, (1).

— The optics include a variable beam splitter which permits a continuous adaptation of the fringe distance in the interference pattern (variation range 10:1).

— The optical components are set at an angle of 90 degrees. By means of a square mask and a defined image scale of the observing optics it is then possible to obtain a fairly exact limitation of the measuring volume. This improves the local resolution and is necessary for measurements at higher particle concentrations, because a "counting system" only works if there is only one particle at a time in the measuring volume.

— The electronic equipment should allow individual analysis and recording of the Doppler frequencies of single bursts. The frequency distributions of particle velocities once being registered, one can — by means of a computer — plot the cumulative distributions $H_o(w)$ and calculate any characteristics of the distributions.

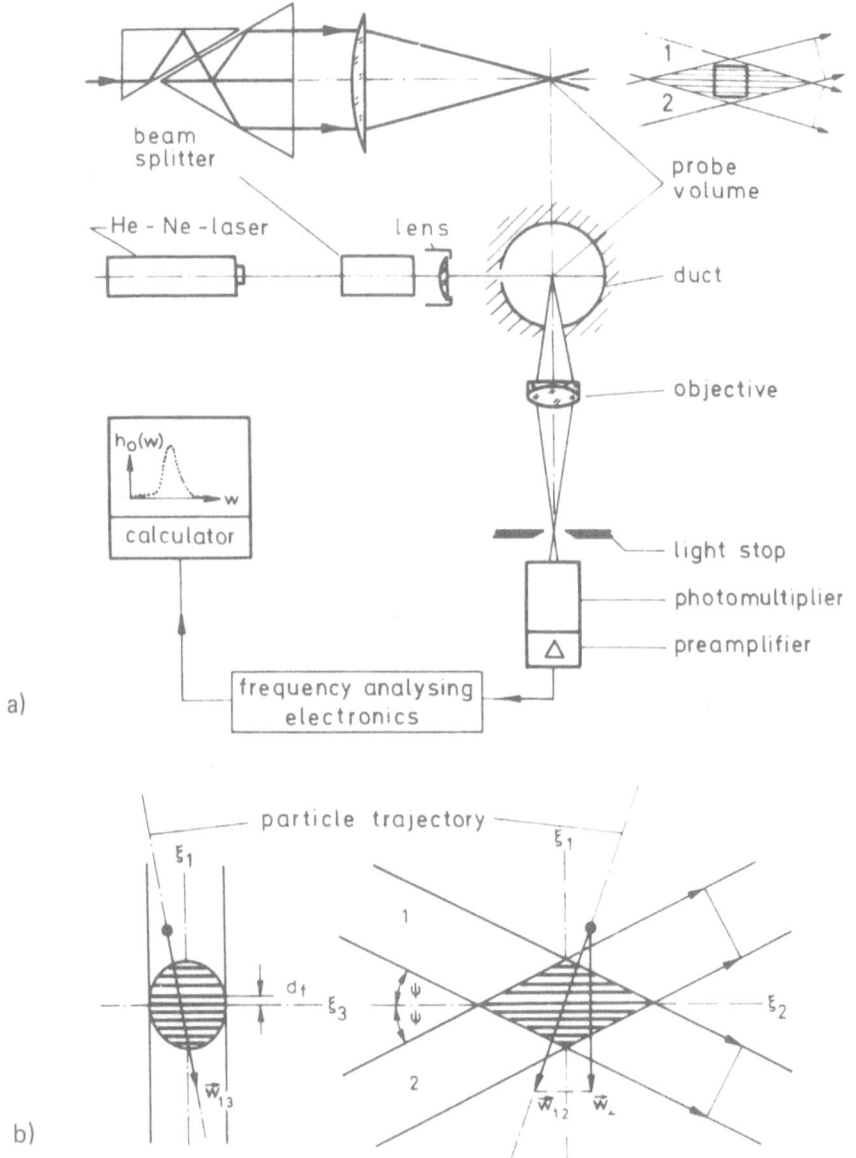

Fig. 3. Laser-Doppler velocimetry (dual-beam method).
a) — arrangement of optical components — example of a variable beam splitter
b) — views of the intersection region, schematic fringe pattern, particle trajec-
tory, components of velocity $\vec{w}$, Cartesian coordinates ξ_i, $i = 1, 2, 3$.

Adaptation, Signal Quality

In order to allow an optimum adaptation of the measuring system, the dependences of the LD-signal quality on the different particle properties for the optical arrangement used have to be known. The authors therefore extensively studied individual particles which had been prepared on rotating glass plates. This method makes it possible purposefully to determine different parameters and the particulate properties. Characteristic dependences of the modulation were established by way of experiment, e.g. the influence of the particle size with different optical material properties, of the surface structure or of the particle shape (1), (4).

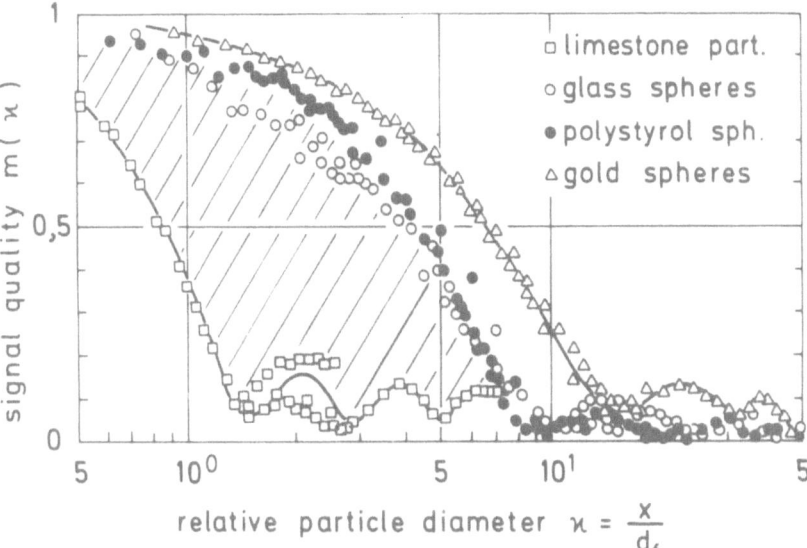

Fig. 4. Signal quality $m(\kappa)$ for different materials.

Fig. 4 shows some typical results; the signal quality m is given as a function of the relative particle diameter $\kappa = x/d_f$ = particle diameter/fringe distance. The modulation of irregular, diffusely scattering limestone particles shows approximately the theoretical decrease at $\kappa = 1.22$ (3); with the spherical glass, polystyrol and gold particles, however, a shift by the factor 8 to 11 occurred. Specially treated glass beads with diffuse scattering centres on their reflecting surfaces yielded modulation values in the hatched area according to whether the diffusely scattered or the reflected light prevailed in the observed signal. The shift of the $m(\kappa)$-curves for the reflecting spheres can be explained satisfactorily by a simple optical model connecting particle size, fringe distance and the aperture of the receiving optics; this could be verified by experiments.

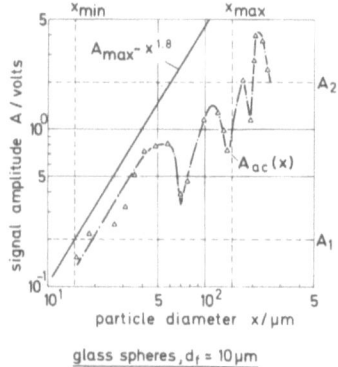

glass spheres, $d_f = 10 \mu m$

Fig. 5. Signal amplitudes $A_{max}(x)$ and $A_{ac}(x)$, particle size range x_{min} to x_{max}, dynamic range of the analyzing electronics A_1 to A_2.

It should be mentioned that the actual optimization of the system has to follow the changes of the signal strength $A_{ac}(x)$ which is influenced by both $A_{max}(x)$ and $m(x)$; *Fig. 5.* The adaptation parameters — fringe distance d_f, aperture of observing optics, amplification/attenuation by optical filters or by the adjustment of photomultiplier and pre-amplifier — have to be chosen such that the amplitude of the actual Doppler-signal A_{ac} in the given particle size range (x_{min} to x_{max}) is always within the dynamic range of the processing electronics (A_1 to A_2).

Measurements in gas-solid flows

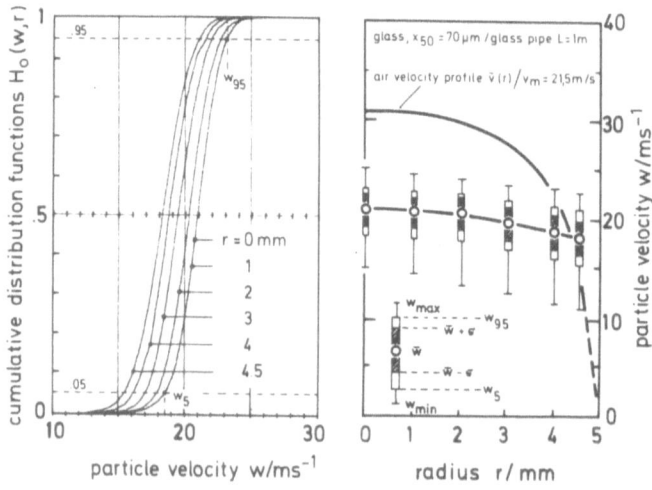

Fig. 6. Axial particle velocities w in a vertical pipe of diameter 10 mm; downward flow.

Taking account of these basic studies, the authors started to do measurements in a vertical pipe flow (diameter D = 10 and 20 mm) to assess the flow behavior of the dispersed phase with different parameters. An analysis of various flow phenomena indicated that measuring the complete velocity distributions gives a lot of additional information and therefore represents a considerable advantage, which is apparent in *Fig. 6.* There the cumulative distributions H_o (w) of the axial particle velocity components w at radius r are shown on the left side. From these curves special values may be derived and plotted as profiles above radius r (right side). The mean velocity $\bar{w}(r)$ and the standard deviation σ_w (r), which corresponds to the rms-value and is symmetrical to $\bar{w}$, can be calculated as well as further characteristics, e.g. ranges w_{95} to w_5, w_{max} to w_{min}, or higher moments of the distribution function. These measurements and numerous additional results will be discussed elsewhere (5).

In conclusion it may be stated, however, that the determination of the local particle velocity distributions according to the method suggested in this paper will be of advantage whenever the motion or the interaction of individual particles are of importance.

References

(1)   Stümke, A. Umhauer, H., paper in preparation
(2)   Durst, F., Melling, A. and Whitelaw, J.H. *Principles and Practice of Laser-Doppler Anemometry,* Academic Press, London, N.Y., San Franc., 1976, pp. 112
(3)   Farmer, W. M., *Appl. Optics,* **11** (1972), 2603
(4)   Stümke, A., Umhauer, H., "Particulate Sources of Error in the Measurement of Local Particle Velocity Distributions Using Laser-Doppler Velocimetry", paper presented at CHISA '78, B 3.3, Prague, 1978
(5)   Stümke, A. et al., paper to be published.

Transducer Techniques

Transducer Techniques

by

P. Buchhave
DISA Elektronik A/S
Mileparken 22, DK-2740 Skovlunde, Denmark

1. Introduction

The label "Transducer Techniques" in fluid dynamic measurements covers a wide range of topics from transducer design and manufacture through transducer principles and application to signal processing. Since the design and principle of operation of many special transducers for both dynamic velocity and dynamic pressure measurements are described in individual papers in this proceedings volume, this paper will focus on the general aspects of transducer response in local, time resolved velocity measurements in fluid flows. The treatment of these problems will center on the two most commonly used instruments in fluid flow research, the hot-sensor anemometer (in particular the hot-wire and the hot-film anemometers) and the laser Doppler anemometer (the LDA). This also provides an opportunity to compare, feature for feature, these two instruments which are too often considered separately and used by people exclusively devoted to one or the other. The two systems will be seen to have much in common, but also some distinct differences. We shall try to identify these differences and thereby define the flow measuring problems in which the specific characteristics of the two systems are most profitably utilized.

The range of physical parameters of interest in dynamic flow measurements is very large: Velocities range from mm/s in creeping flows, lubrication problems, blood flows etc. to km/s in high speed wind tunnels; time scales cover the range from microseconds to hours and the spatial scales of interest extend from fractions of millimeters in high speed turbulence to kilometers in meteorology. Obviously no single transducer can cover this entire range, and even in applications to problems in a specific area it is necessary to look closely at transducer response and transducer-flow interaction. In dynamic flow measurements it is rare to find the trivial application; most measurements require that the user of the equipment evaluates the transducer response in his particular situation and assesses the validity of the measured data or possibly applies the necessary corrections.

To illustrate this process of "experiment design" two specific examples of flow measurements in which the author has recently been involved will be used: Example A, the investigation of the large scale structure of the mixing layer of a 10" d. round jet in air and example B, measurements in a 3" d. fully developed jet in air.

Ideally we would like a transducer to provide information about a physical quantity in a single, fixed, well defined point in the flowing medium (local measurement) and faithfully reproduce the dynamic fluctuations of the physical quantity at that point without attenuation of particular frequencies (time resolved measurement). Also we would like an output (e.g. voltage), that varies proportionally to the input or at least as an algebraic function of the input. Furthermore, noise should be negligible and there should be no interference between transducer and flow.

Even as an approximation these requirements are rarely fulfilled in dynamic flow measurements. To describe the behavior of the non-ideal transducer it is necessary to introduce a number of transducer characteristics: The static transfer function describes the steady state input-output response of the transducer (the calibration curve). The dynamic transfer function describes the response of the transducer to sinusoidal variations of the input as a function of frequency. The resolution is defined as the minimum detectable input variation. The spatial and temporal resolution are concepts that describe the transducer as a function of spatial and temporal wavelengths. In addition it is often necessary to discuss multi-variate sensitivity, i.e. whether the transducer responds to more than one physical parameter at a time. Other factors of importance are noise and error sources other than multi-variate sensitivity, transducer interaction with the flow, and as a final point it is often necessary to consider the signal processing of the transducer output, since the transducer performance may be affected by the particular mode of operation and method of signal processing employed. This is especially true in the case of the LDA for which angular characteristics and spatial and temporal response depend strongly on the signal processing. A further discussion of signal and data processing may be studied in the paper in this volume by W.K. George.

In any particular application of a transducer all these characteristics should be known and compared to the expected flow characteristics before the output is interpreted. To state the situation paradoxically: Ideally the flow should be well known before reasonable measurements can be made! In practice the flow parameters are estimated or guessed before the measurement is performed, and the estimate later compared to the measured results. The particular sensor or instrumentation is chosen in accordance with the expected flow parameters and the possibility of applying suitable correction methods is assessed. Fig. 1.1 illustrates the problem with reference to spectral measurements in air using four different instruments. To be able to distinguish high frequency turbulence from instrumental noise or to assess the influence of high frequency attenuation knowledge of both the flow (spatial and temporal scales) and the instrument (shape of noise spectrum at high frequencies) is required.

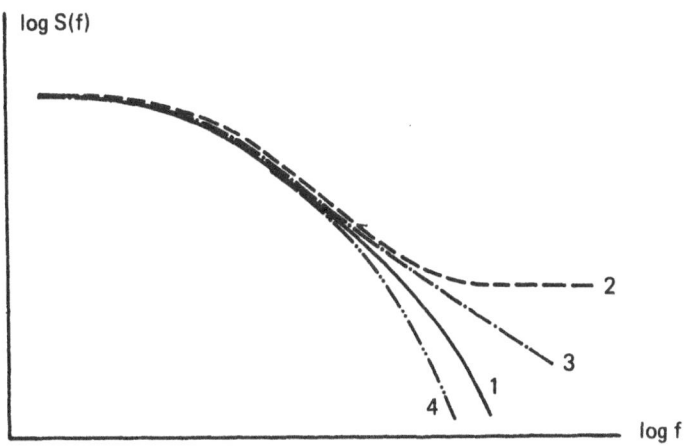

Fig. 1.1. Measurement of power spectrum with different instruments.
1. Correct spectrum
2. LDA tracker (ambiguity noise)
3. LDA counter analog out (drop-out noise)
4. Hot-wire (wire length attenuation)

In the following we shall discuss in more detail the principle of operation of the LDA and the hot-sensor anemometer as well as the important characteristics of these instruments. Then the spatial and temporal resolution of the LDA and the hot-sensor anemometer are described and compared to the requirements imposed by the estimated flow properties in the two specific examples mentioned above.

The problems posed by interference between material probes and the flow are well known, but a detailed discussion will not be attempted within the limitations of this paper. The LDA is generally considered non-interfering, but the irradiance on small particles in e.g. a focused Ar-ion laser beam is very high and effects such as radiation pressure and local heating is likely to influence the measurement at low velocities. However, this problem has not yet been fully investigated.

2. The Laser Doppler Anemometer (LDA)

2.1 Principle of Operation

The laser Doppler anemometer is based on three physical effects or principles:
1. The Doppler shift (rate of change of phase) of light scattered from moving particles.
2. Optical heterodyning of two incident light fields on a square law photodetector.
3. Frequency demodulation of the photocurrent to derive the velocity, which is proportional to the measured frequency.

The implementation of the method is based on the following hardware components:
1. The laser, which provides a light source of high temporal and spatial coherence.
2. The optical system, which directs and focuses the laser light into a small volume, the probe volume.
3. The optical receiver, which collects the scattered light and performs the optical heterodyning on a square-law optical detector (photomultiplier, PIN-diode or Si-avalanche diode).
4. The optical frequency shift, which can add a carrier frequency to the Doppler shift and through that modify the signal processing characteristics.

The signal processing is basically a frequency demodulation, but as we shall see the specific method of extracting the Doppler frequency has a profound influence on the transducer characteristics of the LDA.

The LDA technique has evolved during the last 15 years from the first experiments on light beating spectroscopy by Cummins et al. (1964) and Yeh and Cummins (1964) and has later been documented by a large number of publications and through a series of workshops and symposia (e.g. the International Workshop on laser Doppler anemometry at Purdue University, eds. Thomson and Stevenson (1974) and the LDA-Symposium, Copenhagen, eds. Buchhave et al. (1976)). For further information the reader is referred to a number of recent monographs on LDA (e.g. by Durst et al. (1976) and Durrani and Greated (1977)).

Components

The laser itself is the heart of the LDA. The usefulness of this device in comparison to conventional light sources is due to its extreme stability and coherence. A good reference on gas lasers is a book by Bloom (1968). The output from gas lasers used in LDA work is of the lowest order transverse electromagnetic mode (TEM_{oo}), but may contain several axial modes. The presence of several axial modes does not disturb the LDA measurement as long as the two beams in the heterodyning process travel the same optical path length and have the same polarization. An important property of TEM_{oo} gas laser beams is the Gaussian intensity profile of a cross section of the beam at any axial position. Special expressions are valid for the propagation and imaging of Gaussian laser beams (see e.g. Bloom (1968)), and in some cases these deviate from intuitive conclusions based on geometrical optics. A Gaussian beam is characterized by two parameters only: The width and location of the narrowest part of the beam, the so-called beam waist. An important design goal in LDA optics is to position the beam waist at the location of the probe volume (Hanson, 1973). These concepts are illustrated in Fig. 2.1, which shows a simplified LDA optical system consisting of a laser, a beam splitter and a focusing lens. In order for the beam waist to be located at the intersection point of the two focused beams, i.e. at the focal plane after the lens, the laser should be placed so that the beam waist of the beam directly from the laser is placed at the focal plane before the lens. Special optical "beam waist displacers" may be inserted in the laser beam to transform the beam waist position if it is impossible or impractical to make the correct positioning of the laser directly.

The optical system which focuses and directs the laser beams into the measuring volume defines the transducer characteristics of the instrument.

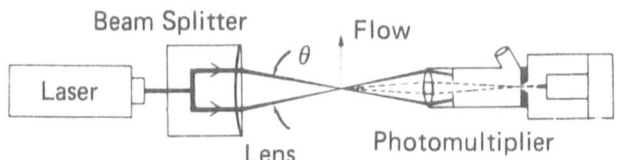

Fig. 2.1. Typical arrangement of optics in laser anemometry.

Most systems today are of the dual-beam type, where the probe volume is defined by the intersection of two beams focused into a common intersection volume by a common lens or optical transmitter. Such a system in a modular design can be built up from the simplest one-component forward scatter LDA with fixed parameters to more sophisticated multi-component systems with adjustable optical parameters and facilities for backscatter operation. As an example of such a system the DISA 55X-optical system is shown in Fig. 2.2.

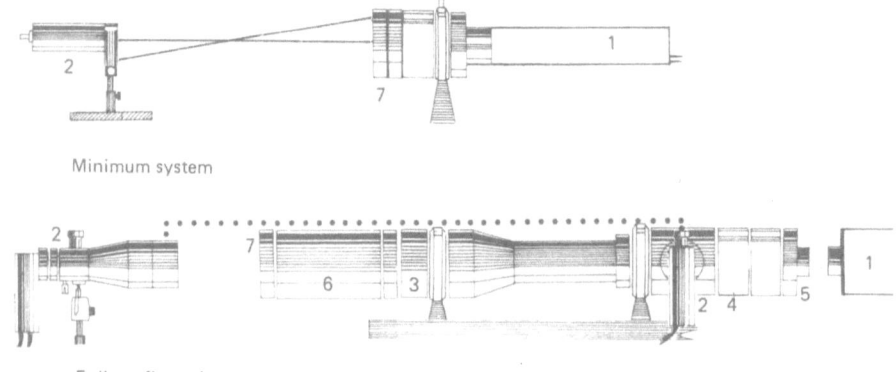

Minimum system

Full configuration

Fig. 2.2. Modular LDA optical transducer.
1. Laser
2. Detector
3. Beam translator
4. Bragg cell
5. Beam waist displacer
6. Beam expander
7. Front lens

The most important controls in the optical system for modification of the transducer properties of the instrument are indicated. The beam translator allows adjustment of the beam separation, which in turn influences a number of parameters as will be explained in the subsequent sections. The Bragg cell introduces an optical frequency shift which in turn influences the signal processing. Also shown are the beam waist displacer, the beam expander and the exchangeable front lens, all of which are important in defining the properties of the probe volume.

The technological sophistication required of the optical and mechanical components in an LDA system depends very much on the specific measuring problem. In a simple forward scatter, one-component LDA, an uncoated simple lens may function satisfactorily as an optical transmitter. In more complex backscatter systems aberration corrected lens systems must be used and the optical elements must have hard, multi-layer anti-reflection coatings. In two-color systems the optical system must also be corrected for chromatic aberration. Fig. 2.3 shows a sophisticated backscatter, two-component LDA with frequency shift in each channel.

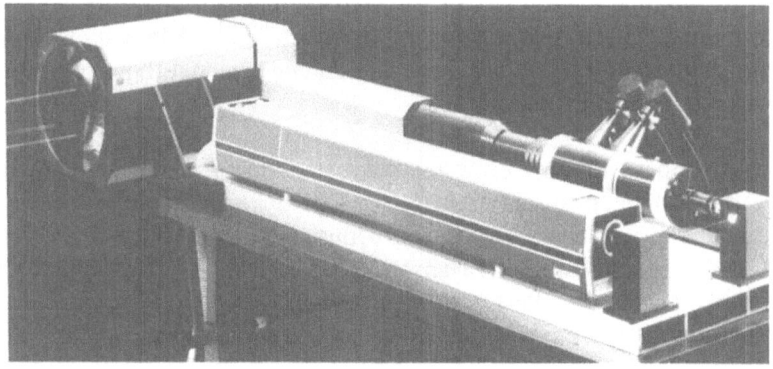

Fig. 2.3. Optical system with traversing feature (zoom optics).

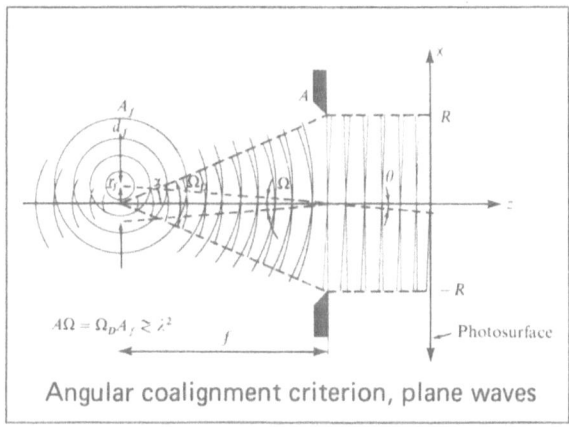

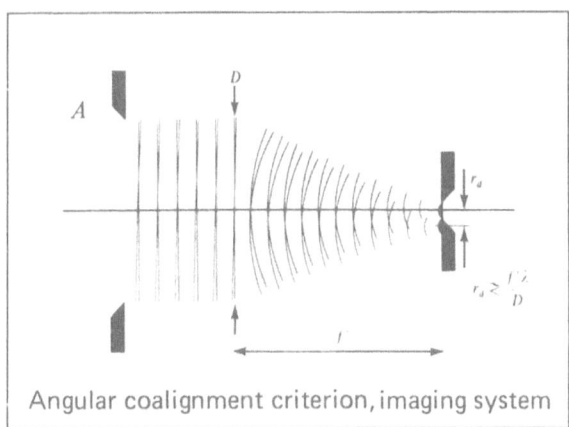

Fig. 2.4. Coalignment criterion for optical heterodyning.

The probe volume can be scanned axially by a zoom-type optical system from 2 to 6 meters. The optics is designed to maintain important transducer characteristics such as probe volume, intersection angle, signal-to-noise ratio etc. constant during the optical scan.

The third important part of the optical transducer is the detector optics. In general optical heterodyning requires the two incident optical wave trains to have parallel wave fronts across the active detector surface (Siegman (1966), Drain (1972), Lading (1973), Hanson (1974)). Fig. 2.4 (from Buchhave 1973) illustrates the case for plane waves and also indicates that the coalignment may be transformed to a coincidence or overlap criterion in the focal plane of the detector lens. The figure also illustrates the use of a pinhole in the imaging plane of the detector optics as a spatial filter to separate ambient light and light received from the probe volume.

In general the photocurrent may be represented by the expression:

$$i_d = i_1 + i_2 + \epsilon \cdot \sqrt{i_1 \, i_2} \, \cos \left(2\pi f_D \, t + \phi \right) \qquad [2.1]$$

where i_1 and i_2 are the currents which would occur with only one beam incident on the detector. f_D is the Doppler frequency shift and ϕ is a phase depending on the initial location of the particles. The degree of coalignment enters through the parameter ϵ which in a well adjusted system should approach one.

A modern optical system as described above is designed so that the heterodyning criterion is automatically fulfilled. It is possible to use one of the two incident beams as a reference beam and operate the system in the so-called reference beam mode. In this case only scattered light within the solid angle defined by the reference beam divergence is active in the heterodyning process, and the detector aperture A shown in Fig. 2.4 may be stopped down to the reference beam diameter. In the dual beam (or fringe mode) with only one particle present in the measuring volume, both scattered beams originate from the same particle and the wave fronts at the photodetector are consequently parallel over a much larger aperture, thus allowing the use of a large collecting lens in this case.

Due to the quantum nature of the photo detection process the photocurrent is a shot noise process with a mean square fluctuating current i'^2 equal to

$$i'^2 = 2 \, e \, \Delta f \, \bar{i} \qquad [2.2]$$

where $\bar{i}$ is the mean photocurrent, Δf is the electronic band width and e is the electron charge. In addition some thermal electronic noise is added from the amplifier input stage of the signal processor. Normally it is possible to operate an LDA in the so-called shot noise limited condition, where the photocurrent noise is dominated by the signal shot noise, and such that noise due to ambient light is negligible. In this case the signal-to-noise ratio of the photocurrent in the reference beam mode and the fringe mode respectively may be expressed by:

$$(S/N)_{R.B.M.} = \frac{\eta \epsilon^2}{h\nu K \Delta f} \, P_s \qquad [2.3]$$

and

$$(S/N)_{F.M.} = \frac{\eta \epsilon^2}{2 h\nu K \Delta f} \, P_s \qquad [2.4]$$

Here η is the quantum efficiency, $h\nu$ the energy of a photon, K the intrinsic gain in the photodetector and P_s the power of the scattered light. Note that in the reference beam mode P_s is collected by a small aperture (coalignment criterion), but that light fields from many scattering particles add coherently, whereas in the fringe mode P_s may be collected over a large aperture, but that light fields from many particles do not add coherently. Thus the reference beam mode is superior in the many-particle case and the fringe mode is superior in the single particle case. More recent information on signal-to-noise in LDA detection may be found in the papers by Durst and Heiber (1977), Adrian and Orloff (1977), Ballik and Shan (1977) and Mayo (1977).

Finally a few words about optical frequency shifting: Optical shift is a simple addition of a carrier frequency f_s to the Doppler shift: $f_d = f_D + f_s$. However, the shift allows detection of flow direction and modifies the behavior of the signal processor increasing the measurable dynamic range of the velocity fluctuations or even changing the angular characteristics of the probe volume as will be described later. The Bragg cell has the advantage that no change occurs in the basic optical parameters of the laser beam, which is the case with frequency shifting from e.g. rotating devices like the rotating diffraction grating. Combined with electronic shift, the Bragg cell allows stable, controlled, variable shift of the Doppler frequency over a range from kHz to many MHz. The combined Bragg cell electronic shift is described by Buchhave (1975). Other methods are described in the literature or commercially available e.g. rotating diffraction gratings by Oldengarm (1977), Kerr cells by Drain and Moss (1972),

optical phase shifting by Foord et al. (1974), and a dual-Bragg cell system by
Brayton et al. (1973).

2.2 Static Transfer Function

The static transfer function for the laser Doppler anemometer is close to ideal.
For a given beam intersection angle θ, the detector current beat frequency, or
Doppler frequency, is given by (referring to the coordinate system in Fig. 2.5):

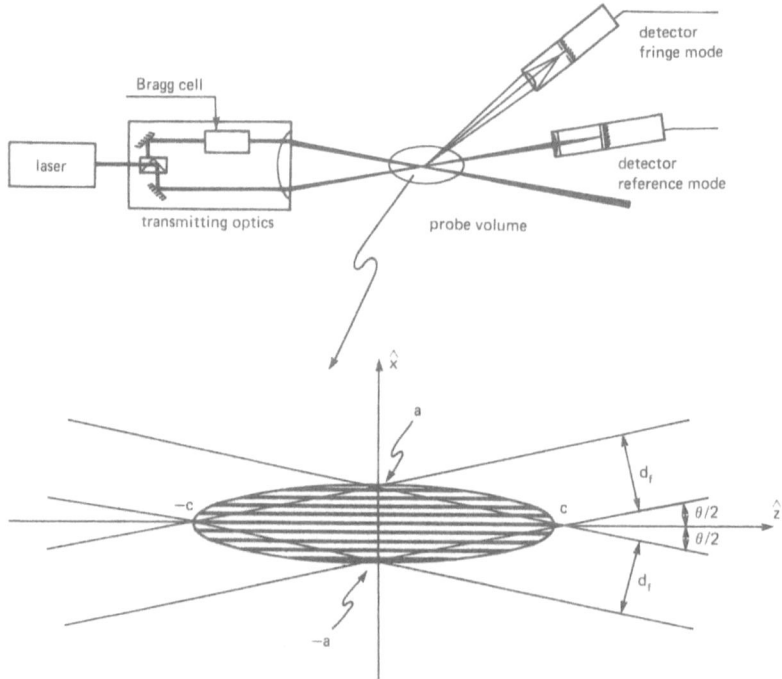

Fig. 2.5. Optical configuration of typical LDA.

$$f_d = f_D + f_s = \frac{2U}{\lambda} \sin \frac{\theta}{2} + f_{s'} \tag{2.5}$$

where U is the component of the velocity vector $\underline{U}$ along the normal to the bisector of the beams in the plane of the two beams (the x-axis in Fig. 2.5), and
f_s is the added shift. Thus the detected frequency is linearly dependent on the
x-component of the velocity. This relation holds over the whole range of
velocities encountered in practice. The LDA is insensitive to other velocity
components and for a given θ the temperature and pressure only enters through
λ, the wavelength of the laser light.

However, due to the random dispersion of scattering particles in space the phase and amplitude of the Doppler signal will fluctuate randomly, which gives rise to the so-called Doppler ambiguity noise, when the continuous Doppler signals are FM demodulated. Properly detected the mean frequency is still proportional to the mean velocity, but the instantaneous frequency fluctuates around the mean. The effect is closely linked to the size of the measuring volume and will be described in the section on spatial resolution.

2.3 The Measuring Volume

The LDA measuring volume is often defined as the volume within the $1/e^2$-boundary of the optical fringe modulation. However, we shall denote this volume the probe volume and reserve the term measuring volume to the region in space from which Doppler signals are received and detected by the system. For Gaussian laser beams the probe volume is an ellipsoid defined by its axes along the coordinates given in Fig. 2.5:

$$2a = 4\sigma_x = \frac{d_f}{\cos \theta/2}$$

$$2b = 4\sigma_y = d_f \tag{2.6}$$

$$2c = 4\sigma_z = \frac{d_f}{\sin \theta/2}$$

d_f is the beam waist of the focused laser beam expressed by:

$$d_f = \frac{4}{\pi} \frac{f\lambda}{d_\varrho} \tag{2.7}$$

where d_ϱ is the beam waist diameter of the unfocused laser, f the focal length of the lens and λ the wave length of the laser light.

The fringe distance is:

$$\delta_f = \frac{\lambda}{2\sin \theta/2} \tag{2.8}$$

and the number of fringes along the x-axis is:

$$N_f = \frac{4}{\pi} \frac{D_s}{d_\varrho} = \frac{8}{\pi} \frac{f}{d_\varrho} \tan (\theta/2) \tag{2.9}$$

where D_s is the separation between the parallel incident beams.

The measuring volume as defined above is influenced by many system parameters and usually differs from the probe volume. Most important is the in-

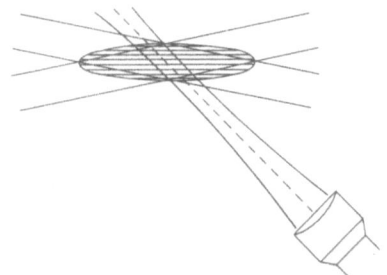

Fig. 2.6. Measuring volume forward scatter.

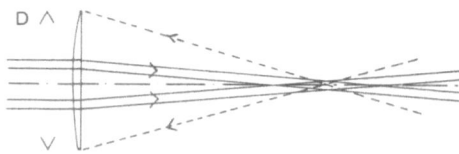

Fig. 2.7. Measuring volume backscatter.

fluence of the detector optics. Referring to Fig. 2.6, it is evident that a detector placed at an angle to the transmitter optical axis cuts out a section of
the probe volume given by the intersection of the probe volume and the cone
defined by the field of view of the detector. More subtle is the effect of the receiver optics in the coaxial backscatter optics shown in Fig. 2.7. The length of
the probe volume is determined by the diameter of the focused beam and the
intersection angle, whereas the actual measuring volume is determined by the
focal region ℓ_f of the receiving lens aperture D, which is proportional to the
square of the receiver F-number and may be much smaller than the probe volume length:

$$\ell_f \cong 8F^2\lambda \qquad\qquad\qquad\qquad\qquad [2.10]$$

with F = f/D.

Furthermore, the total gain of the system enters the definition of the measuring volume. To discriminate against low power shot noise from background
light or thermal noise most LDA signal processors have a minimum signal
threshold, which must be exceeded before the processor starts functioning. Increased signal amplitude by increased photomultiplier gain, preamplifier gain
or by larger particle size allows detection of particles farther away from the
center of the probe volume and thus increases the measuring volume. A still
more subtle influence comes from seeding particle concentration. In some

applications it is desired to maintain a certain mean sampling rate. If the particle concentration varies during a measurement a constant mean sampling rate can nevertheless be maintained by adjusting the gain, e.g. the PM high voltage. But as explained above this leads to a change of measuring volume size, which may be a critical parameter in small scale turbulence measurement or in the presence of large velocity gradients.

2.4 Angular Characteristics

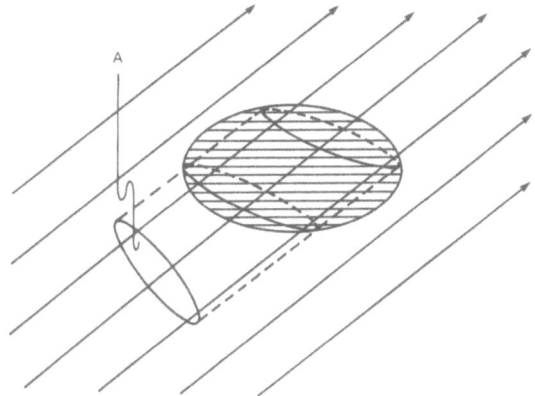

Fig. 2.8. Measuring volume cross section, A.

As mentioned above the LDA is only sensitive to the component of velocity normal to the fringe planes. In the processing of continuous Doppler signals no other angular effects occur as long as the processor is able to function normally. Frequency trackers only function within a limited dynamic range of input frequencies, which results in certain "dead angles" — directions of flow velocity resulting in frequencies below the tracker range. The situation is easily remedied by adding a sufficient frequency shift to insure that the detected frequency always occurs within one tracker range.

A similar problem occurs in burst-type LDA signal processing with the LDA-counter. In this case a gradual decrease in sampling rate is observed for increasing angles between flow direction and the direction of measurement (the x-axis in Fig. 2.5). This effect may be illustrated with reference to Fig. 2.8 and explained in terms of the measuring volume cross section (Buchhave 1976).

As is evident from Fig. 2.8, a measurement of a particle with velocity vector $\underline{U} = (U, V, W)$ can only occur within the section of the probe volume indicated by the dashed lines. This section is defined by a contour in the plane normal to $\underline{U}$, and the measuring cross section is the area within this contour. The cross section defined this way is simply the area normal to $\underline{U}$ within which a particle of a given size produces a Doppler burst of enough periods above the signal processor trigger level to produce a measurement. This cross section depends on the number of fringe planes available along the x-axis, N_f, as well as on the number of zero crossings needed by the counter to complete a measurement, N_e.

Assuming an ellipsoidal measuring volume of half axes a, b, and c the cross section may be written (Dimotakis, 1976):

$$A(\underline{U}) = \pi abcP(1-a^2Q^2P^2)$$ [2.11]

where

$$P = \frac{1}{|\underline{U}|}\sqrt{\frac{U^2}{a^2}+\frac{V^2}{b^2}+\frac{W^2}{c^2}}$$ [2.12]

and

$$Q \equiv N_e/N_f$$ [2.13]

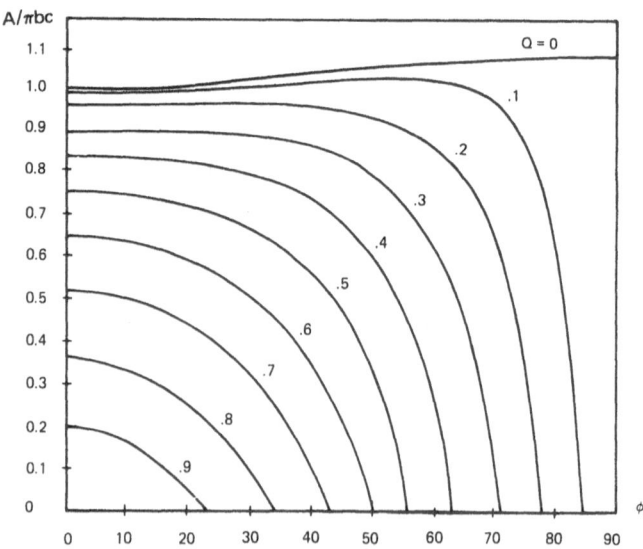

Fig. 2.9. Measuring volume cross section as a function of angle ϕ in the x-y plane ($\theta = 45°$; no frequency shift).

The cross section for flow directions in the x-y plane is shown in Fig. 2.9, as a function of angle between $\underline{U}$ and the x-axis and with Q as a parameter. Note that for $Q \rightarrow 0$, the cross section approaches the geometrical cross section of the ellipsoidal volume. The curves in Fig. 2.9 are valid for a monodisperse particle size distribution. With a range of particle sizes present angular characteristics show a less pronounced dead angle effect.

As in the case of continuous signals the angular characteristics may be improved by the addition of frequency shift. With sufficient shift to allow detection for any flow velocity vector $\underline{U}$ the cross section reduces to the geometrical cross section (corresponding to the curve for $Q = 0$ in Fig. 2.9).

3. The Hot-Sensor Anemometer

3.1 Principle of Operation
This section summarizes the principle of operation of the hot-sensor anemometer. For more detailed discussions the reader is referred to the excellent review articles by Kowasznay (1954), Corrsin (1968) and Compte-Bellot (1976) and the book by Sandborn (1972).

The physical process utilized in the hot-sensor anemometer is the heat transfer from a small, electrically heated sensor (most commonly a thin wire or film) to the surrounding fluid. The velocity dependent heat loss from the wire is detected by measuring the electrical power input to the sensor, i.e. from the current through the sensor and the voltage across the sensor.

The energy balance equation for the element may be written:

$$I^2 R_w = \phi_{conv} + \phi_{cond} + \frac{dQ}{dt} \qquad [3.1]$$

where the power input $I^2 R_w$ (R_w = wire resistance and I = current) is equated to the heat loss by convection plus the heat loss by conduction to the supports plus the rate of heat storage in the element. Other heat transfer mechanisms such as radiation may usually be neglected.

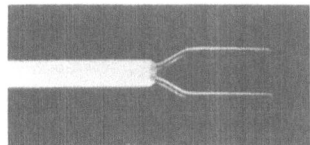

Fig. 3.1. Hot-wire probe (Pt-coated W-wire).

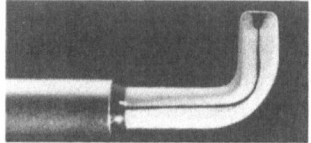

Fig. 3.2. Hot-film probe (Ni-film on quartz substrate, quartz coated).

Two types of sensor elements dominate: 1) The thin wire supported by prongs and 2) The thin metallic coating (film) supported by a non-conducting substrate. Fig. 3.1 shows the design of a typical hot-wire sensor consisting of a platinum-coated tungsten wire welded to two stainless steel supports and Fig. 3.2 shows a film probe consisting of a thin (2000Å) nickel film coated with quartz deposited on a wedge shaped quartz substrate.

TABLE 3.1

Requirements to hot-sensor probes:

1. Large specific resistivity ($\sigma^{-1} = R_w A/\ell$)   }   For good sensitivity
2. Large temperature coefficient, α }   and large signal
3. Large heat conductivity   }   For good
4. Small specific mass   }   frequency
5. Small specific heat   }   response
6. Good mechanical strength at elevated temperature

TABLE 3.2

Common sensor configurations:

wire	coating	prongs	probe body	application
25µ W	—	stainless steel	ceramic	<2kHz
5µ W	Pt	stainless steel	ceramic	air
10 µ Pt-Rd	—	stainless steel	ceramic	air
1 µ Pt	—	stainless steel	ceramic	temp. in air

Film	coating	substrate		application
Ni	100Å Quartz	Quartz cylinder		air
Ni	1000Å Quartz	Quartz cylinder		water
Ni	100Å Quartz	Quartz wedge		air
Ni	1000Å Quartz	Quartz wedge		water

The requirements to the sensor material is summarized in Table 3.1. With the present state-of-the-art in sensor manufacturing and sputtering technology, the most common sensor types are those shown in Table 3.2. The thickness of the sputtered quartz protection coating depends on the application. In water a coating of 1000Å is normal whereas in air only a few hundred Å is necessary.

Electronic systems for heating of the sensor and detection of the power input have in practice been limited to the two well-known cases of either constant current through the sensor (CCA) or constant resistance of the sensor, i.e. constant mean temperature of the sensor (CTA). These two methods are shown in principle in Fig. 3.3.

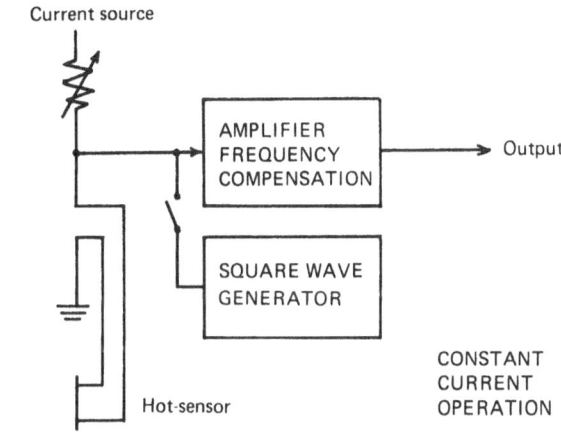

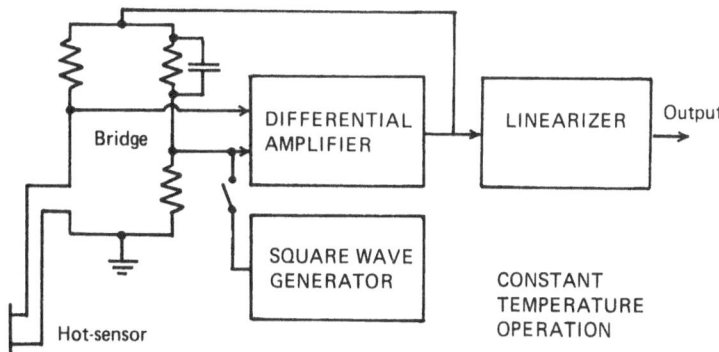

Fig. 3.3. Electronic circuit for hot-sensor measurements.

In both cases the primary output from the instrument is a voltage. Ideally, the CTA operates at a fixed temperature difference between sensor and fluid (i.e. at a fixed operating point). This leads to simplifications in the analytical treatment of the instrument response and results in a fast response because of the constant temperature. The operating point for the CCA varies with flow velocity and the response is inherently slower than for the CTA. In practice the CTA response is more complicated because of the non-uniform temperature distribution along the wire and the heat loss to the prongs, and no simple solution exists for the transfer function.

3.2 Static Transfer Function (Calibration Curve)
In the case of the hot-wire and hot-film anemometer the static transfer function is non-linear. The basic heat transfer relation for heat convected from the sensor may be expressed by the Nusselt number (see e.g. Hinze (1975)):

$$\Phi_{conv} = \pi \ell K_g \, (T_w - T_g) \cdot Nu \qquad\qquad [3.2]$$

where

$$Nu = h \cdot d / K_g \qquad\qquad [3.3]$$

and the constants are:
ℓ, d = wire length and diameter
K_g = conduction coefficient for the fluid
h = convection coefficient for the wire, and
T_w, T_g = wire and fluid temperature respectively.

The physical processes influencing the convection process may be summarized in the following form:

$$Nu \qquad = Nu(Re, Gr, Ma, Pr, Kn, (\ell/d), \varphi, a)$$

where the constants represent various physical processes:
Re $\equiv \rho_g Ud/\mu_g$ (Reynolds No.) — fluid velocity
Gr $\equiv g\rho_g^2 \, d^3 \beta(T_w - T_g)/\mu_g^2$ (Grasshof No.) — bouyancy effects
Ma $\equiv U/c$ (Mach No.) — compressibility effects
Pr $\equiv c_p \mu_g / K_g$ (Prandtl No.) — ratio of viscous to thermal diffusivity
Kn $\equiv \lambda_m /d$ (Knudsen No.) — non-continuum effects
d/ℓ = (length-to-diameter ratio) — effects of finite wire length
φ $\equiv$ (wire orientation) — angular characteristics
a $\equiv (T_w - T_g)/T_g$ — overheat ratio.

The remaining undefined constants are:

ρ_g = specific mass of fluid
μ_g = viscosity of fluid
g = acceleration of gravity
β = coefficient of thermal expansion
c = velocity of sound in fluid
c_p = specific heat of fluid
λ_m = mean free path of molecules of fluid

For discussion of specific effects mentioned above the reader is referred to the cited review articles. For our purpose we shall limit the discussion to aspects of the heat transfer equation relevant to sensor response in incompressible flow at normal temperatures and pressures. The heat transfer equation then reduces to:

$$Nu = Nu(Re, Pr, (\ell/d), \varphi, a) \qquad [3.4]$$

Further simplifications are introduced in the case of hot-wires in gases by considering separately the angular dependence (by expressing the cooling as a function of an effective cooling velocity $U_{eff} \equiv U_{eff} (Re, \varphi)$), the effects of finite wire length and the over-heat dependence. Even so it is not possible to derive a simple, analytical representation of wire cooling as a function of velocity, but recourse must be taken to empirical heat transfer correlations.

A widely used empirical law for wire probes, valid for both gases and liquids, is the so-called Kramer's law:

$$Nu_f = 0.42\, Pr_f^{0.20} + 0.57\, Pr_f^{0.33} \sqrt{Re_f}, \qquad [3.5]$$

where the material parameters are evaluated at the "film temperature", $T_f = (T_w + T_g)/2$. Kramer's law is valid for any Newtonian fluid for $0.7 < Pr < 1000$.

A number of empirical relations for wires in air where Pr may be considered constant $= 0.71$, are similar in form to the relation originally derived by King (1914) for the heat transfer from an infinite cylinder in a potential flow and are therefore often referred to as King's law:

$$RI^2 = (A_K + B_K \sqrt{U_{eff}})(T_w - T_g) \qquad [3.6]$$

A more general expression (the "generalized King's law") is now often used:

$$Nu = a + b\, Re^n \qquad [3.7]$$

or

$$RI^2 = (A+B \, U_{eff}{}^m)(T_w - T_g) \qquad [3.8]$$

where different values of a, b and n or A, B and m have been suggested by different authors depending on the Re-number range. A widely used expression was put forward by Collis and Williams (1959):

$$Nu_f \left(\frac{T_f}{T_g}\right)^{-0.17} = \begin{cases} 0.24 + 0.56 \, Re^{0.45} & \text{for } 0.02 < Re < 44 \\ 0.48 \, Re^{0.50} & \text{for } 44 < Re < 140 \end{cases} \qquad [3.9]$$

The temperature "loading factor", $(T_f/T_g)^{-0.17}$, makes this relation valid over a range of temperatures, $30°C < T_w - T_g < 300°C$. The change at Re = 44 occurs supposedly at the onset of vortex shedding from the wire at that particular Re-number.

In most cases of practical applications of hot-wires the probe cannot be considered an infinitely long cylinder. The conduction to the wire supports is appreciable. Likewise for film probes conduction to the substrate is appreciable. For a wire immersed in a constant, uniform flow normal to the wire the temperature profile along the wire may be derived assuming a fixed temperature T_g at the wire ends:

$$\frac{T - T_g}{T_\infty - T_g} = 1 - \frac{\cosh(z/\ell^*)}{\cosh(\ell/2\ell^*)} \qquad [3.10]$$

T_∞ is the temperature which would result for an infinitely long wire under the same current and flow conditions and ℓ^* is the so-called cold length introduced by Betchow (1948):

$$(\ell^*)^2 = \frac{d^2}{4} \frac{\pi K_w}{R_{w,0} (A + B\sqrt{U} - I^2)} \qquad [3.11]$$

where $R_{w,0}$ is the wire resistance at a reference temperature, and K_w is the thermal conductivity of the wire. Measurements by Champagne et al. (1967) have confirmed this temperature distribution qualitatively, but also shown the appreciable asymmetry resulting for wire orientations other than normal to the flow direction.

In most cases in the forced convection regime the conduction loss is small relative to the convection loss, and the calculated temperature distribution may be used to estimate the end conduction loss and thus link the calibration law for the actual probe to that of the infinite wire (Hinze (1975)). The correction may be written:

$$\varphi_{cond} = \varphi_{conv} \frac{d}{\ell} \left(\frac{R_w}{R_g} \frac{K_w}{K_g} \frac{1}{Nu} \right)^{\frac{1}{2}} \qquad\qquad [3.12]$$

After application of the conduction correction, the generalized King's law may be applied to finite wires as well with the condition that the power term is now only the convection part:

$$Nu_c = a + b\, Re^n \qquad\qquad [3.13]$$

or

$$(I^2 R_w)_c = \varphi_{conv} = (A + BU^m)(R_w - R_g) \qquad\qquad [3.14]$$

The generalized King's law is often applied to finite wires as well (without end loss correction). The constants A, B and m are then determined directly by calibration.

The wider accessability of computing facilities has made it feasible to simply obtain the best least squares fit of the constants in eq. [3.13] and [3.14]. Koppius and Trines (1976) report on an investigation of the calibration of 5μ hot wires in air flows. Least squares fit to the calibration data (corrected for end loss), when both a, b and n are varied gave the best overall fit, but the corresponding A and B coefficients (from eq. [3.14]) were temperature dependent. When a fixed value of n was chosen and only a and b varied, a slightly less accurate fit was obtained, but the A and B coefficients were now approximately independent of temperature. In practical situations the raw calibration data (without end correction) may be used directly and the constants A and B and m found from a least squares fit to eq. [3.14]. Larsen and Busch (1974) used calibration functions of the form eq. [3.14] for fixed m with coefficients A and B depending on temperature, pressure and humidity in a linear approximation. Drubka et al. (1977) evaluated various temperature compensating analog circuits and analyzed their performance on the basis of a similar form for the calibration curve. Other methods of representing the static transfer function for hot-sensor anemometers are reported in Wood (1975), Bellhouse and Bellhouse (1968), Perry and Morrison (1971), Davis and Davies (1972) and Bruun (1971).

3.3 Measuring Volume
The measuring volume of a single hot-sensor is the volume occupied by the sensor itself. The single hot-wire may be represented by a straight line section.

The more complicated geometry of some hot-film probes precludes an analytic representation of the measuring volume. For x-wires or 3-wire probes it is necessary for purposes of computing the spatial and temporal resolution to consider the whole volume between the wires. A 3-wire probe occupies an approximately spherical volume, which for 1 mm long wires may well be approximately 3 mm in diameter. Even special miniature 3-wire probes occupy a 1 mm d. spherical volume. Fig. 3.4 shows the arrangement of the wires in a 3-wire probe.

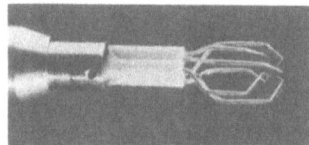

Fig. 3.4. 3-wire probe.

3.4 Angular Dependence

As mentioned above the velocity entering the heat transfer relation for a straight wire probe is the so-called effective cooling velocity. For wires of finite length the effective cooling velocity is not exactly the component of the instantaneous velocity normal to the wire, which is expected from the analysis of infinitely long, straight wires (the cosine law) partly because of the three-dimensionality of the short sensor, partly because of the uneven temperature distribution. An often used empirical expression, which takes into account both the cooling effect of the velocity component along the length of the wire and the effects of prongs and probe body is given by (Jorgensen (1971)):

$$U_{eff}{}^2 = (U_n^2 + K_1^2 U_\ell^2 + K_2^2 U_t^2) \qquad [3.15]$$

where U_n, U_ℓ and U_t are the components of the instantaneous velocity vector $\underline{U}$ along the probe axis, normal to the probe axis and along the wire respectively. Jorgensens measurements show that the best results (i.e. most constant values of K_1 and K_2) are obtained if the yaw factor K_1 is measured at the extreme pitch angle expected to occur during the measurements. For small pitch angles $K_2 \to 0$, and the expression simplifies to that used by Champagne and Sleicher (1967):

$$U_{eff}{}^2 = |U|^2 \ (\cos^2\alpha + K_1^2 \sin^2\alpha) \qquad [3.16]$$

where α is the yaw angle.

For the 3-wire probe shown in Fig. 3.4 a system of equations based on the yaw and pitch coefficients relates the effective cooling velocities for the three wires to the velocity components in the probe coordinate system. For wires mounted in a cube corner configuration the system of equations becomes (with matched wires with same yaw and pitch factors):

$$U_{eff,x}{}^2 = K_1^2 \, U_x^2 + U_y^2 + K_2^2 \, U_z^2$$

$$U_{eff,y}{}^2 = K_2^2 \, U_x^2 + K_1^2 \, U_y^2 + U_z^2 \qquad\qquad [3.17]$$

$$U_{eff,z}{}^2 = U_x^2 + K_2^2 \, U_y^2 + K_1^2 \, U_z^2$$

This system of equations is valid for flow directions within a cone of half top angle of about $37°$ about the probe axis.

4. Spatial Resolution

The fact that any probe has a finite spatial extend results in an integration of the flow field across the transducer, and as a consequence an attenuation of the sensitivity of the probe to high wave number spatial fluctuations (small scale motions). The effect of the attenuation can be quantified for particularly simple probe geometries and for a known spatial structure of the flow field, in particular for isotropic turbulence. In general, however, it must be recommended to use probes of spatial extend smaller than the smallest desired velocity scales.

The spatial resolution problem was first considered for hot-wire probes (see e.g. Uberoi and Kovasznay, 1953 and Wyngaard, 1968) and later similar methods were applied to the LDA by George and Lumley (1973). The attenuation depends on the type of probe, but in general the measured quantity can be expressed as a weighted integral of the actually occurring quantity over a region of space, the measuring volume. Following the method of Uberoi and Kovasznay the measured velocity may be written:

$$U_m \, (\underline{x},t) = \frac{1}{V} \iiint U(\underline{s},t) \, W(\underline{x} - \underline{s},t) \, d^3\underline{s}, \qquad\qquad [4.1]$$

where $W(\underline{x})$ is a weighting function describing the response (or sensitivity) of the transducer at location $\underline{x}$. The spatial averaging is best described in terms of the wave number spectrum. The spatial distribution of velocities is expressed by:

$$\Phi_{ij}(\underline{k}) = \iiint \overline{U_i(\underline{x}) \, U_j(\underline{x}-\underline{r})} \; e^{i \, \underline{k} \cdot \underline{r}} \, d\underline{r} \qquad [4.2]$$

$$= \iiint R_{ij}(\underline{r}) \, e^{-i \, \underline{k} \cdot \underline{r}} d\underline{r} \qquad [4.3]$$

where R_{ij} is the covariance tensor.

The consequences of the above representation may be investigated for a given weighting function and a known flow field.
The general result for the measured one-dimensional spectrum is:

$$F_m(k_1) = \iint \Phi_{11}(\underline{k}) \, \hat{W}(\underline{k}) \, \hat{W}^*(\underline{k}) \, dk_2 dk_3 \qquad [4.4]$$

where

$$\hat{W}(\underline{k}) = \iiint e^{-i \, \underline{k} \cdot \underline{x}} \, W(\underline{x}) \, d^3\underline{x} \qquad [4.5]$$

One can then express the ratio between the measured and actual spectrum, the measuring volume transfer function:

$$\frac{F_m(k_1)}{F_{11}(k_1)} = \frac{\iint \Phi_{11}(\underline{k}) \, \hat{W}(\underline{k}) \, \hat{W}^*(\underline{k}) \, dk_2 dk_3}{\iint \Phi_{11}(\underline{k}) \, dk_2 dk_3} \qquad [4.6]$$

We now evaluate the transfer function for the single hot-wire and for the LDA and consider the implications of the results in the specific flow examples A and B.

4.1 Hot-wire
For a thin, straight wire of length ℓ and diameter $d \ll \ell$ we may write:

$$U_{i,m} = \int_{-\frac{\ell}{2}}^{\frac{\ell}{2}} U_i(s) \, W(y-s) \, ds; \; W(y-s) = \begin{cases} 1 \text{ for } |y-s| \leqslant \dfrac{\ell}{2} \\[2mm] 0 \text{ for } |y-s| > \dfrac{\ell}{2} \end{cases} \qquad [4.7]$$

Contributions from regions outside the wire are zero. The Fourier transform of the weighting function is:

$$\hat{W}(\underline{k}) = \frac{\sin^2(k_2 \ell/2)}{(k_2 \ell/2)^2} \qquad [4.8]$$

And the ratio of the measured to the actual spectrum (the spectral transfer function) is then:

$$\frac{F_m(k_1)}{F_{11}(k_1)} = \frac{\displaystyle\iint \Phi_{11}(\underline{k}) \, \frac{\sin^2(k_2 \ell/2)}{(k_2 \ell/2)^2} \, dk_2 dk_3}{\iint \Phi_{11}(\underline{k}) \, dk_2 dk_3} \qquad [4.9]$$

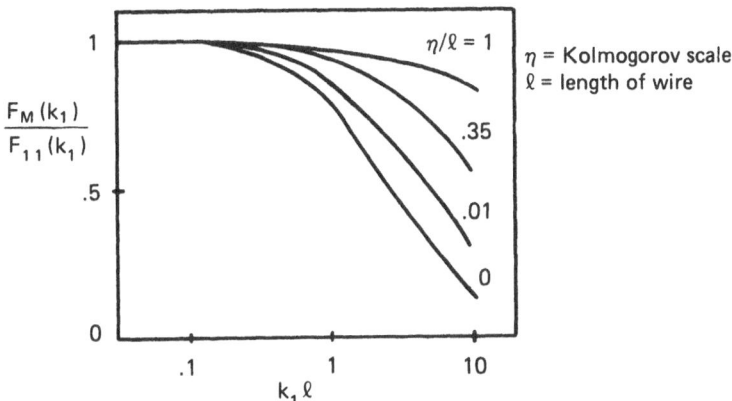

Fig. 4.1. Hot-wire spectral transfer function. (Acc. to Wyngaard (1968)).

This expression has been evaluated numerically for isotropic turbulence with a spectrum according to Pao's model (Wyngaard (1968)). The resulting attenuation of the spectrum is shown in Fig. 4.1 for various ratios between the Kolmogorov length scale η and the wire length.

4.2 Spatial Resolution, LDA
For the LDA the weighting function depends on the particular signal processing method used. For the continuous many-particle Doppler signal particles contribute to the scattered light field according to the light intensity in the measuring volume. George and Lumley (1973) used the following form for the measured velocity:

$$U_o(t) = \frac{1}{\mu} \iiint W(\underline{x})\, U(\underline{x},t)\, g(\underline{x})\, d^3\,\underline{x} \qquad [4.10]$$

where

$$W(\underline{x}) = \frac{1}{(2\pi)^{3/2}\, \sigma_x \sigma_y \sigma_z} \exp\left\{ -\left(\frac{x^2}{2\sigma_x^2} + \frac{y^2}{2\sigma_y^2} + \frac{z^2}{2\sigma_z^2}\right) \right\} \qquad [4.11]$$

$g(\underline{x})$ is a function, which accounts for the presence or absence of a particle at location $\underline{x}$. With this representation and assuming statistically uniformly distributed scattering particles, George and Lumley computed the measuring volume transfer function for isotropic, Gaussian turbulence with a spectrum according to Pao's model. The result is shown in Fig. 4.2. The result is useful for the understanding of the spatial resolution problem of the LDA, but as described above, in most cases the measuring volume is truncated by the detector field of view, and exact corrections are difficult to compute.

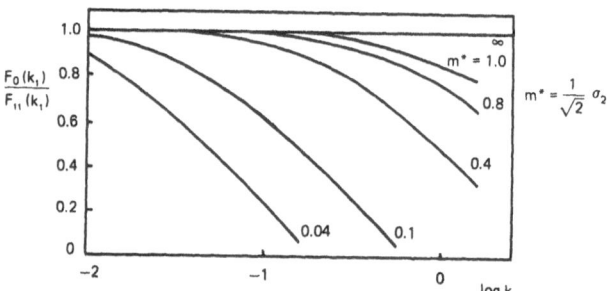

Fig. 4.2. LDA spectral transfer function. (Acc. to George and Lumley (1973)).

A further effect of the finite size of the measuring volume in LDA measurements with continuous Doppler signals is the so-called Doppler ambiguity noise. The effect is caused by the random dispersion of particles in the fluid and the ensuing random composition of the phase of the scattered light. The Doppler ambiguity has been described by many authors. Edwards et al. (1971) derived the shape of the Doppler spectrum for a uniform flow field. Durrani (1973) derived the noise spectrum of the Doppler current for different intensity distributions in the measuring volume using results from communications theory. George and Lumley (1973) combined similar results for the ambiguity noise with the LDA turbulence transfer function to derive expressions for the total broadening of the Doppler spectrum in turbulent flow. Mean gradients within the measuring volume result in additional noise and in a bias of the measured mean velocity. Numerical values of gradient noise and bias have been computed for various cases. Kreid (1974) computed the errors occurring in laminar Poiseuille flow and in turbulent flow with a 7th power wall function. Similar results were published by Owen and Rogers (1976). Berman et al. (1973) compared computed and measured values of ambiguity and gradient noise and found good agreement.

Many LDA signal processors function only when the signal exceeds a preset threshold or trigger level. This is the case for the burst LDA counter. If furthermore only one particle is present in the measuring volume at any time, the weighting function should represent the on-off situation occurring when a particle enters or leaves the volume. In this case a more appropriate representation of the measured velocity is (George, 1976 and Buchhave, George and Lumley, 1978):

$$U_o(t) = \frac{1}{\mu} \iiint W(\underline{x}) \, U(\underline{x},t) \, g(\underline{x},t) \, d^3 \underline{x} \qquad\qquad [4.12]$$

with

$$W(\underline{x}) = \begin{cases} 1 \text{ within the measuring volume} \\ 0 \text{ outside} \end{cases}$$

The actual boundaries of the measuring volume is determined by the optical system, the particle size, the overall gain in the system and the trigger level. The resulting signal is an intermittent function $u_o(t)$, which is "on" only while a particle is inside the measuring volume and zero otherwise. The "on-times", the so-called residence times, are important for the correct signal processing of burst-type signals (see Buchhave, George and Lumley (1978)).

Computations of the measuring volume transfer function has to the best knowledge of the author not been carried out for the burst type LDA. Results very similar to the continuous LDA can be expected, but a more complex transfer function may result from the weight function given above.

Ambiguity noise from random phase jumps does not occur in burst-type LDA as long as only one or zero particle is present in the measuring volume, since the phase in a single particle Doppler burst is well defined (except for the phase jitter resulting from electronic noise). However, gradients within the measuring volume cause noise in a fashion similar to the continuous signal LDA. For non-bias corrected LDA measurements the gradient bias error is proportional to the square of the gradient (see Kreid, 1974 and Owen and Rogers, 1976). However, for residence time weighted signal processing the gradient bias and noise terms are equivalent to those occurring in continuous signal LDA processing (Buchhave, George and Lumley, 1978).

4.3 Examples. Requirements to Spatial Resolution
Example A: The large scale structure in the mixing region of a round jet in air. Theoretical models and empirical relations are used to estimate the spatial scale of the velocity fluctuations in the mixing layer. The goal in this specific example is to measure the energy spectrum at a number of grid points throughout the mixing region to provide data for the reconstruction of the large scale structure of the flow by the method of orthogonal decomposition, Lumley (1967) (see also the paper by S. Herzog in this volume). The wave number k required in order to define the large structure is of the order of $k \cdot \ell \cong 10$, where ℓ is a typical length scale of the energy containing eddies. Empirical relations for the mixing region are:

$$\Delta x \equiv \lambda = \frac{1}{10} \pi \ell \cong \frac{\pi}{100} x \qquad (\ell \cong 0.1 \, x, \, u \cong 0.16 \, U_E) \qquad [4.13]$$

where U_E is the exit velocity at the orifice and u the fluctuating velocity. Thus at x = D we find:

$\Delta x \cong 0.5''$

Thus the spatial resolution requirement in this case is not a problem for either a hot-sensor anemometer or an LDA.

Example B: The small scale structure of the fully developed jet in air. The goal of the experiment was to confirm algorithms for bias-free estimators for mean, mean square, autocorrelation function and spectrum in burst type LDA measurements. The requirements are 1) to be able to resolve the small scale structure (Kolmogorov scale) and 2) to make sure no directional bias is introduced by the large scale fluctuations.

Again we base the estimates on empirical relations valid for the fully developed jet. The Kolmogorov scale is given by:

$$\eta = \left(\frac{\nu^3}{\epsilon} \right)^{\frac{1}{4}} \qquad\qquad\qquad [4.14]$$

where ν is the kinematic viscosity and ϵ the dissipation rate. The dissipation rate is estimated as:

$$\epsilon \cong \frac{u^3}{\ell} \qquad\qquad\qquad [4.15]$$

for air $\nu = 1.5 \cdot 10^{-5}$ m^2/s. Thus at x = 20 D and for an exit velocity of U_E = 3 m/s we find:

$\epsilon \cong 0.45$ m^2/s^2

and

$\eta = 0.29$ mm

Thus, even at this relatively low Re-number flow, only special miniature probes and a well defined LDA measuring volume can resolve the small scales.

The large scale fluctuations approach 25 % of the mean at the center line and even higher relative values at the boundary of the jet. To eliminate directional effects due to the finite number of fringes in the measuring volume it is necessary to add a frequency shift of the order of the mean Doppler frequency at the center of the jet.

5. Temporal Resolution

There are primarily two sources of high frequency fluctuations in the signal from a stationary velocity probe: the small scale turbulent velocity fluctuations in the fluid surrounding the measuring point and the small scale spatial velocity pattern, which is convected past the probe by the mean motion. The distinction is of course artificial and sometimes it can not be made at all.

"Turbulence in a box" or "stationary homogeneous turbulence" are examples approaching the former model, whereas low intensity wind tunnel turbulence is an example of the latter. If the structure of the turbulence does not change appreciably in the time it takes to sweep past the measuring volume, we may use "Taylor's hypothesis", which regards the turbulent velocity structure fixed while it is swept past the volume; this is also called the "frozen turbulence" hypothesis.

5.1 Temporal Resolution, LDA

The inherent temporal resolution of the laser Doppler method is virtually unlimited as far as the optical response time is concerned. In practice the temporal response of the LDA is determined by the limited ability of the seeding particles to follow fast fluctuations in flow velocity and by the electronic response time of the signal processing equipment. The seeding particle response has been studied by e.g. Melling and Whitelaw (1973), Al-Taweel (1975), Melling and Whitelaw (1976) and Gregor (1977). The particle lag results in a spectral transfer function with high frequency attenuation.

The useful electronic bandwidth is determined by the signal-to-noise ratio of the detector current. However, the spatial resolution also enters the considerations. The maximum coherence time of the detected signal is the time it takes a particle to traverse the measuring volume, the residence time. This is the time available for the determination of the frequency in both multi- and single particle Doppler signals. If the processor bandwidth is adjusted accordingly fluctuations within the measuring volume and times faster than the transit time can not be detected. However, if the signal-to-noise ratio allows there is no reason why greater bandwidth can not be used, which would allow measurement of fluctuations during a transit. In special cases as e.g. measurement of acoustic waves this can be useful. The fluctuations should be coherent across the measuring volume otherwise the limited spatial resolution attenuates the transducer output.

5.2 Temporal Resolution, Hot-Sensor Anemometer

The hot-sensor instrument exhibits a time lag between the time for the occurrence of a step input to the time the output has stabilized to within a certain fraction of its final value. The temporal resolution may also be presented in the form of a frequency transfer function given as the output response to a sine wave input as a function of frequency.

The inherent sources of time lag in the heat transfer from sensor to flow are boundary layer transients, thermal storage in the sensor and the electronic lag. In most situations the second effect dominates. Thus the heat loss in the energy balance equation (eq. [3.1]) for the non-steady case is determined by the steady state heat transfer properties. However, because the coefficients in eq. [3.1] are in general temperature dependent the equation is non-linear and not amenable to analytic solution.

Under certain restrictions the equation may be simplified to a 1st order differential equation, and it is then possible to derive a time constant in the usual sense for the probe response. The restrictions are:

— the heat transfer is given by an analytic expression (the static calibration law, e.g. $\Phi = (R_w - R_g)(A + BU^m)$)

— the sensor resistance is assumed linear in temperature ($R_m = R_0(1 + \alpha T)$)

— there is a fixed temperature distribution along the sensor

— small resistance and current changes:
$$R_w = \bar{R}_w + r_w \qquad\qquad r_w \ll R_w; \bar{r}_w = 0$$
$$I = \bar{I} + i \qquad\qquad i \ll I; \bar{i} = 0$$

— small flow fluctuations:
$$U = \bar{U} + u \qquad\qquad u \ll U; \bar{u} = 0$$

— constant material parameters (α, c_p, μ)

Solution of the equation then leads to the well known hot-wire time constants for the constant current anemometer:

Hot-wire

$$\tau_{CCA} = \frac{R}{R_0} \frac{c_w}{\alpha R_0} \frac{1}{A + BU^m} \qquad\qquad [5.1]$$

The feed-back in the constant temperature anemometer with gain G reduces the temperature excursions of the wire and leads to a reduction in the time constant:

$$\tau_{CTA} = \tau_{CCA} \cdot \frac{1}{2RGa} \qquad [5.2]$$

For the hot-film sensor on a relatively large substrate a similar expression results (Bellhouse and Schultz, 1967):

Hot-film

$$\tau_{CCA} = \left(\frac{R}{R_o}\right)^2 \frac{F^2 \rho c K}{(A+BU^m)^2} \qquad [5.3]$$

$$\tau_{CTA} = \tau_{CCA} \frac{1}{(2RGa)^2} \qquad [5.4]$$

At very low frequencies the heat conduction into the substrate and the heat storage in the substrate cause a deviation from the expected flat frequency response.

The small signal theory outlined above is not valid at most overheating ratios used in practice. The most significant deviations occur for the constant current configuration because of the larger relative temperature changes occurring in the sensor without constant temperature feed-back. However, both systems are highly non-linear for large relative flow fluctuations. The non-linear behavior of the constant current system has been analyzed by i.a. Compte-Bellot and Schon, 1969). P. Freymuth (1977) also analyzed the content of higher harmonics of the frequencies occurring in the flow velocity and showed that to keep these terms reasonably low ($<$2 % of the fundamental) the frequencies of the input must be limited to values less than 20 % of the upper frequency limit corresponding to the time constant computed from the simplified, linearized theory: Thus a severe limitation is imposed on the bandwidth of the input for a given anemometer when large signal fluctuations are present.

The influence of a finite wire length on the hot-wire frequency transfer function has recently been investigated by Beljaars (1976). This analysis showed that the temperature re-distribution along the wire results in an additional pole in the equivalent diagram of the system comprising bridge, amplifier (assumed ideal) and sensor. The result is an additional drop in sensitivity at a characteristic frequency, which was found to be of the same order of magnitude as the cut-off frequency for the same wire in the constant-current configuration. This drop in sensitivity could amount to 10-15 %. At higher frequencies the response leveled out until the usual constant temperature cut-off frequency was reached. The author also investigated the effect of finite wire length and the effect of heat storage in the supports around the wire ends on the temporal response of the wire. The author showed the existence of a further drop in sensitivity at a frequency corresponding to the thermal inertia of the wire and support taken together.

All of these effects introduce higher order terms in the response equations, and it becomes impossible to describe the system frequency-and phase characteristics in terms of a single pole circuit even if the electronic bandwidth is so high that phase lag in the electronics itself is negligible. Freymuth (1977) has recently investigated the effect of a non-ideal electronic feed-back amplifier, in particular in the CTA-configuration. He finds that a third order differential equation adequately describes the transient behavior of the CTA in the frequency range of interest. However, experiments with a CTA circuit with a fully compensated bridge show evidence of higher order terms, which can not be explained by the third order theory. Similar investigations of the dynamic response of hot-sensor anemometers have been carried out by i.a. Davis (1970), Perry and Morrison (1971) and Wood (1975).

In conclusion, the hot-sensor anemometer has a quite complicated dynamic transfer function. Under rather restricted circumstances the anemometer response may be described by a first order differential equation, and a time constant may be defined and computed. The CTA best adheres to this description because of the feed-back, which attempts to keep the mean resistance constant, but at the usual operating temperatures higher order effects such as temperature re-distribution along the wire and heat storage in the supports modify the transient behavior of the anemometer and complicate the dynamic response. If accurate measurements with films or short wires are needed a dynamic calibration in situ is recommended. The dynamic response to current and temperature modulation can be shown to be equivalent (Baker 1976, Freymuth 1977 and 1978). This simplifies the practical problems in connection with dynamic calibration.

5.3 Requirements to Temporal Resolution

Example A: Consider the requirement that we want to measure scales of the order of $k \cdot \ell \leqslant 10$, where ℓ is the typical energy containing (large scale) eddy size. This led to a minimum spatial resolution of $\Delta x \cong \pi/100x$. The convection velocity is known empirically: $U_c \cong 0.6 \cdot U_E$. Using Taylor's hypothesis at $x=D$ we find, for $U_E = 30$ m/s and $x = D$:

$$f \cong \frac{U_c}{\Delta x} \cong 2 \text{ kHz}$$

Again we see that the specifications are not very strict in this case.

Example B: The small scale velocity structure in the fully developed jet may be estimated by the Kolmogorov scales: The Kolmogorov time scale is:

$$\tau = \left(\frac{\nu}{\epsilon}\right)^{\frac{1}{2}} \cong 0.2 \text{ kHz}$$

The length scale is $\eta = 0.29$ mm. The axial velocity of the jet is:

$$U(x) \cong 6.5 \ U_E / (\frac{x-x_c}{D}) \qquad\qquad [5.5]$$

Using Taylor's hypothesis at $x = 20$ D and $U_E = 3$ m/s we find:

$$U(x) = 0.99 \text{ m/s}$$

and

$$f_{max} \cong \frac{U(x)}{\eta} = 3.4 \text{ kHz}$$

Thus the required temporal resolution is determined by the convection of the small scale structure past the probe.

6. Conclusion

In dynamic measurements in fluid mechanics it is important to know the transducer characteristics and compare to the resolution required by the specific flow problem. It is necessary to consider the transducer as a whole including the mode of operation of the electronic signal processing equipment. In most hot-sensor circuits and in tracker-based LDA's the output is presented in the form of an analog voltage, and data processing is performed in the conventional analog manner or by sampling and analog-to-digital conversion. The burst-type LDA with counter signal processor is inherently digital, and digital data output is presented directly after each burst (particle passage). When the full information is presented (velocity, residence time and time since last measurement), all the statistical properties of the velocity field can be computed (see e.g. Buchhave et al. 1978).

More detailed studies of some of the transducer characteristics of the LDA are needed. In particular it seems that the spatial and temporal resolution of the burst-type LDA have not been fully explored. The effects of strongly focused, high intensity laser beams on particle motion and local fluid properties also need further study.

References

Adrian, R.J., Orloff, K.L.: Laser anemometer signal: visibility characteristics and application to particle sizing. *Applied Optics*, **16**, 3, 677-684, March 1977

Al-Taweel, A.: Limits of applicability of tracer particles for measuring fluid turbulence. *Canadian J. of Chem. Eng.*

Baker, C.B.: Hot-film probe frequency response. Internal Memorandum, Applied Research Laboratory, Penn. State University, File No. 76-174

Ballik, E.A., Chan, J.H.C.: SNR in optical velocimeters: effects of detection angle. *Appl. Optics*, **16**, 3, 596-600, March 1977

Bellhouse, B.J. and Schultz, D.L.: The determination of fluctuating velocity in air with heated thin-film gauges. *J. Fluid Mech.*, **29**, 289-295, 1967

Bellhouse, B.J. and Bellhouse, F.H.: Thin-film gauges for the measurement of velocity or skin friction in air, water or blood. *J. Phys. E.*, Ser. 2, Vol. **1**, 1211-1213, 1968

Beljaars, A.C.M.: Dynamic behaviour of the constant temperature anemometer due to thermal inertia of the wire. *Appl. Sci. Rev. (Netherlands)*, Vol. **32**, No. 5, p. 509-18, 1976

Berman, N.S. and Dunning, J.W.: Pipe flow measurements of turbulence and ambiguity using laser-Doppler velocimetry. *J. Fluid Mech.*, **61**, part 2, 289-99, 1972

Betchow, R.: L'influence de la conduction thermique sur les anémomètres à fil chauds *Proceedings Koninkl. Ned. Akad. v. Wetenschapen*, **51**, 721-730, 1948

Bloom, A.L.: Gas lasers. New York, London and Sydney: John Wiley and Sons, 1965

Brayton, D.B., Kalb, H.T. and Crosswy, F.L.: Two-Component Dual-Scatter Laser Doppler Velocimeter with Frequency Burst Signal Readout. *Appl. Optics*, **12**, 6, 1145-1155, June, 1973

Bruun, H.H.: Interpretation of a universal calibration law. *J. Phys. E.*, **4**, 225-231, 1971

Buchhave, P.: Light Collecting System and Detector in a Laser Doppler Anemometer. *DISA Information*, **15**, p 15, Oct. 1973

Buchhave, P.: Laser Doppler velocimeter with variable optical frequency shift. *Opt. and Laser Technol.*, **7**, 11-16, 1975

Buchhave, P.: Biasing errors in individual particle measurements with the LDA-counter signal processor. In *Proceedings of the LDA-Symposium Copenhagen, 1975*, 258-278, 1976

Buchhave, P. et al. (eds.) *Proceedings of the LDA-Symposium Copenhagen, 1975*, 1976

Buchhave, P., George, W.K. Jr. and Lumley, J.L.: The measurement of turbulence with the laser Doppler anemometer. Prepared for publication in Annual Review of Fluid Mechanics, Annual Review Inc., Palo Alto Cal. 94306 (Van Dyke and Wehausen, co-editors). Also available as TRL-101 Report from Turbulence Research Laboratory, State University of New York at Buffalo

Champagne, F.H., Sleicher, C.A. and Wehrmann, O.H.: Turbulence measurements with inclined hot-wires; Pt. 1. Heat transfer experiments with inclined hot-wire. *J. Fluid Mech.* Vol. **28**, pt. 1, pp 153-175, 1967

Collis, D.C., Williams, M.J.: Two-dimensional convection from heated wires at low Reynolds number. *J. Fluid Mech.*, **6**, 357-384, 1959

Comte-Bellot, G. and Schon, J.P.: Harmoniques crées par exitation parametrique dans les anémomètres à fil chaud à intensité constante. *Int. J. Heat Mass Transfer*, **12**, 1661-1677, 1969

Comte-Bellot, G.: Hot-Wire Anemometry. *Annual Review of Fluid Mech.,* Vol. **8**, 209-231, 1976

Corrsin, S.: Turbulence: Experimental methods. *Handbook Der Physik,* Vol. **8**, 524-590, 1968, ed. S. Flugge

Cummins, H.Z., Knable, N. and Yeh, Y.: Observation of diffusion broadening of Raleigh scattered light. *Phys. Rev. Letters,* **12**, 6, p 150, Feb. 1964

Davis, M.R.: The dynamic response of constant resistance anemometers. *J. Phys. E.,* **3**, 15-20, 1970

Davis, M.R. and Davies, P.O.A.L.: Factors influencing the heat transfer from cylindrical anemometer probes. *Int. J. Heat Mass Transfer,* **15**, 1659-1677

Dimotakis, P.E.: Single scattering particle laser Doppler measurements of turbulence. *AGARD Conference Proceedings No. 193 on Applications of Non-Intrusive Instrumentation in Fluid Flow Research,* Saint-Louis, France, May 1976

Drain, L.E.: Coherent and non-coherent methods in Doppler optical beat velocity measurement. *J. Phys. D.,* **5**, 481-495, 1972

Drain, L.E. and Moss, B.C.: The frequency shifting of laser light by electro-optic techniques. *Opto-electronics,* **4**, 429-439, 1972

Drubka, R.E., Tan-atichat, J. and Nagib, H.M.: Analysis of temperature compensating circuits for hot-wires and hot-films. *DISA Information,* **22**, 5-14, 1977

Durrani, T.S.: Noise analysis for laser Doppler velocimeter systems. *IEEE Transactions, COM-20,* 296-307, June 1973

Durrani, T.S. and Greated, C.A.: Laser systems in flow measurement. New York and London Plenum Press, 1977

Durst, F., Melling, A., Whitelaw, J.H.: Principles and Practice of laser-Doppler anemometry. New York: Academic Press, 1976

Durst, F., Heiber, K.F.: Signal to noise ratio of laser Doppler signals. *Opt.' Acta (GB),* Vol. **24**, No. 1, 43-67, Jan. 1977

Edwards, R.V., Angus, J.C., French, M.J. and Dunning, J.W., Jr.: Spectral analysis of the signal from the laser Doppler flowmeter: Time-independent systems. *J. Appl. Phys.,* **42**, No. 2, 837-850, Feb. 1971

Foord, R. et al.: A solid state electro-optic phase modulator for laser Doppler anemometry. *J. Phys. D: Appl. Phys.,* Vol. **7**, 136-139, 1974

Freymuth, P.: Further investigation of the nonlinear theory for constant-temperature hot-wire anemometers. *J. Phys. E.,* **10**, 710-713, 1977

Freymuth, P.: Frequency response and electronic testing for constant temperature hot-wire anemometers. *J. Phys. E.,* **10**, 705-710, 1977

Freymuth, P.: Theory of frequency optimization for hot-film anemometers. *J. Phys. E.,* **11**, 177-179, 1978

George, W.K. and Lumley, J.L.: The laser Doppler velocimeter and its application to the measurement of turbulence. *J. Fluid Mech.,* **60**, pt 2, 321-362, 1973

George, W.K.: Limitations to measuring accuracy inherent in the laser Doppler signal. In *Proceedings of the LDA-Symposium Copenhagen, 1975,* 19-63, 1976

Gregor, W.: Limits imposed on the concentration and size of tracer particles in laser-Doppler anemometry. *DISA Information,* **22**, 39-41, 1977

Hanson, S.: Broadening of the measured frequency spectrum in a differential laser anemometer due to interference plane gradients. *J. Phys. D.*, **6**, 164-171, 1973

Hanson, S.: Coherent detection in laser Doppler velocimeters. *Opto-electronics*, **6**, 263, 1974

Hinze, J.O.: Turbulence. McGraw-Hill, Inc., 1975

Jorgensen, F.E.: Directional sensitivity of wire and fiber film probes. *DISA Information*, **11**, 31-37, May 1971

King, L.V.: On the convection of heat from small cylinders in a stream of fluid: Determination of the convection constants of small platinum wires with application to hot-wire anemometry. *Phil. Trans. Roy. Soc. London*, **214A**, 373, 1914

Koppius, A.M. and Trines, G.R.M.: The dependence of hot-wire calibration on gas temperature at low Reynolds numbers. *Int. J. Heat Mass Transfer*, **19**, 967-947, 1976

Kovasznay, L.S.G.: Hot Wire Method. In turbulence measurements, section F of physical measurements in gas dynamics and combustion (ed. R.W. Ladenburg, B. Lewis, R.N. Pease and H.S. Taylor), pp 213-285, Vol. **9** of *High Speed Aerodynamics and Jet Propulsion.* Princeton: Princeton Uni. Press, 1954

Kreid, D.K.: Laser Doppler velocimeter measurements in nonuniform flow: error estimates. *Appl. Optics*, Vol. **13**, 1872-1881, 1974

Lading, L.: Analysis of signal-to-noise ratio of the laser-Doppler velocimeter. *Opto-electrinics*, **5**, 175-187, 1973

Larsen, S.E. and Busch, N.E.: Hot-wire measurements in the atmosphere. Part 9: Calibration and response characteristics. *DISA Information*, **16**, 15-34, 1974

Lumley, J.L.: The structure of inhomogeneous turbulent flows. *Atmospheric Turbulence and Radio Wave Propagation.* (A.M. Yaglom and V.I. Tatasski, eds.). Publishing House Nauka, Moscow, pp 166-188, 1967

Mayo, W.T. Jr.: Photon counting processor for laser velocimetry. *Applied Optics*, **16**, No. 5, 1157-1162, May 1977

Melling, A. and Whitelaw, J.H.: Seeding of gas flows for laser anemometry. *DISA Information*, **15**, 5-14, 1973

Melling, A. and Whitelaw, J.H.: Optical and flow aspects of particles. In *Proceedings of the LDA-Symposium Copenhagen 1975*, 382-402, 1976

Oldengarm, L.: Development of rotating diffraction gratings and their use in laser anemometry. *Opt. & Laser Technol. (GB)*, Vol. **9**, No. 2, 69-71, April 1977

Owen, J.M. and Rogers, R.H.: Velocity biasing in laser Doppler anemometers. In *Proceedings of the LDA-Symposium Copenhagen 1975*, 89-115, 1976

Perry, H.E. and Morrison, G.L.: A study of the constant-temperature hot-wire anemometer. *J. Fluid Mech.*, **47**, Part 3, 577-99, 1971

Perry, A.E. and Morrison, G.L.: Static and dynamic calibrations of constant temperature hot-wire systems. *J. Fluid Mech.*, **47**, part 4, 765-777, 1971

Sandborn, V.A.: Resistance temperature transducers. Fort Collins, Colorado, Metrology Press, 1972

Siegman, A.E.: The antenna properties of optical heterodyne receivers. *Applied Optics*, **5**, 10, pp 1588-1594, Oct. 1966

Thompson, H.D. and Stevenson, W.H. (eds.). *Proceedings of the Second International Workshop on Laser Velocimetry*, March 27-29, 1974. (Bulletin No. 144), Purdue University, W. Lafayette, Ind. 47907

Uberoi, M.H. and Kovasznay, L.S.G.: On mapping and measurement of random fields. *Quart. of Appl. Mathematics,* **10**, 375-393, 1953

Wood, N.B.: A method for determination and control of the frequency response of the constant-temperature hot-wire anemometer. *J. Fluid Mech.,* **67**, pt 4, 769-786, 1975

Wyngaard, J.C.: Measurement of small-scale turbulence structure with hot-wires. *J. Phys. E.,* Ser. 2, Vol. **1**, 1105-8, 1968

Yeh, Y. and Cummins, H.Z.: Localized Fluid Flow Measurements With a He-Ne Laser Spectrometer. *Appl. Phys. Letters,* **4**, 10, p 176, May 1964

Hot-Wire Measurements in Low Speed Heated Flow

by

Louis Fulachier
Institut de Mécanique Statistique de la Turbulence
Laboratoire Associé au C.N.R.S., No. 130
Université D'Aix-Marseille II

1. Introduction

A number of experiments in low speed and heated turbulent flows have been developed over the last few years. These heated flows are indeed very important, particularly in studies dealing with turbulent dispersion of pollutants as well as the structure itself of turbulent flows. When heat can be used as a passive contaminant, the temperature fluctuations can themselves be used to describe the turbulent field (e.g., FULACHIER et al. 1978). And most notably, in the intermittent regions, heat — as compared to velocity fluctuations — enables an easier and more meaningful distinction to be drawn between turbulent and non-turbulent zones (SUNYACH, 1971, DUMAS et al. 1972).

The present paper is concerned with the one experimental technique which seems the most suitable for analysis of turbulent non-isothermal low speed flows: hot-wire anemothermometry. Low speed is here understood to be flows whose compressibility velocity-linked effects are negligible. What is more, temperatures considered are less than about 330 K, and temperature differences are smaller than 30 K (although temperature differences reaching about 100 K are taken into consideration). Also, numerous points of importance in the case of even larger temperature differences (as is the case of combustion) are under-·lined.

In hot-wire anemothermometry the sensor is a metallic wire, whose diameter is of the·order of a micron. This wire is heated by an electric current. Placed in a flow, it is cooled and its response depends essentially on the velocity and temperature of the flow.

Regarding the hot-wire anemometer under normal operating conditions, i.e. when the temperature T_w of the wire is significantly higher than the temperature T of the fluid, numerous studies have been undertaken in the case of isothermal flows. And these studies have dealt with both theoretical and practical levels. In reference, see the didactic works of CORRSIN (1963) and COMTE-

BELLOT (1974, 1976) and, in particular, the works of CHAMPAGNE (e.g. CHAMPAGNE et al. 1967).

The principal accent is here placed upon the determination of temperature fluctuations themselves. In this case, the experimenter arranges for the wire temperature to be only slightly superior to the fluid temperature. The sensor, for the most part made of platinum, acts as a thermometer and thereby reassumes its original use in thermodynamics. Actually, as will be discussed later, the wire remains very often sensitive to velocity and works as a thermoanemometer, now commonly called a "cold-wire".

Therefore, after having written and discussed cold-wire heat balance equations for small and large amplitude fluctuations in both steady and unsteady regimes, this paper insists most particularly upon the precautions needed in order to effectively measure these temperature fluctuations. Secondly, the velocity fluctuations and their correlations with temperature fluctuations in the case of small amplitudes are considered. Emphasis is also placed on the main advantages and inconveniences of the different methods used; and finally, some remarks are made concerning temperature-velocity measurements in the case of high amplitude fluctuations.

2. Determination of the Temperature Fluctuations Themselves: Cold Wire

We propose the undertaking of a theoretical and practical study of cold-wire behaviour with the purpose of estimating and minimizing the different sources of errors encountered by its use in the presence of small or large amplitude fluctuations.

2.1 Cold-Wire Heat Balance Equations and Analysis of Different Parameters in Both the Steady and Unsteady Regimes

We consider a single wire normal to the mean velocity in a given turbulent flow and heated by an electric current of intensity $I(t)$. At every instant t, the electric power, $r_w I^2$, dissipated in this wire, is equal to the thermal power, $mc (\partial T_w/\partial t)$, absorbed by the wire because of its thermal capacity, plus the thermal power, Φ, lost into the fluid. To these two terms are added the thermal power, B, due to conduction by the supports, the "end effects", and the thermal power, R, due to radiation:

$$r_w I^2 = mc \frac{\partial T_w}{\partial t} + \Phi(u_n, T, r_w) + B + R \qquad [1]$$

r_w is the wire resistance at temperature T_w, u_n the velocity component normal to the wire, m the mass of the wire and c the specific heat of the metal.

We suppose for the moment (Cf. 2.2.3) that the wire "aspect ratio" ℓ/d (ℓ, length and d, diameter) is sufficiently large so that the wire may be considered infinitely long; thus, B is negligible. In fact, small diameters are used for the cold wire (d≤1μ). Thus a relatively large ℓ/d ratio can be obtained without the space resolution being necessarily critical. Yet, the problem of "end effects" remains primordial. In particular when ℓ/d is fairly weak, (ℓ/d<600); this corresponds generally to the use of a wire whose diameter is large. The problem is then even more delicate, and this is in particular underlined by BREM-HORST et al. (1977). It is the case for example in experiments with combustion (e.g., MING HO et al. 1976).

Concerning the radiative effects, even in the case of a strongly heated wire ((r_w-r)/r≅0.8, r being the resistance at T), the power loss of the wire is negligible (CORRSIN 1963). This is likewise the case for the power the wire could receive given the aforementioned temperatures.

In the further consideration of eq. [1], two anemothermometer types are possible: the constant-current or the constant-temperature (i.e., in fact the constant-resistance anemometer). However, it can be seen that, when the overheating is reduced, the feedback amplifier intervening in this constant-resistance mode can no longer play its role (e.g., COMTE-BELLOT 1974). The constant-current setup is absolutely necessary in order to try to isolate the temperature fluctuations. This is only logical: T_w being very near to T, the temperature T of the flow can be deduced from the wire voltage (e=r_w I) in a practical, direct way, since we know that:

$$r_w = r_o(1+\chi(T_w-T_o) + \chi_1(T_w-T_o)^2), \qquad [2]$$

where r_o is the wire resistance at a reference temperature T_o and χ and χ_1 are the temperature coefficients of the electric resistivity of the wire. For the platinum wire we have χ≅3.5 10^{-3} (K^{-1})≅1/273 and χ_1≅-5.5 10^{-7} (K^{-2}). Concerning the already mentioned temperature differences we show later on (Cf. 2.1.1) that the linear law suffices:

$$r_w = r_o(1+\chi(T_w-T_o)) \qquad [3]$$

2.1.1 Steady or "Static" Response

Expression of Φ

First of all we consider here the response of a wire to fluctuations sufficiently slow that its thermal lag does not come into play. Equation [1] is thus written:

$$r_w^+ \, |^2 = \Phi, \tag{4}$$

where r_w^+ is the "ideal" wire resistance without thermal lag (CORRSIN 1963). The expression of Φ is deduced from the wire-fluid exchange laws given by the expression of the Nusselt number, Nu:

$$Nu = hd/k = \Phi/\pi k \ell (T_w - T). \tag{5}$$

h represents heat transferred per unit surface area and degree temperature difference from solid to fluid; k is the thermal conductance of the fluid at temperature T. We have adopted the COLLIS and WILLIAMS law (1959), valid for air and for $0.02 \leqslant Re \leqslant 44$, where $Re = u_n d/\nu$ (ν = kinematic viscocity of air):

$$Nu_f = (0.24 + 0.56\, Re_f^n)((T_w - T)/2T)^{0.17} \tag{6}$$

The subscript f indicates that the physical properties are evaluated at the "film temperature" $T_f = (T_w + T)/2$. The exponent n varies in fact very little with the Reynolds number, but remains about 0.45. This empirical law has been confirmed by BRADBURY and CASTRO (1972).

As we shall see (Cf. 2.2.1), the "overheat ratio" $(r_w - r)/r$ must be smaller than 10^{-3}, which allows important simplifications. If we furthermore note, as does HINZE (1975), that k/ν^n is nearly independent of temperature, while taking into consideration [3], [5], [6] and the SUTHERLAND formula (CHAPMAN and COWLING 1960, BRUN et al. 1968), written for small temperature differences $k = k_o(1 + 0.0028(T - T_o))$, it follows that:

$$\Phi = (r_w - r)(A' + Bu_n^n) \quad \text{with} \quad A' = A(1 + 0.0028(r - r_o)/r_o\chi) \tag{7}$$

$$A = 0.24\pi\ell k_o/r_o\chi \quad \text{and} \quad B = 0.56\pi\ell k_o(d/\nu_o)^n/r_o\chi \tag{8}$$

the subscript o corresponds to the reference temperature T_o, often taken as equal to 273 K for simplicity of calculation, but which may very well be chosen equal to the mean temperature of the flow at the measurement point.

Remarks concerning the linear law [3] linking resistance and temperature
Since the difference between the temperatures T_w and T is very small, the value of T can, in the steady regime, be estimated by the measurement of r. This measurement gives a reasonable accuracy, which is in fact linked to velocity contamination. Some possible errors are now estimated, namely those made in the determination of $\overline{T}$, $\overline{T'^2}$, $\overline{T'^3}$ by means of the linear equation [3] rather

than equation [2]. (All throughout is written: $a = \bar{a}+a'$, where $\bar{a}' = 0$). The instantaneous voltage can be written:

$$e^+(t) = e^+(\bar{T}) + \beta(T'/\bar{T}) + \beta c'(T'^2/\bar{T}^2) \qquad [9]$$

where $\beta = \bar{T}(\partial e^+/\partial T)_{\bar{T}}$ is the temperature sensitivity coefficient and $c' = r_o x_1 |\bar{T}^2/\beta$, a coefficient which is zero if [3] is adopted. It can be deduced that the error which affects mean temperature measurements is:

$$(\bar{e}^+ - e^+(\bar{T}))/\beta = (\bar{T}_{meas.} - \bar{T})/\bar{T} - c'(\overline{T'^2}/\bar{T}^2)$$

In the case of relatively moderate temperatures and temperature differences, i.e. $\bar{T} \cong 400$ K and $\Delta\bar{T} \cong 100$ K, $\underline{c'} \cong -0.06$ is obtained; then, even for important turbulent intensities, i.e. $(\overline{T'^2})^{1/2}/\Delta\bar{T} \cong 1$, the maximum error is less than -0.5 %.

For the fluctuations we have:

$$e^{+\prime} = \beta T'/\bar{T} + (\beta c'/\bar{T}^2)(T'^2 - \overline{T}^2)$$

It can be concluded:

$$\overline{e^{+\prime 2}} = \beta^2(\overline{T'^2}/\bar{T}^2)(1+\Psi)$$

with

$$\Psi = 2c'(\overline{T'^2}/\bar{T}^2)^{1/2}S_T + c'^2(\overline{T'^2}/\bar{T}^2)(F_T-1)$$

where $S_T = \overline{T'^3}/(\overline{T'^2})^{3/2}$ and $F_T = \overline{T'^4}/(\overline{T'^2})^2$ are skewness and flatness factors of temperature fluctuations respectively. The relative error in the variance of the turbulent temperature fluctuations is then:

$$[((\overline{T'^2})^{1/2}/\bar{T})_{meas.} - (\overline{T'^2})^{1/2}/\bar{T}]/((\overline{T'^2})^{1/2}/\bar{T}) = (1+\Psi)^{1/2} - 1$$

If we consider the previous experimental case, but with a turbulence intensity such as $(\overline{T'^2})^{1/2}/\Delta T \cong 0.2$, an error is found of the order of -1 %, for $S_T = 3$.

For the skewness factor, the error is given by

$$\overline{e^{+\prime 3}}/(\beta^3(\overline{T'^2})^{3/2}/\bar{T}^3) - S_T \cong 3c'((\overline{T'^2})^{1/2}/\bar{T})(F_T - 1 + c'((\overline{T'^2})^{1/2}/\bar{T})(Q_T - S_T))$$

with $Q_T = \overline{T'^5}/(\overline{T'^2})^{5/2}$. In the previous conditions a relative error less than 3 % is found in S_T for $S_T = 3$ and $F_T = 9$ (the case of the intermittent zone of the boundary layer (DUMAS et al. 1972)).

It must therefore be pointed out that when the temperature differences are slight, say of the order of 30 K, the linear equation [3] is more than sufficient. On the other hand, if these differences were to become important, say of the order of a few hundred degrees, the approximation would be likely to carry along with appreciable errors, in particular when the turbulence intensities are strong, such as in the case of jets. These errors would be even more serious in the case of combustion for instance.

Cold-Wire Response Equations – Sensitivity Coefficients

The cold-wire perpendicular to the mean velocity is in fact not only sensitive to temperature T but also to velocity; we are assuming the wire to be sensitive only to the longitudinal component u'. This assumption is fairly close to reality given the fact that the other components are only corrective terms. The general case of strong amplitude fluctuations is dealt with here. The voltage fluctuation may be written as:

$$e^{+'} = -\alpha \frac{u'}{\overline{u}} + \beta \frac{T'}{\overline{T}} + \gamma_1 \frac{T'u' - \overline{T'u'}}{\overline{T}\,\overline{u}} + \gamma_2 \frac{u'^2 - \overline{u'^2}}{\overline{u}^2} + \gamma_3 \frac{T'^2 - \overline{T'^2}}{\overline{T}^2} \qquad [10]$$

where $\alpha = -\overline{u}(\partial e^+/\partial u)_M$ is the sensitivity coefficient for u-velocity fluctuations (the subscript M indicates values taken at $\overline{u}$ and $\overline{T}$). The coefficients γ can be written:

$$\gamma_1 = \frac{\overline{T}\,\overline{u}}{2}\left(\frac{\partial^2 e^+}{\partial T \partial u}\right)_M, \; \gamma_2 = \frac{\overline{u}^2}{2}\left(\frac{\partial^2 e^+}{\partial u^2}\right)_M \text{ and } \gamma_3 = \frac{\overline{T}^2}{2}\left(\frac{\partial^2 e^+}{\partial T^2}\right)_M.$$

We note here (DUMAS 1978a) that it would be physically more meaningful to introduce $T'/\Delta\overline{T}$ and u'/u_e (where $\Delta\overline{T}$ is the maximum temperature difference of the flow and u_e the velocity reference), instead of $T'/\overline{T}$ and $u'/\overline{u}$. Indeed, the first two are of the same order of magnitude. But we have used the second expressions, which are commonly used.

Considering [4] and [7], it follows:

$$\alpha = \frac{nB\,\overline{u}^n\,(r_w^+ - \overline{r})^2}{\overline{r}\,l} = \frac{nB\,\overline{u}^n\,l^3\,\overline{r_w^+}^2}{\overline{r}(A' + B\,\overline{u}^n)^2} \qquad [11]$$

$$\beta = -\frac{\overline{T}\,r_o\chi l\,\overline{r_w^+}}{\overline{r}}\left(1 + \frac{0.0028A(r_w^+ - \overline{r})^2}{r_o\chi\,\overline{r_w^+}\,l^2}\right) \qquad [12]$$

$\beta \cong \overline{T} \, r_o \chi \overline{l r_w^+}/\overline{r}$ with a relative error of 0.001 for a platinum wire whose characteristics are: $d = 1 \, \mu m$, $\ell = 0.6$ m, $(\overline{r_w^+ - r})/\overline{r} \cong 0.0005$, $I \cong 0.2$ mA (Cf. 2.2.1). The ratio of velocity and temperature sensitivities is:

$$s = \frac{\alpha}{\beta} \cong \frac{nB\overline{u}^n \overline{r_w^+} \, I^2}{\overline{T}(A'+B\overline{u}^n)^2 \, r_o \chi} \cong \frac{nI^2 \, \overline{r_w^+} \, 0.56 \, Re^n}{\overline{T} \pi k_o \ell (0.24 + 0.56 \, Re^n)^2} \qquad [13]$$

This is the relation suggested by WYNGAARD (1971), 0.45 being chosen for n.

Also we obtain:

$$\frac{\gamma_1}{\beta} \cong \frac{nB\,\overline{u}^n}{2(A+B\overline{u}^n)} \cdot \frac{\overline{r_w^+} - \overline{r}}{\overline{r}} \cong \frac{s}{2}$$

$$\frac{\gamma_2}{\beta} \cong \frac{nB\,\overline{u}^n I^2 \, \overline{r_w^{+2}}}{2\overline{T} \, r_o \, \chi \overline{r}(A'+B\,\overline{u}^n)^3} \left((1-n)(A-I^2) + (n+1)B\,\overline{u}^n\right)$$

$$\frac{\gamma_3}{\beta} = 8 \cdot 10^{-6} \, \frac{\overline{T} \, A^2 \, (\overline{r_w^+} - \overline{r})^3}{I^4 \chi \, r_o \, \overline{r} \, \overline{r_w^+}} = 8 \cdot 10^{-6} \, \frac{\overline{T} \, A^2 \, I^2 \overline{r_w^{+2}}}{(A'+B\,\overline{u}^n)^3 \chi \, r_o r}$$

Considering the previous wire at a certain point in the flow, ($u_e \cong 10$ ms^{-1}, $(\Delta T)m \cong 20$ K) where $u \cong 1$ ms^{-1}, $\overline{T} \cong 300$ K and where the turbulence intensity is strong ($(\overline{u'^2})^{1/2}/\overline{u} \cong 0.3$ from which $(\overline{T'^2})^{1/2}/\overline{T} \cong 0.002$), it follows: $s \cong 10^{-4}$, $\gamma_1/\beta \cong \gamma_2/\beta \cong 5 \cdot 10^{-5}$, $\gamma_3/\beta \cong 2 \cdot 10^{-4}$. In comparing the various orders of magnitudes of the different terms of equation [10], it can be noted that only the first two terms remain:

$$\frac{e^{+\prime}}{\beta} = -s\frac{u'}{\overline{u}} + \frac{T'}{\overline{T}} \qquad [14]$$

This is also emphasized by estimating the terms of $\overline{e^{+\prime 2}}/\beta^2$ up to the third order. The only possible contaminant term is the one relative to $\overline{T'u'}$ (Cf. 2.2.1.); but its influence is effectively reduced by using the aforementioned characteristics for the cold-wire (these characteristics are commented upon in section 2.2.).

The measured skewness is given by:

$$(S_T)_m \cong S_T - 0.04 \, (F_T - 1)^{1/2} \, R + 7 \cdot 10^{-4} \, (F_u - 1)^{1/2} \, R' + 3 \cdot 10^{-5} S_u$$

R and R' being the correlation coefficients:

$$R = \overline{u'T'^2} / ((\overline{T'^4} - \overline{T'^2}^2)\overline{u'^2})^{1/2}, \quad R' = \overline{u'^2 T'} / ((\overline{u'^4} - \overline{u'^2}^2) \, \overline{T'^2})^{1/2}$$

In the most pessimistic of cases, $S_T \cong 0$, $F_T \cong F_u \cong 3$, and keeping in mind the fact that the triple correlation coefficients generally remain smaller than 10 %, we can state: $(S_T)_m \cong S_T$.

Thus, if the necessary precautions are taken, there is no cold-wire "static" non-linearity. Nevertheless, it should not be forgotten: a nonlinearity may remain for high moments due to the electronics used (DUMAS 1978b). On the other hand, when the turbulence intensity is strong, the velocity field can be perturbed by the wake of the supports (Cf. 3.3.) and the same may be true for the thermal field.

2.1.2. Unsteady or "Dynamic" Regime

For this case we postulate (CORRSIN 1963) that the heat-loss term can be approximated by the steady heat-loss term Φ (equation [7]). This is especailly justified by the very small diameter of the wires. Equations [1], [3], and [7] and the hypothesis concerning B, lead to the following statement:

$$r_w I^2 = (mc/r_o\chi)(dr_w/dt) + (r_w - r)(A' + B\, u_n^n) \qquad [15]$$

It is to be noted here that not only r_w depends on the time, but so does r, which is the wire resistance at the fluid temperature.

To establish the governing equation for the fluctuations, we may proceed as follows (see e.g. DUMAS 1978a): From [15] we have:

$$r_w' I^2 = (mc/r_o\chi)(dr_w'/dt) + \Phi(u,T,r_w) - \overline{\Phi}$$

A Taylor expansion of Φ is then carried out in the region of mean value $\overline{u}$, $\overline{T}$, $\overline{r_w}$. In particular, this method gives directly sensitivity coefficients, such as α and β, which we have already established ([11] and [12]). This is one of the reasons we have preferred to use the classical notion of an "ideal" wire (Cf. 2.1.1.). In this case equation [15] becomes:

$$r_w^+ I^2 = (r_w^+ - r)(A' + B\, u_n^n)$$

and we obtain:

$$\frac{1}{rI^2}\frac{mc}{r_o\chi}\frac{dr_w}{dt} + \frac{r_w - r}{r_w^+ - r} = 1 \qquad [16]$$

This equation holds true for the case of a cold-wire. It differs to a slight extent from the equation established for the case of a hot-wire in an isothermal flow with the hypothesis of small overheats (COMTE-BELLOT 1974, 1976). The problem here is to determine the "small parameter" in equation [16]. It may be written:

$$r_w^+ - r = (\overline{r_w^+ - r})(1 + \epsilon) \text{ with } \epsilon = (r_w^{+\prime} - r')/(\overline{r_w^+ - r}) \tag{17}$$

From this point onward it should be noted that $r_w^{+\prime}/\overline{r_w^+ - r})$ is not to be considered: this term is very large by the definition itself of the cold wire. Taking into account equations [3] and [14], and regardless of the amplitude of fluctuations, it follows:

$$r_w^{+\prime}| = - \alpha(u'/\overline{u}) + (\overline{r_w^+/r})r'|$$

and, according to [17]:

$$\epsilon = -(1/(\overline{r_w^+ - r})|)\alpha(u'/\overline{u}) + (r'/\overline{r})$$

By introducing the expression for α, A' and B (equations [11] and [8]) we obtain, regardless of the amplitude of fluctuations:

$$\epsilon = \frac{0.56n \text{ Re}^n \overline{r_w^+}}{(0.24 + 0.56 \text{ Re}^n)\overline{r}} \frac{u'}{u} + \frac{r'}{r} = a\frac{u'}{u} + \frac{T'}{T} \tag{18}$$

The factor a tends to zero with decreasing velocity and to $n\overline{r_w}/r$, with increasing velocity or Re; for a cold-wire, a is smaller than approximately 0.5. $r'/\overline{r}$ is nearly equal to T'/T for a platinum wire and $T_o = 273$ K. In the case of a hot-wire with overheat coefficient of the order of 0.8, the value of a is nearly the triple of its previous value. Thus, ϵ is certainly a small parameter. Cold-wire behaviour can then be deduced from [16] in the presence of small or large amplitude fluctuations in the "dynamic" regime.

Small amplitude fluctuations

With first order expansions [16] yields:

$$\frac{mc}{r_o \chi l^2} \frac{dr_w^+}{dt} + \overline{r}\frac{\overline{r_w - r}}{r_w^+ - r}(1 + \frac{r'}{r} + \frac{r_w' - r'}{r_w - r} - \frac{r_w^{+\prime} - r'}{r_w^+ - r}) = \overline{r} + r' \tag{19}$$

And, using the statistical average of [19], we deduce: $\overline{r_w} = \overline{r_w^+}$.

Equation [19] then becomes

$$M\frac{dr_w'}{dt} + r_w' = r_w^{+\prime} \tag{20}$$

M is the so-called time constant of the wire, and is given by:

$$M = (mc/r_o \chi I^2)(\overline{r_w^+ - r})/\overline{r} \quad (\cong (c\delta/k_o)d^2/(1+2.3\ Re^n)\ \text{if}\ I \to 0) \tag{21}$$

where δ is the specific mass.

The differential equation [20] has now constant coefficients, and is identical to the hot-wire equation (CORRSIN 1963).

Large amplitude fluctuations

Taking into account equation [21], equation [16] reads:

$$M \cdot \frac{\overline{r}}{r_w^+ - \overline{r}} \ \frac{r_w^+ - r}{r} \ \frac{dr_w}{dt} + r_w = r_w^+ \tag{22}$$

It is the factor $(r_w^+ - r)/r$ which introduces a "dynamic" nonlinearity. Following equations [17] and [18], equation [22] may be written:

$$M \left(1 + a \frac{u'}{u}(1 - \frac{T'}{\overline{T}})\right) \frac{dr_w'}{dt} + r_w = r_w^+ \tag{23}$$

For a hot-wire in an isothermal flow, using the expression of $e'^+(t)$ (for example, see COMTE-BELLOT 1974), we obtain:

$$M \left(1 + a_u \left(\frac{u'}{\overline{u}} - \frac{1}{2}(\frac{1}{2} - 2a_u) \frac{u'^2 - \overline{u'^2}}{\overline{u}^2} + \frac{1}{2} \frac{v'^2 - \overline{v'^2}}{\overline{u}^2} + \cdots\right)\right) \frac{dr_w'}{dt} + r_w = r_w^+ \tag{24}$$

As already stated herein, a_u is nearly three times larger than a. On the other hand, the terms which multiply a_u can become much more important than the terms which multiply a.

Thus, the "dynamic" nonlinearity is much less critical for the cold-wire than for the hot-wire. It would be interesting to continue these cold-wire calculations as CORRSIN (1963) and COMTE-BELLOT and SCHON (1969) have done for the hot-wire.

2.2 Precautions in Measuring Temperature Fluctuations

In section 2.1. the different cold-wire behaviour laws have been emphasized. In particular the relations established therein allow the prediction of the behaviour of various quantities. But we should always refer to the experiment itself, e.g. direct calibrations.

We now examine the three main problems raised with the cold wire: velocity contamination, time constant, and "end effects".

2.2.1 Longitudinal Velocity Contamination

Equation [14] pointed out that the cold-wire always remains sensitive to u-velocity. We therefore try to reduce as much as possible the sensitivity ratio s. This is done to minimize contamination, but it must be kept in mind that the signal/noise ratio is crucial. For $0.02 < Re < 44$, the ratio s is given by equation [13]. For $Re < 0.5$, corresponding to very small diameter wires, COLLIS and WILLIAMS (1959) propose an equation slightly different from [6]. Their expression is:

$$s = 0.48 \, \overline{r_w^{+2}} \, I^2 / \overline{T} \pi \ell k \overline{r} \qquad [25]$$

with k considered at ambient temperature $\overline{T}$. By introducing the wire resistivity $\overline{\rho_w^{+}}$, which differs very little at $\overline{T}$ and $\overline{T}_w$, and since $k_f \cong k$, the equation [25] becomes equivalent to the equation proposed by LARUE et al. (1975). So, s varies as I^2 for given ℓ and d, according to [13] or [25].

The possible contamination error concerning the desired quantity ($\overline{T'^2}$, $\overline{T'^3}$, $\overline{T'^4}$ ...) may be determined by measuring $(\overline{e^{+'p}})^{1/p}/\beta$ (where p = 2,3,4 ...) for different values of I tending to zero. This "fluctuation-diagram" method was first established for determination of second order moments in supersonic flow (KOVASZNAY 1953). And it was then developed by VEROLLET (1969, 1972) and FULACHIER (1972) in low speed heated flows. From [14] we obtain the statistical equation:

$$\overline{e^{+'2}}/\beta^2 = s^2 (\overline{u'^2}/\overline{u}^2) - 2s(\overline{T'u'}/\overline{T}\,\overline{u}) + (\overline{T'^2}/\overline{T}^2) \qquad [26]$$

The curve, giving $(\overline{e^{+'2}})^{1/2}/\beta$ as a function of current I, enables us to obtain the contamination relative to $\overline{T'^2}$: this is the difference between the ordinate of the point corresponding to I and the ordinate of the origin.

In fact, there exists a minimum current below which the wire has no longer correct behaviour (GIOVANANGELI 1975, FULACHIER et al. 1977). Indeed, considering for various currents the wire resistance $\overline{r_w^{+}}$ as a function of $\overline{r_w^{+}}I^2$, a straight line should be found (see [4] and [7]).

Fig. 1 presents results obtained with a platinum cold-wire. The variation is no longer linear below $a_w = (\overline{r_w^{+}} - \overline{r})/\overline{r} \cong 0.0006$, i.e. $I \cong 0.15$ mA. This anomaly of the wire has been confirmed by PASCAL (1978). It seems linked to thermoelectric effects due to soldering of the wire onto the prongs. Each cold-wire used necessitates determination of this critical minimum current.

476   Louis Fulachier

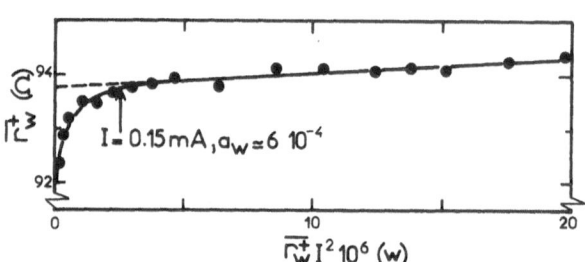

Fig. 1. Evidence of minimum current for cold-wire (d = 1 μ, ℓ = 0.6 mm, ū = 0, T̄ ≅ 295 K).

For a current slightly above this minimum value the term $s^2\overline{u'^2}/u^2$ of equation [26] is always completely negligible relative to the two other terms. So the relative error concerning $\overline{T'^2}$ reads:

$$\epsilon_r \cong -2s\frac{(\overline{u'^2})^{½}/u_e}{(\overline{T'^2})^{½}/\Delta\overline{T}}\frac{\overline{T}}{\Delta\overline{T}}\frac{u_e}{\overline{u}}R \qquad [27]$$

R is the correlation coefficient between T' and u'. An estimate of $|\epsilon_r|$ may then be deduced:

$$(\epsilon_r)_m = 2s(\overline{T}/\Delta\overline{T})/(\overline{u}/u_e)$$

For the previously considered wire ($s\cong10^{-4}$) and a flow such as $\Delta T \cong 20$ K, $T \cong 300$ K, $\overline{u}/u_e \cong 0.5$, we obtain $(\epsilon_r)_m \cong 0.006$.

In respect to moments higher than second order, particularly those introducing derivatives as $(\partial T'/\partial x)^3$, WYNGAARD (1971) underlines the importance of the error caused by the u-velocity contamination in isotropic turbulence. Adopting his proposed hypotheses, we find that the skewness of $\partial T'/\partial x$ — measured with the above mentioned wire — would be less than 0.1, thus acceptable.

The influence of contamination of T' spectrum measurements must also be considered. By writing [14] for the frequency-filtered fluctuations, the following spectra relationship is deduced:

$$\frac{\overline{e^{+'2}}}{\beta^2}F_{e^+} = s^2\frac{\overline{u'^2}}{\overline{u}^2}F_u - 2s\frac{(\overline{T'^2}\,\overline{u'^2})^{½}}{\overline{T}\,\overline{u}}E_{T,u}+\frac{\overline{T'^2}}{\overline{T}^2}F_T \qquad [28]$$

F_T, F_u, F_{e^+}, are respectively the Taylor spectra normalized to unity of T', u', $e^{+'}$, $E_{T,u}$ is the cospectrum relating to T' and u' normalized to the corresponding correlation coefficient. On the other hand, F_u and $E_{T,u}$ are generally shifted towards the low frequencies with respect to F_T (FULACHIER 1971, 1972,

FULACHIER et DUMAS 1971, 1976). Therefore, from [28] — since the term depending on s^2 is negligible — velocity contamination shifts the measured spectrum F_T towards the low frequencies if $\overline{T'u'}$ is negative and towards the high frequencies if $\overline{T'u'}$ is positive. This contamination can thereby result in erroneous values of the temperature-variance "dissipation", if it is calculated from the F_T spectrum, by postulating a local isotropy hypothesis (e.g. FULACHIER 1972).

Remarks Concerning Qualitative Estimation of Contamination
In order to qualitatively determine the u' contamination, the following procedure would be tempting: flow heated, measure $\overline{e^{+'2}}$; flow non-heated, measure $\overline{e_u'^{+2}}$, the wire working under the same conditions. The fact that $\overline{e^{+'2}}$ is very much larger than $\overline{e_u'^{+2}}$ is only one necessary condition for negligible contamination. It is not a sufficient condition: in cold flow, the principal contamination term linked to $\overline{T'u'}$ (see [26]) does not respond to this test. Therefore, this procedure is not an adequate test.

Thus, only if previous precautions with regard to velocity contamination are heeded, we can write from [14] :

$$e_T^{+'} = (\partial \overline{e^+}/\partial \overline{T})T' \tag{29}$$

$\partial \overline{e^+}/\partial \overline{T}$ is constant, as indicated by [3], and as shown by Fig. 2 for a 1 μ, 0.15 mA wire (e.g. BEGUIER et al. 1978).

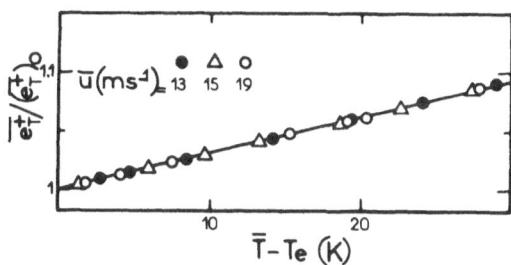

Fig. 2. Cold-wire mean output voltage as a function of mean temperature (d = 1 μ, I $\cong$ 0.15 mA).

2.2.2 Time Constant
From [21] we can see, for the important Reynolds numbers around 44, that the time constant M is varying as $d^{2-n} \cong d^{1.55}$. For small Reynolds numbers (Re<0.5), corresponding in general to the use of very thin wires (d<0.5μ), LARUE et al. (1975) established a variation as $d^{1.6}$. This is the variation which can be adopted.

In order to avoid the necessity of taking into account the wire thermal lag, the experimenter minimizes it by reducing d. Then, if the same sensitivity s is desired, it must not be forgotten to reduce the current as follows: for Re about 44, I^2/d^{2+n} must be constant (equation [13]), and for Re<0.5 I/d must be constant (equation [25]). When the thermal lag of the wire is not compensated, a spectral reduction results for each frequency f; this is deduced from [20] :

$$F'_e = F_e + /(1+(2\pi fM)^2) \qquad\qquad [30]$$

F'_e and F_e+ are voltage fluctuation spectra, the one obtained with an uncompensated wire, the other with a compensated wire. This attenuation often results in non-negligible errors. Let us take as an example a 1μ, 0.15 mA platinum wire in a boundary layer where $u_e \cong 12$ ms^{-1}, $\Delta T = T_w - T_e \cong 20$ K (FULACHIER 1972). At the distance $y^+ = u^+ y/\nu \cong 200$ (u^+ = friction velocity), the M value, measured by the square wave high frequency method (GAVIGLIO 1962) is about 50 μs (ELENA 1977). We can show that with an uncompensated wire an error of only approximately 4 % is made on $\overline{T'^2}$. But the error concerning temperature "dissipation" is about 30 %. Specifically speaking, for a frequency f = 3 kHz a lack of wire compensation means we neglect 1 before 1!.

2.2.3. "End Effects"
These effects depend essentially upon the "aspect ratio" (CORRSIN 1963). For a strongly heated wire, when ℓ/d is of the order of 600, they seem negligible (COMTE-BELLOT 1974). Unfortunately, wires generally used are of the order of several microns (d $\cong 5\ \mu$), which would cause a very important spatial resolution problem. It can thus be understood why experimenters often work with wires with $\ell/d \cong 200$.

For a cold-wire, spatial resolution problems are less critical, given the fact that their diameters are very small (d $\lesssim 5\ \mu$). Regarding "end effects" we can quote the works of MAYE (1970), HOJSTRUP et al. (1976), MILLON (1977), MILLON et al. (1978), BREMHORST and GILMORE (1978).

One idea stands out from the recent studies of PARANTHOEN et al. (1978), namely that end effects have little influence on the high frequencies (the prong heat transfer not having enough time to be established) but, on the other hand, reduce the low frequency response. When the wire is soldered directly onto supports — i.e. when thermal lag of supports can be considered infinite with regard to the wire thermal lag — the measured variance is:

$$\overline{T_m'^2} = \overline{T'^2} \int_0^\infty H(f) F_T(f) \, df \qquad\qquad [31]$$

with

$$H = 1-2 \, (\ell(b+(1/2aM))^{\frac{1}{2}} -1)/b\ell^2$$

and

$b = (1+(2\pi fM)^2)^{\frac{1}{2}} /2aM$ where "a" is wire thermal diffusivity.

According to PARANTHOEN et al. (1978), an ℓ/d ratio of 600 seems to be a minimum value, and it would be preferable to attain values about 1000.

3. Measurements of Velocity Fluctuations and of Their Correlations With Temperature Fluctuations

In as much as detection of temperature fluctuations with a single wire is, on a practical level, possible (Cf. 2.2.1), direct detection of velocity fluctuations is impossible. Indeed, the sensitivity ratio s cannot become much larger than one because the overheat cannot exceed unity. CHAMPAGNE (1978) has recently reiterated this point.

On the whole there are two types of measurement methods: the "fluctuation-diagram" method and those using several wires. The final choice of method depends essentially on the spatial scales of the flow studied and also on the quantities to be measured. We shall firstly study the small amplitude fluctuations.

3.1 "Fluctuation-Diagram" Methods

The aim of these methods is to reduce as much as possible the numbered wires needed. If a single wire normal to the mean velocity works successively at three different overheats, we can see from [26] that $\overline{u'^2}$, $\overline{T'^2}$ and $\overline{T'u'}$ may be determined (CORRSIN 1947). However, a better solution is to make use of more overheats, thereby assuring a more accurate determination of these quantities. It is the so-called "fluctuation-diagram" method, already mentioned in section 2.2.1. Let us recall here the application of this method to spectral measurements (Cf. equation [28], FULACHIER 1971, 1972). Fig. 3 gives an example of diagrams obtained in a boundary layer at the distance $y/\delta = 0.113$ ($s' = s\overline{T}/(T_w -\overline{T})$ where T_w = wall temperature). A 2.5 μ wire and a constant current anemometer were used. The curves are really portions of hyperbolae (relation [28]) fitted to the data. Each curve yields for a given frequency f the corresponding values of F_u, F_T and $E_{T,u}$.

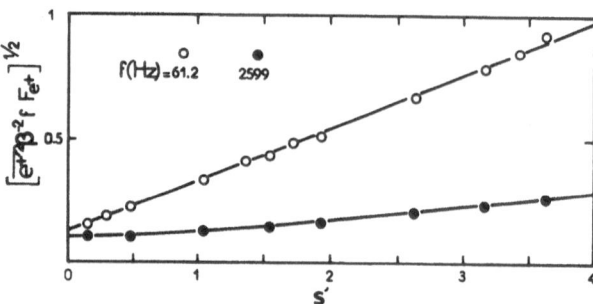

Fig. 3. Fluctuation-diagram of the filtered signals (single wire).

This method assures <u>even</u> more accuracy when the hyperbola has an apparent minimum, i.e. when $\overline{T'u'}$ is positive. This is especially the case for a boundary layer on a cold wall, for a cold cylinder wake, or for a heated jet.

For determining the variance of the transversal component $\overline{v'^2}$ and the correlation $\overline{T'v'}$ an X-wire probe and two constant current anemometers are used. The "fluctuation-diagram" method may be applied in a similar way (notation e^+ reminds us that the thermal lag of the wire must be compensated):

$$(\overline{e_{\text{I}}^{+'2}} - \overline{e_{\text{II}}^{+'2}})/4\epsilon\beta = -s(\overline{u'v'}/\overline{u^2}) + (\overline{T'v'}/\overline{T}\ \overline{u}) \qquad (\overline{e_{\text{I}}^{+'} - e_{\text{II}}^{+'}})^2/4\epsilon = \overline{v'^2}/\overline{u^2} \qquad [32]$$

where $e_{\text{I}}^{+'}$ and $e_{\text{II}}^{+'}$ are the outputs of the two wires of the X-probe; ϵ is its v-sensitivity. Equation [32] reads in the spectral form:

$$(\overline{e_{\text{I}}^{+'2}}F_{e_{\text{I}}^+} - \overline{e_{\text{II}}^{+'2}}F_{e_{\text{II}}^+})/4\epsilon\beta = -s((\overline{u'^2}\ \overline{v'^2})^{1/2}/\overline{u^2})E_{u,v} + ((\overline{T'^2}\ \overline{v'^2})^{1/2}/\overline{T}\ \overline{u})\ E_{T,v}$$

Fig. 4 gives as an example diagrams obtained in a boundary layer at $y/\delta = 0.113$ (FULACHIER 1971, FULACHIER and DUMAS 1976). The X-probe is made up of two wires of 2.5 μ; the distance separating them is about 0.4 mm. Each straight line gives, for a certain frequency f, the corresponding values of co-spectra $E_{u,v}$ and $E_{T,v}$.

In the previous experiment heat can be considered as a passive contaminant. A testing of the diagram method's accuracy has shown that F_u and E_{uv} were each the same for heated or non-heated flows (FULACHIER 1972).

The diagram method's main advantages are: reduction of spatial resolution, use of a single constant-current anemometer or of two similar setups (I = constant), and finally, accuracy in attaining the quantities we want to measure.

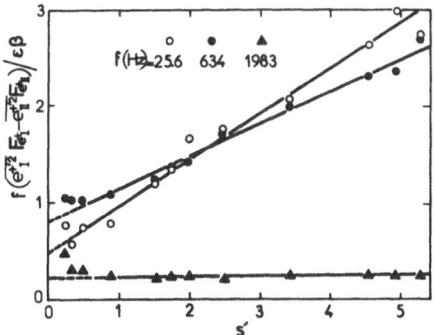

Fig. 4. Fluctuation-diagram of the filtered signals (X-wires).

On the other hand, the main inconveniences are: separation of the fluctuations themselves (T', u', v') is impossible and results concern only 2nd order moments; besides, the difficulty of the method is important. However, this inconvenience could become almost irrelevant by using a data acquisition and digital processing system.

Concerning the direct determination of various sensitivity coefficients for each current, see the works of VEROLLET (1969, 1972).

3.2 Multi-Wire Probes
The main purpose is to obtain temperature and velocity fluctuations at the same time. The temperature fluctuation, T', is given by a cold-wire (Cf. section 2.). Velocity component fluctuations are deduced from responses given by one or several hot-wires which work in general at constant resistance r_w. Unfortun tely, these wires are also sensitive to temperature. For example, in the case of a constant resistance hot-wire, the mean output voltage may be written as:

$$(\overline{e^+})^2 = A''(\overline{T}) + B''(\overline{T})\, \overline{u}^n \qquad [33]$$

Fig. 5 shows that A'' is a decreasing linear function of ambient temperature $\overline{T}$, but that B'' is constant. This experiment is carried out with a 5 μ wire and an overheat coefficient of 0.8; the n-value is 0.45 (BEGUIER et al. 1978). Concerning the A'' and B'' variation laws with temperature T, see in particular the works cf CHEVRAY and TUTU (1972) and ALI (1975).

To simultaneously obtain T' and a longitudinal velocity fluctuation u', two single wires are used. This yields, besides [29], the following equation:

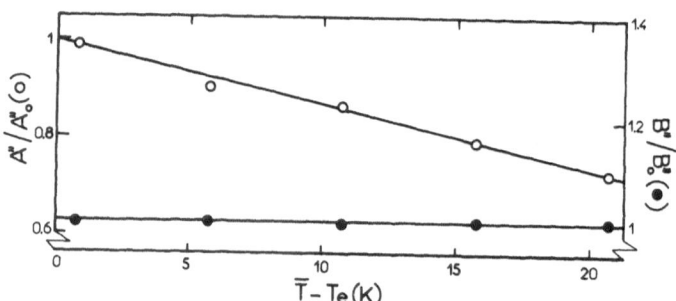

Fig. 5. A'' and B'' as functions of temperature.

$$e_1^{+\prime} = -\alpha_1 \frac{u'}{\overline{u}} + \beta_1 \frac{T'}{\overline{T}} \qquad\qquad [34]$$

Then, the variables T' and u' are separated using an analog device (e.g., CHE-VRAY and TUTU 1972, ALI 1975, BEGUIER et al. 1978), or the digital technique (e.g., CHAMPAGNE et al. 1977, KEFFER et al. 1978). Regarding wire positions, it is always necessary to locate the cold-wire upstream from the hot-wire in order to avoid contamination of the former by the latter. This is imperative mainly for large turbulence intensities. By using two cross-wires — one horizontal, the other vertical — the spatial resolution is reduced, but the wakes of the prongs of the cold-wire can sometimes perturbe the hot-wire. Two parallel single-wires are often employed, but the spatial resolution may be critical. For these two kind of probes it is not possible to perform measurements close to walls. In order to avoid this inconvenience, two single-wires located on a straight line parallel to the wall can be used, each one separated by the same prong (FULACHIER 1972); in this instance, the spanwise spatial resolution is fairly critical.

In order to simultaneously obtain T', u', and v' fluctuations, a cold-wire and a hot X-wire are often used (JOHNSON 1955). This yields, besides [29] the following equations:

$$e_1^{+\prime} = -\alpha_1 (u'/\overline{u}) + \epsilon(v'/\overline{u}) + \beta_1 (T'/\overline{T}) \qquad\qquad [35]$$

$$e_2^{+\prime} = -\alpha_2 (u'/\overline{u}) - \epsilon(v'/\overline{u}) + \beta_2 (T'/\overline{T}) \qquad\qquad [36]$$

The variables T', u', and v' are separated using an analog device (e.g., SCHON and BAILLE 1972, CHARNAY et al. 1973, ANTONIA et al. 1977, BEGUIER et al. 1978) or the digital technique (e.g. CHARNAY et al. 1977). The cold-wire must here also be located upstream from the geometric center of the

X-wire. The distances between the two wires of the X-wire is generally at least equal to 0.8 mm. The spatial resolution of these probes does not allow measurements close to walls. The minimal distances which can be reached are of the order of 1 mm.

With the purpose of carrying out more exact measurements, BEGUIER et al. (1975) proposed the use of a four parallel wire probe P_4 (Fig. 6).

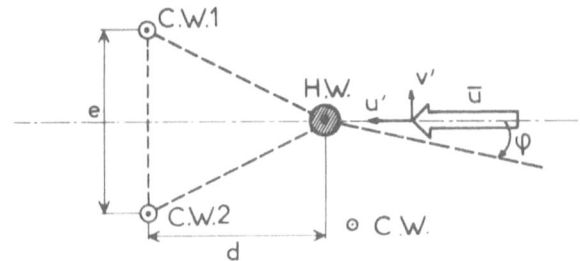

Fig. 6. Four parallel wire probe configuration.

This probe is built with a three parallel wire probe P_3 (BEGUIER et al. 1973, REY 1973, REY and BEGUIER 1977) and a cold-wire C.W. located upstream from P_3. The P_3 probe is configured with an upstream hot-wire, H.W., and two downstream cold-wires, C.W. 1 and C.W. 2. The H.W. is 5 μ in diameter and works as a constant-temperature anemometer. The C.W. 1 and C.W. 2 wires are located as shown on Fig. 6 ($e \cong 0.1$ mm, $d \cong 0.1$ mm) in the wake of the hot-wire H.W. The difference between the C.W. 1 and C.W. 2 output voltages gives a signal proportional to the v' component. T' and v' are then directly obtained. The main advantage of this P_4 probe is its good spatial resolution. Moreover, the influence of the transverse component w' is very weak contrary to the X--wire. But unfortunately, given the fragility and small dimensions of the P_4, C.W. 1 and C.W. 2 must have relatively large diameters (2.5 μ) and their thermal lag compensation is crucial. This may explain why its high frequency response is somewhat erroneous.

FABRIS (1977) has perfected a four wire probe for simultaneously obtaining T', u', v' and w'. The spatial resolution of this probe is about 1 mm^3. Besides, FABRIS uses a tractable digital technique. However, it seems that such a probe has at the very least the same inconveniences as the X-probe. Also it is difficult to suppress interactions between these four wires and the prong wakes.

The main advantages of multi-wire probes are: direct measurement of fluctuations T', u', v' and even sometimes w' — all at the same time — thus enabling measurements of moments of any order and correlation. Furthermore these measurements can be effectuated in an easy and quick way. The main inconveniences are the following ones: limitations of spatial resolution, which can be crucial; prong wakes; use of different setups for cold-wire and hot-wire, which can introduce phase shifts.

Concerning spatial resolution, it must be noticed that when the equations [35] and [36] are considered we assume the two wires A and B to be located at the same point. Indeed, in most flows the correlation coefficient R_u between u'_A and u'_B is effectively near unity. But the correlation coefficient R_v, between v'_A and v'_B or R_T, between T'_A and T'_B — mainly in wind tunnels where spatial scales are relatively small — decrease strongly with respect to distance AB. This is linked to the fact that the F_v and F_T spectra present a more important contribution than F_u at high frequencies. For example, in a heated wall boundary layer ($u_e = 12$ ms^{-1}, $\Delta T \cong 20$ K, $\delta = 62$ mm), at the distance $y/\delta = 0.04$ and for $\overline{AB} = 0.35$ mm, we have $R_u \cong 0.98$, $R_T \cong 0.90$ (FULACHIER 1972). When $\overline{AB} \cong 1$ mm we can expect to obtain $R_T \cong 0.7$!

3.3 Large Amplitude Fluctuations

These large fluctuations are found mainly in the following flows: free jets, mixing layers, boundary layers very close to the wall, etc.. As TUTU and CHEVRAY (1975) point out for isothermal flows, it seems that most of the published measurements, $\overline{u'v'}$, for example, have been done in a conventional manner without any correction. In heated flows, the problem of large fluctuations is even more complicated.

Regarding isothermal flows, we can mention in particular the works of TUTU and CHEVRAY (1975) and those of ACRIVLELLIS (1977) about $\overline{u}$, $\overline{u'^2}$, $\overline{v'^2}$, $\overline{w'^2}$, $\overline{u'v'}$ measurements. It must be noticed, however, that we cannot obtain the true instantaneous u' and v' values. And also, it must be noticed that the prong wake effects which are in practice not easily estimated, can become much more important than nonlinearity effects when the turbulence intensity $(\overline{u'^2})^{1/2}/\overline{u}$ is about 30 % (DUMAS 1978a). Indeed, in this case the instantaneous fluctuation $u'/\overline{u}$ can reach more than 100 %.

In the case of a heated jet flow, FARCY and LEUCHTER (1978) have used the "fluctuation-diagram" method (Cf. 3.1.) for single wires in order to measure $\overline{T'^2}$, $\overline{u'^2}$ and $\overline{T'u'}$. Additional terms are introduced because of large fluctuations. However, when the overheat ratio is increased the wire becomes mainly sensitive to velocity; and since the anemometer is a constant current one we know its response to be erroneous.

Therefore, in heated flows with large fluctuations, use of digital processing seems almost obligatory and might well help solving the problem by applying iterative processes. Nevertheless, given these inconveniences and especially the wake effects, the simultaneous utilization of a laser anemometer and a cold-wire could perhaps give the solution. But at first glance, the two techniques seem incompatible (e.g. the dust problem).

4. Conclusion

The thermal turbulent field of slightly heated flows can be accurately analysed from cold-wire measurements. Indeed, if precautions are taken, temperature fluctuations can be practically isolated. That is to say, the following is necessary: effectively negligible velocity contamination, wire thermal lag correctly compensated and an "aspect ratio", ℓ/d of about 1000. "Static" nonlinearity does not exist under such conditions. As for "dynamic" nonlinearity, it plays a much less prominent role for cold-wires than for hot-wires. A thorough study of this whole question would nevertheless be useful.

With regard to the velocity field and temperature-velocity correlations, the choice between possible methods depends essentially upon the flow spatial scale. The "fluctuation-diagram" method is toilsome and limited to second order moments, but it never fails. The multi-wire method is on the contrary quick and easy to handle and does allow the determination of high order moments. However, in spite of its apparent advantages, the experimenter can only wonder at the certainty of his findings — i.e. if the quantities obtained are really those he was effectively seeking! Concerning large fluctuations, still other new studies using digital techniques are necessary.

Acknowledgements

I should like to thank Dr. R. DUMAS for his valuable suggestions and comments throughout the course of this study. Thanks are also due to Drs. E. AR-ZOUMANIAN and G. TAVERA for their help in carrying our certain calculations. I am also grateful to M. ASTIER for his efforts in constructing special electronic circuits, certain ones of which are used in this paper.

References

ACRIVLELLIS, M., 1977 *DISA Information,* **22**, 15

ALI, S.F., 1975 PH.D., Johns Hopkins University, Baltimore, Maryland

ANTONIA, R.A., DANH, H.Q. and PRABHU, A., 1977 *J. Fluid Mech.,* **80**, 153

BEGUIER, C., FULACHIER, L. and DUMAS, R., *1975 Euromech,* **63**, Copenhagen

BEGUIER, C., FULACHIER, L. and KEFFER, J.F., 1978 *J. Fluid Mech.,* **89**, 561

BEGUIER, C., REY, C., DUMAS, R. and ASTIER, M., 1973 *C.R. Acad. Sci., Paris,* **277A**, 475

BRADBURY, L.J.S. and CASTRO, I.P., 1972 *J. Fluid Mech.,* **51**, 487

BREMHORST, K. and GILMORE, D.B., 1978 *Int. J. Heat Mass Transfer,* **21**, 145

BREMHORST, K., KREBS, L. and GILMORE, D.B., 1977 *Int. J. Heat Mass Transfer,* **20**, 315

BRUN, E.A., MARTINOT-LAGARDE, A. and MATHIEU, J., 1968 *Mécanique des Fluides,* Dunod, Paris

CHAMPAGNE, F.H., 1978 *J. Fluid Mech.* **86**, 67

CHAMPAGNE, F.H., FRIEHE, C.A., LARUE, J.C. and WYNGAARD, J.C., 1977, *J. Atmos. Sci.* **34**, 515

CHAMPAGNE, F.H., SCHLEICHER, C.A. and WEHRMANN, O.H., 1967 *J. Fluid Mech.* **28**, 153

CHAPMAN, S. and COWLING, T.G., 1960 *The Mathematical Theory of Non Uniform Gases* Cambridge Univ. Press

CHARNAY, G., SCHON, J.P., ALCARAZ, E. and MATHIEU, J., 1977 *Proc. Symp. Turbulence Shear Flows,* Univ. Park, Pennsylvania, 15-47

CHARNAY, G., SCHON, J.P. and SUNYACH, M., 1973 *J. Entropie,* **50**, 24

CHEVRAY, R. and TUTU, N.K., 1972 *Rev. Sci. Instr.* **43**, 1417

COLLIS, D.C. and WILLIAMS, M.J., 1959 *J. Fluid Mech.* **6**, 357

COMTE-BELLOT, G., 1974 *Techniques de Mesures dans les Écoulements,* Eyrolles, Paris, 117

COMTE-BELLOT, G., 1976, *An. Review Fluid Mech.,* **8**, 209

COMTE-BELLOT, G. and SCHON, J.P., 1969 *Int. J. Heat Mass Transfer,* **12**, 1661

CORRSIN, S., 1947 *Rev. Sci. Instr.,* **18**, 469

CORRSIN, S., 1963 *Handbuch der Physik,* Springer Verlag, **8**, 524

DUMAS, R. 1978a, Anémothermométrie à fils chauds. 3rd Cycle, Lect. Rep. I.M.S.T. Univ. Aix-Marseille, France

DUMAS, R., 1978b *Proceedings of the Dynamic Flow Conference 1978,* Marseille and Baltimore

DUMAS, R., FULACHIER, L. and ARZOUMANIAN, E., 1972 *C.R. Acad. Sci. Paris,* **274A**, 267

ELENA, M., 1977 Rep. C.E.A.-R-4843, C.E.N. Saclay, Thesis 1975, I.M.S.T., Univ. Aix-Marseille II, France

FABRIS, G., 1977 *Proc. Symp. Flow Measurement in Open Channels and Closed Conduits,* NBS Sp. Publ. 484

FARCY, A. and LEUCHTER, O., 1978, *Dynamic Flow Conference* 1978, Marseille, France

FULACHIER, L., 1971 *C.R. Acad. Sci. Paris,* **272A**, 1022
FULACHIER, L., 1972 Thesis I.M.S.T., Univ. Provence, France
FULACHIER, L. and DUMAS, R., 1971 *Meet. AGARD Turbulent Shear Flows,* No. 93, 4
FULACHIER, L. and DUMAS, R., 1976 *J. Fluid Mech.,* **77**, 257
FULACHIER, L., ARZOUMANIAN, E. and DUMAS, R., 1977, *3e Cong. Franc. Mec.,* Grenoble
FULACHIER, L., ARZOUMANIAN, E. and DUMAS, R., 1978, Proc. Symp. Structure and Mechanisms of Turbulence, *Lecture Notes in Physics,* Springer Verlag, **76**, 46
GAVIGLIO, J., 1962, *P.S.T., Minist. Air* 385
GIOVANANGELI, J.P., 1975, Thesis I.M.S.T., Univ. Aix-Marseille II, France
HINZE, J.O., 1975 *Turbulence,* McGraw-Hill
HOJSTRUP, J., RASMUSSEN, K. and LARSEN, S.E., 1976, *DISA Information,* **20**, 22
KEFFER, J.F., BUDNY, R.S. and KAWALL, J.G., 1978 *Rev. Sci. Instr.,* To be published
KOVASZNAY, L.S.G., 1953, N.A.C.A. Tech. Note, No. 2939
LARUE, J.C., DEATON, T. and GIBSON, C.H., 1975 *Rev. Sci. Instr.,* **46**, 757
MAYE, J.P., 1970 *DISA Information,* **9**, 22
MILLON, F., 1977, Thesis Lab. Thermodyn. Fac. Sci. Univ. Rouen, France
MILLON, F., PARANTHOEN, P. and TRINITE, M., 1978 *Int. J. Heat Mass Transfer,* **21**, 1
MING HO, C., JAKUS, K. and PARKER, K.H., 1976, *Combust. Flame,* **27**, 113
PARANTHOEN, P., MILLON, F., and TRINITE, M., 1978 *Dynamic Flow Conference 1978,* Marseille, France
PASCAL, A., 1978 Thesis I.M.F.T., Inst. Nat. Polytech. Toulouse, France
REY, C., 1973 Thesis I.M.S.T., Univ. Provence, France, Bul. Dir. et Rech. E.D.F. A,3
REY, C. and BEGUIER, C., 1977 *DISA Information,* **21**, 11
SCHON, J.P. and BAILLE, A., 1972 *C.R. Acad. Sci. Paris* **274A**, 116
SUNYACH, M., 1971, Thesis E.C.L. Univ. Lyon, France
TUTU, N.K. and CHEVRAY, R., 1975 *J. Fluid Mech.,* **71**, 785
UBEROI, M.S. and KOVASZNAY, L.S.G., 1953 *Q. Appl. Math.,* **10**, 375-93
VEROLLET, E., 1969 P.S.T. Minist. Air 449, Thesis I.M.S.T., Univ. Aix-Marseille, France 1962
VEROLLET, E., 1972, Thesis I.M.S.T., Univ. Provence, France, Rep. C.E.A.-R-4872, C.E.N. Saclay 1977
WYNGAARD, J.C., 1971, *J. Fluid Mech.,* **48**, 763

Examples of the Use of the Pulsed Wire Anemometer in Highly Turbulent Flow

by

L.J.S. Bradbury
Mechanical Engineering Department,
University of Surrey, Guildford, England

Summary

This paper describes further applications of the pulsed wire anemometer to velocity measurements in highly turbulent flows. The measurements were all made in the near wake of the flow downstream of a normal flat plate. The near wake is characterised by a reverse flow region in which the velocity fluctuations contain both a random component and also a near periodic contribution arising from the rolling up of the vortex sheets shed from the sharp edges of the plate. Previous measurements of normal turbulent stresses by Bradbury (1976) are extended to include shear stress measurements using an inclined pulsed wire probe. Although it is not possible, at this stage, to establish the precise accuracy of these measurements, the internal consistency of the results is satisfactory and suggests that worthwhile studies involving shear stress measurements can be carried out using inclined pulsed wire anemometers.

Examples of the use of the pulsed wire anemometer in conditional sampling experiments are also described. Signals from a hot wire anemometer placed outside the wake of the flat plate were used to trigger the pulsed wire at a fixed phase in the vortex shedding cycle. The results show clearly that the complete development of the vortex shedding process in the near wake region could be studied using this technique.

Finally, the problems of obtaining spectral information from a pulsed wire anemometer are discussed. Experiments using random sampling and a sequential sampling technique are briefly described. A difficulty with both techniques is that autocorrelation estimates cannot be obtained for time lags less than about 10 milliseconds and an application of the autocorrelation function theorem to estimate these unknown autocorrelation values is described. This theorem is used in the development of maximum entropy power spectral analysis but its application here represents a new area of application of the theorem.

1.0 Introduction

Although a number of papers have been published on both the development and application of the pulsed wire anemometer, the potential of the technique has not yet been fully explored and there remain many interesting applications that can be considered. However, the principle area of application remains the study of highly turbulent flows and this paper presents some further examples of the use of the instrument in an investigation of the near wake of a two-dimensional flat plate. The principle characteristic of this flow is the strong periodic vortex shedding that takes place and it is a flow of considerable intrinsic interest. Moreover, the presence of a highly turbulent reverse flow region behind the plate with velocity fluctuations containing both a random broad band element and a near periodic vortex shedding component make it a particularly difficult environment in which to obtain accurate and reliable velocity measurements. In the present context, it is for precisely these characteristics that it is an admirable flow in which to examine the application of the pulsed wire anemometer.

Bradbury (1976) discussed the use of the pulsed wire anemometer in measurements in highly turbulent flows and presented some results for both the mean velocity field and the three normal components of the turbulent intensity in the wake of a normal flat plate. In section 3 of this paper, this work is extended to the use of a sloping pulsed wire anemometer probe to obtain shear stresses in a highly turbulent flow. In the earlier work, it was possible to make comparisons with hot wire anemometer results by the rather tortuous process of using the pulsed wire data to compute the response of a linearized hot wire anemometer in this highly turbulent flow. In the case of the present shear stress measurements, it is not so simple to make such comparisons and intimations of accuracy can only be obtained by comparisons between mean velocity and normal stresses results obtained from sloping pulsed wire anemometer data and those obtained from normal probe measurements.

In section 4, the possible application of the pulsed wire anemometer to conditional sampling experiments is discussed. The signals from a hot wire anemometer placed outside the wake of the flat plate are used to provide a synchronising signal to trigger the pulsed wire anemometer at a fixed phase in the vortex shedding cycle. The results of two traverses in the reverse flow region show clearly that it would be possible to undertake studies of the vortex shedding process as a function of phase angle using this technique although more comprehensive on-line computer facilities would be required than were available in these experiments.

Finally, section 5 describes preliminary results of experiments to obtain spectral estimates from pulsed wire anemometer measurements. An inherent restriction with the pulsed wire anemometer is that velocity estimates cannot be obtained at intervals less than about twice the time constant of the pulsed wire anemometer. This restricts the maximum sampling rate to between about 100 Hz and 200 Hz so that spectral estimates obtained in this way would be limited only to about 50 Hz to 100 Hz and subject to serious aliasing errors. As a result of the work of Gaster & Roberts (1975), some of these difficulties may be avoided by using a random sampling technique (Gaster & Bradbury (1976)) but, for a variety of reasons, the use of random sampling is inconvenient if the data is to be collected and processed by a mini-computer. An alternative possibility is to use a sequential sampling technique in which the sampling interval increases in a simple arithmetic progression. Some early results of the use of this technique are described but a difficulty with both random sampling and sequential sampling when applied to the pulsed wire anemometer is that it is not possible to obtain autocorrelation estimates at lag times of less than about 10 milliseconds. In order to obtain autocorrelation estimates in this "dead" time, a technique using the autocorrelation function theorem is described. This theorem forms the cornerstone of maximum entropy power spectral analysis but its application to the present problem represents a new application of the theorem.

2.0 Experimental Arrangements

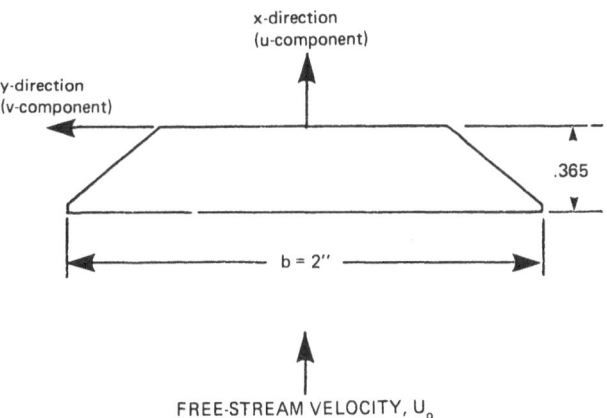

Fig. 1. Cross-section of flat plate.

The experiments were carried out in the 20" × 30" wind tunnel in the Mechanical Engineering Department. The cross-section of the flat plate used in the experiments is shown in Fig. 1 along with a definition of the coordinate system to be used in discussing the experimental results. The width of the flat plate was 2″ and it spanned the 30″ side of the wind tunnel working section. In all respects, the model set-up was the same as in the earlier experiments of Bradbury (1976). However, in order to rotate the probe to carry out the shear stress measurements, the probe was also mounted with its axis parallel to the span of the plate as shown in Fig. 2. A simple mechanism at the base of the probe support enabled the probe to be set accurately at 0°, ±45° or 90° to the free-stream direction. In the earlier experiments, the probe was mounted with its axis normal to the plate span and, using a dummy probe, it was not possible to observe any probe interference effects. However, with the present arrangement, clear evidence of probe interference effects on the flow were found (see section 3.0) and, indeed, this is hardly surprising in view of the relative sizes of the probe support to the model width. There would be no difficulty in constructing a less obtrusive probe support but, within the time scale for these preliminary experiments, it was necessary to use the original design.

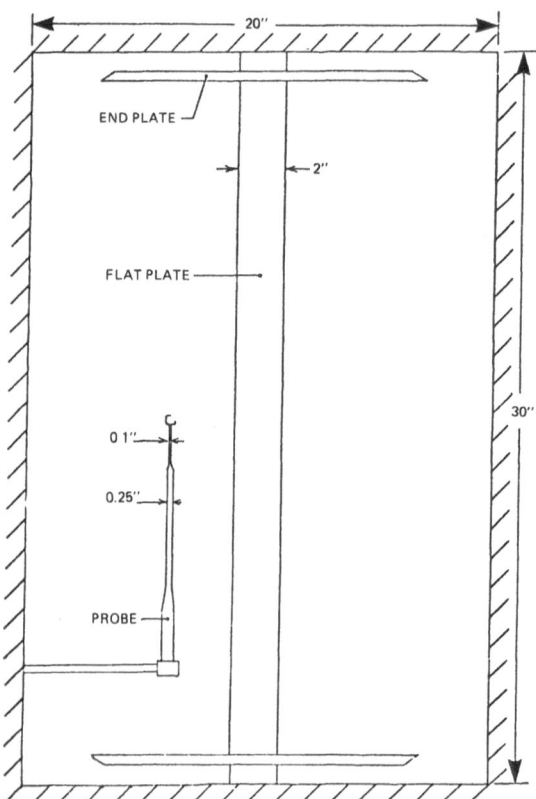

Fig. 2. Cross-section of flat plate.

Examples of probe calibration results have been given in earlier papers in which some of the effects of departures from the ideal response have been discussed. However, further studies are required in order to assess the errors that might arise in shear stress measurements of the type reported in this paper.

In the experiments discussed in sections 3 and 4, the pulsed wire anemometer was connected on-line to a programmable desk calculator. This restricted the sampling rate to about four samples per second and, whilst this was just acceptable for the straight forward mean velocity and turbulence intensity measurements, faster sampling rates are required for extensive conditional sampling experiments of the sort described in section 4 and also for the spectra measurements reported in section 5. For these latter experiments, the anemometer was interfaced to a Hewlett Packard 21MX minicomputer and the sampling rate was then only restricted by the time constant of the pulsed wire. In these experiments, pulsed wires of 5 micron diameter Tungsten were used with 2.5 micron diameter sensor wires. With these probes, it was possible to continuously sample at a rate of about 150 samples per second.

3.0 Measurements with an Inclined Pulsed Wire Anemometer
In order to examine the possibility of using the pulsed wire anemometer to measure the shear stress in a highly turbulent flow, two traverses were made downstream of the flat plate corresponding to values of x/b = 1 and x/b = 2. In the former case, the mean flow on the centre-line is in the reverse direction to the free-stream whereas two plate widths downstream corresponds closely to the downstream extent of this reverse flow region (Bradbury (1976)). At each position, measurements were made with the plane of the probe set at ±45° to the free stream direction in addition to measurements with the plane of the probe both normal and tangential to the free-stream direction. Two thousand samples were recorded in each measurement and the period for recording and analysing this number of samples occupied about 8 minutes.

If the departures of the probe yaw response from the ideal cosine law are ignored then the two mean velocity estimates, U_1 and U_2, recorded by the two inclined probe measurements are given simply by

$$U_{1,2} = \frac{U}{\sqrt{2}} \pm \frac{V}{\sqrt{2}}$$ [3.1]

where U and V are the mean velocities in the free-stream x-direction and the transverse y-direction respectively. The variances of σ_1^2 and σ_2^2 are simply

494   L.J.S. Bradbury

$$(\sigma_{1,2})^2 = \frac{\overline{u^2}}{2} + \frac{\overline{v^2}}{2} \pm \overline{uv}$$ [3.2]

where $\overline{u^2}$, $\overline{v^2}$ are the x and y components of the normal turbulent stresses respectively and $\overline{uv}$ is the shear stress. Thus

$$U = \frac{U_1 + U_2}{\sqrt{2}}$$ [3.3a]

$$V = \frac{U_1 - U_2}{\sqrt{2}}$$ [3.3b]

$$\overline{uv} = \frac{\sigma_1^2 - \sigma_2^2}{2}$$ [3.3c]

$$\overline{u^2} + \overline{v^2} = \sigma_1^2 + \sigma_2^2$$ [3.3d]

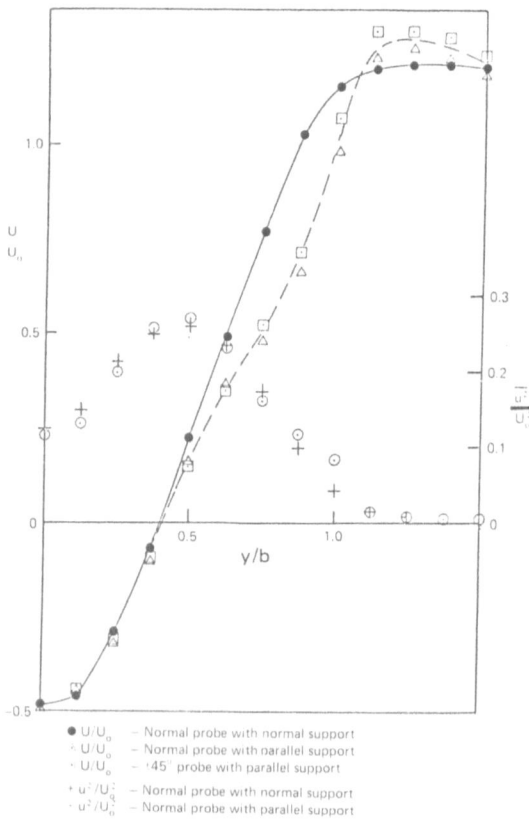

Fig. 3. Mean velocity and normal stress profiles at x/b = 1.

The similarities with the use of inclined hot wires are obvious. The measurements with the plane of the pulsed wire anemometer set normal to the freestream gave independent estimates of U and $\overline{u^2}$ whereas the tangential measurements yielded independent estimates of V and $\overline{v^2}$. The effect of the finite extent of the yaw response and the departure from ideal cosine law yaw response have already been discussed by Bradbury (1976) for normal and tangential probe measurements and no further comment will be made here except to note that satisfactory measurements with a tangential probe can only be made once the local turbulent intensity exceeds about 50 %.

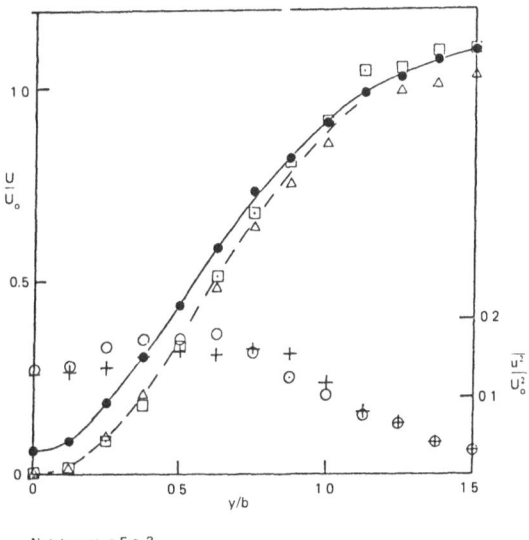

Notation as in Fig 3

Fig. 4. Mean velocity and normal stress profiles at x/b = 2.

Figs. 3 and 4 show results of the mean velocity in the free-stream direction obtained from the inclined probe measurements using equation [3.3a] and the normal probe measurements. The two sets of results are generally in good agreement with one another. In addition, the blocked symbols show results obtained previously with a normal probe mounted with its axis at right angles to the plate span. It is clear that there is some significant distortion in the mean velocity profiles due to probe interference in the present test results particularly in the results at x/b = 1 in the region corresponding to about 0.5 < y/b < 1.0. From flow visualisation studies, this is the region in which the most obvious rolling up of the vortex sheets takes place and, as mentioned in section 2, it is clear that it is necessary to make use of a far less obtrusive probe support than the one currently in use. However, this does not effect the conclusions about the consistency of the inclined probe results when they are compared with normal probe results using the same probe mounting geometry.

Also shown in Figs. 3 and 4 are the results for the normal turbulent stresses, $\overline{u^2}/U_o^2$, obtained from the present normal probe measurements compared with previous measurements with the probe axis at right angles to the model span. The agreement between the two sets of data is good and probe interference effects are not obviously apparent in these results.

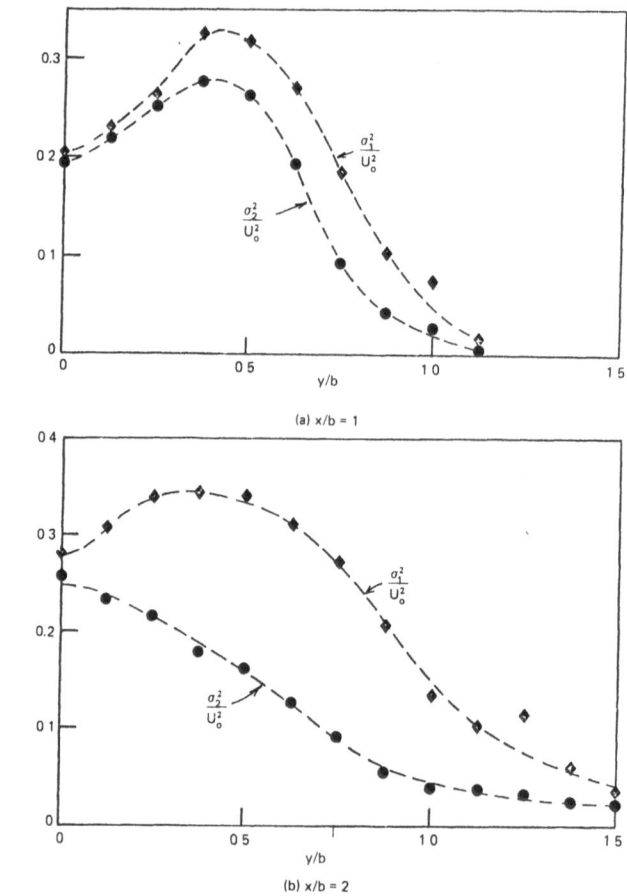

Fig. 5. Profiles of the variances from inclined probe measurements.

Fig. 5 shows the results for the variances obtained with the probe at ±45° to the free-stream direction. Before discussing the shear stress results obtained from these measurements, it is perhaps worthwhile commenting briefly on the confidence intervals of these measurements. For large sample sizes, the confidence interval $\Delta\sigma$ on the standard deviation σ for a Gaussian signal is given by

$$\frac{\Delta\sigma}{\sigma} = \frac{t(\alpha)}{\sqrt{2n}}$$

where n is the sample size. $t(\alpha)$ is the Student t-distribution and is a function of the probability α that the true value of the standard deviation lies within the range $\sigma \pm \Delta\sigma$. If we consider the difference between the variances of two independently recorded Gaussian random sequences then the probability that one of the estimates is outside one extreme of the confidence interval is $(1-\alpha)/2$ and the probability that the other estimate is at the same time outside the other extreme of its confidence interval is therefore apparently $[(1-\alpha)/2]^2$. Therefore, the confidence interval of the difference between the variances of two Gaussian random sequences would seem to be $\pm(4t(\alpha)/\sqrt{2n})(\sigma_1^2 + \sigma_2^2)$ but with a probability that the true value lies within this interval given by $1 - [(1-\alpha)/2]^2$. Although the above heuristic argument applies strictly only to Gaussian signals, it gives a guide to the confidence interval of shear stress measurements and emphasises the rather obvious point that whereas the shear stress is proportional to the difference between the variances, the confidence interval is more likely to depend on the sum of the variances. If we require the 95 % confidence interval on the shear stress, the above argument indicates that we should use the value of $t(\alpha)$ appropriate to $\alpha = 55.3$ %, namely $t(\alpha) = 0.76$. The shear stress results are plotted in Fig. 6 showing the upper and lower limits of the 95 % confidence interval. Of course, sampling size errors are only one of

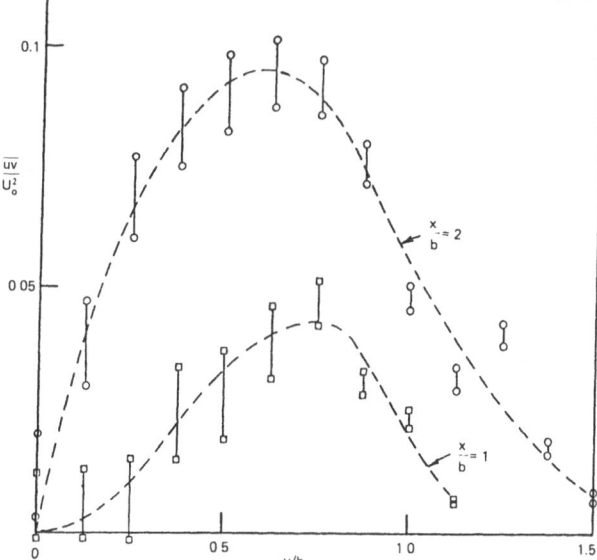

Fig. 6. Shear stress profiles showing the 95 % confidence intervals.

many sources of error and, at this stage, it is not possible to assess the precise accuracy of these shear stress results. However, the data seems reasonably consistent and this suggests that the pulsed wire can certainly be used to carry out studies involving shear stress measurements. Further work is planned in which comparisons with hot wires will be made in flows at lower turbulence levels and, in high turbulence level flows, using a photon correlation laser Doppler anemometer.

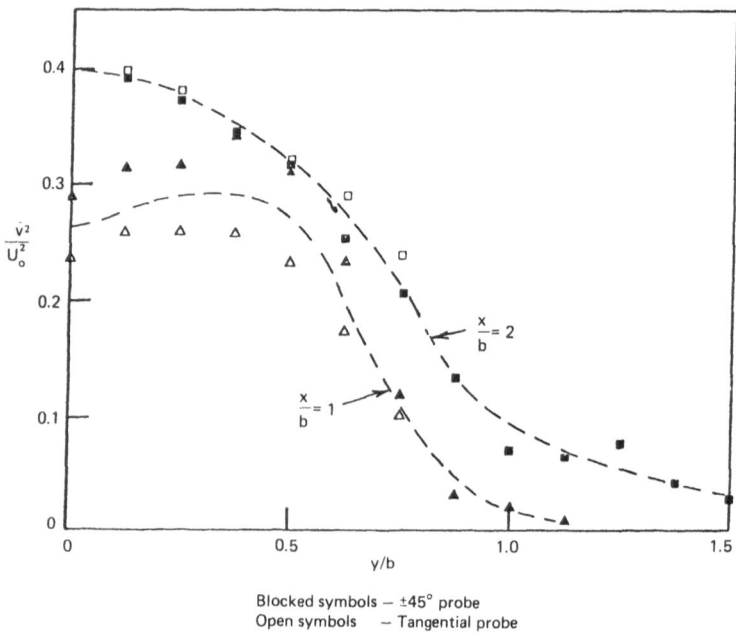

Blocked symbols — ±45° probe
Open symbols   — Tangential probe

Fig. 7. v-component normal stress profiles.

As a final indication of the accuracy of the inclined probe measurements, Fig. 7 shows values of $\overline{v^2}/U_o^2$ obtained from the inclined probe and normal probe measurements using equation [3.3d]. These are compared with direct measurements of $\overline{v^2}/U_o^2$ obtained from a tangential probe. These latter measurements only extend out to y/b = 0.75 because, beyond this position, the turbulence level was too low for this technique to be satisfactorily used. As with the shear stress measurements, the confidence intervals on the results with the inclined probe are significantly larger than the individual variances but, as Fig. 8 shows, reasonable agreement with the tangentially mounted probe results is obtained. For clarity, the confidence intervals are not shown in this occasion.

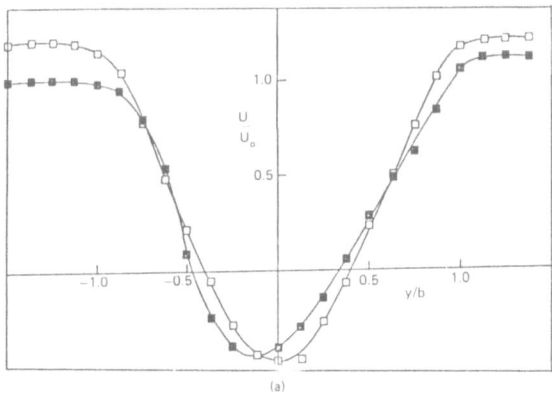

(a)

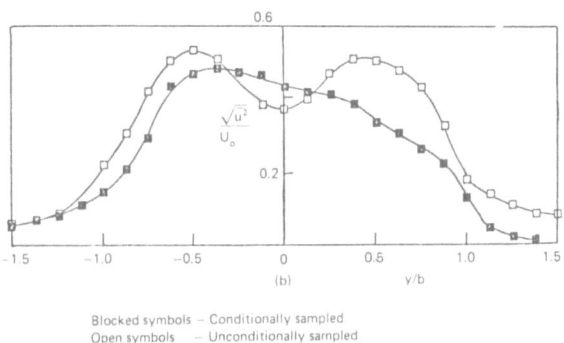

(b)

Blocked symbols — Conditionally sampled
Open symbols   — Unconditionally sampled

Fig. 8. Conditionally samples mean velocity and $\sqrt{\overline{u^2}}/U_o$ profiles at $x/b = 1$.

4.0 Examples of Conditional Sampling Measurements

To obtain a proper understanding of the flow downstream of a normal flat plate, it is necessary to study the almost periodic rolling up of the vortex sheets shed from the sharp edges of the plate. This clearly requires a conditional sampling experiment in which measurements are recorded only at particular phases of the vortex shedding cycle. Davies (1975) carried out such experiments using conventional hot wire anemometers in the wake of a circular cylinder using the output from a hot wire outside the wake as a synchronising signal for measurements within the wake. However, these measurements were restricted to regions of lower turbulence level some way downstream of the cylinder and the interesting flow in the immediate wake could not be included in the investigation. Recently, Cantwell (1976) has extended hot wire measurements to the near wake region using an ingenious flying wire technique. In this section, some examples of the use of the pulsed wire anemometer to undertake conditional sampling experiments are described.

In the present experiment, a hot wire anemometer was mounted at a position two plate widths from the centre-line of the flow and two plate widths downstream. This position was well outside the turbulent wake and the velocity fluctuations were essentially irrotational and almost periodic. After narrow pass filtering at the vortex shedding frequency, this hot wire signal was used to trigger the pulsed wire anemometer only at the zero crossing points. The technique was therefore essentially the same as that used by Davies (1975).

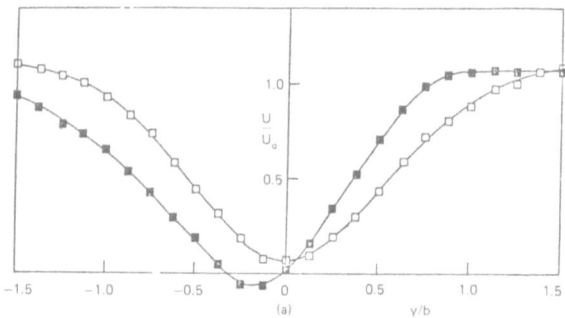

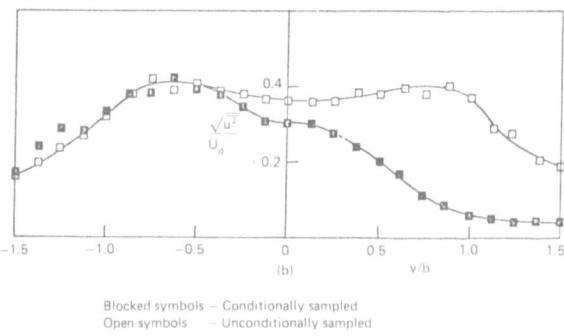

Blocked symbols – Conditionally sampled
Open symbols   – Unconditionally sampled

Fig. 9. Conditionally samples mean velocity and $\sqrt{\overline{u^2}}/U_o$ profiles at $x/b = 2$.

Figs. 8 and 9 show conditionally sampled mean velocity and $\sqrt{\overline{u^2}}$ intensity profiles at x/b = 1 and x/b = 2 compared with conventionally measured values. As would be expected, the conditionally measured mean velocity profiles are not now symmetrical about the flow centre-line and, in principle, they could be used to construct the stream functions for the flow at various stages of the vortex shedding cycle. However, the integration would be complex because the y/b = 0 axis is not now a plane of symmetry and the integration would have to use an external reference streamline sufficiently far from the plate to remain undisturbed by the vortex shedding process.

As far as the intensities are concerned, it should be noted that the synchronising hot wire was mounted on the side of the plate corresponding to positive values of y/b. At both x/b = 1 and x/b = 2, the conditionally sampled intensities are significantly smaller than the conventionally measured values for y/b in the region corresponding to about $0.5 < y/b < 1.5$. However, on the other side of the plate in the region $y/b < -1.0$, the conditionally and unconditionally sampled intensities seem to approach one another. It is not obvious how to interpret these results since if the velocity fluctuations consisted of a simple harmonic wave and an uncorrelated random signal then the conditionally sampled variance would arise simply from the random component and would be everywhere less than the unconditional variance which would be the sum of the variance of the simple harmonic signal and the random signal. Clearly this is not the situation that occurs in this flow and, indeed, even casual observation of hot wire signals at the edge of the flow shows that the occurrence of "bursts" of turbulence are phase related to the vortex shedding. There is also, of course, the likelihood that phase related harmonics are present in the signals. However, even in such cases, it is difficult to understand how the two variances can approach one another unless the flow on one side of the plate becomes completely uncorrelated from the flow on the other side. However, there is evidence to show that this is not so and therefore the interpretation of the intensity results remain puzzling.

Although the above results are very limited, they do demonstrate the plausibility of undertaking conditionally sampled studies of vortex shedding with the pulsed wire anemometer. However, the quantity of data involved in a thorough study of this sort is very great and the storage and computing facilities available at the stage at which these initial measurements were made were not adequate for such a task.

5.0 Examples of Measurements of the Turbulence Spectrum
With an instrument such as a frequency tracking laser Doppler anemometer or

a conventional hot wire anemometer, it is possible to obtain estimates of the energy spectrum of the turbulence by periodic sampling of the data provided that aliasing effects are avoided by pre-filtering the signal to remove frequencies above the Nyquist frequency. With the pulsed wire anemometer, it is not possible to pre-filter the signal and, moreover, to avoid destroying the pulsed wire, it is not possible to periodically sample at an interval amounting to less than, say, twice the time constant of the pulsed wire. Even with pulsed wires of 5 micron diameter, this limits the sampling frequency to a maximum of about 200 Hz so that aliasing would result from frequencies above only 100 Hz. Two possibilities have so far been explored to overcome this limitation. The first possibility arises directly from the work of Gaster & Roberts (1975), in which it is shown that an alias free estimate of the autocorrelation of a continuous stationary signal can be obtained by sampling at random intervals provided the sampling intervals are Poisson distributed. In practice, the time lag axis of the autocorrelation function is divided into a number of equi-spaced slots and the cross-products of the randomly sampled signal are accumulated in the time lag slots nearest to the appropriate random interval. However, with the pulsed wire, it is still necessary to avoid pulsing the probe at an interval of less than about twice the pulsed wire time constant and, therefore, in the exploratory experiments of Gaster & Bradbury (1976) using random sampling, it was necessary to eliminate random pulses that occurred within 10 milleseconds of one another. As a result, a "dead" zone was introduced so that no autocorrelation estimates were available for lag times less than this period. Fig. 10 shows one example of the autocorrelation function obtained by Gaster & Bradbury for the v-component velocity fluctuations on the centre-line of the plate flow at x/b = 2. The time lag slot width in these experiments was two milliseconds and autocorrelation estimates up to two seconds were obtained. However, in Fig. 10, the results up to a lag time of 200 milliseconds only are shown. The presence of a strong periodic component to the velocity fluctuations is obvious from the results. However, in order to Fourier transform the autocorrelation function, it is necessary to assign values to the unknown autocorrelation coefficients in the "dead" zone. In the absence of any more reliable technique, Gaster & Bradbury simply sketched in these values in what seemed to be a plausible manner, However, the resultant spectrum at low frequencies contained negative values and it proved impossible to avoid these by simple adjustments to the unknown autocorrelation values. One possible technique for interpolating these unknown values in a more soundly based manner is based on the autocorrelation function theorem. This theorem — whose proof is remarkably simple — states that the determinant of the autocorrelation matrix

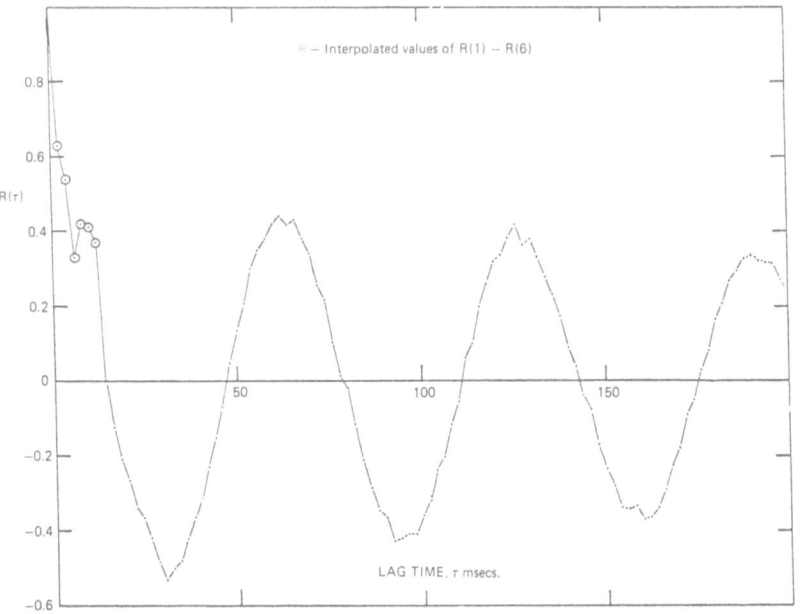

Fig. 10. Autocorrelation of v-component turbulence at y/b = 0, x/b = 2 using random sampling.

$$
T = \begin{bmatrix}
R(0) & R(1) & R(2)\cdots\cdots\cdot R(N) \\
R(1) & R(0) & R(1)\cdots\cdots\cdot R(N-1) \\
R(2) & R(1) & R(0) \\
\vdots & \vdots & \ddots \\
R(N) & R(N-1) & R(0)
\end{bmatrix}
\qquad [5.1]
$$

where R(0) to R(N) are the first (N+1) coefficients of the discrete autocorrelation functions, must be non-negative definite i.e. $T \geqslant 0$. This theorem has been used in the past to develop a means of minimising the effects of truncation errors - Burg (1967). In this type of application, the problem is that values of the autocorrelation coefficient from R(0) to R(N−1) are assumed known but R(N) is unknown. In this case, expanding the determinant T and equating it to zero results in a quadratic expression in R(N) and thus sets upper and lower bounds on R(N) in order to satisfy the condition that $T \geqslant 0$. In maximum entropy analysis, it is the mid-range value of R(N) that is chosen. This is equivalent to maximising the value of the determinant and, having obtained this estimate, this value may then be used to obtain an estimate of R(N+1) and so on. This extrapolation process is continued until the autocorrelation function has decayed sufficiently to avoid significant truncation errors and it is then possible to Fourier transform to obtain the power spectrum in the normal way. The use of this extrapolation process has been studied in detail by Stone (1978) and it should be noted that, because of symmetric nature of the auto-correlation matrix, the extrapolation process does not explicitly involve the evaluation of the determinant.

Another important feature of extrapolation using the autocorrelation function theorem is that the theorem is equivalent to setting upper and lower bounds on the unknown values of the autocorrelation function in order to ensure that the power spectrum resulting from the infinitely extended autocorrelation function is everywhere positive. The great attraction of this approach to the problem of truncation errors is that it leads to estimates of the power spectrum that are entirely consistent with the known values of the autocorrelation function. This is not true of techniques such as Hanning which, for this reason, are extremely unsatisfactory.

We now consider the problem in which, say, values of R(0) to R(N) are known with the exception of an unknown value of R(1). An application of the autocorrelation function theorem would seem to yield a polynomial in R(1) and R(1) may lie within a number of allowable ranges. However, the autocorrelation function theorem applies to the autocorrelation matrix of any order N.

Thus

$$
\begin{vmatrix} R(0) & R(1) \\ \\ R(1) & R(0) \end{vmatrix} \geqslant 0 \text{ and }
\begin{vmatrix} R(0) & R(1) & R(2) \\ R(1) & R(0) & R(1) \\ (R(2) & R(1) & R(0) \end{vmatrix} \geqslant 0
$$

and so on.

The first determinant limits R(1) to lie between ±1 and the second also yields a quadratic in R(1) such that (1) must lie within the range

$$
\pm \sqrt{\frac{1+R(2)}{2}} .
$$

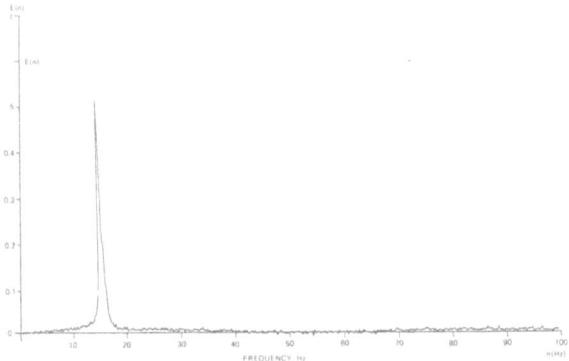

Fig. 11. Power Spectrum of randomly sampled signal using interpolated values of R(1) to (R6).

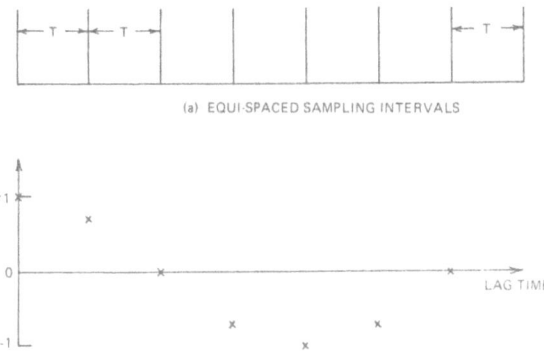

Fig. 12. Equi-spaced sampling of a sine wave.

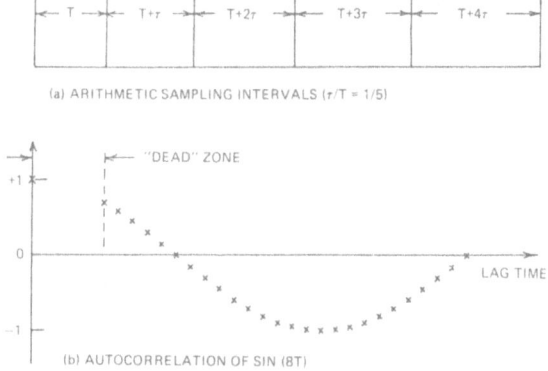

Fig. 13. Sequential sampling of a sine wave.

As this process is continued, only one overlapping range is obtained from different order matrices and this diminishes in extent with the increasing order of the matrices. But the difficulty remains of selecting a value within this range. By analogy with the maximum entropy analysis, it might seem that the value that maximises the determinant is a reasonable choice but Stone (1978) has demonstrated that this is not so. The equivalent criterion is to select a value $R(1)$ within the allowable range such that it gives a resultant matrix that if extended as described before from $R(N-1)$ would give a value of $R(N)$ equal to the known value of $R(N)$. It should again be emphasised that the symmetric nature of the autocorrelation matrix can be made use of to reduce this interpolation technique to a comparatively simple process not involving the explicit evaluation of the determinants.

In the example shown in Fig. 10, values $R(1)$ to $R(6)$ are unknown. In the first instance, the above process is used to generate $R(4)$ using the known values $R(8)$, $R(12)$ and so on. $R(5)$ and $R(6)$ may similarly be estimated. Using these values, $R(2)$, $R(3)$ and $R(1)$ may then be found in that order. Fig. 10 shows the result of this interpolation process using the first 100 known values of the autocorrelation function and Fig. 11 shows the resultant power spectrum using the full 1000 lag values and the interpolated values for $R(1)$ to $R(6)$. Unlike the spectrum obtained by Gaster & Bradbury, there are no significant regions in which this power spectrum has negative values.

Of course, it is impossible to know whether the interpolated values are really accurate estimates of the unknown values but Stone (1978) has considered many examples in which comparisons between correlation values obtained both by extrapolation and interpolation can be made with known values and the technique does invariably produce estimates of surprisingly high accuracy and detail.

Although spectral estimates can be obtained from the pulsed wire anemometer using random sampling, the requirements for data storage and the complexity of the software is such that it is difficult to undertake measurements using an on-line minicomputer. In consequence, recent efforts have been directed to a simple sampling technique in which the sampling interval increases in a simple arithmetic progression. As a simple illustration of the technique, Fig. 12 shows the result of sampling a simple harmonic signal at equi-spaced intervals T. If, however, the interval between samples is increased in an arithmetic progression by an amount τ then the autocorrelation obtained in Fig. 13 is obtained. The choice of T and τ in these examples has no significance. The important feature of this sequence is that although the frequency of sampling is less than in the periodic sampling case, the Nyquist frequency is apparently increased from 1/2T to 1/2τ. In practice, a finite length sequence should be repeated with possibly a random interval between each sequence or, alternatively, at an interval exceeding the period over which there is any significant correlation. It should also be noted that the number of samples contributing to the autocorrelation estimates differs for different lag times. Obviously, there are many questions that arise over the use of an arithmetic sampling sequence but it does seem to offer the possibility of using a comparatively slow sampling frequency with the pulsed wire anemometer to obtain spectral information beyond the conventional Nyquist frequency.

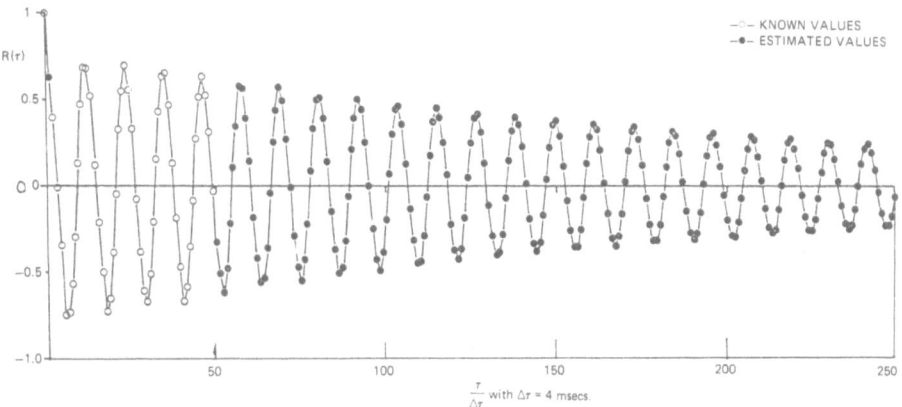

Fig. 14. Autocorrelation of u-component turbulence at x/b = 2, y/b = 1 using sequential sampling.

Work on the development of an efficient software package for obtaining autocorrelation estimates from an arithmetic sampling sequence is still not complete but Fig. 14 shows the results of an early experiment in which the initial sampling interval was 81 milliseconds with the period increasing by 41 milliseconds at each sample. The sequence length consisted of 48 samples and this was repeated 500 times. Only one estimate was required in the "dead" zone but, of more interest in this case, are the extrapolation results which extend the known autocorrelation function from 48 values to 250 values. Fig. 15 shows the results of Fourier transforming the original truncated autocorrelation function compared to the spectrum obtained from the extended autocorrelation function. Truncation errors are still present with the spectrum obtained from the extended autocorrelation function but these cannot be observed on the scale of Fig. 15. By contrast, the truncation errors in the raw transform are obvious.

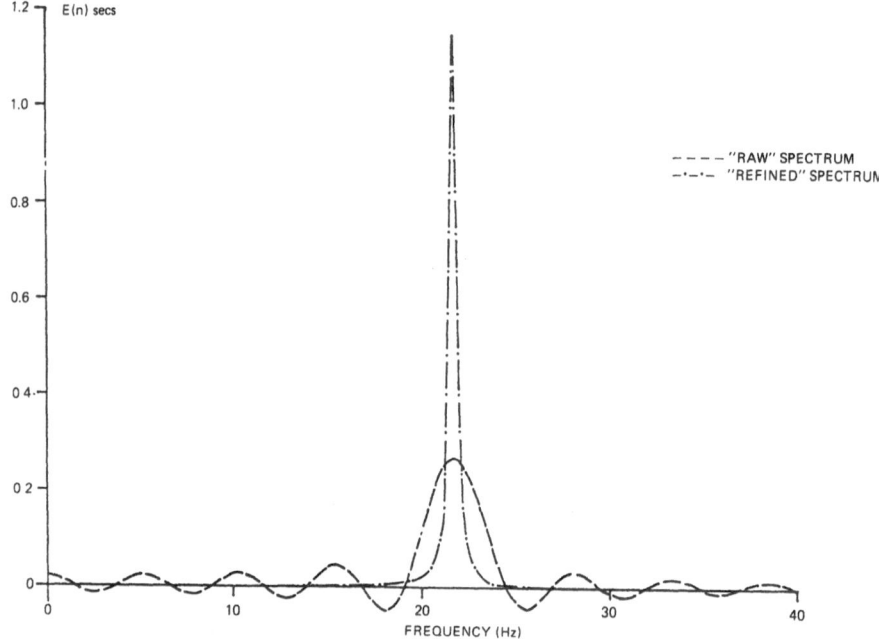

Fig. 15. Power spectrum of sequentially sampled signal using interpolated & extrapolated values of R(τ).

It should be emphasised that the discussion and results presented in this section represent only the briefest outline of the work that has been undertaken by Stone (1978) on the use of the autocorrelation function theorem and by Dr. N. Toy in the Mechanical Engineering Department on the use of sequential sampling.

6.0 Concluding Remarks

The results contained in this paper are intended to illustrate applications of the pulsed wire anemometer to the study of highly turbulent flows. The emphasis has been placed on the measurement technique and no attempt has been made to discuss any significance that the results might have in understanding the flow about the normal flat plate.

References

Bradbury, L.J.S., 1976 Measurements with a pulsed-wire and a hot-wire anemometer in the highly turbulent wake of a normal flat plate. *J. Fluid Mech.*, **77**, 473

Bradbury, L.J.S. and Castro, I.P., 1971 A pulsed-wire technique for velocity measurements in highly turbulent flows. *J. Fluid Mech.*, **49**, 657

Bradbury, L.J.S. and Moss, W.D., 1975 Pulsed wire anemometer measurements in the flow past a normal flat plate in uniform and in shear flow. *Proc. 4th Int. Conf. Wind Effects on Buildings & Structures,* London

Burg, J.P., 1967 Maximum entropy spectral analysis. *37th Annual Int. Soc. Explor. Geophys. Meeting,* Oklahoma

Cantwell, B.J., 1976 A flying hot wire study of the turbulent near wake of a circular cylinder at a Reynolds number of 140,000. Ph.D. thesis, Graduate Aeronautical Laboratories, California Inst. of Tech.

Davies, M.E., 1975 Wakes of oscillating bluff bodies. Ph.D. thesis, Aeronautics Dept., Imperial College

Gaster, M. & Roberts, J.B., 1975 Spectral analysis of randomly sampled signals. *J. Inst. Math. Appl.*, **15**, 195

Gaster, M. & Bradbury, L.J.S., 1976 The measurement of spectra of highly turbulent flows by a randomly triggered pulsed-wire anemometer. *J. Fluid Mech.*, **77**, 499

Gaster, M. & Roberts, J.B., 1977 On the spectral analysis of randomly sampled records by a direct transform. *Proc. Roy. Soc. A.*

Stone, P.P., 1978 The use of the autocorrelation function theorem in spectral analysis. Ph.D. thesis, University of Surrey.

Laser Doppler Anemometer Diagnostics in Unsteady Flows

by

Kenneth L. Orloff

Ames Research Center, NASA, Moffett Field, CA 94035

Abstract

The application of the laser Doppler anemometer (LDA) to unsteady flows is discussed with respect to (1) necessary features of the signal processor, (2) properties of the optical system, and (3) character of the flow under investigation. The discussion of signal processors includes consideration of frequency trackers, counter-type processors, particle properties, data rates, and statistics. Secondly, diffraction limitations for an optical system are viewed with respect to spatial resolution. Finally, the total velocity field is decomposed into its subfields and the feasability of, criteria for, and possible types of conditional sampling are defined. Several reported LDA experiments using conditional sampling are presented to demonstrate the different techniques that may be used.

1. Introduction

The successful application of the laser Doppler anemometer (LDA) to the measurement of the velocity field of steady or quasi-steady fluid flow fields has prompted consideration of the application of LDA to unsteady flows that are more complex in character. These more complex flow fields may include those wherein the physical insertion of a measuring probe is impossible because the environment is too hostile or because such a probe might disturb the flow parameters being measured by too great an extent. If the LDA is to be used to measure the time-dependent fluid velocities in unsteady flow fields, those factors must be considered that influence the accuracy of the measurement and that force the compromises that must often be made in the LDA optical system to optimize the instrument for a particular investigation. Accordingly, this paper is concerned with factors that influence the performance of the LDA when applied to unsteady flows. Since the design of the LDA instrument involves a wide variety of factors too numerous to cover in this paper, the topics to be discussed here are limited to: (1) signal processors and the required seed particle density as they are related to the LDA data rate; (2) particle flow tracing fidelity; and (3) optical diffraction limitations. These factors have been encountered by various investigators and are considered to be among those that limit LDA performance for unsteady flows.

With the performance limitations of the LDA in mind, the LDA is next considered in combination with conditional sampling, where appropriate, to provide a feasible technique for studying the fluid mechanics of time-dependent motions. For steady turbulent flows, LDA measurements have generally been concerned with mean velocity, turbulence intensity (variance), and turbulent shear stress (derived from variances or coincident measurements). If a flow (turbulent or laminar) is also unsteady in the sense of being periodic, then these statistical quantities can still be computed by time or ensemble averaging over a sufficient number of characteristic periods to ensure reproducibility of the statistical mean and variance. Whereas such data will not reflect the periodicity of the motion, they may prove useful when, for example, the mean velocities are integrated over an appropriate area, thereby yielding the mass flow rate. In the case of rotating machinery, such information is useful for evaluating machine performance (e.g. thrust, drag, swirl). On the other hand, if the intention is to *improve* the performance of say a turbine engine or a helicopter rotor, then statistical results cannot provide enough detail of the time-dependent interactions involved, and conditional sampling should be considered.

Indeed, to use averaged velocity data to infer the dynamics of an unsteady process is to risk being misled; that is, to bypass many aspects of the fluid mechanics of the situation and to invite erroneous conclusions. An illustrative, humorous example of the pitfalls of using statistical measures was given some years ago by Mollo-Christensen in an informative paper advocating the use of conditional sampling to enhance our understanding of turbulent boundary layer and shear flows (Mollo-Christensen 1971). The illustration suggests that a blind man uses a road bed sensor in an attempt to understand what motor vehicles look like. He chooses a road that is only traveled by limosines and motorcycles. He concludes that the average vehicle is a car with 2.4 wheels. He then formulates an elaborate theoretical model of the physics of this vehicle and attains fame for his model that looks like a motorcycle with a sidecar whose wheel is only in contact with the ground 40 % of the time. This illustration, although slightly absurd, makes a strong case for incorporating conditional sampling techniques into LDA investigations of unsteady flows. Accordingly, the second section of this paper discusses the varied types of fluid motions that may be encountered in LDA applications in unsteady flows and considers the degree to which conditional sampling is feasible.

Unsteady flow fields may be divided into a group wherein the time-dependent character is somewhat regular and a group wherein the motion is random. When the unsteadiness is random, the flow field is defined as being turbulent. When the motion can be described as periodic motion or to have a somewhat regular eddy pattern, it is labeled a *conditional* unsteady flow. This is not to say, however, that the instantaneous structure of this conditional flow cannot itself be turbulent in nature, with irregular, random, perhaps small-scale structure.

A review of several successful LDA investigations of conditional unsteady flows was compiled by Runstadler (1976); his review is updated here and examples of recent LDA applications techniques for conditional flows are presented.

2. LDA Performance For Unsteady Flows

Trackers, counters, realizations, and statistics

Frequency tracking signal processors have been successfully applied to transient and periodic LDA signals. Respectively, the unsteadiness can be measured with (or by) either rapid optical scanning through a complex flow (Orloff et al. 1975) or by fixed-point measurements in a time-dependent flow (Orloff and Biggers 1974). Frequency trackers are appropriate only when the occurrence rate of realizations (signals generated by the passage of single particles through the LDA focal volume) is high enough to guarantee that: (1) signal drop-out is not greater than the tracker hold time, and (2) the frequency difference between sequential realizations is not greater than the tracking bandwidth (Fridman et al. 1975). These criteria are met when the particle density is sufficiently high to yield many realizations during each characteristic velocity fluctuation time. However, for high particle densities there may be a significant number of overlapping realizations, thereby generating random phase shifts in the optical signal (Lading and Edwards 1975) that are demodulated as rapid frequency fluctuations with varying amplitudes. Unfortunately, it is difficult to differentiate between these "apparent" velocity fluctuations and the true unsteady fluid fluctuations.

Because of these limitations of frequency trackers, period-counting LDA signal processing electronics have been used nearly exclusively for unsteady flow studies. In combination with acousto-optic frequency biasing, the time required to obtain a single realization is of the order of, at most, several microseconds, which, for most flow studies, can be considered instantaneous. Hence, it is the *number* of realizations that occur during a given time interval that will limit the real-time response of the LDA to unsteady flows. However, it very well may be that the time to collect enough realizations to compute an accurate mean velocity is several orders of magnitude greater than the characteristic fluctuation time of the unsteady motion, and the details of the unsteady flow may not be properly represented. Therefore, when the unsteadiness in the flow is found to be correlated with some other detectable, conditional event, the electronics can be gated by this event and realizations accumulated as a statistical ensemble; the realizations are only collected when they occur during a short "data window" of duration δt. Depending on the duration and repetition rate of this data window, the *total* time required to acquire the statistical ensemble may become substantial. As a result, the exact repeatability of the motion is of paramount importance, because the ensemble average is implicitly assumed to be an accurate representation of the instantaneous conditions at some chosen phase of the conditional event.

Particle flow-tracing fidelity

The phrase "particle flow-tracing fidelity" is used to denote the accuracy with which a particle pathline follows a mean pathline of a fluid element as it moves through an unsteady flow field. Additionally, if the flow is turbulent, it is necessary to quantify the manner in which the particle responds to small-scale, high-frequency motions. The size of the seeding particle is therefore of paramount importance in the evaluation of its flow-tracing fidelity. Its diameter must be large enough to yield a scattering cross section that is acceptable for the particular optical configuration of the LDA (i.e., backscatter or forwardscatter, high or low power, etc.). On the other hand, the particle must be small enough to follow, to within the desired accuracy, (1) the convective accelerations of the fluid where there is sharp curvature of the mean flow, (2) the turbulent fluctuations, and (3) the local accelerations resulting from an unsteady flow.

The dynamical equation that describes the particle motion has been studied by many authors. To obtain a tenable solution the assumption is usually made that during the motion of the solid particle the same fluid element remains in its neighborhood. Such an assumption is highly restrictive because it is only satisfied if the pathline of the particle and the fluid streamline nearly coincide. This constraint can be easily violated when the particle is followed over the characteristic time scale of an unsteady flow. In this case, the unsteady flow problem should be treated over a series of shorter time intervals for which the assumption is valid. This approach lends itself well to the time-dependent or unsteady flow because it allows specification of an updated fluid velocity field as the initial conditions for each short time interval. Such an analysis was carried out by Base (1975) to study the difference between the particle pathline and the fluid streamline for a two-dimensional, time-dependent, vortex-modelled turbulent flow without mean flow curvature. More work needs to be directed toward similar studies that compute the particle pathlines for a two-dimensional time-dependent mean flow with significant curvature.

The conventional approach to particle fidelity for turbulent flows is to simplify the analysis by Fourier transforming the dynamical equation into the frequency domain and examining the particle's response to turbulent fluctuations at discrete frequencies. However, such a frequency-domain analysis does not provide the information required for considering the particle in an unsteady flow (the time-domain problem). A case in which the time-domain analysis has been successfully applied is that of the dynamical behavior of micron-sized particles entrained in a two-dimensional gas flow in the interblade region of a circular stationary cascade of turbine stator blades. Maxwell (1974) determined the particle velocity lag and angular deviation relative to the gas as a function of particle diameter and mass density. The extension of his work to the two-dimensional time-dependent case would be useful for evaluating the particle trackability errors for unsteady flows.

Diffraction limitations
The spatial resolution requirements of the LDA are determined by the scale of the flow structure being studied. The size of the LDA focal volume is generally not critical for large-scale flows with unconditional diagnostics. However, conditional measurements of smaller scale periodic structure often demand a significant improvement in the spatial resolution of the LDA.

An important property of a spatially coherent laser beam is the fact that the curvature of its wavefront at any point in space can be specified precisely in terms of the beam's history since it left the laser and passed through various optical elements. A beam that is divergent as it leaves the laser results in nonplanar wavefronts at the focus of even a perfect lens. The interference of two such focused beams (i.e. LDA dual-scatter) generates a nonuniform fringe spacing that introduces additional uncertainty into the ensemble average of the realizations. This phenomenon has been covered in depth by Durst and Stevenson (1975). In the discussions that follow, this divergence is taken to have been corrected; the laser light incident upon lenses is assumed to be collimated with planar wavefronts and Gaussian intensity distributions. Further complications due to a Gaussian beam profile and higher order laser modes have been adequately covered by Marshall (1971), among others. Only those uncertainties introduced by diffraction are considered herein.

Spherical aberration
The optical components used in LDA systems are subject to aberrations such as astigmatism, coma, curvature of field, distortion, and spherical aberration. Because LDA optics generally deal with laser light propagating along or parallel to the optic axis, the only aberration that is usually serious is spherical aberration.

Third order spherical aberration (Jenkins and White 1957) in a thin lens is illustrated in Fig. 1. This shows the path through a positive (converging) thin lens of light rays from an infinitely distant point source on the optical axis of the lens. Such parallel rays are equivalent to the rays in a collimated beam from a laser. In Fig. 1, it can be seen that the paraxial rays (those nearest to the optical axis) are focused at one point (called the paraxial focus), while the marginal rays (those furthest from the optical axis) are focused at a point (called the marginal focus) nearer the lens. The longitudinal spherical aberration L is defined as the distance of the marginal focus from the paraxial focus. The smallest blur spot lies between these two points.

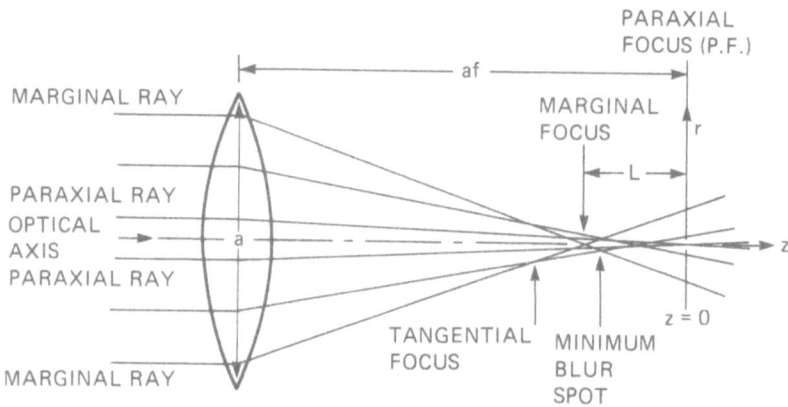

Fig. 1. Diagram showing phenomenon and nomenclature for spherical aberration.

Fig. 1 demonstrates longitudinal spherical aberration by means of ray tracing. More correctly, Innes and Bloom (1966) have used the exact Kirchhoff diffraction integral to calculate intensity contour diagrams for the focusing of a Gaussian beam with varying amounts of spherical aberration. Fig. 2(a) shows three levels of spherical aberration for a single laser beam focused by a positive lens. In order to avoid having to draw separate intensity diagrams for every possible f-number, Innes and Bloom define a system of "normalized" or dimensionless coordinates that can be used for all f-numbers. These coordinates are u and v, where u is the coordinate along the optic axis and v is the coordinate in the plane perpendicular to the axis. If z and r are the actual physical dimensions parallel and perpendicular to the optic axis, respectively, then the definitions of u and v are:

$$u = \pi z/2\lambda f^2 \qquad v = \pi r/\lambda f$$

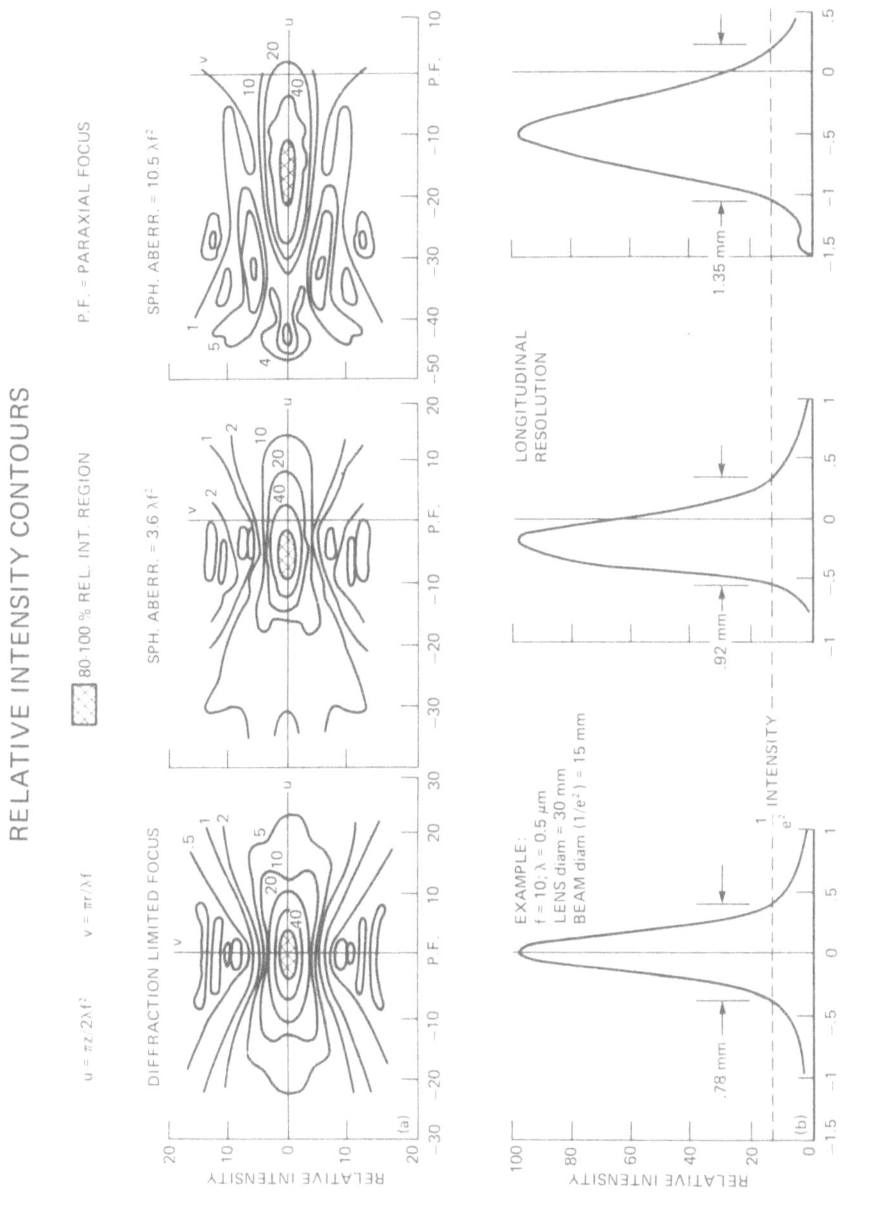

Fig. 2. Spherical aberration degrades spatial resolution and reduces density at focus.

(a)   Intensity contour diagrams for positive spherical aberration.

(b)   Intensity variation along the optical axis showing loss of longitudinal spatial resolution due to spherical aberration.

Fig. 2(a) presents the computations for a single laser beam whose $1/e^2$ intensity diameter is one-half the diameter of the focusing lens. The aberration has the effect of elongating the focal region and redistributing the available energy over a much larger region. As an example, Fig. 2(b) shows the variation of the intensity along the optical axis for an f= 10 lens. Using the $1/e^2$ intensity points, the longitudinal resolution (the length of the central maximum) varies from 0.78 mm for a perfect lens to 1.35 mm for a lens with a nominal amount of spherical aberration. Clearly, the spatial resolution is degraded and the energy density reduced.

This diffraction effect also influences the performance of an LDA operated in the crossed-beam configuration. For a "bundle" of rays in the marginal zone of the lens (see Fig. 1), the tangential focus is noted to occur at a location other than along the optical axis. As a result, the spherical aberration degrades the spatial resolution of the LDA probe volume and introduces uncertainty into the velocity measurement due to overlapping spherical rather than planar wavefronts at the junction of the two bundles of rays (Durst and Stevenson 1975).

The spherical aberration of a lens varies greatly with its shape. However, for a given focal length, there will be a best lens shape for which the aberration blur spot diameter is a minimum (Jenkins and White 1957). Wilson (1968) has shown that the best-form lens for a laser application is determined by equating this minimum blur spot diameter to the diameter of the Airy disc of a diffraction limited lens. This provides the minimum allowable focal length in order not to exceed this minimum aberration. Although spherical aberration cannot be entirely eliminated for a single spherical lens, it is possible to do so for a combination of two or more lenses of opposite sign. The amount of aberration introduced by one lens of such a combination must be equal and opposite to that introduced by the other. This strongly suggests the use of diffraction limited multielement lenses for LDA applications in which spatial resolution is critical (e.g., boundary layer studies).

Chromatic aberration
The advent of ion lasers and the development of LDA optical systems using more than a single color have given rise to situations where the correction for chromatic aberration is important. Since the index of refraction is a function of the wavelength of the light, the properties of the optical elements also vary with wavelength. In general, the index of refraction of optical materials is higher for short wavelengths than for long wavelengths; this causes the short wavelengths to be more strongly refracted at each surface of a lens so that, in a simple positive lens for example, the blue light rays are brought to a focus

closer to the lens than the red rays. The distance along the axis between the two focal points is called the longitudinal or axial chromatic aberration. To correct for this aberration, it is common practice to combine optical elements with axial aberration of opposite sign so that the aberrations contributed to the system by one element are cancelled out, or corrected, by the other. Such a lens combination is called an achromatic doublet.

The technique of using doublet lenses can also be employed to correct spherical aberration so that both chromatic and spherical aberration may easily be corrected simultaneously with one optical unit. Unfortunately, a perfect correction for these aberrations can only be accomplished at a single design focal distance for a given lens combination. An example will be given in a later section of the paper that demonstrates the reduction in spatial resolution due to these aberrations for LDA operation at other than design focus.

3. Conditional Sampling

The degree to which conditional sampling is feasible in any particular LDA investigation of an unsteady flow can be determined by decomposing the total velocity field into three subfields,

$$\vec{Q}(\vec{x},t) = \vec{U}(\vec{x},t) + \underset{\sim}{\vec{U}}(\vec{x},t) + \vec{u}'(\vec{x},t),$$

with respective length scales L, ℓ, and λ, where $L \gg \ell \gg \lambda$, and with respective dissimilar time scales, T, t and τ, where $T \gg t \gg \tau$. This decomposition is presented schematically in Fig. 3. An assumed velocity variation over two periods of this hypothetical motion is also depicted in Fig. 3.

The procedures for ensemble averaging LDA data have been studied by several authors (Durao and Whitelaw 1975; Buchhave 1975, McLaughlin and Tiederman 1973) and are not herein reviewed. It suffices here to denote the appropriately weighted ensemble average by angular brackets, and the interval over which the LDA realizations are accepted by a subscript. Then, with reference to Fig. 3:

unconditional mean: $\langle Q_i(\vec{x},t) \rangle_{n\underset{\sim}{t}} = U_i(\vec{x},t); \ \langle \underset{\sim}{U}_i(\vec{x},t) \rangle_{n\underset{\sim}{t}} = \langle u_i'(\vec{x},t) \rangle_{n\underset{\sim}{t}} = 0;$

conditional ensemble mean: $\langle Q_i(\vec{x},t) \rangle_{\delta t} = \underset{\sim}{U}_i(\vec{x},t); \ \langle u_i'(\vec{x},t) \rangle_{\delta t} = 0.$

Now, each of the velocity subfields is considered and several successful LDA applications are briefly reviewed to exemplify these subfields.

The unconditional mean velocity, $\vec{U}(\vec{x},t)$

The unconditional mean velocity $\vec{U}(\vec{x},t)$ is the average velocity determined *unconditionally* over a long time interval T (usually $T \gg \underset{\sim}{t}$ for a periodic flow). The time-dependence is dropped for steady-state periodic or random flows. However, for developing flows, such as the decay of an isolated vortex, the spin-up of a turbine engine, or the forward acceleration of a helicopter from hover to 50 knots, the time-dependence in $\vec{U}(\vec{x},t)$ must be retained, and the characteristic time for a significant change in the mean flow is taken to be of the order of T. Over the data acquisition time, $\vec{U}(\vec{x},t)$ is treated as quasi-steady as if it were a nondeveloping flow and becomes $[\vec{U}(\vec{x})]_t$ where the subscript t indicates that $\vec{U}(\vec{x})$ is slowly changing with time.

Bein and Penner (1970) have used an LDA to investigate the unsteady spin-up and spin-down characteristics for confined rotating flows. The characteristic decay time for their flows was $T \cong 17$ sec; the typical data acquisition time for a single velocity value was of the order of 0.3 sec, during which the flow structure underwent negligible change. The flow could, therefore, be treated as quasi-steady during the signal averaging process.

Using an LDA with optical scanning, Ciffone and Orloff (1975) and Luebs et al. (1976) studied the decay characteristics of the vortex wake generated by towing an aircraft model through a towing tank. In that case, to obtain a meaningful time history of the decay process and to assign an "age" to each measured profile, the time required for an optical traversal of the vortex had to be short compared with the vortex decay time. The average decay time in their studies was $T \cong 60$ sec and the time for an LDA traversal was about 1 sec, thereby meeting the above criteria for quasi-equilibrium during the velocity profile measurement.

The conditional mean velocity, $\underset{\sim}{\vec{U}}(\vec{x},t)$

The conditional mean velocity $\underset{\sim}{\vec{U}}(\vec{x},t)$ is a fluctuation whose occurrence is either random, periodic, or some combination thereof, and whose time-dependence is highly correlated with the phase of one or more other discrete, measurable events (e.g., surface pressures, periodic timing pulses).

Periodic conditional.

The shedding of a Von Kármán vortex street from a circular cylinder, the flow associated with rotating machinery, and Tollmien-Schlicting waves in the transition zone of a boundary layer are all examples of periodic motions. The instantaneous velocity at a point within such flows can be determined at a particular phase of the motion by restricting the measurement time to a short interval, δt, at a fixed phase of the correlated event (see Fig. 3). At this fixed phase, the inertial properties of the fluid are expected to be repeatable within a spatial region in which the velocity is highly correlated with the chosen gating event. In fact, for periodic flows, the detection and analysis of the influence of this time-dependent fluid inertia may be precisely what the measurements are intended to reveal.

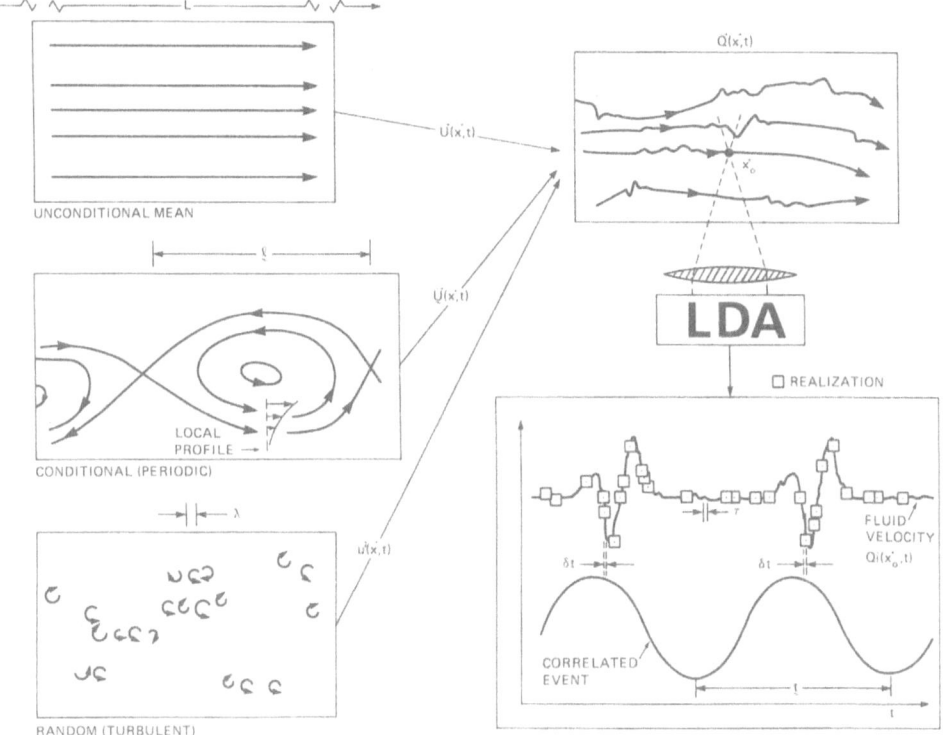

Fig. 3. Schematic representation of the total velocity field $\vec{Q}(x,t)$, the resulting LDA signal, and the correlated event for gating the LDA. The velocity subfields are: $\vec{U}(x,t)$ — the unconditional mean; $\underset{\sim}{U}(x,t)$ — the conditional mean; $\vec{u}'(x,t)$ — the turbulent fluctuation.

A few examples of periodic conditional LDA measurements reported to date follow.

Denison and Stevenson (1970) used a reference-beam LDA to study the laminar oscillatory flow of an incompressible Newtonian fluid within a ridig cylindrical tube. The mean flow velocity was zero; the desired oscillations were induced by imposing a periodic pressure gradient on the fluid. The streamwise velocity structure within the tube was measured for several phases of the axial pressure gradient. They report the period of the imposed pressure fluctuation to be $t \cong 6$ sec; the use of frequency tracking electronics was therefore found to be adequate to effectively document the instantaneous flow.

Walker et al. (1975) used a two-color confocal backscatter LDA to study the velocity patterns and shock locations in the interblade region of a transonic fan by traversing the optical package and varying the time delay (gated to blade locations) of the velocity samples. A typical time duration of the velocity sampling was 1/20 of the time (about 10 μsec) required for one interblade gap to pass the probe volume of the LDA. Using a variable time delay, successive measurements were made from blade to blade, thereby enhacing the data rate.

Seegmiller et al. (1978) studied the unsteady transonic flow over an 18 % thick circular arc airfoil at $0°$ angle of attack. Their LDA is of the two-color forward-scatter type with an ellipsoidal spatial resolution of 0.3 mm diameter by 3 mm length in the spanwise direction. For a wing chord of 20.32 cm, the resolution is $0.3/20.32 \cong 1.5$ % chord, yielding high spatial resolution in the vicinity of the wing. Over a narrow range of Mach and Reynolds numbers, the flow surrounding the airfoil was found to be very nearly periodic and highly correlated with the signal from an upper surface pressure transducer, as indicated in Fig. 4. The difference in the signal between consecutive cycles of the pressure influences the repeatability of the velocity realizations for a given phase of the pressure. These variations can be attributed to inaccuracies in: (1) determining the onset (zero phase) of the pressure signal or; (2) to random external perturbations (discussed in more detail in the next section) induced by wind tunnel unsteadiness.

The influence of the seed particle trackability (through the shock wave) on the accuracy of the LDA measurements was analyzed by Seegmiller and his co-workers. They used polystyrene particles with a diameter of 0.4 μ as the tracer particles. They note that the results of Maxwell and Seasholtz (1974) indicate that downstream of the shock location these particles should have a relaxation length of less than 1 mm; this length corresponds to 0.5 % of the wing chord. Hence, the particle flow-tracing fidelity was deemed adequate for the spatial resolution of the experiment.

Owen and Johnson (1978) surveyed the wake flow of a circular cylinder using a forward scatter LDA. The real-time signal from a surface hot-film gauge was used to provide a correlated event. In this manner the unsteady vortex shedding process was conditionally investigated. They compared both constant-phase,

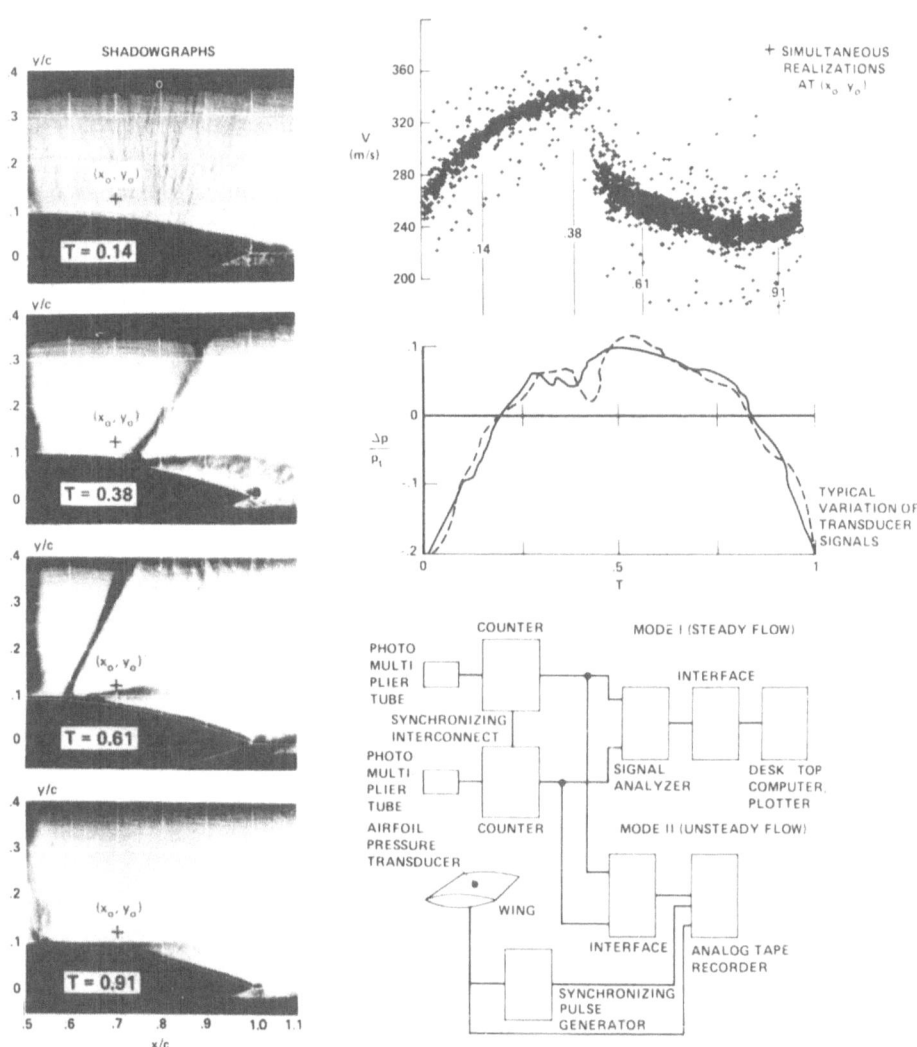

Fig. 4. Conditionally sampled periodic unsteady transonic flow over a circular arc airfoil by Seegmiller et al. (1978).

ensemble-averaged data and time-averaged data to demonstrate how the detailed features of the flow may be either revealed or hidden, respectively. By conditionally (constant-phase) excluding the organized, periodic motion, they show that the measured local turbulence intensity can be considerably lower than would be concluded using time-averaged (unconditional) information; in other words, they have attempted to isolate $\sqrt{\overline{u'^2}}$ from $\sqrt{\overline{|\underset{\sim}{U}+\vec{u'}|^2}}$.

Biggers and Orloff (1974), Biggers et al. (1975), and Biggers et al. (1977) used a two-color, dual-beam, on-axis backscatter LDA with optical scanning capability to investigate the instantaneous flow around a model helicopter rotor (2.13 m diameter) operated in a forward flight configuration in the NASA-Ames 7- by 10-Foot Wind Tunnel. A sample of the results for the downward component of the velocity along a survey line below the rotor disk plane is presented in Fig. 5. The velocities have been nondimensionalized by the rotor tip speed, $\Omega R = 35$ m/sec ($\Omega = 300$ rpm).

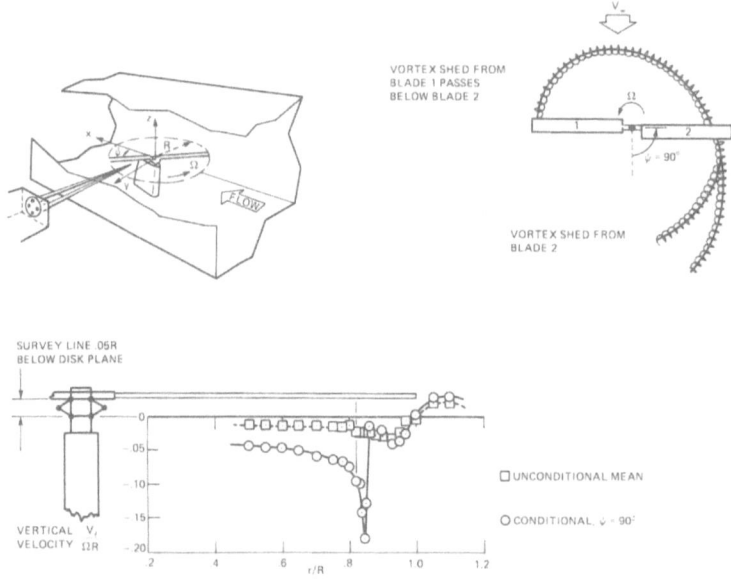

Fig. 5. Periodic conditional and unconditional mean velocity profiles below a model helicopter rotor in the forward flight configuration by Biggers and Orloff (1974).

The unconditional mean velocities were obtained using a spectrum analyzer set to a wide bandwidth and a long integration time. The median frequency for the distribution was visually estimated from the CRT display. These values are equivalent to those that would be measured by a device with poor frequency response, such as a pitot tube, located at the same point. It is noted in Fig. 5 that near the blade tip ($y/R = 1.0$) the unconditional mean velocities indicate the presence of an *average* tip vortex that results from the rotor acting as a lifting device in forward flight. However, the unconditional mean does not even suggest the presence of *discrete* vortices passing, periodically, by the LDA focal volume.

The conditional velocities shown in Fig. 5 were obtained by gating the counter-type LDA processor according to the rotor periodicity as the conditional event. As indicated, the conditional structure below the advancing blade discloses the influence of the vortex that was shed by the preceding blade. These data were obtained with the processor enabled for a short time ($\delta t = 150$ μsec) only when the blade was exactly at the $90°$ azimuth. During this time the blade motion was 0.0006 revolutions; typical data rates during this interval for substantial seeding of the flow were approximately two realizations per second. The conditional structure clearly shows the details of the tip vortex shed from the preceding blade; also, the instantaneous downward velocity is, as expected, instantaneously greater beneath the blade than it is (unconditionally) on the average.

The accuracy of the results obtained by Biggers and his coworkers are compromised by several properties of the LDA used for these measurements. Uncertainties are introduced due to the effects of spherical and chromatic aberration as suggested in the earlier general discussion of these phenomena.

The scanning optical system used in the above experiment provides diffraction-limited operation only at a specified design focus distance (Orloff and Biggers 1974). As one optically traverses to a scan range other than design, the spherical aberration becomes either over-corrected or under-corrected, resulting in a tangential focus that does not coincide with the junction of the central ray and the optical axis. The image of the crossed-beam interference volume is therefore elongated vertically. This "vertical blur" and the ensuing increased length of the focal volume not only reduce the LDA spatial resolution, but they also reduce the laser energy density within the interference fringe pattern. The loss of spatial resolution is small compared to the overall size of the rotor radius and as such should not influence the accuracy of unconditional data. On the other hand, conditional measurements of the vortex passing beneath the blade indicate the vortex core radius to be less than 1 cm; hence, the spatial resolution of the LDA is of great concern. Fig. 6 quantifies these relationships for more recent measurements by Biggers et al. (1977) using improved scanning optics.

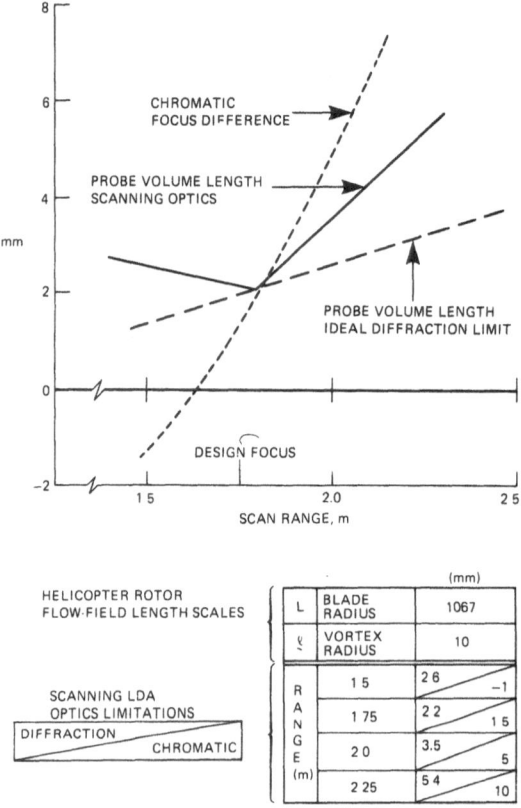

Fig. 6. Relationship between length scales for the flow beneath the helicopter rotor and the spatial resolution of the LDA. L is the unconditional length scale; ℓ is the conditional length scale.

Additionally, this scanning optical system is perfectly achromatized for two-color operation only at one design distance. Hence, as the scan range is varied, the crossover locations for the two colors separate longitudinally. As before, this error is relatively unimportant when only unconditional results are desired, but the error may become significant when conditional data regarding the tip vortex are desired, as indicated in Fig. 6.

Random external conditional.
A fluid motion that is unsteady in a more random sense may still be conditional upon a correlated event (e.g., visual or optical positioning, surface pressure detection, or force measurement). The velocity field of interest may or may not be time-dependent; it may be immersed within a randomly fluctuating flow

of larger scale. As a result, the unsteadiness subsequently detected by a measuring instrument (for a fixed probe position) is generated in part by the motion of the external flow rather than by the internal (conditional) flow. Since the unsteadiness is neither associated with nor produced by the embedded conditional flow, the dynamic influence that these external fluctuations have on the embedded flow must be considered. When the length and time scales of the external fluctuations are large compared to ℓ and t, respectively, the effects are probably negligible. Conversely, if the random motion has a characteristic time scale on the order of t, then the influence of the unsteadiness on the fluid inertia of the conditional structure should be taken into account. Random, externally imposed motions of this type can be induced by fluctuations in the free-stream velocity of a wind tunnel facility. Examples of LDA measurements in the presence of random external perturbations follow.

East (1976) investigated the interaction region of a normal shock wave and a two-dimensional turbulent boundary layer using a two-color backscatter LDA optical system with photon correlation signal processing. The streamwise location of the normal shock wave was unsteady due to small disturbances propagating downstream in the supersonic section and upstream in the subsonic section. Random flow variations such as these can compromise the quality of the data by introducing uncertainty as to the exact location of the LDA test point within the conditional flow field. In particular, sharp changes in the flow conditions across the shock wave can be undesirably smoothed out by these motions. To reduce these effects, the correlator processor was only enabled when the shock location was determined to be constant (to a specified accuracy). A pressure transducer was connected to the tunnel floor at a location where the steep pressure rise at the start of the shock-wave boundary-layer interaction occurred most frequently. The output of the transducer was then used to gate the correlator, activating it only when the measured static pressure was between set values. The fluid inertia imparted to the conditional flow as a result of this shock motion was determined to be small, thereby validating the results as a description of the steady-state velocity structure.

The LDA transducer effect of greatest concern in the above investigation is the particle flow-tracing fidelity. The flow is reported to have been seeded with oil particles with diameters of about 1 μ. Evidence (Maxwell and Seasholtz 1974; Yanta 1973) suggests that the velocity lag of these particles must be considered. East (1976) found that unless the seeding tube was regularly cleaned, larger particles were formed which did not accurately follow the flow. When these larger particles were present, low mean values of the velocity were measured upstream of the shock wave, indicating that the particles had lagged behind the accelerating flow through the tunnel nozzle. Downstream of the shock wave, the measured velocity was too high, indicating that the larger particles had not responded rapidly enough to the change in the flow conditions through the shock wave.

The vortex wakes shed from lifting wings have been studied using both hot-wire and LDA techniques. A problem that hinders wind-tunnel vortex measurements is the lateral, nearly random motion of the vortex within the wind-tunnel test section; this motion has been termed "vortex meander". One of the earliest wind-tunnel studies of these vortices was conducted by Chigier and Corsiglia (1972). They used a hot-wire anemometer to measure the velocity structure of the vortex generated by a rectangular semispan wing installed in the NASA-Ames 7- by 10-Foot Wind Tunnel. The hot-wire probe was held fixed at selected points across the vortex diameter and average values were obtained over long measuring times. This resulted in a loss of information regarding the detailed vortex structure, particularly near the central or core region. A sample of their results is presented in Fig. 7.

Also shown in Fig. 7 are the LDA results obtained by Orloff (1974) for the same wing at the same wind tunnel conditions; here, an LDA with high-speed optical scanning was used to overcome the problem of vortex meander by obtaining the velocity scan across the vortex during a time interval so short that the vortex did not move or meander appreciably. Particulate scattering material was introduced into the diffuser section of the wind tunnel by means of a mineral oil aerosol generator. Recirculation of the air around the wind tunnel loop provided a light concentration of the aerosol throughout the test section, and LDA data were obtained at all points except very near the core of the vortex where all but the smallest particles were centrifuged outward. This centrifugal action distinguished the core region as a much dimmer illumination of the outgoing laser beams, allowing a conditional determination of when an "on-diameter" traversal had been accomplished. Such a traversal was only considered valid if, during the optical traversal, the vertical location of the vortex core remained aligned with the outgoing laser beams that defined a horizontal plane. A small amount of core motion was acceptable when the probe volume of the LDA was well outside the central region where the tangential velocity gradients were low. The vortex was, however, required to remain nearly stationary as the probe volume traversed the core region.

The scale wavelength of the wind-tunnel turbulence for the NASA-Ames 7- by 10-Foot Wind Tunnel, operating under the conditions indicated above, was determined by Reed (1973) to be nominally 10 m. If the turbulence is con-

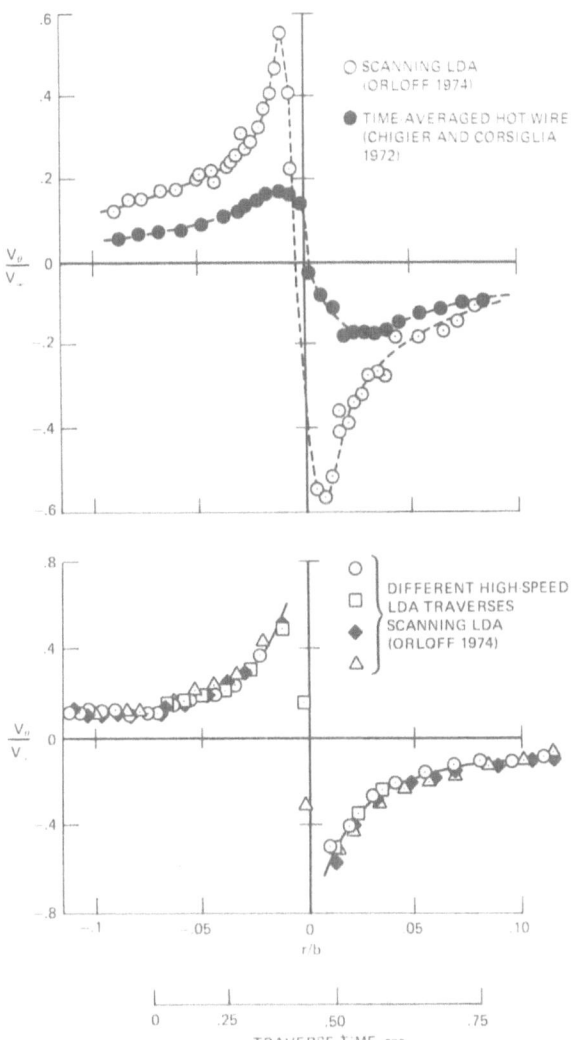

Fig. 7. Random external conditional (Orloff 1974) and unconditional (Chigier and Corsiglia 1972) mean rotational velocity structure of a wing tip vortex in a wind tunnel.

vected at a mainstream velocity of 25 m/sec, then the characteristic meander period is 0.4 sec. However, the time for the LDA to optically traverse the vortex region is shown in Fig. 7 to be typically 0.3 sec. Since this time is not a great deal smaller than the typical meander time, some variation between different optical traverses is expected and is apparent in Fig. 7.

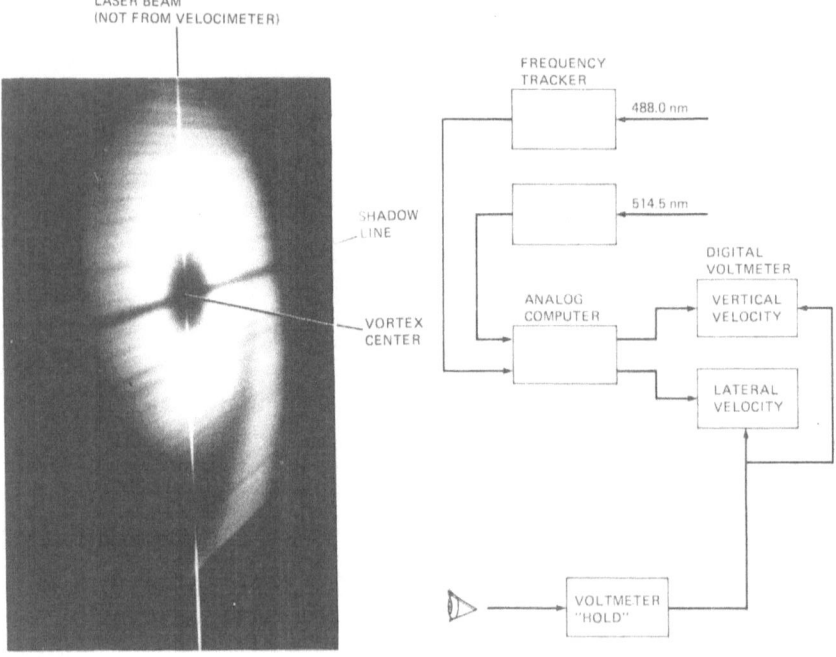

Fig. 8. LDA experiment of Corsiglia et al. (1978) to study interacting vortices
in an open-circuit wind tunnel in the presence of random external disturbances.

More recently, the interaction and merger of two free vortices in an open-circuit wind tunnel has been studied by Corsiglia et al. (1978) with an LDA designed to measure the two cross-stream components of the velocity (Fig. 8). High-speed optical traverses were not used to overcome the problem of vortex meander because the wind tunnel had a very low level of free-stream turbulence. Nevertheless, the small amount of meander present was compensated for by taking data only when the embedded vortex core flow was visually observed to be in a prescribed location. To accomplish this, the meander was monitored with a crossed target comprised of a light slit shadow and a laser beam mutually orthogonal to each other as shown in Fig. 8. Data were continuously averaged by the frequency trackers with a time-constant of 0.3 sec to remove high-frequency fluctuations. The voltmeter "hold" was then used to sample the velocity several times at each location when the vortex core was observed to be "in position". In this way, repeatability of the results was generally within 4 %.

The turbulent fluctuation, $\vec{u}'(\vec{x},t)$

The velocity $\vec{u}'(\vec{x},t)$ is characterized by the smallest time and length scales, τ and λ, respectively. The scale of these fluctuations is such that high spatial and temporal resolution are required of the velocity measuring instrument. A mechanical probe inserted into the flow to sense these fluctuations might itself generate disturbances greater than those it is intended to measure. The LDA overcomes these problems. For most flows it is possible to spatially resolve to scales much smaller than λ when diffraction-limited optics are used. Also, the measurement is not intrusive, and the flow can be seeded with particles that are sparsely distributed and small enough so that the turbulent structure is not altered.

The turbulent fluctuations generally result from instabilities in the instantaneous velocity profile of the conditional flow (see Fig. 3). The production of these secondary fluctuations is dominated by their local environment and is assumed to be uncorrelated with the gating condition used to obtain conditional measurements. At each point within $\vec{U}(\vec{x},t)$, the influence of the turbulent fluctuations on the LDA measurement is to increase the standard deviation associated with the ensemble average for the conditionally sampled velocity. As a result, the time-dependent mechanics of the turbulent flow process remain undetected and $\vec{u}'(\vec{x},t)$ must be analyzed by statistical means (e.g. rms value, turbulent shear stress).

An accurate measurement of $\overline{u'^2}$ (variance of u') for a conditional unsteady flow is difficult. Because δt is generally quite small, an extremely long elapsed time is required to collect enough samples to provide an accurate variance.

Also, if the correlation between the "correlated event" and the velocity $\vec{Q}$ is not close to unity, then a high "apparent" variance will be generated because $\underset{\sim}{U}$ is not reproducible at times t and t + nt. Seegmiller (private communication) suggests that some uncertainty may exist in the phase-averaged measurement of the turbulence intensity for a given phase of the shock motion (Seegmiller et al. 1978) due to cycle-to-cycle variations that may have been induced by mainstream fluctuations or other as yet unidentified factors; however, the mean values are still quite accurate. Owen and Johnson (1978) report distributions of the conditionally averaged turbulence intensity but have not considered to what degree random external unsteadiness may have influenced the measurements. Indeed, the reported real-time signals (Fig. 4 of Owen and Johnson 1978) indicate larger variations per cycle between the correlated event (hot film output) and the LDA signal than would be expected due to turbulence alone.

4. Concluding Remarks

The LDA has proved to be a powerful tool for providing data that can augment our understanding of unsteady flows. Where these flows are conditional and a correlated event can be accurately monitored, this understanding can be enhanced even further. However, it has been shown that care must be taken to clearly define the character of the conditional flow. When the conditional flow is periodic, the dynamics of the unsteady motion are inherent in the results; this is generally a desirable situation and no special precautions are necessary. On the other hand, the unsteadiness may be due to external influences that are unrelated to the flow field of interest (random external conditional). In this case, even though a correlated event can be identified, there must be some assurance that the externally imposed motion does not significantly influence the accuracy of the results.

Conditional sampling often reduces the spatial scale at which the unsteady motion is diagnosed. This can impose constraints on the LDA optical system that would not be necessary in order to examine the same flow field unconditionally. Hence, it is important to understand the geometrical arrangement and the optical aberrations of a particular LDA instrument before conducting such experiments. As indicated, due consideration must also be given to the properties of the tracer particles and to the signal processing electronics to determine that conditional sampling is practical.

This paper has presented only a few examples of successful conditional sampling investigations using LDA. The author acknowledges that there have been many other noteworthy LDA investigations of unsteady flows and apologizes for the lack of space to include them all.

References

Base, T.E., 1975 The motion of aerosol particles in a computed turbulent flow model — to determine the accuracy of an LDV system. Proceedings of the Minnesota Symposium on Laser Anemometry, pp. 277-292

Bien, F. and Penner, S.S., 1970 *Phys. Fluids,* Vol. **13**, pp. 1665-1671

Biggers, J.C. and Orloff, K.L., 1974 Laser velocimeter measurements of the helicopter rotor induced flow field. Proceedings of the 30th Annual National Forum of the American Helicopter Society

Biggers, J.C., Chu, S. and Orloff, K.L., 1975 Laser velocimeter measurements of rotor blade loads and tip vortex rollup. Proceedings of the 31st Annual National Forum of the American Helicopter Society

Biggers, J.C., Lee, A., Orloff, K.L. and Lemmer, O.J., 1977 Measurements of helicopter rotor tip vortices. Proceedings of the 33rd Annual National Forum of the American Helicopter Society

Buchhave, P., 1975 Biasing errors in individual particle measurements with the LDA-counter signal processor. *The Accuracy of Flow Measurements by Laser Doppler Methods.* Proceedings of the LDA-Symposium, Copenhagen, Denmark, pp. 258-278

Chigier, N.A. and Corsiglia, V.R., 1972 *J. Aircraft,* Vol. **9**, pp. 820-825

Ciffone, D.L. and Orloff, K.L., 1975 *J. Aircraft,* Vol. **12**, pp. 464-470

Corsiglia, V.R., Iversen, J.D. and Orloff, K.L., 1978 Laser-velocimeter surveys of merging vortices in a wind tunnel. AIAA 16th Aerospace Sciences Meeting, Huntsville, Alabama

Denison, E.B. and Stevenson, W.H., 1970 *Rev. Sci. Instrum.* Vol. **41**, pp. 1475-1478

Durao, D.F.G. and Whitelaw, J.H., 1975 The influence of sampling procedures on velocity bias in turbulent flows. *The Accuracy of Flow Measurements by Laser Doppler Methods.* Proceedings of the LDA-Symposium, Copenhagen, pp. 138-149

Durst, F. and Stevenson, W.H., 1975 Properties of focused laser beams and the influence on optical anemometer signals. Proceedings of the Minnesota Symposium on Laser Anemometry, pp. 371-388

East, L.F., 1976 The application of a laser anemometer to the investigation of shock-wave boundary-layer interactions. NATO-AGARD Symposium on Applications of Non-intrusive Instrumentation in Fluid Flow Research, St. Louis, France

Fridman, J.D., Young, R.M., Seavey, R.E. and Orloff, K.L., 1975 Modular high accuracy tracker for dual channel Laser Doppler velocimeter. Proceedings of the Minnesota Symposium on Laser Anemometry, pp. 485-503

Innes, D.J. and Bloom, A.L., 1966 Spectra-Physics Laser Technical Bulletin No. 5

Jenkins, F.A. and White, H.E., 1957 *Fundamentals of Optics,* McGraw-Hill, pp. 131ff

Lading, L. and Edwards, R.V., 1975 The effect of measurement volume on laser Doppler anemometer measurements as measured on simulated signals. *The Accuracy of Flow Measurements by Laser Doppler Methods.* Proceedings of the LDA-Symposium, Copenhagen, pp. 64-80

Luebs, A.B., Bradfute, F.G. and Ciffone, D.L., 1976 NASA TM X-73, 197

Marshall, L., 1971 *Laser Focus* (April, pp. 26-28)

Maxwell, B.R., 1974 NASA CR-134543

Maxwell, B.R. and Seasholtz, R.G., 1974 NASA TN D-7490

McLaughlin, D.K. and Tiederman, W.G., 1973 *Phys. Fluids*, Vol. **16**, pp. 2082-2088

Mollo-Christensen, E., 1971 *AIAA J.*, Vol. **9**, pp. 1217-1228

Orloff, K.L. and Biggers, J.C., 1974 Laser velocimeter measurement of developing and periodic flows. Proceedings of the Second International Workshop on Laser Velocimetry, Vol. II, pp. 143-168

Orloff, K.L., Corsiglia, V.R., Biggers, J.C. and Ekstedt, T.W., 1975 Investigating complex aerodynamic flows with a laser velocimeter. *The Accuracy of Flow Measurements by Laser Doppler Methods*. Proceedings of the LDA-Symposium, Copenhagen, pp. 624-643

Orloff, K.L., 1974 *J. Aircraft*, Vol. **11**, pp. 477-482

Owen, F.K. and Johnson, D.A., 1978 Measurement of unsteady vortex flowfields. AIAA 16th Aerospace Sciences Meeting, Huntsville, Alabama, Paper No. 78-18

Reed, R.E., 1973 TR 47, 1973, Nielsen Engineering and Research, Inc., Mountain View, California

Runstadler, P.W. Jr., 1976 Creare, Inc. TN-241

Seegmiller, H.L., Marvin, J.G. and Levy, L.L., 1978 Steady and unsteady transonic flow. AIAA 16th Aerospace Sciences Meeting, Huntsville, Alabama, Paper No. 78-160

Walker, D.A., Williams, M.C. and House, R.D., 1975 Intrablade velocity measurements in a transonic fan utilizing a laser Doppler velocimeter. Proceedings of the Minnesota Symposium on Laser Anemometry, pp. 124-145

Wilson, D.K., 1968 *Optical Spectra* (April, pp. 52-55)

Yanta, W.J., 1973 Measurements of aerosol-size distributions with a laser Doppler velocimeter. AIAA 6th Fluid and Plasma Dynamics Conference.

Application of Laser Doppler Anemometry to Aeroacoustic Research

by

J. Haertig

Institut Franco-Allemand de Recherches de Saint-Louis (ISL)

12, rue de l'Industrie, 68301 Saint-Louis, France

Introduction

It is the purpose of this paper to demonstrate the interest of Laser Doppler Anemometry (LDA) in the field of experimental aeroacoustics, the emphasis thereby being placed on jet noise research.

For a long time a number of laboratories throughout the world have been concerned both experimentally and theoretically with the problem of jet noise in investigating the jet flow itself and the acoustic far field as well. In general, the local measurements inside the jet chosen as a model are conducted with the aid of hot wire anemometers or pressure transducers (1, 2). These measuring techniques are well adapted in the case of low flow velocities. However, they are difficult to be put into operation and the analysis of the measured results becomes rather doubtful as soon as the flow attains higher velocities (in the order of 100 m/s). Finally, in the case of very high velocity flows (several hundreds of m/s) with strong turbulences such as they are encountered in the mixing zones of transonic free jets, the above measuring techniques are practically of no use. This does not apply to LDA which can be put into operation in all types of flow. This will be demonstrated in the present paper in reporting on a few typical experiments performed at ISL in the framework of an investigation of jet noise. These experiments were conducted on a cold jet at relatively high velocities which, without attaining those of real engine jets, are nevertheless not very far below.

The cold round free jet facility of ISL will be described first. Thereafter, we shall report on our experimental studies which, based on simultaneously performed theoretical research work, can be classified as follows:
1. Causality correlations in order to find the apparent distribution of noise sources in a transonic free jet.
2. Relations of velocity field coherence with generated noise. Effects of upstream excitation on both the flow and acoustic far field of a free jet.
3. Resonance phenomenon in transonic jet.

The Cold Round Free Jet Facility of ISL

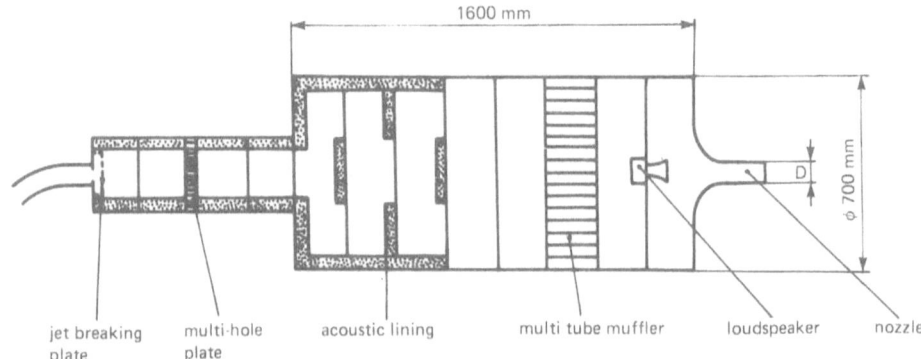

Fig. 1. Settling chamber of the ISL cold round free jet facility.

It consists essentially of a nozzle exhausting into a room with walls covered with sound absorbing material. Compressed air at approximately room temperature is supplied to the nozzle via a control valve and a silencer settling chamber. This chamber (Fig. 1) is designed to reduce the noise generated by the supply system and to weaken its resonance eigen modes. Two nozzles are used with diameters D = 40 mm and 80 mm. In any case, the area contraction ratio is larger than 60 so that the flow speed in the settling chamber is very low. The jet exit velocity can be adjusted between 100 m/s and more than 300 m/s. The settling chamber, and the jet itself, are movable vertically (Oy axis) and along the axis of the jet (Oz axis) so that it is possible to probe the flow without moving the optical set-up which is, therefore, very stable.

The LDA consists of a 2 W laser beam delivered by an Argon laser which is split in two partial beams. The latters are passed through a Bragg cell allowing an optical frequency shift and then focused in the flow making a probe volume whose diameter is close to 200 μm, the distance between the fringes being 15 μm. A photomultiplier (PM) collects the front scattered light and the Doppler signal issuing from the PM is processed by a two-counter LDA processing unit.

The flow is seeded with particles of D.O.P. whose mean diameter is roughly 0.2 μm in order to have accurate measurements at high rates (more than 10^5 measurements/s). The analog output of the LDA processing unit as well as the signals delivered by the microphones placed in the acoustic field and by other control probes are recorded with a 14 tracks tape recorder (FM mode-bandwidth 0 - 40 kHz) and processed.

1. Causality Correlation

Theoretical Background

LIGHTHILL has shown (3) that it is possible to express the wave operator acting on the pressure p using momentum and continuity equations. By means of some hypotheses (homoentropic flow, viscous effects neglected . . .) the fluctuating part $\tilde{p}$ of the pressure satisfies the following non homogeneous wave equation:

$$\frac{\partial^2 \tilde{p}}{\partial x_i \partial x_i} - \frac{1}{a_\infty^2} \frac{\partial^2 \tilde{p}}{\partial t^2} = -\frac{\partial^2}{\partial x_i \partial x_j} (\rho U_i U_j)$$ [1]

x_i = space coordinates
U_i = velocity component in the x_i direction
ρ = density
a_∞ = sound velocity in the ambient air

Outside the flow one assumes that the right hand side (the "source term") of equation [1] is small enough to be dropped out. From the acoustical point of view the whole space is thus divided into two regions:

1. the flow which is represented by the "source term" $-(\partial^2/\partial x_i \partial x_j)\rho U_i U_j$ and
2. the acoustic field where this term is dropped out (homogeneous wave equation).

Using KIRCHHOFF's solution of a non homogeneous wave equation and by means of additional hypotheses PROUDMAN (4) has shown that the acoustic pressure $\tilde{p}(\vec{x},t)$ in the far field is given by

$$\tilde{p}(\vec{x},t) = \frac{\rho_\infty}{4\pi a_\infty^2} \int_{jet} \left\{ \frac{\partial^2}{\partial t'^2} U_x^2 (\vec{y},t') \right\}_{t'=t-r/a_\infty} \cdot dy^3/r$$ [2]

ρ_∞ is the density outside the jet, $r = |\vec{x}-\vec{y}|$ and $U_x (\vec{y},t')$ are the components of the local flow velocity in the direction of observation $\overrightarrow{Ox}$.

Using equation [2] we can express the acoustic intensity in the far field $\overline{\tilde{p}^2} (\vec{x},t)$ or, more generally, the autocorrelation function of the pressure (5, 6):

$$\overline{\tilde{p}(\vec{x},t)\tilde{p}(\vec{x},t+\tau)} = \frac{\rho_\infty}{4\pi a_\infty^2} \int_{jet} \left\{ \frac{\partial^2}{\partial t'^2} [U_x^2(\vec{y},t)\tilde{p}(\vec{x},t+\tau')] \right\}_{\tau'=\tau+r/a_\infty} dy^3/r$$ [3]

Causality Correlation Experiments
Equation [3] exhibits the correlation function

$$R_{U_x^2\tilde{p}}(\vec{y},\tau') = \overline{U_x^2(\vec{y},t)\tilde{p}(\vec{x},t+\tau')} = R(\vec{y},\tau')$$

of the acoustic pressure $\tilde{p}(\vec{x},t)$ with the square of the flow speed component $U_x(\vec{y},t)$ in the direction of observation $\vec{Ox}$ which is easy to measure directly using LDA with inclined fringes. Equation [3] suggests to measure the correlation coefficient $R(\vec{y},\tau')$ throughout the flow. This has been made for more than 60 stations within the jet and various flow conditions and directions of observation. We present here results obtained by SCHAFFAR (7) in a jet whose Mach number is close to M = 0.98 and for sound radiated in direction $\vec{Ox}$ at angles $\theta = 20°$ and $30°$ with the jet axis.

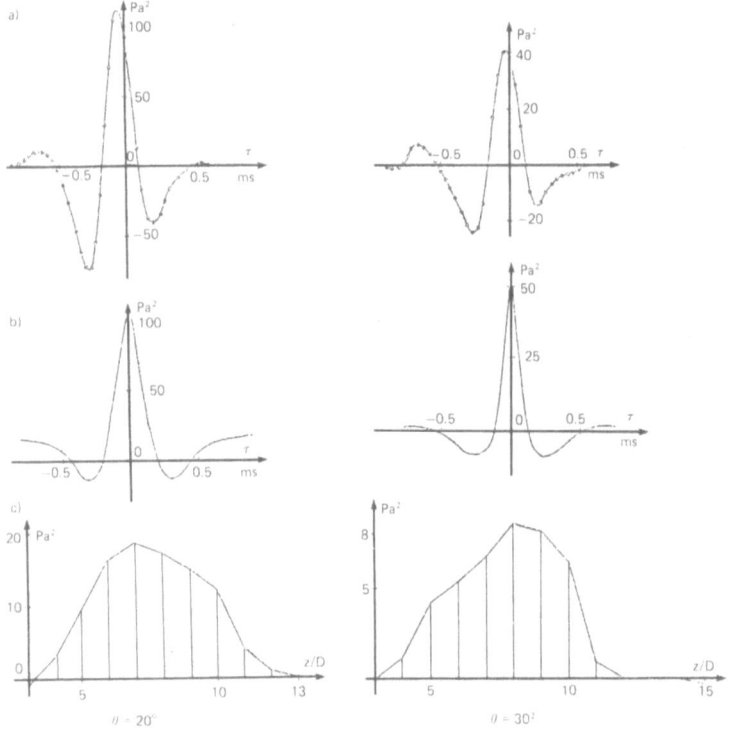

a) Far field pressure autocorrelation function obtained by integration (equation [3])
b) Directly measured far field pressure autocorrelation function
c) Source strength distribution along the jet

Fig. 2. Causality correlations.

Fig. 2a shows the autocorrelation functions of the pressure obtained by integration of the intercorrelation function $R(\vec{y},\tau')$ over the jet (integrated curve) and
the directly measured autocorrelation of the pressure (2b).

The "integrated curve" is similar in shape to the direct correlation curve except
for a small time shift of the maximum.

Quantitatively the sound intensity given by the maximum of the "integrated
curve" is in good agreement with the directly measured sound velocity which
was found to be 104 Pa^2 and 50 Pa^2 for 20° and 30° respectively.

Fig. 2c illustrates the contribution of different regions along the jet to the
whole radiated noise. This is obtained by integration of the correlation function $R(\vec{y},\tau')$ over slices of jet whose length is one diameter. It is clear that the
transition region ($5 \leqslant z/D \leqslant 10$) is the most active part of the jet as regards jet
noise generation for these angles.

These results agree with those obtained by other authors at lower jet exit
velocities and this kind of study is being continued with noise measurements
performed at larger angles with jet axis. Under those conditions, LDA is a quite
convenient method to achieve direct measurements of the velocity component
at large angles with the mean flow direction.

2. Relation of Velocity Field Coherence with Sound Radiated by a Free Jet

Theoretical Background
If the velocity field of the source region is known, we have seen that it is
theoretically possible, with LIGHTHILL analogy, to determine the acoustic
far field.

However, this is in general not the case and we can only use a model of the turbulent flow which is, therefore, defined by some empiric parameters: mean
velocity, mean shear, turbulence intensity, correlation length . . .

From an analytical point of view it is of interest to consider the jet flow and
its acoustic field as a whole described with the same unique physical quantity.
Following POWELL (8), we have focused our attention on the velocity vector
field $\vec{U}(\vec{x},t)$ because it is the easiest measurable quantity in the flow and, in the
acoustic field, velocity is directly related to pressure.

By means of classical assumptions we have established (9) an equation governing the velocity field taking into account the local mean value of the sound

velocity $a(\vec{x})$ and entropy $\bar{s}(\vec{x})$:

$$\frac{\partial}{\partial t}\frac{d\vec{U}}{dt} - a^2 \overrightarrow{grad} \, div \, \vec{U} + \overrightarrow{grad}\frac{dU^2/2}{dt} + a^2 \, \overrightarrow{grad} \, (\vec{U}\cdot\overrightarrow{grad} \, \bar{s}/cp)$$

$$+(\gamma-1)\frac{d\vec{U}}{dt} \cdot div \, \vec{U} - \frac{dU^2/2}{dt} \cdot \overrightarrow{grad} \, \bar{s}/cp - (\gamma-1)(\vec{U}\cdot\overrightarrow{grad} \, \bar{s}/cp) \cdot \frac{d\vec{U}}{dt} = 0$$

[4]

We have then applied this equation to an undimensional axisymmetric shear flow modeling a round free jet and have looked for wave-like solutions in the direction of the axis of the flow Oz (axial wave number k_z, angular frequency ω). We have found that slow stable disturbances can propagate in such a flow, these particular disturbances being defined by an azimuthal wave number n = 0 (axisymmetric disturbances) an axial wave number k_z and an angular frequency ω which depend mainly on the mean velocity profile. In the next section these stable slow disturbances are shown to be the orderly structures observed by many authors in round free jets.

Mean Velocity Profile — Orderly Structures

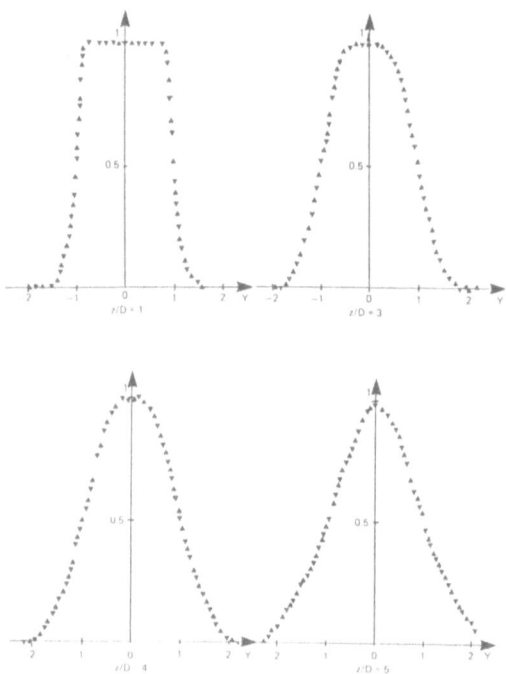

Fig. 3. Mean velocity profile at different stations along the jet (exit jet velocity 200 m/s).

The axial wave number and frequency i.e. the phase velocity of stable slow disturbances are deduced from mean velocity profile. We have, therefore, measured such profiles at different stations along the jet whose exit speed ranges from 100 m/s beyond 300 m/s. Fig. 3 shows a few examples of the profiles we have obtained in a 200 m/s jet.

Using a suitable analytical form of these experimental curves, computation gives the frequency and phase velocity of the stable disturbances propagating in a theoretical model jet corresponding to different slices of actual jet.

Space time correlation of the fluctuating axial component u of the velocity has been performed on the axis of the quiet and the excited jet as well.

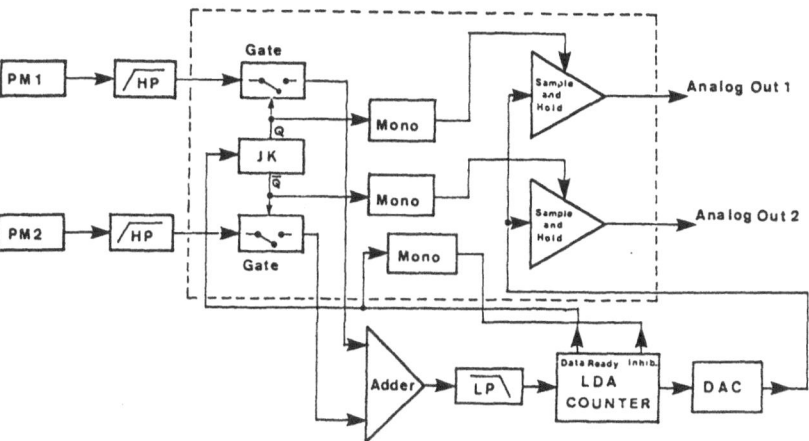

Fig. 4. Double analogical synchronous multiplexer.

These measurements are made with a simple optical set-up (10) allowing two probe volumes to be formed on the axis of the jet. Light front scattered by the particles is collected on two photomultipliers (PM1 and PM2). In order to process the Doppler signals delivered from these PM's we have developed a double analogical synchronous multiplexer (Fig. 4) which allows two LDA channels to be processed using only one LDA processing unit (11).

This device has been connected to a DISA 55L90 LDA processing unit which was slightly modified to increase the slew-rate of the digital to analog converter (DAC). We have obtained a measurement rate larger than 6×10^4/s for each channel which is high enough to perform intercorrelation with a standard real time correlator. These measurements have been made between 1 and 8 D down-

J. Haertig

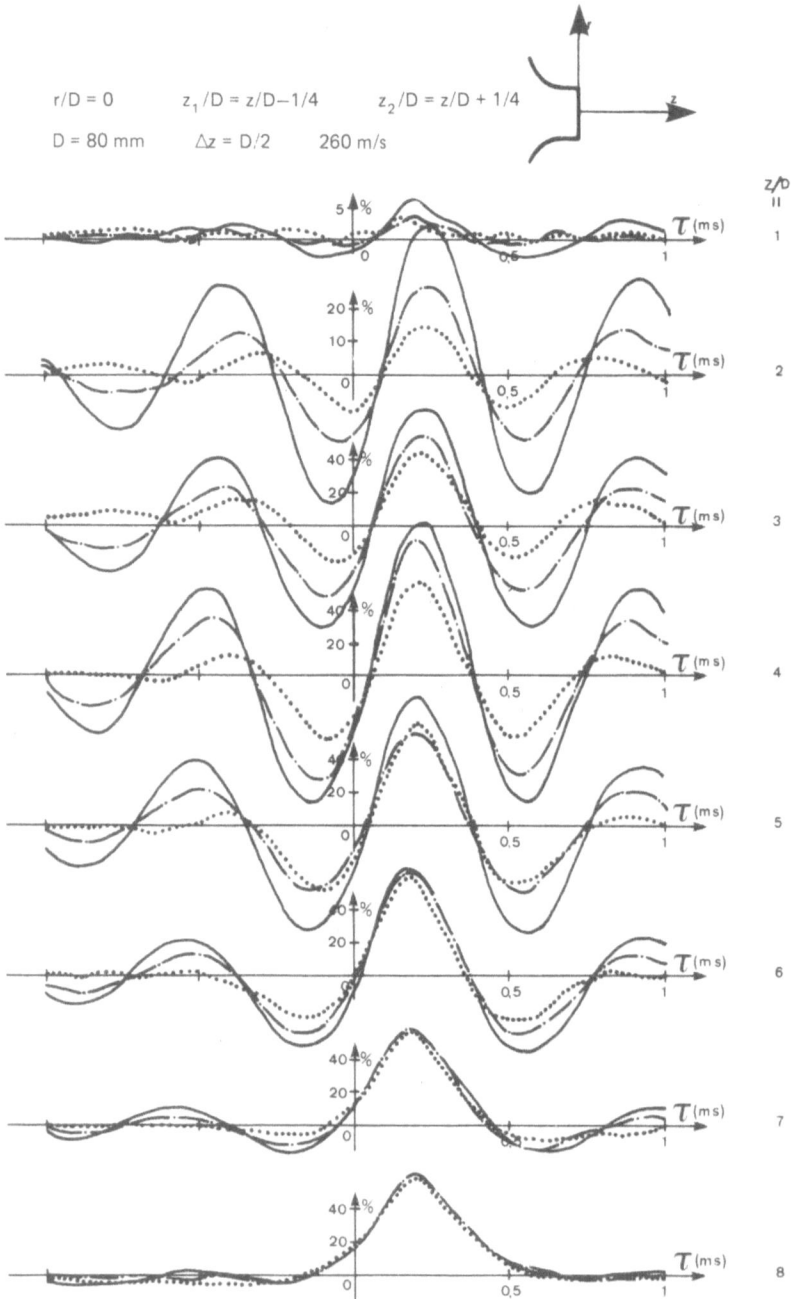

$r/D = 0$ $z_1/D = z/D-1/4$ $z_2/D = z/D + 1/4$

$D = 80\ mm$ $\Delta z = D/2$ $260\ m/s$

*Fig. 5. Space time velocity correlation at different locations along the jet axis.
Excitation levels: 0 V: $\cdots\cdots$; 2.5 V: $-\cdot-$; 5 V $\longrightarrow$*

stream the nozzle (D = 80 mm) at jet exit velocities of 130 m/s, 260 m/s and 300 m/s in quiet jets and excited jets as well. The excitation was made with a loudspeaker set into the settling chamber in front of the nozzle. The loudspeaker was supplied with a pass-band filtered white noise ($\Delta f/f \approx 0.3$), the center frequency of which corresponds to a Strouhal number ST ≈ 0.5.

Fig. 5 shows examples of correlation coefficients obtained for a jet exit speed of 260 m/s.

We observe a large increase in coherence of the velocity field with higher excitation level and we see that the convection velocity of these coherent structures is independent of the excitation.

Fig. 6 collects convection velocity data inferred from correlation measurements on the axis of the jet. Furthermore, this figure shows the computed phase velocity of stable disturbances deduced from measured mean velocity profiles. These data are very close to each other and we find also a good agreement for the frequencies: It is then concluded that slow stable disturbances are locally coincident with the orderly structures observed by several authors (1, 2, 10), even in case of excitation.

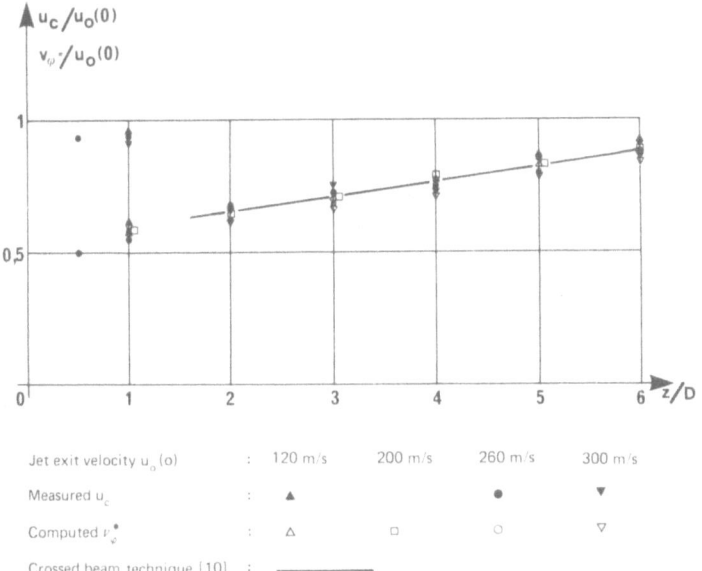

Fig. 6. Measured convection velocity u_c of orderly structures and computed phase velocity of slow stable disturbances along the jet.

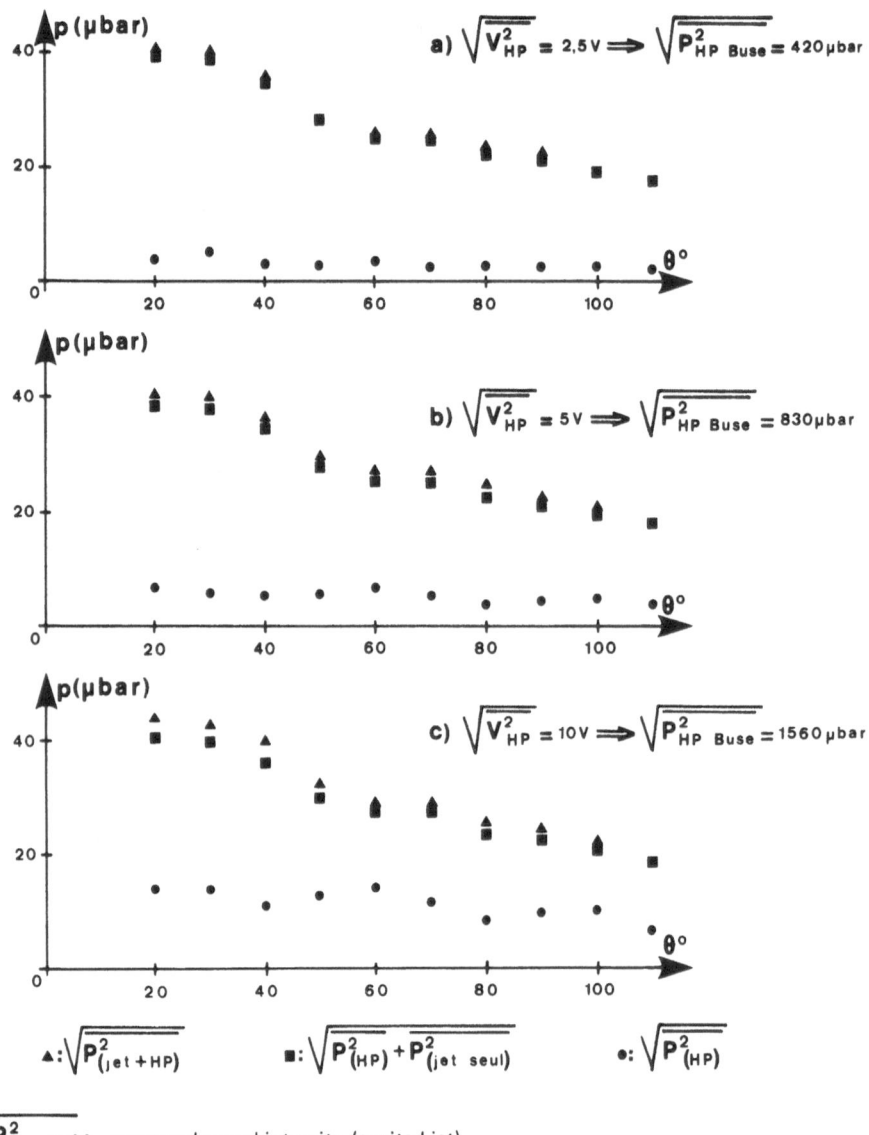

$\blacktriangle: \sqrt{\overline{P^2_{(jet+HP)}}}$ $\blacksquare: \sqrt{\overline{P^2_{(HP)}} + \overline{P^2_{(jet\ seul)}}}$ $\bullet: \sqrt{\overline{P^2_{(HP)}}}$

$\overline{P^2_{(jet+HP)}}:$ measured sound intensity (excited jet)

$\overline{P^2_{(HP)}}$: sound intensity radiated by the loudspeaker through the nozzle (without flow)

$\overline{P^2_{(jet\ seul)}}:$ sound intensity generated by the non-excited jet

Fig. 7. Far field acoustic pressure at different angles θ with jet axis and various excitation levels $\sqrt{\overline{V^2_{HP}}}$. Jet exit velocity: 260 m/s; excitation band: 1200 - 1600 Hz.

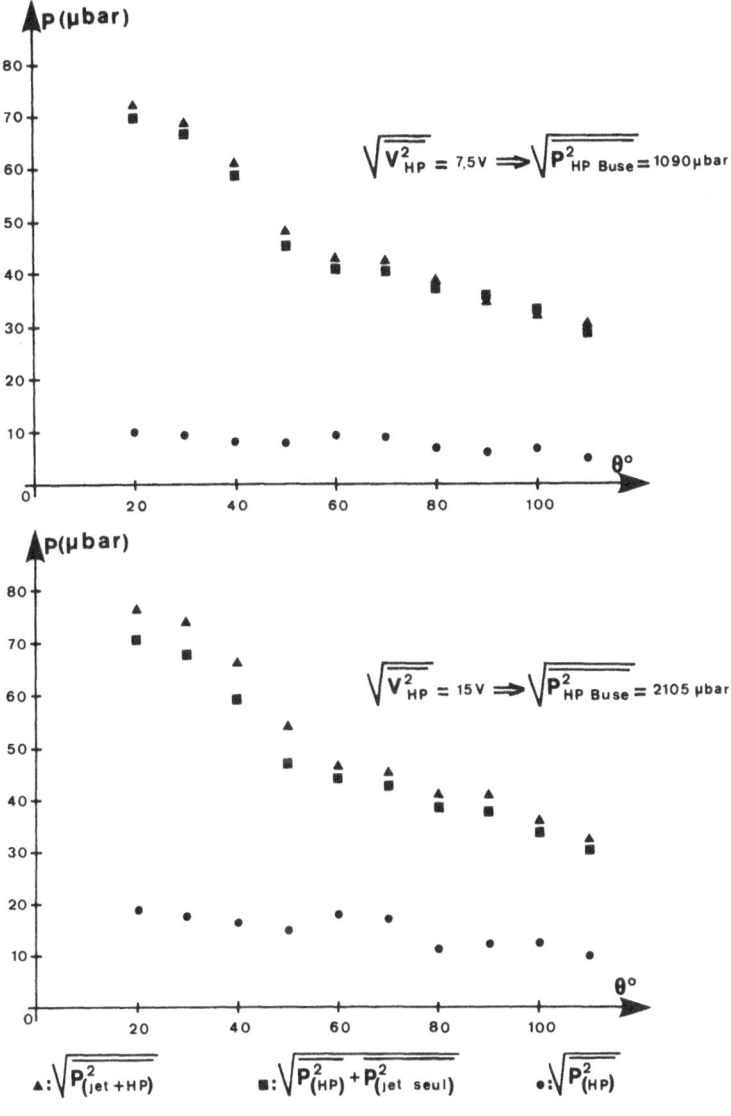

$\blacktriangle : \sqrt{\overline{P^2_{(jet+HP)}}}$ $\blacksquare : \sqrt{\overline{P^2_{(HP)} + P^2_{(jet\ seul)}}}$ $\bullet : \sqrt{\overline{P^2_{(HP)}}}$

$\overline{P^2_{(jet+HP)}}$:   measured sound intensity (excited jet)

$\overline{P^2_{(HP)}}$:   sound intensity radiated by the loudspeaker through the nozzle (without flow)

$\overline{P^2_{(jet\ seul)}}$:   sound intensity generated by the non-excited jet

Fig. 8. Far field acoustic pressure at different angles θ with jet axis and various excitation levels $\sqrt{\overline{V^2_{HP}}}$. Jet exit velocity: 300 m/s; excitation band: 1500 - 2000 Hz.

As the acoustic efficiency of these disturbances is theoretically poor, experimental work has been made to study the acoustic far field of the jet with increasing excitation level. This was made by putting ten microphones (B and K 1/4") on a semi circle (radius 50 D) centered onto the nozzle. Figs. 7 and 8 give examples of the result obtained when comparing the generated noise of the excited jet with the addition of the noise radiated through the nozzle by the loudspeaker and the noise generated by the non-excited jet. One concludes that the noise generated by the excited jet (which exhibits a strong coherence of the flow velocity field) is approximately the addition of loudspeaker noise and quiet jet noise, except for large excitation levels. In the latter case we note that non linear effects become important and give rise to a modification of the mean flow field. These results agree with MOORE's results (12) and subsequent works have been demonstrated that in our experimental conditions, the noise generated by a jet is due to non linear interaction and not directly related to the coherence of the large orderly structures propagating in it.

3. Resonance in Transonic Jet

Theoretical Background
Theoretical work based on equation [4] and applied to shear axisymmetric flow have shown that guided waves appear in a round free jet (9). Fig. 9 shows a few examples of dispersion curves of propagating modes in free jet at different Mach numbers.

It must be noted that in the case of Mach number close to 0.9, the dispension curves are almost horizontal straight lines. This means that such a jet gives rise to resonance phenomena. The lowest resonant frequency corresponds to a Strouhal number ST $\approx$ 0.4. Experimental study of this particular phenomenon indicates disturbances to appear when putting obtrusive probes in nearly transonic jets.

Experimental Investigation
To check experimentally this characteristic of transonic jets, we have measured the intercorrelation coefficient $C_{u\,p_1}\,(\tau)$ relating the axial velocity fluctuation u in the jet to the static pressure p_1 upstream the nozzle. Velocity is measured using LDA three diameters downstream the nozzle (240 mm) and pressure is measured with a microphone p_1 in the settling chamber 270 mm upstream so as to not disturb the flow.

The correlation coefficient $C_{u\,p_1}\,(\tau)$ is generally zero, except if the Mach number of the jet is within 0.9. In this case $C_{u\,p_1}\,(\tau)$ reaches 5 % and exhibits many oscillations whose frequency corresponds to a Strouhal number close to 0.4 (Fig. 10).

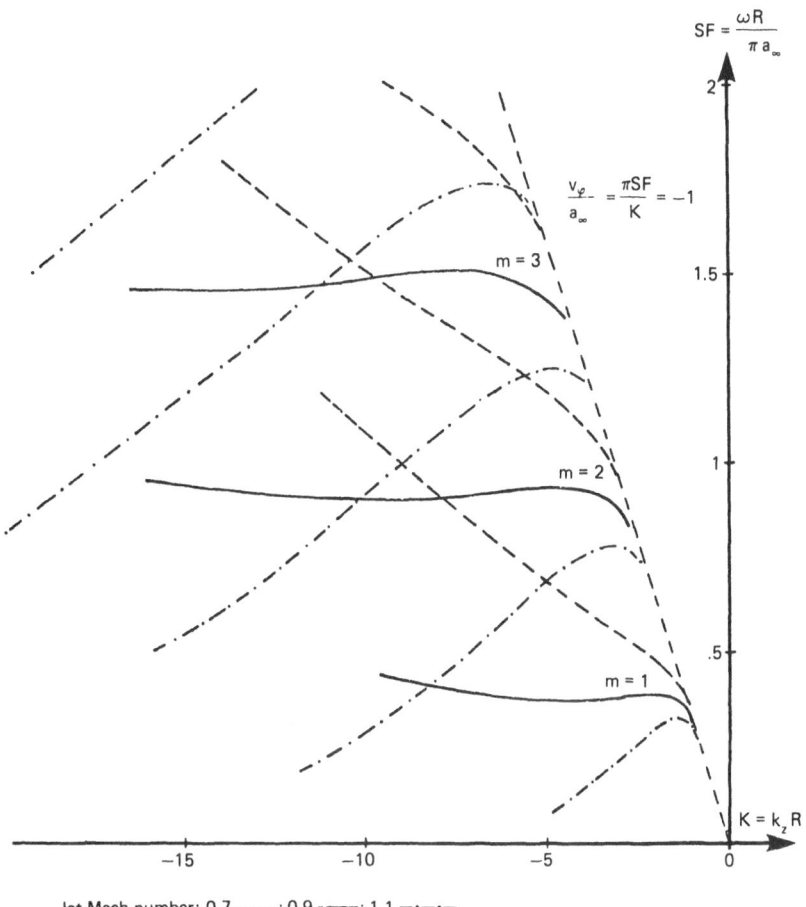

Jet Mach number: 0.7 ———; 0.9 ·——·; 1.1 —·—·—

Fig. 9. Computed dispersion curves* of "upstream" guided waves in jets azimuthal wave number n = 0, radial wave number m = 1, 2, 3.

If we put into the flow (near the probe volume of the LDA, for example) an obtrusive probe and even if it is of small size, the correlation coefficient $C_{u\,p_i}\,(\tau)$ increases strongly and reaches more than 50 % showing oscillations to occur at the same frequency as before (ST ≈ 0.4).

Taking into account the large distance separating the LDA probe volume and the microphone, one concludes that a very large structure oscillating at a well defined frequency (close to the theoretical value) is present in the jet flow and eventually radiates in the acoustic field. If we insert a probe into the flow, the probe acts as an antenna which radiates white noise exciting the jet resonator

---

* These curves have symmetrical branches with respect to the coordinates origin 0.

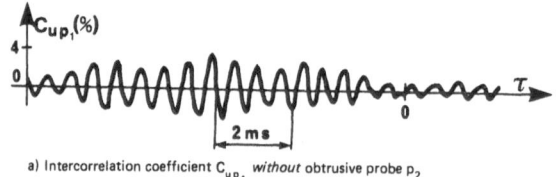

a) Intercorrelation coefficient C_{up_1} *without* obtrusive probe p_2

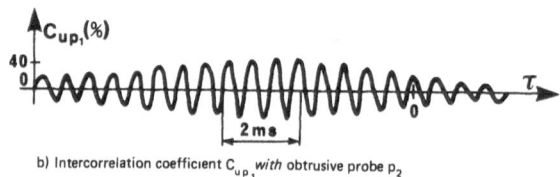

b) Intercorrelation coefficient C_{up_1} *with* obtrusive probe p_2

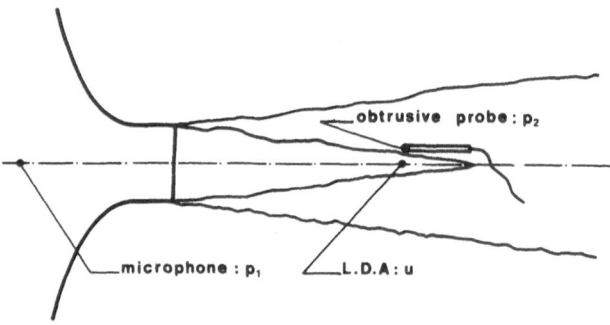

Fig. 10. Resonance in transonic jet.

which acts, therefore, as a filter at the resonance frequency. This effect is different from those observed by other authors (13) because, here, the frequency is practically independent of both the position and geometry of the solid probe. The experiment shows disturbances to occur when putting a probe in a high speed jet and the advantage of non obtrusive methods to perform measurements in such flows.

Conclusion

In aeroacoustic research and especially in jet noise studies LDA is easily put into operation because there are no windows, no walls, and consequently no seeding troubles. We have demonstrated this method to give reliable data without disturbing neither the flow nor the acoustic fields so that this technique can be considered to be a very convenient one in experimental jet noise research.

References

(1) CROW, S.C. and CHAMPAGNE, F.H. "Orderly Structures in Jet Turbulence" *JFM*, Vol. **48**, part 3, 1970, pp. 547-591

(2) FUCHS, H.V. "Über die Messung von Druckschwankungen mit umströmten Mikrofonen im Freistrahl", D.L.R. F.B. 70-22 (1970)

(3) LIGHTHILL, M.J. "Sound Generated Aerodynamically. I. General Theory", *Proc. Roy. Soc. (London),* **A211** (1952) pp. 564-587

(4) PROUDMAN, I. "The Generation of Noise by Isotropic Turbulence", *Proc. Roy. Soc. (London),* **A214** (1952) pp. 119-132

(5) RIBNER, S. "The Generation of Sound by Turbulent Jets", *Adv. Appl. Mech,* **8** (1964) Academic Press, pp. 103-182

(6) SIDDON, T.E. "Noise Sources Diagnostics Using Causality Correlations", AGARD C.P. 131 - On Noise Mechanism — Bruxelles 1973

(7) SCHAFFAR, M. "Relation de cause à effet entre les fluctuations de vitesse d'un jet froud et le bruit émis près de l'axe" Rapport ISL R 109/78 (1978)

(8) POWELL, A. "Theory of Vortex Sound", *J.A.S.A.,* Vol. **36**, No. 1 (1964), pp. 177-195

(9) HAERTIG, J. "Etude théorique de l'évolution d'une perturbation acoustique dans un écoulement non uniforme. Application au bruit des jets libres". Thèse: Institut National Polytechnique de Lorraine, Nancy 1978

(10) KOCH, B. and PFEIFER. H.J., "Detection of Large Scale Coherent Structures in Free Jet by Laser Anemometry and Crossed Beam Schlieren Techniques" Rapport ISL CO 210/77 (1977)

(11) HAERTIG, J. and WIETRICH, F. "Unité de multiplexage analogique pour anémometrie laser: Application à la mesure d'intercorrelation" Rapport ISL R 103/78 (1978)

(12) MOORE, C.J. "The Role of Shear Layer Instability Waves in Jet Exhaust Noise" *JFM,* Vol. **80**, part 2 (1977), pp. 321-337

(13) EVERTZ, E., KLÖPPEL, V., NEUWERTH, G. and QUICK, A.W., "Noise Generating by Interaction between Subsonic Jets and Blown Flaps", D.L.R. F.B. 76-20 (1976)

Sonic Anemometer Measurement of Atmospheric Turbulence

by

J.C. Kaimal
Wave Propagation Laboratory
NOAA/ERL, Boulder, Colorado 80302

1. Introduction

The study of atmospheric turbulence and its role in transporting momentum, heat, water vapor and other constituents within the layers close to the ground has been a central goal in micrometeorology. Essential to such a study is detailed information on the fluctuations of all relevant parameters. The frequencies of interest cover a range of nearly five decades, corresponding to scales of motion ranging from a few centimeters to several kilometers, presenting a serious challenge to the experimentalist. Over the last 50 years a number of different measurement techniques have been tried with varying degrees of success. Particular attention was given to fluctuations in the velocity field because of their importance to turbulent transport in the atmosphere.

The earliest attempt to estimate momentum flux through direct measurements of velocity fluctuations was made by F.J. Scrase (1930), a British meteorologist. He made his measurements in 1962 over Salisbury Plain using an airmeter, a swinging plate anemometer, and a simple bivane attached to a recording pen. His instrumentation was replaced by more sensitive bivanes and hot-wire anemometers in later experiments at other sites, but a clear understanding of the structure and dynamics of turbulences in the first 50 m or so of the atmosphere, usually referred to as the surface layer, did not emerge until the late 1960's. To a large extent this breakthrough was possible because of advances in sonic anemometry which, coupled with developments in the rapidly growing data processing and recording field, enabled scientists to analyze large quantities of turbulence data in a relatively short time.

The sonic anemometer offered several advantages over the bivane and hot-wire systems. It has no moving parts to come into dynamic equilibrium with the flow, so its frequency response is limited only through the attenuation in the spatial response imposed by line averaging along the path. It responds linearly to wind velocity and is relatively free of contamination from other velocity components or temperature. As an absolute instrument, its calibration is established by its design parameters.

The use of sonic anemometers in micrometeorological studies was pioneered by Suomi (1957). In his instrument, acoustic pulses were propagated in opposite directions along a 1-m path to determine wind velocity along the path. He demonstrated the feasibility of sonic anemometry for turbulence measurement but encountered substantial difficulties in detecting the leading edge of the received pulses. This problem led later workers in the United States (Kaimal and Businger, 1963) and the USSR (Gurvich, 1959; Bovsheverov and Voronov, 1960) to turn to continuous wave techniques, which avoided some of the difficulties of the pulse technique. Measuring transit time differences with phase comparators proved more dependable, so the early field instruments used the continuous-wave technique. However, inherent drifts in the zero-wind calibration due to thermal drifts in the transducers and the need for a multiplicity of frequencies in any three-axis configuration led investigators to reexamine the pulse approach and overcome some of the problems encountered in the earlier versions. The instruments described by Mitsuta (1966) and Kaimal et al. (1974) use the pulse approach to advantage. In the USSR continuous-wave systems still remain in use (Koprov and Sokolov, 1973), and a three-dimensional version utilizing one transmitter and four receivers is described by Bovsheverov et al. (1973).

Recent developments in sonic anemometry have been aimed at adapting it for continuous operation on towers and reducing its cost. The author's orthogonal, three-dimensional array used on the Boulder Atmospheric Observatory's 300-m tower is designed for wider azimuth coverage of wind directions than was possible in earlier versions. Kaijo Denki Co. of Tokyo, Japan, offers several array configurations including the type mentioned above. A single-axis prototype of a low-cost phase locked loop system has been tested successfully by the Atmospheric Sciences Department at the University of Washington, but further development work is needed to convert it into an operational three-axis system.

2. Principle of Operation

The sonic anemometer measures wind velocity components from arrival times (or phase) of acoustic signals transmitted in opposite directions across a fixed path. Fig. 1 illustrates the effect of wind V on the sound ray vectors for a single-axis, dual-path sonic anemometer. The paths are parallel and closely spaced to minimize errors arising from velocity and temperature differences between the two paths. Assuming a uniform wind and temperature field within the array the transit times for two opposing pulses traveling from T_1 to R_1 and T_2 to R_2 can be approximated by

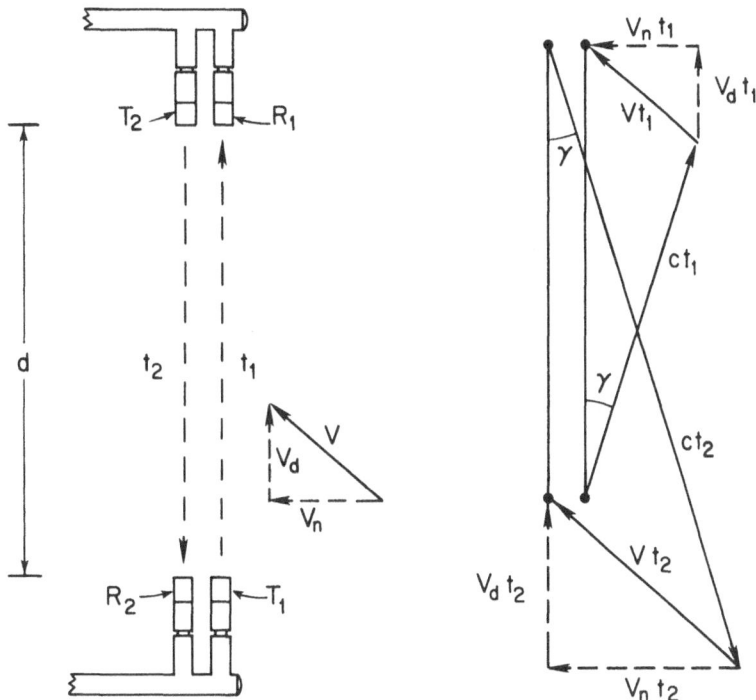

Fig. 1. Sound ray vectors for single-axis sonic anemometer showing principle of operation. Symbols T and R represent the transmitter and receiver, respectively.

$$t_1 = \frac{d}{c\cos\gamma + V_d} \qquad [1]$$

$$t_2 = \frac{d}{c\cos\gamma - V_d} \qquad [2]$$

where t_1 and t_2 are the transit times for the two pulses, d is the path length, V_d is the velocity component along the path, c is the velocity of sound in air, and $\gamma = \sin^{-1}(V_n/c)$, V_n being the velocity component normal to the path. The transit time difference is therefore

$$\Delta t = t_2 - t_1 = \frac{2V_d d}{c^2 - V^2} \, , \qquad [3]$$

where $V^2 = V_d^2 + V_n^2$. Assuming $V^2 \ll c^2$,

$$\Delta t = \left[\frac{2d}{c^2}\right] V_d . \qquad [4]$$

The speed of sound is a function of temperature and water vapor content and can be expressed in the form (Kaimal and Businger, 1963)

$$c^2 = 403 \ T(1+0.32 \ e/P) \tag{5}$$

where T is the absolute temperature, e is the vapor pressure of water and P is the atmospheric pressure with all terms expressed in SI units. The vapor pressure term is small and usually negligible, as are fluctuations in T compared to the mean absolute temperature. From [4] and [5] we have

$$V_d = [\frac{201.5 \ \overline{T}}{d}] \ \Delta t. \tag{6}$$

When $\overline{T}$ (mean temperature) is known, the measurement of V_d reduces to a simple measurement of transit time difference whereas the sign of Δt indicates the direction of the velocity component. Therefore the accuracy of the measured velocity depends on the precision of the Δt measurement. For typical values of d = 0.2 m and $\overline{T}$ = 25°C, a minimum resolution of 0.1 μs (3 cm/s^{-1}) is needed. The effective resolution can be improved by averaging 10 or 20 successive acoustic transmission so that each data point is a block average over the preceding sampling interval (Kaimal et al. 1974). This prefiltering has the added advantage of reducing distortions at the high frequency end of spectra due to aliasing.

Sonic anemometers using only one transducer at each end of the path, switching alternately between transmitter and receiver modes, are now available from Kaijo Denki Co. The benefits include reduced interference to airflow and smaller temperature drifts in the measurements. The drawbacks are a lower sampling rate (<10 Hz) to allow for settling times in the transducers and consequent inability to block average data by transmitting more frequently than is needed.

It is instructive to examine the error introduced by the simplifying assumption $c^2 \cong 403 \ \overline{T}$ which led to the expression in Eq. [6]. Separating the terms in Eq. [5] into their mean and fluctuating parts (denoted by overbars and primes, respectively), and neglecting the higher order terms,

$$\frac{c'}{c} \cong \left(\frac{T'}{2\overline{T}}\right) + 0.16\left(\frac{e'}{P}\right). \tag{7}$$

The humidity fluctuation term is usually negligible. The effect of the temperature fluctuations on the velocity measurement can then be expressed as

$$(V_d)_m \cong V_d (1 - \frac{T'}{\overline{T}}), \qquad\qquad\qquad [8]$$

where the subscript m indicates the measured value. The error is less than 1 %, but could be large if a significant temperature trend occurs during the observation period. Hourly updating of $\overline{T}$ is essential especially during the transition periods. Friehe (1976) has shown that error introduced by the temperature fluctuation term in Eq. [8] is negligibly small in the variance and heat flux calculations but can be as large as -3.6 % in the momentum flux.

An alternate approach avoids the velocity error caused by fluctuations in c^2. If instead of $(t_2 - t_1)$, the difference between their reciprocals $(1/t_1 - 1/t_2)$ is computed, the resulting term $2V/d$ is independent of c^2. The difficulty of measuring small differences between the reciprocals of two relatively large numbers with sufficient accuracy diminishes its attraction to the instrument designer. However, Kaijo Denki offers this option in the digital version of its sonic anemometer.

In addition to wind measurement some sonic anemometers (Mitsuta, 1974) provide temperature measurements by computing the sum of the transit times. But the temperature information is contaminated by the crosswind V_n and by humidity (to a lesser extent) since they have the same effect on t_1 and t_2. Under convective conditions in the first 30 m or so of the boundary layer, the contamination is negligible, but in neutral and stable air where the actual temperature fluctuations are small, the fluctuations in V_n appear prominently in the measured temperature signal (Kaimal, 1969).

3. Measuring the Vertical Component of the Wind
The geometry of the single-axis sonic anemometer array is particularly well suited for measuring the vertical wind component w since the transducers interfere little if at all with airflow along the path when the path is oriented normal to the flow. This is significant since in most other wind sensors w is the hardest component to measure accurately. Shifts in azimuth wind directions do not affect the accuracy of the measurement except when the acoustic path is in the wake of the supporting frame. Bivanes and hot-wire x-probes do not measure the w component directly, but provide inclination angle measurements from which w must be computed.

Fig. 2 is one of the first records of simultaneous w and temperature measurements. It shows a convective plume at 3.5 m, with pronounced updraft within the core of the plume (as defined by the sawtooth temperature burst), and downdrafts in the thermally quiescent regions upwind and downwind of it.

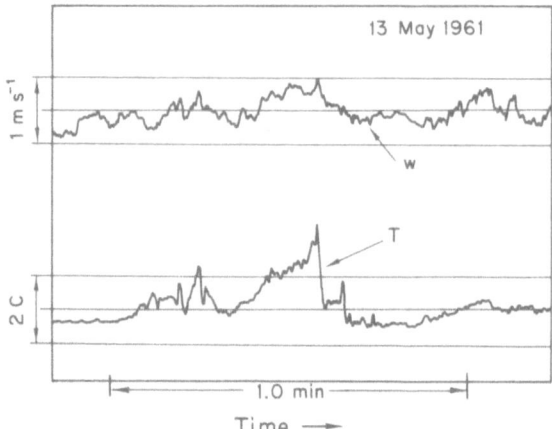

Fig. 2. Simultaneous records of vertical velocity and temperature fluctuations in a convective plume obtained with a sonic anemometer-thermometer.

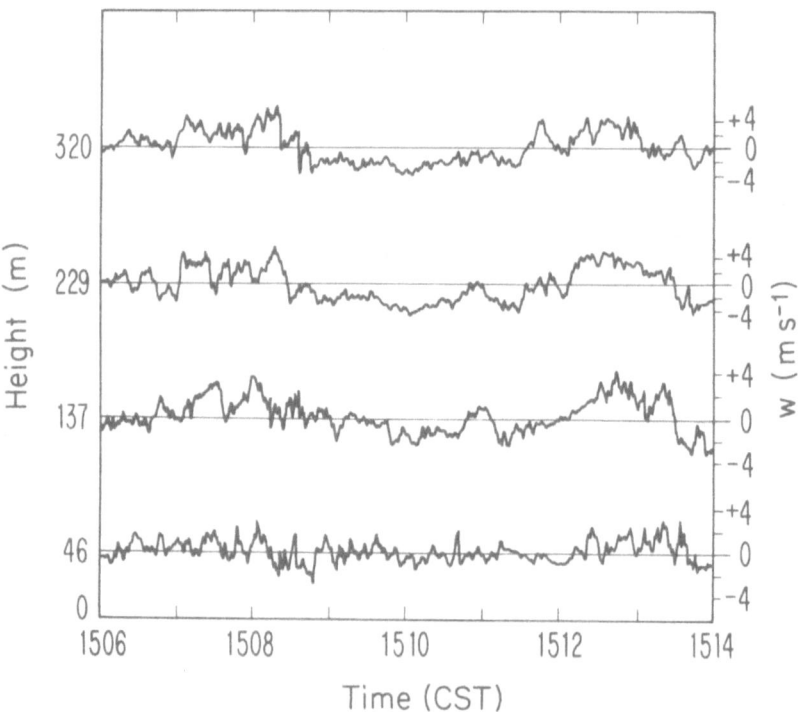

Fig. 3. Vertical velocity fluctuations measured on a 430-m tower show large-scale convective plume formed by merging of smaller ground-based plumes.

These individual updrafts merge into larger plumes as they ascend above 100 m as seen in sonic anemometer records in Fig. 3 obtained from four levels on a 430-m tower (Kaimal and Haugen, 1967). Peak vertical velocities observed here are of the same order as the horizontal wind speeds (5 to 6 ms^{-1}) during that period. Thus, sonic anemometers intended for measuring w on tall towers should have a range comparable to that needed for horizontal velocity measurements.

4. Measurement of Horizontal Velocity Components

The integration of two horizontal axes to the vertical array should, in principle, provide the full three-dimensional information on the wind field needed for turbulence studies. But the problem of designing a three-dimensional array with unobstructed exposure for all the axes is not a trivial one. A non-orthogonal array which has evolved in recent years (Kaimal et al. 1974; Mitsuta, 1974) has its horizontal axes separated by a 120-deg. angle to provide a reasonably wide (90 deg) azimuth coverage (see Fig. 4). Even with such a coverage the array often must be reoriented once every hour to ensure that the natural variations in wind direction do not exceed the desired range. Such reorientation of the array can be achieved remotely if the array is mounted on an antenna rotor (see Fig. 5). The wind outputs from the horizontal axes are monitored on a strip-chart recorder to determine the orientation of the array with respect to the wind.

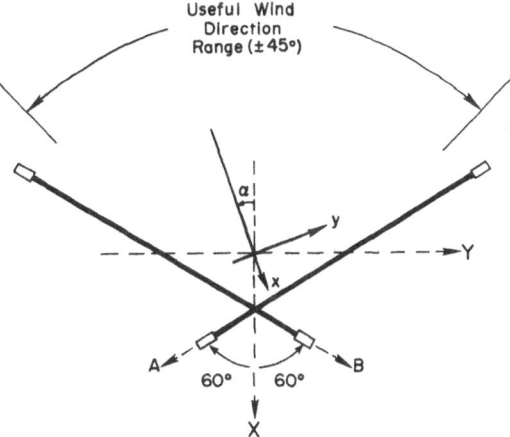

Fig. 4. Relationship between the horizontal axes of a three-axis sonic anemometer and the vector-mean axes of the wind. Coordinate transformation for converting wind measurements to velocity components along the vector-mean axes are given in Eq. [9].

In the array of Fig. 5 the vertical component (w) is obtained directly from the vertical axis, but the streamwise (u) and lateral (v) components must be computed from velocities V_A and V_B measured along the horizontal paths A and B (see Fig. 4). If α is the angle between the vector-mean wind directions and the axis of symmetry for the array,

$$u = V_A \left(\cos\alpha - \frac{1}{\sqrt{3}}\sin\alpha\right) + V_B \left(\cos\alpha + \frac{1}{\sqrt{3}}\sin\alpha\right); \qquad [9a]$$

$$v = -V_A \left(\frac{1}{\sqrt{3}}\cos\alpha + \sin\alpha\right) + V_B \left(\frac{1}{\sqrt{3}}\cos\alpha - \sin\alpha\right); \qquad [9b]$$

$$\alpha = \tan^{-1}\left[\frac{\overline{V}_B - \overline{V}_A}{\sqrt{3}(\overline{V}_B + \overline{V}_A)}\right] \qquad [9c]$$

The overbar denotes time average over a suitably long interval. In the instance of $\alpha = 0$ we have $u = V_A + V_B$ and $v = (1/\sqrt{3})(V_B - V_A)$

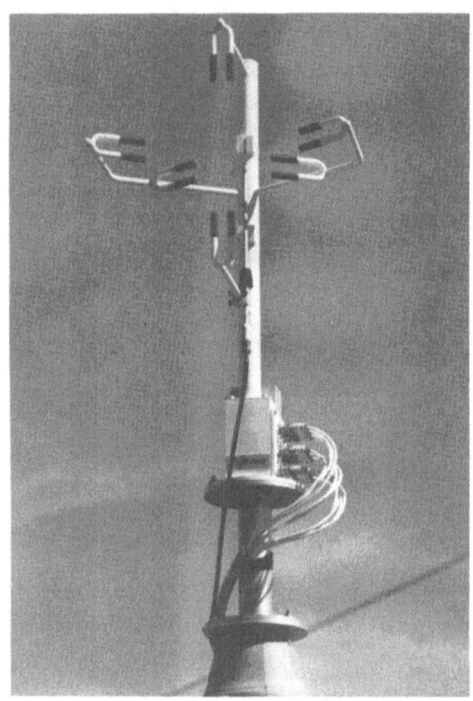

Fig. 5. Three-axis sonic anemometer array (EG&G Model 198-3) using the geometry of Fig. 4 avoids the need for support members upwind of acoustic paths. Array is mounted on a remotely-controlled, antenna rotor and periodically oriented into the mean wind. Path length in each axis is 20 cm.

Although this arrangement is ideal for surface layer measurements where observation periods can be conveniently broken up into hourly segments it is not well-suited for continuous, unattended operation on tall towers. A different configuration of the axes is needed. One such array, designed for use on the Boulder Atmospheric Observatory (BAO) 300-m tower, is shown in Fig. 6. The axes are orthogonal, with the vertical axes kept upwind of the other axes to optimize the vertical velocity measurement. The acceptable azimuth range for the vertical axis is clearly much larger in this array than in the array in Fig. 5; the unacceptable segment would be useless anyway because of shadowing caused by the tower. However, the horizontal wind measurements fare less well in this arrangement. An underestimation in the wind (Mitsuta, 1974) can be expected when the mean wind direction is along either axis. This error is corrected in the data processing routine by a first order approximation of the wind direction (see Section 8). Interference from supports for the crosswind axis is also minimized by vertically separating the horizontal paths. The main disadvantage for surface layer measurements is the large spatial separation between the horizontal axes and the vertical axis.

5. Some Consequences of Array Geometry

The dimensions of the array and the configuration of the acoustic paths have pronounced effect on the spectral responses at wavelengths (λ) approaching the size of the array.[*] Line-averaging along the acoustic path attenuates wavelengths smaller than $2\pi d$. The shape of the transfer function varies with orientation of the path relative to the mean wind direction (Kaimal et al., 1968; Horst, 1973). In the streamwise direction the response drops as $\sin^2 x/x^2$, approaching zero at wavelength d. In the crosswind direction the roll-off is more gradual, with the half-power point at wavelength d. When measurements along the different axes are combined to resolve the wind fluctuations along the streamwise (x) and lateral (y) directions (Fig. 4), the response is more complex because of the spatial separation between the axes. Transfer functions for u, v, and w measured by the array of Fig. 5 are presented in Fig. 7. The term transfer function is used rather loosely here for the ratio of measured to true spectral density. Strictly speaking the curves are not transfer functions since they require an assumption of the spectral form for the velocity field. The curves were computed for the specific instance of d = 0.2 m and a horizontal separation distance, s = 0.6 d, between the midpoints of the paths.

---

[*]Wavelength λ is approximated by $\overline{V}_H/n$ where $\overline{V}_H$ is mean horizontal wind vector and n is the cyclic frequency.

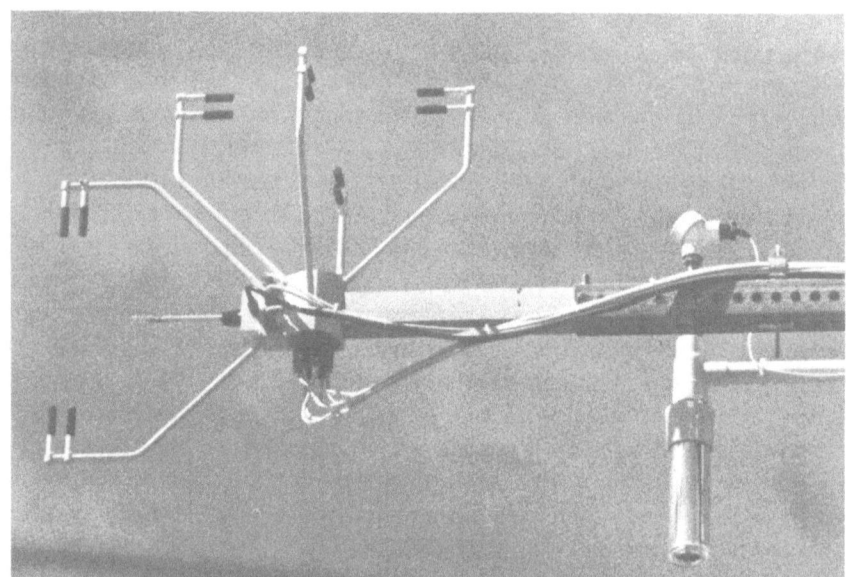

Fig. 6. Three-axis sonic anemometer array (Ball Bros. Model 125-197, 198) designed for fixed-boom operation uses an orthogonal configuration with 25-cm path length on each axis. Array provides wide azimuth coverage for vertical axis and for fast-response temperature sensor mounted in vertical probe. Horizontal velocity measurements need correction for blockage from transducers.

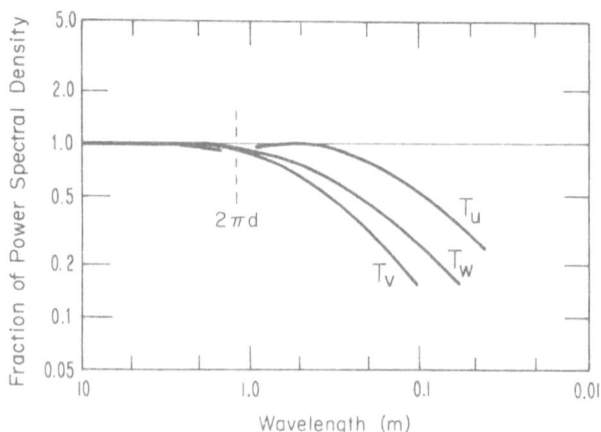

Fig. 7. Transfer functions for u, v, and w for three-axis array in Fig. 5. Distortions in spectral response become apparent at wavelengths smaller than 2πd (1.26 m for this array).

The effects of spatial separation merit further discussion. Distortions in the spectral response of u and v arise because wavelengths smaller than $2\pi s$ are not properly transformed by Eq. [9]. As the wavelength decreases, the Fourier components of velocity observed along A and B become increasingly uncorrelated. In the limit when no correlation exists, the u spectrum is overestimated by a factor of 2.5 and the v spectrum is underestimated by a factor of 0.625 (Kaimal et al., 1974).

Spatial separation between sensors (e.g. the w path and the fast-response temperature and humidity sensors) also introduces a limitation on the high-frequency response in flux cospectra. A diminishing correlation between the sensors at wavelengths smaller than 2π times the separation distance can be expected, which leads to the simple requirement that the separation distance be made no larger than the length of the w path in the sonic anemometer.

In the orthogonal array of Fig. 6, partial shadowing of the acoustic path by the transducers causes the velocity readings to be underestimated. Results of wind tunnel and atmospheric tests indicate a linear drop in response for angles between 0 and 75 deg relative to the acoustic path (see Fig. 6). For the path length-to-transducer diameter ratio of 25 appropriate to this array (heavy line in Fig. 8), the measured wind component $(V_d)_m$ can be approximated by

$$(V_d)_m = \begin{cases} V_d\,(0.87 + 0.13\,\theta/75), & \text{for } 0 \leqslant \theta \leqslant 75 \\ V_d & \text{for } 75 \leqslant \theta \leqslant 90 \end{cases} \qquad [10]$$

where

$$\theta = \tan^{-1} \frac{|V_n|}{|V_d|} . \qquad [11]$$

A first order approximation to the instantaneous angle θ using wind components measured along the two horizontal axes is found to be adequate for correcting the two wind components. With such a correction, the effect of transducer shadowing is hardly apparent in the mean wind profiles and in the second moments involving u and v.

Another type of error can result from small offsets in the array geometry and in the leveling of the array. The parameter most affected is the covariance of u and w, which represents the vertical momentum flux. Kaimal and Haugen (1969, 1971) have shown that the error can be very large in unstable air (typically 25 % for 1-deg tilt) and recommend a 0.1-deg accuracy in the mechanical

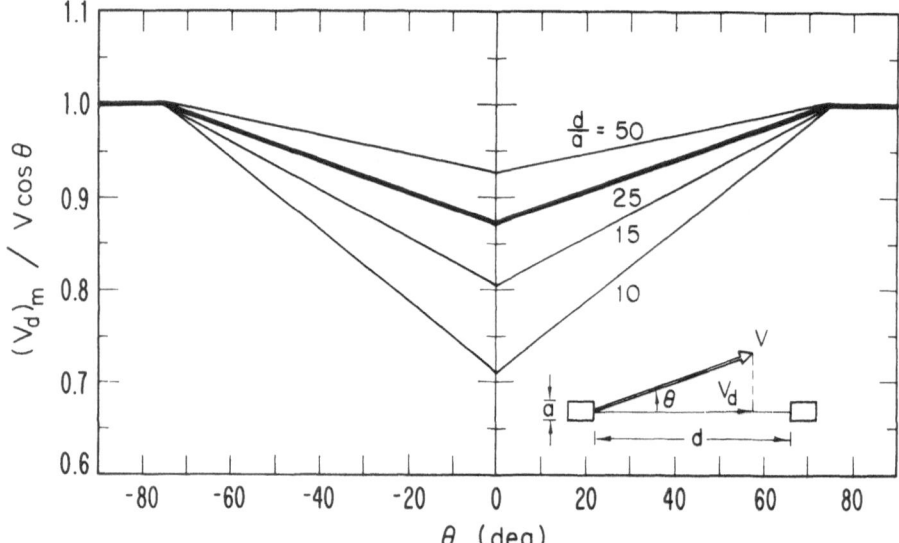

Fig. 8. Sonic anemometer response to wind component along acoustic path for various path length to transducer diameter ratios shown as a function of wind direction. Heavy line represents velocity error in the sonic anemometer of Fig. 6 (see Eq. [10]).

alingment of the probe and in the leveling. This requirement is stringent but worth pursuing for accuracy in the momentum flux measurement. Heat flux and variance measurements, on the other hand, are not as sensitive to alignment and tilt errors.

6. Sampling Rates and Run Lengths
The upper frequency limit dictated by the array dimensions determines the maximum sampling rate for sonic anemometer data. Allowing an additional octave at the high-frequency end for aliasing effects, this sampling rate n_s may be specified in terms of the path length and the mean horizontal wind vector (V_H) as

$$n_s = 2V_H/\pi d. \tag{12}$$

For typical 5 to 6 ms^{-1} winds and a path length of 0.2 m the sampling requirement can be rounded off to 20 Hz. This restriction on bandwidth imposed by array size is not a serious limitation for most turbulence studies at heights above 4 m since the inertial subrange is usually well established at 10 Hz in that height range and no significant contributions to the turbulent fluxes are expected from higher frequencies.

The relation between observation height and high-frequency spectral behavior is highlighted in Fig. 9 by a line marking the frequency at which λ equals z, the height above ground. The spectra shown here are typical for velocity fluctuations over flat, unobstructed terrain. Their behavior is predictable for wavelengths $\lambda(\cong V_H/n)$ smaller than $z/2$. The $-2/3$ slope at the high-frequency (or short-wavelength) end corresponds to the Kolmogorov $-5/3$ power law predicted for a one-dimensional velocity spectrum in the inertial subrange. The 4/3 ratio between the transverse (v and w) and longitudinal (u) spectral intensities predicted for local isotropy is also apparent in that frequency range. Thus, with the spectral curves established down to at least an octave into the inertial subrange, their behavior through the rest of the inertial subrange can be obtained by simple extrapolation.

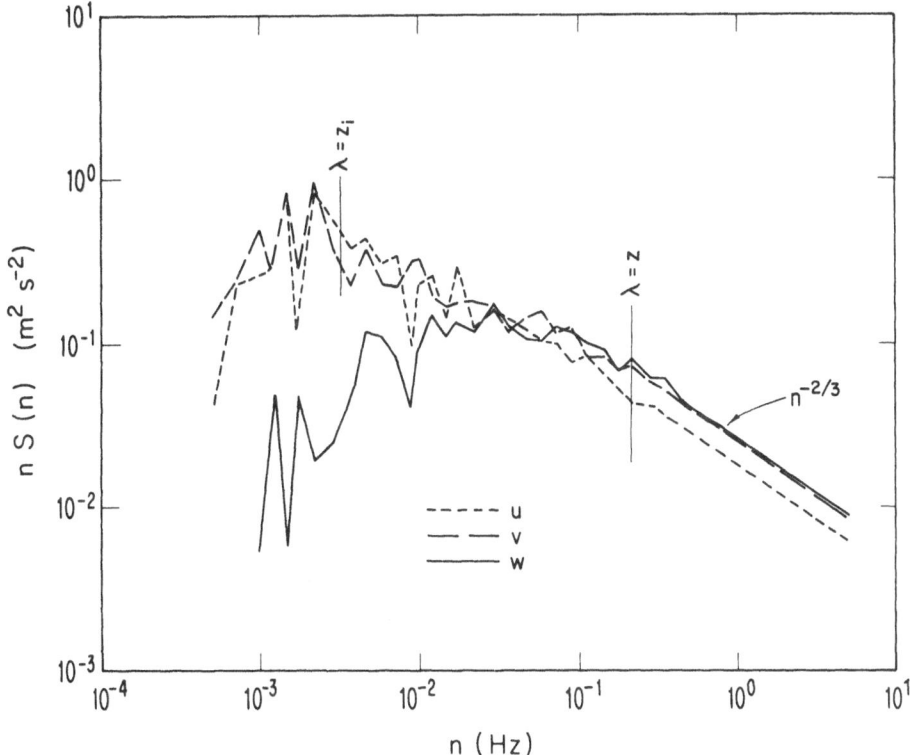

Fig. 9. Typical surface layer velocity spectra for unstable daytime conditions obtained with the sonic anemometer shown in Fig. 5 at z = 32 m. The high-frequency ends of spectra have been corrected for distortions introduced by the anemometer's spatial response and by aliasing. The boundary layer depth z_i for this run was 2095 m.

On the low-frequency side, the day-time w spectrum rolls off as n^{+1}, peaking at a wavelength roughly 5.9 times z (Kaimal et al., 1972) which suggests that in principle a height-dependent, low-frequency cutoff can be specified for w. This is not possible with u and v since they tend to scale with z_i, the convective boundary layer depth, rather than z. The wavelength at the peak corresponds to roughly $\lambda = 1.5z_i$ (Kaimal, 1978) as seen in Fig. 9. In contrast to w, the u and v spectra often show a tendency to rise again in response to the large-scale, two-dimensional fluctuations in the velocity field (not shown in figure). An optimum cutoff point for boundary layer studies would be the spectral minimum which occurs roughly a decade below the peak frequency. Record lengths of 1-hr duration provide spectral information down to approximately 0.0003 Hz, close to the expected u and v spectral minima. For most surface layer work the bandwidth offered by data collected at 20 Hz rate over 1-hr observational periods should prove adequate.

7. Concluding Remarks

The sonic anemometer with its rapid response, linear output and stable calibration has become a valuable tool for atmospheric research. It has provided high-quality turbulence data in a number of field experiments conducted during the last decade. A new three-axis configuration is currently being used for continuously monitoring the wind at eight levels on the Boulder Atmospheric Observatory's 300 m tower.

Because of its complexity and high cost, the sonic anemometer is likely to remain a research instrument. Attempts are being made to reduce cost and to simplify   its operation. Even if these attempts are successful, the sonic anemometer has an inherent limitation which precludes its use in adverse weather conditions. The instrument fails to respond in rain, wet snow, and heavy fog. Formation of water drops on the transducer temporarily affects its operation. Thus the sonic anemometer's main contribution will be in fair-weather observations of atmospheric turbulence, where it is still the instrument of choice amoung boundary-layer physicists.

References

Bovsheverov, V.M. and Voronov, V.P., 1960 Acoustic anemometer. *Izv. Geophys. Series,* **6,** 882-885

Bovsheverov, V.M., Koprov, B.M., and Mordukhovich, M.I., 1973 Spatial correlation functions of velocity and temperature components in the surface layer of the atmosphere. *Izv. Atmos. Oceanic Phys.,* **9,** 434-437

Friehe, C.A., 1976 Effects of sound speed fluctuations on sonic anemometer measurements. *J. Appl. Meteorol.,* **15,** 607-610

Gurvich, A.S., 1959 Acoustic microanemometer for investigation of the microstructure of turbulence. *Acoust. J.* (USSR), **5,** 368-369

Horst, T.W., 1973 Spectral transfer functions for a three-component sonic anemometer. *J. Appl. Meteor.,* **12,** 1072-1075

Kaimal, J.C., 1969 Measurement of momentum and heat flux variations in the surface boundary layer. *Radio Sci.,* **4,** 1147-1153

Kaimal, J.C., 1978 Horizontal velocity spectra in an unstable surface layer. *J. Atmos. Sci.,* **35,** 18-24

Kaimal, J.C. and Businger, J.A., 1963 A continuous wave sonic anemometer-thermometer. *J. Appl. Meteorol.,* **2,** 156-164

Kaimal, J.C. and Haugen, D.A., 1967 Characteristics of vertical velocity fluctuations observed on a 430-m tower. *Q.J.R. Meteorol. Soc.,* **93,** 305-317

Kaimal, J.C. and Haugen, D.A., 1969 Some errors in the measurement of Reynolds stress. *J. Appl. Meteorol.,* **8,** 460-462

Kaimal, J.C. and Haugen, D.A., 1971 Comments on "Minimizing the levelling error in Reynolds stress measurement by filtering". *J. Appl. Meteorol.,* **10,** 337-339

Kaimal, J.C., Wyngaard, J.C., and Haugen, D.A., 1968 Deriving power spectra from a three-component sonic anemometer. *J. Appl. Meteorol.,* **7,** 827-837

Kaimal, J.C., Wyngaard, J.C., Izumi, Y., and Coté, O.R., 1972 Spectral characteristics of surface layer turbulence. *Q.J.R. Meteorol. Soc.,* **98,** 563-589

Kaimal, J.C., Newman, J.T., Bisberg, A. and Cole, K., 1974 An improved three-component sonic anemometer for investigation of atmospheric turbulence. *Flow – Its Measurement and Control in Science and Industry,* Vol. **1** (Instrum. Soc. Amer.), 349-359

Koprov, B.M. and Sokolov, D. Yu., 1973 Spatial correlation functions of velocity and temperature components in the surface layer of the atmosphere. *Izv, Atmos. Oceanic Phys.,* **9,** 178-182

Mitsuta, Y., 1966 Sonic anemometer-thermometer for general use. *J. Meteorol. Soc., Japan,* **44,** 12-24

Mitsuta, Y., 1974 Sonic anemometer-thermometer for atmospheric turbulence measurements *Flow – Its Measurement and Control in Science and Industry,* Vol. **1,** (Instrumen. Soc Amer.), 341-347

Scrase, F.J., 1930 Some characteristics of eddy motion in the atmosphere. Meteorological Office Geophysical Memoirs, No. 52. His Majesties Stationary Office, London, England

Suomi, V.E., 1957 Sonic Anemometer. *Exploring the Atmosphere's First Mile,* Vol. **1,** Pergamon, New York, 356-266.

Computer Analysis of Flow Visualization Records Obtained by the Smoke-Wire Technique

by

H. Nagib[*], T. Corke[*], K. Helland[**] and J. Way[*]
[*]Illinois Institute of Technology, Chicago, Illinois, USA
[**]University of California, San Diego, California, USA

Summary

A technique utilizing a "smoke-wire" for introducing controlled sheets of smoke streaklines has been developed over the last two years for flow visualization in wind tunnels. The smoke-wire, which is mounted on a portable probe, has been used to generate vertical or horizontal sheets of smoke streaklines in several regions of interest in various flow fields. Recently, smoke-wire visualization records of the flow downstream of cylinders and turbulence generating grids, and in turbulent boundary layers have been studied with the aid of digital image processing. The photographic records were digitized on an optical drum scanner and stored on digital tape for analysis. Extensive software packages have been developed by our group for studies of turbulence and are available for array processing, statistical computation, and two- and three-dimensional plotting of results on drum and CRT media. Typical processing of the photographs includes multiple image averaging, thresholding, edge and texture detection, streakline following and spectral analysis of the intensity distribution. This paper outlines the first stages of a long range program and reports on preliminary results obtained from computer analysis of records of grid turbulence with mesh Reynolds numbers ranging from 1,500 to 4,000, the far and near wakes of circular cylinders for the Reynolds number ranges 50 to 150 and 5,000 to 10,000 and turbulent boundary layer flow with Reynolds numbers based on momentum thickness from 600 to 3,300.

Introduction

A flow visualization technique (1) utilizing a "smoke-wire" for introducing controlled sheets of smoke streaklines into wind tunnels has been recently developed at IIT and used to provide global and local information about diverse problems in fluid mechanics. The smoke wire which can be mounted on a portable probe in the traversing mechanism of a wind tunnel, is capable of generating vertical or horizontal sheets of smoke streaklines in various regions of

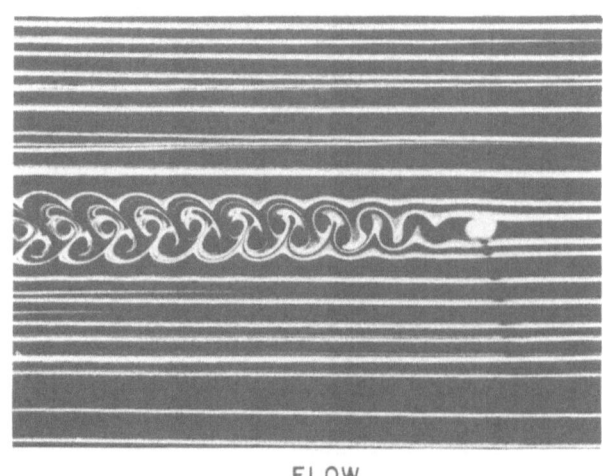

FLOW

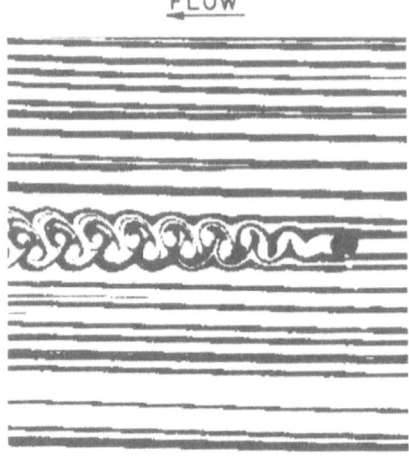

Fig. 1. Photograph and Digital Plot of the Laminar Wake Behind a Circular Cylinder.

interest in the flowfield. The surface of a 0.1 mm diameter wire is coated with uniformly spaced minute droplets of oil which are vaporized by resistive heating, resulting in sheets of discrete streaklines similar to those shown in Figs. 1 and 2. A synchronization circuit controls the duration of time the wire is supplied with the heating current and triggers the camera and lights after an adjustable delay. The synchronization circuit can be triggered manually or by an external input signal which can be provided from the output of a transducer sensing the flowfield, such as a conditioned anemometer output. The wire is operated from outside the wind tunnel without interruption of experiments.

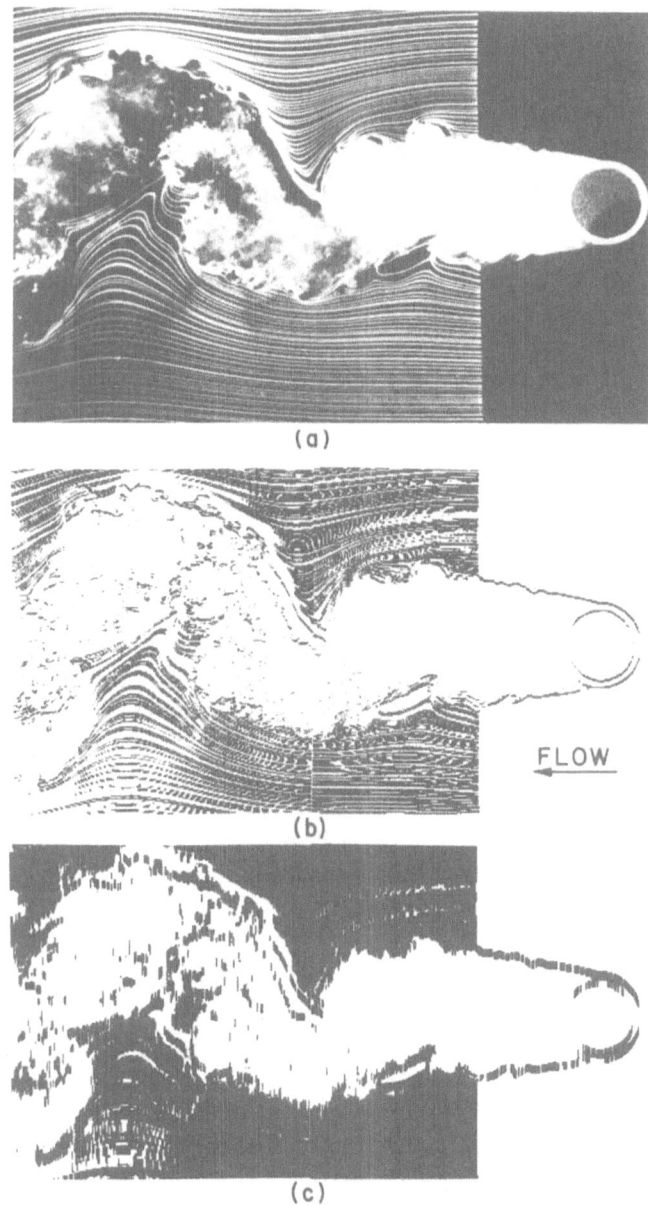

(a)

(b)

FLOW

(c)

Fig. 2. Photograph (a), and Digital Edge (b) and Texture (c) Representations of Flow in the Wake of a Circular Cylinder at High Reynolds Number.

Applications of this technique to study flows near building models in thick turbulent boundary layers and to investigate wakes of complex bluff bodies have been presented by Nagib et al. (1,2,3).

Utilizing the improved visualization capabilities of this technique, our overall objectives were to obtain global and local information on simple as well as complex flow processes in a manner which is far less subjective than previously possible through flow visualization. This was to be accomplished by combining flow visualization techniques, such as the smoke-wire or hydrogen bubble, with the power of digital computers in a manner similar to that used for some years in such areas as earth science and bio-medical imaging (4,5,6). This combination can objectively extract desired information from an image while suppressing extraneous information that often masks the sought after features when no enhancement is done. Other advantages of this method are the quantitative nature of the processed information available from the computer, the capability of processing and storing large numbers of flow visualization records, and the standardization of data gathered through various visualization techniques or by different investigators.

In order to facilitate the digitization of the flow visualization records at this stage in our program, the images recorded on negatives at IIT were sent to San Diego for optical scanning in the UCSD Visibility Laboratory. This also permitted the processing of the data, recorded on digital magnetic tapes, with the Image Processing Software System of the facility. This system permits versatile and immediate operations to be performed on the digitized record of the acquired picture. Included in these operations are averaging, filtering, enhancement and various non-linear gray level reconstruction. Automatic printing and movie devices are available for advanced graphic output of the processed images.

Copies of the digital magnetic tapes were also shipped back to IIT. Further processing of the tapes was done on a UNIVAC 1108 in the university computer center, and by the Data Acquisition and Processing System (utilizing a PDP-11) in the Fluids Group laboratories (7). Extensive software packages have been developed by this group for their studies of turbulence and are available on these machines for array processing, statistical computation, and two- and three-dimensional plotting of results on drum and CRT media. Typical processing includes multiple image averaging, thresholding, edge and texture detection, streakline following and spectral analysis of intensity distributions. As an example, several of the unsteady smoke-wire photographic images which are synchronized to the same instantaneous flow characteristics can be averaged and the resulting "mean" and "fluctuating" images reconstructed. In addition to the existing software already mentioned, pattern recognition routines are being developed to extract specific features, such as large-scale coherent structures and instabilities, from the image data.

Facilities and Techniques

The processes necessary for obtaining digital representations of the visually recorded images began in the darkroom. There, the film taken with a 35 mm SLR camera was developed and used to produce prints. The techniques utilized in the photographing of the flowfield have been described in our previous publications, e.g. reference (1). Frequently in the printing process, various "dodging" techniques were used to compensate for any non-uniform lighting of the smoke streaklines caused by the wind tunnel arrangement. The final prints, made on high contrast F-6 paper, were then photographed with a 4X5 camera (actual negative size is 100 mm X 125 mm) and the film was developed for use in the digitization. In cases where multiple-realization averaging would be required, the negatives were overlayed on a master and aligned with two or more fixed reference points. Two edges of the negatives were then trimmed to facilitate accurate registration of the negatives during the digitization at the Visibility Laboratory. The negatives were then flown to San Diego with the time outlay at this point being on the order of 10 hours per negative for a high-quality realization.

The Visibility Laboratory located at UCSD consists of an IBM 360/44 computer with a 384k byte (8 bits/byte) core memory, four IBM 2311 disk drives, two nine track tape drives, and associated scanning and picture viewing equipment. Negatives are scanned with an Optronics drum scanner which under computer control digitizes the film density and transmits the scanned data to the computer at a rate of 8000 16-bit words/sec. After a negative has been digitized, the picture can be displayed on a CRT device and written out to digital tape for processing at some later time on either the Visibility Laboratory computer or on other computers such as the IIT UNIVAC 1108 and PDP-11 machines. The software system on the IBM 360 consists of many individual programs stored on a disk which can be called into the core overlay structure by the image processing supervisor that remains in core memory at all times. Included in this software system are subroutines for basic arithmetic operations, filtering, Fourier transforming, and input-output operations on various combinations of the two dimensional matrices that contain the computer representation of the images. The image processing system can operate in a batch mode or in an interactive CRT oriented mode. In the batch mode, a card deck written in the language of the image processing system consists of operation codes followed by their required parameters. In the interactive mode, the computer executes each operation code keyed in by the user and then waits for the next instruction. In this interactive mode, intermediate results of a series of operations can be displayed automatically on the CRT display device. The batch mode is used primarily for repetitive computing jobs, or with calculations on very large arrays which are especially time consuming such as a two-dimensional Fourier transform of a 512 X 512 image.

In our application, the negatives required a few preliminary steps before being digitized. Since some groups of pictures were to be "averaged", proper registration of successive images had to be maintained. The edges of the negatives were trimmed to remove slight differences in the placement of the image on the film stock. Using a common edge as a reference, alignment holes were punched in the negatives. These holes served to position the negative over a reference grid on a light table and on the Optronics scanning drum itself. The area to be scanned was determined from measurements made on the light table. The scanner has a switchable resolution of 50, 100 or 200 microns. For the 10 cm X 12.5 cm negatives used in this study, we chose a 200 micron scanning resolution and an array size of 256 X 256. The array size was chosen so that the resulting digital images would be compatible with memory capabilities of the UNIVAC 1108 Computer at IIT. With these parameters we were able to cover most of the negatives with two or three 5.12 cm X 5.12 cm scans. The 200 micron pixel (picture element) size appeared to be consistent with the finest spacing of the smoke lines that could be resolved by ordinary photographic reproduction and enlargement of the negatives.

The negatives presented here were scanned with the computer operating in the interactive mode. The scanning coordinates and storage disk file number were keyed into the computer and the resulting image was automatically displayed on the CRT screen. This process permits quick judgments on the quality of the scan being made before removing the negative from the scanning drum. Some difficulty was encountered when the negatives were not placed in the proper plane of the drum and a number of pictures were recorded out of focus. These pictures were redigitized and carefully checked for sharp focus on the CRT display screen and for proper negative mounting. After scanning the negatives, the computer was switched to the card input mode and a program was run to copy the digitized pictures on standard 9-track digital tape in an integer format suitable for reading on the IIT UNIVAC 1108 and PDP-11 computers. It should be noted that this method requires about 5 to 10 minutes per scan, including mounting the negative on the scanning drum, setting the initial starting coordinates of the scanning head, and keying in the information necessary for the computer to scan the negative. Although this technique is reasonably efficient, it is not suitable for a very large number of negatives. Future digital processing of images obtained in unsteady or turbulent flows would be greatly enhanced by one of the direct photodigitizing cameras available commercially for interfacing to various digital computers, such as the I.I.T. Data Acquisition and Processing System (DAPS).

Copies of the digitized images stored on digital magnetic tape were flown back to IIT for analysis. Various software routines were written for use on the UNIVAC 1108 and PDP-11 systems. In addition, routines to handle the format conversion between the two machines were necessary. All communication between the machines was made via magnetic tape. The documentation of the results was always made by the DAPS through its CRT and hard copy units or via analog signals provided by the digital to analog conversion channels of the system.

The software routines were broken up into two groups: Univac Software and PDP-11 Software. The criteria for their distribution was purely one of allocatable computer memory. Since the digital representations were comprised of two or three files of 256 records each containing 256 pixel intensities, 65k words of memory were necessary to store each portion of the photographs. In the case of mean and variance calculations, two data blocks, 65k words each in size, were necessary. The PDP-11 system has 8k words of memory, so that its use was limited to those computations which required only a few successive records to be in core simultaneously, thereby preventing excessive magnetic tape movement. As a result of the large number of required data manipulations (wherever possible on the Univac and in all cases on the PDP-11), the software routines were written in machine assembly language to hold the computations and plotting time to a minimum.

The Univac routines consisted mainly of array manipulations such as shifting rows and columns by prescribed increments n and m

$$F(j,k) \leftarrow F(j-n, k-m)$$

where $F(j,k)$ is the pixel intensity at point (j,k) in the array; taking j as the record number (or column) and k as the intrarecord pixel location (or row). To rotate the image by $\pi/2$, the rows and columns of the pixel arrays were interchanged through

$$F(j,k) \leftarrow F(k,j)$$

Similar techniques were used to flip the digitized image about an axis passing through any row or column. These routines were useful for correcting small registration errors and reorienting the digital representations that were to be averaged.

The mean and variance of N numbers of related digital images were computed through the relations

$$\frac{1}{N} \sum_{n=1}^{N} F_n(j,k) \equiv M(j,k)$$

and

$$\frac{1}{N} \sum_{n=1}^{N} [F_n(j,k) - M(j,k)]^2$$

where $F_n(j,k)$ is the pixel intensity at point (j,k) in photo n. Here, $M(j,k)$ is the mean value of the pixel intensities of N digital representations at point (j,k).

The PDP software package written for the digital picture analysis consisted of computational routines with two and three dimensional plotting. The simplest of the three-dimensional computation-plotting routines lead to the reconstruction of the 128 gray levels on a video monitor, e.g., an oscilloscope with a controllable z axis intensity. Here, the display intensity $I(j,k)$ for the entire digital file is plotted as the third dimension with

$$I(j,k) = F(j,k)$$

The output from the D/A converters was used in this type of application and the results were recorded photographically. Remarkable similarity to the original photographs was achieved by this approach. Often this method was used to check the quality of the digital records.

A one bit representation of the digital image of the type shown in the lower portion of Fig. 1 was accomplished through the relation

$$I(j,k) = \begin{cases} 0: F(j,k) < T_p \\ 1: F(j,k) \geqslant T_p \end{cases}$$

where T_p is a threshold value. Those pixel points having intensities greater than or equal to the threshold value are plotted as bright screen vectors on the CRT. Note that the hard copy reproductions presented in the figures are photographic negatives of the CRT display and thus bright points appear dark in these representations. Therefore, white smoke streaklines appear black in these hard-copy displays. For the high contrast photographs of the present work the resulting images were reasonably insensitive to the particular choice of T_p. However, future applications may require careful selection procedures for T_p and could benefit from some standardization techniques. In view of the good reproduction achieved by this approach for reconstructing the original photographs we will refer to its output as the "digital plot" of the recorded image.

The edges of adjoining regions of different intensities were computed using the relation

$$I(j,k) = \begin{cases} 0: |F(j,k) - F(j,k+1)| < T_e \\ 1: |F(j,k) - F(j,k+1)| \geq T_e \end{cases} \equiv G(j,k; T_e)$$

where T_e is again a threshold for the difference between adjoining pixel intensities. In the above representation, the edges were computed along each record. However, they could just as well be computed across records (in the flow direction for our manner of digitization) or across both. Changes in the texture of the photographs were detected by simply counting edges within any particular region. That is

$$I(j,k) = \begin{cases} 0: \sum_{n=k-w}^{k+w} G(j,n; T_e) < T_t \\ 1: \sum_{n=k-w}^{k+w} G(j,n; T_e) \geq T_t \end{cases}$$

where T_t is a threshold value for the number of edges encountered in a region of width w. Again, the texture can also be computed across records. The three parameters T_e, T_t and w were varied to select "optimum" values prior to documenting the results by the hard copy unit. While the choice is often somewhat subjective, various techniques for the selection or standardization of these parameters can be employed. Although so far we have used simple selection criteria for the parameters, we plan on utilizing more sophisticated and objective techniques in the future.

Plots of the pixel intensity values taken along selected records comprising the digital image were represented by the two-dimensional form

$$y_j(k) = F(j,k)$$

The power spectral density of the pixel intensity along these records was represented as

$$Y_j(k) = |\Phi_j[F(j,k)]|^2$$

where Φ_j is the one-dimensional Fourier transform taken along each record. The running variable is not temporal frequency but rather spacial frequency, i.e., per length in the plane of visualized flow.

Typical times required for processing of the digital images on the UNIVAC 1108 and the DAPS are 0.5 and 1 minute per negative, respectively. The use of machine assembly language was very instrumental in achieving such efficient processing.

Discussion of Results

The results presented here represent only a small portion of our efforts to date in developing capabilities for digital picture analysis with special applications in turbulence research. To this end, 31 smoke-wire visualization photographs were digitized and stored in 59 digital-tape files for analysis. The flow conditions are comprised of the laminar steady flow through a two-dimensional contraction; the laminar periodic wake of a circular cylinder, $50 < Re_D < 130$; grid generated turbulence, $1.5 \times 10^3 < Re_M < 4 \times 10^3$; turbulent boundary layer flow, $600 < Re_\theta < 3300$; and the phase conditioned photographs of the turbulent unsteady ("periodic") wake of a circular cylinder, $5 \times 10^3 < Re_D < 10^4$, in both high and low free-stream turbulence. Such a variety of cases is intended to provide a "test bed" for different image processing techniques as we increase our sophistication in developing pattern recognition routines to extract specific features from the image data, e.g., identification and documentation of large-scale coherent structures, detection of turbulent and non-turbulent zones, and discovery of various instabilities.

In Fig. 1 a photograph and its digital representation are presented for the laminar wake downstream of a circular cylinder at a Reynolds number of 70. This figure is intended to demonstrate the excellent degree of reproduction obtainable from a one-bit plot of the digital image for such smoke-wire photographs. The wake behind a circular cylinder at a Reynolds number of 10,000 is documented in the top portion of Fig. 2 by a smoke-wire placed downstream of the cylinder. The middle and lower plots in this figure are the respective edge and texture representations of the flowfield visualized in part (a) of the figure. The computations in these plots were done along the records, i.e., in the cross-flow direction. Such representations may be useful in marking turbulent or nonturbulent regions.

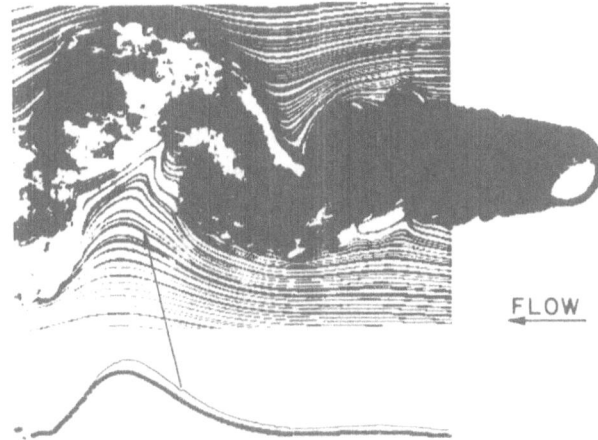

Fig. 3. Digital Plot and Plot of Two Selected Streaklines of the Flow in the Wake of a Circular Cylinder at a High Reynolds Number.

Fig. 4. Photograph, Digital Plot and Plot of Two Selected Streaklines in a Turbulent Boundary Layer Flow.

In Fig. 3, the digital plot of the photograph in Fig. 2 is reproduced along with two selected streaklines. The streaklines were plotted separately by a special streakline following routine which searches for a local maximum or an edge, as in this case as well as in Fig. 4, while advancing in the flow direction from record to record. As the search proceeds from one record to the next, the co-ordinates of the identified streakline are stored in memory for future recall. The local pressure gradient can be subsequently computed, if the flow is two-dimensional, by calculating the slopes between adjacent pixel locations. The smoke-wire visualization photograph of the flow in a turbulent boundary layer and its digital representation are presented in Fig. 4, along with the reproduction of two selected streaklines.

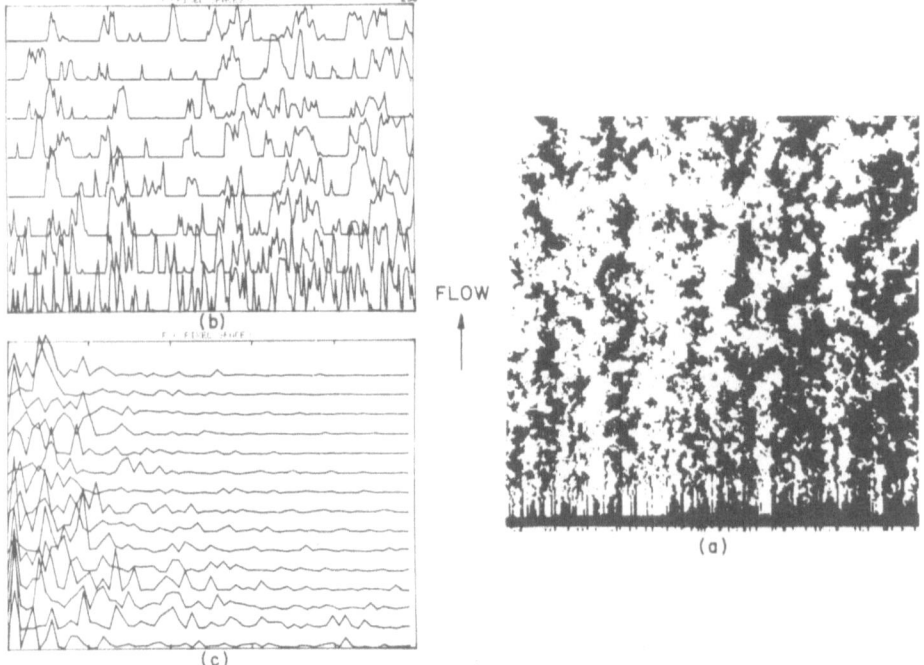

FLOW

Fig. 5. Digital Plot of Visualized Flow Downstream of a Grid (a), Intensity Plots (b), and Power Spectral Density of these (c) taken Along Cuts Normal to the Flow Direction.

The digital plot of the flow visualized immediately downstream of a grid is shown in the right portion of Fig. 5. The pixel intensities taken along selected cuts normal to the flow direction for this flowfield are presented in the upper left portion of the same figure. In such representations, we plot the distribution with its datum at a position corresponding to the location of the selected

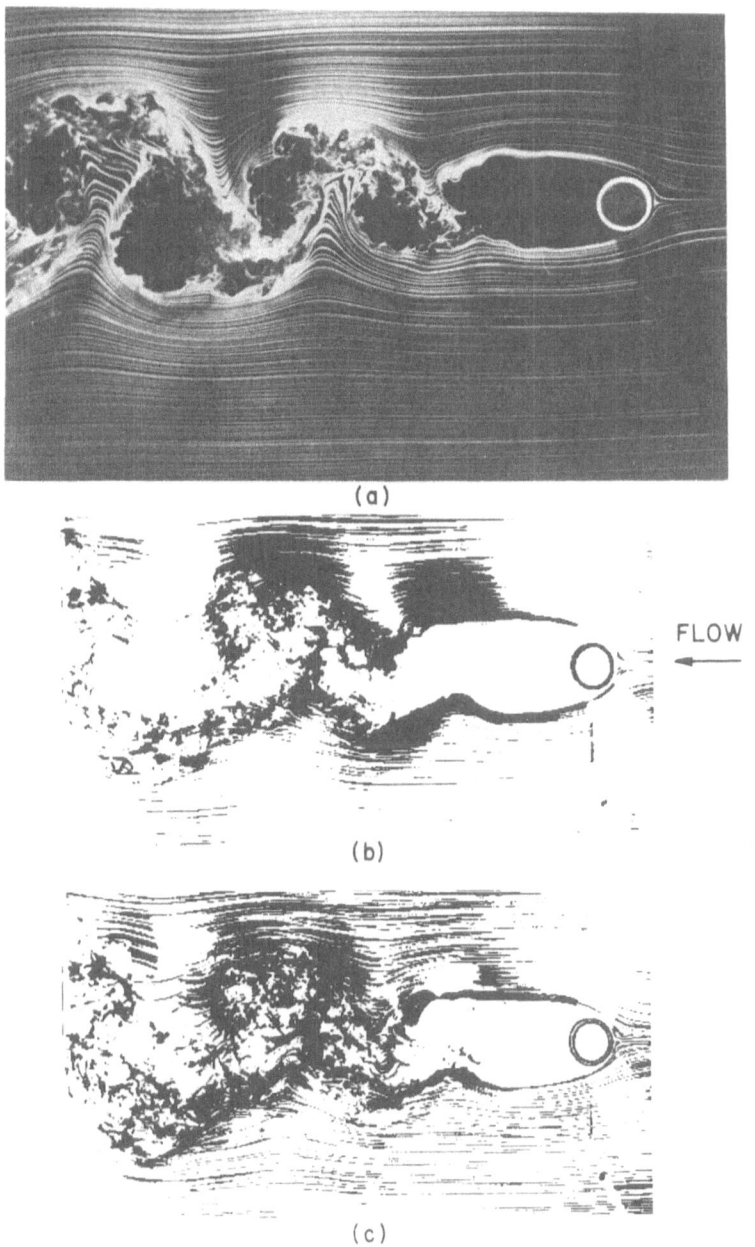

(a)

FLOW

(b)

(c)

Fig. 6. Photograph of Flow Past a Circular Cylinder (a) and Digital Plot of the Mean (b) and Variance (c) of Four Such Representations Taken at a Relative Phase Equal to Zero.

cut in the photograph. It is apparent from these intensity distributions that the scales of turbulence are increasing as the flow moves downstream of the grid. This is further demonstrated by the power spectral density plots in the lower left of the figure. The spectral plots of this figure were obtained for a larger number of intensity distributions and, hence, do not correspond exactly to those shown above them. Here, we see a definite shift in spectral energy to the right as the analysis is taken closer to the grid. This indicates more closely spaced clumps of smoke (smaller scales) in the vicinity of the grid. Note that these are only one-dimensional Fourier transforms presented without any digital filtering. We expect two-dimensional transforms and digital filters to give us much better results, including length scale estimates.

The flow past a circular cylinder at a Reynolds number of 10,000 is document-ed with the aid of smoke-wire visualization in the top portion of Fig. 6. Also included in this figure are digital plots of the mean and variance of four such realizations taken at a relative phase of the wake shedding cycle equal to zero (the photograph in part (a) was not taken at the same exact relative phase). The phasing was provided by a hot-wire probe placed in the potential flow just outside of the separated shear layer. The average of the four realizations, ob-tained digitally, indicates that a fairly coherent wake exists at this relative phase in the cylinder shedding cycle. In the plot of part (c) of this figure, large variance values are exhibited in the upper shear layer. This presumably is a re-sult of some flapping of the shear layer which occurs during the formation of a rolled-up eddy on the opposite side of the wake.

Concluding Remarks

The preliminary computer analyses outlined in the preceding sections have demonstrated not only the feasibility, but also the strong potential of digital processing of images obtained by flow visualization. In addition to the samples presented in Figs. 1 - 6, numerous other records of flowfields visualized by the smoke-wire method have been analyzed by these techniques, and the results suggest that this approach can be an effective tool for future turbulence re-search. For example, image enhancement techniques similar to those presented in Fig. 2 can be employed in marking turbulent or non-turbulent regions.

All of the image processing methods used here have been limited to one-dimen-sional analysis, primarily because of computer system limitations. Preliminary tests utilizing two-dimensional techniques, not discussed in this paper, have indicated substantially better results, as well as a more expanded scope of appli-cations. These techniques would lead to more sophisticated image analysis, such as feature detection and extraction. For example, one goal of our work deals with the correlation of events, detectable by sensors in a turbulent boundary layer, with some visually recognizable flow structures which are recorded simultaneously with the sensors' output. Repeating this type of correlation for a large sample of events would lead to less-biased probabilistic interpretations and better understanding of the mechanisms of the flowfield.

In view of this, steps have been taken at I.I.T. to provide an in-house capability for digitizing the photographs and to even by-pass the darkroom processes. Monitoring the flowfield directly, or through any photographic records of it, by a photodigitizing camera interfaced with the DAPS will lead to this improvement. In addition, the computation capabilities of the DAPS (7) are being expanded to handle the requirements of this project. In this way we hope to make more efficient use of the strong points that flow visualization and the smoke wire technique have to offer to the turbulence experimentalists.

Acknowledgement
The authors are indebted to Valerie Mattioli for the typing of this manuscript and to Yann Guezennec for the many long hours he spent in the darkroom. This work was sponsored under NSF Grant ENG 76-04112 and AFOSR Contracts F44620-76-C-0062 and F49620-78-C-0047.

References

(1) Corke, T., Koga, D., Drubka, R. and Nagib, H., 1977, A New Technique for Introducing Controlled Sheets of Smoke Streaklines in Wind Tunnels. *Proceedings of International Congress on Instrumentation in Aerospace Simulation Facilities,* IEEE Publication 77 CH1251-8 AES, p. 74

(2) Nagib, H.M., 1977, Visualization of Turbulent and Complex Flows Using Controlled Sheets of Smoke Streaklines. *Proceedings of the International Symposium on Flow Visualization,* Tokyo, Japan

(3) Nagib, H.M., Merati, P. and Crawford, A.C., 1978, Visualization of Turbulent Atmospheric Flows. *Proceedings of the Third U.S. National Conference on Wind Engineering Research,* Gainesville, Florida

(4) Pratt, W.K., 1978, *Digital Image Processing,* Wiley, New York

(5) Aggarwal, J.K., Duda, R.O. and Rosenfeld, A., editors 1977, *Computer Methods in Image Analysis,* Wiley, New York

(6) Rosenfeld, A. and Kak, A., 1976, *Digital Picture Processing,* Academic Press, New York

(7) Way, J., 1975, Applications in Fluid Mechanics Research of a Portable Data Acquisition and Processing System. *Proceedings of International Congress on Instrumentation in Aerospace Simulation Facilities,* IEEE Publication 75 CHO 993-6 AES, 1975, p. 50.

Simultaneous Measurement of Flow Velocities in Multipoint by the Laser Doppler Velocimeter

by

Noboru Nakatani, Ryoichi Yorisue and Tomoharu Yamada
Department of Precision Engineering, Faculty of Engineering
Osaka University; Yamada-Kami, Suita, Osaka, Japan

1. Introduction

An important shortcoming of the conventional laser Doppler method is that it furnishes the instantaneous velocity at a single point in the fluid. To obtain the complete flow pattern, the experiment has to be repeated for different regions of the flow field (1) ~ (6).

In this study some new optical systems for the simultaneous measurement of flow velocity distribution by making one- or two-dimensional intersection of the two incident laser beams in flow field and using optical fibers for receiving light are described. To measure unsteady flow, frequency shift using acousto-optic modulators (4) ~ (6) is combined with these systems. Using these velocimeters unsteady flow in a branch tube is measured.

2. Some Optical Systems for the Simultaneous Measurement of Flow Velocity Distribution

Optical systems of the differential mode are used. In these systems light scattered from the measuring volume can be collected in a larger solid angle than in the optical systems of the reference mode or the dual scattered mode.

The two thin beams with a width obtained by expanding laser beams with some spherical lenses and some cylindrical lenses are used as the incident beams into flow field. The two line beams obtained by focusing laser beams with some spherical lenses are also used as the incident beams. The one- or two-dimensional intersection of the two incident beams defines the measuring region in the flow field. Optical fibers are used for receiving intensity of light in image plane and leading it to photodetectors, i.e. for non-coherent detection. Therefore both multimode fibers and single mode fibers can be used, but multimode fibers of large numerical aperture had better be used for collecting light. In this study the optical fibers, made of plastics, of step index type are used. Their numerical aperture is about 0.5 and their diameter, 100 μm. Avalanche photodiodes are used as photodetectors.

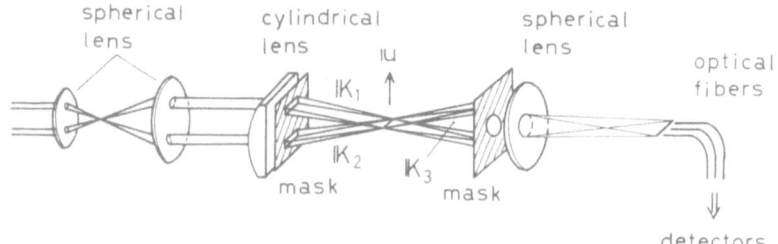

Fig. 1. Optical system A of the laser Doppler velocimeter for the measurement of flow velocity distribution.

In the optical system A shown in Fig. 1 the Doppler frequency is given by $f = (|K_1 - |K_2) \cdot |u/2\pi = (2n||u|/\lambda)\sin(\theta/2)$, where $|K_1$ and $|K_2$ are the wave vectors, $|u$ is the velocity vector, n is the refractive index of the measuring medium, λ is the laser wavelength, and θ is the angle between $|K_1$ and $|K_2$ beams. The velocity components perpendicular to the direction of line intersection of the two incident beams can be measured. To measure the velocity components parallel to the direction of the line intersection, the two thin incident laser beams are intersected by the use of some mirrors (in the optical system A'). As the forward scattered beams are observed, the high intensity of light on the photodetectors can be obtained.

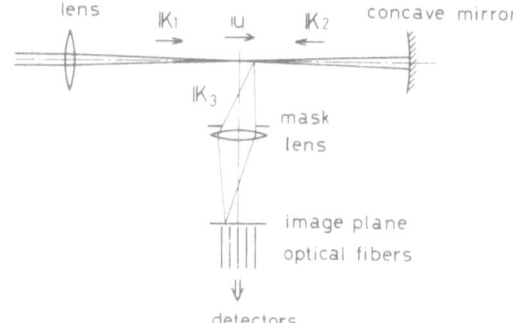

Fig. 2. Optical system B of the laser Doppler velocimeter for the measurement of flow velocity distribution.

In the optical systems B and C shown in Figs. 2 and 3 the Doppler frequency is given by $f = (|K_1 - {}_|K_2) \cdot |u/2\pi \approx 2n||u|/\lambda$. The velocity component parallel to the direction of line intersection of the two incident line beams can be measured. By changing the line beams for the thin beams with a width two-dimensional intersection can be obtained. The observed beams make it necessary to have equal intensities for obtaining high signal to noise ratio (SNR). Hence the scattered beams should be observed from the direction perpendicular to the incident beams. Then the intensity of the observed light becomes low. As the laser beam, however, is not expanded, the intensities of light observed in the optical systems B and C become little weaker than that in the

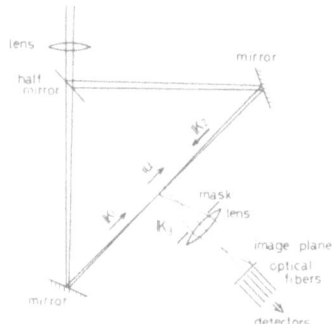

Fig. 3. Optical System C of the laser Doppler velocimeter for the measurement of flow velocity distribution.

system A. In the optical system B SNR depends on the transmittance T of the flow field as given by SNR = 2T/(1+T). The SNR little decreases with the increase of T. For example, in the case of T = 0.7, the SNR is 0.8. And it also decreases with the increase of the optical path difference between k_1 beam and k_2 beam to the observation point due to the axial modes of a laser used. To exclude the influence of the axial modes, in the optical system C the laser beam is divided into two incident beams with the half mirror and the optical path difference of the two incident laser beams is made smaller. The intensities of the incident beams in the optical system C become half of those in the optical system B. It had better use the optical system B for the measurement of small flow field without the influence of the axial modes.

For the measurement of velocity components in two dimensions, the combinations of the optical systems A and A', or A and either B or C can be used.

3. Measurement of Unsteady Flow in a Branch Tube
After a step input of pressure gradient into a branch tube, the variations of flow velocity distributions and flow rates for elapsed times are measured. The branch tube shown in Fig. 4 is used. The tube is placed on the horizontal plane without the influence of gravitation. The head tank of large diameter is used for obtaining a constant pressure. The step input of the pressure gradient is obtained by opening the exits of the tube with two magnetic valves of two ports.

The Doppler signal from the avalanche photodiodes are recorded by a multi-channel wave memory or data recorder. The records of the Doppler signals are limited by their recordable frequency range. To make the range wider, the mean frequency of the Doppler signals are shifted by the optical systems,

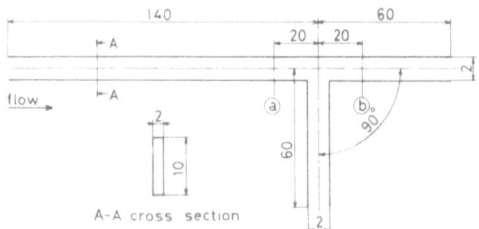

Fig. 4. Dimension of the branch tube and the measured points ⓐ and ⓑ.

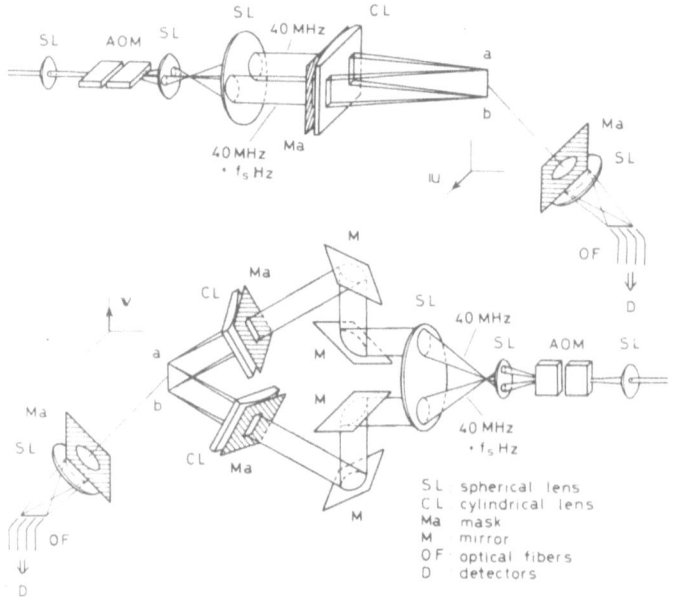

Fig. 5. Optical systems of the laser Doppler velocimeters for the measurement of flow velocity vector distribution by the use of the two acousto-optic modulators.

shown in Fig. 5, using the two acousto-optic modulators. The Doppler signals reconstructed from the wave memory or the data recorder are processed by the methods shown in Fig. 6. In the method (a) the flow velocities at an elapsed time after the step input can be measured stroboscopically. Hence it is required to repeat the experiment for obtaining the flow velocities at some elapsed times. As digital signals output with slow speed can be obtained, the signals can easily be processed by the use of a minicomputer and the flow velocity distri-

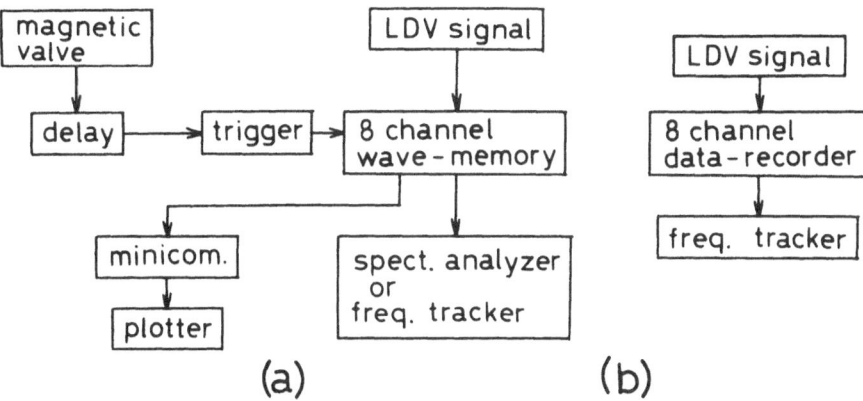

Fig. 6. Block diagrams of Doppler signal processing. (a): using a wave memory, (b): using a data recorder.

butions can be displayed on a CRT or a plotter. And the flow rate can be obtained by integrating the flow velocities. In the method (b) the velocities at each elapsed time can be obtained by one experiment. The signals reconstructed from the data recorder are processed with a tracker to obtain the instantaneous velocities. It is necessary to use the optical system, shown in Fig. 5, with an up-shifted frequency, to keep the width and frequency of the signal frequency variation within the measurable range of the frequency tracker.

4. Results

The flow velocity distributions can be obtained as shown in Fig. 7. At the early stages the velocity near the center axis in the main tube above the branching part is approximately constant. In the latter stages the velocity profile tends asymptotically to the parabolic distribution for steady flow. However, in the main tube below the branching part, the velocity profile does not tend to the parabolic distribution. This is because the flow becomes turbulent in the branching part. Curves of the flow rate as a function of elapsed time are obtained by integrating the velocities along the axis at each elapsed time and are shown in Fig. 8. At early stages the ratio Q_2/Q_1 of the flow rate in the main tube below the branching part to that in the main tube above the branching part decreases. And the ratio Q_3/Q_1 of the flow rate in the branch to that in the main tube above the branching part increases. These phenomena are considered to be caused to the variation of the flow velocity profiles in the branching part shown in Fig. 9. The two ratios of the flow rates tend asymptotically to a constant at about 0.2 s.

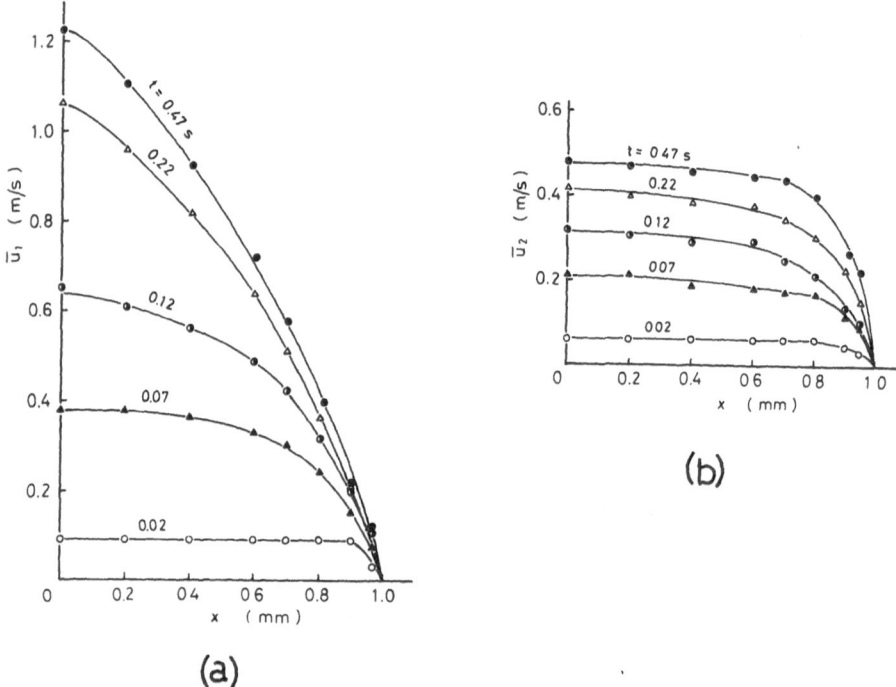

Fig. 7. Mean flow velocity profiles $\bar{u}$ at various elapsed times t. (a): measured points Ⓐ, (b): measured points Ⓑ. x: the distance from the center axis of the tube in the direction perpendicular to the centerline.

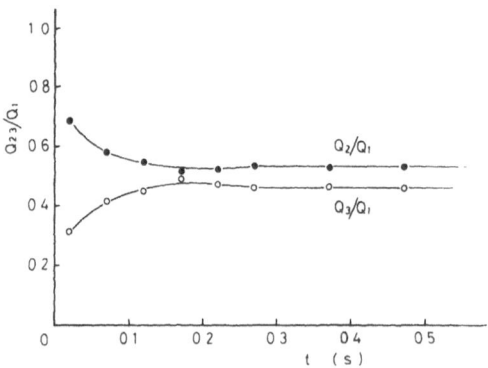

Fig. 8. Ratios of the flow rates Q_2/Q_1 and Q_3/Q_1 versus elapsed time t. Q_1: the flow rate in the main tube above the branching part; Q_2: in the main tube below the branching part, Q_3: in the branch.

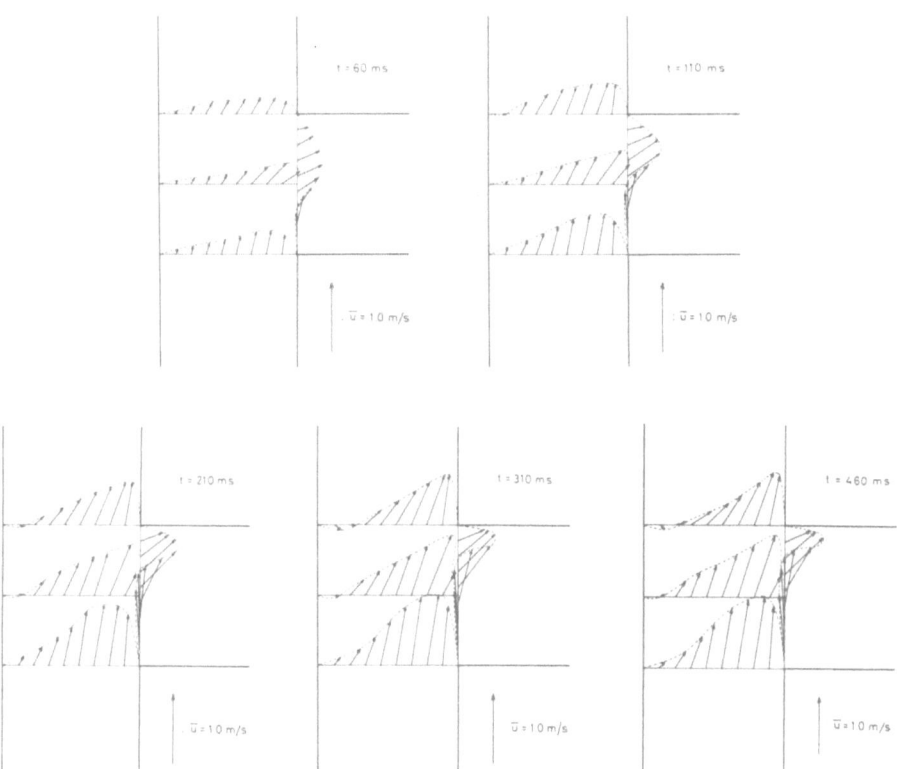

Fig. 9. Mean velocity vector profiles at various elapsed times in the branching part.

5. Conclusions

Some new optical systems for the simultaneous measurement of flow velocity distribution are developed by making one- or two-dimensional intersection of the two incident laser beams in flow field and using optical fibers for receiving light. The optical systems of the differential mode are used for collecting light scattered from the scattered regions through a large solid angle. The Doppler signals are processed after recording with a multichannel data recorder or wave memory. To make their recordable frequency range wider and to give more allowance to the limitation of the dynamic response of the frequency tracker, the Doppler signals are shifted by the use of the two acousto-optic modulators. It is demonstrated using the unsteady flow in a branch tube that these veloci-meters can be applied to the measurement of simultaneous flow velocity distribution and flow rate. Application of these velocimeters to the measurement of space correlation will be reported in the near future.

References

(1)　Nakatani, N., Ono, A. and Yamada, T.: *Proc. of the 2nd Inter. JSME Symposium Fluid Machinery and Fluidics,* **3** (1972), 141, Tokyo, JSME

(2)　Nakatani, N., Fujiwara, K., Morimoto, S., Ono, A., and Yamada, T.: *Fluidics Quarterly* **5**-3 (1973), 49

(3)　Nakatani, N., Yatomi, S. and Yamada, T.: *Fluidics Quarterly,* **7**-1 (1974), 47

(4)　Nakatani, N. Hanioka, N., Konishi, T. and Yamada, T.: *Proc. of the 10th Fluidics Symposium* (1975), 105, Tokyo SICE

(5)　Nakatani, N., Konishi, T., Yorisue, R. and Yamada, T.: *Proc. of 1977 Joint Gas Turbine Congress,* (1977), 461, Tokyo, GTSJ, JSME and ASME

(6)　Nakatani, N., Yorisue, R. and Yamada, T.: *Tech. Rep. of the Osaka Univ.,* **27**-1382 (1977), 463

Analysis of Data from 3-Dimensional Hot-Wire Probes using Comparison with Profile Instrumentation for Calibration

by

Sören E. Larsen, Olaf Mathiassen and Niels E. Busch
Risö National Laboratory, Denmark

1. Introduction

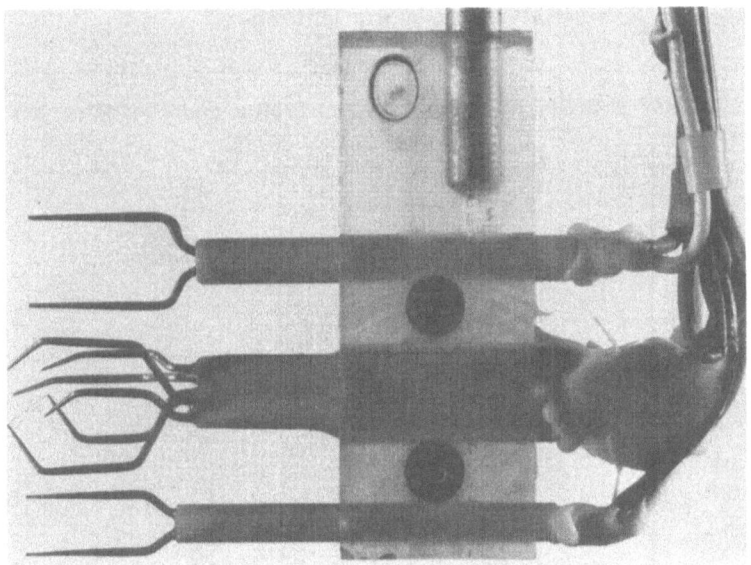

Fig. 1. Probe head for mounting on wind vane. From above is seen: The vertical wire, the 3D-sensor, and a temperature sensor.

During the JONSWAP air-sea interaction experiment in 1975, we measured turbulent velocity fluctuations by means of simple vertically aligned hot wires and 3-dimensional hot-wire sensors mounted on a wind vane. The sensor configuration is shown on Fig. 1.

All probes were calibrated in wind tunnels before the experiment. During the experiment linearization was performed in accordance with these calibrations. However, due to contamination and corrosion of the wires during the experiment the calibrations became unreliable. Instead we have assumed that the linearized anemometer outputs are linear for small fluctuations around a mean value, and have determined the parameters describing this linear behaviour by comparing simultaneous 10 minute averages of the hot wire outputs with calibrated outputs from a cup-anemometer located at the same measuring height, but at a horizontal distance of 300 m.

This comparison gave a rough estimate of the calibration only. For "fine tuning" of the calibration we have assumed that the sensor system to a certain degree is self-calibrating due to the combination of linear transformation of the signal from the vertical wire and quadratic transformation of signals from the 3D probe.

2. Comparison between Averaged Sensor Outputs

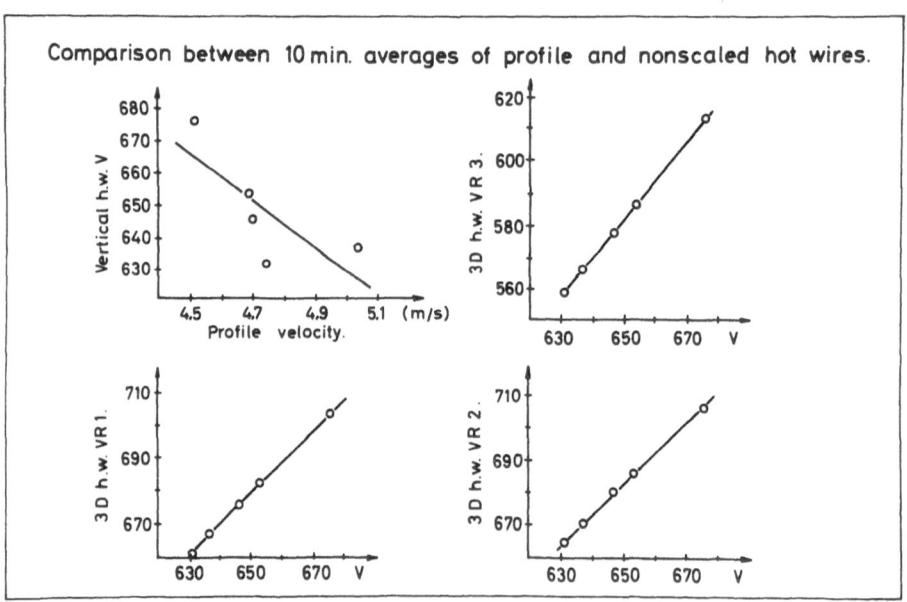

Fig. 2. Comparison between 10 min. averages of outputs from a cup-anemom-eter and the vertical sensor V and the 3D-sensor. VR 1, 2, 3. The hot-wire out-puts are noncalibrated.

Fig. 2 shows a typical comparison between noncalibrated hot-wire outputs and the cup-anemometer readings. The correlation between the linearized output from the vertical wire, V, and the cup-anemometer can be considered reason-ably good, leading to considerable uncertainty on the scaling of the fluctuating velocities.

On the other hand, the correlations between the three linearized outputs $VR1$, $VR2$, and $VR3$ from the 3D probe and V are excellent. The following set of calibration equations is suggested

$$\overline{V} = a + G\ \overline{V}cup \tag{1}$$

$$\overline{VRi} = b_i + G_i\ \overline{Vh} = b_i + G_i a + G_i G\ \overline{V}cup \tag{2}$$

where overbar denotes 10 min. averages.

The calibration factors a and G in [1] must be understood as least squares fit parameters. The scatter in the $\overline{V} - \overline{V}cup$ comparison in Fig. 2 reflects statistical uncertainty on the 10 minute averages. Owing to the horizontally homogeneous conditions over the ocean, we should for usual atmospheric conditions expect the scatter to reduce and become negligible when the averaging time increases to about an hour.

Based on such reasoning we postulate that [1] must be identically true, when averaged over all 10 min. sections, here five. Thus, letting double overbar denote total average

$$\overline{\overline{V}} = a + G\ \overline{\overline{Vcup}} \tag{3}$$

which establish a bond between a and G.

To relate $\overline{VRi}$ to the actual cooling velocities, we must consider the transformation equation for the 3D-probe.

3. Response of 3D-Probes

We shall follow the discussion in Larsen and Busch (1974) and Gaulier (1977). The wire configuration is shown on Fig. 3. In a coordinate system moving with the vane mounted probe, the attacking velocity is written $(U_1 + u_1, u_2, u_3)$, where $U_1 + u_1$ is along the horizontal probe axis, u_3 is vertical and u_2 is lateral, $\overline{u}_1 = \overline{u}_2 = \overline{u}_3 = 0$.

The velocity components parallel to the three wires in the sensor are V_1, V_2, V_3, where the sensor is oriented such that V_3 is in the vertical plane spanned by $U_1 + u_1$ and u_3. The relations between u_i and V_j are:

$$\left\{ \begin{array}{c} V_1 \\ V_2 \\ V_3 \end{array} \right\} = \begin{bmatrix} \frac{1}{3}\sqrt{3} & -\frac{1}{2}\sqrt{2} & -\frac{1}{6}\sqrt{6} \\ \frac{1}{3}\sqrt{3} & \frac{1}{2}\sqrt{2} & -\frac{1}{6}\sqrt{6} \\ \frac{1}{3}\sqrt{3} & 0 & \frac{1}{3}\sqrt{6} \end{bmatrix} \cdot \left\{ \begin{array}{c} U_1 + u_1 \\ u_2 \\ u_3 \end{array} \right\} \tag{4}$$

The effective cooling velocities VR_i are related to V_i through

$$
\left\{ \begin{array}{c} V^2 R_1 \\ V^2 R_2 \\ V^2 R_3 \end{array} \right\} = \left| \begin{array}{ccc} k_1^2 & 1 & k_2^2 \\ k_2^2 & k_1^2 & 1 \\ 1 & k_2^2 & k_1^2 \end{array} \right| \cdot \left\{ \begin{array}{c} V_1^2 \\ V_2^2 \\ V_3^2 \end{array} \right\}
\tag{5}
$$

where k_1 and k_2 are yaw and pitch response coefficients, respectively. We have used $k_1 = 0.15$ and $k_2 = 1.02$ from Gaulier (1977).

By use of [4] and [5] and the definition equation for Vh, the total horizontal velocity,

$$
Vh^2 = (U_1 + u_1)^2 + u_2^2,
\tag{6}
$$

we can derive the relation between $\overline{VR}_i$ and $\overline{Vh}$ through a first order expansion of turbulence intensity terms, e.g. for $\overline{VR}_1$

$$
\overline{VR}_1 = \alpha^{\frac{1}{2}} \overline{Vh} (1 + \overline{Vh}^{-2} [0.74\,\overline{Vh'^2} - 0.13\,\overline{u_2^2} + 0.64\,\overline{u_3^2} + 0.59\,\overline{u_1 u_2}
\tag{7}
$$
$$
+ 0.57\,\overline{u_1 u_3} - 0.42\,\overline{u_2 u_3}]),
$$

where $\alpha = 1/3(1 + k_1^2 + k_2^2)$ and the numbers in the square bracket result from various combinations of the terms in the matrices in [4] and [5].

To estimate the terms in the square bracket in [7] we recall that the probe is placed on a fast responding wind vane, for which reason we neglect all terms involving u_2.

Estimates of the remaining terms are well described in the literature about the atmospheric surface layer above the ocean, e.g. Busch (1977). We have used, with $\overline{u_1 u_3} = -u_*^2$

$$
\overline{Vh}/u_* \cong 24, \quad \overline{u_1^2}/u_*^2 \cong 4, \quad \overline{u_3^2}/u_*^2 \cong 1.44
\tag{8}
$$

We now obtain

$$
\overline{VR}_i = \alpha^{\frac{1}{2}} \overline{Vh} [1 + \rho_i], \quad i = 1,2,3
\tag{9}
$$

where ρ_i is given in table 2 and the smallness of ρ_i justifies both the first order expansion and the nonsubtle method of estimation.

Equations [9], [2], and [1] now yields the resulting calibration for VR_i and Vh

$$V = a + GVh$$

[10]

$$VRi = b_i + G_i a + G_i G\alpha^{-\frac{1}{2}} (1 + \rho_i)^{-1} VR_i$$

where we have let $\overline{Vh} = \overline{V}cup$ and dropped the averaging symbols and where we by virtue of [3] have one uncertain parameter only, say G.

4. "Fine Tuning" of G

With G estimated by e.g. the least squares fit on Fig. 2 and [3], we can estimate $(U_1 + u_1, u_2, u_3)$ and thereby Vh from [4] and [5] and we can determine Vh directly from the vertical wire.

Sensor	Variable	$G_o \cdot 0.85$	G_o	$G_o 1.087$	$G_o \cdot 1.15$	Units
3D	$\overline{u}_3$	−0.036	−0.029	−0.029	−0.028	m/s
3D	$\overline{u}_2$	−0.054	−0.047	−0.045	−0.037	−
3D	$\overline{Vh}$	4.44	4.46	4.47	4.47	−
Vertical wire	$\overline{Vh}$	5.74	4.89	4.48	4.22	−
3D / Vertical wire	$\dfrac{\overline{Vh^2}}{\overline{Vh^2}}$	0.58	0.83	1.005	1.12	
Vertical wire	$\sqrt{\overline{Vh'^2}}$	0.564	0.481	0.441	0.415	m/s
3D	$\sqrt{\overline{u_1^2}}$	0.560	0.476	0.437	0.413	−
Vertical wire 3D	$\overline{Vh'u_3}$	−0.012	−0.008	−0.007	−0.007	m^2/s^2
3D	$\overline{u_1 u_3}$	−0.011	−0.009	−0.007	−0.006	−
Vertical wire 3D	$\dfrac{\overline{Vh'u_1}}{\sqrt{\overline{Vh'^2}\,\overline{u_1^2}}}$	1.000	0.996	0.996	0.998	

Table 1. Variation of estimates of the velocity of the vertical wire and the 3D-sensor with the estimate of G. G_o is based on a least squares proximation against the cup-anemometer values. $G = G_o$ 1.087 is our best estimate of G. The table is based on a 400 sec. data section.

We notice that the linear calibration of the straight vertical wire means that uncertainty on G corresponds, in a simple way, directly to uncertainty in estimates of turbulence quantities. The quadratic transformation [5] for the 3D-probe means that the probe uncertainty on G (in a much more complicated way) gives rise to uncertainty on both mean value and turbulence quantities, see e.g. Larsen and Rasmussen (1978). This means that we are here much more critically dependent on a good estimate of G.

The "true" G value can now be estimated as the one giving the same value of Vh from both the vertical wire, where G enters linearly, and the 3D-probe, where G enters nonlinearly. This method is illustrated in Table 1 for a short section of the same data series as is used in Fig. 2.

With G determined, the final data transformation can be executed. Aside from the transformations described above, it involves transformation from the moving wind vane based coordinate system to a fixed coordinate system. This transformation is trivial, however, Larsen and Busch (1974), and shall not be described here.

5. Discussion
The most important assumptions in the present analysis are that all wires were properly aligned and that the k-values used in [5] are correct.

Given the complexity of an error analysis concerning these assumptions, we have avoided such an analysis and rather considered a number of control quantities such as $\bar{u}_2$ and $\bar{u}_3$ (see Table 1) but also the spectral behaviour of the signals, being prepared to accept deviation from ideal behaviour on the 1 % level, and to discard the runs in case of more serious deviations. Fortunately, this has so far been necessary only once.

$$\rho_i$$

i	Estimated	Calculated
1	$5.61 \ 10^{-3}$	$7.94 \ 10^{-3}$
2	$5.64 \ 10^{-3}$	$7.68 \ 10^{-3}$
3	$6.4 \ 10^{-3}$	$8.01 \ 10^{-3}$

Table 2. Values of the ρ_i parameter that enters the equation [9] for the mean value of the outputs of the 3D-sensor, as estimated in the text and as calculated from the data used in Table 1.

Concerning determination of ρ_i in [9], the method may include a feed back of experimentally determined values into the calculation scheme. Due to the smallness and rather accurate estimate of ρ_i for the here considered low turbulence intensity studies, see Table 2, this has so far not been necessary. In studies of large turbulence intensity flows such a scheme is probably necessary.

References

Busch, N.E. (1977), Fluxes in the Surface Boundary Layer over the Sea. In *Modelling and Prediction of the Upper Layers of the Ocean.* Ed. E.B. Kraus. Pergamon Press, 72-91

Gaulier, C. (1977), Measurement of Air Velocity by Means of a Triple Hot-wire Probe. *DISA Information* No. 21, 16-20

Larsen, S.E. and Busch, N.E. (1974). Hot-wire Measurements in the Atmosphere, Part 1, Calibration and Response Characteristics, *DISA Information* No. 16, 15-34

Larsen, S.E. and Rasmussen, K.R. (1978). Comment on: Measurement of Air Velocity by Means of a Triple Hot-wire Probe. *DISA Information* No. 23, 4-5.

Directional Sensitivity of Cylindrical Hot-Film Probes in Liquids

by

H. Klages
Max-Planck-Institut für Strömungsforschung
D 34 Göttingen, Fed. Rep. of Germany

Abstract
Hot-film probes are mainly used to measure mean and fluctuating velocities in liquids. Until now, however, the directional sensitivity has not been investigated for all probes of interest. Only for wedge-shaped probes this has been done (see F.J. Resch, DISA Inf. No. 14, 1973). The present paper completes the knowledge about the directional sensitivity for cylindrical hot-film probes in liquids.

The effective cooling velocity, u_{eff}, of a hot-film sensor can be expressed by

$$u_{eff} = U(\cos^2\alpha + k^2 \sin^2\alpha)^{\frac{1}{2}}$$

Here $U\cos\alpha$ is the normal component of the velocity vector U with respect to the hot-film sensor axis and $U\sin\alpha$ the tangential velocity component. The influence of the tangential component is described by the so-called k-factor, which is a function of the velocity U and the yaw angle α.

The following results were obtained for two types of cylindrical DISA hot-film probes:
The k-factor for the standard probe 55F06 and for the slanted probe 55F07 is independent of the velocity U and approximately 0.7. For yaw angles $\alpha \lesssim 50°$ the k-factor does not depend on the yaw angle for both types of probes investigated. Only the slanted probe showed a dependence for yaw angles $\alpha \gtrsim 50°$.

1. Introduction
For the measurement of mean and fluctuating velocities in liquids, hot-film probes are generally used. For example, a probe having two cylindrical sensors arranged in an X-configuration is used to measure the streamwise and normal velocity components. The two sensors of these probes are inclined at an angle of 45° to the main direction of the flow. The directional sensitivity of hot-film probes with cylindrical sensors in liquids has not yet been investigated. The

directional sensitivity of the DISA probes 55F06 (standard probe) and 55F07 (slanted probe) was investigated in a laminar oil flow.

2. Method of Measurement

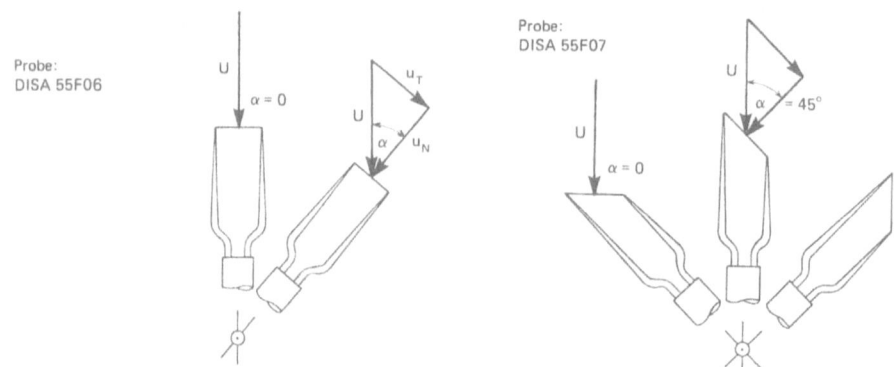

Fig. 1. Definition of yaw angle.

The experiments were carried out in an open flow channel with oil as the flow medium. The probes were investigated in a one-dimensional flow field. *Fig. 1* shows the geometrical relationships for the investigated probe types. Here U is the velocity vector, $u_T = U \sin\alpha$ is the tangential velocity component and $u_N = U \cos\alpha$ is the normal component of the velocity vector U. The yaw angle between the velocity vector U and the normal of the axis of the sensor is designated as α.

It is known, from hot-wire anemometry, that the effective cooling velocity u_{eff} is equal to the normal component of the velocity vector u_N and is in part dependent upon the tangential component u_T. The influence of the tangential component is the so-called k-factor which is defined by the equation

$$u_{eff} = U(\cos^2\alpha + k^2\sin^2\alpha)^{\frac{1}{2}} \qquad [1]$$

The directional sensitivity of the probes was investigated by determining the k-factor by two different measuring methods. The yaw angle α was measured optically in both methods. It was possible to measure the yaw angle α to within $1°$ accuracy by using a microscope with a $360°$ dial.

In the first method, the hot-film probe is investigated in a constant temporal and spatial velocity field. The k-factor is calculated by using the generalized King's law

$$E^2 - E_0^2 = B\ U^n \tag{2}$$

The experimentally determined exponent was found to be $n = \frac{1}{2}$. The graph of $(E^2 - E_0^2) = f(\sqrt{U})$ was found to be straight lines for the yaw angles $\alpha = 0°$ and $\alpha \neq 0°$; the slopes of these lines $B(\alpha = 0)$ and $B(\alpha \neq 0)$ are shown in *Fig. 2.* Combining Equation [1] and the ratio of the two slopes yields

$$\left[\frac{B(\alpha \neq 0)}{B(\alpha = 0)}\right]^2 = (\cos^2 \alpha + k^2 \sin^2 \alpha) \tag{3}$$

The k-factor is calculated from Equation [3], taking a given yaw angle α.

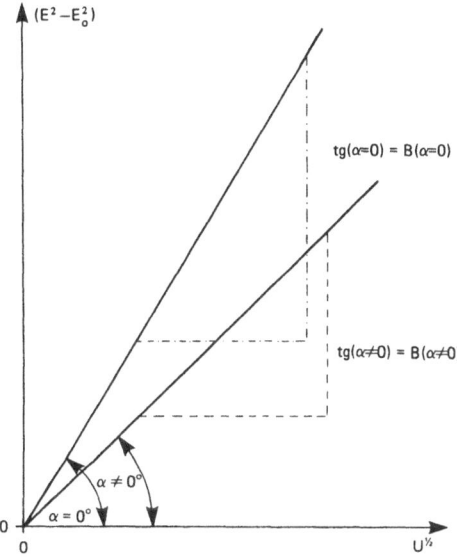

Fig. 2. Calibration curves for two different yaw angles α.

In the second method the probe is sinusoidally moved with an alternating velocity component $u_\sim$ in a one-dimensional flow field with constant velocity $U(u_\sim \ll U)$. This movement is produced mechanically by a sine generator in which the amplitude A and the angular frequency ω are set. The alternating velocity component is given by

$$u_\sim = -A\omega \sin\omega t \tag{4}$$

By measuring the r.m.s. values of the anemometer output voltage $\sqrt{\overline{E'^2}}$ and by taking the sensitivity $dE/du(\overline{u})$ derived from the static calibration curve of the investigated probes, the k-factor can be calculated according to the following relation

$$\left[\frac{\sqrt{2\overline{E'^2}}}{dE/du(\overline{u})}\right]^2 = (A\omega)^2 (\cos^2\alpha + k^2\sin^2\alpha) \qquad [5]$$

The k-factor can be determined by rewriting relation [5].

The second method will not be discussed in more detail here. However, this method is also well suited for investigating the directional sensitivity of cylindrical hot-film probes. A comparison of the two methods of measurement showed good agreement in the values obtained for the k-factor. For more details, see (1).

3. Results

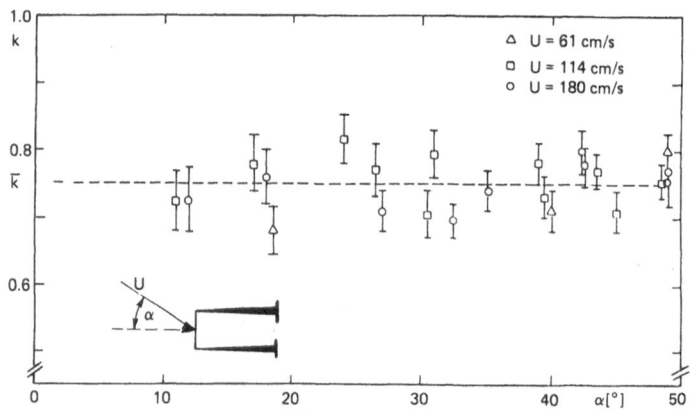

Fig. 3. k-factor as a function of yaw angle α for three different flow velocities for DISA probe 55F06.

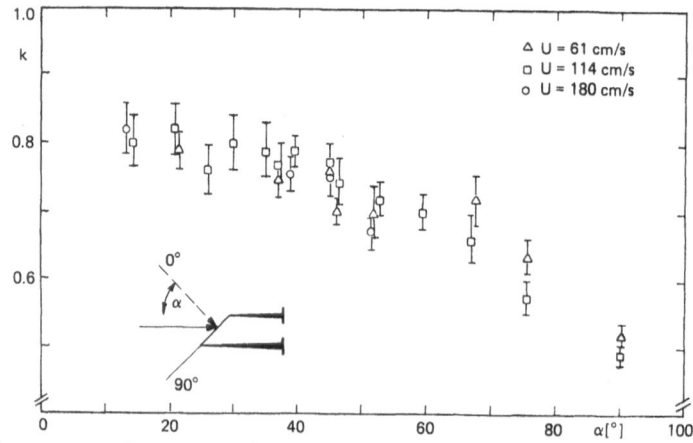

Fig. 4. k-factor as a function of yaw angle α for three different flow velocities for DISA probe 55F07.

The results shown in *Figs. 3 and 4* were obtained by the first method. Fig. 3 shows the k-factor for the standard probe 55F06 as a function of the yaw angle α for three different velocity vectors. The mean value k is approximately 0.75. From these measurements no dependency of the $\bar{k}$-factor can be observed for angles α up to approximately 50°. Nor can a dependency on the velocity vector be deduced from these measurements. The measured values for the k-factors lie in the range of 0.3 to 0.9 given by J.C. Mill and C.A. Sleicher (2). F.E. Jorgensen (3) found, for a cylindrical hot-film probe (DISA 55F26) in air, a value of approximately 0.45 at flow velocities between 12-31 m/sec.

For the slanted probe 55F07 (see Fig. 4), the k-factor was measured for angles from 0° - 90°. In the range of α = 0° up to approximately 45°, the k-factor is constant at approximately 0.78. In the regular position of the probe (α = 45°), the k-factor is 0.75, as it is with the standard probe. For $\alpha \gtrsim 45°$, the k-factor decreases as was also found by F.E. Jorgensen (3). This decrease can be explained by the influence of the sensor prongs. As the yaw angle increases, the sensors enters the wake of its prongs, which changes the flow characteristics at the sensor.

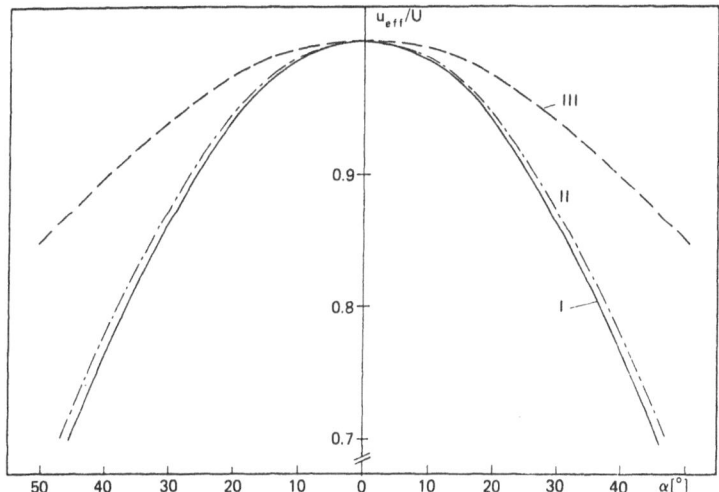

Fig. 5. Directional characteristic for hot wire (II) and a DISA Type 55F06 probe (III); for comparison curve (I) shows a cosine function.

With known k-factor, a directional characteristic (see *Fig. 5*) can be plotted. Here the relation u_{eff}/U is shown as a function of the yaw angle. Curve I shows the Cosin Function (k = 0). Curve III shows the directional characteristic for the DISA standard probe.

Curve II is a typical directional characteristic for a hot-wire with k = 0.2.

With cylindrical hot-film probes, the part of the cooling by the tangential component of the velocity vector u_T is considerably greater than it is with a hot-wire. One reason for this is the very much smaller length-to-diameter ratio (ℓ/d). Characteristic values can be regarded here as $\ell/d \approx 20$ for a hot-film and $\ell/d \approx 200$ for a hot-wire.

4. Summary

In a one-dimensional flow, DISA probes of types 55F06 and 55F07 were investigated. For the standard probe, a k-factor of 0.75 was found in the range of the velocity vector U from 0.6 - 1.8 m/sec. It is independent of U, and in the angle range $0° \leqslant \alpha \leqslant 50°$ it is independent of α. For the slanted probe a value of 0.78 was found for the range $0° \leqslant \alpha \leqslant 40°$. At $\alpha = 45°$ (regular position of the probe), the k-factor is 0.75, as it is with the standard probe. For $\alpha \geqslant 45°$, it decreases continuously until it reaches a value of $k \approx 0.5$ at approximately 90°.

5. References

(1)   Klages, H.: Beiträge zur experimentellen Untersuchung und Anwendung von Heißfilm-sonden. Max-Planck-Institut für Strömungsforschung, Bericht Nr. 14/1977, Göttingen, Juli 1977

(2)   Hill, J.C. and Sleicher, C.A.: Equations for errors in turbulence measurements with inclined hot wires. *Phys. Fluids,* **12** (1969) 1126-1127

(3)   Jorgensen, F.E.: Directional sensitivity of wire and fiber-film probes. *DISA Information,* Nr. **11** (1971).

Electrochemical Method for Dynamic Measurements in Two-Phase Flow

by

M. Souhar, G. Cognet

LEMTA - INPL, Nancy Cedex, France

Two phase flows have a great importance in engineering sciences for which transport phenomena are essential such as in petroleum, nuclear and chemical engineering, so a more precise study of these flows is very needed. Knowledge of wall shear stress is of great interest for elaboration of valuable models particularly in pressure drop prediction. This work gives results of friction obtained in bubble and plug flow using an electrochemical technique: the polarography (1), (2).

I - Experimental Conditions

1. Apparatus (Fig. 1a, Fig. 1b)

The test column is a vertical plexiglass pipe (internal diameter D = 44 mm, length 7300 mm) with a profiled entrance at its lower part and elements including probes for friction measurement in its upper part with temperature control and a device for pressure drop measurement in two-phase flow (7).

The circuit of liquid essentially contains:

— a stock vessel with temperature regulation
— a pump insuring a flow up to 30 m^3/h and a flow meter in the coming duct
— a vessel before the test column and a vessel for gas-liquid separation at the exit

Gas is provided from high pressure bottles of Nitrogen (200 Atm). After a double expansion the flow rate is measured through a sonic nozzle. Feeding of the column is made axially through a conical injector (37 holes, ϕ 0.5 mm).

2. Experimental Technics

The use of double wall probes (3) allows local and instantaneous measurement of friction: τ (sign and modulus).

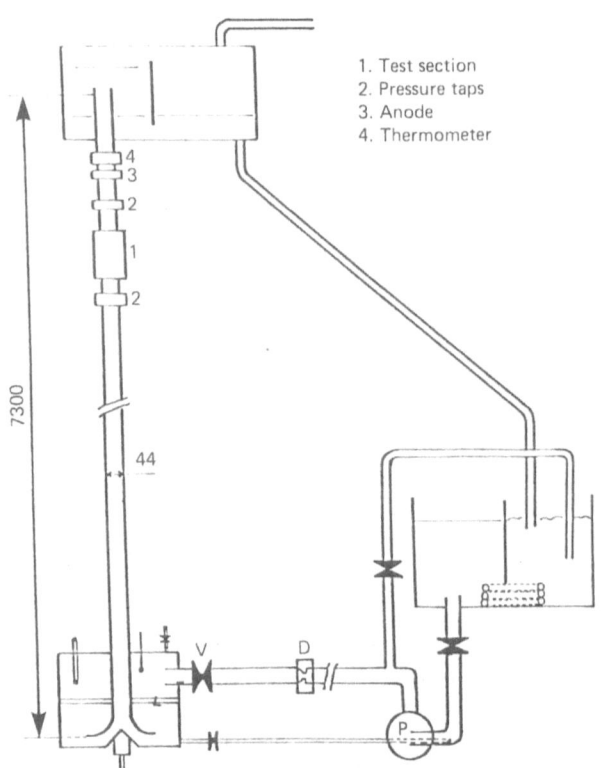

1. Test section
2. Pressure taps
3. Anode
4. Thermometer

Fig. 1a. Liquid circuit

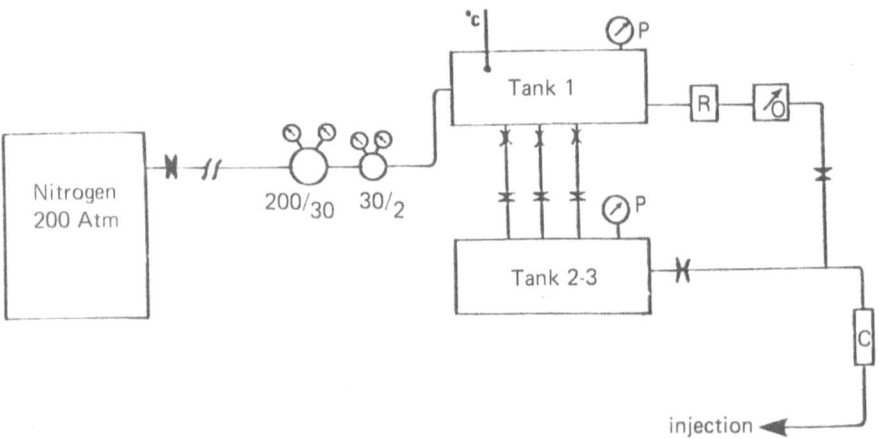

Fig. 1b. Gas circuit

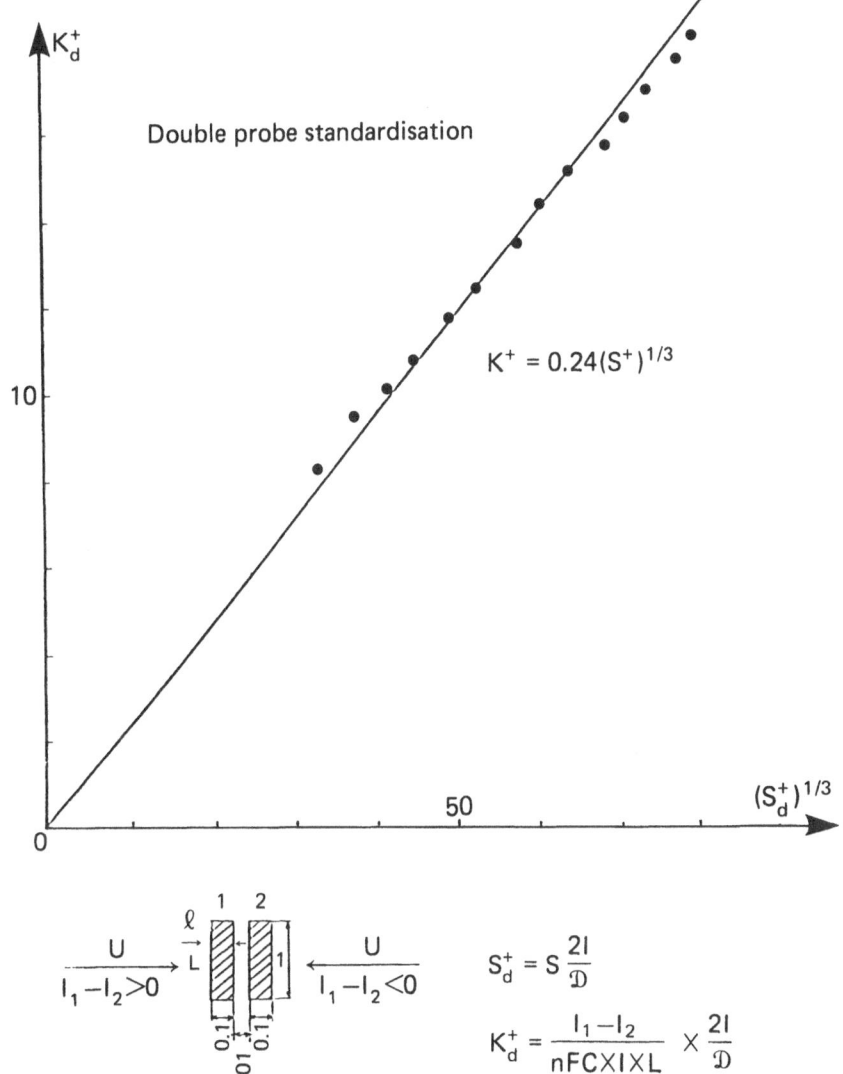

Fig. 2.

The transducer law is: $\tau \sim (I_1 - I_2)^3$, $I = I_1 - I_2$ being the difference of intensity measured on each element of the probe.

The liquid used is a water solution of potassium ferricyanide ($C_9 = 3.10^{-6}$ mole/cm^3) with an excess of potassium chloride (3.10^{-3} mole/cm^3). At 18°C $\rho = 1.016$ g/cm^3, $\sigma = 59$ dynes/cm, $\nu = 1.06 \ 10^{-2}$ cm^2/s. The probe sizes and standardisation curve are given in Fig. 2.

The flow velocities for liquid and gas can vary respectively from 15 to 135 cm/s and from 2 to 375 cm/s corresponding to bubble and plug flows. The direct observation of different flows is in good agreement with the map proposed by Griffith and Wallis (4).

II - Results

1. Flow Characterisation

a) Bubble flow
This flow is obtained for low rates of gas, the signal received from a double probe (Fig. 3a) is near from that observed for a turbulent flow, as it can be seen by comparison of the spectras. However, the energy level seems to be more important for the low frequencies in bubble flow. The histogram of amplitude (Fig. 3c) shows always a direct flow with a little exception for $U_L = 6.80$ cm/s.

b) Plug flow
This flow is obtained for higher flow rates of gas. The signal given by the double probe shows alternative direct and counter flows characteristic of each plug passage (Fig. 4a - 4b). The flow in the film appears to be essentially counter current. The spectra (Fig. 4c) gives the plug frequency.

2. Wall Friction Measurements
An experiment is made with a constant liquid flow rate Q_L and variable gas flow rate Q_G. In each case mean value $\overline{I^3} \sim \overline{\tau}$ is measured and compared to $\overline{I_0^3} \sim \overline{\tau}_0$ mean value of the corresponding one phase flow (Q_L, $Q_G = 0$). The curves $\tau/\tau_0 = f(U_{GS})$, with U_L parameter, are presented in Fig. 5 with $U_{GS} =$ mean gas velocity in the test section, $U_L =$ mean liquid velocity, $\overline{\tau}$ and $\overline{\tau}_0$: mean friction, respectively in two phase and one phase flow for the same U_L. The curves have different shapes according to the liquid flow rate. For $U_L > 40$ cm/s the mean friction can be negative for low gas flow rates.

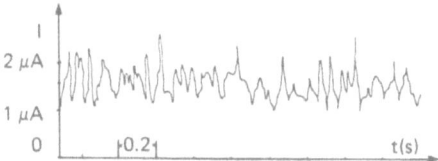

Fig. 3a.

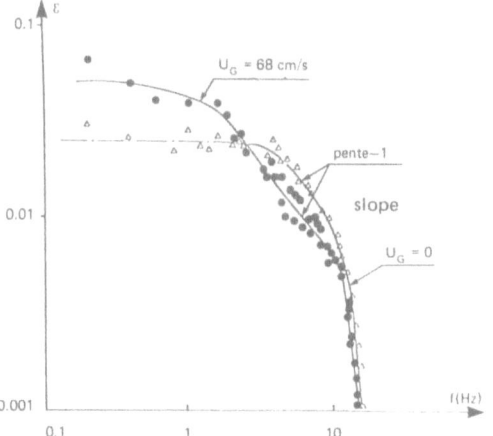

Fig. 3b.

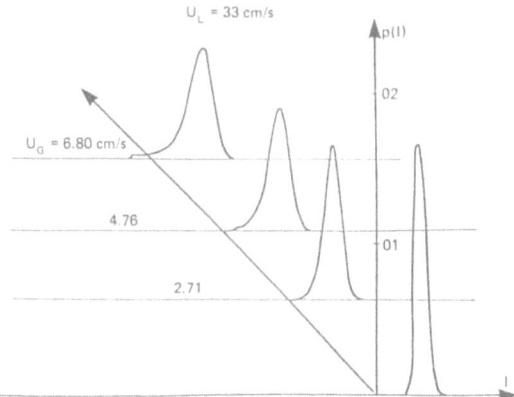

Fig. 3c.

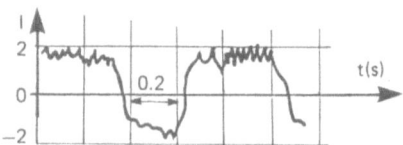

Fig. 4a.

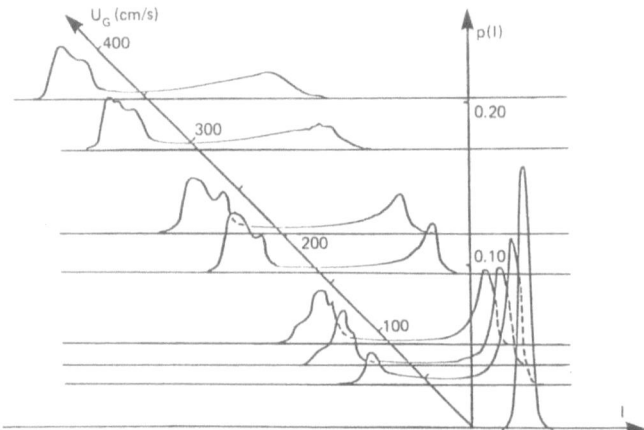

Fig. 4b.

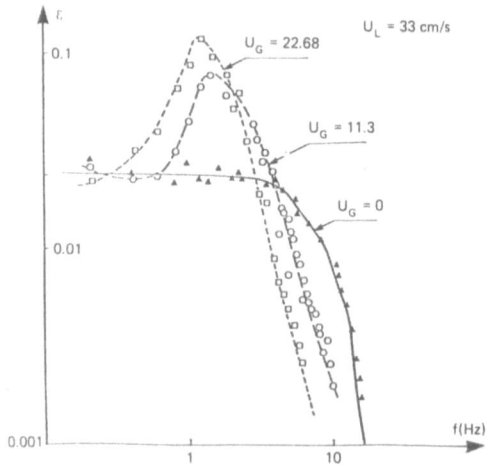

Fig. 4c.

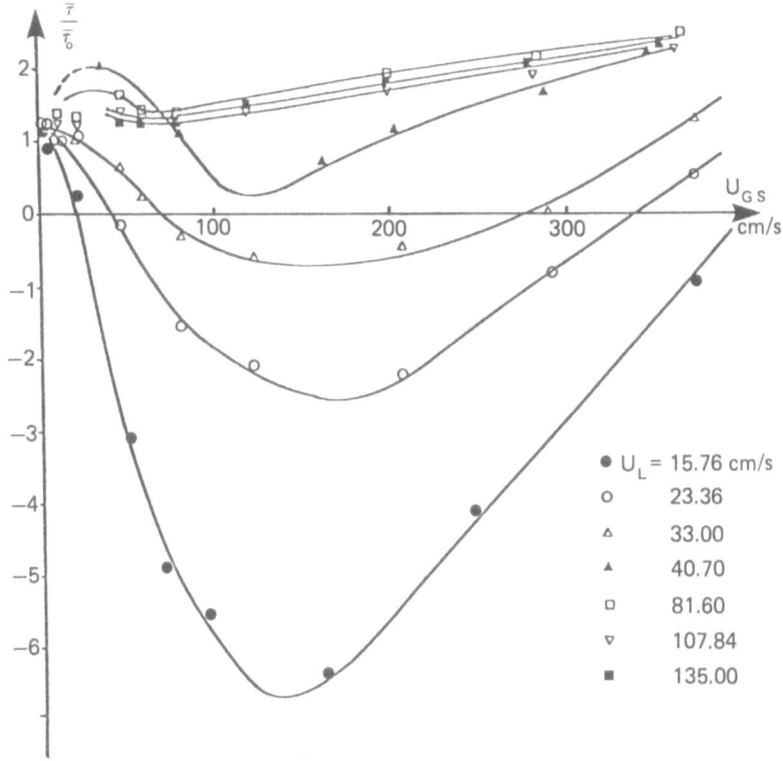

Fig. 5a.

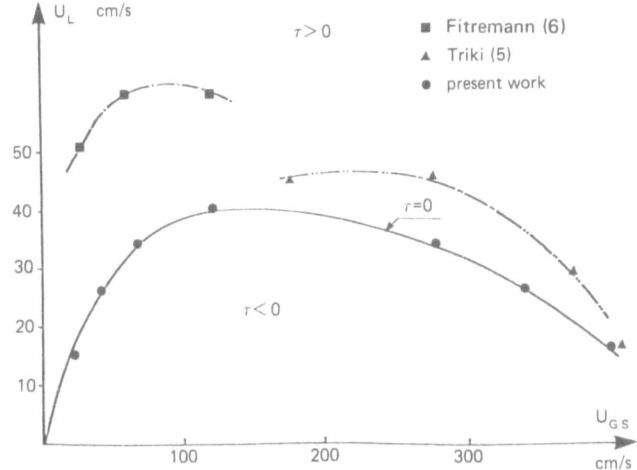

Fig. 5b.

For $U_L > 40$ cm/s, the mean friction is always positive, with relative maximum and minimum according to the appearance and growth of the plug flow. All the curves are increasing for large values of U_{GS}.

For $U_L < 40$ cm/s each curve has a zero friction point (Fig. 5b) defining, in the plane $(U_{GS}; U_L)$ two regions characterised by positive or negative value of $\bar{\tau}$. The agreement with recent results (5) is rather good.

III - Balance of Momentum

The balance of momentum calculation for evaluating the validity of the measurements can be written (7):

$$-\frac{\partial p}{\partial z}\bigg|_t = \frac{4\bar{\tau}}{D} + [\alpha \rho_G + (1-\alpha)\rho_L]g + (\frac{\partial p}{\partial z})_a + (-\frac{\partial p}{\partial z})_{int}$$

pressure gradient   friction   gravity   acceleration   interface

For the velocities considered in the work, the acceleration term can be neglected. The direct total pressure gradient measurement $(-\partial p/\partial z)_t$ and the mean friction, from one part, and evaluation of gravity forces from the other part $(\alpha = U_{GS}/(1.2(U_L + U_{GS}) + 0.35 \sqrt{gD})$sic Nicklin (6)) satisfy the balance of momentum better than 2.5 % in bubble flow. The interface term in that case is of the order of the measurement uncertainty. For example: for $U_L = 135$ cm/s; $U_{GS} = 23$ cm/s we obtain:

$$-\frac{\partial p}{\partial z}\bigg|_t = 9230 \text{ P/m}; \quad -\frac{\partial p}{\partial z}\bigg|_a \cong 0; \quad \frac{4\bar{\tau}}{D} = 430 \text{ P/m}; \quad -\frac{\partial p}{\partial z}\bigg|_{gravity} = 8900 \text{ P/m}$$

$$-\frac{\partial p}{\partial z}\bigg|_{interface} = -100 \text{ P/m}$$

In plug flow such a verification is difficult due to the inaccuracy in the total pressure gradient measurement.

These results show that the use of double electrochemical probes gives correct values for the wall friction in two phase flow, in spite of the changes in flow direction observed in many cases.

For elaboration of more complete models of flow the results can be improved by more precise measurement of the total pressure gradient with void fraction determined by probes placed in the column.

References

(1)   COGNET: Thèse (1968 NANCY)
(2)   LEBOUCHE - COGNET: Chimie Industrie — génie chimique Vol 97 no12 (1967)
(3)   LABBE: Thèse de docteur ingénieur (1975 NANCY)
(4)   HSU-GRAHAM: Transport Processes in Boiling and Two-Phase Systems (1976)
(5)   TRIKI: Thèse de docteur ingénieur (1978 PARIS)
(6)   FITREMANN: Thèse de doctorat d'état (1977 PARIS)
(7)   TRUONG QUANG MINH: Thèse de Docteur Ingénieur (1965 GRENOBLE)

Special Problems

Comments on the Statistical Equations of Turbulent Flows and Experiments

by

A. Favre
I.M.S.T., Marseille, France

Let us recall briefly the methods which can be used concerning the analytical expressions of the statistical equations of fluids in turbulent flows in the general case when the density is randomly variable, and in relation to the experimental magnitudes of the terms which are making the differences of the formulations in various fields of applications.

1. Survey of the Equations

1.1 Separation Methods
In the case when the specific mass, or density ρ, is constant, the statistical equations have been derived by BOUSSINESQ and REYNOLDS by averaging the Navier-Stokes equations, using the method which we shall call "A", by separation of each random quantity w into a mean value $\bar{w}$ and a fluctuating part w'':

$$\text{"A"} \quad w \equiv \bar{w} + w'' \Rightarrow \overline{w''} \equiv 0, \quad \overline{\rho w''} \neq 0, \tag{1}$$

w being the average of w, by averaging in time for application to stationary flows.

For the extension of applications to new fields of turbulent flows, where ρ is random, some authors wrote the equations by the same method "A", but it led to complicated equations that are difficult to use for analytical treatment and difficult to interpret. In addition, because of these difficulties it happens that some terms are neglected which are not always negligible.

Another method which we shall call "B", has been employed by meteorologists HESSELBERG (1926), VAN MIEGHEM (1949), BLACKADAR (1950), wherein the velocity is separated into a mass-weighted average macroscopic velocity and a fluctuating part. This method has been extended by FAVRE (1958) to the case of a compressible gas experiencing turbulent fluctuations of velocity, density, pressure, internal energy and temperature as well as viscosity heat conductivity and specific heats. Each turbulent quantity w is separated into a mean mass-weighted quantity $\widetilde{w}$ and a fluctuating part w', such that:

"B" $w \equiv \widetilde{w} + w'$ with $\widetilde{w} \equiv \overline{\rho w}/\overline{\rho} \Rightarrow \overline{\rho w'} \equiv 0, \quad \overline{w'} \equiv -\dfrac{\overline{\rho' w'}}{\overline{\rho}},$ [2]

and an extension to nonstationary flows is made by using ensemble averages, more general than time averages, by FAVRE (1958).

The resulting equations have a simpler form and a clearer physical interpretation.

At that moment the question was: the method "B" seems to be better than the method "A", but is it the best one? Then I developed the equations FAVRE (1965, 1969, 1972, 1975), and the most complete text is given by FAVRE (1977), by a more general method called "G" in which each turbulent quantity w is also separated into a macroscopic part, $\overset{*}{W}$, and a fluctuating part, $\overset{*}{w}$

$$w \equiv \overset{*}{W} + \overset{*}{w},$$ [3]

but in which the second equation for the complete definition of the macroscopic part $\overset{*}{W}$:

$$\overset{*}{W} \equiv \ldots ?,$$ [4]

is retained for a later stage. The only requirement is that $\overset{*}{W}$ must not be a random variable, then the equation [4] must fulfill the condition:

$$\overset{*}{W} = \overline{\overset{*}{W}}$$ [5]

This condition implies that, if h is a random quantity:

$$\overline{\overset{*}{W}h} = \overline{\overset{*}{W}h} = \overline{\overset{*}{W}}\,\overline{h} = \overset{*}{W}\,\overline{h},$$ [6]

which permits the macroscopic quantities to be extracted from the averages and to write the statistical equations in a generalized form "G".

These equations then can be used as guides for the discussion and the choice of the definitions of the macroscopic parts $\overset{*}{W}$, with the available equations [4].

For instance the velocity components u_k are separated into:

$u_k \equiv \overset{*}{U}_k + \overset{*}{u}_k$ with the condition $\overset{*}{U}_k = \overline{\overset{*}{U}}_k$ and the equation $\overset{*}{U}_k \equiv \ldots?$ has to be chosen later.

Now, by definition the macroscopic material derivative is:

$$\frac{\overset{*}{D}}{Dt}(\ ) \equiv \frac{\partial}{\partial t}(\ ) + \overset{*}{U}_k \frac{\partial}{\partial x_k}(\ ).$$

1.2 Reference Volumes

We use first a reference volume (Ω) bounded by a fixed closed surface Σ (Fig. 1), with $\mathbf{n}$ as the unit vector normal to the surface having the components ℓ_k.

We use also a reference volume $(\overset{*}{\tau})$, bounded by a closed surface $\overset{*}{C}(t')$, which is moving at the macroscopic velocity $\overset{*}{U}$ of the fluid and not at the velocity $\overset{*}{u}$ because the turbulent diffusion would give the surface a very complicated shape $C'(t')$ and make it impracticable to integrate over either that surface or the volume bounded by it. The moving surface $\overset{*}{C}$ has a unit normal vector $\overset{*}{m}(\lambda_k^*)$ (Fig. 2). The macroscopic material derivative of the volume integral is then:

$$\frac{\overset{*}{D}}{Dt} \int_{\overset{*}{\tau}} (\ ) \, d\overset{*}{\tau}.$$

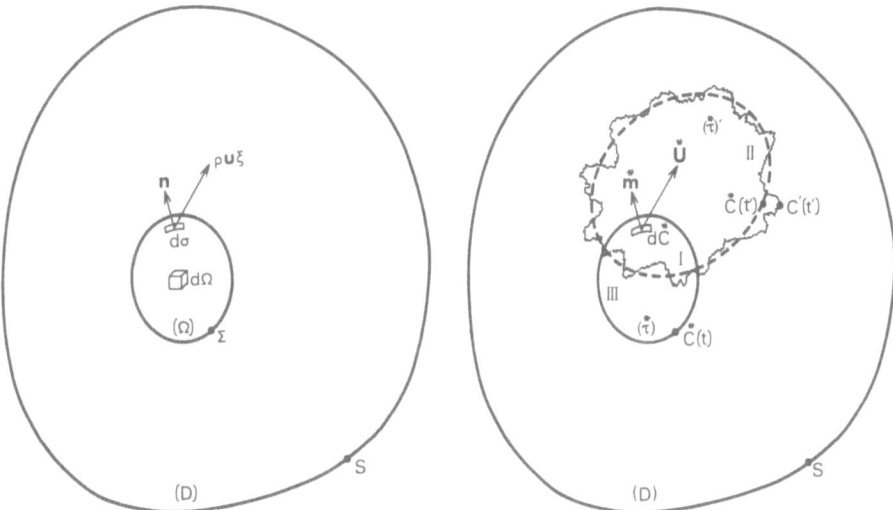

Fig. 1. Fixed reference volume.   Fig. 2. Moving reference volume.

1.3 Statistical Balance of Transportable Quantities

1.3.1. Transportable Quantities
The statistical equations are expressing the mean balance of mass, momentum and energy into the reference volumes.

It can be expressed by the unique form of the balance of a transportable quantity $\xi(x,t)$ per unit mass, ξ being a scalar of a component of a vector.

Then it must be emphasized that *the fundamental quantity for transportation of ξ by the fluid flow is not the velocity* u, *but the mass flux* ρu

The flux of ξ through the surface element is proportional to the mass flux, i.e.:

$$\rho u \xi n d\sigma \tag{7}$$

1.3.2. Conservation of Mass
When $\xi \equiv 1$, we obtain the balance of mass. Into the fixed infinitesimal volume element we have the equation:

$$\frac{\partial \bar{\rho}}{\partial t} + \frac{\partial}{\partial x_k} (\bar{\rho}\overset{*}{U}_k + \overline{\rho u_k}) = 0, \tag{8}$$

or into the moving volume $(\overset{*}{\tau})$ the equation:

$$\frac{\overset{*}{D}}{Dt} \int_{\overset{*}{\tau}} \bar{\rho} d\overset{*}{\tau} = \boxed{-\int_C \overline{\rho u_k} \overset{*}{\lambda}_k d\overset{*}{c}} \tag{9}$$

The equation [9] shows that, in general *the mean mass is not constant* into the volume $(\overset{*}{\tau})$ moving at a macroscopic velocity $\overset{*}{U}$, because there is a *mean mass flux* through the surface $\overset{*}{C}$ *due to turbulence* and expressed by the term $\overline{\rho u_k}$.

As a consequence, the surfaces defined by the macroscopic stream lines, tangent to the macroscopic velocity $\overset{*}{U}$ are not impenetrable, even with respect to the mean value. Then it is loosing the usual physical meaning of streamlines.

The only way to obtain the conservation of mass in mean value into the moving volume, and to have significant macroscopic streamlines impenetrable in the sense of the mean value, is to make the choice of three equations [4] such that:

$$\overline{\rho u'}_k \equiv 0 \Rightarrow \overline{\rho u_k} \equiv \bar{\rho}\tilde{u}_k, \tag{10}$$

which is the method "B".

Then the surface C is moving at the mean mass-weighted $\tilde{u}$ and equation [9] becomes

$$\frac{D}{Dt}\int_\tau \bar{\rho}\,d\tau = 0, \tag{11}$$

thus the mean mass is constant into the moving volume τ, bounded by surface C.

1.3.3. Balance of Transportable Quantities, Generalized Statistical Equations

Let us separate the transportable quantity ξ into a macroscopic part $\overset{*}{Z}$ and a fluctuating part $\overset{*}{\xi}$:

$$\xi \equiv \overset{*}{Z} + \overset{*}{\xi} \text{ with the condition } \overset{\tilde{*}}{Z} = \overset{\tilde{*}}{Z}. \tag{12}$$

We obtain the balance of the macroscopic part $\overset{*}{Z}$ at one point:

$$\bar{\rho}\frac{D\overset{*}{Z}}{Dt} + \boxed{\frac{D}{Dt}\overline{\rho\overset{*}{\xi}}} + \boxed{\overline{\rho\overset{*}{\xi}\frac{\partial\overset{*}{U}_k}{\partial x_k}}} + \boxed{\overline{\rho\overset{*}{u}_k\frac{\partial\overset{*}{Z}}{\partial x_k}}} + \frac{\partial}{\partial x_k}\overline{\rho\overset{*}{u}_k\overset{*}{\xi}} = \mathfrak{D}(\rho\xi). \tag{13}$$

$\mathfrak{D}(\rho\xi)$ being the rate of local creation or destruction of $\rho\xi$ by unit volume. And the balance in the moving volume $(\overset{*}{\tau})$ is:

$$\frac{\overset{*}{D}}{Dt}\int_{\overset{*}{\tau}} (\bar{\rho}\overset{*}{Z} + \boxed{\overline{\rho\overset{*}{\xi}}})\,d\overset{*}{\tau} = \int_{\overset{*}{\tau}}\overline{\mathfrak{D}(\rho\xi)}\,d\overset{*}{\tau} - \int_C (\boxed{\overline{\rho\overset{*}{u}_k}}\overset{*}{Z} + \overline{\rho\overset{*}{u}_k\overset{*}{\xi}})\lambda_k\,d\overset{*}{C}. \tag{14}$$

We obtain also the equations for variances:

$$\frac{\overset{*}{D}}{Dt}\overline{\rho\overset{**}{\xi\xi}} + \overline{\rho\overset{**}{\xi\xi}\frac{\partial\overset{*}{U}_k}{\partial x_k}} + \frac{\partial}{\partial x_k}\overline{\rho u_k\overset{**}{\xi\xi}} + 2\overline{\rho u_k\overset{*}{\xi}\frac{\partial\overset{*}{Z}}{\partial x_k}} + 2\overline{\rho\overset{*}{\xi}\frac{\overset{**}{DZ}}{Dt}} = 2\overline{\overset{*}{\xi}\mathfrak{D}(\rho\xi)}. \tag{15}$$

and the equations for turbulent fluxes, taking into account the gravity forces g_i and the effects of rotation T_k of the axis:

$$\frac{\overset{*}{D}}{Dt}\overline{\rho\overset{*}{u}_i\overset{*}{\xi}} + \overline{\rho\overset{*}{u}_i\overset{*}{\xi}\frac{\partial\overset{*}{U}_k}{\partial x_k}} + \overline{\rho\overset{*}{u}_k\overset{*}{\xi}\frac{\partial\overset{*}{U}_i}{\partial x_k}} + \overline{\rho\overset{*}{u}_k\overset{*}{u}_i\frac{\partial\overset{*}{Z}}{\partial x_k}} +$$

$$\frac{\partial}{\partial x_k}(\overline{\rho\overset{*}{u}_k\overset{*}{u}_i\overset{*}{\xi}} - \overline{\overset{*}{\xi}f_{ik}} + \overline{\overset{*}{\xi}p'\delta_{ik}}) + \boxed{\overline{\rho\overset{*}{\xi}\frac{\overset{**}{DU}_i}{Dt}}} + \boxed{\overline{\rho\overset{*}{u}_i\frac{\overset{**}{DZ}}{Dt}}} = \tag{16}$$

$$\boxed{\overline{\rho\overset{*}{\xi}g_i}} - 2\epsilon_{ik\ell}T_k\overline{\rho u_\ell\overset{*}{\xi}} + \overline{p'\frac{\partial\overset{*}{\xi}}{\partial x_i}} - \overline{f_{ik}\frac{\partial\overset{*}{\xi}}{\partial x_k}} + \overline{\overset{*}{u}_i\mathfrak{D}(\rho\xi)} - \overline{\overset{*}{\xi}\frac{\partial p}{\partial x_i}},$$

where f_{ik} is the viscous stress tensor.

Looking now at these equations, we see that the method "B" applied to velocity [10] is again making simplifications by cancelling by definition the terms
in the frames which include $\overline{\rho \overset{*}{u}_k}$.

In addition, for the momentum equations, where $\rho \xi \equiv \rho u_i$ the method "B"
will cancel all the terms in the frames of equations [13], [14], [15], [16].

This will make strong simplifications in the equations of motion which have
then the same form when ρ is variable and constant. Also the equations for
kinetic energy of macroscopic motion, kinetic energy of turbulence, Reynolds
stresses, vorticity, entropy, helicity, and concentration of a binary mixture
are simplified. If we apply now the same method "B" to internal energy, temperature, enthalpy, total enthalpy, entropy, all the terms in the frames of
equations [13], [14], [15], [16], will vanish by definition because, for
balance in a volume of such quantities, it is more logical to retain the *values by
unit volume* $\rho \xi$ than the values by unit mass ξ. For total enthalpy these simplifications permit the integration of the equations along each mass-weighted
streamline.

Then the method "B" is making great simplifications in the analytical expressions of all the transport terms of the statistical equations. It is also clearing
up the physical meaning of the terms, for instance the equation [13] for the
macroscopic quantities becomes:

$$\bar{\rho} \frac{D\tilde{\xi}}{Dt} + \frac{\partial}{\partial x_k} \overline{\rho u'_k \xi'} = \overline{\mathcal{D}(\rho \xi)}. \tag{17}$$

These terms represent the average rates of variation of $\rho \xi$ per unit time and
volume: (I) by the variation of $\bar{\rho} \tilde{\xi} \equiv \overline{\rho \xi}$ following the mean mass-weighted
motion and taking into account the macroscopic changes of volume and the
mass conservation, (II) by turbulent diffusion which represents then *all* the
transport effects of turbulence, (III) the rate of local creation or destruction
of $\rho \xi$.

In the case of entropy, there is a special difficulty because when separating
the entropy per unit mass s into a macroscopic part $\overset{*}{S}$ and a fluctuating part
$\overset{*}{s}$:

$$s \equiv \overset{*}{S} + \overset{*}{s}, \text{ with } \overset{*}{S} = \overline{\overset{*}{S}}, \tag{18}$$

it is desirable that the macroscopic part $\overset{*}{S}$ of the entropy has the physical
properties of the entropy s.

Now, the equation [14] gives for entropy:

$$\frac{\overset{*}{D}}{Dt} \int_{*\tau} (\bar{\rho}\overset{*}{S} + \boxed{\overline{\overset{*}{\rho s}}})d\overset{*}{\tau} - \int_{*C} \left(\overline{\frac{\lambda}{\theta}\frac{\partial\theta}{\partial x_k}} - \boxed{\overline{\rho\overset{*}{u}_k}}\overset{*}{S} - \overline{\rho\overset{*}{u}_k\overset{*}{s}} \right) \overset{*}{\lambda}_k d\overset{*}{C} = \qquad [19]$$

$$\int_{*\tau} \left(\overline{\left(\frac{\bar{\varphi}}{\theta}\right)} + \overline{\frac{\lambda}{\theta^2}\frac{\partial\theta}{\partial x_k}\frac{\partial\theta}{\partial x_k}} \right) d\overset{*}{\tau} \geqslant 0,$$

where θ is the temperature, λ the conductivity, φ the dissipation. This is the expression of the second law of thermodynamics concerning the mean entropy

$$\overline{\rho s} \equiv \bar{\rho}\overset{*}{S} + \overline{\rho\overset{*}{s}}.$$

But the evolution of the terms $\overline{\rho\overset{*}{s}}$ is unknown. Thus the only solution for the condition that the macroscopic part of entropy has the property expressed by that second law is that the unknown part $\overline{\rho\overset{*}{s}}$ is zero by definition. This leads again to the "B" method:

$$s \equiv \tilde{s} + s' \text{ with } \overline{\rho s} \equiv \overline{\rho\tilde{s}} \Rightarrow \overline{\rho s'} \equiv 0, \qquad [20]$$

which also cancels the terms in frames in equation [19].

Then the mean mass-weighted entropy $\tilde{s}$ times $\bar{\rho}$, which is equal to the mean entropy into the moving volume τ (which contains a constant mean mass) minus the mean flux of entropy through the surface C by heat conduction and by turbulent diffusion, cannot decrease.

1.4 Equation of State

As for the equation of state for a perfect gas, without phase change, and within its domain of validity, let us separate the temperature θ into a macroscopic part $\overset{*}{\Theta}$ and a fluctuating part $\overset{*}{\theta}$, with $\overset{*}{\Theta} = \overset{*}{\Theta}$. The statistical equation in general form "G" reads

$$\bar{p} = R(\bar{\rho}\overset{*}{\Theta} + \overline{\rho\overset{*}{\theta}}). \qquad [21]$$

where R is the gas constant.

One can see that, in order to keep for the statistical equation of state the same form as for the macroscopic equation of state, the solution is again the method "B"

$$\theta \equiv \tilde{\theta} + \theta' \text{ with } \overline{\rho\theta} \equiv \overline{\rho\tilde{\theta}} \Rightarrow \overline{\rho\theta'} \equiv 0, \qquad [22]$$

the equation of state then reads:

$$\bar{p} = R\bar{\rho}\tilde{\theta}.$$ [23]

2. Experimental Comparisons Between the Methods "A" and "B"

One can of course use either the methods "A" or "B", which are both correct. But when the method "A" is used, there are many terms like $\overline{\rho'u_k''}$, $\overline{\rho'\xi}$, and so on, which are not zero by definition, and which cannot be neglected unless a check is made in *each case* proving that these terms are negligible in the equations. This must take into account their magnitude relative to the other analogous terms and the fact that their gradients have to be considered.

This is why I gave (FAVRE (1975, 1977)) formulas helping to compute the differences between the main terms of the equations by methods "A" and "B", i.e. the relative errors in the case when, using method "A" the terms like $\overline{\rho'u_k''}$, $\overline{\rho'\xi}$ and others would be neglected.

There is also in Fig. 3 an indicative list of numerical values measured or estimated in the case of different types of flows in air: isovolumetric, incompressible with small heat transfer, incompressible with strong heat transfer, supersonic, and atmospheric surface layer.

In this paper are given in addition some values, following R.W. BILGER (papers and private communication: BILGER (1975, 1977)) and others concerning binary mixtures of helium air (STANFORD and LIBBY (1974), helium-nitrogen (BROWN and ROSHKO (1974) and ROSHKO (1976)); and also combustion of hydrogen by diffusion flame into air (GLASS and KENT (1977)) and premixed flames (BRAY and LIBBY (1976)). Experiments have also been made by BONNIOT, C. and BORGHI, R. (1977).

One can notice first that even for low speed flows with small heat transfer i.e. air boundary-layer at 11.9 ms^{-1}, $\Delta\bar{\theta}$ = 21°K, Fig. 4 (VEROLLET (1972)), there is an *important difference* between the mean velocity $\bar{u}_2$ and the mean mass-weighted velocity $\tilde{u}_2$ for the components *in the gradient direction*, because it is the same as for the longitudinal direction *multiplied by the ratio* $||\bar{u}||/|\bar{u}_2|$ of the total velocity to the transverse component, which is *very large in quasi parallel flows*.

The real mean mass flux $\overline{\rho u_2}$ is given by $\bar{\rho}\tilde{u}_2$ by method "B". With method "A" one should take account the mean mass flux due to turbulence $\overline{\rho'u_2''}$

Quantities	Isovolume $\rho = c^{te}$	Air incompressible flow heat transfer		Air supersonic Mach 1.5 to 4	Atmospheric surface layer stable-unstable	Binary Mixture		Combustion	
		$\Delta\overline{\theta}/\overline{\theta}\sim$0 to 0.1	$\Delta\overline{\theta}/\overline{\theta}\sim$-0.1 to 0.5			Porous pipe helium-air	Mixing layer helium-nitrogen	Diffusion flames, hydrogen-air	Premixed flames, hydrogen-air
Velocity $(\overline{u_k'^2})^{1/2}/|\overline{u}|$	0,30	0,30	0,30	0,10	0,30	0,07 to 0,22	0,30	0,50 to 0,70	0,50 to 0,70
Density $(\overline{\rho'^2})^{1/2}/\overline{\rho}$	0	0 to 0,03	0,03 to 0,15	0,02 to 0,15	0,003 to 0,01	0,18	0,44	~1,30	~1,30
Temperature $(\overline{\theta'^2})^{1/2}/\overline{\theta}$	0	0 to 0,03	0,03 to 0,15	0,02 to 0,15	0,03 to 0,01	0	0	—	—
Concentration $(\overline{c'^2})^{1/2}/c$	0	0	0	0	0	0,60	0,53 to 1,00	~1,10	~2,00
Scalar fluctuations $\|\overline{\xi''}\xi'\|/(\overline{\xi'^2})^{1/2}=\overline{\rho'\xi'}/\overline{\rho}(\overline{\xi'^2})^{1/2}$	0	0 to 0,03	0,03 to 0,15	0,02 to 0,15	0,003 to 0,01	0,17	0,32 to 0,80	1,00	1,00
Scalar turbulent fluxes $(\rho\overline{u_k'\xi'} - \overline{\rho'u_k''\xi''})/\overline{\rho u_k''\xi''}$	0	0 to 0,001	0,001 to 0,02	0,0004 to 0,02	10^{-5} to 10^{-4}	0,03	~0,10	0,20	0,20
Concentration $\overline{c'c}/c \equiv \overline{\rho'c'}/\overline{\rho}c$	0	0	0	0	0	0,10	0,16 to 0,80	1,10	2,00
Temperature $\|\overline{\theta}-\overline{\theta}\|/\overline{\theta} \equiv \|\overline{\rho'\theta'}\|/\overline{\rho}\overline{\theta}$	0	0 to 0,001	0,001 to 0,02	0,004 to 0,02	10^{-5} to 10^{-4}	0	0	—	—
Longitudinal velocity component $\|\overline{u}_1-\overline{u}_1\|/\|\overline{u}_1\| \equiv \|\overline{\rho'u_1'}\|/\overline{\rho}\overline{u}_1$	0	0 to 0,01	0,01 to 0,05	0,002 to 0,015	0,001 to 0,003	0,01	0,01 to 0,20	0,30	0,30
Transversal velocity component $\|\overline{u}_k-\overline{u}_k\|/\|\overline{u}_k\| \equiv \|\overline{\rho'u_k'}\|/\overline{\rho}\overline{u}_k$ with k = 2 or 3	0 to $0,01\|\overline{u}\|/\overline{u}_k$	$0,01\|\overline{u}\|/\overline{u}_k$ to $0,05\|\overline{u}\|/\overline{u}_k$		$0,002\,\overline{u}/\overline{u}_k$ to $0,015\,\overline{u}/\overline{u}_k$	$0,001\|\overline{u}\|/\overline{u}_k$ to $0,003\|\overline{u}\|/\overline{u}_k$	$0,01\|\overline{u}\|/\overline{u}_k$ to ~0,50	$0,01\|\overline{u}\|/\overline{u}_k$ to $0,20\|\overline{u}\|/\overline{u}_k$	$0,30\|\overline{u}\|/\overline{u}_k$	$0,30\|\overline{u}\|/\overline{u}_k$

Fig. 3. Magnitudes of differences between terms by methods "A" and "B".

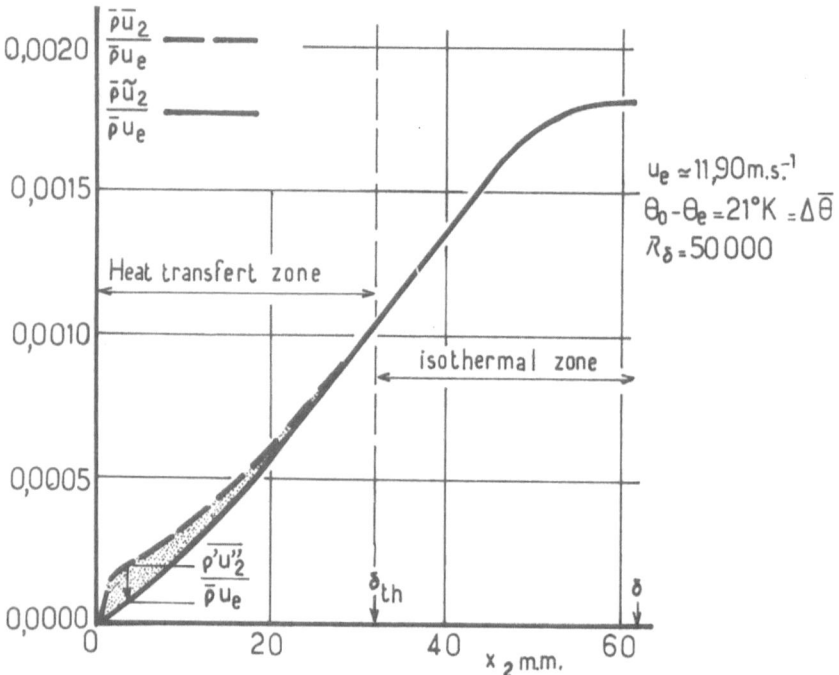

Fig. 4. Boundary layer at low speed with small heating on internal layer. (E. Verollet).

$$\overline{\rho u_2} \equiv \overline{\rho \tilde{u}_2} \equiv \overline{\rho \tilde{u}_2} + \overline{\rho' u_2''}$$

O(1)   O(1)   O(2)   O(−1)

Using the method "A" and neglecting the term $\overline{\rho' u_2''}$ would make result in a significant error. In practice it does not usually appear, because one is measuring the longitudinal component $\overline{u}_1 \cong \tilde{u}_1$, and computing the transversal component through the two dimensional continuity equation, which gives the value of $\tilde{u}_2$ and not of $\overline{u}_2$, that is applying in fact the method "B".

Another remark is that of course the relative importance of terms like $\overline{\rho' u_2''}$, $\overline{\rho' \xi}$ and others, is increasing with the differences of temperature, density, velocity, concentration, and that such terms are important in supersonic flows, strong heat transfer, heavy mixtures and reach the order of 100 % or 200 % for combustion.

References

BILGER, R.W. (1975) *Combustion and Technology,* **11**, 215-217
BILGER, R.W. (1977) *A.I.A.A. Journal,* **15**, 1056
BLACKADAR, A.K. (1950) New York Univ., *Meteo. Papers,* 1,4
BONNIOT, C. and BORGHI, R. (1977) *6th Int. Coll. on Gasdynamics of Explosions and Reactive Systems*
BRAY, K.N.C. and LIBBY, P.A. (1976) *The Physics of Fluids,* **19**, 1687-1701
BROWN, G.L. and ROSHKO, A. (1974) *Journal of Fluid Mech.,* **64**, 775-816
FAVRE, A. (1958)   *C.R.A.S., Paris,* **246**, pp. 2576-2579, pp. 2723-2725, pp. 2839-2842, pp. 3216-3219
FAVRE, A. (1965) *Journal de Mécanique,* **4**, pp. 361-390, 391-421
FAVRE, A. (1969) *Nauka Phys. and Math. Moscou,* 483-511, Cont. Mech. Soc. for Ind. and Appl. Math. Philadelphia, 231-266 (English version)
FAVRE, A. (1972) Instituto Naz. Alta Matematica. *Symposia Matematica 9*, 371-390. Acad. Press, New York and C.R.A.S., A, **273**, 1087-1289 (1971) (English version)
FAVRE, A. (1975) C.R. Fifth Canadian Congr. Appl. Mech., New Brunswick, Univ. Fredericton, Canada
FAVRE, A. (1977) La Turbulence en Mécanique des Fluides, by A. FAVRE, L.S.G. Kovasznay, R. DUMAS, J. GAVIGLIO, M. COANTIC, Gauthiers-Villars, Paris
GLASS, M. and KENT, J.H. (1977) *Second Australian Conference on Heat and Mass Transfer,* The University of Sydney, pp. 445-452
HESSELBERG, Th. (1926) *Beits. Phys. Freien Atmosph.,* **12**, pp. 141-160
ROSHKO, A. (1976) *A.I.A.A. Journal,* **14**, 1349-1357
STANFORD, R. and LIBBY, P.A. (1974) *The Physics of Fluids,* **17**, 1353-1361
VAN MIEGHEM, J. (1949) Mem. Inst. Roy. Meteo., Belgique, 34
VEROLLET, E. (1972) Thesis, Université de Provence, I.M.S.T.

LDA Applications to Internal Combustion Engines

by

M.L. Yeoman
AERE, Harwell, Oxfordshire, England

Abstract

Significant improvement in efficiency, market competitiveness and pollutants control of petrol and diesel engines will only be achieved with the aid of accurate quantitative information of the total engine function. Laser Doppler anemometry is currently employed to provide precise data on the quality and distribution of the air-fuel mixture prior to combustion. The technique monitors spatial and temporal profiles of mean velocity vectors and variance with the prospect of measuring fuel droplet size distributions. The problem of applying optical methods to pulsed systems with limited access, pressure fluctuations, engine vibrations, scattering centres with diameters 1-500 microns and windows covered by moving films of fluid have, to a large extent, been successfully overcome.

The Doppler signal must be matched to a point in tne engine cycle. Several possible methods are available but the usual technique is to synchronize the pulse trains from an optical encoder on the crankshaft to the gate of a data retrieval system. Histograms of velocity information are stored in multichannel analysers or computerised systems for a number of windows per cycle. Large data banks can rapidly be filled and the problem becomes one of on-line processing and data storage.

Most engine research and development laboratories rely heavily on steadily blown rigs and static pressure/high temperature cells for testing volumetric efficiencies, swirl, spray distributions etc. LDA is established in these areas as a complementary approach to conventional monitoring procedures. Velocity components normal to the beam are sampled with differential Doppler in the forward or backscatter mode. On-axis components require less conventional methods. The technique has been applied successfully to manifolds and cylinders of firing engines in limited situations. Here backscatter is obligatory and the behaviour of the visibility function for particles much larger than the fringe spacing is important for a number of beam polarisations. Large particles will not follow the air movement in turbulent systems and droplet and air-flow characteristics cannot be resolved. A system developed to avoid this problem will measure simultaneously the velocity and particle size.

Some progress has been made towards integrating and packaging equipment for use by inexperienced personnel. In this context the application of light fibres gives extra flexibility to scanning arrangements. To date most measurements have accumulated velocity and variance information averaged over a large number of engine cycles. However, cycle to cycle and in-cycle behaviour is important to a complete understanding of the engine function. Future developments will incorporate microprocessor systems to store and relate velocity and particle size information to specific engine cycles.

Introduction
At least a dozen establishments in Europe, the USA and Japan are using laser Doppler anemometry to look inside petrol and diesel internal combustion engines. The aim is to provide the engine designer with precise quantitative information on the composition and distribution of air-fuel mixtures throughout the total engine function. Much of the work is funded by the motor manufacturer and the motivation for their interest in quantitative data on the combustion process can be traced to three reasons which have come suddenly together, namely the fourfold increase in the price of oil since 1974, the realisation that oil reserves are declining and the availability of the present composition of petrol and diesel fuels are severely limited, and thirdly the application of government legislation in many countries which strictly limits the permitted pollutants produced by internal combustion engines. Current programmes of research are designed to reduce pollution and increase the flexibility of permitted fuel compositions with improved efficiency through a greatly improved understanding of the combustion process.

The parameters sampled with LDA are the spatial and temporal dependence of mean velocity and turbulence with the prospect more recently of using the amplitudes of the Doppler signal to monitor fuel droplet size.

Type of Engine
The area of engine to which the LDA technique has been applied varies with the type of engine. In the Otto cycle or petrol engine of Fig. 1 the carburrettor, inlet manifold, valve and combustion chamber are all of major importance in the production and delivery of the fuel/air mixture to the point of ignition.

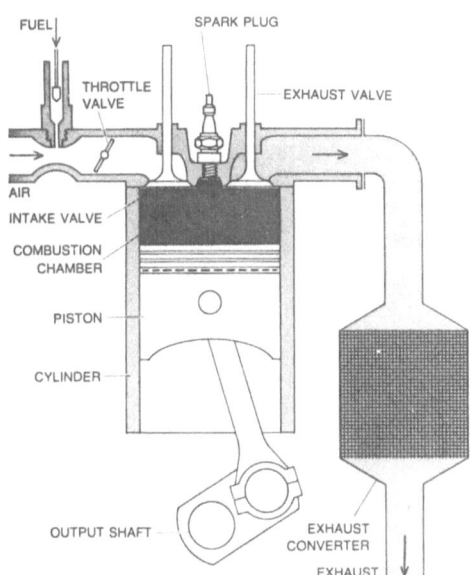

Fig. 1. Petrol Engine

Up to the present time measurements in this type of engine have been made near the carburrettor and throttle plate and in the plenum and arms of the inlet manifold with the engine in the firing mode.

The diesel engine shown in Fig. 2 dispenses with a spark plug and ignites the fuel mixture in the cylinder by the heat of compression with a glow plug to facilitate starting. The fuel air mixing in the diesel engine occurs inside the cylinder with the inlet valve designed to produce a strongly swirling air intake to encourage good mixing. The places of interest to the engine designer are the regions just after the inlet valve where air turbulence and swirl are of crucial importance and in the bowl of the piston at TDC where the firing process occurs.

Up to the present time measurements have been made of the air movement after inlet valves and, by placing a window in the injector port and firing the engine on three cylinders, detailed velocity and turbulence profiles have been mapped in the piston bowl of the fourth cylinder. An account of some of the Harwell work in this area was given in Open Forum Session F.

The stratified charge engine of Fig. 3 is presently of great interest to the motor industry as a promising alternative to catalysts for reducing emissions from spark ignition systems. It employs a modification of the combustion chamber

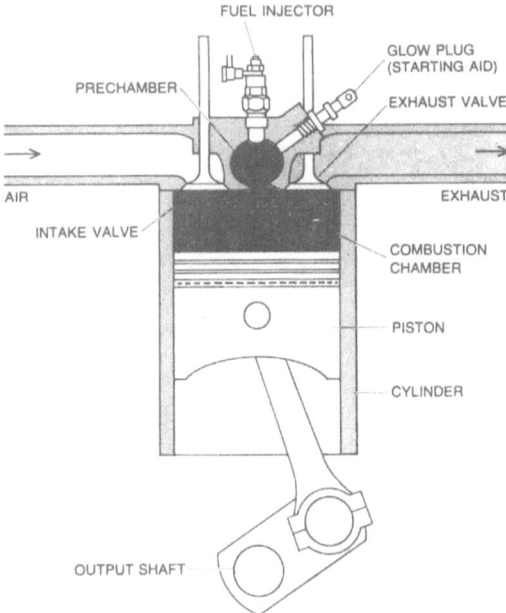

Fig. 2. Diesel Engine

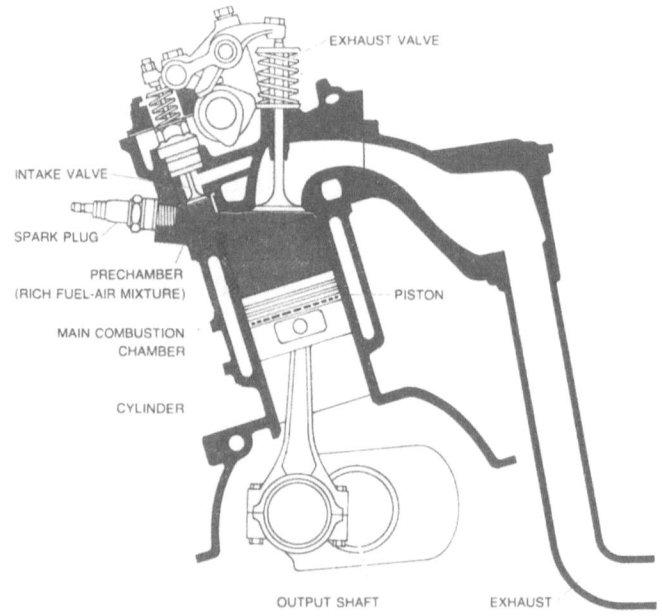

Fig. 3. Stratified-Charge Engine

and fuel injection which results in a two stage combustion process. A very rich fuel-air mixture is ignited first and is then used to burn a very lean mixture. In this way the maximum combustion temperature is lowered with a consequent reduction in the production of NO_x emissions. Here the region between the pre-chamber and main chamber and the propagation of the flame front into the lean main chamber mixture is of great interest to the motor manufacturer. LDA measurements have been made on engines of this type in the firing mode.

Component Testing

Research programmes on the application of LDA to internal combustion engines initially used the technique as a more sophisticated form of anemometer to fingerprint given systems in a comparative approach to engine component behaviour.

Fig. 4. Monitoring Inlet Valve Characteristics

Fig. 4 illustrated a system utilising forward scattered light to examine the air movement after different types of inlet valve in a static continuously blown air rig operating at flow rates equal to those obtained through the ports on a firing engine. The results of this measurement confirmed earlier hot wire studies and showed the existence of a vortex pattern which differed from one valve type to another. The vortex behaviour was later correlated with results obtained on a motored diesel engine.

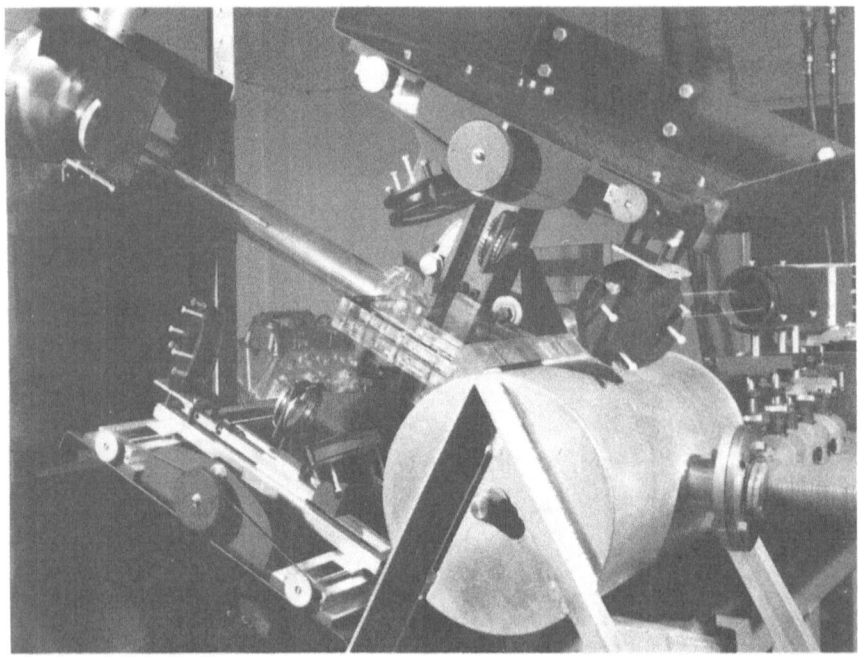

Fig. 5. Inlet Manifold Testing

The forward scatter mode was used to measure air flows in a caricature perspex manifold system shown in Fig. 5. The optical scanning system was a fixed non-portable arrangement for scanning in three dimensions without adjustment of illuminating and collection optics.

Measurements were made with air drawn through one of the arms at flow rates corresponding to 30, 50 and 70 mph. Turbulence intensity profiles were shown to follow the gradients in mean velocity and it is clear from Fig. 6 that the maximum mass transfer of air occurred near the outer wall of the manifold arm.

Although engine research and development laboratories rely heavily on steadily blown rigs the real potential of LDA in engine research lies in the application directly to firing reciprocating engines. Here one can identify the problems as those of significant background vibrational levels, windows in carburrettors, manifolds and cylinders which are covered with static or moving films of fuel, scattering centres ranging in diameter from a few microns to several hundred microns and, in the case of incylinder measurements, severe and spatially limiting access problems.

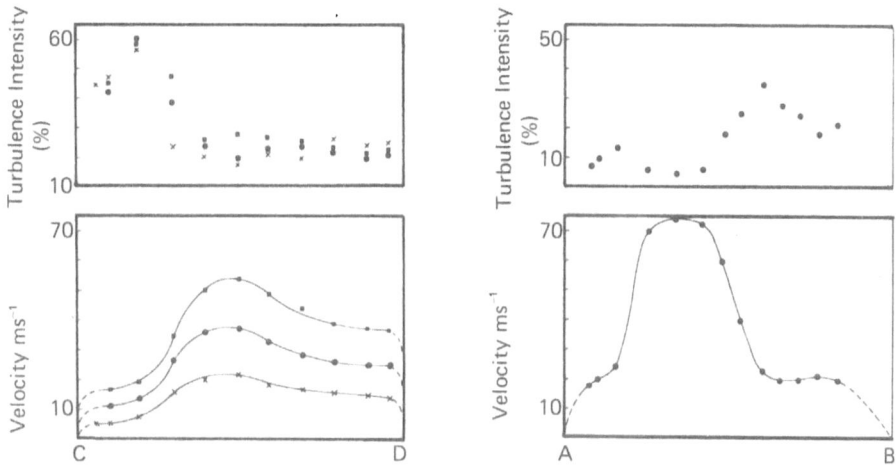

Fig. 6. Mean velocities and turbulence intensities in the manifold plenum (AB) and the entry to arm 4 (CD). Gas flows corresponding to 30, 50 and 70 rnph are represented by crosses, circles and squares respectively.

Fig. 7. LDA Measurements in a Perspex Manifold on a Motored Engine.

In most of the investigations to date the backscatter differential Doppler mode
of operation has been utilised. Because of the high vibration levels and temperature fluctuations found in engine test cells the optical unit should be self-
aligning, stable and suitable for use with severe problems of flare from windows, oil films, pistons etc.

An optical unit designed for this situation using a polarising beam splitter and
movable prism collection is shown in Fig. 7 as part of a scanning system installed in an engine test cell at an Esso Research Laboratory. Frequency shifting of either sign was here provided by an electro-optic Kerr cell placed in front
of the laser. A perspex manifold was attached to a motored Triumph Dolomite
two litre engine and the air movement monitored using reflected forward scattered light.

Timing Systems

The accessing of velocity information must be synchronised with a given interval of engine rotation relative to a known point in the engine cycle. In principle
the laser output could be gated with a pocket cell switch but most researchers
have preferred to monitor Doppler signals continuously and to gate the data
processing capability. An early systems is shown in Fig. 8.

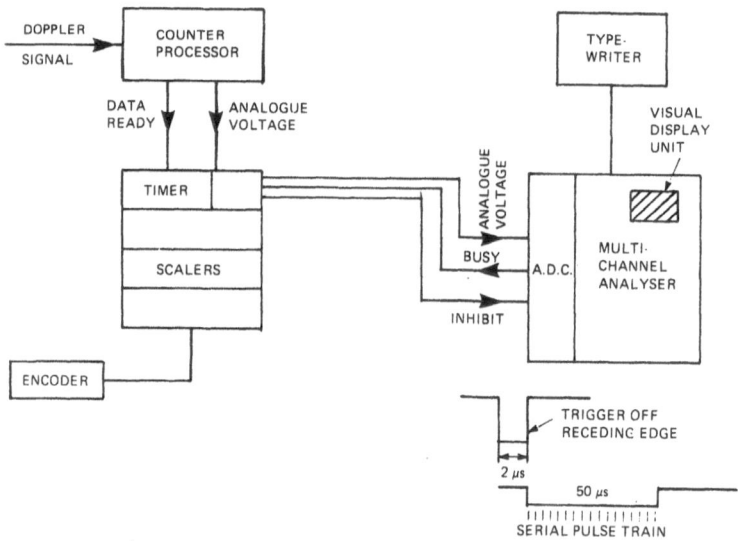

Fig. 8. An early timing system with gated data processing.

An encoder attached directly to the engine crankshaft outputs a single pulse
and two 360 pulse trains per revolution. The single pulse was used to trigger a

delay unit and the two 360 pulse trains were used for timing purposes and to drive a revolution rate counter. When a Doppler burst has been analysed and a voltage is about to appear as an analogue voltage proportional to the Doppler frequency the Counter Processor outputs a 'data read' signal which primes the Timer to receive the analogue signal from the Counter. The Timer selects a 2 μs sample of the signal and feeds the sample to the TDC of a Laben multichannel analyser, but the analyser is inhibited by the Timer until the required point in the cycle relative to the TDC pulse is reached when the analyser is allowed to collect data for a predetermined window period. As the 2 μs pulse begins to turn over a 'busy signal' is produced by the Laben which is fed to the Timer inhibiting the system from accepting data from the Counter. When, after about 50 μs, the 'busy signal' disappears the Timer waits for the next 'data ready' signal from the Counter and the process is repeated.

The 512 channels of the multichannel analyser were split into eight subgroups and the data displayed, as in Fig. 8, as velocity histograms for a predetermined interval of crankshaft rotation. The peak of the histogram for a symmetric distribution gives the mean velocity and the width of the curve is related to the variance of the velocity fluctuations.

Researchers in the USA have preferred to use a minicomputer to store data until the memory is full. Typically 10^6 raw velocity/crank angle pairs are recorded and transferred to a main frame computer for subsequent processing. An absolute encoder is attached to the crankshaft and the pulse train used to select the window interval and crank angle after the data has been taken. The current processor and data storage system in use at Harwell is shown in Fig. 10. The system is designed for easy transport to engine manufacturer test cells with a facility for viewing and checking the data as it is taken.

The system consists of an interface which takes in data, a floppy disc storage capability and a visual display unit. The front end estimates live time of the Doppler burst for particle biasing corrections and the unit records 64 contiguous velocity histograms per engine cycle with selected window intervals of from one-sixth of a degree of crankshaft rotation. The processor estimates the mean velocity and variance of the histogram which is available immediately as a hard copy or can be stored on disc for further analysis. A second channel is included for drop sizing measurements and for processing single cycle operation.

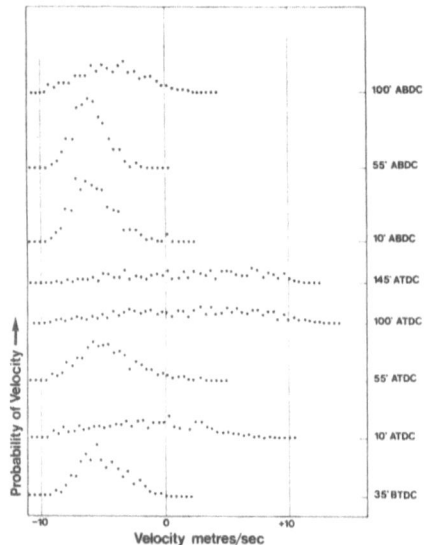

Fig. 9. Velocity spectra in a manifold attached to a motored Triumph Dolomite engine

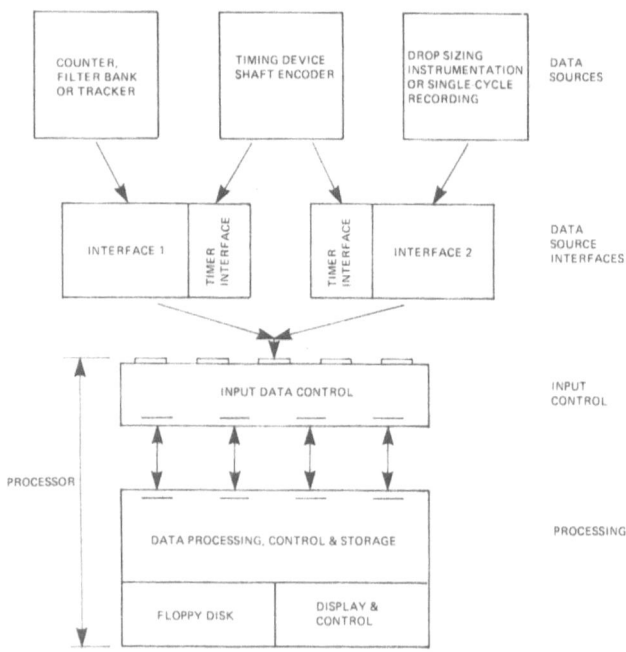

Fig. 10. Harwell LDA Data Processor and Storage

Velocity and Turbulence Measurements in a Firing Engine

Engine test cells are frequently small, highly congested and not at all suitable for setting up laser systems with a three dimensional scanning capability. To meet this situation a Universal Laser Mount has been designed and manufactured for general purpose use in engine environments. The argon laser is located below control units on a base tank which can be filled with water for mechanical stability and acts as a reservoir and heat exchanger for cooling the laser tube and power supply. A Kerr cell electro-optic frequency shifter is placed just after the laser as shown in Fig. 11, or a Bragg cell is attached to the beam splitter under the main arm. A second arm directs the beam through the focussing lens into a final mirror and down into the engine. Backscattered or reflected forward scattered light is collected by the same mirror-lens-mirror arrangement and enters the beam splitting unit through a scanning prism which periscopes the scattered light into the focussing system of a fast photomultiplier. For forward scattering applications a fibre optic bundle transmits the scattered light to the collection prism of the beam splitting unit. The mount is mobile, self-contained and can be wheeled into test cells after alignment for direct application to engine measurements.

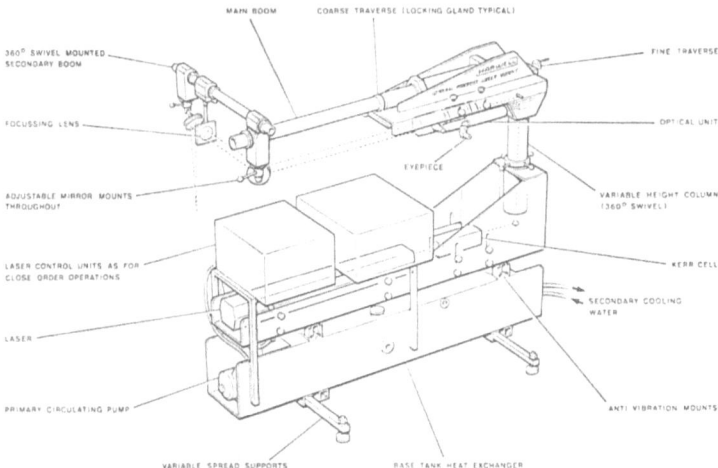

Fig. 11. Universal Laser Mount

Fig. 12 shows the Universal Laser Mount installed in a British Leyland engine test cell at Rover Cars in Solihull, Birmingham, U.K. In this application the optical unit has a balanced twin photodetector system for removal of the pedestal component of the signal when scattering directly from petrol droplets. Backscattered light was collected by the beam splitting optical unit and the

Fig. 12. LDA Measurements in a Manifold on a Firing Production Engine

Doppler signal processed with a digital counter. Mean and fluctuating velocity measurements were synchronised with a measurement of instantaneous pressure by locating a piezo-resistive pressure transducer below the probe volume in one of the arms of the manifold. The pressure fluctuation as a function of engine cycle for three different engine speeds is shown in Fig. 13.

The distance between the start of the marker pulses is one engine rotation. Each stroke of the piston has a characteristic pressure fluctuation pattern which is repeated in the next cycle. The pressure fluctuates by about 15 % above a base pressure of typically 0.5 bar. Fig. 14 illustrates a plot of mean velocity against pressure with the inlet valve sequence included. When the inlet valve to the cylinder is fully open the mean velocity is high and the pressure is low. As the valve begins to close the pressure increases and the velocity decreases.

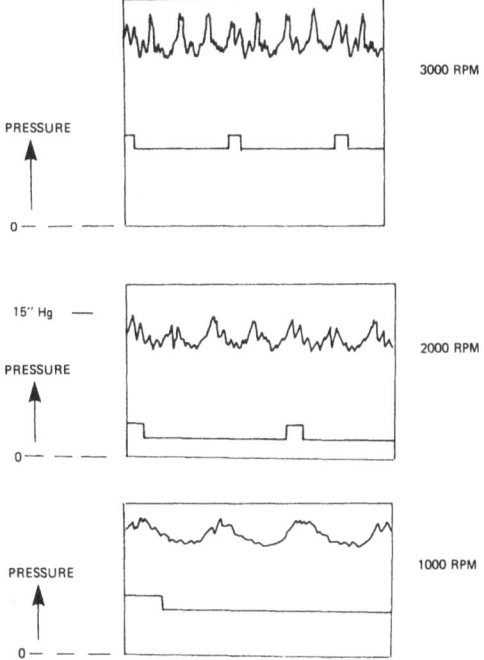

Fig. 13. Pressure Fluctuations in Arm 4 of Inlet Manifold

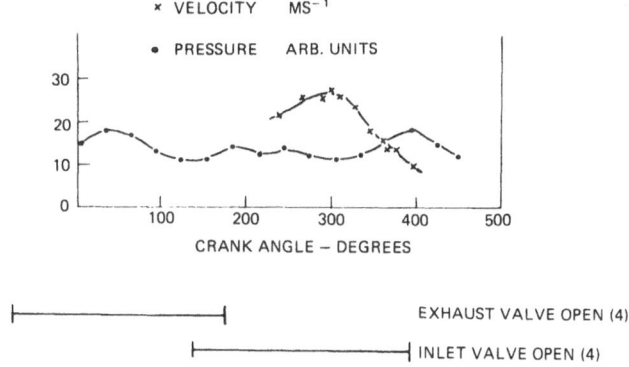

Fig. 14. Crank Angle Dependence of Mean Velocity and Pressure

The mean velocity is approximately universely dependent on pressure with, as shown in Fig. 15, only a weak turbulence/pressure dependence.

Laser anemometry as applied in the measurements described above monitors the droplet velocity which may not be representative of the air velocity as the carburrettor/baffle valve arrangement on a standard petrol engine produces a

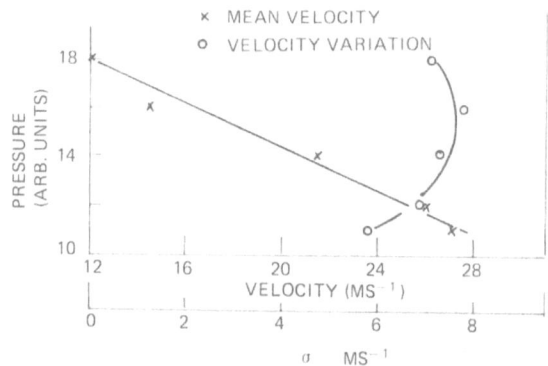

Fig. 15. Mean Velocity and Variance as a Function of Pressure

*Fig. 16. Engine Firing on Three Cylinders with Arm 4 of Manifold Replaced by
Plastic Ram Pipe*

range of droplet size. The influence of the droplet size distribution on the measured mean velocity and variance was examined by replacing one of the arms
of the inlet manifold with a perspex ram pipe of variable length, as indicated in
Fig. 16, and firing the engine on the remaining three cylinders.

Fuel was introduced to the ram pipe in two ways. The first method employed a small tube positioned half-way down the section which was connected to a seeding generator producing droplets of 1 μm diameter. The second arrangement employed an identical carburrettor to that used on the firing cylinders to function in its normal mode of operation at the end of the ram pipe. The first mode should correspond to seeding with very small droplets, much more likely to follow flow than the droplets produced in the second mode. Mean velocities and variance averaged over a large number of engine rotations were measured for contiguous window intervals of five degrees of crankshaft rotation. Fig. 17 shows that the measured value are up to thirty percent higher with the small droplets than with the range of droplet sizes produced by carburrettor.

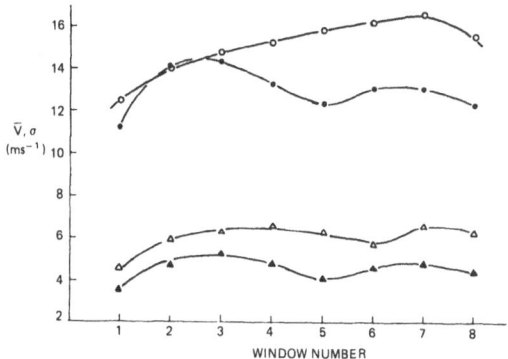

Fig. 17. Dependence on Droplet Distribution of $\overline{V}$ and σ
Seeding upstream of throttle plate: $\overline{V}$■, $σ^{½}$▲. Seeding downstream of throttle plate: $\overline{V}$○, $σ^{½}$△

It is clear from this type of measurement that in order to measure air flow velocities in firing engines with laser anemometry it is necessary to measure simultaneously droplet size and velocity. A unit working in backscatter is currently under development at Harwell for this application. The smaller droplets present in the flow will be used to estimate the air flow velocity.

Summary
Velocity and turbulence information on the flow of fuel droplets and seeding particles inside the component parts of petrol and diesel engines has been obtained with laser Doppler anemometry. Data has been averaged over many thousands of engine rotations because of the low arrival data and small sampling intervals relative to the period of crankshaft rotation. Timing systems have been developed which gate the data storage capability and synchronize the measurement interval to a known engine sequence. Accessing the in-cylin-

der locations continues to be a difficult problem, but problems of flare, vibration and piston beam interference have largely been overcome. Data is presently being collected which is of direct importance to the engine designer.

The technique is however capable of providing information on the particle or droplet size distribution and concentration for a direct measurement of air flow and droplet velocity. Future developments will lead to techniques for monitoring cycle to cycle variations and for measuring simultaneously two velocity components in firing engines.

Specific Problems in Cardiovascular Fluid Dynamics Measurements

by

C. Oddou

Laboratoire d'Hydrodynamique, Université Paris 7

75221 Paris Cedex 05

Abstract

The subject of this paper is confined to some applications of the physics of fluids to the study of characteristic hemodynamical features such as vortices dynamics, wave propagation and hydromechanical instabilities. Useful techniques in cardiovascular fluid dynamics measurements are reviewed with relation to the physical aspects of these topics with emphasis upon the unsteady nature of the flow and its interactions with moving boundaries of distensible vessel walls.

One of the main objectives in hemodynamics research is the determination of unsteady flow patterns generated by cardiac pressure waves. Focus has to be placed on accurate measurements of wall shear stresses, due to the high sensitivity of endothelial cells in respect of mechanical events which are localised in the flow boundary layer. Flow field microstructure at various sites inside the flow may be also very important with regard to thrombogenic mechanisms, dynamics of blood constituants and eventual blood cells trauma. In area of branchings, stenoses, aneurysms and at sudden enlargement, appearance of recirculation zone and transient formation of vortex are frequently occurring. The dynamics of such flow patterns and their related interactions with wall boundaries and blood components are of fundamental importance for pathological investigations.

One gives some detailed examples of flow patterns measured by ultrasound probings either inside hydromechanical models or during in vivo experiments concerning animal models and human cardiovascular system. It is underlined that such obtained information need to be extended and interpreted by means of complex numerical models.

1. Introduction

Modelling and diagnostics are two areas in which physics of fluids can efficiently contribute to the understanding of physiological phenomena and in return can gain some new insight in fluid dynamics. For instance, the behavior of cardiovascular system provides a field rich with flow dynamics problems. Such a system presents a complex vascular network in which blood flows in a pulsatile way at a wide range of velocity. In this respect, very different flow field structures can be encountered going from creeping motion and lubrication plasmatic layers inside very narrow capillaries or non newtonian suspension plug flow in venules and arterioles to high Reynolds number, unsteady and "disturbed" flows within large vessels such as arteries, veins and cardiac cavities. Our main concern in this topic will be to stress some aspects of physics of fluids related to the characterization of blood flows inside large vessels.

Main characteristics of such flows are their constant interaction with moving distensible walls whose mechanical properties are very unusual (viscoelastic, orthotropic and non linear rheological properties). It results from this that flow measurements in cardiovascular system are rather difficult due not only to the unsteady nature and physiological changes of the flow but also to the very complex motion of the physiological frame of reference, all major blood vessels and organs moving with various magnitude. From a theoretical standpoint the investigation of such phenomena requires that the mechanics of the flow should be closely coupled to the mechanics of the vessel walls. Therefore, the rheological properties of the vessel material must be known with particular emphasis on nonlinear and viscoelastic effects and accurate measurements of parietal displacements have to be performed.

Such an interaction is at the origin of the pressure wave propagation through the arterial system. Examples of local flow rate patterns associated with these waves are given on Fig. 1 (on the right part, note the phase delay corresponding to the effect of wave propagation between two points). The transient pressure gradients related to the waves are locally generating unsteady flow fields whi are generally not fully developed. Furthermore, due to the complex geometi , (arch, branch and tapering), local dynamics of fluids under such circumstances is far from to be fully understood. For instance, transient generation of recirculation zone in stenosed vessel and of secondary flow in branching tubes, unsteady interactions between recirculation vortex and primary entry flow, stability of such complex flow structure are fluid mechanics problems that require more study when applied into the physiological context.

One aspect which has recently been of particular interest for both physicists and physicians, is the influence of flow dynamics upon the transport of blood elements inside the arterial wall. Evidence for the predilection of athero-sclerosis plaque to develop at certain sites of the vasculature where the flow is in a "perturbed" state has arised very rapidly (Cf. Fig. 1). It has resulted a pressing

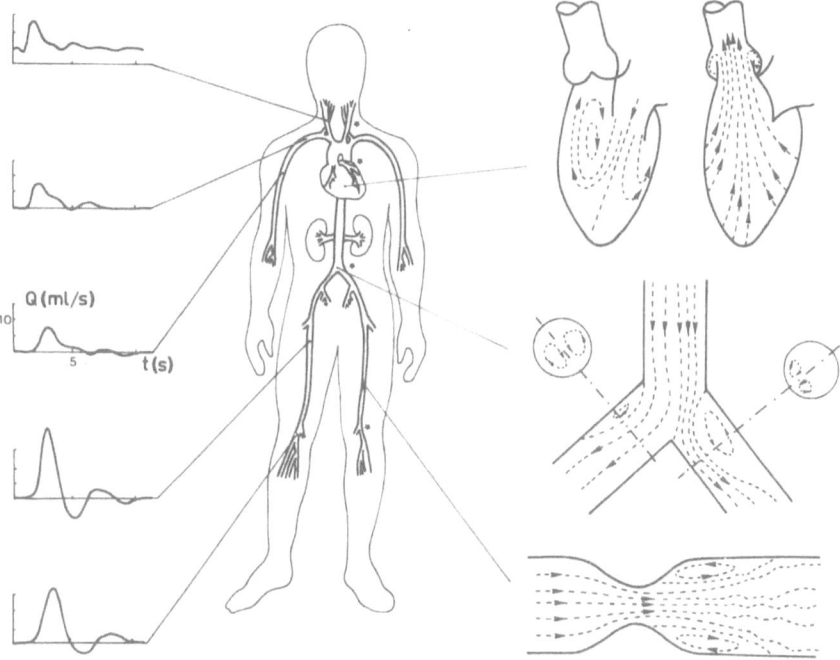

*Fig. 1. The fluid filled dynamic system of the human circulation is illustrated schematically with quotation of some of the favored sites of atherosclerosis**. When an artery (such as coronary artery) is narrowed by atherosclerosis (that is deposition of cholesterol and other fats inside the vessel wall), it may reach such a degree of constriction that the blood supply becomes inadequate to sustain nourishment of tissues and organs (causing heart failure). Note the tendency for the disease to begin at points where the flow geometry is complex such that bends and branches (from SPAIN, 1966).*

For a more comprehensive and accurate characterization of the fluid mechanical effects affecting disease of the cardiovascular system (i.e. atherosclerosis and rheumatic heart diseases) a detailed instantaneous mapping of flow field inside cavities, branches and stenoses is required. As noted on the right, particular attention should be paid on presence of recirculation zone and secondary flows with stagnant blood (thrombolytic factor) or part of flow with high strain rate (hemolytic factor) and enhanced wall shear stresses (endothelial layer injuries).

Are also shown, on the left, some characteristic flow rate patterns inside different arteries (carotid, subclavian, axillary, femoral and popliteal) of healthy subject, recorded by means of a pulsed ultrasound Doppler velocimeter (from ANLIKER et al., 1977).

need of designing method for accurately measuring details in parameters of the local flow fields both in human and in models (animals and physical models). Other factors such as acceleration of the blood flow in great vessels may be very useful to assess heart performance. These goals, when applied to human physiology have to be carried out by means of non invasive methods for velocimetric measurements such as transcutaneous ultrasound sensing.

Some recent techniques concerning vessel wall and flow dynamics characterization are briefly reviewed with their applications to some problems in physics of fluids raised by cardiovascular flow dynamics.

2. Dynamics of the Vessel Walls Cardiovascular System

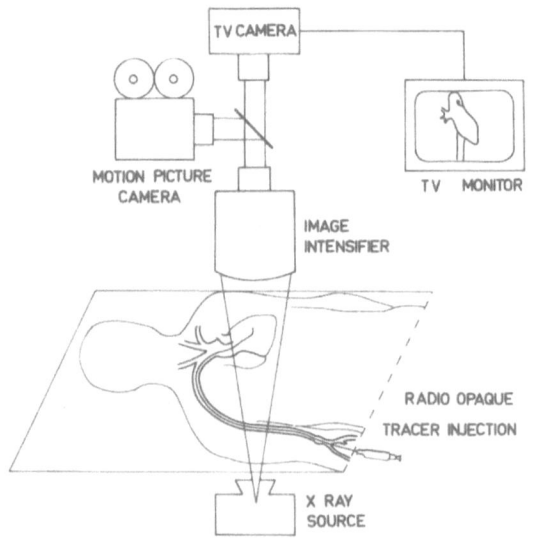

Fig. 2. Much of the evaluation of the human cardiovascular system performance has been based on data collected from x-ray technics using contrast media injected into the blood and thus obtaining images of large blood and heart vessels (cineangiographic methods). Visual diagnosis of the ventricules and related structures requires traumatic insertion of a flexible catheter through an artery in the patient's arm or leg up to a point inside the heart cavities and deliverance of a relatively large bolus of contrast material in a relatively short time (for details, cf. GROSSMAN, 1976 and WOOD, 1976). The main disadvantages of such a method are perturbations in sinus rythm and myocardial function following injection. Cardiac structure images formed by x-rays are displayed on a cathodic ray tube monitor after image intensification and recorded by motion picture camera (high speed cinematography of about 100 frames/s). Recently, quantitative imaging have been performed using information content of multiplanar x-ray cineangiograms (biplane anteroposterior and lateral views, for instance).

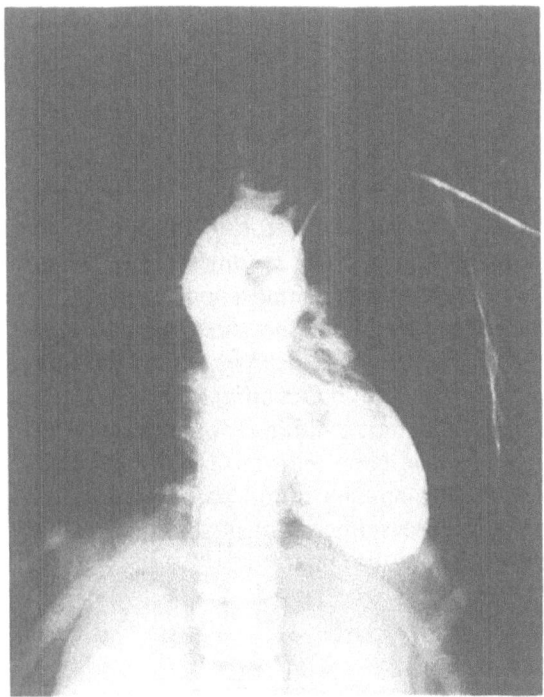

Fig. 3. Example of a left ventricular cineangiographic antero-posterior view (reproduced by courtesy of Dr. A. NITENBERG, Hopital H. Mondor, Creteil). Quantitative analysis of film images provides useful information on cardiac performance and patient clinical status. Computerized processing of cineangiographic frames for detection of left ventricular contour and volume rebuilding are the object of numerous studies (Cf. session on cardiac imaging and myocardial performance in Biosigma, 1978, and workshop on imaging of cineroentgenograms in computers in cardiology, 1977).

Human cardiovascular mechanics is now being extensively studied in hospital clinical investigation laboratories by means of quantitative imaging methods such as cineangiographic techniques. Excellent dynamic x-ray projection images of radioopacified heart cavities and related structures are commonly produced. They provide quantitative information about shapes and dimensions of cardiac silhouettes and their rates of change. Progresses in injection catheters manufacture and injection technique in order to produce a rapid delivery of an adequate bolus of contrast material, associated with recent developments of filming equipments (x-ray tubes and image intensifier as schematically shown on Fig. 2) have permitted to achieve high resolution dynamic imaging and relatively good definition of anatomic details. An example of such a cine frame given on Fig. 3 shows the complexity in heart shape which hardly varies during the cardiac cycle.

Rapid developments in computer based quantitative imaging, during the last years, have allowed to process cineangiographic images for the detections of left ventricular contours, taking into account distortion errors that are introduced by the image intensifier. Related calculations, on the basis of cross sectional shapes assumption, give cavity volumes, circumferential lengths and wall thicknesses. When associated with simultaneous intraventricular pressure measurements, wall mechanical parameters such as myocardial stresses, parietal compliance and cardiac work output can be estimated. Moreover, under special conditions concerning the blood opacification and the observation time associated to such a visualization technique, qualitative information about intracardiac fluid dynamics have been reported (ODDOU et al., 1978a); streaklines from entrained non radio-opaque blood particles during the filling phase of human left ventricle are revealing an unsteady vortical flow field structure in interaction with the mitral valve apparatus.

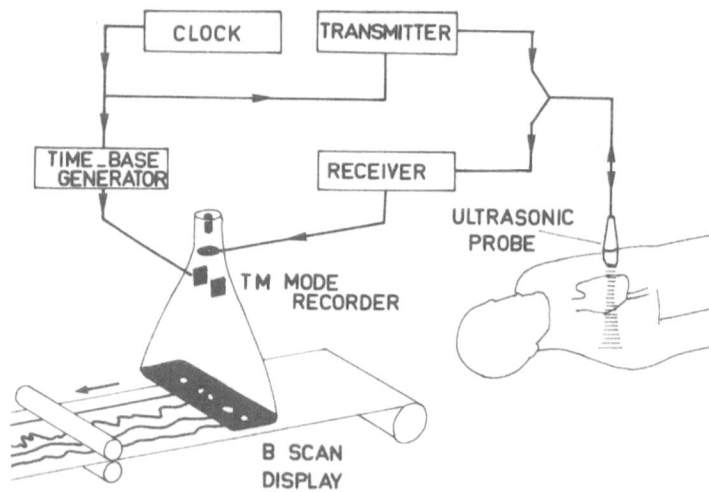

Fig. 4. Echocardiographic technics use time resolution scanning system (B scan, display) to monitor heart structure dynamics. Ultrasound probe excited by high-frequency (~ 5 MHz) emitter, which is timed by a clock, transmits a short acoustic pulse sequentially into the tissues. Simultaneously the clock drives through a time-base generator, the plates of a fiber-optic cathode-ray tube. Echoes signals after amplification and synchroneous detection in the receiver modulates the brightness of the display. Records of the echograms are obtained on ultraviolet sensitive paper which is fed past the display at a constant speed (from DEWEY G.B. and P.N.T. WELLS, 1978).

Such an invasive and traumatic method gives, in its monoplane version, a useful evaluation of left ventricular function. Nevertheless, due to the high complexity of cardiac cavities shapes, detailed analysis of segmental wall motion cannot be actually performed without improvements. Synchroneous multiplanar cineangiography associated with computerized dynamic spatial reconstruction are future possibilities for studying the kinetics of these anatomic structures.

Another method for measurements of vessels dynamics that has recently become a major tool in clinical investigations is the ultrasound echography. This non invasive and non harzardous ultrasonic imagery is based upon the reflection phenomenon of ultrasound pulses at the boundaries between media of different acoustic impedance (blood and soft tissues of the vessel walls, for instance). It uses ultrasonic waves of a few megacycle frequency which give good enough resolution and penetration depth. Using a 2 MHz transducer, for example, about 20 cm of soft tissues depth can be explored with a 1 mm resolution. The mode of display currently used in cardiovascular investigations is the time motion mode whose principle has been recalled on Fig. 4. These investigations are often improved by real time visualization using sector scan or multi-scan imaging system. When applied to cardiac structure echography, this graphic technique (recent recordings being made using fiber optic stripchart recorders) provides a lot of characteristic motion patterns, few of them being schematically given on Fig. 5.

One of the landmarks in echographic examination, due to its easiness to be recorded and to be recognized, is the signature of the mitral valve motion. The flow dynamics of normal human mitral valve has been recently well documented by ultrasound echotomographic acquisition of instantaneous mitral leaflet shape profiles (BRUN et al., 1977) processing data from an ultrasound multi transducer probing with a small computer system (Cf. Fig. 6).

The pressure gradient pulse, coming from the heart with each contraction, travels along the arteries much more rapidly than the induced bulk movement of the blood. This rapid propagation of the pressure pulse occurs primarily because of the distributed elasticity in the arterial walls. Due to their compliance (which is a function of the anatomic location, pathologic state and age), the arteries expand at each heart beat. In order to characterize mechanical behavior of arterial wall tissues, this amount of expansion has to be measured without direct exposure of the artery. Ultrasound echography can be used for non invasively recording arterial wall motion.

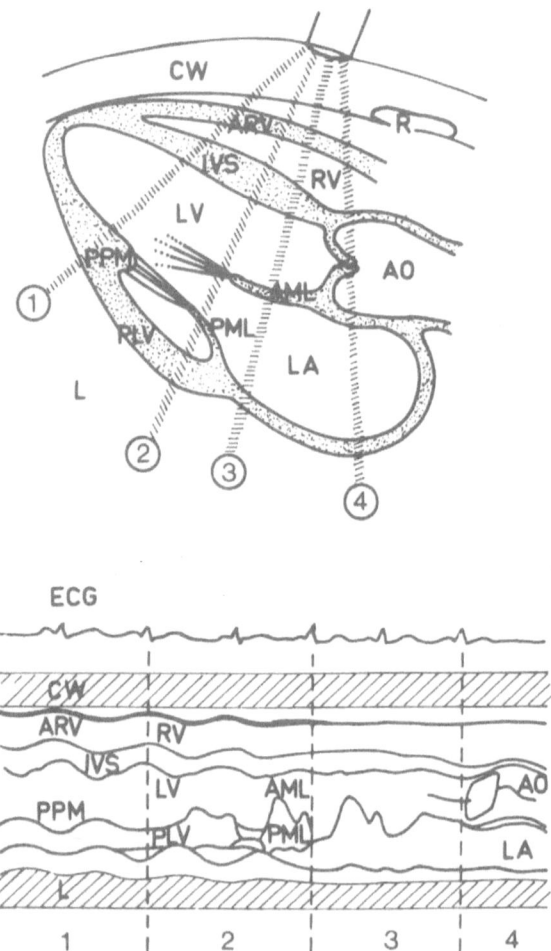

Fig. 5. Schematic heart cross section, showing the structures through which
the ultrasonic passes as it is aimed in four successive directions from the apex
(position 1) towards the base of the heart (position 4). Diagrams of the different echocardiograms thus obtained are shown at the bottom with the various
structures motions traces which are synchronized on electrocardiogram (ECG).
C.W.: Chest Wall; R.: Rib; A.R.V.: Anterior Right Ventricular wall; R.V.:
Right Ventricular cavity; I.V.S.: Intraventricular Septum; L.V.: Left Ventricular cavity; A.M.L.: Anterior Mitral Leaflet; P.M.L.: Posterior Mitral Leaflet;
AO: Aorta; P.P.M.: Posterior Papillary Muscle; P.L.V.: Posterior Left Ventrical
wall; L.A.: Left Atrium. (From FEINGENBAUM, H., 1976).

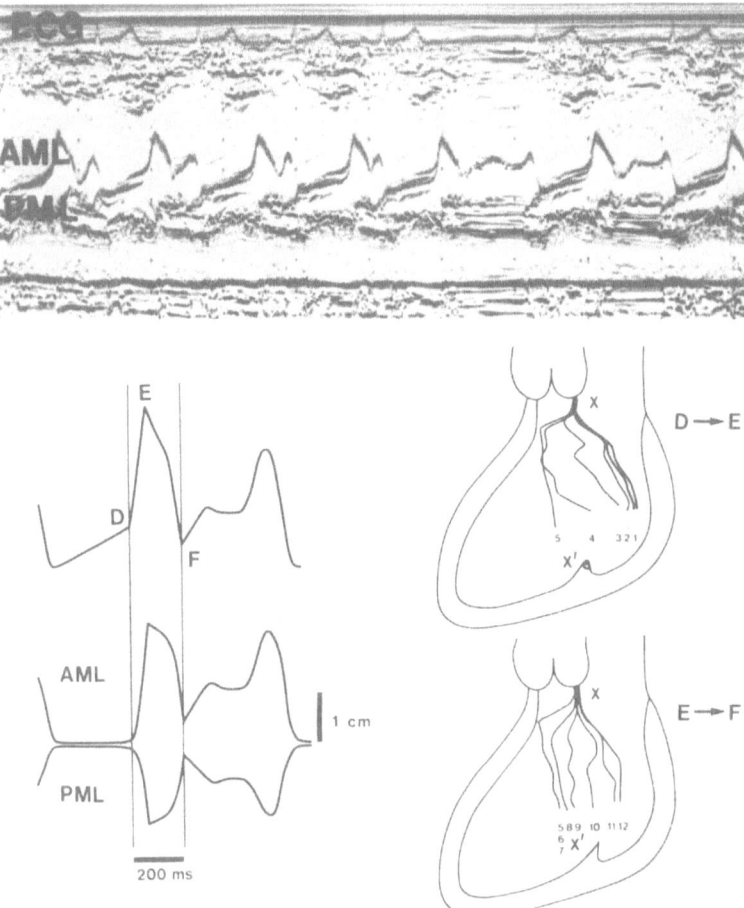

Fig. 6. Echocardiographic data: on top, an original record of an echogram on ultrasensitive paper showing traces of the mitral valve leaflets motion. On middle left is schematically shown details of the anterior mitral leaflet echogram (D,E,F: standard echocardiographic nomenclature) and on bottom left the corrected leaflet motion obtained by subtraction of the overall heart cavity motion. On right, are instantaneous profiles of the A.M.L. and chordae. Multiscan signals from high performance multiprobe time motion echocardiographic system have been synchroneously analysed in order to record the time evolution (1 to 12, 16 ms time increment) of the leaflet shape between the mitral annulus (X) and papillary muscle (X'). Notice the sigmoid instantaneous echographic shape of the leaflet during the opening phase in contrast with a concave aspect during the closing motion (from C. ODDOU et al., 1978a).

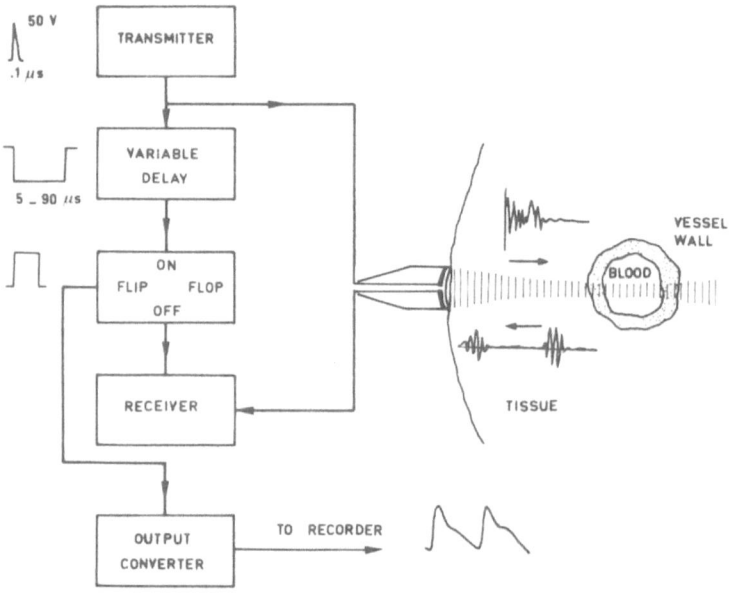

Fig. 7. Diagram of an ultrasonic echo tracking system device specifically developed for recording arterial wall motion (from HOKANSON et al., 1970). Repetitive voltage pulses at a rate of 10 kHz are exciting a piezoelectric crystal of 5 MHz resonant frequency. The received echoes are processed in order to convert transit time in an analog output that corresponds to movement of the arterial wall. This instrument is particularly designed for atraumatic studies of wall displacement of peripheral and external arteries in human.

Principle of an echo-tracking system is shown on Fig. 7. It is based on the use of a flip-flop element which is turned on at a set time after transmission and turned off by the echo. The time this element is on, changing with the motion of the arterial wall, is converted to an analog output and recorded. With such a system, arterial wall in its unexposed state has been shown to have an axisymmetric motion, the arterial diameter change ($\sim$ 1 mm) having large values in comparison with measurements made in exposed state.

Nevertheless, a large uncertainty in the anatomic location of the reflecting acoustic interfaces (particularly the proximal wall blood interface) prevents adequate definition of ultraluminal dimensions. New ultrasonic technique for arteriography, producing images of arteries which are comparable with those obtained by means of roentgenography, have been recently developed, based upon Doppler blood flow visualization (MOZERSKY et al., 1971; POURCE-LOT, 1978).

3. Blood Flow Dynamics in Large Blood Vessels

Thin film anemometers have been used to determine the velocity distribution in major blood vessels such as aorta of humans and animals (SCHULTZ, 1972). This highly traumatic method provided information on arterial blood flow dynamics which has also been studied by hydromechanical models associated with laser velocimetric measurements (ODDOU et al., 1978b). Nevertheless, pulsed ultrasound emission and Doppler shift detection of the echo during selected gated time remains the only non invasive way for in vivo measurements of both the blood velocity and position of moving walls.

As previously seen, ultrasound echo imaging is basically a specular process originating at the tissues interface whereas flow detection requires to approach parallelism between blood vessels and sound axis, waves back-scattered by moving blood cells being frequency shifted in proportion to the velocity component along the beam direction. The choice of the working frequency in order to achieve signal to noise ratio optimization has to take into account different physical processes, yet not well interpreted, such as ultrasound absorption in tissues and backscattering by red cells: under the assumption of (i) tissues absorption varying at a rate proportional to the frequency (attenuation coefficient of the order of 1 dB/MHz/cm) and (ii) Rayleigh type diffusion mechanism for ultrasound backscattering (whose intensity is varying with the fourth power of the frequency) and (iii) the Doppler shift frequency receiver bandwidth is proportional to the emitter frequency, the optimal frequency for maximum signal to noise ratio is found to be inversely proportional to the penetration depth (about 3 MHz for a 5 cm depth). Thus, the performance of such an apparatus depends of the emission power, the average power acceptable for biological safety being about 50 mW/cm^2.

As already mentioned, characteristics of biological flows are the dimensional variability of boundary structures (cross sectional area change and translational motion) and variations in blood flow fields (including smooth and highly disturbed velocity patterns). Therefore it appears necessary to accurately discriminate distance as well as velocity using a combination of echographic and Doppler method whose operating principle is shown on Fig. 8. Measurements accuracy both in velocity and position is a function of the sample volume (which depends of transducer size, ultrasonic beam shape, emission and detection durations and is generally of the order of few mm^3) and of the beam angulation and probe positioning precisions. For instance, transducer axis misalignment in the vessel meridian plane and finite length of the sample volume affect both the velocity profile measurement and the computed volume flow rate.

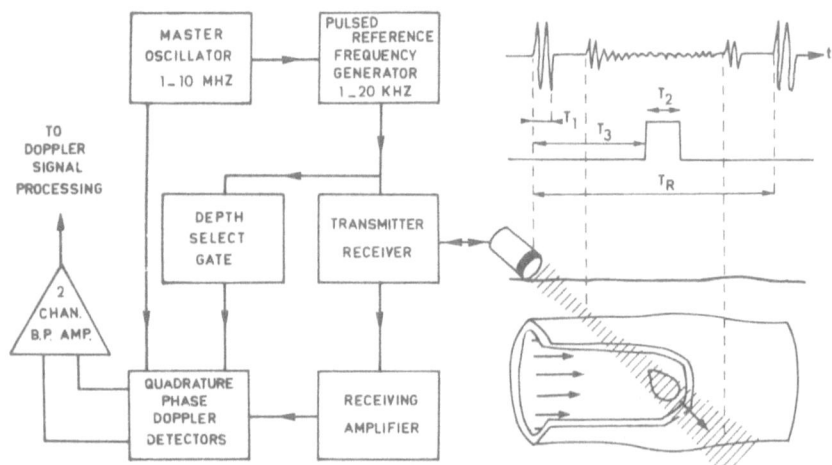

Fig. 8. Basic principles and functional block diagram of the pulsed Doppler system showing different steps of detection process: Based on a combination of echographic and Doppler methods, the pulsed Doppler technics uses repetitive emission of coherent bursts of ultrasound followed by reception and electronic selection of signals reflected by red cells within a particular sample volume. The depth position along the ultrasound beam of this volume is being fixed by the reception time. Doppler shifted component in the echoes due to the moving blood are detected by phase comparison between the raw echoes and the emitter reference signal. Various spectral analysis methods have been proposed in order to quantify the Doppler signal and to characterize the flow behavior being tested (for details in the different technics, cf. BAKER and DAIGLE, 1977 and PERONNEAU, 1977).

Such a phase coherent pulsed Doppler velocimeter has basic limitations if one wishes to measure velocity and position unambiguously. The maximum range of depth is inversely proportional to the pulse repetition frequency and the maximum Doppler shift that can be detected is equal to the pulse repetition rate divided by 2 (sampling theorem). Both basic principles, definition of performances and specific apparatus designs have been reviewed by several authors (Cf. in RENEMAN, 1973 and PERONNEAU, 1974). Particularly one of the principal characteristics of such a device is the fact that a number of spectral broadening effects occur including transit time effects, velocity gradient and flow disturbances fluctuations broadening.

Thus, the Doppler spectrum contains velocity information relative to all the streamlines within the sample volume and flow structure characterization requires a quantitative spectral analysis in order to obtain both local mean velocity and fluctuations mean intensity. Different Doppler signal processing methods have been developed. They are based on either spectral analysis methods involving fast Fourier transformation, use of set of filters and computers or zero crossing histogram technics which provide the distribution of characteristic frequencies (defined as the inverses of the time intervals between flow patterns and spectrum features by means of definitions of perturbation indices (PERONNEAU, 1977)).

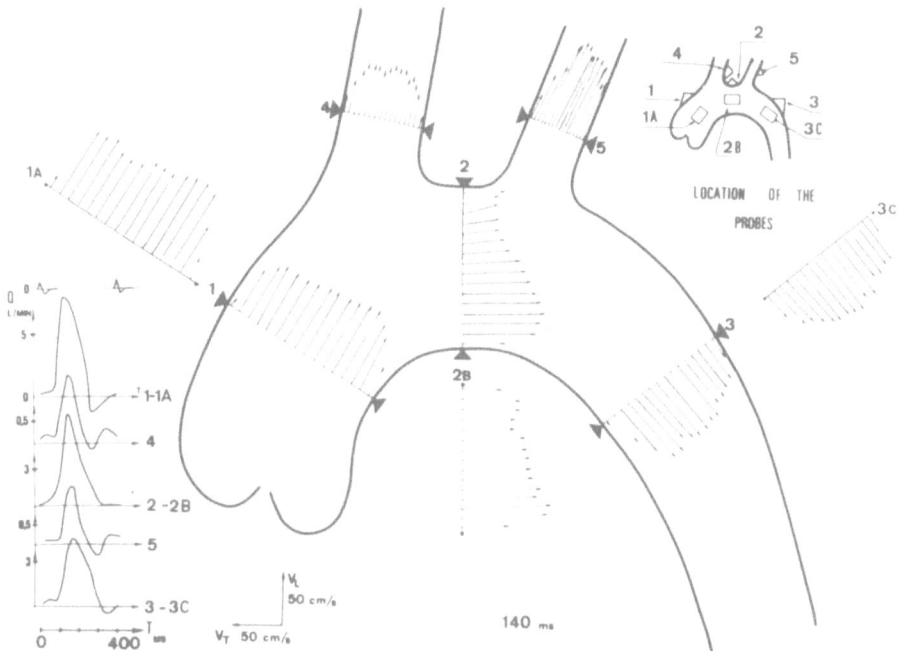

Fig. 9. An example of velocity vector profiles in the dog aortic arch obtained by multigated Doppler velocimetry at the time of systolic peak velocity (from PERONNEAU et al., 1977). Such a 16 channel pulsed Doppler velocimeter permits the determination of quasi-instantaneous spatial velocity profiles within human large blood vessels near the skin or within animal deep arteries by perivascular probe approach. In this example, special double probes were used in order to get the components of the velocity both in the plane of curvature of the arch and in perpendicular planes. Patterns of flow fields thus recorded, are showing characteristics of non fully developed entry-flows in a curved pipe. These patterns are very sensitive to the branched geometry and inlet conditions of the flow. Their boundary layers have a small thickness particularly on the interior wall side and profiles in the perpendicular plane generally present notches in their middle part (effects of inertial secondary flows).

Recent developments in the pulsed Doppler velocimetry technics offers new ways of obtaining information on in vivo experimental hemodynamics based on instantaneous recording of velocity profiles by multichannel processing. Example of such measurements in the dog aortic arch is given on Fig. 9. They yield quantitative data on instantaneous flow patterns in large blood vessels near the skin and their variations during the cardiac cycle. Moreover, ultrasound velocimetry have been successfully applied in hydromechanical models of local unsteady flow patterns induced by vascular stenosis, examples of which are given on Fig. 10. It yields to a detailed analysis of recirculating zone pattern and vortex shedding phenomenon.

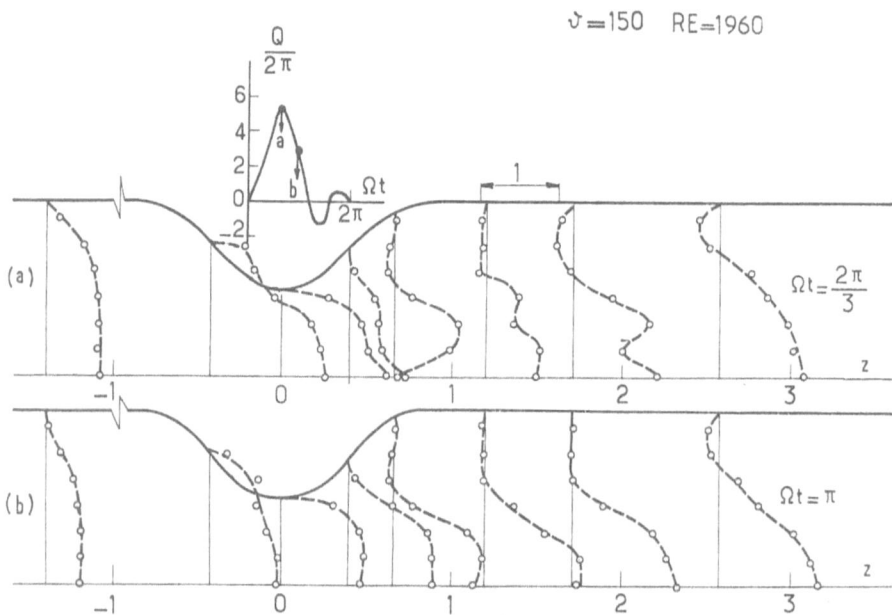

Fig. 10. Flow patterns recorded with pulsed Doppler ultrasonic velocimetry in a model of vascular stenosis (from SIOUFFI et al., 1977). In such in vitro experiments on cylindrical and stenosed test section of an hydrodynamical bunch, fairly accurate measurements permit a detailed study of local flow structure, vortex shedding and hydrodynamical instabilities generation behind a constricted unsteady flow. Notice the observation of unsteady recirculation zone behind the stenosis and its breaking in a periodic wake in the downstream part of the flow (skewed profiles). Combining studies based on flow visualization experiments, quasi local and instantaneous ultrasound velocimetric measurements and numerical models data, a very precise description of the complex dynamics of such flow under physiological conditions can be given.

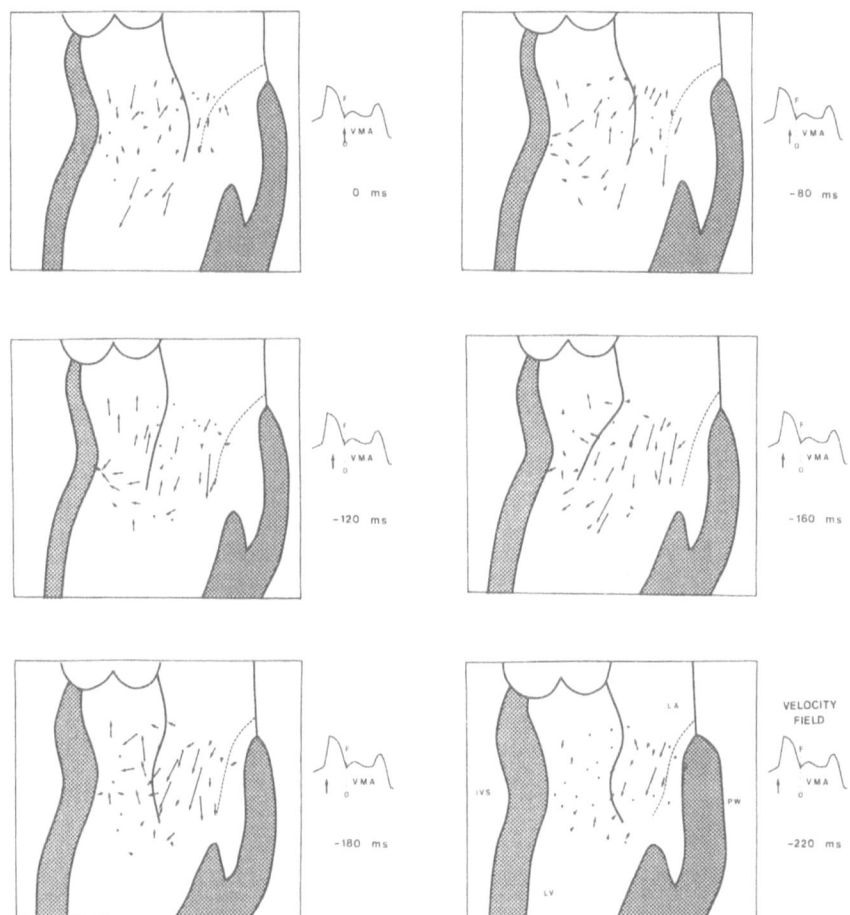

Fig. 11. Illustrations of instantaneous velocity field in human left ventricle recorded by pulsed Doppler ultrasound velocimetry during the cardiac filling phase (from BRUN et al., 1978). By optical technic for sensing the positions of the transducer with respect to a fixed reference and by post synchroneous accordance of velocity components waveforms mapping of part of the flow field in a diametral plane has been achieved. It can be noted the birth and development of a recirculation zone inside the cavity as the entry jet flow decelerates. Interaction between such a vortex and the anterior mitral leaflet is playing an important role in the closure motion of the valve. During early stage in the acceleration phase whipping motion of the leaflet generates at its free edge a very localized small vortex.

Another example of vortices dynamics generated behind mitral valve as the flow enters inside the human left ventricular cavity is illustrated by pulsed anemometric data shown on Fig. 11. It is shown that recirculation zone and leaflet continuous interaction is playing a dominant role in the control of valve movement.

4. Discussion and Conclusion

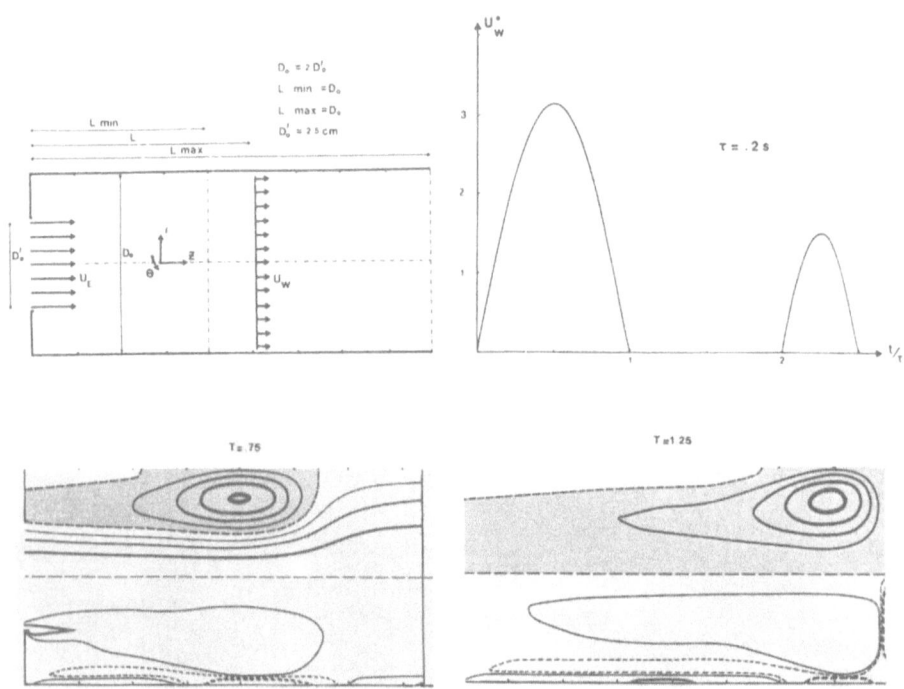

Fig. 12. Example of numerical crude model designed to study flow dynamics in the human left ventricle (from C. ODDOU et al., 1977): Numerical computations based on finite difference formulations of the fluid dynamics equations have been performed in an idealized cylindrical cavity, the entry flow being controlled by varying the length of the cavity. The flow rate thus given, simulates initial flow rate during the filling phase of the heart (Cf. upper right). Interactions between captive eddies generated at the rigid wall corner and the main entry flow through the axisymmetric expansion of the cavity are illustrated by the output data of the numerical model that gives stream and vorticity lines at different times within the filling phase (lower part of the figure).

From this brief review concerning quantitative and non invasive flow measurements in large blood vessels, it has been shown that numerous problems of physics of fluids in medical environment are not yet fully solved. For instance, our knowledge of the performance of the heart as a pump or of the peripheral vessels as a blood distribution system is yet very unperfected. Particularly, it is now generally felt that purely mechanical effects could be major factors in the generation of pathologic situations such as atheromatous or thrombotic deposits.

Dynamic measurements of velocity field under in vivo conditions characterized by complexes unsteadiness and geometry have only been recently performed using ultrasonic technics. They had to face problems of non-toxic intervention on living material in systems where the physical frame of reference is continuously moving. The results of such measurements have to be interpreted by continual interacting comparison with, on one side, data from hydromechanical models (under well defined and reproducible conditions) and on the other side results of theoretical computations including numerical models.

Progress in this field are only possible if is respected such a pluridisciplinary approach bringing together engineering, physical, physiologic and clinical point of view. As an example of this type of study, vortices dynamics generated behind mitral valve annulus, well documented by pulsed ultrasound anemometric measurements (Cf. Fig. 11), has been also illustrated by cineangiographic observations. The motion of valve leaflets has been measured by ultrasonic echotomography with associated data processing using small computer (Cf. Fig. 6). Data from these clinical investigations have been compared and partially interpreted by numerical model of unsteady flow field inside a growing cylindrical cavity (Cf. Fig. 12). Instantaneous valve leaflet profiles have been calculated under the assumptions of massless, infinitely thin and neutrally buoyant material.

Such studies in the field of cardiovascular research have been recently performed owing to new designs in ultrasound techniques which offer the opportunity of non invasively obtaining detailed information on both the flow structures and vessel wall motions.

References

ANLIKER, M., CASTY, M., FRIEDLI, P., KUBLI, R., and KELLER H., 1977 Non-invasive measurement of blood flow. *In Cardiovascular flow dynamics and measurements,* N.H.C. HWANG and N.A. NORMAN Eds., University Park Press, Baltimore, pp. 43-61

BAKER, D.W. and DAIGLE, R.E., 1977 Non invasive ultrasonic flowmetry. *In Cardiovascular flow dynamics and measurements,* N.H.C. HWANG and N.A. NORMAN, Eds., University Park Press, pp. 151-189

BIOSIGMA, 1978 *Proceedings of the International Conference on signals and images in medicine and biology.* Paris April 24-28th

BRUN, P., ODDOU, C., KULAS, A., LAURENT, F., 1977 Small computer development of echographic information related to left ventricle and mitral valve in diastole. *Computers in Cardiology.* pp. 267-273

BRUN, P., ODDOU, C., DANTAN, Ph., PLAPORTE, J., LAURENT, F., PERROT, P., 1978 Blood flow dynamics during human left ventricular filling phase. *3rd International Conference on Cardiovascular System Dynamics,* Leiden, Netherlands

COMPUTERS IN CARDIOLOGY, 1977. I.E.E.E. Catalog No. 77 CH 1254 - 2 C

DEVEY, G.B. and WELLS, P.N.T., 1978 Ultrasound in Medical Diagnosis, Scientific American

FEIGENBAUM, H., 1976 *Echocardiography,* Lea and Febiger, Philadelphia

GROSSMAN, W., 1976 *Cardiac catheterization and angiography,* Lea & Febiger, Philadelphia

HOKANSON, D.E., STRANDNESS, D.E. and MILLER, C.W., 1970 An echo-tracking system for recording arterial wall motion. *I.E.E.E. transaction on sonics and ultrasonics S.U.,* 17, 3, 130

MOZERSKY, D.J., HOKANSON, D.E., BAKER, D.W., SUMNER, D.J. and STRANDNESS, D.E., 1971 *Ultrasonic arteriography. Arch. surg.,* 103, pp. 663-667

ODDOU, C., BRUN, P.,DANTAN, Ph., BERALDO, E., KULAS, A., de VERNEJOUL, F., 1977 Fluid mechanics in the human left ventricle during cardiac filling phase, *in Cardiovascular and pulmonary dynamics.* Ed. M.Y. Jaffrin Editions INSERM Paris

ODDOU, C., BRUN, P., DANTAN, Ph. et KULAS, A., 1978a Relation entre mecanique du fluide intracardiaque et dynamique de la valve mitrale. *J. Fr. Biophys. et Med. Nucl.,* 1,61-67

ODDOU, C., DANTAN, Ph., FLAUD, P. and GEIGER, D., 1978b Hydrodynamics in cardiovascular research, in *Engineering Principles in Cardiovascular Research,* University Park Press, Baltimore, N.H.C. HWANG and D.R. GROSS (Eds.)

PERONNEAU, P., 1974 *Velocimetrie ultrasonore Doppler. Application à l'étude de l'écoulement sanguin dans le gros vaisseaux.* INSERM Vol. 34, Paris

PERONNEAU, P., 1977 Analyse de l'écoulement sanguin dans les gros vaisseaux par methode ultrasonore - Thèse de Doctorat d'Etat, Orsay

PERONNEAU, P., SANDMANN, W. and XHAARD, M., 1977 Blood flow patterns in large arteries in *"Ultrasound in Medicine"* WHITE, D.N. and BROWN, R.E., Eds., Vol. 3B, p. 1193 Plenum Press Publishing, New York

POURCELOT, L., BESSE, D., PEJOT, C., PLANIOL, Ph., 1978 Visualisation du sang circulant par effet Doppler. In *Biosigma,* 78, pp. 407-411

RENEMAN, R.S., ed., 1973 *Cardiovascular applications of ultrasound.* North Holland Publishing Company, Amsterdam

SIOUFFI, M., PERONNEAU, P., WILDT, E. and PELISSIER, R., 1977 Modification of flow patterns induced by a vascular stenosis. In *cardiovascular and pulmonary dynamics.* Euromech 92. M.Y. Jaffrin Ed., Editions INSERM, Paris, Vol. **71**, pp. 73-88

SPAIN, D.M., 1966 Atherosclerosis, Scientific American

WOOD, E.H., 1976 Cardiovascular and pulmonary Dynamics by quantitative imaging, Circulation Research 38 (3), pp. 131-139

Instantaneous Measurements of Flame Temperature and Density by Laser Raman Scattering

by

M. Lapp and C.M. Penney
General Electric Company
Corporate Research and Development
P.O. Box 8, Schenectady, NY 12301

Introduction

Combustion science has substantial need for spatially and temporally well-resolved, non-intrusive and non-perturbing measurement techniques. Demanding requirements are put upon candidate methods by the necessity to probe rapidly fluctuating, hot flows which can be out of thermal and chemical equilibrium.

One of these methods, which we discuss here, is based upon vibrational Raman scattering (VRS). This method possesses useful characteristics for probing major variables over a substantial range of flame conditions. Among these virtues are: possibility to probe small three-dimensional test zone; good time resolution; molecular species specificity; clear connection between experimental data and desired values of temperature and density; accessibility of temperature and density information from the same general type of data; capability for probing systems not in thermal or chemical equilibrium; and relative lack of interferences. The major disadvantage for VRS diagnostics is its general weakness in intensity, which severely restricts its applicability for brightly luminous and multi-phase flows. However, for many important experimental investigations, VRS diagnostics can be utilized to obtain essential information not otherwise available from conventional techniques.

Raman scattering has been used in a number of studies during the past several years which have probed unsteady flows. (See, for example, Hartley, 1974; Birch et al. 1975 and 1978; Brown et al. 1975; Boiarski, 1975; Lederman, 1976; Eckbreth, 1976; Setchell, 1976; Pealat et al. 1977; Lapp and Penney, 1977; Bridoux et al. 1978; Black and Chang, 1978; Bailly et al. 1978; Chabay et al. 1978; and Lapp, 1978). This body of work has included applications ranging from slowly-varying fluid conditions to turbulent flows and to shock-wave gasdynamics. Although we concentrate here upon the characterization and analysis of turbulent combustion by VRS measurements of fluctuation *values* of temperature and major species densities, the specific techniques are applicable to a wide range of fluid problems.

Raman Scattering Techniques

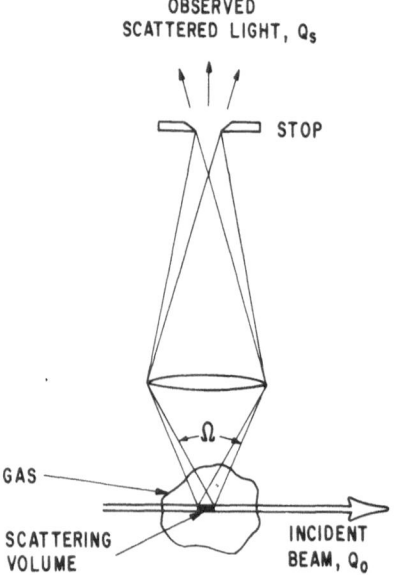

Fig. 1. Geometric configuration for light scattering diagnostics.

A commonly used experimental arrangement for light scattering flow measurements is shown in Fig. 1. An intense monochromatic light beam is focused at the measurement volume. Some of the light scattered from this beam by molecules and particles is collected and analyzed by the receiving optics. This configuration provides excellent spatial resolution because the collected light is restricted to that which comes from the intersection of the incident beam and receiver field of view. Time resolution is obtained by using a pulsed laser as the light source; pulses shorter than one microsecond can be obtained without difficulty.

Most of the collected light in a scattering experiment will be at the same wavelength as the incident light. The unshifted scattered light from molecules is called Rayleigh scattering. However, some of the scattered light will be substantially shifted in wavelength. These shifts result from the internal motion — rotation and vibration — of the molecules. In a classical sense, the shifts are sidebands arising from modulation of the scattered light by this internal molecular motion. The scattered light shifted in wavelength by this process is called Raman scattering.

Raman scattering is useful for flow measurements because different molecules produce different wavelength shifts, and the intensity of scattering at a particular wavelength is directly proportional to the number density of the cor-responding molecule. There is also a characteristic temperature dependence in the scattered light spectrum. Thus, Raman scattering can be used to measure absolute concentrations of different species and temperature. Data reduction to obtain desired information is relatively straightforward and free of uncertain corrections. In fact, analysis of Raman spectra does not require the assumption of equilibrium, so this technique is also applicable to non-equilibrium systems.

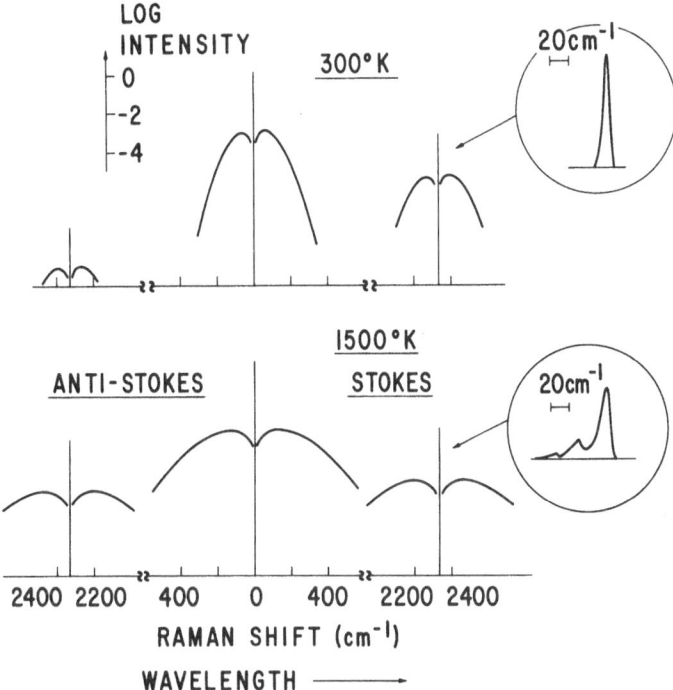

Fig. 2. Raman and Rayleigh scattering from N_2 at ambient (300°K) and elevat-ed (1500°K) temperatures for an exciting laser line in the mid-visible. The cen-tral unshifted peak corresponds to Rayleigh scattering, which is flanked by ro-tational Raman scattering represented here by wing envelopes of the rotational line peak intensities. The vibrational Q-branches on the Stokes and anti-Stokes sides are shown at the characteristic Raman shifts for N_2 of 2331 cm^{-1}. These Q-branches are surrounded by weaker vibrational bands called the O- and S-branches, shown also by wing envelopes. Note that relative intensities are drawn on a logarithmic scale and that large breaks occur along the wavenumber and wavelength axes. The spectral contours of the Q-branches shown in the two inset diagrams are presented on a linear scale, and have been calculated using a triangular spectrometer slit function with 6 cm^{-1} (~ 0.18 nm) FWHM.

As mentioned previously, the major disadvantage to be overcome in Raman scattering flow measurements is the weakness of the scattering. In most experiments, these measurements are presently limited by this weakness to major species densities and excitation temperatures for these species in relatively clean flows. High power lasers and sensitive detection apparatus are often employed to provide the required time ($\leqslant 1$ μs) and space ($\leqslant 1$ mm^3) resolution, as well as to discriminate against undesired flow luminosity. A typical Raman spectrum from nitrogen at ambient and elevated temperatures is shown in Fig. 2. The shifted sharp peaks are vibrational Raman scattering Q-branches. Rotational Raman scattering takes the form of wings — actually composed of numerous sharp lines — around the unshifted Rayleigh scattering. Less intense wings also appear about the vibrational Q-branches.

Although we focus our consideration upon VRS in order to provide reasonable bounds for this discussion, other Raman approaches also have strong merit. Progress has been achieved in the application to combustion measurements of both rotational Raman scattering (RRS) and non-linear optical processes such as coherent anti-Stokes Raman spectroscopy (CARS).

The prime advantage of RRS over VRS is its increased strength — roughly an order of magnitude — and its ability to produce sensitive data at much lower temperatures. However, because RRS corresponds to small molecular rotational energy increments and because the scattered light wavelength shifts from the exciting incident laser beam are therefore very small (see envelope of RRS transitions in Fig. 2), it is relatively difficult to separate the corresponding spectral signatures of spectroscopically-similar constituent flame molecules (such as N_2, O_2, CO, ...) from each other. Additionally, the presence of very strong Rayleigh and particle scattering spectrally nearby can present severe problems in signal discrimination. That these problems can be overcome in particular cases is now clearly documented; thus, RRS has been utilized for combustion gas mixtures in the very difficult environment of a rocket plume with constraints imposed upon the data reduction by flame composition considerations (Williams et al. 1977).

Additionally, the RRS spectrum of a particular molecule can be extracted from that of a complicated gas mixture by an interferometric technique (Barratt and Myers, 1971). In effect, the interferometer establishes a "comb" pattern through which only the specific RRS line spectrum of the desired molecule passes. This method becomes more complicated to apply as the temperature is increased, since more rotational lines become important and nonrigid rotor effects cause these lines to depart from precisely even spacings. (Of course, in the event that the flow is steady, the RRS spectrum of a gas mixture can be clearly analyzed and used for accurate diagnostics from careful spectrometer scans (Drake and Rosenblatt, 1976 and 1978).

In many cases the CARS process and other related non-linear scattering processes provide much stronger signals than the spontaneous Raman effect that we have been discussing. Application of these processes for measurements in turbulent flames is now being investigated by several groups (see, for example, Tolles et al. 1977, and Eckbreth et al. 1978). Among the disadvantages off-setting the stronger signals from CARS are its possible sensitivity to beam intensity fluctuations produced by unsteady fluid flow, and its sensitivity to collision processes. In some ways this latter sensitivity is similar to that which complicates quantitative interpretation of fluorescence measurements. Successful CARS contour fits for flame gases have been made with simplified assumptions concerning linewidth (such as a value constant with rotational quantum number and with temperature), but the ultimate limitations to CARS imposed by these assumptions are not yet known (Pealat et al. 1978). However, CARS has proven utility for qualitative imagery of flows, and it is likely to be the method of choice for temperature measurements in bright flames.

Raman Scattering Characteristics for Density and Temperature Measurements
The quantitative characteristics of Raman scattering can be explained in terms of the energy levels of a molecule, such as those shown in Fig. 3. If a molecule gains energy during a scattering event, going to a higher energy level, then conservation of energy requires that the scattered light lose energy, shifting to a longer wavelength λ and lower frequency ν according to the relationship for light photon energy $E = hc/\lambda = h\nu$. Light shifted to longer wavelengths, i.e. to the red, is called Stokes Raman scattering. On the other hand, if the molecule loses energy, the corresponding scattered light is shifted to shorter wavelengths. This process, which can occur only with excited molecules, is called anti-Stokes scattering. The number of possible lines is tightly limited by selection rules (viz., $\Delta v = 0, \pm 1$ and $\Delta J = 0, \pm 2$ for diatomic molecules). The intensity of each line is proportional to the number of molecules in the initial state leading to that line.

The strong central peak characteristic of vibrational Raman scattering is frequently called a Q-branch, using the standard spectroscopic terminology for a transition in which $\Delta J = 0$. For the vibrational Stokes Q-branch (i.e., $\Delta v = +1$, $\Delta J = 0$) shown in Fig. 2 at room temperature and at $1500°K$, the envelopes of the allowed rotation-vibration lines are presented in the inset diagrams as they would appear if viewed with a spectrometer slit function of about 0.18 nm full width at half-maximum (FWHM). In the $1500°K$ inset diagram, the strongest peak, called the ground state band, corresponds to the vibrational transition

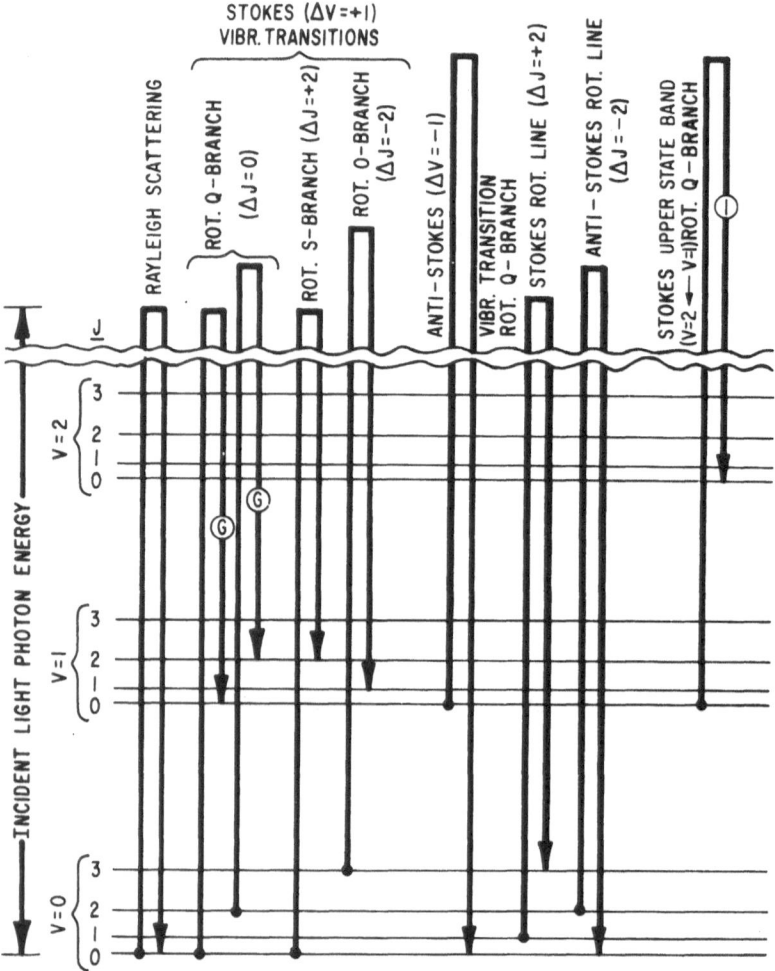

Fig. 3. Schematic of vibrational and rotational molecular energy levels and corresponding indications of Raman and Rayleigh scattering.

0 → 1 for Stokes scattering, whereas adjacent peaks, called hot bands because of their prominence at higher temperature, arise from the vibrational transitions 1 → 2, 2 → 3, etc. On the anti-Stokes side, the corresponding fundamental band series arises from the transitions 1 → 0, 2 → 1, etc.

Density measurements can be obtained from vibrational Raman scattering observations by comparing the intensity of the observed line with that of a known reference, such as ambient air. Temperature measurements can be obtained from the ratio of two bands or lines, such as a Stokes and anti-Stokes pair, from the ratio of the fundamental to a hot band of a single vibrational band, or from a band contour fit. (See, for example, Lapp and Penney, 1977).

Such a measurement of temperature is extremely basic in its nature, since temperature is defined on a molecular basis from the ratio of populations of adjacent states, and this ratio is directly proportional to the ratio of a corresponding pair of Raman lines or bands.

Temperature Measurements for Turbulent Flames

We now discuss, in particular, measurements that apply to turbulent flows and, wherever possible, to flames. To obtain fluctuation measurements, significant information must be obtained from the scattered light that can be related to short time intervals — say, to times approaching 10 μs or less. This information can be obtained directly through use of short-time-duration pulsed lasers or longer-time-duration pulsed lasers of rather high power (so that the laser energy per 10 μs-time interval is appreciable). Such data can also be obtained through use of cw laser sources. An outline of major current effort, illustrating these various approaches, is shown in Table 1.

The work performed in this laboratory has been focused upon the first of these two pulsed laser alternatives (Lapp and Penney, 1977). In Fig. 4 we show a schematic of our experimental configuration, which is composed of Stokes/ anti-Stokes (SAS) diagnostics applied to a turbulent diffusion flame produced in an open-throat coaxial jet burner. (See Lapp, 1978, for further details of the Raman experiments, and Wang and Gerhold, 1977, for characterization from turbulent velocity profiles obtained from a similar closed-throat burner by laser velocimetry). The air flow was produced in a 100 mm diameter pipe at about 0.1 m^3/s, while the hydrogen fuel flowed at about 585 cm^3/s from a tube of 2.7 mm diameter. The laser source used was a modified Phase-R flashlamp-pumped dye laser which can produce submicrosecond pulses in the mid-visible with energies of 1 J within a spectral width of $\sim$ 0.15 nm. All laser shots were monitored for laser spectral position and lineshape with a TV camera coupled to a small grating monochromator, and for laser pulse energy. Also shown in Fig. 4 is a schematic of the relationship between SAS intensity ratios and temperature, and expected forms of the probability distribution functions (pdf's) or histograms of temperature at various flame positions.

In Fig. 5 we show the results of one particular experiment with the apparatus just described. Each of the pdf's shown on the left-hand side corresponds to runs of 170 shots. The probed volume corresponds approximately to a cylinder of 0.7 mm height, with a volume less than 0.1 mm^3.

As will be seen in the next section, analytical estimates of accuracy for the SAS method are poor below about 800°K because the number of detected anti-Stokes photons becomes very small. It is nonetheless, clear under these condi-

Table 1.

Character of Laser Source	Experimental Capabilities			Comments	References
	pdf	Frequency Spectra	Spatial Mapping		
Short energetic laser pulses, low rep rate (ex. dye, YAG, or ruby lasers)	Yes	No	Yes	Can use variable wavelength laser.	a, b, c, d,
Long strongly energetic laser pulses (ex. free-running ruby laser, intracavity experiment)	Yes	Mid (kHz) to high (~ 50 kHz) frequencies	Difficult	Can probe somewhat luminous, particulate flows.	e
cw laser: • operated cw • modulated (chopped or cavity dumped) [• Fourier transform data analysis • autocorrelation function analysis • moments of photon count distribution]	Yes	Low (Hz) to mid (kHz) frequencies	Difficult	Restricted to low luminosity flows. pdf's obtained approx. from finite sequence of photon count moments	f, g

a. This work — pdf's
b. Lederman (1976) — Average values and stand. deviations
c. Bridoux et al. (1978) — Spatial mapping in flame
d. Black and Chang (1978) — Spatial mapping in open jets

e. Bailly et al. (1978) and Pealat et al. (1977) — Time history shown for ~ 300 μs
f. Chabay et al. (1978) — Fourier transform
g. Birch et al. (1978 and 1975) — Autocorrelation function and pdf's

Table 1. Comparison of VRS fluctuation measurement capabilities for methods using different laser sources. Here, the column for frequency spectra is the same as that for time history capability, over the time duration of the laser emission (i.e., total time interval for cw laser, or time duration of one free-running ruby pulse). Spatial mapping refers to one-dimensional spatial multiplexing of Raman data, best accomplished through use of spectral-multiple-channel detectors, but also possible with difficulty utilizing multiple photomultipliers, optical fibers, etc.

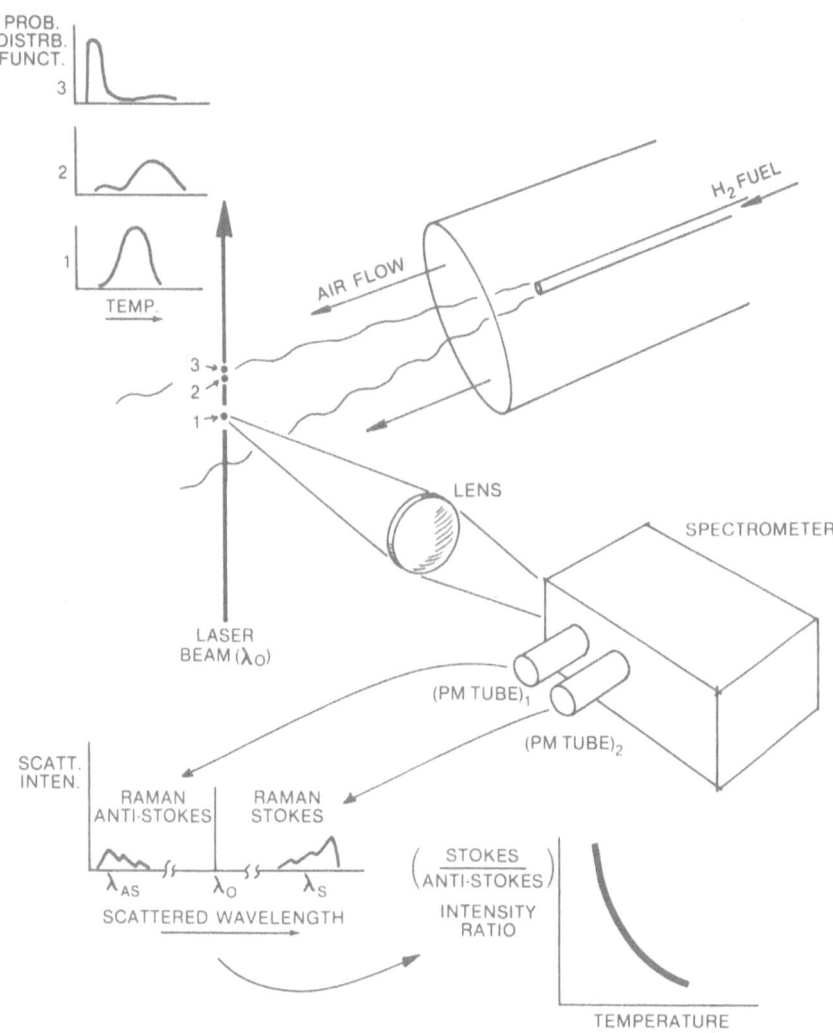

Fig. 4. Schematic of turbulent combustor geometry and optical data acquisition system for vibrational Raman scattering temperature measurements using Stokes/anti-Stokes intensity ratios. Also shown are the expected Raman contours viewed by each of the photomultiplier detectors, the temperature calibration curve, and several expected pdf's of temperature at different flame radial positions.

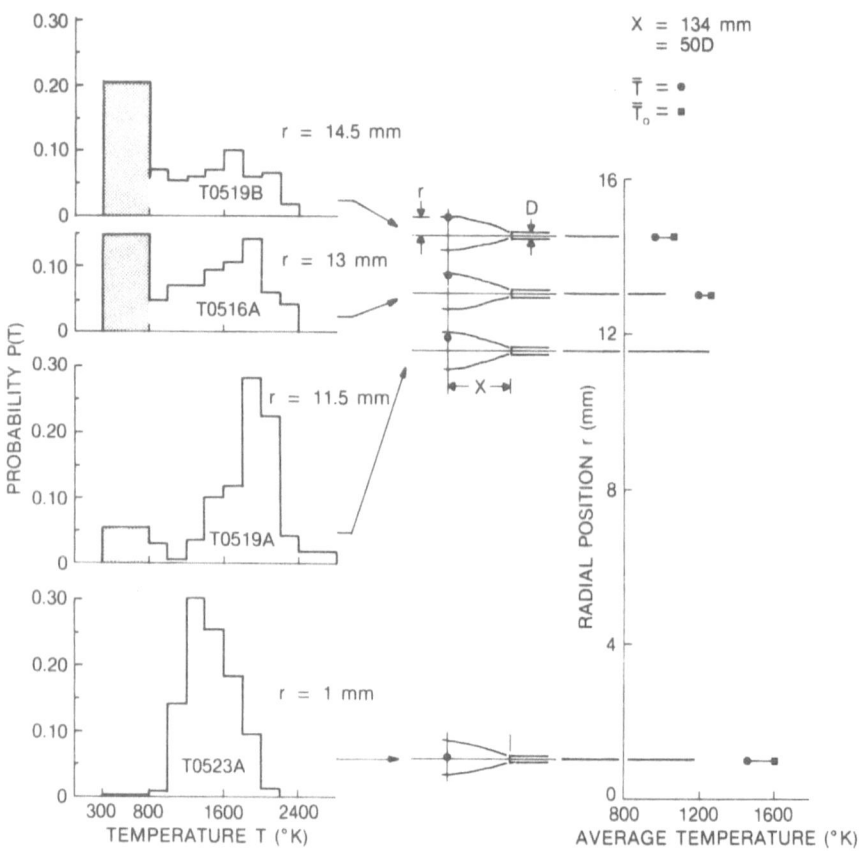

Fig. 5.Probability distribution functions of temperature for H_2-air turbulent flame determined at various radial positions 134 mm downstream of the fuel line tip according to procedures indicated in Fig. 4. The measurement positions are drawn schematically in the center of the figure to correspond to the radial positions r on the ordinate of the average temperature vs r plot at the right-hand side. (The different values of temperature on this plot correspond to either complete subtraction of measured apparent optical background in the data analysis (T), or to no background subtraction (T_0). The values of $\bar{T}$ and $\bar{T}_0$ bound the true temperature).

tions, that the values of accuracy obtained degrade rapidly as temperature is decreased further. Thus, below $800°K$ we do not treat the fluctuation data in the same detail as at higher temperatures, but rather integrate all data for $T < 800°K$ into one large bin for the pdf's — the "cold" portions shown as shaded areas in Fig. 5. These areas provide a clear estimate of the entrainment of cool ambient gas into the flame, and thus provide an excellent measure of flow intermittency.

The mean temperatures $\overline{T}$ and $\overline{T}_0$ for the four measurement positions shown here, plotted as a function of radial position in the right-hand part of Fig. 5, correspond to two different treatments of the optical data. Values of T (and $\overline{T}$) correspond to full subtraction from each Raman scattering datum of background radiation, measured from subsidiary runs at adjacent spectrometer wavelengths. Values of T_0 (and $\overline{T}_0$) ignore this potential correction. Since methods for measuring background corrections are not yet demonstrated to be completely correct, it is believed that the true value of temperature for each datum lies somewhere between T and T_0. Further work in reducing the effect of background radiation corrections wherever possible is clearly to be preferred (Lapp and Penney, 1977). It is believed that worthwhile improvements can be made in this direction, and also that unambiguous background correction procedures can be developed. Work to be reported elsewhere will detail our current efforts in this direction.

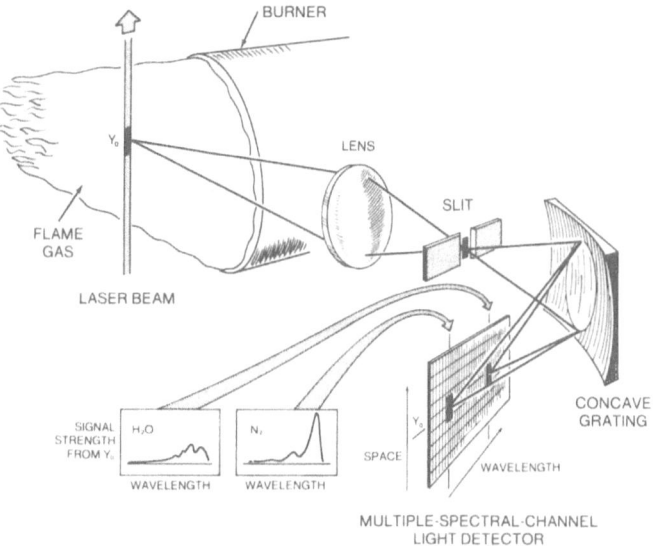

Fig. 6. Experimental schematic for planned Raman scattering measurement configuration suitable for obtaining spatially-multiplexed set of temperature and species concentration data.

Spatial multiplexing of the experimental configuration just described can be accomplished with a two-dimensional multiple-spectral-channel detector, such as the Optical Multichannel Analyzer OMA® detector* (see Fig. 6). Thus a spatial map of fluctuation values of major chemical components in the flame, and temperature along the incident laser beam, can be obtained with each laser shot. It appears practical, for example, to obtain data for ten to twenty positions over a 1 to 2 cm linear dimension in a flame in this fashion. Such data have been obtained recently for flames (Bridoux et al. 1978) and for open jets (Black and Chang, 1978).

The second pulsed laser technique (Pealat et al. 1977, and Bailly et al. 1978) is based upon use of a high power (free-running ruby) laser which can provide a time history of composition and temperature fluctuations over the duration of its relatively long laser pulse (typically 0.1 to 1 msec). About 1.5 J of light energy at the ruby wavelength (694.3 nm) is required to provide a statistically significant measurement of major constituent densities and temperature from a test zone of 1 mm length in flame gases at $1500°K$. These workers used a laser providing 100 kW of circulating power within the laser cavity averaged over 0.7 msec. Thus, they obtained roughly fifty distinct measurements in that short period. Within the limits of short pulse duration, this time history can be used to provide pdf's, correlation functions over short times, and the mid to high frequency portion of the frequency spectrum of turbulence fluctuations. However, it is not easy to obtain simultaneous measurements at several spatial points (spatial multiplexing) with this technique.

Up to this point, we have described approaches to VRS unsteady flow diagnostics based upon use of energetic laser pulses at relatively low repetition rates (e.g., 0.1 or less to perhaps an upper limit of 10 per second), in order to provide time-resolved information in the form of pdf's. Information concerning fluctuations in low luminosity environments can also be obtained from cw laser sources of sufficient power by direct observation of the scattering time history, but time resolution is necessarily limited in this case. (For example, a flame measurement of N_2 density at $1500°K$ can be made with roughly 0.5 J of light energy with the argon laser blue line at 488 nm. If one utilizes a strong laser source of 10 W power, one could obtain a time history with resolution of about 50 ms. Multiple-reflection optical cells can be used to enhance the available laser power, as can intracavity geometries, but even another factor of 50 in available laser power leads to only a 1 msec time resolution).

---

*OMA® is a registered trademark of the Princeton Applied Research Corporation, Princeton, New Jersey.

Alternatively, one can use photon correlation techniques to obtain density correlation functions (Birch et al. 1975, and 1978). This information can be extended to at least mid frequencies, because it can be accumulated over long times. Information about pdf's can be obtained from treatment of the moments of the photon count distribution (Birch et al. 1978). The technique does not appear to be easily extendable to temperature measurements, and spatial multiplexing is difficult in this case. It is interesting that the data analysis can use the same equipment employed for laser Doppler velocimetry. Fourier transform methods applied to cw laser-induced VRS have also been exploited recently to obtain fluctuation data for a turbulent jet flow (Chabay et al. 1978).

Experimental Accuracy of Fluctuation Data: Absolute Calibration and Temperature Measurement Spread Function
Up to this point, we have shown how fluctuation data for fluids have been obtained from Raman scattering data, but we have not yet commented upon experimental accuracy, except for background corrections. Even in the ideal case of a gas at constant elevated temperature, measurement errors due to the statistical fluctuations in photon detection introduce a spread of measured temperature values around the true value. This spread function can be described by some representative characteristic of its contour, such as the standard deviation σ. (The standard deviation is related to the probable temperature error ϵ through the relation $\epsilon = 0.6745\ \sigma$. The value of ϵ describes the temperature range within which lies half the spread function area and thus within which half the measurements can be expected to fall). In our work we have calculated spread functions and corresponding probable errors, and compared these to the measured values obtained. This work will be documented later in this section.

Temperature is determined from the ratio of integrated signals from the photomultipliers monitoring the Stokes and anti-Stokes vibrational bands. The photomultiplier signals can be represented by the total numbers of photons N_S and N_{AS} detected by these photomultipliers during the laser pulse. (We assume background can be neglected for this discussion). Thus a measured value T of gas temperature is obtained from

$$T = \theta_V / \ln(RN_S/N_{AS}). \tag{1}$$

Here θ_V is the characteristic vibrational temperature for the observed molecule ($3374°K$ for N_2) and R is a calibration factor which takes into account the wavelength dependence of the scattering cross section and the variation in transmission and detection efficiency across the spectral range from the anti-Stokes to Stokes wavelengths.

Thus

$$R = \left[\frac{\lambda_S}{\lambda_{AS}}\right]^3 \left|\frac{\text{anti-Stokes detector quantum efficiency}}{\text{Stokes detector quantum efficiency}}\right| \times \left|\frac{\text{anti-Stokes spectrometer transmission}}{\text{Stokes spectrometer transmission}}\right|$$

For our experimental configuration, we estimate R to be approximately 0.96.

For use of Eq. [1] in determining temperature from experimental data, an additional calibration factor must be evaluated to connect the detected photon numbers to recorded signal voltages for the Stokes and anti-Stokes signals, V_S and V_{AS}. The most practical method for completing this absolute calibration is simply to measure temperature by the SAS method for known conditions, and thereby to determine an overall constant K such that we now have

$$T = \theta_V / \ln(K V_S / V_{AS}). \tag{2}$$

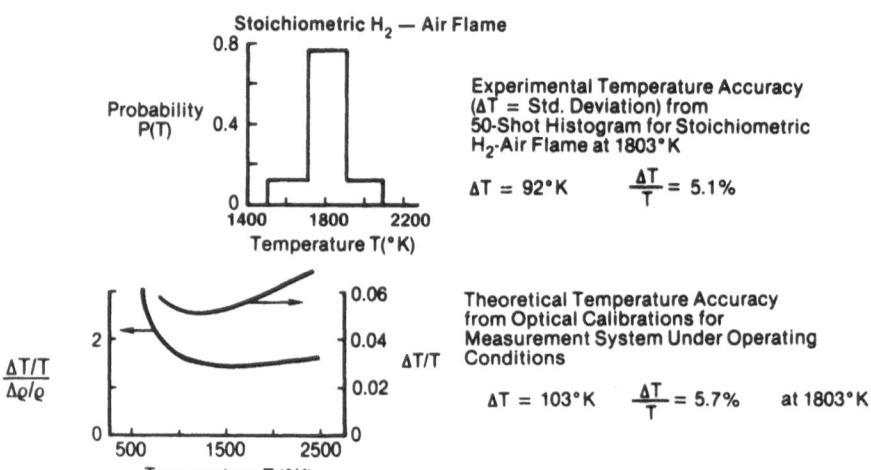

Stoichiometric H_2 — Air Flame

Experimental Temperature Accuracy (ΔT = Std. Deviation) from 50-Shot Histogram for Stoichiometric H_2-Air Flame at 1803°K

$\Delta T = 92°K \qquad \frac{\Delta T}{T} = 5.1\%$

Theoretical Temperature Accuracy from Optical Calibrations for Measurement System Under Operating Conditions

$\Delta T = 103°K \qquad \frac{\Delta T}{T} = 5.7\% \qquad$ at 1803°K

Fig. 7. *Experimental and analytical determinations of fluctuation temperature measurement accuracy. In the upper part of the figure, a pdf for calibration flame is shown. The standard deviation ($\sigma \equiv \Delta T$) for this experimental measurement can be compared with the value found from the analytical treatment illustrated below. The results of this analysis of vibrational Stokes/anti-Stokes temperature measurement accuracy $\Delta T/T$ vs T is presented here both as absolute values and relative to Stokes density measurement accuracy $\Delta\rho/\rho$. These data have been computed using detailed optical calibrations of the laboratory system in use for the experimental calibration shown in the upper part of the figure. The close agreement of the accuracies (with experiment being better than theory!) is fortuitous because of the difficulties of performing highly accurate experimental or theoretical optical calibrations of this sort.*

We show in the upper part of Fig. 7 such a calibration obtained with 50 shots on a laminar stoichiometric H_2/air flame produced on a horizontal porous plug burner. The temperature of this flame had been carefully determined in previous experiments both by thermocouple probes and detailed Raman scans with computer-fits of the vibrational contours (Lapp, 1974). By setting the average temperature for this run to the previously measured value, K was determined.

By presenting the data for this essentially isothermal calibration run in the form of a pdf and by determining its standard deviation, we have also obtained an indication of the fluctuation measurement capability of our apparatus. The shape of the pdf in Fig. 7 can be likened qualitatively to a spectrometer slit function for interpretation of the instrument smearing of the optical data. No attempts were made to deconvolute the resultant fluctuation data because of uncertainties which would have been introduced into the results; the narrowness of this calibration pdf in comparison to those shown for the turbulent combustor in Fig. 5 does show, however, that the broad distributions shown by the pdf's in Fig. 5 are real, and not simply introduced by experimental non-reproducibility.

In order to determine whether or not the accuracy found from calibration pdf's of the type shown in Fig. 7 was reasonable, theoretical estimates of accuracy were first made using a propagation of errors analysis. For a temperature measurement, σ can be calculated approximately from

$$\sigma \approx \left[(\frac{\partial T}{\partial N_S} \sigma_S)^2 + (\frac{\partial T}{\partial N_{AS}} \sigma_{AS})^2 \right]^{\frac{1}{2}} \tag{3}$$

where $\sigma_S = \sqrt{\overline{N}_S}$ and $\sigma_{AS} = \sqrt{\overline{N}_{AS}}$ are the standard deviations in the individual photon counts. Here $\overline{N}_S$ and $\overline{N}_{AS}$ are average values of detected photon numbers. The approximation is good if $\sigma_S/\overline{N}_S$ and $\sigma_{AS}/\overline{N}_{AS}$ are small compared to unity.

From Eqs. [1] and [2],

$$\sigma \approx \frac{T^2}{\theta_V} \left[\frac{1}{\overline{N}_S} + \frac{1}{\overline{N}_{AS}} \right]^{\frac{1}{2}} . \tag{4}$$

The number of photons detected in each channel was determined by measuring the pulse height distribution of individual photon pulses when the photomultiplier was illuminated by a low level light source, then determining the average pulse height, and dividing this value into the integrated pulse height obtained from the burst of photons detected during a measurement at known temperature. In this way we found, for example, with incident laser pulses of about 0.6 J that $\bar{N}_{AS} = 77$ and $\bar{N}_S = 759$ detected photons for scattering from N_2 in a steady ambient pressure hydrogen-air flame at a point where the gas temperature is 1500°K. Values corresponding to other temperatures can be calculated from the known dependence of the Stokes and anti-Stokes intensities on temperature. Using these values we have calculated the standard deviation $\sigma \equiv \Delta T$ as a function of temperature for our experimental configuration. The results are shown in the lower part of Fig. 7. Note the very good agreement between theoretical predictions for σ and the experimental results. The fact that the latter are actually slightly better is clearly fortuitous, since the values of σ cannot be determined to great accuracy from such a limited sample.

There are several reasons why a theoretical calculation of the temperature spread function is desirable in addition to this approximate characterization in terms of standard deviation. First, the propagation of errors analysis is inaccurate at low temperatures, where $\sigma_{AS}/\bar{N}_{AS}$ is no longer small. Second, that analysis does not yield the shape of the spread function. Third, we have not taken into account the additional noise, beyond the ideal lower limit imposed by quantum statistics, introduced by statistical fluctuations in the charge amplification process within the photomultiplier.

For these reasons, we have calculated temperature spread functions using a Monte Carlo program, including both the effects of photon statistics and amplification fluctuations. The latter were represented in terms of the pulse height distribution function measured from the photomultipliers. The Monte Carlo simulation of the temperature spread function works in the following manner: The function is built up from a large number of individual simulated measurements ($\sim 10^4$). In an individual simulation, a random number generator is modified to yield a Poisson-distributed variable to represent the number of photons detected in the anti-Stokes channel. For an individual measurement, a second random number yields an independent result for the Stokes channel. The Poisson distributions are chosen to yield the correct average value and standard deviation for each channel, where the average values are derived from our experimental results. Two more random variables yield photomultiplier amplification factors distributed according to the measured pulse height distributions of the photomultipliers. The result is two simulated voltages representing the Stokes and anti-Stokes photomultiplier signals. From these a temperature is calculated using Eq. [2]. The process is repeated many times, as mentioned previously, to build up the temperature distribution which represents the spread function. Results of several of these calculations are shown in Fig. 8.

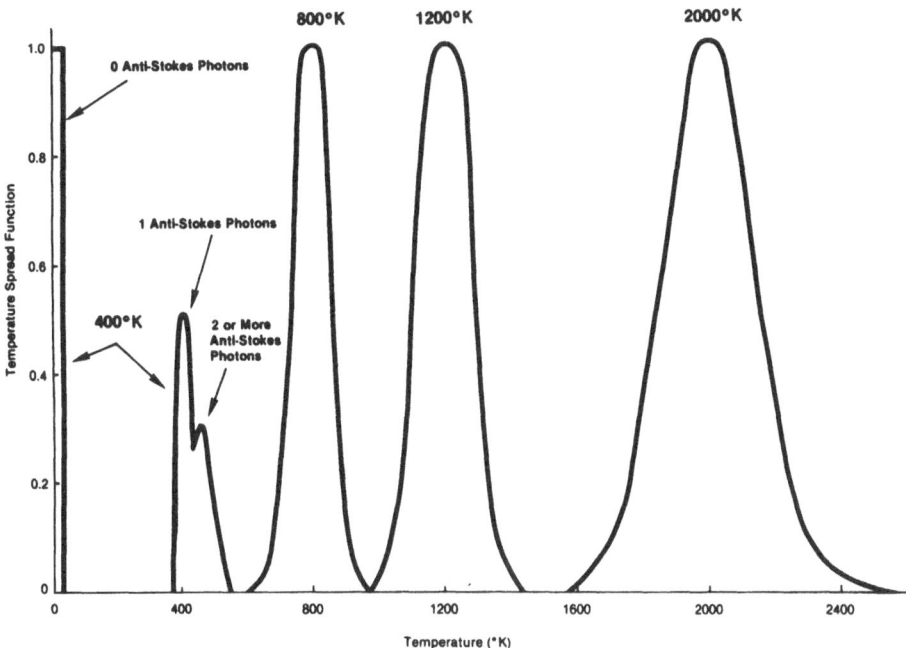

Fig. 8. Temperature spread function vs temperature for various mean flame temperatures, based upon Monte Carlo calculations. The effect of photon statistics and amplifier fluctuations have been included in this analysis.

The calculated shapes of these functions reveal several interesting features. First, statistical fluctuations in the amplification process add little to the probable errors calculated from the simpler analysis. Thus, values of σ measured from these experimental spread functions (σ = half-width at half-maximum/$\sqrt{2\ln 2}$) are roughly 7 % for T $\geqslant$ 800°K, in comparison with the approximate figure of 5 to 6 % shown in Fig. 7.

At temperatures below roughly 600°K, the spread function breaks up into several peaks, representing 0, 1, 2, . . . photons detected in the anti-Stokes channel. The 0-photon peak is located at 0°K, and for our configuration and laser pulse energy, the 1-photon peak is located slightly above 400°K. Finally, significant wings extend out from the center to several times the standard deviation. This effect is particularly noticeable at the higher temperatures. Of course, the probable error and wing significance can be reduced by increasing the laser pulse energy or the detection efficiency.

Conclusion

We have reviewed briefly the current status of fluctuation measurements in unsteady flows — and, especially, in flames — by Raman scattering techniques. Although these techniques have clear limitations in scope, they are already providing new data otherwise unobtainable from more conventional probe systems. Development and utilization of these methods, based upon both pulsed and cw lasers, are leading to data in the form of pdf's, frequency spectra, and spatial mapping.

Acknowledgement

The authors are indebted to S. Warshaw for his valuable contributions to the experimental program and to M.C. Drake for his critical reading of the manuscript. One of the authors (ML) is grateful to I. Chabay and G.J. Rosasco of NBS for illuminating discussions concerning the material presented in Table 1. We are pleased to acknowledge the support of Project SQUID (Office of Naval Research) and the Air Force Office of Scientific Research for portions of this work.

References

Bailly, R., Pealat, M., & Taran, J.P.E., 1978 Real Time Measurement of Temperature and Density in Hot Flows by Raman Scattering. In *Proceedings of the Sixth International Conference on Raman Spectroscopy,* Vol. 2 (ed. E.D. Schmid, R.S. Krishnan, W. Kiefer, & H.W. Schrötter) pp. 256-257. Heyden and Son Ltd.

Barratt, J.J. & Myers, S.A., 1971 New Interferometric Method for Studying Periodic Spectra Using a Fabry-Perot Interferometer. *J. Opt. Soc. Am.,* **61,** 1246-1251

Birch, A.D., Brown, D.R., Dodson, M.G., & Thomas, J.R., 1975 The Determination of Gaseous Turbulent Concentration Fluctuations Using Raman Photon Correlation Spectroscopy. *J. Phys. D: Appl. Phys.,* **8,** L167-L170

Birch, A.D., Brown, D.R., Dodson, M.G., & Thomas, J.R., 1978 The Turbulent Concentration Field of a Methane Jet. *J. Fluid Mech.,* **88,** 431-449

Black, P.C. & Chang, R.K., 1978 Laser-Raman Optical Multichannel Analyzer for Transient Gas Concentration Profile and Temperature Determination. *AIAA J.,* **16,** 295-296

Boiarski, A.A., 1975 Shock-Tube Diagnostics Utilizing Laser Raman Spectroscopy. Naval Surface Weapons Center (White Oak, Silver Spring, Maryland) Technical Report NSWC/WOL/TR 75-53

Bridoux, M., Crunelle-Cras, M., Grase, F., & Sochet, L.R., 1978 Space Resolved Analysis of Flames Stabilized in a Flat-Flame Burner by Multichannel Pulsed Raman Spectroscopy. In *Proceedings of the Sixth International Conference on Raman Spectroscopy.* Vol. 2, (ed. E.D. Schmid, R.S. Krishnan, W. Kiefer, & H.W. Schrötter) pp. 256-257. Heyden and Son Ltd.

Brown, D.R., Thomas, J.R., Pomeroy, W.R.M., & Vaughan, J.M., 1975 The Application of Laser Raman Spectroscopy to Transient Mixture and Combustion Measurements. British Gas Corp., Research and Development Division Report MRS E 250

Chabay, I., Rosasco, G.J., & Kashiwagi, T., 1978 FFT Analysis of Raman Intensities: A Probe of Concentration Fluctuations in Turbulent Flow. In *Proceedings of the Sixth International Conference of Raman Spectroscopy.* Vol. 2 (ed. E.D. Schmid, R.S. Krishnan, W. Kiefer, and H.W. Schrötter) pp. 516-517. Heyden and Son Ltd.

Drake, M.C. & Rosenblatt, G.M., 1976 Flame Temperature from Raman Scattering. *Chem. Phys. Lett.,* 44, 313-316

Drake, M.C. & Rosenblatt, G.M., 1978 Rotational Raman Scattering from Premixed and Diffusion Flames. *Comb. Flame,* 33, 179

Eckbreth, A.C., 1976 Laser Raman Thermometry Experiments in Simulated Combustor Environments. AIAA Paper No. 76-27

Eckbreth, A.C., Bonczyk, P.A., & Verdieck, J.F., 1978 Laser Raman and Fluorescence Techniques for Practical Combustion Diagnostics. *Appl. Spectrosc. Rev.,* 13, 15-164

Hartley, D.L., 1974 Application of Laser Raman Scattering to the Study of Turbulence *AIAA J.,* 12, 816-821

Harvey, A.B., 1978 Chemical Applications of Coherent Anti-Stokes Raman Spectroscopy (CARS). In *Proceedings of the Sixth International Conference on Raman Spectroscopy.* Vol. 1 (ed. E.D. Schmid, R.S. Krishnan, W. Kiefer, & H.W. Schrötter) pp. 355-356i. Heyden and Son Ltd.

Lapp, M., 1974 Flame Temperature from Vibrational Raman Scattering. In *Laser Raman Gas Diagnostics.* (ed. M. Lapp and C.M. Penney) pp. 107-145. Plenum Press.

Lapp, M. & Penney, C.M., 1977 Raman Measurements on Flames. In *Advances in Infrared and Raman Spectroscopy.* Vol. 3 (ed. R.J.H. Clark & R.E. Hester). Chapt. 6. Heyden and Son Ltd.

Lapp, M., 1978 The Study of Flames by Raman Spectroscopy. In *Proceedings of the Sixth International Conference on Raman Spectroscopy.* Vol. 1 (ed. E.D. Schmid, R.S. Krishnan, W. Kiefer, & H.W. Schrötter) pp. 219-232. Heyden and Son Ltd.

Lederman, S., 1976 Some Applications of Laser Diagnostics to Fluid Dynamics. AIAA Paper No. 76-21

Pealat, M., Bailly, R. and Taran, J.P.E., 1977 Real Time Study of Turbulence in Flames by Raman Scattering. *Opt. Comm.* 22, 91-94

Pealat, M., Taran, J.P., & Schnepp, O., 1978 Analytical Applications of CARS in Combustion Diagnostics. In *Proceedings of the Sixth International Conference on Raman Spectroscopy.* Vol. 2 (ed. E.D. Schmid, R.S. Krishnan, W. Kiefer, & H.W. Schrötter) pp. 260-261. Heyden and Son Ltd.

Setchell, R.E., 1976 Time-Averaged Measurements in Turbulent Flames Using Raman Spectroscopy. AIAA Paper No. 76-28

Tolles, W.M., Nibler, J.W., McDonald, J.R. & Harvey, A.B., 1977 A Review of the Theory and Application of Coherent Anti-Stokes Raman Spectroscopy (CARS). *Appl. Spectrosc.* 31, 253-271

Wang, J.C.F. & Gerhold, B.W., 1977 Measurements on Turbulent Hydrogen Flames in a Circular Air Duct., AIAA Paper 77-48

Williams, W.D., Power, H.M., McGuire, R.L., Jones, J.H., Price, L.L., & Lewis, J.W.L., 1977 Laser-Raman Diagnostics of Temperature and Number Density in the Mixing Region of a Rocket Engine Exhaust and a Coflowing Air Stream. AIAA Paper No. 77-211.

Diagnostic Techniques in Combustion MHD Flows

by

Sidney A. Self
High Temperature Gasdynamics Laboratory
Stanford University

1. Introduction

The U.S. national program in MHD electric power generation is predicated on the direct firing of pulverized coal, using MHD as a topping cycle for a regular steam plant. Extensive system studies predict the thermal efficiency of a large combined MHD/Steam plant will be in the range 50 - 55 % compared with 35 - 40 % for a conventional steam plant. The basic idea is to burn a fossil fuel at a very high temperature ($\sim$ 3000 K), seed it with an easily ionizable material (K, Cs) and expand the resulting combustion products plasma through a nozzle into a generator channel at high (usually subsonic) velocity u $\leqslant c_s \sim$ 1000 m/s. The generator channel is immersed in a strong transverse magnetic field B, and is fitted with electrodes in the walls for electric power take off. The induced electric field in the plasma is $\underline{E} = \underline{u} \times \underline{B}$, the induced current density is $\underline{J} \sim \sigma \underline{E}$, where σ is the electrical conductivity, and the generated power density is $P \sim \underline{J} \cdot \underline{E} \sim \sigma u^2 B^2$.

To be economically competitive the power density should be of the order of 100 Mw/m^3. Since u is fixed at $\leqslant$1000 m/s, and the practical limit of large scale superconducting magnets is B$\sim$5T, this leads to a requirement for a conductivity of $\sim$10 mho/m, and corresponding fields and currents of 5 kV/m and 5$\times$10^4 A/m^2.

The electrical conductivity $\sigma = e\, n_e\, \mu_e$ is determined by the product of the concentration n_e of electrons and their mobility μ_e. The electron concentration can be calculated from Saha equilibrium in terms of the concentration of easily ionizable seed, while for typical combustion MHD conditions at a pressure of p$\sim$1 Atm, the effective electron mobility is $\mu_e \sim$0.5 m^2/V $\cdot$ s and is a slow function of temperature. This leads to a requirement of electron concentration $n_e \sim$10^{20} m^{-3}. At practical levels of seed concentration, e.g. $\sim$1 % by mass of K, Saha's equation then leads to a requirement for temperatures in the range 2800-3000 K at the generator input.

To achieve fossil fuel combustion temperatures of this order requires a high degree of air preheat ($\sim$ 1500-1800 K) which is provided by regenerative, pebble-bed type preheaters. The exhaust temperature, 2000-2200 K, is determined by the fact that at this temperature the electrical conductivity has fallen to such a low value that further MHD power extraction is uneconomic. The high combustion temperature leads to formidable materials problems for the walls of the flow train.

While a considerably body of experience has been accumulated with cleanfuel MHD, the additional problems associated with coal-firing have received less attention until recently. The principal new problems relate to the presence of mineral ash in the pulverized coal — typically 10 % by mass with principal constituents in order SiO_2, Al_2O_3, Fe_2O_3, CaO, MgO, as well as SO_3.

Since the combustion temperature is much higher than in conventional coal-fired plants, a higher fraction of the ash is vaporized, and the particulate ash droplets appear to be distributed over a smaller size range (perhaps $\sim$ 1 μm instead of 10 μm mass mean diameter). It is anticipated that a multistage cyclone combustor will be used to remove a large fraction of the ash, but still a significant fraction ($\sim$ 10 %) of the mineral matter will be carried over into the generator in the form of vapor and particulates. This has two important implications: first as regards the effects on the plasma conductivity in the volume of the channel; and second with regard to the operation of the generator channel with slagging walls, including the mechanism of current transfer through the slag layer.

The performance of an MHD generator depends not only on the characterization of the plasma as regards to species concentrations and electrical properties, but also on the fluid mechanics and heat transfer associated with the turbulent channel flow. The latter determine the spatial distribution of temperature in the channel, and hence the distribution of composition, and electrical properties.

Turbulent channel flow computer codes have been developed which calculate the flow and temperature field by matching a two dimensional boundary layer model to the core flow, and include the electromagnetic interaction terms. The latter comprise the $\underset{\sim}{J} \times \underset{\sim}{B}$ force term in the momentum conservation equation and the $\underset{\sim}{J} \cdot \underset{\sim}{E}$ Joule dissipation term in the energy conservation equation. The inclusion of these terms can result in considerable modifications of the ordinary gasdynamic flow excluding electromagnetic interaction, and produce significant changes in the wall friction and heat transfer coefficients.

The turbulent boundary layers can have a profound effect on MHD generator performance because the wall temperature is necessarily appreciably lower than the core plasma temperature. The resulting thermal boundary layer has a lower electrical conductivity and can give rise to significant internal resistance (or voltage drop) in the generator. Consequently detailed understanding of the boundary layers, on both the electrode wall [1] and insulating sidewall [2] is very important in calculating generator performance.

Finally, mention should be made of the significance of temporal fluctuations of conductivity on the performance of MHD generators. Calculations indicate that they produce an effective conductivity which is lower than the time-average value, which results in a degradation of generator performance. The effect is stronger the higher the fluctuation amplitude and increases with magnetic field strength. Such fluctuations may be induced by combustion instabilities or possibly from inherent magnetoplasma instabilities. Thus the study of fluctuations of pressure, temperature and electrical conductivity in MHD generator channels is of vital concern to the evaluation of MHD power generation.

II. Stanford MHD Research Facility

The High Temperature Gasdynamics Laboratory's MHD facility is a sub-scale system designed for flexibility in research into the basic physics of MHD phenomena rather than the demonstration or development of efficient power generation systems. The latter can only be effectively carried out in facilities of pilot plant scale, say $\sim$ 100 MW (thermal), because of the inevitable effects of unfavorable surface to volume ratio in sub-scale systems.

The present hardware centers around either of two combustors, the M-2 system rated at 2 MW (th) and the M-8 system rated at 8 MW (th). These are liquid fuel spray combustors, stabilized by counter swirling injection of oxygen and nitrogen (used as a diluent to control the temperature). Simulation of coal-firing is achieved by admixture of a pulverized coal slurry to the ethanol fuel. Potassium seed, up to $\sim$ 2 % by mass of K, is added by dissolving KOH in the fuel.

The combustor normally feeds into a MgO brick-lined plenum chamber to provide adequate residence time for mixing and to reduce fluctuations of plasma properties. The exit of the plenum is fitted with a nozzle to accelerate the flow into various experimental channel section which are constructed to suit the purpose of particular experimental studies. Sometimes, especially for preliminary development of diagnostic techniques, use is made of the free jet issuing from the nozzle.

The channels are usually constructed in a water cooled metal body lined with insulators and electrodes, depending on the configuration required. The insulators are normally of MgO firebrick (occasionally dense Al_2O_3) and the electrodes are stainless steel and water cooled to run at a surface temperature $\sim$ 1200 K. Below this temperature seed material (K_2CO_3) condenses on the surface, while much higher temperatures result in corrosion. With coal firing, the walls are usually of water-cooled metal, grooved and filled with castable ceramic to ensure attachment of the slag layer, which does not adequately wet plain metal surfaces.

The channels are sized to fit the available electromagnet. This is a conventional iron yoke, water cooled copper magnet producing fields up to 2.7 Tesla (27 k Gauss) in a working volume between pole-pieces which allow channel internal dimensions of 5 cm wide (B field direction) X 10 cm high X 70 cm long.

The channel exit attaches to a diffuser and exhaust section which includes a water spray quench and scrubber system for removing particulates.

Storage capacity of fuel, oxygen and nitrogen are sufficient to allow tests of several hours duration. Most tests are run under subsonic conditions (<1000 m/s), but occasionally supersonic tests have been run. Total mass flow rates may range up to 1 lb/sec.

In addition, a small superconducting magnet capable of 6 T is available and for the future a large superconducting magnet with a field of 7T, a bore of 55 cm and a working length of 1.5 m will be available.

Metering of fuel, oxygen, nitrogen and potassium seed is routinely measured, as are wall temperatures and pressures. The outputs from the various transducers are coupled into a minicomputer for on-line display. A patch panel allows the electrodes to be connected to appropriate loads in a variety of configurations and currents and voltages are recorded via the minicomputer. To enhance the currents and MHD interaction to simulate full-scale generator operation, the electrodes are often powered from a battery pack.

A great variety of diagnostic techniques have been developed and used to measure the properties of the plasma channel flow and to study slag layer dynamics and electrode current transfer and breakdown effects.

A major problem with any diagnostic technique is set by the difficulty of access and the hostility of the environment. For optical diagnostics, ports and windows are normally shuttered and purged with dry N_2 to prevent condensation of water and potassium carbonate (dew point $\sim 1000°C$). The bulk of the surrounding magnet particularly makes for difficult access problems. In addition, the stray field precludes the use of photomultipliers, oscilloscopes, etc. in close proximity to the rig.

III. Diagnostic Techniques
A major component of the Stanford program in MHD research is the development and use of diagnostic techniques for characterizing the plasma flow inside an operating generator, with the purpose of studying the basic physical processes, and providing engineering data useful in the design of large scale systems.

Both the plasma properties, especially the electrical conductivity, and the flow properties are important in determining generator behavior, and the boundary layers, in particular, are of interest since they play an important role in determining generator performance. This places a premium on measurement techniques capable of high spatial resolution. Moreover, fluctuations also have an important effect on generator performance, so that temporal resolution, (to $\sim 1\text{-}10$ kHz) is also highly desirable.

A considerable variety of techniques are available for measuring the various quantities of interest, which include velocity (including turbulence characterization), temperature, electrical conductivity, electron concentration, electron mobility, and seed (K) concentration. In addition, for coal-fired flows, the ash loading and particulate size distribution is important, as is the thickness and surface temperature of the surface slag layers. The majority of the techniques are non-intrusive optical ones, though some material probe methods are valuable, in spite of the fact that they must be either heavily cooled or restricted to very short insertion times.

An extensive review of the status of diagnostic techniques for combustion MHD flows, as of two years ago, was given in reference [3]; A more recent review, including Soviet contributions in this area, appears in reference [4]. In the following, the status of the various techniques will be summarized according to the quantity measured; for further details the cited references should be consulted.

(i) Velocity
Boundary layer profiles of average streamwise velocity and turbulence intensity have been measured using laser anemometry [5] in two distinct configurations, the configurations being determined primarily by the access possibilities.

690   Sidney A. Self

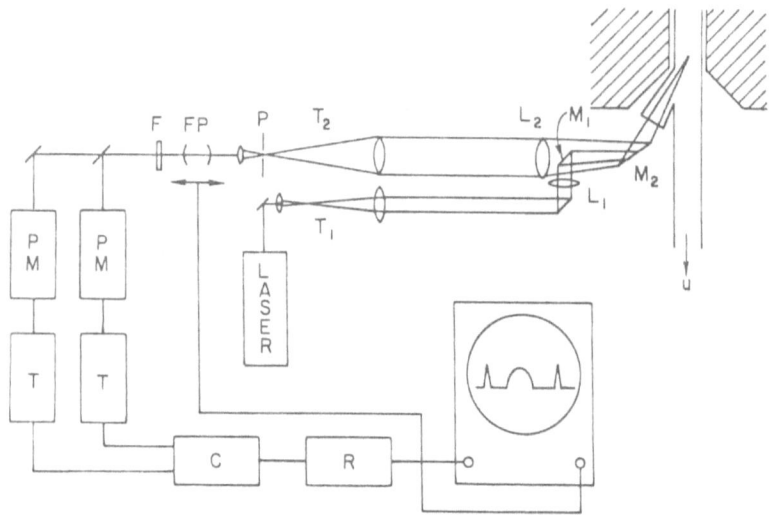

Fig. 1. Single beam backscatter anemometer for electrode wall boundary layer.

In the first configuration, used to measure the electrode wall boundary layer
[1], the only possibility for access was from a port in the downstream channel
sidewall looking upstream into the magnet, as shown in Fig. 1. This single-
beam backscatter configuration, aligned at a small angle to the flow axis, combined with the high velocities, results in a Doppler frequency shift (several
GHz) which is too high to measure by heterodyne techniques. Consequently it
was measured as a wavelength shift using a scanning Fabry-Perot interferometer. In this system, and in that described below, micron-sized ZrO_2 powder was
added to the flow to act as scattering centers. Boundary layer profiles with a
spatial resolution of $\sim$ 0.1 mm were made to within $\sim$ 0.3 mm of the electrode
surface, at both subsonic and supersonic core velocities. An example of a measured profile at a free stream velocity of 1600 m/s is shown in Fig. 2.

In the second configuration, used to measure the insulating sidewall boundary
layer [2], access was possible between the magnet pole pieces through the top
and bottom electrode walls. This allowed the use of the more conventional
dual beam, real fringe technique in (near) forward scatter, as shown in Fig. 3,
which shows how access and scanning of the measurement point was achieved
using a series of mirrors. Fig. 4 compares the measured and computed profiles
of average velocity with and without electromagnetic interaction. The marked
change in profile is due to the fact that the $\underset{\sim}{J} \times \underset{\sim}{B}$ braking force is strongest in
the core flow, because the current is concentrated there due to the lower con-
ductivity in the sidewall thermal boundary layer. This results in significant

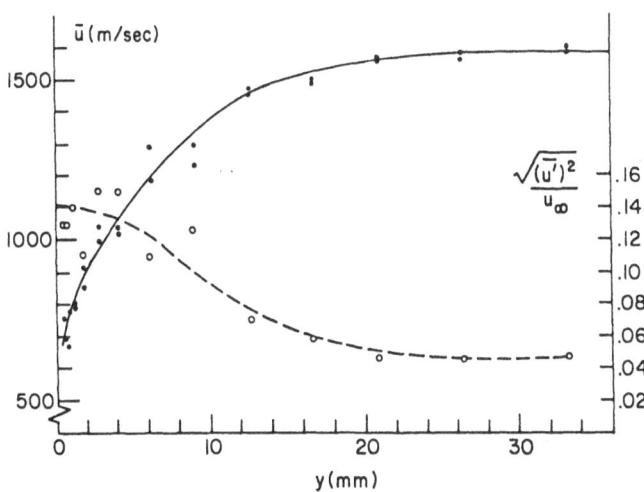

Fig. 2. Electrode wall boundary-layer velocity profiles from single beam anemometer. Mean velocity — solid line and dots; turbulence intensity — broken
line and circles.

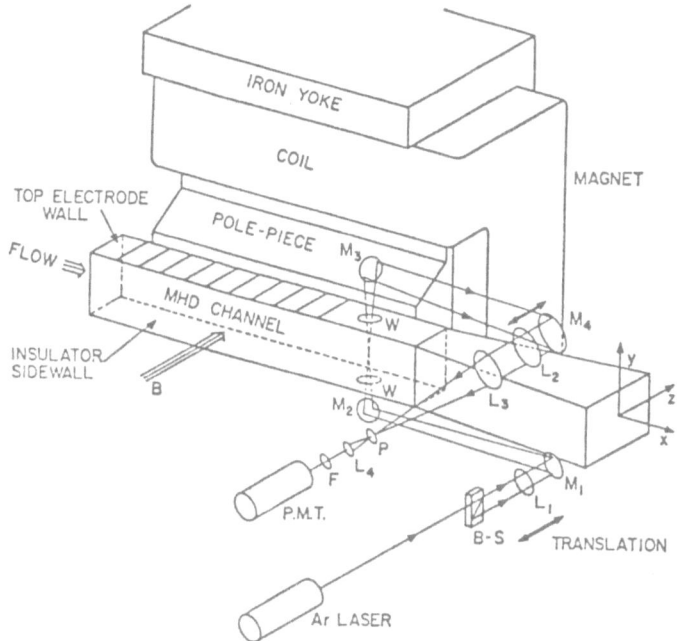

Fig. 3. Dual beam forward scatter anemometer for side wall boundary layer.

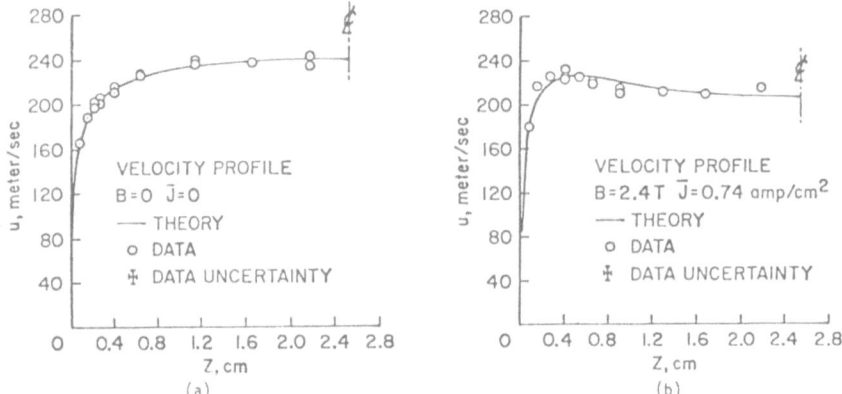

Fig. 4. *Sidewall boundary layer profiles of average streamwise velocity: (a) without MHD interaction, (b) with MHD interaction.*

increases in the friction and heat transfer coefficients. A reduction of turbulence intensity was observed in the presence of magnetic field due to magnetic damping of turbulence.

(ii) Temperature

The gas temperature is among the most critical parameters for operation of MHD generators. Since thermochemical equilibrium prevails, except in near-electrode arcs, the temperature is a measure of the enthalpy and composition of the gas.

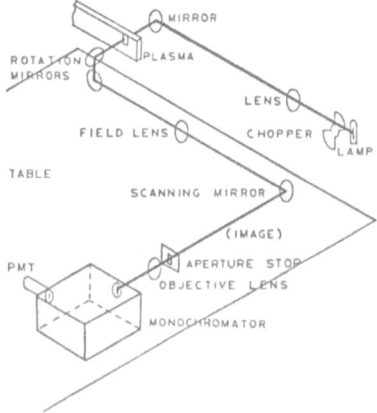

Fig. 5. *Optical system for spectroscopic measurements of temperature and electron concentration.*

(a) **Resonance Line Reversal Method.** The old-established resonance line reversal method, which yields a line-of-sight average temperature, has been adapted to measure boundary layer temperature profiles in MHD channels, [1, 6]. Modification of the standard reversal procedure is necessary for two reasons. First, because the plasma temperature is higher than the temperature available from tungsten ribbon lamps, the true reversal condition cannot be attained, and some sort of extrapolation technique is necessary. Second, under typical conditions using small concentrations of Sodium seed, the temperature and seed concentration gradients in the boundary layers along the line of sight, cause the center of the emission line to be self-reversed. To overcome this, measurements are made in the line wing where the optical depth is $\leqslant 0.5$, and the radiative transfer equation is solved, using an assumed boundary layer temperature profile to obtain the core temperature [7]. The effects of fluctuations and nonuniformities has also been taken into account [8]. Fig. 5 shows a schematic of the optical system used to make electrode wall boundary layer profile measurements with a spatial resolution of 0.3 mm to within 0.4 mm of the electrode. Fig. 6 compares experimental and theoretical profiles with and without current drawn to the electrode. The temperature rise due to Joule heating is clearly evident.

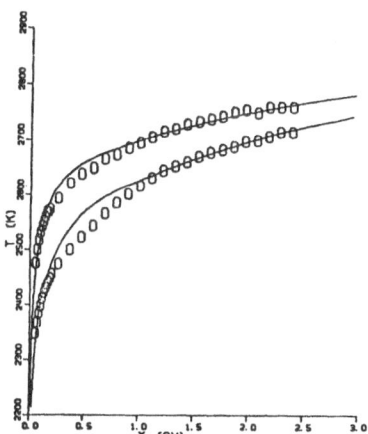

Fig. 6. Comparison of measured and computed electrode wall boundary layer temperature profiles: lower curve J = 0; upper curve J = 1.54 A/cm². (u ~ 350 m/s).

(b) **Laser Fluorescence Technique.** As an alternative to line-of-sight emission/ absorption techniques, a "point" measurement technique has recently been developed at Stanford [9] using laser fluorescence. A tunable dye laser is used to pump transitions from an excited electronic level of Na or K to a higher level. Since the population of the absorbing level is exponentially dependent on temperature, the resulting fluorescence re-radiation is a sensitive function of relative temperature and temperature fluctuations. Observations of this signal along a path orthogonal to the exciting beam provides a spatial resolution of $\sim$ 1 mm^3. Spatially resolved measurements, relative temperature profiles and temperature fluctuations, as well as two point fluctuation correlations, should be feasible in MHD plasma using this technique.

(iii) Electrical Conductivity
There is no known non-intrusive technique for directly measuring the electrical conductivity. However, four-wire conductive probes have been used at Stanford. While reliable measurements can be made in seeded Argon flows from an arc-jet [10], their use in combustion MHD flows poses serious problems due to seed condensation, probe heating and electrode corrosion.

Soviet workers have developed rf inductive probes [11] which avoid such problems by enclosing the coils in a protective dielectric housing of airfoil shape. The change in Q of the coil due to eddy currents induced in the surrounding plasma as the probe is rapidly inserted and withdrawn from the plasma, is used to measure the plasma conductivity. The probe is calibrated by immersion in conducting salt solutions. Corrections must be applied for the conductivity profile in the boundary layer surrounding the probe, and the absolute accuracy is probably no better than $\pm$50 %.

(iv) Electron Concentration
Methods for the measurement of electron concentration that have found use in combustion MHD flows include electric (Langmuir) probes, a spectroscopic line intensity method, and submillimeter laser interferometry.

(a) **Electric (Langmuir) Probes.** The use of Langmuir probes in flames and MHD type plasmas has been reviewed in reference [12], and the utility of such probes in combustion MHD flows was demonstrated by Clements. The appropriate theory predicts the ion current collected by a negative biased spherical probe as (MKS units):

$$I \approx 12(\mu_i \, \epsilon_o)^{\frac{1}{4}} \, (n_e e \, u)^{\frac{3}{4}} \, V^{\frac{1}{2}} \, r_p^{\,5/4}.$$

Here μ_i is the ionic mobility, ϵ_o the vacuum permittivity, e the electron charge, u the gas velocity, V the probe potential and r_p the probe radius. The probe actually measures the ion flux $(n_e u)$ and a separate knowledge of u is required to determine n_e.

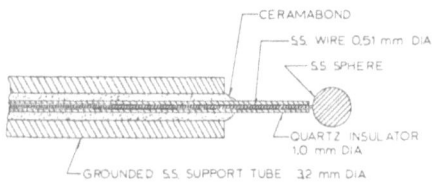

Fig. 7. Spherical Langmuir Probe: Construction details.

The practical problems caused by probe heating due to the high temperatures and velocities is overcome by using very short residence times (< 100 ms). An incidental problem resulting from the use of a cold probe is the deposition of potassium carbonate on the surface. However, this is very soluble and can be washed off between insertions. The typical probe construction is shown in Fig. 7. Recently probe measurements have been made [14] inside a combustion MHD channel using a fast reciprocating probe driven by an air cylinder as shown in Fig. 8. While the absolute accuracy of the probe is not high (~ ±50 %) it is simple and inexpensive, and gives good spatial and temporal resolution for relative measurements.

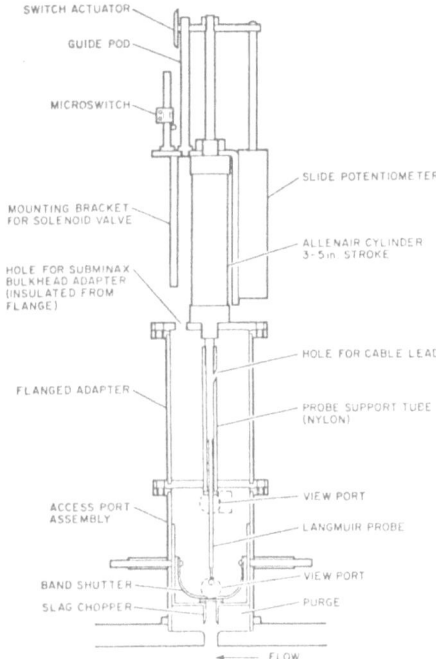

Fig. 8. Probe access port and sweep mechanism.

(b) Spectroscopic Method. A method for measuring the electron concentration has been developed [1, 6] using the absolute emission intensity of high lying electronic levels of potassium seed atoms. The measurement yields line of sight averages which must be unfolded to yield core values, and while the absolute accuracy is not high the technique offers high spatial resolution (~ 0.3 mm) for boundary layer measurements. The apparatus is essentially the same as that used for line-reversal temperature measurements (Fig. 5). A feature of the technique is that the high lying electronic levels should be in thermal equilibrium with the electron temperature, even though the lower levels may not be. This allows it to be used to explore non-equilibrium effects in MHD boundary layers. Fig. 9 compares measured n_e profiles with those calculated and with those deduced from Saha equilibrium using the measured temperature. Within 2 mm of the surface a small electron concentration non-equilibrium is suggested, but the increase of n_e relative to equilibrium is considerably less than that predicted by kinetic considerations using conventional three-body electron recombination coefficients.

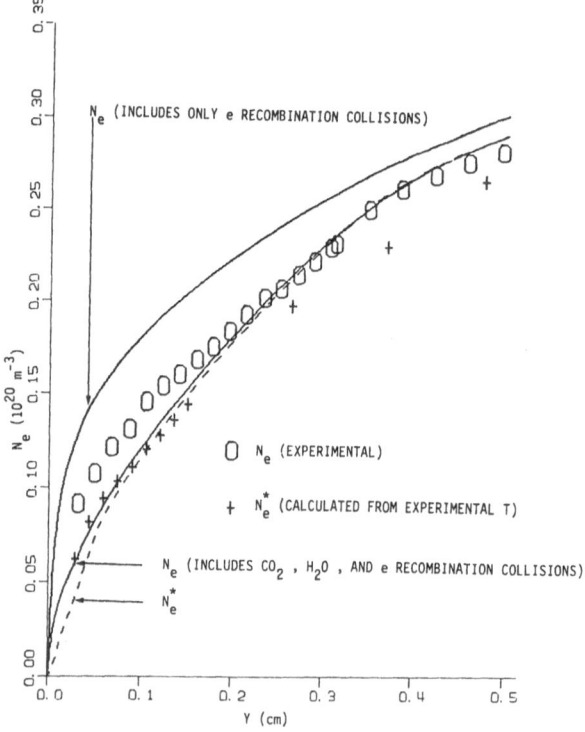

Fig. 9. Comparison of experimental and calculated profiles of n_e in the electrode wall boundary layer.

(c) Submillimeter Laser Interferometry. The utility of submillimeter (far infrared) laser wave propagation as a non-intrusive technique for accurate measurements of electron concentration, n_e, and mobility, μ_e, in combustion MHD flows has been established by preliminary measurements [15]. A laser operating at 394 μm and 496 μm, optically pumped by a CO_2 laser, was used with a Michelson interferometer (Fig. 10) to measure the contribution to the refractive index by free electrons from the phase shift observed with the addition of potassium seed to the free jet from the M-8 combustor. The mobility was obtained from a measurement of the absorption of the submillimeter radiation by the plasma. From the separate measurements the electrical conductivity $\sigma = e\,n_e\,\mu_e$ is given directly.

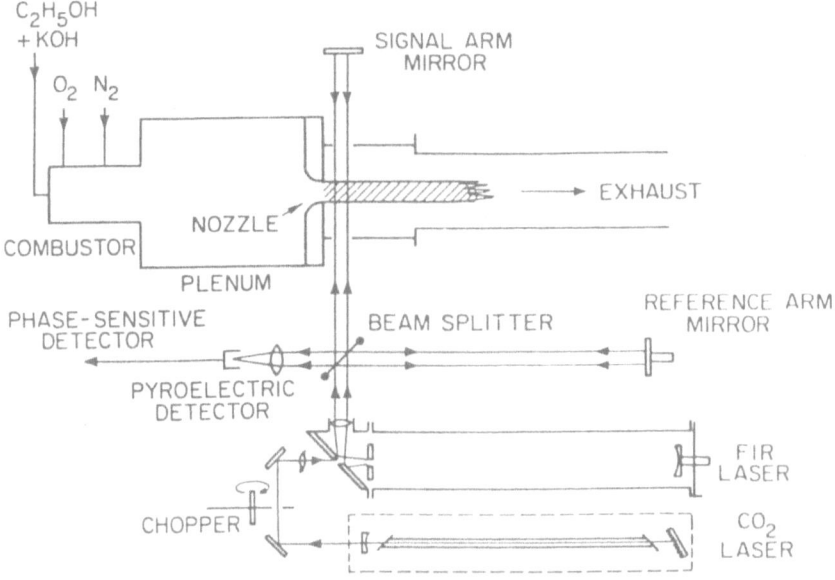

Fig. 10. Schematic of submillimeter laser interferometer system.

For a Michelson interferometer the number of fringe shifts N is given by

$$N = \frac{1}{n_c \lambda} \int_0^L n_e \, dx$$

where L is the plasma width, λ is the free-space wavelength and (MKS units)

$$n_c = \frac{\epsilon_o m_e c^2}{4\pi^2 e^2 \lambda^2} \approx \frac{1.12 \times 10^{15}}{\lambda^2}$$

is the critical density above which free propagation ceases.

The double pass transmission is given by

$$Tr = \exp - \frac{2}{n_c c} \int_0^L n_e \, \nu \, dx$$

where ν is the effective electron collision frequency for momentum transfer. ($\mu_e = e/m_e \, \nu$)

The choice of working frequency is set by the requirement that the number of fringe shifts should be neither too large nor too small, and that the attenuation be not too large. For typical combustion MHD plasma conditions

$$n_e \sim 10^{20} m^{-3}, \nu \sim 3 \times 10^{11} \, s^{-1},$$

and for small installations this leads to a requirement for submillimeter waves. For instance for L = 50 mm and λ = 0.5 mm, we find N $\approx$ 2 and Tr $\approx$ 0.14. In practice the shift can be determined to $\sim$ 1/50 fringe, leading to an accuracy of $\sim$ 1 % in the determination of $\overline{n}_e$. For larger systems, the use of shorter wavelengths is indicated to maintain reasonable transmission.

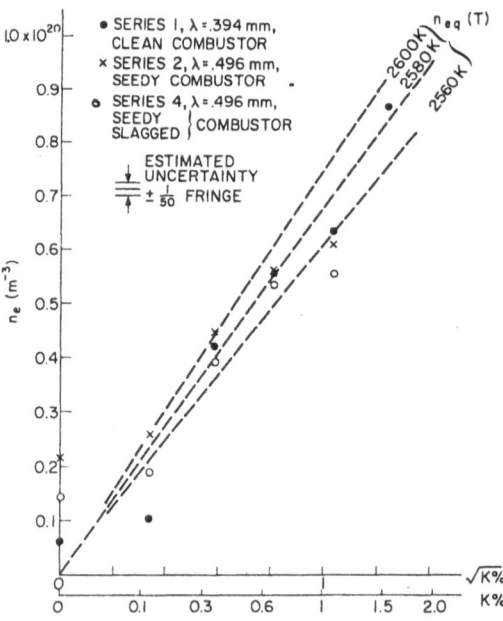

Fig. 11. Measurements of electron concentration in free jet by submillimeter laser interferometer versus square root of potassium seed concentration: broken lines show values computed from equilibrium for temperatures shown.

Fig. 11 shows measured n_e values plotted against $\sqrt{K}$, where K is the input potassium seed concentration to the combustor. The broken lines show the equilibrium n_e values calculated from an equilibrium properties program for three temperatures at intervals of 20 K. Simultaneous measurements were made of the temperature by line reversal and the temperature was also calculated from the metered combustor inputs and measured heat losses. The measured n_e values agreed with those calculated from equilibrium for both the measured and calculated temperatures all to within their respective estimated uncertainties. Because the electron concentration is such a strong function of temperature, it is preferable to deduce T from a measurement of n_e than vice versa.

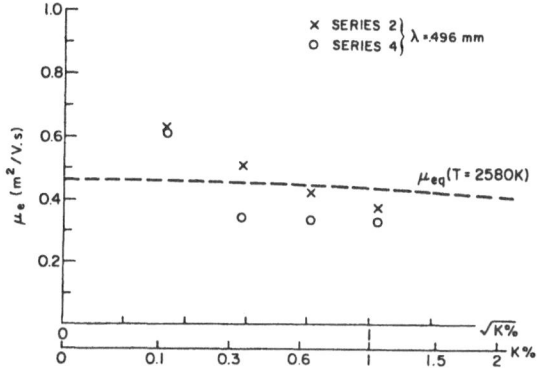

Fig. 12. Mobility measured by submillimeter wave alternation compared with theoretical estimate.

Measured values of the electron mobility versus $\sqrt{K}$ are shown in Fig. 12 where they are compared with those calculated from the best estimates of collision cross sections for the species present. It is anticipated that more accurate measurements will be possible with refinement of the technique.

A considerable advantage of this technique is that it should be unaffected by the presence of ash droplets, since the Rayleigh scattering should be negligible at these long wavelengths. This should be particularly valuable on large coal-fired systems which are optically thick due to scattering and thermal radiation at visible wavelengths.

(v) Particulates
In coal-fired MHD flows, it is important to know the ash droplet loading and size distribution. Assuming a cyclone type combustor is used to reject ~ 90 % of the ash, the mass fractional loading at the input to the generator channel

may be of the order of 1 % or less. In view of the higher combustion temperature and rejection of large particles in the cyclone, the mean droplet size entering the channel may be expected to be much smaller (say ~ 1 μm) than in conventional pulverized coal fired boilers (~ 10 μm).

Because the problems of using sampling probes to capture high velocity, high
temperature liquid droplets are formidable, our attention has focused on non-
intrusive optical techniques, using lasers to facilitate the rejection of background radiation by spectral filtering. The simplest of these is the transmissometer which measures the extinction of a collimated light beam resulting from
scattering and absorption by the particles. This is a well-established technique
for cold flows, which effectively measures the total projected area of the particles in the beam. To deduce a mean particle diameter, a separate measure is
required of the fractional loading of ash.

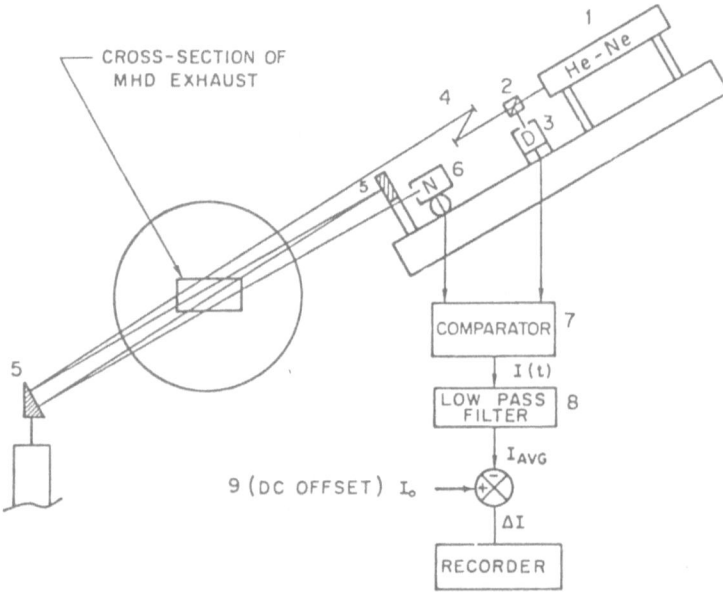

Fig. 13. Schematic of He-Ne transmissometer for ash droplet measurements.

(a) **Laser Transmissometer.** Fig. 13 shows a two-pass transmissometer, employing a He-Ne laser which has been used [16] to measure the average particle
size in the plasma issuing from a channel (5×10 cm) attached to the M-8 combustor fired with a slurry of pulverized coal in ethanol, at input ash mass fractions of ~ 0.50 %. The ash mass fraction at the measurement station was
estimated as ~ 0.23 %, taking account of loss by evaporation, and particle

deposition in the combustor. Under these conditions the extinction is small, ~ 1 %, and considerable care must be taken to obtain reliable measurements. The Sauter mean diameter obtained from these measurements was 1.5 μm with an estimated uncertainty of ±50 %.

(b) Multiple Wavelength Laser Transmissometer. To avoid the necessity of estimating the ash mass fraction, a new two-wavelength laser transmissometer system is being constructed, which should allow the measurement of smaller extinction values. By using two widely separated wavelengths, it is possible to determine the mean particle size and ash loading.

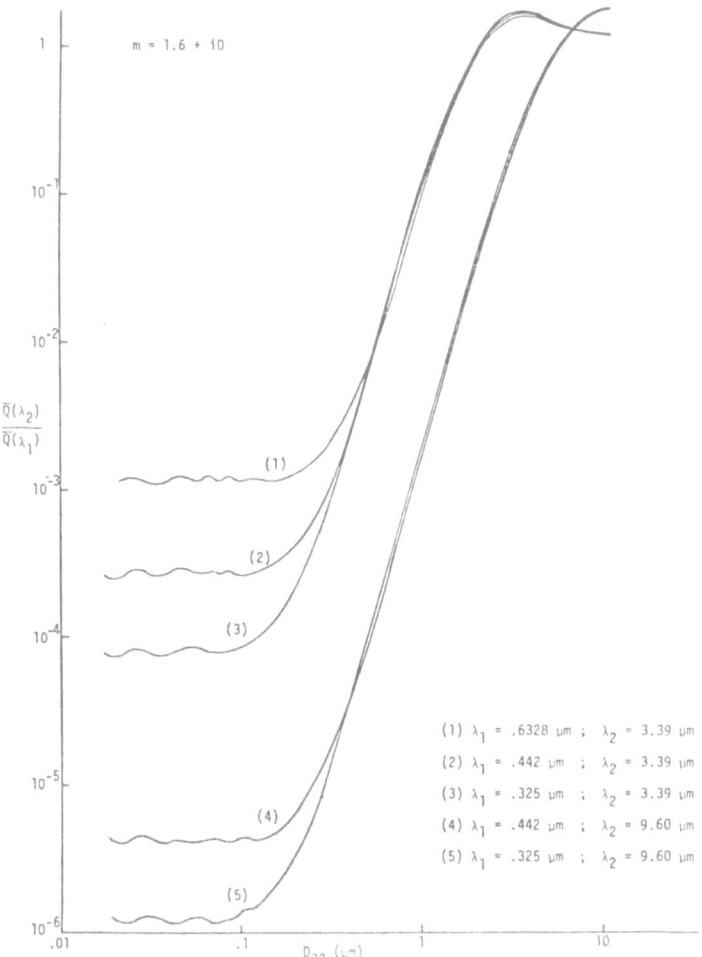

Fig. 14. Ratio of average extinction efficiencies for selected pairs of laser wavelengths versus Sauter mean diameter.

Fig. 14 shows the calculated ratios of the extinction coefficients as a function of the Sauter mean diameter D_{32} for a dispersion of droplets of refractive index m = 1.6 + i 0, for various pairs of readily available laser wavelengths. It can be seen that the combination of λ_1 = 0.325 μm (He-Cd laser) and λ_2 = 3.39 μm (He-Ne laser) should allow D_{32} values in the range 0.1 - 3.0 μm to be determined, while the combination of λ_1 = 0.442 μm (He-Cd laser) and λ_2 = 10.6 μm (CO_2 laser) should cover the range 0.2 - 10 μm. However, the lower limit in either case may be excluded by the fact that the extinction for the longer wavelength may be too small on subscale systems at ash loadings of interest. On larger scale systems the extinction would be correspondingly greater and should facilitate the measurement.

(c) **In-Situ Particle-Sizing Counter.** In separate work, a laser scattering device has been developed which determines the particle size distribution from the amplitude distribution of scattered light pulses as individual particles pass through the focus of a laser beam. It is essentially an extension of the commercially available sampling type of optical sizing counter to allow in-situ measurements, and is reported in more detail in reference [17].

As presently configured, the collection axis is at 12° to the forward direction, and allows measurements of particles in the size range 2 - 30 μm. The measurement volume is then $\delta V \sim 10^{-5}$ cm^3, which allows concentrations as high as 10^5 cm^{-3} to be measured without coincidence problems.

Compared with a sampling type optical counter, where the sampled and diluted particles are passed through a uniformly illuminated jet, a major problem with the in-situ device is that the scattered signal amplitudes depend not only on particle size but also on their trajectory through the measurement volume. Thus a monodispersion will yield a signal pulse height distribution with a sharp upper cut-off corresponding to particles passing through the center of the measurement volume, and extending down to small values, corresponding to particles passing off the center.

A novel and vital feature of the present instrument is the use of an on-line computer-based inversion technique, which unfolds the measured count distribution of pulse heights to yield the true particle size distribution. The ability of the inversion technique to unfold the fine structure of a polydispersion artificially created by the addition of four separate monodispersions was demonstrated.

Measurements have also been made of the size distribution of glass beads introduced into a bench scale flat flame burner, to test the instrument's immunity to background radiation and refractive index fluctuations.

At present, the instrument has been demonstrated to be capable of measuring particle size distributions in the range 2 - 30 μm at concentration up to 10^5 cm^{-3} in small laboratory flames. By modifying the configuration, the range can probably be extended from 0.5 to 50 μm or more. Its extension to the high velocity, high temperature conditions of coal-fired MHD plasma flows poses a number of problems, primarily the need for very much faster pulse-processing electronics to handle the very short pulses resulting from the high velocity.

IV. Conclusion
The particular diagnostic measurement needs for understanding and monitoring the performance of combustion MHD generator systems have been described. Despite the hostile environment and access difficulties, a considerable variety of techniques have been developed for measuring velocity, gas temperature, electron concentration and mobility, electrical conductivity, alkali seed atom concentration and ash droplet loading and size distribution. Many of these techniques have been used as research tools in sub-scale laboratory systems. Their extension to use for monitoring important performance-determining parameters in pilot scale systems is, in principle, straightforward, but will present some challenging technical problems.

Acknowledgments
It is a pleasure to record the vital contributions and active support of many members of the High Temperature Gasdynamics Laboratory to the work described herein. Particular mention should be made of Professor R.H. Eustis, Laboratory Director, for constant encouragement, of Professor C.H. Kruger who has supervised the spectroscopic work, of Dr. D. Holve who was instrumental in the particulate measurements, and of Prof. R.M. Clements who pioneered the Langmuir probe measurements and assisted with the submillimeter laser technique. Last, but by no means least, the contributions of numerous graduate students and of the technical staff, especially Mr. F.O. Reigel, to the diagnostic measurement efforts are gratefully acknowledged.

This work has been supported by the Department of Energy under Contract EX 76-C-01-2341, by the National Science Foundation under Grant ENG-76-04116-A01, and by the Office of Naval Research through Project SQUID.

References

1. Daily, J.W., Kruger, C.H., Self, S.A. and Eustis, R.H., "Boundary Layer Profile Measurements in a Combustion-Driven MHD Generator", *J. AIAA,* **14**, 997-1005 (1976)
2. Rankin, R.R., Self, S.A. and Eustis, R.H., "A Study of the MHD Insulating Wall Boundary Layer", *16th Symposium on Engineering Aspects of MHD,* Pittsburgh, May 1977
3. Self, S.A. and Kruger, C.H., "Diagnostic Methods in Combustion MHD Flows", *AIAA J. Energy,* **1**, 25-43 (1977)
4. "Joint US-USSR Report on the Status of Open Cycle MHD Power Generation", Argonne National Laboratory, 1978, Chapter 14 "Diagnostics" by S.A. Self, I.A. Vasil'eva and A.P. Nefedov
5. Self, S.A., "Boundary Layer Measurements in Combustion MHD Channels", *3rd International Workshop on Laser Velocimetry,* W. Stevenson and D. Thompson, Editors, Purdue University, July 1978
6. James, R.K. and Kruger, C.H., "Boundary Layer Profile Measurements in the Electrode Wall of a Combustion Driven MHD Channel", *17th Symposium on Engineering Aspects of MHD,* Stanford, March 1978
7. Daily, J.W. and Kruger, C.H., "Effect of Cold Boundary Layers on Spectroscopic Temperature Measurements in Combustion Gas Flows", *J. Quant. Spectrosc. Radiat, Transfer,* **17**, 327-338 (1977)
8. Kowalik, R.M. and Kruger, C.H., "The Effects of Fluctuation and Non-uniformities on Line-Reversal Temperature Measurements", *Ibid.* **18**, 627-636 (1977)
9. Kowalik, R.M. and Kruger, C.H., "Laser Fluorescence Temperature Measurements", Combustion and Flame (to be published)
10. Hower, N., "Measurements of Electrical Conductivity of MHD Plasmas with Four-Pin Probes", HTGL Report 108, Stanford University, February 1978
11. Gapanov, I.M., Poberezhshky, L.P. and Chernov, Yu. G., "Study of the Electrical Conductivity of Plasma of Combustion Products with Seeding in the U-02 MHD Generator Channel and on a Laboratory Installation", *Combustion and Flame,* **23**, 29 (1974)
12. Clements, R.M. and Smy, P.R., "Collection of Ions by Electric Probes in Combustion MHD Plasmas: An Overview", *AIAA J. Energy,* **2**, 53-58 (1978)
13. Clements, R.M. and Smy, P.M., "Ion Current to a Spherical Probe in a Flowing High Pressure Plasma under Thin-Sheath Conditions", *Proc. IEEE,* **117**, 1721 (1970)
14. Reigel, F.O. and Self, S.A., Unpublished Work, 1978
15. Self, S.A., Reigel, F.O., Clements, R.M. and James, R.K., "Electron Concentration Measurements in Combustion MHD Flows by Submillimeter Laser Interferometry", *AIAA J. Energy,* **1**, 206-211 (1977)
16. Holve, D. and Self, S.A., "Optical Measurements of Mean Particle Size in the Exhaust of a coal-Fired MHD Generator", *Western States Meeting of the Combustion Institute,* San Diego, October 1976
17. Holve, D. and Self, S.A., "An Optical Particle-Sizing Counter for In-Situ Measurements", Technical Report SU-2-PU, Project SQUID, Purdue University, January 1978. Also, Applied Optics, to be published.

Disorder in Pulsatile Flow:
Biomedical Measurements with Hot Film,
Laser and Ultrasound Anemometers

by

Don P. Giddens
School of Aerospace Engineering
Georgia Institute of Technology, Atlanta, Georgia 30332

Summary

Biomedical measurements of flow and velocity may require a variety of instrumentation methods. The case of disordered velocities occurring in pulsatile flows downstream of localized atherosclerotic plaques is selected as an example for consideration. The hot film, laser and ultrasound anemometers are brought to bear upon studies in laboratory models, animals and humans to illustrate the advantages, limitations and complementary nature of these tools for cardiovascular flow measurements. Example results from each method are given and their interrelationship is discussed.

Introduction

Cardiovascular diseases account for over one-half of all deaths in the United States annually. Since the prime function of the cardiovascular system is to circulate blood, it is not surprising that fluid dynamicists have turned increasing attention to this intricate network and its associated complex flow patterns. Accompanying this heightened involvement was the demand for improved flow measurement methods which could operate reliably and accurately in a living system or in representative models. Techniques employed in vivo at present include thermal anemometry, Doppler ultrasound, electromagnetic flowmetry, nuclear magnetic resonance, and in very limited situations, laser methods. Because flow measurements in experimental animals and in humans are quite difficult to control, investigators frequently resort to model studies in vitro. The choice of flow or velocity measurement technique depends upon several factors: the level of detail required, which relates to the present state of knowledge of the problem; the availability of animals or humans for the experiments and the accompanying safeguards for these subjects; the possibility of alternate or complementary model studies; the type of measurement required; and, of course, the availability and expense of instrumentation and associated data analysis equipment. In bio-fluid dynamics it is essential to begin with an appreciation of the first factor listed. For example, the noninvasive measurement of blood velocity in an artery even though accurate to only 30 per cent is a vast improvement over the inability to make the measurement at all or the necessity of using invasive means which pose risk to the subject.

Rather than attempt a general review of bio-fluid dynamic measurements, this paper will take the approach of selecting a particular problem in cardiovascular flow and describe how our research group approached its study by experimental methods. Hopefully, this will serve both as an illustrative example and a vehicle for discussing some of the primary considerations for selecting bio-fluid dynamic instrumentation and assessing the accuracy of the measurement results.

Background

The situation to be considered is the flow field distal to stenoses, or constrictions, in major arteries. Atherosclerosis, the leading cause of cardiovascular disease, results in an accumulation of lipid-laden material among the interior of arterial walls giving rise to protrusions into the vessel lumen. In earlier stages of development these plaques may be the source of small fragments or emboli which are convected downstream to occlude smaller vessels. As the disease progresses the likehood as well as the size of fragments increases; and, for example, the occurrence of a cardiovascular accident arising in the coronary arteries or a cerebrovascular episode resulting from carotid or intracranial disease becomes a distinct risk. Finally, in very advanced stages total occlusion of a major artery can occur.

The insidious nature of atherosclerosis is such that it has usually progressed to a rather mature state before recognizable clinical symptoms are evident. Frequently, the first symptoms are also the last. For this reason there is considerable interest in developing safe, nontraumatic, noninvasive methods of diagnosing atherosclerosis at presymptomatic stages. Aiding in this effort is the fact that the nature of the disease is such that there are certain sites of the vascular tree which show predilection for localized plaque deposition and thus narrow the field of search and of surgical correction, if required. Our group has become particularly interested in extracranial carotid artery disease since this accounts for 34 per cent of all strokes (Mohr, 1978) and these arteries are close to the body surface and therefore more accessible to noninvasive instrumentation than the coronary arteries which supply blood to the heart.

The questions asked by our research are twofold: (i) what are normal flow patterns in these arteries and (ii) how are these patterns altered in the presence of localized atherosclerosis? The first question has bearing on the interesting puzzle of atherogenesis while the second relates directly to the problem of early disease detection.

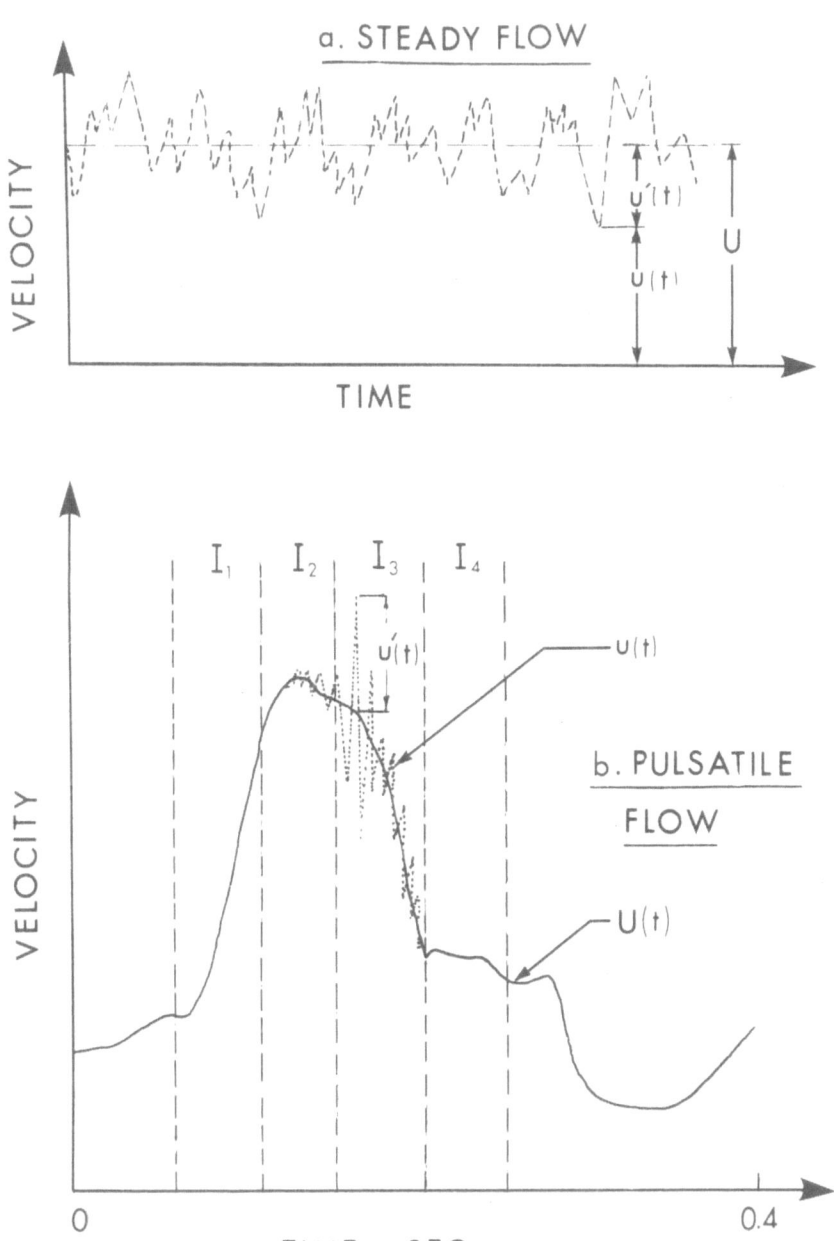

Fig. 1. Illustration of disorder in steady and in pulsatile flow.

Fluid Dynamic Complexities

The poststenotic flow field in a major artery poses an exceedingly complex fluid dynamic problem. The arterial walls are viscoelastic and thus respond to and interact with the fluid motion. The geometry of the conduits is not simple since regions of interest often are sites of vessel branching or bifurcation, giving rise to secondary flows. The flow is pulsatile and at low to moderate Reynolds numbers. When stenoses are present, instabilities can exist which either decay or transit into turbulence depending upon the conditions present. The field may include vortical, transitional, or fully turbulent phenomena — and these may be intermittent. Poststenotic flow separation and recirculation are possible and severe stenoses may yield intensely turbulent jets. The fluid is a non-Newtonian particulate suspension, although this is not a major factor in large arteries in most cases, and, fortunately, incompressibility usually may be assumed.

Biological Complexities

There is considerable biologic variability, ranging from anatomical variations among subjects to heart rate variability in a given subject. Control of parameters in experiments is extremely difficult, particularly when dealing with humans. Animal models are frequently useful but may not be representative of man. For animal experimentation surgical methods may often be required to introduce probes or to gain access to vessels, thus necessitating a team effort unless the fluid dynamicist is qualified in surgical techniques. Human studies require noninvasive, noninjurious methods except in rare cases where fluid dynamic studies can be performed secondary to surgical intervention.

The choice of instrumentation and the expected nature of results are often dictated by biologic constraints.

Approach

Clearly, the poststenotic flow field may be studied at a variety of levels — from a detailed, fundamental, investigation of the nature of instability in a pulsatile shear layer to a relatively crude, but quite useful, measurement of velocity waveforms in an awake human in a clinical setting. A comprehensive study should cut across much of this spectrum for it is difficult to interpret the clinical measurement without some basic knowledge of the flow field; and it is likewise difficult to properly construct a relevant fundamental study without appreciation of the "real-life" problem.

The logic of our experimental approach can be summarized as follows, the first two steps of which were available from work by others:

(i)   Clinical observation that stenosis occurs in atherosclerosis
(ii)   hypothesis that flow is disturbed distal to stenosis
(iii) animal experiments to determine whether flow disorder is a significant factor for moderate and mild constrictions in a living system
(iv)   development of methods to characterize disorder in a pulsatile flow
(v)   well-controlled in vitro studies to aid in interpreting animal results
(vi)   parallel development of noninvasive instrumentation to measure flow disorder in humans
(vii) selection of an anatomical site (carotid complex) and subsequent model and human studies.

At the outset we had no velocity-measuring instrumentation at hand. The first step was to obtain thermal anemometry equipment since it had been established earlier by several investigators that hot films could be successfully employed in blood flow studies (e.g. Ling et al. (1978), Bellhouse and Bellhouse (1968), Schultz et al. (1969), Seed and Wood (1970), Nerem and Seed (1972)). Although this was the only device available for high frequency velocity measurements in vivo, it soon became clear in our model studies that the hot film was inadequate for gaining the data required due to the presence of flow separation, recirculation, and large velocity fluctuations. We thus turned to laser Doppler anemometry for the in vitro experiments. Fortunately, both of these instrumentation methods were highly developed already. Since we deemed human studies as necessary and hoped for eventual clinical application, the question of making noninvasive measurements arose; and in this area we were not so fortunate. Although Doppler ultrasound methods for measuring velocity have existed for approximately two decades (Kalmus (1954) and Baldes et al. (1957)), there was no noninvasive instrumentation to measure turbulent flows reliably. Consequently, some development was required to be able to realize our objectives.

Thus, for the poststenotic flow field study we bring to bear three types of velocity-measuring instrumentation, each with its own advantages and handicaps, and attempt to incorporate these into a coherent mosaic of fluid dynamics experimentation.

Data Analysis
Before discussing the measurement methods, it is appropriate to briefly consider techniques for treating the data. A more detailed description is found in Khalifa and Giddens (1978a) along with other references in this area.

Fig. 1 illustrates the cases of disorder in two classes of flows:

the first has a steady mean velocity and represents the familiar situation of a stationary turbulent flow, while the second depicts a flow which is basically pulsatile with disturbances superimposed during part or all of the cycle. We are, of course, interested in the latter case for which the mean flow is periodic. Time-varying ensemble average waveforms may be computed if a reliable "clock" is available, according to the equation

$$U(t) = \frac{1}{N} \sum_{n=0}^{N-1} u(t+nP) \qquad [1]$$

where $u(t)$ is the instantaneous velocity, U is the periodic ensemble average, P is the period, and N is a "large" number of cycles. Disturbance velocities may then be determined from the relation

$$u'(t) = u(t) - U(t) \qquad [2]$$

and various analyses performed on this component. Since $u'(t)$ may well be a nonstationary random variable for the poststenotic flow, it is useful to consider evolutionary energy spectra and turbulence intensities (e.g. Priestley (1965) and Khalifa and Giddens (1978a)) as a method for describing the development of flow disorder during the cycle.

Hot Film Anemometer
The hot film anemometer has proven particularly useful in arterial flow studies and has been utilized by investigators too numerous to list here.

Advantages
Principal advantages are

(i)   It is an "accepted" tool — much is known concerning its frequency response and accuracy
(ii) It can be used for turbulence measurements due to good high frequency characteristics
(iii) Probes for use in blood are available and can be employed for in vivo studies
(iv) Cost of anemometer is relatively inexpensive.

Disadvantages
Principal disadvantages are
(i)   The probe is invasive, it must be surgically inserted for in vivo experiments making it unsuitable for most human studies
(ii) Calibration is both difficult and tedious, particularly for in vivo work

(iii) Probes for blood flow use must be exceptionally sturdy since insertion into the artery is required

(iv) The invasive nature precludes reliable measurement in separated or reversed flows and in flows with large angularity, including highly turbulent regions

(v) Probes for blood velocity measurements are expensive to purchase and difficult to manufacture.

Probes

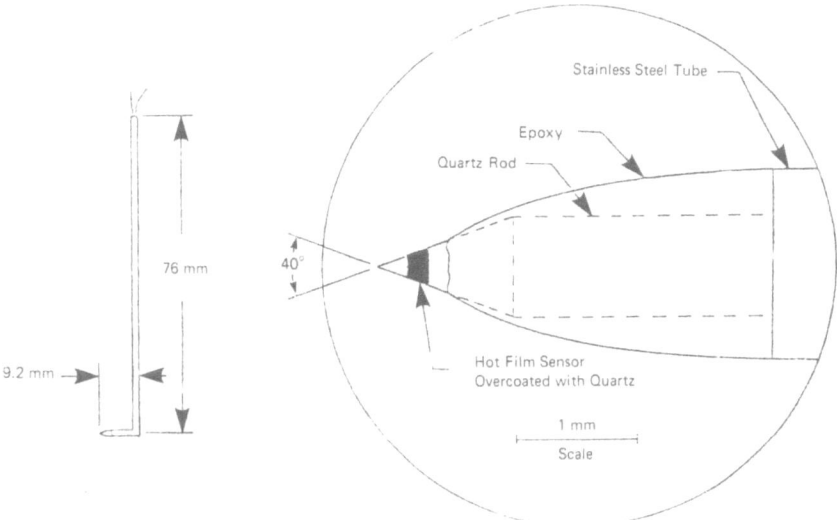

Fig. 2. Sketch of conical-tipped hot film probe for blood velocity measurement.

Various probe designs have been employed for blood flow studies. Typically, a thin film is deposited near the tip of a hypodermic needle or mounted at the end of a catheter and then covered with a protective quartz coating. The probes we have employed most successfully were conical-tipped, right-angle needles manufactured by DISA. A typical one is shown in Fig. 2. Also, catheters are available with conical-tipped probes mounted at the end. These have the advantages of entering the artery at an easily accessible site remote from the region of study and of freedom of motion in the axial direction. The radial position within the lumen cannot, however, be controlled. The right-angle type can be traversed radially across a vessel with a micrometer mounting but requires surgical entry at the site of interest and, once in place in an artery, is not easily relocated due to problems of hemorrhage at the puncture wound.

One of the troublesome features of the hot film is that calibration of each probe is required and this calibration must be performed in the fluid being studied. This is particularly problematical for blood flow experiments since ideally the calibration should be obtained from the blood of the animal under investigation. Another aspect is that usually only limited quantities of blood are available. Our procedure to overcome this was the following.

(i)   We employed a slotted turntable-type device with a variable speed motor. Approximately 200 ml of fluid were used to fill the slot and the probe was placed into the rotating liquid. Numerous studies with water in this device and in a separate apparatus in which a large volume of water in a cylinder underwent rigid body rotation assured us that the calibration in the turntable was accurate.

(ii) For each new probe an anesthetized dog was killed by potassium chloride injection and sanguinated to obtain the required blood sample. The blood was heparinized to prevent coagulation. Calibration was obtained over a velocity range from zero to approximately 100 cm/sec.

(iii) During the actual experiment (with another dog) the linearizer was set to the initial calibration; however, both linearizer output and the (unlinearized) anemometer output were recorded on FM tape.

(iv) At the conclusion of the experiment blood was drawn from the animal being studied and the probe was recalibrated. If this was in agreement with the presurgical study, the recorded linearized data was accepted; if not, the recorded anemometer signal could be used as input to the linearizer with new settings at the time of data analysis.

Experimental Methods

Details of our stenosis studies with anesthetized dogs have been given by Giddens et al. (1976) and Khalifa and Giddens (1978a). Briefly, a variable stenosis was imposed by placing a flexible band about the descending thoracic aorta of the dog and attaching this band to a device which could give constriction in measured increments. The hot film probe was inserted at a desired location distal to the band and measurements were recorded at various radial positions and for various degrees of stenosis. Also recorded were the ECG, to be used as a reference "clock" in data analysis, and the volume flow rate as measured by an electromagnetic flowmeter with a cuff-type probe located several vessel diameters downstream of the hot film. The hot film anemometer and linearizer were DISA models 55D01 and 55D10. The EM flowmeter was manufactured by Carolina Medical Electronics.

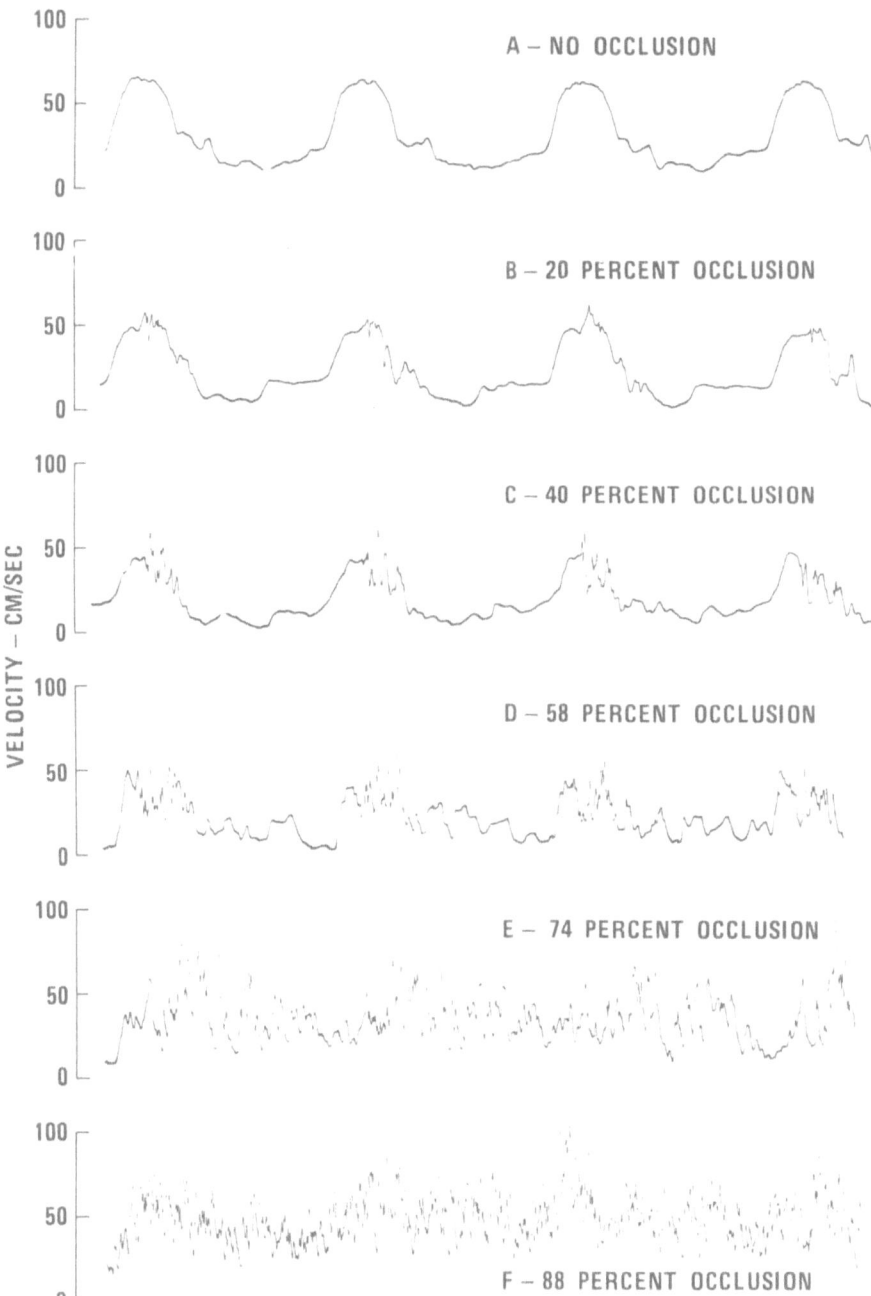

Fig. 3. Velocity recorded at centerline of dog aorta with hot film probe located 2 cm distal to throat of constriction.

Results

Since results of these studies have been reported previously, we include here
only data which gives a representation of the capability of the hot film anemometer. Fig. 3 gives an example of centerline velocity waveform measurements as a function of the degree of stenosis, measured by cross-sectional area
reduction, at a location 2 cm distal to the constriction. The very good frequency response is a notable feature in the figures. Flow disturbances can be
seen even in the 20 percent occlusion case beginning in the deceleration phase
of the cycle. As the stenosis becomes more constrictive, this disturbed flow
region extends in duration and becomes more violent until, for the 88 per cent
case, it is difficult to distinguish pulsatility features in the instantaneous velocity waveforms.

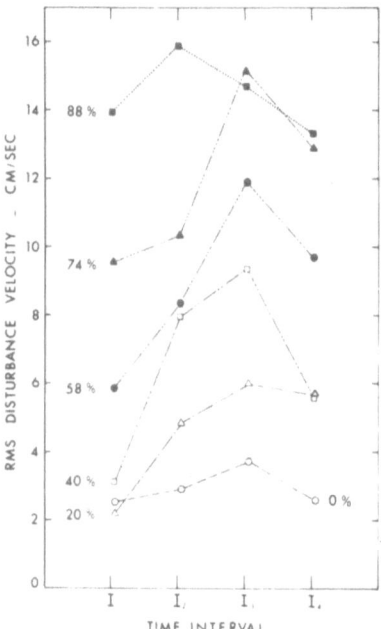

*Fig. 4. Variation of r.m.s. disturbance velocity with time interval of cycle. Data
is from velocity measurements of Fig. 3 and time intervals noted in Fig. 1.*

The velocity cycle can be divided into intervals for purposes of analysis. Fig. 1b
gives four such intervals for this data based upon the ensemble average of the
zero occlusion waveform and employing the ECG signal as a timer. The period
of time between I_4 in one cycle and I_1 in the next has been excluded from the
analysis. For low degrees of stenosis it is possible that reverse velocities occur in

the dog aorta and these would be undetectable with our probe. For higher degrees of occlusion this would not be the case due to the capacitance of the vessel upstream of the constriction. Fig. 4 gives the rms disturbance velocity as a function of interval in the cycle and percentage stenosis. Dramatic increases are apparent even for mild stenosis, particularly in I_3, the deceleration phase. Another way of viewing this is with energy spectra computed as a function of interval. Fig. 5 gives a similarity plot of the spectra for I_3 and compares the results with the in vitro data of Clark (1976), also obtained with a hot film. The coordinates are the energy function $E^* = F\overline{U}/(2\pi d)$ and Strouhal number $N_s^* = 2\pi f d/\overline{U}$. Here, d is the throat diameter, $\overline{U}$ is the average velocity over I_3, f is frequency in Hertz, and F is the normalized spectrum, $F(\omega) = E(\omega)/u'^2$ where $E(\omega)$ is the velocity-squared content per unit frequency.

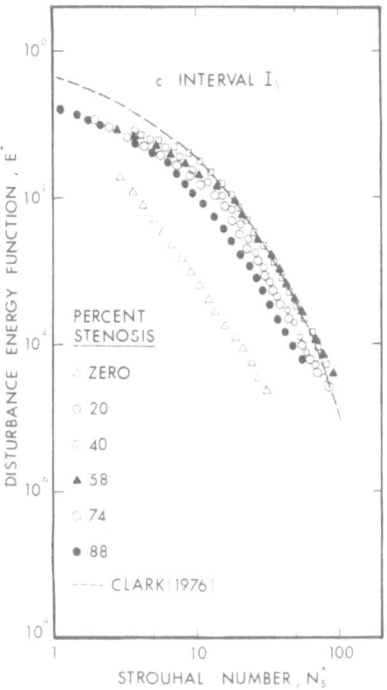

Fig. 5. Energy spectra for turbulence velocity during interval I_3; variation with degree of stenosis.

Although the 20 and 40 percent stenoses give readily apparent velocity disturbances under these conditions (the peak upstream Reynolds number is approximately 2000 and the frequency parameter is 13.8), it is unlikely that these constrictions would yield any clinically observable symptoms in humans.

Reductions in volume flow rate do not occur until approximately 75 per cent reduction in area, although creation of emboli can occur at earlier stages.

Measurements in several dogs yielded essentially the same results. In each case the nearfield of the stenosis exhibited obvious flow disturbances for mild constrictions. Thus, this series of experiments demonstrated that flow disorder might be employed as an early indicator of plaque formation provided we could understand more about the conditions under which it occurs and could develop methods for its noninvasive measurement.

Laser Doppler Anemometer
The results of the hot film studies in dogs were sufficiently interesting to entice us into more detailed in vitro experiments under better controlled conditions. Although the hot film anemometer provided useful data in vitro (Cassanova and Giddens (1978)), we would not locate separation and reattachment points distal to the constriction, nor were we confident in measured velocity profile data in certain regions for cases of large turbulence intensity or flow angularity. This led to consideration of laser Doppler anemometry for in vitro studies.

Advantages
For our purpose the primary advantages of the LDA are:

(i)   It is noninvasive
(ii)   The frequency response to velocity fluctuations is reasonably adequate to allow turbulence measurements
(iii) The sample volume of measurement is acceptably small
(iv)   One component of the flow velocity is isolated
(v)   With a moving fringe pattern measurements in separated and recirculating flows are possible.

Disadvantages
There are, however, several notable disadvantages:

(i)   The LDA requires a good optical path and cannot be used in arteries
(ii)   Considerable corrections are required for curved surfaces and/or careful matching of indices of refraction among various interfaces must be achieved
(iii) Ambiguity affects frequency response and correction may be required
(iv)   Equipment is expensive.

Experiments

In keeping with the theme of this conference we report here a few results obtained with the LDA in two sets of experiments with the objective being to illustrate the versatility and suitability of the instrument to modeled arterial flows as opposed to concentration on the fluid dynamic results. The system employed was a DISA Mark II LDA with frequency tracker. The laser was a 15 mW helium-neon Spectra-Physics Model 120 laser operating in the forward scatter mode. Data reduction was accomplished with a Hewlett-Packard 5451A Fourier Analyzer. A block diagram of the system is given in Fig. 6.

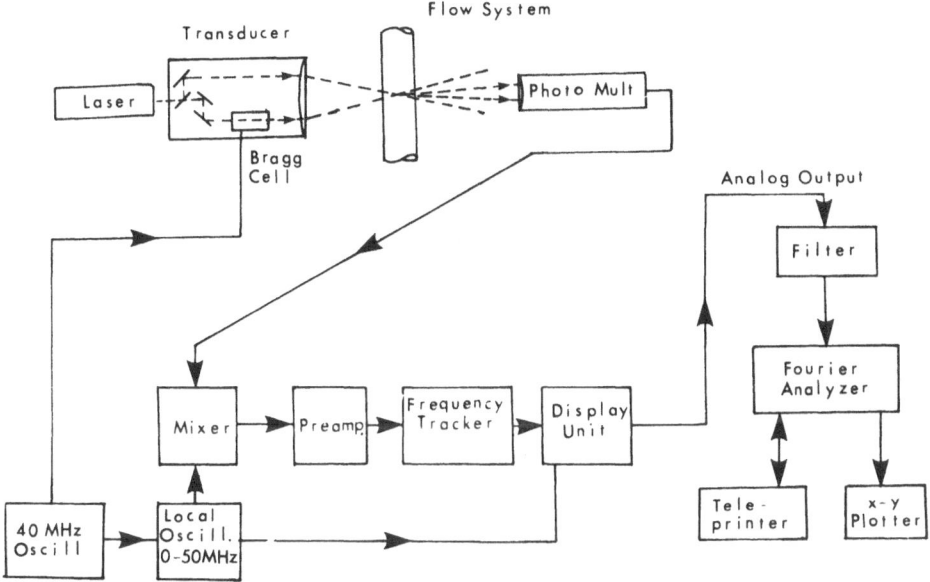

Fig. 6. Block diagram of LDA system.

Steady Flow

It was decided to depart somewhat from the purely biomedical motivation and to study first a fully turbulent steady flow through a contoured 75 per cent constriction at upstream Reynolds numbers ($Re = UD/\nu$) of 5,000, 10,000, and 15,000. We, and others, have interest in turbulent modeling for computational fluid dynamics and there is little available in the way of detailed measurements in flows containing recirculation zones with which the theoretician may compare his results. These experiments served the two-fold purpose of providing some insight into flows at (and beyond) the upper limit of conditions expected in the major arteries and of giving detailed measurements for comparison with turbulence computations.

The flow system details are given by Deshpande (1977) along with a complete set of the experimental data. Briefly, the stenosis followed a cosine contour in a plexiglas tube. The upstream unoccluded diameter was 2 inches and the entrance plane of the six inch long stenosis was located 74 inches from the tube inlet. The fluid was water and a rectangular water bath surrounded the region of the tube where LDA measurements were obtained. Light seeding with one micron silicon carbide particles was employed. Many initial experiments to compare hot film and LDA measurements gave confidence in the Doppler results and are described by Deshpande (1977). Also, methods for correcting for the optical path are described in that reference.

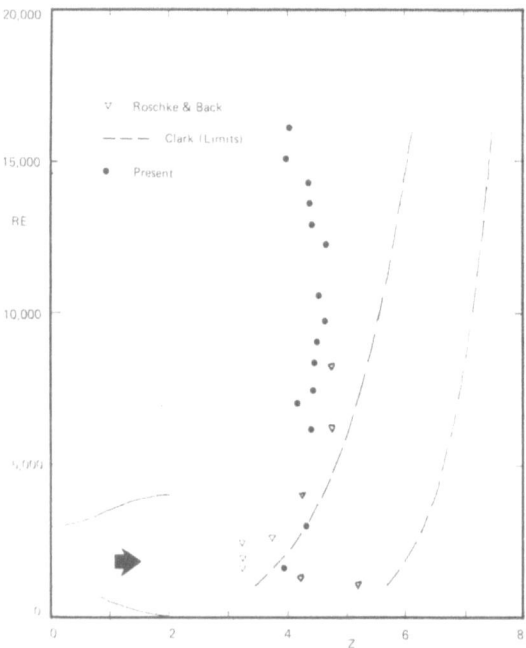

Fig. 7. Reattachment points for flow distal to contoured stenosis; variation with Reynolds number.

One of the important uses of the LDA was in locating the reattachment "point' of the separated flow. The poststenotic field is highly turbulent and dye injection methods are virtually impossible to interpret. The approach taken here was to measure the velocity as close to the wall as possible and at several points 0.10 inch apart in the axial direction. Mean velocity values were found at these points by integrating the signal over sufficiently long time periods. Very near reattachment it was necessary to average for as long as 500 seconds before a satisfactory velocity was obtained. These values were then plotted versus axial location and the reattachment "point" was taken to be that at which the mean velocity changed sign. Results for various Reynolds numbers are graphed in Fig. 7 along with the experiments of Roschke and Back (1976) for the recirculation length in a pipe with an abrupt expansion. The empirical relations proposed by Clark (1976) for the upstream and downstream limits of reattachment are also shown. Our data differ substantially from Clark's predictions based on dye injection observations and, of course, it is our opinion that the LDA results are the more reliable.

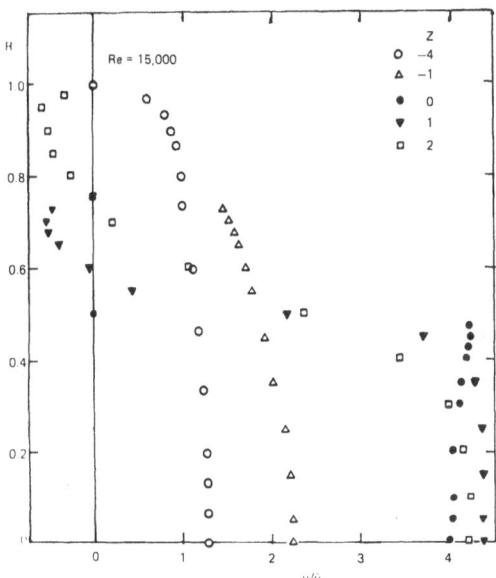

Fig. 8. Axial velocity profiles in stenosis at Reynolds number of 15,000.

Fig. 8 gives an example of axial velocity profile measurements at Re = 15,000. The Z-coordinate is the axial distance measured from the throat plane and nondimensionalized by the unoccluded tube radius. The R-variable is the radial coordinate similarly nondimensionalized. Interesting features which can be noted are the off-axis peak in the velocity profile at the throat and the negative velocities of the recirculation region.

Turbulence fluctuations in u' and w' were recorded at various locations. Fig. 9 gives a graph of the maximum values of u'_{rms} and w'_{rms} in a given cross-sectional plane as a function of axial position. These are nondimensionalized by the bulk velocity. The maximum values tend to occur in the shear layer in the

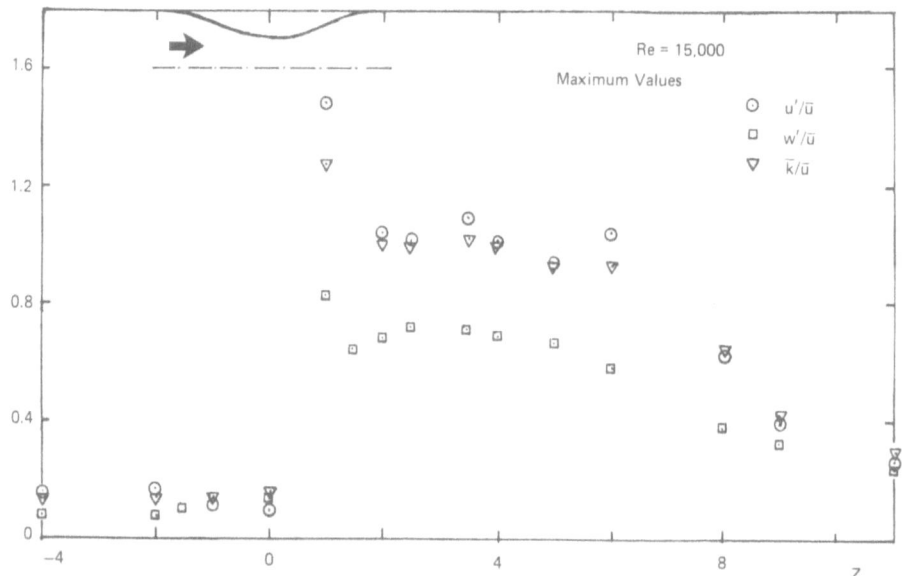

Fig. 9. Maximum r.m.s. disturbance velocities as a function of axial location;
Re = 15,000.

near poststenotic field and at the centerline farther downstream. Interestingly
enough, there is a substantial region of very high intensities and further studies
at more representative Reynolds numbers may give clues to potential "target
areas" when clinical search is made for signs of stenotic disease.

Pulsatile Flow

In an effort to understand more about the initiation and evolution of poststenotic flow disturbances a series of pulsatile flow studies was begun. The first
experiments were directed toward obtaining an overview of the phenomena
occurring under conditions reasonable representative of flow in the dog aorta.
The tube diameter was 1.27 cm, upstream waveform period was one second,
peak and minimum velocity were 68 cm/sec and 13 cm/sec, respectively. With
a glycerin/water mixture of viscosity 3.6 centipoise, this gave a pulsatile flow
whose peak Reynolds number was 2400.

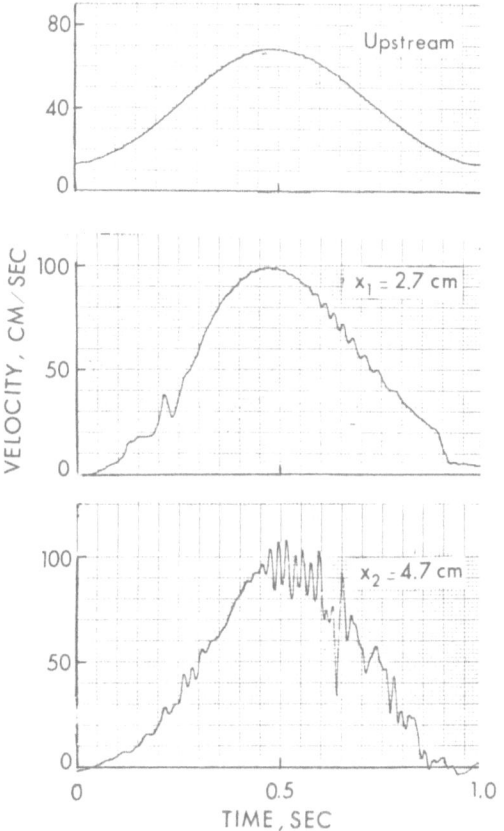

Fig. 10. Centerline velocity waveform measurements for pulsatile flow through a 50 percent contoured constriction.

Measurements were made with the LDA at various axial locations along the centerline. Fig. 10 gives several examples of velocity waveforms taken for a 50 percent contoured occlusion. The upstream waveform is included for reference. At position X_1 two phenomena are apparent. The first is a large velocity fluctuation caused by a passing vortex at low Reynolds number as the flow accelerates. Later, during the deceleration phase following the peak velocity, a periodic instability can be seen. At a position X_2 farther downstream it can be seen that this instability has increased in amplitude and fragmented into a more random pattern. The vortex patterns in the acceleration phase can also be seen to be moving downstream. Another way of viewing the flow behavior is by employing the disturbance velocity.

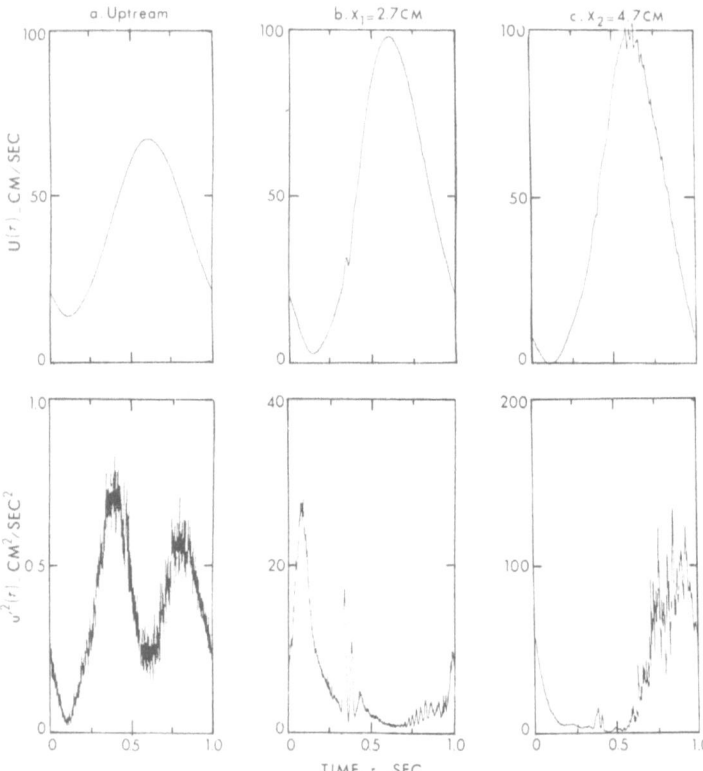

Fig. 11. Ensemble average and disturbance velocities for the data of Fig. 10.

Fig. 11 gives three examples of ensemble average waveforms and the corresponding variation of u'^2 as a function of time. The driving sinusoidal signal for
the pulsating pump was used as the triggering signal. If a vortex formation was
perfectly reproducible there would be no corresponding value for u'^2. However, slight non-reproductibilities in the shedding frequency occur and these
can be readily identified as sharp spikes in the u'^2 curves. The magnitude of
these peaks has no physical meaning, but their occurrence in the graphs gives
a clear identification for the vortex location. Note that the upstream u'^2
values are quite low, indicating a laminar waveform. At X_1 the first peak in u'^2
is an indication of residual disorder at the end of each cycle. The next two
peaks can be identified with vortex formation, while the series of peaks between $\tau \cong 0.75$ and 1.0 correspond to the instability seen in the deceleration
phase (Fig. 10). At X_2 the vortex formation can again be identified at $\tau \cong 0.4$
and an intensely turbulent field is seen in the deceleration phase.

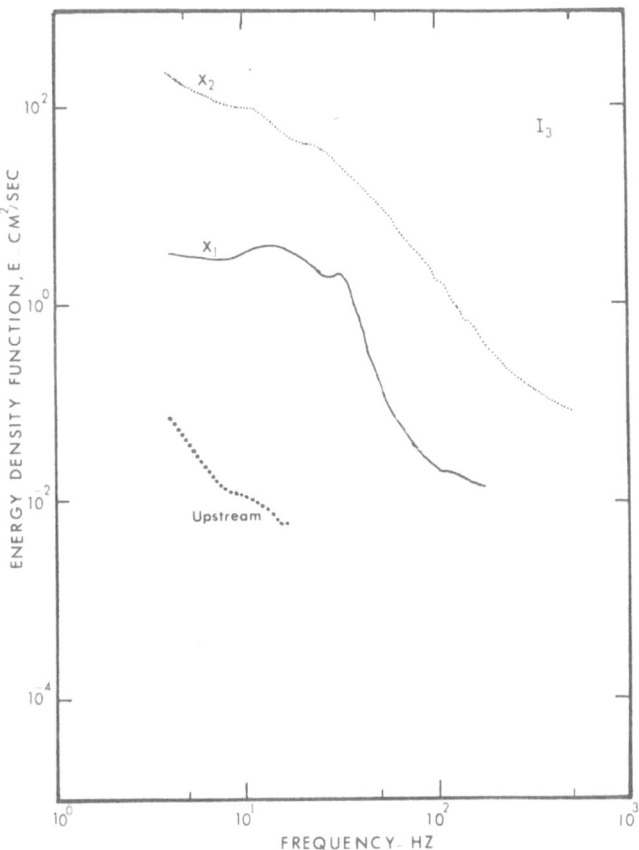

Fig. 12. Energy spectra for the disturbance velocity for three axial positions (upstream, X_1 and X_2) during time $0.5 < \tau < 0.75$ sec.

Evolutionary energy spectra again provide useful information. Fig. 12 gives spectra for upstream, X_1, and X_2 positions during the time interval $0.5 < \tau < 0.75$. Note the distinct peaks present, corresponding to a frequency of approximately 50 Hertz. (Again, the amplitude itself is not relevant but the frequency agrees with that measured directly from the waveform curves). Also, the overall spectral level has risen considerably, consistent with the u'^2 increase from X_1 to X_2 during this time interval.

The data are given as examples of the detail afforded by an LDA possessing good frequency response. Further results and interpretations may be found in Khalifa (1978). The LDA appears capable of allowing study in vitro, under careful controlled conditions, much of the phenomena observed in vivo.

Pulsed Doppler Ultrasound

Although the hot film and laser Doppler anemometer provide valuable data on the poststenotic flow field, the eventual clinical applicability depends wholly on the ability to make the required measurements noninvasively since very few people are prone to undergo catheterization or direct arterial puncture with a hot film as a routine diagnostic procedure for presymptomatic disease. For this reason our interests turned to pulsed Doppler ultrasound as a promising method to achieve measurements on human subjects.

Advantages

Ultrasound is ideally suited for certain blood flow measurements.

Principal advantages are:
(i)   Ultrasound is noninvasive to the flow
(ii)   Power levels required are sufficiently low that it can be safely employed on humans
(iii) By pulsing and gating the transducer, range information is possible
(iv)   Ultrasound works very well with heavy seeding and nature has provided almost ideal scattering particles in the form of red blood cells
(v)   Velocity measurement depends upon the Doppler principle which reduces the complexity of calibration.

Disadvantages

As with the other methods discussed there are serious disadvantages

(i)   The sample volume for velocity measurement is relatively large, despite the use of range-gating
(ii)   The frequency response to velocity fluctuations has been very poor so that reliable turbulence data were not possible
(iii) The relationship between scattering phenomena and fluid dynamic events in the sample volume are not as well understood as for the LDA
(iv)   A good acoustic path is necessary
(v)   Pulsed systems are not readily available commercially.

Before ultrasound can be satisfactorily used for the purpose desired here, two needs must be met: an improvement in the signal processing and a better understanding of the relationship between acoustic scattering and fluid dynamics within the sample volume.

A Pulsed Doppler Ultrasound System

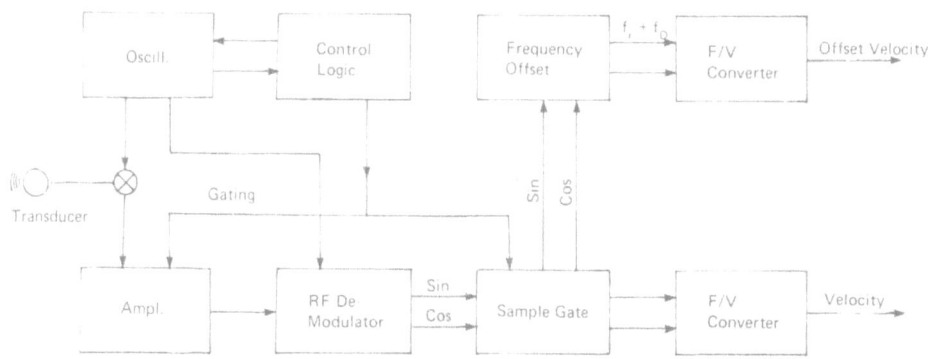

Fig. 13. Block diagram of pulsed Doppler system of McLeod.

The pulsed Doppler (PD) system employed in our laboratory was designed by F.D. McLeod (1974). A block diagram is given in Fig. 13. The heart of the system is a master oscillator from which the transmission frequency, pulse repetition frequency, gating and demodulation reference frequency are derived. The carrier in this particular system is 7.7 MHz. A lead zirconate piezoelectric crystal is stimulated by the master oscillator for a desired number of cycles, giving rise to an acoustic burst which has an intensity distribution resembling a teardrop (e.g. Jorgensen et al. (1973)). The transducer is acoustically coupled to the skin with a gel and transmission through tissue occurs at a speed of sound very close to that through water (approximately 1500 m/sec). Attenuation occurs as $(frequency)^4$ and biological safety considerations limit the input power. Bone or airspace cannot be tolerated in the acoustic path. Despite these restrictions several major arteries are readily accessible to the PD, including the carotid complex.

Sound is scattered from interfaces and the red blood cells act as particles for blood flow studies (e.g. Reid et al. (1969)). The concentration is quite high, being on the order of 5 to 6×10^6 particles per mm^3 and 40 to 47 percent by volume. The return signal is gated for range and the transducer assumes the alternate role of receiver. Demodulation with the carrier is accomplished employing a phase shift technique which distinguishes the sign of the Doppler shift, allowing separation of forward and reverse velocities. It is useful to provide a constant frequency offset, analogous to the Bragg cell frequency shift utilized in the LDA, so that zero velocity corresponds to a reference frequency, f_r. Various filters are required throughout the system to remove harmonics of the pulse repetition frequency (PRF). PRF settings of approximately 12.5, 25, and 50 kHz are available; and the maximum range is limited by, among other considerations, the requirement that there be no ambiguity in determining which pulse is being received at a given time. Listening times up to 3 μsec are possible, and the transmission burst may be set for 4, 8 or 16 cycles for this system.

After demodulation the Doppler frequency must be determined, and several methods are either available or under development. For this case of backscatter the Doppler equation is

$$\frac{f_D}{f_o} = \frac{2u}{c} \tag{3}$$

where f_D is the Doppler frequency, f_o is the carrier frequency, u is the velocity component in the beam direction, and c is the speed of sound in the medium. There are several sources of spectral broadening just as with the LDA (e.g. George and Lumley (1973) and Brody and Meindl (1974)). Furthermore, the sample volume is larger than that for the laser and consequently there is likely to be a greater influence of velocity variations within this volume, such as mean gradients and fluctuating velocity, upon the return signal. In passing it may be noted that for a 7.7 MHz carrier and flow velocities in the range 0 to 100 cm/ sec, the corresponding Doppler frequency varies from 0 to 10 kHz so that virtually all of the signal arising from an arterial flow measurement is in the hearing range, and the signal is frequently referred to as an "audio" signal. (This also has a clinical usefulness but it has the unfortunate side effect that a physician may tend to rely overly on the "sounds" as opposed to seeking quantitative results).

Signal Processing Methods
It is generally assumed in the literature that the power spectrum of the return signal is representative of the probability distribution of particle velocities in the sample volume (e.g. Peronneau et al. (1974); Brody and Meindl (1974)). If there are no nonuniformities of acoustic intensity particle distribution, and particle size, the velocity is given by the normalized first moment of the power spectrum (in principle, one can account for "negative" frequencies)

$$u = K \frac{\int_{-\infty}^{\infty} \omega \, P(\omega) \, d\omega}{\int_{-\infty}^{\infty} P(\omega) \, d\omega} \tag{4}$$

where K is the Doppler constant, $P(\omega)$ is the power spectral density, and ω is the frequency. However, spectral broadening occurs, primarily due to transit time, mean velocity gradient, and turbulence fluctuations. If these effects are symmetric in their distribution about the mean frequency, then the first moment calculation of Eq. [4] gives an accurate result (Gaussian contributions are an example). However, this is not necessarily the case and, unless the tedious task of deconvolution of the measured P(ω) is successful, use of Eq. [4] is subject to some error. Further, since this equation is based on the assumption of incoherent scattering, it is natural to question whether its application to obtain *instantaneous* velocities is appropriate. For the present, we assume that this is, in fact, suitable although further study of the scattering process and its relation to fluid dynamic behavior is in progress.

Several methods of signal processing will be briefly discussed.

Zero Crosssing Detector

The zero crossing detector (ZCD) has been the most popular method of frequency-to-voltage conversion in Doppler ultrasound systems. Statistically, the ZCD detects the square root of the second moment of the power spectrum (Rice (1954))

$$u_{ZCD} = K \left[\frac{\int_{-\infty}^{\infty} \omega^2 \, P(\omega) \, d\omega}{\int_{-\infty}^{\infty} P(\omega) \, d\omega} \right]^{\frac{1}{2}} \qquad [5]$$

If the distribution is narrowband, this is close to the first moment value; however, as the bandwidth increases so does the difference between Eqs. [4] and [5]. In practice, the ZCD has not proven satisfactory for quantitative velocity measurements. Statistical fluctuations in the rate of zero crossing lead to noisy output if the averaging period of the frequency meter is too short; while for large averaging time, the response to velocity fluctuations suffers. We are not aware of any highly reliable results reported using zero-crossing detectors.

Spectrum Analyzer

Several investigators have taken the approach of employing analog spectrum analyzers (either off-line of real time (e.g. Coghlan et al. (1974)). This method of processing has the advantage of giving a visual image of the spectral information in the Doppler signal as a function of time and, in fact, has been employed advantageously in clinical studies (e.g. Strandness et al. (1977)).

To the fluid dynamicist this is a somewhat unsatisfying method of presentation, however, since he is accustomed to dealing with fluctuating velocities rather than spectral broadening. Ideally, the frequency-to-velocity transformation should be performed.

Phase-Lock Loop
This analog method of frequency tracking has been employed successfully in LDA applications. It performs best on a relatively narrow band signal. The ability to reject "noise", to maintain "lock" and to recapture the signal is strongly dependent upon the specific circuitry design. As previously discussed the ultrasonic Doppler spectrum is rather complex and presents some difficulty to the phase-lock loop (PLL). The basic PLL system consists of a phase detector, low pass filter, and VCO. The low pass filter time constant should be relatively small so that rapid changes in frequency (hence particle velocity) can be followed. If, however, the constant is too short, the loop will tend to jump from one frequency to another of a complex spectrum thus giving a "noisy" output. This situation is analogous to that of the ZCD. However, the capture and lock behavior are sensitive to the relative amplitudes of the frequency components of the input signal and so there is a tendency to ignore low amplitude components. Although there is a somewhat "artistic" approach to applying the PLL to Doppler ultrasound signal analysis, we shall present results of turbulence measurements which are quite encouraging in a subsequent section.

Digital First Moment Method
The proliferation of minicomputers and advanced state of Fast Fourier Transform techniques makes it attractive to consider digital processing of the Doppler signal. Digital methods, although relatively expensive, offer a great deal of flexibility in data analysis and manipulation such as examining effects of windows and sampling parameters. The Digital First Moment (DFM) method is based on the assumption that Eq. [4] is valid for small times and we write

$$u_{DFM}(t) = K \frac{\int_{-\infty}^{\infty} \omega\, P(\omega,t)\, d\omega}{\int_{-\infty}^{\infty} P(\omega,t)\, d\omega} \qquad [6]$$

This inherently assumes a separation of time scales; that is, the velocity cannot vary greatly over the time required to compute the "instantaneous" power spectrum of the Doppler signal. Although there are clearly limits on velocity resolution and frequency response, the sampling theory is relatively straightforward so that these limits can be predicted for a given set of conditions.

The DFM method has been tested on simulated turbulence Doppler signals and performance agrees well with design estimations. Brief examples will be given in the section dealing with results. The primary question remaining is that of performance with real Doppler signals since, again, spectral broadening and the relatively large sample volumes pose problems. This latter aspect is under study.

The advantage of this method is that it is straightforward in concept and thus has a well-defined theoretical model with which one can experiment.

Sample Results

Phase-Lock Loop
A series of flow experiments was run to compare the pulsed Doppler/phase-lock loop measurements with those of the laser Doppler anemometer (Khalifa and Giddens (1978b)).

An apparatus was constructed to provide steady or pulsatile flow through a 1/2 inch diameter tube. The test section was designed to accommodate a 1/16 inch diameter ultrasonic probe aligned at 45° with the tube axis. The outer walls of the section were machined flat, facilitating the introduction of the laser beams of the LDA. The sample volumes of the PD and LDA were adjusted to be centered at the same point along the tube centerline. Simultaneous measurements with the two instruments were not possible due to conflicting seeding requirements. One micron silicon carbide particles were used for the laser and a greater concentration of 10 μm particles was employed for the ultrasonic measurements.

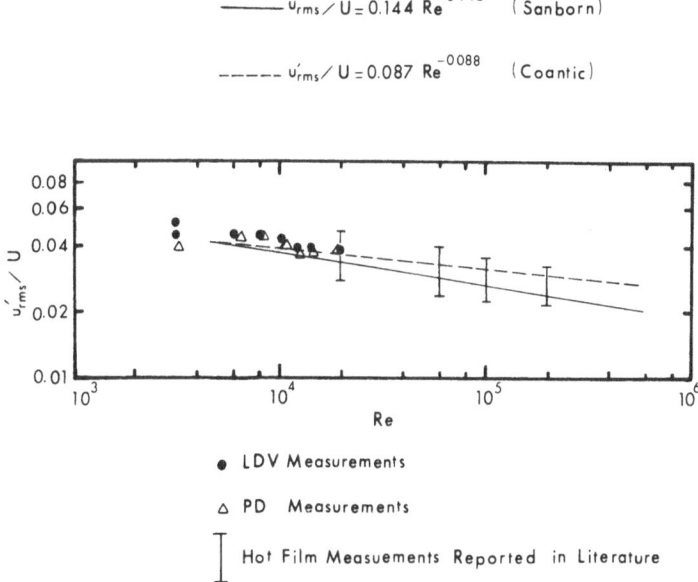

$$\text{------} \quad u_{rms}/U = 0.144 \, Re^{-0.146} \quad \text{(Sanborn)}$$

$$\text{- - - - -} \quad u'_{rms}/U = 0.087 \, Re^{-0.0088} \quad \text{(Coantic)}$$

- • LDV Measurements
- △ PD Measurements
- ⊥ Hot Film Measuements Reported in Literature

Fig. 14. Turbulence intensity measurements in fully developed pipe flow.

Steady flows at Reynolds numbers ranging from approximately 6000 to 15,000 were studied at this single centerline location. For the PD system a transmission length of 8 cycles and receiving times of 0.5 to 1.0 μsec were employed. The length of the sample volume was thus approximately 1.5 to 3.0 mm.

Fig. 14 gives the results of turbulence intensity measurements and compares the present results with hot film data reports by Resch (1970) and with the empirical correlations of Sanborn (1955) and Coantic (1966). In determining u'_{rms} for the ultrasound data with probe inclined at 45° to the flow, it was assumed that the turbulence was isotropic at the centerline.

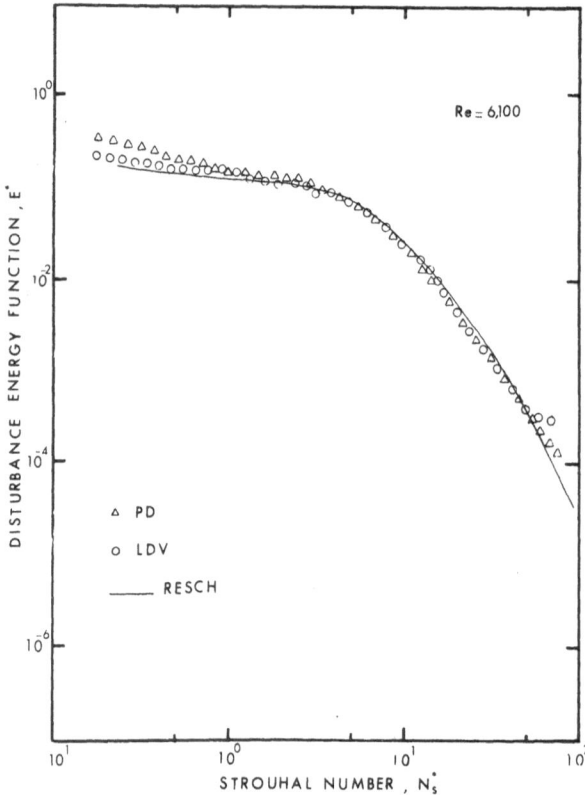

Fig. 15. Turbulence energy spectra measurements with PD ultrasound compared with LDA and hot film data at Re = 6,100.

Fig. 15 gives a comparison of hot film (Resch), LDA and PD/PLL measurements of turbulence energy spectra at a Reynolds number of 6100. Additional data at other Reynolds numbers may be found in Khalifa and Giddens (1978b) along with results in pulsatile flow and in flows distal to a 50 percent contoured occlusion. For this latter case the assumption of turbulence isotropy is not valid and only the shape of the spectra for LDA and PD/PLL can be compared since different turbulence components were measured.

Further studies with the system are required, however. We still experience signal drop-out during portions of the cycle in some of the pulsatile data, and it is necessary to further define the effects of the filter on the turbulence fluctuations, particularly for cases of large turbulence intensity. Despite these remaining uncertainties, results have been sufficiently encouraging to undertake a series of experiments on animals in which hot film and PD/PLL measurements will be compared in the dog aorta. Preliminary clinical studies are also in progress.

Digital First Moment

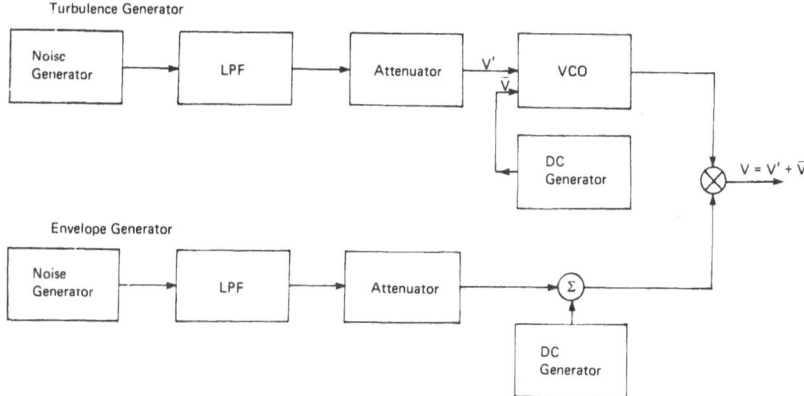

Fig. 16. Block diagram of Doppler signal simulator.

To date we have applied the DFM processing only to simulated turbulence signals. A block diagram of the simulation system is shown in Fig. 16. The input signal to the VCO is produced by sequential low pass filtering of the output of a noise generator. Fig. 17 shows the resulting "turbulence" spectrum of the input for a particular case. The output of the VCO is mixed with the filtered output of a second noise generator which simulates the amplitude modulations found in a real Doppler signal recorded in previous steady flow studies. The resulting simulation is both amplitude and frequency modulated. This signal is introduced into the DFM processing method and the results compared

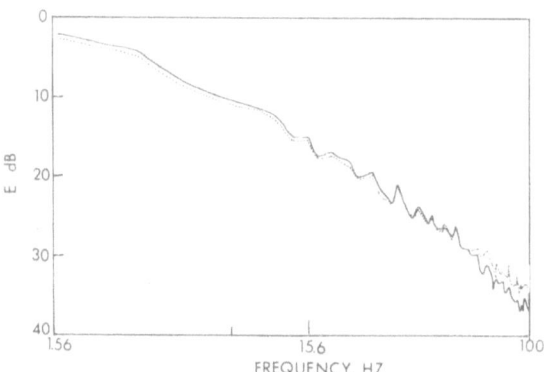

Fig. 17. Comparison of simulated turbulence energy spectrum with that obtained from DFM technique.

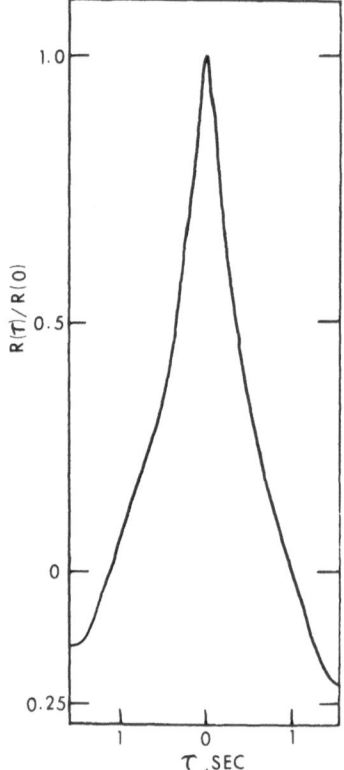

Fig. 18. Cross correlation of simulated turbulence velocity with that constructed by DFM approach.

with the input to the VCO. The sampling criteria employed for this particular study give a maximum expected frequency response of 100 Hz to velocity fluctuations. The DFM-produced energy spectrum is also shown in Fig. 17 for comparison with that of the VCO input. The corresponding cross-correlation curve is given in Fig. 18 as an alternate manner of presenting the data. Additional results for simulated laminar, oscillating laminar, and other turbulent flows are given by Saxena (1978).

The DFM operation is not presently accomplished in real time since we are in the concept-proving phase, nor has it been attempted on "real" Doppler signals, which we anticipate will be somewhat more complex than the simulated signal. Also, we again emphasize that the measurement accuracy and interpretation is clearly dependent upon the sample volume characteristics so that considerable work remains to be done in this study.

Discussion

It is hoped that the example of disorder in a pulsatile flow has illustrated many of the advantages, limitations and complementary interrelationships of hot film, laser and ultrasound anemometers. The advantages of one instrument can often be employed profitably in a case where the other two are totally inadequate. And yet, at least for this example, it is difficult to work the problem without the use of all three instruments. Again, we stress the fact that the level of useful information must be considered in biomedical studies. For instance, although the fluid dynamicist would like to measure turbulence at frequencies to several hundred Hertz for many applications, there may be a wealth of clinically relevant information contained below one hundred Hertz. Also, a laboratory model may tolerate a triple-wire thermal probe, but a dog or human aorta cannot.

It is important to play the in vitro against the in vivo, the invasive against the noninvasive, and the fundamental study against the clinical one in approaching biofluid dynamic problems. Both the experimentalist and theoretician will profit by keeping this as a basic tenet.

Acknowledgments

The work of the following co-investigators is gratefully acknowledged: R.F. Mabon, R.A. Cassanova, M.D. Deshpande, F.D. McLeod, J.I. Craig, R. Rubenstein, A.M.A. Khalifa, and Vijay Saxena. Research was supported in large part by the National Science Foundation and the National Institutes of Health.

Bibliography

Baldes, E.J., Farral, W.R., Haugen, M.C. and Herrick, J.F. (1957). A forum on an ultrasonic method for measuring the velocity of blood. In *Ultrasound in Biology and Medicine* (E. Kelly, ed.), 165-176, A.I.B.S., Washington

Bellhouse, B.J. and Bellhouse, F.H. (1968). Thin film gauges for the measurement of velocity or skin friction in air, water or blood. *J. Sci. Inst.,* 1, 1211-1213

Brody, W.R. and Meindl, J.D. (1974). Theoretical analysis of the CW Doppler ultrasonic flowmeter. *IEE Trans. Biomed. Eng.,* **BME**-21, 1830192

Cassanova, R.A. and Giddens, D.P. (1978). Disorder distal to model stenoses in steady and pulsatile flow. *J. Biomechanics* (in press)

Clark, C. (1976). Turbulent velocity measurements in a model of aortic stenosis. *J. Biomechanics,* 9, 677-687

Coantic, M. (1966). Contribution a l'Etude de la Structure de la turbulence dans une conduite de section circulaire. These Doct. es Sc., Marseille, France

Coghlan, B.A., Taylor, M.G. and King, D.H. (1974). On-line display of Doppler shift spectra. In *Cardiovascular Applications of Ultrasound* (R.S. Reneman, ed.), 55-65, North-Holland (Elsevier, Amsterdam-London, New York)

Deshpande, M.D. (1977). Steady laminar and turbulent flow through vascular stenosis models. Ph.D. Thesis, Georgia Institute of Technology, Atlanta, Georgia

George, W.K. and Lumley, J.L. (1973). The laser Doppler velocimeter and its application to the measurement of turbulence. *J. Fluid Mech.,* 60, 321-362

Giddens, D.P., Mabon, R.F. and Cassanova, R.A. (1976). Measurements of disordered flows distal to subtotal vascular stenoses in the thoracic aortas of dogs. *Circulation Res.* 39, 112-119

Jorgensen, J.E., Campau, D.N. and Baker, D.W. (1973). Physical characteristics and mathematical modeling of the pulsed ultrasonic flowmeter. *J. Biomechanics,* 6, 701-

Kalmus, H.P. (1954). Electronic flowmeter system. *Rev. Scient. Instrum.,* 25, 201-206

Khalifa, A.M.A. (1978). The role of flow disorder in the noninvasive detection of atherosclerosis. Ph.D. Thesis, Georgia Institute of Technology, Atlanta, Georgia

Khalifa, A.M.A. and Giddens, D.P. (1978a). Analysis of disorder in pulsatile flows with application to poststenotic blood velocity measurement in dogs. *J. Biomechanics,* 11, 129-141

Khalifa, A.M.A. and Giddens, D.P. (1978b). Turbulence measurements with pulsed Doppler ultrasound. In *Proceedings of the First Mid-Atlantic Conference of Biofluid Dynamics* (D.P. Schneck, ed.), 229-238, V.P.I. and S.U. Press, Blacksburg, Virginia

Ling, S.C., Atabek, H.B., Fry, D.L., Patel, D.J. and Janicki, J.S. (1968). Applications of heated film velocity and shear probes to haemodynamic studies. *Circulation Res.,* 23, 789-801

McLeod, F.D. (1974). Multichannel pulse Doppler techniques. In *Cardiovascular Applications of Ultrasound* (R.S. Reneman, ed.), 85-107, North-Holland/Elsevier, Amsterdam, London, New York

Mohr, J.P. (1978). Transient ischemic attacks and the prevention of stroke. *New England J. Medicine* (editorial) 299, 93-94

Nerem, R.M. and Seed, W.A. (1972). An in vivo study of the nature of aortic flow disturbances. *Cardiovasc. Res.* **6**, 1-14

Peronneau, P.A., Bournat, J.P., Bugnon, A., Barbet, A. and Xhaard, M., (1974). Theoretical and practical aspects of pulsed Doppler flowmeter: real-time application to the measurement of instantaneous velocity profiles in vitro and in vivo. In *Cardiovascular Applications of Ultrasound* (R.S. Reneman, ed.), 60-84, North-Holland/Elsevier, Amsterdam, London, New York

Priestley, M.B. (1965). Evolutionary spectra and nonstationary processes. *J.R. Statist. Soc.,* **B27**, 204-237

Reid, J.M., Sigelmann, R.A., Nassar, M.G. and Baker, D.W. (1969). The scattering of ultrasound by human blood. *Proc. 8th ICOMBE,* Chicago, Ill.

Resch, F.J. (1970). Hot-film turbulence measurements in water flow. J. Hydraulics Div., *Proc. A.S.C.E. HY3,* 787-800

Rice, S.O. (1954). Mathematical analysis of random noise. Bell System Tech. J. 23-24, 1-162

Roschke, E.J. and Back, L.H. (1976). The influence of upstream conditions on flow reattachment lengths downstream of an abrupt circular channel expansion. *J. Biomechanics,* **9**, 481-483

Sandborn, V.A. (1955). Experimental evaluation of momentum terms in turbulent pipe flow. NACA TN 3266, Washington

Saxena, Vijay (1978). Signal analysis of turbulence measurements using pulsed Doppler ultrasound. Ph.D. Thesis, Georgia Institute of Technology, Atlanta, Georgia

Schultz, D.L., Tunstall-Pedoe, D.S., Lee, G. deJ., Gunning, A.J. and Bellhouse, B.J. (1969). Velocity distribution and transition in the arterial system. In *Ciba Foundation Symposium on Circulatory and Respiratory Mass Transport* (G.E.W. Wolstenholme and J. Knight, eds.), Churchhill, London

Seed, W.A. and Wood, N.B. (1970). Development and evaluation of a hot film velocity probe for cardiovascular studies. *Cardiovasc. Res.,* **4**, 253-263

Strandness, D.E., Ward, K.J., Phillips, D.J. and Harley, J.D. (1977). Recent aspects of ultrasonic techniques in clinical angiology. In *Proceedings, Conference on Ultrasound and Angiology,* 82-89, Munich, Germany

Wall Shear Stress Measurements in Artificial Heart Valves with Hot-Film Probes[*]

by

W. Tillmann, G. Häußinger

Helmholtz-Institut für BMT, Aachen, Germany

For blood traumatization after the implantation of artificial organs, such as artificial heart valves, the shear field generated by the prosthesis is believed to be a prime factor. Therefore, it is of interest to determine the shear stress experimentally in heart valve prostheses in a mock circuit, especially since the results of several research groups indicate that shear stresses above a critical level lead to irreversible damage of blood components. Hot film anemometry with flush mounted probes is used as measurement technique. Since the flow in the human systemic arteries is pulsatile, the hot film wall shear probes are to be calibrated under steady and unsteady flow conditions in order to determine their dynamic characteristics.

Fig. 1. Device for calibration in steady flow.

[*]The study was supported by the "Deutsche Forschungsgemeinschaft SFB 109".

For the calibration in steady flow a new calibration apparatus (1) was developed. The device consists of two concentric cylinders with the outer one rotating. Laminar Couette flow is established within the gap. Thus, for absolute calibration, determination of the shear stress at the wall of the inner cylinder, which holds the probes, is possible. Temperature control of the test liquid during the calibration procedure is provided by two separate circuits, since exact temperature control is essential for a measurement technique, which is based on a heat transfer measurement. The calibration device together with the electric motor for driving the outer cylinder by a flat belt is depicted in Fig. 1.

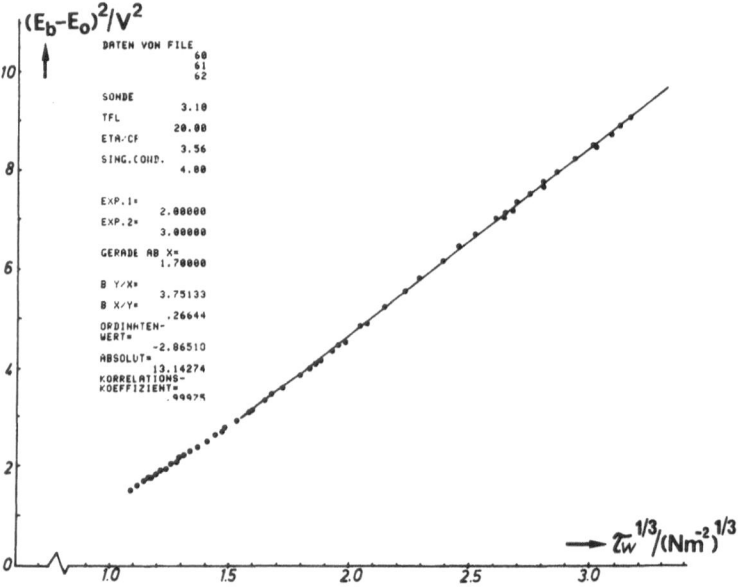

Fig. 2. Steady calibration plot.

A typical calibration curve for a hot film wall shear probe as recorded in the above apparatus is shown in Fig. 2, where the output voltage is plotted as a function of the wall shear stress. The measured bridge voltage as indicated by E_B, E_O is the voltage read at the voltmeter at zero wall shear stress. The calibration curve can be divided into a free and forced convection regime. The calibration curve in the free convection zone (dashed line) is characterized by a parabolic behaviour. A linear relationship of the calibration plot is obtained for the forced convection zone. This linear behaviour of the wall shear probe output is predicted by theory for the heat transfer due to forced convection of a heated film in a two dimensional boundary layer on a flat pate (2). The dimensionless heat transfer is given by the equation

$$Nu = A' + B'\, \tau_W{}^{1/3}$$

or in electrical quantities

$$(E_B - E_O)^2/V^2 = A + B\, \tau_W{}^{1/3}/(Nm^{-2})^{1/3}$$

From the computed regression curve (solid line), which is based on the measuring points of three different runs, the constants A and B can be determined, leading to the following calibration formula:

$$(E_B - E_O)^2/V^2 = -2.865 + 3.751\, \tau_W{}^{1/3}/(Nm^{-2})^{1/3}$$

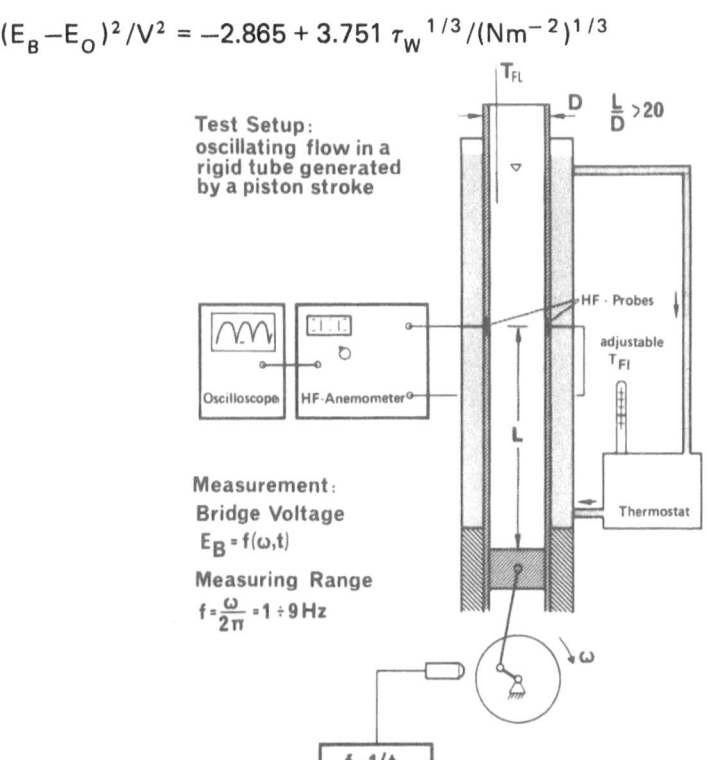

Fig. 3. Test set-up for calibration in oscillating flow.

The probes were also calibrated in oscillating flow for evaluation of amplitude dependence and phase distribution of the probe signal. Sinusoidal flow in a rigid tube was used for these experiments (Fig. 3), since the analytical solution of the wall shear stress distribution for this type of flow is available (3). The study was restricted to the evaluation of the peak values, since with respect to the blood damage-shear stress correlation, only the maximum stresses are of interest.

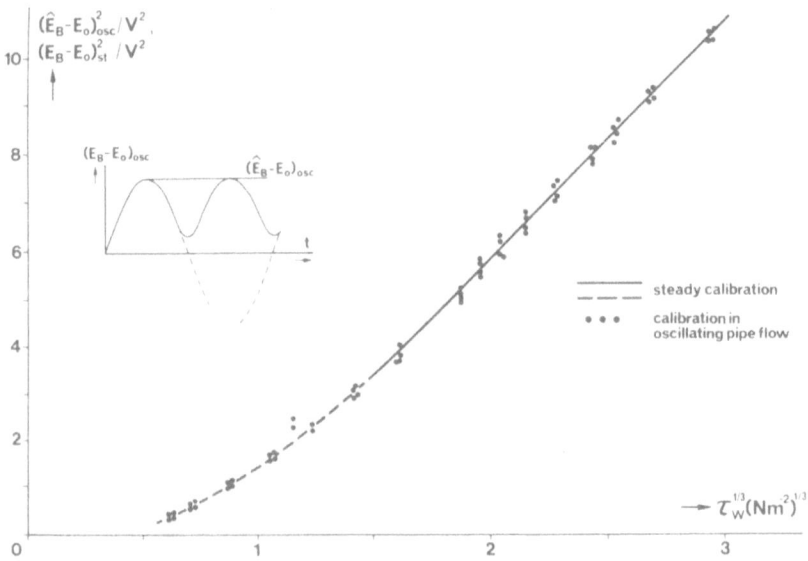

Fig. 4. Oscillating calibration plot.

The results of these comparative studies for the amplitude ratio are shown in Figs. 4 and 5. Amplitude ratio, in this context is defined as the ratio of probe output in oscillating flow to steady flow at the same theoretically predicted instantaneous shear stress level.

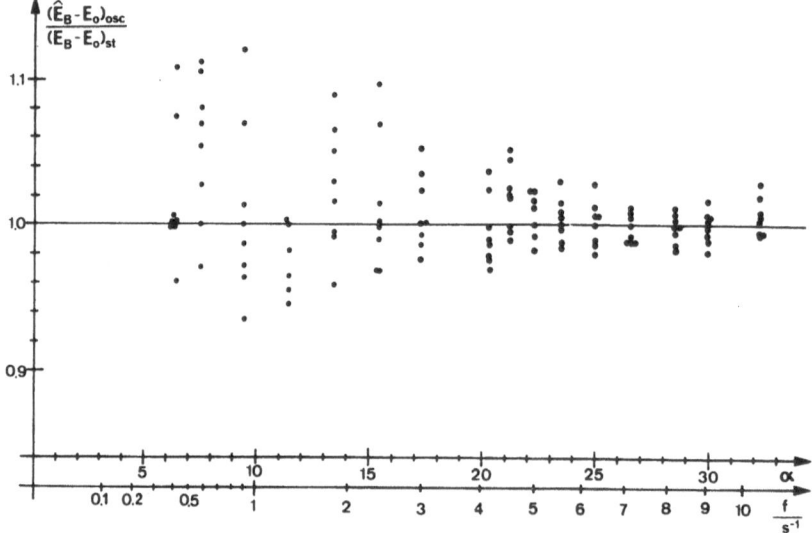

Fig. 5. Amplitude ratio of steady and oscillating calibration as computed from Fig. 4.

Fig. 4 shows the comparison between an oscillating and steady calibration plot for one probe under the same conditions with respect to overheat ratio, viscosity and fluid temperature. In Fig. 5 the amplitude ratio of this probe is plotted for the corresponding frequency range. The comparison of the probe voltages shows good agreement of the steady and oscillating calibration. In the measured frequency range no amplitude attenuation could be observed. The results of the investigation of the phase lag between the shear stress at the wall and the mean flow wave, which corresponds to the piston velocity, is shown in Fig. 6. The measured data points are compared with the analytical solution for the phase lag as a function of the dimensionless frequency parameter $\alpha = R\sqrt{\omega/\nu}$. The regression curve computed for the indicated range of the frequency parameter shows slightly lower values than those predicted by theory. This indicates a phase lag of the probe signal, relating to the actual wall shear stress.

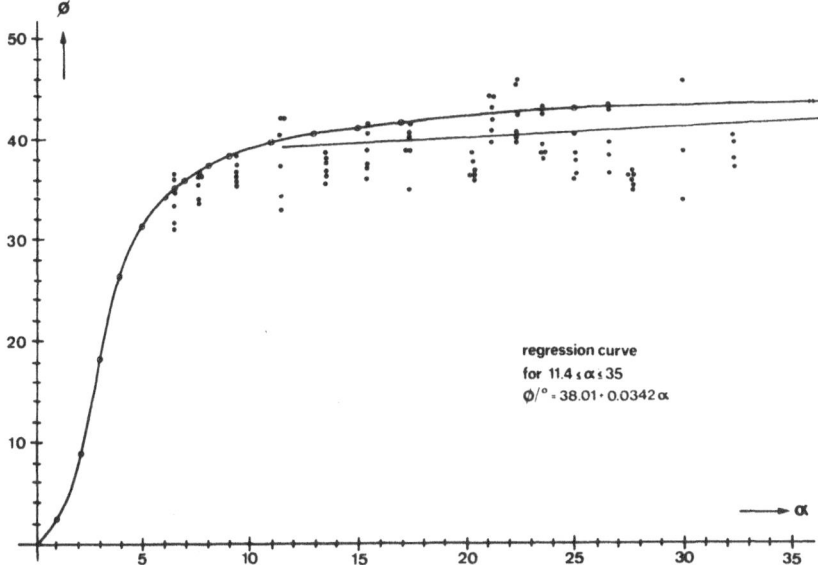

Fig. 6. Theory (solid line) and measurement (full circles) for the phase lag between wall shear stress and mean flow wave for sinusoidal flow in a rigid tube.

The discrepancy in the order of $2°$, however, is in the order of the overall accuracy which can be achieved in the calibration apparatus. Thus, the results of the calibration in an oscillating flow field show that at least for the maximum values of the cycle and the measured frequency range the calibration in steady flow is sufficient. Hence, the measured wall shear distribution in three different heart valve prostheses, shown in Fig. 7, are based on the calibration in steady Couette flow.

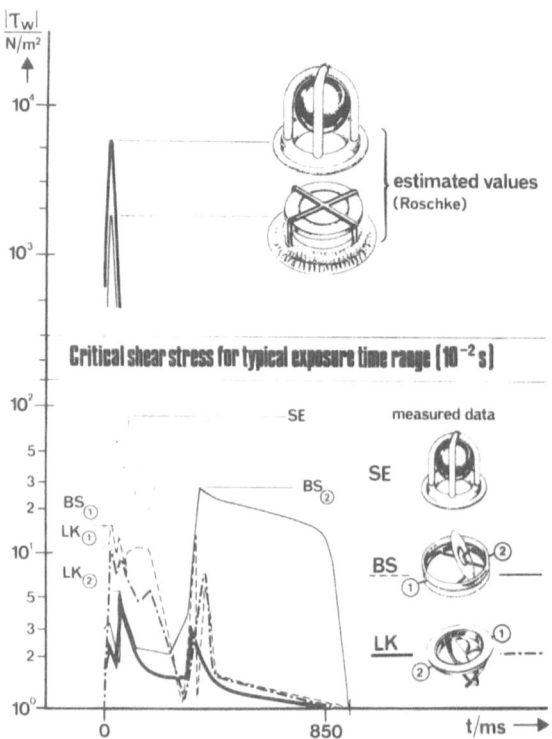

Fig. 7. Wall shear stress measurements at different heart valve prostheses.

Summarizing the experimental results the following can be stated: the wall shear stresses measured at the valve rings of Björk-Shiley, Lillehei-Kaster and Starr-Edwards prostheses are subcritical with respect to the threshold shear stresses reported by other authors for the same exposure time range. This is in contrast to estimations previously published by Roschke (4) for two investigated valve types. These estimations lead to errors, since their results are very sensitive on the input-assumptions, which in this case are based on published data of blood flow and poppet displacement in different heart valves. From the measured valve prostheses only the Starr-Edwards valve generates, under special conditions, shear stresses which can reach the critical range. This supports the better hemodynamical characteristics of the disc-valve over the ball valve. Comparing the two measured disc-valve types, it can be found that the produced maximum stresses are comparable, but that in the Björk-Shiley valve this level is maintained over one half of the cardiac cycle. This effect is due to the special valve design, since in this type of disc valve a narrow ring slit is left open between the disc and the valve ring in closed state. Through this narrow slit flow reversal occurs during the filling phase of the heart chamber due to the pressure gradient between aorta and ventricle, resulting in high shear stresses.

By means of the wall shear stress measurement with hot film anemometry it is possible to detect flow induced stresses and stimulate the development of new and — from the hydro-dynamical point of view — better heart valve prostheses.

Literature

(1)   Tillmann, W., Schlieper, H.: "A Device for the Calibration of Hot Film Wall Shear Probes in Liquids" *Journal of Physics E: Scientific Instruments,* Vol. **12,** 1979
(2)   Ludwieg, H.: "Ein Gerät zur Messung der Wandschubspannung turbulenter Reibungsschichten", *Ingenieurarchiv,* Vol. **17,** p. 207, 1949
(3)   Ling, S.C., Atabeck, H.B. et al.: "The Application of Heated-Film Velocity and Shear Probes to Hemodynamic Studies", *Circulation Research,* Vol. **23,** p. 789, 1968
(4)   Roschke, E.J., Harrison, E.C.: "Fluid Shear Stresses in Prosthetic Heart Valves" *Journal of Biomechanics,* Vol. **10,** p. 299-311, 1977

Simultaneous Measurement of Velocity and Concentration in Fiber Suspension Flow

by

R. Ek, K. Moller and B. Norman
STFI, Stockholm, Sweden

Abstract

The interaction between the state of flocculation and turbulence in a sheared paper pulp suspension is important in many of the unit operations in pulp and papermaking.

Combined Laser Doppler Anemometry and light reflection techniques have given promising initial results in the simultaneous measurement of local velocity and local concentration in pulp suspensions flowing through narrow glass tubes.

Some results are presented together with details of the measuring techniques, signal processing equipment and the particular problems associated with measurements in fiber suspensions.

Introduction

The shear flow behaviour of pulp suspensions in narrow channels is important in many unit operations in papermaking such as beating, screening and forming. However, little is known about the interaction between the fluid and the fiber movements and the resulting fiber distribution due to the lack of satisfactory techniques for measuring local velocities and concentrations. Various types of impact probe and light-guide probes for measuring these basic variables have been afflicted with a bad resolution, fiber stapling problems, and produce flow disturbances due to their presence in the flow field (1).

That a fiber suspension has a complex and in some respects a unique flow behaviour in pipes has been known for many years. In Fig. 1 typical curves of pipe friction loss versus mean flow velocity for a sulfate pulp have been compared to that of water (2).

The curves exhibit maxima and minima as well as drag reduction at high velocities.

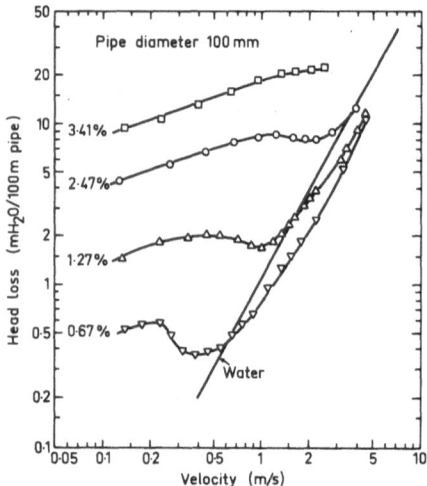

Fig. 1. Pipe friction loss versus mean velocity for an unbeaten sulfate pulp with the concentration as parameter (2).

This paper describes how to complement the LDA to enable the simultaneous measurement of local fiber velocity and fiber concentration. It is not possible to measure the water velocity in such a suspension since the fibers themselves act as scattering particles.

That this new technique offers a great potential for new information is manifested in some of the results from the initial measurements presented later in this paper.

Experimental
The fiber suspension was recirculated around a closed flow loop where the measuring section was located approximately 90 diameters downstream of the inlet to a 1 m long smooth glass tube with an inner diameter of 10 mm. The arrangement is shown schematically in Fig. 2.

A paper pulp suspension is composed of particles with a very broad size distribution, from fines with the characteristic dimensions of 1 μm to fibers about 3 mm long. The fines have a very large specific surface area and since the amount of scattered light is proportional to the scattering surface, the optical beam is greatly attenuated by even a few percent fines in the suspension. To a first approximation it may be assumed that the flow characteristics are influenced only by the large fibers, so for these measurements the fines were removed from the pulp.

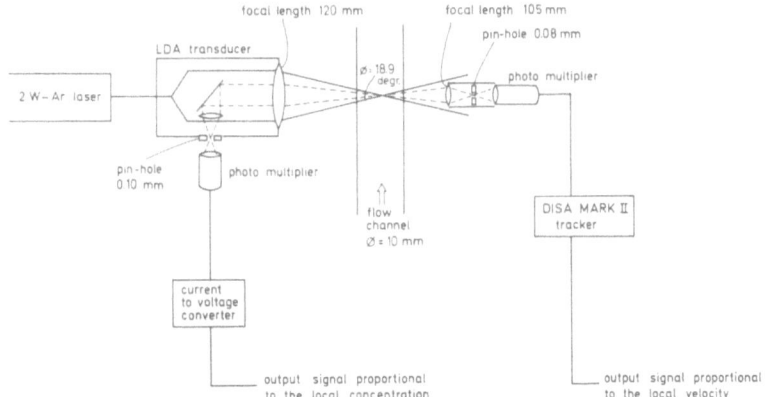

Fig. 2. Optical arrangement and signal flow diagram for LDA-system.

Velocity — Forward Scatter

The Doppler bursts were detected in the forward scatter mode since paper-making fibers are strongly forward scattering. The dual beam mode was used for low concentrations but the reference beam mode proved to be more effective for higher concentrations.

A large portion of the detected signal is composed of Doppler bursts because of the large number of particles in the fluid and it is therefore a straightforward procedure to track the signal. Fig. 3 shows a typical filtered (0.75 - 5 MHz) photomultiplier signal from 0.5 % pulp suspension flowing with a mean velocity of 2.1 m/s. The laser was operated at the green (514.5 $\times$ 10^{-9} m) line and the angle of intersection of the incident beams was 18.9 degrees.

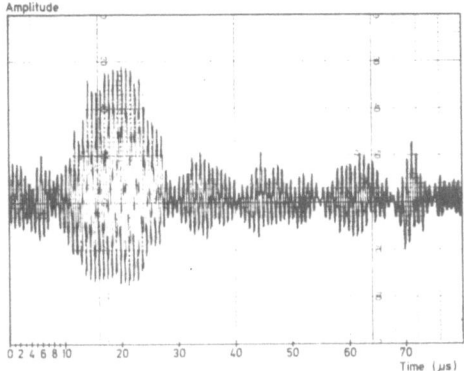

Fig. 3. Doppler bursts for fiber/water suspension flow.

The corresponding velocity signal with a drop out of 10 % is shown in Fig. 4.

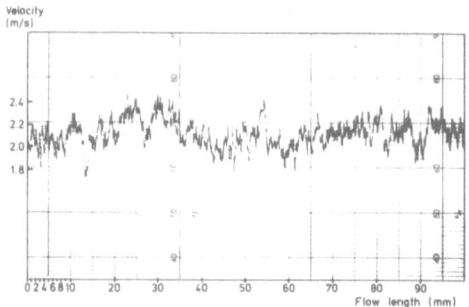

Fig. 4. Fiber velocity signal from the LDA-tracker unit.

The probability density function for the velocity signal is shown in Fig. 7. The agreement with Gaussian behaviour is good.

Concentration — Back Scatter

The optical system used (DISA Mark II) includes a back-scattered light-detecting device in which the light is detected through the same lens which focusses the two laser beams. The photomultiplier current of low frequency is converted to an analog voltage which is related to the local concentration in the measuring volume. This signal is shown in Fig. 5.

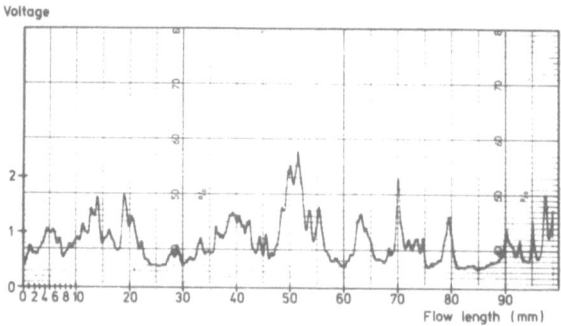

Fig. 5. Concentration signal proportional to the amount of back scattered light.

In Fig. 6 the mean voltage is plotted against the mean concentration in the pipe for a number of penetration depths. Obviously the voltage is only proportional to the concentration within certain limits. Such calibrations must be made as an initial step in any investigation since the linearity depends on the type of fiber and in particular on the amount of fines in the suspension.

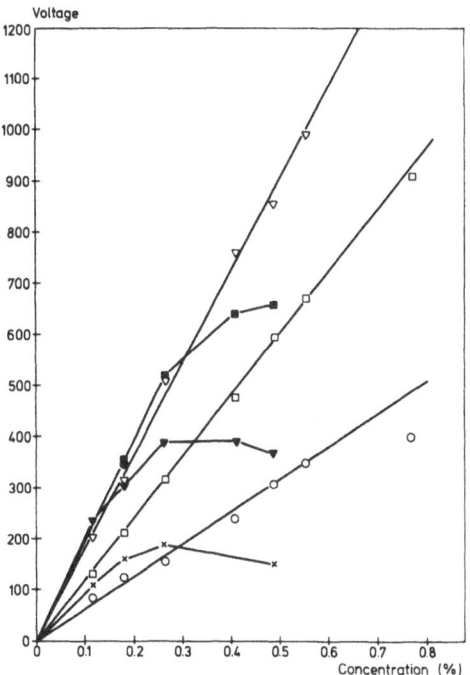

Fig. 6. Time-smoothed mean voltage (light reflection) as a function of mean concentration for different penetration distances.
(○, 1 mm; □, 2 mm; △, 3 mm; ■, 5 mm; ▼, 6 mm; x, 8 mm)

Fig. 7 shows the probability density function for the concentration signal. It is considerably skewed, indicating that concentration fluctuations do not follow Gaussian behaviour.

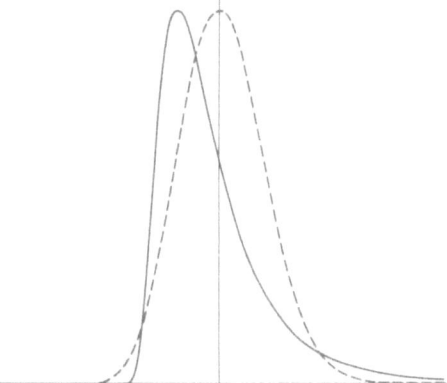

Fig. 7. Normalized probability density distribution for velocity (– – –) and concentration (——) fluctuations, 2 mm from the wall in a 10 mm pipe at 2 m/s flow velocity.

Signal Analysing Equipment

The basic properties of the velocity and concentration signals are described by three statistical functions, namely the mean value, the probability density distribution (amplitude domain) and the spectral density function (frequency domain).

Mean values were obtained with a DISA 52B30 true integrator, RMS values with a Brüel and Kjaer 2409 electronic voltmeter, probability density functions with a Hewlett Packard 3721A Correlator, power spectra with a General Radio GR 1925/26 frequency analyser, and coherence functions with a Hewlett Packard 3582A spectrum analyser.

The frequency spectra were transformed to wavelength spectra in order to make it easier to interpret the results. The following transformations were used (3):

$$\ell = \frac{\overline{u}}{f} \qquad [1]$$

$$E(\ell) = \frac{f^2}{\overline{u}} E(f) \qquad [2]$$

where $\overline{u}$ = local mean flow velocity
ℓ = wavelength
f = frequency
E = spectral density function

Results

The intention here is to demonstrate the potential of the modified LDA system for measuring local fiber velocities and fiber concentrations in a flowing fiber suspension.

Velocity Measurements

The LDA has enabled the accurate measurement of time-mean velocities close to the pipe wall. For the first time it has been possible to show that the normalised fiber velocity profile crosses the normalised water profile at small distances from the wall, as shown in Fig. 8.

This is also accompanied by a crossing of the turbulence intensity profiles for fiber suspension and water respectively, see Fig. 9.

It is advisable to measure the turbulence energy spectrum because it indicates within which frequency range a valid velocity signal is obtained. Turbulence

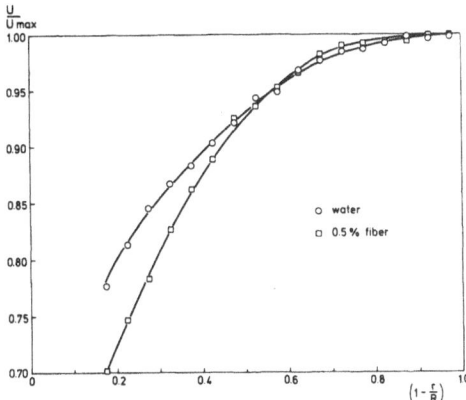

Fig. 8. Velocity profiles for fibers in a 0.5 % suspension and for pure water in a 10 mm pipe at 2 m/s.

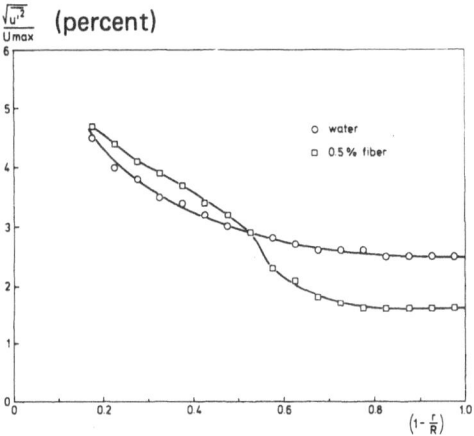

Fig. 9. Turbulence intensity profiles for fibers in a 0.5 % suspension and for pure water in a 10 mm pipe at 2 m/s.

spectra for fibers in a 0.5 % suspension and for pure water are plotted in Fig. 10 for two different radial positions at the flow velocity of 2 m/s.

The following conclusions can be drawn from the turbulence spectra. The water turbulence is lower at the pipe centerline than 2 mm from the wall. At the centerline the fiber velocity fluctuations are much less intense than the fluctuations in pure water. But 2 mm from the wall the reverse is true, an effect which has not previously been observed.

At small wavelengths the spectra approach the −2 slope for random variations indicating that the signal has been drowned in noise. Therefore the signal was low pass filtered below 1000 Hz (corresponding to 2 mm wavelength) for the turbulence intensity profile measurements reported above.

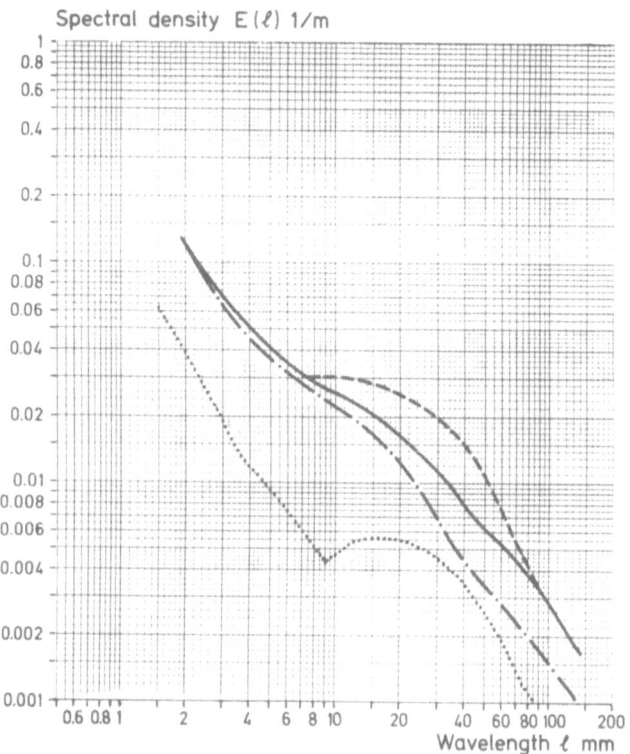

Fig. 10. Turbulence spectra for water and 0.5 % fiber suspension at two different radial positions at 2 m/s, in a 10 mm pipe.
5 mm from the wall, − · − · − water; . . . . 0,5 % fiber suspension
2 mm from the wall, —— water; −−− 0,5 % fiber suspension.

Concentration Measurements

With the modified LDA system the resolution at the concentration measurements is better than 1 mm. Some typical spectra are shown in Fig. 11 for measurements 2 mm from the wall at three different velocities. Apparently the number of fiber flocs with wavelengths less than 8 mm decrease as the velocity increases. Larger flocs do not seem to be affected by a change of velocity from 2 to 5 m/s.

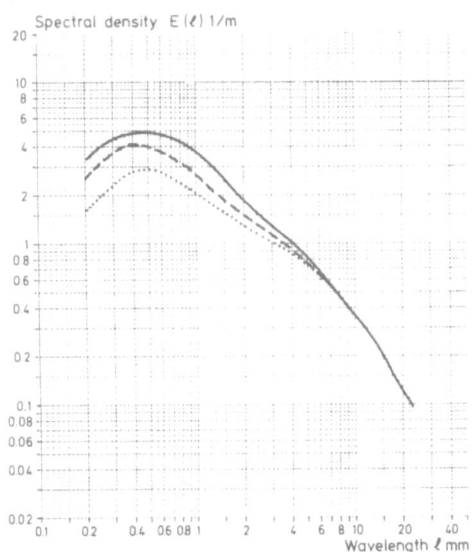

Fig. 11. Wavelength spectra for the concentration fluctuations 2 mm from the wall at 3 different velocities (——— 2 m/s, ——— 3 m/s, ... 5 m/s) for a 0.5 % fiber suspension.

Simultaneous Measurement of Fiber Velocity and Fiber Concentration

The local fluctuations in fiber concentration and fiber velocity are not independent of one another. This can be seen by measuring the coherence function, which is a normalised measure of the degree of causality between the two signals, ranging from 0 for uncorrelated to 1 for fully correlated signals. The coherence function $\gamma^2(f)$ is shown in Fig. 12 for the frequency range from 60 to 1000 Hz and for the flow velocities 1, 2 and 5 m/s.

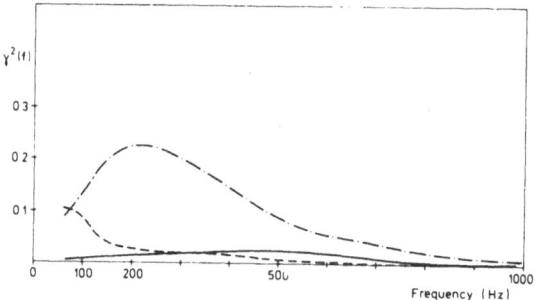

Fig. 12. Coherence spectra at the centerline in a 10 mm tube for a 0.5 % fiber suspension at three different velocities (——— 1 m/s; —·—·— 2 m/s; ——— 5 m/s)

At 2 m/s there is clearly some connection between the fiber movements and the concentration, with a maximum at about 200 Hz (10 mm). However, at 1 m/s and 5 m/s there is very little coherence between local fiber velocity and fiber concentration.

References

1.   Norman, B., Moller, K., Ek, R., Duffy, G.G. *Fibre-water Interactions in Paper-Making* Vol. **1**, pp 195-249, Published by Technical Division, The British Paper and Board Industry Federation, London, 1978
2.   Moller, K., Norman, B. *Svensk Papperstidning,* **78** (16): 582, 1975
3.   Norman, B., Wahren, D. *Svensk Papperstidning,* **75** (20): 807, 1972

Signal and Data Processing

Processing of Random Signals

by

William K. George, Jr.
State University of New York at Buffalo

with Appendices by

Paul D. Beuther
State University of New York at Buffalo
and
John L. Lumley
Cornell University

Introduction

The experimentalist who is involved in dynamic flow measurement is almost always interested in some type of random signal processing. This interest can arise because the flow itself is random, as in turbulent flow. Or it can arise from concern about whether measured data represent real flow phenomena or merely statistical fluctuations in the data. These interests can occur simultaneously. For example, in the problem of determining periodicities in the turbulent flow behind a moving blade row the question of distinguishing between fluctuations in the data and the true spectral peaks is crucial to the experiment.

In this review article, attention will be confined to the problem of measuring and interpreting statistics (to the second moment) of a stationary random signal. After an introduction to the nature of stationary processes, we shall review the great strides of the last two decades in the processing of discretely sampled signals. The advances of the last decade in the understanding of randomly sampled data will be reviewed and these techniques will be compared to the even more recent advances in understanding the burst-processor laser-Doppler anemometer. Finally, some new ideas for measurement using both continuous and random sampling will be introduced.

No attempt will be made to be comprehensive, either in the sense of covering all topics of interest or in the sense of providing detailed references to the material. Rather, the approach adopted is tutorial and is intended to both introduce the material and provide a framework for subsequent learning. This article should be viewed in the context of the other articles in this volume which also address topics relevant to signal processing and interpretation — especially the review articles by Buchhave, Kovasznay and Van Atta.

The historical roots of this presentation are many, but particular credit must be given to the excellent articles on the subject found in references (1) - (5). In no sense should this article be viewed as a substitute for careful reading of these and the other references.

The Ensemble Average
The concept of an *ensemble average* is familiar in some sense to every layman. To compute such an average we simply add the individual realizations of the process (supposed random) and divide by the number of realizations; that is,

$$X_N = \frac{1}{N} \sum_{i=1}^{N} x_i \qquad [1]$$

where X_N is the "average" computed on the basis of N realizations and the x_i denote the individual realizations.

By the very act of using such an algorithm, the layman is expressing a belief that a *true average* exists and that the computed number X_N is representative of it. He even has a primitive idea of convergence in that he believes that the more realizations he has, the better will be his estimate of the true value. If $\bar{x}$ denotes the true average, we express this convergence formally as

$$\lim_{N \to \infty} X_N = \bar{x} \qquad [2]$$

The fact that these concepts are widespread is probably an expression of a basic belief (conditioned or inherited) that there is some underlying order in the universe; in effect, we insist that even random events are orderly.

We can quantify the above ideas by noting that since our estimator X_N given by equation [1] is the sum of random variables, it is itself a random variable. Therefore, we can define the variance of X_N as

$$\text{var} \left\{ X_N \right\} = \overline{(X_N - \bar{x})^2} \qquad [3]$$

where the overbar denotes the true ensemble average and is assumed to exist. The question of convergency can now be expressed as: Does the variance of estimator, X_N, become vanishingly small as the number of realizations, N, becomes large?

$$\lim_{N\to\infty} \text{var}\{X_N\} \overset{?}{\to} 0 \tag{4}$$

It is easy to show (and is well-known) that if the samples are identically distributed and are statistically independent, then

$$\text{var}\{X_N\} = \frac{1}{N}\,\text{var}\,\{x\} \tag{5}$$

where $VAR\{x\}$ is the mean square fluctuation in the ensemble given by

$$\text{var}\,\{x\} = \overline{(x-\overline{x})^2} = \sigma_x^{\,2} \tag{6}$$

If an acceptable fluctuation in our estimate is given by ϵ where

$$\epsilon^2 \equiv \frac{\text{var}\{X_N\}}{\overline{x}^2}\,, \tag{7}$$

then for N realizations,

$$\epsilon = \frac{1}{\sqrt{N}} \cdot \left[\frac{\sigma_x}{\overline{x}}\right] \tag{8}$$

Thus the variability of our estimator is proportional to the relative fluctuation of the random variable itself $(\sigma_x/\overline{x})$ and is inversely proportional to the square root of the number of *independent* samples. Clearly the estimator given by equation [1] converges as $N \to\infty$ if the samples are identically distributed and independent.

We have assumed to this point that the estimator converges to the *true* mean. When this is not true the estimator is said to be *biased*. Examples of biased estimators will be introduced later.

Time Series, the Time Average, and the Ergodic Hypothesis
Let us now imagine a continuous signal u(t) which is random. By this we mean both that the time evolution of a single signal can not be predicted by a deterministic function of time nor can the value at a particular time be predicted for an ensemble of realizations. If we further imagine a number of different experiments all evolving under the same physical constraints, we can immediately apply our concept of an ensemble average and speak of the ensemble average at a particular instant in the evolution of the process. The conditional and periodic averages discussed elsewhere in this volume are simply variations on this ensemble average approach.

In many situations in which flow measurements are to be undertaken it is inconvenient, or impossible, to carry out a number of independent experiments.
In such situations a time average defined by

$$U_T = \frac{1}{T} \int_0^T u(t)dt \qquad [9]$$

is often employed with the hope that U_T will approach the *true* average as the
length of the integration interval increases. That is, we hope that

$$\lim_{T \to \infty} U_T = \bar{u} \qquad [10]$$

It is obvious that this operation of time averaging makes no sense unless the
averages themselves are independent of time — at least in the measurement
interval. When $\bar{u}$ is time-independent, we say that the process is *statistically
stationary* and have some reason to believe that the operation of time averaging
might make sense. Our hopes can be summarized in the *ergodic* hypothesis
which for our purposes can be stated as:

The Ergodic Hypothesis

Time averages converge to a mean value as the averaging time becomes large.
Moreover, the time average always converges to the same value regardless of
when the averaging process is initiated. Volumes have been written about the
ergodic hypothesis. For the physical scientist, however, it is simply a statement
that the world is working as we think it should.

The *ergodic hypothesis* has a number of implications about the nature of
random time series. Of particular interest to us are the following:

(i) Random variables must become uncorrelated (and, in fact, statistically
    independent) at large time delays.
    Symbolically,

$$\overline{u(t) \, u(t+\tau)} \to 0 \text{ as } \tau \to \infty \qquad [11]$$

(ii) An integral scale exists and is a measure of the memory of the process
     (time to lose correlation).

For our purposes, the integral scale can be defined from the autocorrelation as:

$$T_u = \int_0^\infty \rho_u(\tau) \, d\tau \qquad [12]$$

where $\rho_u(\tau)$ is the autocorrelation coefficient defined by

$$\sigma_u^2 \ \rho_u(\tau) = \overline{[u(t) - \bar{u}] \ [u(t') - \bar{u}]} = B_u(\tau) \qquad [13]$$

where

$$\sigma_u^2 = \overline{[u(t) - \bar{u}]^2} = \text{var}\{u\} \qquad [14]$$

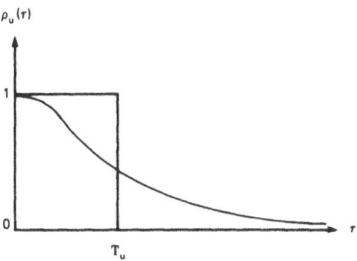

Fig. 1. Typical autocorrelation showing integral scale from equation 12

From Fig. 1, T_u is seen to be the intersection with the time delay axis of a rectangle having unit height and area equal to that under autocorrelation coefficient curve. The figure makes it clear that T_u is a measure of the time over which the signal is correlated, as required.

We note that other integrals scales can be defined; all of these will be about the same. Also, if we generate a new process from u(t), say F[u(t)], the integral scale of F will be no larger than that of the original process.

The Convergence of Time Averages
We are now prepared to answer the two major questions of random signal processing:

1. Do my time averages converge to the correct value?
2. At what rate do my time averages converge?

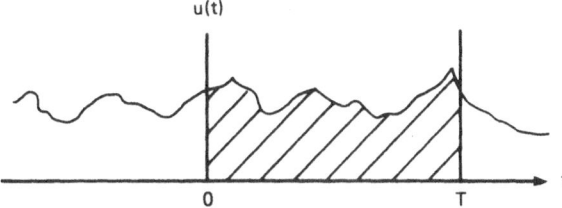

Fig. 2. Illustration that the time average is a random variable (shaded area).

The operation of time averaging which we have defined is illustrated in Fig. 2 by the area bounded by the random signal and the averaging interval. It is clear that our estimator U_T is also a random quantity. The answer to the first question is easily obtained by ensemble averaging U_T. Since the signal is assumed stationary the operations of ensemble averaging and integration commute and we have

$$\overline{U}_T = \frac{1}{T} \int_0^T \overline{u(t)} \, dt = \overline{u}$$

Thus in our previously accepted terminology: U_T is an unbiased estimator of $\overline{u}$. We answer the question of convergence as before by examining the variance of U_T.

$$\text{var}\{U_T\} = \overline{[U_T - \overline{u}]}^2 = \frac{1}{T^2} \iint_0^T \overline{[u(t) - \overline{u}] [u(t') - \overline{u}]} \, dt dt' \qquad [15]$$

The integrand is just the autocorrelation of the signal $u(t)$ given by equation [13]. After a partial integration (see ref. 3) we have

$$\text{var}\{U_T\} = \frac{2 \, \text{var}\{u\}}{T} \int_0^T \rho(\tau) [1 - \frac{\tau}{T}] \, d\tau \qquad [16]$$

Since $\rho \to 0$ as T becomes large, this reduces to

$$\text{var}\{U_T\} \cong \frac{2 \, \text{var}\{u\}}{T} \int_0^T \rho_u(\tau) \, d\tau = \frac{2 \, T_u}{T} \, \text{var}\{u\} \qquad [17]$$

The relative error in the estimator is then given by

$$\epsilon^2 = \frac{\text{var}\{U_T\}}{\overline{u}^2} \cong \frac{2 T_u}{T} \, \frac{\text{var}\{u\}}{\overline{u}^2} \quad [18] \quad \text{or} \quad \epsilon = \sqrt{\frac{2 T_u}{T}} \cdot \left[\frac{\sigma_u}{\overline{u}}\right] \qquad [19]$$

The analogy between equations [19] and [8] is obvious. In effect, the number of independent samples in our average is

$$N = \frac{T}{2 T_u} \qquad [20]$$

Thus segments of our time record of two integral scales in length contribute to the average as though they were statistically independent. This is illustrated in Fig. 3.

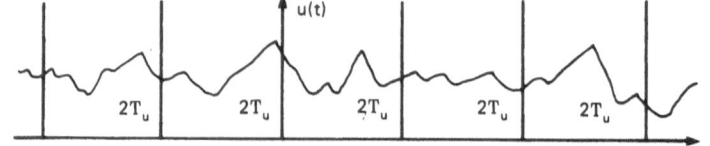

Fig. 3. Segments of stationary random process illustrating pieces of the record which contribute to the mean as though they were statistically independent.

The results given in equations [19] and [20] occur over and over again in the analysis of time series. We only assumed that the signal $u(t)$ was stationary.

Another example which leads to a similar result is the time average estimator for the autocorrelation defined by

$$B_T(\tau) = \frac{1}{T} \int_0^{T-|\tau|} u(t)u(t+\tau)\, dt \qquad [21]$$

Note at the outset that this estimator is biased since

$$\overline{B_T(\tau)} = \frac{1}{T} \int_0^{T-|\tau|} \overline{u(t)u/t+\tau)}\, dt = \frac{T-|\tau|}{T}\, B(\tau) \qquad [22]$$

Clearly this bias could have been avoided by dividing by $T-|\tau|$ instead of T. It will be seen in the following paragraphs that the unbiased estimator does not converge for large time lags and is therefore unsuitable; hence, the widespread use of the biased estimator.

The variance of our estimator(s) can be computed in the following manner:

$$\text{var}\{B_T\} = \overline{[B_T(\tau) - B(\tau)]}^2 = \frac{1}{T^2} \iint_0^{T-|\tau|} \overline{u'(t)u'(t+\tau)u'(t_1)u'(t_1+\tau)}\, dt\, dt_1 \qquad [23]$$

where the primes denote only the fluctuating values. By assuming that the fourth order moments of $u(t)$ are jointly Gaussian

$$\overline{uu'u''u'''} = \overline{uu'}\cdot\overline{u''u'''} + \overline{uu''}\cdot\overline{u'u'''} + \overline{uu'''}\cdot\overline{u'u''} \qquad [24]$$

and modelling the autocorrelation of $u(t)$ as a simple exponential

$$\rho_u(\tau) = \exp\{-|\tau|/T_u\} \qquad [25]$$

it can be shown that (see ref. 1)

$$\text{var}\{B_T\} \cong \frac{2T_u}{T}\, \text{var}\{u\} \qquad [26]$$

The relative error is then

$$\epsilon^2 = \frac{\text{var}\{B_T\}}{B^2(\tau)} = \frac{2T_u}{T} \cdot \left[\frac{1}{\rho_u(\tau)}\right]^2 \qquad [27]$$

since var $\{u\}$ = B(o). Thus the relative accuracy to which we have determined the autocorrelation with this estimator decreases with increasing time lag since the autocorrelation goes to zero. Alternately, the larger the lag at which we want to compute the autocorrelation, the longer we must average.

A similar derivation reveals the reason for abandoning the unbiased estimator (see ref. 1). The occurrence of a term $T-|\tau|$ in the denominator of equation [27] increases the relative error with time lag. As long as the largest time lag of interest is much less than the averaging time, there is no significant difference between the biased and unbiased estimators.

Spectral Estimates From Finite Record Lengths

We first consider the problem of Fourier decomposing a stationary random signal (of infinite record length). Originally this subject was approached through the use of Fourier-Stieltjes integrals which, to the uninitiated, appeared strange and complicated. The subject has now been considerably simplified, thanks to the introduction of generalized functions into the Fourier analysis. For a complete review, the interested reader is referred to the comprehensive monograph by Lumley (ref. 4).

The Fourier transform of a signal $u(t)$ in the usual sense is defined by

$$\hat{u}(\omega) = \frac{1}{2\pi} \int_{-\infty}^{\infty} e^{i\omega t} \, u(t) dt \qquad\qquad [28]$$

For this transform to exist, $u(t)$ must satisfy a number of conditions including smoothness and vanishing at infinity (see any standard calculus text). The inverse transform is given by

$$u(t) = \int_{-\infty}^{\infty} e^{-i\omega t} \, \hat{u}(\omega) \, d\omega \qquad\qquad [29]$$

Stationary random functions of time do not, in general, satisfy the necessary conditions for their Fourier transform to exist in the above sense. However, if one agrees to work in the domain of generalized functions so that our signal of infinite length is approached as the limit of a sequence of functions whose Fourier transforms exist, then we can define the Fourier transform of the stationary random signal as the limit of the Fourier transforms of the members of the sequence. Under a set of conditions which need not concern us here, this limit can be assumed to exist. Hence, we can write the Fourier transform of $u(t)$ as in equation [28] and denote it by $\hat{u}(\omega)$, if we agree to say that this is the Fourier transform of $u(t)$ *in the sense of generalized functions*. This seems a small price to pay since once we have made this qualification we can treat $\hat{u}(\omega)$ as though it were an ordinary Fourier transform (for our purposes, at least).

It should be obvious that for any given realization of $u(t)$, there will be a particular realization of $\hat{u}(\omega)$. In other words, if $u(t)$ is random, so must be $\hat{u}(\omega)$.

Since u(t) is assumed stationary (eg. $\overline{u(t)\,u(t+\tau)}$ is function of τ only), it is reasonable to expect some constraints on the expected values of the Fourier transforms. It can be shown that a consequence of stationarity is that the Fourier coefficients are uncorrelated at different frequencies; that is

$$\overline{\hat{u}(\omega)\hat{u}(\omega')} = 0 \;\; ; \omega \neq \omega' \tag{30}$$

In fact, we can write

$$\overline{\hat{u}(\omega)\hat{u}(\omega')} = S(\omega)\delta(\omega-\omega') \tag{31}$$

where $S(\omega)$ is the spectrum of the signal u(t) and $\delta(\omega-\omega')$ is the familiar Dirac delta-function.
It is well-known and follows immediately that the spectrum is the Fourier transform of the autocorrelation

$$S(\omega) = \frac{1}{2\pi} \int_{-\infty}^{\infty} e^{i\omega\tau}\, B(\tau)\, d\tau \tag{32}$$

and that the autocorrelation is the inverse Fourier transform of the spectrum

$$B(\tau) = \int_{-\infty}^{\infty} e^{-i\omega\tau}\, S(\omega)\,d\omega \tag{33}$$

By evaluating the last equation at $\tau=0$ we have

$$B(0) = \sigma^2 = \int_{-\infty}^{\infty} S(\omega)\,d\omega \tag{34}$$

Thus, as is well known, the integral of the spectrum over all frequencies yields the variance (or mean square) of the signal.
Now that we are assured (in principle, at least) that a spectrum exists, we turn to the problem of its estimation from a finite length of record. Equation [28] suggests that an appropriate estimator for the Fourier transform $\hat{u}(\omega)$ might be

$$\hat{u}_T(\omega) = \frac{1}{2\pi} \int_{-\frac{T}{2}}^{\frac{T}{2}} e^{i\omega\tau} u(t) dt \tag{35}$$

The interval $(-\frac{T}{2}, \frac{T}{2})$ has been chosen instead of (0,T) for symmetry and simplicity with no loss of generality.

That this is a good choice is confirmed by the fact that the average value of the spectral estimator defined by

$$S_T(\omega) = 2\pi \frac{|\hat{u}_T(\omega)|^2}{T} \tag{36}$$

does reproduce the correct value of spectrum as T becomes large. This is easily seen by carrying out the operations implied by equations [35] and [36] to obtain

$$\overline{S_T(\omega)} = \frac{1}{2\pi} \int_{-T}^{T} e^{i\omega\tau} B(\tau) \cdot \left[1 - \frac{|\tau|}{T} \right] d\tau \qquad [37]$$

Clearly from equation [37],

$$\lim_{T \to \infty} \overline{S_T(\omega)} = S(\omega) \qquad [38]$$

and the estimator is unbiased.

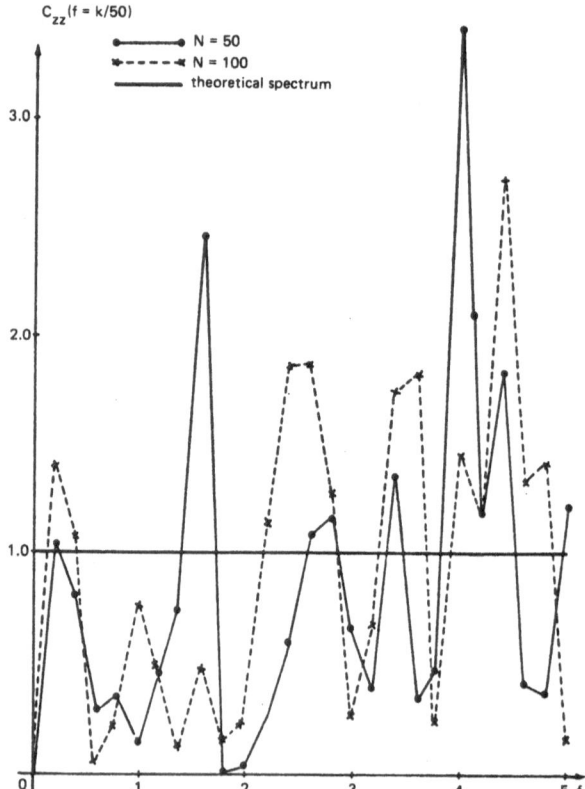

Fig. 4. Spectral estimates for two different record lengths of white noise signal illustrating 100 % relative error (adapted from ref. 1).

Unfortunately, unlike our previous estimators we have a problem with the variability of $S_T(\omega)$. This is illustrated in Figs. 4 and 5 taken from ref. (1) which show both the true spectrum and the estimated spectrum. It is clear that the scatter is unacceptable and that things do not improve with increasing record length.

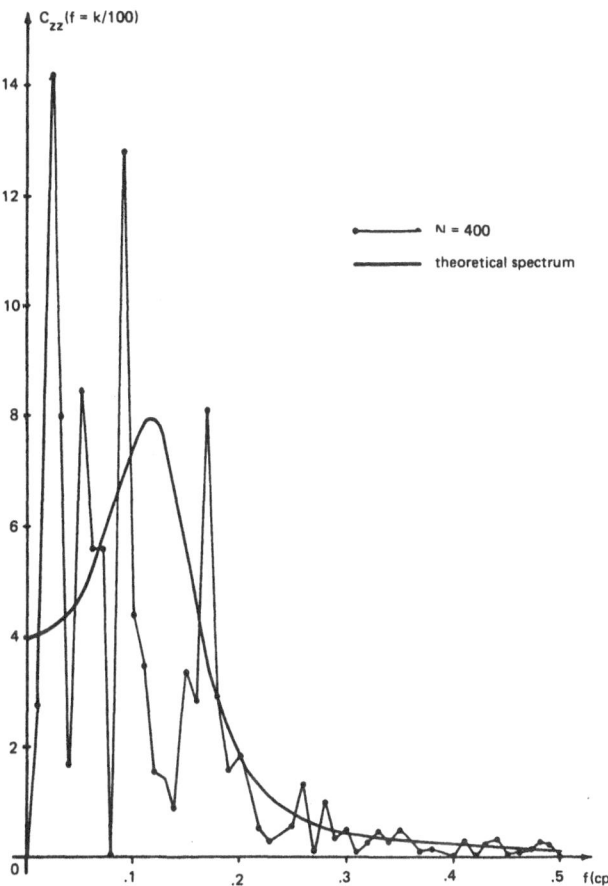

Fig. 5. Spectral estimate for filtered noise signal illustrating 100 % relative error (adapted from ref. 1)

To see why this is so we compute the variance of our estimator. By assuming that fourth-order moments of u(t) are jointly normal, we can obtain (see ref. 1)

$$\text{var} \left\{ S_T(\omega) \right\} = [S(\omega)]^2 \tag{39}$$

It immediately follows that the relative error is unity!

$$\epsilon^2 = \frac{\text{var}\left\{ S_T(\omega) \right\}}{[S(\omega)]^2} = 1 \tag{40}$$

Thus the rms fluctuations are as big as the spectrum!

The reason for the above can be seen by examining equation [37]. We define a new function $h_T(\tau)$ by

$$
h_T(\tau) = \begin{cases} 1 - \dfrac{|\tau|}{T} & ; \ |\tau| < T \\[2mm] 0 & ; \ |\tau| > T \end{cases}
\tag{41}
$$

and denote its transform by $H_T(\omega)$. Equation [37] can be rewritten as

$$
\overline{S_T(\omega)} = \frac{1}{2\pi} \int_{-\infty}^{\infty} e^{i\omega\tau} \; \overline{u(t)u(t+\tau)} \; h_T(\tau) \, d\tau
\tag{42}
$$

or using Parseval's theorem

$$
\overline{S_T(\omega)} = \int_{-\infty}^{\infty} S(\omega_1) \, H_T(\omega-\omega_1) \, d\omega_1
\tag{43}
$$

Thus, our spectral estimator is viewing the spectrum through a window whose width is determined by the averaging time T ($\Delta\omega_T = 2\pi/T$). As T increases, $h_T(\tau)$ gets wider and $H_T(\omega)$ gets narrower. In the limit as $T \to \infty$, $H_T(\omega)$ simply selects a single realization of the spectrum. While this explains the lack of bias, it also accounts for the variance since it means that we have used only a single independent sample for our estimate!

There are two ways in which we can make our spectral estimator converge with increasing averaging time:

1. We can ensemble average independent estimates of the spectrum based on non-overlapping record segments. This technique can be viewed as either doing the experiment many times or as subdividing a long record.

2. We can use the fact that the estimates at different frequencies are uncorrelated when separated by more than $\Delta\omega_T = 2\pi/T$, and average over different frequencies. This process is called smoothing. The fact that adjacent estimates are correlated only over $\Delta\omega_T$ follows immediately from the fact that the true Fourier coefficients (which are uncorrelated at different frequencies) are seen through the window $H_T(\omega)$ by virtue of the finite record length.

The rate of convergence due to method (1) is simply the same as that for any ensemble average ($\epsilon \sim 1/\sqrt{N}$). For method (2) it should be clear that the number of independent samples making the smoothed estimate depends on the relative value of the effective window width $\Delta\omega$ and the correlation width $\Delta\omega_T$.

Thus, $N \sim \Delta\omega/\Delta\omega_T$ and the relative spectral error is given by

$$\epsilon^2 = \frac{2\pi}{\Delta\omega T} \qquad [44]$$

This does go to zero as $T \to \infty$ as required.

There are many ways in which the averaging (or smoothing) processes above can be implemented. All involve the choice of a window in either the time domain $D(\tau)$ or the frequency domain, $W(\omega)$. The smoothed spectrum is then given by

$$S_M(\omega) = \frac{1}{2\pi}\int_{-\infty}^{\infty} e^{i\omega\tau}\, B_q(\tau)\, D(\tau)d\tau \qquad [45]$$

where

$$B_q(\tau) = B(\tau)h_T(\tau) \qquad [46]$$

or

$$S_M(\omega) = \int_{-\infty}^{\infty} S_q(\omega_1)\, W(\omega-\omega_1)\, d\omega_1 \qquad [47]$$

where $S_q(\omega)$ is the Fourier transform of $B_q(\tau)$.

TABLE I   Lag and Spectral Windows (from Figure 6)

	Lag Window	Spectral Window	Effective Bandwidth ($\Delta\omega$)
Rectangular	$D_R(\tau) = \begin{cases} 1, & \|\tau\| \leqslant M \\ 0, & \|\tau\| > M \end{cases}$	$W_R(\omega) = \frac{M}{\pi}\left\{\frac{\sin \omega M}{\omega M}\right\}$	$\dfrac{\pi}{M}$
Bartlett	$D_B(\tau) = \begin{cases} 1-\frac{\|\tau\|}{M}, & \|\tau\| \leqslant M \\ 0, & \|\tau\| > M \end{cases}$	$W_B(\omega) = \frac{M}{2\pi}\left\{\frac{\sin \omega M/2}{\omega M/2}\right\}^2$	$\dfrac{3\pi}{M}$
Tukey	$D_T(\tau) = \begin{cases} 1/2\left\{1+\cos\frac{\pi\tau}{M}\right\}, & \|\tau\| \leqslant M \\ 0, & \|\tau\| > M \end{cases}$	$W_T(\omega) = \frac{M}{2\pi}\left\{\frac{\sin \omega M}{\omega M}\right\}\left[\frac{1}{1-(\omega M/\pi)^2}\right]$	$\dfrac{8\pi}{3M}$
Parzen	$D_P(\tau) = \begin{cases} 1-6\left\{\frac{\tau}{M}\right\}^2 +6\left\{\frac{\|\tau\|}{M}\right\}^3, & \|\tau\| \leqslant \frac{M}{2} \\ 2\left\{1-\frac{\|\tau\|}{M}\right\}^3, & \frac{M}{2} \leqslant \|\tau\| \leqslant M \\ 0, & \|\tau\| > M \end{cases}$	$W_P(\omega) = \frac{3M}{8\pi}\left\{\frac{\sin \omega M/4}{\omega M/4}\right\}^4$	$\dfrac{3.77\pi}{M}$

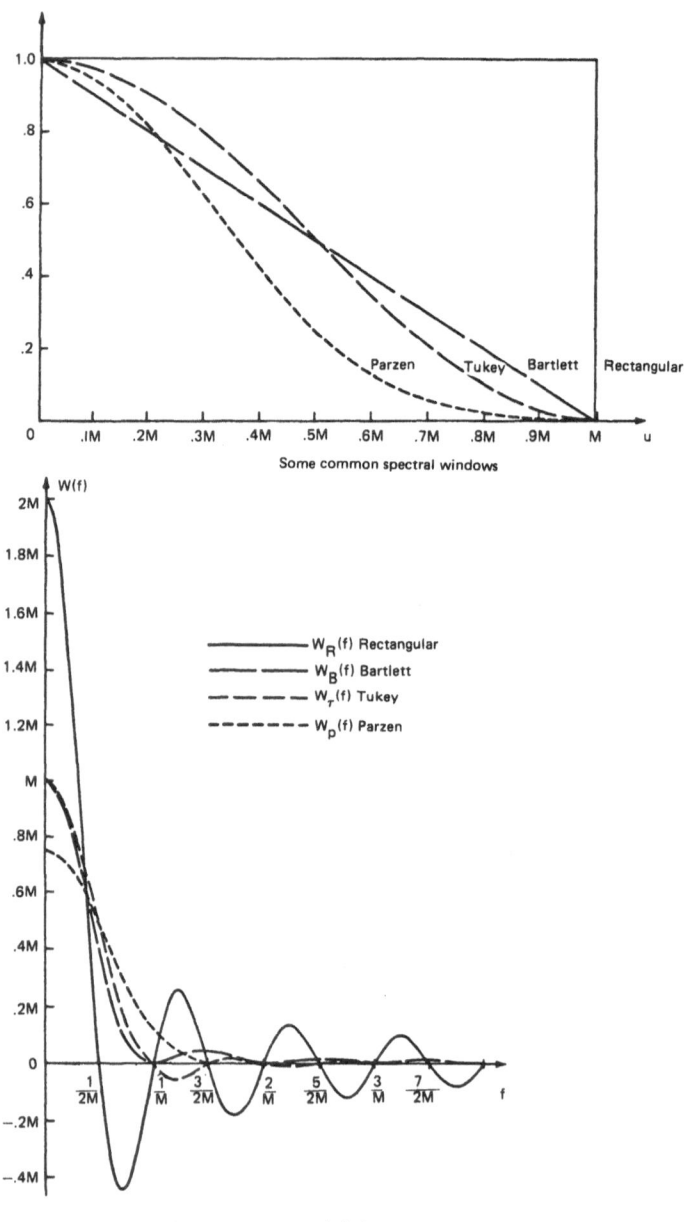

Fig. 6. Lag and spectral windows commonly in use (adapted from ref. 1). For definitions and effective bandwidths see Table I.

Some sample windows are shown in Fig. 6 in the form in which they are usually applied to digital data. For analog estimation the window is determined by the characteristics of the band-pass filter used in the analysis.

That we have not solved all of our problems is illustrated by Fig. 7 from ref. 1 which shows the effect of various window widths on a spectrum estimated from a fixed length of record. For the narrowest window, the spectrum cannot be determined because of the variability of the estimator (which peaks are real?), while for the widest window the spectrum, although smooth, cannot be believed. Unfortunately, *the smoothing process introduces a bias.*

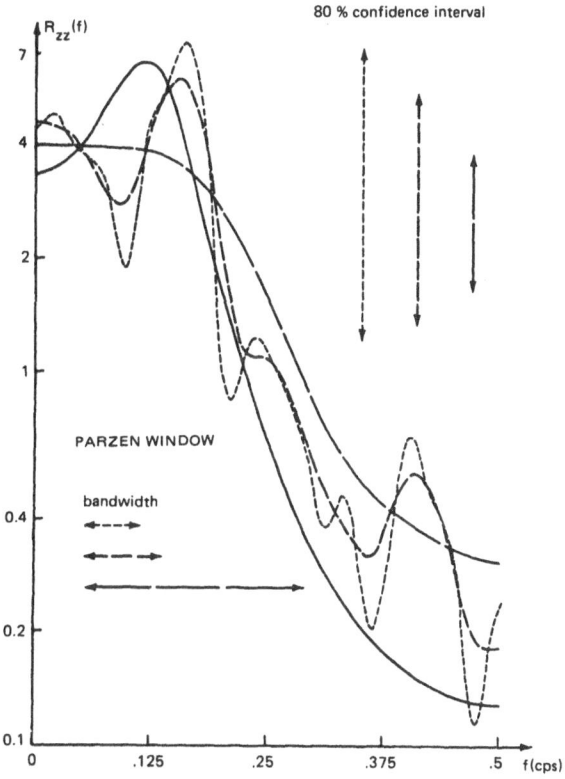

Fig. 7. Spectral estimates illustrating dependence of relative error and bias on window width (from ref. 1). Solid line is true spectrum.

It is not difficult to show that this bias can be related to the curvature of the spectrum (ref. 1) and simply places a constraint on the maximum window width (and minimum averaging time) which can be used. Approximately we must have

$$\Delta\omega << \left[S(\omega) \middle/ \frac{d^2}{d\omega^2} S(\omega) \right]^{\frac{1}{2}} \quad \text{to avoid bias.} \tag{48}$$

Discretely Sampled Signals

The emergence over the past 30 years of the digital computing machine as a major experimental tool and especially the rapid advances in mini- and micro-computer technology of the last decade have made this mode of signal process-ing the most common. Even on-line hard-wired correlators and spectrum ana-lyzers are no longer analog devices but rather dedicated microprocessors opera-ting on digitized inputs.

In the preceding section we have examined the basic character of random signals and the limitations imposed on our ability to measure them. In this sec-tion we shall look at the additional problems introduced by digitizing this sig-nal with a sampling device which takes samples at a fixed rate. We should not expect that the process of digitization will eliminate any of the considerations of the preceding section. Rather, the important question is: how much in-formation is lost in the conversion process?

We begin by. defining a sampling function g(t) which simply selects values of the process being sampled at programmed times. If we agree that g(t) can be represented as a generalized function and that we can work with such functions, then the sampled signal [denoted by $u_o(t)$] is most easily represented as

$$u_o(t) = u(t) g(t) \tag{49}$$

It is clear that $u_o(t)$ only has meaning in the sense of generalized functions.

The question we can now ask is: How much information about u(t) is left in $u_o(t)$? To answer this we need to select a form for g(t). An appropriate choice is easily shown to be

$$g(t) = \Delta t \sum_{n=-\infty}^{\infty} \delta(t-n\Delta t) \tag{50}$$

Whether we sum or simply regard g(t) as a sequence is irrelevant since the delta functions never overlap. The factor Δt is introduced to keep g(t) dimension-less.

We first compute the time averaged mean value of $u_o(t)$.

$$U_{o_T} = \frac{1}{T} \int_0^T u_o(t)dt = \frac{1}{T} \int_0^T u(t)g(t)dt = \frac{\Delta t}{T} \sum_{n=1}^{N} u(n\Delta t) \tag{51}$$

But $T/\Delta t$ is simply the number of samples and we have

$$U_{o_T} = \bar{u} = \frac{1}{N} \sum_{n=1}^{N} u(n\Delta t) \qquad [52]$$

which is reminiscent of our ensemble average.

It is immediately obvious that our sampled function indeed retains all the information on the moments of $u(t)$. It is also obvious from our previous considerations that we can simultaneously minimize the number of data to be handled while maximizing the convergence rate of the estimator by insuring that the samples are, in effect, statistically independent by choosing

$$\Delta t = 2T_u \qquad [53]$$

Thus the optimal sampling rate is one sample for every two integral scales in time.

If we analyze our estimator for the autocorrelation we find that its digital counterpart is

$$B_{o_T}(n\Delta t) = \frac{1}{N^2} \sum_{i=1}^{N-n} \sum_{j=1}^{N-n} u(i\Delta t)u([j+n]\Delta t) \qquad [54]$$

where the time lag is given by $n\Delta t$.

A detailed analysis of the convergence of this estimator will reveal what we could have guessed; namely, that all of these time lag products do not contribute to the convergence of the estimator, only those separated by two integral scales. Thus a much more efficient estimator which would converge at the same rate would be

$$\tilde{B}_{o_T}(n\Delta t) = \frac{1}{N'} \left\{ u(o)u(n\Delta t) + u(2T_u)u(2T_u + n\Delta t) + \right. \qquad [55]$$

$$\left. \ldots + u(2[N'-1]T_u)\, u(2[N'-1]T_u + n\Delta t) \right\}$$

where $N' = T/2T_u$ and is, in effect, the number of independent samples. The algorithm computes a time-lag pair for the appropriate time delay, then leaps down the record $2T_u$ before computing another. This is illustrated in Fig. 8.

This method of computing the autocorrelation was suggested to this author by Lumley in 1972 who also showed that a very efficient spectral estimator could be derived from it. Since this suggestion was unpublished, it has been included here as Appendix II with the appropriate credit given.

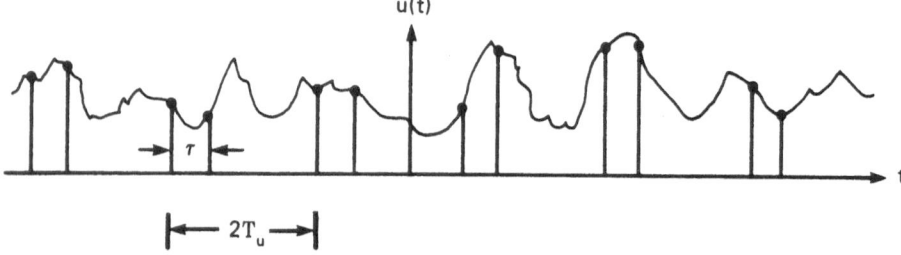

Fig. 8. Illustration of skip method for computing autocorrelations in which pairs of samples are taken only every two integral scales.

Spectral Analysis of Discretely Sampled Signals

It is straightforward to compute the Fourier transform (in the sense of generalized functions) of the signal $u_o(t)$.

$$\hat{u}_o(\omega) = \frac{1}{2\pi} \int_{-\infty}^{\infty} e^{i\omega\tau} u(t)g(t) \, dt \qquad [56]$$

Substitution for $g(t)$ and some manipulation yields the spectrum of $u_o(t)$ as

$$S_o(\omega) = \sum_{n=-\infty}^{\infty} S(\omega - 2\omega_n) \qquad [57]$$

where

$$2\omega_n = \frac{2\pi}{\Delta t} \qquad [58]$$

Thus our discretely sampled signal produces not only the desired result, $S(\omega)$, but an infinity of similar spectra all shifted by frequency ω_n. The phenomenon is illustrated in Fig. 9 from ref. (1).

It is clear that if the signal $u(t)$ has spectral components above the frequency ω_n, these components are aliased to lower frequencies with the result that the desired spectrum can no longer be distinguished. The phenomenon is called aliasing and ω_n, the folding frequency, is called the *Nyquist frequency.*

The *Nyquist criterion* can be summarized as follows: The spectral information in a discretely sampled time signal can be retained only if the sampling rate is *at least* twice the highest frequency present in the original signal.

Thus for digital processing of random signals we have conflicting requirements. We must sample at small time intervals to avoid aliasing, yet the additional data contribute little to the convergence of our estimators. The "waste factor" is

$2T_u/\Delta t$ which for wide-band signals can be very large indeed.

The problem of processing the large data blocks necessitated by the Nyquist and convergence criteria has received a large amount of attention since the infancy of digital processing techniques. As we shall see, there are several clever algorithms and techniques that have been developed to reduce the computational effort. A number of the commonly used and new approaches are summarized below.

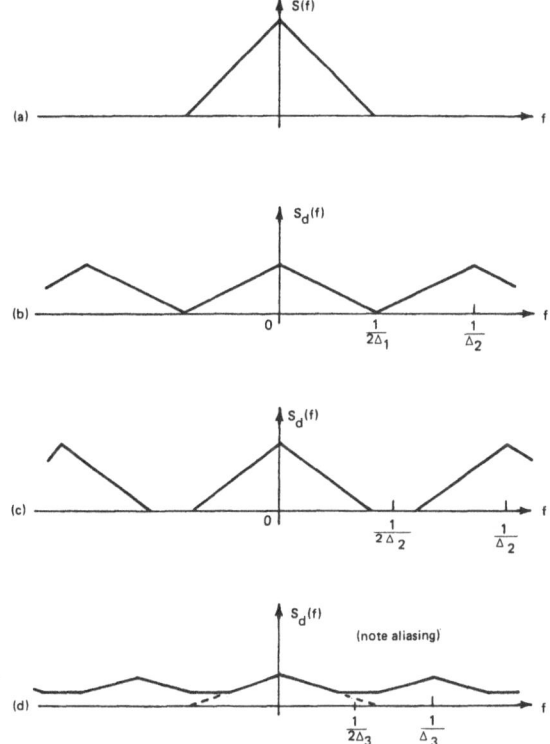

Fig. 9. Effect of sampling rate on measured spectra. Original spectrum (uppermost); aliased spectrum (lowermost) (adapted from ref. 1)

1. The time-lag product approach

This is a direct approach based on a discrete version of the time average estimator for the autocorrelation, Equation [54]. The spectrum is then computed by multiplying this autocorrelation by the appropriate lag window and taking its discrete transform. Reference (6) provides a complete discussion of the application of this approach.

2. The Fast Fourier Transform

This is a very fast and efficient algorithm for computing the Fourier co-efficients of the signal by utilizing a bit-reversal of the binary address of the time-series data. The smoothed spectrum is then computed by squaring the coefficients and convolving them with the appropriate window. The savings over the direct approach is approximately $\ln N/N$ where N is the number of data points (see ref. (7) for complete discussion).

3. The Low-Pass FFT

This is a variation on the FFT in which the incoming signal is divided by frequency into band-limited blocks. The first block might be low-pass filtered at 0.1 ω_n, the second at 0.2 ω_n, the third at 0.4 ω_n, etc.
Each block has a length inversely proportional to the cutoff frequency of the preceding block; that is, the higher blocks have shorter records. Each block is processed by the F.F.T. and smoothed by filters whose width increases with block width. Not only does this technique allow substantially larger spectral samples to be computed on a fixed core machine, but a substantial savings in computational costs is achieved, at some increases in experimental complexity. For a complete discussion see reference (8).

4. The Auto-Regressive Spectral Technique

Although not in common use in the fluid dynamics community, these techniques are evolving rapidly elsewhere. Because of this a detailed summary by Beuther of how these techniques work is included as Appendix I. In brief, an autoregressive series is fitted to the incoming time-data which optimizes the ability to predict the next data point. The spectrum is then simply determined by the coefficients of the autoregression and the mean-square error in the fit. A bonus is that the spectrum is already smoothed in a manner which minimizes both bias and relative error and is thus optimally filtered in some sense.

5. The Skip Technique

We have already mentioned this efficient estimator due to Lumley which utilizes only *independent samples of the time-lag products.* This method is described in detail in Appendix II. For most applications to turbulence measurement it appears to be the most efficient and is particularly well-suited to small computers.

To conclude this section we note that all spectral estimates must be smoothed in the same manner as for the continuous signal. The rate of convergence of the smoothed estimator is again proportional to inverse product of the window width in frequency space and the averaging time. As for the continuous signal, bias can result from this smoothing operation.

Randomly Sampled Data

There is no reason that the process of converting a continuous signal to a digital one has to be carried out at uniformly spaced time intervals. One of the most significant advances in signal processing over the past ten years has been the understanding and appreciation of the merits of random sampling. That is, instead of sampling at uniformly spaced intervals in time, the intervals are themselves random and are governed by some statistical process.

The development of the theory of random sampling is due to a number of independent investigators; among them are Gaster and Roberts, Mayo, Shapiro and Silverman, and others. For a comprehensive review the reader is referred to the article in this proceedings by Mayo. Our purpose here is to briefly review the basic principles and place the theory into the context of the overall problem of signal processing.

We begin by insisting that our random sampling process be *statistically independent of the process being sampled*. As will be seen later this *excludes* the burst-mode LDA from consideration. We choose a sampling function g(t) so that it selects values of u(t) at random instants in time. The sampled signal is then given by

$$u_o(t) = u(t)\, g(t) \qquad\qquad [59]$$

as before.

A suitable choice for g(t) is

$$g(t) = \delta(t-t_i),\ i = 1, \ldots, \infty \qquad\qquad [60]$$

where the t_i are uncorrelated. If the average sample rate is denoted by ν, it is easy to show that

$$\overline{g} = \nu \qquad\qquad [61]$$

and

$$\overline{g(t)\, g(t')} = \nu^2 + \nu\delta(t'-t) \qquad\qquad [62]$$

and

$$\overline{(g-\overline{g})(g'-\overline{g})} = \nu\delta(t'-t) \qquad\qquad [63]$$

Since, by hypothesis, u(t) and g(t) are statistically independent, it follows immediately that

$$\overline{u_o(t)} = \nu\overline{u} \tag{64}$$

$$\overline{u_o'(t)\,u_o'(t')} = \nu^2\,\overline{u'(t)u'(t')} + \nu\,\overline{u^2}\,\delta(t'-t) \tag{65}$$

An immediate consequence is that the autocorrelation and spectrum of $u_o(t)$ are given by:

$$B_o(\tau) = \nu^2\,B(\tau) + \nu\,\overline{u^2}\,\delta(\tau) \tag{66}$$

and

$$S_o(\omega) = \nu^2\,S(\omega) + \frac{\nu}{2\pi}\,\overline{u^2} \tag{67}$$

Thus with the exception of the spike at the origin in the autocorrelation and the corresponding white noise added to the spectrum, all of the information contained in the original signal has been retained.

In practice, the nuisance due to the spike at the origin in the autocorrelation can be eliminated by simply agreeing that we will never include self-products when we multiply signals together. It is easy to show that this corresponds to writing

$$\overline{g(t)\,g(t')} = \nu^2 \tag{68}$$

from which it follows that

$$B_o(\tau) = \nu^2\,B(\tau) \tag{69}$$

$$S_o(\omega) = \nu^2\,S(\omega) \tag{70}$$

Thus, by agreeing to remove self-products from any algorithms, we need only the sample rate to recover the desired information.

To derive practical estimators we turn to time-averaged value of $u_o(t)$ and its moments. For the mean value we have simply

$$U_{o_T} = \frac{1}{T}\int_o^T u_o(t)dt = \frac{\nu}{T}\sum_{i=1}^{N} u_i \tag{71}$$

where N is the number of samples which arrive in the interval (o, T) and the u_i

are the randomly sampled realizations of the signal. For T large enough, $\nu T \cong N$ and an unbiased estimator for the mean can be written as

$$U_{TR} = \frac{1}{N} \sum_{i=1}^{N} u_i \tag{72}$$

An immediate question is: How does this estimator converge? It is easy to show in the same manner as before that

$$\text{var}\{U_{TR}\} = \frac{2T_u}{T} \left[1 + \frac{1}{2\nu T_u}\right] \cdot \text{var}\{u\} \tag{73}$$

from which it follows that the relative error is

$$\epsilon^2 = \frac{\text{var}\{U_{TR}\}}{\bar{u}^2} = \frac{2T_u}{T} \left[1 + \frac{1}{2\nu T_u}\right] \cdot \left[\frac{\sigma_u}{\bar{u}}\right]^2 \tag{74}$$

Note that as $\nu T \to \infty$, this reduces to our previous result for continuous signals; while for $\nu T \to 0$ we obtain

$$\lim_{\nu T \to 0} \epsilon \cong \sqrt{\frac{1}{\nu T}} \cdot \frac{\sigma_u}{\bar{u}} \cong \frac{1}{N} \frac{\sigma_u}{\bar{u}} \tag{75}$$

These limits are reasonable since as the sampling rate becomes very high relative to the time scale of the signal, we reproduce the continuous signal. As the sampling rate becomes very low, all the samples are, in effect, statistically independent.

A simple estimator for the mean square value of the original signal is given by

$$\sigma^2_{TR} = \frac{1}{N} \sum_{i=1}^{N} [u_i - U_{TR}]^2 \tag{76}$$

It is straightforward to show that the relative error is analogous to that in equation [74].

The best estimator for estimating the autocorrelation and spectrum from the randomly arriving samples is still a matter of debate (see the article by Mayo in this volume). The approach adopted here will be to first obtain the spectrum, and then compute the inverse transform to obtain the autocorrelation.

The finite time estimate for the Fourier coefficients of $u_o(t)$ is given by

$$\hat{u}_{o_T}(\omega) = \frac{1}{2\pi} \int_0^T e^{i\omega\tau} u'_o(t)dt = \frac{1}{2\pi} \sum_{i=1}^{N} e^{i\omega\tau} u'(t_i) \tag{77}$$

It follows that a spectral estimator for $S_o(\omega)$ is

$$S_{o_T}(\omega) = 2\pi \frac{|\hat{u}_{o_T}(\omega)|^2}{T} = \frac{1}{2\pi T} \sum_{\substack{i=1 \\ i \neq j}}^{N} \sum_{j=1}^{N} e^{i\omega (t_i - t_j)} u'(t_i) u'(t_j) \qquad [78]$$

where we have employed our prohibition on self-products.

From equation [78] we can immediately obtain an estimator for $S(\omega)$, the desired spectrum, as

$$S_{TR}(\omega) = \frac{1}{2\pi \nu^2 T} \sum_{\substack{i=1 \\ i \neq j}}^{N} \sum_{j=1}^{N} e^{i\omega (t_i - t_j)} u'(t_i) u'(t_j) \qquad [79]$$

This estimator is unbiased *and unaliased.* The lack of aliasing is a consequence of the random sampling and is independent of the mean sample rate. Thus random sampling would appear to have significant advantages over the traditional discrete sampling technique.

The absence of aliasing is not without price, however. The price is the increased relative error due to the random sampling. It can be shown by assuming that fourth order moments are jointly Gaussian and by noting that   [80]

$$\overline{gg'g''g'''} = \nu^4 + \nu^3 \left[\delta(t''-t) + \delta(t'''-t) + \delta(t''-t') + \delta(t'''-t') \right] + \nu^2 \left[\delta(t''-t)\delta(t'''-t') + \delta(t''-t')\delta(t'''-t) \right]$$

that the relative error of the estimator $S_{TR}(\omega)$ is given by

$$\epsilon^2 = \frac{\text{var}\{S_{TR}(\omega)\}}{[S(\omega)]^2} = \left\{ 1 + \frac{S(o)}{S(\omega)} \cdot \left[\frac{1}{2\nu T_u} \right] \right\}^2 \qquad [81]$$

Since the spectrum is assumed to fall as frequency increases, the effect of the random sampling is to increase the relative error with frequency. Note that as the sample rate becomes large, the relative error reduces to the result for continuous signals. Clearly the relative error is never less than unity and is therefore unacceptable. Hence we must smooth the estimator as before.

There are several techniques by which the estimator above can be smoothed to make it converge. These are:

(i) **Block Averaging**
      This approach is certainly the most straightforward to implement and simply involves averaging the spectral estimators obtained from independent record lengths of the process. As in the continuous case, the relative error is decreased by the inverse square root of the number of independent samples.

(ii) **Spectral Window**

As for the continuous case this can be done either in the time or frequency domain. The simplest approach is to insert a lag-window into the estimator of equation [79] to obtain

$$S_{TRM}(\omega) = \frac{1}{2\pi\nu^2 T} \sum_{\substack{i=1 \\ i \neq j}}^{N} \sum_{j=1}^{N} e^{i\omega (t_i - t_j)} \, u'(t_i)u'(t_j)D(t_i - t_j) \qquad [82]$$

where $D(\tau)$ can be any of the windows already introduced. It is tedious but straightforward (see ref. 9 or proceed as above) to show that the relative error is the error given by equation [81] reduced by a factor of $2\pi/\Delta\omega T$ where $\Delta\omega$ is the effective window width in the frequency domain. We have

$$\epsilon^2 = \frac{2\pi}{\Delta\omega T} \left\{ 1 + \frac{S(o)}{S(\omega)} \cdot \frac{1}{2\nu T_u} \right\}^2 \qquad [83]$$

A similar result is obtained if spectrum estimated by equation [79] is convolved with a filter of width $\Delta\omega$. In view of the increased error as frequency increases, this approach can have significant advantages if the window varies with the frequency at which the spectrum is estimated. Gaster and Roberts (10) suggest that for power law spectrum a filter whose width increases logarithmically with frequency would be desirable. Clearly combinations of all the above approaches could be used to advantage.

(iii) **The Time-Slot Approximation**

This is best illustrated by first taking the inverse Fourier transform of $S_{TR}(\omega)$ to obtain

$$B_{TR}(\tau) = \frac{1}{\nu^2 T} \sum_{\substack{i=1 \\ i \neq j}}^{N} \sum_{j=1}^{N} u'(t_i)u'(t_j)\delta(t_i - t_j - \tau) \qquad [84]$$

This represents a collection of random realizations of the autocorrelation. We now group these realizations into slots of width $\Delta\tau$ by defining the time-slot approximation to the autocorrelation as

$$B_s(n\Delta\tau) = \frac{1}{\Delta\tau} \int_{(n-\frac{1}{2})\Delta\tau}^{(n+\frac{1}{2})\Delta\tau} B_{TR}(\tau)d\tau \qquad [85]$$

which yields the "smoothed" autocorrelation as

$$B_s(n\Delta\tau) = \frac{1}{\Delta\tau} \left(\frac{1}{\nu^2 T} \right) \sum_{\substack{i=1 \\ i \neq j}}^{N} \sum_{j=1}^{N} u(t_i)u(t_j) \qquad [86]$$

where

$$(n-\tfrac{1}{2})\Delta\tau < |t_i-t_j| < (n+\tfrac{1}{2})\Delta\tau$$

This algorithm can be directly applied to the incoming data to compute the autocorrelation which can, in turn, be Fourier transformed to yield a discrete spectrum. This transform can even incorporate a lag window to improve convergence in the usual manner.

There is some confusion in the literature as to what constitutes the proper relative error for spectra computed from the time-slot approximation. The operation defined by equation [85] is, in fact, a convolution of the autocorrelation with a time window. In frequency space this means that the spectrum is simply filtered by the window function which is assumed broadband since the time window $\Delta\tau$ is narrow. Since the variance is proportional to the spectrum squared, it is reduced by this low-pass filtering, but only at the expense of bias. Thus the primary cause for convergence is the window used in computing the smoothed spectrum from the time-slot correlation, and the convergence is identical to equation [83]. This is, in fact, the conclusion of Mayo from empirical evidence.

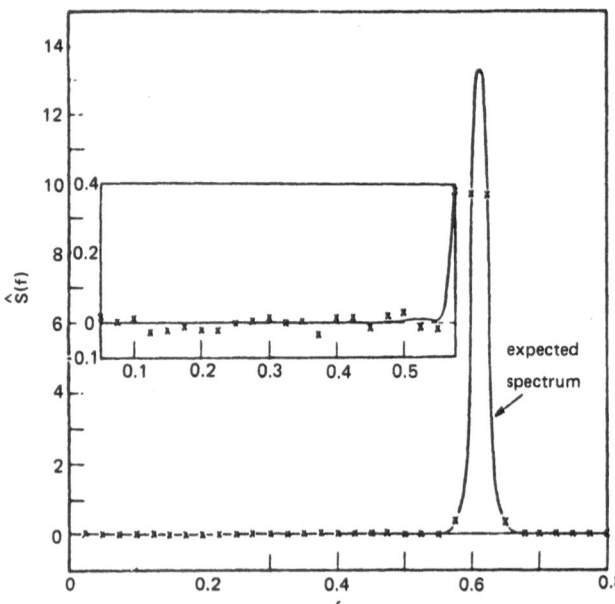

Fig. 10. Spectrum of sine wave from random samples $(2\pi v/\omega = 1.7)$ (from ref. 10).

Examples of spectra computed by Gaster and Roberts from randomly sampled data are shown in Figs. 10 and 11. In Fig. 10, a sine wave is sampled so that the mean sample rate is less than the Nyquist frequency. Clearly the expected peak is reproduced. In Fig. 11, a low-passed noise is sampled so that the mean sample rate corresponds to the breakpoint in the spectrum. Although the increased variability at the high frequencies is obvious, the spectrum is accurately reproduced.

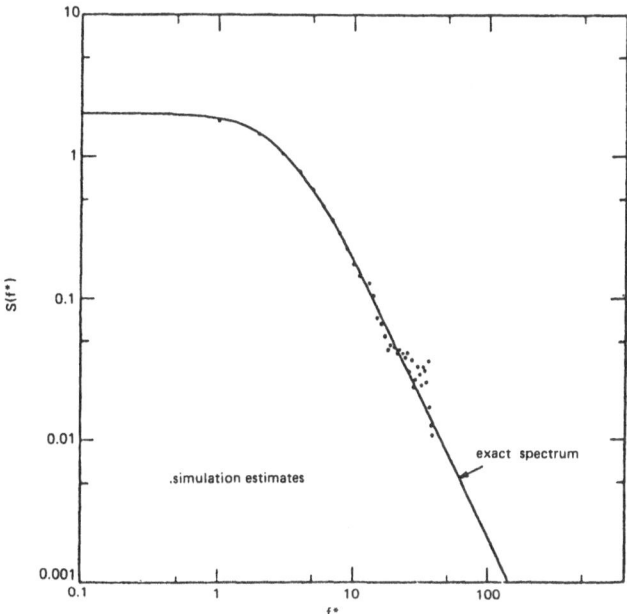

Fig. 11. Spectrum of low-passed noise $(v/f_o = 1)$ *(from ref. 10).*

The Burst-Mode Laser Doppler Anemometer

The purpose of this section is not to discuss the burst-mode LDA as a tool for flow measurement but rather to concentrate on the interesting signal analysis problems it presents. The phrase "burst-mode LDA" is used to refer to an LDA operating in such a way that for most of the time there are no scattering particles in the measuring volume and there is never more than one particle at a time. As the particles arrive randomly in time, the processor measures their velocity and makes this data available as a digital word for further processing. As we shall see, the processor must also provide other information if the signal is to be interpreted correctly. For a complete description of this instrument the reader is referred to the review article by Buchhave in this proceedings and to references (11) and (12).

The randomly arriving particles do, in fact, randomly sample the flow velocity. Unlike the sampling scheme just discussed, however, the sampling process is *not* independent of the process being sampled. This is easily seen if one considers that the particles are assumed randomly distributed at statistically independent locations *in space* and carried to the measuring volume by the flow. Thus, arrival time and flow are correlated.

We can correctly analyze this instrument if we design a sampling function which samples the velocity at the spatial location of the particle. This was done in references (12) and (13) with the following result: the sampled velocity is given by

$$u_0(t) = \int\limits_{\text{all space}} U(\underline{a},t)g_1(\underline{a})w[\underline{x}(\underline{a},t)]\,d^3\underline{a} \qquad [87]$$

where $U(\underline{a},t)$ is the Lagrangian velocity of the flow which the particles are assumed to follow, the function $w(\underline{x})$ accounts for the fact that the measuring volume is of finite extent, and $g_1(\underline{a})$ is a random function which accounts for the presence or absence of a particle at a particular Lagrangian coordinate.

Because of the motion the particle is moved by the flow to a location given by

$$x = x(\underline{a},t) = \int_0^T U(\underline{a},t_1)\,dt_1 \qquad [88]$$

Thus, when the particle wanders into the measuring volume, the function $w(\underline{x})$ "turns on" the signal $u_0(t)$. A typical scattering volume is shown in Fig. 12 and the signal $u_0(t)$ resulting from the randomly arriving particles is shown in Fig. 13.

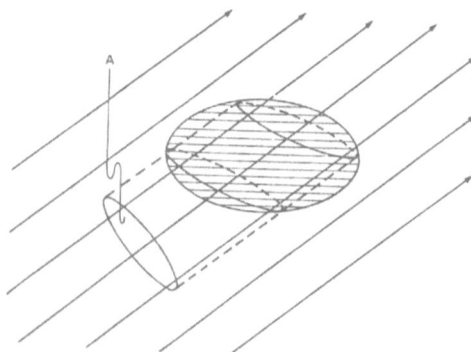

Fig. 12. Schematic showing measuring volume defined by laser Doppler ane-mometer.

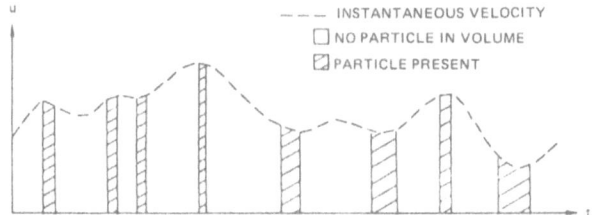

Fig. 13. Typical velocity signal as sampled by individual and randomly arriving particles.

By a series of arguments (see reference (13)), we can transform the dependence on Lagrangian coordinates to Eulerian (or spatial) ones and write

$$U_0(t) = \int\limits_{\text{all space}} u(\underline{x},t)\, w(\underline{x})\, g(\underline{x},t) d^3\underline{x} \tag{89}$$

where $u(\underline{x},t)$ is the velocity at the point $\underline{x}$ and $g(\underline{x},t)$ has the following statistical properties:

$$\overline{g(\underline{x},t)} = \mu \tag{90}$$

$$\overline{g(\underline{x},t)\, g(\underline{x}',t')} = \mu\delta(\underline{x}-\underline{x}' - \underline{U}\cdot[t'-t]) + \mu^2 \tag{91}$$

and μ is the average number of particles per unit volume.

We compute the moments of $u_0(t)$ directly as

$$u_0 = \int\limits_{\text{all space}} \overline{u(\underline{x},t)}\ \overline{g(\underline{x},t)}\ w(\underline{x}) d^3\underline{x} \tag{92}$$

$$B_0(\tau) = \overline{u_0(t)u_0(t')} = \iint\limits_{\text{all space}} \overline{u(\underline{x},t)u(\underline{x}',t')}\ \overline{g(\underline{x},t)g(\underline{x}',t')}\cdot w(\underline{x})w(\underline{x}')d^3\underline{x}d^3\underline{x}' \tag{93}$$

where we have used the fact that velocity and the occurrence of a particle in space are independent. It follows immediately that (see ref. (11))

$$\overline{u_0(t)} = (\mu V)\,\overline{u} \tag{94}$$

and

$$B_0(\tau) = (\mu V)^2 B(\tau) + (\mu V)\,\overline{u^2}\rho_1(\tau) \tag{95}$$

where we have assumed that the effect of averaging the velocity and its space-time correlation over the scattering volume is negligible. The function $\rho_1(\tau)$ is easily shown to be a "spike-like" function of width proportional to the mean transit time of particles crossing the volume.

Since μV corresponds to the expected number of particles in the volume and is assumed small, the "spike-like" term will dominate the spectrum in the same manner the self-product terms dominated the random time sampling process discussed earlier. The "cure" here is the same as before: we agree to never include self-products in any computation. This corresponds to eliminating the delta function in equation [91] yielding

$$\overline{g(\underline{x},t)\, g(\underline{x}',t')} = \mu^2 \qquad\qquad [96]$$

from which it follows that

$$B_0(\tau) = (\mu V)^2 B(\tau) \qquad\qquad [97]$$

and

$$S_0(\omega) = (\mu V)^2 S(\omega) \qquad\qquad [98]$$

Thus the moments and spectra are completely recoverable, in principle, from the information available from the particles. That complete flow statistics could be obtained from the randomly arriving particles was suspected long before it was proven in 1975 (see for example ref. (14)). Unfortunately it was not until after this (and from this) that correct estimators were formulated (see ref. (11), (12) and (13)).

The mean value can be calculated from the time integral as before:

$$\begin{aligned}
U_{0_T} &= \frac{1}{T}\int_0^T u_0(t)dt \\
&= \frac{1}{T}\left\{\int_{t_1}^{t_1+\Delta t_1} u_0(t)dt + \int_{t_2}^{t_2+\Delta t_2} u_0(t)dt + \ldots + \int_{t_N}^{t_N+\Delta t_N} u_0(t)dt\right\}
\end{aligned} \qquad [99]$$

where the t_i are the arrival times and the Δt_i are the *residence times* or times that the signal is on.

If we assume that the velocity is relatively constant while the particle is traversing the volume, this immediately yields a practical estimator for the mean velocity as

$$U_{TBP} = \frac{\displaystyle\sum_{i=1}^{N} u_i \, \Delta t_i}{\displaystyle\sum_{i=1}^{N} \Delta t_i} \qquad [100]$$

where in the denominator we have used the fact that as the number of realizations becomes large, the "on-time" is given by

$$\mu VT = \sum_{i=1}^{N} \Delta t_i \qquad [101]$$

A similar analysis for the mean square fluctuating velocity yields

$$\sigma^2{}_{TBP} = \frac{\displaystyle\sum_{i=1}^{N} (u_i - U)^2 \Delta t_i}{\displaystyle\sum_{i=1}^{N} \Delta t_i} \qquad [102]$$

It is easy to show that the relative error in the estimator is the same as for the
randomly sampled case discussed earlier if the mean sampling rate ν is interpreted as the mean data rate given by

$$\nu_{eff} = (\mu V)/T_p \qquad [103]$$

where T_p is the average particle residence time and is approximately given by
$d/\bar{u}$ where d is the effective width of the measuring volume.

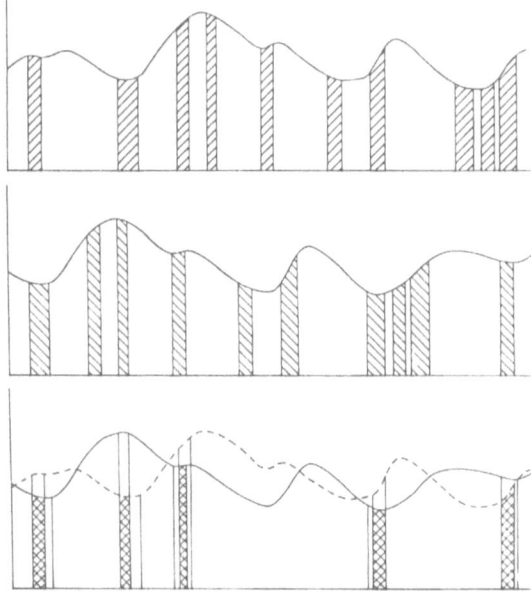

Fig. 14. Computation of the autocorrelation from burst processors using only
the overlap times. The middle trace is the upper trace displaced by amount τ as
shown

An estimator for the autocorrelation is more difficult to interpret since both $u_o(t)$ and $u_o(t+\tau)$ must on "on" to contribute to the time integral. This is illustrated in Fig. 14. A time-slot approximation to the autocorrelation is easily shown to be given by

$$B_T(n\Delta\tau) = \frac{1}{N_{\Delta\tau}(\sum_{i,j} \Delta t_{ij})} \sum_{i=1}^{N} \sum_{\substack{j=1 \\ i \neq j}}^{N} u_i u_j \, \Delta t_{ij} \qquad [104]$$

where Δt_{ij} is the overlap time and only those $N_{\Delta\tau}$ realizations satisfying

$$(n-\frac{1}{2})\Delta\tau < |t_i - t_j| < (n+\frac{1}{2})\Delta\tau$$

can contribute. It can be shown (most directly by transforming the spectrum derived below) that an estimator which is easier to implement is given by

$$B_{TS}(n\Delta\tau) = \frac{1}{\Delta\tau} \left[\frac{T}{\sum_{i=1}^{N}(\Delta t_i)^2} \right] \sum_{i=1}^{N} \sum_{\substack{j=1 \\ i \neq j}}^{N} u_i u_j \, \Delta t_i \Delta t_j \qquad [105]$$

A direct Fourier transform can be defined from equation [35] as

$$\hat{u}_{o_T}(\omega) = \frac{1}{2\pi} \int_{-T/2}^{T/2} e^{i\omega\tau} u_o(t) dt \qquad [106]$$

which is approximately given by

$$\hat{u}_{o_T}(\omega) = \frac{1}{2\pi} \sum_{i=1}^{N} e^{i\omega t_i} u_i \Delta t_i \qquad [107]$$

(The neglected variation of t in the interval $(t_i, t_i + \Delta t_i)$ can be shown to be equivalent to low-pass filtering at the inverse residence time, which introduces no further approximations than have already been made).

From this a spectral estimator is readily calculated to be

$$S(\omega) = \frac{T}{2\pi(\sum_{i=1}^{N}\Delta t_i)^2} - \sum_{i=1}^{N}\sum_{\substack{j=1 \\ i\neq j}}^{N} e^{i\omega(t_i-t_j)} u_i u_j \Delta t_i \Delta t_j \qquad [108]$$

It is tedious but straightforward to show that the relative error of this estimator is also given by equation [81] with ν given by equation [103]. As might be expected the convergence of the smoothed estimator is given by equation [83].

The preceding analysis of the LDA along with experimental evidence is given in a series of recent papers (ref. (12), (15), (11)). Also included in the latter is a detailed description of the hardware and software for implementing these algorithms and an assessment of the errors resulting from ignoring the need for residence time information.

Mixed Mode Data Processing

Fig. 15. Photograph showing simultaneous use of LDA and hot-wire anemometer

Our success with randomly sampled data leads as to inquire whether a mixed-mode signal processing system might be possible. An example of where such a mode might be desirable is illustrated in Fig. 15 which shows the simultaneous use of a laser Doppler system and a hot-wire anemometer to measure cross-correlations. The primary advantage of such a system over conventional hot-wire techniques is that there is no wake generated by the upstream probe. Moreover, the use of a downstream hot-wire instead of another LDA considerably reduces experimental complexity and cost.

To analyze this problem we assume the upstream LDA to be operating in the burst mode analyzed earlier. We note that any signal generated at the hot-wire probe by the scattering particles will be uncorrelated with the velocities and can be treated as noise (for example, water drops will be rapidly evaporated causing random spikes in the signal).

The mode of data analysis is the following: We continuously monitor the hot-wire signal. When the LDA receives a burst, we measure the particle's velocity and residence time, and send a delayed "freeze" command to the hot-wire monitor. We then compute an instantaneous realization of the space-time

cross-correlation at *all* time lags of interest by multiplying the single LDA realization by the appropriately delayed hot-wire signal. Each succeeding particle
arrival is treated in the same manner and the accumulated realizations of the
correlation are averaged to yield the time-averaged cross-correlation.

For simplicity, we consider only the case where the LDA and hot-wire are measuring the same velocity. Extension of the analysis to spatially separated signals is straightforward. The analytical statement of the procedure described
above is then

$$C_0(\tau) = \frac{1}{T} \int_0^T u_0(t) \, u(t+\tau) dt \qquad [109]$$

where $u_0(t)$ is the LDA burst signal given by equation [89], $u(t)$ is the hot-wire
signal, and the interval (o,T) is determined by the first and last burst, (Note:
$t+\tau > T$ and $t+\tau < o$ are allowed).
It is easy to show that

$$C_T(\tau) = \frac{1}{\mu VT} \, C_0(\tau) \qquad [110]$$

and that this is an unbiased estimator of the cross-correlation. Hence an algorithm consistent with the approximations in $u_0(t)$ of the preceding section is

$$C_T(n\Delta\tau) = \frac{\displaystyle\sum_{i=1}^{N} u_0(t_i) u(t_i + n\Delta\tau) \, \Delta t_i}{\displaystyle\sum_{i=1}^{N} \Delta t_i} \qquad [111]$$

It remains to examine the convergence of our estimator C_T. Proceeding as before we write

$$\text{var}\{C_T(\tau)\} = \frac{1}{(\mu VT)^2} \iint_0^T \overline{u_0(t) u_0(t_1) u(t+\tau) u(t_1+\tau)} \, dt dt_1 - [C(\tau)]^2 \qquad [112]$$

By substituting for $u_0(t)$, the integrand is readily obtained as

$$\overline{u_0(t) u_0(t_1) u(t+\tau) u(t_1+\tau)} \qquad [113]$$

$$= \iint_{\text{all space}} \overline{u(\underline{x},t) u(\underline{x}',t_1) u(\underline{x}_0,t+\tau) u(\underline{x}_0,t_1+\tau)} \; \overline{w(\underline{x}) w(\underline{x}')}, \overline{g(\underline{x},t) g(x_1,t_1)} \cdot d^3\underline{x} \, d^3\underline{x}'$$

where $\overline{g(\underline{x},t) \, g(\underline{x}',t')}$ is given by equation [91].

After considerable manipulation, it can be shown that the relative error is given
by

$$\epsilon^2 = \frac{\text{var}\{C_T(\tau)\}}{[C(\tau)]^2} \cong \frac{2T}{T} \left[\frac{1}{\rho_c^2(\tau)} + \frac{1}{\nu_{eff}T} \right] \qquad [114]$$

where ν is given by equation [103]. Comparison of this with equation [27] shows that the added variability is measured by the second term which is independent of the lag. Thus for modest sampling rates relative to the integral scale, the estimate converges nearly as well as for the continuous case.

A similar convergence criterion can be shown to exist if the randomly sampled signal is randomly sampled in time instead of a burst LDA signal. Moreover, it is straightforward to derive convergence criteria for smoothed spectral estimators using the previously illustrated techniques.

A Final Word
In this paper we have reviewed the development of our current understanding of random signal analysis. We have concentrated on the estimation of mean quantities by time average with particular attention to the problems of bias and convergence. Of particular interest because of its recent development, has been the randomly sampled process which holds considerable promise for application as digital transducer components become more common.

We have entirely avoided the practical and theoretical problems which arise from questions relating to frequency response, dynamic range, quantization error and noise. All of these play a significant role in real processing. For more information on this subject the interested reader is referred to refs. (5) and (7).

Bibliography

1. Jenkins, G.M. and D.G. Watts (1968) *Spectral Analysis and Its Applications.* Holden-Day, San Francisco.

2. Lumley, J.L. and H.A. Panofsky (1964) *The Structure of Atmospheric Turbulence,* Interscience, New York.

3. Tennekes, H. and J.L. Lumley (1972) *A First Course in Turbulence,* MIT Press, Cambridge.

4. Lumley, J.L. (1970) *Stochastic Tools in Turbulence,* Academic Press, New York.

5. Van Atta, C.W. (1974) Sampling Techniques in Turbulence Measurements, in *Annual Review of Fluid Mechanics,* Vol. **6**, Academic Press, Palo Alto (Van Dyke, Vincenti, and Wehausen, coeditors).

6. Blackman, R.B. and J.W. Tukey (1959) *The Measurement of Power Spectra,* Dover Publications, Inc., New York.

7. Enochson, L.D. and R.K. Otnes (1968) *Programming and Analysis for Digital Time Series Data,* The Shock and Vibration Monograph Series, SVM-3, Naval Publication and Printing Office, Washington.

8. Pierce, R.E. (1972) Spectral Analysis by Large Block Transforms, Ph.D. Dissertation, Department of Aerospace Engineering, The Pennsylvania State University.

9. Gaster, M. and J.B. Roberts (1975) Spectral Analysis of Randomly Sampled Signals, *J. Inst. Maths. Appl.,* **15**, 195-216.

10. Gaster, M. and J.B. Roberts (1977) The spectral analysis of randomly sampled records by a direct transform, *Proc. R. Soc. Lond A.,* **354**, 27-58.

11. Buchhave, P., W.K. George and J.L. Lumley (1979). The measurement of turbulence with the laser Doppler anemometer, *Annual Reviews in Fluid Mechanics,* vol. **11** (Van Dyke and Wehausen, editors).

12. Buchhave, P. (1979) Errors and Correction Methods in Turbulence Measurements with the LDA, Ph.D. Dissertation, Department of Mechanical Engineering, State University of New York at Buffalo.

13. George, W.K. (1976) Limitations to measurement accuracy inherent in the laser Doppler signal, in *The Accuracy of Flow Measurements by Laser Doppler Methods,* Copenhagen, (P. Buchhave et al., editors).

14. Stevenson, W. (ed) (1974) *Proceedings of the Second International Workshop on Laser Velocimetry,* Purdue University, W. Lafayette, Ind.

15. Buchhave, P. and W.K. George (1978) Bias corrections in turbulence measurements by the laser Doppler Anemometer, *Proceedings of the Third Workshop on Laser Doppler Anemometry,* Purdue University, W. Lafayette, Ind. (W. Stevenson, ed.).

16. Akaike, H. (1972) "Use of an Information Theoretic Quantity for Statistical Model Identification. *Proceedings of the Fifth Hawaii International Conference on System Sciences,* Western Periodicals, Inc. pp.249-250.

17. Parzen, E. (1974) "Some Recent Advances in Time Series Analysis: *IEEE Trans, Auto. Control,* Vol. **AC-19** pp. 723-730.

18. Parzen, E. (1974) "Some Recent Advances in Time Series Analysis: SUNY/B Department of Statistical Science Technical Report No. 10.

19. Parzen, E. (1975) "Multiple Time Series: Determining the Order of Approximating Autoregressive Schemes" SUNY/B Department of Statistical Science Technical Report No. 23.
20. Pagano, M. (1976) Time Series Analysis Course Notes, Department of Statistical Science, SUNY/B.
21. Parzen, E. (1976) Multivariate Time Series Analysis Course Notes, Department of Statistical Science, SUNY/B.

Appendix I: Autoregressive Spectral Estimation

by

Paul D. Beuther
State University of New York at Buffalo

Introduction

One of the major difficulties in estimating the spectrum of a turbulent process is reducing the variance in the spectral level. This variance is a statistical consequence of the Fourier transformation from time domain to frequency domain. Consider a stationary process, Y(t), which has a "true" spectrum, $F(\omega)$. It can be shown that for a finite time segment, Y(t), t=O,T, the sample spectral density $F_T(\omega)$ is unbiased $(\overline{F_T(\omega)} \to F(\omega))$, but it is not consistent (VAR $F_T(\omega) \neq 0$). In fact, $VAR(F_T(\omega)) \sim [F(\omega)]^2$! Without some type of smoothing, a typical spectrum appears to be mostly noise.

Since the expected value for $F_T(\omega) \to F(\omega)$, one could certainly average many independent spectral realizations to obtain a better estimate for $F(\omega)$. However, in many cases it is impractical to obtain and analyze the necessary amount of data. In some experiments many realizations of the same event are not possible. In these situations other techniques are needed. Due to recent advances, autoregressive techniques are some of the most powerful methods available.

Theory of Autoregressive Spectra

A time series, Y(t), can be represented as a sum of its weighted past plus some random "shock", $\epsilon(t)$. This can be represented mathematically as

$$Y(t) = a^*(1)\, Y(t-1) + a^*(2)\, Y(t-2) + \ldots + \epsilon(t). \tag{1}$$

However, it is more common to define the process in the following manner:

$$Y(t) + \sum_{j=1}^{\infty} Y(t-j)\, a(j) = \epsilon(t) \text{ where } a(j) = -a^*(j). \tag{2}$$

In practice there is an order p such that, for $j > p$, $a(j) \cong 0$. This is called a pth order autoregression and is written as:

$$Y(t) + \sum_{j=1}^{p} Y(t-j)\, a(j) = \epsilon(t). \tag{3}$$

Techniques for determining p will be discussed in detail later. Since $a(j) = 0$ for $j < 0$ and $j > p$, equation 3 can be rewritten as:

$$\sum_{j=-\infty}^{+\infty} a(j)\, Y(t-j) = \epsilon(t) \text{ where } a(0) \equiv 1. \qquad [4]$$

The LHS is merely a convolution of a and Y which can be transformed to a frequency domain to read

$$F_a(\omega) \cdot F_Y(\omega) = F_\epsilon(\omega). \qquad [5]$$

$F_a(\omega)$, $F_Y(\omega)$, and $F_\epsilon(\omega)$ are the spectral density functions for $a(j)$, $Y(t)$ and $\epsilon(t)$ respectively. From this it is easy to obtain an expression for the spectrum of Y:

$$F_Y(\omega) = \frac{F_\epsilon(\omega)}{F_a(\omega)}. \qquad [6]$$

$\epsilon(t)$ is a random process, so $F_\epsilon(\omega)$ is a constant equal to the residual variance $\hat{\sigma}^2$. By substituting for $F_a(\omega)$ and $F_\epsilon(\omega)$ equation 6 becomes:

$$F_y(\omega) = \frac{\hat{\sigma}^2}{\left| \sum_{j=0}^{p} a(j)e^{ij\omega} \right|^2} \qquad [7]$$

Calculating the a's for a pth Order Autoregression

The set of coefficients, $a(j)$, can be calculated from the covariance function $R(\tau)$. Transforming equation 4 into time-lag products results in the set of equations known as the Yule-Walker equations.

$$[8]$$

$$\sum_{j=0}^{p} R(\tau-j)\, a(j) = \hat{\sigma}^2 \delta(\tau) \qquad \tau = 0, 1, 2, \ldots. p \text{ where } R(\tau) \equiv \frac{1}{T} \sum_{t=0}^{T=\tau} Y(t)\, Y(t-\tau).$$

This is a set of p linear equations with p unknowns, and can be solved by one of several means in a straightforward manner.

Deciding the Order p

There are two methods for determining the order p. The first was developed by Akaike (16), the second by Parzen (17). Both yield approximately the same result, but the theory behind each is quite different in approach. The Akaike method assumes that a given time series is a product of an autoregressive process with a "true" order p. Choosing p reduces to minimizing the function AIC(m), defined by

$$AIC(m) = Ln(\hat{\sigma}_m^2) + \frac{2m}{T}. \qquad [9]$$

m is order, T the total number of points, and $\hat{\sigma}^2_m$ the residual variance for order m. The Parzen approach assumes that the given time series is actually the product of an infinite autoregressive process. The goal is to find the best estimate of this infinite series, which can be reduced to minimizing the function CAT (m), defined by

$$CAT(m) = 1 - \frac{\hat{\sigma}^2_\infty}{\sigma^2_m} + \frac{m}{T}. \qquad [10]$$

m is the order, $\sigma^2_m = \frac{T}{T-m} \hat{\sigma}^2_m$ is the "unbiased" estimator of the residual variance and $\hat{\sigma}^2_\infty$ is the residual variance of an infinite order autoregression. Substituting for $\hat{\sigma}^2_\infty$ equation [10] can be rewritten as:

$$CAT(m) = \frac{1}{T} \sum_{j=1}^{m} \left(\frac{T-j}{T} \frac{1}{\hat{\sigma}^2_j} \right) - \frac{T-m}{T} \frac{1}{\hat{\sigma}^2_m}. \qquad [11]$$

In practice the two methods give approximately the same order. It can be shown that the Akaike method is an upper bound of the Parzen criterion.

Extension of autoregressive techniques to multidimension time series is straightforward (see refs. (16) - (21)). However, the solution to the regression equations is more involved and may require more computer memory than some laboratory mini- and micro-computers have.

Appendix II:
Computation of Spectra and Cross Spectra
by a Skipping Technique

by

John L. Lumley
Cornell University

It is straightforward to show (Lumley (4), for example) that if the mean of a quantity is computed by integrating in time over an interval of length T, the mean square relative error is given by (asymptotically)

$$\epsilon^2 \cong \frac{2T}{T} a^2 \qquad [1]$$

where a^2 is the relative fluctuation level of the quantity being measured, and T is the integral scale. On the other hand, if N *independent* estimates of the same quantity are made, the mean square relative error is given by

$$\epsilon^2 \cong \frac{a^2}{N} \qquad [2]$$

Comparison of [1] and [2] indicates that the interval T may be considered to contain T/2T = N independent estimates. Evidently one point in each interval of length 2T will produce just as fast a convergence of the statistics. The extra points included in the integral over 2T do not improve the convergence, since they are not independent — they do not give new information.

This concept may be immediately applied to the measurement of spectra. Suppose it is desired to measure the spectrum of a process having no energy above a frequency $1/\eta$. The lowest frequency desired is $1/\tau$, where it is presumed that $\tau \geqslant T$. A tape of length T is taken, sufficient to provide the required accuracy by [1] and [2], and digitized at the Nyquist frequency $2/\eta$, providing $2T/\eta$ points. At each lag, T/2T multiplications are required, followed by T/2T additions, or T/T operations; there are $2\tau/\eta$ lags, for a total of $2\tau T/T\eta$ operations to produce the correlation. This must now be transformed by a fast Fourier transform of $2\tau/\eta$ points. If we indicate the number of operations required for this by $FFT(2\tau/\eta)$, the total number of operations is

$$2\tau T/T\eta + FFT(2\tau/\eta) \qquad [3]$$

This may be compared with other ways of obtaining the spectrum. For example, each block of $2\tau/\eta$ points could be transformed by FFT, requiring

$(T/\tau)FFT(2\tau/\eta)$ operations; then each Fourier coefficient must be squared, requiring $2T/\eta$ operations; then the squared coefficients at each frequency must be added together, requiring $(2\tau/\eta)(T/\tau) = 2T/\eta$ operations. This gives a total of

$$4T/\eta + (T/\tau)FFT(2\tau/\eta) \qquad [4]$$

operations.

Still another way would involve taking the FFT of the entire tape, $FFT(2T/\eta)$, squaring the coefficients obtained, for $2T/\eta$ operations, and adding each group of T/τ together to form $2\tau/\eta$ block averages, requiring $2T/\eta$ operations, for a total of

$$4T/\eta + FFT(2T/\eta)$$

We may compare these by writing directly

$$\frac{\tau}{2T} \cdot \frac{4T}{\eta} + FFT(2\tau/\eta) \qquad [3]$$

$$\frac{4T}{\eta} + \frac{T}{\tau} FFT(2\tau/\eta) \qquad [4]$$

$$\frac{4T}{\eta} + FFT((T/\tau)(2\tau/\eta)) \qquad [5]$$

Now, ordinarily $\tau/2T \sim 1$, while $T/\tau \sim 10^2$, so that it is evident that method [3] is the cheapest. The ratio can be estimated from the expression $FFT(n) \sim n \log_2 n$, from which approximate expressions for [3], [4] and [5] may be derived; setting $\tau/2T = C$, we have

$$C\frac{4T}{\eta} + \frac{2\tau}{\eta} \log_2 \frac{2\tau}{\eta} \sim C\frac{4T}{\eta} \qquad [3]'$$

$$\frac{4T}{\eta} + \frac{T}{\tau} \cdot \frac{2\tau}{\eta} \log_2 \frac{2\tau}{\eta} \qquad [4]'$$

$$\frac{4T}{\eta} + \frac{T}{\tau} \frac{2\tau}{\eta} \log_2 \frac{2\tau}{\eta} + \frac{2T}{\eta} \log_2 \frac{T}{\tau} \qquad [5]'$$

If $C = 1$, the relative cost differences are

$$\frac{[4]' - [3]'}{[3]'} = 1/2 \log_2 \frac{2\tau}{\eta} \qquad [6]$$

$$\frac{[5]' - [3]'}{[3]'} = 1/2 \log_2 \frac{2T}{\eta} \qquad [7]$$

In turbulent flows τ/η ranges from 10 to 10^3; T/τ will ordinarily be at least 10^2 (for 10 % accuracy); thus $2 \times 10^3 \geqslant 2\tau/\eta > 20$, $2 \times 10^5 \geqslant 2T/\eta \geqslant 2 \times 10^3$. Thus roughly,

$$2 \leqslant \frac{[4]' - [3]'}{[3]'} \leqslant 5.5 \qquad [8]$$

$$5.5 \leqslant \frac{[5]' - [3]'}{[3]'} \leqslant 9 \qquad [9]$$

and the right-hand side of [9] may be even higher if greater accuracy is desired.

Hence, under the worst circumstances, the proposed method, leading to [3], is 1/3 the cost of the next best method, and may be far cheaper. It should be noted that tape reading times and storage requirements are the same for all methods; the proposed method gives a bonus, in the form of the correlation. This would only be obtained at extra cost in the other methods.

It must be noted that there may be other reasons for computing the Fourier coefficients than obtaining the spectrum, such as determining higher order statistics of the Fourier coefficients. In such a case, there is, of course, no substitute for the FFT. In addition, if it is desired to compute correlations to very long lags ($C \gg 1$), corresponding to very low frequencies★, the relative costs change, and [4]' may be less expensive than [3]' (although it is clear that [5]' will always be more expensive than [4]').

★ a typical need in calculating derivative spectra.

Acknowledgements

The author is grateful to P. Buchhave and P. Beuther for many helpful discussions and for their assistance in preparing the manuscript, to the organizing committee of DFC 78 for the invitation to participate, to DISA Electronics for their contributions to a stimulating and pleasureable experience, and to E. Graber for typing the manuscript.

The support of the U.S. National Science Foundation, Engineering, Fluid Dynamics Program and the U.S. Air Force Office of Scientific Research, Aerospace Sciences Branch, is gratefully acknowledged.

Processing of Laser Anemometry Signals

by

Lars Lading
Risø National Laboratory
DK-4000 Roskilde, Denmark

Abstract
A number of signal processing schemes are discussed for the Laser Doppler Anemometer and the Time-of-Flight Laser Anemometer. Special emphasis is placed on time-resolved measurements with a tracking processor.

The essential statistical properties of the detector signal(s) is (are) reviewed. The concept of Maximum Likelihood Estimation is discussed in relation to the implementation of processors.

1. Introduction
The velocity of a scattering particle(s) is determined by estimating some quantity of the temporal signal(s) of one (or several) photodetector(s). The intensity distribution in the measuring volume, as seen from the detector, defines the quantity that gives the velocity. We shall call this intensity distribution the code of the laser anemometer. The simplest code is given by a beam of known spatial width: measuring the temporal width or correlation time of the detector signal yields the velocity. A slightly more complicated code is given by two displaced peaks: the velocity is determined here from the time lag between two signals of the photodetector(s). The most commonly used code is a spatial wave packet: the velocity is given by the frequency of the temporal signal.

In the analysis and/or design of a laser anemometer, it must be realized that the system is based on stochastic processes: the velocity fluctuations (turbulence) are usually random; the scattering particles are supposed to follow the flow exactly, but are randomly distributed and will thus give a random sampling of the flow; the number of detected photons within a given time is random, but the mean (or expected) value is modulated by the movements of the particles. The way in which the expected photocurrent (or photon count-rate) is modulated — the encoding — is given by the code of the anemometer.

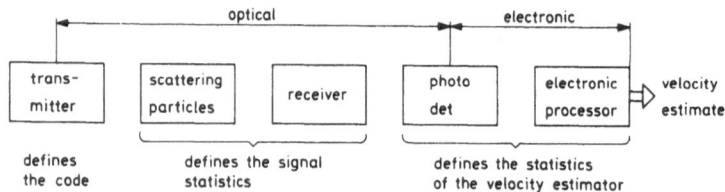

Fig. 1. Structure of a laser anemometer.

In Fig. 1 is shown the essential components of a laser anemometer, and their impact on the information flow is indicated.

The purpose of the electronic processor is to give the best possible estimate of the desired velocity information, i.e. the randomness of the processes involved should have the smallest possible impact on the measurement.

We shall in the following discuss the synthesis of electronic processors on the basis of simple optimization criteria, assuming that a measurement is to be performed within a time shorter than the temporal microscale of the turbulence. Two laser anemometers are considered, namely the Laser Doppler Anemometer (LDA) where the code is given by a wave-packet, and a Time-of-Flight Laser Anemometer (TFLA) where the code is given by two displaced peaks.

2. Basic Optical Model

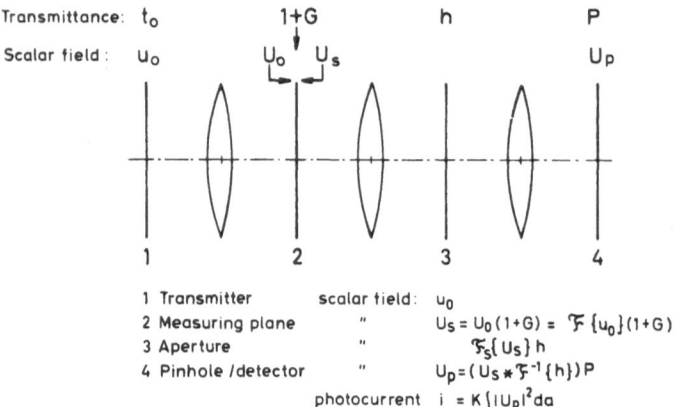

Fig. 2. Basic optical model. Each lens acts as a Fourier transforming device. The field in the right focal plane U_r is related to the field of the left focal plane by $U_r = 1/\lambda F \cdot \mathcal{F}\{U_o\}$ is evaluated at the spatial frequencies $f_x = x/\lambda F$ and $f_y = y/\lambda F$.

Our considerations are based on a Fourier optical model as shown in Fig. 2 and the general expressions for the photocurrents as given in references 1 and 2.

The photocurrent will in general contain three components: (1) a dc-current, caused only by directly transmitted light, which carries no information and only generates noise; (2) a component caused only by scattered light, and (3) a component caused by heterodyning between scattered and directly transmitted light.

The incoherent mode is given by the following: The directly transmitted beams are blocked; the scattered light is collected by an aperture so large that the width of its spatial Fourier transform in the measuring plane is much smaller than the average particle separation. The photocurrent is then given by

$$i = K \frac{A}{F^2} \sum_p \sigma_p |U_o(\underline{r}_p)|^2 \qquad\qquad [1]$$

where K is the detector sensitivity (amps/watt), A the receiver aperture area, F the focal length, σ_p the scattering cross section of particle p, and r_p its position vector, and $U_o(r)$ the scalar field in the focal (measuring) plane.

The coherent mode is identical to the incoherent except for the fact that the aperture is now so small that its Fourier transform is larger than the focal beam diameters. The photocurrent is then given by

$$i = K \frac{A}{F^2} \left| \sum_p \sqrt{\sigma_p} \; U_o(\underline{r}_p) \right|^2 , \qquad\qquad [2]$$

i.e. the summation has to be performed before the squaring.

Directly transmitted light is also detected in the reference beam mode. Let the aperture be larger than the directly transmitted beam. Then we get a dc-current

$$i_{dc} = K \int |U_o(r)|^2 \; dA \qquad\qquad [3]$$

and a dynamic component

$$i = 2K\lambda \sum_p \sqrt{\sigma_p} \; |U_o(\underline{r}_p)|^2 . \qquad\qquad [4]$$

The distinction between coherent and incoherent mode (or detection), as given here, is general and therefore in agreement with the way in which these terms were introduced for laser Doppler anemometers [3].

3. Statistics of the Detector Signal, no Photon or Electronic Noise, Constant Velocity

The statistical properties of the expected photo-current $<i>$ are investigated in this paragraph. The photo-current itself is considered to be a deterministic quantity, i.e. photon and electronic noise is neglected. The velocity is also assumed to be constant. These assumptions confine the problem to the statistics of the photo-current given by the random distribution of the particles.

In the case of single particle detection, the velocity measuring problem is reduced to estimating a parameter in a deterministic signal, which — in principle — is trivial. We shall therefore concentrate on the "many particle case".

a. The Doppler Signal

The temporal signal of a narrow-band process can be written as

$$s(t) = A(t)\cos(\omega_0 t + \varphi(t)) \tag{5}$$

where $A(t)$ is a slowly varying function (slow compared with $1/\omega_0$) with a Rayleigh distribution, $\varphi(t)$ is evenly distributed between π and $-\pi$, and the width of the spectrum of $\dot{\varphi}$ is of the order of the signal-bandwidth. The statistics of the derivative of the phase $\dot{\theta} = \omega_0 + \dot{\varphi}$ have been derived by Rice (4). One of the important results of Rice's analysis is the fact that the variance of $\dot{\theta}$ is infinite (any physical detector will, of course, give a finite value). George and Lumley (5) showed that equation [5] is a very good approximation to an LDA signal if there are many particles in the measuring volume. They also analyzed some of the implications of the rather "unpleasant" statistics of $\dot{\theta}$ for turbulence measurements and showed that the intrinsic fluctuations of $\dot{\theta}$ set a limit to turbulence spectral measurements that cannot be overcome by any linear operation on $\dot{\theta}$. Some of the same reults have been obtained by computer simulation (6) (7).

However, if the measurement of $\dot{\theta}$ is conditioned by properties of the envelope $A(t)$, then the variance of $\dot{\theta}$ can be reduced to any desired level as shown below.

Let the normalized amplitude be $\alpha = A/\psi_0^{1/2}$, where ψ_0 is the mean signal power, and the normalized phase derivative be $\mu = \dot{\varphi}/\Delta$, where Δ is the signal-bandwidth. Appendix I shows that, on the basis of Rice's calculation (4), we find for the conditional variance

$$\text{var}\{\mu|\alpha>\alpha_0\} = \tfrac{1}{2}E_1(\tfrac{1}{2}\alpha_0^2)/e^{-\frac{1}{2}\alpha_0^2} \tag{6}$$

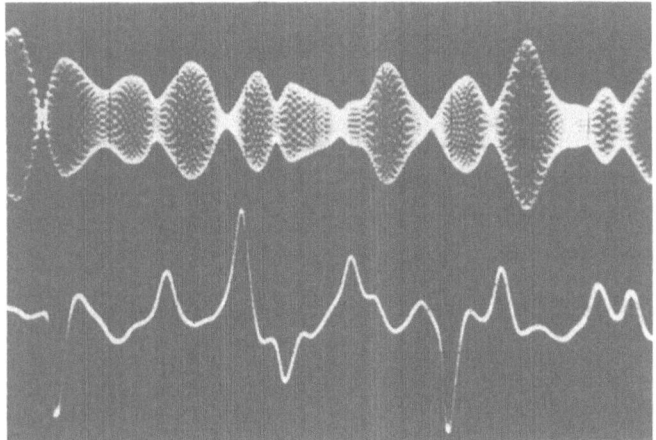

Fig. 3. Simulated Doppler signal (only the envelope is visible) and the corresponding fluctuations in the derivative of the phase averaged over a time roughly one third of the correlation length.

where $E_1(z)$ is the exponential integral   [9]

$$\int_z^\infty \frac{e^{-t}}{t}\, dt.$$ [7]

The expression in equation [6] is a decreasing function of α_o and it has a singularity for $\alpha_o = 0$.

For very large z, $E_1(z)$ converges towards e^{-z}/z, which substituted in equation [6] yields

$$\text{var}\{\mu | \alpha > \alpha_o\} \to 1/\alpha_o^2.$$ [8]

A consequence of introducing a threshold condition is that the process will be randomly sampled: both the times and the lengths of the samples will be randomly distributed. By computer simulation with a system as that described in reference 7, it was found that the distribution function for the samples will approximately follow a Poisson distribution, if a maximum of one sample is taken within a time equal to the coherence time of the signal. It has also been verified that the "ambiguity noise" is a decreasing function of the sampling threshold.

Now, the mean sampling rate decreases with increasing sampling amplitude, so that for a given turbulence "bandwidth" an optimum sampling rate − and thus threshold — must exist. The optimum condition can be defined as that which maximizes the ratio between the turbulence signal and the "noise" uncorrelated with the turbulence.

Amplitude conditioned sampling will essentially have the same effect as a finite particle concentration; this has been verified by computer simulation. For a given turbulence, there must therefore also be an optimum particle concentration.

The displaced part of the power spectrum for a constant velocity can be written as

$$S(\omega) = K \exp\left\{-\tfrac{1}{2}\left(\frac{\omega-\omega_o}{\Delta_d}\right)^2\right\}. \tag{9}$$

where Δd is the signal bandwidth and K a constant. The probability distribution for the estimated power spectrum $\hat{S}(\omega)$ will be a X^2-distribution (9). In general, $E\left\{\hat{S}(\omega)\right\} \neq S(\omega)$. For a finite averaging time and a given analyzing filter bandwidth, $\hat{S}(\omega_i)$ and $\hat{S}(\omega_j)$ will be correlated. However, if the analyzing filter bandwidth is much smaller than the signal bandwidth, and the averaging time is long compared with the coherence time of the reciprocal filter bandwidth (signal coherence time), then (1) the probability distribution for $\hat{S}(\omega)$ will be Gaussian (this follows from the central limit theorem; see also ref. 10); (2) $E\left\{\hat{S}(\omega)\right\} = S(\omega)$ and the individual estimated spectral components can be considered as statistically independent (the correlation length in ω-space is equal to the width of the analyzing filter).

b. The Time-of-Flight Signals

Let us assume two detectors. This is not essential but will simplify the discussion (this assumption is equivalent to discarding the zero frequency and "minus" frequency peaks in the Doppler spectrum). The photocurrents are $i_1(t)$ and $i_2(t)$, and

$$i_2(t) = i_1(t-\tau_o). \tag{10}$$

One detector would give a photocurrent equal to the sum of i_1 and i_2.

By comparing equation [10] with equation [5], it is noticed that τ_o is apparently not obscured by a term equivalent to $\dot{\varphi}$. This may indicate an absence of "ambiguity noise" for the TFLA. However, to the knowledge of the author this question has not yet been fully resolved. It must be remembered, though, that no meaningful measurement can be performed within a time shorter than the time-of-flight.

The cross correlation function is

$$R_{12}(\tau) = R_{11}(o) \exp\left\{-\tfrac{1}{2}\left(\frac{\tau-\tau_0}{t_t}\right)^2\right\}. \tag{11}$$

The probability distribution is rather complicated because a product of two random variables is involved [11], but for an averaging time much longer than the coherence time of the signals it will approach a Gaussian distribution (the central limit theorem). The variance is given by [12]

$$\text{var}\left\{\hat{R}_{12}(\tau)\right\} \cong \frac{1}{T}\int_{-\infty}^{\infty} R_{11}(\xi)R_{22}(\xi) + R_{12}(\xi+\tau)R_{21}(\xi-\tau)\,d\xi. \tag{12}$$

4. Photon Statistics

The expressions given in the preceding section are expected values of the photon count-rate without background. The probability distribution for the number of photon counts within a time interval t, which is small compared with the signal coherence time, is a Poisson distribution,

$$p(n) = \frac{\langle n\rangle^n}{n!}\, e^{-\langle n\rangle}, \tag{13}$$

where $\langle n\rangle = E\{n\}$, the expected value at a given time. For high count-rates, the distribution will approach a Gaussian.

In general the expression

$$n(t) = (s(t) + b(t))\,\Delta t \tag{14}$$

holds, where $s(t)$ and $b(t)$ are the signal and background count-rates, respectively. It is our experience that the limiting factor in the performance of a given system for practical measurements is the fluctuations associated with the background $b(t)$. In principle, $b(t)$ could be zero; the limitations would then be given by the fluctuations (noise) associated with the signal itself. In "difficult" situations, such as boundary layer measurements or measurements at long ranges, $\langle b(t)\rangle \gg \langle s(t)\rangle$ for any t. This will be assumed in deriving estimators for ω_0 or τ_0. The cases where the intrinsic signal noise dominates are computationally very difficult.

5. Maximum Likelihood Estimation

There are many possible criteria for optimizing a processor. Some give identical results and some are of a very fundamental nature, but it may be difficult or impossible to devise a scheme which fulfils the optimizing criteria. We shall here make use of the maximum likelihood estimator (MLE). The MLE for a given parameter α is that value which maximizes the conditional probability [13]

$$p(\{r\}|\alpha) \qquad\qquad\qquad [15]$$

where $\{r\} = (r_1, r_2, \ldots \ldots r_n)$ is the data-set of a given measurement. In the present context, $\{r\}$ may be the photon-counts $\{n\} = (n(t), n(t+\Delta t), n(t+2\Delta t), \ldots \ldots, n(t+n\Delta t))$ and α the Doppler frequency ω_0 or the time-of-flight τ_0. The MLE makes no use of pre-knowledge about the parameter to be estimated, although this usually exists. We shall see later how we can combine the MLE with some *a priori* knowledge about the probability distribution for α in the design of "trackers".

A lower limit for the variances of a MLE is given by [13]

$$\mathrm{var}\{\hat{\alpha}\} \geqslant \left(E \; \frac{\partial^2}{\partial \alpha^2} \{ \ln p(\{r\}|\alpha) \} \right)^{-1}. \qquad\qquad [16]$$

For cases where background noise is the primary noise contribution and the signal-to-noise ratio is high, the variance of $\hat{\alpha}$ is given by equation [16] (14).

A comparison of laser anemometers has been performed in reference 15 on the basis of equation [16].

a. Single Particle Detection
Let the number of photon-counts in a time interval Δt at the time t be given by equation [14], and the probability distribution for n by equation [13]. Let $\langle n(t) \rangle = \langle n(t, \alpha) \rangle$, where α is the parameter to be estimated (ω_0 or τ_0). The count numbers at different times are statistically independent:

$$p(\{n\}|\alpha) = \Pi p(n(t_i)|\alpha). \qquad\qquad [17]$$

The conditional probability $p(\{n\}|\alpha)$ will obtain its maximum value for the same value of α as $\ln p(\{n\}|\alpha)$ does. The expression $\ln p(\{n\}|\alpha)$ is in general easier to handle.

From equations [13] and [17], we find that a MLE is the value that maximizes the correlation function

$$\sum n(t_i) \; \ln \langle n(t_i, \alpha) \rangle, \qquad\qquad [18]$$

i.e. the measured signal has to be correlated with the logarithm of the expected signal. Substituting equation [14] into [18] and assuming $\langle b(t) \rangle \gg \langle s(t) \rangle$ yields by expansion of the logarithmic term

$$\sum n(t_i) \langle s(t_i, \alpha) \rangle. \tag{19}$$

If there is only one maximum, $\hat{\alpha}$ is the value of α for which

$$\sum n(t_i) \frac{\partial}{\partial \alpha} \left\{ \langle s(t_i, \alpha) \rangle \right\} \text{ is maximum.} \tag{20}$$

Now, in single particle detection it must be necessarily so that no particle is present in the measuring volume for most of the time, but the background noise will imply that equation [20] is quite often fulfilled anyway. A maximum seeking device may be rather slow or complicated. The way to solve this problem is to detect whether a particle is present or not (hypothesis testing (16)) on the basis of the magnitude of the product (correlation in equation [19]; if and only if this product exceeds a given threshold level, the estimator can be found from equation [20].

A complication in arriving at a MLE may be the fact that the expected signal is not fully known even for a given α: it may contain unknown parameters, e.g. the phase of a detected wave-packet. The conditional probability distribution (equation [15]) for such a case is given by

$$p(\{n\} | \omega_0) = \int p(\{n\} | \omega_0, \varphi) p(\varphi) d\varphi \tag{21}$$

where the probability for φ in the present case is evenly distributed between $-\pi$ and π. For a discussion of this topic, see e.g., (16). We shall later make use of some of the results given in this reference.

b. Many Particles in the Measuring Volume

In a laser anemometer signal a "parameter" may not only be unknown, it may also be a variable. In the many-particle Doppler case, the phase of the signal contains such a term: $\varphi(t)$ in equation [5]. Two ways of handling this problem are possible:

In the first method the frequency $\dot{\theta}$ is estimated within a time shorter than the coherence time of $\varphi(t)$. This is done many times. ω_0 is then estimated from the "new" data set by simple averaging.

The second method is based on some statistical quantity that still contains the parameter to be estimated, but where the undesired variable is averaged out.

In the present cases the estimated correlation function or power spectrum will be the basis on which a MLE is derived. It is again assumed that the background dominates.

For the Doppler case, the power spectrum is

$$S_d(\omega) = \frac{A_d}{\sqrt{2\pi}\Delta_d} \exp\left\{-\frac{1}{2}\left(\frac{\omega-\omega_0}{\Delta d}\right)^2\right\} + N_d \qquad [22]$$

Let the measured spectrum be represented by a set of discrete values $(S_1, S_2, \ldots, S_n) = \{S\}$ and $S_{io} = E\{\hat{S}(\omega_i)\}$. The spacing between these sprectral values is equal to the analyzing filter bandwidth; this gives an almost complete but non-redundant determination of $S(\omega)$ (10). Using the same procedure as in the single particle case and changing from sums to integrals, we find $\hat{\omega}_0$ as the value of ω_0 for which

$$\int \hat{S}(\omega)\, S_0(\omega, \omega_0)\, d\omega \qquad [23]$$

is maximum, or as the solution to the equation

$$\int \hat{S}(\omega)\, \frac{\partial}{\partial\omega_0} S_0(\omega, \omega_0)\, d\omega. \qquad [24]$$

For the time-of-flight case, we find equivalently that $\hat{\tau}_0$ is given as the solution to the equation

$$\int \hat{R}(\tau)\, \frac{\partial}{\partial\tau_0} R_0(\tau, \tau_0)\, d\tau = 0. \qquad [25]$$

6. Trackers

In the previous section it was shown that a MLE is usually obtained by maximizing the correlation between the measured signal and the expected signal. This can in principle be performed in a multichannel analyzer as shown in Fig. 4. In measuring the temporal evolution of turbulence, the averaging time of the processor should be shorter than the micro scale — or preferably the inner scale. If this is the case, we can utilize the fact that the probability for a large difference between two consecutive measurements is much smaller than the probability for a small difference. The basic principle is then the use of only one channel, but this channel is continuously up-dated to maximize the correlation function that gives the MLE. Let us consider some examples of tracking processors.

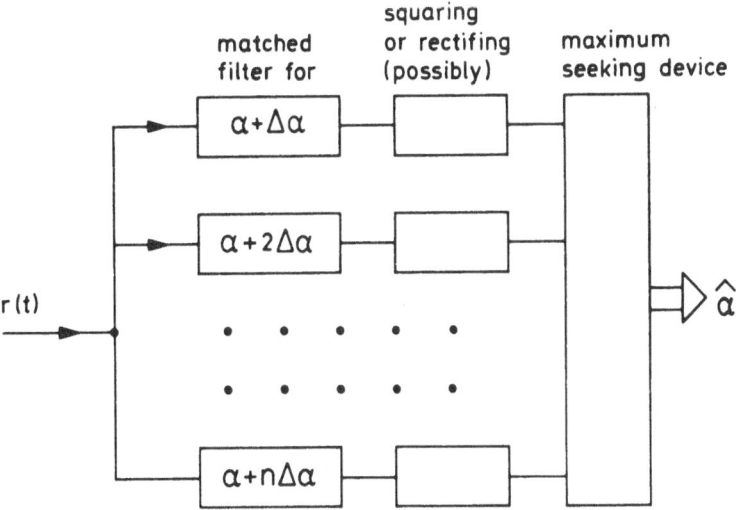

matched
filter for

squaring
or rectifing
(possibly)

maximum
seeking device

$\alpha + \Delta \alpha$

$\alpha + 2\Delta \alpha$

r(t)

$\hat{\alpha}$

$\alpha + n\Delta \alpha$

Fig. 4. Structure of a multichannel processor. Each of the channels incorporates a filter with an impulse response equal to the expected signal for $\hat{\alpha}_i$. An envelope or square law detector may be incorporated to overcome an unknown parameter, such as e.g., the phase.

a. A Phase-locked Loop

Let the signal be

$$r(t) = s(t) + n(t) = A(t)\cos(\omega_o t + \varphi(t)) + n(t) \qquad [26]$$

where n(t) is the background noise. the correlation between r(t) and $\sin\omega_r t$ obtained with an averaging time T is

$$R(\tau, \omega_r, T) = \frac{1}{T}\int_0^T (A(t)\cos(\omega_o t + \varphi(t)) + n(t))\sin(\omega_r(t+\tau))dt. \qquad [27a]$$

For an infinite averaging time T this average is zero, indendedently of ω_r and τ. This reflects the fact that the spectral power has a continuous distribution and is not concentrated at any single frequency. Let us assume $\omega_r = \omega_o$. For an averaging time so short that A(t) and $\varphi(t)$ can be considered constant, we get

$$R(\tau,T) = A(t_i)\sin(\omega_o \tau + \varphi(t_i)) + \frac{1}{T}\int_{t_i}^{t_i+T} n(t)\sin(\omega_o(t+\tau))dt. \qquad [27b]$$

Let us for the moment neglect the effect of the noise n(t). It is seen that we have obtained information on the relative phase of the Doppler signal. This can be used to correct the phase $\omega_o \tau$ of the reference oscillator, so that $\omega_o \tau$ tends to follow the phase of the Doppler signal. It is important to realize that the averaging time has to be shorter than the correlation time ($\sim$ transit time) of the Doppler signal.

Let us consider the noise, and assume that $n(t)$ is uncorrelated with $A(t) \cdot \cos(\omega_o t + \varphi(t))$. This is not always the case, but it is adequate for the present discussion.

The quantity $\overline{n(t) \cdot \sin(\omega_o (t+\tau))}_T$ is random with estimated mean of zero. The variance of the fluctuations is given by equation [12], which yields

$$\text{var}\left\{ R_{nr}(0) \right\} \cong \frac{1}{T} \int R_n(\tau) R_r(\tau) d\tau \qquad [28]$$

where $R_n(\tau)$ and $R_r(\tau)$ are the autocorrelation functions for the noise and for the reference signal, respectively. Assuming white noise gives

$$R_n(\tau) = N \, \delta(\tau) \qquad [29a]$$

$$R_r(\tau) = \cos(\omega_o \tau), \qquad [29b]$$

which substituted in [28] yields

$$\text{var}\left\{ \hat{R}_{nr}(0) \right\} = N/T. \qquad [30]$$

These fluctuations reduce our ability to derive the desired phase information: they should be as small as possible, which implies a large averaging time T.

Concluding this discussion we can say that owing to the stochastic nature of the Doppler signal (without noise), a short averaging time is needed, whereas the photon/amplifier noise implies a long averaging time. The optimum time must be somewhere around the correlation time or transit time of the Doppler signal.

The lay-out for such a tracking correlator may be as shown in Fig. 5.

For such a setup some of the tools of automatic control theory can be used (17) assuming that the loop is locked. For the closed loop transfer function we get (17) (Laplace transforms)

$$\frac{\theta_r(s)}{\theta_d(s)} = \frac{A \cdot K_o \, F(s)}{s + A \, K_o \, F(s)} \qquad [31]$$

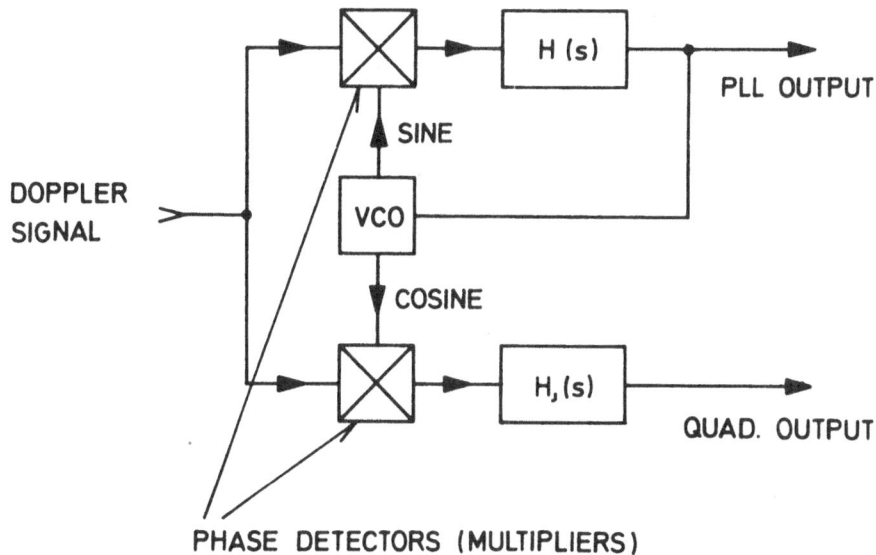

Fig. 5. A phase lock loop with a quadrature output. A display of the loop versus the quad. outputs gives useful information about the operational status of the loop.

where A is the signal amplitude, K_o the VCO gain, and $A \cdot K_o \cdot F(s)$ the open loop transfer function. In order to be able to track the signal, $F(s)$ has to contain an integrator. The gain of the loop is proportional to A, which is a (slowly) varying positive quantity (Rayleigh distribution) as appears from Section 3. Since A can have any positive value, the loop must be stable for all gains. A very convenient way to show this is by root-locus plots. Fig. 6 shows the root-locus plot for a second-order loop where

$$F(s) = \frac{s\tau_2 + 1}{s\tau_1}.$$ [32]

The loop should be adjusted so that for A = <A> it has a bandwidth equal to the Doppler bandwidth and slightly undercritical damping (17).

The discussion so far provides some general ideas of how a PLL will perform as a demodulator for Doppler signals. However, an exact general analysis is hardly possible. It is noted that if the quadrature output is at its maximum theoretical value, the PLL will give a MLE of ω_o.

In order to investigate the actual performance of a PLL, we simulated the Doppler signal as well as signal detection and processing on a hybrid computer.

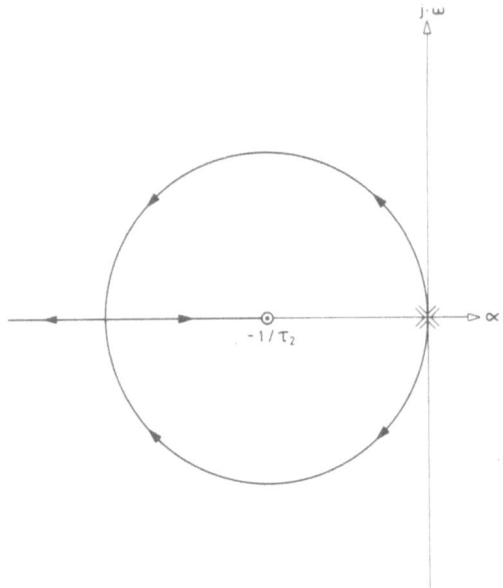

Fig. 6. Root-locus plot for the PLL. Note that because the amplitude of the signal fluctuates, the poles are not at constant positions, but always in the left half-plane.

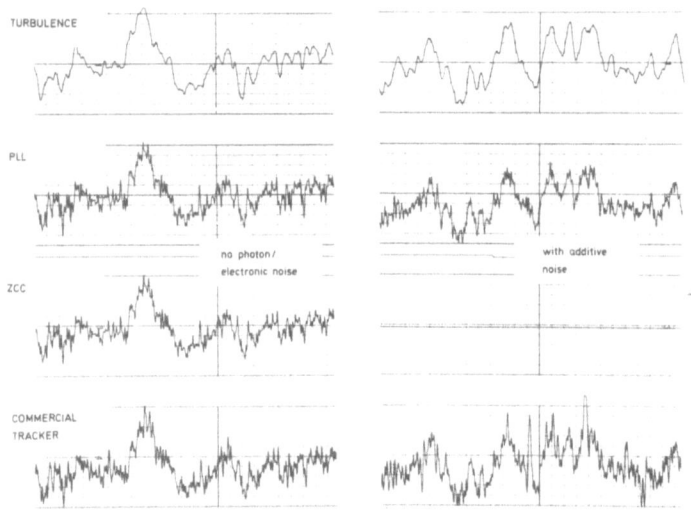

Fig. 7. a. Input turbulence. b. Output of the PLL. c. Output of a zero-crossing counter. d. Output of a commercial tracker. Case I is without noise, case II with additive noise. The zero-crossing counter could not operate with any significant noise added.

The simulated setup is described in reference (7); here only some of the results (or confirmations) are pointed out. A second-order phase-locked loop with a bandwidth larger than the Doppler bandwidth is essentially able to give an output that is proportional to the derivative of the phase in equation [26], averaged over a time given by the loop bandwidth. An example of a true turbulence signal and the "tracked" output is shown in Fig. 7. It is seen that the detector signal looks like the turbulence with some noise added — the so-called "ambiguity noise" given by the random phase fluctuations.

The level of the ambiguity noise was measured both absolutely and relative to the low-frequency turbulence intensity for different numbers of fringes in the measuring volume (7). These measurements showed excellent agreement with the (ideal) predictions of reference (5).

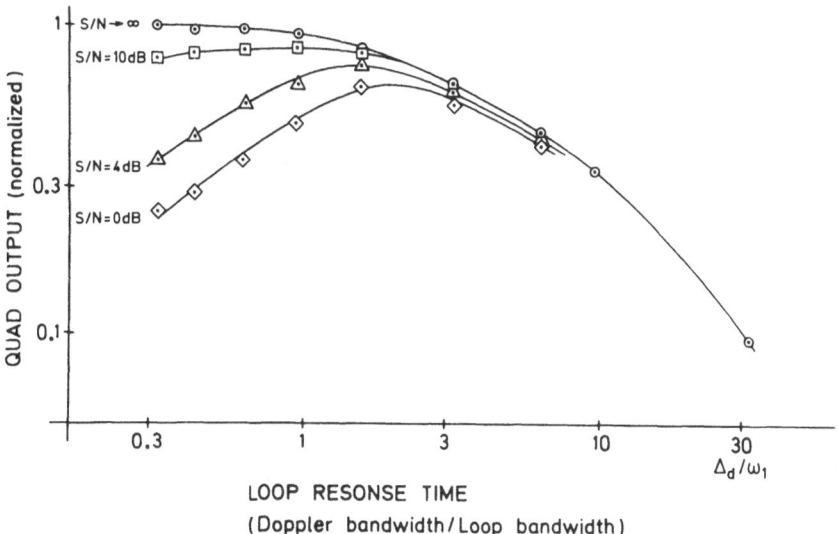

Fig. 8. Quadrature output (lock indication) versus the normalized reciprocal loop bandwidth Δ_d. The signal-to-noise ratio is taken as a peak-to-peak ratio.

We studied the effect of the bandwidth and found that the loop ceased to perform if the loop bandwidth was considerably narrower than the Doppler spectrum (Fig. 8). The mean of the quadrature or QUAD output was in this case taken as a measure of loop performance: If the loop is locked with no phase error, then the output of the phase detector is zero. The output of a detector giving the phase difference between the input signal and a signal 90° out of phase with respect to the VCO, will then be at its maximum. The mean of this QUAD output can be taken as a measure of lock quality (17).

Actual flow measurements with a PLL processor seem to confirm the results of the simulation experiments (18). The conditional sampling of the Doppler signal described in section 3 has been performed with a phase lock loop. If the quadrature output exceeded a threshold level, a sample was taken. The simulation experiments proved that both the "ambiguity noise" level and the sampling rate decrease with increasing threshold. It has also been verified that decreasing the particle concentration has the same effect as increasing the threshold. If the noise had any significant effect on the output of the PLL, conditional sampling would not work.

b. A Frequency-locked Loop

Let the estimated power spectrum be $\hat{S}(\omega)$. An MLE of ω_o from $\hat{S}(\omega)$ is given by equation [24]. Substituting equation [9] into [24] yields

$$\hat{\omega}_o \cong \frac{\int \omega\, \hat{S}(\omega)\, S_o(\omega - \hat{\omega}_o)\, d\omega}{\int \hat{S}(\omega)\, S_o(\omega - \omega_o)\, d\omega}. \qquad [33]$$

If the relative variance of the denominator is small, $S_o(\omega)$ is the expected signal spectrum with $\omega_o = 0$. Let us estimate $\omega_o - \omega_r$ and assume $|\omega_o - \omega_r|$ smaller than the Doppler bandwidth, i.e.

$$\hat{\omega}_o - \omega_r = \frac{\int \omega \hat{S}(\omega + \omega_r)\, S_o(\omega)\, d\omega}{\int \hat{S}(\omega + \omega_r)\, S_o(\omega)\, d\omega}. \qquad [34]$$

Using the moment theorem and the fact that
$\mathcal{F}\{\hat{S}(\omega + \omega_r) S_o(\omega)\} = \hat{R}(\tau) \exp j\omega_r \tau \,\star\, R_o(\tau)$ yields

$$\hat{\omega}_o - \omega_r = \frac{-j}{\hat{P}} \frac{\partial}{\partial \tau} \{\hat{R}(\tau) \exp\{j\omega_r \tau\} \star R_o(\tau)\}|_{\tau = 0} \qquad [35]$$

where $\hat{P}$ is the numerator in equation [34]. An optimum tracker for $\hat{\omega}_o$, given $\hat{S}(\omega)$, is obtained if ω_o is controlled so that $\hat{\omega}_o - \omega_r = 0$, which would fulfil equation [33]. The estimator given by equation [35] can be implemented as shown in Fig. 9.

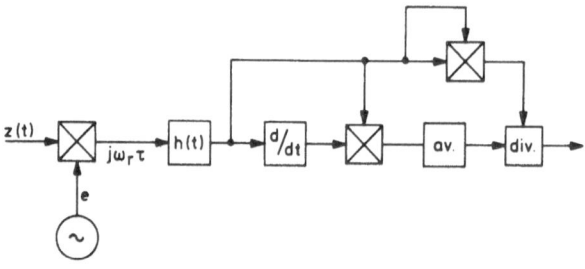

Fig. 9. Estimating $(-j/\hat{P})\partial/\partial \tau \{\hat{R}(\tau)\exp\{j\omega_r\tau\} \star R_o(\tau)\}|_{\tau = 0}$.

However, normally only the real part $r(t)$ of $z(t)$ is accessible and therefore we have to generate the analytic signal $z(t) = r(t) + j\check{r}(t)$ from $r(t)$. $\check{r}(t)$ is the Hilbert transform of $r(t)$. Let $r(t)$ be given by

$$r(t) = a(t)\cos\omega t - b(t)\sin\omega t; \qquad\qquad [36a]$$

then it follows that (19)

$$\check{r}(t) = a(t)\sin\omega t + b(t)\cos\omega t. \qquad\qquad [36b]$$

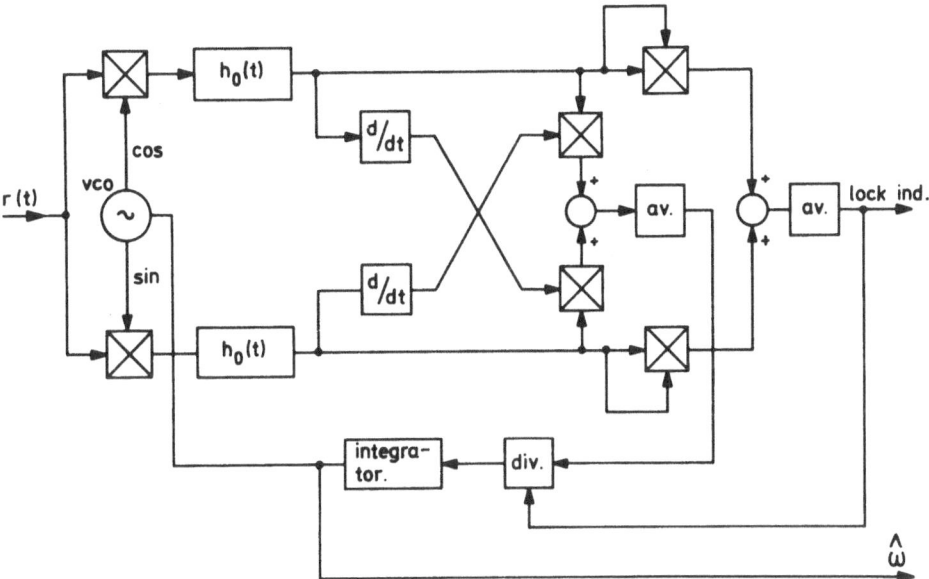

Fig. 10. A frequency lock loop. The input to the division circuit is also used for monitoring the track quality.

On the basis of equations [36a] and [36b], the processing scheme of Fig. 9 can be modified to cope with real signals as shown in Fig. 10. It is noted that $h(t)$ is the impulse response of a filter for which $\overline{h(t)h(t+\tau)} = R_o(\tau)$. This filter will in general be non-causal, but it can be approximated by a physically realizable filter.

An important property of the FLL as compared to the PLL is that, for the FLL, the averaging time can be much longer than the coherence time of the signal. In principle a poor S/N can always be counterbalanced by a longer averaging time, which, of course, means a slower response.

An FLL as described here is also described in reference (20). The basic concept for this tracker was to design a processor that would give as output the derivative of the phase (see equation [5]); and in fact

$$\bar{\dot{\theta}}_T = \left(\frac{a(t)\,\dot{b}(t) + \dot{a}(t)\,b(t)}{a^2(t) + b^2(t)}\right)_T$$

which is the control output.

c. The Delay Lock Loop

The Delay Lock Loop (DLL) is a device which can track the temporal displacement between two signals. Here it is used as a processor for the TFLA in the case of many particles in the measuring volume.

Let the cross-correlation function be

$$R_{12}(\tau) = A\,\exp\left\{-\tfrac{1}{2}\left(\frac{\tau-\tau_0}{t_t}\right)^2\right\} + N\delta(\tau) \tag{37}$$

and let $N \gg At_t$. An MLE of τ_0 implies

$$\hat{G}(\tau_0) = \int R_{12}(\tau)\,R'_{12}(\tau-\tau_0)\,d\tau = 0. \tag{38}$$

The integral is a convolution between the measured correlation function and the expected correlation function. The expresssion on the left of equation [38] can be obtained as shown in Fig. 11. Note that $d\,R(\tau)/d\tau = \overline{r(t)r(t+\tau)}$ and that $\int h(t)h(t+\tau)$ is proportional to $R_{11}(\tau)$.

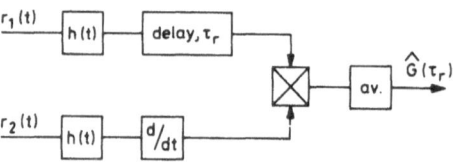

Fig. 11. Estimating $\int \hat{R}_{12}(\tau)\,R'_{12}(\tau-\tau_r)\,d\tau.$

Let τ_r be the actual delay in Fig. 11. If $\tau_0-\tau_r < t_t$ then the expected output will be proportional to $\tau_0-\tau_r$. This output can be used to control the delay so that $\tau_0-\tau_r \cong 0$. The variance of the output is (by equation [12])

$$\text{var}\{\hat{G}(\tau_0)\} = \frac{\sqrt{\pi}}{2}\frac{N^2}{t_t^2}. \tag{39}$$

It is noted that if N = 0, the variance is zero (!) indicating the absence of "ambiguity noise". However, equation [12], and thus [39], is only valid if $T_{av} \gg t_t$, and note that no measurement can be performed within a time shorter than τ_o.

The design of the feedback circuitry is the same as for the PLL and the FLL. In this case also the loop will have to be stable for a gain range larger than that given by the variance on $\hat{G}(\tau)$.

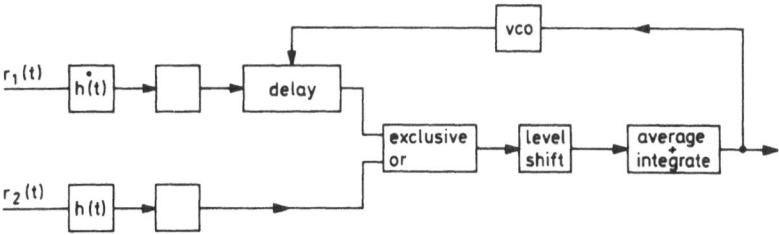

Fig. 12. A delay lock loop. This DLL in a TFLA has been compared with the PLL in an LDA. In the case of high mono-disperse seeding density, the PLL was slightly superior. With very large fluctuations in the signal level, the DLL was far superior.

A DLL has been built as shown in Fig. 12. In order so simplify the system, the actual signals (without the dc-component) are represented by a two level signal

$$x(t) = \begin{cases} 1 \text{ if } r(t) > 0 \\ 0 \text{ if } r(t) < 0. \end{cases}$$

This implies that the delay can be accomplished with a simple shift register. The multiplication is then performed with an exclusive-or unit.

The dynamic range is essentially limited by the input filters. Ideally, these filters should track the delay, too, but this would complicate the implementation.

If the filters do not track the delay, filter functions other than h(t) may provide for a better control signal.

d. The Coincidence Tracker
In connection with work on remote measurement of wind velocity, a special processor has been developed: a coincidence tracker (Fig. 13). The processor is designed for cases where the mean number of particles in the measuring

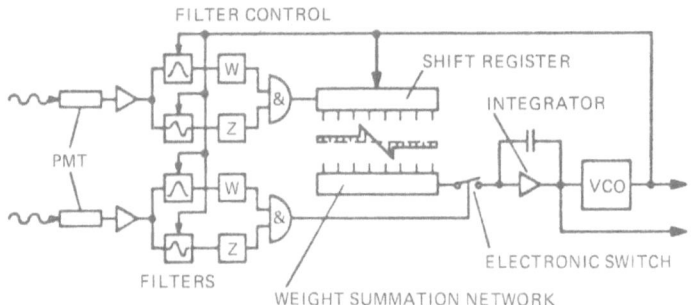

Fig. 13. A coincidence tracker.

volume is much smaller than one, but the sampling rate is still high enough for time resolved measurements.

The filter performs an (near) MLE of the temporal position of the detected pulses. The pulses are fed to a shift register, which is tapped with the shown weight function. Each time a time-of-flight measurement is performed, the clock frequency is changed so that the "new" τ_0 corresponds to the zero-crossing of the tapped shift register. However, this will only occur if the difference between two consecutive measurements is smaller than the "lock-in" range.

A coincidence tracker has been (1) tested on an analog computer, (2) built, and (3) found to perform as expected in actual flow measurements.

Conclusion
To conclude, the table lists a number of processors classified according to their suitability for different situations in which time-resolved measurements are to be performed.

The burst correlator and the zero-crossing counter have not been dealt with in this paper. However, a few comments on them may be in order.

The photon-counting burst correlator is a very powerful processor in the case of very low photon-count rates (21), but it has two shortcomings: (1) it does not use the previous measurements in the evaluation of a new measurement, and (2) as it has a digital design, it cannot cope with a very high background.

A counter will, in general, be very far from an MLE. However, in cases where background noise is no problem a good processor does not have to perform like an MLE. The amplitude conditional sampling described in section 3 can be carried out with a counter.

Acknowledgement

The author wishes to thank R.V. Edwards, A. Skov Jensen, C. Fog and R.L. Schwiesow for useful discussions and suggestions on various matters related to the subject of this paper. Most of the work on the PLL was done together with R.V. Edwards. C. Fog was responsible for the implementation of the coincidence tracker and the photon burst correlator.

TABLE

	Doppler				Time-of-Flight	
	Counter	PLL	FLL	Burst corr. (photon count)	DLL	Coinc. Tr.
single burst det. high S/N	xxx	xx				xxx
single burst det. low S/N high background		x		x		xxx
single burst det. low S/N low background				xxx		xxx
many particles in the m.v. high S/N	x	xxx	xx		xxx	
many particles in the m.v. low S/N		xx	xxx		xxx	

Table: The performance of different processors under different conditions. The number of x marks gives the relative performance under the conditions given. It is assumed that the optical configuration is matched to the processor. The transition from a low S/N to a high is in the range 1-10. In cases where the intrinsic signal noise dominates, the S/N concept is not well defined. Here a low S/N means that the total number of detected photons during one transit is — say — lower than 100.

Appendix I

The conditional probability for the derivative of the phase $\dot{\varphi}(t)$ for a given amplitude A_0 and a narrowband process is evaluated in this section.

From ref. 4 the joint-probability for $\dot{\varphi}$ and A is obtained

$$p(\varphi,A) = \left(\frac{2\pi}{\Delta^2 \psi_0^3}\right)^{\frac{1}{2}} \frac{A^2}{2} \exp\left\{-\frac{1}{2}\left(\frac{1}{\psi_0} + \frac{\dot{\theta}^2}{\Delta^2 \psi_0} A^2\right)\right\} \qquad \text{[AI-1]}$$

The probability distribution for A is

$$p(A) = \frac{A}{\psi_0} \exp\left\{-\frac{A^2}{2\psi_0}\right\}, \text{ where} \qquad \text{[AI-2]}$$

φ_0 is the mean signal power and Δ the bandwidth of the process in radians/sec. A Gaussian-shaped power spectrum has been assumed, but this is not essential. Normalizing, so that $A/\psi_0^{\frac{1}{2}} = \alpha$ and $\dot{\varphi}/\Delta = \mu$, yields

$$p(\mu, \alpha) = (2\pi)^{-\frac{1}{2}} \alpha^2 \exp\left\{-\frac{1}{2}(1+\mu^2)\alpha^2\right\} \qquad \text{[AI-3]}$$

$$p(\alpha) = \alpha \exp\left\{-\frac{1}{2}\alpha^2\right\} \qquad \text{[AI-4]}$$

Applying Bayes' rule to equations [AI-3] and [AI-4], gives for the conditional probability

$$p(\mu|\alpha) = \frac{p(\mu, \alpha)}{p(\alpha)} = (2\pi)^{-\frac{1}{2}} \alpha \exp\left\{-\frac{1}{2}\mu^2 \alpha^2\right\} \qquad \text{[AI-5]}$$

which is a Gaussian distribution for μ with a variance $1/\alpha^2$

The conditional probability for μ, given that $\alpha > \alpha_0$, is

$$p(\mu|\alpha > \alpha_0) = \frac{\int_{\alpha_0}^{\infty} p(\mu, \alpha)d\alpha}{\int_{\alpha_0}^{\infty} p(\alpha)d\alpha} \qquad \text{[AI-6]}$$

The variance is

$$\text{var}\{\mu|\alpha > \alpha_0\} = \frac{\int_{-\infty}^{\infty} \mu^2 \int_{\alpha_0}^{\infty} p(\mu, \alpha)d\alpha}{\int_{\alpha_0}^{\infty} p(\alpha)d\alpha} \qquad \text{[AI-7]}$$

Substituting [AI-3] and [AI-4] into [AI-7] yields (after doing some arithmetic)

$$\text{var}\{\mu|\alpha > \alpha_0\} = \frac{\frac{1}{2}E_1(\frac{1}{2}\alpha_0^2)}{e^{-\frac{1}{2}\alpha_0}} \qquad \text{[AI-8]}$$

where $E_1(z)$ is the exponential integral (9) $\int_z^{\infty} \frac{e^{-t}}{t} dt$.

References

1. Lading, L.: A Fourier Optical Model for the Laser Doppler Velocimeter, *Opto-electronics,* **4**, 385-398 (1972)
2. Lading, L.: The Time-of-Flight Laser Anemometer, *AGARD Conference Proceedings No. 193,* paper No. 23
3. Drain, L.E.: Coherent and Non-Coherent Methods in Optical Beat Velocity Measurements, *J. Phys. D: Appl. Phys,* **5**, 481-495 (1972)
4. Rice, S.O.: Statistical Properties of a Sine Wave Plus Random Noise, *Bell Syst. Tech. J.* **27**, 109-157 (1948)
5. George, W.K. and Lumley, J.L.: The Laser Doppler Velocimeter and its Application to the Measurement of Turbulence, *J. Fluid Mech.,* **60**, 321-362 (1973)
6. Durrani, T.S. and Grated, C.A.: Statistical Analysis and Computer Simulation of Laser Doppler Velocimeter Systems, *IEEE Trans. Instr. and Measurements,* **IM-22**, 23-24, (1973)
7. Lading, L. and Edwards, R.V.: The Effect of Measurement Volume on Laser Doppler Anemometer Measurements as Measured on Simulated Signals, *Proceedings of the LDA-Symposium, Copenhagen,* 64-80, (1975)
8. Abramowitz, M. and Stegun, I.A. (ed): Handbook of Mathematical Functions, New York, Dover, 1972, pp. 228-231
9. Whalen, A.D.: Detection of Signals in Noise, New York, 1971, pp. 109-112
10. Papoulis, A.: Probability, Random Variables, and Stochastic Processes, New York, 1965, pp. 461-467
11. Epstein, B.E.: Some Applications of the Mellin Transforms in Statistics, *Ann. Math. Statist.,* **19**, 370-379 (1948)
12. Bendat, J.S. and Piersol, A.G.: Random Data: Analysis and Measurement Procedures, New York, Wiley, 1971, pp. 182-183
13. (Reference 9) chapter 10
14. Van Trees, H.L.: Detection, Estimation and Modulation Theory, New York, Wiley, 1968, chapter 4
15. Lading, L.: The Time-of-Flight Laser Anemometer versus the Laser Doppler Anemometer (to be published in the *Proceedings of the Third International Workshop on Laser Velocimetry, Purdue University,* July 11-13, 1978)
16. (Reference 9) chapters 5, 6, and 7
17. Gardner, F.M.: Phaselock Techniques, New York, Wiley, 1966, pp. 7-16 and p. 52
18. Edwards, R.V., Lading, L. and Coffield, F.: Design of Frequency Trackers for Laser Anemometer Measurements (to appear in the *Proceedings of the Fifth Biennial Symposium on Turbulence, Univ. of Missouri-Rolla,* Oct. 3-5, 1977
19. (Reference 10) p. 375
20. Wilmshurst, T.H. and Rizzo, J.E.: An Autodyne Frequency Tracker for Laser Doppler Anemometry, *J. Phys. E.: Sci. Instr.,* **7**, 924-930 (1974)
21. Fog, C.: A Photon-Statistical Correlator for LDA Applications, *Proceedings of the LDA-Symposium, Copenhagen,* 25-28, (1975).

LDA Signal Analysis

by

Aldo Coghe

C.N.P.M., C.N.R.-Politecnico di Milano, Italy

and

Umberto Ghezzi

Istituto di Macchine, Politecnico, Milano, Italy

Abstract

It is the purpose of this paper to review some fundamental contributions related to the problem of understanding the fundamental properties of the Doppler signal produced by a single scattering particle and to present typical results obtained by a computational procedure developed at the Politecnico di Milano. The analysis is based on Mie scattering theory for spherical particles and provides useful and reliable information in any experimental situation, without restricting to asymptotic cases. Only by accounting for this information, mainly the signal visibility and the S/N, the optimum optical configuration and the proper signal processing system can be selected. It is shown that some general criteria can be inferred in order to optimize the LDA system design, including the possibility of selecting particle size and type by optical means, in the presence of a polydisperse distribution of scattering particles.

1. Introduction

In recent years considerable effort has been made on the development of the laser Doppler anemometer (LDA), for which several different optical configurations and theoretical models have been reported (Durst et al. 1976; Pike & Cummins 1976; Durrani & Greated 1977; Lading 1972; George & Lumley 1973). Despite the numerous useful velocity information that has been obtained with LDA systems, there are still many uncertainties in the estimate of the Doppler signal dependence from the geometrical and scattering parameters, resulting mainly from lack of adequate experimental data and from oversimplifications in the theoretical models.

The correct choice of the LDA signal processing system requires a considerable progress in understanding the fundamental properties of the Doppler signal produced by scattering particles. In this work we shall attempt to present a rigorous analysis of the Doppler signal generated by an individual particle with a differential mode LDA. We take as its starting point the Mie theory for scattering by spherical particles.

We shall first briefly discuss some background on the scattering of light by small spherical particles and on the differential mode LDA.

2. Differential Laser Doppler Anemometer

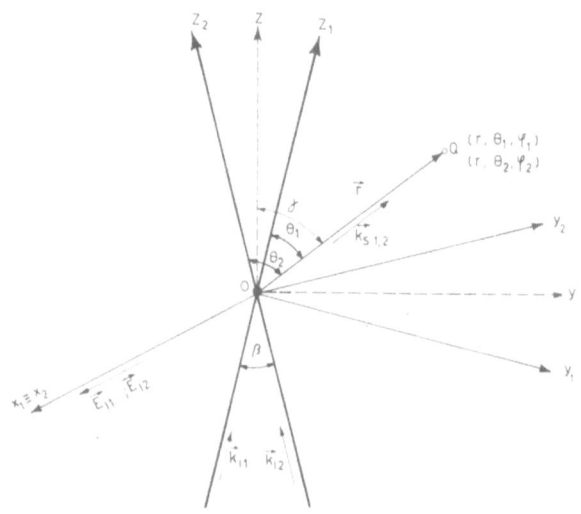

Fig. 1. Geometry of the LDA system. Origin of all coordinate systems is chosen at the center of the scattering volume.

The typical differential laser Doppler anemometer (DLDA) employs two plane polarized and coherent beams which are focused to a common point. The receiving optics consists of a converging lens which images the cross-beam region on the photodetector. The geometry of a DLDA system is schematically shown in Fig. 1.

The main advantage of the DLDA is that it is quite simple to align and is not sensitive to small vibrations. Also, the frequency is independent of the detection angle γ. This is an important feature since it allows a large collection angle to be used without a spread of Doppler frequency.

A picture which is often used to describe this system is the so-called "fringe model" that involves visualizing a set of interference fringes produced by the two incident beams. While the "fringe model" has a useful value in many practical situations, the veracity of this model has been questioned. In fact, we must remember that scattering is a linear process and that no fringes exist in reality in the cross-beam region. The Doppler signal is produced by light scattered from particles that cross the intersection of the illuminating beams when the scattered light, collected through a receiving aperture, reaches the photodetector.

In many practical situations (high speed flows, reactive flows, etc.) the Doppler signal that is available for processing is produced by no more than one particle in the cross-beam region at a time. Discrete signal processors accept individual signals and determine the correct Doppler frequency and hence the velocity. Proper use of the instrument would require rejection of multiple particle signals because of random phase fluctuations which will lead to incorrect velocity measurements. Further information could also be obtained by individual realization DLDA: particle size and concentration. Individual realization anemometers do not provide a continuous record of the velocity. Moreover, since the particles cross randomly the probe volume with an approximately Poisson distribution of time instants, they provide a randomly sampled time series of the fluctuating velocity. Time resolution could be limited by the above mentioned characteristic, but it has been shown that random sampling with a Poisson distribution eliminates aliasing (Shapiro & Silverman 1960) and enables spectral estimates to be obtained at frequencies up to many times the mean sampling rate (Gaster & Roberts 1977).

The rate of particle arrival at the probe volume can be dependent on the velocity field and this phenomenon has serious implications on the types of averaging that can be used (McLaughlin & Tiederman 1973; Kreid 1974). It has been shown (George 1975) that unbiased velocity averages are available from discrete LDA. The signal processing, however, must keep track of the particle residence time. This can be accomplished with commercial counter processors. Note that individual Doppler signals can be easily obtained when seeding with a suitable concentration is used and a high spatial resolution is required. In the following sections we will show that the properties of individual Doppler signals can be accurately predicted on the basis of Mie scattering theory.

3. Theory

Consider two linearly polarized illuminating beams with the same direction of polarization and directions of propagation forming an angle β, whose electric vectors are denoted by $\underline{E}_{i1}$ and $\underline{E}_{i2}$ (see Fig. 1). The two beams are focused to a common intersection where they are plane parallel. The electric vector of the

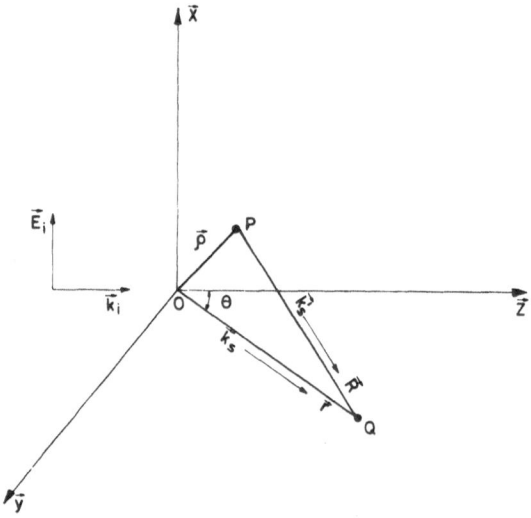

Fig. 2. Coordinates used in the scattering model.

ζ-th incident wave at a point ρ is given by

$$\underline{E}_{i\zeta}(\underline{\rho},t) = \underline{E}_{0\zeta}(\underline{\rho})e^{\,j(\omega t - \underline{k}_{i\zeta}\cdot\underline{\rho} - \Phi_{\zeta})} \;;\; (\zeta = 1,2) \tag{3.1}$$

where $\omega = 2\pi\nu$ is the circular frequency, $\underline{k}_i$ and $\underline{\rho}$ are defined by Fig. 2, λ is the wavelength in the medium, $|\underline{k}_i| = 2\pi/\lambda$ and Φ_{ζ} is the phase of the wave at the origin of the coordinate system.

In the "far field" approximation, the waves scattered by a particle P from each incident beam may be treated as spherical waves emitted by the geometric center of the beam intersection. The radial components of the field vectors fall off as $(\lambda/r)^2$ and, hence, they may be neglected in the "far field" zone. In the spherical coordinates defined by Fig. 3 the other two components of the two scattered fields are given by

$$\underline{E}_{\phi\zeta}(\underline{r},t) = -j\,\frac{E_{0\zeta}(\underline{\rho})\,S_{1\zeta}(\theta_{\zeta})}{k\,r}\,e^{\,j(\omega t - kr + \underline{K}_{\zeta}\cdot\underline{\rho} - \Phi_{\zeta} + \chi_{\zeta})}\,\sin\phi_{\zeta}\,\hat{e}_{\Phi\zeta} \tag{3.2}$$

$$\underline{E}_{\theta\zeta}(\underline{r},t) = j\,\frac{E_{0\zeta}(\underline{\rho})\,S_{2\zeta}(\theta_{\zeta})}{k\,r}\,e^{\,j(\omega t - kr + \underline{K}_{\zeta}\cdot\underline{\rho} - \Phi_{\zeta} + \psi_{\zeta})}\,\cos\phi_{\zeta}\,\hat{e}_{\theta\zeta} \tag{3.3}$$

where

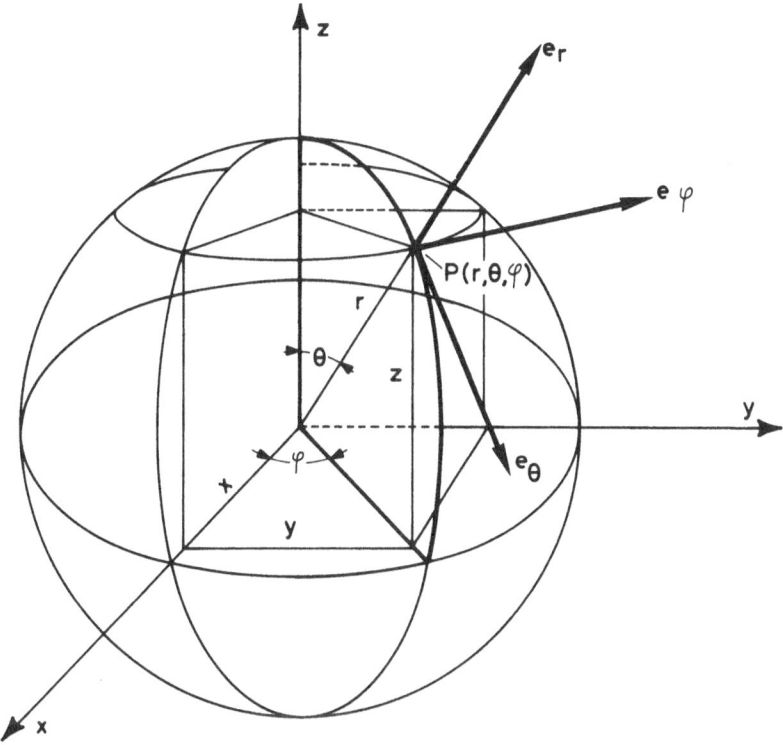

Fig. 3. Spherical coordinates used in Mie scattering theory.

$$\chi_\varsigma(\theta_\varsigma) = \text{arctg}\left[\frac{\text{Im}(S_1)}{\text{Re}(S_1)}\right] \quad \text{and} \quad \psi_\varsigma(\theta_\varsigma) = \text{arctg}\left[\frac{\text{Im}(S_2)}{\text{Re}(S_2)}\right] \tag{3.4}$$

$k=|\underline{k}_s| \cong |\underline{k}_i|$; $\underline{K} = \underline{k}_s - \underline{k}_i$ and $\underline{K}_\varsigma \cdot \underline{\rho}$ gives the instantaneous phase of the field scattered from a single particle due to its position ρ relative to the coordinate origin 0. $S_1(\theta)$ and $S_2(\theta)$ denote the amplitude functions of the particle and are dimensionless factors. They are complex quantities and determine the phase of the scattered wave as well as its amplitude and polarization. They depend on two scattering parameters, $\alpha = \pi d/\lambda$, where d is the particle diameter and $m = m_R - im_I$, the complex refractive index (Born & Wolf 1975). The complex amplitudes depend also upon the scattering angle θ (see Fig. 1). The exact expressions for S_1 and S_2 are given by Mie theory (Van de Hulst 1957; Kerker 1969) and consist of a series of Legendre polynomials which require to be computed by numerical procedures, with "double precision" routines to avoid unacceptable truncation errors.

The total scattered field from each beam may be written

$$\underline{E}_{s\varsigma}(\underline{r},t) = j \frac{E_{0\varsigma}}{kr} \underline{S}(\theta_\varsigma, \phi_\varsigma) e^{j(\omega t - kr + \underline{K}_\varsigma \cdot \underline{\rho} - \Phi_\varsigma)}$$ [3.5]

where

$$\underline{S}(\theta, \phi) = S_1(\theta)\sin\phi \ \hat{e}_\phi + S_2(\theta)\cos\phi \ \hat{e}_\theta$$ [3.6]

At any point $Q(r, \theta, \phi)$ on the photodetector surface the electric vector of the scattered light field is the sum of the two scattered wave electric vectors

$$\underline{E}_T = \underline{E}_{s1} + \underline{E}_{s2}$$ [3.7]

Averaging over a time interval much larger than the reciprocal of the wave frequency yields the intensity

$$I(r,\theta,\phi) = \frac{1}{2T} \int_0^T Re(\underline{E}_T \cdot \underline{E}_T^*) \ dt$$ [3.8]

The integrated intensity i.e. the power over the collection solid angle Ω is

$$I_\Omega = \int_\Omega r^2 \ I(r,\theta,\phi) \ d\Omega$$ [3.9]

Under the assumption that all of the energy passing through the receiving lens impinges upon the photodetector, the photocurrent generated by this field is given by

$$i = \eta I_\Omega$$ [3.10]

where η is the sensitivity of the photodetector which is usually a function of the wavelength of the incident radiation. Lenses can generally be regarded as thin, in which case their only effect is to cause a phase delay in the wavefront proportional to the thickness of the lens at the position where the ray enters.

Taking into account the above relation, the general expression of the photocurrent generated by a single particle can be written

$$i = \frac{\eta}{k} \left\{ I_{01} P_1(\theta,\phi) + I_{02} P_2(\theta,\phi) + (I_{01} I_{02})^{1/2} \times D(\theta,\phi)\cos(\underline{K} \cdot \underline{\rho} - \Psi(\theta,\phi)) \right\}$$ [3.11]

where $\underline{K} = (\underline{k}_{i1} - \underline{k}_{i2})$; P_1, P_2, D and Ψ are integral quantities defined in terms of the complex amplitudes given by the Mie scattering theory (cf. Adrian & Early 1976); I_{01} and I_{02} denote the intensities of illuminating beams. The term Ψ determines the phase of the scattered intensity, due to the scattering process; the term $\underline{K}\cdot\underline{\rho}$ denotes the phase difference between the two incident waves due to the position $\underline{\rho}$ of the scattering particle.

A moving scatterer generates a time-varying scattered intensity as it crosses the beam intersection region. In fact, for a uniform velocity $\underline{U}$ in the $\underline{y}$ direction of Fig. 1, the position $\underline{\rho}$ may be defined as

$$\underline{\rho} = \underline{\rho}_0 + \underline{U} t \tag{3.12}$$

If we assume $\underline{\rho}_0 = 0$, the phase term $\underline{K}\cdot\underline{\rho}$ can be rewritten

$$\underline{K}\cdot\underline{\rho} = (\underline{k}_{i1} - \underline{k}_{i2})\cdot\underline{U} t = \omega_D t \tag{3.13}$$

where

$$\omega_D = 2 k U \sin(\beta/2) = 2\pi U/\delta \tag{3.14}$$

is the frequency of the Doppler signal, and

$$\delta = \lambda/(2\sin(\beta/2)) \tag{3.15}$$

indicates the spacing between fringes.

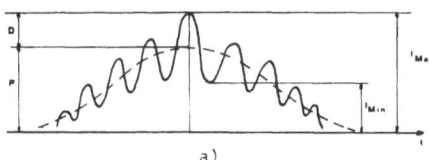

a)

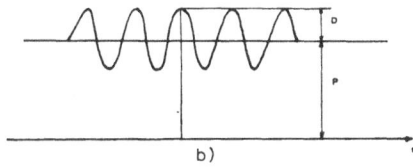

b)

Fig. 4. Doppler signal generated by a single moving particle. (a) Gaussian cross-section beams; (b) present model assumption.

Substitution of [3.13] into [3.11] yields

$$i(t) = \frac{\eta I_0}{k} \left\{ P(\theta,\phi) + D(\theta,\phi)\cos(\omega_D t - \Psi(\theta,\phi)) \right\} \tag{3.16}$$

where $P=P_1+P_2$ and we assumed $I_{01}=I_{02}=I_0$. The term P represents the "pedestal" amplitude of the current, while D represents the Doppler amplitude, as shown in Fig. 4. The usual shape of the Doppler signal of Fig. 4a is determined by the Gaussian cross-section of laser beams in the TEM_{00} mode.

It should be noted that Mie scattering theory is valid strictly for plane incident waves having constant intensity. Consequently, it is necessary to assume that the intensity of laser beams does not vary appreciably over the particle surface. In the present model a constant intensity over the beam intersection region was assumed, hence the Doppler signal has a constant "pedestal" amplitude as in Fig. 4b. Results are expected to be reasonably accurate for the peak portion of a Doppler signal generated by a particle crossing the geometric center of the probe volume. The same assumption was made by Adrian & Orloff (1977).

The computational procedure that we used needs as input parameters: m, d, r, $\alpha = \pi d/\lambda$, the polarization direction of incident waves and the cross angle β. The axis of the collecting aperture can be chosen at any angle γ with respect to the forward direction, i.e. the bisector of the two incident beams in Fig. 1. The first step is the evaluation of the complex amplitude functions $S_{1\zeta}$ and $S_{2\zeta}$ at each grid point on the collecting aperture, by means of a Mie scattering routine. In the second step, the total scattered field is determined at each point by taking into account polarizations and amplitudes of the two scattered waves. Finally, P,D and Ψ of [3.16] are evaluated by numerical integration over a specified solid angle Ω. Full details of the computational procedure are reported by Coghe et al. (1978).

4. Doppler Signal Visibility and S/N

The quantities P,D and Ψ depend only on the scattering properties of the particle and the geometry of the LDA system, whereas the photocurrent is a time-varying function of the particle position. Indeed a moving fringe patterns is seen by the detector due to two beam interference and particle motion. Following Michelson (Born & Wolf 1975) we take as measure of the contrast of the fringes, hence, of the quality of the Doppler signal of Fig. 4 the visibility V defined by

$$V = \frac{i_{max}-i_{min}}{i_{max}+i_{min}} = \frac{D}{P} \qquad\qquad [3.17]$$

where i_{max} and i_{min} correspond to the maximum and minimum intensities of the peak portion of the Doppler signal.

In general, V is a function of the illuminating beam parameters, $I_{0\xi}$ and $\underline{k}_{i\xi}$, the scattering parameter α, the refractive index m, the parameter d/δ, the detection angle γ and the collecting solid angle Ω. In addition, V depends on the position $\underline{\rho}$ of the scattering particle, but in this work the particle is supposed as lying at the center of the beam intersection, where $I_{01} = I_{02}$. Then V has a functional form (Adrian & Orloff 1977)

$$V = V(\pi d/\lambda, m, d/\delta, \Omega) \qquad [3.18]$$

The exact dependence of V on the above parameters can be numerically obtained only by models based on Mie scattering theory. The first derivation of the visibility function was made by Farmer (1972). He obtained a first-order Bessel function of ($\pi d/\delta$), under the assumption of paraxial scattering, which implied that scattering properties of the particles were ignored. Robinson & Chu (1975) used scalar diffraction theory to determine the forward scattered fields and showed the aperture size effect, confirmed by experimental measurements. But their analysis was only valid for particle diameter much larger than the wavelength and for forward scatter. More recently Durst & Eliasson (1975), Chu & Robinson (1976) and Adrian & Earley (1976) developed computer programs based on Mie scattering theory, and reported experimental results of the visibility in comparison with the theoretical predictions.

Visibility is an important parameter because its computed values can be easily compared with experimental ones; moreover, it has been shown (Adrian & Earley 1976) that the S/N is directly proportional to the ratio $D^2/P = V \, D$. In fact, if the photodetector shot noise is the only significant noise source in the Doppler signal, the signal-to-noise ratio is given by

$$S/N = C \frac{\langle I_{01} I_{02} \rangle}{\langle I_{01} \rangle} \frac{D^2}{P} \qquad [3.19]$$

where it is assumed that $\langle I_{01} \rangle = \langle I_{02} \rangle$, and the constant C depends on the quantum efficiency of the photodetector, the wavelength of the laser light and the noise band-width of the signal processing electronics.

Lading (1973) derived a similar expression for an ensemble of scattering particles

$$S/N = C' \frac{S}{r^2} I_0 \frac{\langle C_{sc}^2 \rangle}{\langle C_{sc} \rangle} \qquad [3.20]$$

where S is the receiver aperture area and $\langle C_{sc} \rangle$ is the ensemble average of the scattering cross-sections of the particles. When the index of refraction is real and there is no absorption, C_{sc} is equivalent to the extinction cross-section C_{ext} and is related to amplitude functions (Van de Hulst 1957) evaluated at zero scattering angle

$$|S_1(0°)|^2 = |S_2(0°)|^2 = (\pi/\lambda^2) C_{ext} = (\pi/\lambda^2) C_{sc} \qquad [3.21]$$

For an ensemble of scattering particles, if we assume that the LDA system is operating with incoherent scattering, the total photocurrent is simply the sum of the currents generated by each individual scatterer and C_{sc} is a random variable. In this case the relation [3.20] can be very useful, but, it is equivalent to the [3.19] only for small values of the scattering angle.

In theoretical models of LDA it is common practice to make assumption [3.21] (paraxial approximation). According to this hypothesis, Durrani & Greated (1977) obtained for the general expression of the photocurrent generated by a single moving particle, traveling with the uniform velocity U across the probe volume,

$$i(t) = A W(t) (1 + \cos\omega_D t) \qquad [3.22]$$

where A is a time-independent term which denotes the amplitude of photocurrent at the center of the observation volume. It varies with the scattering cross-section C_{sc} and depends on the optical geometry of the LDA, laser power and efficiency of the detector. The quantity W(t) is a low-frequency term which depends on the Gaussian intensity profile of the two laser beams within the intersection region and on the particle velocity. The cosine term defines the high-frequency component of the current produced by particle motion. There is a close similarity between the signal of [3.22] and classical noise signal. Indeed, the statistics of the LDA photocurrent has been determined starting from this expression (George & Lumley 1973; Durrani & Greated 1977).

However, the assumption [3.21] is probably valid only for particles less than 0.1 μm diameter and/or for small scattering angles. For large particles the scattering cross-section exhibits a rapid increase with d (see Fig. 5), becoming equal to twice the geometric cross-section $\pi(d/2)^2$, in the geometric optics limit $\alpha \gg 1$. At the same time, amplitude functions exhibit a marked and rapidly fluctuating dependence on the scattering angle as it is shown in Figs. 6, 7 and 8 for three different particle diameters. For d $\geqslant$ 5 μm we can note a sharp decrease of the scattered amplitudes up to two orders of magnitude or more in the range 0° - 5° of the scattering angle. In the same range, dielectric (m_I=0) and absorbing (m_I=1) particles denote a similar structure. In back-

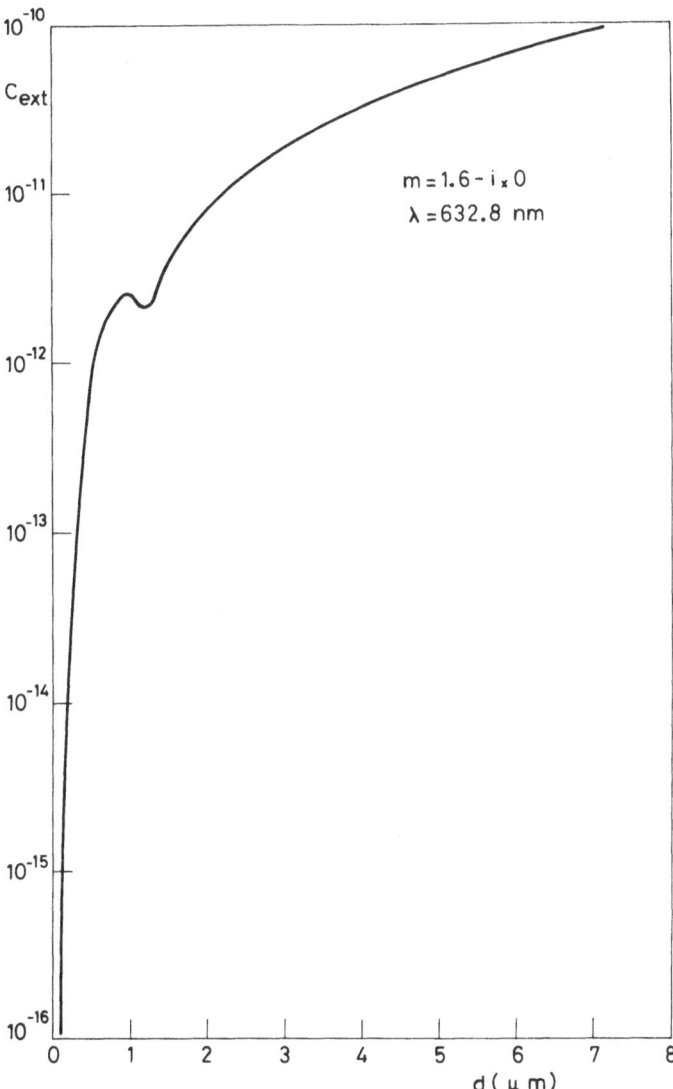

Fig. 5. Scattering cross-section for non absorbing spheres.

scatter, however, absorbing particles show a nearly constant value of amplitude fucntions over a wide range of θ, and hence, in this case the assumption
[3.21] could be valid. It should be noted that results of Figs. 5 to 8 have been
obtained by the computational procedure presented in this work and based on
Mie scattering theory.

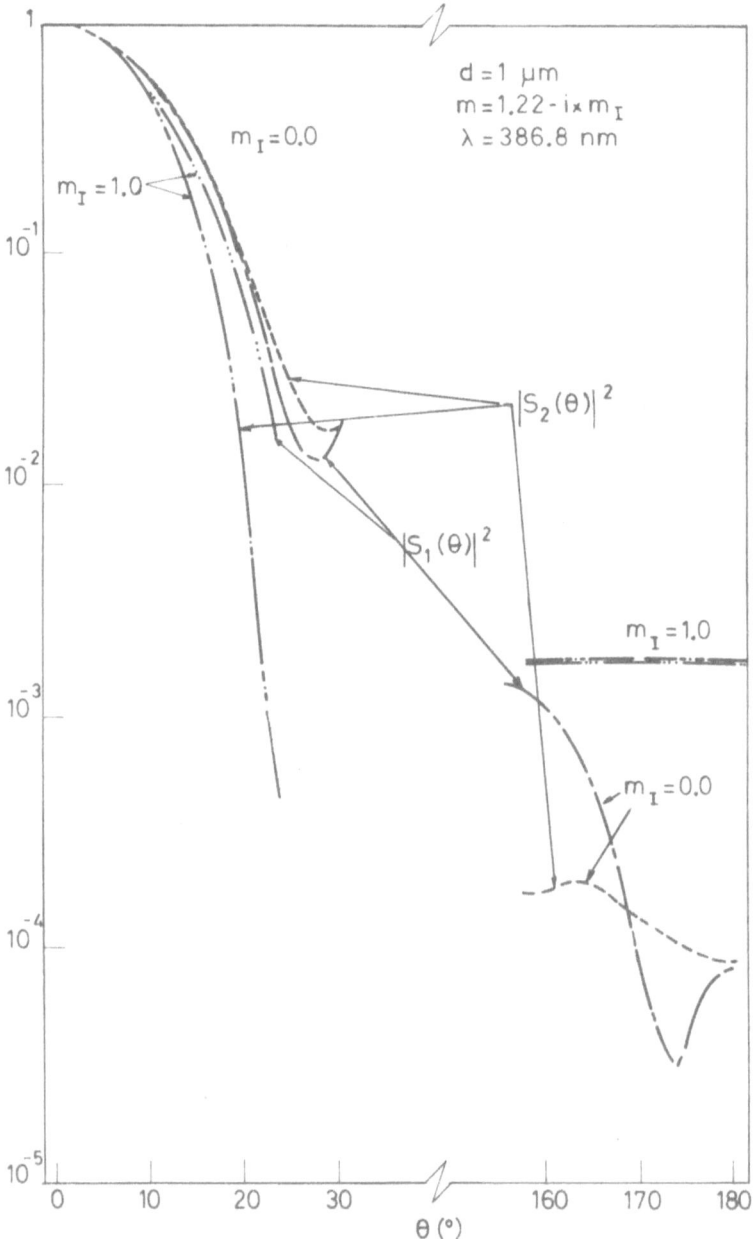

Fig. 6. Amplitude functions, $|S_1(\theta)|^2$ and $|S_2(\theta)|^2$ for absorbing ($m_1=1.$) and non absorbing ($m_1=0.$) particles. θ is the scattering angle in Fig. 2.

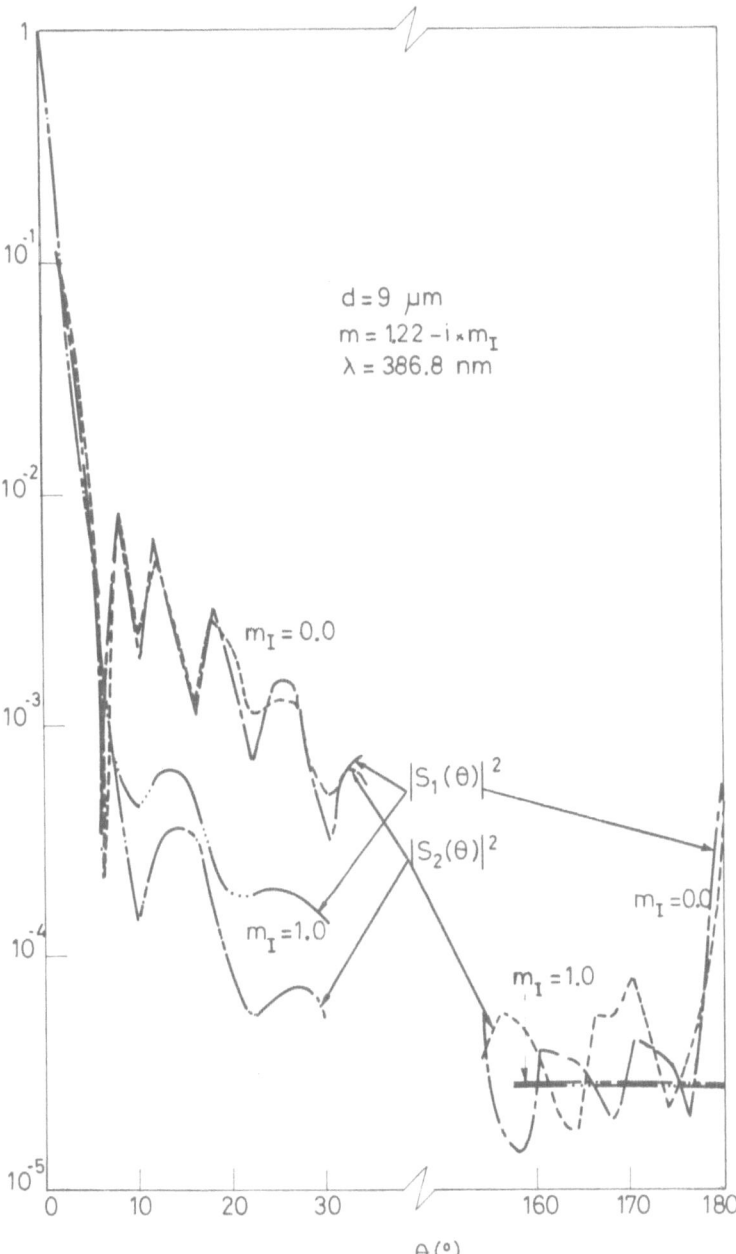

Fig. 7. Amplitude functions, $|S_1(\theta)|^2$ and $|S_2(\theta)|^2$ for absorbing ($m_1=1$.) and non absorbing ($m_1=0$.) particles. θ is the scattering angle in Fig. 2.

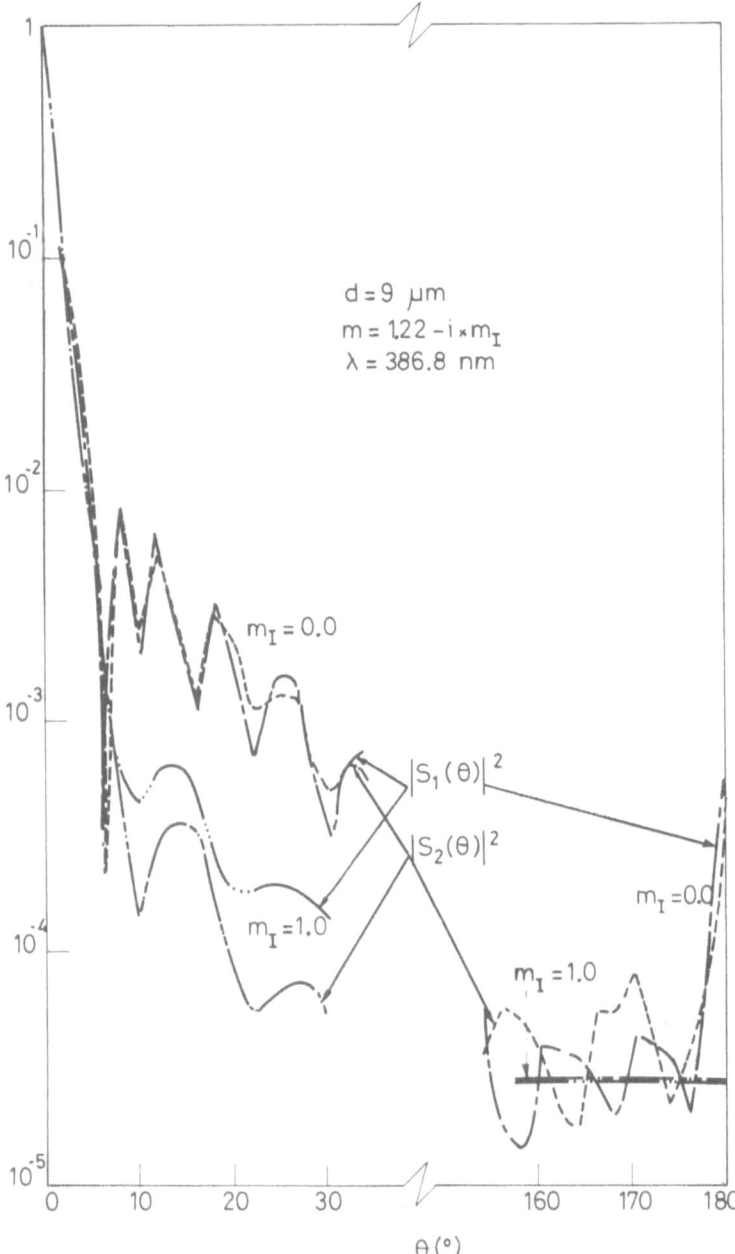

Fig. 8. Amplitude functions, $|S_1(\theta)|^2$ and $|S_2(\theta)|^2$ for absorbing $(m_1=1.)$ and non absorbing $(m_1=0.)$ particles. θ is the scattering angle in Fig. 2.

It can be inferred from the above results that in a DLDA system with large β and Ω angles at any point on the photodetector surface it will be, in general

$$|S(\theta_1, \phi_1)|^2 \neq |S(\theta_2, \phi_2)|^2 \qquad\qquad [3.23]$$

where (θ_1, ϕ_1) and (θ_2, ϕ_2) are the angular coordinates of the same point in the two coordinate systems of Fig. 1. As a consequence, $D \leqslant P$, from equations [3.6] and [3.11], hence, fringe contrast can be reduced also if $I_{01} = I_{02}$. Fig. 9 shows typical visibility curves vs particle diameter for two collecting solid angles and two cross-beam angles. They refer to non absorbing particles and detection angle $\gamma = 5°$ with respect to the forward direction. It should be noted that the visibility exhibits a marked dependence on Ω and β for $d > 2$ μm. These results are in contrast with assumption [3.21]. In fact, a comparison of [3.22] with [3.16] shows clearly that in the former expression of the photo-current the signal visibility is always equal to unity. In general this is not the case and we must remember that the S/N of Doppler signal is strongly related to the visibility function.

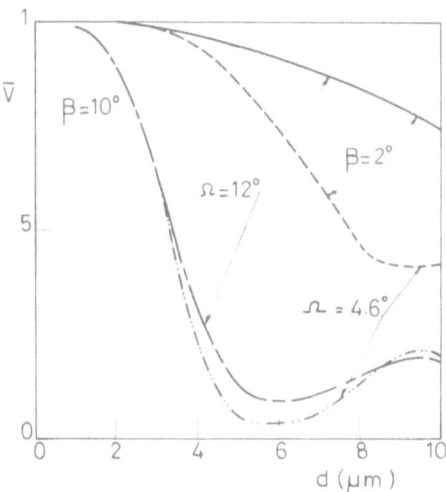

Fig. 9. *Theoretical visibility V vs particle diam. for two collection solid angles Ω and two cross-beam angles β. $m = 1.6-ix0.$, $\lambda = 632.8$ nm and detection angle $\gamma = 5°$.*

In practical applications of LDA the situation is complicated because the seeding particles are seldom monodisperse and solid particulates are commonly of irregular shape, so it is difficult to make any precise prediction. However, useful information can be obtained by theoretical models based on Mie scattering theory.

5. Numerical Results

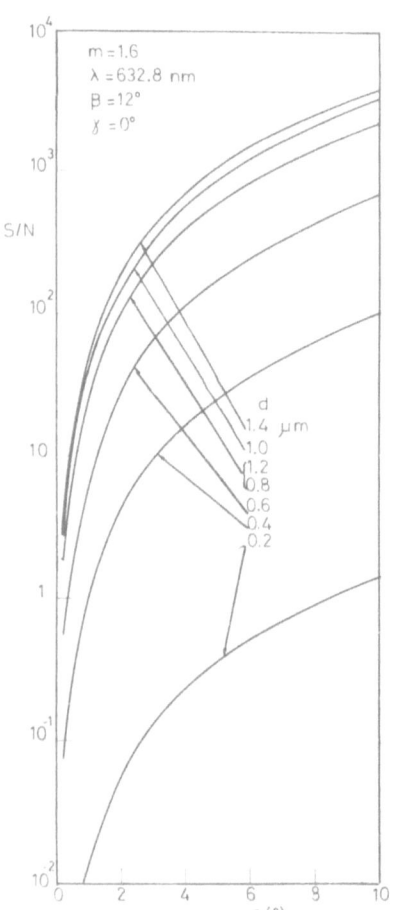

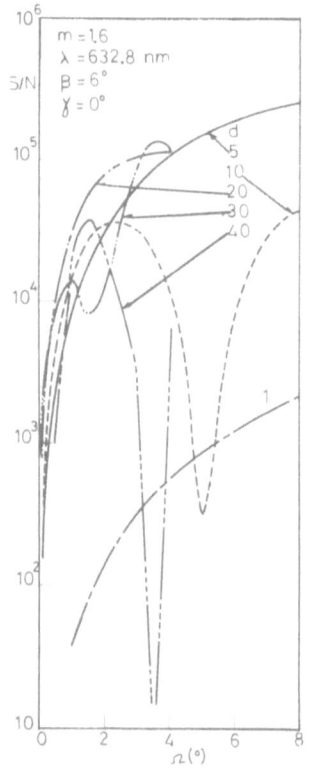

Fig. 10. S/N vs collection angle computed from present model for various particle diam. d. Forward scattering

Fig. 11. S/N vs collection angle computed from present model for various particle diam. d. Forward scattering.

All the numerical results reported in this section refer to the S/N of an individual Doppler signal and have been computed from equation [3.19]. For simplicity we assumed $I_{01} = I_{02} = 1$ and C = 1. In practical situations it will be possible to evaluate the exact value of C or to determine it by calibrations. It should be noted that the functional form of the S/N is similar to that of the visibility, so we analyzed the dependence on m, d, β, γ and Ω.

Figs. 10 and 11 show typical S/N curves vs collecting solid angle for various particle diameters in forward scattering. It is useful to note the rapid increase of the S/N with d for diameters less than 1 μm at any Ω. For d approaching unity and increasingly for d>1, the S/N increases in an irregular way and exhibits maxima and minima. Results of calculations based on the present model are different from those, for example, of Lading (1973) obtained in the assumption that the particles have scattering cross-sections independent on the scattering angle. Consequently, it appears that the simplified models based on assumption [3.21] are not always adequate, also for small collecting apertures. However, for small particles the S/N increases monotonically with collecting apertures as predicted by [3.20]. Experimental results reported by Lading (1973) are not in contrast with present results, but probably they show only one of the possible behaviour of the S/N.

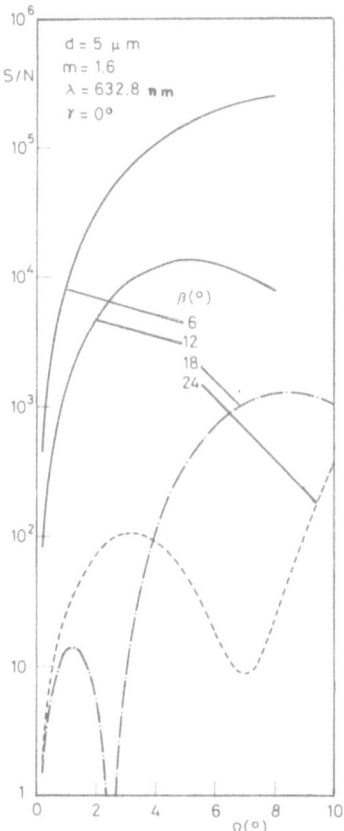

Fig. 12. S/N vs collection angle computed from present model for various cross angles. Forward scattering.

Fig. 12 shows the influence of the cross-beam angle on the S/N for one particle diameter, d = 5 μm. Again, the S/N exhibits maxima and minima when the
angle β increases and, hence, radiation scattered at large θ angles is collected.
The behaviour of the S/N curves of the above figures is very similar to the
theoretical results reported by Adrian & Earley (1975) for a symmetric backscatter LDA system.

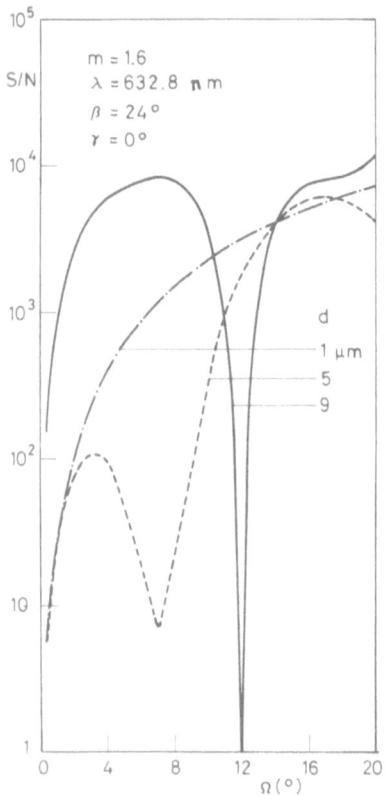

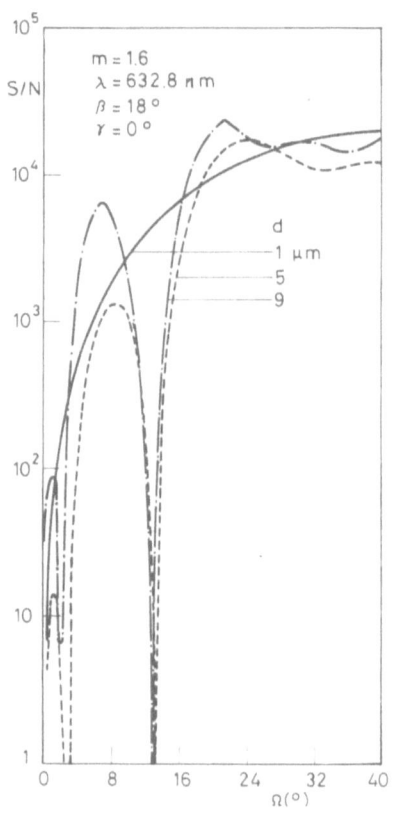

Fig. 13. The effect of large collection angles on S/N for three particle diam. Forward scattering.

Fig. 14. The effect of large collection angles on S/N for three particle diam. Forward scattering.

Figs. 13 and 14 show the effect of very large collecting apertures for two
values of β and three particle diameters in forward scattering. It can be seen
that at large Ω the S/N curves exhibit an asymptotic behaviour for all the
considered particle sizes. In fact, as Ω increases, the infinite aperture limit
(Ω → 2π) is sufficiently approximated. It should be noted that we did not consider the influence of a beam stop or of the laser beam noise when one of the
undeflected laser beams falls into the receiver aperture.

For many LDA systems it is often impractical and unnecessary to use so large an aperture that the condition $\Omega \to 2\pi$ is valid. In this case it can be seen from the above reults that, generally, the aperture of the receiving optics is very critical with respect to the S/N for particles larger than about 1 μm. Note also that the assertion that with the DLDA the S/N can be increased by having a large collecting solid angle is not always satisfied. Although this is the general trend, large oscillations in the S/N curves become dominant at sufficiently large particle sizes.

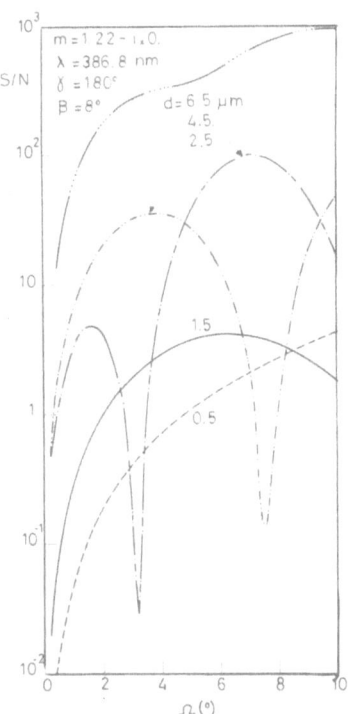
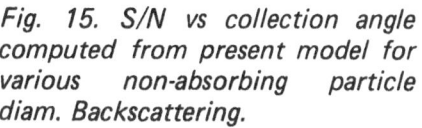
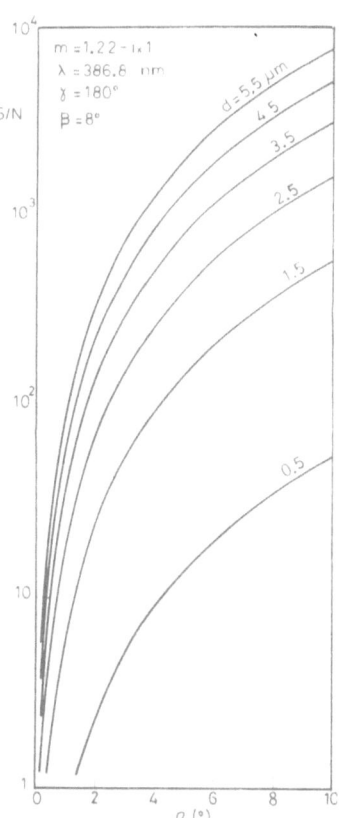

Fig. 15. S/N vs collection angle computed from present model for various non-absorbing particle diam. Backscattering.

Fig. 16. S/N vs collection angle computed from present model for various absorbing particle diam. Backscattering.

As far as the difference between absorbing and non absorbing particles is concerned, some examples are reported in Figs. 15 and 16. Two limiting cases were analyzed, referring to a backscattering geometry as suggested by curves of Figs. 6, 7 and 8. Firstly, (Fig. 15) we assumed m_R = 1.22 and m_I = 0, $\beta = 8°$, λ = 386.8 nm and S/N curves were calculated vs Ω for different particle sizes. In the second case (Fig. 16), we changed the imaginary part of the refractive index and assumed m_I = 1. Numerical results show that the S/N curves have a more regular shape for absorbing particles, and in this case they increase with the collecting angle. This is expected since the C_{sc} is independent of the scattering angle (see Figs. 6, 7 and 8).

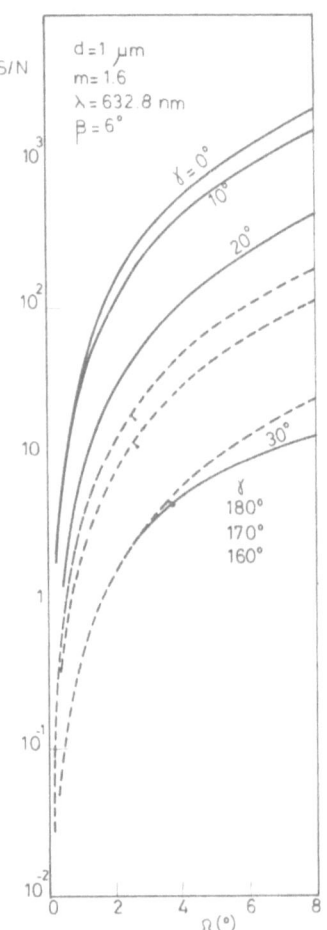

Fig. 17. The effect of detection angle on the S/N in forward and backscattering. d = 1 μm.

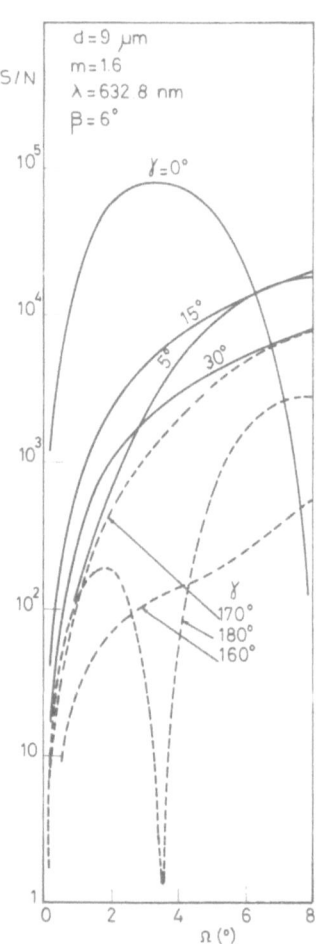

Fig. 18. The effect of detection angle on the S/N in forward and backscattering. d = 9 μm.

Investigation of the effect of aperture location has been extended to various detection angles in the forward and backscattering geometry. From the results reported in Figs. 17 and 18 it can be seen that another assertion generally accepted is not completely valid. Indeed, the optimum S/N is not always obtained when the direction of detection is along the bisector of the two incident beams. It would seem that the S/N dependence on detection angle is complicated because of the simultaneous dependence on Ω and d. As long as the particle size is small, the S/N is not strongly unfluenced by the detection angle (up to about 20°), but for large particles the forward detection set-up seems to be better for small Ω, but this could not be the case when the collecting aperture is increased. More difficult to explain are the results obtained in the same conditions but with a backscattering detection. All results refer to non absorbing polystyrene spheres and cross-beam angle $\beta = 6°$.

6. Experimental Results

Theoretical signal visibility results can be easily compared with the experimental data. We performed experimental investigations of the relationship between the visibility function and the geometric parameters of a DLDA. The optical arrangement, the experimental procedure and preliminary results were reported by Ghezzi et al. (1978). Here we note that the usual optical configuration of the DLDA was employed and that Doppler signals of individual particles crossing the center of the probe volume were selected visually by means of a storage oscilloscope. Measurements were made on an air flow seeded with polystyrene particles, 5.7 μm mean diameter and 26 % standard deviation, at a suitable concentration. The visibility of each accepted signal was determined by relation [3.17]. In Fig. 19 we reported experimental results referring to four series of measurements, each at a given solid collecting angle in the forward scattering. All other geometric parameters were maintained constant. Experimental points represent mean values over several hundreds of particles.

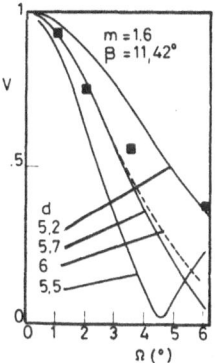

Fig. 19. Comparison of theoretical and experimental results of signal visibility vs Ω for polystyrene spheres, 5.7 μm diam. Forward scattering.

Comparison with the theoretical curve for d = 5.7 μm shows good agreement, except at the point corresponding to Ω = 6°, where the mean of experimental results seems to correspond to d = 5.2 μm. A totally satisfactory explanation of this discrepancy has not been found, but the theoretical curves of Fig. 20 could explain it. In fact, the visibility function shows a strong oscillatory behaviour that corresponds closely to the similar behaviour denoted by the single scattering cross-section C_{sc} for non absorbing particles (Kerker 1969). This large scale amplitude oscillation is caused by anomalous diffraction (Van de Hulst 1957) and was also found by Chu & Robinson (1977). When α is sufficiently large, the geometrical optical rays incident upon a scattering sphere do undergo a significant phase shift in passing through, because of the long path length through the large sphere. This leads to an interference between the original field and the forward scattered field. In order to resolve this large-scale amplitude oscillation we have computed the visibility function vs d in increments of 0.1 μm. Superimposed there would be a ripple structure which becomes increasingly irregular at higher refractive indices, but to observe it we need increments of 0.01 μm.

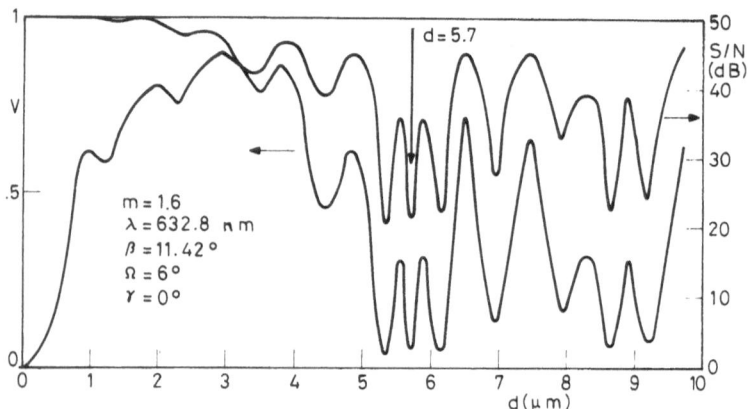

Fig. 20. Theoretical visibility and S/N (dB) curves vs particle diam. for poly-styrene spheres. Forward scattering.

The S/N curve of Fig. 20 exhibits a similar behaviour, as was already shown by Durst & Eliasson (1975). From observation of this figure, it can be noted that a large oscillation characterizes the diameter range around 5.7 μm, and, hence, experimental imprecisions could be justified. A better agreement of theoretical and experimental results of signal visibility was found by changing the parameter d/δ, i.e. the cross-beam angle. Results reported in Fig. 21 refer to other four series of measurements that were made with a detection angle γ = 5° with respect to the forward direction.

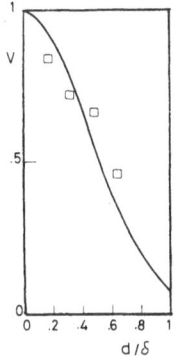

Fig. 21. Comparison of theoretical and experimental results of signal visibility vs the parameter d/δ for polystyrene spheres, 5.7 μm diam. Detection angle γ = 5°.

More difficult is to obtain reliable measurements of the S/N. Also in the literature a few results were reported. We attempted to compare our theoretical model with experimental results reported by Ballik & Chan (1977) that show the effect of detection angle for two apertures and low particle concentrations (Fig. 22). It can be seen that the behaviour of the theoretical results obtained by the present model shows some agreement, although we did not consider the volume correction and the influence of a beam stop as made by the authors. Also their theoretical model is not adequate and the discrepancy for the large (f/2.6) aperture case is not completely understood.

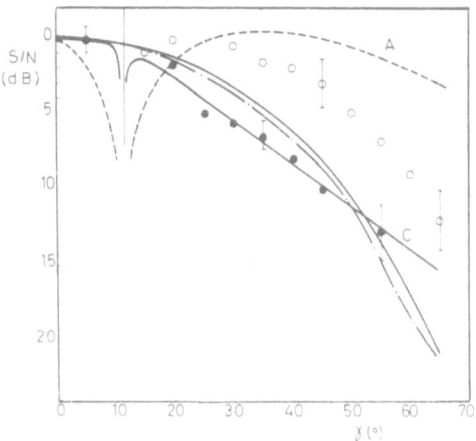

Fig. 22. Comparison of S/N results of the present model (—— f/2.6, —·— f/22) and theoretical (A,C) and experimental results of Ballik & Chan (1977).

7. Conclusions

Application of Mie scattering theory to the analysis of Doppler signals shows that the inherent characteristics of the signal produced by an individual spherical particle can be accurately predicted. However, in view of the complex diversity of effects which influence the S/N of Doppler signals, together with its sensitivity to modest changes of the optical system parameters, it is only possible to infer some general criteria, but it would be better to define the optimum particle size and system configuration by means of numerical evaluations for any specific case.

The present analysis shows that by varying the cross angle of the incident beams, the size of the receiving aperture and its angular position, together with the sensitivity of the electronic signal processing system, the LDA instrument could be optimized so that only a narrow range of particle size and/or type (absorbing or non absorbing) is accepted for signal processing and velocity measurements. This procedure, based on correlations of particle size with the signal magnitude, the S/N and the visibility, could be very useful in flow systems with polydisperse particle distributions, in two phase flows and, more generally, in situations in which we are forced to use flows with natural seeding particles of different sizes and compositions.

As a final result, parallel measurements of particle sizes and velocities could be performed.

Theoretical visibility functions obtained by the present model are in excellent agreement with experimental data reported in this work and in the literature. Further researches involving visibility and S/N measurements in practical experimental situations will aid in testing theoretical models and developing more sophisticated LDA signal processing systems conceived to extract all possible information or to discriminate in favour of some specific demand.

Acknowledgements

The authors wish to thank Dr. V. Bergamaschi, Dr. G.L. Sensalari and Dr. A. Vismara for their assistance in obtaining the numerical and experimental data.

References

ADRIAN, R.J. & EARLEY, W.L., 1976, in Proc. Minnesota Symposium on Laser Anemometry, Minnesota University, Minneapolis, p. 426

ADRIAN, R.J. & ORLOFF, K.L., 1977, *Applied Optics,* **16**, 3, 677-684

BALLIK, E.A. & CHAN, J.H.C., 1977, *Applied Optics,* **16**, 3, 674-676

BORN, M. & WOLF, E., 1975, *Principles of Optics,* Pergamon Press

CHU, W.P. & ROBINSON, D.M., 1977, *Applied Optics,* **16**, 3, 619-626

COGHE, A., GHEZZI, U. & PASINI, S., 1978 *XXXIII Congresso Nazionale ATI,* Ancona, Italy

DURRANI, T.S. & GREATED, C.A., 1977, *Laser Systems in Flow Measurement,* Plenum Press, New York

DURST, F. & ELIASSON, B., 1975 in *Proc. LDA-Symposium, Copenhagen,* p. 457

DURST, F., MELLING, A. & WHITELAW, J.H., 1976, *Principles and Practice of Laser-Doppler Anemometry,* Academic Press

FARMER, W.M., 1972, *Applied Optics,* **11**, 11, 2603-2612

GASTER, M. & ROBERTS, J.B., 1977, *Proc. R. Soc. London A.,* **354**, 27-58

GEORGE, W.K. & LUMLEY, J.L., 1973, *J. Fluid Mech.,* **60**, 2, 321-362

GEORGE, W.K., 1975 in *Proc. LDA-Symposium, Copenhagen,* p. 20

GHEZZI, U., COGHE, A., BERGAMASCHI, V., GIORGETTI, A., SENSALARI, G.L. & VISMARA, A., 1978, *XXXIII Congresso Nazionale ATI.* Ancona, Italy

KERKER, M. 1969, *The Scattering of Light,* Academic Press

KREID, D.K., 1974, *Applied Optics,* **13**, 8, 1872-1881

LADING, L., 1972, *Opto-electronics,* **4**, 385-398

LADING, L., 1973, *Opto-electronics,* **5**, 175-187

McLAUGHLIN, D.K. & TIEDERMAN, W.G., 1973, *The Physics of Fluids,* **16**, 12, 2082-2088

PIKE, E.R., 1977, in *Photon Correlation Spectroscopy and Velocimetry,* (eds. H.Z. Cummins & E.R. Pike), p. 246, Plenum Press

ROBINSON, D.M. & CHU, W.P., 1975, *Applied Optics,* **14**, 9, 2177-2183

SHAPIRO, H.S. & SILVERMAN, R.A., 1960, *J. Soc. Ind. Appl. Math.,* **8**, 225-248

VAN de HULST, C.H., 1957, *Light Scattering by Small Particles,* J. Wiley.

Spectrum Measurements with Laser Velocimeters

by

W.T. Mayo, Jr.
Spectron Development Laboratories, Inc.
Costa Mesa, California 92626

Introduction

Several different theoretical researchers have shown that alias-free spectral estimation is possible for stationary random processes if randomly timed instantaneous samples of the random process are available. There is no such thing as a physically realizable alias-free spectrum estimator because the information concerning the higher frequencies of the measured process is largely contained in the subset of the random samples sufficiently close together to measure them and in the absence of uncertainty of the exact time instants of the samples. Physical measuring devices always exhibit a small, but finite, dead time following a measurement instant during which no additional measurement can be made and/or there is always finite time jitter, or uncertainty, of the measurements. In spite of these effects, which produce bias errors, useful practical estimators have been developed over the past ten years which do not require the Nyquist criterion with respect to the mean sample rate. Such techniques have been developed by ourselves and others for computer analysis of burst-counter laser velocimeter data.

Spectron Development Laboratories has recently conducted a study[30] for the NASA Langley Research Center to review and extend, if necessary, the presently available theory concerning the required length of time, or number of samples, over which a laser velocimeter experiment must be run in order to obtain a specified variability error level. The study did not emphasize bias errors due to time jitter and digitization or due to correlations between instantaneous mean sample rate and instantaneous velocity. The time errors have been previously treated[10,15,16]. The latter bias error effect is suspected by some but may be nonexistent for certain estimators[10].

This paper introduces the currently available techniques for estimation of turbulence point spectra using a Poisson impulse sampling signal formalism, and provides an extensive historical bibliography[1-32] for a thorough introduction. Some of the literature is couched in much more 'rigorous' mathematical language which obscures physical practicalities. In the contract report[30], we provide additional simplifying background theory which will be of assistance in understanding the basis for the techniques described in this paper. Space limitations require us to omit this material here.

The major significant contributions of this paper are the determination of the best available variability error formula for one of the estimators based on extension of existing theory by a somewhat ad hoc approach and excellent confirmation with data from previously performed experiments. This provides the basis for experiment planning, as we discuss in more detail elsewhere[30].

Classical Techniques

Infinite Time Statistics

The terminologies commonly found in aerodynamics, mathematical statistics, and electrical engineering differ. The following definitions and assumptions are discussed at length by Papoulis[1]. The random velocity processes which are considered will be assumed to be statistically stationary with interchangeable statistical and infinite time averages denoted by $<>$.

The *autocorrelation* is

$$R(\tau) = <U(t)U(T+\tau)>. \qquad [1]$$

The *autocovariance* is

$$C(\tau) = <u(t)u(t+\tau)> = R(\tau) - <U(t)>^2 \qquad [2]$$

where u(t) is the time varying part of U(t)

$$u(t) = U(t) - <U(t)>.$$

The two-sided *power spectral density* or, more simply *power spectrum* of U(t) is defined as the Fourier transform* of $R(\tau)$ in a manner which includes an impulse at the frequency origin if $<U(t)> \neq 0$

$$S(\omega) = \int_{-\infty}^{\infty} R(\tau)e^{-j\omega\tau} \, d\tau. \qquad [3]$$

This quantity is real, even, and positive for all ω. There are two standard approaches to power spectrum estimation**.

* Many authors include a factor of $1/2\pi$ in the definition of $S(\omega)$; care must be used in using the results in the literature.

**The sample mean is generally removed first to avoid 'leakage' from the zero frequency spike. For a general introduction, the reader is advised to see[2].

The first general approach is to compute the Fourier transform of a weighted finite-time autocovariance estimate. The second general approach uses a modified periodogram estimate, where the raw periodogram is defined as

$$\hat{S}_T(\omega) = \frac{1}{T}\left|\int_{-T/2}^{T/2} \hat{u}(t)e^{-j\omega t}\,dt\right|^2 \tag{4}$$

where $\hat{u}(t)$ has an estimate of the mean subtracted from $U(t)$. The modified periodogram includes a weighting function $\omega(t)$ multiplying $\hat{u}(t)$. It is known[3] that the periodogram is not a consistent estimator: the variance of $\hat{S}_T(\omega)$ does not decrease as T is extended to infinity,

$$\lim_{T\to\infty} \hat{S}_T(\omega) \neq S(\omega) \tag{5}$$

However, Papoulis[1] shows that in the limit of large T, the expected value of $\hat{S}_T(\omega)$ is $S(\omega)$. The reason the raw periodogram does not converge is that there exists a fundamental inverse relation between frequency resolution of the estimate and the length of the data, T. Increasing T decreases the distance between independent frequency estimates without decreasing the variance of the estimate. One presently favored approach to power spectrum analysis with uniformly spaced samples is to use modified periodogram estimates[4] in which the variance is reduced by averaging many estimates with small T or by smoothing the periodogram with a weighted running average which trades resolution for reduced variance. The many subtle variations of smoothing techniques are theoretically similar. The specific smoothing methods favored, as well as the relative merits of the autocovariance approach (mean lagged products) and the modified periodogram approach (also direct approach, and complex demodulation approach) vary with application and generally vary historically according to the capabilities of digital computers. In particular, the autocovariance route was selected over the direct route to save computer time. The advent of the fast Fourier transform has reversed this.

Classical Analysis of Sample and Hold Data
Classically, a band-limited signal may be sampled periodically at a rate greater than twice the highest signal frequency (Nyquist criterion) and the continuous signal may be completely recovered using a linear interpolation filter. It is tempting to try to reconstruct the continuous time history in the randomly sampled case and then perform classical spectral analysis. Any linear filter which passes the signal spectrum also passes a white noise spectrum which results from the random sampling. The 'signal-to-noise' power ratio in such case can be shown to be no better than unity for sampling at a mean rate equal to twice the equivalent (integral) signal bandwidth[30].

The topic of the error in reconstruction of the time history waveform from random samples has been considered in detail by Leneman and Lewis[7]. They have shown that the theoretically optimum linear filter performs much worse than non-linear interpolation filters. They investigated sample-and-hold and polygonal (straight line between sample points) filters and found that the mean-square error becomes very small for the polygonal filter when the mean sample rate is greater than *six times* the equivalent signal bandwidth. Thus, using a nonlinear filter, the error in reconstruction became small at three times the 'Nyquist rate' in a loosely defined sense.

These Leneman and Lewis results may have serious consequences for classical CW spectrum analysis of either laser velocimeter tracker data or burst-counter with sample and hold output. The implication is that serious errors may occur for mean data rates of less than three times the effective Nyquist rates in some loosely defined sense. Although this may not be a serious concern for some applications, typical backscatter measurements from transonic air flow does not often afford one the luxury of such high data rates. We have not pursued the classical analysis approach for this reason.

Point Sample Techniques
In all that follows we assume the availability of a set of instantaneous point samples of the measured velocity component at a single point in space. The sampling is assumed to be 'random' in the sense that individual sample times are random variables which are uncorrelated with each other and with the instantaneous velocity. All the techniques are valid for uniform Poisson sampling. Estimator 3 appears to be useful for inhomogeneous (nonstationary) Poisson sampling as well[10], even if the instantaneous mean sample rate is correlated with velocity; but this latter assertion may be still open for debate.

Alias-Free Spectral Estimation Without Sample Times
Spectrum estimation based upon classical techniques requires the knowledge of sampling times as well as the sample values of the function. It has been shown mathematically[5,6] that with random sampling there is no aliasing, except for a white level; there is no bandwidth restriction; and that *theoretically* the spectrum exists in terms of the sample values and the sampling probabilities without the time instants at which the samples occur. None of several investigators of experimental estimators have found it practical to attempt spectral estimation without the sample times.

Estimator 1: Exact Lag Product Approach
In all practical spectrum estimation problems, the mean value of the process is

estimated and removed first. Usually, the autocovariance estimate is then multiplied by a smoothing window function $D(\tau)$, which is unity at $\tau = 0$ and nonzero on the interval $(-\tau_{max} \leqslant \tau \leqslant \tau_{max})$, prior to Fourier transformation to reduce spectrum leakage due to finite time truncation effets. Thus, a conventional continuous time approximation of equation [3] is given by

$$\hat{S}(\omega) = \int_{-\tau_{max}}^{\tau_{max}} D(\tau) \, \hat{C}(\tau,T) \cos(\omega\tau) \, d\tau \qquad [6]$$

where $\hat{C}(\tau,T)$ is the autocovariance estimate for data duration T:

$$\hat{C}(\tau,T) = \frac{1}{T-\tau} \int_{-T/2}^{T/2-\tau} \hat{u}(t)\hat{u}(t+\tau)dt, \; \tau \geqslant 0 \qquad [7]$$

$$= \hat{C}(-\tau) \qquad\qquad , \tau \leqslant 0 \quad \text{where } \hat{u}(t) \text{ is an estimate of } u(t):$$

$$\hat{u}(t) = U(t) - \frac{1}{T} \int_{-T/2}^{T/2} U(t)dt. \qquad [8]$$

One sees, of course, that T much be much larger than the maximum delay τ_{max} and also much greater than the signal coherence time in order for a useful estimate to result. The $\cos(\omega\tau)$ term in equation [6] is used instead of $\exp(-j\omega t)$ because of the definition of $\hat{C}(\tau)$ as an even function about $\tau = 0$. Now substituting equation [7] into [6] gives a double integral in the variable time t and delay τ:

$$\hat{S}(\omega) = \int_0^{\tau_{max}} \frac{1}{T-\tau} \int_{-T/2}^{T/2-\tau} \hat{u}(t)\hat{u}(t+\tau)D(\tau)\cos(\omega\tau)dtd\tau, \; \tau > 0 \qquad [9]$$

+ corresponding term for $\tau < 0$.
From equation [9] it is a somewhat believable leap of faith to obtain Estimator 1, which has been rigorously analyzed[*] and shown to be asymptotically unbiased and alias-free

$$\hat{S}_1(\omega) = \frac{2}{\lambda^2 T} \sum_{R(j,k)}\sum D(t_k - t_j)\hat{u}(t_k)\hat{u}(t_j)\cos\omega(t_k - t_j) \qquad [10]$$

where R(j,k) is the range of the summation. The form[**] given in equation [10] is that of Gaster and Roberts[20] if R(j,k) is given by $(1 \leqslant k \leqslant N)$, $(j < k)$, where N is the total number of samples, which for stationary random sampling at mean rate λ is given approximately by its expectation:

---

[*]Gaster and Roberts[20] and Masry and Lui[25,26]. Adegebola[24] described a similar estimator much earlier. We disclaimed such an approach[16] as being impractically slow, as did Gaster and Roberts[20].

[**]Both Gaster and Roberts[20] and Masry and Lui[25] include a factor of $1/2\pi$ in the definition of power spectrum that Papoulis does not include. We have omitted this factor here for consistency.

$$N \approx \lambda T. \tag{11}$$

The estimator given by Masry and Lui[25,26] is similar but with differences in the premultiplier and limits of summation.

In principle, Estimator 1 is alias-free for any mean sample rate. In experimental practice, the sample instants t_j can never be measured exactly, and there is always some 'dead' time or minimum interval between sample instants. These two effects do actually limit the bandwidth over which the signal spectrum can be reconstructed without severe attenuation and distortion. As a matter of practicality, one might just as well digitize the time measurements of the instants t_k and accept the restrictions (now known precisely)[15] which this imposes, instead of taking what you get without knowing and having a slower algorithm besides. For these reasons, or others, Jones; Gaster and Roberts; and Mayo, Shay and Riter all adapted Estimator 3, described below.

Estimator 2: Direct Spectrum Estimation
To obtain the second estimator we use equation [4], but use it to obtain a direct estimate, without approximation, of the spectrum $S_r(\omega)$ of the Poisson sample process:

$$\hat{S}_r(\omega) = \frac{1}{T} \left| \int_0^T U_n \delta(t-t_n) e^{-j\omega t} dt \right|^2 = \frac{1}{T} \left| \sum_{n=1}^{N} U_n e^{-j\omega t} {}_n \right|^2. \tag{12}$$

As described in Reference 30, assuming temporarily that we know R(0) and λ, where λ is the mean sample rate, we form the estimate of $S(\omega)$ as

$$\hat{S}_2(\omega) = \frac{1}{\lambda^2} \hat{S}_r(\omega) - \frac{R(0)}{\lambda}. \tag{13}$$

Unfortunately, λ and R(0) are not known, but we do have the data for estimating these quantities:

$$\hat{\lambda} = \frac{N}{T} \tag{14}$$

$$\hat{R}(0) = \frac{1}{N} \sum_{n=1}^{N} U_n^2. \tag{15}$$

In order to obtain a practical estimator, we insert these estimated quantities with the sample mean subtracted to obtain[9,16]:

$$\hat{S}_2(\omega) = \frac{T}{N^2} \left[\left| \sum_{n=1}^{N} \hat{u}_n e^{-j\omega t} {}_n \right|^2 - \sum_{n=1}^{N} \hat{u}_n^2 \right]. \tag{16}$$

As discussed by Gaster and Roberts[23], a smooth data window W(t) may be added to obtain:

$$\hat{S}_2(\omega) = \frac{T}{N^2} \left[\left| \sum_{n=1}^{N} \hat{u}_n W(t_n) e^{-j\omega t}{}_n \right|^2 - \sum_{n=1}^{N} \hat{u}_n^2 W^2(t_n) \right]. \qquad [17]$$

We comment that there is no limit on how many values of ω one may use evaluating $\hat{S}_2(\omega)$. However, given N data values, there are at most N independent values of $\hat{S}_2(\omega)$ which may be calculated. With uniform sampling, it is known that the values of ω should be periodically chosen with $\Delta\omega = \pi/T$ and that the N values of $\hat{S}(\omega)$ which result are independent. There is no guarantee for the random sampling situation, but we assume that π/T is still a useful guide for the available radian frequency resolution.

Estimator 2 was compared with Estimator 1 by Gaster and Roberts[23] and found to be faster computationally. However, Estimator 1 produced less variance (both theoretically and experimentally) than Estimator 2 for the same amount of data. Mayo, Shay and Riter[9,16] did not even consider Estimator 1 due to the obvious computational slowness; they compared Estimators 2 and 3 (described below) experimentally and found Estimator 3 to be both faster *and* to produce less estimate variance for exactly the same data set. New work by Roberts and Gaster[29] has shown the potential for even faster techniques for evaluating Estimator 2. Perhaps an optimum procedure would be to use such fast techniques for on-line preliminary checks of data as it is recorded with the more precise lag product computations made later with the recorded raw data.

Estimator 3: Discretized Lag Products
The method we have found most useful for spectrum analysis[9,10,14-21] begins with a discretized autocovariance estimate similar to that reported by Jones[8,11-13]. The exact time differences between pairs of samples are represented approximately as integral binary numbers with a time increment $\Delta\tau$ which satisfies the Nyquist criterion. For each pair of samples for which the discretized time difference does not exceed the preselected maximum lag number M, the product of the two samples (lag product) is accumulated in a $1 \times M$ matrix SUM(k) and 1 is added to a corresponding histogram H(k). We assume that the sample mean has been previously subtracted from the data. After N samples are processed, the resulting estimate $\hat{C}(k)$ of the value of $C(\tau)$ at $\tau = k\Delta\tau$, $k \neq 0$, is

$$\hat{C}(k) = \frac{1}{H(k)} \text{SUM}(k) \qquad [18]$$

where the summation includes only the H(k) products for which

$$\left| \frac{|t_i - t_j|}{\Delta\tau} - k \right| < 0.5. \qquad [19]$$

For k = 0, the practical advisability of including products for which $|t_i - t_j|$ < $\Delta\tau/2$ along with the $\hat{u}_i^2$ terms depends on the implementation; without these terms, H(0) = N and

$$\hat{C}(0) = \frac{1}{N} \sum_{i=1}^{N} \hat{u}_i^2 \qquad\qquad [20]$$

We note, however, that equation [20] includes all measurement errors as false turbulence which may be eliminated if only $i \neq j$ terms can be utilized.

A very similar autocovariance estimate may also be expressed in the alternate notation which assumes assignment of t_i to the next value of $j\Delta\tau$ prior to computation as

$$\hat{C}'(k) = \frac{\sum_{j=0}^{N'-k} \hat{u}(j)\hat{u}(j+k)I(j)I(j+k)}{\sum_{j=0}^{N'-k} I(j)I(j+k)} \qquad\qquad [21]$$

where $N \cong \lambda T = \lambda N'\Delta\tau$ and I(j) is a binary indicator sequence of ones and zeroes. The estimator is denoted with a prime to remind us that equation [21] differs from equations [19] and [20] because $\pm\Delta\tau/2$ errors in the discretization of t_i and t_j prior to subtraction allows maximum errors in time differences of $\pm\Delta\tau$. The maximum time error for $\hat{C}(k)$ is $\pm\Delta\tau/2$. Also $\hat{C}(k)$ uses a fixed number of samples, while $\hat{C}'(k)$ uses a fixed length of time. The relation N = λT is only true on the average.

Jones dismisses the analysis of the discretized autocovariance estimate by reference to his analysis of periodic sampling with random omissions[12]. If, in fact, $\Delta\tau$ is assumed small enough to make the time errors negligible in both $\hat{C}(k)$ and $\hat{C}'(k)$ then the application of conditional expectation techniques to the random omission model shows that the estimates are unbiased provided that N is sufficiently large so that each lag value has at least one lag product. We note, however, that practical limitations will sometimes require $\Delta\tau$ to be as large as is permissible without large bias error. Bias error due to excessive width includes both the conventional aliasing of periodic sampling at rate $1/\Delta\tau$ and a low-pass filter effect due to time smearing. These effects are discussed by Scott[10], Mayo et al.[15,16], and Shay[17].

The histogram H(k) of the number of lag products obtained at each lag value k plays a significant role in practical application of the estimator $\hat{C}(k)$. In Reference[16] we have evaluated its expectation approximately under the assumption that $\lambda\Delta\tau$ is small and N is large and with only $\hat{u}_i^2$ terms used in $\hat{C}(0)$:

$$H(k = 0) = N$$

$$<H(k)> \cong N\lambda\Delta\tau - k(\lambda\Delta\tau)^2 \cong (\frac{T}{\Delta\tau} - k)\ (\lambda\Delta\tau)^2. \tag{22}$$

The result shows that the number of products at each lag is generally much less than the number of $\hat{u}_i^2$ terms. The expected number of products at each lag decreases linearly with k, but if the data collection time, T, for each data segment is much greater than the maximum delay, i.e., if

$$T \gg M\Delta\tau, \ N \gg M\lambda\Delta\tau \tag{23}$$

then the value of $<H(k)>$ is nearly constant. This condition is an important consideration in implementation which must be observed when segmented data methods are used.

The variance of H(k) has not been theoretically evaluated, but we expect it to behave in the manner of a Poisson process for acceptably large values; i.e.,

$$<(H(k) - <H(k)>)^2> = <H(k)>. \tag{24}$$

Once the discrete autocovariance estimate $\hat{C}(k)$ has been computed, the two-sided power spectrum estimate as a function of frequency follows using a discrete Fourier transform as

$$\hat{S}_3 (i\Delta\omega) = \Delta\tau \left[\hat{C}(0) + 2\sum_{k=1}^{M-1} W(k)\hat{C}(k)\cos(\frac{ik\pi}{M}) \right] \tag{25}$$

$$i = 1, 2, \ldots M$$

where

$$M = \frac{\tau_{max}}{\Delta\tau}$$

$$\Delta\omega = \frac{\pi}{M\Delta\tau} = \frac{\pi}{\tau_{max}} \tag{26}$$

$$\omega = i\Delta\omega$$

and W(k) is a smoothing window function which is unity at k = 0 and which decreases to zero at k = M. The effect of the window function is to reduce side band leakage and variability error at the expense of some loss of resolution[32]. The discrete transform may be implemented as a fast Fourier transform if desired.

Comments on Estimators

Optimum selection of the shape of the window function is an art involving knowledge of the spectrum shape (either *a priori* or by trial and error procedure) and the measurement objectives. This art has been discussed in the classical literature[2,3]. For smooth broadband turbulence spectra without spikes due to periodic components, the simple Bartlett window given by

$$W(k) = 1 - \frac{k}{M}, k < M \qquad [27]$$

$$= 0 \text{ otherwise}$$

may be adequate. This window has the advantage of simplicity and, therefore, computational speed over others which have lower side lobes. The use of no window is equivalent to using a rectangular window with sinx/x side lobes which are usually not acceptable.

We note that Gaster and Roberts[20] have also developed a *logarithmic* slotting technique and a corresponding Fourier cosine transform approach for obtaining logarithmically-spaced frequency bands. This variation may be quite valuable in many experimental situations; or somewhat equivalently, one may wish to perform non-uniform smoothing of the spectral estimate directly in the frequency domain with wider bands at the higher frequencies[18].

Another observation concerns the temptation on the part of some to compute $\hat{S}(f)$ at frequency intervals less than $1/(2\Delta\tau)$ because the result is smoother and more pleasing looking than the somewhat jagged appearance of the result of using equation [25]. Giving into this temptation does not produce more useful information or resolution since the width of the convolving window $W(\omega)$ generally exceeds $1/(2M\Delta\tau)$. On the other hand, natural spectra are usually 'smooth', and the jagged appearance resulting from errors is not disguised when the frequency spacing is maintained large enough for the estimates to remain nearly independent.

For completeness, we point out that there are some additional estimators that one should avoid. A likely candidate might seem to be a straightforward numerical approximation to the continuous periodogram integral. Such an estimator was defined and evaluated experimentally by Mayo et al. [9,16]. This estimator which was called 'Estimator 1' in the given references failed to produce any useful spectral estimates. Gaster and Roberts[23] also mention this estimator (their equation [2]) as a logical and often used discrete time version of the periodogram integral which is well known for uniformly-spaced samples. They then neglect it and go on to define what we have identified as 'Estimator 2' in this report.

In searching for possible additional estimators, one could consider single or multiple regression techniques which attempt to best fit sinusoidal wave forms to a set of sample data. Such techniques were discovered and discarded during a literature search in 1973 as being very useful only for spectra composed of a few discrete frequencies. Similar conclusions have recently been obtained independently by Norsworthy[31] who found such techniques to be inadequate for broadband spectra.

Finally, we wish to point out that the (1976) paper by Masry and Lui[26] mentions the fact that the spectral estimates from randomly-timed samples are not necessarily positive. This is unfortunate since the true power spectrum is positive by definition. They, however, at the end of the paper[26] propose a new modified periodogram estimator which is always positive to eliminate the problem. Their new estimator is none other than that prescribed by Thompson[22] in 1971 and shown by both Gaster and Roberts[23] and Mayo et al.[9,16] to be inadequate without the subtraction of the estimate of the white level. Once this correction is made, the estimator is no longer positive definite and, in fact, is more subject to error due to sample rate fluctuations than is either of the lag-product estimators. In two new papers by Masry, he analyzes another version of Estimator 2 and its minor differences from earlier similar estimators[27,28].

Estimator Equivalences
The literature makes distinctions over even the smallest nuances of differences between different spectral estimators. After many years of computing spectral estimates with uniform sampling, the near equivalence mathematically of various lag product and direct algorithms became clear, and the significant issues became computational speed in most cases. It is almost embarrassing to realize that history appears to have repeated itself. Masry[27] seems to be first to note the fact that without the window functions and neglecting the differences in limits of summation, we obtain the form of Estimator 1 from Estimator 2 simply by expanding the sum of complex exponentials, taking the products, and collecting the terms as cosines. As had been discussed by Gaster and Roberts[23], the spectral resolution of the direct transform method (Estimator 2) can be made worse while decreasing the variance by summing many spectral estimates formed with short blocks of data in a manner analogous to that commonly done in conventional spectral analysis[4]. When this is done, the variance is still approximately twice as much as with Estimator 1, probably

due to the lag products which are missed between blocks. This could perhaps be partially overcome by using overlapping blocks. The computational time remains slightly faster for the block averaged version of Estimator 2.

We may summarize by saying that the various forms of Estimators 1 and 2 are exact alias-free estimators which have minor differences when used with block averaging of Estimator 2. In comparison, Estimator 3 appears distinct in that it trades predictable digitization effects for increased computational speed. Furthermore, Estimator 3 offers the elimination of effects due to nonstationarity of the sampling process. It thus has some distinct practical advantages. It is also less well characterized theoretically and will receive most of the attention in what follows.

Variance of Estimator 3

In our NASA report[30] we review the various formulas for the variability error of Estimators 1, 2, and 3, and derive the variability of the intermediate autocovariance estimate included in Estimator 3 in agreement with a different earlier derivation by Scott[10]. We are forced by length to briefly summarize the results here. Unfortunately, there has been no complete derivation of the variability error of spectral Estimator 3. Mayo et al.[16] and Scott[10] have obtained comparable formulas based on the assumption of low mean sample rate with statistically independent lag products. Gaster and Roberts[20] obtained a complete theoretical expression for Estimator 1 and used it without proof as an approximate formula for Estimator 3 (which they call the 'slotted' technique). Masry and Lui[25,26] also obtained a variability formula for their version of Estimator 1, which is functionally similar (if certain quantities are exchanged with their expected values) and differs only by a constant factor near 2.7 from that of Gaster and Roberts. The difference is presumed to be due to the use of a window function by Gaster and Roberts, although a more detailed comparison may be needed.

We have used the following deductive process concerning Estimator 3 variability error. The first term in the Estimator 1 formula by Gaster and Roberts is identical in form to that obtained by Mayo and Scott for Estimator 3 with low mean sample rates. The second, and only other term in the Estimator 1 variance formula has been identified by Masry and Lui[25,26] as that due to the finite duration of the unsampled continuous random process under observation. It is obvious by physical reasoning (assuming mathematically well-behaved spectra) that the second term must be the high-data-rate asymptotic limit of any consistent estimator, including Estimator 3. This follows because at the limit of high sample rates, new samples cannot add new information, and the information content of the signal sample depends only on the length of the sample. This logic does not exclude additional positive or negative terms in the formula for Estimator 3 variability, but it does offer support for the simple adoption of the Estimator 1 formula, as was done by Gaster and Roberts. The result of this logic in the case of using a Bartlett window (see Reference (16) for derivation of premultiplying constant) for frequencies other than zero is

$$\text{Var}\hat{S}_3(\omega) \approx \frac{2\tau_m}{3T}\left[S(\omega) + \frac{C(0)}{\lambda}\right]^2 \qquad [28]$$

where the $\approx$ refers to our substitution of $N \approx \lambda T$ under the assumption of large N and the notational equivalences

$$\tau_m = M\Delta\tau \qquad [29]$$

$$C^2(0) = \sigma^4$$

where σ^2 is the variance of the sampled process and $C(\tau)$ is the autocovariance as previously defined.

For a low-pass velocity spectrum, we may define an integral effective energy bandwidth B from the true power spectrum as

$$2B = \frac{\dfrac{1}{2\pi}\displaystyle\int_{-\infty}^{\infty} S(\omega)\,d\omega}{S_{peak}} = \frac{C(0)}{S_{peak}} \qquad [30]$$

where S_{peak} is the peak value of the spectrum, or any other chosen value, as was done in Reference (16). When this is done, the variance equation becomes

$$\text{Var}[\hat{S}_3(\omega)] = \rho\left(\frac{\tau_m}{T}\right)\left|S(\omega) + S_{peak}/\frac{\lambda}{2B}\right|^2 \qquad [31]$$

where ρ is a constant near unity defined by the lag window function used.

The above form of the expression makes the functional dependence of the estimator variability most intuitively clear. For the same asymptotic assumption of large T and $N \approx \lambda T$, we obtain approximately

$$\text{Var}[\hat{S}_3(\omega)] \approx \rho\left(\frac{\tau_m \lambda}{N}\right)\left|S(\omega) + S_{peak}/\frac{\lambda}{2B}\right|^2 . \qquad [32]$$

In this form we see that when the number of data words is fixed, the variance depends on the ratio of the number of data within the lag window time divided by the total number.

Examination of the equation shows that for fixed N, there is a separate optimum data rate which minimizes the variance for each specified frequency, and this optimum depends on the spectrum being measured. This means that generally speaking one must have *a priori* knowledge of both the experiment objectives and the general spectrum shape before detailed experiments can be correctly conducted. If such *a priori* knowledge is not available, preliminary experiments and/or data processing may be required to obtain it.

Experimental Confirmation

Again due to length limitations, we omit here all of the details concerning a computer simulation study of Estimator 3 variability error which is reported in Reference (30). Those results also support the use of equation [32]. Only a brief comparison of equation [32] with previously reported experimental data is given here since it is more appropriate.

A series of careful experiments performed in 1973 by Mayo et al.[16] at Texas A&M University did not agree well with the available variability error theory for the pass band portion of the spectrum; absolute agreement was obtained for the stop band portion of the spectrum where $S(\omega) \equiv 0$. The discrepancy is now easy to explain, and the data directly supports the variability formula given above in equation [32]. In order to show this, we rewrite the variability formula in the form of normalized RMS error used in Reference (16) as

$$\epsilon'_{\ell p} = \frac{\sqrt{\text{Var}\hat{S}(\omega)}}{S_{peak}} = A\sqrt{2B\alpha}\left(\frac{S(\omega)}{S_{peak}} + \frac{1}{\alpha}\right) \qquad [33]$$

where A is a constant which collects terms, and $\alpha = \lambda/2B$. Now in Reference (16), the $S(\omega)/S_{peak}$ term was not included and the corresponding formula was plotted in Fig. 5 of that reference (with B fixed) as

$$\epsilon_{\ell p} = \frac{A\sqrt{2B}}{\sqrt{\alpha}} = \frac{0.06}{\sqrt{\alpha}} \qquad [34]$$

where the value 0.06 was determined absolutely from the experiment parameters. If we now consider that in the pass band the value of $S(\omega)/S_{peak} \approx 1$, we may evaluate the newer results for comparison using the formula equation [35] below*.

$$\epsilon'_{\ell p} = 0.06\sqrt{\alpha}\left[1 + \frac{1}{\alpha}\right]. \qquad [35]$$

* This formula does not apply when bandwidth is a variable, because B is included in the constant 0.06. Equation [32] is the correct form to use.

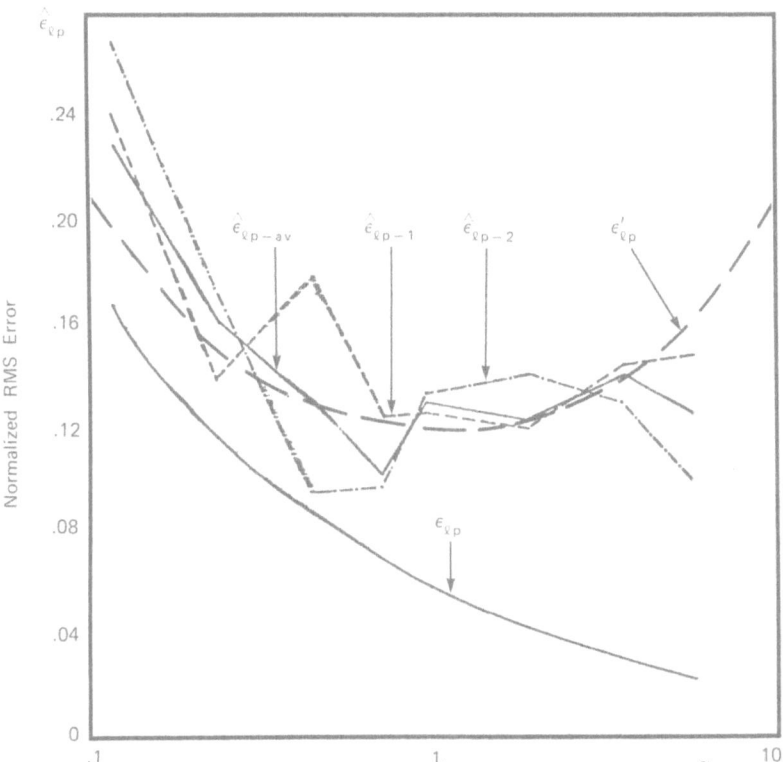

Fig. 1. Reproduction of Fig. 5, Reference 16, with Addition of Corrected Error Prediction Equation $\epsilon'_{\ell p}$.

In Fig. 1 this formula is plotted on a reproduction of Fig. 5 from Reference (16). The agreement is excellent. This does not detract from the previous excellent agreement with the theory in the stop band where $S(\omega) \equiv 0$, since the $S(\omega)/S_{peak}$ term vanishes in that region, and the formula collapses asymptotically back to that which was used in Reference (16).

In their 1975 paper[20] Gaster and Roberts reported computer simulation studies of Estimator 1. They determined that the variability formula, which we have now adopted for both Estimator 1 and Estimator 3, was valid in the stop band of simulated narrow band signals. They do not appear to have directly tested the 'slotting' technique.

In reviewing the literature, there does not appear to be any other experimental confirmation of the variability formula that we have found.

W.T. Mayo, Jr.

Discussion and Conclusions

A review of the available methods of spectrum analysis from randomly occuring samples as occurs with laser velocimeter data has been presented. The discretized lag product technique (Estimator 3) still appears to be the most data efficient practical spectral estimator, although computationally faster methods (derivatives of Estimator 2) are being developed[29]. Considering the expense of data acquisition time in certain situations, the need exists for the development of special-purpose high-speed computational hardware for Estimator 3, since software is still slow on even the fastest minicomputers.

A comparison with other derivations and deductive logic and with previously conducted experiments indicates that the variability error theory derived by Gaster and Roberts[20] for their exact lag product algorithm is a useful practical formula for the discretized lag product algorithm. We have expanded this variability error formula in a form which shows a strong dependence on mean sample rate and that there is an optimum smoothing window width and mean sample rate which depends on the value of the spectrum being measured. Thus, experiment objectives may dictate particle seeding or long data collection times in certain cases, and the desired level of variability error may not be practically possible in certain other situations.

Acknowledgements

The author wishes to thank NASA for financial support and J.F. Meyers and W.W. Hunter, Jr., NASA Langley Research Center, for their encouragement and technical support.

References

1. Papoulis, A., *Probability, Random Variables, and Stochastic Processes,* McGraw-Hill, New York, 1965
2. Blackman, R.B. and Tukey, J.W., *The Measurement of Power Spectra from the Viewpoint of Communications Engineering,* Dover Publications, Inc., New York, 1959
3. Jenkins, G.M. and Watts, D.G., *Spectral Analysis and Its Applications,* Holden Day, San Francisco, 1968
4. Welch, P.D., "The Use of Fast Fourier Transform for the Estimation of Power Spectra: A Method Based on Time Averaging over Short, Modified Periodigrams", *IEEE Trans. on Audio and Electro-Acoustics,* p. 70-73, June 1967
5. Shapiro, H.S. and Silverman, R.A., "Alias-Free Sampling of Random Noise", *S. Joc. Indust. Appl. Math.,* Vol. 8, June 1960, pp. 225-248
6. Beutler, F.J., "Alias-Free Randomly Times Sampling of Stochastic Processes", *IEEE Trans. on Information Theory,* Vol. IT-16, No. 2, March 1970, pp. 147-152
7. Leneman, O.A.Z. and Lewis, J.B., "Random Sampling of Random Processes: Mean Square Comparison of Various Interpolators", *IEEE Trans. on Automatic Control,* p. 396-403, July 1966
8. Jones, R.H., "Spectrum Estimation with Unequally Spaced Observations", *Proceedings of the Kyoto International Conference on Circuit and System Theory,* September 9-11, 1970, AFOSR-70-2747R
9. Mayo, W.T. Jr., Riter, S. and Shay, M.T., "An Introduction to the Estimation of Power Spectra from Single Particle LV Data", *Proceedings of the LDA Workshop,* Oklahoma State Univ., June 11-13, 1973
10. Scott, P.F., "Distortion and Estimation of the Autocorrelation Function and Spectrum of a Randomly Sampled Signal", Ph.D. Dissertation, Rennsselaer Polytechnic Institute and General Electric Corporate Research and Development, Schenectady, New York, Report No. 76CRD180, Sept. 1976
11. Jones, R.H., "Spectral Analysis with Regularly Missed Observations", *Annals of Mathematical Statistics,* Vol. 33, pp. 445-461, June 1962
12. Jones, R.H., "Spectrum Estimation with Missing Observations", *Annals of the Institute of Statistical Mathematics,* Vol. 23, pp. 387-398, 1971, AFOSR-TR-72-0513
13. Jones, R.H., "Aliasing with Unequally Spaced Observations", *Journal of Applied Meteorology,* Vol. 11, No. 2, pp. 245-254, March 1972, AFOSR-TR-72-1548
14. Mayo, W.T. Jr., Shay, M.T. and Riter, S., "Digital Estimation of Turbulence Power Spectra from Burst Counter LDV Data", *Proceedings of the Second International Workshop on Laser Velocimetry,* Purdue Univ., W. Lafayette, Indiana, March 1974
15. Mayo, W.T. Jr., "A Discussion of the Limitations and Extensions of Power Spectrum Estimation with Burst Counter LDV Systems", *Proceedings of the Second International Workshop on Laser Velocimetry,* Purdue Univ., W. Lafayette, Indiana, March 1974
16. Mayo, W.T. Jr., Shay, M.T. and Riter, S., "The Development of New Digital Data Processing Techniques for Turbulence Measurements with a Laser Velocimeter", USAF Arnold Engineering Development Center Report No. AEDC-TR-74-53, August 1974

17. Shay, M.T., "Digital Estimation of Autocovariance Functions and Power Spectra from Randomly Sampled Data Using a Lag Product Technique", A Ph.D. Dissertation, Texas A&M University, College Station, TX, Aug. 1976

18. Smith, D.M. and Meadows, D.M., "Power Spectra from Random-Time Samples for Turbulence Measurements with a Laser Velocimeter", *Proceedings of the Second International Workshop on Laser Velocimetry*, Purdue Univ., W. Lafayette, Indiana, March, 1974

19. Scott, P.F., "Theory and Implementation of Laser Velocimeter Turbulence Spectra Measurements", *Proceedings of the Second International Workshop on Laser Velocimetry*, Purdue Univ., W. Lafayette, Indiana, March 1974

20. Gaster, M. and Roberts, J.B., "Spectral Analysis of Randomly Timed Signals", *J. Inst. Maths. Applics.*, **15**, pp. 195-216, 1975

21. Meyers, J.F. and Clemens, J.I. Jr., "Processing Laser Velocimeter High-Speed Burst Counter Data", *Proceedings of the Third International Workshop on Laser Velocimetry*, Purdue Univ., W. Lafayette, Indiana, July 11-13, 1978

22. Thompson, R.O., "Spectral Estimation from Irregularly Spaced Data", *IEEE Trans. Geoscience Electronics*, **GE-9**, pp. 29-35, 1971

23. Gaster, M. and Roberts, J.B., "The Spectral Analysis of Randomly Sampled Records by Direct Transform", *Proc. R. Soc. Lond. A.*, **354**, pp. 27-58, 1971

24. Adegebola, M.O., *Alias-Free Spectral Estimation of Stochastic Processes*, Ph.D. Dissertation, California Institute of Technology, 1971 (available from University Microfilms)

25. Masry, E. and Lui, M.C., "A Consistent Estimate of the Spectrum by Random Sampling of the Time Series", *SIAM J. Appl. Math.*, **28**, No. 4, pp. 793-810, 1975

26. Masry, E. and Lui, M.C., "Discrete-Time Spectral Estimation of Continuous-Parameter Processes — A New Consistent Estimate", *IEEE Trans. Information Theory*, **IT-22**, No. 3, pp. 298-312, 1976

27. Masry, E., "Poisson Sampling and Spectral Estimation of Continuous-Time Processes", *IEEE Trans. Information Theory*, **IT-24**, No. 2, p. 173, (March 1978)

28. Masry, E., "Alias-Free Sampling: An Alternative Conceptualization and Its Applications", *IEEE Trans. Information Theory*, **IT-24**, No. 3, p. 317 (May 1978)

29. Roberts, J.B. and Gaster, M., "Rapid Estimation of Spectra from Irregularly Sampled Records", *Proc. IEE*, **125**, No. 2, pp. 92-96, 1978

30. Mayo, W.T. Jr., "Error Prediction for LV Turbulence Power Spectra", Final Report NASA Langley Research Center Contract NAS1-15353 (to be published)

31. Norsworthy, K.N., "Fourier Transformation and Spectrum Analysis of Sparsely Sampled Signals", *Proceedings of the Third International Workshop on Laser Velocimetry*, Purdue Univ., W. Lafayette, Indiana, July 11-13, 1978

32. Harris, F.J., "On the Use of Windows for Harmonic Analysis with the Discrete Fourier Transform", *Proceedings IEEE*, **66**, No. 1, pp. 51-83 (January 1978).

Determination of Large Eddy Structures in the Viscous Sublayer: A progress report

by

S. Herzog
Applied Research Laboratory
The Pennsylvania State University
University Park, PA 16802, USA[*]

and

J.L. Lumley
Sibley School of Mechanical and Aerospace Engineering
Cornell University, Ithaca, NY 14853, USA

Abstract

An experiment to determine the dynamics of the large eddies of the viscous sublayer is described. Lumley's orthogonal decomposition for partially homogeneous and stationary flow is to be used to determine the eddy structures from the experimental cross-spectral data. The experiment is carried out in a glycerine tunnel designed specifically for this investigation. The extensive cross-spectra (16,000) are measured with miniature split film probes. Included in the paper are a description of the facility, a detailed analysis of the probe response, a description of the data processing procedure, and finally an interpretation of the resulting decomposed eddies.

1. Introduction

1.1 General Approach

The purpose of this investigation is to contribute to a better understanding of the viscous sublayer and buffer layer of a turbulent boundary layer over a smooth surface. To this end, the streamwise and lateral velocity fluctuations are measured simultaneously at two different movable locations in the near-wall region. Altogether 1728 different spacial constellations between the two probes are realized, covering a volume of 580 viscous lengths in the streamwise direction and 130 in the circumferential direction, covering the region normal to the wall from $y^+ = 1.25$ to $y^+ = 40$ (where y^+ is the Reynolds number based on the friction velocity and the distance to the wall). The configuration

[*]Present address: Sibley School of Mechanical and Aerospace Engineering, Cornell University, Ithaca, NY 14853, USA.

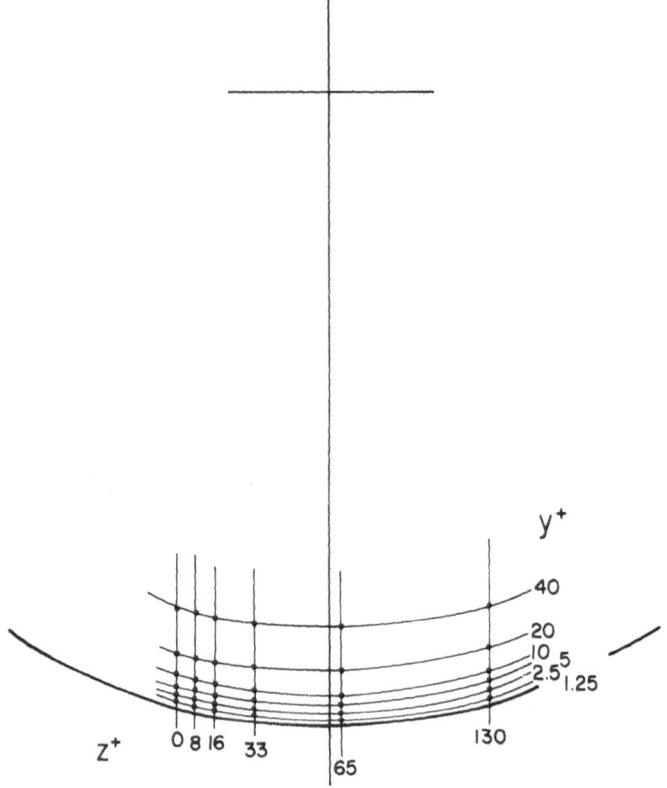

Fig. 1. Plane of positions of 'movable' probe. Plane of 'fixed' probe is at $X^+ = 0.9, 18, 36, 72, 145, 291,$ and 581 upstream with fixed probe at $Y^+ = 1.25, 2.5, 5, 10,$ and 40.

of points measured is shown in Fig. 1. It can be seen that the spacing of the points is approximately logaritmic in the three directions. This is motivated by a desire to reduce as far as possible the number of point pairs to be measured, and is permissable because only larger eddies contribute to the correlation at the larger spacing; hence, progressively fewer points are required to define the correlation as the spacing increases.

This investigation is still in the data-acquisition stage. We give, however, some details of the method by which we propose to analyse the data. Briefly, the data will be analog/digital converted on line and recorded on digital tape to permit calculation of the complete space-time correlation tensor. From this tensor the complete orthogonal decomposition scheme introduced by Lumley (1967) will be used to identify the large-scale structure.

1.2 Use of Continuity

Again, in an effort to reduce as far as possible the number of point pairs to be measured, we measure only the streamwise and lateral fluctuations, and omit the component normal to the wall. Bakewell (1967) measured only the streamwise component, and were forced to use an *ad hoc* assumption to complete the velocity field. We wish to avoid this. Fortunately, the incompressibility condition is sufficient to allow us to obtain the entire space-time cross-correlation tensor from measurements of the components containing the streamwise and lateral components. We can see this easily in a Cartesian coordinate system. Suppose that the field is homogeneous in the 1 and 2 directions, and inhomogeneous in the 3 direction. Continuity gives

$$u_{i,i} = 0 \tag{1}$$

By multiplying by u_j at a different place and time we have

$$\overline{u_j(\underline{x}',t')\, u_{i,i}(\underline{x},t)} = 0 \tag{2}$$

If the correlation is given by $R_{ij}(\underline{x},\underline{x}',t,t')$, equation [2] produces the three equations

$$(\partial/\partial x_3)R_{31} = (\partial/\partial x_1)R_{11} + (\partial/\partial x_2)R_{21}$$

$$(\partial/\partial x_3)R_{32} = (\partial/\partial x_1)R_{12} + (\partial/\partial x_2)R_{22} \tag{3}$$

$$(\partial/\partial x_3)R_{33} = (\partial/\partial x_1)R_{13} + (\partial/\partial x_2)R_{23}$$

When the correlations involving 11, 12, 21 and 22 have been measured, the right hand sides of the first two equations will be known; they can then be integrated (using various interpolation and smoothing techniques) to obtain the 31 and 32 components; the right-hand side of the third equation will then be known, and the 33 component can then be obtained. While differentiation of experimental data is somewhat dangerous, very good results have been obtained by Huber (1974) by fitting a Fourier series to the measured points, discarding Fourier coefficients at the high wavenumber end that clearly correspond to noise, and obtaining the derivative from the series. This amounts to spacial filtering of the measurement noise.

1.3 Computational Efficiency

As shown by Lumley (see George 1978), it is more efficient to compute the correlations directly, using a sequence of point pairs, the members of a pair being separated by the lag desired, but successive pairs being displaced by two integral scales, the correlation so obtained being transformed by fast Fourier transform. The alternative customarily used, direct transforming by Fast Fourier Transform of the signals, followed by formation of the spectrum, can be shown to use considerably more (at least a factor of four) computer time. Hence, we will use the former technique. The explanation for the difference lies in the fact that points less than two integral scales apart do not contribute to the convergence of the quantity in question (correlation or spectrum) because they are statistically dependent.

1.4 The Orthogonal Decomposition Scheme
The orthogonal decomposition has been described in detail by Lumley 1970. Briefly, in any inhomogeneous direction a random function can be represented by a series of deterministic functions with uncorrelated coefficients. The functions are orthonormal, and represent the extraction from the fluctuating velocity of such random recurrent structural patterns as exist. The functions are obtained as eigensolutions of an integral equation involving the correlation tensor (or in certain cases, the cross-spectral tensor). The series is optimal, in the sense that it converges as rapidly as possible; the maximum possible energy is in the first eigenfunction, the maximum of the remainder is in the next one, and so forth. The eigenvalues of the equation are the energies in the various terms. In the homogeneous directions a different decomposition has been suggested by Lumley 1970, analogous to the shot effect decomposition; the random function is represented by deterministic functions sprinkled at random, with random strengths. The deterministic functions are obtained as the transforms of the square-roots of the spectra. Here there is an ambiguity, since the phase information is lost in the spectrum; a family of these deterministic functions is possible, having phases which vary with frequency (or wavenumber) in various ways, and the variation cannot be determined from the spectrum. It is possible, however, to determine the phases from the higher order statistics, and an effort will be made to do this.

Of course, in our flow, we have a situation which combines these two possibilities: the flow is homogeneous in the streamwise and lateral directions, and stationary in time, but inhomogeneous normal to the wall. This situation has been treated by Lumley 1970. It is necessary first to transform the velocity covariance tensor to obtain the velocity spectrum tensor in the two homogeneous directions and time; the spectrum tensor will be a function of the two distances from the wall and the cross-stream and streamwise wave numbers and frequency. The eigenvalues and eigenfunctions of the cross-stream wavenumbers and frequency are parameters. A function is then constructed from the product of the eigenfunction and the square root of the eigenvalue. At this point, if an appropriate phase can be determined from higher order statistics, it can be inserted. This function is then transformed back again to give the deterministic function in which the velocity field is expanded.

The experience of Bakewell 1967 indicated that in these low Reynolds number flows, a large percentage of the energy is in the first eigenfunction, so that an excellent representation of the velocity field is obtained using only the first.

Proceeding in this way, we obtain a deterministic function in which the fluctuating velocity field can be represented, which is in a sense more like (Lumley 1967) the randomly occurring eddies, including here all three velocity components, and three spacial dimensions and time; hence we will have a compact eddy structure which evolves in time, to replace the subjective impressions from flow visualization of sweeping, lifting, etc.

2. Experimental Facility

2.1 Summary

In most of the ordinary occurring flow situations the viscous boundary layer region is confined to such small physical dimensions that detailed measurements are extremely difficult to perform. For turbulent pipe flow the viscous length can be expressed in terms of the pipe-diameter D, the kinematic viscosity ν, and the bulk velocity $\bar{U}_m$ via the friction-law of Blasius:

$$\ell_\nu = 5.03 \cdot D^{1/8} \cdot \left(\frac{\nu}{\bar{U}_m}\right)^{7/8}$$

Except for changes in test-pipe diameter over two or three order of magnitudes, changes in D have no significant influence of ℓ_ν which we want to have as large as possible. High values of kinematic viscosity and/or low values of mean-velocities are left as significant parameters, but the latter one certainly has lower limits, especially for gases as bouyancy effects come into play. The pipe diameter then finally serves to control the Reynolds-number $Re = D \cdot \bar{U}_m / \nu$ in order to establish fully developed turbulent pipe flow.

For our experiments we have obtained a viscous length of 0.56 mm by taking almost pure glycerine, a highly viscous and clear liquid, as working fluid. The experimental facility is described elsewhere in detail (Bakewell 1966); some important gross-data are summarized in Table 1. The closed-circuit facility consists of a 100 hp centrifugal pump, a settling section with a built-in honeycomb followed by a 16 to 1 contraction to the test-pipe diameter of D = 0.284 m. The test section is located twenty-five diameters downstream of the pipe-inlet which is equipped with a serrated trip-ring to fix the location of transition to a turbulent boundary layer.

Working Fluid		Glycerine, 2.2 % Water
Operating Temperature	T	34.85 ±0.1°C
Fluid Density	ρ	1246 kg/m
Kinematic Viscosity	ν	216.10^{-6} m²/sec
Pipe Diameter	D	0.285 m
Mean Velocity	$\bar{U}_m$	6.63 m/sec
Reynolds Number	$R_e = D \cdot U_m / \nu$	8750
Wall Shear Stress	τ_w	184.9 N/m²
Friction Velocity	U_τ	0.385 m/sec
Viscous Length	$\ell_\nu = \nu/U_\tau$	0.561 mm

Table 1. Pertinent data of pipe flow

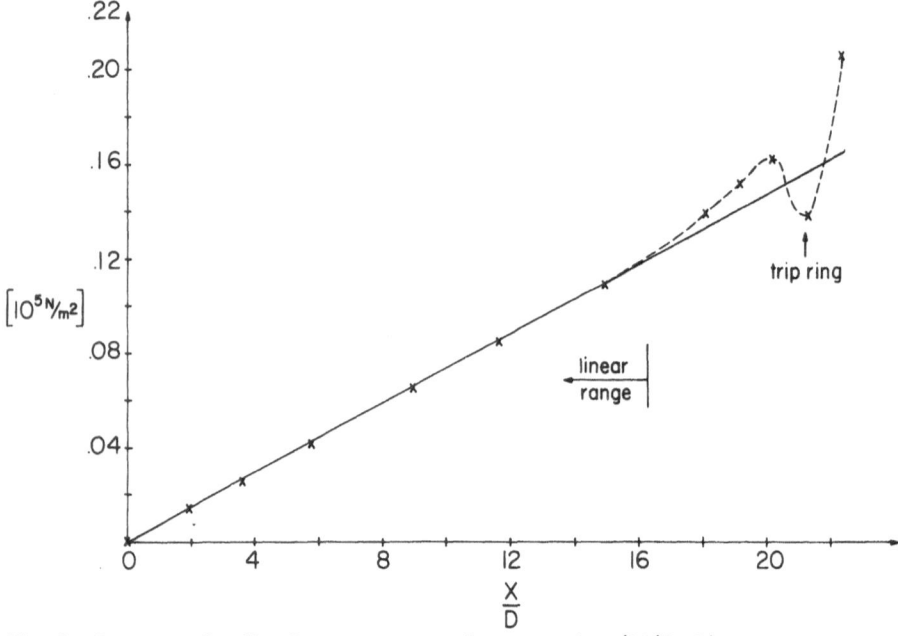

Fig. 2. Pressure distribution upstream of test section (X/D=0).

From pressure measurements along the pipe axis, see Fig. 2, we conclude that this boundary layer then grows rapidly into the core region of the pipe. The linearity of the pressure drop over 17 pipe-diameters upstream of the test-section convinces us that we indeed have a fully developed turbulent flow condition inside the test-section, so far as mean velocity and drag are concerned. Since we are interested only in the near-wall region, where the time scale is short, we may expect higher order moments also to come to equilibrium (Comte-Bellot 1963). These considerations are fully discussed by Bakewell 1966.

2.2 Glycerine as Working Fluid

Kinematic Viscosity	ν	$216.2 \cdot 10^{-6}$ m^2/sec
Density	ρ	1246 kg/m^3
Specific Heat Capacity	c_ρ	2430 Nm/(kg·°C)
Thermal Conductivity	λ	0.280 N/(°C·sec)
Thermal Diffusivity	$\alpha=\lambda/(\rho \cdot c_\rho)$	$9.25 \cdot 10^{-8}$ m^2/sec
Prandtl-Number	Pr	2340

Table 2. Physical Properties of Glycerine at 35°C.

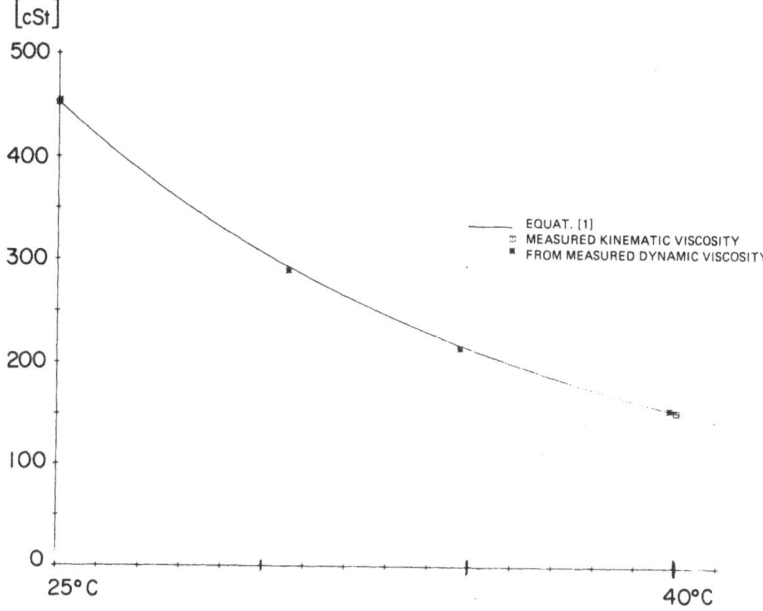

Fig. 3. Kinematic viscosity of glycerine vs. temperature.

Table 2 and Fig. 3 summarize values of the physical properties of glycerine. Except for the viscosity all properties change only insignificantly with temperature. In addition to the strong temperature-dependency of the viscosity of glycerine its hygroscopic character has to be mentioned, because of the capture of moisture from the ambient air.

Like many liquids the viscocity-temperature dependency can be empirically described by a simple exponential law. From independent measurements of dynamic and kinematic viscocity and density we found:

$$\nu[\text{cSt}] = -0.7 + \exp\left\{\left(\frac{T[^\circ C] + 273}{471.28}\right)^{-3.951}\right\} \tag{4}$$

with a gradient of

$$\frac{1}{\nu} \cdot \frac{d\nu}{dT} = 0.069\ 1/^\circ C \tag{5}$$

at 35°C.

Because the viscosity directly influences the size of the viscous sublayer a temperature control is obviously necessary, which is done by means of a by-pass loop with heat-exchanger. The cooled glycerine is re-introduced into the tunnel upstream of the driving pump. Since the operating temperature is about 10 to 15°C above ambient air-temperatures a temperature and viscosity variation across the boundary layer would occur. In addition the wall-temperature would vary with changes in ambient temperature conditions and influence the size of the viscous sublayer.

Fiberglass insulation of the entire tunnel-circuit keeps the temperature difference between pipe-centerline and wall to less than 0.3°C, (Bakewell 1966), which is considered to be sufficient.

3. Velocity Sensors

3.1 General Description
In order to evaluate the complete cross-correlation tensor as needed for Lumley's orthogonal decomposition scheme, it is sufficient to measure only two velocity components.

Presently we measure with commercially available, single-ended hot-splitfilm sensors the axial and circumferential velocity component simultaneously. The sensor consists of a cylindrical, single-ended quartz body which carries two separate hot-films, each operated by a separate constant-temperature anemometer unit.

Some physical data of these sensors are collected in Table 3. The axis of the sensor is oriented roughly perpendicular to the pipe wall.

The actually measured signal is the response to a spacial average over one half of the viscous length perpendicular to the wall and one fourth parallel to the wall. Related to these errors of spacial averaging are the velocity disturbances

Sensor Diameter	D	0.153 mm
Length of Hot-Film	L	0.254 mm
Thickness of Hot-Film		0.001 mm
Typical Electric Resistance	R	6 - 8 Ω
Temperature Gradient of R $\dfrac{1}{R}\dfrac{dR}{dT}$		0.002°C^{-1}
Substrate Material		Fused Quartz
Density of substrate	ρ_s	2100 kg/m^3
Heat-Capacity of s.	C_{ps}	1.65 · 10^6 Nm/(m^3·°C)
Thermal Conductivity of s.	λ_s	1.38 W/(m·°C)
Thermal Diffusivity of s.	α_s	8.36 · 10^{-7} m^2/sec

Table 3. Physical Data of typical Hot-Splitfilm Sensor as encountered during our experiments.

introduced by the very presence of the probe. For measurements within the viscous sublayer the probe Reynolds number based on sensor diameter and local mean velocity simply reads:

$$Re_{probe} = \frac{d}{\ell_\nu} \cdot y^+ \qquad [6]$$

with ℓ_ν the viscous length as defined by the ratio of kinematic viscosity and friction velocity, and y^+ the dimensionless wall distance.

For $y^+ = 1.25$, our closest wall distance we obtain

$$Re_{probe} = 0.34$$

so that the velocity disturbances surely propagate several times the probe dimensions into the flow. It should be mentioned that this is not a unique feature of our experimental set-up but holds for all sublayer investigations.

3.2 Steady-state Response of Hot-film Sensors
Several potential problems have to be discussed to insure proper working conditions of the sensors.

Since glycerine is electrically non-conductive uncoated hot-film probes can be used in order to avoid electrostatic build-up. On the other hand, glycerine is chemically slightly aggressive to most metals. If air or oxygen have access to a surface wetted with glycerine we observed the products of chemical reactions with aluminum, brass, steel and the platinum-alloy of the hot films.

From equation [6] we infer that the expected Reynolds number of the probe is of order unity. For flow around cylindrical bodies several flow situations occur at different Reynolds numbers, e.g. attached viscous flow up to Re = 5, separation with two steady, symmetrically situated vortices at the rear of the cylinder for up to Re = 40, and regularly shedding vortices in the region of Re up to 150. The highest occurring Re-number in our investigations is 5, but based on the 'free-stream' fluid temperature. Neither in our steady-state probe-behavior investigation in a special calibration pipe nor through wave-analysis of obtained turbulent signals could we detect any hint of signals induced by the probe's own shedding vortex-street.

Because relevant data about split hot-film sensors were completely unknown we built a small calibration station inserting the probe into a stationary, laminar pipe flow.

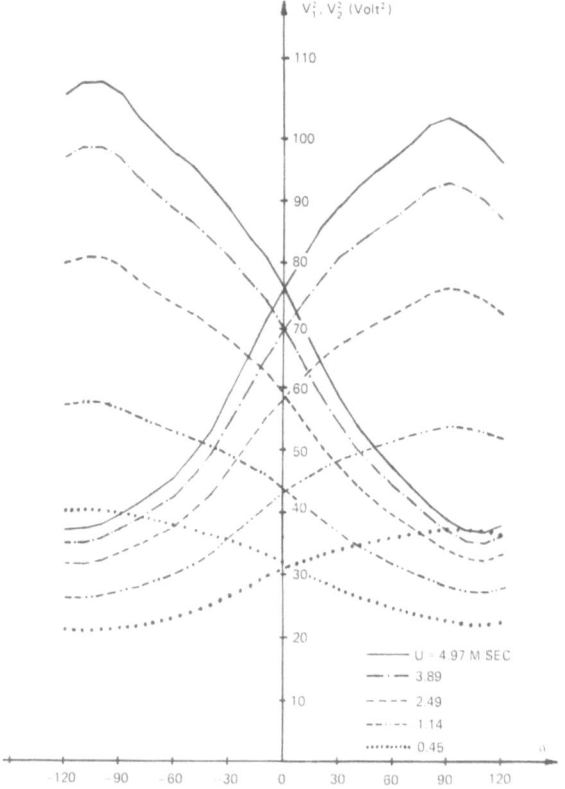

Fig. 4a. Energy output of the two films of adjusted sensor vs. angle of attack at various velocities.

Fig. 4a shows the squared voltage-output of the two anemometer units which form one probe as function of angle of attack with the normal velocity as parameter.

Like most high-viscosity liquids glycerine has an extremely high Prandtl-number, see Table 2, so that the temperature field around the probe is (in comparison to the velocity field) confined to the immediate vicinity of the probe. This explains the irregularities of the calibration curves as immediate response to the physical splits in the carrier body separating the individual film-segments. Specifically the angular dependency of the total energy output of the probes, as shown in Fig. 4b, at higher velocities is not really satisfying.

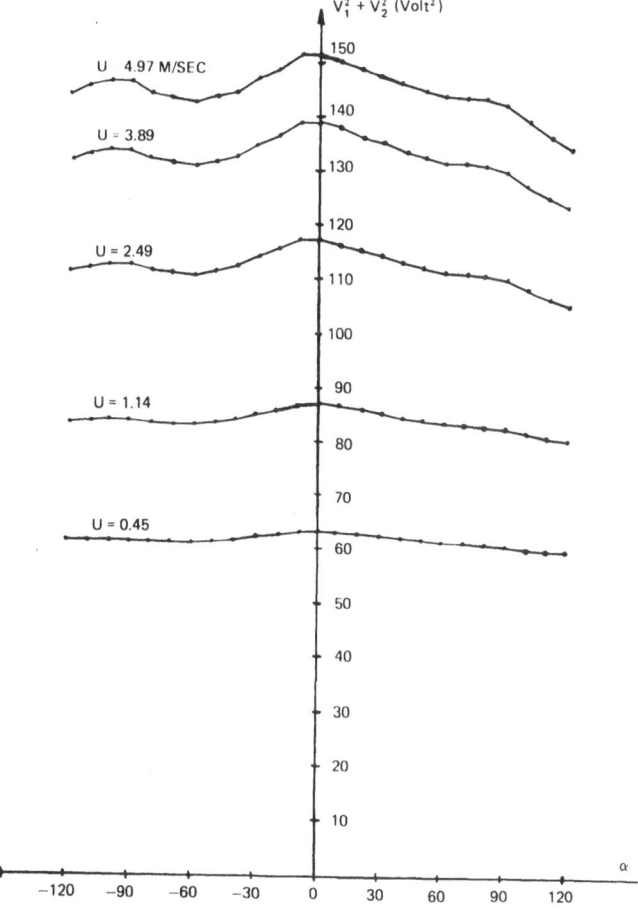

Fig. 4b. Total energy output of sensor vs. angle of attack at various velocities.

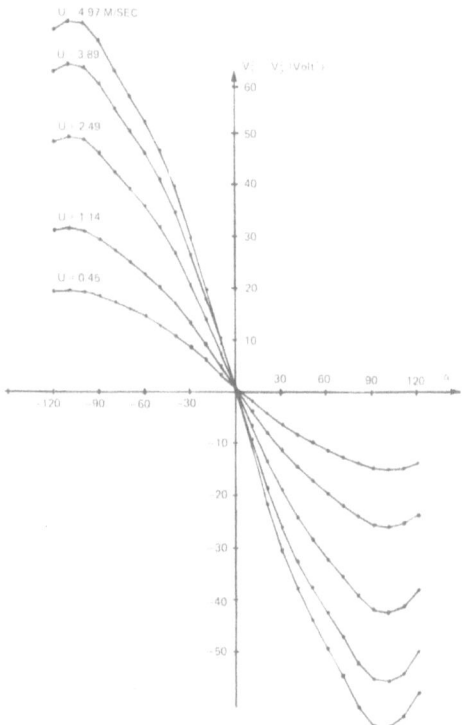

Fig. 4c. Difference in energy-output of the two films of a sensor vs. angle of attack at various velocities.

Generally speaking we found it extremely difficult to operate the hot-splitfilm sensors. The adjustment of the operational resistances of the two film-units with respect to each other is a very time-consuming task. Fig. 4d shows the voltage-output of the two films at constant velocity for varying angle of attack with two different operational resistances for one of the films. The overheat-temperature of the two films is roughly 35°C. Clearly visible is the effect of the thermal coupling between the two films via the substrate body.

Chemical leaching, altering effects and changes in the ambient temperature conditions of the anemometer units unfortunately never allowed us to reproduce calibration curves over an extended time, say a few hours.

During the main experiments we constantly have to readjust the hot-film sensors.

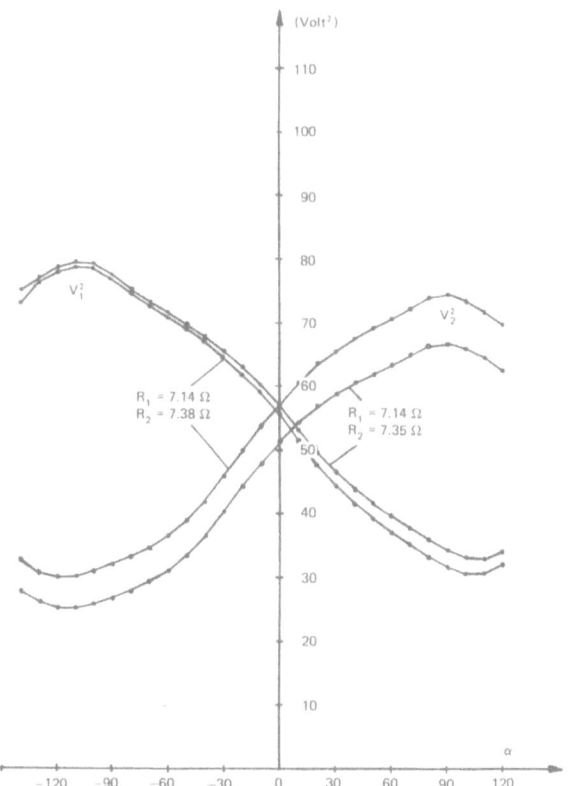

Fig. 4d. Influence of change in operational resistance onto sensor response at constant velocity.

3.3 Theoretical Investigation of Probe-Frequency Response

In the field of anemometry the term constant-temperature anemometer (CTA) can be seriously misleading because the very function of such an anemometer unit is only to hold the overall electric resistance of the hot-film or wire constant, leaving the actual temperature distribution subject to the local, instantaneous heat-fluxes from film to surrounding fluid and into the substrate body. Although this is generally known, detailed investigations of probe frequency response are scarce and somewhat discouraging.

In order to evaluate the frequency response of our probes we are presently developing a computer code to simulate the thermophysical and electrical properties of a hot-splitfilm sensor. Preliminary results convinced us that our probes have a sufficiently flat frequency response. Although a separate report is in the preparation we would like to bring attention to some ideas and results of our theoretical investigations.

The starting point of our interest was some basic calculations of a simple thought-model. Two electric resistors (i = 1,2), the resistances of which vary with temperature

$$R_i = R_o + S \cdot T_i$$

are connected electrically parallel and operated upon by a CTA-unit holding the overall resistance constant. For each resistor the internal electric heat-generation is balanced by losses to its surrounding with a certain heat-transfer coefficient H_i and temperature difference T_i. For not too large overheat-ratios $(ST_i/R_o \ll 1)$ one roughly obtains

$$\frac{H_i \cdot T_i}{H_2 \cdot T_2} = \text{constant.}$$

Transferred to our cylindrical hot-film sensors this means that the circumferential distribution of temperature difference, film to fluid, shows variation to the same degree as the distribution of the local heat-transfer coefficient does.

Because for our sensors we estimate a characteristic frequency of propagation of thermal events to be $\omega = \alpha/d^2 = 36$ Hz more detailed investigations were appropriate.

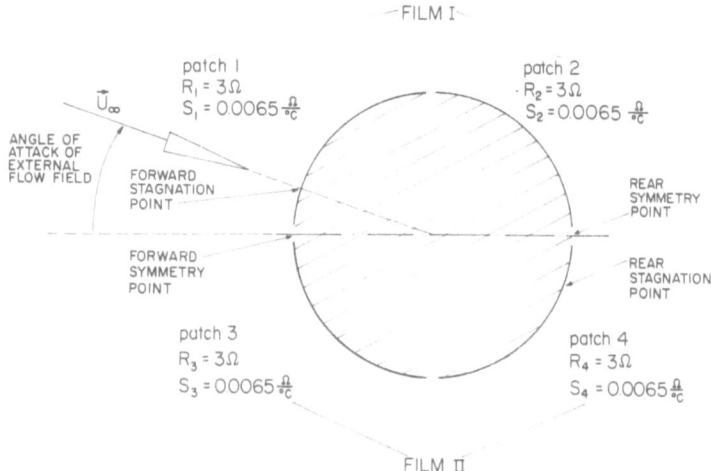

Fig. 5a. Cross-section view of hot-splitfilm sensor.

On the basis of numerical calculations of the flow field around circular cylinders at low Reynolds numbers we were able to develop a corresponding, simple heat-transfer model for high Prandtl number fluids. Representing the heat-transfer within the substrate body's cross-section by a simple lumped-node model and degrading the anemometer units to black boxes, the function of which simply is to hold the overall resistance of their respective films constant, a relatively small computer program is able to simulate thermal events in a plane perpendicular to the sensor axis, neglecting for the time being any axial effects. With the physical properties of the sensor according to Table 3 (cross-section view of sensor see Fig. 5a) we obtained a circumferential temperature distribution as shown in Fig. 5b.

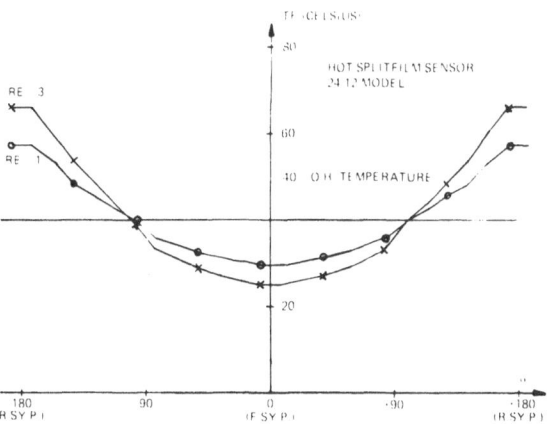

Fig. 5b. Circumferential temperature distribution as obtained with 2-dimensional computer model.

The two distributions shown are stationary solutions obtained at constant angle of attack for Reynolds Numbers Re = 1 and 3 and are representative for distributions at other Reynolds numbers and angles as well. The operational electric resistances of the two equal films were selected to give a nominal mean temperature difference, film to fluid, of 40°C. The heat fluxes inside the substrate body are of equal order of magnitude as those from the film to the fluid, damping considerably the temperature maximum at the rear symmetry point and the minimum at the forward symmetry point. Associated with variations in temperature distribution is a thermal enertia. The transient response of our sensor-model to stepfunctions in external flow conditions (Reynolds number or angle of attack) is shown in Fig. 5c, 5d. VF(1) and VF(2) are the voltage responses of the two films. The asymptotic approach to the steady-

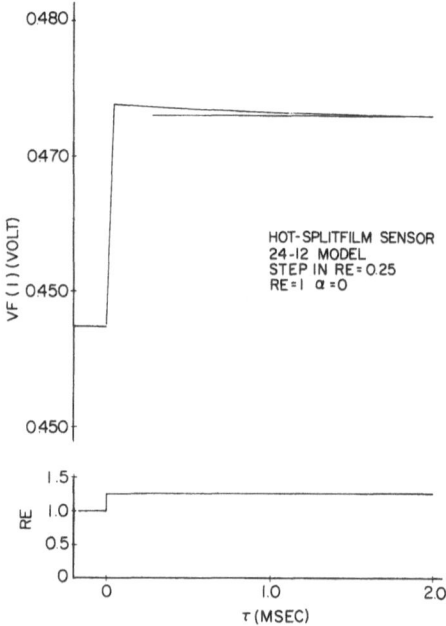

Fig. 5c. Sensor response to step in Reynolds number. (Theoretical).

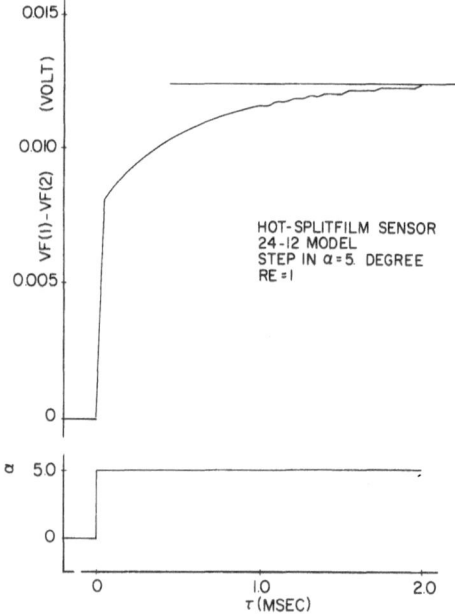

Fig. 5d. Sensor response to step in angle of attack.

state solution is exponential and the time-constants associated with it are roughly

$$\tau^{-1} = 50 \cdot \alpha/d^2 \qquad \text{for rotation}$$
$$= 30 \cdot \alpha/d^2 \qquad \text{for step in Reynolds number.}$$

Whether the sensor response yields over- or undershooting in the first moment and how strong this is is hard to predict at this stage of our investigations and depends very much on the assumed external heat-transfer coefficient distribution.

3.4 Summary of Theoretical Sensor Response Calculations
As was said earlier our computer model neglects any axial effects of the heat-transfer inside the substrate body.

Axial heat-conduction will damp the magnitude of the thermal inertia effects but will shift them to much lower frequencies as associated with the length scale of the axial sensor dimensions.

Acknowledgements
This work was supported in part by the U.S. Office of Naval Research through the General Hydrodynamics Research Program of the David W. Taylor Naval Ship Research and Development Center (Bethesda, MD), and in part by the Fluids Engineering Unit of the Applied Research Laboratory, The Pennsylvania State University (University Park, PA). The formulation of the investigation benefitted from many helpful discussions with W.K. George, Jr.

References
1. Bakewell, H.P., 1966. An experimental investigation of the viscous sublayer in turbulent pipe flow. Ph.D. Thesis, The Pennsylvania State University
2. Bakewell, H.P and Lumley, J.L., 1967. Viscous sublayer and adjacent wall region in turbulent pipe flow. *The Physics of Fluids*, 10, 1880
3. Comte-Bellot, G. 1963. Contribution à l'étude de la turbulence de conduite. Ph.D. Thesis, L'Université de Grenoble
4. George, W.K., Jr., 1978 *Proceedings of the Dynamic Flow Conference, Marseille and Baltimore, 1978*
5. Huber, A. 1974. Calculations of transport coefficients for a numerical model of turbulence applied to a plane wake. M.S. Thesis, The Pennsylvania State University
6. Lumley, J.L., 1967. The structure of inhomogeneous turbulent flows. *Atmospheric Turbulence and Radio Wave Propagation* (A.M. Yaglom and V.I. Tatasski, eds.) Publishing House Nauka, Moscow, pp. 166-188
7. Lumley, J.L. 1970. Stochastic Tools in Turbulence. Academic Press, NY.

Data Analysis of Atmospheric Measurements

by

N.E. Busch, S.E. Larsen and D.W. Thomson[*]
Risø National Laboratory, Roskilde, Denmark

I. Introduction

Dynamic measurements in unsteady atmospheric flows are greatly complicated, principally, by the fundamental properties of the characteristically inhomogeneous and nonstationary natural atmosphere. Since meteorological processes of everyday human interest range from evolution of major weather systems to short range diffusion of anthropogenic pollutants, the dynamic range of required atmospheric measurements extends essentially from global to atomic spatial and temporal scales.

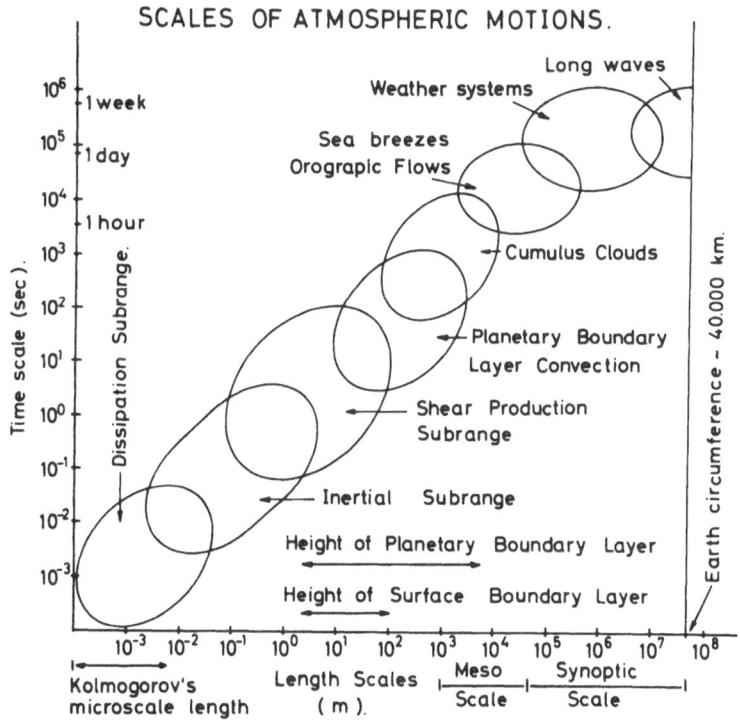

Fig. 1. Spatial and temporal scales for different atmospheric processes.

[*]Present affiliation: Pennsylvania State University, University Park, P.A., USA

Fig. 1 shows length and time scales associated with various atmospheric phenomena. In addition to the range of scales involved, it is important to note the
overlap between the scales for the different phenomena. This overlap results in
a continuous spectrum of the basic internal thermodynamic and dynamic
variables.

We have not attempted in this paper to review the problems of large-scale
meteorological measurements for analysis and prediction of "weather" phenomena. Although there exist many unsolved details regarding the fluid dynamic behaviour of the mixed, radiation absorbing, moist atmosphere (on the rotating earth), its general behaviour can be sufficiently well measured so that
little impetus appears to exist for making major changes in the existing weather observation system. In fact on the other hand, it may be the sheer logistic
inertia of the international network rather than the lack of recognition of de-
ficiences in its combination of regular (1 per hour to 2 per day) in-situ measure-
and satellite observations that discourages adoption of improved measurement
techniques.

However, this is certainly not the case in ongoing studies of the atmosphere's
planetary boundary layer (PBL) and the surface layer (SL) within it. Our
present understanding of the physical structure of and processes occurring
within these regions of the lower atmosphere has been extraordinarily dependent upon technological developments in measurements and data processing.
Furthermore, continued progress toward the goal of being able to analyse and
predict motions and structure in the lower atmosphere for the "real" situation
of normal weather conditions above typical complex terrain locations will
depend upon how clearly we can extract heretofor unavailable information
from both existing and new in-situ and remote measurement systems.

In this paper we review and contrast the observational and data processing requirements and procedures for several types of micrometeorological experiments. These include: a surface layer experiment at a location with the simplest possible underlying ground surface during "steady-state", readily analyzed
atmospheric conditions; a boundary layer experiment also conducted in the
simplest possible location and conditions and, finally, a boundary layer experiment performed at a simple "complex" terrain location.

II. Regarding Interpretation of Atmospheric Measurements
Even the simplest micrometeorological measurements, time series of wind
speed, temperature and humidity from which mean values can be deduced require painstaking care if they are to be of sufficient quality for estimation of

fundamental quantities such as the vertical fluxes of momentum, heat and moisture. In order to obtain spatially and temporally invarient parameter values, it is, first of all, necessary that the instrument site be in uniform, homogeneous terrain. For SL measurements this means *flat* with either a bare ground surface or uniform vegetative cover of at most a few cm height. Observations are best made in at least "quasi-stationary" conditions, but low frequency variations in SL wind speeds and temperatures occur continuously as a part of the natural diurnal and weather-system cycles. Hence, stationarity in an absolute sense is virtually impossible to attain. In order to achive quasi-stationarity, measurements are normally made during unchanging weather conditions at times of the day where changes in the diurnally varying heat flux is least pronounced.

Since the theoretically relevant equations are easily established from the governing equations by separating the flow field into a constant mean and fluctuating part, much research is directed toward relating different "mean" quantities of variables to each other within the above mentioned framework of assumed quasi-stationarity and homogeneity.

But even for those conditions where the assumption of stationarity/homogeneity can be defended, the experimenter must determine which time/space averages he will use to approximate the theoretical concept of an ensemble average.

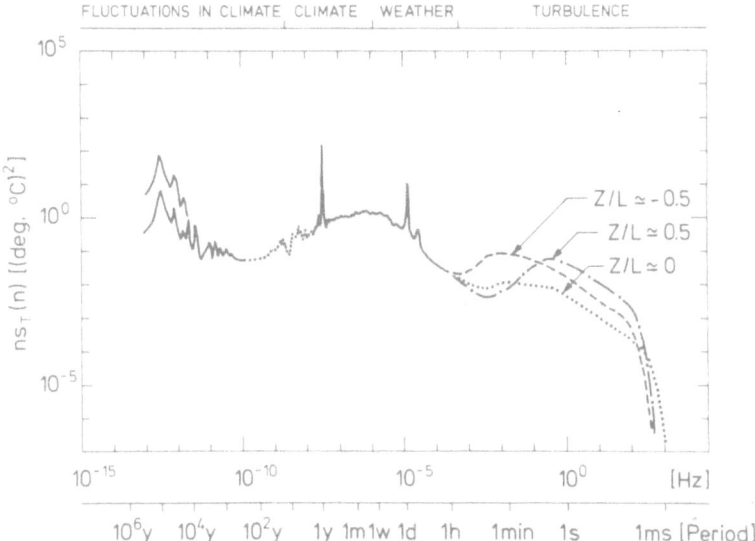

Fig. 2. Spectrum of atmospheric temperature fluctuations at mid latitude close to the surface. z/L is the Monin-Obuchov stability index.

As guidance for such decisions, the spectrum of the involved variable is often used. Consider the example shown in Fig. 2, which summarizes the frequency spectrum of temperature variations close to the surface at mid latitude. In this example the frequency scale ranges from ice-age periods of 10^5 years to the dissipation cut-off at around 1 ms. In the time scale range relevant to most micrometeorological experiments, the spectrum has been split according to thermal stability into three parts.

In stable situations ($z/L > 0$) a spectral gap exists between the turbulence and lower frequency regions. If such a gap is reasonably broad and one chooses averaging time scales within it, mean values will vary so slowly with respect to the fluctuations that the assumption of quasi-stationarity will be well fulfilled. Furthermore, the variance of the fluctuations (in Fig. 2 the temperature variance) will be relatively insensitive to the precise value of the averaging time used, a fact which facilitates comparisons between independent experiments.

It should be mentioned, however, that the above neat picture is somewhat disturbed by the fact that turbulence spectra shift towards lower frequencies with increasing height. Hence, in principle different averaging times should be used for different heights.

In thermally unstable situations ($z/L < 0$) analysis is complicated by the fact that the gap fills up due to convective activity. Now the concept of stationarity is ambiguous because one must either extend the averaging time so far towards the daily cycle that conditions obviously will be changing during an averaging period, or else let the averaging time decrease to a time scale interval which may be associated with the process of interest. Experimenters normally choose to use the shorter averaging times. This, however, introduces all kinds of statistical analysis problems because of the high energy content in the low frequency part of the measured time series. Obviously, estimates of the various statistics of the fluctuations can become rather sensitive to the selected averaging times.

In spite of such problems the scheme of separating the flow into mean and fluctuating parts has proven useful for analysis of unstable conditions. One reason for this is that not all of the observed variables show as bad a behaviour in the gap in unstable situations as the illustrated temperature variations. Some of the best behaved spectra have been those pertaining to important quantities such as the vertical fluxes of momentum, heat and moisture. Hence, the basic philosophy in choosing averaging times has been to get the best possible estimates of those fluxes, and then deal with the problems associated with other quantities as they appear.

III. Surface Layer Techniques and Experiments

The existence of ideal experimental conditions and optimized analytic techniques does not in any way reduce the performance demanded of the meteorological sensors which are used in the field. Since the wind speed varies approximately logarithmically with height above the surface from 0 to, typically, three to five $msec^{-1}$ at three to ten m height, individual anemometers must have a threshold velocity of a few $cm \cdot sec^{-1}$, be calibrated to within less than $0.01\ msec^{-1}$, and at least four must be used in order to adequately define the linearity (on a log-linear plot) of the vertical speed profile.

In neutral conditions, the vertical temperature gradient is less than $0.01°C \cdot m^{-1}$. Hence, mean surface layer temperature differences must be resolved to within a few hundredths of a degree C in an environment in which the radiative heating error of a poorly designed sensor may be several 10th's of a °C and the rms value of natural temporal variations may be nearly one °C.

Accurate atmospheric humidity measurements are difficult in the best of circumstances. In the surface layer the vertical humidity gradient is normally determined by directly sensing wet-bulb temperature differences.

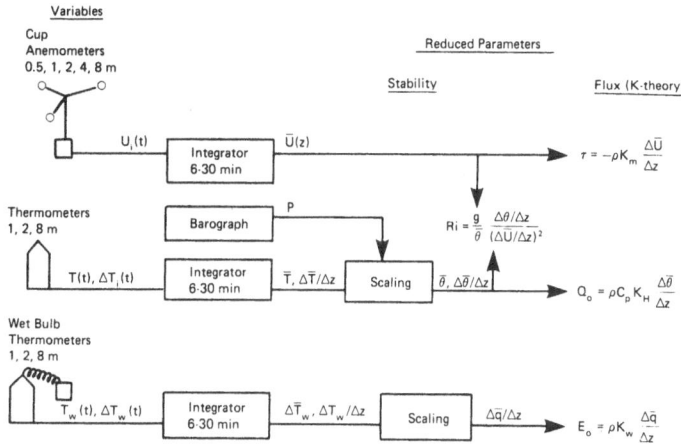

Fig. 3. Surface layer mean profiles and derived parameters.

One of the first definitive surface layer experiments which included most of the components necessary for analysis of surface layer energetics and vertical momentum, heat and moisture fluxes was the so-called Great Plains experiment conducted at O'Neill Nebraska in 1951 (Lettau and Davidson, 1957). Fig. 3 summarizes the basic "K-theory" method which was used to guide and interpret most of the Great Plains measurements. Observations from this program are still used for evaluating new analytic and numerical surface layer models.

The next level of complexity in a surface layer experiment is to modify or replace the basic sensors so that parameter fluctuations may also be sensed and recorded. For example, cup anemometers may be replaced by 2 or 3-d ultrasonic anemometers, hot-wire anemometers, sensitive propellers or drag anemometers. Aspirated mean-value thermometers may be replaced with fine-wire resistance of thermocouple probes of sufficiently small cross-section ($\leqslant 25~\mu m$) so that radiative heating errors are negligible. Although the increased costs of using delicate, fast response sensors include their propensity for catastrophic failure and the necessity for more sophisticated, wider bandwidth data processing and recording, the advantages, particularly for research applications, of being able to sense and record individual parameter fluctuations are easily justified. Fig. 4 clearly illustrates some of the advantages insofar as reduced parameters such as fluxes are concerned. There is simply no way, for example, in which problems associated with counter-gradient fluxes can be studied on the basis of mean-profile observations only.

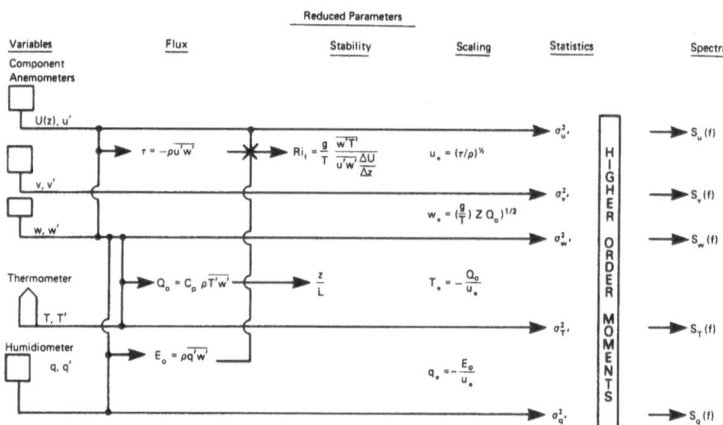

Fig. 4. Surface layer turbulence measurements and derived parameters.

With few exceptions our classic knowledge of turbulent processes in the surface layer is based on interpretation of measurements such as those shown in Figs. 3 and 4. The processing of the real time series from which the various reduced parameters are derived is in principle straightforward although it requires processing of extensive data sets. Obviously the difficulties lie in the evaluation of the statistical significance of the results and in their interpretation.

One of the last intensive field measurement programs which was limited to surface layer measurements was the 1968 Kansas experiment (Izumi, 1971, Kamial et al., 1972). It was specifically designed to study the behaviour of the dry atmospheric SL, especially with the aim of verifying, and establishing the limits for, the validity of the Monin-Obuchov hypothesis.

The Monin-Obuchov hypothesis is based on the observation that for stationary and horizontally homogeneous conditions, there exists a layer close to the surface, where momentum and heat flux ($\overline{u'w'}$ and $\overline{w'\theta'}$) may be considered independent of height, z. In this layer local statistics of the flow, such as vertical mean-gradients, cross- and co-variances and so forth, will be universal-functions of z/L, when properly nondimensionalized by $\overline{w'\theta'}$, $\overline{u'w'}$ and z. The so-called Monin-Obuchov length L is given by

$$L = \frac{T \, |\overline{u'w'}|^{3/2}}{kg \, \overline{w'\theta'}}$$

where g is acceleration due to gravity and k is the universal v. Kármán constant.

KANSAS 68   SURFACE   LAYER   EXPERIMENT.

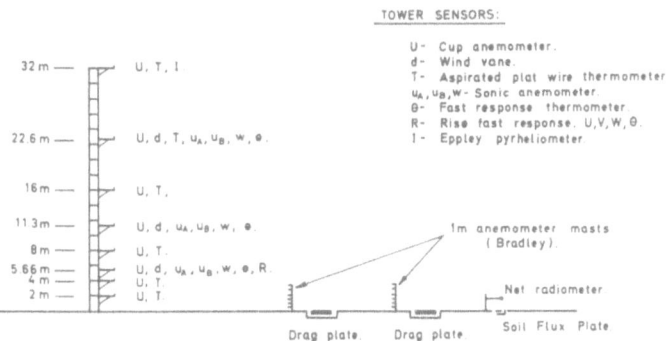

Fig. 5. Experimental set-up for the Kansas 1968 experiment.

On Fig. 5 the experimental set-up is shown. As seen, it was centered around one 32 m instrumented tower, which means that the data taken consisted of time-series at different z-levels. The experiment was, therefore, totally dependent upon Taylor's hypothesis to obtain relevant length scales.

From the description of the Monin-Obuchov hypothesis it is clear that an experiment to test its validity must provide independent measurements of $\overline{u'w'}$, $\overline{w'\theta'}$ and mean gradients. Because of the need for establishing accurate estimates of vertical gradients of temperature and velocity along the tower, mean values were measured fairly densely, while turbulent fluctuations were measured with sonic anemometers, hot wires and fast thermometers at only three heights. One reason for fewer levels with turbulence instrumentation was the much greater difficulty of running such instrumentation not only in terms of keeping them properly functioning, but also in terms of handling the much larger data flow. A second reason for more levels for mean value measurements was that the Monin-Obuchov hypothesis, in relation to the mean flow, concerns vertical mean-flow gradients, which experimentally are generated by differencing data from two levels. On the other hand, estimates of quantities like $\overline{u'w'}$ and $\overline{w'\theta'}$ need only one level of turbulence instrumentation. In the Kansas experiment three levels with turbulence instrumentation were used because three is one more than the absolute minimum required to establish height variations of turbulence statistics, and because it is just enough to estimate height variations of turbulent transport terms like $\partial/\partial z \; \overline{w'u'}$.

Due to the overwhelming importance of the $\overline{u'w'}$ correlation in the Monin-Obuchov hypothesis, surface drag was also measured directly by means of drag-plates.

Table I gives an impression of the kind of statistics which are output from such experiments, the number of sensors involved, and the kind of spatial and temporal resolution which is needed.

Let us relate these columns concerning averaging time and digital filtering to our discussion of averaging times in section II. The basic averaging time is 1 hour. For quantities such as fluxes and mean values, however, 15 min. mean values are considered first, thus making it possible to discard non-stationary runs. For those statistics which are computed on the basis of 1 hour straight averaging, the same digital bandpass filter is applied to all data. This, of course, makes the computed variances and cross variances somewhat artificial, since they reflect the position of the spectrum relative to the band pass region as well as true energy content. But at least the computational scheme is well defined. Note further that the fluxes are computed both by averaging 4, 15 min. averages and by the bandpass method, hence the sensitivity of these critical parameters to the computational method can be checked. Finally, the spectra themselves are computed for selected runs, not only because they are of physical interest themselves, but also because one thus obtain further checks on the variance and cross-variance estimates. That the spectra are computed in two overlapping frequency regions is due merely to computational problem, namely that even the Fast Fourier Transform method is costly when used to resolve data covering a large frequency interval.

TABLE I.	Raw Analysis Variables	Instruments	No. of Measuring Heights	Spatial Resolution Instruments	Pre-recording Filtering	Recording Rate	Digital Filtering	Avg. Time	Spectral Analysis	No. of 1 h runs Analyzed
Mean Values	$u(z)$	cups	13	—	none	1 Hz	none	4x15 min.		32
	$T(z)$	Pt-wire	8	—	—	—	—	—		—
2nd order Moments	$\overline{uw}, \overline{w\theta}, \overline{u\theta}$	Dragplates	1	D.P. 1 m	Low Pass	20 Hz	Band pass $3.3 \cdot 10^{-3}$ –5 Hz [none]	4x15 min. 1 h		32
	$\sigma^2(u,v,w,\theta)$	Sonic ane.	3	20 cm	Dragplate	—		1 h		—
	$CO(u,w,\theta)(n)$	Pt-wires			5 Hz else 10 Hz			1 h	FFT $3 \cdot 10^{-3}$ – 0.6 Hz	15
	$S(u,w,\theta)(n)$							—	$5 \cdot 10^{-3}$ – 10 Hz	—
3rd Order Moments	Skewness (u,v,w,θ)	Sonic ane.	3	20 cm	Low Pass 10 Hz	20 Hz	Band Pass $3.3 \cdot 10^{-3}$ – 5 Hz	1 h		32
	$\overline{u^2 w}, \overline{v^2 w}, \overline{w^2\theta}, \overline{uw^2}, \overline{w\theta^2}$	Pt-wires	—	20 cm	—	—		—		—
4th Order Moments	Kurtosis (u,v,w,θ)	Sonic ane.	3	20 cm	Low Pass 10 Hz	20 Hz	Band Pass $3.3 \cdot 10^{-3}$ –5 Hz	1 h		32
		Pt-wires	—	—	—	—		—		—
High Frequency	$\epsilon = 15\,\nu(\partial u/\partial x)^2$	Hot-wire linearized	3	1.5 mm	Differentiation Low Pass, 2 kHz	Analog digitizing 3 kHz	none	1 h		52

Table 1. Summary of measurements during the Kansas 1968 experiment. $Co(u,w,\theta)(n)$ means Co-spectra involving two of the signals in the first paranthesis as function of the frequency, n. The spectra, S, are described correspondingly.

Another important category of SL experiments is that in which measurements are restricted essentially only to one level but in which an array of sensors is arranged such that the spatial characteristics of mean and turbulent parameter fluctuations can be studied. Spatial properties from single point observations can be inferred only through application of the Taylor hypothesis and transformation using the mean velocity from the time to space domain. One important exception is, of course, aircraft measurements in which the platform velocity is so high with respect to most micrometeorological velocity scales that the turbulence may be regarded as "frozen".

Contemporary separated sensor experiments may be roughly divided into two groups. In the first, commonly referred to as "coherence" experiments the sensors such as anemometers are distributed along and across the mean flow at separations ranging from a minimum of a few meters to a maximum array dimension of one to two kilometers. From the reduced coherence data it is possible to extract many details regarding not only the spatial characteristics of eddies and, thus, their effect on structures such as buildings and bridges but also relations such as those between the Lagrangian and Eulerian time scales (Norman et al., 1976).

A second important group of separated sensor experiments are the measurements, normally at relatively small separations ($\leqslant 1$ m) of parameter fluctuations from which the so-called structure parameters may be deduced. If, for example, we measure the temperature difference fluctuations across a distance r, they may be used to evaluate the 1-dimensional temperature structure parameter

$$C_T^2 = \frac{\overline{(T(x) - T(x+r))^2}}{r^{2/3}}$$

For isotropic turbulence $C_T^2(r)_{1\text{-}d}$ may be related to $C_T^2(\lambda/2)_{3\text{-}d}$ which will be sensed by for example, a monostatic indirect sensor sodar system, operating at wavelength λ (Thomson et al., 1978).

Table II summarizes the important functional relationships which determine the meteorological parameter sensitivity of selected remote probing systems. If the appropriate parameter differential sensors are not available, structure parameters may also, of course, be deduced from the power spectrum evaluated from a single sensor time series.

With the exception of some optical anemometers, remote probing systems are normally employed as "virtual" towers for PBL and other tropospheric mea-

Remote Sounding System	Output Signal	Relevant Refractive Index Parameter	Meteorological Variable(s)
Line-of-sight optical	$A(t)$	$C_{N_o}^2$, C_T^2	$N_o \cong 77.6\ P/T$
Optical anemometer	$A(x,t)$, $A(x+\Delta x,t)$	''	Transverse wind
Lidar	$A(t)$	Particulate Scattering	Particulate Properties
Monostatic sodar	$A(t)$	$C_{N_{ac}}^2$, C_T^2	$N_{ac} \cong 10^6\ (\frac{T_o - T}{2T_o})$ ★
Bistatic sodar	$A(t)$	$C_{N_{ac}}^2$, C_V^2, C_T^2	$N_{ac} \cong 10^6\ (\frac{T_o - T}{2T_o} - \frac{V\cos\phi}{C_o})$ ★
Doppler sodar	$\phi(t)$, $f_d(t)$	''	u, v, w
Radar	$A(t)$	$C_{N_{mw}}^2$, C_T^2, C_q^2	$N_{mw} \cong \frac{77.6}{T}\ (P+4810\ e/T)$
Doppler radar (incl. FM-CW)	$\phi(t)$, $f_d(t)$	$C_{N_{mw}}^2$	u, v, w

★ Neglects small humidity contribution 0.14 e/p

Table II. Remote sensing of selected meteorological variables.

surements. However, the interpretation of the observations, particularly of turbulence, made with them is based upon our knowledge of parameters such as the structure functions, which have been most extensively studied in the surface layer environment. Consequently, the application of many remote probes, such as sodars, for quantitative PBL measurements depends upon how confidently we can extrapolate our knowledge of selected surface layer processes to those which occur at higher altitudes.

IV. Reaching into the PBL

Historically, as well as logically, the 1973 Minnesota experiment was an extension of the Kansas-experiment. As may be seen on Fig. 6, it is basically an extension of the Kansas experiment to greater heights (Izumi and Caughey 1976, Kaimal et al., 1976).

The purpose of this experiment was to study the processes in the unstable boundary layer. The recording equipment and the ground-based sensor system were with minor modifications the same as used in Kansas. For measurements at greater heights specially designed sensors were mounted on the tethering cable of a large captive balloon.

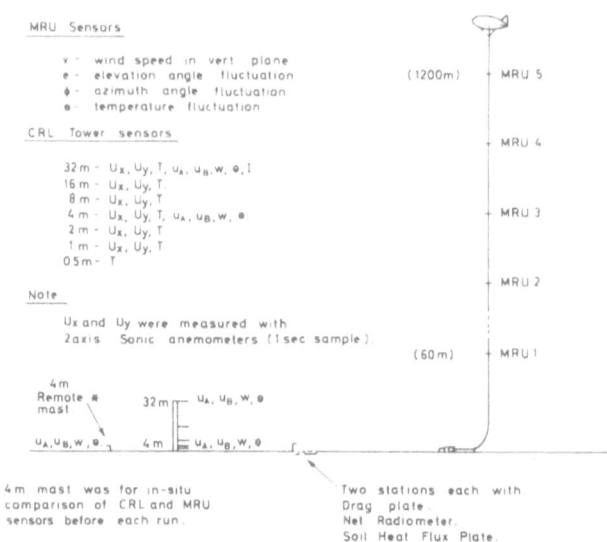

Fig. 6. Experimental set-up for the 1973 Minnesota experiment.

Fig. 6 also shows a small mast which was used for periodic intercomparison between the balloon-based and the tower-based sensor system. This was of crucial importance since the sensors on the two platforms were of significantly different design. The problem of using different sensor systems to measure spatial variations of the same variables was, of course, also complicated by the fact that the balloon data inevitably was influenced by the motion of the tethering cable. For this reason several years with intercomparison experiments preceded the final experiment.

In comparing the Kansas and Minnesota experiments, we note first that more mean variables have appeared. The turning of the mean wind with height, induced by the Coriolis force, makes it important to measure the components of the mean velocity (in this experiment, the mean wind direction at the lower level was taken as reference). The same aspect showing up in the stress computations now containing two component also, $\overline{u'v'}$ and $\overline{u'w'}$.

A new scale height appears. The height z_i of the unstable boundary layer becomes an important parameter because it constitutes the vertical scale of limitation for the flow motion throughout a large part of this layer. In this experiment z_i was taken as the height to the capping temperature inversion.

TABLE III.	Raw analysis variables	Sensors		Measuring Levels		Recording rate	Electronic filtering	Averaging time	Digital filtering	Spectral bandwidth	No. of runs
		Tower	Balloon cable	Tower	Cable						
Mean quantities	$u_x(z)$ $u_y(z)$ $w(z)$ $T(z)$ z	sonic anem. quartz cryst.	cup anem. hot-wires wind vane Pt-wire radio tracking Theodolite	6 7	5	1 Hz		75 min	high pass 10^{-3} Hz		
	$z_i(z)$	Rawinsonde				every 2 hours					
Second order moments	$\overline{uv}, \overline{uw}, \overline{w\theta}$ $\sigma^2(u,v,w,\theta)$ $S(u,v,w,\theta)(n)$ $Co(u,v,w,\theta)(n)$	sonic anem. Pt-wires	cup anem. hot-wires wind vane	2	5	10 Hz	5 Hz	75 min	high pass 10^{-3} Hz	FFT $2\cdot10^{-4} - 4\cdot10^{-2}$ Hz $2.5\cdot10^{-3} - 5$ Hz Corrections for HP filtering	11
Third order moments	$\overline{w^2\theta}$ $\overline{w\theta^2}$	sonic anem. Pt-wires	cup anem. hot-wires wind vane	22	5	10 Hz	5 Hz	75 min	high pass 10^{-3} Hz		

Table III. Summary of measurements during the 1973 Minnesota experiment. The notation is described in the text and in the legend to table I.

As compared to the surface boundary layer motions in the unstable boundary layer, we have larger associated time scales. In Table III this shows up in slower recording rates and larger averaging times, as compared to Table I. Also note that the experimenters here found it necessary to high pass filter also the data processed for spectral computations, as opposed to the strategy in Table I. The high pass filtering here works as a detrending procedure, of importance because spectra based on data with trends and calculated with fast Fourier transform technique will be deformed in the whole frequency region considered.

The difficulties associated with even a limited objective PBL experiment are manyfold larger than those for an SL experiment. With about the same preparation time and effort the Kansas experiment resulted in a definitive description of the horizontally homogeneous surface layer. However, the Minnesota experiment was able to provide insight into only limited aspects of the PBL's behaviour.

V. A "Simple" Complex Terrain Experiment

"Risø 78" was a comprehensive micrometeorological experiment designed to study the effects of a terrain and surface roughness change on the mean and fluctuating wind components and temperature. In addition to the diverse set of SL profile and turbulence measurements, the following other related experiments were concurrently performed: A series of remote probing sodar measurements of PBL turbulence and winds, a sequence of optical mirage observations from which overwater SL temperature profiles could be deduced and, finally, several tracer diffusion experiments for evaluation of the dispersion parameters σ_y and σ_z at the shoreline escarpment site.

The field program, which took place in the summer of 1978, was the third in a series of terrain and surface roughness change experiments conducted at Risø. The first was performed at the same location in 1974-75 with the aim of studying the flow response to change of roughness, when the flow moved from the water inland, see Fig. 7. A result of that experiment was that even the quite modest escarpment at the shoreline, Fig. 7, did seriously influence the flow on the scales considered (Peterson et al., 1976). The flow response to the combined roughness and terrain change is shown on Fig. 9. The second experiment was therefore conducted at another site, where elevation differences between land and water was virtually absent, and, hence the effect of water-land roughness change could be better isolated and evaluated. Based on the experience from these two experiments, plans could now be made for a new experiment at the first site. Hence, the principal objective of Risø 78 was to obtain a set of turbulent velocity component and temperature fluctuations at selected heights upwind, and at downwind of the shoreline escarpment, and a

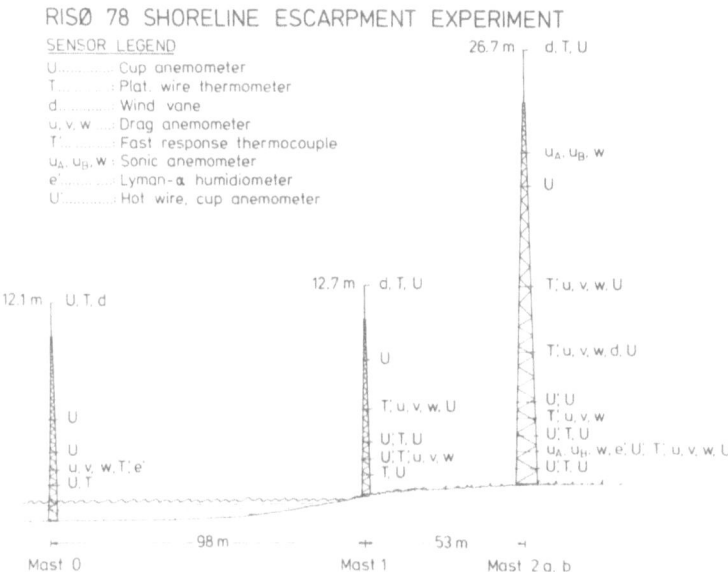

RISØ 78 SHORELINE ESCARPMENT EXPERIMENT

SENSOR LEGEND
U............. Cup anemometer
T.............. Plat. wire thermometer
d.............. Wind vane
u, v, w ..... Drag anemometer
T'............. Fast response thermocouple
u_A, u_B, w : Sonic anemometer
e'............. Lyman-α humidiometer
U'............. Hot wire, cup anemometer

Fig. 7. Set-up of experimental masts in the Risø 78 experiment.

sufficient set of high tower and indirect sensing measurements, so that the "background" planetary boundary layer structural characteristics could be specified.

Table IV summarizes the kind of measurements which were made during the experiment. It also includes some interesting numbers on the data bandwidth and recording requirements for such an experiment. For the terrain-change experiments a necessary condition for operation included winds within $\pm20°C$ of the bearing $328°$ mast line, see Fig. 8. Thus, the extensive surface layer turbulence measurements could be performed only on those few days with suitable wind directions which occurred in the six week field program.

The turbulent, u,v,w velocity measurements were made using seven fast response drag anemometers which were designed, constructed and tested at The Pennsylvania State University (PSU) (Norman et al., 1976).

In addition to the drag anemometer turbulent velocity measurements six hot wires were used in conjunction with fast-sampled cup anemometer signals to sense the turbulent longitudinal velocity fluctuations. In-situ calibration of the hot wire signals were performed using the same cup-anemometers as were used for the 10 min. mean profile values. However, to facilitate calibration of and comparison with the hot wire signals, the cup signal outputs were samples using the same high speed digital data logger which was recording the hot wire signals, see Table IV.

Parameter	Sensor	No. of Units	No. of Data Channels	Samp. Freq.	Samples per 45 min run
MASTS					
Windspeed	Cup	18	18	$(10 \text{ min mean})^{-1}$	90
Temperature	Resistance	7	7	$(10 \text{ min})^{-1}$	35
Wind Direction	Vane	4	4	$(10 \text{ min})^{-1}$	20
u,v,w	Drag	7	21	10 Hz	567,000
T'	Thermocouple	7	7	10 Hz	189,000
q'	Lyman-α	2	2	10 Hz	54,000
u,v,w,T	Sonic	3	12	10 Hz	324,000
p	Aneroid	1	1	10 Hz	27,000
Overwater T					
Profile	Thermocouples	5	5	10 Hz	135,000
u'	Hot wire	6	6	200 Hz	3,240,000
u'	Cup	18	18	200 Hz	9,720,000
RISO TOWER					
Windspeed	Cup	5	5	$(10 \text{ min mean})^{-1}$	25
Wind Direction	Vane	4	8	$(10 \text{ min})^{-1}$	40
(and Variance)		4	8	$(10 \text{ min})^{-1}$	40
Dewpoint	Hygrometer	1	1	$(10 \text{ min})^{-1}$	5
SODARS					
Aerovironment	—	1	2	150 Hz	810,000
Penn State	—	1	10	150 Hz	4,050,000
Sensitron	—	1	5	150 Hz	2,025,000

TOTAL 21,141,255

Table IV. Summary of measurements during a run in the Risø 78 experiment.

In order to obtain estimates of the turbulent water vapor fluctuations from which the vertical latent heat flux could be estimated, Lyman-α humidiometers were operated at 2 m on masts 0 and 2a. Each was located as close as possible, without introducing flow interference, to one of the PSU drag anemometers. Thus, at 2 m on the fjord and inland masts a complete time series of u', v', w' and e' (or q') was available. The Lyman-α signals were sampled at 10 Hz. Hourly ventilated psychrometer readings were taken to facilitate extraction of the calibration drift which is well known and characteristic of these units.

Two ultrasonic anemometer-thermometer units were also used to acquire independent u, v, w and T measurements. The sonic measurements had several purposes. One was to obtain a comparison between the sonics, which are regularly used by Risø for field measurements, and the present version of the PSU drags.

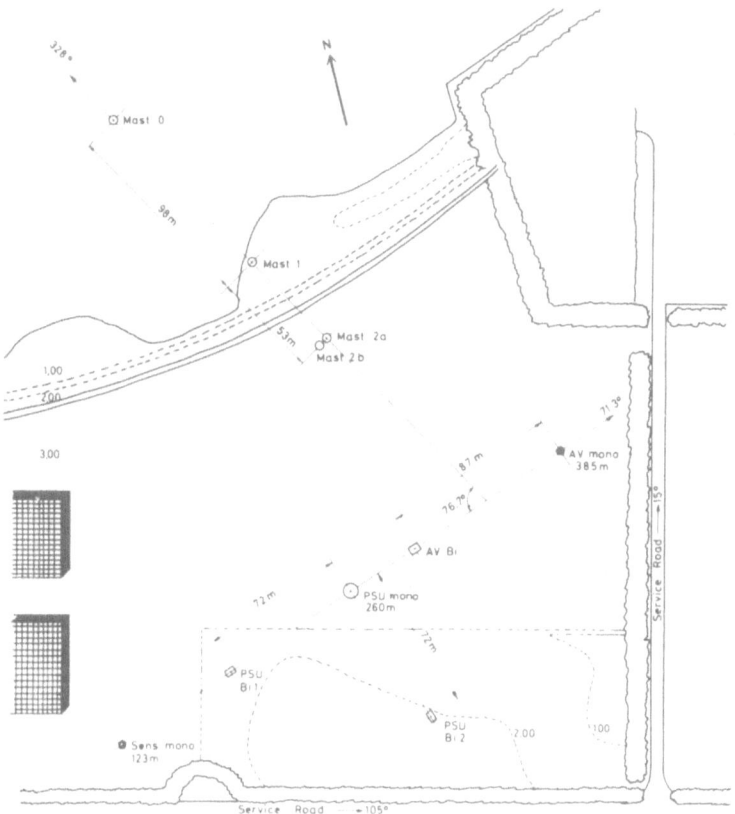

Fig. 8. Positioning of the sodars relative to the masts in the Risø 78 experiment.

The second component of Risø 78 consisted of an array of three sodar (sonic detection and ranging) systems which were configured for a variety of remote PBL wind and turbulence measurements. The various sodars were operated in conjunction with the 123 m Risø tower. In order of increasing distance from the Risø tower (Figs. 8 and 10) the sodars were:

— At 123 m: A monostatic Sensitron SR101A operating at 1800 Hz for profile measurements of $C_T^2(z)$ and $w(z)$.
— At 260 m: The PSU tristatic Doppler system operating at 1000 Hz for profile measurements of $\vec{V}(z)$, $C_T^2(z)$ and $C_V^2(z)$.
— At 385 m: An Aerovironment 300 operating at 1600 Hz for profile measurements of $C_T^2(z)$.

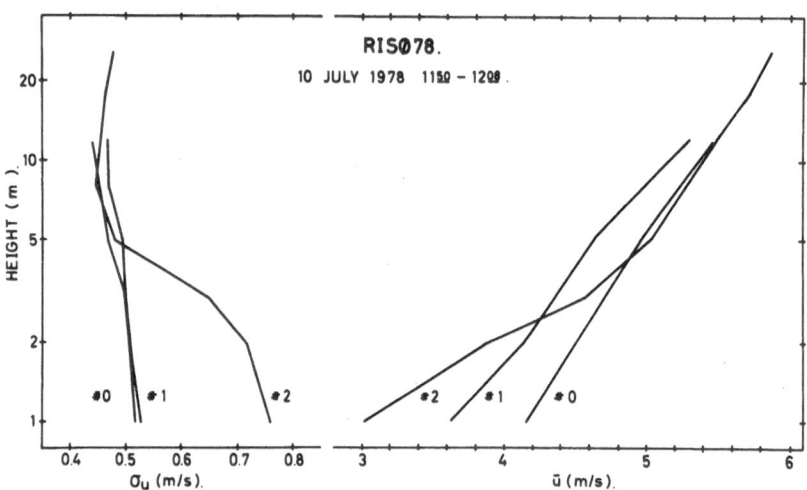

Fig. 9. *Flow response to roughness change and terrain experiment as measured in the Risø 78 experiment. ★ 0, 1 and 2 refers to the masts positioned in the water, on the shoreline and inland, respectively.*

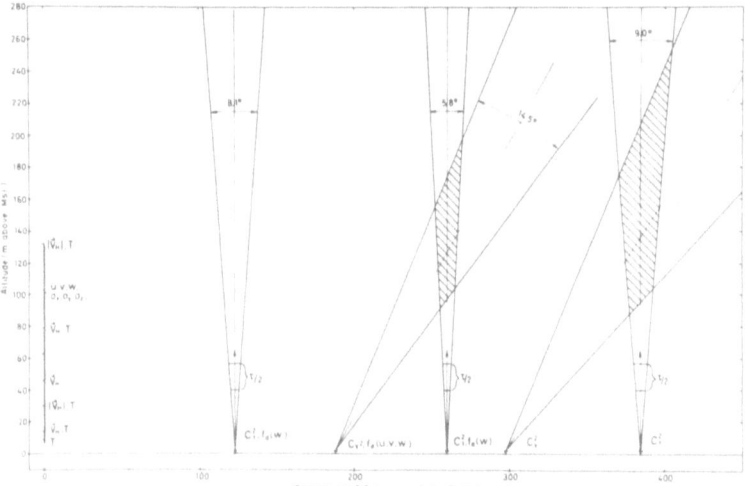

Fig. 10. *Sodar Target volumes and the Risø tower during the Risø 78 experiment.*

Finally, three diffusion experiments to measure the horizontal and vertical diffusion of sulphurhexaflouride (SF_6) tracer were conducted. The concentration measurements enabled a direct determination of the lateral σ_y and vertical σ_z spread parameters at the shoreline escarpment location.

Platform or System	Type of Measurement	Variable(s)	Principal Advantage(s); Disadvantage(s)
High (>50 m) Tower	Fixed point time series, carriage vertical profiles	State parameters and winds (mean and turbulence)	Scope of possible sensors; fixed position, "Longitudinal" time series only
Low (≤50 m) Mast Array	Fixed point time series	as above	Improved definition of lateral and longitudinal structure; cost
Tethered Balloon	"Fixed" point time series, vertical profiles	as above	Extends vertical range; balloon motions, not all weather system
Tetroon	Lagrangian time series	as above	Lagrangian; limited positional control
Radio or Drop-sonde	Quasi-lagrangian vertical profile	State parameters and winds (mean only)	Attitude range: single shot only
Aircraft	"Spatial" time series	State parameters and winds (mean and turbulence)	Mobility; cost of removing aircraft motions
Sodar	Vertical profile volume integral	Winds, 3-d Structure $C_{N_{ac}}^2$ parameter	Low cost, "longitudinal" time series only
FM-CW Radar	Vertical or slant range profile	Winds, 3-d Structure $C_{N_{mw}}^2$ parameter	Spatial resolution; cost
Lidar	Vertical or slant range profile	Particulate distribution and properties, winds	Resolution; operational maintenance
Bistatic Optical	Line integral	Transverse wind $C_{N_o}^2$	Low cost; signal interpretation in complex terrain

Table V. Measuring systems for boundary layer experiments.

The basic field instrumentation and techniques used were identical to those reported by Gryning et al. (1978). The release point was located at mast 0, 150 m upwind of samplers in a vertical and horizontal array at mast 2b. In order to best center the SF_6 plume on the sampler array, the lateral position of the upwind release point was adjusted according to the best (subjective) estimate of the 10 min. mean wind direction prior to starting the sampling units. In each experiment three sequential 10 min. samples of gas for analysis were "grabbed" by clock-controlled sampling units.

Concluding Remarks

Although it is still possible for a single laboratory group to assemble the equipment for and conduct a scientifically productive single tower or mast array experiment, the scope of effort required for a PBL study with objectives beyond those of the Minnesota experiment virtually dictates that it be a cooperative multigroup effort. Boundary layer processes include the evolution of structural features such as thermal plumes, helical rolls, and buoyancy waves having characteristic scales which may be comparable to the boundary layer dimensions. In comparison to surface layer processes which clearly fall within the turbulence regime, many boundary layer phenomena are definitely locally non-isotropic, and may occur as either intermittent or periodic "events", Hence, definition of the three-dimensional structure is essential (Warhaft, 1973).

Even a cursory examination of the observational requirements demonstrates the enormity of the measurement problems. The depth of the planetary boundary layer ranges, depending upon the time of day and ambient weather and terrain conditions, from about 50 to 2500 m. The costs of a single, well instrumented 300 m tower places it in the *major national facility* category (Hall, 1977). A single tower will provide at best only a "slice" or time-cross section through time-series measurements at selected heights. Alternatively (or in addition), a carriage may be used to obtain vertical profiles between fixed instruments. Since both the wind speed and direction normally vary with height, detailed analysis of structural features can be exceedingly difficult or in some cases impossible without additional independent observations. In practice, the possibilities for acquiring additional measurements are limited to the techniques summarized in Table V. It includes some of the principal advantages and limitations of the various sensor platforms and systems.

In order to resolve 3-d structure, it is generally necessary to synchronously operate several of the above referenced systems. The particular ones used in a given experiment will depend upon its principal objectives. A modern PBL turbulence experiment will certainly include a sodar or FM-CW radar; a particulate diffusion study, tetroons and lidar, aircraft, etc.

Although remote probing systems are probably the only viable long term solution to PBL measurement problems, once they are incorporated into a measurement program both the scope and logistics of data analysis problems can greatly change. Lidar systems require multi-MHz capability transient digitizers, radars include video bandwidth analog and digital processors, Doppler sodars may include digital processors for real-time evaluation of the power spectrum of a complex signal, or for any of the remote probes signal amplitude variations may be routinely evaluated in terms of a variety of statistical parameterizations

— many of which are more familiar to electronic engineers than to meteorologists. As an example of the signal processing and recording demands made by indirect probing systems, one may note in Table IV the data channels and recording demands for the sodars used during the Risø 78 experiment. Without careful planning a single remote probe can easily suffocate a small research group in data tapes.

Finally, no discussion of the basic conditions for boundary layer experiments is complete without mentioning the practical difficulties in carrying out field work. Some of them are:

a) The experimental conditions, in terms of weather, are neither controllable nor repeatable. The latter means that one, irrespectively of on-line analysis facilities, tends to record as many raw data as possible.
b) The many and varied instruments and instrument carrying devices must involve many people from different groups, a fact that tends to place narrow time limits on the experiment. Furthermore, the hazards of transporting sophisticated, delicate equipment ensures a fair amount of malfunction.
c) The preference for "simple" and therefore often remote experimental sites means that the experiments must be carried out without laboratory facilities.

As a consequence, a successful field experiment will, at most, result in only a few hours worth of data although it takes years to prepare it and months to carry it out. This is clearly illustrated by the experiments described in the present paper. They were all preceded by several years of planning and test experimenting, each of the experiments lasted 2 - 3 months, and a few tenths of hourly runs of data were finally obtained.

Acknowledgement
We wish to thank J.C. Kaimal for supplying us with Figs. 5 and 6.

References

Gryning, S.E., Lyck, E. and Hedegaard, K., 1978, Short range diffusion experiments in unstable conditions over inhomogeneous terrain, *Tellus,* **30**, No. 5, 392-403, 1978

Hall, F.F. Jr., 1977, The Boulder Atmospheric Observatory and its Meteorological Research Tower, *Optics News*

Izumi, Y., 1971, The Kansas 1968 Field Program Data Report, Environmental Research Papers No. 379, AFCRL-72-0041. Air Force Cambridge Research Laboratories, Bedford, Mass.

Izumi, Y. and Changhey, S.J., 1976, Minnesota 1973, Atmospheric Boundary Layer Experiment Data Report, AFCRL Research Papers No. 547, AFCRL-TR-76-0038, Cambridge Research Laboratories, Bedford, Mass.

Kaimal, J.C., Wyngaard, J.C., Izumi, Y. and Coté, O.R., 1972, Spectral characteristica of surface-layer turbulence, *Quart. J.R. Met. Soc.,* **98**, 563-589

Kaimal, J.C., Wyngaard, J.C., Haugen, D.A., Coté, O.R, and Izumi, Y., 1976, Turbulence structure in the convective boundary layer *J. Atmos. Sci.,* **33**, 2152-2169

Lettau, H.H. and Davidson, B., 1957, *Exploring the Atmosphere's first mile,* Pergamon Press, N.Y.

Norman, J.M., Perry, S.G., Panofsky, H.A., 1976, Measurement and theory of horizontal coherence at a two meter height, *Third Symp. Atmos. Turb. and Air Quality, Amer. Meteor. Soc.,* Oct. 19-22, 1976, Rayleigh, N.C.

Peterson, E.W., Kristensen, L. and Chang-Chun Su 1976, Some observations and analysis over non-uniform terrain, *Quart. J.R. Met. Soc.,* **102**, 857-869

Thomson, D.W., Coulter, R.L. and Warhaft, Z., 1978, Simultaneous measurements of turbulence in the lower atmosphere using sodar and aircraft, *J. Appl. Meteor.,* **17**, 723-734

Warhaft, Z., 1973, The relation between temperature and humidity in the free atmosphere under conditions of stable stratification and strong thermal intermittency — a case study, *Quart. J.R. Met. Soc.,* **99**, 89-104.

Digital Analysis of LDA Counter Signals in a Separated Boundary Layer

by

J.P. Melinand and G. Charnay

Laboratoire de Mécanique des Fluides de l'École Centrale de Lyon
(L.A. C.N.R.S. No. 04263), 69130 Ecully, France

Abstract

The data obtained in a separated boundary layer by means of LDA is discussed. The measurements are performed using an LDA, Bragg cell and counter. The data validated by the LDA system are stored on the disk of a computer. Statistical moments up to order four, power spectra, external and directional intermittency factor of the longitudinal velocity component are computed. The effect of false data is reduced by digital processing. The results are compared with those obtained by hot wire anemometry.

1. Introduction

This work is a part of a general study of the characteristics of turbulent boundary layers developing in an adverse pressure gradient. The description of the experimental device and the flow in the longitudinal equilibrium state has been given in an earlier paper (1). In the present note it is our purpose to determine the characteristics, with some details, of the turbulent field in the region where the flow separates. In such a region the direction of the flow is random and the conventional hot wire is not well adapted to determine the turbulent field. Therefore, measurements are made with a Laser Doppler Anemometer (LDA) system with a Bragg cell and a counter, and the validity of the data is discussed.

It is possible to distinguish between two regions downstream of the separation. In the first one located near the wall the flow is recirculating and the direction of the mean velocity is not constant. In the second one, named here the "external region", the direction of the flow is always downstream. In the first region only LDA measurements can be considered as valid (2), in the second one, both LDA and hot wire measurements can be performed, and so comparisons can be made between the two methods.

Simpson et al. (3) have already given a detailed description of a similar flow using LDA techniques. However, the signal from the photomultiplier is processed in a different way: they use sampling spectrum analysis, we use a counter associated with a computer.

2. Experimental Arrangement

2.1 Description of the Flow
As already quoted the description of the flow is given in reference (1). The width of the wind-tunnel is 1.8 m, so the flow is two-dimensional. Before the separation the value of the external pressure gradient is +290 Pa/m. The data are taken at 1.9 m from the leading edge of the plate, where the thickness of the whole boundary layer is .2 m and the thickness of the recirculating region is .06 m. The free stream velocity is 16 m/s at this point.

2.2 Measuring Equipment
A two color DISA LDA system with type 55X02-06 and 07-13 optics is used. The laser is a SPECTRA-PHYSICS model 164-09 Argon laser, the power of which is 2 W at 514.5 nm. The front lens focal length is 1.5 m and the forward scatter mode is used. Only one channel is equipped, so only the longitudinal velocity component is considered. The effective Bragg cell frequency is −0.6 MHz in the external region and +0.5 MHz in the recirculating region.

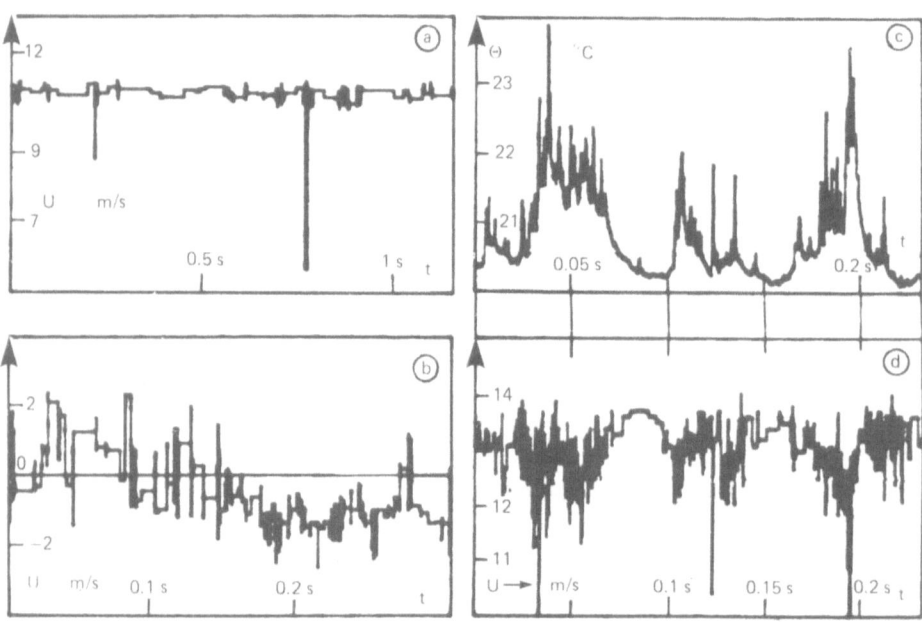

Fig. 1. Records of LDA velocity versus time. (a): external region. (b): recirculating region. (c) and (d): simultaneous records; (c) temperature signal, thermal tagging by heating of the plate; (d) LDA velocity signal. Seeding, (a) and (b): upstream of the test section, (c): in the boundary layer.

The velocity is obtained through a DISA 55L90 counter. All the measurements were performed with a band pass filter in the range .256 MHz - 1 MHz and a 1.5 % validation when counting on 5 and 8 fringes. A typical frequency observed is .5 MHz and the accuracy on the velocity is of about 4 cm/s. The validated data (velocity u and time τ between two successive values) obtained from the counter are digitally stored on the disk of a H.P. 2108 computer. The electronic circuitry limits the sampling rate at $3 \cdot 10^4$ samples per second. The maximum data frequency is 4 kHz and the usual recording time is 40 s. The data are then processed through Fortran IV programs to obtain the turbulent quantities with or without correction. The averages are data averages denoted $\bar{Q}$.

The smoke generator is a T.E.M. apparatus which produces oil smoke. The smoke particles are introduced at the intake of the wind tunnel or through lateral holes located 2 cm downstream of the leading edge of the flat plate.

3. Statistical Moment of the Velocity Fluctuation

In this section the measurement of the statistical moments $\overline{u^m}$ ($m \leqslant 4$) are discussed. Some of these moments are of importance in the budget of the turbulent kinetic energy and allow the explanation of some turbulent mechanisms.

Some obviously false data appear on the records of the velocity although they have been validated by the counter. This is mainly visible in the external region (Fig. 1a) and less visible in the recirculating region (Fig. 1b). Noise in general, as well as the validation method (4) used in the counter, can be regarded as responsible of these errors. The rate of false data is lower than 5 per thousand for optimum adjustment. This rate increases with the bandwidth Δf of the filters and counting accuracy. The maximum value of the false data is 5 m/s for our Δf.

When the mean velocity is considered, the effect of false data can be neglected, but for higher order moments the effect can be important. We propose a correction method of these false data. Two steps are used, first we eliminate false data through:

$$\left. \begin{array}{l} D_f \text{ is false if } |D_f - \bar{M}| > S_1 \\[2mm] D_j \text{ is a good value if } |D_j - \bar{M}| < S_1 \end{array} \right\} \tag{1}$$

$M \equiv \bar{D}_i$ where the D_i are all the data (D_f and D_j). Then we calculate the new mean and fluctuation by

$$M' = \overline{D}_j \text{ and } d'_j = D_j - M' \tag{2}$$

When the r.m.s. fluctuation is higher than 0.4 m/s, this method does not work, so we make a new condition:

$$\left. \begin{array}{l} D_f \text{ is false if } |2D_f - D_{f+1} - D_{f-1}| > S_2 \\[2ex] D_j \text{ is a good value if } |2D_j - D_{j+1} - D_{j-1}| < S_2 \end{array} \right\} \tag{3}$$

D_{j-1}, D_j, D_{j+1} being successive data. On Fig. 2 the influence of the corrections is shown. For $\sqrt{\overline{u^2}} \leqslant 0.4$ m/s, the two methods give the same result with $S_2 \sim 2S_1$. The values obtained with $S_2 = 1.5$ m/s are in agreement with those obtained by the hot wire anemometer (5).

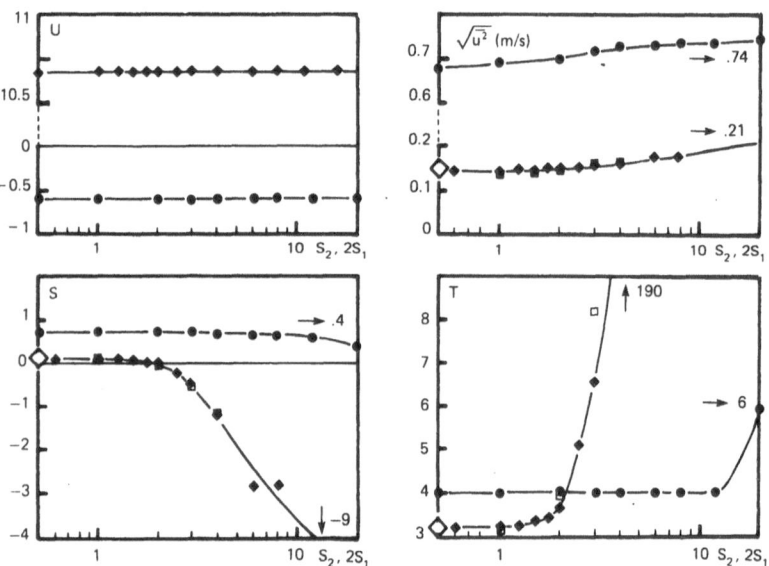

Fig. 2. Correction of false data. LDA measurements: ● *recirculating region, correction [2] + [3], threshold S_2;* ◆ *external region, correction [2] + [3], threshold S_2;* □ *external region, correction [2] + [1], threshold S_1. Hot wire measurements:* ◇ *external region; $S = \overline{u^3}/(\overline{u^2})^{1/2}$. $T = \overline{u^4}/(\overline{u^2})^2$. (Seeding: upstream of the test section).*

Without correction the errors which occur on U, $(\overline{u^2})^{1/2}$, S and T are respectively 0.1 %, 17 %, > 200 % and > 200 % in the external region. In the recirculating region the relations [2] + [3] are used and the error without correction is then respectively 0.1 %, 8 %, 50 % and 40 % on the same quantities.

4. Intermittency Factors

A detailed description of a turbulent flow includes the analysis of intermittent phenomena. The signal of an LDA is not continuous, so the measurements are fairly delicate. We present here some preliminary results with a high density of particles.

4.1 Free Boundary Intermittency in the External Region

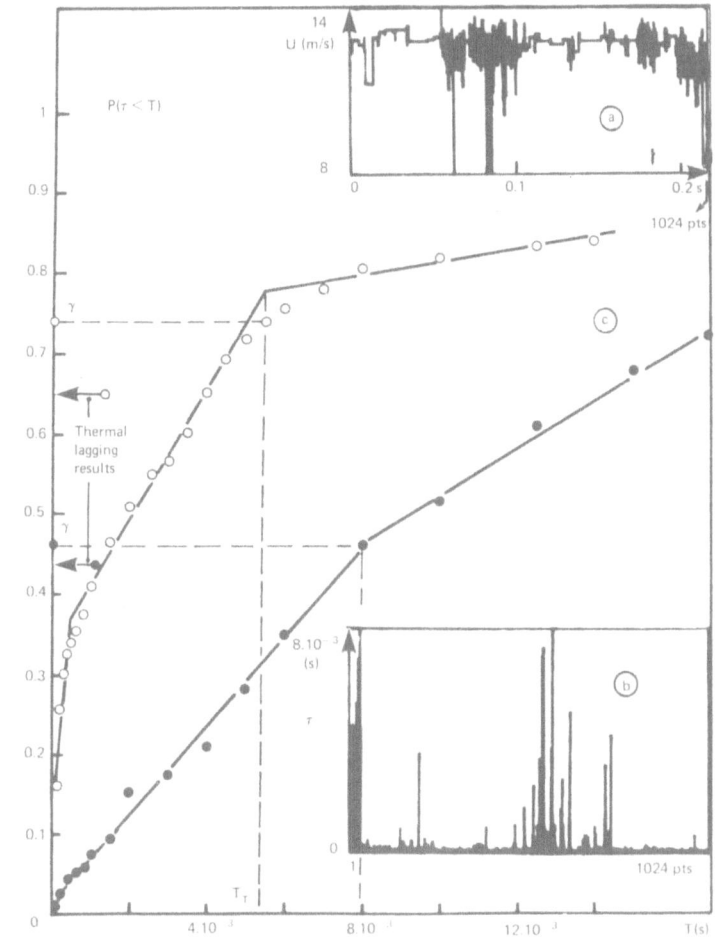

Fig. 3. Free edge intermittency factor (external region). LDA Method. Seeding in the boundary layer, downstream of leading edge of the plate. (a) LDA velocity versus the time, 1024 data validated by counter. (b) time τ between two successively validated data, same record as (a), x axis: data number (linear scale). (c) $P(\tau < T)$, T is variable threshold in time, $\circ$ y = .12 m, $\bullet$ y = .16 m.

This intermittency is usually detected by a thermal tagging of the boundary layer (6). The temperature signal (Fig. 1c) shows then a succession of periods with high frequencies and amplitudes and of more calm periods. If we assume that the boundary layer can be tagged in a similar way by injecting the particles at the beginning of the boundary layer, the density of particles will be much higher in the boundary layer than outside. In effect, the LDA signal u(t) has an obvious intermittent aspect (Fig. 1d) and the intermittency factor γ can be obtained by classical treatment (7). The measurement of γ can be improved by using the distribution of the time τ between two successively validated data, γ is then given by:

$$\gamma = \text{Prob}(\tau < T_T) = P(T_T)$$

The threshold T_T is chosen at the point where P(T) shows a break (Fig. 3). The values for γ given by LDA are comparable to the values obtained by thermal tagging, but always somewhat higher. This is already visible on the signals (Fig. 1c, d). Different reasons can be claimed; however, it seems that the particles do not follow the flow perfectly but migrate away from their original fluid particle control volume (Crossing-Trajectories Effect, (8), (9)).

4.2 Directional Intermittency in the Recirculating Region

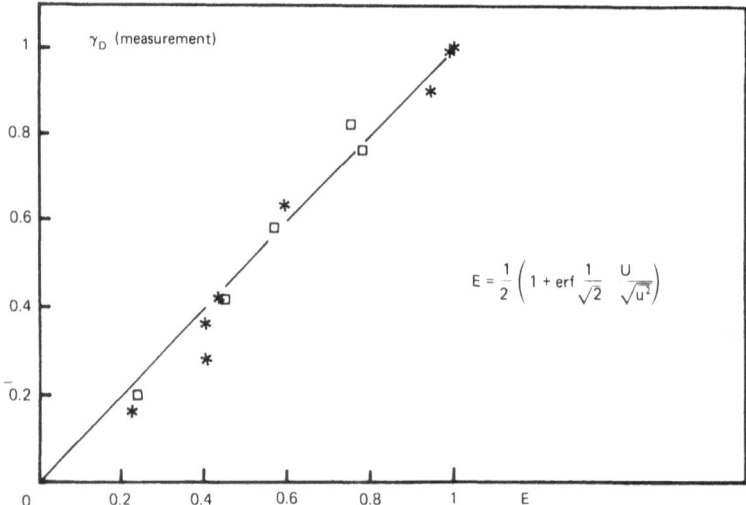

Fig. 4. Directional intermittency factor. (Recirculating region). γ_D: LDA measurements, E: Gaussian estimation. *: present experiments, □: Simpson et al. results (3). Seeding upstream the test section.

The factor γ_D is the factor of time during which the flow moves downstream (3). In regions where the u signal can be considered Gaussian, γ_D can be estimated with the formula

$$\gamma_D = \frac{1}{2}\left(1 + \text{erf}\ \frac{1}{\sqrt{2}}\ \frac{U}{(\overline{u^2})^{\frac{1}{2}}}\right)$$

In the region where $S \cong 0$ and $T \cong 3$ this appears to be verified (Fig. 4).

5. Power Spectra

The power spectra of u are computed from the u(t) signal. This signal is first converted so that it corresponds to a constant samplerate. In Fig. 5(a) the LDA spectra are compared to the hot wire spectra. According to the initially validated data, the mean frequency of which is 500 Hz, the spectra are in good agreement for frequencies f lower than 200 Hz. For higher values of f the discrepancy is probably due to the accuracy on the velocity measurement and the frequency of the whole LDA system.

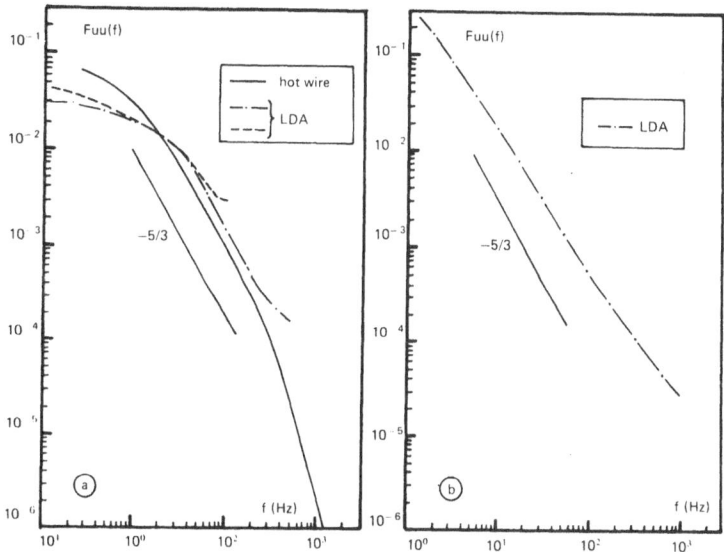

Fig. 5. Power spectra of u component, normalized by $\overline{u^2}$: $\int_0^\infty Fuu(f)\ df = \overline{u^2}$. a: external region, ———: hot wire, — · —: LDA $(f_D = 500$ Hz, $\tau max = 0.025$ sec), — — —: LDA $(f_D = 200$ Hz, $\tau max = 0.08$ sec). b: recirculating region, — · —: LDA $(f_D = 2300$ Hz, $\tau max = 0.01$ sec). Seeding upstream the test region.

The spectra in the recirculating region show a slope different from −5/3, further investigations appear to be necessary to explain this fact.

6. Conclusion

The main results of the studies are:

1. The LDA technique with counter is well adapted to separated boundary layer studies,
2. Mean velocity, higher order statistical moments, external and directional intermittency factors and power spectra have been measured,
3. In the external region, measurements with LDA are in good agreement with hot wire measurement,
4. False data may be removed by digital processing, their importance is large in the external region and low in the recirculating region.

Acknowledgements

We would like to thank Mr. J.P. SCHON for his helpful assistance and Mrs. M. Allier for typing the paper and treating the digitalized signals.

References

1. CHARNAY, G. and BARIO, F.: Structure d'une couche limite turbulente en équilibre longitudinal dans un gradient de pression. Association aéronautique et astronautique de France, Note Technique 77-18, 1977
2. DURST, F., MELLING, A. and WHITELAW, J.H.: *Principles and practice of Laser-Doppler Anemometry,* Academic Press, p. 368, 1976
3. SIMPSON, R.L., STRICKLAND, J.H. and BARR, P.W.: Features of a separating turbulent boundary layer in the vicinity of separation, *J. Fluid Mech.,* Vol. 77, Vol. 3, pp. 553-594, 1977
4. STEENSTRUP, F.V.: Counting techniques applied to Laser Doppler anemometry. *DISA Information,* No. 18, pp. 21-25, 1975
5. HAERTIG, J., KOERBERG, G. and BOUIS, X.: Comparaison de mesures de vitesse instantanée faites par anémomètre à fil chaud et anémomètre laser. Rapport I.S.L. 117/75, 1975
6. DUMAS, R., FULACHIER, L. and ARZOUMANIAN, E.: Facteurs d'intermittence et de dissymétrie des fluctuations de température et de vitesse dans une couche limite turbulente. C.R. Acad. Sc. Paris, t. 274, série A, 1972
7. SCHON, J.P. and CHARNAY, G.: Conditional sampling. In measurement of unsteady fluid dynamic phenomena, Edited by RICHARDS, B.E., Hemisphere P.C., pp. 291-325 1977
8. LUMLEY, J.L.: Two-phase and Non-Newtonian Flows. In *Turbulence* edited by BRADSHAW. Springer-Verlag, 1976, pp. 289-324
9. SCHON, J.P., SERRES, E., KREISS, J., REY, C. and RIBON, M.: Feasibility of the simulation of atmospheric boundary layer, application to the study of diffusion on complex terrain. Euromech 109, Delft, 1978.

Multichannel Measurements and
High Order Statistics

Multi-Channel Measurements and High-Order Statistics

by

C.W. Van Atta
Department of Applied Mechanics and Engineering Sciences
and
Scripps Institution of Oceanography
University of California, San Diego
La Jolla, California 92093

Abstract

Multi-channel measurements in turbulent flows include those using probe arrays to simultaneously measure velocity or scalar field fluctuations at a number of points in the flow field and those to measure a more complex flow property or combinations of several variables at a single point.

The first category includes probe arrays of sensors spanning lateral and longitudinal directions in the interior of the flow or on the boundaries. Such measurements provide information on instantaneous spatial patterns or profiles of vector and scalar fields, allowing assessment of the detailed structure, method of generation, convection, mutual interaction, and eventual degeneration or annihilation of identifiable flow structures contributing to the turbulence. Consideration must be given to the optimal placement of a finite number of sensors with regard to information content of the array, mutual aerodynamic interference, and spatial resolution. A large number of sensors usually dictates economy of design in associated electronics, the necessity of digital sampling and recording of the data, frequently controlled by a computer, and computer processing of the resulting digital data. Various applications of spatial probe arrays will be discussed, especially those employed for studies of the turbulent boundary layer, where their use is the most highly developed.

Multi-channel measurements of the second type have included those using arrays of several hot-and-cold-wires to simultaneously measure combinations of several velocity components of their spatial gradients, vorticity, temperature, or other scalar variables in the neighborhood of a single point. Observations of several kinds of intense vortices which can explosively grow into larger scale organized motions in turbulent shear flows has led to further refinement of vorticity probes to measure both longitudinal and transverse components of the fluctuating vorticity. Because the highly organized motion making substantial contributions to the dynamics of turbulent flows includes internal shear layers that are vortical in nature, "vorticity meters" furnish a useful tool for defining important "events" in the flow and for extracting their principal signatures for use in pattern recognition schemes to elucidate the contributions of different kinds of events. Use of conditional sampling discriminating simultaneously on the instantaneous components of vorticity and velocity can in some cases unambiguously identify the allowable topology of fluid motions associated with a particular type of event, allowing more decisive measurements and interpretation. Recent experiments using such ideas and techniques will be summarized.

Higher-order correlations, i.e., those involving third- or higher-order products of variables or measurements at more than two points in the flow have considerable physical importance both in shear flows and in nonsheared homogeneous turbulence. This survey will conclude with a discussion of measurements to determine the behavior of such correlations and their relationships to one another in connection with closure problems in turbulence theory, spectral energy transfer, intermittency of the fine scale structure and energy dissipation of turbulence, and in studies of the large scale features of turbulent shear flows.

Multi-Channel Measurements
Multi-channel measurements using hot-wires or other sensors can provide

1.   Simultaneous information at several points in the flow field for the measurement of instantaneous profiles of velocity or other flow variables, phase velocities, and multi-point correlations.

2.   Measurements of local vorticity, simultaneous measurements of two or more velocity components or several different variables, e.g., velocity and temperature, at a single point.

In the first case a spatial array of sensors globally spanning the flow may be employed, while in the second the attempt is made to confine the sensors to the smallest possible volume of space consistent with probe interference effects in order to approach as closely as practical to a pointwise measurement.

Two-channel measurements, such as those used for determining two velocity components, phase speeds, or two-point spatial correlations of u have been employed extensively, and will for the most part not be discussed here, with the exception of space-time correlations. For the present discussion, the term multi-channel will refer to measurements which require three or more sensors and the associated processing of three or more channels of information. The main components of a multi-channel measurement are the probe array and the associated electronics needed to operate the elements of the array, a method to calibrate the array, and the data acquisition and processing operations applied to the resulting signals.

Spatial Arrays

Most applications of probe arrays to date have been in measuring the vertical and lateral structure of turbulent shear flows. Table I summarizes the main characteristics of some studies of this kind. The work by Kovasznay, Komoda, and Vasudeva (1962) on breakdown of periodic waves in a boundary layer undergoing transition was an early example of a probe array and the use of transistorized multi-channel, D.C. coupled, linearized hot-wire anemometers, permitting the simultaneous measurement of velocity components using as many as 10 channels of anemometry. In this way maps could be drawn of instantaneous spatial fluctuation patterns. For the simultaneous display of 10 traces a display oscilloscope was built around a commercial television receiver, producing a sampling rate of 15 kHz. Photographic recording was used and the data were sampled visually. Faced with similar problems of handling large amounts of data from experiments, during this period some workers developed other special purpose sampling systems designed to process and average information from multi-channel experiments. In another study on the nonlinear development of periodic disturbances in a laminar boundary layer Komoda (1967) used a traversing four wire-array to measure u, v, w and three components of vorticity. In order to simultaneously measure the instantaneous distributions of u, v, and w, and $\partial u/\partial y$, a ten-channel synchronized sampler described by Komoda and Handa (1968) was used, taking advantage of the fact that the fluctuations were periodic and controlled by the exciting current of the vibrating ribbon. An analog computing circuit was used to preprocess the signals and to obtain instantaneous values of u, v, w and $\partial u/\partial y$. Spatial gradients were measured by traversing the probe in z (or y) with the sampling phase held constant and taking continuous records of u, v, w and $\partial u/\partial y$ versus z (or y). In this way, for various values of the phase, records of the instantaneous values of the variables were obtained for continuous values of z (or y) and at several y (or z) positions. Instantaneous space and time derivatives of u, v, and w were evaluated graphically from the records, except for the x derivative which was evaluated in the usual way from the time derivative. However, the adoption of digital sampling and computer analysis of turbulence data in the early sixties (e.g. Coles and Van Atta (1966, 1967)) made it possible to analyze very large amounts of data and contributed to the subsequent increase in measurements using large probe arrays performed in the last ten years.

Kaplan and Laufer (1969) used a vertical array of ten hot-wires operated at constant temperature measuring the u component of velocity to study the intermittently turbulent region of the boundary layer ($0.7 \leqslant y/\delta \leqslant 1.1$).

The turbulence and wire calibration data were first FM tape recorded, then later digitized and analyzed with a computer, using disk for intermediate storage of data. For hot-wire calibrations, data was recorded with the array moved vertically into the free stream. For the turbulence data, a turbulence detector function was generated from each of the converted hot-wire voltages for all ten channels and used to compute conditionally sampled averages for each wire. A small time sharing system enabled the overlapping of data conversion and calculations. No permanent records of the raw digital data were generated. The conditional zone averages of the u component obtained generally showed good agreement with the results obtained at about the same time by Kovasznay, Kibens, and Blackwelder (1970) using a single x-wire probe and on-line non-computerized data sampling and averaging. The simultaneous data from the ten hot-wires was also used to generate two-dimensional intermittency maps representing the extent of the turbulent zones in both the x and y directions. This study demonstrated the usefulness of probe arrays and computer analysis for generating global information about the instantaneous state of the turbulent boundary layer. Such information would prove to be very illuminating, especially in comparison with different global measurement techniques such as the photographic tracking of hydrogen bubbles and other tracker particles.

In the next study, Blackwelder and Kaplan (1976) examined the wall structure of the turbulent boundary layer using vertical and lateral hot-wire rakes and conditional sampling techniques. The ten-wire vertical rake, which spanned a vertical distance of about $100y^+$, was fixed to a plug mounted flush with the wall of the wind tunnel and could not be placed in the free stream for calibration. The rake was therefore calibrated in situ assuming that the velocity profile was the same as that measured by a single calibrated hot-wire probe for several values of the free-stream velocity. A single hot-wire sensitive to u was located at $y^+ = 15$ and used as a detector probe for bursts.

As previously observed in flow visualization studies, a high degree of correlation between velocities was found in the region below $y^+ \leqslant 20$, associated with the bursting process.

In a comprehensive study of the topography and motion of the turbulent interface in a two-dimensional wall jet Paizas and Schwarz (1974) used a vertical linear array of twenty hot-wire probes coupled to analog turbulence detectors

Table I

Summary of Some Multi-Channel Shear Flow Measurements Using Probe Arrays

Authors	Flow	Variables Measured	Type of Array	Number of Wires	Wire Spacing (cm)
Kovasznay, Komoda, & Vasudeva (1962)	Periodic disturbance in boundary layer	u,v,w ω_x, ω_z	Vertical Horizontal (Sting mounted)	5 6	0.07
Komoda (1967)	Periodic disturbance in boundary layer	u,v,w	Vertical, Sting mounted	4	0.25
Kaplan & Laufer (1969)	Turbulent boundary layer	u	Vertical, Sting mounted	10	0.25
Gupta, Laufer, & Kaplan (1971)	Turbulent boundary layer	u τ_w	Lateral (z), flush wall mounted	9 1	Irregular over 10.7 mm, $z^+ \sim 101$ or 324
Sunyach (1971)	Heated 2-D mixing layer	θ (temperature)	Vertical Sting mount	6	1.5 mm
Paizis & Schwarz (1974)	Turbulent wall jet	I (intermittency detector function)	Vertical tube mount	20	0.48
Fiedler (1975)	Heated 2-D mixing layer	θ (temperature)	Vertical Spanwise	10 10	1.0 variable
Chen (1975)	Heated boundary layer	θ (temperature)	Vertical Sting mount	10	0.635
Blackwelder & Kaplan (1976)	Turbulent boundary layer	u u	Vertical, flush wall mount Lateral (z), Sting mount	10 12	logarithmic over 0.96 cm, $x_2^+ = 5$ to $x_2^+ = 100$ $\Delta x_3 = 0.254$
Haritonidis, Kaplan, & Wygnanski (1978)	Turbulent spot interaction with turbulent boundary layer	u	Vertical Sting mount	10	logarithmic over 24 mm

whose outputs were in turn coupled to digital logic circuits. The outputs of the constant-temperature anemometers were fed into turbulence (intermittency) detector circuits whose function was to produce an output which was 'on' (1) if the probe was in a turbulent field and 'off' (0) otherwise. These outputs were processed to determine properties of the interface. In one operation all these outputs were added in a summing amplifier to give the total width of the turbulent region passing the rake. Other logic circuitry used to determine the widths of segments of the multiple-valued interface counted the number of consecutive probes that 'saw' turbulence, the counting being started from the probe closest to the fully turbulent region. The number of times that the interface intersected the probe rake, or the degree of multiple-valuedness of the interface, was determined by a crossing circuit which counted the number of changes in state that occurred along the rake. Consecutive detector outputs were compared in 'exclusive or' gates whose outputs were 'on' only if the inputs differed, i.e. only if an interface crossing was detected. The number of 'on' signals occurring at any instant was equal to the number of times that the interface intersected the probe rake. Calibration of the hot-wire probes was not attempted. The gate levels of the twenty intermittency detectors were set by matching the intermittency profile and probability density of the interface position measured by the rake with that determined by a single vertically traversing hot-wire. In the wall jet, the interface position, expressed as the height of the interface from the wall, was found to be multiple-valued for as much as 40 % of the time, but the quantity of turbulent fluid carried in the folded regions was small, about 10 % of the turbulent fluid in the intermittent region. The expected valuedness, or the average number of values taken on by the interface position, was found to be 2.35, with a ratio of surface area to projected area of 7.2.

Vorticity Measuring Arrays

The fluid vorticity, like the fluid velocity, is a vector (more exactly, a pseudo-vector), but its measurement is much more difficult than for the velocity because each component of vorticity involves spatial gradients, in two different directions, of two different velocity components. Vorticity measurements require multi-channel measurements utilizing configurations of several hot-wires. The full three-dimensional vector vorticity at a point in a turbulent flow has never been measured. Usually, only a single component or one of the terms contributing to a single component is measured, but some recent measurements have attempted to determine two out of three of the vector components of the fluctuating vorticity simultaneously. Table II summarizes the main characteristics of some studies of this kind.

Table II

Summary of Some Vorticity Fluctuation Measurements

Author	Turbulent Flow	Vorticity Components Measured	Type of Array	Number of Wires	Cross-Stream Spatial Resolution
Kovasznay Kistler (1952)		ω_x	"pyramidal"	4	≈ 1 mm
Corrsin & Kistler (1954)	Boundary Layer, Grid	ω_x	"pyramidal"	4	≈ 1 mm
Uberoi (1957)	Boundary Layer	$\omega_\sigma, \omega_\sigma'$	principal strain axis (45°)	6	≈ 2 mm
Kastrinakis, Wallace & Willmarth (1978)	Channel Flow	ω_x	"pyramidal"	4	≈ 0.1 mm, $\ell^+ \approx 5$
Eckelmann, Nychas, Brodkey, & Wallace (1977)	Channel Flow (Oil)	ω_y, ω_z	V-probe, X-probe & u-probe	5	≈ 4 mm, $d^+ \approx 7$
Foss (1978)	Axisymmetric Jet, Vortex Rings	ω_z	X-probe & parallel u-probes	4	4 mm×1 mm
Blackwelder & Eckelmann (1978)	Channel Flow	$(\omega_x)_{wall}$ $(\omega_z)_{wall}$	Surface V-films	4 or more films	$z^+ \approx 10$
Kastrinakis, Eckelmann, Wallace & Nychas (1978)	Boundary Layer	ω_x	"pyramidal" probe	4	≈ 0.1 mm

Kovasznay (1954) described an ingenious technique for measuring the stream-wise component of vorticity using an array of four hot-wires in a "pyramidal" array. Under assumptions of linearized voltage response of the wires to small fluctuations in all three coordinate directions, the array has the simple proper-ty that for matched hot-wires having the same resistances the voltage difference generated between two of the diagonally opposite points of the bridge of hot-wires is proportional to the streamwise vorticity fluctuation. In principle, the probe can be calibrated by rotating it about the x-axis in the free stream, but other calibration methods are normally used. Corrsin and Kistler (1954) com-pared the readings in a decaying isotropic turbulence with the values of vortici-ty fluctuation levels computed from turbulence levels and microscale measure-ments. Estimates of the parasitic sensitivities, especially to the three compo-nents of turbulent velocity, were made by measuring the steady-state yaw and speed sensitivities in a low-turbulence stream, and found to be negligible for their particular array. No corrections were made for finite wire length or for the ratio of wire spacing to turbulence microscale, a characteristic giving para-sitic sensitivity to the second derivatives of velocity fluctuations. Wyngaard (1969) has given an analysis for the spatial resolution of the Kovasznay vortici-ty meter and of arrays for measuring velocity derivatives in isotropic turbu-lence, assuming a form for the three-dimensional spectrum. In all cases it was found that spectral response deteriorates rapidly for array sizes significantly larger than the Kolmogorov microscale η of the turbulence. The response of the streamwise vorticity meter was found to be relatively insensitive to wire angle but strongly dependent on wire length ℓ and spacing. Spectral attenua-tion at arbitrarily low wave numbers was found. For small $k_1 \ell$ the measured and true spectra do not agree if η/ℓ is much less than 1.0, in contrast to the result for velocity spectra where there is always a $k_1 \ell$ value below which there is no spectral error. A physical explanation is that the dominant wave number of the streamwise vorticity fluctuations is near $1/\eta$ in magnitude. These modes cut the x_1 axis at all possible angles giving contributions to mean-square vor-ticity at all k_1 down to zero. From Wyngaard's results it appears than an η/ℓ value not smaller than 0.3 is a reasonable design goal. This is difficult to achieve in typical high Reynolds number laboratory flows, but should be within reach for application in atmospheric turbulence. All of these results are sensitive to the form of the three-dimensional spectrum, on which there is not general agreement, but the calculations provide a basis for probe design. The mean square vorticity in the directions of the principal axes of the rate of deforma-tion was measured by Uberoi (1957) using an array of six constant-current hot-wires. Local isotropy in shear flows was tested by measuring the ratio of the mean square vorticity in the two principal directions. For local anisotropy the vorticity should be highest in the direction in which fluid elements are being stretched and lowest in the direction of contraction. Since the directions are at 45° to the mean flow direction for parallel or nearly parallel flows, the probe could be rotated by 90° about its x-axis to allow measurements in both principal directions. The probe was not directly calibrated for absolute vorticity, but relied for its accuracy on the matching of hot-wire resistances and careful attention to the angular orientation of the hot-wires in the array. Near the wall of a duct the mean-square vorticity was found to be larger (by about 30 %) in the direction of rate of elongation than in the direction of rate of contraction. Towards the center of the channel where the rate of deformation decreases the ratio approached unity, the value for local isotropy.

In principle, a vorticity meter measuring one of the components of the fluctuating vorticity would be an ideal way of measuring the position of the turlent-non-turbulent interface or superlayer. However, because of the relatively large size of the vorticity meter and its complexity, interface detection has usually been done using velocity components or their time derivatives, a scalar variable like temperature, or single spatial gradients like $\partial u/\partial y$, obtained with two closely spaced hot-wires. Considerable miniaturization and consequent substantial improvement in spatial resolution of the streamwise vorticity probe has been achieved by Kastrinakis, Wallace, and Willmarth (1978), and a report on progress to data by Willmarth (1978) will be included in the proceedings of this conference. The results show that the probe responds correctly to streamwise vorticity if the cross-flow velocity is zero. In this case the sensitivity is a function of streamwise velocity u, and can be corrected using digital measurements. If the cross-flow velocity is not zero the vorticity response of the probe is contaminated by the lateral velocity components v and w, and this effect cannot be instantaneously corrected, although some time-averaged quantities can still be measured.

While interest in the bursting process associated with intense streamwise vortices near the wall in turbulent boundary layers spurred the development of miniature streamwise vorticity probes, visual evidence for intense spanwise (z) or y-oriented vorticity in turbulent shear flows motivated Eckelmann, Nychas, Brodkey, and Wallace (1977) and Foss (1978) to develop hot-wire arrays for the measurement of the other two vorticity components ω_y and ω_z.

The five-sensor, commercially built, probe of Eckelmann et al. used five quartz coated platinum films operated at constant temperature and linearized to measure ω_y, ω_z, u, v, w, and $\partial u/\partial z$ simultaneously. In the x-z plane, parallel to the wall, two sensors in a V configuration measured the u and w components of the velocity. Directly below this V probe, in the x-y plane, was an X-array used to

928 C.W. Van Atta

obtain the u and v components of velocity. The centers of the X and V-arrays were separated by about 2 mm (two Kolmogorov length scales). At 1 mm in the positive y direction from the V-array a single u sensor aligned in the z direction was used to determine the gradient $\partial u/\partial z$. The probe thus measured the gradients $\partial u/\partial y$ and $\partial u/\partial z$, and by assuming equivalence of time and x derivatives, the gradients $\partial u/\partial x$, $\partial v/\partial x$, and $\partial w/\partial x$. None of the gradients composing the streamwise vorticity were obtained, just as no components of the non-streamwise vorticity are obtained with the usual streamwise vorticity probe. Calibration consisted of towing the probe and assuming cosine-law response calculated from known geometrical factors. Apparently, no direct calibration of the probe in pitch and yaw was attempted to verify this procedure. The calibration and turbulence signals from the five sensors were digitized and processed with a computer. Considerable difficulties were found in accurately measuring the true gradients, because the sensors should be within a Kolmogorov scale of each other for gradient measurements, but at these separations one is subtracting the differences of two large numbers (the two u velocities at two arrays), and calibration errors have large effects. To test the accuracy of the gradient measurements, several sets of data were taken using a gradient probe consisting of two single sensors separated by a distance of one mm and then repeating the measurements using the five sensor probe. At positions beyond $y^+ \sim 10$, the error in measuring the mean velocity gradient was very large, but the instantaneous fluctuating gradients were very similar.

A similar array using four hot-wires to measure the lateral component of vorticity $\omega_z = \partial v/\partial x - \partial u/\partial y$ has been developed by Foss (1978, proceedings of this conference). In this probe, an X array at z with wires parallel to the x-y plane is placed adjacent to a parallel wire array at $z + \delta z$ with wires parallel to the x-z plane. The term $\partial v/\partial x$ is evaluated from the X-array using the time derivative of v. The adjacent parallel array and the pitch angle ($\gamma = \tan^{-1} v/u$) from the X-array are used to infer $\partial u/\partial y$, viz.

$\partial u/\partial y = \cos\gamma[V(y+\delta y) - V(y-\delta y)]/2\,\delta y$

where V is the velocity magnitude inferred from the voltage magnitude. The probe is calibrated for various pitch angles γ, which defines unknown functions of γ in the analytical hot-wire response equations. Another function of γ, $G(\gamma)$ is defined by eliminating V from the X-array expressions, and the difference between this analytical form and its measured value is treated as an error measure. Evaluation of u, v or V, γ from the X-array proceeds iteratively by first assuming the error measure to be zero to obtain an initial γ, then evaluating V from the X-array equations, improving the value of γ using the error measure expression, etc. The computing scheme successfully recovers the velocity vector information over a range of $-40° \leqslant \gamma \leqslant +40°$, and has been implemented both in a batch mode on a digital computer and also in a 50 kHz rate, real time, hard-wired processor.

Space-Time Correlation
Experiments in turbulent shear flows have demonstrated the crucial importance of persistent large scale motions and coherent structure in turbulent flows. These structures cannot be described in terms of local Eulerian power spectra and space correlations, but are more naturally described in terms of a Lagrangian approach, which is not possible to implement using customary fixed probe arrays. Space-time correlations and conditional sampling techniques can be used to approximate true Lagrangian measurement techniques and to follow the generation, evolution, and ultimate decay or organized flow structures. Table III lists some of the many experiments carried out to measure space-time correlations in various flows. There have been many others which will not be discussed here, and those listed are offered only as a representative example. Favre, Gaviglio, and Dumas (1953) originally applied analog techniques to the measurement of space-time correlations in grid turbulence and in turbulent boundary layers. The signals from probes at different spatial locations were recorded on analog tapes, and time differences were generated during playback by running the tape over two different playback heads separated by a known distance along the tape. Various convection velocities associated with the filtered and unfiltered turbulent structure were computed from the isocorrelation curves. The lower wavenumber part of the space-time correlations was found to be much more persistent than the high wavenumber part. The larger scale motions were found to be convected with roughly the average mean velocity across the boundary layer. By displacing the movable probe in x, y, or z the measurements were extended to measure the general space-time velocity correlations $R_{1,1} (\xi_1, \xi_2, \xi_3; \tau)$, and isocorrelation curves for optimum delay were calculated with different positions x_2 of the fixed hot-wire. The isocorrelation curves are very elongated in the streamwise direction, and not symmetric with respect to $\xi_1 = 0$.

Willmarth and Wooldridge (1962, 1963) and Willmarth and Tu (1967) also used the playback head displacement method to measure space-time correlations between wall-pressure fluctuations p_w at two points on the wall in a turbulent boundary layer and between p_w at one point and u, v, or w at another point near the wall. The pressure transducer was a lead-zirconate-titanate disk of 1.5 mm diameter and 0.5 mm thickness mounted flush with the wall. An X-array was used to measure u, v, and w, and the wall shear stress was measured with

Table III

Some Space-Time Correlation Measurements

Authors	Turbulent Flow	Quantities used for Space-Time Correlations	Number of Spatial Points	Spatial Separation Coordinate
Favre, Gaviglio, & Dumas (1953)	Boundary Layer	u,v	2	ξ_1, ξ_2, ξ_3
Willmarth & Wooldridge (1962)	Boundary Layer	p,u,v,w	2	ξ_1, ξ_2, ξ_3
Frenkiel & Klebanoff (1966)	Grid Turbulence	u	2	ξ_1
Kovasznay, Kibens, & Blackwelder (1970)	Boundary Layer	u,v,uv	2	ξ_1
Champagne, Harris, & Corrsin, (1970)	Uniform Shear Flow	u,v,w	2	ξ_1, ξ_2, ξ_3
Comte-Bellot & Corrsin (1971)	Grid Turbulence	u	2	ξ_1
Fulachier, Giovanangeli, Dumas, Kovasznay, & Favre (1974)	Boundary Layer	u	3	ξ_1, ξ_3
Harris, Graham, & Corrsin (1977)	Uniform Shear Flow	u	2	ξ_1

a u-wire placed 0.05 mm from the wall. Kovasznay, Kibens and Blackwelder (1970) combined conditional sampling and space-time correlation measurements to study the extent and lifetime of organized structure in the turbulent boundary layer. They observed that the signal in the streamwise direction remains correlated over three boundary layer thicknesses (3δ) and in the spanwise direction over approximately only one δ. Blackwelder and Kaplan (1976) in studying the "bursting" phenomenon near the wall estimated the average trajectory of the burst using a two point correlation method, and found that the coherence of the fluctuations did not extend beyond 3 δ even along the path of maximum correlation. One cannot deduce the scale of the large eddies from these correlation measurements because the eddy structure may be distorted by interaction with other eddies. In order to reduce the apparent dispersion effect of the "random walk" of an eddy about some mean trajectory, Fulachier, Giovanangeli, Dumas, Kovasznay, and Favre (1974) measured three-point space-time correlations in the boundary layer, with two points equally spaced laterally from a third one upstream. The triple correlation contours are much narrower in the spanwise direction than the double correlation contours suggesting that the large eddy does not extend beyond 0.4 in the spanwise direction. The triple correlation eliminates one effect contributing to the smearing out of the eddies, but, like other unconditioned correlation measurements, does not help much in defining the shape and size of individual large eddies.

High Order Moments
The replacement of analog techniques by digital computing in the reduction of turbulence data has made it practical to examine the behavior of higher-order statistical properties of turbulence. Digital computation allows the accurate computation of powers and products of fluctuations, relatively easy application of accurate calibration data and compensation for spatial and frequency response. Digital Fourier analysis techniques also allow the computation of higher-order spectra, which depend upon several different wave-numbers and play a central role in describing the dynamical interactions occurring in the flow.

Higher-Order Correlations, Spectra, and Structure Functions
Despite the central importance of the closure problem in theories of turbulence, measurements of third- and fourth-order correlations are rare. Reported analog measurements include those of third order by Stewart (1951), Mills et al. (1958), Favre et al. (1962), and Bradshaw & Ferriss (1965), and measurements of fourth order by Uberoi (1953). Higher-order correlations were not measured until more recently, when digital recording and analysis helped make it practical to perform such measurements.

Using digital data sampling and high-speed computing techniques, Frenkiel & Klebanoff (1967a, b) measured (by means of constant-current anemometry) two-point time correlation up to eighth order and associated skewness and flatness factors of the streamwise component u of the fluctuating velocity in grid turbulence. The correlations were obtained by direct calculation of mean time-lagged products, using 160,020 digital data samples (corresponding to 12.5 sec of real time) for each sample calculation of the complete correlation function. A number of these sample calculations were then averaged to obtain the final measured correlations. They found that a non-Gaussian joint-probability density of Gram-Charlier type described relations between their measured odd-order correlations fairly well, while the even-order correlations were related according to a Gaussian joint distribution, except for very small time separations.

Van Atta and Chen (1968, 1969a, b) obtained significantly different results for third-order moments, using constant-temperature anemometry, digital sampling, and computer analysis to measure two-point time correlations of third order in a study of the dynamics of spectral energy transfer in grid turbulence under nearly the same conditions as those of Frenkiel and Klebanoff. This led them to extend their measurements to higher orders to obtain a full comparison with the results of Frenkiel and Klebanoff. To reduce computer time, while at the same time using longer samples of data than were previously employed, they developed a method for very fast and efficient digital-computer calculation of two-point correlations of arbitrary order, using the FFT-algorithm. Their digital data were Fourier transformed in records each containing 2048 velocity samples (the number of discrete frequencies in each transform was also 2048), corresponding to a time interval about 20 times longer than the integral time scale of the turbulence. The appropriate products of the discrete transforms were averaged over 150-200 records, and these averaged products, each consisting of 2048 complex numbers, were then inverse fast-Fourier transformed to obtain the correlation functions. Excellent agreement was found between higher-order correlations computed this way and those obtained by conventional averaging of mean lagged products. The results for all even-order correlations were in very close agreement with those of Frenkiel and Klebanoff, illustrating the power of digital techniques as an accurate tool for turbulence measurements. However, all measured odd-order correlations were strikingly different from those obtained by Frenkiel and Klebanoff, a difference that has been traced to the different response of constant-current and constant-temperature hot-wire anemometers by Helland and Van Atta (1976), but which is not fully understood at present. However, the multi-variate Gram-Charlier distribution was again found to describe relations between higher- and lower-order odd-order correlations very well.

Yeh and Van Atta (1973) and Helland, Van Atta, and Stegen (1977) employed digital Fourier techniques to measure higher-order cross spectra, which were used to directly compute energy transfer spectra without recourse to intermediate computations of third-order correlation functions.

Van Atta and Yeh (1970) extended the digital-Fourier-analysis grid-turbulence measurements to three- and four-point time correlations using a modification of the one-dimensional fast-Fourier-transform method. As computer memories became larger and computational techniques improved, it became feasible to perform these and other computations via multi-dimensional fast-Fourier transforms and associated higher-order spectra.

The third-order spectrum (bispectrum) was of immediate interest because of its central importance in the problem of spectral energy transfer between different wavenumber components in turbulence (Yeh and Van Atta (1973)). The bispectrum also gives the distribution of contributions to the mean cube of a signal as a function of two interacting wavenumbers of arbitrary magnitude.

Inertial subrange measurements of bispectra of turbulent velocity fluctuations in the atmospheric boundary layer were made by Lii, Rosenblatt, and Van Atta (1976). A recent examination of this data by Van Atta (unpublished) shows that the dependence of the measured bispectrum on wavenumber is consistent with the spectral form predicted by the usual type of inertial subrange dimensional arguments, and that for fixed wavenumber magnitude the variation with angle in wavenumber space is in good agreement with the functional form predicted by Herring (1978, unpublished) from numerical calculations for high Reynolds number using the test field model. The atmospheric measurements were extended to higher frequencies using the same data by Helland, Lii, and Rosenblatt (1977), who also computed the bispectrum for grid turbulence. In examining the interpretation of these bispectra in terms of spectral energy transfer, Van Atta (1978) found that within the limitations of a method of interpretation originated by Wilson (1974) the measured behavior of wavenumber triplet interactions was consistent with a cascade of energy from lower to higher wavenumbers.

Although Kolmogorov's original inertial subrange formulation as well as the work of some other authors have been framed in terms of structure functions rather than correlation functions, relatively few measurements of structure functions have been made. Structure functions involve only differences in velocity or other variables and very accurate determination of the mean velocity may not be essential, thus providing distinct advantages over correlation

934   C.W. Van Atta

measurements in geophysical flows such as the atmospheric boundary layer. With digital techniques, it became easy to generate the necessary time difference over a wide range and to obtain measurements of higher-order moments, for which there are simple but widely conflicting theories. Some measurements of this type are summarized in Table IV. Park's (1976) measurements of higher-order structure functions of temperature in the atmospheric boundary layer exhibited a nearly linear dependence on separation distance unexpected for the inertial range. Suspecting that this dependence might be caused by the organized large scale structure of the temperature field, Van Atta (1977) developed a simple model based on the observed ramp-like events in the temperature signal which yields behavior for the higher order moments similar to that measured. These results have been extended and applied to laboratory flows by Antonia and Van Atta (1977). For sufficiently high-order moments and small enough separation distances, such measurements can suffer from the dynamic-range problems encountered for moments of derivatives, discussed in the next section on small-scale structure.

Small-Scale Structure

Much of the motivation for recent measurements of the fine structure of turbulence, characterized by statistical properties of derivatives of the velocity and other variables that are heavily weighted by the finer scales of the flow, came from the work of Kolmogorov (1962), who modified the earlier hypothesis (Kolmogorov (1941)) that the small-scale structure of turbulence should be of universal form for all turbulent flows at sufficiently high Reynolds number. This theoretical development led to predictions that high order moments should depend strongly on the value of the turbulence Reynolds number R_λ, whereas the original Kolmogorov theory predicted that such quantities would have constant values independent of R_λ. Direct digital sampling and computer analysis of fine-structure data is the best means available for measuring these quantities, because the dynamic range achievable with A-D converters is considerably larger than that obtained with analog tape recorders and other analog devices. However, the dynamic range required for obtaining acceptably accurate measurements of higher moments is a strongly increasing function of the order of the moment and of the turbulent Reynolds number. The spread of the all-important tails of the probability densities of derivatives of velocity and other variables increases with Reynolds number. The problem becomes so severe for large R_λ that Tennekes and Wyngaard (1972) conclude that measurements of moments of velocity derivatives higher than the fourth are practically impossible, with present instrumentation, for some cases of geophysical turbulence. Because atmospheric turbulence records much longer than an hour are likely to exhibit trends, it will be necessary to obtain ensemble averages, repeating the same experiment many times until sufficient statistical stability in the extreme tails of the probability distribution is achieved. In lower-R_λ laboratory measurements these problems are not as severe, and higher moments can be measured, but careful attention must be given to such considerations.

A number of measurements of third-and fourth-order moments (skewness and flatness) of time derivatives of velocity or scalar variables in the atmospheric boundary layer, most of which employed digital sampling and analysis, have been reported (see Van Atta (1974) for a summary). It would be useful to find out how much of the scatter in the data may be attributed to the effects discussed by Tennekes and Wyngaard (1972) and how much is due to physical differences in the large scale structure of the flows studied, since the latter differences are one of the main points of contention in the modified Kolmogorov theory. Future experiments and analysis should attempt to sort out these effects.

Acknowledgements
The present paper was written in the course of research funded by NSF Grant ENG 76-13147, from the Fluid Mechanics Program of the Engineering Mechanics Section of the National Science Foundation.

Table IV

Some Measurements of Higher-Order Correlations, Structure Functions, and Spectra

Author	Turbulent Flow	Higher-Order Quantities Measured	Highest Order Measured
a) Correlations			
Stewart (1951)	Grid Turbulence	$R_{m,n}(\xi_1) = \langle u^m(x)u^n(x+\xi_1)\rangle$ (spatial correl.) for u	$m+n=3$ (triple correl.)
Uberoi (1953)	Grid Turbulence	$R_{m,n}(\xi_1,\xi_2)$ for u,v, and w	$m+n=4$
Mills, Kistler, O'Brien & Corrsin (1958)	Heated Grid Turbulence	$R_{m,n}(\xi_1)$ for u and θ (temperature)	$m+n=3$
Favre, Gaviglio, & Dumas (1962)	Boundary Layer	$R_{m,n}(\xi_1)$, $R_{m,n}(\tau)$	$m+n=3$
Frenkiel & Klebanoff (1967)	Grid Turbulence	$R_{m,n}(\tau)$ (time correl.) for u	$m+n=8$
Van Atta & Chen (1968)	Grid Turbulence	$R_{m,n}(\tau)$ for u	$m+n=8$
Van Atta & Yeh (1973)	Grid Turbulence	$R_{\ell,m,n}(\tau_1,\tau_2,\tau_3)$ for u	$1+m+n=5$
Yeh & Van Atta (1973) and Yeh	Heated Grid Turbulence	$R_{m,n}(\tau)$ for u and θ	$m+n=3$ $m+n=8$
b) Higher-Order Structure Functions			
Stewart, Wilson, & Burling (1970)		$S_n(\tau) = \langle(u(t+\tau)-u(t))^n\rangle$ time structure functions	$n=4$
Van Atta & Chen (1970)	Atmospheric Boundary Layer	$S_n(\tau)$ for u	$n=4$

Van Atta & Park (1972)	Atmospheric Boundary Layer	$S_n(\tau)$ for u	n = 8
Mestayer (1975)	Heated & Cooled Boundary Layers	$S_n(\tau)$ for u and θ	n = 6
Park (1976)	Atmospheric Boundary Layer	$S_n(\tau)$ for u and θ	n = 8
Antonia & Van Atta (1978)	Atmospheric & Laboratory Boundary Layers, Jet	$S_n(\tau)$ for θ	n = 8

c) Higher-Order Spectra

Dutton & Deaven (1972)	Atmospheric Turbulence	Spectra of u^n	n = 4
Wilson (1974)	Atmospheric Boundary Layer	$B(k_1, k_2)$ Bispectrum of $\partial u/\partial t$	Third
Van Atta & Wyngaard (1975)	Atmospheric Boundary Layer	Spectra of u^n	n = 9
Lumley & Takeuchi (1976)	Jet, Pipe	$S_4(\omega_1, \omega_2, \omega_3)$ Cumulant Spectrum of u	Fourth
Lii, Rosenblatt, & Van Atta (1976)	Atmospheric Boundary Layer	$B(k_1, k_2)$ Bispectrum of $\partial u/\partial t$	Third
Champagne, Pao & Wygnanski (1976)	2-D Mixing Layer	Spectra of u^n	n = 4
Helland, Lii, & Rosenblatt (1977)	Atmospheric Boundary Layer, Grid Turbulence	$B(k_1, k_2)$ for $\partial u/\partial t$	Third

References

Antonia, R.A. and Van Atta, C.W. (1978) "Structure functions of temperature fluctuations in turbulent shear flows", *J. Fluid Mech.*, **84**, 561-580

Blackwelder, R.F. and Kaplan, R.E. (1976) "On the wall structure of the turbulent boundary layer", *J. Fluid Mech.* **76**, 89-112

Blackwelder, R.F. and Eckelmann, H. (1978) "The spanwise structure of the bursting phenomenon" in *Structure and mechanisms of turbulence I, Lecture Notes in Physics, 75,* Springer-Verlag, H. Fiedler, ed., p. 190-204

Blackwelder, R.F. and Eckelmann, H. (1978) "Streamwise vortices associated with the bursting phenomenon", *Bericht 101,* Max-Planck Institut für Strömungsforschung

Bradshaw, P. and Ferriss, D.H. (1971) National Physical Laboratory Aero Report 1144

Champagne, F.H., Harris, V.G. and Corrsin, S.C. (1970) "Experiments on nearly homogeneous turbulent shear flow", *J. Fluid Mech.*, **41**, 81-139

Champagne, F.H., Pao, Y.H. and Wygnanski, I.J. (1976) "On the two-dimensional mixing region", *J. Fluid Mech.* **74**, 209-250

Chen, P.Y. (1975) "The large scale motion in a turbulent boundary layer: A study using temperature contamination", Ph.D. Thesis, University of Southern California

Coles, D.E. and Van Atta, C.W. (1966) "Progress report on a digital experiment in spiral turbulence", *AIAA J.* **4**, 1969-71

Coles, D.E. and Van Atta, C.W. (1967) "Digital experiment in spiral turbulence", *Phys. of Fluids,* **10**, Supplement, Boundary Layers and Turbulence, S120-121

Comte-Bellot, G. and Corrsin, S. (1971) "Simple Eulerian time correlation of full- and narrow-band velocity signals in grid-generated 'isotropic' turbulence", *J. Fluid Mech.* **48**, 273-337

Corrsin, S.C. and Kistler, A.L. (1954) "The free stream boundaries of turbulent flows" NACA Tech. Note 3133

Dutton, J.A. and Deaven, (1972) "Some properties of atmospheric turbulence" in *Statistical models and turbulence, Lecture Notes in Physics,* Vol. **12** (ed. M. Rosenblatt and C.A. Van Atta) Springer, 402-426

Eckelmann, H., Nychas, S., Brodkey, R. and Wallace, J. (1977) "Vorticity and turbulence production in pattern recognized turbulent flow structures", *Phys. Fluids,* **20**, S225-S231

Favre, A., Gaviglio, J. and Dumas, R. (1953) "Quelques mesures de corrélation dans le temps et l'éspace en soufflerie", *Rech. Aeron,* **32**, Transl. as NACA Tech. Memo (1955) 1370

Favre, A., Gaviglio, J. and Dumas, R. (1962) "Corrélations spatio-temporelles en écoulements turbulents", *Mécanique de la turbulence,* CNRS, Paris, 419-445

Fiedler, H. (1975) "On turbulence structure and mixing mechanism in free turbulent shear flows", in Turbulent mixing in nonreactive and reactive flows, Plenum, ed. S.N.B. Murthy

Foss, J., (1978) " Transverse vorticity measurements", *Proceedings of Dynamic Flow Conference,* Marseille and Baltimore

Frenkiel, F.N. and Klebanoff, P.S. (1966) "Space-time correlations in turbulence" in *Dynamics of fluids and plasmas,* Academic Press Inc., New York 257-274

Frenkiel, F.N. and Klebanoff, P.S. (1967a) "Higher-order correlations in a turbulent field", *Phys. Fluids,* **10**, 507-520

Frenkiel, F.N. and Klebanoff, P.S. (1967b) "Correlation measurement in a turbulent flow using high-speed computing methods", *Phys. Fluids,* **10**, 1737-1747

Fulachier, L., Giovanangeli, J.P., Dumas, R., Kovasznay, L.S.G., and Favre, A. (1974) "Structure des perturbations dans une couche limité turbulente: Zone interne" *Comtes Rendus de l'Academie des Sciences,* Paris, t. 278 Serie B.

Komoda, H. (1967) "Non-linear development of a disturbance in a laminar boundary layer", *Phys. Fluids,* **10**, Suppl. Boundary Layers and Turbulence S87-S94

Komoda, H. and Handa, N. (1968) "Applications of a synchronized sampling technique to the measurement of quasi-periodic velocity fluctuations" in Advances in hot-wire anemometry, *Proc. International Symposium on Hot-Wire Anemometry,* Univ. of Maryland, ed. W.L. Melnik and J.R. Weske

Kovasznay, L.S.G. (1954) "Turbulence measurements" in Physical measurements in gas dynamics and combustion, *High Speed Aerodynamics and Jet Propulsion,* Vol. IX (R.W. Ladenburg, ed.) Princeton Univ. Press, 227

Kovasznay, L.S.G., Komoda, H. and Vasudeva, B.R., (1962) "Detailed flow field in transition", *Proc. 1962 Heat Transfer and Fluid Mechanics Institute* (F. Ehlers, et al. eds.) Stanford Univ. Press, 1-26

Kovasznay, L.S.G., Kibens, V. and Blackwelder, R.F. (1970) "Large-scale motion in the intermittent region of a turbulent boundary layer", *J. Fluid Mech.,* **41**, 285-320

Lii, K.S., Rosenblatt, M. and Van Atta, C.W. (1976) "Bispectral measurements in turbulence", *J. Fluid Mech.,* **77**, 45-62

Lumley, J.L. and Takeuchi, K. (1976) "Applications of central-limit theorems to turbulence and higher-order spectra", *J. Fluid Mech.,* **74**, 433-468

Mestayer, P.G. (1975) "Étude de certaines caractéristiques statistiques locales d'une couche limite turbulente à grand nombre de Reynolds", Thèse Docteur-Ingénieur, Université d'Aix Marseille, France

Mills, R.R., Kistler, A.L., O'Brien, V. and Corrsin, S. (1958) N.A.C.A. Tech. Note 4288

Paizis, S.T. and Schwarz, W.H. (1974) "An investigation of the topography and motion of the turbulent interface", *J. Fluid Mech.,* **63**, 315-343

Park, J.T. (1976) "Inertial subrange turbulence measurements in the marine boundary layer" Ph.D. Thesis, University of California, San Diego

Stewart, R.W., (1951) "Triple velocity correlations in isotropic turbulence", *Proc. Camb. Phil. Soc.,* **47**, 146-157

Gupta, A.K., Laufer, J. and Kaplan, R.E. (1971) "Spatial structure in the viscous sublayer", *J. Fluid Mech.* **50**, 493-512

Haritonidis, J.H., Kaplan, R.E. and Wygnanski, I. (1978) "Interaction of a turbulent spot with a turbulent boundary layer", in *Structure and mechanisms of turbulence II, Lecture Notes in Physics,* **76**, Springer-Verlag, 234-247

Harris, V.G., Graham, J.A.H. and Corrsin, S.C., (1977) "Further experiments in nearly homogeneous turbulent shear flow", *J. Fluid Mech.,* **81**, 657-687

Helland, K.N. and Van Atta, C.W. (1976) "Response of constant-current and constant-temperature anemometers to artificial turbulence, *Phys. Fluids,* **19**, 1109-1117

Helland, K.N., Van Atta, C.W. and Stegen, G.R. (1977) "Spectral energy transfer in high Reynolds number turbulence", *J. Fluid Mech.,* **79**, 337-361

Helland, K.N., Lii, K.S. and Rosenblatt, M. (1977) "Bispectra of atmospheric and wind tunnel turbulence", in *Proc. Symp. on Application of Statistics,* P.R. Krishaniah, ed.

Kaplan, R.E. and Laufer, J. (1969) "The intermittently turbulent region of the boundary layer", *Proc. 12th Int. Cong. Appl. Mech.,* Stanford, Springer 236-245

Kastrinakis, E.G., Eckelmann, H., Wallace, J.M. and Nychas, C.S.G. (1978) "The vorticity probe in turbulent flow", *Proceedings of Dynamic Flow Conference,* Marseille and Baltimore

Kastrinakis, E.G., Wallace, J.M. and Willmarth, W.W. (1978) "On the mechanism of bounded turbulent shear flows" in *Structures and Mechanisms of Turbulence I, Lecture Notes in Physics,* **75**, Springer-Verlag, 175-189

Kistler, A.L. (1952) M.Sc. Thesis, Johns Hopkins University

Kolmogorov, A.N. (1941) "The local structure of turbulence in incompressible viscous fluid for very large Reynolds numbers", *C.R. Acad. Sci. USSR,* **30**, 301-305

Kolmogorov, A.N. (1962) "A refinement of previous hypotheses concerning the local structure of turbulence in a viscous incompressible fluid at high Reynolds number", *J. Fluid Mech.,* **13**, 82-85

Stewart, R.W., Wilson, J.R. and Burling, R.W. (1970) "Some statistical properties of small scale turbulence in an atmospheric boundary layer", *J. Fluid Mech.,* **41**, 141-152

Synyach, M. (1971) "Contribution à l'étude des frontieres d'écoulements turbulents libres", Sc.D. Thesis, École Centrale de Lyon

Tennekes, H. and Wyngaard, J.C. (1972) "The intermittent small-scale structure of turbulence: data processing hazards", *J. Fluid Mech.,* **55**, 93-103

Uberoi, M.S. (1953) "Quadruple velocity correlations and pressure fluctuations in isotropic turbulence", *J. Aeronaut. Sci.* **20**, 197-204

Uberoi, M.S. (1957) "Equipartition of energy and local isotropy in turbulent flows", *J. Appl. Phys.,* **28**, No. 10, 1165-1170

Van Atta, C.W. and Chen, W.Y. (1968) "Correlations measurements in grid turbulence using digital harmonic analysis", *J. Fluid Mech.,* **34**, 497-515

Van Atta, C.W. and Chen, W.Y. (1969a) "Correlation measurements in turbulence using digital Fourier analysis", *Phys. Fluid,* **12**, Suppl. II-264-269

Van Atta, C.W. and Chen, W.Y. (1969b) "Measurements of Spectral energy transfer in grid turbulence", *J. Fluid Mech.,* **38**, 743-764

Van Atta, C.W. and Chen, W.Y. (1970) "Structure functions of turbulence in the atmospheric boundary layer over the ocean", *J. Fluid Mech.* **44**, 145-160

Van Atta, C.W. and Yeh, T.T. (1970) "Some measurements of multi-point time correlations in grid turbulence", *J. Fluid Mech.* **41**, 169-178

Van Atta, C.W. and Park, J.T. (1972) "Statistical self-similarity and inertial subrange turbulence", In *Lecture Notes in Physics,* Vol. **12**, Statistical Models and Turbulence (ed. M. Rosenblatt and C.W. Van Atta), Springer-Verlag, 402-426

Van Atta, C.W. (1974) "Sampling techniques in turbulence measurements", *Ann. Rev. Fluid Mech.,* **6**, 75-91

Van Atta, C.W. and Wyngaard, J.C. (1975) "On higher-order spectra of turbulence", *J. Fluid Mech.,* **72**, 673-694

Van Atta, C.W. (1977) "Effect of coherent structures on structure functions of temperature in the atmospheric boundary layer", *Archives of Mechanics,* **29**, 161-171.

Van Atta, C.W. (1978) "Bispectral measurements in turbulence computations", *Proc. 6th International Conf. on Numerical Methods in Fluid Mechanics, Lecture Notes in Physics,* Springer Verlag (in press).

Willmarth, W.W. and Tu, B.J. (1967) "Structure of turbulence in the boundary layer near the wall", *Phys. Fluids,* Suppl. **10**, Part II, S134-137

Willmarth, W.W. and Wooldridge, C. (1962) "Measurements of the fluctuating pressure at the wall beneath a thick turbulent boundary layer", *J. Fluid Mech.,* **14**, 187-210

Wilson, J.R. (1974) "Some small scale properties of atmospheric turbulence", Ph.D. Thesis, University of British Columbia

Wyngaard, J.C. (1969) "Spatial resolution of the vorticity meter and other hot-wire arrays", *J. Sci. Instr. (Journal of Physics E),* **1** (ser. 2) 983-987

Yeh, T.T. and Van Atta, C.W. (1973) "Spectral transfer of scalar and velocity fields in heated-grid turbulence", *J. Fluid Mech.* **58**, 233-261

Yeh, T.T. (1971) "Spectral transfer and higher-order correlations of velocity and temperature fluctuations in heated grid turbulence", Ph.D. Thesis, University of California, San Diego.

Multi-Probes and Higher Moments

by

H.H. Bruun
Whittle Laboratory, Cambridge University

Abstract
This paper discusses the application of multi-wire probes and the utilization of higher order moments in fluid dynamic probes. Multiprobe techniques are considered for conventional hot-wire probes and special probes are presented for flow structure detection. The use of higher order moments is related to the statistical description of turbulence.

1. Introduction
The output from a hot-wire anemometer depends on the heat transfer from the hot-wire element to the surrounding fluid. It therefore follows that a hot-wire probe can be used for the measurements of velocities, temperatures, densities and also for concentration measurements in gas mixtures. In this paper we shall discuss the probes and techniques related to velocity and temperature measurements.

2. Velocity Component Evaluation
Hot-wire probes can provide information of one or more of the velocity components at fixed points in a flow field. Several probe configurations are available. To demonstrate the relative merit and associated data evaluation for the different probe types, we shall in this section of the paper consider single normal and yawed hot-wire probes, an X hot-wire probe, a 3 wire probe and a vorticity probe.

2.1 Single Normal Hot-Wire Probe
When the hot-wire element is placed perpendicular to the mean flow direction, then the hot-wire probe provides information on the instantaneous velocity component in the mean flow direction.

For most velocity measurements the hot-wire anemometer is operated in the constant temperature mode at a high overheat ratio. The calibration function

$$E = f(V) \tag{1}$$

between the direct anemometer voltage E and the flow velocity V is highly non-linear. A large number of analytical approximations to [1] have been suggested in the literature as summarized by e.g. Bruun (1977a). The most frequently used 'calibration law' is

$$E^2 = A + BV^n \tag{2}$$

as suggested by King (1914) with n = 0.5. Collis and Williams (1959) have since shown that a better approximation is obtained with n = 0.45. However, to obtain accurate results using equation [2] a detailed calibration is required for *each* hot-wire probe. To avoid this problem an accurate 'universal shape' calibration method has been introduced by Bruun (1971, 1977a) in the form

$$E^2 - E_o^2 = Kf(V) = K_1 V^{n(V)} \tag{3}$$

with E_o being the voltage output at zero velocity and f(V) and n(V) being universal functions for a given probe type (and support orientation).

In low turbulence intensity flows the mean velocity $\bar{V}$ is obtained from the measured mean voltage $\bar{E}$ by inverting equations [2] or [3], while the fluctuating velocity and voltage signals are related by the probe sensitivity $\partial E/\partial V$, obtained from equation [2] and [3] as

$$\frac{\partial E}{\partial V} = \frac{n(V)}{2} \frac{E^2 - E_o^2}{EV} \cong \frac{n}{2} \frac{E^2 - A}{EV} \tag{4}$$

In many applications the anemometer signal is 'linearized' *before* the data processing by feeding the direct anemometer signal through a 'linearizer' which inverts the relationship E = f(V). The linearized hot-wire signal

$$E_L = CV \tag{5}$$

has the obvious advantage that the velocity information can be obtained directly from the linear signal. However, for measurements of the mean velocity and turbulence intensity the linearized signal does usually *not* provide more accurate results than the direct non linear signal. Both evaluation procedures are influenced by higher order correction terms as discussed by Bruun (1976), and illustrated by comparative measurements in tables 1 and 2.

Table 1 Mean velocity $\bar{U}$

η	$\bar{U}^{nl}$ (m s^{-1})	$\bar{U}^{l}$ (m s^{-1})				$\bar{U}$ (m s^{-1})	$\bar{U}^{l}-\bar{U}^{nl}$ (m s^{-1})		$\dfrac{\bar{U}^{l}-\bar{U}^{nl}}{U}$ (%)
		eq (2)	eq (4)	eq (6)	eq (8)		Measured	Theory	
-0.165	28·0	28·0	28·07	28·03	28·03	28·0	0	0·01	—
-0.098	27·82	27·83	27·84	27·78	27·76	27·80	-0.01	0·02	0·1
-0.062	26·29	26·32	26·42	26·40	26·38	26·23	0·03	0·05	0·2
-0.031	22·86	22·96	22·99	22·93	22·96	22·72	0·10	0·14	0·5
0·037	11·45	11·76	11·83	11·75	11·75	11·15	0·30	0·30	2·6
0·098	3·88	4·26	4·33	4·32	4·19	3·53	0·38	0·35	9·8

Table 2 RMS value $(\overline{u^2})^{1/2}$

η	$(\overline{u^{2\,nl}})^{1/2}$ (m s^{-1})	$(\overline{u^{2\,l}})^{1/2}$ (m s^{-1})				$(\overline{u^2})^{1/2}$ (m s^{-1})	Tu	$\overline{u^{2\,nl}}-\overline{u^{2\,l}}$ (m s^{-1})^2		$\dfrac{\overline{u^{2\,nl}}-\overline{u^{2\,l}}}{u^2}$ (%)
		eq (2)	eq (4)	eq (6)	eq (8)			Measured	Theory	
-0.165	0·739	0·742	0·702	0·736	0·726	0·74	0·026	0	0	—
-0.098	1·06	1·06	1·01	1·06	1·05	1·07	0·038	0	0·01	—
-0.062	2·19	2·28	2·19	2·31	2·28	2·30	0·088	0·04	0·36	8·0
-0.031	3·42	3·27	3·18	3·27	3·26	3·34	0·147	1·0	0·60	8·7
0·037	3·71	3·75	3·75	3·76	3·80	3·58	0·32	-0.25	-0.46	-1.9
0·098	2·24	2·52	2·57	2·52	2·58	2·17	0·62	-1.3	-1.8	-26

Table 1 and 2

However, when probes with more than one wire is used, then the data analysis is usually much simpler for the linearized signals. In the remaining discussion it will therefore be assumed that the hot-wire signals have been linearized.

To obtain information on the velocity components perpendicular to the mean flow direction, it is necessary to place one or more hot-wires at a yawed angle relative to the mean flow direction. For accurate results it is necessary first to determine the hot-wire yaw response.

2.2 Yaw Response

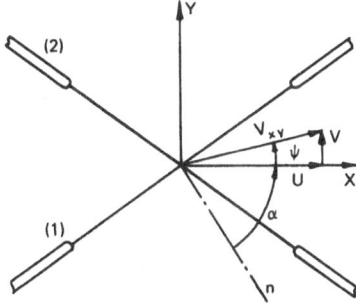

Fig. 1. Notation for X Hot wire probe

The effective cooling velocity $\overline{V}_e$ for a hot-wire probe placed at an angle to the mean flow direction is usually expressed as

$$\overline{V}_e = \overline{U}f(\alpha) \qquad [6]$$

where $\overline{U}$ is the mean velocity and α the angle between the normal n to the hot-wire probe and the mean velocity direction (see Fig. 1). $f(\alpha)$ deviates from a simple cosine law, and yaw calibration has branched into either direct calibration of the velocity component sensitivities $S_u = \partial E/\partial u$ and $S_v = \partial E/\partial v$ (e.g. Bradshaw (1971), p. 123), or into the determination of analytical approximations for the yaw dependence $f(\alpha)$ (e.g. Champagne et al. (1967) and Bruun (1971, 1972))

$$f(\alpha) = \begin{cases} (\cos^2\alpha + k_\alpha^2 \sin^2\alpha)^{\frac{1}{2}} & [7a] \\ \\ \cos^m\alpha & [7b] \end{cases}$$

In the following sections the data evaluations related to a single yawed hot-wire probe, an X hot-wire probe and a 3 wire probe are considered. Both of the relationships [7a] and [7b] have been shown by Bruun (1972) to give adequate yaw approximations. Both functions are therefore used in the following discussion, which apply to all hot-wire probes and support orientations. The relevant values of m (or k) for different probe configurations and support orientations are discussed in Bruun (1972).

2.3 Single Yawed Hot-Wire Probe

The linear low turbulence intensity equation for a single yawed hot-wire probe, using $f(\alpha) = \cos^m\alpha$, is (Bruun 1972))

$$V_e = \cos^m\alpha(\overline{U}+u - m\tan\alpha\, v) \qquad [8]$$

It is well known that measurement of the rms values corresponding to the hot-wire positions $\pm\alpha$ in the XY plane and one measurement with a single normal hot-wire probe, provide the values of $\overline{u^2}$, $\overline{v^2}$ and $\overline{uv}$. An alternative procedure is to place the yawed hot-wire probe at three different values of α and to solve the three simultaneous equations for $\overline{u^2}$, $\overline{v^2}$ and $\overline{uv}$ (Fujita and Kovasznay (1968)). If the hot-wire is placed in the XZ plane, then the values obtained are $\overline{u^2}$, $\overline{w^2}$ and $\overline{uw}$, but to obtain the shear stress $\overline{vw}$ a more complex method is necessary.

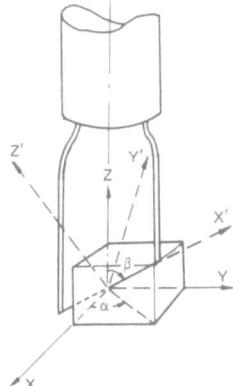

Fig. 2. Single yawed hot-wire probe with spatially fixed co-ordinate system (XYZ) and probe fixed co-ordinate system (X', Y', Z').

If a hot-wire at a fixed point is placed at six different orientations, then it is possible to calculate all the six Reynolds stresses, $\overline{u^2}$, $\overline{v^2}$, $\overline{w^2}$, $\overline{uv}$, $\overline{uw}$ and $\overline{vw}$, as well as the mean velocity components $\overline{U}$, $\overline{V}$ and $\overline{W}$. The velocity components U, V and W are evaluated in a spatially-fixed co-ordinate system XYZ, while the effective cooling velocity is specified in an X'Y'Z' co-ordinate system fixed relative to the hot-wire probe (Fig. 2). (The hot-wire and the Z axis are in the X'Z' plane). The relationship for the effective cooling velocity expressed in terms of the velocity components U, V and W in the spatially-fixed co-ordinate system XYZ is obtained as follows: The effective cooling velocity in the X'Y'Z' co-ordinate system is most conveniently expressed in the form (Jorgensen (1971))

$$V_e^2 = V_z'^2 + k_\alpha^2 \, V_x'^2 + k_\beta^2 \, V_y'^2 \qquad [9]$$

where k_α and k_β are the yaw and pitch corrections.

The hot-wire orientation in the spatially fixed co-ordinate system XYZ (Fig. 2) is specified by the angles α and β. The co-ordinate system transformation is therefore related to angle rotations of α and β, giving the following relationship between the velocity components in the two co-ordinate systems

$$V_x' = \sin\beta(\cos\alpha \ U + \sin\alpha \ V) + \cos\beta \ W \qquad [10a]$$

$$V_y' = -\sin\alpha \ U + \cos\alpha \ V \qquad [10b]$$

$$V_z' = -\cos\beta(\cos\alpha \ U + \sin\alpha \ V) + \sin\beta \ W \qquad [10c]$$

By combining equations [9] and [10a-c] the required equation is obtained. The values of α and β are arbitrary, but the simplest solution is obtained by rotating the hot-wire probe around the Z axis (β constant) (Fig. 2). Selecting six values of α and measuring the corresponding mean and rms values of the effective velocity gives the necessary simultaneous equations for the evaluation of the three mean velocity components and the six Reynolds stresses. The yaw and pitch factors k_α and k_β make the solution of the six simultaneous equations [9] somewhat lengthy, but the data evaluation can be reduced to simple matrix operations if a digital computer is available.

A similar six orientation method for a single yawed hot-wire probe, using the calibration function

$$E^2 = A + B(\alpha,\beta) \, |\vec{V}|^n \tag{11}$$

has been developed by Hoffmeister (1972). The slope parameter B contains the yaw and pitch dependence, which was determined by direct calibration of the hot-wire probe. Hoffmeister also used a digital computer to simplify the data reduction.

The above calculations have shown that it is possible to obtain the mean velocity and Reynolds stress components from a single yawed hot-wire probe. The single wire measurements, however, do not enable the relationship between the instantaneous velocity fluctuations to be derived. To obtain this information two or more hot-wires must be used simultaneously.

2.4 X Hot-Wire Probe

An X hot-wire probe (Fig. 1) provides information about the instantaneous U and V velocity components in the XY plane of the X hot-wires. The W velocity component in the Z direction introduces interpretation errors, but if the relative mean velocity component $\overline{W}/|\vec{V}|$ is small, then these errors can normally be ignored (Bruun (1975a)).

A simple sum and difference solution exists if a cosine yaw dependence is valid. Using the notation on Fig. 1, the effective cooling velocity of the two hot-wires 1 and 2, can for small values of $\overline{W}/|\vec{V}|$ be written as

$$V_{e,1} = V_{XY} \cos(\alpha+\psi) \tag{12a}$$

$$V_{e,2} = V_{XY} \cos(\alpha-\psi) \tag{12b}$$

and the instantaneous velocity components U and V are obtained as

$$U = \frac{V_{e,1} + V_{e,2}}{2\cos\alpha} \tag{13a}$$

$$V = \frac{V_{e,2} - V_{e,1}}{2\sin\alpha} \tag{13b}$$

However, the yaw dependence of a hot-wire probe, deviates from a cosine law (section 2.2), and for practical hot-wire applications, first order, low turbulence intensity equations have been derived. When the mean flow direction and the X hot-wire probe are aligned ($\overline{V} = 0$), then the deviation from a cosine law can be included in a simple sum and difference analysis (Champagne and Sleicher (1967) and Bruun (1972)).

The first order solution for the yaw function [7a] is

$$\frac{V_{e,1} + V_{e,2}}{2} = (\cos^2\alpha + k_\alpha^2 \sin^2\alpha)^{\frac{1}{2}} (\overline{U} + u) \tag{14a}$$

$$\frac{V_{e,2} - V_{e,1}}{2} = (\cos^2\alpha + k_\alpha^2 \sin^2\alpha)^{\frac{1}{2}} A\tan\alpha \, v \tag{14b}$$

with the yaw factor A being

$$A = \frac{\cos^2\alpha(1 - k_\alpha^2)}{\cos^2\alpha(1 - k_\alpha^2) + k_\alpha^2} \tag{14c}$$

The similar sum and difference equations corresponding to the yaw function [7b], and the higher order correction terms applicable in high turbulence intensity flows, have been derived by Bruun (1972).

However, a simple solution (equations [13a] and [13b] exists for a cosine yaw dependence even when misalignment occurs, implying that a similar simple method should exist for other yaw dependence functions. If the misalignment is included in the usual series expansion method, then rather complicated equations are obtained. However, evaluation of the values of the coefficients to the velocity components show these to be nearly independent of the misalignment angle ψ (Fig. 1), permitting a simple sum and difference method to be used even when misalignment occurs (Bruun (1975a)).

2.5 3 Wire Probe
Three hot-wires placed at different orientations at a point 0 enable all three velocity components to be evaluated simultaneously. If the hot-wires, as well as their prongs, are placed perpendicular to each other (Fig. 3), then a vector notation similar to equation [9] can be used for all three hot-wires. Specifying

H.H. Bruun

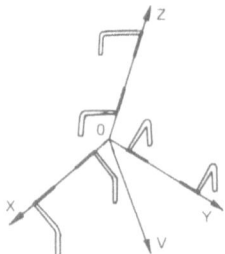

Fig. 3. Co-ordinate system for 3 wire probe

the velocity vector V by the velocity components U, V and W in the XYZ co-ordinate system of the three hot-wires (Fig. 3), the equations for the effective cooling velocities of the three hot-wires become:

$$V_{e,1}^2 = V^2 + k_\beta^2 W^2 + k_\alpha^2 U^2 \qquad [15a]$$

$$V_{e,2}^2 = W^2 + k_\beta^2 U^2 + k_\alpha^2 V^2 \qquad [15b]$$

$$V_{e,3} = U^2 + k_\beta^2 V^2 + k_\alpha^2 W^2 \qquad [15c]$$

By solving the equations [15a-c] one obtains the squared values U^2, V^2 and W^2 of the instantaneous velocity components of the velocity vector. To be able to deduce the velocity components from these values, it is necessary to know the octant the flow is in, since identical readings can be obtained from the eight octants.

2.6 Vorticity Probe

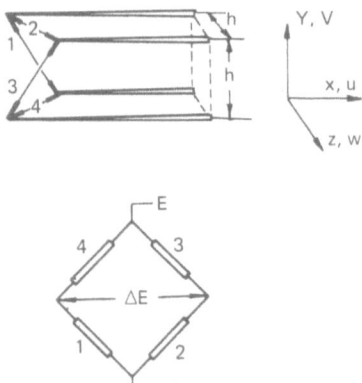

Fig. 4. Vorticity probe

A four wire probe has been developed for the measurement of the vorticity component in the mean flow direction (see e.g. Bradshaw (1971)).

The three vorticity components in a rectangular X,Y,Z co-ordinate system are

$$\omega_x = \frac{\partial W}{\partial y} - \frac{\partial V}{\partial z} \; ; \; \omega_y = \frac{\partial U}{\partial z} - \frac{\partial W}{\partial x} \; ; \; \omega_z = \frac{\partial V}{\partial x} - \frac{\partial U}{\partial y} \qquad [16]$$

To evaluate ω_x we must measure the transverse variations in the V and W velocity components. The 'streamwise' vorticity probe (Fig. 4) therefore consists basically of two 'X hot-wire' probes: the first 'A' parallel to the XY plane for the measurement of (u,v) and a second 'B' parallel to the XZ plane to obtain (u,w). To evaluate the gradient $\partial V/\partial z$ the two wires 1 and 3 in 'probe A' must be separated a distance h in the z direction, and similarly the two wires 2 and 4 in 'probe B' are separated a distance h in the y direction.

Placing the hot-wires at angles of ±45°, and assuming that a cosine yaw law is valid (see section 2.2), it follows from a first order Taylor expansion of the velocity components that the 'summation signals' from the two 'X hot-wire' probes A and B are

$$V_{e1} + V_{e3} \sim 2u - h\frac{\partial v}{\partial z} \qquad [17a]$$

$$V_{e2} + V_{e4} \sim 2u - h\frac{\partial w}{\partial y} \qquad [17b]$$

Therefore connecting the four wires from the vorticity probe to the wheatstone bridge as shown in Fig. 4, the bridge balance signal ΔE

$$\Delta E = \tfrac{1}{2}(e_1 + e_3 - e_2 - e_4) \sim h(\frac{\partial w}{\partial y} - \frac{\partial v}{\partial z}) \qquad [18]$$

will be proportional to the streamwise vorticity component ω_x.

Accurate measurements of the two transverse vorticity components ω_y and ω_z are much more difficult. The first term $\partial u/\partial z$ in ω_y can be obtained from the 'difference signal' from two normal hot-wires with a spatial separation in the Z direction. However, the direct measurement of $\partial w/\partial x$ requires: a) the 'summation signal' from the two normal hot-wires = 2u, and b) the 'summation signal' from two yawed hot-wires (±45°) placed in the XZ plane and separated a distance h in the X direction. Mathematically such a probe arrangement should provide the correct value of $\partial w/\partial x$. However, in this wire arrangement the downstream wire is placed directly in the thermal wake from the upstream yawed wire. An approximate solution to this problem can be obtained provided a 'frozen pattern convection' relationship:

$$\frac{\partial}{\partial t} = -\bar{U}\frac{\partial}{\partial x} \qquad [19]$$

can be applied to the existing turbulence. If equation [19] is an accurate approximation, then we can replace $\partial w/\partial x$ by $-\bar{U}\,\partial w/\partial t$, and $\partial w/\partial t$ is obtainable from an X hot-wire probe placed in the XZ plane. This technique has recently been used in a special probe design (see section 5).

3. Temperature Measurements

In the previous section the temperature T_a of the air flow has been assumed to be constant. There are, however, many flow situations where either the mean temperature changes during the experiment, or where the velocity fluctuations are accompanied by temperature fluctuations. In both cases the dependence of the anemometer output voltage on the fluid temperature must be known.

When simultaneous velocity and temperature fluctuations occur, then both the velocity and temperature sensitivities must be known. To obtain accurate temperature results it is necessary to work in terms of the wire voltage E_w, instead of the direct anemometer output voltage E. Accurate temperature measurements are therefore considerably more involved than velocity measurements. The velocity and temperature dependent of the wire voltage E_w can, to first order, be expressed by the equation

$$\frac{E_w^2}{R_w} = (A + BV^n)(T_w - T_a) \qquad [20]$$

where A, B and n are constants independent of the velocity and temperature. The necessary second-order corrections to equation [20] are discussed in Bruun (1975b).

Introducing the notation $\theta = \Delta T$, the velocity and temperature sensitivities corresponding to equation [20] become

$$S_u = \frac{\partial E_w}{\partial u} = \frac{nBV^{n-1}}{2}\sqrt{\frac{R_w(T_w - T_a)}{A + BV^n}} \qquad [21a]$$

$$S_\theta = \frac{\partial E_w}{\partial \theta} = -\tfrac{1}{2}\sqrt{\frac{R_w(A + BV^n)}{T_w - T_a}} \qquad [21b]$$

giving the following equation for the fluctuating voltage

$$e = S_u\, u + S_\theta\, \theta \qquad [22]$$

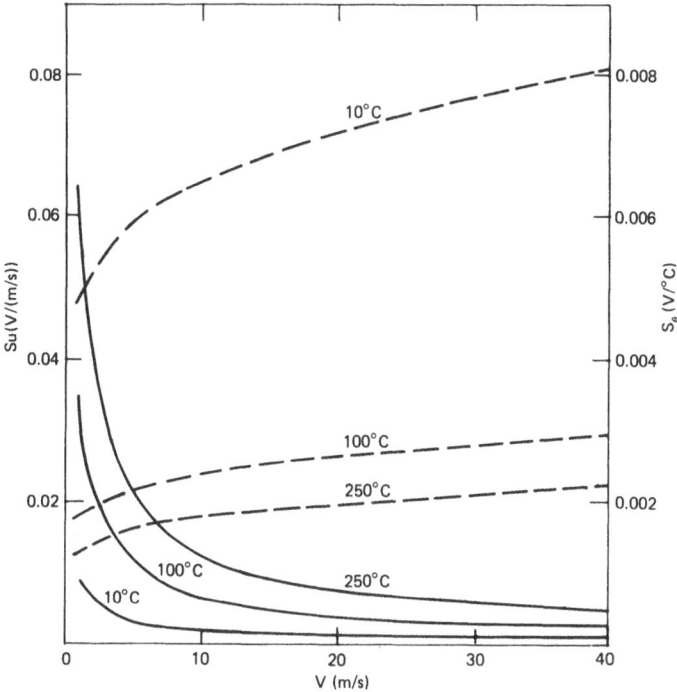

Fig. 5. Variation in velocity and temperature sensitivities with velocity and overheat ratio. —— S_u, – – – S_θ.

Equations [21a] and [21b] demonstrate that both the temperature and velocity sensitivities depend on the velocity V, the temperature difference $(T_w - T_a)$ and the wire temperature T_w (or the hot resistance R_w). For a standard normal ISVR, typical values for A, B and n in equation [20] are quoted in Bruun (1975b). The author has evaluated the related velocity and temperature sensitivities S_u and S_θ, and typical values corresponding to $T_w - T_a = 10°, 100°$ and $250°C$ are plotted in Fig. 5 for the velocity range 1 - 40 m/s. For a given velocity V, the sensitivities S_u and S_θ are seen to vary inversely with the temperature difference $T_w - T_a$, the value of S_u increasing and the value of S_θ decreasing with increasing values of $T_w - T_a$. A high overheat ratio is therefore recommended for velocity fluctuation measurements. However, even at the highest overheat ratio we observe a significant temperature sensitivity, and it is therefore necessary to control the temperature even in velocity experiments. To obtain a high temperature sensitivity a low overheat ratio must be used (see Fig. 5). Consequently for temperature measurements the hot-wire is often operated at a very low overheat ratio as a resistance thermometer.

The sensitivities S_u and S_θ have different overheat ratio dependence, and several different techniques have been developed for the evaluation of simultaneous velocity and temperature fluctuations.

Operating a single normal hot-wire probe at three different overheat ratios and measuring the corresponding voltage rms values (equation [39]), enables $\overline{u^2}$, $\overline{u\theta}$ and $\overline{\theta^2}$ to be evaluated (see e.g. Corrsin (1947)).

If a single yawed hot-wire is placed in a flow with simultaneous velocity and temperature fluctuations, then the equation for the fluctuation voltage e becomes

$$e = S_u\, u + S_v\, v + S_\theta\, \theta \qquad\qquad\qquad [23]$$

where S_u and S_v are the u and v sensitivities of the yawed hot-wire probe.

By applying equation [23] to an X hot-wire probe, and combining the voltage outputs from the two wires, it is possible to evaluate $\overline{v^2}$ $\overline{uv}$ and $\overline{v\theta}$ (Corrsin (1947)). The complete u, v and θ solutions can be obtained by placing a resistance wire close to the X hot-wire probe. However, due to the large difference in the values of the velocity and temperature sensitivities S_u and S_θ, a considerable uncertainty exists in the evaluated temperature terms, as discussed by Arya and Plate (1969) for both single and X hot-wire probes.

If two normal hot-wires are placed close to each other and operated at different overheat ratios, then it is possible to separate the velocity and temperature fluctuations. In the investigations by Chevray and Tutu (1972) and Ali (1975), one of the hot-wires was operated in a constant-current low-overheat-ratio mode giving a signal directly proportional to the temperature fluctuations. By using this signal to compensate the high-overheat-ratio wire signal for the fluctuation temperature, the fluctuating velocity signal was obtained. In the investigation by Sakao (1973) the hot-wires were operated at different overheat ratios in the constant temperature mode. Using the difference between the squared voltages from the two hot-wires, a compensation technique which separates the velocity signal from the temperature fluctuations was developed.

4. Special Probes
The previous sections have described the evaluation of hot-wire probe signals in 'simple' flow fields (nearly uniform mean velocity field, low turbulence intensity etc.). Most practical flow situations are very complex, and additional anemometer problems may occur including, aerodynamic disturbance effects, non

linear frequency response, mean-shear corrections, spatial resolution and wire length errors, wall effects, high turbulence intensity corrections etc. The magnitude of these effects have been evaluated in many papers as outlined in e.g. Comte-Bellot (1976) and Bruun (1977a).

Special probes have been designed to overcome or minimize these problems. Miniature probes are available for measurements in confined spaces, and multi-probes are now designed with special precautions to avoid thermal wake interference effects.

Recently some new probe techniques, based on the detection of a thermal wake, have been developed to overcome some of the limitations of conventional hot-wire probes. X hot-wire probes are used to measure lateral velocity components, but due to their finite size they cannot be used close to a wall. A different technique based on an array of three parallel hot-wires overcomes this problem. One heated wire is placed upstream and two 'cold' detection wires set downstream symmetrically with respect to the first wire are used for the measurement of the lateral velocity component (Rey and Beguier (1977)).

Hot-wire probes are sensitive to all three velocity components. In low and moderate turbulence intensity flows it is usually possible to evaluate the necessary correction terms, as illustrated in e.g. Bruun (1975b). However, in highly turbulent flows, flow reversal will often occur for a considerable period of the time. A conventional hot-wire probe cannot detect flow reversal, and large interpretation errors occur in such cases. To overcome this problem, thermal wake probes have been developed. The 'pulsed wire' technique (Bradbury and Castro (1971)) utilize one heated (pulsed) wire and two cold detection wires placed perpendicular on either side of the heated wire. A shielded dual-sensor hot-wire probe has recently been developed by Cook and Redfearn (1976), using two conventional hot-wire anemometers. In this design the sign of the backward sensor was inverted, and a simple electronic circuit was used continuously to select the signal with the largest magnitude. Due to the thermal wake effect of the upstream or downstream sensor on the other sensor, this probe provides both the magnitude and direction of the flow velocity.

5. Probes for Flow Structure Detection
Special probes have been developed during the last 5-10 years for the study of specific turbulent quantities.

A recent five sensor film probe (Brodkey et al. (1975)) is shown in Fig. 6. This probe consists of a) a single sensor sensitive to u, b) a V probe sensitive to u,w

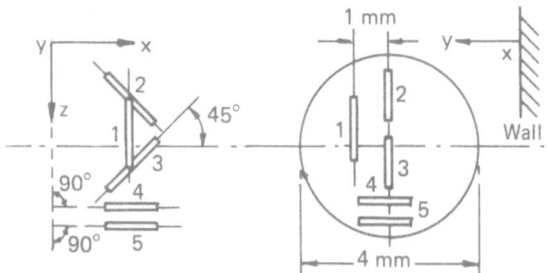

Fig. 6. Arrangement of the sensors in the 5-sensor probe. 1: Single probe; 2 & 3: V probe; 4 & 5: X probe. (From Brodkey et al. (1975)).

and c) an X probe sensitive to u, v. Measuring the simultaneous output from all 5 sensors, it is possible to calculate the simultaneous instantaneous signals of u, v, w and also $\partial u/\partial y$ and $\partial u/\partial z$. These measurements enable the evaluation of the instantaneous turbulence production $-\rho uv[\partial U(t)/\partial y]$ and estimates of the two vorticity components $\omega_y(t)$ and $\omega_z(t)$ as outlined in section 2.6:

$$\omega_y(t) = \frac{\partial u}{\partial z} - \frac{\partial w}{\partial x} \cong \frac{\partial u}{\partial z} + \frac{1}{U}\frac{\partial w}{\partial t} \qquad [24a]$$

$$\omega_z(t) = \frac{\partial v}{\partial x} - \frac{\partial u}{\partial y} \cong -\frac{1}{U}\frac{\partial v}{\partial t} - \frac{\partial u}{\partial y} \qquad [24b]$$

With the advent of fast digital computers conditional sampling techniques have been developed, which enable separate flow phenomena to be isolated and studied. These techniques have provided considerable insight into the physical nature of turbulent flows, as illustrated by some of the other conference papers. Only a few of these multi-probe techniques shall therefore be mentioned.

Betchov (1974) studied the occurrence of positive spikes in the time derivative signal of the turbulent streamwise velocity component $\partial u/\partial t$. Recording the simultaneous signal from both 3 and 4 wire probes, he concluded from probability evaluations that the occurrence of these spikes were related to large scale coherent flow structures.

A considerable number of experiments have recently been carried out in turbulent boundary layers using 'hot-wire rakes', illustrated on Fig. 7 (from Blackwelder and Kaplan (1976)). A typical set of their results, shown in Fig. 8, for both the instantaneous and conditional sampled signals, clearly demonstrates the coherent nature of the related large scale flow processes. The tem-

Fig. 7. Photograph of the ten-wire rake used for measuring the instantaneous streamwise velocities at different positions in the normal direction.

poral and spatial relationship of these large flow structures can be determined by a new time domain technique developed by Bruun (1977b).

6. Higher Order Moments

The evaluation of higher order moments is usually related to the statistical description of turbulent flow. The relationship between the different statistical properties of turbulence are governed by the Navier-Stokes equations, and the central problem in statistical turbulence theory, is to obtain solutions to the time mean Navier-Stokes equation. It can be shown that in a turbulent flow the equation for the *mean* velocity flow field can be written as

$$\rho \frac{d\overline{V}_i}{dt} = \rho V_j \frac{\partial \overline{V}_i}{\partial x_j} = -\frac{\partial \overline{P}}{\partial x_i} + \mu \frac{\partial^2 \overline{V}_i}{\partial^2 x_j} - \rho \frac{\partial \overline{v_i v_j}}{\partial x_j} \qquad [25]$$

Therefore, to describe the mean velocity motion of a turbulent fluid, an additional term related to the Reynolds stresses $\overline{v_i v_j}$ must be included in the equation for the mean motion.

The describing equation for the shear stress term $\partial \overline{v_i v_j}/\partial x_j$ can be obtained from the Navier-Stokes equation [25] as shown by Townsend (1956):

$$\overline{V}_j \frac{\partial \overline{v_i v_\varrho}}{\partial x_j} = -(\overline{v_i v_j} \frac{\partial \overline{V}_\varrho}{\partial x_j} + \overline{v_\varrho v_j} \frac{\partial \overline{V}_i}{\partial x_j}) - \frac{\partial \overline{v_i v_\varrho v_j}}{\partial x_j}$$

$$- (\overline{v_i \frac{\partial p}{\partial x_\varrho}} + \overline{v_j \frac{\partial p}{\partial x_i}}) + \nu (\overline{v_i \frac{\partial^2 v_\varrho}{\partial^2 x_j}} + \overline{v_j \frac{\partial^2 v_i}{\partial^2 x_j}}) \qquad [26]$$

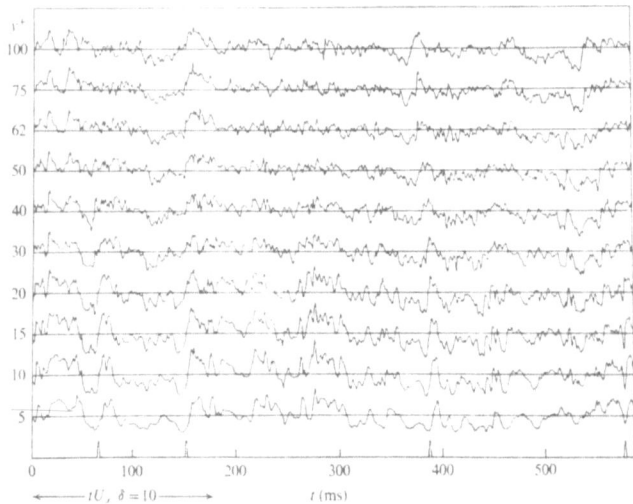

Instantaneous streamwise velocities as a function of the normal direction.
The detector function obtained at $y^+ = 15$ is shown between the tick marks.

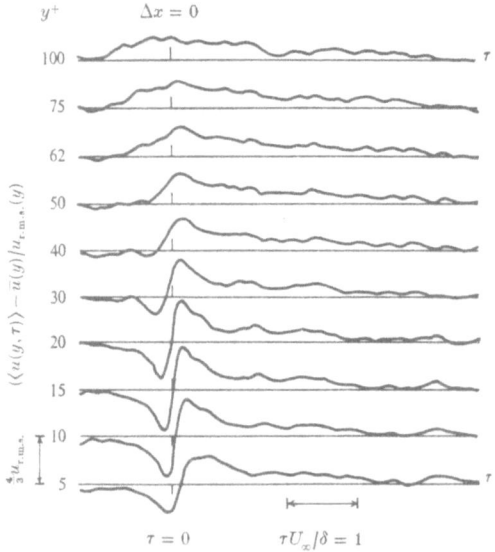

Conditionally averaged streamwise velocities at ten y^+ locations.
The detection criterion was applied at $y^+ = 15$. $R_\theta = 2550$.

Fig. 8. (From Blackwelder and Kaplan 1976).

This (stationary) equation for the Reynolds stresses $\overline{v_i v_\varrho}$ demonstrates that the shear stresses are a balance between the gain by advection by the mean flow, production by interaction between the turbulence and the mean flow, gain by convective movements of the turbulence and by the action of the pressure gradient, and destruction by viscous forces (Townsend (1956)).

Much of the turbulence work is concerned with the transfer of turbulent energy from low wave numbers to high wave numbers. Various flow processes may be identified by considering the governing equation for the turbulent energy $E_o = \frac{1}{2}\overline{q^2}$, with $\overline{q^2}$ being the sum of the normal stresses

$$\overline{q^2} = \overline{v_i^2} = \overline{v_1^2} + \overline{v_2^2} + \overline{v_3^2} \qquad [27]$$

The corresponding turbulent energy equation, obtainable from the Navier-Stokes equations, can be written as:

$$\underbrace{\frac{d\overline{E}_o}{dt}}_{I} = \underbrace{\frac{\partial \overline{E}_o}{\partial t}}_{II} + \underbrace{\overline{V}_j \frac{\partial \overline{E}_o}{\partial x_j}}_{III} = -\underbrace{\frac{\partial}{\partial x_i}\left(\overline{v_i \frac{P}{\rho}}\right)}_{IV} - \underbrace{\frac{\partial}{\partial x_j}(\overline{v_j E_o})}_{V} - \underbrace{\overline{v_i v_j}\frac{\partial \overline{V}_i}{\partial x_j}}_{VI}$$

$$-\underbrace{\nu \frac{\partial}{\partial x_j}\,\overline{v_i\left(\frac{\partial v_j}{x_i} + \frac{\partial v_i}{x_j}\right)}}_{VII} - \underbrace{\nu\,\overline{\left(\frac{\partial v_j}{\partial x_i} + \frac{\partial v_i}{\partial x_j}\right)\frac{\partial v_i}{\partial x_j}}}_{VIII} \qquad [28]$$

This equation, which is known as the turbulent energy equation, states that

I: The rate of change in the turbulent energy (per unit mass) in a "Lagrangian" description which follows the *mean* flow can be related to the following fixed point measurements:

II: The rate of change in the turbulent energy as observed at the fixed point X. In a *stationary* turbulent flow this term is zero.

III: The advection of this energy past the fixed point by the mean velocity field.

IV and V: are inertial terms, which represent convective diffusion of the (total) energy by the turbulence. These terms neither create nor destroy turbulence.

VI: Production of turbulent energy from the mean motion by the turbulent shear stresses.

VII: Represent the work done per unit mass and time by the viscous shear stresses of the turbulent motion.

VIII: is the viscous energy dissipation (ϵ).

6.1 Basic Statistical Relationships in Isotropic Turbulence

The most significant theoretical contributions to the statistical description of turbulence has occurred in the areas of homogeneous and isotropic turbulence. In the *statistical* description of turbulence the flow field is assumed to be random, and we can consequently only obtain a probalistic description of the flow field. In this description of turbulence, the conditions of *homogeneous* turbulence exist when the averaged properties are independent of the position in the fluid. The condition of *isotropic* turbulence is obtained when furthermore the averaged properties are identical in all directions.

In isotropic turbulence, the following basic statistical relationships are valid

$$\overline{u^2} = \overline{v^2} = \overline{w^2} \qquad\qquad\qquad [29a]$$

$$\overline{v_i v_j} = 0 \qquad\qquad\qquad [29b]$$

$$S_u = \overline{u^3}/\overline{u^2}^{3/2} = 0; \; F_u = \overline{u^4}/(\overline{u^2})^2 = 3 \qquad\qquad\qquad [29c]$$

The measurements of these terms in a turbulent flow, will often give a simple estimate of the deviation from isotropic turbulence.

References

Ali, S.F., 1975, *Rev. Sci. Instr.*, **46**, 185-191

Aray, S.P.S. and Plate, E.J., 1969, *Instr. Contr. Syst.*, **42**, 87-90

Betchov, R., 1974, *Phys. Fluids,* **17**, 1509-1512

Blackwelder, R.F. and Kaplan, R.E., 1976, *J.F.M.*, **76**, pp. 89-112

Bradbury, L.J.S. and Castro, I.P., 1971, *J.F.M.*, **49**, 657-691

Bradshaw, P., 1971, *'An Introduction to turbulence and its measurements'*, Pergamon Press

Brodkey, R.S., Eckelmann, H., Nychas, S.G. and Wallace, J.M., 1975, Euromech 63, Copenhagen, Denmark

Bruun, H.H., 1971, *J. Sci. Instr.*, **4**, pp. 225-231

Bruun, H.H., 1972, *J. Sci. Instr.*, **5**, pp. 812-818

Bruun, H.H., 1975, *DISA Information,* No. **18**, pp. 5-10

Bruun, H.H., 1975, *J. Sci. Instr.*, **8**, pp. 942-951

Bruun, H.H., 1976, *J. Sci. Instr.*, **9**, pp. 53-56

Bruun, H.H., 1977, VKI Lecture serie 96

Bruun, H.H., 1977, *J.F.M.*, **83**, pp. 641-671

Champagne, F.H., Sleicher, C.A. and Wehrman, O.H., 1967, *J.F.M.*, **28**, 153-176

Champagne, F.H. and Sleicher, C.A., 1967, *J.F.M.*, **28**, 177-182

Chevray, R. and Tutu, N.K., 1972, *Rev. Sci. Instr.*, **43**, 1417-1421

Collis, D.C. and Williams, M.J., 1959, *J.F.M.*, **6**, 357-389

Comte-Bellot, G., 1976, *Ann. Rev. Fluid Mech.*, **8**, 209-231

Cook, N.J. and Redfearn, D., 1976, *J. Ind. Aerod.*, **1**, 221-232

Corrsin, S., 1947, *Rev. Sci. Instr.,* **18**, 469-471

Hoffmeister, M., 1972, *DISA Information* No. **13**, 26-28

Fujita, H. and Kovasznay, L.S.G., 1968 *Rev. Sci. Instr.,* **39**, 1351-1355

Jorgensen, F.E., 1971, *DISA Information,* No. **11**, 31-37

King, L.V., 1914, *Phil. Trans. Roy. Soc.,* **214A**, 373-432

Rey, C. and Beguier, C., 1977, *DISA Information* No. **21**, 11-15

Sakao, F., 1972, *J. Sci. Instr.,* **6**, 913-916

Townsend, A.A., 1956, *'The structure of turbulent shear flow',* Cambridge Univ. Press

Corrélations spatiotemporelles d'ordre élevé

par

R. Dumas
I.M.S.T., Marseille, France

1. Introduction

Les corrélations du deuxième ordre dépendent principalement des fluctuations dont le niveau est de l'ordre de leur écart type σ. Dans le cas d'un signal qui ne s'écarte pas trop d'une distribution Gaussienne, les fluctuations de niveaux inférieurs à 0,2 σ et supérieurs à 3,5 σ n'influent pratiquement pas sur la covariance. Il n'en est pas de même si l'on considère des corrélations d'ordres supérieurs. A titre indicatif la Figure 1 récapitule les contributions statistiques aux moments $\overline{t^2}$, $\overline{t^4}$ et $\overline{t^8}$ d'une variable Gaussienne t centrée, réduite. Les maxima correspondent à $\sqrt{n}$.

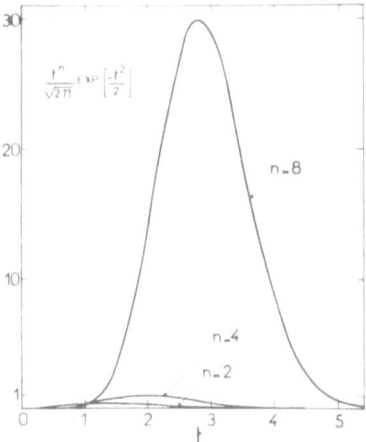

Figure 1. Distributions statistiques: $\overline{t^2} = 1$, $\overline{t^4} = 3$, $\overline{t^8} = 115$.

La difficulté de mesurer par exemple $\overline{t^8}$ est qu'il s'agit d'effectuer des moyennes de signaux de très forts niveaux relatifs, se produisant très rarement par rapport à l'échelle de temps caractéristique du phénomène. Par ailleurs les hypothèses de fermeture des équations statistiques des écoulements turbulents s'arrêtent pratiquement aux termes du 4ème ordre, aussi nous limiterons-nous aux corrélations d'ordre 3 et 4.

Le fait de considérer des corrélations spatiotemporelles (C.S.T.) ne modifie pas sensiblement les remarques précédentes, bien que les distorsions des pointes de grandes amplitudes du signal soient moins critiques, celles-ci étant peu correlées lorsque le décalage d'espace est notable. Par contre, il s'introduit dans le cas des C.S.T. les effets de sillages aérodynamique et thermique du support situé en amont sur la, ou les, sondes en aval. Nous nous plaçons en effet dans le contexte de l'utilisation du "fil chaud", car il ne semble pas que des mesures systématiques de C.S.T. aient été faites avec l'anémomètre à laser-Doppler, bien que cet appareil présenterait de grands avantages du fait de l'absence de sondes intrusives.

Nous rappelerons successivement les causes d'erreurs qui affectent les mesures des C.S.T., les principes des méthodes de calculs utilisées, et enfin les différentes interprétations physiques qui ont été données des C:S.T. d'ordre 3 et 4, notamment par FRENKIEL et KLEBANOFF, VAN ATTA et à l'I.M.S.T.

Nous ne considérerons pas le cas des C.S.T. pression-vitesse qui pourraient à certains égards être classées parmi les corrélations triples en deux points. On se reportera en premier lieu aux travaux de WILLMARTH. Rappelons que ces expériences ont suggéré à WILLMARTH (BO-TUNG TU et WILLMARTH, 1966) qu'il existait dans la région de paroi une structure tourbillonnaire bien définie dont il a proposé un modèle. Ces structures sont liées au processus d'éjections visualisées par KLINE (Cf. WILLMARTH, 1975); elles seront évoquées aux paragraphes 4.1 et 4.2.

2. Causes d'erreurs dans les mesures de corrélations d'ordre élevé

Un certain nombre d'erreurs interviennent déjà dans les mesures des corrélations du deuxième ordre, certaines qui deviennent plus critiques avec les ordres 3 et 4 sont rappelées dans ce chapitre.

2.1 Réponse de l'anémomètre à fil chaud aux fluctuations de grandes amplitudes

Nous ne reprendrons pas l'étude des effets de la réponse non-linéaire du fil chaud aux fluctuations de vitesse. Elle est traitée en détail dans de nombreux articles, en particulier par COMTE-BELLOT (74) et dans l'article de FULACHIER (78). On sait que cet effet peut être plus ou moins compensé par un linéarisateur analogique, ou même par linéarisation numérique. A titre d'exemple la figure 2 présente des résultats obtenus par FRENKIEL et KLEBANOFF, (1967b) sur les valeurs non corrigées et corrigées de l'effet de la non-linéarité d'un anémomètre à fil chaud, des corrélations du troisième ordre:

$$R^{1,2} = \frac{\overline{u_1'^2(t)\, u_1'(t+\tau)}}{(\overline{u_1'^2})^{3/2}} \qquad R^{2,1} = \frac{\overline{u_1'(t)\, u_1'^2(t+\tau)}}{(\overline{u_1'^2})^{3/2}}$$

u_1' est la composante longitudinale de la vitesse, $\overline{u}$ la vitesse moyenne, τ le temps retard, M la maille de la grille.

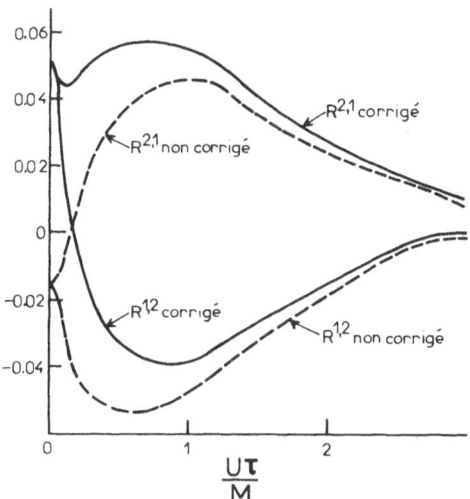

Figure 2. Effets de la correction de non-linéarité sur $R^{1,2}$ et $R^{2,1}$. M maille de la grille. (FRENKIEL et KLEBANOFF).

Il s'agit de mesures de turbulence en aval d'une grille. On voit que, malgré le faible niveau de turbulence $\sqrt{\overline{u_1'^2}}/\overline{u} \cong 0{,}018$ les corrections *changent le signe* des corrélations triples.

On sait qu'il existe une autre non-linéarité qu'il est impossible de corriger sans hypothèses simplificatrices: elle est due à ce que le fil chaud, détecteur thermique, ignore par principe le sens de la vitesse. On considère que la vitesse effective de refroidissement peut se mettre sous la forme (HINZE 1959):

$$u_{eff} = [u_n^2 + k^2 u_t^2]^{1/2} \qquad\qquad [1]$$

u_n est l'intensité de la projection de $\overrightarrow{u}$ sur le plan normal au fil, u_t sur le fil lui-même. En fait, $k \ll 1$ tend vers zéro pour un fil assez long, environ 600 diamètres (CHAMPAGNE et al. 1967) la relation [1] s'écrit (DUMAS, 1978):

$$u_{eff} = [|\overrightarrow{u}\times\overrightarrow{f}|^2 + k^2 |\overrightarrow{u}\cdot\overrightarrow{f}|^2]^{1/2} \qquad\qquad [2]$$

où $\vec{f}$ est le vecteur unitaire porté par le fil. On a dans le cas le plus simple d'un fil "droit", perpendiculaire à la vitesse moyenne, $\vec{u}(\overline{u}+u'_1, u'_2, u'_3)$, $\vec{f}(0,1,0)$, en négligeant le terme de produit scalaire:

$$u_{eff} \approx \overline{u}[1+2\frac{u'_1}{\overline{u}} + \frac{u'^2_1}{\overline{u}^2} + \frac{u'^2_3}{\overline{u}^2}]^{\frac{1}{2}} \qquad\qquad [3]$$

Il n'est possible de négliger u'^2_1 et u'^2_3 que si les fluctuations sont d'amplitudes modérées. On peut dans la linéarisation de débarrasser de l'exposant 1/2, mais il demeure que la fluctuation de tension électrique $e'(t)$, après linéarisation, est entachée des termes parasites précités:

$$e'(t) = e(t)-\overline{e}(t) \propto 2\frac{u'_1}{\overline{u}} + \frac{u'^2_1}{\overline{u}^2} + \frac{u'^2_3}{\overline{u}^2} - (\frac{\overline{u'^2_1}}{\overline{u}^2} + \frac{\overline{u'^2_3}}{\overline{u}^2}) \qquad\qquad [4]$$

Des calculs de corrections sur $\overline{u'^2_1}$, $\overline{u'^3_1}$, etc... ont été faits en élevant [4] à la puissance correspondante. Il est toutefois nécessaire de se donner plus ou moins arbitrairement les divers moments et corrélations entrant dans les termes correctifs. On simplifie en général les calculs avec des hypothèses d'isotropie et de lois normales.

Notons d'ailleurs que ZARIC (1969) a donné et calculé pour quelques valeurs des paramètres l'expression de la distribution en probabilité du module $|u_n|$ dans le cas de *distributions normales* correlées pour u'_1 et u'_2. A partir de cette expression on pourrait construire des abaques qui permettraient de déterminer les erreurs de mesure du fil chaud en fonction de l'intensité élevée de turbulence.

FRENKIEL et KLEBANOFF (1967b) ont explicité les termes correctifs déterminés à partir de relations analogues à [4] dans le cas plus complexe de correlations entre les fluctuations en deux points pour diverses composantes jusqu'à l'ordre 4. Les expressions obtenues contiennent de très nombreux termes; ces auteurs ont montré que les termes correctifs étaient négligeables dans le cas des mesures derrière grille. Mais il conviendrait, dès que l'intensité de turbulence est élevée, supérieure à 10 %, d'examiner chaque pas particulier.

Ces divers effets de non-linéarité sont donc particulièrement critiques dans le cas des corrélations d'ordre élevé; il faut rappeler à ce propos que ces difficultés n'existent pratiquement pas lorsque l'on considère les fluctuations de température, le fil fonctionnant en thermomètre (Cf. FULACHIER 1978).

2.2 Sillages
Notons tout d'abord que si les fluctuations sont de grandes amplitudes il serait

nécessaire de tenir compte du terme $k^2 u_t^2$ dans [1]. Dans le cas de fils droits, ce terme correctif est en général négligeable même pour des intensités fortes, de l'ordre de 20 %; par contre, il peut devenir important dans des mesures avec des fils inclinés sur la direction de la vitesse moyenne (BAILLE 1973).

Par ailleurs, l'angle instantané du vecteur vitesse avec la normale au fil ne doit pas dépasser en pratique 60° pour éviter des interactions support-fil. Si l'on admet que cet angle limite correspond à un écart de 4σ où σ est l'écart type rapporté à la vitesse moyenne de la composante u_2', l'intensité relative maximale sans distorsion est dans le cas d'un fil "droit" $\sigma = $ Arctg 15° $\cong$ .42 et dans le cas d'un fil incliné à 45° $\sigma = $ Arctg (15°/4) $\cong$ 0.07 seulement. Les fils inclinés en x sont d'ailleurs impropres à fonctionner correctement avec une intensité de turbulence excédant 15 % par suite des interactions entre fils. Les performances des sondes à trois fils parallèles sont nettement supérieures de ce point de vue (REY, BEGUIER 1977).

Dans le cas des C.S.T. il existe de plus le risque d'interactions entre les fils. En principe le sillage d'une sonde en amont augmente l'intensité de turbulence en aval, ce qui sert d'ailleurs de test. Toutefois, il peut exister des déformations du champ de vitesse telles que l'intensité de turbulence à l'aval soit diminuée (FAVRE et al. 1958). De même, des perturbations de l'écoulement de type à potentiel de vitesse peuvent être induites par un support sur un fil en amont (COMTE-BELLOT 1974).

Pour éviter le sillage de la sonde en amont il est indiqué de décaler dans la direction où la turbulence est homogène, le fil aval par rapport à la ligne de courant moyenne passant par l'amont. En diminuant progressivement ce décalage on peut espérer déterminer par extrapolation la valeur de la C.S.T. avec décalage nul (COMTE-BELLOT, CORRSIN 1971).

2.3 Distorsions d'origine électroniques
Une première difficulté bien connue de mesurer les C.S.T. d'ordre 3 et 4 est d'amplifier ou d'acquérir sans distorsion de non-linéarité des signaux de grandes amplitudes relatives par rapport à leur écart type. La figure 1 est révélatrice à cet égard.

Une deuxième difficulté est due à l'étalement spectral. Considérons par exemple le spectre de puissance F_e (f) de $e'(t)$:

$$F_e(f) = \int_{-\infty}^{\infty} \overline{e'(t)e'(t+\tau)} \exp(-2\pi i f \tau) \, d\tau \qquad [5]$$

Si l'on suppose que $e'(t)$ est Gaussian on trouve que le spectre de e'^2, $F_e{}^2$ (Cf. BLANC-LAPIERRE et PICINBONO 1961):

$$F_e(f) = (e'^2)^2\ \delta(f) + 2 \int_{-\infty}^{\infty} F_e(f')\ F_e(f-f')\ df' \qquad [6]$$

$\delta(f)$ opérateur de Dirac. De l'énergie est projétée en composante continue, aux fréquences basses et élevées. Le coefficient d'autocorrélation de $e'^2(t)$, $r_2(\tau)$:

$$r_2(\tau) = \frac{\overline{e'^2(t)\ e'^2(t+\tau)} - \overline{(e'^2)^2}}{\overline{(e'^2 - e'^2)^2}} \qquad [7]$$

est lié au coefficient d'autocorrélation $r(\tau)$ de $e'(t)$ par la relation [8] qui sert d'ailleurs à établir [6];

$$r_2(\tau) = [r(\tau)]^2 \qquad [8]$$

$$\left(\frac{\partial^2 r_2}{\partial \tau^2}\right)_{\tau=0} = 2\left(\frac{\partial^2 r}{\partial \tau^2}\right)_{\tau=0} \qquad [9]$$

La relation [9] indique une énergie spectrale relative plus élevée aux hautes fréquences. Il s'ensuit que pour avoir une même définition relative au voisinage de $\tau = 0$ de $r_2(\tau)$ que pour $r(\tau)$, il faut doubler la bande passante, et la fréquence d'acquisition. La composante continue de e'^2 peut être aussi une source d'erreurs, aussi bien en calculs analogiques que numériques. En effet, considérons le produit $\overline{e'^2 s'}$; si $s'(t)$ est légèrement décentré $s' = s'' + \epsilon$, $\overline{s''} = 0$, on a: $\overline{e'^2 s'} = \overline{e'^2 s''} + \overline{e'^2}\epsilon$. Le terme parasite $\overline{e'^2}\epsilon$ peut fausser le résultat, car la corrélation $\overline{e'^2 s''}$ est en général faible.

Enfin, S. CORRSIN a montré que la constante de temps d'un fil chaud en fonctionnement dynamique est en fait variable car elle dépend de l'amplitude du signal. Ceci introduit des harmoniques spectraux qui semblent peu importantes même par des amplitudes très fortes dans le cas d'un fonctionnement à température constante (COMTE-BELLOT 1974). Par contre il n'en est pas de même dans le cas du fonctionnement à intensité constante avec la compensation usuelle de l'inertie thermique. La figure 3 est relative à des résultats obtenus par calculs par COMTE-BELLOT et SCHON à l'aide d'un calculateur analogique, montrant l'erreur considérable qui pourrait être faite sur le facteur de dissymétrie $\overline{u'^3}/(\overline{u'^2})^{3/2}$ dans le cas de très fortes intensités simulées de turbulence de l'ordre de 25 %.

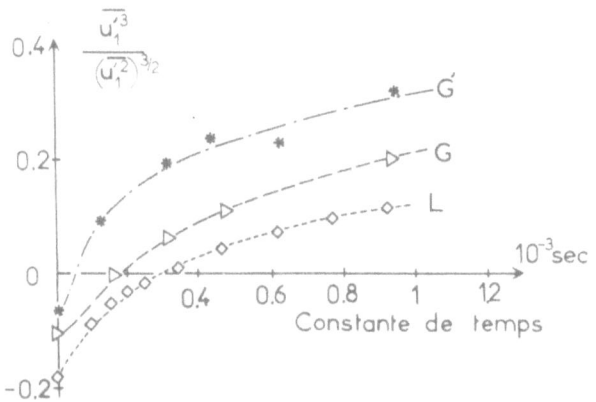

Figure 3. Effets de la non-linéarité dynamique de la constante de temps M. G, G' turbulence derrière grille, L de couche limite. (COMTE-BELLOT et SCHON).

3. Méthodes de Calculs

A partir des signaux analogiques délivrés par les anémomètres à fils chauds, on peut soit calculer directement les corrélations, soit les déterminer à partir des transformées de FOURIER de ces signaux.

3.1 Méthodes Directes

Les premières mesures de corrélations dans le temps ont été effectuées à l'I.M.S.T. dès 1948 (FAVRE 1948, FAVRE et al. 53) par des méthodes analogiques. Le décalage de temps entre les deux signaux était obtenu par décalages des têtes d'enregistrement et de lecture sur deux rubans magnétiques. Cette méthode garde d'ailleurs son intérêt actuellement, si l'on désire introduire un décalage de temps très important. Par la suite on utilisa assez généralement des appareils hybrides analogique-numériques introduisant des retards de temps incrémentaux. Ces appareils présentent des limitations de bande passante (Cf. GIOVANANGELI 1975).

Les méthodes numériques qui sont maintenant très largement utilisées, ont été introduites pour les C.S.T. par FRENKIEL et KLEBANOFF (1967 a et a). La méthode numérique employée à l'I.M.S.T. avec un calculateur de 16 K et une unité de ruban magnétique (800 bpi) est décrite en détail dans les notes de LACHARME (1975) et ARZOUMANIAN (1973). Les données essentielles sont les suivantes:

Un produit moyen entre deux variables quelconques x et y est estimé par:

$$\overline{xy}^N (k\Delta\tau) = \frac{1}{N-k} \sum_{1}^{N-k} x_n y_{n+k} \tag{10}$$

N est le nombre de couples $k\Delta\tau$ et le retard de temps. $\overline{xy}^N (k\Delta\tau)$ est un estimateur correct de la valeur moyenne $x(t)y(t+\tau)$ $\tau = k\Delta\tau$ obtenue pour $N \to \infty$. Toutefois, l'estimation, en considérant l'ensemble de tous les points obtenus à partir d'un échantillon n'est pas correcte (Cf. par ex., BENDAT et PIERSOL 1966). La figure 4 illustre ce fait, les blocs sont de 1000 couples (x,y). On voit que la courbe se stabilise à 100000 couples. Un échantillonage insuffisant se traduit par des ondulations caractéristiques sur la figure 4.

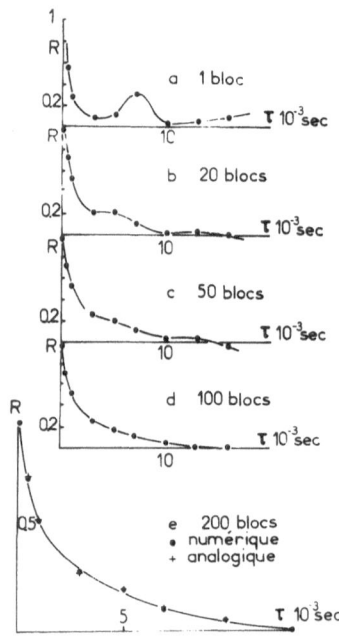

Figure 4. Convergence d'une courbe de corrélation. (ARZOUMANIAN).

En ce qui concerne la cadence d'échantillonnage, $f_e = 1/\Delta\tau$, elle n'introduit pas d'erreur par elle-même. Seul importe le *temps de conversion analogique-numérique*, $\delta\tau'$. Dans notre cas $\delta\tau' \approx 9$ μs ce qui correspond à une fréquence discernable maximale $1/2\ \delta\tau' \sim 55,000$ Hz.

En pratique dans le cas des C.S.T., le décalage de temp étant important, il n'est pas nécessaire d'avoir une définition de la corrélation avec des incréments $\Delta\tau \cong$ petit. La cadence d'échantillonnage f_e est surtout choisie pour que le temps T, durée de l'échantillonnage, $T = N/f_e$, avec $N > 100.000$, soit très supérieure aux échelles de temps les plus basses significative pour la C.S.T. considérée. En fait, il est nécessaire de faire des essais, préliminaires à N fixe et f_e variable. On peut aussi effectuer deux mesures successives avec deux fréquences f_e' et f_e'' voisines. Les décalages de temps sont intercalés. On peut arriver à la limite de définition (sauf pour les premiers incréments). $1/f_e' - 1/f_e'' \leqslant \delta\tau'$ même si la cadence maximale d'échantillonnage, toujours limitée par le temps d'accès à la mémoire de masse, est très inférieure à $\delta\tau$.

Finalement, les coefficients de corrélations sont estimés par la formule:

$$r_{xy}(k\Delta\tau) = \frac{\overline{xy}^N(k\Delta\tau) - \overline{x}^N\,\overline{y}^N}{[\overline{x^2}^N - (\overline{x}^N)^2]\,[\overline{y^2}^N - (\overline{y}^N)^2]^{\frac{1}{2}}} \qquad [11]$$

Chaque moyenne est déterminée sur le même ensemble d'échantillonnage.

3.3 Méthode Spectrale

Les corrélations dans le temps sont les transformées de FOURIER des spectres et cospectres. Si l'on définit par Sx(f) la transformée de FOURIER, T.F., de x(t):

$$S_x(f) = \int_{-\infty}^{\infty} x(t)\exp[-i2\pi ft]\,dt; \quad S(f) = S^*(-f) \qquad [12]$$

x(t) étant une function stationnaire dans le temps l'intégrale $\int_{-\infty}^{\infty} x^2\,dt$ diverge, et cette définition n'est applicable que pour une durée finie T, avec x(t) pris nul en dehors de cet intervalle. On a alors:

$$\overline{x(t)y(t+\tau)} = \int_{-\infty}^{\infty} \frac{S_x(f)S_y^*(f)\exp(i2\pi f\tau)\,df}{T} \qquad [13]$$

Dans les calculs pratiques on considère les séries de FOURIER; elles sont équivalentes aux intégrales sur un même intervalle t, t+T. L'allogarithme utilisé est en général la F.F.T. (par ex. RESCH et ABEL 1975). Le calcul est effectué sur p blocs de n valeurs, avec $n = 2^r$, r entier. Si f_e est la fréquence d'échantillonnage, on a: $T = pn/f_e$. Toute l'énergie spectrale doit être concentrée entre $f_{min} = f_e/n$ et $f_{max} = f_e/2$ ou les effets d' "aliasing" apparaissent. Pour que l'énergie mesurée à la fréquence f_{min} ait un sens physique, il faut $p \ggg 1$.

L'intérêt de la méthode spectrale est qu'elle est rapide; l'inconvénient est que l'intégrale [13] met en jeu tout le spectre. Notons que l'on obtient aussi la fonction de cohérence:

$$coh(f) = \frac{|S_x S_y^*|^2}{|S_x|^2\,|S_y|^2} \qquad [14]$$

et le déphasage, ou temps retard τ_o:

$$S_x S_y^* = [\cosh(f)]^{\frac{1}{2}} \exp[-i2\pi f \tau_0] \qquad [15]$$

Le coefficient de corrélation avec filtrage en fréquence s'écrit (FAVRE et al. 1977)

$$r(f,\tau) = [\cosh(f)]^{\frac{1}{2}} \cos[2\pi f(\tau-\tau_0)] \qquad [16]$$

VAN ATTA a appliqué les T.F. aux corrélations d'ordres supérieurs par la méthode suivante (VAN ATTA et YEH 1970). Soit à calculer par exemple la corrélation du troisième ordre:

$$R^{1,1,1} = \frac{\overline{u'(t)\, u'(t+\tau_1)\, u'(t+\tau)}}{[u'^2]^{3/2}}$$

On génère dans l'intervalle de mesure la série:

$$x(t,\tau_1) = u'(t)\, u'(t+\tau_1)$$

et l'on est ramené à calculer une corrélation du deuxième ordre:

$$\overline{x(t,\tau_1)\, u'(t+\tau)}$$

Le temps τ_1 étant un paramètre, le calcul doit être recommencé pour chaque valeur de τ_1. Cette façon de procéder est aussi applicable au calcul par la méthode directe (§3.2.) dans le cas où l'on veut déterminer une corrélation en trois temps $\overline{x(t)\, y(t+\tau_1)\, y(t+\tau_2)}$.

4. Interprétations des corrélations spatiotemporelles d'ordre élevés

A notre connaissaince, peu de mesures de C.S.T. d'ordres élevés ont été faites. On pourrait ranger avec l'hypothèse de TAYLOR les mesures avec décalage en trois temps et plus, de VAN ATTA et YEH (1970) parmi les C.S.T. Toutefois, on sait que le champ turbulent même en turbulence homogène et isotrope en aval d'une grille n'est pas figé. Le coefficient de corrélation

$$\frac{\overline{u'(o,t)\, u'(x_1,t+\tau)}}{u'^2}$$

entre deux-points P_0 et P passe par un maximum pour $t = x_1/\overline{u}$ (FAVRE et al. 1953, FRENKIEL et KLEBANOFF 1966). Ceci est une conséquence de la diffusion de l'énergie cinétique turbulente par la turbulence elle-même, comme le montrent les équations aux C.S.T. du deuxième ordre (Cf. § 4.1.). Par contre, on trouve en turbulence homogène avec une bonne approximation que les coefficients de corrélations spatiaux et temporels sont égaux; soit:

$$\overline{u'(o,t)\, u'(x_1,t)} \approx \overline{u'(o,t)\, u'(0,t+\tau)}$$

si $\tau = x_1/\bar{u}$

Aussi, la corrélation en trois temps fixe et un temps variable

$$R^{1,1,1,1} = \frac{\overline{u'(t)\, u'(t+\tau_1)\, u'(t+\tau_2)\, u'(t+\tau)}}{(\overline{u'^2})^2}$$

donnée sur la Figure 5 pourrait s'interpréter comme des mesures en trois points, $x_1 = 0, \bar{u}\tau_1, \bar{u}\tau_2$ fixes et un point mobile $x_1 = \bar{u}\tau$.

En fait, VAN ATTA et YEH dans les expériences précitées avaient pour but de tester des hypothèses sur les lois de distribution en probabilité de la turbulence qui permettent de calculer les moments d'ordres supérieur en fonction des moments d'ordres inférieurs. Par exemple, dans le cas de la Figure 5, $R^{1,1,1,1}$ mesuré, est comparé aux valeurs calculées avec l'hypothèse Gaussienne de MILLIONSHCHIKOV. L'accord dans ce cas est bon, (voir aussi Fig. 6), il n'en est pas de même en turbulence de couche limite (FRENKIEL et KLEBA-NOFF 1973). D'une manière générale il est nécessaire d'ajuster des distributions de GRAM-CHARLIER qui permettent d'avoir des moments impairs non nuls (FRENKIEL et KLEBANOFF 1967a). Toutefois, le moment d'ordre 3 qui joue un rôle essentiel dans les modélisations de la turbulence ne peut être calculé et doit être déterminé par l'expérience, ce qui enlève de l'intérêt pratique aux considérations précitées.

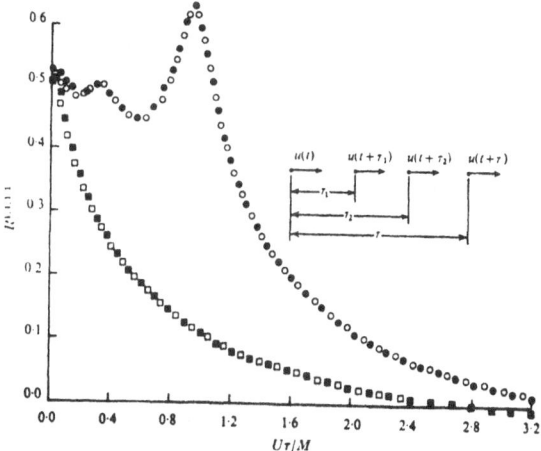

Figure 5. Corrélations en quatre temps test hypothèse MILLIONSHCHIKOV ● ■ positif-négatif. Calculés ○ positif □ négatif. (VAN ATTA et YEH).

4.1 Mesures en deux points de l'expace

Citons tout d'abord les expériences de FRENKIEL et KLEBANOFF (1966) qui avaient pour but de tester l'hypothèse de TAYLOR et les lois statistiques dans l'écoulement turbulent en aval d'une grille. La Figure 6 donne la C.S.T.

$$R^{2,2} = \frac{\overline{u'^2(o,t)\, u'^2(x_1\,, t+\tau)}}{(\overline{u'^2})^2}$$

et les valeurs qui peuvent être calculées par l'hypothèse de MILLIONSHCHI-KOV précitée.

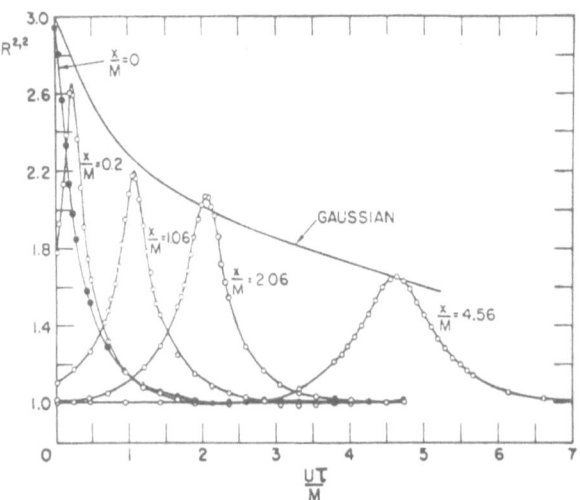

Figure 6. Corrélations spatiotemporelles quadruples. Tests hypothèses de TAYLOR et de MILLIONSHCHIKOV (FRENKIEL et KLEBANOFF).

Les mesures faites à l'I.M.S.T. ont été faites dans un tout autre ordre d'idée, puisqu'il s'agissait de mettre en évidence l'influence à travers la couche limite de perturbations en provenance de la paroi dans une couche limite turbulente (DUMAS et al. 1973, ARZOUMANIAN et al. 1978). A titre d'exemple, la figure 7 donne les courbes d'iso—C.S.T.

$$r_{1,11} = \frac{\overline{u'_1(x_0,t)\, u'^2(\underline{x}, t+\tau)}}{[\overline{u'^2_1(x_0)}\,(\overline{u'^2(\underline{x})} - \overline{u'^2(\underline{x})})]^{\frac{1}{2}}} = \text{cste}$$

obtenues dans une couche limite turbulente. Le point en amont $P_o(\underline{x_o})$ est fixe, situé à y_o = 0.056 δ proche de la paroi, le point en aval $P(\underline{x})$ est à distance longitudinale x_1 = 1.41 δ. Il est déplacé au sein de la couche limite selon la direction y perpendiculaire à la paroi, latéralement x_3 = 0. Les figures 8a et b donnent les ISO-CST $r_{1,13}$ et $r_{1,33}$ faisant intervenir la composante u'_3 normale à la paroi, au point en aval.

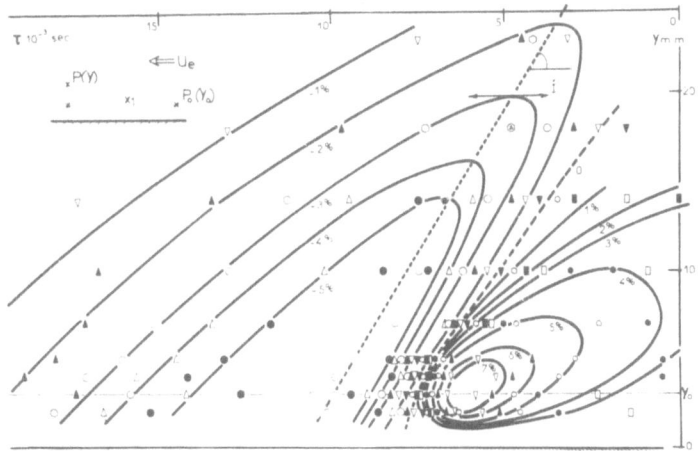

Figure 7. Corrélations spatiotemporelles triples $r_{1,11}$. Modes dominants de perturbations dans une couche limite.

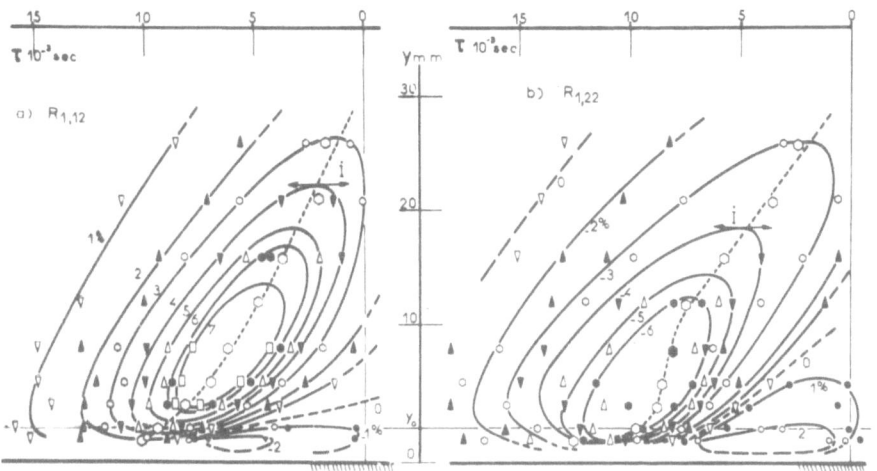

Figure 8 a et b. Corrélations spatiotemporelles triples a) $r_{1,13}$ b) $r_{1,33}$. Modes dominants de perturbations dans une couche limite.

En considérant les figures 7, 8 a et b, on peut en conclure que ces C.S.T. sont en accord avec l'existence de deux modes dominants de perturbations de fortes intensités. Le premier mode ($u_1' < 0$, $u_3' > 0$), connu comme "éjections", est décelé à travers toute l'épaisseur de la couche limite ($r_{1,11} < 0$, $r_{1,13} > 0$). Le deuxième mode ($u_1' > 0$, $u_3' < 0$) appelé souvent "sweep flows" a une influence dominante au long de la paroi ($r_{1,11} > 0$, $r_{1,13} < 0$); les temps retards correspondants étant plus courts, il a une célérité plus forte que le premier.

Des déductions de cet ordre ne peuvent pas être obtenues à partir des C.S.T. doubles $r_{1,1}$ et $r_{1,3}$, car elles ne distinguent pas les produits de deux valeurs négatives de celles de deux valeurs positives, par exemple. De plus, elles ne donnent pas une pondération suffisante aux fluctuations de fortes amplitudes.

Notons encore que les C.S.T. triples interviennent directement dans les équations donnant le taux de perte au long du mouvement moyen des C.S.T. doubles (FAVRE et al. 1968), tout comme les corrélations spatiales triples interviennent dans les équations de KARMAN et HOWARTH des corrélations doubles (Cf. STEWART 1950, VAN ATTA et CHEN 1969). Les mesures des C.S.T. triples montrent que la perte de corrélation est due à la diffusion turbulente et à l'action de la pression (DUMAS et al. 1976).

4.2 Mesures en trois points de l'espace

La structure du champ turbulent a été aussi étudiée en mesurant des C.S.T. en trois points de l'espace, ce qui permet de privilégier les perturbations de grandes échelles spatiales et de fortes intensités, comme le feraient des C.S.T. en deux-points, mais de plus de déterminer les dimensions de ces perturbations du point de vue statistique, comme on le verra ci-après. Cette méthode, préconisée par L.S.G. KOVASZNAY, a été appliquée à l'I.M.S.T. pour déterminer l'envergure des perturbations en provenance de la paroi dans une couche limite turbulente (FULACHIER et al. 1974a et b). Les trois points P_0, P_1, P_2 étaient disposés comme l'indique le schéma ci-contre, dans un plan parallèle à la paroi. Les points P_1 et P_2 sont placés symétriquement par rapport à la ligne de courant moyenne P_0P.

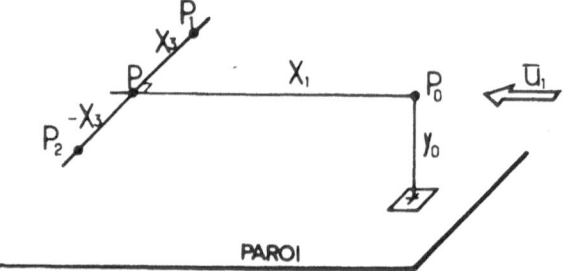

Pour éviter les effets de non-linéarité, signalés en §2.1., on a utilisé les fluctuations de température θ_0', θ_1', θ_2' aux trois points considérés. En effet, si l'on chauffe légèrement un écoulement, par exemple dans le cas considéré par la paroi, la chaleur se comporte comme un contaminant pratiquement passif transporté par le fluide. La température marque alors le fluide à la manière d'un colorant. Le signe de la fluctuation de température, par rapport à la température moyenne, indique l'origine du fluide à l'instant considéré.

Considérons alors le coefficient spatiotemporel en trois points:

$$r_{\theta_0\theta_1\theta_2} = \frac{\overline{\theta_0'\,(t)\,\theta_1'\,(t+\tau)\,\theta_2'\,(t+\tau)}}{[\overline{\theta_0'^2}\,(\overline{\theta_1'\,\theta_2'} - \overline{\theta_1'\,\theta_2'})^2]^{\frac{1}{2}}}$$

Si, pour un certain décalage de temps t, $r_{\theta_0\theta_1\theta_2}$ est notable, ceci peut s'interpréter comme étant dû à des perturbations passées au temps t en P_0, et atteignant en aval *simultanément* en t + τ, les points P_1 et P_2' Du point de vue statistique la largeur de la zone de cohérence des perturbations est au moins de l'ordre de P_1P_2. Un renseignement analogue ne peut être obtenu à partir d'une C.S.T. double, par exemple entre P_0 et P_1, car du fait de son *trajet aléatoire* une perturbation peut passer en P_0 puis P_1 sans que sa largeur soit PP_1. Nous dirons que les C.S.T. doubles définissent des domaines *d'influence,* alors que les C.S.T. triples définissent des domaines de *cohérence*. Notons encore que pour définir la largeur du domaine de cohérence on aurait pu mesurer la corrélation purement spatiale en P_1P_2, mais, dans ce cas, on considère l'ensemble du champ turbulent. En entroduisant un point en amont P_0, la distance P_0P étant relativement importante de l'ordre de l'épaisseur δ au moins, on ne prend en compte que les perturbations de fortes cohérences, de nature peut-être pseudo-déterministes; on élimine l'effet des fluctuations à petites échelles participant à la dissipation par viscosité.

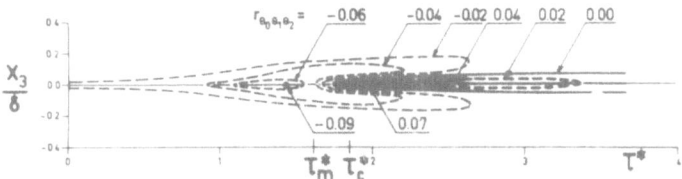

Figure 9. Corrélations spatiotemporelles en trois points $r_{\theta_0\theta_1\theta_2}$. Domaines de cohérence en envergure. $\tau^* = \overline{u}\tau/\delta,\ \tau_m^*\ (r_{\theta_0\theta})_{max},\ \tau_c^* = x_1/\delta,\ x_1 = 1{,}86\ \delta,\ y_0 = 0.034\ \delta$

La figure 9 donne un exemple des courbes d'isocoefficients $r_{\theta_0\theta_1\theta_2}\ (x_1,\tau) =$ cste obtenues. On distingue deux zones principales selon les signes positifs et négatifs de $r_{\theta_0\theta_1\theta_2}$. Les zones positives correspondent à du fluide plus chaud en provenance de région plus proche de la paroi; au contraire des zones négatives correspondant à des arrivées de fluide plus froid. En ordonnées on note la largeur du domaine de cohérence de l'ordre de 0,2 δ à cette position des perturbations en provenance de la paroi.

A la même position de P_0, P_1 et P_2 étant confondus en P, la figure 10 donne les C.S.T. triples conditionnelles entre les points P_0 et P.

$$R_{m,nn} = \frac{\overline{J(t)\,\theta'_0(t)\,\theta'^2(t+\tau)}}{[\overline{\theta'^2_0}\,\overline{(\theta'^2-\overline{\theta'^2})^2}]^{\frac{1}{2}}}$$

où m et n ont les signes + ou −. J(t) = 1 quand $\theta'_0(t)$ et $\theta'(t+\tau)$ ont les signes m et n respectivement. (J(t) = 0 autrement (DUMAS et al. 1977). On a la relation: $r_{\theta_0\theta^2} = R_{+++} + R_{+--} + R_{-++} + R_{---}$

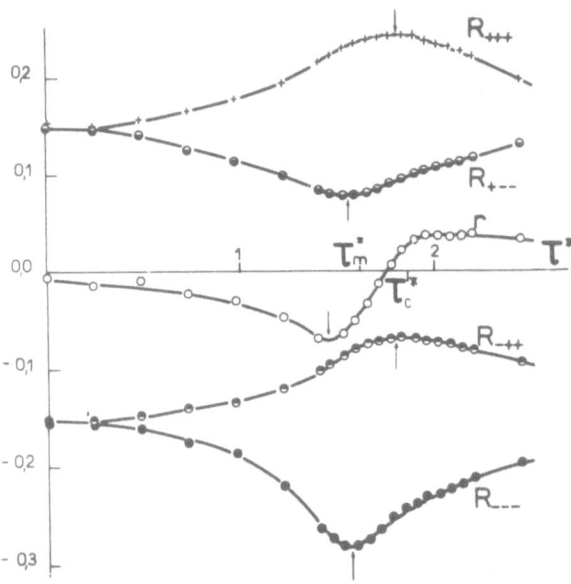

Figure 10. Corrélations spatiotemporelles. Triples: Décomposition en corrélations conditionnelles.

On voit qu'une corrélation triple d'aspect antisymétrique, comme il est usuel en écoulement pleinement turbulent, et de faible niveau, est la composition de corrélations conditionnelles d'aspect symétrique et de forts niveaux lorsqu'il s'agit de séquences conservatives, R_{+++} et R_{---}. On note que le temps retard correspondant au maximum des séquences de fluide plus chaud R_{+++} est plus grand que le temps correspondant au minimum des séquences de fluide froid R_{---}, ce qui était attendu, étant donné les provenances du fluide. Il est plus suprenant de constater que le temps correspondant au maximum de R_{+++} est plus petit que $\tau_c = \bar{u}\tau/x_1$ donné par la vitesse moyenne locale. Ceci montre que le processus d'éjection est lui-même accéléré par rapport au mouvement moyen.

Enfin, des résultats obtenus à diverse distances de la paroi sont regroupés sur la figure 11.

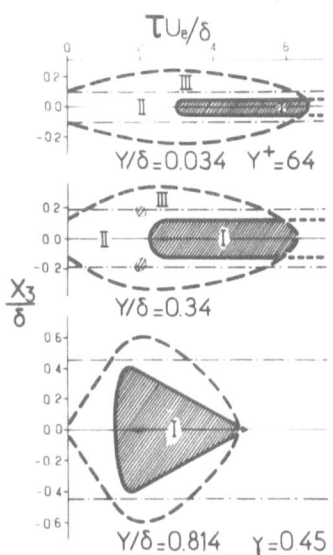

*Figure 11. Evolution des perturbations en provenance de la paroi zone I
— — C.S.T. doubles r — · — corrélations spatiales r.*

On a conservé la ligne de zéros des C.S.T. $r_{\theta_0 \theta_1 \theta_2}$, délimitant la zone positive correspondant aux perturbations en provenance de la paroi, ainsi que la ligne de zéros de la C.S.T. double $r_{\theta_0 \theta_1}$, ou $r_{\theta_0 \theta_1} = 0$, et de la corrélation purement spatiale $r_{\theta_1 \theta_2} = 0$. On constate que la largeur des perturbations en provenance de la paroi augmente au fur et à mesure qu'elles s'éloignent de la paroi. Dans la zone d'intermittence à $y/\delta = 0.814$, où le facteur $\gamma = 0.45$, la largeur moyenne de ces perturbations est de l'ordre de δ; elles forment pratiquement les protubérances de turbulence (FULACHIER et al. 1974b).

En tenant compte de la liaison statistique à travers la couche limite mise en évidence sur les figures 7, 8 a et b, on peut en déduire que des perturbations dimensions fixes, mais de très fortes intensités, nées près de la paroi de la frontière de la sous-couche visqueuse, se diffusent dans la couche limite jusqu'à former les protubérances d'intermittence.

Par ailleurs, on voit que la largeur du domaine d'influence donnée par les C.S.T. doubles $r_{\theta_0 \theta_1}$, est nettement supérieure à celle des C.S.T. triple, car d'une part les C.S.T. doubles ne distinguent par les séquences positives m=n=+ des séquences négatives m=n=− et que, d'autre part, elle prend en compte l'agitation des perturbations. Dans la publication de FULACHIER et al. (1974b) une discussion approfondie de ce type de résultats est donnée concernant les régions I, II et III délimitées par les lignes des trois corrélations.

5. Conclusions

5.1. Les C.S.T. d'ordres élevés 3 et 4, et à fortiori plus, du fait qu'elles mettent surtout en jeu les grandes amplitudes du signal, sont plus affectées par les divers effets non-linéaires que les C.S.T. doubles. On note aussi un étalement spectral. Ces effets sont particulièrement sensibles pour les CST impaires de faibles valeurs, mais qui sont en fait (Cf. figure 10) la différence de corrélations de fortes valeurs.

5.2. En ce qui concerne l'interprétation des C.S.T., on sait que les C.S.T. doubles sont essentielles pour déterminer les paramètres de conservation ou des mémoires du champ turbulent. D'une autre façon elles permettent de déterminer la différence avec le champ gelé de l'hypothèse de TAYLOR due à des mécanismes fondamentaux de diffusion et d'action de la pression de la turbulence. Les C.S.T. en deux points d'ordre 3 interviennent dans les équations des C.S.T. doubles précitées; elles sont essentielles si l'on veut "modeler" ces équations pour une résolution numérique. De plus, les C.S.T. d'ordre 3 peuvent s'interpréter en termes de perturbations dominantes, distinguant les signes des fluctuations, ce que ne permettent pas les C.S.T. doubles.

Les C.S.T. en trois points de l'espace permettent de délimiter, au moins dans une direction, la dimension du point de vue statistique du domaine de cohérence des perturbations. On pourrait envisager, grâce aux systèmes d'acquisition multivoies, des C.S.T. en multipoints donnant les trois dimensions de domaine de cohérence d'une perturbation passée en un point amont P_0. Notons que dans les sens longitudinal on peut envisager de remplacer deux points de l'espace par deux temps, en admettant que $R(X_1 \ldots)$ $R(\tau \bar{u}, \ldots)$. Il est indiqué d'utiliser les fluctuations de température, le fil froid thermomètre présentant beaucoup d'avantages sur le fil chaud anémomètre (Cf. 2.1. et FULACHIER 1978), en chauffant légèrement le fluide. De plus, on obtient d'après la source de chaleur, l'origine de la masse fluide considérée.

Références
ARZOUMANIAN, E. 1973, N.T. I.M.S.T.
ARZOUMANIAN, E., DUMAS, R., FAVRE, A., 1978 *C.R. Acad. Sc.* **286**, B, 113
BAILLE, A., 1971 Bulletin DER-EDF A No. 3
BENDAT, J.S., PIERSOL, A.G., 1966 *Measurement and analysis of random data*, John Wiley and Sons
BLANC-LAPIERRE, PICINBONO, B. 1961, *Propriétés statistiques du bruit de fond*, Masson

BO-JANG TU, WILLMARTH, W.W., *tech. Rep. 02920-3-T* Dpt. Aerospace Eng. Univ. Michigan

CHAMPAGNE, F.H., SCHLEICHER, C.A., WEHRMAN, O.H., 1967 *J. Fluid Mech.* **28**, 153

COMTE-BELLOT 1974 DER-EDF-21, *Techniques de Mesure dans les Écoulements,* Eyrolles

COMTE-BELLOT, G., CORRSIN, S., 1971 *J. Fluid Mech.,* **48**, 273

COMTE-BELLOT, G., SCHON, J.P., 1969 *Int. J. Heat Mass Transfer,* **12**, 1661

DUMAS, R., 1978 Anémothermométrie à fils chauds - Cours D.E.A., I.M.S.T.

DUMAS, R., ARZOUMANIAN, E., FAVRE, A., 1973, C.R. Acad. Sc. A 759

DUMAS, R., ARZOUMANIAN, E., FULACHIER, L., FAVRE, A., 1977 *13th Biennal Fluid Dyn. Symp. Olsztyn Proceedings,* à paraître

DUMAS, R., FULACHIER, L., ARZOUMANIAN, E., FAVRE, A., 1976 *Jour. Physique Sup.* No. 1, 37, CI-181

FAVRE, A., *7ème Cong. Inter. Meca. Appl. Londres,* 1948

FAVRE, A., DUMAS, R., VEROLLET, E., 1968 *Proc. 12th Cong. Inter. Meca. Stanford* Springer-Verlag

FAVRE, A., GAVIGLIO, J., DUMAS, R., 1958 Onera, Publ. No. 92

FAVRE, A., GAVIGLIO, J., DUMAS, R., 1953 *La Recherche Aéronautique No. 32*

FAVRE, A., KOVASZNAY, L.S.G., DUMAS, R., GAVIGLIO, J., COANTIC, M., 1976, *La Turbulence en Mécanique des Fluides,* Gauthier-Villars

FRENKIEL, F.N., KLEBANOFF, P.S., 1966, *Dyn. of Fluids and Plasmas,* S.I. Pai ed. Acad. Press

FRENKIEL, F.N., KLEBANOFF, P.S., 1967, a — *The Phy. of Fluids,* **10**, 1737 b— Idem., **10**, 507

FRENKIEL, F.N., KLEBANOFF, P.S., 1973 *The Phy. of Fluids,* **16**, 725

FULACHIER, L., 1978, *Proceedings of the Dynamic Flow Conference Marseille & Baltimore*

FULACHIER, L., GIOVANANGELI, J.P., DUMAS, R., KOVASZNAY, L.S.G., FAVRE, A., 1974 a — *C.R. Acad. Sc.,* **278** B 683 b — Idem. **278** B, 994

GIOVANANGELI, J.P., 1975, Th. Doct. Spé. I.M.S.T.

LACHARME, J.P., 1975, N.T., I.M.S.T.

HINZE 1959, *Turbulence,* McGraw-Hill Book

REY, C., BEGUIER, C., 1977, *DISA Information,* **21**, 11

RESCH, F., ABEL, R., 1975, *Inter. J. Num. Meth. Eng.,* **9**, 869

STEWART, R.W., 1950, *Proc. Camb. Ph. Soc.,* **47**, 146

VAN ATTA, C.W., 1974 *Annual Review Fluid Mech.*

VAN ATTA, C.W., CHEN, W.Y., 1969, *J. Fluid Mech.,* **38**, 793

VAN ATTA, C.W., YEH, T.T., 1970, *J. Fluid Mech.,* **41**, 169

WILLMARTH, W.W., 1975, *Advances Appl. Mech.,* **15**, 159

ZARIC, Z., 1969, *C.R. Acad. Sc.,* **269**, A, 987.

Transverse Vorticity Measurements

by

J.F. Foss
Michigan State University

Abstract

For a flow in which the mean velocity distribution can be described in terms of the streaming flow direction (x) and the direction of the principal velocity gradient (y), the predominant vorticity ($\vec{\omega}$) will exist in the transverse direction (z). It was the purpose of the present study[*] to develop an experimental technique to form a time series representation of the instantaneous ω_z values. The motivation for this study was to develop a diagnostic tool for those flow fields where a knowledge of ω_z(t) would be valuable, such flows include (i) laminar and turbulent vortex rings, (ii) unsteady aerodynamics, and (iii) sound producing flows. The specific objectives of the present study were to develop the hardware and support software to make instantaneous ω_z measurements in typical laboratory shear flows with maximal accuracy and allowance for the large variation in pitch angles associated with free shear flows.

The measurement technique for ω_z follows directly from its defining equation.

$$\omega_z = \partial v/\partial x - \partial u/\partial y \qquad [1]$$

An x-array (at z) is placed adjacent to a parallel array (at $z + \delta z$). The derivative ($\partial v/\partial x$) is evaluated from the u, v time series obtained from the x-array; viz.,

$$\partial v/\partial x \doteq u_a(t)^{-1} \left[v(t+\delta) - v(t-\delta t) \right]/2\delta t. \qquad [2]$$

The "instantaneous Taylor hypothesis" [$u_a = (2u(t) + u(t-\delta t) + u(t+\delta t))/4$] is considered to properly represent the instantaneous spatial derivative. The adjacent parellel array and the pitch angle ($\gamma \equiv \tan^{-1} v/u$) from the x-array are used to infer $\partial u/\partial y$, viz.,

$$\partial u/\partial y \doteq \cos\gamma \left[V(y+\delta y) - V(y-\delta y) \right]/2\delta y \qquad [3]$$

where V is the velocity magnitude inferred from the voltage magnitude.

[*]Supported by NASA Grant NGR. 23-004-091

The algorithm to infer (u,v) from the instantaneously sampled, x-array voltage magnitudes, is considered to be an important by-product of the present development work. This computing scheme successfully recovers the velocity vector information from a wide range of velocity magnitudes (V) and a wide range of pitch angles ($-40° \leqslant \gamma \leqslant 40°$). It is readily implemented in a batch process mode on a digital computer and it will be implemented in a 40 kHz rate, real time, hard-wired procesor (The VORCOM).

The algorithm allows an analytical evaluation of the sensitivity (e.g. $\partial u/\partial E_1$) and uncertainty effects associated with the computation of (u,v) from the x-array voltages. For the uncertainty, a pitch angle difference ($\delta\gamma$), a velocity magnitude difference (δV) or the presence of a transverse velocity at the two wires will cause the computed (u,v) values to be in error. The influence of these effects upon the E_1 and E_2 values are analytically determined and the sensitivity analysis is used to infer the resulting perturbations in u and v. A representative set of (δu, δv) values are presented.

The error associated with the transverse velocity creates a bias in the stochastic value of u and, if the p.d.f. of v is skewed, in v as well. Suitable assumptions allow the "most probable" correction for the influence of w to be applied to the instantaneous measurements. Instantaneous time series of u, v, $\partial v/\partial x$, $\partial u/\partial y$, and ω_z from x/d = 10 in an axisymmetric jet are also presented.

Introduction
The vorticity, $\vec{\omega}$, in a flow is defined in terms of the velocity, $\vec{V}$, as

$$\vec{\omega} \equiv \nabla \times \vec{V}. \tag{1}$$

The vorticity vector field, $\vec{\omega}(x,y,z,t)$, has several properties which make it attractive as a primary variable for the description of a flow field. Its transport equation

$$\frac{D\vec{\omega}}{Dt} = \vec{\omega} \cdot \nabla\vec{V} + \nu\nabla^2\vec{\omega} \tag{2}$$

involves a balance of local effects (unlike the velocity which responds to the gradient of the ubiquitously controlled pressure[*]) and this localized character is associated with several distinctive phenomena. Prominent among these is the viscous super layer, Corrsin and Kistler (1955), which forms the outer "skin" separating the vortical inner flow from the external inviscid flow in jets, wakes, boundary layers, etc. The sharp demarcation between the vortical and non-

[*]$p(\vec{x}) = -\frac{1}{4\pi}\int \frac{\partial^2}{\partial x'_\varrho \partial x'_m} [u_\varrho u_m] \frac{1}{|r|} d\tau (x')$, for an isothermal flow, see Townsend (1976).

vortical flow states is easily rationalized from [2]; the external fluid can only gain vorticity through the direct action of viscocity ($\nu\nabla^2\vec{\omega}$) and hence the length scale of the ∇^2 operator must be sufficiently small to counteract the small value of the kinematic viscocity. (Implicitly, the superlayer occurs for large turbulence Reynolds numbers, ($\tilde{u}\lambda/\nu$)).

Coherent motions, such as vortex rings, e.g. Maxworthy (1977), and large scale structures in shear flows, e.g., Roshko (1976), are logically evaluated in terms of vorticity since it is this property, and its integral — the circulation[*], which are associated with the long lifetimes of these structures. The relationship of the vorticity to the acoustic noise producing effects in a flow field has been given an Eulerian reference frame description by Hardin (1973). Specifically, the density fluctuation at the point $\vec{x}$ in the far field near a noise producing flow is

$$\rho_a(x,t) = \frac{-\rho_0}{4\pi a_0{}^4} \frac{x_i x_j}{x^3} \frac{d^2}{dt^2} \int y_i L_j d\vec{y} \qquad [3]$$

where $\rho_a = \rho - \rho_0$, a_0 = sound speed, and $\vec{L} = \vec{\omega} x \vec{v}$. Equation 3 suggests that the acoustic fluctuations of the far field are related to the time dependence of the vorticity field; hence $\vec{\omega}(\vec{y},t)$ measurements would serve a useful diagnostic function in the investigation of noise producing effects.

For flows with plane or axial symmetry, the time mean vorticity vector can be defined in terms of a single component; viz, with x as the streaming flow direction and x-y as the plane of the time mean velocity gradients,

$$\omega_z = \frac{\partial v}{\partial x} - \frac{\partial u}{\partial y} \qquad [4]$$

The other two components, ω_x and ω_y, can be expected to have large fluctuating values, albeit a zero time mean value, in a turbulent flow.

Measurements of ω_x are made difficult by the inherent difficulties of measuring v and w. A four-wire array, operating in the constant current mode was suggested by Kovasznay (1950), developed by Kistler (1952) and has been examined and utilized by numerous investigators; a current account of such measurements is provided by Willmarth in the present conference proceedings.

The measurement of the transverse vorticity, ω_z, has been the subject of rather less development. The relatively direct approach of measuring: i) the velocity magnitude at two locations $V(y-\delta y)$, $V(y+\delta y)$ to form $\delta u/\delta y$ and ii) a time

[*]$\Gamma \equiv \int \vec{\omega} \cdot \hat{n}\, dA$

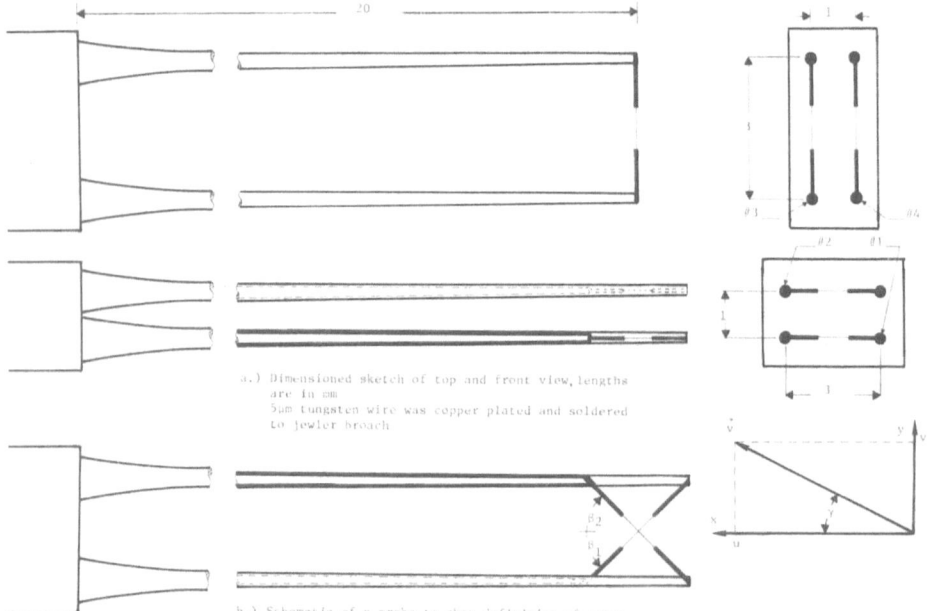

a.) Dimensioned sketch of top and front view, lengths are in cm
5μm tungsten wire was copper plated and soldered to jewler broach

b.) Schematic of x probe to show definition of terms

Fig. 1. The four-wire array vorticity probe.

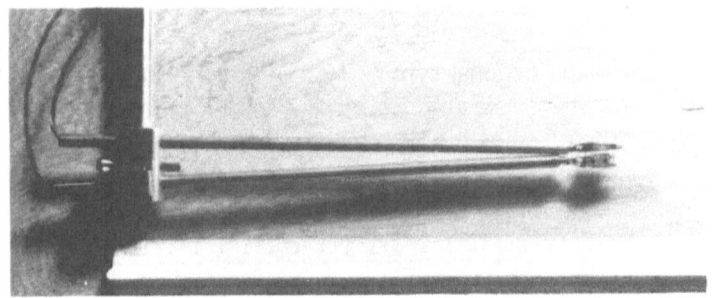

a) Probe support, design of probe support head allows proper positioning of wires.

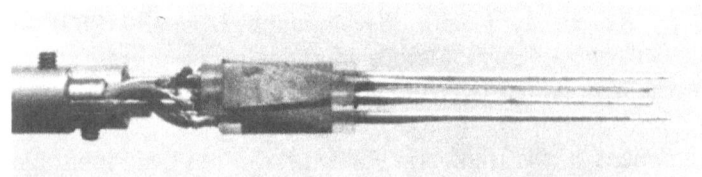

b) Detail view of four wire array.

Fig. 2. The Vorticity Probe.

series for u and v at (y−δz), from an adjacent x-array, to form an approximation for ∂v/∂x as

$$\frac{\partial v}{\partial x} \cong - \frac{1}{u(t)} \frac{v(t+\delta t) - v(t-\delta t)}{2\delta t} ,$$ [5]

was presented by Foss (1976). A schematic representation of this technique is provided in Fig. 1; Fig. 2 shows a photograph of the probe used in the present study.

This approach was made more general by Eckelmann, Nychas, Brodkey and Wallace (1977), in their investigation of the near-wall region of a low speed turbulent channel flow. The basic probe construction was as described above with the exceptions: a (horizontal) "V" array in the x-z plane was located at (y, Δz − δz/2) and (y, Δz + δz/2) and a single wire to record V was placed at (y + δy, Δz). By making use of various "smoothing" assumptions, the authors extracted ∂u/∂z and w, as well as ∂u/∂y. They identified the experimental difficulties associated with executing measurements of a spatial derivative with small δy and δz displacements; similar observations are made in the results section of the present communication. The total sampling cross section of their probe was quite small (4 mm); however, the closely spaced prongs and short prong lengths are in violation of the suggestions of Strohl and Comte-Bellot (1973). Boundary layer measurements, especially in the near-wall region, may not be as sensitive as free shear flows to probe/prong interference effects. The competing requirements of a small sample volume and a probe which can be physically fabricated and is free of probe/prong effects represent a major problem associated with the transverse vorticity measurements. (The streamwise vorticity probe is spatially more compact but its signals have a more tenous relationship to the desired property of the flow field).

The transverse vorticity can be approximated by the single measurement of δu/δy; the data acquisition/processing and the spatial resolution are all favorably effected by this approximation. The validity of the approximation is effected by the corresponding magnitude of ∂v/∂x; this point is addressed in the results section for the present experiments. This approximation was used by Kovasznay, Komoda and Vasudeva (1962) in their study of boundary layer transition and by Kovasznay, Kibbens and Blackwelder (1970) for their conditionally sampled measurements in the outer region of a turbulent boundary layer. The wire separation, δy = 5 mm, of the latter study would seem to provide considerable spatial smoothing; analog processing with linearized hot-wire signals was used in both studies.

Wyngaard (1969) has evaluated the effects of the spatial displacement of the probe sensors for (i) the Kovasznay type streamwise vorticity meter, (ii) the two-wire $\delta u/\delta y$ measurement and (iii) the frozen flow $\partial u/\partial x$ and $\partial v/\partial x$ measurements given the isotropic turbulence field described by the Pao form of $E(k)$. These results show that the ratio of

$$\frac{<(\partial u/\partial y)^2 \overline{>}_{\text{meas}}}{<(\partial u/\partial y)^2 >}$$

depends strongly on the ratios of the Ii) displacement and (ii) the wire length to the Kolmogorov Scale. Ratios much larger than 1 result in a pronounced reduction of the $<(\partial u/\partial y)^2 >$ value. It can be intuitively expected that these effects are not as severe if the vortical structures are anisotropically oriented by the axial or plane symmetry of the gross flow field. A theoretical investigation of such an effect would add greatly to the interpretation of the response of the present probe.

The developmental work reported herein has concentrated on the creation of calibration/processing techniques to recover, with the maximum possible accuracy, the time series representations of u, v, ω_z from the four hot-wire signals. (Accuracy considerations are paramount since ω_z is formed from the difference of differences). To this end, extant response equations have been incorporated into a new computing algorithm in order to extract (u,v) from the x-wire voltages (E_1, E_2). This algorithm and the optimal technique to evaluate $\delta u/\delta y$ are presented in the next section; the sensitivity (dV/dE) and uncertainty analytical consideration which are based upon the algorithm are presented in the subsequent section. A representative set of u, v, $\partial u/\partial y$, $\partial v/\partial x$ and ω_z measurements from the intermittent region of an axisymmetric jet are presented to demonstrate the vorticity measurement capability.

Calibration and Computational Algorithms
The vorticity calculations are based upon a time series of instantaneous and simultaneous samples of the E_1, E_2, E_3 and E_4 voltages (Fig. 1 defines 1, 2, 3, 4). Digital calculations are favored because of their accuracy and flexibility; they are well suited to deal with the non-linear response of the basic anemometer transfer function. The signal-to-noise ratio of the nonlinear bridge voltage is generally superior to that of the linearized signal commonly incorporated into analog processing schemes; a discrete, versus the more desirable continuous record of u, v, etc., is a necessary penalty of the digital processing. Digital techniques may be implemented using a table-look-up scheme; the alternative of a computing algorithm was adopted because: i) it provides a smoothing and an interpolation scheme for the use of the discrete calibration data and ii) it can

be readily used as the basis of sensitivity and uncertainty calculations. (The sensitivity and uncertainty matters are covered in the next section).

The computing algorithm is based upon the (slightly modified) Collis and Williams (1959) relationship

$$E_i^2 = E_{0i}^2 + K_i(\gamma)\, V_{eff}^m \qquad [6]$$

and the pitch angle relationship due to Friehe and Schwarz (1968)[*]

$$V_{eff} = V\left\{1-b\,[1-\cos^{1/2}(\beta-\gamma)]\right\}^2 \qquad [7]$$

Fig. 1 presents a definition sketch for the terms of the analysis. The individual voltages E_1 and E_2 from the wires of the x-array depend upon the velocity magnitude V and pitch angle γ. However, an appropriately defined function, $G(E_1, E_2)$ is (ideally) independent of V; viz,

$$\underbrace{\frac{[(E_1^2-E_{01}^2)/K_1(0)]^{1/m_1}}{[(E_2^2-E_{02}^2)/K_2(0)]^{1/m_2}}}_{G_{meas}} = \underbrace{\frac{[K_1(\gamma)/K_1(0)]^{1/m_1}\left\{1-b_1[1-\cos^{1/2}(\beta_1-\gamma)]\right\}^2}{[K_2(\gamma)/K_2(0)]^{1/m_2}\left\{1-b_2[1-\cos^{1/2}(\beta_2-\gamma)]\right\}^2}}_{<G>} \qquad [8]$$

The two functions G_{meas} and $<G>$ are introduced to acknowledge (i) their independent evaluation and (ii) that the two functions are not identical. The seemingly redundant division by $K_i(0)$, (i=1,2) is related to the structure of the calibration/acquisition procedures. The $[K(\gamma)/K(0)]$ ratio in $<G>$ is determined from the master[**] calibration data set for a given probe; these ratios are used in the computation of $<G>$. Because this ratio primarily reflects the geometric properties of the probe, a master data set is acquired only once during the lifetime of the probe. The minor changes in the probe response, which reflect the aging of the wire, ambient temperature and humidity changes, etc., are accounted for in the calibration data set at $\gamma=0$ which is acquired at the time of the data acquisition. (Pre and post $\gamma=0$ calibration sets are acquired to ensure reliable interpretation of the data). When the calibration and data days differ, the exponents m_1 and m_2 are set to the Collis and Williams suggested

[*]This relationship has been generalized and extensively examined by Drubka, Nagib and Tan-atichat (1978). They allow "best-fit" evaluations of b and m and show improved fits to the experimental data. The $K(\gamma)$ of [6] partially achieves the same result.

[**]A typical master calibration data set will include γ values of (−40, −35, . . . 40 degrees) and V=(3,6,9. . . 30 mps). Nominally, a three hour time period is required for the acquisition of these data.

value of 0.45[*] in order to provide an invariant exponent for the E_0^2 and $K(0)$ evaluations. If the master calibration data and the flowfield data are acquired on the same day, then a "best fit" m is used.

The computing algorithm is implemented by evaluating G_{meas} (E_1, E_2), from the data and using the inverse function $\gamma = \gamma(<G>)$ to evaluate the pitch angle; this initial estimate of the pitch angle is termed γ_i. The velocity is calculated using an appropriate form of equation [6] ; viz.,

$$V = \left\{ [E_i^2 - E_{0i}^2]/K_i(0) \right\}^{1/m_i} F_i^{-1} \qquad [9]$$

where

$$F_i = [K_i(\gamma)/K_i(0)]^{1/m_i} \left\{ 1 - b[1 - \cos^{\frac{1}{2}}(\beta_i - \gamma)] \right\}^2 \qquad [10]$$

The choice of i=1 or 2 is made to minimize the sensitivity of the calculated V to the first estimate of γ. If β_i and γ_i have the same sign, then $\partial K_i/\partial\gamma \cong 0$; hence this similarity of signs selects the i value (1 or 2) to use in [9]. The value of V from [9] is used to estimate a more precise value for γ; specifically, the original calibration data are used to define $G_{meas} - <G>$ as:

$$G_{meas} - <G> = A(\gamma_i) + S(\gamma_i) [V - <V>] \qquad [11]$$

where $<V>$ is the midpoint of the calibration velocities. Equation [11] reflects the observation that G_{meas} is weakly dependent upon the velocity magnitude whereas $<G>$, by definition, is independent of $|\vec{V}|$. The magnitude of this correction can be seen in the master calibration data of Fig. 3.

The error that would be incurred in γ without this correction can be estimated from this figure by using:

$$\delta\gamma = \frac{d\gamma}{dG} \delta G. \qquad [12]$$

The final estimate for γ is taken from $<G_{jj}>$ where

$$<G_{jj}> = G_{meas} - A(\gamma_j) - S(\gamma_j) [V - <V>] \qquad [13]$$

This final estimate for $<G>$ is used to define the final estimate for γ and, using $F(\gamma_{jj})$, the final estimate for V. A block diagram representation of the algorithm is presented in Fig. 4.

[*]A logical improvement is to use the Drubka et al. (1978) procedures to define the best m for a given probe.

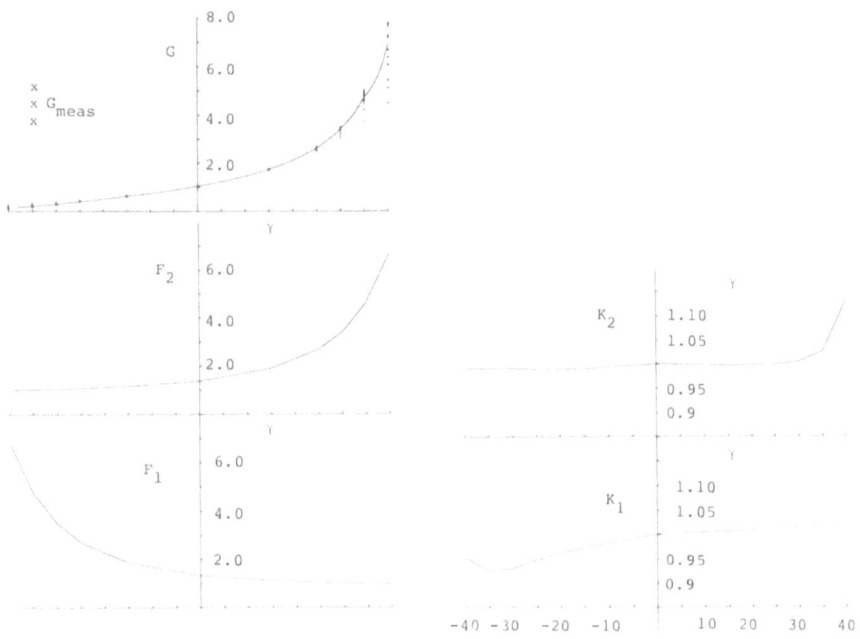

Fig. 3. Pitch dependent functions, used in the computing algorithm, as evaluated for the present x-array. Note: d = 5 μm tungsten wires, 1/d = 200, R_H = 1.8 R_c.

The computation of $\partial v/\partial x$ employs a smoothed "instantaneous" Taylor hypothesis as the most reasonable conversion between the available time derivative and the desired spatial derivative. Specifically,

$$\frac{\partial v}{\partial x} \cong -\frac{1}{u_{avg}} \frac{v(t+\delta t) - v(t-\delta t)}{2\delta t} \tag{14}$$

where $u_{avg} = \left\{ 2u(t) + u(t-\delta t) + u(t+\delta t) \right\}/4$.

The single wires are assumed to respond to the magnitudes of the velocity; further, the pitch angle γ (from the x-array) is assumed to describe the pitch angle at the adjacent parallel array. Hence,

$$\frac{\partial u}{\partial y} \cong \cos\gamma \left[\frac{V_3 - V_4}{\delta y} \right] \tag{15}$$

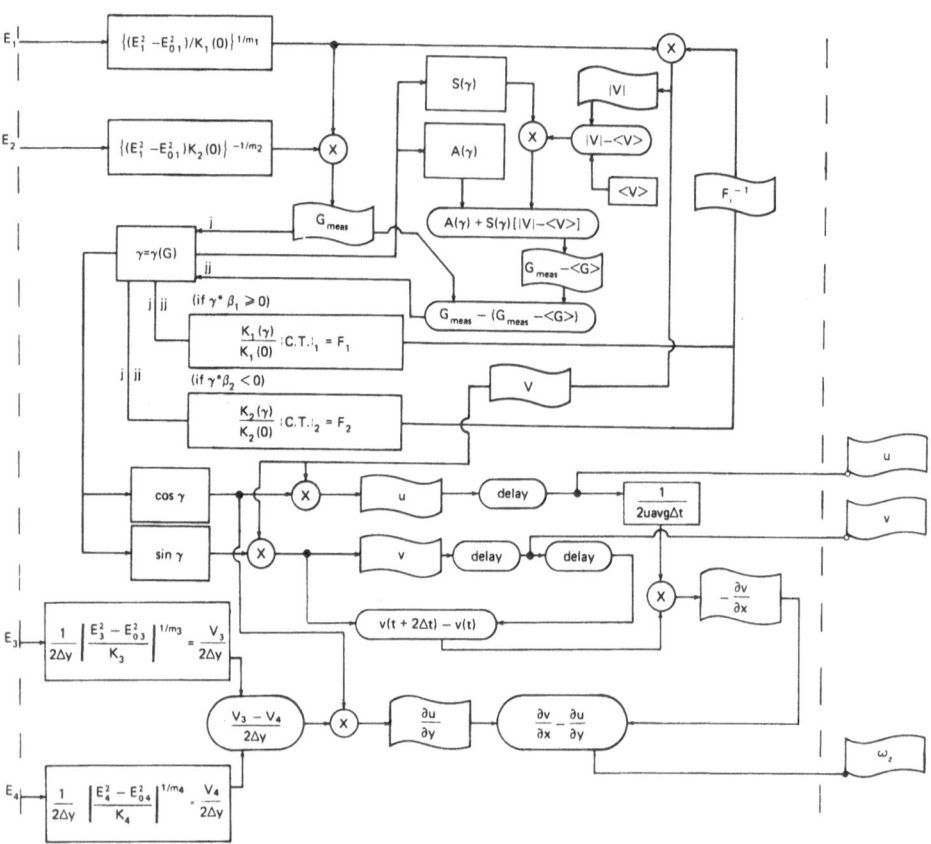

Fig. 4. The VORCOM-flow chart for $E_i^* \to u,v,\omega_z$ computations. Note: $(C.T.)_i =$ $(1-b[1-\cos^{1/2}(\beta_i-\gamma)])^2$. auvg $= (u(t)+2u(t+\delta t)+u(t+2\delta t))/4$.

The following procedure was implemented to ensure maximum accuracy in the evaluation of the velocity difference $[V_3-V_4]$. Wire 3 was selected as the "standard" and the best fit of E_{03}^2, K_3 and m_3 were evaluated. E_3 was then used to calculate V_3 ... using E_{03}^2, K_3, and m_3. This velocity (V_3) was used as the "calibration" velocity to evaluate E_{04}^2, K_4 and m_4. The standard deviation between V_4 and V_3 was typically one order of magnitude smaller than the standard deviation between V_3 and the measured velocity. (For the present experiments, the former value is typically 0.015 mps, the latter is $\cong 0.15$ mps. It is estimated that the nonlinearity of the pressure transducer and the A/D resolution combine to create an uncertainty in the measured velocity for the calibration data of the order 0.15 mps; hence the latter value is as small as can be expected).

Sensitivity, Uncertainty and the Bias Effect of w
Sensitivity Analysis

The sensitivity (dV/dE) of the parallel wires is known in terms of the calibration constants: $dV/dE = 2EV^{1-m}/mK$; the sensitivity of the x-wire response $(\partial u/\partial E_1$, etc.) cannot be stated as simply. Using $u = V\cos\gamma$ and $v = V\sin\gamma$, the normalized equations for du and dv are

$$\frac{du}{V} = \cos\gamma\,\frac{dV}{V} - \sin\gamma\,d\gamma; \quad \frac{dv}{V} = \sin\gamma\,\frac{dV}{V} + \cos\gamma\,d\gamma \qquad [17a,b]$$

The logical steps to describe dV/V and $d\gamma$ in terms of dE_1 and dE_2 are summarized below; the details are presented in the final report to the NASA, Foss (1978).

The analytical expression for $d\gamma\,(dE_1, dE_2)$ is based upon the dual nature of the G function. Specifically

$$d\gamma = (d\gamma/dG)dG \qquad [18]$$

where $d\gamma/dG$ is evaluated as $d\gamma/d\langle G\rangle$ using the functional form on the r.h.s. of [8] and the differential dG is evaluated from the l.h.s. of [8]. Namely,

$$dG = G\left[\frac{2E_1 dE_1}{m_1(E_1^2 - E_{01}^2)} - \frac{2E_2 dE_2}{m_2(E_2^2 - E_{02}^2)}\right] \qquad [19]$$

The differential, dV, is based upon equation [9]; the basic expression is:

$$\frac{dV}{V} = \frac{2E_i dE_i}{m_i(E_i^2 - E_{0i}^2)} - \frac{1}{F_i}\frac{dF_i}{d\gamma}\,d\gamma \qquad [20]$$

Since $F_i(\gamma)$ is known from [10] and since $d\gamma$ is known from the above, the sensitivities $(\partial u/\partial E_1$, etc.) may be calculated.

Uncertainty Considerations[*]

Consider a three-dimensional, spatially variable (e.g., turbulent) flow field in which the instantaneous velocity vector is decomposed into a transverse component, w (in the z direction), and its projection, V, in the x-y plane. The direction of the velocity vector is prescribed by the angular displacement, γ, of V from the x-axis[**]. A properly oriented x-array will not, in general, provide E_1 and E_2 values which can be used to compute the true values of V and γ because

[*]A limited portion of the complete (and complex) uncertainty considerations is considered herein; uncertainties caused by the spatial displacements used to define the x and y derivatives and those associated with single wire sampling length errors are not specifically considered.
[**]V and γ, so defined, represent the same physical quantities as before; however, they are here defined as a part of a more general flow field.

the three-dimensional and spatially variable conditions of the flow field will influence these output voltages. Specifically, the transverse velocity, w, will contribute to the heat transfer from the wires with the same effect as if V increased or $(\beta_i-\gamma)$ decreased. The spatial variability condition can create an instantaneous pitch angle and a velocity vector at wire 2 which is different from γ and V at wire 1. The spatial-variability conditions will be designated as $\Delta\gamma = (\gamma_2-\gamma_1)$ and $\Delta V = V_2-V_1$. The purpose of the uncertainty analysis is to allow quantitative estimates of the effects on u and v of the perturbation factors: $\Delta\gamma$, $\Delta V/V$, w/V.

The calculation procedure for these estimates makes use of the sensitivity analysis presented above. The perturbation voltage for wire "i", δE_i, can be evaluated for given magnitudes of $\Delta\gamma$, $\Delta V/V$, and w/V and the δE_i values can be input to the sensitivity analysis for the calculation of δu and δv. The uncertainty analysis will allow the validity of each individual u,v calculation to be expressed in terms of the true, $(\ )_t$, and measured, $(\ )_m$, values. That is,

$$\delta u = u_m - u_t; \quad \delta v = v_m - v_t \tag{21}$$

The perturbation voltage, δE_i, represents the difference between the measured voltage E_{im} and a hypothetical voltage, E_{it}, which would exist if the perturbations: $(\Delta\gamma, \Delta V/V, w/V)$ were equal to zero. The response equations presented above, and the assumption that $V_{eff}^2 = V^2\left\{1+b\left(1-\cos^{\frac{1}{2}}(\beta-\gamma)\right)\right\}^4 + w^2$ allow E_{im} and E_{it} to be calculated for prescribed values of V and γ. Namely,

$$E_{1m}^2 = E_{01}^2 + K_1(\gamma)V^{m_1}\left[\left\{1-b_1\left(1-\cos^{\frac{1}{2}}(\beta_1-\gamma)\right)\right\}^4 + (\frac{w}{V})^2\right]^{m_1/2} \tag{22}$$

and

$$E_{2m}^2 = E_{02}^2 + K_2(\gamma+\Delta\gamma)V^{m_2}\left[(1+\frac{\Delta V}{V})^2\left\{1-b\left(1-\cos^{\frac{1}{2}}(\beta_2-\gamma-\Delta\gamma)\right)\right\}^4 + (\frac{w}{V})^2\right]^{m_2/2} \tag{23}$$

δE_1 and δE_2 have been computed using the indicated procedures; the corresponding values of δu and δv are presented in the Results Section.

The w-bias Effect*

The perturbations $\Delta\gamma$ and $\Delta V/V$ are expected to have a zero net effect on E_1 and E_2 since they are symmetrically distributed and they have a skew symmetric effect on the voltages. The symmetric effect of (w/V) leads to an average

*Instantaneous, as well as stochastic, corrections for w are possible; the latter are discussed in this section.

change in the voltages and hence a bias in the calculated value of u and a possible bias in v. The bias effect can be expressed as

$$\langle \delta u(\gamma, V) \rangle = \int_{-\infty}^{\infty} \delta u \left(\frac{w}{V} ; \gamma, V \right) p_{w/V} (\xi) \, d\xi \qquad [24]$$

with a similar expression for $\langle \delta v \rangle$. Using $\Delta\gamma = \Delta V = 0$ in [22] and [23] and using the two leading terms of the expansion:

$$[1+(w/V_*)^2]^{m_i/2} = 1 + (\frac{m_i}{2}) (\frac{w}{V_*})^2 + \frac{(\frac{m_i}{2}) (\frac{m_i}{2} - 1)}{2!} (\frac{w}{V_*})^4 + \ldots$$

where $V_* = V \left\{ 1-b(1-\cos^{1/2} (\beta - \gamma) \right\}^2$,

an approximate relationship for δE_i can be developed as

$$\delta E_i = \frac{m_i K_i V^{m_i} \{C.T.\}^{m_i - 2}}{4 E_i} (\frac{w}{V})^2 ; \quad C.T._i = \left\{ 1 - b[1 - \cos^{1/2} (\beta_i - \gamma)] \right\}^2 \qquad [25]$$

When [25] is combined with [17a,b], [18], [19] and [20], expressions of the form:

$$\delta u = (f_1^* + f_2^*) (w/V)^2 ; \quad \delta v = (g_1^* + g_2^*) (w/V)^2 \qquad [26]$$

result. (Note that f and g are functions of γ and V). From [24],

$$\langle \delta u(\gamma, V) \rangle = (f_1^* + f_2^*) (\tilde{w}/V)^2 ; \quad \langle \delta v(\gamma, V) \rangle = (g_1^* + g_2^*) (\tilde{w}/V)^2 \qquad [27]$$

Hence, the *instantaneous* velocities may be corrected for the effect of (w/V) and the statistical quantities of interest may be formulated using these corrected values.

Analog data processing schemes can also be "corrected" by suitable procedures; recent examples are the work of Chevray (1975) and Rodi (1975) (the latter utilizes the unconventional, but well motivated procedure of using e^2 to infer the higher order statistics). A significant difference between the analog techniques and the present one is that the former execute statistical operations on the voltages and then make conversions/corrections. For the present technique, all corrections and statistical processing are executed with the velocities.

Results and Discussion

Analytical evaluations of the sensitivity and uncertainty effects have been
made, and experimental data from an axisymmetric jet have been acquired, to
evaluate the proposed vorticity measurement technique. These results are presented and discussed in this section.

The sensitivity analysis was used to assess the influence of the digital resolution
on the u and v calculations. The present experimental equipment (DISA
55M10 anemometer, 5 μ tungsten wire, $R_H = 1.8\ R_c$, 10 bit A/D converter,
0-100 fps velocity range) provides an active range of 2.5 - 5.5 volts, and hence
a 3 mV resolution of the suppressed and amplified bridge voltage. Various
combinations of (3, 0, −3) mV were input to the equations of the sensitivity
analysis; the results are presented in Table I. Similar results are obtained for
the evaluation of the uncertainty caused by the velocity magnitude ($\Delta V/V$),
pitch angle ($\Delta \gamma$) and transverse velocity (w/V) effects; these results are presented in Table II. The uncertainty calculations support the plausible conjec-
ture that large perturbation voltages are produced (for a given $\Delta V/V$, $\Delta \gamma$, or
w/V perturbation) when large pitch angles are present, i.e., for the condition
where the perturbation creates a substantial change in the relatively small
effective cooling velocity.

The sensitivity of the calculations to the pitch angle is an intrinsic feature of
extracting V and γ from an x-array. It is, however, interesting to note that the
well known summation technique ($E_1 + E_2$) provides a relatively reliable estimate for u for sufficiently small γ and sufficiently similar and properly aligned
wires. Unfortunately, the use of the difference ($E_1 - E_2$) to recover v (or γ) is
quite sensitive to the wire geometry as well as to resolution (noise) and the perturbation effects identified above. These effects are more fully explored in the
companion report, Foss (1978). The observed sensitivity to γ suggests that the
computing algorithm for the vorticity calculation be modified. Specifically,
let $V_N = (V_3 + V_4)/2$, as extracted from the parallel array, and let $u = V_N \cos\gamma$
and $v = V_N \sin\gamma$ where γ is determined from the x-array. With this change, the
uncertainty effects only influence the trigonometric functions and the spatial
resolution of the 4 wire probe is "improved" since only $(\partial\gamma/\partial x)\delta z$ need be
assumed small. Similarly, if an optimal scheme to evaluate time series for (u,v)
is considered, a four wire array — where one of the parallel wires is yawed to
provide an estimate of w and hence an instantaneous (w/V) correction to the
x-array voltages — is suggested.

High rate, time series data were collected at selected y locations. These data
were used to prepare plots like that shown in Fig. 5 (the appearance of a con-

TABLE I

(DU/U)★100. DE1 = .000; DE2 = .003

γ	V=10	V=30	V=50	V=70	V=90
−40	.61	.43	.36	.33	.31
−25	.49	.34	.29	.26	.25
−10	.42	.29	.25	.22	.21
0	.38	.26	.22	.20	.19
10	.34	.24	.20	.18	.17
25	.28	.19	.16	.14	.13
40	.12	.08	.07	.06	.06

(DV/VELOCITY)★100. DE1 = .000; DE2 = .003

γ	V=10	V=30	V=50	V=70	V=90
−40	−.53	−.37	−.32	−.29	−.27
−25	−.52	−.37	−.31	−.28	−.26
−10	−.49	−.34	−.29	−.26	−.24
0	−.44	−.31	−.26	−.24	−.22
10	−.39	−.27	−.23	−.21	−.19
25	−.28	−.19	−.16	−.15	−.13
40	−.10	−.07	−.06	−.05	−.05

(DU/U)★100. DE1 = .003; DE2 = −.003

γ	V=10	V=30	V=50	V=70	V=90
−40	−.40	−.29	−.25	−.23	−.21
−25	−.16	−.12	−.10	−.09	−.09
−10	−.03	−.02	−.02	−.02	−.02
0	.04	.03	.02	.02	.02
10	.11	.08	.07	.06	.06
25	.25	.18	.15	.14	.13
40	.53	.37	.32	.29	.27

(DV/VELOCITY)★100. DE1 = .003; DE2=−.003

γ	V=10	V=30	V=50	V=70	V=90
−40	.69	.48	.40	.36	.34
−25	.84	.58	.49	.44	.41
−10	.89	.62	.53	.47	.44
0	.89	.62	.53	.48	.44
10	.88	.61	.52	.47	.43
25	.79	.55	.47	.42	.39
40	.61	.42	.36	.33	.30

(DU/U)★100. DE1 = .003; DE2 = .000

γ	V=10	V=30	V=50	V=70	V=90
−40	.20	.14	.11	.10	.09
−25	.33	.22	.19	.17	.16
−10	.38	.26	.22	.20	.19
0	.41	.29	.24	.22	.20
10	.45	.31	.27	.24	.22
25	.52	.37	.31	.28	.26
40	.65	.45	.39	.35	.33

(DV/VELOCITY)★100. DE1 = .003; DE2 = .000

γ	V=10	V=30	V=50	V=70	V=90
−40	.16	.10	.09	.08	.07
−25	.31	.21	.18	.16	.15
−10	.41	.28	.24	.21	.20
0	.45	.31	.27	.24	.22
10	.48	.34	.29	.26	.24
25	.51	.36	.31	.28	.26
40	.51	.36	.30	.28	.26

(DU/U)★100. DE1 = .003; DE2 = .003

γ	V=10	V=30	V=50	V=70	V=90
−40	.81	.56	.48	.43	.40
−25	.82	.57	.48	.43	.40
−10	.80	.55	.47	.43	.39
0	.79	.55	.47	.42	.39
10	.79	.55	.47	.42	.39
25	.80	.55	.47	.43	.39
40	.77	.53	.46	.41	.38

(DV/VELOCITY)★100. DE1 = .003; DE2 = .003

γ	V=10	V=30	V=50	V=70	V=90
−40	−.38	−.27	−.23	−.21	−.20
−25	−.21	−.16	−.13	−.12	−.12
−10	−.08	−.06	−.05	−.05	−.04
0	.01	.00	.00	.00	.00
10	.09	.07	.06	.05	.05
25	.23	.17	.14	.13	.12
40	.41	.29	.25	.22	.21

Table I. Results of the sensitivity analysis. Note: V is in feet/sec; γ is in degrees

J.F. Foss

TABLE II

(DV/VELOCITY)★100. Δγ = .00; ΔV/V = .00; W/V = .25

γ	V=10	V=30	V=50	V=70	V=90
−40	5.10	5.06	5.02	5.01	5.00
−30	3.17	3.17	3.14	3.15	3.15
−20	1.71	1.71	1.70	1.70	1.71
−10	.66	.66	.67	.66	.66
0	− .23	− .22	− .22	− .23	− .23
10	−1.15	−1.15	−1.14	−1.15	−1.14
20	−2.30	−2.29	−2.30	−2.29	−2.29
30	−3.96	−3.93	−3.93	−3.92	−3.93
40	−5.40	−5.32	−5.27	−5.25	−5.22

(DV/VELOCITY)★100. Δγ = .00; ΔV/V = .05; W/V = .00

γ	V=10	V=30	V=50	V=70	V=90
−40	−3.69	−3.69	−3.69	−3.68	−3.68
−30	−3.64	−3.64	−3.64	−3.64	−3.64
−20	−3.47	−3.47	−3.46	−3.47	−3.46
−10	−3.18	−3.17	−3.18	−3.18	−3.18
0	−2.78	−2.78	−2.78	−2.78	−2.78
10	−2.28	−2.28	−2.28	−2.28	−2.28
20	−1.71	−1.70	−1.71	−1.70	−1.71
30	−1.06	−1.06	−1.06	−1.06	−1.06
40	− .36	− .36	− .36	− .36	− .36

(DU/U)★100. Δγ = .00; ΔV/V = .00; W/V = .25

γ	V=10	V=30	V=50	V=70	V=90
−40	11.61	11.55	11.49	11.49	11.48
−30	8.24	8.21	8.18	8.20	8.19
−20	6.44	6.43	6.43	6.42	6.43
−10	5.68	5.67	5.67	5.67	5.67
0	5.49	5.49	5.49	5.49	5.48
10	5.75	5.74	5.74	5.73	5.73
20	6.59	6.59	6.59	6.59	6.58
30	8.54	8.52	8.51	8.51	8.50
40	11.24	11.13	11.07	11.04	11.01

(DU/U)★100. Δγ = .00; ΔV/V = .05; W/V = .00

γ	V=10	V=30	V=50	V=70	V=90
−40	3.07	3.08	3.08	3.07	3.08
−30	2.92	2.93	2.93	2.92	2.93
−20	2.81	2.81	2.81	2.81	2.81
−10	2.64	2.63	2.64	2.64	2.64
0	2.36	2.36	2.36	2.36	2.35
10	1.99	1.99	1.99	1.99	1.99
20	1.59	1.59	1.59	1.59	1.59
30	1.10	1.11	1.10	1.11	1.10
40	.43	.43	.43	.43	.43

(DV/VELOCITY)★100. Δγ = 1.00; ΔV/V = .00; W/V = .00

γ	V=10	V=30	V=50	V=70	V=90
−40	.13	.13	.13	.13	.13
−30	.34	.34	.34	.34	.34
−20	.55	.55	.55	.55	.55
−10	.75	.75	.75	.75	.75
0	.92	.92	.93	.92	.92
10	1.07	1.07	1.07	1.07	1.07
20	1.19	1.19	1.19	1.19	1.19
30	1.27	1.27	1.27	1.27	1.27
40	1.33	1.33	1.33	1.33	1.33

(DV/VELOCITY)★100. Δγ = 1.00; ΔV/V = .10; W/V = .40

γ	V=10	V=30	V=50	V=70	V=90
−40	.78	.63	.58	.50	.42
−30	− .29	− .35	− .38	− .40	− .40
−20	−2.08	−2.05	−2.10	−2.11	−2.05
−10	−3.32	−3.28	−3.35	−3.31	−3.31
0	−4.20	−4.19	−4.18	−4.19	−4.18
10	−4.97	−4.95	−4.92	−4.90	−4.90
20	−5.94	−5.87	−5.87	−5.85	−5.86
30	−7.30	−7.19	−7.13	−7.12	−7.10
40	−6.63	−6.45	−6.35	−6.26	−6.22

(DU/U)★100. Δγ = 1.00; ΔV/V = .00; W/V = .00

γ	V=10	V=30	V=50	V=70	V=90
−40	− .11	− .11	− .11	− .11	− .11
−30	− .27	− .27	− .27	− .27	− .27
−20	− .45	− .45	− .45	− .45	− .45
−10	− .62	− .62	− .62	− .62	− .62
0	− .78	− .78	− .79	− .78	− .78
10	− .94	− .93	− .93	− .94	− .94
20	−1.11	−1.11	−1.11	−1.11	−1.11
30	−1.32	−1.32	−1.32	−1.32	−1.32
40	−1.60	−1.60	−1.60	−1.61	−1.61

(DU/U)★100. Δγ = 1.00; ΔV/V = .10; W/V = .40

γ	V=10	V=30	V=50	V=70	V=90
−40	26.84	26.57	26.43	26.35	26.35
−30	22.78	22.66	22.59	22.59	22.50
−20	19.40	19.37	19.36	19.33	19.28
−10	17.42	17.37	17.39	17.31	17.29
0	16.31	16.29	16.26	16.26	16.25
10	15.88	15.83	15.80	15.79	15.79
20	16.47	16.38	16.38	16.37	16.37
30	18.48	18.35	18.29	18.28	18.26
40	19.64	19.38	19.26	19.16	19.10

Table II. Results of the Uncertainty Analysis. Note: V is in feet/sec; γ is in degrees

tinuous signal is from the linear connection of the discrete data points) and the values shown in Table III. The quite small value of the $(\partial v/\partial x)(\partial u/\partial y)$ correlation suggests that a single measurement of $\partial u/\partial y$ to indicate the vorticity may be quite deceiving.

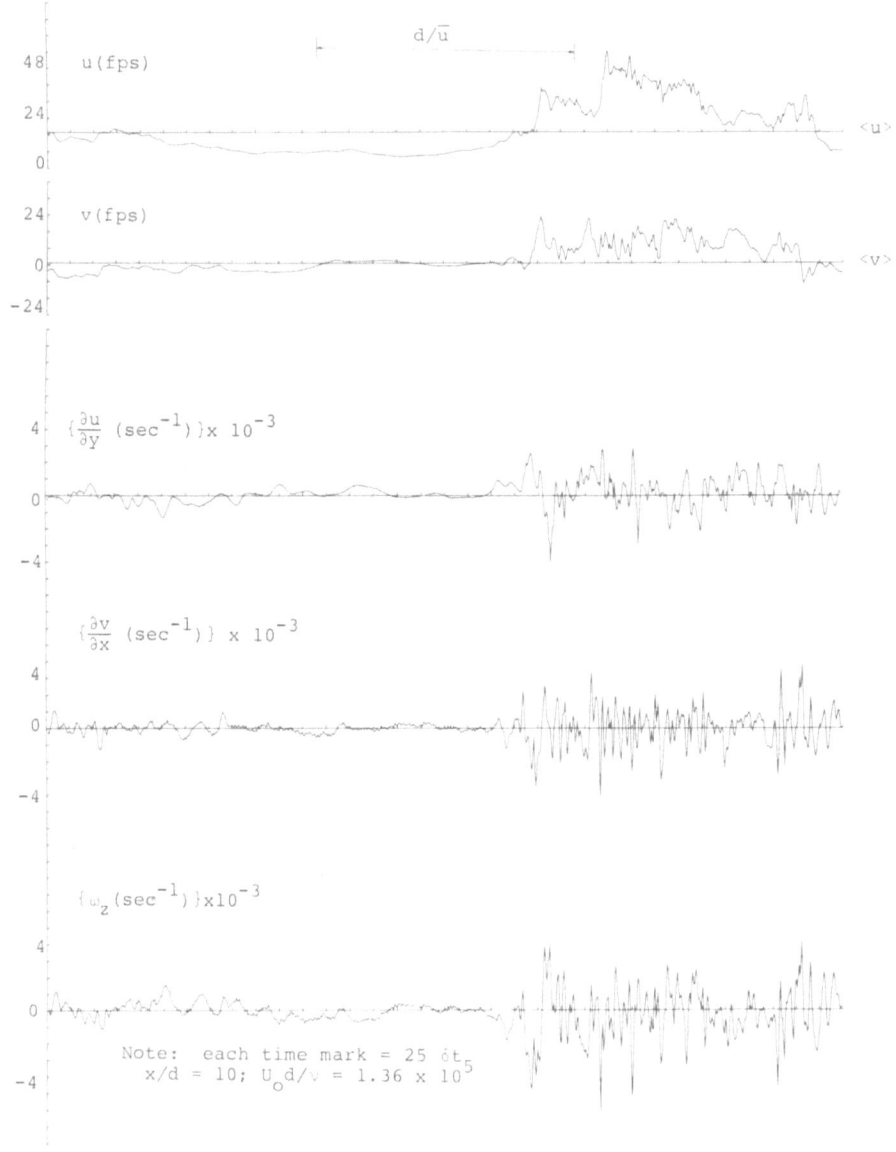

Fig. 5. *Discretely evaluated time series of the vorticity $y/d = 1.32$, $\delta t = 0.0435$ ms*

Table III. Stochastic Results from the axisymmetric jet: $x/d = 10$, $u_o d/\nu = 1.36 \times 10^{-5}$

y/d	u/u_o	$\dfrac{<(\partial v/\partial x)(\partial u/\partial y)>}{(\text{S.D.: } \partial v/\partial x)(\text{S.D.: } \partial u/\partial y)}$	$\tilde{\omega} d/u_o$
1.32	0.16	−0.0094	2.42
−0.88	0.35	−0.017	3.67
0.02	0.71	−0.033	4.02

Some decrease in this correlation can be attributed to the transverse dimension between the two probe arrays.

The average values from the strongly vortical segment (480→850) of Fig. 5 were used to evaluate the turbulence field characteristics. The dissipation ϵ was estimated from: $\epsilon \cong 15\nu<(\partial u_i/\partial x_j)^2>$ for i=1, j=1; i=1, j=2; i=2, j=1. The average answer of 1.8×10^3 ft/sec provided a Kolmogorov scale of 0.066 mm. The analysis of Wyngaard would suggest that the probe does not fully respond to the fluctuating vorticity.

The validity of the small scale variations in the $(\partial v/\partial x)$ signal can be evaluated by considering the sensitivity and uncertainty analyses. The sensititity analysis shows that an unfavorable combination of 3 mV resolution jumps could lead to a $\partial v/\partial x$ change of the order 150 sec^{-1}; hence the variation from zero is significant with respect to the resolution. (A "favorable" combination of the 3 mV changes would provide an order of magnitude reduction in this estimate). An examination of the uncertainty results suggests that viable combinations of $\Delta\gamma$, ΔV and W effects could be responsible for the non-zero vorticity: time marks 14-19.

Conclusions

A comprehensive technique for the evaluation of the transverse vorticity from an array of four hot-wire probes has been developed. The technique involves digital calculations which are implemented with appropriate programs.

The technique includes a new computing algorithm to evaluate (u,v) from an x-array. Analyses of the sensitivity, uncertainty, and the bias effect of w are presented; the bias effect is shown to be significant in the calculation of mean quantities. The demonstrated sensitivity of the velocity (V) to the pitch-angle (γ)-calculation provides a stimulus to modify the (u,v) calculation given the presence of an x-array and at least one straight wire. The proposed modifications will provide improved accuracy and spatial resolution for the vorticity measurements.

Representative measurements in an axisymmetric jet are presented to demonstrate the technique.

References

Collis, D.C. and Williams, M.J. (1959) "Two-dimensional convection from heated wires at low Reynolds numbers", *Jour. Fluid Mech.,* Vol. **51** part 3, p. 487

Corrsin, S. and Kistler, A.L., (1955), "Free stream boundaries of turbulent flows", NACA Report 1244

Drubka, R.E., Nagib, H.M. and Tan-atichat, J., (1977) "On temperature and yaw dependence of hot-wire probes", IIT Fluids and Heat Transfer Report R77-1, August

Eckelmann, H., Nychas, S.G., Brodkey, R.S. and Wallace, J.M., "Vorticity and turbulence production in pattern recognized turbulent flow structures", *Physics of Fluids,* Vol. **20**, No. 10, Pt. 11, Oct.

Foss, J.F. (1978), "Transverse Vorticity Measurements", Final Report, NASA Grant NGR-23-004-091, July. (Revised 1979)

Foss, J.F., (1976), "Accuracy and uncertainty of transverse vorticity measurements", *Bull. Am. Phys. Soc.,* Series II, Vol. **21**, No. 10, p 1237, November

Friehe, C.A. and Schwarz, W.H. (1968), "Deviations from the cosine law for yawed cylindrical anemometer sensors", *Trans ASME, Jour. Appl. Mech.,* No. **35**, p 655

Hardin, J.C. (1973), "Analysis of noise produced by an orderly structure of turbulent jets", NASA TND-7242

Kistler, A.L. (1952), The vorticity meter, M.S. Thesis, The Johns Hopkins University

Kovasznay, L.S.G., (1950), Quarterly progress report of aeronautics department, contract No. rd-8036-JHB-3D. The Johns Hopkins University, Jan. 1 - March 31.

Kovasznay, L.S.G., Kibbens, V. and Blackwelder, R.F., (1970), "Large-scale motion in the intermittent region of a turbulent boundary layer", *Jour. Fluid Mech.,* Vol. **41**, Part 2, pp. 283-325

Kovasznay, L.S.G., Komoda, H. and Vasudeva, B.R. (1962), "Detailed flow field in transition", *Proc. Heat Trans. and Fluid Mech. Inst.,* Stanford Univ. Press

Maxworthy, T. (1977), "Some experimental studies of vortex rings", *Jour. Fluid Mech.,* Vol. **81**, p. 3, pp. 465-495

Rodi, W. (1975), A new method of analyzing hot-wire signals in highly turbulent flow, and its evaluation in a round jet, *DISA Information* No. **17**, p 9

Roshko, A. (1976), "Structure of turbulent shear flows: A new look" *AIAA Jour.* Vol. **14**, pp. 1348-1357

Strohl, A. and Comte-Bellot, G. (1973), "Aerodynamic effects due to configuration of x-wire anemometers", *Jour. Applied Mech.,* pp. 661-666

Townsend, A.A. (1976), *The structure of turbulent shear flow,* Cambridge University Press

Tutu, N.K. and Chevray, R. (1975), "Cross wire anemometry in high intensity turbulence", *Jour. Fluid Mech.,* Vol. **71**, part 4, pp. 785-800

Wyngaard, J., (1969), "Spatial resolution of the vorticity meter and other hot-wire arrays", *Jour. of Sci. Instruments* (Jour. Phys. E), Ser. 2, Vol. **2**.

Nonsteady Vorticity Measurements: Survey and New Results

by

William W. Willmarth
Department of Aerospace Engineering
The University of Michigan, Ann Arbor, Michigan 48109

Abstract

The paper consists of a brief description of our experience in making measurements of streamwise vorticity. The measurements were made using a probe consisting of an array of four hot wires with precise geometrical and electrical characteristics. This type of probe was developed by Kovasznay in 1954 but the influence of the flow velocity on the vorticity signal had not been investigated in detail. We describe the probe construction, calibration, spatial resolution and the influence of the flow velocity on the vorticity signal. The later work has been done at the Max-Planck-Institut für Strömungsforschung by E. Kastrinakis and H. Eckelmann. It is concluded that in a turbulent flow field this probe cannot be used without simultaneous knowledge of the instantaneous transverse velocity components.

Introduction

Various research groups using different methods of investigation have proposed that vortex structures in bounded turbulent shear flows are an important feature of such flows. For example Kim et al. (1971) have visually observed vortices in a turbulent boundary layer and Willmarth and Tu (1967) were able to explain their velocity and wall pressure-velocity correlation measurements in a boundary layer by means of a horseshoe-vortex model.

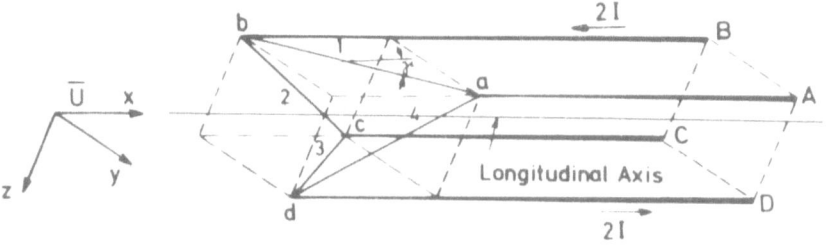

Fig. 1. Sketch of the vorticity probe.

The hot wire probe shown in Fig. 1, which was originally proposed by Kovasznay (1954), was developed to measure the streamwise component of the vorticity $\omega_x = \partial w/\partial y - \partial v/\partial z$ as well as the streamwise velocity component, $\bar{U} + u$. This hot wire probe will be called a vorticity probe from now on. The hot wire array is heated using the constant current method (Fig. 2). The voltage between the input terminals b,d of the circuit is a measure of the longitudinal velocity component, $\bar{U} + u$, and the voltage between the diagonal points a,c is a measure of the streamwise component of the vorticity, ω_x.

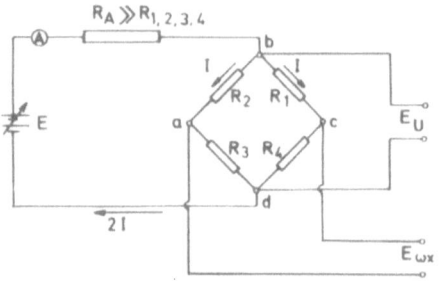

Fig. 2. Circuit of the vorticity probe.

With a vorticity probe, measurements have been conducted in an isotropic turbulent flow by Uberoi and Corrsin (1951) and Kistler (1952). Wyngaard (1969) analyzed the probe response to varying scales of isotropic turbulence. Recently, Kastrinakis (1977) and Kastrinakis et al. (1978a) carried out correlation measurements in a turbulent channel flow with two calibrated vorticity probes.

In this paper I will discuss my personal experience with vorticity probes. In 1966 Dr. Bo Jang Tu and I began working with them. You could not purchase a vorticity probe commercially at that time and I don't believe they are commercially available at the present time. Therefore, I will first describe the method of construction of the probes. This is followed by a discussion of various applications of the probe taken from my experience with them. Initially Dr. Bo Jang Tu and I used them in a high Reynolds number boundary layer. Unfortunately the probe response was not very useful owing to poor spatial resolution in this situation. Our observations in the boundary layer and Wyngaard's (1969) calculations of the probe response to isotropic turbulence will be discussed.

I did no further work with the vorticity probe until 1975. In 1974 work began at the Max-Planck-Institut für Strömungsforschung in Göttingen on the construction of small vorticity probes and a calibration device. At Jim Wallace's suggestion I visited Göttingen and worked with E. Kastrinakis, H. Eckelmann and J. Wallace during the summer of 1975. Numerous vorticity probes were constructed by E. Kastrinakis for use in a channel flow with a very large viscous length scale in an attempt to make measurements in which spatial resolution was not a problem. In the course of this work it was determined that the probes were also sensitive to fluctuations in the flow velocity. Corrections for streamwise fluctuations were easily made but corrections for the affect of transverse velocity fluctuations were not possible. The result is, as will be shown, that the application of vorticity probes to turbulent flow measurements is severely limited.

Methods of Probe Construction

The prongs of our first probes were constructed using four equally spaced sewing needles. The needles must be very straight and precisely spaced to form a square when viewed looking downstream along the probe axis. Furthermore, each pair of needles, diagonally opposite one another, must be of equal length; one pair being shorter than the other. It is convenient to make the difference in length equal to the needle spacing along the sides of the square. Then, when the hot wires are fastened to the tips of the needles, the wires make an angle of 45° with the flow along the probe axis. The precision required is of the order of 1 or 2 % and can be checked using an optical comparator. It was usually necessary to bend the needles and examine the result many times before the required dimensional precision could be attained.

In order to attach hot wires to the prongs the wire was positioned using a microscope. To attain a precise 45° alignment of each wire with the probe axis the wire was suspended from a support with a small weight attached to the end. The probe was then placed in a fixture aligned at precisely 45° to the vertical. The probe could be rotated about its axis in the fixture and positioned so that the tips of two of the needles just contacted the hanging wire.

In our earlier work, in 1966, Dr. Tu constructed the probes using copper plated tungsten wire which could be soft soldered to the needles. In our recent work, Dr. E. Kastrinakis used platinum plated tungsten wire which was spot welded to the tips of the needles. Both operations must be viewed using a microscope since one must attempt to construct as small a probe as possible.

In addition to precise geometrical arrangement of the hot wires it is necessary that the electrical resistance of each wire and the sensitivity of each wire to the normal (cooling) velocity be the same. This can be attained with the precision of 2 - 3% if wire of very uniform diameter is used and if the length of each wire is the same. In order to make the length of the active portion of the wires the same, in our early work, the copper plating process was done using a specially constructed fixture. The bare wire was immersed in two drops of copper sulphate solution formed at the ends of two closely spaced glass tubes. I believe this method originated at Johns Hopkins, at any rate I learned about it from M.S. Uberoi. In spite of the very careful plating and construction methods described above the probes were usually not electrically symmetrical. Dr. Tu devised a method using reverse plating for removing additional copper from a hot wire whose resistance was too low. The probe resistance was continuously monitored while a very small drop of plating solution held in a tiny loop of platinum wire was carefully applied to the active region of the offending wire. Reverse plating current was then applied until the Wheatstone bridge was balanced.

In the more recent work we depended upon very accurate needle spacing to obtain equal length and therefore resistance of each wire. After spot welding the wires to the prongs it was often possible to make small adjustments in the wire resistance by rewelding the ends of the wire with the largest resistance. This often repaired a faulty weld and/or slightly reduced the hot wire length. In general, the wire resistances were the same with a precision of the order of 2 - 3 %.

Spatial Resolution of Vorticity Probes

In our early (1966) work two vorticity probes were used in a thick (13 cm) turbulent boundary layer at high Reynolds number. The vorticity signal was observed to contain much more "energy" at high frequencies than was observed in the spectra of the velocity fluctuations at the same point in the boundary layer. Our interest was focused on the vorticity field near the wall. We used two probes in an attempt to study the space-time correlations of the streamwise vorticity. We were not successful because all the correlations measured were zero even when the probes were placed as close together as possible.

In this situation the viscous length scale ν/u_τ in the wall region of the boundary layer was of the order of 8×10^{-3} mm but the smallest possible spacing of the probes was slightly larger than the prong spacing, d, and d was 1.9 mm. As discussed by Kim et al. (1971), streamwise vortices in the wall region were observed with an average spacing of 100 viscous lengths or 0.8 mm. Thus, the vorticity probes were too far apart to resolve the pairs of streamwise vortices in the sublayer.

After reducing the free stream velocity from 61 to 6 m/sec and moving the probes well downstream, we were able to increase the viscous length by a factor of the order of 10. A very small maximum correlation coefficient, of the order of 0.05, between the vorticity signals for two probes placed side by side was then measured. However, this correlation was so small that further study of the vorticity correlation field was abandoned. In this case the spacing of streamwise vortices, 100 ν/u_τ, was of the order of 8 mm or 4 times the probe separation. However, the prong spacing, d, for a single probe was still large relative to the sublayer thickness, δ_s. If we assume $\delta_s \cong 10 \, \nu/u_\tau$ then $d/\delta_s \cong 2.5$. This suggests that the vorticity probes were much too large to be used in the highly sheared boundary layer flow near the wall.

Further information about the spatial resolution of a vorticity probe when exposed to isotropic turbulence is contained in Wyngaard's (1969) paper. In this case the mean flow is uniform. Wyngaard computed the vorticity probe response when the scale of the turbulence relative to the prong spacing, d, the hot wire length, ℓ, and the wire inclination angle, θ, were varied. He found that the change in the probe reponse as a function of wire angle θ in the range $30° \leqslant \theta \leqslant 60°$ was relatively small. Vorticity probes with wires more highly inclined to the flow were somewhat more sensitive at high wave numbers. At low wave numbers θ had very little affect.

Both the ratio d/ℓ and η/ℓ had a relatively strong influence on the probe response at all wave numbers, even very low wave numbers. Here, η, is the Kolmogorov length. The usual result for hot wires used to measure a turbulent velocity component (or for sensors used to measure the turbulent pressure) is that at very low wave numbers the sensitivity is the same and is not reduced when the probe size is increased. Wyngaard gave a physical explanation for this unexpected result for the vorticity probe. He reasoned that the dominant wave number of streamwise vorticity fluctuations is near $1/\eta$ in magnitude and therefore small. But all modes of oscillation can cut the streamwise probe axis at all possible angles giving contributions to streamwise vorticity down to zero wave number. The probe must be as small as the dominant scale of the fluctuations in order to resolve them. Let us assume that a valid model for the vorticity field is like a random field of spaghetti of various diameters. A vorticity probe must then be small enough so that all four prongs are covered by the smallest diameter strands of spaghetti before wave number contributions particularly those of low number in which the spaghetti lies along the probe axis can be accurately measured.

Table 1 below, taken from Wyngaard's paper (1969), gives the reduction in the response of the vorticity probe at zero wave number as a function of prong spacing. It can be seen that the ratio η/ℓ must not be less than 0.3 in order that 5 % accuracy be attained.

η/ℓ	Measured Vorticity / Actual Vorticity
1.000	1.0
0.320	0.95
0.100	0.78
0.032	0.53

Table 1. Reduction in response of a vorticity probe to isotropic turbulence at zero wave number; $\theta = \pi/4$, $d/\ell = 1.0$. From Wyngaard (1969).

Calibration of the Vorticity Probe

In 1974 J. Wallace and H. Eckelmann at the Max-Planck-Institut für Strömungs-forschung were instrumental in starting a new program of streamwise vorticity measurements. The new measurements were to be made in a low speed turbulent channel flow with a relatively large viscous length of the order of 0.63 mm so that good spatial resolution could be obtained. The prong spacing of the probes was planned to be of the order of 0.5 mm. However, the construction of such small probes was found to be very difficult. The first probes were made with a prong spacing of 2 mm giving a ratio of viscous length to prong spacing of the order of 0.3. The probes were calibrated by spinning them about their streamwise axes in a laminar flow using a specially constructed mechanism fitted with mercury slip rings (for technical details please see the thesis of Kastrinakis (1977)).

The probe construction and calibration was done primarily by E. Kastrinakis (1977). Fig. 3 shows a typical vorticity probe calibration measured at two different speeds in a laminar flow. The streamwise component of the vorticity is given by twice the angular velocity of the spinning probe since the flow rotates relative to the probe as a solid body. The magnitude of the vorticity signal (measured between points a and c of Fig. 2) decreases with increasing flow speed. The decrease is caused by the usual decrease in hot wire sensitivity as the mean speed increases.

Fortunately, one can in principle correct for this change in sensitivity of the vorticity probe since the velocity u is available. One must simply measure the voltage between the points b and d of Fig. 2. (The probe sensitivity to u can be obtained by calibration as for an ordinary hot wire).

During the calibration process it was discovered that the vorticity signal between points a and c of Fig. 2 was a function of u even when the probe was not

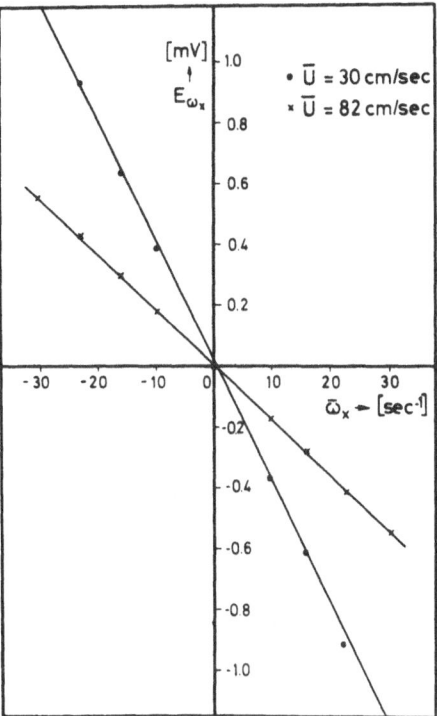

Fig. 3. Calibration curve of the vorticity probe with respect to $\overline{\omega}_x$ ($E_{\omega x}$ is the voltage across the diagonal points a-c in Fig. 2).

spinning. This behavior is discussed in detail in a forthcoming paper, Kastrinakis et al. (1978b), and is caused by slight geometrical or electrical asymmetries of the probe or by differences in the calibration curves for the individual hot wires. For the calibration curves of Fig. 3, the bias voltage at zero streamwise vorticity has been removed.

During the calibration with respect to ω_x, a periodic signal was observed superimposed on the mean output voltage $E_{\omega x}$. It was found that this oscillatory signal was dependent upon the alignment of the axis of rotation of the probe with respect to the direction of the main flow. Fig. 4 (top) is an example of the vorticity signal observed with imperfect alignment of the probe. This effect is a result of the influence of the two transverse velocity components v and w on the vorticity signal. After more exact alignment of the probe with respect to the mean flow, the amplitude of the periodic signal could be brought to a *minimum* (Fig. 4, bottom). Furthermore, misalignment of the probe axis with

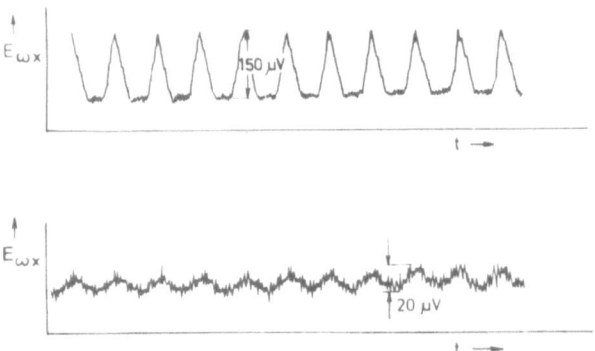

Fig. 4. Output voltage of the vorticity probe $E_{\omega\,x}$ across the diagonal points a-c (Fig. 2) during the calibration procedure. Top: the longitudinal axis of the probe is not parallel to the main stream. Bottom: after alignment of the probe, the periodic signal can be minimized.

respect to the mean flow resulted in a nonlinear calibration curve (i.e. $E_{\omega\,x} = f(\omega_x)$ was nonlinear). The amplitude of the oscillatory signal was found to be a measure of the quality of the probe (uniformity of the wires) and its alignment with the flow.

An experimental study of the influence of the transverse velocity on the vorticity signal has been made by yawing a stationary probe in a uniform flow and measuring the vorticity signal as a function of the yaw angle, Kastrinakis et al. (1978b). The erroneous vorticity signal caused by the transverse velocity depends upon the orientation of the probe about the streamwise axis. When the probe orientation is such that the transverse velocity is parallel to the x,y or x,z plane of Fig. 1, the erroneous vorticity signal is a maximum. When the transverse velocity is oriented at $45°$ to the x,y and x,z planes of Fig. 1 (i.e. the transverse velocity is directed parallel to a diagonal formed by prongs on opposite sides of the array when viewed along the x-axis) the erroneous signal is very small. A mathematical analysis of the response to transverse velocity yielded results in good agreement with the measurements, Kastrinakis et al. (1978b).

In the analysis it was necessary to calculate the change of the resistance of each wire caused by the transverse velocity to terms of second order. This was done by computing the velocity component normal to each wire to terms of second order and then using the hot wire calibration data. In the computation it was also necessary to account for the redistribution of the current in the Wheatstone bridge caused by the change in resistance of each wire. The resulting vorticity signal (i.e. the voltage) between points a and c of Fig. 1 was then computed. The results of the computations and measurements are shown in Fig. 5. In the analysis the hot wire calibration data for the total probe was used. It was assumed that each wire was the same. The results shown in Fig. 5 are in good qualitative agreement. The determination of the calibration data for each wire would have been extremely difficult.

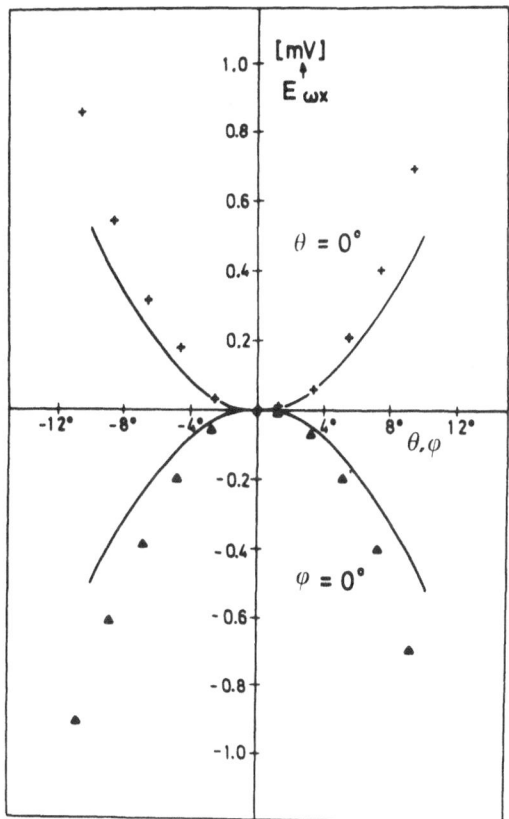

Fig. 5. Comparison of the calculated (solid lines) and measured (symbols) vorticity signal, $E_{\omega x}$, as a function of the pitch and yaw angles, ϕ and θ. The probe is oriented so that the transverse velocity is parallel to planes formed by adjacent prongs a,b or b,c.

Conclusions
The use of a streamwise vorticity probe for measurements in turbulent flows has been discussed. Wyngaard (1969), analyzed the vorticity probe response to isotropic turbulence. He concluded that the probe must be small enough to resolve the smallest scales of the vorticity field before accurate measurements can be made, even at low wave number. This is a result of the spatial structure of the vorticity field.

Methods for constructing small vorticity probes have been described. The problem of making small vorticity probes that can be used to give good spatial resolution in ordinary laboratory flows or in the boundary layer on an airplane (for example) has not been solved.

The calibration of vorticity probes has been discussed and the influence of the transverse velocity components investigated, Kastrinakis et al. (1978b). The result of this work is that the influence of the transverse velocity fluctuations in a turbulent channel flow on the vorticity signal is of the same order of magnitude as the vorticity signal. The affect of the transverse fluctuations cannot be corrected, since the simultaneous knowledge of these fluctuations is unavailable. Thus, the measurement of instantaneous values of the longitudinal vorticity component with a single vorticity probe in flows with high turbulence levels is impossible. However, when the probe can be oriented properly with the flow mean values of the vorticity can be measured.

Part of this work was supported by the Deutsche Forschungsgemeinschaft. Additional support of the author from the Office of Naval Research and the National Science Foundation is gratefully acknowledged.

References

Kastrinakis, E.G. (1977), Max-Planck-Institut für Strömungsforschung, Bericht 5/1977

Kastrinakis, E.G., Wallace, J.M., Willmarth, W.W., Ghorashi, B., and Brodkey, R.S. (1978a) *Lecture Notes in Physics,* **75**, 175-189

Kastrinakis, E.G., Eckelmann, H., and Willmarth, W.W., (1978b), "The Influence of the Flow Velocity on a Kovasznay Type Vorticity Probe". Submitted to Review of Scientific Instruments

Kim, H.T., Kline, S.J., and Reynolds, W.C. (1971), *J. Fluid Mech.,* **50**, 133-160

Kistler, A.L. (1952). M.S. Thesis, The Johns Hopkins University

Kovasznay, L.S.G. (1954), *Physical Measurements in Gasdynamics and Combustion,* Princeton University Press

Uberoi, M.S. and Corrsin, S. (1951). Progress report on the propagation of turbulence into a "non turbulent" flow, Department of Aeronautics, The Johns Hopkins University

Willmarth, W.W. and Tu, B.J. (1967), *Phys. Fluids,* **9**, Suppl., S134-137

Wyngaard, J.C. (1969), *J. Sci. Instrum.,* **2**, (Ser. 2), 983-987.

Real-Time Measurements of Turbulence Quantities With a Triple Hot-Wire System

by

R.J. Moffat
Department of Mechanical Engineering, Stanford University
Stanford, California 94305

S. Yavuzkurt
Department of Mechanical Engineering, Israel Institute of Technology
Haifa, Israel

M.E. Crawford
Department of Mechanical Engineering, Mass. Inst. of Technology
Cambridge, Massachusetts 02139

Abstract

This paper describes a technique for measuring turbulence quantities in highly turbulent flows whose mean-flow direction is not well known. Three orthogonal hot-wire anemometers are used, with individual linearizers. Their signals are simultaneously processed, in real time, by a network of high-speed analog devices. Instantaneous values of u, v, and w (velocity components in laboratory coordinates) are calculated, using the directional characteristics of each wire and its orientation within the test tunnel. The instantaneous velocity signals can be either processed by a digital computer or introduced into a second analog network which separates the signals into mean and fluctuating components and forms any desired double or triple correlation, again in real time.

Real-time processing avoids some of the ambiguities which arise in time-averaging methods: those which are a consequence of the non-zero values of the time-averaged product of the fluctuating terms. Those ambiguities remain which arise as a consequence of non-zero wire length or lack of perfect knowledge of the directional characteristics of the wires. The present work does not address either of those problems, and they remain as the principal hindrances to further improvements in hot-wire anemometry.

Electrical qualification tests of the analog system have shown the output amplitude to be independent of frequency to within 0.1 % up to 20 kHz and the phase shift between components to be less than $1°$ up to 7 kHz.

Fluid dynamic qualification tests were conducted in a two-dimensional channel flow whose mean and turbulence quantities are believed to be well known. A standard DISA triple-wire probe was used and traverses made with various combinations of pitch and roll on the probe.

Both in the central region of the channel and near the wall, the measured streamwise component of mean velocity was independent of roll and independent of pitch up to 20°. Small spurious transverse components of mean velocity were indicated. In the central region, turbulence quantities were generally within ±10 %, so long as the velocity vector was within 20° of the probe axis. Near the wall, in the region of high velocity gradient (on the order of 100 m/sec/m), the acceptance cone for turbulence quantities became smaller, and roll angle became a more important factor. This latter behavior is attributed to the effect of probe size.

Progress is being made in the development of a compact triple-wire probe, which is expected to improve the near-wall behavior.

1. Introduction

Current research programs in turbulent boundary layer heat transfer involve situations in which the turbulence levels are very high (perhaps 25-50 %) and the direction of the mean flow is either unstable or uncertain (within ±15°). Examples would be flows over surfaces cooled by injection through discrete holes (turbine blade cooling), or flows past large roughness elements, or flows over backward-facing steps. In each of these cases, large amounts of data are desired to map the turbulence fields, and, since no predictive model yet exists, it is not known which one or two properties are most relevant. One would like to document the entire structure and seek out the most important properties.

These are all laboratory flows, of relatively large scale, operating at essentially ambient conditions of temperature and pressure and at velocities between 5 and 50 m/s. The high turbulence levels and the partially unknown flow direction make it impossible to use conventional techniques in interpreting the hot-wire signals. The usual approach is to measure the time-averaged output and the mean square of the fluctuating output from a hot wire at each of several consecutive positions (or from two or more wires at the same time) and then deduce the turbulence components by solving a set of simultaneous equations involving these time-averaged outputs.

The ambiguities which arise when the flow direction is unknown or the turbulence is large are brought on by the fact that the signal from the hot wire is sensitive to all three components of the velocity relative to the wire.

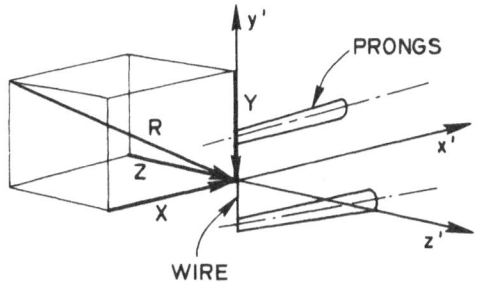

Fig. 1. Velocity components in wire coordinates

To illustrate the nature of the problem, it is instructive to examine the equations necessary for the usual treatment of a flow of unknown direction. The relation between the effective velocity and the velocity components will be taken as in Eqn. [1] from Jorgensen (1971), with the nomenclature as shown in Fig. 1.

$$U_{eff}^2 = X^2 + k_1^2 Y^2 + k_2^2 Z^2 \qquad [1]$$

In the above equation X, Y, Z are the velocity components in the directions of the wire coordinates x', y', and z', respectively, and k_1 and k_2 are pitch and yaw sensitivities of the wire.

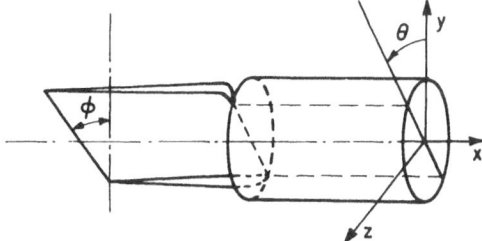

Fig. 2. Orientation of the wires in laboratory coordinates

Fig. 2 shows the wire orientation with respect to the laboratory coordinates, x, y and z. In Fig. 2 ϕ is the slant angle (angle in the plane containing the prongs, between the wire and y direction), and θ is the roll angle in the (yz) plane. The value of θ is zero when the plane containing the prongs is in the (xy) plane. U, V, and W are the velocities in the laboratory coordinates in the x, y, and z directions, respectively. With this nomenclature, Eqn. [1] can be written in terms of velocities in laboratory coordinates, as follows:

$$U^2_{eff} = A(U^2) + B(V^2) + C(W^2) + D(UV) + E(VW) + F(UW) \tag{2}$$

where

$A = \cos^2\phi + k_1^2\sin^2\phi,$

$B = (\sin^2\phi + k_1^2\cos^2\phi)\cos^2\theta + k_2^2\sin^2\theta,$

$C = (\sin^2\phi + k_1^2\cos^2\phi)\sin^2\theta + k_2^2\cos^2\theta,$

$D = (1 - k_1^2)\sin2\phi\cos\theta,$

$E = (\sin^2\phi + k_1^2\cos^2\phi - k_2^2)\sin2\theta,$

$F = (1 - k_1^2)\sin2\phi\sin\theta.$

In Eqn. [2], each velocity can be written as the sum of a mean and a fluctuating part:

$U = \bar{U} + u'$

$V = \bar{V} + v'$

$W = \bar{W} + w'$

and

$U_{eff} = \bar{U}_{eff} + u'_{eff}$

If these definitions are substituted in Eqn. [2], a very complicated equation results. In the classical time-averaged method, the quantities recorded are $\bar{U}_{eff}$ and U'^2_{eff}. Eqn. [2] can be brought to a theoretically solvable form by a Taylor's series expansion of both sides. This requires the assumptions that the flow has a strongly preferred direction and has low fluctuations ($\bar{V}$, $\bar{W}$, and u', v', and w' are all one order of magnitude smaller than $\bar{U}$). The simplified equations are given below:

$$\bar{U}_{eff} = \frac{1}{2\sqrt{A}\,\bar{U}}\left[(2A\bar{U}^2 + B\bar{V}^2 + C\bar{W}^2) + (D\overline{UV} + E\overline{VW} + F\overline{UW})\right.$$

$$\left. + (A\overline{u'^2} + B\overline{v'^2} + C\overline{w'^2}) + (D\overline{u'v'} + E\overline{v'w'} + F\overline{u'w'})\right] + 0(3) \tag{3}$$

$$u'^2_{eff} = \frac{1}{4A}\left[\left(4A^2 + D^2\frac{\bar{V}^2}{\bar{U}^2} + F^2\frac{\bar{W}^2}{\bar{U}^2} + 4AD\frac{\bar{V}}{\bar{U}} + 4AF\frac{\bar{W}}{\bar{U}} + 2DF\frac{\bar{V}\bar{W}}{\bar{U}^2}\right)\overline{u'^2}\right.$$

$$+\left(D^2 + 4B^2\frac{\bar{V}^2}{\bar{U}^2} + E^2\frac{\bar{W}^2}{\bar{U}^2} + 4BD\frac{\bar{V}}{\bar{U}} + 2DE\frac{\bar{W}}{\bar{U}} + 4BE\frac{\bar{V}\bar{W}}{\bar{U}^2}\right)\overline{v'^2}$$

$$+\left(F^2 + E^2\frac{\bar{V}^2}{\bar{U}^2} + 4C^2\frac{\bar{W}^2}{\bar{U}^2} + 2EF\frac{\bar{V}}{\bar{U}} + 4CF\frac{\bar{W}}{\bar{U}} + 4CE\frac{\bar{V}\bar{W}}{\bar{U}^2}\right)\overline{w'^2} \qquad [4]$$

$$+\left(4AD + 4BD\frac{\bar{V}^2}{\bar{U}^2} + 2EF\frac{\bar{W}^2}{\bar{U}^2} + (8AB+2D^2)\frac{\bar{V}}{\bar{U}} + (4AE+2DF)\frac{\bar{W}}{\bar{U}} + (2DE+4BF)\frac{\bar{V}\bar{W}}{\bar{U}^2}\right)\overline{u'v'}$$

$$+\left(4AF+2DE\frac{\bar{V}^2}{\bar{U}^2} + 4CF\frac{\bar{W}^2}{\bar{U}^2} + (4AE+2DF)\frac{\bar{V}}{\bar{U}} + (8AC+2F^2)\frac{\bar{W}}{\bar{U}} + (2EF+4CD)\frac{\bar{V}\bar{W}}{\bar{U}^2}\right)\overline{u'w'}$$

$$\left.+\left(2DF+4BE\frac{\bar{V}^2}{\bar{U}^2} + 4CE\frac{\bar{W}^2}{\bar{U}^2} + (2DE+4BF)\frac{\bar{V}}{\bar{U}} + (4CD+2EF)\frac{\bar{W}}{\bar{U}} + (8BC+2E^2)\frac{\bar{V}\bar{W}}{\bar{U}^2}\right)\overline{v'w'}\right] + 0(3)$$

Even with these assumptions, the resulting equations are quite complicated. In two-dimensional flows where the flow direction is known, aligning the probe with the flow direction sets $\bar{V}$ and $\bar{W}$ to zero, allowing the simplified equations to be used. In cases where there is symmetry, ($\overline{u'w'} = 0$ and $\overline{v'w'} = 0$), the equations reduce to the following form, which is generally used for rotatable slant wires:

$$\bar{U}_{eff} = A\,\bar{U} + 0(2) \qquad [5]$$

$$\overline{u'^2_{eff}} = A\,\overline{u'^2} + \frac{D^2}{4A}\,\overline{v'^2} + \frac{F^2}{4A}\,\overline{w'^2} + D\,\overline{u'v'} + 0(3) \qquad [6]$$

One measures $\overline{U_{eff}}$ and $\overline{u'^2_{eff}}$ in each of four different positions, thus obtaining enough equations to solve for these unknowns.

There are several ways to make measurements in 3-D turbulent flows using these classical time-averaged equations. The most important problem in 3-D flows is that the direction of the mean flow is unknown. This direction can be found with one of the following ways. Johnston (1970) measured the local pitch and yaw angles for the mean velocity with a Conrad probe. Reynolds stresses were then obtained from a horizontal wire and a rotatable cross-wire aligned according to the known flow direction. The low-fluctuation assumption was invoked. Moussa and Eskinazi (1975) tried to measure the mean flow direction by using a rotatable slant wire. Making use of the directional properties of hot wires, they calibrated the probe for all possible angles and prepared detailed charts which included the flow angles as functions of four mean voltages obtained at different rotations of the probe. Delleur (1966) used a cross-wire to measure the flow direction. He argues that the cross-wire technique is twice as accurate as the single-wire technique. Both methods require the use of a hot-wire calibration curve for flow-direction measurement and frequent calibration of the hot wire to renew the calibration charts.

Some other 3-D hot-wire methods were developed which do not require the flow direction to be known *a priori*. Majola (1974) gives a hot-wire method (rotatable slant wire or cross-wire) for measuring the three mean components of the velocity and six Reynolds stresses without knowing the flow direction. His equations are valid when there is a strongly preferred mean flow direction and for low fluctuations. Hoffmeister (1972) describes a scheme which employs a single rotatable slant wire to obtain three mean velocities and six Reynolds stresses. In this scheme the interpretation of the anemometer voltages is based on calibrations of the probe in the entire range of angles between the wire and the flow existing during measurements.

None of the preceding methods is practical for taking large amounts of data. Further, the accuracy with which the higher-ordered terms can be measured is seriously limited. As the number of terms retained from Eqn. [4] increases, the number of independent realizations required increases; thus the number of probe rotational positions increases, and the strength with which the equations converge diminishes.

The problem lies, basically, in the time-averaged approach to turbulence measurement. With one or two wires, one does not have enough information to solve for the instantaneous velocities, and, hence, time-averaging is required.

2. The Present Approach
Three orthogonal hot wires provide enough information to solve for the instantaneous velocity without time averaging. Commercial circuitry is available for this purpose. Such a scheme was used by Zimmerman and Abbott (1975) to measure the Reynolds stresses in a 3-D boundary layer created by a flat plate skewed with respect to the oncoming flow. A triaxial probe was used with an analog processor, yielding the mean components and Reynolds stresses.

The present approach differs mainly in the frequency response capabilities of the circuity and the fact that the directional properties of the individual wires can be used, instead of a global assumption. The present system provides more output information, but that is of secondary importance.

The first requirement is a triple-wire probe whose wires form an orthogonal set. For such a probe, the x component for one wire is the y component for another and the z component for the third. Such a probe is commercially available from the DISA Corporation.

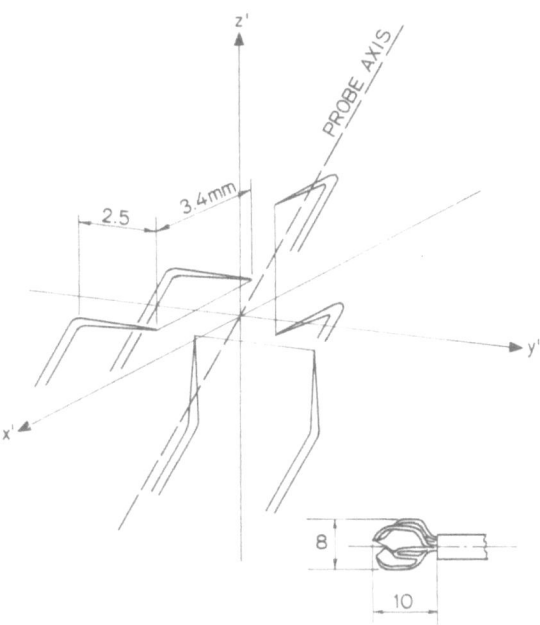

Fig. 3. The triple-wire coordinate system

The DISA triaxial wire probe has three wires with separate ground leads, driven by three separate anemometers. The wires are mutually orthogonal, forming a right-angled coordinate system. The sensors provided by DISA are 3.2 mm long, with 5 micron Pt-plated tungsten wires, plated at the ends with copper and gold to leave a sensitive length of 1.25 mm. In the present work, bare Pt-plated tungsten wires were used without gold plating, giving 3.2 mm active length. The three wires form a cone of apex angle $70.6°$ around the axis of the probe stem. The wire coordinate system can be seen in Fig. 3. With the special prong structure of the probe, the effective velocity indicated by each wire is related to the velocity components in the wire coordinates in the following manner.

$$U^2_{eff_1} = k^2_{11} X^2 + Y^2 + k^2_{21} Z^2$$

$$U^2_{eff_2} = k^2_{22} X^2 + k^2_{12} Y^2 + Z^2 \qquad [7]$$

$$U^2_{eff_3} = X^2 + k^2_{23} Y^2 + k^2_{13} Z^2$$

The linearized effective voltages (linearizer outputs) are related to the effective velocities as follows:

$$U_{eff_1} = A_1 + B_1 E_{eff_1}$$

$$U_{eff_2} = A_2 + B_2 E_{eff_2} \qquad\qquad [8]$$

$$U_{eff_3} = A_3 + B_3 E_{eff_3}$$

In Eqns. [8] the A's and B's are constants, obtained from calibrations of the wires.

Equations [7] have three unknowns — the instantaneous velocities in the wire coordinates — which can be obtained from the equations shown below:

$$
\begin{bmatrix} X^2 \\ Y^2 \\ Z^2 \end{bmatrix} =
\begin{bmatrix} k_{11}^2 & 1 & k_{21}^2 \\ k_{22}^2 & k_{12}^2 & 1 \\ 1 & k_{23}^2 & k_{13}^2 \end{bmatrix}^{-1}
\begin{bmatrix} U_{eff_1}^2 \\ U_{eff_2}^2 \\ U_{eff_3}^2 \end{bmatrix} \qquad [9]
$$

or

$$
\begin{bmatrix} X^2 \\ Y^2 \\ Z^2 \end{bmatrix} =
\begin{bmatrix} k_{11}^2 & 1 & k_{21}^2 \\ k_{22}^2 & k_{12}^2 & 1 \\ 1 & k_{23}^2 & k_{13}^2 \end{bmatrix}^{-1}
\begin{bmatrix} (A_1 + B_1 E_{eff_1})^2 \\ (A_2 + B_2 E_{eff_2})^2 \\ (A_3 + B_3 E_{eff_3})^2 \end{bmatrix} \qquad [10]
$$

Once the instantaneous velocities in the wire coordinates are obtained, the instantaneous velocities in the laboratory coordinates can be obtained easily with a transformation of coordinates.

$$
\begin{bmatrix} U \\ V \\ W \end{bmatrix} = N
\begin{bmatrix} X \\ Y \\ Z \end{bmatrix} \qquad [11]
$$

Where N is the coordinate transformation matrix from wire coordinates to laboratory coordinates. The solution to Eqns. [11] is a set of three instantaneous velocities in the laboratory coordinate system. With these in hand, it is a simple matter to find the mean value of each, and the fluctuating component. The three individual fluctuating components can then be multiplied by one another and averaged to yield $\overline{u'v'}$, $\overline{v'w'}$, $\overline{u'w'}$, $\overline{u'^2}$, $\overline{v'^2}$, $\overline{w'^2}$, q^2 and any desired triple correlations.

The question is whether this task is better performed by analog or by digital means. Both methods were investigated, and the final selection was strongly influenced by cost.

We sought to develop a low-cost, stand-alone unit with which we could process the signals from three linearized anemometers. Our objective was to obtain information concerning both the amplitude and relative phase of the three components of the instantaneous velocity, over the frequency range from D.C. up to 10 kHz. The "low-cost, stand-alone" requirement was important since it was our intent to make these units available to each of several experiments which might all be running at the same time.

The system finally adopted uses analog processing, although digital techniques were considered. The principal argument which led to selection of the analog scheme was that we could obtain acceptable accuracy (1 % to 2 %) over an acceptable frequency range (4 Hz to 10 kHz) at much lower cost than by using digital processing. It should be pointed out that this choice reflects the state of the art as we saw it late in 1976. The trend in digital systems has since been towards higher speed and lower cost, and it may soon be economically feasible to do these calculations digitally. The factor which raised the projected cost of the digital processing system was the rate of acquisition and processing required in order to preserve the waveform and relative phase between the three components. The triple-wire processor would require a sampling rate of about 150 kHz for 10 kHz variations on each wire. Steady-state computation would require about 30 algebraic operations per data bit, or about 4.5×10^6 calculations per second. Burst mode and buffer type computations would soften this requirement, but would require storage. It did not seem possible to duplicate the projected analog performance without going to a digital system costing much more than the analog (about \$5200 parts and 1 man-month labor).

3. The Three-Dimensional Turbulent Flow Analyzer
An analog device was built to solve Eqns. [10] and [11] using high-speed analog components. All quantities were magnitude-scaled using $(U_{eff})_{max}$, the value which resulted in the maximum output of the linearizer. The coefficients A and B in Eqn. [10] are the same for each wire, since A and B are functions of the minimum and maximum velocities of the calibration range for the triple-wire probe. Different values of k_1 and k_2 can be set for each wire, although an

uncertainty analysis has shown that these coefficients are not critically important. For example, a change of 1 % in the value of k_1 causes a change of 0.02 % in the indicated velocity, while a change of 1 % in k_2 yields a change of 1.04 %.

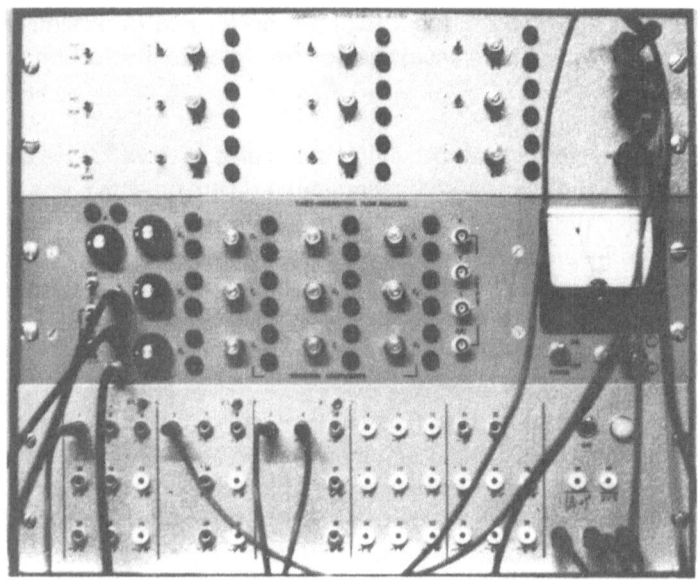

Fig. 4. The control panel of the Real-Time Flow Analyzer

Fig. 4 shows the control panel of the Flow Analyzer (center panel), the coordinate rotation unit (upper panel), and the post-processor unit (lower panel). On the Flow Analyzer panel, there are three potentiometers for setting the B coefficients, even though the three values are the same, because of the circuit requirements. The nine coefficient potentiometers for the Jorgensen matrix are shown, each with its own pair of test points for checking its value. The meter provides a continuous display of R/C, the time-averaged value of the magnitude of the velocity vector. The outputs are the normalized velocity components in wire coordinates X, Y, Z and a value of R/C, the root sum square of the components. The upper panel accepts the X, Y, Z inputs and contains nine potentiometers whose values can be set according to the wire position to calculate the values of U, V, and W, the normalized velocity components in laboratory coordinates. The lower unit (the post-processor) contains low-pass and high-pass filters for separating the mean and fluctuating values, and the multipliers and summers necessary to display all of the Reynolds stress components. There are no adjustments possible on the post-processor.

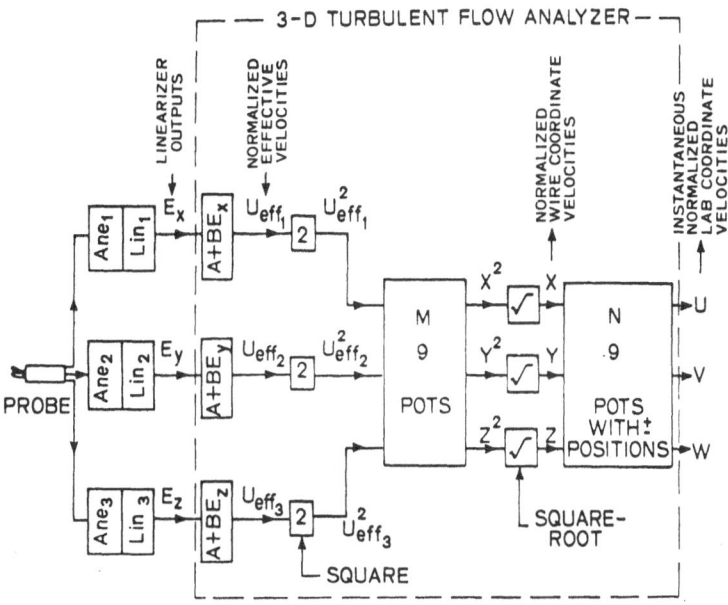

Fig. 5. Electrical flow diagram of the Flow Analyzer and Coordinate Rotation units

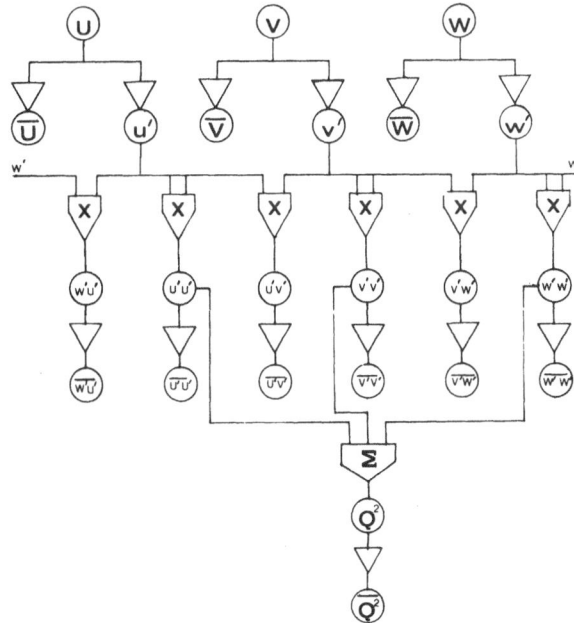

Fig. 6. Electrical flow diagram of the post-processor unit

The flow diagrams of the Flow Analyzer and the coordinate rotation unit are shown in Fig. 5, while the post-processor diagram is shown in Fig. 6.

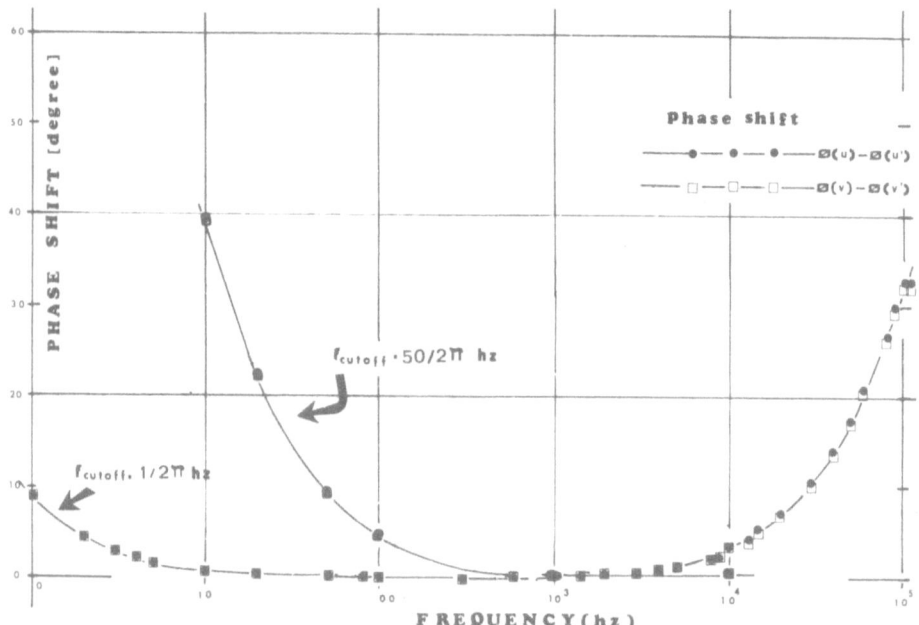

Fig. 7. Phase shift characteristics of the high-pass filter units of the post-processor

Electrical performance tests were conducted on each unit, to ensure that no phase shift or attenuation occurred. The Flow Analyzer and coordinate rotation unit show a phase shift less than 2° up to 20 kHz, with signal attenuation of about 0.1 % (maximum). The post-processor, since it involves filter units, warrants separate study. Fig 7 shows the measured phase shift characteristics: less than 3° phase shift between 4 Hz and 10 kHz. This low-frequency performance was made possible by selecting the low-frequency cutoff frequency to be 0.16 Hz, the purpose being to preserve phase relationships in studies of large-scale, low-frequency structures. The second line, labeled (50/2π), shows the performance which would have been achieved using an 8 Hz cutoff instead of 0.16. Our interest in large-scale structures prevailed.

Fig. 8 shows a sample of the attenuation test results. The RMS value of u'(out) is compared to the RMS value of u'(in) as a function of the frequency, the AC voltage level, and the DC level. Between 1 Hz and 10 kHz, there is no measurable roll-off for any combination of voltages and frequencies.

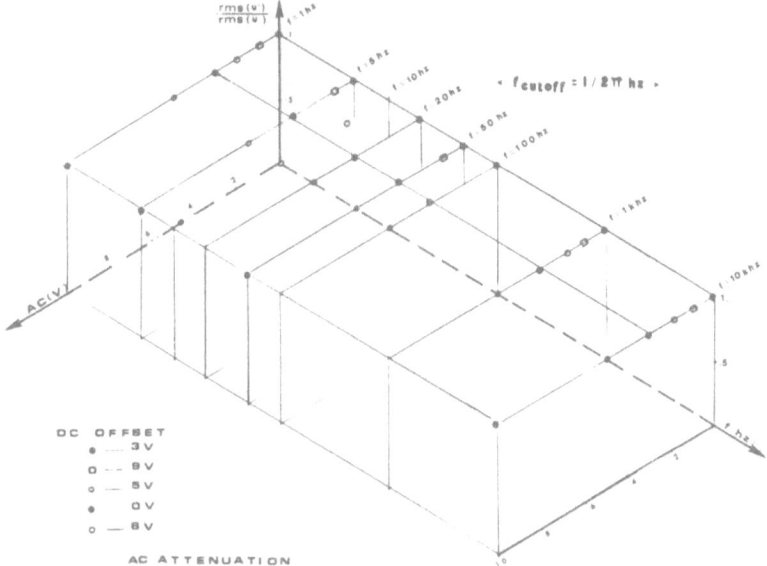

Fig. 8. Typical results of attenuation tests on the post-processor circuit: atten-uation of a single fluctuation component, μ'

4. Fluid Mechanic Qualification Tests

The Thermosciences Laboratory in the Mechanical Engineering Department of Stanford University has a two-dimensional channel which gives fully developed mean velocity and turbulence profiles up to at least the second-order turbu-lence quantities. The performance characteristics of this channel have been fully explored; it has been used by almost all recent experimenters to calibrate their hot-wire technique, their probes, and their systems. The Flow Analyzer and coordinate rotation unit were tested in this channel to qualify their per-formance and to explore their limitations for turbulence measurements.

The post-processor was added after these tests had been completed, and only spot checks have been run, aside from the electrical check-out. In every case checked, the post-processor results agreed with the results presented here within about 2 %. The values of the turbulence quantities reported in the fol-lowing figures were calculated digitally from tape-recorded outputs of the Flow Analyzer and coordinate rotation unit.

For the qualification tests, the probe was mounted in a two-axis probe holder so that it could be streamwise rotated around the axis (roll angle a) and also tilted (pitch angle ω) against the approaching flow, as well as traversed to several different distances from the wall. By measuring at several distances from the wall, the system performance was recorded both for high shear regions (near the wall) and zero shear regions (at the centerline of the channel). The outputs were compared with the outputs of the other acceptable methods of measurement in the channel (single horizontal wire, pitot tube measurements, and linear shear stress distribution calculated from the pressure gradient along the channel).

The two-dimensional channel is 6.35 cm wide and 117 cm high. The experiments were made with air flow at ambient conditions and with a centerline speed of 11.2 m/sec.

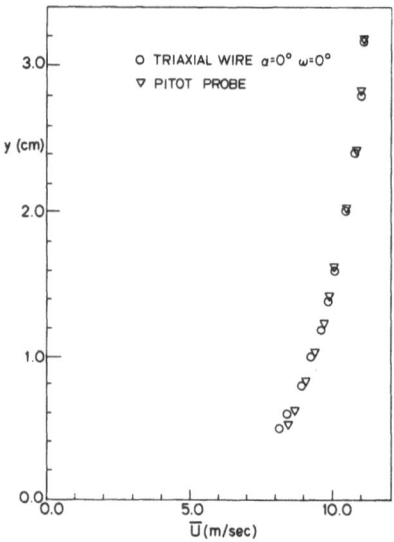

Fig. 9. Comparison of triaxial probe and pitot probe measurements of $\bar{U}$ in the 2-D channel

In Fig. 9 the values of $\bar{U}$ (the streamwise mean velocity) obtained from a pitot probe and from the triaxial probe are compared. The pitot probe was modified to create the same kind of stem blockage effect in the channel as did the triaxial probe. In this test the probe was set to zero roll ($a = 0°$) and the axis was aligned with the flow ($\omega = 0°$). The readings of the pitot probe were corrected for shear displacement effect and for turbulence level. The maximum difference between the two probes occurs near the wall — about 2.4 %. The difference diminishes rapidly as the distance from the wall increases. The difference near the wall may be due to the finite size of the triaxial probe, as was explained earlier, interacting with the velocity gradient.

Next, the effect of the rotation of the probe around its axis was investigated. When the probe axis is aligned with the flow direction ($\omega = 0°$), rotation around its streamwise axis should not affect the result if the velocity is uniform, but may affect the result in a shear flow. To investigate this, the probe was aligned with the flow direction ($\omega = 0°$), and for each transverse position across the tunnel, the probe was rotated around its axis to the values of $a = 0°$, $90°$, $180°$, and $270°$. This angle range covers the extreme positions for the wires and exposes different configurations of wires to the shear at different angles.

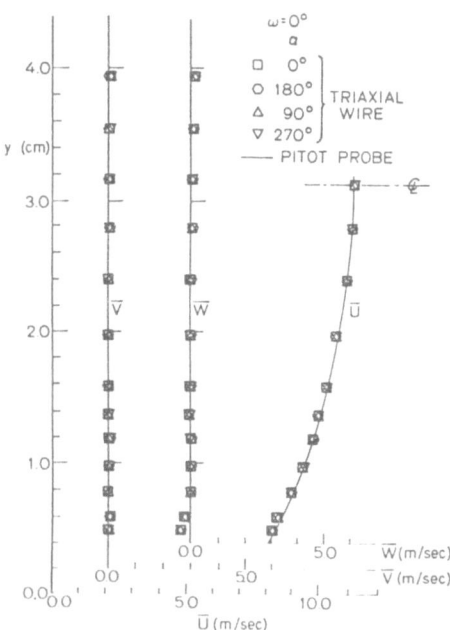

Fig. 10. The effect of roll angle about the probe axis on the indicated mean velocity components

Fig. 10 shows the three mean velocity components ($\overline{U}$, $\overline{V}$, and $\overline{W}$) as a function of the distance from the wall for several values of the roll position, a. Roll around the probe axis does not affect the $\overline{U}$ values. The effect on $\overline{V}$ and $\overline{W}$ is small, but not negligible. In this figure, $\overline{V}$ and $\overline{W}$ values are also plotted to the same scale as $\overline{U}$. The values of $\overline{V}$ and $\overline{W}$ should be zero, but due to the probe size, some deviation from zero is observed within the shear region. The most meaningful comparison for error in $\overline{V}$ and $\overline{W}$ is to compare them to the $\overline{U}$ at the same location. The largest deviations occur at the point near the wall for $a = 0$, $\overline{V}/\overline{U} = 1$ % and for $a = 270°$, $\overline{W}/\overline{U} = 4.5$ %. These deviations from zero become smaller as the distance from the wall increases. In the zero gradient region at the centerline there is no deviation.

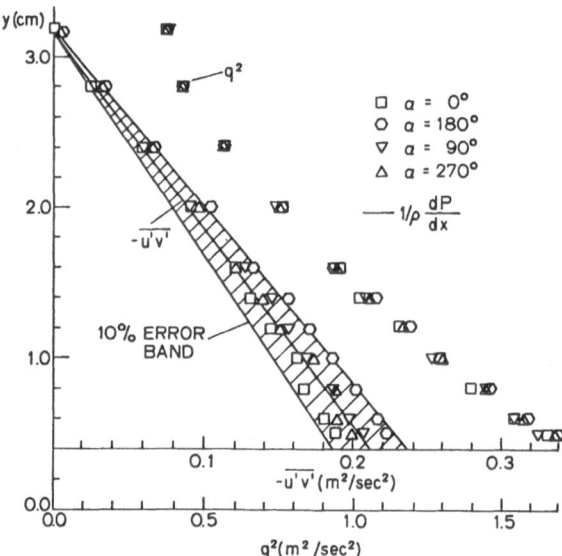

Fig. 11. The effect of roll angle about the probe axis on indicated shear stress and turbulent kinetic energy

Fig. 11 shows the turbulent kinetic energy and shear stress distributions as a function of the distance from the wall for several values of the roll angle. The shear stress ($-u'v'$) measurements are compared with the linear shear stress distribution obtained from the pressure gradient along the channel (dp/dx). As is seen, all the experimental data lie inside the 10 % error band, but at angles $a = 90°$ and $a = 270°$ the deviations are much smaller; therefore, one would like to measure $-u'v'$ at these angles. The measured turbulent kinetic energy is not much affected by the roll. The largest difference between results occurs in the high shear region near the wall; it is about 3.5 %. In the zero shear region there is no effect of roll.

Fig. 12 shows the diagonal Reynolds stress components ($\overline{u'^2}$, $\overline{v'^2}$, $\overline{w'^2}$) as a function of the distance from the wall, normalized with centerline velocity. The streamwise normal stress ($\overline{u'^2}$) does not seem to be affected much by roll around the probe axis, even in the high shear regions near the wall. On the same figure, the $\overline{u'^2}$ distribution obtained with a conventional single horizontal wire is also given, and its agreement with the triaxial wire data is not bad. The other normal Reynolds stresses ($\overline{v'^2}$ and $\overline{w'^2}$) are affected by the roll angle, especially in the high shear region, but the data collapse on each other quickly as the shear decreases. In the zero shear region on the centerline there are no deviations. One important observation is that, at a certain a value, if $\overline{v'^2}$ reads high compared to the value at $a = 0°$, then $\overline{w'^2}$ reads low, or vice versa, while $\overline{u'^2}$

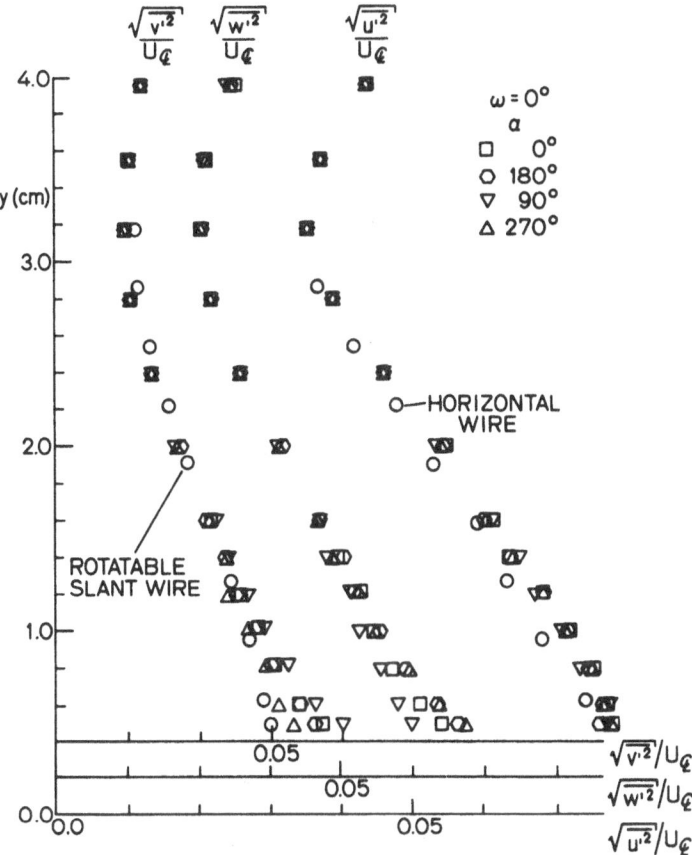

$$\frac{\sqrt{\overline{v'^2}}}{U_{\mathbb{C}}} \quad \frac{\sqrt{\overline{w'^2}}}{U_{\mathbb{C}}} \qquad \frac{\sqrt{\overline{u'^2}}}{U_{\mathbb{C}}}$$

$\omega = 0°$

α
□   0°
○   180°
▽   90°
△   270°

○—HORIZONTAL
    WIRE

ROTATABLE
SLANT WIRE

0.05 $\sqrt{\overline{v'^2}}/U_{\mathbb{C}}$

0.05 $\sqrt{\overline{w'^2}}/U_{\mathbb{C}}$

0.05 $\sqrt{\overline{u'^2}}/U_{\mathbb{C}}$

Fig. 12. The effect of roll angle about the probe axis on the indicated values of $\overline{u'^2}$, $\overline{v'^2}$, and $\overline{w'^2}$

does not change much with a. This combination leads to q^2 values which are quite insensitive to the changes in a, a fortuitous result for the measurement of q^2 — the main interest of the general research.

The data discussed above are enough to qualify this system for measurements when the probe axis is aligned with the flow direction. But one of the most important objectives of this research was to find a method which would work in a flow of unknown direction. To investigate this, the probe axis was tilted against the approaching flow direction (ω), again in the 2-D channel. Some rotations around the probe axis (a) were also tested to see the combined effect of both a and ω. The results are discussed in the following paragraphs.

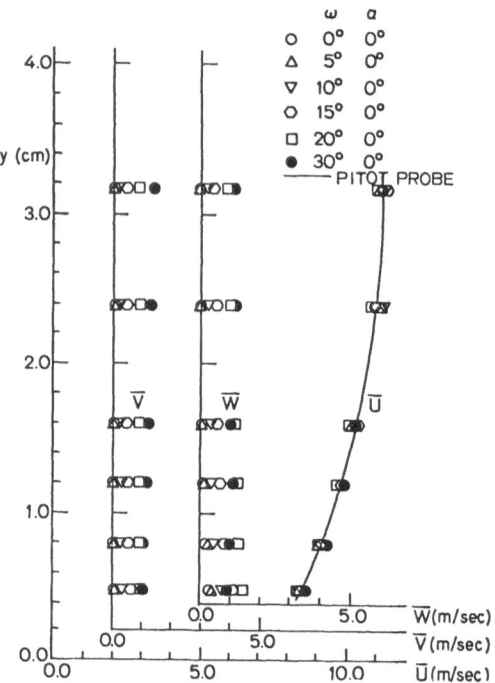

Fig. 13. The effect of pitch angle on the indicated values of the mean velocity components

Fig. 13 shows the three mean velocity components as a function of distance from the wall for several values of the angle between the flow and the probe axis (ω). There the value of a was held constant, because it was seen above that the mean velocities were not much affected by the roll angle. For $\bar{U}$ up to $\omega = 20°$ the data collapse on top of each other and the deviation for $\omega = 30°$ is not very large. The largest deviation at this angle is about 3.5 % in the high shear region near the wall, and about 2 % in the zero shear region. Deviation is calculated as the difference between two extremes, not from the pitot probe data. This result means that mean velocity can be measured with good accuracy if the approaching flow direction is within ±30° of the probe axis; i.e., one does not have to know the flow direction better than within a cone of half apex angle 30° around the probe axis to measure the mean velocity with acceptable accuracy. (If $\bar{V} = 0.1\ \bar{U}$, this will give an angle of about ±6°). As the angle between the flow direction and the probe axis increases, the errors in $\bar{V}$ and $\bar{W}$ also increase (ideally, their values should be zero). Some of this may be the effect of the probe size, as was explained earlier, and some may be due to the probe blockage effect in the channel.

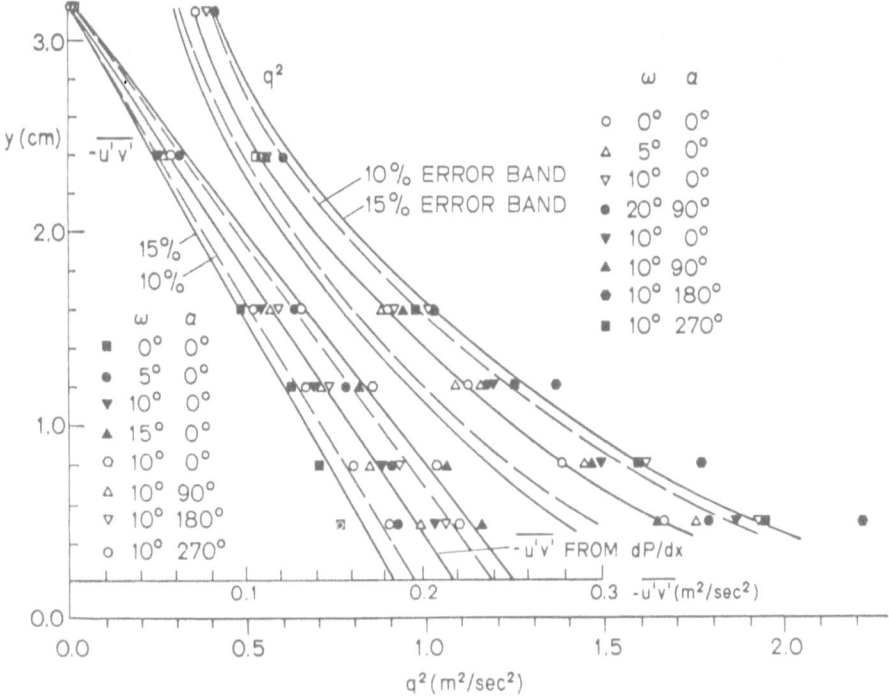

Fig. 14. The effect of combined roll and pitch angle on indicated shear stress and turbulent kinetic energy

In Fig. 14 the turbulent kinetic energy (TKE) and shear stress are plotted for several values of roll angle (a) and pitch angle (ω). In the TKE plot the line at the center is faired through the data at $\omega = 0°$, $a = 0°$ (this measurement should be the one closest to reality). The other lines denote the ±10 % and ±15 % error bands around the reference. Again, the data points converge rapidly as the distance from the wall increases. Deviations are much smaller in the zero shear region. In the same figure, the q^2 distribution for $a = 90°$, $\omega = 20°$ is shown, to demonstrate the increase in deviation as ω increases. Another important point to observe from this figure is that for $a = 90°$ and $\omega = 10°$ the data lie very close to the center profile. This shows that, depending on the quantity being measured, there are angle combinations a and ω for which the measurement cone can be enlarged. For example, for $a = 90°$ and $\omega = 10°$ and 20°, it appears that even in the highest shear region q^2 can be measured within 12 % inside a cone of 15° half apex angle. Generally, q^2 can be measured within 10 % or 12 % inside a cone of 10° half apex angle around the probe axis. The deviations in q^2 are not like uncertainty scatters, but rather have a preferred direction. It may be possible to devise a scheme to correct the data based on the first estimate of the flow direction, to improve the accuracy. In the shear stress part of Fig. 14, the straight line in the middle of the figure is the shear stress distribution obtained from the pressure gradient in the streamwise direction. The other straight lines are the boundaries for 10 % and 15 % error. Most of the data up to the angle ω = 10° lie within 10 % error band, except a few points near the wall for angles a = 270° and a = 90°. Almost all the data, including ω = 15 % and a = 0°, lie within the ±15 % error band. In conclusion, it can be said that the shear stress $-\overline{u'v'}$ can be measured within 10 % within a cone of half apex angle 10° and within 15 % inside a cone of half apex angle 15°, except very near the wall.

In all of the qualification data which were discussed above, it is seen that, depending on the quantity measured, the size of the measurement cone changes. Some quantities, such as $\overline{v'^2}$ and $\overline{w'^2}$ are affected by rotation around the probe axis, especially in high-velocity gradient regions. The reason for these changes may be the large probe size.

An important point shown by the qualification data is that the major mean velocity component $\overline{U}$ can be measured quite accurately within a cone of 30° half apex angle around the probe axis. In three-dimensional flows where the probe axis makes a large angle with the unknown flow direction, the measured turbulence quantities may have large errors. In critical cases, one can find the flow direction approximately (within 3-4°), since the mean components are accurately measured, even when the probe is misaligned. Then the probe could be approximately aligned with this flow direction. In this position all of the turbulence quantities could be measured quite accurately.

5. Operating Experience With The Present System

The Flow Analyzer and Coordinate Rotation units have been used with the DISA triple-wire probe in an extensive experimental program covering several months of intensive data acquisition. Of particular importance is the stability of the electrical circuitry. None of the 18 set point potentiometers which adjust the matrix coefficients was adjusted during the test program. The system was shut down for a period of six months, yet upon re-energizing it and recalibrating, the same values were recorded from the same flow situation within about 2 %.

Wire-breakage problems with hot wires seems to arise much more often during the first few months of operator experience than after that, providing that the tunnel air is reasonably well filtered. A triple-wire probe is not recommended for a novice, but once the operator skill has reached a level where a single wire can be kept alive, the triple-wire probe poses no significant new hazard. The same probe has now been in use for over a year, with no wire breakage.

6. Miniaturizing The Triple-Wire Probe

The standard DISA triple-wire probe head is 8 mm in diameter, which limits the use of the probe in boundary layers. In part, the size is necessary in order that the prongs, as well as the wires, form an orthogonal set in satisfaction of the Jorgensen criteria for the directional sensitivity.

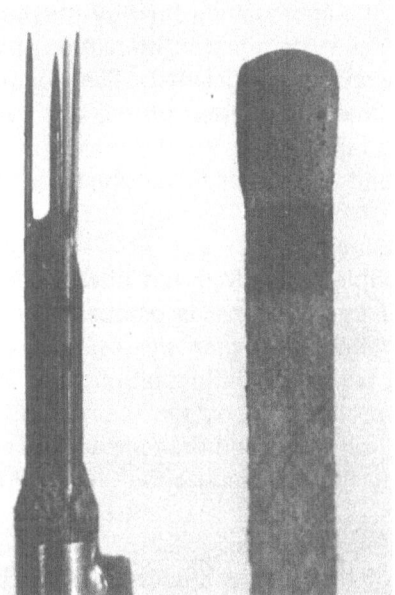

Fig. 15. The compact triple-wire probe

A compact triple-wire probe has been built whose outside diameter is 2 mm, as shown in Fig. 15. There are three significant differences between the compact probe and the DISA probe: size, prong orthogonality, and number of prongs. The prongs on the compact probe are parallel to one another, rather than being orthogonal. A new directional matrix will have to be determined for this arrangement. The general form can be deduced by analysis based upon the Jorgensen decompositions, but the values of the new directional sensitivity coefficients will have to come from experiment. The use of four support prongs instead of six introduces the possibility of electrical cross-talk between the channels, either by virtue of ground potential differences between the chassis, or because the shared length of wire will generate IR potentials in every channel which reflect the sum of the currect flows in the individual branches. The ground-potential problem can be resolved at the chassis location, by proper grounding. The signal cross-talk can be eliminated by making the shared resistance very small. Tests have shown that if the common lead wire has a resistance of 0.001 ohm or less, the cross-talk is not significant: less than 3 micro-volts passed to the Y and Z channels as a consequence of a 3-volt signal from the X bridge.

It is believed that the spurious values of $\overline{V}$ and $\overline{W}$ displayed by the present system when used at large angles of pitch and yaw are partly a system-disturbance effect (diversion of the approaching flow by the pressure field surrounding the probe stem) and partly system-sensor interaction error (incorrect assessment of the directional sensitivity coefficients). The compact triple-wire probe will offer less blockage than the present probe, and the directional sensitivity co-efficients chosen to represent it will be biased in favor of minimum artifact generation in pitch and yaw rather than using an arithmetic mean.

7. Temperature Measurements
Fine-wire thermocouple technology has now advanced to the point where a 5μ dia, butt-welded thermocouple is practical. Such a thermocouple has the same frequency response characteristics as a 5μ cold-wire, as is frequently used for measuring temperature fluctuations, and its signal is free of velocity contamination. Time-constant compensation circuitry can be applied to im-prove the frequency response, and the temperature signal then used in conjunc-tion with the velocity signal to deduce the turbulent heat transfer rates.

8. Higher-Ordered Measurements
Triple correlations of the velocity fluctuations can be evaluated in the present apparatus using additional multiplier modules and filter modules of the same design as already tested. Of particular interest are two measurements aimed at estimating the turbulence production and dissipation.

Circuits have been designed to measure:

$$D^* = \overline{\frac{d}{dt}(u'^2)}$$
$$P^* = \overline{u'v'U}$$

The quantity D^* is related to the dissipation, while P^* is related to the produc-tion, though the connection is not as straightforward. It will be necessary to acquire considerable experience with this measurement capability, before its usefulness can be assessed.

9. Conclusions
Real-time data reduction from an orthogonal triple-wire anemometry system offers significant advantages over time-averaged data reduction from single or

x-wires. Processing by analog means is both practical and economical, though digital processing may be a better choice in the future if the processing speed goes up and the cost comes down significantly.

The present system uses a commercially available (DISA) triple-wire probe. Mean velocity is measured within 3.5 % so long as the velocity vector is within 30° of the probe axis, in a velocity gradient up to 1600 m/sec/m. Turbulence kinetic energy can be measured within 12 % so long as the instantaneous velocity vector is within 15° of the probe axis. The turbulent shear stress can be measured within 15 %, inside the same cone. Miniaturization of the probe is expected to further improve the pitch and yaw tolerance of the system.

With the present analog processor, the waveform and relative phase are preserved for components between 4 Hz and 10 kHz, with less than 3° phase shift and less than 1 % amplitude attenuation.

Some problems remain, in that spurious signals of several percent are generated in the $\bar{V}$ and $\bar{W}$ components. These are believed to be induced partly by probe blockage in the yawed position and partly by use of less-than-optimum coefficients for directional sensitivity. A compact triple-wire probe with different directional characteristics is under development, in an effort to remedy this deficiency.

References

Delleur, J.W., 1966, "Flow Direction Measurement by Hot-Wire Anemometry", *Journal A.S.C.E., Engineering Mechanics Division*, Vol. **92**, p. 45

Hoffmeister, M., 1972, "Using a Single Hot-Wire Probe in Three-Dimensional Turbulent Flow Fields", *DISA Information No. 13*, May 1972, pp. 26-28

Johnston, J.P., 1970, "Measurements in a Three-Dimensional Turbulent Boundary Layer Induced by a Swept Forward-Facing Step", *J. Fluid Mechanics*, Vol. **42**, Part 4, pp. 823-844

Jorgensen, F.E., 1971, "Directional Sensitivity of Wire and Fiber Film Probes, An Experimental Study", *DISA Information No. 11*, May 1971, pp. 31-37

Majola, O.O., 1974, "A Hot-Wire Method for Three-Dimensional Shear Flows", *DISA Information No. 16*, July 1974, pp. 11-14

Moussa, Z.M., and Eskinazi, S., 1975, "Directional Mean Flow Measurements Using A Single Inclined Hot Wire", *Physics of Fluids*, Vol. **18**, No. 3, pp. 298-305

Zimmerman, D.R., and Abbott, D.E., 1975, "An Experimental Investigation of a Three-Dimensional Turbulent Boundary Layer", Technical Report CFMTR-75-1, May 1975